1989 Recommended Dietary Allowances (RDA)

Age (yr)	Energy (kcal)	Protein (g)	Vitamin A (µg RE)	Vitamin K (µg)	Iron (mg)	(mg)	(µg)
Infants							
0.0–0.5	650	13	375	5	6		
0.5–1.0	850	14	375	10	10		50
Children							
1–3	1300	16	400	15	10	10	70
4–6	1800	24	500	20	10	10	90
7–10	2000	28	700	30	10	10	120
Males							
11–14	2500	45	1000	45	12	15	150
15–18	3000	59	1000	65	12	15	150
19–24	2900	58	1000	70	10	15	150
25–50	2900	63	1000	80	10	15	150
51+	2300	63	1000	80	10	15	150
Females							
11–14	2200	46	800	45	15	12	150
15–18	2200	44	800	55	15	12	150
19–24	2200	46	800	60	15	12	150
25–50	2200	50	800	65	15	12	150
51+	1900	50	800	65	10	12	150
Pregnancy	+300	60	800	65	30	15	175
Lactation							
1st 6 mo.	+500	65	1300	65	15	19	200
2nd 6 mo.	+500	62	1200	65	15	16	200

In addition to the values that serve as goals for nutrient intakes (presented in the adjacent tables), the Dietary Reference Intakes include a set of values called Tolerable Upper Intake Levels—the maximum amount of a nutrient that appears safe for most healthy people to consume on a regular basis.

Tolerable Upper Intake Levels for Selected Nutrients (per day)

Age (yr)	Vitamin D (µg)[a]	Niacin (mg)	Vitamin B_6 (mg)	Folate (µg)	Choline (mg)	Vitamin C (mg)	Vitamin E (mg)	Calcium (mg)	Phosphorus (mg)	Magnesium (mg)	Fluoride (mg)	Selenium (mg)
Infants												
0.0–0.5	25	—[b]	—[b]	—[b]	—[b]	—[b]	—[b]	—[b]	—[b]	—[b]	0.7	45
0.5–1.0	25	—[b]	—[b]	—[b]	—[b]	—[b]	—[b]	—[b]	—[b]	—[b]	0.9	60
Children												
1–3	50	10	30	300	1000	400	200	2500	3000	65	1.3	90
4–8	50	15	40	400	1000	650	300	2500	3000	110	2.2	150
9–13	50	20	60	600	2000	1200	600	2500	4000	350	10.0	280
14–18	50	30	80	800	3000	1800	800	2500	4000	350	10.0	400
Adults												
19–70	50	35	100	1000	3500	2000	1000	2500	4000	350	10.0	400
>70	50	35	100	1000	3500	2000	1000	2500	3000	350	10.0	400
Pregnancy	50	35	100	1000	3500	2000[c]	1000[d]	2500	3500	350	10.0	400
Lactation	50	35	100	1000	3500	2000[c]	1000[d]	2500	4000	350	10.0	400

Note: An Upper Limit was not established for thiamin, riboflavin, vitamin B12, pantothenic acid, and biotin because of a lack of data, not because these nutrients are safe to consume at any level of intake; all nutrients can have adverse effects when intakes are excessive.

[a] To convert µg to IU, multiply by 40. For example, 50 µg x 40 = 2000 IU.

[b] Upper Levels were not established for many nutrients in the infant category because of a lack of data.

[c] The vitamin C upper level for women 18 years and younger is 1800 mg during pregnancy and lactation.

[d] The vitamin E upper level for women 18 years and younger is 800 mg during pregnancy and lactation.

Source: Adapted from the first two of the *Dietary Reference Intakes* series, National Academy Press. Copyright 1997 and 1998, by the National Academy of Sciences. Courtesy of the National Academy Press, Washington, D.C. Search for Dietary Reference Intakes online at: http//www.nap.edu/readingroom.

NUTRITION
FOR
HEALTH
AND
HEALTH
CARE

SECOND
EDITION

NUTRITION

FOR HEALTH AND HEALTH CARE

Eleanor N. Whitney

Corinne B. Cataldo

Linda K. DeBruyne

Sharon R. Rolfes

SECOND EDITION

WADSWORTH

THOMSON LEARNING

Australia • Canada • Mexico • Singapore • Spain
United Kingdom • United States

PUBLISHER:	Peter Marshall
DEVELOPMENT EDITOR:	Laura Graham
MARKETING MANAGER:	Becky Tollerson
PROJECT EDITOR:	Sandra Craig
PRINT BUYER:	Barbara Britton
PERMISSIONS EDITOR:	Joohee Lee
PRODUCTION:	The Book Company
TEXT AND COVER DESIGNER	Norman Baugher
PHOTO RESEARCHER:	Myrna Engler
COPY EDITOR:	Patricia Lewis
ILLUSTRATORS:	Impact Publications, Greg Gambino, J & B Woolsey and Associates
COVER IMAGES:	Nurse: © Bob Rowan, Progressive Image/CORBIS; tropical fruits and spices from Grenada: © Wolfgang Kaehler/CORBIS; steaming vegetables: © John Heseltine/CORBIS; beans and eggplants: © PhotoDisc, Inc.
COMPOSITOR:	Parkwood Composition
PRINTER:	R. R. Donnelley & Sons / Willard

Wadsworth/Thomson Learning
10 Davis Drive
Belmont, CA 94002-3098
USA

For more information about our products, contact us:
Thomson Learning Academic
Resource Center
1-800-423-0563
http://www.wadsworth.com

International Headquarters
Thomson Learning
International Division
290 Harbor Drive, 2nd Floor
Stamford, CT 06902-7477
USA

UK/Europe/Middle East/South Africa
Thomson Learning
Berkshire House
168-173 High Holborn
London WC1V 7AA
United Kingdom

Asia
Thomson Learning
60 Albert Street, #15-01
Albert Complex
Singapore 189969

Canada
Nelson Thomson Learning
1120 Birchmount Road
Toronto, Ontario M1K 5G4
Canada

Printed in the United States of America
6 7 04

For permission to use material from this text, contact us by
Web: http://www.thomsonrights.com
Fax: 1-800-730-2215
Phone: 1-800-730-2214

Library of Congress Cataloging-in-Publication Data

Nutrition for health and health care/Eleanor N. Whitney . . . [et. al.].
    p. cm.
Includes bibliographical references and index.
ISBN 0-534-51552-5
1. Nutrition. 2. Nutritionally induced diseases. 3. Diet therapy.
I. Whitney, Eleanor Noss.

QP143.N8942001
612.3—dc21   00-036513

DEDICATION

To the memory of my parents, Edith Tyler Noss,

and Henry H.B. Noss, who supported me with love,

discipline, and pride.

ELLIE

To my son, Cory, a uniquely independent and determined

individual, who has filled my life with precious memories,

challenges, and pride. My love and best wishes go with you

as you enter college and begin your life as an adult.

MOM (CORKIE)

To my son, Zak, whose smiles, love, and determination fill

me with happiness and wonder, and whose future looks as

bright as the sun. Go for it baby.

MOM (LINDA)

To my son Lyle, whose smile makes my heart smile. The

early chapters of your life have been delightful, and I look

forward to seeing how your knowledge grows and your

talents unfold as you enter your college years.

MOM (SHARON)

ABOUT THE AUTHORS

ELEANOR NOSS WHITNEY, PH.D., R.D., received her B.A. in biology from Radcliffe College in 1960 and her Ph.D. in biology from Washington University in St. Louis in 1970. She is a registered dietitian and a member of the American Dietetic Association. Formerly on the faculty at Florida State University, she now devotes full time to research, writing, and consulting. Her earlier publications include articles in *Science, Genetics,* and other journals. Her textbooks include *Understanding Nutrition, Understanding Normal and Clinical Nutrition, Nutrition Concepts and Controversies, Life Span Nutrition: Conception through Life, Nutrition and Diet Therapy,* and *Essential Life Choices* for college students and *Making Life Choices* for high school students. Her most intense interests currently include energy conservation, solar energy uses, alternatively fueled vehicles, and ecosystem restoration.

CORINNE BALOG CATALDO, M.M.SC., R.D., C.N.S.D., received her B.S. in community health nutrition from Georgia State University in 1976 and her M.M.Sc. in clinical dietetics from Emory University in 1979. She has worked in private practice in Atlanta, as a clinical dietitian and metabolic support nutritionist at Georgia Baptist Medical Center in Atlanta, as a faculty member and dietetic internship coordinator at Emory University, and as a nutritionist for the Infant Formula Council. She has made numerous presentations, and in addition to this book, she has written a manual on tube feedings and the books *Understanding Normal and Clinical Nutrition, Nutrition and Diet Therapy,* and *Understanding Clinical Nutrition.* She maintains professional memberships in the American Dietetic Association, the American Society for Parenteral and Enteral Nutrition, the American Diabetes Association, and Dietitians in Nutrition Support.

LINDA KELLY DEBRUYNE, M.S., R.D., received her B.S. in 1980 and her M.S. in 1982 in nutrition and food science at Florida State University. She is a founding member of Nutrition and Health Associates, an information resource center in Tallahassee, Florida, where her specialty areas are life cycle nutrition and fitness. Her other publications include the textbooks *Nutrition and Diet Therapy, Life Span Nutrition: Conception through Life, Health: Making Life Choices,* and a multimedia CD-ROM called *Nutrition Interactive.* As a consultant for a group of Tallahassee pediatricians, she teaches infant nutrition classes to parents. She maintains a professional membership in the American Dietetic Association.

SHARON RADY ROLFES, M.S., R.D., received her B.S. in psychology and criminology in 1974 and her M.S. in nutrition and food science in 1982 from Florida State University. She is a founding member of Nutrition and Health Associates, an information resource center that maintains an ongoing bibliographic database that tracks research in over 1000 nutrition-related topics. Her other publications include the textbooks *Understanding Nutrition, Understanding Normal and Clinical Nutrition, Understanding Clinical Nutrition,* and *Life Span Nutrition: Conception through Life,* and a multimedia CD-ROM called *Nutrition Interactive.* In addition to writing, she also lectures at universities and at professional conferences and serves as a consultant for various educational projects. She maintains her registration as a dietitian and is a member of the American Dietetic Association.

CONTENTS IN BRIEF

PART ONE: NUTRITION FOR HEALTH

CHAPTER 1
Perspectives on Health and Nutrition   1
Finding the Truth about Nutrition   28

CHAPTER 2
Carbohydrates   31
Nutrition and Dental Health   50

CHAPTER 3
Lipids   52
Vegetarian Diets   71

CHAPTER 4
Proteins and Amino Acids   74
Supplements and Ergogenic Aids Athletes Use   90

CHAPTER 5
Digestion and Absorption   95
Food Safety 113

CHAPTER 6
Metabolism and Energy Balance   118
Inborn Errors of Metabolism   133

CHAPTER 7
Overweight, Underweight, and Weight Control   137
Eating Disorders   160

CHAPTER 8
The Vitamins 166
Vitamin Supplements   196

CHAPTER 9
Water and the Minerals   199
Replacing Fluid Losses: Sports Drinks and Rehydration
Formulas   226

CHAPTER 10
**Nutrition through the Life Span: Pregnancy
and Infancy**   229
Encouraging Successful Breastfeeding   258

CHAPTER 11
**Nutrition through the Life Span: Childhood
and Adolescence**   261
Childhood Obesity and the Early Development of Chronic
Diseases   284

CHAPTER 12
Nutrition through the Life Span: Later Adulthood   289
Food for Singles   309

PART TWO: NUTRITION FOR HEALTHCARE

CHAPTER 13
Nutrition and Nursing Assessments   312
Nutrition and Mental Health   338

CHAPTER 14
Nutrition Intervention   341
Community Nutrition   353

CHAPTER 15
Nutrition and Upper GI Tract Disorders   356
Helping People with Feeding Disabilities   374

CHAPTER 16
Nutrition and Lower GI Tract Disorders   379
Food and Foodservice in the Hospital   402

CHAPTER 17
Enteral Formulas   406
Nutrition and Cost-Conscious Health Care   424

CHAPTER 18
Parenteral Nutrition   429
Ethical Issues in Nutrition Care   444

CHAPTER 19
Nutrition in Severe Stress   448
Intestinal Immunity and Stress   461

CHAPTER 20
Nutrition and Diabetes Mellitus   464
Mastering Diabetes Control   492

CHAPTER 21
**Nutrition and Disorders of the Heart, Blood Vessels,
and Lungs**   495
Free Radicals, Antioxidants, and CHD   517

CHAPTER 22
Nutrition and Renal Diseases   519
Dialysis and Nutrition 539

CHAPTER 23
Nutrition and Liver Disorders   542
Nutrition and Alcohol Abuse   558

CHAPTER 24
Nutrition, Cancer, and HIV Infection   562
Alternative Therapies   585

APPENDIX A
Table of Food Composition

APPENDIX B
Canada: Recommendations, Choice System, and Labels

APPENDIX C
United States: Recommendations and Exchanges
World Health Organization: Recommendations

APPENDIX D
Nutrition Resources

APPENDIX E
Nutrition Assessment: Supplemental Information

APPENDIX F
Aids to Calculation

APPENDIX G
Enteral Formulas

APPENDIX H
Answers to Self Check Questions

GLOSSARY

INDEX

INSIDE FRONT COVER, LEFT
Dietary Reference Intakes (DRI)

INSIDE FRONT COVER, RIGHT
Recommended Dietary Allowances (RDA)
Tolerable Upper Intake Levels for Selected Nutrients

INSIDE BACK COVER, LEFT
Daily Value for Food Labels

INSIDE BACK COVER, RIGHT
Body Mass Index (BMI)

CONTENTS

PREFACE   xvii

PART ONE: NUTRITION FOR HEALTH

CHAPTER 1
Perspectives on Health and Nutrition   1
Food Choices   2
The Nutrients   8
Nutrition Standards and Guidelines   10
Dietary Reference Intakes   10
Dietary Guidelines   13
Fitness Guidelines   13
Diet-Planning Principles   14
Food Group Plans   17
Nutrition Surveys   20
Food Labels   21
Nutrition in Practice: Finding the Truth about Nutrition   28

CHAPTER 2
Carbohydrates   31
The Chemist's View of Carbohydrates   32
Monosaccharides (Single Sugars)   32
Disaccharides (Double Sugars)   33
Starch and Glycogen (Energy-Yielding Polysaccharides)   34
The Fibers   35
Health Effects of Sugars and Alternative Sweeteners   36
Sugars   36
Alternative Sweeteners: Sugar Alcohols   39
Alternative Sweeteners: Artificial Sweeteners   39
Health Effects of Complex Carbohydrates   41
Fibers   42
Carbohydrates: Food Sources and Recommendations   45
Energy Nutrients in Perspective   47
Nutrition in Practice: Nutrition and Dental Health   50

CHAPTER 3
Lipids   52
The Chemist's View of Lipids   54
Triglycerides   54
Fatty Acids   54
Phospholipids   57
Sterols   57
Fats and Health   58
Fats and Fatty Acids   58
Fat Substitutes   60
Fats in Foods   62
Finding the Fats in Foods   62
Cutting Fat Intake and Choosing Unsaturated Fats   64
Nutrition in Practice: Vegetarian Diets   71

CHAPTER 4
Proteins and Amino Acids   74
The Chemist's View of Proteins   75
The Structure of Proteins   75
Essential Amino Acids   76
Proteins in the Body   77
Protein and Health   82
Protein-Energy Malnutrition   82
Protein Excess   84
Protein in Foods   86
Nutrition in Practice: Supplements and Ergogenic Aids Athletes Use   90

CHAPTER 5
Digestion and Absorption   95
Anatomy of the Digestive Tract   96
The Digestive Organs   96
The Involuntary Muscles and the Glands   97
The Process of Digestion   102
Digestion in the Mouth   102
Digestion in the Stomach   103
Digestion in the Small and Large Intestines   103
The Absorptive System   105
The Small Intestine   105
Release of Absorbed Nutrients   107
Transport of Nutrients   107
The Vascular System   108
The Lymphatic System   109
Transport of Lipids: Lipoproteins   109
The System at Its Best   111
Nutrition in Practice: Food Safety   113

CHAPTER 6
Metabolism and Energy Balance   118
The Organs and Their Metabolic Roles   119
The Principal Organs   119
Energy for Metabolic Work   120
The Body's Energy Metabolism   121
Building Up Body Compounds   121
Breaking Down Nutrients for Energy   121
The Body's Energy Budget   123
The Economics of Feasting   123
The Economics of Fasting   125
Energy Balance   127
Nutrition in Practice: Inborn Errors of Metabolism   133

CHAPTER 7
Overweight, Underweight, and Weight Control   137
Body Weight and Body Composition   138
Defining Healthy Body Weight   138
Body Composition   140
How Much Is Too Much Body Fat?   141
Causes of Obesity   142

Risks of Overweight and Obesity   144
Aggressive Treatments of Obesity   146
 Obesity Drugs   146
 Very-Low-kCalorie Diets   147
 Surgery   148
Reasonable Strategies for Weight Loss   148
 Diet   149
 Physical Activity   151
 Behavior and Attitude   153
Underweight   155
 Health Risks of Underweight   155
 Strategies for Weight Gain   155
Nutrition in Practice: Eating Disorders   160

CHAPTER 8
The Vitamins   166
The Fat-Soluble Vitamins   167
 Vitamin A and Beta-Carotene   168
 Vitamin D   173
 Vitamin E   175
 Vitamin K   176
The Water-Soluble Vitamins   177
 The B Vitamins   178
 Thiamin   181
 Riboflavin   181
 Niacin   182
 Pantothenic Acid and Biotin   182
 Vitamin B_6 183
 Folate   184
 Vitamin B_{12} 184
 Non-B Vitamins   186
 Vitamin C   187
**Nutrition in Practice:
Vitamin Supplements
196**

CHAPTER 9
**Water and the Minerals
199**
Water and Body Fluids   200
 Water Balance   200
 Fluid and Electrolyte Balance   201
 Acid-Base Balance   202
The Major Minerals   203
 Sodium   203
 Chloride   204
 Potassium   205
 Calcium   205
 Phosphorus   209
 Magnesium   210
 Sulfur   210
The Trace Minerals   211
 Iron   211
 Zinc   216
 Selenium   218
 Iodine   218
 Copper   219
 Manganese   219
 Fluoride   220

 Chromium   221
 Other Trace Minerals   221
**Nutrition in Practice: Replacing
Fluid Losses: Sports Drinks and
Rehydration Formulas   226**

CHAPTER 10
**Nutrition through the Life Span:
Pregnancy and Infancy   229**
Pregnancy: The Impact of Nutrition on the Future   230
 Preparing for Pregnancy   230
 Nutrient Needs during Pregnancy   230
 Weight Gain   236
 Physical Activity   237
 Problems in Pregnancy   237
 Practices to Avoid   239
 Adolescent Pregnancy   241
Breastfeeding   241
 Nutrient Needs during Lactation   242
 Contraindications to Breastfeeding   243
Nutrition of the Infant   244
 Nutrient Needs during Infancy   244
 Breast Milk   246
 Infant Formula   247
 The Transition to Cow's Milk   249
 Introducing First Foods   249
 Looking Ahead   252
 Mealtimes   253
**Nutrition in Practice: Encouraging
Successful Breastfeeding   258**

CHAPTER 11
**Nutrition through the Life Span:
Childhood and Adolescence   261**
Early and Middle Childhood   262
 Energy and Nutrient Needs   262
 Malnutrition in Children   264
 Lead Poisoning in Children   266
 Food Allergies   269
 Hyperactivity   270
Food Choices and Eating Habits of Children   271
 Mealtimes at Home   272
 Nutrition at School   273
The Teen Years   275
 Growth and Development during Adolescence   276
 Energy and Nutrient Needs   276
 Food Choices and Health Habits   277
 Problems Adolescents Face   278
**Nutrition in Practice: Childhood Obesity and the Early
Development of Chronic Diseases   284**

CHAPTER 12
**Nutrition through the Life Span:
Later Adulthood   289**
Nutrition and Longevity   290
 Slowing the Aging Process   290
 Nutrition and Disease
 Prevention   292
Nutrition-Related Concerns during Late
Adulthood   294

Cataracts   294
Arthritis   294
The Aging Brain   295
Energy and Nutrient Needs during Late Adulthood   297
Energy and Energy Nutrients   298
Water   299
Vitamins and Minerals   299
Nutrient Supplements for Older Adults   301
The Effects of Medications on Nutrients   302
Food Choices and Eating Habits of Older Adults   303
Nutrition in Practice: Food for Singles   309

PART TWO: NUTRITION FOR HEALTHCARE

CHAPTER 13
Nutrition and Nursing Assessments   312
Nutrition in Health Care   313
Responsibility for Nutrition Care   313
Illness and Nutrition Status   314
Diet-Medication Interactions   316
Identifying Risk for Malnutrition   322
Health History   323
Health Problems and Treatments   323
Methods for Gathering Food Intake Data   325
Physical Examinations   327
Measures of Growth and Development   327
Physical Signs of Malnutrition   331
Biochemical Analysis   332
Follow-Up Care   335
Nutrition in Practice: Nutrition and Mental Health   338

CHAPTER 14
Nutrition Intervention   341
Care Plans   342
Nutrition-Related Nursing Diagnoses   342
Outcome Identification and Planning   343
Special Diets   345
Standard and Modified Diets   345
Routine Progressive Diets   346
Medical Records   350
Types of Medical Records   350
The Medical Record and Nutrition Care   350
Nutrition in Practice: Community Nutrition   353

CHAPTER 15
Nutrition and Upper GI Tract Disorders   356
Disorders of the Mouth and Esophagus   357
Difficulties Chewing   357
Difficulties Swallowing   359
Disorders of the Stomach   360
Indigestion and Reflux Esophagitis   360
Nausea and Vomiting   363
Gastritis   364

Ulcers   365
Gastric Surgery   365
Nutrition in Practice: Helping People with Feeding Disabilities   374

CHAPTER 16
Nutrition and Lower GI Tract Disorders   379
Common Problems of the Lower Intestine   380
Constipation   380
Diarrhea and Dehydration   381
Irritable Bowel Syndrome   383
Lactose Intolerance   383
Fat Malabsorption   385
Medical Nutrition Therapy   386
Pancreatitis   388
Cystic Fibrosis   389
Inflammatory Bowel Diseases   390
Bacterial Overgrowth   391
Short-Bowel Syndrome   392
Celiac Disease   393
Disorders of the Large Intestine   394
Diverticular Disease of the Colon   394
Colostomies and Ileostomies   396
Nutrition in Practice: Food and Foodservice in the Hospital   402

CHAPTER 17
Enteral Formulas   406
Enteral Formulas: What Are They?   407
Types of Formulas   407
Distinguishing Characteristics   408
Enteral Formulas: Who Needs What?   409
Oral Feedings   409
Tube Feedings   409
Tube Feedings: How Are They Given?   414
Safe Handling   414
Initiating and Progressing a Tube Feeding   415
Delivering Medications through Feeding Tubes   417
Addressing Tube Feeding Complications   418
From Tube Feedings to Table Foods   421
Nutrition in Practice: Nutrition and Cost-Conscious Health Care   424

CHAPTER 18
Parenteral Nutrition   429
Intravenous Solutions: What Are They?   430
Intravenous Nutrients   430
Intravenous Solutions   432
Intravenous Nutrition: Who Needs What?   434
Intravenous Solutions: How Are They Given?   435
Insertion and Care of IV Catheters   435
Administration of Complete Nutrient Solutions   436
From Parenteral to Enteral Feedings   438
Specialized Nutrition Support at Home   440
The Basics of Home Programs   440
How Home Enteral and Parenteral Nutrition Programs Work   441
Nutrition in Practice: Ethical Issues in Nutrition Care   444

CHAPTER 19
Nutrition in Severe Stress   448
The Body's Responses to Stress   449
Nutrition Support following Acute Stress   452
   Nutrient Needs   453
   Delivery of Nutrients following Stress   456
Nutrition in Practice: Intestinal Immunity and Stress   461

CHAPTER 20
Nutrition and Diabetes Mellitus   464
What Is Diabetes Mellitus?   465
   Types of Diabetes   465
   Acute Complications of Diabetes   467
   Chronic Complications of Diabetes   469
Treatment of Diabetes Mellitus   470
   Medical Nutrition Therapy for Diabetes   471
   Meal-Planning Strategies   473
   Physical Activity   480
   Drug Therapy for Diabetes   482
   Monitoring Diabetes Management   484
Diabetes Management throughout Life   484
   Diabetes Management in Childhood   485
   Diabetes Management in Pregnancy   486
   Diabetes Management in Later Life   488
Nutrition in Practice: Mastering Diabetes Control   492

CHAPTER 21
Nutrition and Disorders of the Heart, Blood Vessels, and Lungs   495
Atherosclerosis and Hypertension   496
   Atherosclerosis   496
   Hypertension   498
   Risk Factors for Coronary Heart Disease (CHD)   499
Prevention and Treatment of CHD   500
   Medical Nutrition Therapy   502
   Drug Therapy   506
Heart Failure and Strokes   508
   Heart Attacks   508
   Chronic Heart Failure (CHF)   509
   Strokes   510
Disorders of the Lungs   511
   Acute Respiratory Failure   511
   Chronic Obstructive Pulmonary Disease (COPD)   512
Nutrition in Practice: Free Radicals, Antioxidants, and CHD   517

CHAPTER 22
Nutrition and Renal Diseases   519
Kidney Stones   521
   Consequences of Kidney Stones   521
   Prevention and Treatment of Kidney Stones   522
The Nephrotic Syndrome   523
   Consequences of the Nephrotic Syndrome   523
   Treatment of the Nephrotic Syndrome   524
Acute Renal Failure   525
   Consequences of Acute Renal Failure   526
   Treatment of Acute Renal Failure   527
Chronic Renal Failure   528
   Consequences of Chronic Renal Failure   529
   Treatment of Chronic Renal Failure   530

   Kidney Transplants   534
Nutrition in Practice: Dialysis and Nutrition   539

CHAPTER 23
Nutrition and Liver Disorders   542
Fatty Liver and Hepatitis   543
   Fatty Liver   543
   Hepatitis   544
Cirrhosis   544
   Consequences of Cirrhosis   545
   Treatment of Cirrhosis   548
   Medical Nutrition Therapy   549
Liver Transplants   553
Nutrition in Practice: Nutrition and Alcohol Abuse   558

CHAPTER 24
Nutrition, Cancer, and HIV Infection   562
Cancer   563
   How Cancers Develop   563
   Consequences of Cancer   566
   Treatments for Cancer   568
   Medical Nutrition Therapy   570
HIV Infection   575
   Consequences of HIV Infection   575
   Treatments for HIV Infection   578
   Medical Nutrition Therapy   580
Nutrition in Practice: Alternative Therapies   585

APPENDIX A
Table of Food Composition

APPENDIX B
Canada: Recommendations, Choice System, and Labels
RNI
Choice System for Meal Planning
Food Labels
Canada's Food Guide

APPENDIX C
United States: Recommendations and Exchanges
RDA
Exchange Lists for Meal Planning
Diet and Health Recommendations
Nutrition Recommendations from WHO

APPENDIX D
Nutrition Resources

APPENDIX E
Nutrition Assessment: Supplemental Information
Drug History: Nutrition Interactions
Growth Charts and Measures of Body Composition
Laboratory Tests of Nutrition Status

APPENDIX F
Aids to Calculation
Conversion Factors
Percentages
Ratios
Weights and Measures

APPENDIX G
Enteral Formulas

APPENDIX H
Answers to Self Check Questions

GLOSSARY

INDEX

INSIDE FRONT COVER, LEFT
Dietary Reference Intakes (DRI)

INSIDE FRONT COVER, RIGHT
Recommended Dietary Allowances (RDA)
Tolerable Upper Intake Levels for Selected Nutrients

INSIDE BACK COVERS
Daily Values for Food Labels
Body Mass Index (BMI)

CASE STUDIES APPEAR THROUGHOUT THE CHAPTERS

CHAPTER 10:
Pregnant Woman with Weight Problem   248

CHAPTER 11:
Boy with Disruptive Behavior   271

CHAPTER 12:
Elderly Man with a Poor Diet   304

CHAPTER 13:
**Risk Factors for Malnutrition in a Computer Scientist
Following a Car Accident   336**

CHAPTER 14:
Computer Scientist with Car Accident Injuries   345

CHAPTER 15:
Accountant with Reflux Esophagitis   363

CHAPTER 16:
College Student with Crohn's Disease   391

CHAPTER 17:
Graphic Designer Requiring Enteral Nutrition   421

CHAPTER 18:
Mail Carrier Requiring Parenteral Nutrition   439

CHAPTER 19:
Journalist with a Third-Degree Burn   457

CHAPTER 20:
Child with Type 1 Diabetes   486

CHAPTER 21:
History Professor with Cardiovascular Disease   511

CHAPTER 22:
Store Manager with Acute Renal Failure   528

CHAPTER 23:
Carpenter with Cirrhosis   554

CHAPTER 24:
Travel Agent with HIV Infection   581

"HOW TO" FEATURES

CHAPTER 1:
Calculate the Energy a Food Provides   9

CHAPTER 3:
Lower Fat Intake—By Food Group   64

CHAPTER 7:
**Rate Sound and Unsound Weight-Loss Schemes
and Diets   149**
Change Behaviors   154
Combat Eating Disorders   162
Establish Healthy Eating Patterns   164

CHAPTER 9:
Add Calcium to Daily Meals   209
Add Iron to Daily Meals   216

CHAPTER 11:
Protect Children from Lead Exposure   269

CHAPTER 13:
Manage Diet-Medication Interactions   319
Measure Weight   328
Measure Length and Height   329
Estimate % IBW and % UBW   330

CHAPTER 14:
Help Clients Improve Their Food Intakes   351

CHAPTER 15:
Improve Acceptance of Pureed Foods   358
Prevent Indigestion   361
Prevent and Treat Reflux Esophagitis   362
Minimize Nausea   364
Adjust Meals to Prevent the Dumping Syndrome   369

CHAPTER 16:
Add Fiber to the Diet   382
Improve Acceptance of Fat-Restricted Diets   388

CHAPTER 17:
Help Clients Cope with Tube Feedings   412
Plan a Tube Feeding Schedule   416

CHAPTER 18:
Read IV Solution Abbreviations   431
**Calculate Energy and Energy Nutrient Content of IV
Solutions Expressed as Percentages   433**

CHAPTER 19:
**Estimate Energy and Protein Needs Following
Severe Stress   455**

CHAPTER 20:
Plan a Diet for Diabetes Using Exchange Lists   477
Help Clients Count Carbohydrates   481

CHAPTER 21:
Help Clients Implement the DASH Diet   504

CHAPTER 22:
Help Clients Comply with a Renal Diet   533

CHAPTER 23:
Help the Person with Cirrhosis Eat Enough Food   553

CHAPTER 24:
Help Clients Handle Food-Related Problems   572

PREFACE

Health care professionals recognize that optimal health care depends on a thorough understanding of a client's physical, emotional, intellectual, social, and spiritual needs. As the professional with a broad understanding of medical needs and the most consistent and frequent contact with clients, nurses have a unique opportunity to identify these needs. This second edition of *Nutrition for Health and Health Care* aims to provide future nurses with the basic nutrition facts, an understanding of the role of nutrition in disease prevention and treatment, and practical information for using nutrition concepts, addressing nutrition concerns, and incorporating nutrition into care plans. Although primarily written for nursing students, this book can serve as a valuable resource for students of all health-related professions, including nursing assistants, dietary technicians, health educators, health science majors, dietitians, and physicians.

Nutrition is a young and rapidly expanding science, with many questions remaining to be answered and new "facts" surfacing every day. Each chapter of this book has been substantially revised since the first edition to reflect these many changes. To help readers keep abreast of changes that will occur in the future, one of the missions of this text is to teach them how to ascertain and cautiously interpret new information.

The book begins by introducing the basics of nutrition and shows how nutrition supports health. It then describes how nutrient needs change throughout the life cycle. The second half of the book begins with a look at how nurses can use information from nursing assessments to identify and address clients' nutrient needs. The remaining chapters examine medical nutrition therapy and its role in the prevention and treatment of medical conditions.

Within each chapter, definitions and notes in the margins clarify nutrition information, remind readers of previously defined terms, and provide cross-references. "How to" skill boxes help readers work through calculations or give practical suggestions for applying nutrition advice. Case Studies in the later chapters guide readers in applying information to everyday situations. The Self Check at the end of each chapter provides questions to help review chapter information. Clinical Applications encourage readers to practice mathematical calculations, synthesize information from previous chapters, or understand the impact of nutrition care on health care professionals or their clients. Later chapters also include Nutrition Assessment Checklists, which remind readers of assessment parameters particularly relevant to specific stages of the life cycle or groups of disorders.

New to this edition are life cycle and fitness icons, learning links, web site references, nursing diagnoses, and diet-medication interaction boxes. Life cycle and fitness icons highlight age-related and fitness-related information, respectively. Learning links help readers understand the relationships between nutrition concepts by tying together information from various parts of the text. Notations of web site addresses at the end of each chapter offer readers Internet resources for finding additional information on the topic discussed in the chapter. Nursing diagnoses with nutrition implications are provided in the later chapters to help nursing students correlate nutrition care to nursing care. The diet-medication interaction boxes in the later chapters describe interactions relevant to medications described in each chapter.

At the end of each chapter, a Nutrition in Practice section explores current topics, advanced subjects, or specialty areas such as ethical issues in nutrition care. This edition includes new Nutrition in Practice sections examining supplements and ergogenic aids, food safety, fluid replacement, childhood obesity and chronic diseases, community nutrition, nutrition and cost-conscious health care, diabetes control, free radicals and antioxidants, dialysis, and alternative therapies.

The appendixes support the book with a wealth of information on nutrient contents of foods and enteral formulas, Canadian nutrient recommendations and food choices, U.S. nutrient intake recommendations and the exchange system, supplemental information about nutrition assessments, aids of calculations, nutrition resources, and answers to Self Check questions.

We hope that as you discover the many fascinating aspects of nutrition science, you will enthusiastically apply the concepts in both your personal and professional life. For nutrition updates and other resources, we invite you to visit our Web site: http://www.wadsworth.com/nutrition.

ELLIE WHITNEY
CORINNE CATALDO
LINDA DeBRUYNE
SHARON ROLFES

ACKNOWLEDGMENTS

Among the most difficult words to write are those that express the depth of our gratitude to the many dedicated people whose efforts have made this book possible. A special note of thanks to Dale Hugo, R.Ph., for answering many questions about medications and how they are delivered in hospitals.

Special thanks to Kellie Hatcher for her word processing assistance. Thanks also to JooHee Lee for her help in securing permissions, Pat Lewis for copyediting our manuscript, and Elizabeth Hand and Bob Geltz for creating an extremely accurate and practical food composition appendix. We are indebted to our editorial team, Peter Marshall and Laura Graham, and to our production team, Sandra Craig and Dusty Friedman, for seeing this project through from start to finish. We would also like to acknowledge Becky Tollerson for her marketing efforts. To the many others involved in designing, indexing, typesetting, dummying, and marketing, we offer our thanks. We are especially grateful to our associates, family, and friends for their continued encouragement and support and to our reviewers who consistently offer excellent suggestions for improving the text.

REVIEWERS

Ethel Avery, Trentholm State Community College

Beth Ann Batturs, Anne Arundel Community College

Dea Baxter, Georgia State University

Mary Beck, Owens Community College

Roseann Cutroni, St. Elizabeth Hospital

Gloria Ferrito, Maric College

Cindi Fuller, University of North Carolina—Greensboro

Sandra Godwin, Tennessee State University

Christina Hasemann, Broome Community College

Carolyn Hollingshead, University of Utah

Jacklyn Lee, California State University—Long Beach

Samantha Logan, University of Massachusetts—Amherst

Martha Ludemann, University of Connecticut

Marge McNairn, Butte College

Johnnie Nichols, Panola College

Linda Parker, University of Miami

Evelyn L. Parsons, Anne Arundel Community College

Carol Powell, Indiana Wesleyan University

Helen Reid, Southwest Missouri State University

Tammy Sakanashi, Santa Rosa Junior College

Sue Smith, Iowa Western Community College

Tina Speer, Southwest Technical Center

Lynn Thomas, University of South Carolina

NUTRITION
FOR
HEALTH
AND
HEALTH
CARE

SECOND
EDITION

CHAPTER 1

PERSPECTIVES ON HEALTH AND NUTRITION

CONTENTS

Food Choices

The Nutrients

Nutrition Standards and Guidelines

Food Labels

Nutrition in Practice: Finding the Truth about Nutrition

Nutrition is only one of the many factors that influence people's food choices.

health: a range of states with physical, mental, emotional, spiritual, and social components. At a minimum, *health* means freedom from physical disease, mental disturbances, emotional distress, spiritual discontent, social maladjustment, and other negative states. At a maximum, *health* means "wellness."

wellness: maximum well-being; the top range of health states; the goal of the person who strives toward realizing his or her full potential physically, mentally, emotionally, spiritually, and socially.

nutrition: the science of foods and the nutrients and other substances they contain, and of their ingestion, digestion, absorption, transport, metabolism, interaction, storage, and excretion. A broader definition includes the study of the environment and of human behavior as it relates to these processes.

phytochemicals: nonnutrient compounds in plant-derived foods that have biological activity in the body.

Every day, several times a day, you make choices that will either improve your **health** or harm it. Each choice may influence your health only a little, but when these choices are repeated over years and decades, their effects become significant.

The choices people make each day affect not only their physical health, but also their **wellness**—all the characteristics that make a person strong, confident, and able to function well with family, friends, and others. People who consistently make poor lifestyle choices, on a daily basis, increase their risks of developing diseases. Figure 1-1 shows how a person's health can fall anywhere along a continuum, from maximum wellness on the one end to total failure to function (death) on the other.

When health care professionals take responsibility for their own health by making daily choices and practicing behaviors that enhance their well-being, they prepare themselves physically, mentally, and emotionally to meet the arduous demands of their profession. Health care professionals, however, have a responsibility to their clients as well as to themselves. They have unique opportunities to make their clients aware of the benefits of positive health choices and behaviors, to show them how to change their behaviors and make daily choices to enhance their own health, and to serve as role models for those behaviors.

This text focuses on how nutrition choices affect health and disease. The early chapters introduce the basics of nutrition to support good health. The later chapters emphasize clinical nutrition, describing how diseases, their symptoms, and their treatments influence nutrient needs and how meeting those needs affects recovery.

■■ Food Choices

Sound **nutrition** throughout life does not ensure good health and long life, but it can certainly help to tip the balance in their favor. Nevertheless, most people choose foods for reasons other than their nourishing value. Even people who claim to choose foods primarily for the sake of health or nutrition will admit that other factors also influence their food choices. Because food choices become an integral part of people's lifestyles, they sometimes find it difficult to change eating habits. Health care professionals who help clients make diet changes must understand the dynamics of food choices, because people will alter their eating habits only if their preferences are honored.

■■ **PREFERENCE** Why do people like certain foods? One reason, of course, is their preference for certain tastes. Some tastes are widely liked, such as the sweetness of sugar and the zest of salt. Research suggests that genetics may influence people's taste preferences, a finding that may eventually have implications for clinical nutrition.[1] For example, sensitivity to bitter taste is an inheritable trait. People born with great sensitivity to bitter tastes tend to avoid foods with bitter flavors such as broccoli, cabbage, brussels sprouts, spinach, and grapefruit juice.[2] These foods, as well as many other fruits and vegetables, contain compounds called **phytochemicals** that may reduce the risk of cancer.[3] Thus, the role that genetics may play in food selection is gaining importance in cancer research.

■■ **HABIT** Sometimes habit dictates people's food choices. People eat a sandwich for lunch or drink orange juice at breakfast simply because they have always done so.

■■ **ASSOCIATIONS** People also like foods with happy associations—foods eaten in the midst of warm family gatherings on traditional holidays or

Figure 1-1
THE HEALTH LINE
No matter how well you maintain your health today, you may still be able to improve tomorrow. Likewise, a person who is well today can slip by failing to maintain health-promoting habits.

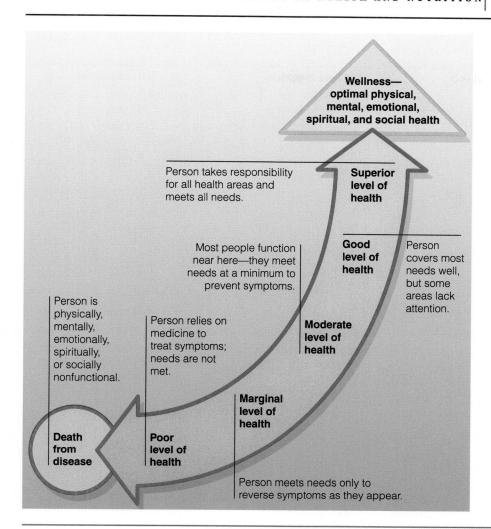

Wellness—optimal physical, mental, emotional, spiritual, and social health

Person takes responsibility for all health areas and meets all needs.

Superior level of health

Most people function near here—they meet needs at a minimum to prevent symptoms.

Good level of health

Person covers most needs well, but some areas lack attention.

Person is physically, mentally, emotionally, spiritually, or socially nonfunctional.

Person relies on medicine to treat symptoms; needs are not met.

Moderate level of health

Marginal level of health

Death from disease

Poor level of health

Person meets needs only to reverse symptoms as they appear.

given to them as children by someone who loved them. By the same token, people can attach intense and unalterable dislikes to foods that they ate when they were sick or that were forced on them when they weren't hungry.

■■ **ETHNIC HERITAGE AND TRADITION** Every country, and every region of a country, has its own typical foods and ways of combining them into meals. The foodways of North America reflect the many different cultural and ethnic backgrounds of its inhabitants. Many foods with foreign origins are familiar items on North American menus: tacos, egg rolls, lasagna, and gyros, to name a few. Still others, such as spaghetti and croissants, are almost staples in the "American diet." North American regional cuisines like Cajun and TexMex blend the traditions of several cultures. Table 1-1 (pp. 4–6) presents profiles of selected **ethnic diets,** together with comments on their nutritional merits and drawbacks.

■■ **VALUES** People's values, environmental ethics, religious beliefs, and political views also influence their food choices. By choosing to eat some foods or avoid others, people make statements about themselves that reflect their values. For example, people may select only foods that come in containers that can be reused or recycled. Some people choose only brands of canned tuna fish that state "dolphin safe" on the label, meaning that dolphins were not killed when the tuna were netted. Religion also influences many people's food choices. Jewish law sets forth an extensive set of dietary rules. Many Christians

Ethnic meals and family gatherings nourish the spirit as well as the body.

ethnic diets: foodways and cuisines typical of national origins, races, cultural heritages, or geographic locations.

Table 1-1

CHARACTERISTICS OF SELECTED ETHNIC DIETS

Staple Foods	Strengths of the Diet	Weaknesses of the Diet
Hispanic Americans from Cuba, Haiti, Puerto Rico		
Include: • Steamed white rice; wheat breads. • Starchy vegetables (beans, cassavas, yuccas); plantains; green peppers; tomatoes; garlic. • Dried, salted fish; chicken; pork. • Lard; olive oil; sugar; jams and jellies; sweet pastries; sugared fruit juices; coffee. *Exclude:* • Green, leafy vegetables. • Milk as a beverage for adults. • Fish other than dried and salted.	Provides adequate protein, many other nutrients, and fiber.	May provide too much fat, especially animal fat; may lack calcium.
Hispanic Americans from Mexico, Central America		
Include: • Steamed rice, corn products such as tortillas. • Many varieties of beans; chili peppers; tomatoes; mangoes; prickly pear fruit; potatoes. • Meat and sausages; fish; poultry; eggs. • Lard; chocolate and coffee drinks; cakes; pastries. *Exclude:* • Green, leafy vegetables; yellow vegetables. • Milk as a beverage for adults.	Most nutrients can be obtained.	Is high in kcalories and fat, especially saturated fat, and high in sugar.
Black Americans from West Indies, Central or South America and Recent African Immigrants		
Include: • Millet, corn, wheat, rice, or barley. • Starchy roots such as cassavas, yams; plantains; bananas; coconuts; peanuts; fresh fruits; hot peppers; tomatoes; onions; okra. • Palm oil; fruit wine; tea; coffee; honey; molasses. *Exclude:* • Milk and milk products (meat and fish limited use).	Is low in fat and salt; is high in fiber.	Is low in calcium, iron, and vitamin B_{12}; is potentially low in protein, depending on availability of foods.
Southern Black Americans from West Africa (Many Generations in United States)		
Include: • Rice; hominy grits; biscuits; cornmeal and cornbread. • Legumes; potatoes; onions; tomatoes; hot peppers; green, leafy vegetables; okra; sweet potatoes; squashes; corn; cabbage; melons; peaches. • Smoked pork; meats and poultry; fish; thick stews. • Pecans; butter, shortening, and lard; sugar; bread puddings, pies, and sweets. *Exclude:* • Milk and milk products. • Yeast breads.	Provides ample nutrients of meat.	Provides excess protein; is high in kcalories; provides excess fat, especially saturated fat; is high in salt; is low in calcium.

Staple Foods	Strengths of the Diet	Weaknesses of the Diet
Chinese Americans from China (Diets Sometimes Vary With Region)		
Include: • Rice and rice gruel; wheat noodles; soybean noodles. • Corn; vegetables from the cabbage family; squashes; cucumbers; eggplant; leafy vegetables; various shoots (bamboo, mung, and soy); sweet potatoes; radishes; onions; peas and pods; mushrooms; roots; local vegetables; pickled vegetables; sea vegetables; plums; peaches; tangerines; kumquats; other citrus fruits; litchis; longans; mangoes; papayas; pomegranates. • Soybean products (tofu and soy milk); meat; fish with bones; poultry; seafood. • Soup or tea as beverage; soy sauce; sugar. *Exclude:* • Milk and most milk products.	Is low in fat; is high in fiber and many nutrients.	Depending on availability of protein-rich foods, protein and iron may be low; high in salt.
Japanese Americans from Japan		
Include: • Rice. • Vegetables (including pickled and sea); fruits; salads. • Soy (miso, tofu, bean paste); fish with bones; seafood. • Sugars as seasoning; ginseng; soy sauce. Exclude: • Milk and milk products.	Provides abundant nutrients with little fat.	Is high in salt.
Korean Americans from South Korea		
Include: • Rice; noodles. • Leafy vegetables; kimchi (hot pickled cabbage); sea vegetables; hot peppers; seasonal fruits; mushrooms. • Small fish with bones; grilled beef; chicken; squid, octopus, and lobster; mussels; eggs. • Lard and vegetable fat for frying; sesame oil; nuts and seeds; ginger; sugar as seasoning. *Exclude:* • Milk and milk products.	Provides adequate protein.	Is high in fat; monotonous in winter (kimchi is served at each meal, to the exclusion of other vegetables); without the traditional small fish with bones, calcium can be lacking.
Vietnamese Americans from Vietnam		
Include: • Rice, rice noodles; french bread and croissants. • Hot peppers; curries of asparagus and potatoes; salads; tropical fruits and vegetables; lemons and limes. • Small portions of poultry; eggs; fish pâtés; nuoc nam (a strong, fermented fish sauce). • Sweets, candies, sweetened drinks; coffee; tea; butter. *Exclude:* • Milk and milk products.	Provides adequate protein and vitamins.	Can be low in iron or calcium.

continued

Table 1-1 (continued)

Staple Foods	Strengths of the Diet	Weaknesses of the Diet
Native Americans		
Include:	Varies with region; may provide adequate protein and fiber.	Varies with region; may be low in calcium.
• *Southeast:* corn; cornmeal; coontie (flour from a palmlike plant); fried breads; pumpkins; squashes; papayas; alligator, snake, wild hog, duck, fish, and shellfish.		
• *Northeast:* blueberries; cranberries; beans; corn; pumpkins; fish; lobster; wild game; maple syrup.		
• *Midwest:* bison; beans; corn; melons; squashes; tomatoes.		
• *Southwest:* corn (many varieties); beans; squash; pumpkins; chili peppers; melons; pinenuts; cactus.		
• *Northwest:* salmon; caviar; other fish; otter; seal; elk; whale; bear; other game; wild fruits, nuts, and greens.		
Exclude:		
• Milk and milk products.		
Italian Americans from Italy		
Include:	Provides adequate protein and most nutrients.	Varies with region; can be high in fat.
• *Northern Italy:* egg-based, ribbon-shaped pastas; cheese; cream; butter; meat; eggs.		
• *Southern Italy:* wheat pastas; artichokes, eggplants, peppers, and tomatoes; beans; olive oil.		

forgo meat during Lent, the period prior to Easter. Other faiths prohibit some dietary practices and promote others. Diet planners can foster sound nutrition practices only if they respect and honor each person's values.

SOCIAL PRESSURE Social pressure is another powerful influence on people's food choices. Such pressure operates in all circles and across all cultural lines. It is often considered rude to refuse food or drink being shared by a group or offered by a host. Often you are accepted as a member of a social gathering only when you "break bread" with the others.

EMOTIONAL STATE People may eat in response to emotional stimuli—for example, to relieve boredom or depression or to calm anxiety. A lonely person may choose to eat rather than to call a friend and risk rejection. A person who has returned home from an exciting evening out may unwind with a late-night snack. Eating in response to emotions can easily lead to overeating and obesity, but may be appropriate at times. For example, sharing food at times of bereavement serves both the givers' need to provide comfort and the receiver's need to be cared for and to interact with others.

In contrast, some people respond to emotions by not eating at all or by eating very little. A depressed person may simply have no appetite for food. A person overcome with grief can easily lose all interest in food (see the Nutrition in Practice in Chapter 13). As long as a person's appetite or interest in food returns quickly, little harm is done.

AVAILABILITY, CONVENIENCE, AND ECONOMY The influence of these factors on people's food selections is clear. You cannot eat foods if they are not available, if you cannot get to the grocery store, if you do not have the time or skill to prepare them, or if you cannot afford them. Convenience plays a major role in many people's food selections today, but along with convenience, consumers want great taste and quality—foods that are fast, delicious, and nutritious.[4] The demand for foods that are ready to eat or can be easily prepared in a microwave oven demonstrates this influence.

AGE Age influences people's food choices. Infants, for example, depend on others to choose foods for them. Older children also rely on others, but become more active in selecting foods that taste sweet and are familiar to them and rejecting those whose taste or texture they dislike. In contrast, the links between taste preferences and food choices in adults are less direct than in children.[5] Adults often choose foods based on health concerns such as body weight. Indeed, adults may avoid sweet or familiar foods because of such concerns.

OCCUPATION Some people have jobs that keep them away from home for days at a time, or require them to conduct business in restaurants or at conventions, or involve hectic schedules that allow little or no time for meals at home. For these people, the kinds of restaurants available to them, and the cost of eating out so often, may limit food choices.

BODY IMAGE Sometimes people select foods that they associate with ideals of body image. The fashion and movie industries, not the medical community, have defined what people believe to be the ideal body—sometimes an excessively thin body for women or an excessively muscular body for men. Both men and women seek "beautiful bodies," and in doing so, they select or avoid foods that they believe will improve or impair their physical appearance. Such intentions are rational when based on sound nutrition and fitness knowledge, but when based on faddism or carried to extremes, they undermine good health.

MEDICAL CONDITIONS Sometimes medical conditions and their treatments (including medications) limit the foods a person can select. For example, a person with heart disease might need to adopt a low-fat diet. The chemotherapy needed to treat cancer can interfere with a person's appetite or limit food choices by causing vomiting. Allergy to certain foods can also limit choices. The second half of this text discusses how diet can be modified to accommodate different medical conditions.

HEALTH AND NUTRITION Consumers today cite nutrition as a primary concern in making food choices, yet the foods they choose do not always reflect this concern. Since sound nutrition promotes health and longevity, it is in every person's own best interest to know more about the nutrients in food and how to apply that knowledge when choosing foods.

You have arrived at your first Learning Link. Learning Links help you understand how the information you have just read applies to later chapters and, in turn, to your career in health care. In this case, understanding the many factors that affect a client's food choices is important in developing nutrition care plans (described in Chapter 14).

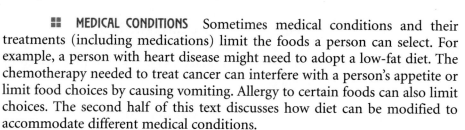

LEARNING LINK

Diet histories can provide clues about a person's food choices, eating habits, and lifestyle. Armed with this information, health care professionals can provide practical suggestions and offer advice tailored to the individual. For example, a dietitian or nurse who notices that the client never drinks milk can offer suggestions for appropriate nonmilk sources of calcium. Chapter 13 provides more details.

■■ The Nutrients

You are a collection of molecules that move. All these moving parts are arranged in patterns of extraordinary complexity and order—cells, tissues, and organs. Although the arrangement remains constant, the parts are continually changing, using **nutrients** and energy derived from nutrients.

Almost any food you eat is composed of dozens or even hundreds of different kinds of materials. Spinach, for example, is composed mostly of water (95 percent), and most of its solid materials are the compounds carbohydrates, fats (properly called lipids), and proteins. If you could remove these materials, you would find a tiny residue of minerals, vitamins, and other compounds. Water, carbohydrates, fats, proteins, vitamins, and minerals are the six classes of nutrients commonly found in spinach and other foods. Some of the other materials in foods, such as the pigments and other phytochemicals, are not nutrients, but may still be important to health. The body can make some nutrients for itself, at least in limited quantities, but it cannot make them all, and it makes some in insufficient quantities to meet its needs. Therefore the body must obtain many nutrients from foods. The nutrients that foods must supply are called **essential nutrients.**

■■ **CARBOHYDRATES, FATS, AND PROTEINS** Four of the six classes of nutrients (carbohydrates, fats, proteins, and vitamins) contain carbon, which is found in all living things. They are therefore **organic** (meaning, literally, "alive"). During metabolism, three of these four (carbohydrates, fats, and proteins) provide energy the body can use. These **energy-yielding nutrients** continually replenish the energy you spend daily. Carbohydrates and fats meet most of the body's energy needs; proteins make a significant contribution only when other fuels are unavailable.

■■ **VITAMINS, MINERALS, AND WATER** Vitamins are organic but do not provide energy to the body. They facilitate the release of energy from the three energy-yielding nutrients. In contrast, minerals and water are **inorganic** nutrients. Minerals yield no energy in the human body, but like vitamins, they help to regulate the release of energy, among their many other roles. As for water, it is the medium in which all of the body's processes take place.

■■ **KCALORIES: A MEASURE OF ENERGY** The amount of energy carbohydrates, fats, and proteins release can be measured in calories (or, more properly, kilocalories or **kcalories**). kCalories are not constituents of foods; they are a measure of the energy foods provide. It is as incorrect to refer to the kcalories in a food as it is to refer to the inches in a person. It is correct to refer to the energy a food provides and to the height of a person. The energy a food provides depends on how much carbohydrate, fat, and protein the food contains.

Carbohydrate yields 4 kcalories of energy from each gram, and so does protein. Fat yields 9 kcalories per gram. If you know how many grams of each energy-yielding nutrient a food contains, you can derive the number of kcalories potentially available from the food. Simply multiply the carbohydrate grams times 4, the protein grams times 4, and the fat grams times 9, and add the results together (the accompanying "How to" box describes steps to calculate the energy a food provides).

nutrients: substances obtained from food and used in the body to provide energy and structural materials, as well as regulating agents to promote growth, maintenance, and repair. Nutrients may also reduce the risks of some diseases.

The six classes of nutrients are water, carbohydrates, fats, proteins, vitamins, and minerals.

essential nutrients: nutrients a person must obtain from food because the body cannot make them for itself in sufficient quantity to meet physiological needs.

organic: carbon containing. The four organic nutrients are carbohydrates, fats, proteins, and vitamins.

Metabolism, the sum total of all chemical reactions that go on in living cells, is described in Chapter 6.

energy-yielding nutrients: the fuel nutrients, those that yield energy the body can use. The three energy-yielding nutrients are carbohydrates, proteins, and fats.

inorganic: not containing carbon or pertaining to living things.

kcalories: units by which energy is measured. One kcalorie (kcal) is the amount of heat necessary to raise the temperature of 1 kilogram (kg) of water 1°C. Most people speak of these units simply as calories, but on paper the word *calorie* is prefaced by a k for *kilocalorie*. We use kcalories and kcal throughout this book.

Food energy can also be measured in kilojoules (kJ). The kilojoule is the international unit of energy. One kcalorie equals 4.2 kJ.

HOW TO:

CALCULATE THE ENERGY A FOOD PROVIDES

The following example shows how to calculate the energy available from 1 slice of bread that has 1 teaspoon of butter on it. From food tables such as Appendix A in this book, you can determine that buttered bread contains 15 grams carbohydrate, 2 grams protein, and 5 grams fat:

$$15 \text{ g carbohydrate} \times 4 \text{ kcal/g} = 60 \text{ kcal.}$$
$$2 \text{ g protein} \times 4 \text{ kcal/g} = 8 \text{ kcal.}$$
$$5 \text{ g fat} \times 9 \text{ kcal/g} = 45 \text{ kcal.}$$
$$\text{Total} = 113 \text{ kcal.}$$

From this information, you can calculate the percentage of kcalories each of the energy nutrients contributes to the total. To determine the percentage of kcalories from fat, for example, divide the 45 fat kcalories by the total 113 kcalories:

$$45 \text{ fat kcal} \div 113 \text{ total kcal} = 0.398 \text{ (round to 0.40).}$$
$$\text{Then multiply by 100 to get the percentage:}$$
$$0.40 \times 100 = 40\%$$

Health recommendations that urge people to limit fat intake to 30 percent of kcalories refer to the day's total energy intake, not to individual foods. Still, if the proportion of fat in each food choice throughout a day exceeds 30 percent of kcalories, then the day's total surely will, too. Knowing that this snack provides 40 percent of its kcalories from fat alerts a person to the need to make lower-fat selections at other times that day.

ENERGY NUTRIENTS IN FOODS Practically all foods contain mixtures of the energy-yielding nutrients, although foods are sometimes classified by their predominant nutrient. To speak of meat as "a protein" or of bread as "a carbohydrate," however, is inaccurate. Each is rich in a particular nutrient, but a protein-rich food such as beef contains a lot of fat along with protein, and a carbohydrate-rich food such as cornbread also contains fat (corn oil) and protein. Only a few foods are exceptions to this rule, the common ones being sugar (which is pure carbohydrate) and oil (which is pure fat).

ENERGY STORAGE IN THE BODY The body first uses the energy-yielding nutrients to build new compounds and fuel metabolic and physical activities. Excesses are then rearranged into storage compounds, primarily body fat, and put away for later use. Thus, if you take in more energy than you expend, whether from carbohydrate, fat, or protein, the result is usually a gain of body fat. Too much meat (a protein-rich food) is just as fattening as too many potatoes (a carbohydrate-rich food).

ALCOHOL, NOT A NUTRIENT When taken in excess of energy need, alcohol, too, is converted to body fat and stored. The body derives energy from alcohol at the rate of 7 kcalories per gram. Alcohol is not a nutrient, however, because it cannot support the body's growth, maintenance, or repair. When alcohol contributes too much energy to a person's diet, the harm it does extends far beyond the problems of excess body fat. Nutrition in Practice 23 discusses alcohol's effects on nutrition.

LEARNING LINK

Diets that control both the total energy and the proportion of carbohydrate, fat, and protein often play an important role in the treatment of many medical conditions, notably obesity (Chapter 7), diabetes mellitus (Chapter 20), cardiovascular diseases (Chapter 21), and renal disease (Chapter 22). To meet the specific energy and nutrient needs of each of these diseases, diet planners often use *exchange lists* that group foods according to the amount of energy, carbohydrate, fat, and protein they contain.

Nutrition Standards and Guidelines

Appendix C presents nutrient recommendations developed by two international groups, the Food and Agricultural Organization and the World Health Organization (the FAO/WHO recommendations).

How well you nourish yourself depends on the many different foods you select at numerous meals over days, months, and years. Several nutrition standards and dietary guidelines have been developed to help you whenever you are selecting foods—whether shopping at the grocery store, choosing from a restaurant menu, or preparing a home-cooked meal.

Dietary Reference Intakes

Defining the amounts of energy, nutrients, and other dietary components that best support health is a huge task. For more than 50 years, nutrition experts produced a set of energy and nutrient standards known as the Recommended Dietary Allowances (RDA) for people in the United States. The Canadian equivalent of the RDA is the RNI (Recommended Nutrient Intakes). Over the years, these standards were revised periodically as new evidence became available, but each revision maintained the original goal of protecting against nutrient deficiencies. Given the abundance of research now linking diet and health, that goal has been broadened to include supporting optimal activities within the body and preventing chronic diseases as well. Previous editions of the recommendations also focused narrowly on nutrients known to be essential. With recent research revealing the health benefits of other dietary components such as phytochemicals, their recommended intakes need to be addressed as well.

Dietary Reference Intakes (DRI): a set of values for the dietary nutrient intakes of healthy people in the United States and Canada. These values are used for planning and assessing diets.

A major revision of the dietary recommendations is currently under way. The revised recommendations will replace both the RDA and the RNI and are called **Dietary Reference Intakes (DRI).**[6] The first in a series of reports was published in 1997 and presents both the framework for developing the DRI and the revised recommendations for the five nutrients that play key roles in bone health—calcium, phosphorus, magnesium, vitamin D, and fluoride. A second report came out in 1998 and features revised recommendations for the eight B vitamins and a related compound, choline. The third report, released in 2000, discusses beta-carotene and the other antioxidant nutrients—vitamin C, vitamin E, and selenium. The report presents recommendations for vitamin C, vitamin E, and selenium, but not for beta-carotene. The recommendations for these 13 nutrients appear on the inside front cover under the heading 1997–2000 Dietary Reference Intakes. For the remaining nutrients and energy, the 1989 RDA will continue to serve health professionals until DRI can be established. The values for these nutrients also appear on the inside front cover under the heading 1989 Recommended Dietary Allowances.

The 1991 Canadian Recommended Nutrient Intakes (RNI) are presented in Appendix B.

Recommended Dietary Allowances (RDA): the average daily amounts of nutrients considered adequate to meet the known nutrient needs of practically all healthy people; a goal for dietary intake by individuals.

Adequate Intake (AI): the average amount of a nutrient that appears sufficient to maintain a specified criterion; a value used as a guide for nutrient intake when an RDA cannot be determined.

SETTING NUTRIENT GOALS FOR INDIVIDUALS: RDA AND AI   One advantage of the DRI is that they apply to individuals. In the past, nutrient standards were appropriate for planning and assessing the diets of populations, but their use for individuals was limited. In contrast, individuals are a prime concern of the DRI committee. The committee offers two sets of values to be used as nutrient intake goals by individuals: a revised set of **Recommended Dietary Allowances (RDA)** and a set called **Adequate Intakes (AI).**

Based on solid experimental evidence and other reliable observations, the RDA are the foundation of the DRI. The AI values are also based on scientific findings as much as possible, but estimating their values requires some educated guesswork. The committee establishes an AI value whenever scientific evidence is insufficient to generate an RDA.[7]

In the last decade, abundant new research has linked nutrients in the diet with the promotion of health and the prevention of chronic diseases. An advantage of the DRI is that they take into account disease prevention where appropriate, as well as nutrient adequacy. For example, the AI for calcium is based on intakes thought to reduce the likelihood of osteoporosis-related fractures later in life.

To ensure that the vitamin and mineral recommendations meet the needs of as many people as possible, the recommendations are set near the top end of the range of the population's estimated average requirements (see Figure 1-2). Small amounts above the daily requirement do no harm, but amounts below the requirement lead to health problems. When people's intakes are consistently **deficient,** their nutrient stores decline, and over time this decline leads to deficiency symptoms and poor health. In contrast to the vitamin and mineral recommendations, the energy recommendation is not generous, but rather is set at the population's estimated average requirement (see Figure 1-2). Although insufficient energy may cause undernutrition, too much energy is as bad for health as too little because excess energy leads to obesity.

■■ **FACILITATING NUTRITION RESEARCH AND POLICY: EAR** In addition to the RDA and AI, the DRI committee has established another set of values: **Estimated Average Requirements (EAR).** These values establish population-wide average requirements that researchers and nutrition policymakers need in their work. Nutrition scientists may use the EAR as standards in research. Public health officials may use them to assess nutrient intakes of populations and to make recommendations.

■■ **ESTABLISHING SAFETY GUIDELINES: UL** The DRI committee also establishes upper limits of intake for nutrients posing a hazard when consumed in excess. These values, the **Tolerable Upper Intake Levels (UL),** are indispensable to consumers who take supplements. (The UL are listed on the inside front cover.) Consumers need to know how much of a nutrient is too much. The UL are also of value to public health officials who set allowances for nutrients that are added to foods and water.

■■ **USING NUTRIENT RECOMMENDATIONS** Each of the four DRI categories serves a unique purpose. For example, the EAR are most appropriately used to develop and evaluate nutrition programs for *groups* such as schoolchildren or

deficient: the amount of a nutrient below which almost all healthy people can be expected, over time, to experience deficiency symptoms.

Estimated Average Requirement (EAR): the amount of a nutrient that will maintain a specific biochemical or physiological function in half the people of a given age and sex group.

Tolerable Upper Intake Level (UL): the maximum amount of a nutrient that appears safe for most healthy people and beyond which there is an increased risk of adverse health effects.

Nutrient recommendations are set high enough to cover nearly everyone's requirements (the boxes represent people).

The energy recommendation is set at the mean so that half the population's requirements fall below and half above it.

Figure 1-2
NUTRIENT RECOMMENDATIONS AND THE ENERGY RECOMMENDATION COMPARED

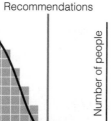

Nutrient recommendations

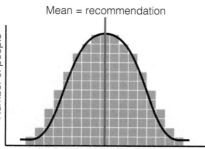

Energy recommendation

military personnel. The RDA (or AI if an RDA is not available) can be used to set goals for *individuals.* The UL help to keep nutrient intakes below the amounts that increase the risk of toxicity. With these understandings, professionals can use the DRI for a variety of purposes. Table 1-2 sums up the names and purposes of the nutrient intake standards just introduced.

In addition to understanding the unique purposes of the DRI, it is important to keep their uses in perspective. Consider the following:

requirement: the lowest continuing intake of a nutrient that will maintain a specified criterion of adequacy.

- Nutrient recommendations are estimates for optimal and safe intakes, not **requirements;** except for energy, they include a generous margin of safety. Figure 1-3 presents an accurate view of how a person's nutrient needs fall within a range, with marginal and danger zones both below and above the range.
- Nutrient recommendations are designed to meet the needs of most *healthy* people. Medical problems alter nutrient needs as later chapters describe.
- Nutrient recommendations are specific for people of both sexes as well as various ages and stages of life: infants, children, adolescents, men, women, pregnant women, and lactating women.

LEARNING LINK

To evaluate the nutrients in a person's diet, the dietitian must first determine what the person is eating. Two of the most common tools used for gathering such data are:

- Twenty-four-hour recalls, which ask clients what they have eaten over the past day.
- Food records, which ask clients to keep complete lists of all the foods and beverages they consume over a specified period of time (often several days or more).

After food intake data have been collected, the energy and nutrients in a person's diet can be calculated by hand or by computer and then compared with nutrient recommendations such as the DRI. Chapter 13 describes the correct techniques for collecting food intake data, the advantages and drawbacks of each technique, and the methods for analyzing the information.

Table 1-2
THE DRI AND THEIR USES

Reference Value	Description
Recommended Dietary Allowances (RDA)	Nutrient intake goals for individuals. RDA values are derived from the Estimated Average Requirements (see below).
Adequate Intakes (AI)	Nutrient intake goals for individuals. AI values are set whenever scientific data are insufficient to allow establishment of an RDA value.
Tolerable Upper Intake Levels (UL)	Suggested upper limits of intakes of potentially toxic nutrients.. Intakes above the UL are likely to cause illness from toxicity.
Estimated Average Requirements (EAR)	Population-wide average nutrient requirements for nutrition research and policy making. The basis upon which RDA values are set.

SOURCE: Adapted from Food and Nutrition Board, Institute of Medicine, National Academy of Sciences, Dietary Reference Intake (DRIs) for calcium, phosphorus, magnesium, vitamin D and fluoride, *Nutrition Today* 32 (1997): 182–190.

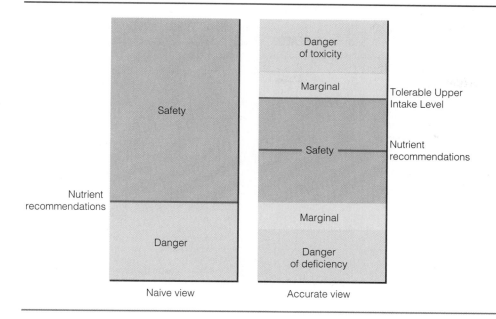

Naive view   Accurate view

Figure 1-3
NAIVE VERSUS ACCURATE VIEW OF NUTRIENT RECOMMENDATIONS

The recommendation for a given nutrient represents a point within a range of appropriate and reasonable intakes that lies between toxicity and deficiency. The recommendation is high enough to provide reserves in times of short-term dietary inadequacies, but not so high as to approach toxicity. Nutrient intakes above or below this range might be equally harmful.

Dietary Guidelines

Nutrient recommendations such as the DRI were developed to ensure adequate nutrient intakes, but they do little to protect people from excess intakes of fat, cholesterol, sugar, salt, and alcohol. Government authorities are now as much concerned about **overnutrition** as they once were about **undernutrition.** Research confirms that dietary excesses, especially of energy, fat, and alcohol, contribute to many **chronic diseases,** including heart disease, cancer, diabetes, and liver disease.[8] The *Dietary Guidelines for Americans* originated from the awareness that overnutrition contributes to disease.

The 2000 *Dietary Guidelines* emphasize the importance of physical fitness to health and answer the question, what should an individual eat to stay healthy? The first guideline encourages people to aim for a healthy weight, while the second guideline encourages people to be physically active each day. The next three guidelines urge people to use the Food Guide Pyramid when making food choices and to choose a variety of grains, fruits, and vegetables each day. The sixth guideline addresses the importance of food safety. The last four guidelines encourage a diet that is low in saturated fat and cholesterol and moderate in total fat, sugars, salt, and alcohol. Together, these ten guidelines point the way toward better health. Table 1-3 (p. 14) presents the 2000 *Dietary Guidelines for Americans.* Table 1-4 (p. 14) presents Canada's *Guidelines for Healthy Eating.*

The *Dietary Guidelines* suggest not only what people should eat but also what they should limit. In addition, the recommendations refer to weight maintenance and physical activity. Some people's diets are close to these recommendations, but the typical U.S. diet falls far short of the recommendations. So does the sedentary lifestyle of many who fail to meet even minimal guidelines for physical activity.[9]

Fitness Guidelines

The American College of Sports Medicine (ACSM) has published guidelines on the types and amounts of physical activity recommended for developing and maintaining **fitness** in healthy adults.[10] The types and amounts of physical activity needed to promote *fitness,* however, may differ from those

overnutrition: overconsumption of food energy or nutrients sufficient to cause disease or increased susceptibility to disease; a form of malnutrition.

undernutrition: underconsumption of food energy or nutrients severe enough to cause disease or increased susceptibility to disease; a form of malnutrition.

chronic diseases: degenerative diseases characterized by deterioration of the body organs; also called chronic, **noncommunicable diseases (NCD).** Examples include heart disease, cancer, and diabetes.

fitness: the characteristics of the body that enable it to perform physical activity; more broadly, the ability to meet routine physical demands with enough reserve energy to rise to a sudden challenge; or the body's ability to withstand stress of all kinds.

Physical activity helps you look good, feel good, and have fun, and it brings many long-term health benefits as well.

Table 1-3
2000 DIETARY GUIDELINES FOR AMERICANS THE ABC'S OF GOOD HEALTH

Aim for fitness.
- Aim for a healthy weight.
- Be physically active each day.

Build a healthy base.
- Let the pyramid guide your food choices.
- Choose a variety of grains daily, especially whole grains
- Choose a variety of fruits and vegetables daily.
- Keep food safe to eat.

Choose sensibly.
- Choose a diet that is low in saturated fat and cholesterol and moderate in total fat.
- Choose beverages and foods that limit your intake of sugars.
- Choose and prepare foods with less salt.
- If you drink alcoholic beverages, do so in moderation.

NOTE: These guidelines are intended for adults and healthy children ages 2 and older.

Table 1-4
CANADA'S GUIDELINES FOR HEALTHY EATING

- Enjoy a variety of foods.
- Emphasize cereals, breads, other grain products, vegetables, and fruits.
- Choose lower-fat dairy products, leaner meats, and foods prepared with little or no fat.
- Achieve and maintain a healthy body weight by enjoying regular physical activity and healthy eating.
- Limit salt, alcohol, and caffeine.

SOURCE: These guidelines derive from *Action Towards Healthy Eating: The Report of the Communications/ Implementation Committee* and *Nutrition Recommendations . . . A Call for Action: Summary Report of the Scientific Review Committee and the Communications/ Implementation Committee*, which are available from Branch Publications Unit, Health Services and Promotion Branch, Department of Health and Welfare, 5th Floor, Jeanne Mance Building, Ottawa, Ontario K1A 1B4.

needed to obtain *health* benefits (see Table 1-5). For health's sake, the ACSM specifies that people should spend an accumulated minimum of 30 minutes in some sort of physical activity on most days of each week.[11] The activity need not be sports. A few minutes spent climbing stairs, another few spent pulling weeds, and several more spent walking the dog all contribute to the day's total. The guidelines for developing fitness are still optimal, though, because improving fitness provides additional health benefits.[12]

Extensive evidence confirms that regular physical activity promotes health and prevents disease.[13] Still, despite an increasing awareness of the health benefits that physical activity confers, as many as 60 percent of adults in the United States are either irregularly active or completely inactive.[14] Physical inactivity is linked to the major degenerative diseases—heart disease, cancer, stroke, and hypertension—that are the primary killers of adults in developed countries.[15] Therefore, one of the most important challenges health care professionals face is to motivate more people to become physically active. To motivate others, health care professionals must first become more physically active themselves, thereby enhancing their own health. Second, they can include regular physical activity as a component of therapy for their clients. Table 1-6 (p. 16) summarizes the benefits fitness confers.

Diet-Planning Principles

How can dietitians, nurses, and other health care professionals help people select foods to create a diet that supplies all the needed nutrients in the appropriate amounts for good health? The principle is simple enough: encourage clients to eat a variety of foods that supply all the nutrients the body needs. In practice, how do people do this? It helps to keep in mind six basic diet-planning principles, which are listed in the margin in alphabetical order for ease in remembering them.

ADEQUACY   The earlier discussion on the DRI already addressed the ideal of **dietary adequacy.** An adequate diet provides enough energy and enough of every nutrient to meet the needs of healthy people.

Diet-planning principles:

- *Adequacy.*
- *Balance.*
- *kCalorie control.*
- *Density (nutrient)*
- *Moderation.*
- *Variety.*

dietary adequacy: the characteristic of a diet that provides all the essential nutrients, fiber, and energy necessary to maintain health and body weight.

Table 1-5
PHYSICAL ACTIVITY GUIDELINES

Guidelines for developing and maintaining *physical fitness*:

- Frequency of activity: 3 to 5 days per week.
- Intensity of activity: 55 to 90% of maximum heart rate.
- Duration of activity: 20 to 60 minutes of continuous activity.
- Mode of activity: Any activity that uses large-muscle groups.
- Resistance activity: Strength training of moderate intensity at least two times per week.
- Flexibility activity: Stretching of the major muscle groups two to three times per week.

Guidelines for obtaining *health* benefits:

- Frequency of activity: Every day.
- Intensity of activity: Moderate.
- Duration of activity: At least 30 minutes total of activity (can be intermittent).
- Mode of activity: Any activity.

NOTE: Duration and intensity are inversely related. To obtain similar fitness benefits, a person may exercise either at a low intensity for a long duration or at a high intensity for a short duration. For example, a person may choose to walk briskly for 40 to 50 minutes (lower intensity and longer duration) or to jog for 20 to 30 minutes (higher intensity and shorter duration).

This cola and bunch of grapes illustrate nutrient density. Each provides about 150 kcalories (mainly from carbohydrate), but the grapes offer a trace of protein, some vitamins, minerals, and fiber along with the energy; the cola beverage offers only "empty" kcalories. Grapes, or any fruit for that matter, are more nutrient dense than cola beverages.

BALANCE   As for **dietary balance,** the essential minerals calcium and iron illustrate its importance. Meats, fish, and poultry are rich in iron but poor sources of calcium. Similarly, milk and milk products are rich in calcium but poor sources of iron. In fact, milk (except breast milk) and milk products are so low in iron that overuse of these foods can actually lead to iron-deficiency anemia by displacing iron-rich foods from the diet. Yet milk is the single most nutritious food for infants and can be an important calcium source for people of all ages.

Advise clients to use some meat and meat alternates for iron and some milk and milk products for calcium. They need to save some space for other foods, too, since a diet consisting only of milk and meat would not be adequate. To obtain the other needed nutrients, people have to eat vegetables, fruits, grains, and other foods. In short, balance in the diet helps to ensure adequacy.

dietary balance: providing foods of a number of types in balance with one another such that foods rich in one nutrient do not crowd out foods that are rich in another nutrient.

KCALORIE (ENERGY) CONTROL   While it takes thought and skill to design an adequate, balanced diet, incorporating **kcalorie control** presents an added challenge—to eat enough without eating too much. Energy balance and weight control are discussed in Chapter 7, but the key to controlling energy intake is to select foods of high nutrient density.

kcalorie (energy) control: management of food energy intake.

NUTRIENT DENSITY   To eat well without overeating, select foods that deliver the most nutrients per kcalorie, a concept known as **nutrient density.** For example, among foods containing calcium, a 1½-ounce portion of cheddar cheese and 1 cup of nonfat milk provide about the same amount of calcium; but the cheese contributes twice as much food energy (kcalories) as the non-fat milk (see Appendix A). The nonfat milk, then, is twice as calcium dense as the cheddar cheese: it offers the same amount of calcium for half the kcalories.

nutrient density: a measure of the nutrients a food provides relative to the energy it provides. The more nutrients and the fewer kcalories, the higher the nutrient density.

MODERATION   Foods rich in fat and sugar provide enjoyment and energy but relatively few nutrients. In addition, they promote weight gain when eaten in excess. A person practicing **moderation** would eat such foods only on occasion and would regularly select foods low in fat and sugar, a practice that automatically improves nutrient density. Returning to the example of cheddar cheese and nonfat milk, the nonfat milk not only offers the same amount of calcium for less energy, but it contains far less fat than the cheese. When working with clients advised to follow a diet that is low in fat and sugar, remind them that the most nutrient-dense foods are the best choices for the sake of both kcalorie control and moderation.

moderation: providing enough, but not too much, of a substance.

Variety helps to ensure an adequate and balanced diet.

Table 1-6
BENEFITS OF FITNESS*

- *Restful sleep:* Rest and sleep occur naturally after periods of physical activity. During rest, the body repairs injuries, disposes of wastes generated during activity, and builds new physical structures.
- *Nutritional health:* Physical activity spends energy and thus allows people to eat more food. If they choose wisely, active people will consume more nutrients and be less likely to develop nutrient deficiencies.
- *Optimal body composition:* A balanced program of physical activity limits body fat and maintains lean tissue. Physically active people have relatively less body fat than sedentary people at the same body weight.[a]
- *Optimal bone density:* Weight-bearing physical activity builds bone strength and protects against osteoporosis.[b]
- *Resistance to colds and other infectious diseases:* Fitness enhances immunity.[†c]
- *Low risks of some types of cancers:* Lifelong physical activity may help to protect against colon cancer, breast cancer, and some other types of cancer.[d]
- *Strong circulation and lung function:* Physical activity that challenges the heart and lungs slows the aging of the circulatory system.
- *Low risk of cardiovascular disease:* Physical activity lowers blood pressure, slows resting pulse rate, and lowers blood cholesterol, thus reducing the risks of heart attack and strokes.[e] Some research suggests that physical activity may reduce the risk of cardiovascular disease in another way as well—by lowering intra-abdominal fat stores.[f]
- *Low risk of diabetes:* Physical activity normalizes glucose tolerance, especially via the secretion of insulin.[g]
- *Low incidence and severity of anxiety and depression:* Compared with sedentary people, physically active people deal better with stress.
- *Strong self-image:* The sense of achievement that comes from meeting physical challenges promotes self-confidence.
- *Long life and high quality of life in the later years:* Active people have a lower mortality rate than sedentary people. Even a 2-mile walk daily can add years to a person's life.[h] In addition to extending longevity, physical activity supports independence and mobility in later life by reducing the risks of falls and minimizing the risk of injury should a fall occur.[i]

*The reference notes for this table are on pp. 26–27.
†Moderate physical activity can stimulate immune function. Intense, vigorous, prolonged activity such as marathon running, however, may compromise immune function. J. A. Smith, Guidelines, standards, and perspectives in exercise immunology, *Medicine and Science in Sports and Exercise* 27 (1995): 497–506.

variety (dietary): eating a wide selection of foods within and among the major food groups (the opposite of monotony).

The reason may not be obvious, but, choosing nutrient-dense, low-fat, low-sugar foods can also help control salt intake. Although added salt does not influence the nutrient density of foods, manufacturers usually add all three—salt, sugar, and fat—to enhance the taste of foods. Thus, pursuing nutrient density helps keep salt intake moderate.

■■ **VARIETY** A diet can have all of the characteristics just described but still lack **variety** if a person eats the same foods day after day. Encourage clients to vary their choices within each food group from day to day, for at least two reasons. First, different foods in the same group contain different arrays of nutrients. Among the fruits, for example, strawberries are especially rich in vitamin C, and peaches are rich in vitamin A. Thus, variety helps ensure adequacy. Second, no food is guaranteed entirely free of substances that in excess could cause harm. Choosing strawberries today, peaches tomorrow, and watermelon the day after, ensures that the total diet will have diluted concentrations of any contaminants that may be present in the foods.

These diet-planning principles offer a framework of excellence to strive for in planning diets; they are dietary ideals. To plan a diet that achieves these ideals, the planner needs to know which foods offer which nutrients and how many daily servings of these foods are recommended.

Food Group Plans

The human body needs about 40 different vitamins and minerals. Each nutrient has its own unique pattern of distribution in foods. Although working all the nutrients into the meals a person eats might seem quite a challenge, people all over the world meet their nutrition needs from an amazing variety of diets.

FOOD GROUPS

For many people, **food group plans** such as the **Daily Food Guide,** serve as the basis for planning adequate, balanced diets. Imagine shopping for groceries in a store with no signs and no organization. You might find canned peaches next to a box of taco shells and find canned pears several aisles over next to a box of cereal. To pick up even a few groceries in such a store would be quite a task. You might even forget which groceries you wanted and, instead, simply grab whatever you could find. To ease the task of shopping, grocers group similar foods together and provide signs so that you can find them. In much the same way, food group plans ease the task of planning nutritious diets by sorting foods according to their key nutrient contributions. For example, cheese and yogurt fall within the milk group because each is notable for calcium, riboflavin, and protein, as well as small amounts of other nutrients. Legumes are included in the meat, fish, and poultry group because these foods are notable for their protein, iron, and zinc.

Food group plans exclude some foods from the main clusters because they make no notable nutrient contributions to people's intakes. These foods, called "miscellaneous" foods, are included in a diet plan only as extras in small quantities once basic nutrient needs have been met by nutritious foods. Examples of miscellaneous foods are soft drinks, jams, salad dressings, and butter.

THE DAILY FOOD GUIDE

Figure 1-4 (pp. 18–19) presents the Daily Food Guide, a food group plan that assigns foods to five major food groups. The figure lists the number of servings recommended, the most notable nutrients of each food group, the serving sizes, and the foods within each group categorized by nutrient density. It also includes the Food Guide Pyramid, which presents the Daily Food Guide in pictorial form. The pyramid shape, with grains at the base, is designed to convey the message that grains should be abundant and form the foundation of a healthy diet. Fruits and vegetables share the next level of the pyramid, indicating that they, too, should have a prominent place in the diet. Meats and milks appear in a smaller band near the top. A few servings of each can provide valuable nutrients, such as protein, vitamins, and minerals, without too much fat and cholesterol. Fats, oils, and sweets occupy the tiny apex, indicating that they should be used sparingly.

The Daily Food Guide offers a strong foundation for a healthy diet, but it fails to specify food energy intakes. Large fat and energy differences exist

food group plans: diet-planning tools that sort foods of similar origin and nutrient content into groups and then specify that people should eat certain numbers of servings from each group.

Daily Food Guide: the USDA's food group plan for ensuring dietary adequacy that assigns foods to five major food groups.

Figure 1-4
THE DAILY FOOD GUIDE

Breads, Cereals, and Other Grain Products: 6 to 11 servings per day.

These foods are notable for their contributions of complex carbohydrates, riboflavin, thiamin, niacin, iron, protein, magnesium, and fiber.
Serving = 1 slice bread; ½ c cooked cereal, rice, or pasta; 1 oz ready-to-eat cereal; ½ bun, bagel, or English muffin; 1 small roll, biscuit, or muffin; 3 to 4 small or 2 large crackers.

♦ Whole grains (wheat, oats, barley, millet, rye, bulgur, couscous, polenta), enriched breads, rolls, tortillas, cereals, bagels, rice, pastas (macaroni, spaghetti), air-popped corn.
◊ Pancakes, muffins, cornbread, crackers, cookies, biscuits, presweetened cereals, granola, taco shells, waffles, french toast.
♦ Croissants, fried rice, doughnuts, pastries, cakes, pies.

Vegetables: 3 to 5 servings per day (use dark green, leafy vegetables and legumes several times a week).

These foods are notable for their contributions of vitamin A, vitamin C, folate, potassium, magnesium, and fiber, and for their lack of fat and cholesterol.
Serving = ½ c cooked or raw vegetables; 1 c leafy raw vegetables; ½ c cooked legumes; ¾ c vegetable juice.

♦ Bamboo shoots, bean sprouts, bok choy, broccoli, brussels sprouts, cabbage, carrots, cauliflower, corn, cucumbers, eggplant, green beans, green peas, leafy greens (spinach, mustard, and collard greens), legumes, lettuce, mushrooms, okra, onions, peppers, potatoes, pumpkin, scallions, seaweed, snow peas, soybeans, tomatoes, water chestnuts, winter squash.
◊ Candied sweet potatoes.
♦ French fries, tempura vegetables, scalloped potatoes, potato salad.

Fruits: 2 to 4 servings per day.

These foods are notable for their contributions of vitamin A, vitamin C, potassium, and fiber, and for their lack of sodium, fat, and cholesterol.
Serving = typical portion (such as 1 medium apple, banana, or orange, ½ grapefruit, 1 melon wedge); ¾ c juice; ½ c berries; ½ c diced, cooked, or canned fruit; ½ c dried fruit.

♦ Apples, apricots, bananas, blueberries, cantaloupe, grapefruit, guava, oranges, orange juice, papaya, peaches, pears, plums, strawberries, watermelon; unsweetened juices.
◊ Canned or frozen fruit (in syrup); sweetened juices.
♦ Dried fruit, coconut, avocados, olives.

Meat, Poultry, Fish, and Alternates: 2 to 3 servings per day.

Meat, poultry, and fish are notable for their contributions of protein, phosphorus, vitamin B_6, vitamin B_{12}, zinc, iron, niacin, and thiamin; legumes are notable for their protein, fiber, thiamin, folate, vitamin E, potassium, magnesium, iron, and zinc, and for their lack of fat and cholesterol.
Servings = 2 to 3 oz lean, cooked meat, poultry, or fish (total 5 to 7 oz per day); count 1 egg, ½ c cooked legumes, 4 oz tofu, or 2 tbs nuts, seeds, or peanut butter as 1 oz meat (or about ⅓ serving).

♦ Poultry (light meat, no skin), fish, shellfish, legumes, egg whites.
◊ Lean meat (fat-trimmed beef, lamb, pork); poultry (dark meat, no skin); ham; refried beans; whole eggs, tofu, tempeh.
♦ Hot dogs, luncheon meats, ground beef, peanut butter, nuts, sausage, bacon, fried fish or poultry, duck.

Key:
♦ Foods generally highest in nutrient density (good first choice).
◊ Foods moderate in nutrient density (reasonable second choice).
♦ Foods lowest in nutrient density (limit selections).

Milk, Cheese, and Yogurt: 2 servings per day.

3 servings per day for teenagers and young adults, pregnant/lactating women, women past menopause; 4 servings per day for pregnant/lactating teenagers. These foods are notable for their contributions of calcium, riboflavin, protein, vitamin B_{12}, and, when fortified, vitamin D and vitamin A.
Serving = 1 c milk or yogurt; 2 oz process cheese food; ½ oz cheese.

◆ Fat-free and 1% low-fat milk (and fat-free products such as buttermilk, cottage cheese, cheese, yogurt); fortified soy milk.

◇ 2% reduced-fat milk (and low-fat products such as yogurt, cheese, cottage cheese); chocolate milk; sherbet; ice milk.

◆ Whole milk (and whole-milk products such as cheese, yogurt); custard; milk shakes; ice cream.

NOTE: These serving recommendations were established before the 1997 DRI, which raised the recommended intake for calcium; meeting the calcium recommendation may require an additional serving from the milk, cheese, and yogurt group.

Fats, Sweets, and Alcoholic Beverages: Use sparingly.

These foods contribute sugar, fat, alcohol, and food energy (kcalories). They should be used sparingly because they provide food energy while contributing few nutrients. Miscellaneous foods not high in kcalories, such as spices, herbs, coffee, tea, and diet soft drinks, can be used freely.

◆ Foods high in fat include margarine, salad dressing, oils, lard, mayonnaise, sour cream, cream cheese, butter, gravy, sauces, potato chips, chocolate bars.

◇ Foods high in sugar include cakes, pies, cookies, doughnuts, sweet rolls, candy, soft drinks, fruit drinks, jelly, syrup, gelatin, desserts, sugar, and honey.

◆ Alcoholic beverages include wine, beer, and liquor.

KEY

● Fat (naturally occurring and added)

▼ Sugars (added)

These symbols show fats, oils and added sugars in foods.

Fats, Oils & Sweets
Use sparingly

Milk, Yogurt & Cheese Group
2–3 servings

Meat, Poultry, Fish, Dry Beans, Eggs & Nuts Group
2–3 servings

Vegetable Group
3–5 servings

Fruit Group
2–4 servings

Bread, Cereal, Rice & Pasta Group
6–11 servings

Food Guide Pyramid
A GUIDE TO DAILY FOOD CHOICES

The breadth of the base shows that grains (breads, cereals, rice, and pasta) deserve most emphasis in the diet. The tip is smallest: use fats, oils, and sweets sparingly.

within a single food group—for example, between nonfat milk and ice cream, baked chicken and bacon, green beans and french fries, apples and avocados, or bread and croissants. Yet, according to the Daily Food Guide, any of these choices would be acceptable. Figure 1-4 provides a nutrient density key for foods within each group. People can therefore control their food energy intakes while using the Daily Food Guide. The foods color-coded with green are the highest in nutrient density; they contribute the most nutrients for the fewest kcalories. Those with yellow diamonds are moderate in nutrient density; and those with red diamonds are lowest in nutrient density. People who choose the smallest recommended number of servings, choose them all from the green-coded foods, and strictly limit foods from the miscellaneous group will obtain the needed nutrients at an energy intake of about 1600 kcalories. People who choose more servings or less nutrient-dense foods can easily boost their energy intakes. Depending on their energy needs, such choices might add to their body weight.

The beauty of the Daily Food Guide is that it is easy to learn and simple to use. It may appear rigid, but it actually offers great flexibility once its intent is understood. For example, cheese can be substituted for milk because both supply the key nutrients for the milk group (protein, calcium, and riboflavin) in about the same amounts. To limit kcalories, choose nonfat milk; to add kcalories, choose cheese. Legumes are alternative choices for meats, so vegetarians can adapt the pattern by using legumes in place of meat selections. Nutrition in Practice 3 shows how to plan healthy vegetarian diets.

▪▪ Nutrition Surveys

Researchers use nutrition surveys to determine which foods people are eating, to assess people's nutritional health, and to measure people's knowledge, attitude, and behaviors about nutrition and how these relate to health.[16] The resulting wealth of information can be used for a variety of purposes. For example, Congress uses this information to establish public policy on nutrition education, assess food assistance programs, and regulate the food supply. Scientists use the information to establish research priorities. One of the first nutrition surveys, taken before World War II, suggested that up to a third of the U.S. population might be eating poorly. Programs to correct **malnutrition** have been evolving ever since.

malnutrition: any condition caused by deficient or excess energy or nutrient intake or by an imbalance of nutrients.

food consumption survey: a survey that measures the amounts and kinds of food people consume (using diet histories), estimates the nutrient intakes, and compares them with a standard such as the DRI.

Nutrition assessment methods are described in Chapter 13.

nutrition status survey: a survey that evaluates people's nutrition status using nutrition assessment methods.

▪▪ FOOD CONSUMPTION SURVEYS **Food consumption surveys** determine the kinds and amounts of foods people eat. Researchers calculate the energy and nutrients in the foods and compare the amounts consumed with a standard such as the DRI. An example of this type of survey is the Nationwide Food Consumption Survey (NFCS).

▪▪ NUTRITION STATUS SURVEYS **Nutrition status surveys** examine the people themselves, using nutrition assessment methods. The data provide information on several nutrition-related conditions, such as growth retardation, heart disease, and nutrient deficiencies. The National Health and Nutrition Examination Survey (NHANES) is an example of a nutrition status survey. Both the NFCS and the NHANES oversample high-risk groups (low-income families, infants and children, and the elderly) in order to glean an accurate estimate of their health and nutrition status.

▪▪ HEALTHY PEOPLE REPORTS Until 1990, findings from the nation's many nutrition surveys were almost impossible to compare and synthesize into a single cohesive report. Then the National Nutrition Monitoring and Related Research Act was enacted to coordinate the many nutrition-related

Grocers often post nutrition-information placards or pamphlets near nonpackaged items such as raw fruits, vegetables, and seafood.

activities that had been under way within 22 different federal agencies. The law mandated that the U.S. Department of Agriculture (USDA) and the Department of Health and Human Services (DHHS) establish and implement a Ten-Year Comprehensive Plan for nutrition monitoring and related research.[17] All major reports that examine the contribution of diet and nutrition status to the health of the people of the United States depend on information collected and coordinated by this national program. These data provided the basis for the mid-1990s report *Healthy People 2000*.[18] This report showed that we are not meeting many of our health goals; in fact, we are not even heading in the right direction for some goals, such as reducing the prevalence of overweight in the United States.[19] Table 1-7 (p. 22) lists the 20 nutrition-related priorities of the new objectives, *Healthy People 2010*.

The Healthy People 2010 *report sets national objectives in health promotion and disease prevention for the year 2010.*

Food Labels

Today consumers know more about the links between diet and disease than they did in the past, and they are demanding still more information on disease prevention. Many people rely on food labels to tell them which substances to select and which ones to limit for health reasons. Figure 1-5 (p. 23) illustrates the requirements for label information. Most food labels must conform to all these requirements. Exceptions include plain coffee, tea, spices, and other foods contributing few nutrients; foods produced by small businesses; packages with fewer than 12 square inches of surface area; and those prepared and sold in the same establishment as long as the foods do not make nutrient or health claims.*[20]

THE DAILY VALUES ON LABELS   The **Daily Values** (inside back cover) are a set of nutrient standards created by the Food and Drug Administration (FDA) for use on food labels. The Daily Values do two things: they set adequacy standards for nutrients that are desirable in the diet such as protein,

Daily Values: reference values developed by the FDA specifically for use on food labels.

---

*For example, restaurants need not provide complete nutrition information unless they make "heart-healthy" claims for menu items.

Table 1-7
PROPOSED NUTRITION-RELATED OBJECTIVES FOR THE NATION, YEAR 2010[a]

Disease-Related Objectives

1. Reduce *coronary heart disease* deaths.
2. Reduce *cancer* deaths.
3. Decrease the incidence of *diabetes* (type 2), increase diagnosis of existing diabetes, and reduce rates of diabetes-related illness and death.
4. Reduce the prevalence of *osteoporosis*.
5. Reduce *dental caries*.

Nutrient Objectives

6. Increase the prevalence of *healthy weight* and decrease the prevalence of *obesity*.
7. Reduce *growth retardation* among low-income children.
8. Increase the proportion of people aged 2 and older who meet the *Dietary Guidelines* for *fat* and *saturated fat* in the diet.[b]
9. Increase intakes of *fruit and vegetables* to at least six servings a day.
10. Increase intakes of *grain products* to at least six servings a day.
11. Increase the proportion of people who meet the recommendation for *calcium*.
12. Increase the proportion of people who meet the Daily Value of 2400 milligrams or less of *sodium* a day.
13. Reduce *iron deficiency* in children, adolescents, women of childbearing age, and low-income pregnant women.
14. Increase the proportion of mothers who *breastfeed* immediately after birth, for the first six months, and preferably, through the infant's first year of life. Increase the proportion of mothers who *breastfeed exclusively*.
15. Increase the proportion of children and adolescents whose intakes of *meals* and *snacks at school* contribute to overall dietary quality.
16. Increase the proportion of schools teaching *essential nutrition topics*.

Food Safety Objectives

17. Reduce the proportion of *infections* caused by foodborne pathogens and *antibiotic-resistant* pathogens.
18. Reduce deaths from *food allergy* (anaphylaxisis).
19. Increase the proportion of *consumers* who practice four critical food safety behaviors when handling foods: washing hands, preventing cross-contamination, cooking meats thoroughly, and chilling foods promptly.[c]
20. Reduce occurrences of improper food safety techniques in *retail food establishments*.

[a]Adapted. *Italic type* has been added to emphasize main areas of concern. Details of the *Healthy People 2010 Objectives* may be viewed at their web site: http://web.health.gov/healthypeople/
[b]The objective specifies an average of 30 percent of kcalories or less from fat, and 10 percent or less from saturated fat.
[c]See Nutrition in Practice 5 for details on these and other food safety concepts.
SOURCE: *Health People 2010: National Health Promotion and Disease Prevention Objectives* (Washington, D.C.: U.S. Department of Health and Human Services, 2000).

vitamins, minerals, and fiber, and they also set moderation standards for other nutrients that must be limited, such as fat, cholesterol, and sodium, according to the *Dietary Guidelines*.

⠿ HEALTH CLAIMS ON LABELS The FDA has established strict guidelines pertaining to claims about health on food labels. Table 1-8 (p. 24) presents the health claims that have been approved to date and the criteria they must meet. A review of the table reveals that fats play a key role in whether a food can make a health claim. Whole milk, even though it is high in calcium, may not make a claim about osteoporosis because it contains too much satu-

Figure 1-5
EXAMPLE OF A FOOD LABEL

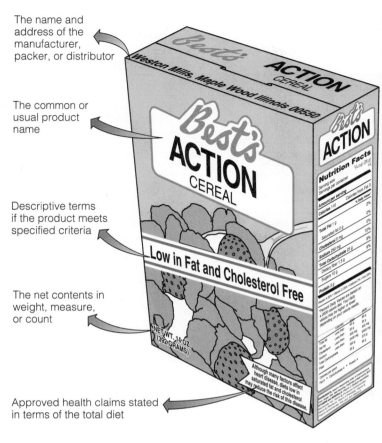

The name and address of the manufacturer, packer, or distributor

The common or usual product name

Descriptive terms if the product meets specified criteria

The net contents in weight, measure, or count

Approved health claims stated in terms of the total diet

The standard serving size expressed in common household and metric measures and the number of servings per container

kCalorie information and quantities of specified nutrients per serving, in actual amounts

Quantities of nutrients as "% Daily Values" based on a 2000-kcalorie energy intake

Daily Values for selected nutrients for a 2000- and a 2500-kcalorie diet*

kCalorie per gram reminder

The ingredients in descending order of predominance by weight

Nutrition Facts

Serving size   ³/₄ cup (28 g)
Servings per container   14

Amount per serving

Calories 110   Calories from fat 9

	% Daily Value*
Total Fat 1 g	2%
Saturated fat 0 g	0%
Cholesterol 0 mg	0%
Sodium 250 mg	10%
Total Carbohydrate 23 g	8%
Dietary fiber 1.5 g	6%
Sugars 10 g	
Protein 3 g	

Vitamin A 25% • Vitamin C 25% • Calcium 2% • Iron 25%

*Percent Daily Values are based on a 2000 calorie diet. Your daily values may be higher or lower depending on your calorie needs.

	Calories:	2000	2500
Total fat	Less than	65 g	80 g
Sat fat	Less than	20 g	25 g
Cholesterol	Less than	300 mg	300 mg
Sodium	Less than	2400 mg	2400 mg
Total Carbohydrate		300 g	375 g
Fiber		25 g	30 g

Calories per gram
Fat 9 • Carbohydrate 4 • Protein 4

INGREDIENTS, listed in descending order of predominance: Corn, Sugar, Salt, Malt flavoring, freshness preserved by BHT. **VITAMINS and MINERALS:** Vitamin C (Sodium ascorbate), Niacinamide , Iron, Vitamin B₆ (Pyridoxine hydrochloride), Vitamin B₂ (Riboflavin), Vitamin A (Palmitate), Vitamin B₁ (Thiamin hydrochloride), Folic acid, and Vitamin D.

*A package with fewer than 40 square inches of surface area may present Daily Values for a 2000-kcalorie diet only.

rated fat to qualify. Low-fat and nonfat milk, however, do qualify to bear the calcium and osteoporosis claim. Table 1-9 (p. 25) defines the descriptive terms consumers may see on food labels. Those who design food labels must proceed carefully. For example, use of the word *healthy* in a name such as "Healthy Start" or the use of a heart-shaped logo may imply that a food is health promoting. Foods bearing such words or logos must not exceed limits set for fat, saturated fat, cholesterol, and sodium contents.

LEARNING LINK

Health claims on food labels offer valuable information for clients concerned with specific conditions. A client with cardiovascular disease, for example, can look for food labels specifying low fat, low saturated fat, and low cholesterol, as well as identify good sources of oat bran and fiber. Later chapters discuss the links between specific nutrients and diseases.

Consumers may wonder whether they are reducing their health risks if they eat only "healthy" foods. They may be, and that is all that a label is

Table 1-8

FOOD LABEL HEALTH CLAIMS AND THEIR CRITERIA

Health Claim	Criteria
Calcium and reduced risk of osteoporosis	• High in calcium (≥20% Daily Value) • No more phosphorus than calcium
Sodium and reduced risk of hypertension	• Low in sodium (≤140 mg/serving)
Dietary saturated fat and cholesterol and reduced risk of coronary heart disease	• Low in saturated fat (≤1 g/serving) • Low in cholesterol (≤20 mg/serving) • Low in fat (≤3 g/serving)
Dietary fat and reduced risk of cancer	• Low in fat (≤3 g/serving)
Fiber-containing grain products, fruits, and vegetables and reduced risk of cancer	• Low in fat (≤3 g/serving) • Good source of dietary fiber (10 to 19% Daily Value)
Fruits, vegetables, and grain products that contain fiber, particularly soluble fiber, and reduced risk of coronary heart disease.	• Low in saturated fat (≤1 g/serving) • Low in cholesterol (≤20 mg/serving) • Low in fat (≤3 g/serving) • Soluble fiber (≥0.6 g/serving)
Fruits and vegetables and reduced risk of cancer	• Low in fat (≤3 g/serving) • Good source of vitamin A, vitamin C, or dietary fiber (10 to 19% Daily Value)
Folate and reduced risk of neural tube defects	• Good source of folate (10 to 19% Daily Value) • Limited in vitamin A and vitamin D (≤100% Daily Value)
Sugar alcohols and reduced risk of tooth decay	• Sugar-free • Cannot lower dental plaque pH below 5.7 by bacterial fermentation
Soluble fiber from whole oats and reduced risk of heart disease; soluble fiber from psyllium seed husk and reduced risk of heart disease	• Soluble fiber from oats (≥0.75 g/serving) or from psyllium seed husk (≥ 1.7 g/serving) • Low in saturated fat (≤1 g/serving) • Low in cholesterol (≤20 mg/serving) • Low in fat (≤3 g/serving)
Soy protein and reduced risk of heart disease	• Soy protein (6 g/serving)

NOTE: With the exception of sugar alcohols and dental caries, all other health claims must also meet two additional criteria. First, a food making a health claim must be a naturally good source (containing at least 10 percent of the Daily Value) of at least one of the following nutrients: vitamin A, vitamin C, iron, calcium, protein, or fiber. Second, foods are disqualified from making health claims if a standard serving contains more than 20 percent of the Daily Value for total fat, saturated fat, cholesterol, or sodium.

allowed to say: that a substance "may" or "might" reduce disease risks. This wording is tentative because scientists are still accumulating evidence concerning the roles of diet in disease. A claim must also state that the development of a disease rests on many factors. A permissible health claim reads like this: "Development of heart disease depends on many factors. A healthful diet low in saturated fat and cholesterol may lower blood cholesterol levels and may reduce the risk of heart disease." Health claims on labels are carefully controlled so that they can be an asset to consumers who would rather not worry about grams, percentages, and other mathematical stumbling blocks as they shop for foods.

Table 1-9
TERMS USED ON FOOD LABELS

General Terms

Free, without, no, zero: Contains no amount or a trivial amount. *kCalorie-free* means containing fewer than 5 kcal per serving; *sugar-free* or *fat-free* means containing less than 0.5 g per serving.
Fresh: Raw, unprocessed, or minimally processed (blanched or irradiated) with no added preservatives.
Good source: Provides 10 to 19% of the Daily Value of a given nutrient per serving.
Healthy: A food that is low in fat, saturated fat, cholesterol, and sodium and that contains at least 10% of the Daily Value for vitamin A, vitamin C, iron, calcium, protein, or fiber.
High: Provides 20% or more of the Daily Value per serving.
Less, fewer, reduced: Provides 25% less of a nutrient or kcalories than a reference food. This may occur naturally or as a result of altering the food. For example, pretzels, which are usually low in fat, can claim to provide less fat than potato chips, a comparable food.
Light: This descriptor has three meanings on labels:
- A serving provides one-third fewer kcalories or half the fat of the regular product.
- A serving of a low-kcalorie, low-fat food provides half the sodium normally present.
- The product is light in color and texture; the label must make this intent clear, as in "light brown sugar."

More: Contains at least 10% more of the Daily Value for a given nutrient than a comparable food. The nutrient may be added or may occur naturally.

Carbohydrates: Fiber and Sugar Terms

High fiber: Provides 5 g or more fiber per serving; a high-fiber claim made on a food that contains more than 3 g fat per serving and per 100 g of food must also declare total fat.
Sugar-free: Provides less than 0.5 g sugar per serving.

Energy Terms

kCalorie-free: Contains fewer than 5 kcal per serving.
Low kcalorie: Contains 40 kcal or less per serving.
Reduced kcalorie: Contains at least 25% fewer kcalories per serving than the comparison product.

Fat Terms (Meat and Poultry Products)

Extra lean: A product contains:
- Less than 5 g fat.
- Less than 2 g saturated fat.
- Less than 95 mg cholesterol per serving.

Lean:[a] A product contains:
- Less than 10 g fat.
- Less than 4.5 g saturated fat.
- Less than 95 mg cholesterol per serving.

Fat and Cholesterol Terms (All Products)

Cholesterol-free: Contains less than 2 mg cholesterol per serving and 2 g or less saturated fat per serving.
Fat-free: Contains less than 0.5 g fat per serving.
Low cholesterol: Contains 20 mg or less cholesterol per serving and 2 g or less saturated fat per serving.
Low fat: Contains 3 g or less fat per serving.
Low saturated fat: Contains 1 g or less saturated fat per serving.
Percent fat free: May be used only if the product meets the definition of low fat or fat-free. Requires disclosure of grams fat per 100 g food.
Reduced or less cholesterol: Contains 25% or less cholesterol than the comparison food and 2 g or less saturated fat per serving.
Reduced or less fat: Contains 25% or less fat than the comparison food.
Reduced or less saturated fat: Contains 25% or less saturated fat than the comparison food and reduced by more than 1 g per serving.
Saturated fat-free: Contains less than 0.5 g saturated fat and less than 0.5 g *trans*-fatty acids.

Sodium Terms

Low sodium: Contains 140 mg or less sodium per serving.
Sodium-free: Contains less than 5 mg sodium per serving.
Very low sodium: Contains 35 mg or less sodium per serving.

---

[a]The word *lean* as part of the brand name (as in "Lean Supreme") indicates that the product contains fewer than 10 grams of fat per serving. *Lean* ground beef can contain up to 22.5 percent fat by weight.
SOURCES: The new food label, *FDA Backgrounder*, December 10, 1992; Nutrition labeling of meat and poultry products, *FSIS Backgrounder*, January 1993.

SELF CHECK

SELF CHECK

1. When people eat the foods typical of their families or geographic region, their choices are influenced by:
   a. occupation.
   b. nutrition.
   c. emotional state.
   d. ethnic heritage or tradition.

2. The energy-yielding nutrients are:
   a. fats, minerals, and water.
   b. minerals, proteins, and vitamins.
   c. carbohydrates, fats, and vitamins.
   d. carbohydrates, fats, and proteins.

3. The inorganic nutrients are:
   a. proteins and fats.
   b. vitamins and minerals.
   c. minerals and water.
   d. vitamins and proteins.

4. Alcohol is not a nutrient because:
   a. the body derives no energy from it.
   b. it is organic.
   c. it is converted to body fat.
   d. it does not contribute to the body's growth or repair.

5. DRI stands for:
   a. Daily Recommended Intakes.
   b. Dietary Requirements for Individuals.
   c. Dietary Reference Intakes.
   d. Daily Recommendations for Individuals.

6. Which of the following is consistent with the *Dietary Guidelines for Americans*?
   a. Choose a diet moderate in saturated fat and cholesterol.
   b. Be physically active each day.
   c. Choose a diet with plenty of milk products and meats.
   d. Eat an abundance of foods to ensure nutrient adequacy.

7. A slice of apple pie supplies 350 kcalories with 3 grams of fiber; an apple provides 80 kcalories and the same 3 grams of fiber. This is an example of:
   a. kcalorie control.
   b. nutrient density.

   c. variety.
   d. essential nutrients.

8. The diet-planning principle that provides enough, but not too much, of a constituent is:
   a. adequacy.
   b. balance.
   c. moderation.
   d. variety.

9. Foods within a given food group are similar in their contents of:
   a. energy.
   b. proteins and fibers.
   c. vitamins and minerals.
   d. carbohydrates and fats.

10. According to the Food Guide Pyramid, which foods should form the foundation of a healthy diet?
    a. vegetables
    b. breads, cereals, rice, and pasta
    c. fruits
    d. milk, yogurt, and cheese

Answers to these questions appear in Appendix H.

CLINICAL APPLICATIONS

1. Make a list of the foods and beverages you've consumed in the last two days. Look at each item on your list and consider why you chose the particular food or beverage you did. Did you eat cereal for breakfast because that's what you always eat (habit), or because it was the easiest, quickest food to prepare (convenience)? Did you put fat-free milk on the cereal because you want to control your energy and fat intake (nutrition)? In going down your list, you may be surprised to discover exactly why you chose certain foods.

2. As a nurse, you can uncover clues about a client's food choices by paying close attention. You may be surprised to discover why a client chooses certain foods, but you can then use this knowledge to serve the best interests of the client. For example, an elderly, undernourished widower may eat the same sandwich for lunch every day. In talking with the client, you discover that is what he and his wife fixed together each day. Consider ways you might be able to help the client learn to eat other foods and vary his choices.

3. Using the list of foods and beverages from exercise #1, compare your day's intake with the Daily Food Guide. Did you vary your choices within each food group? Did your intake match the recommended number of servings from each group? If not, list some changes you could have made to meet the recommendations.

Table 1-6 Notes

[a]J. H. Wilmore, Increasing physical activity: Alterations in body mass and composition, *American Journal of Clinical Nutrition* 63 (1996): S456–S460.

[b]D. Teegarden and coauthors, Previous physical activity relates to bone mineral measures in young women, *Medicine and Science in Sports and Exercise* 27 (1995): i–iv.

[c]D. C. Nieman, Exercise, upper respiratory tract infection, and the immune system, *Medicine and Science in Sports and Exercise* 26 (1994): 128–139.

[d]I. Thune and coauthors, Physical activity and the risk of breast cancer, *New England Journal of Medicine* 336 (1997): 1269–1275; M. M. Kramer and C. L. Wells, Does

physical activity reduce risk of estrogen-dependent cancer in women? *Medicine and Science in Sports and Exercise* 28 (1996): 322–334; J. A. Woods and J. M. Davis, Exercise, monocyte/macrophage function, and cancer, *Medicine and Science in Sports and Exercise* 26 (1994): 147–157.

[e]S. N. Blair, Physical inactivity and cardiovascular disease risk in women, *Medicine and Science in Sports and Exercise* 28 (1997): 9–10; G. B. M. Mensink and coauthors, Intensity, duration, and frequency of physical activity and coronary risk factors, *Medicine and Science in Sports and Exercise* 29 (1997): 1192–1198; NIH Consensus

Continued

NUTRITI**ON**THE**NET**

FOR FURTHER STUDY OF THE
TOPICS IN THIS CHAPTER,
ACCESS THESE WEB SITES.

www.nas.edu
Dietary Reference Intakes

www.usda.gov/fcs/cnpp/guide.htm
1995 Dietary Guidelines for Americans

www.shapeup.org
Shape Up America!

www.nalusda.gov/fnic/Fpyr/ pyramid.html
Food Guide Pyramid

www.usda.gov
U.S. Department of Agriculture

www.os.dhhs.gov
Department of Health and Human Services

web.health.gov/healthypeople

vm.cfsan.fda.gov/~lrd/newlabel.html
Background on Food Labels

Notes

[1] A. Drewnowski and C. L. Rock, The influence of genetic taste markers on food acceptance, *American Journal of Clinical Nutrition* 62 (1995): 506–511.

[2] A. Drewnowski, Taste preferences and food intake, *Annual Review of Nutrition* 17 (1997): 237–253.

[3] Drewnowski and Rock, 1995.

[4] A. E. Sloan, America's appetite '96: The top ten trends to watch and work on, *Food Technology* 50 (1996): 55–71.

[5] M. Nestle and coauthors, Behavioral and social influence on food choice, *Nutrition Reviews* 56 (1998): S50–S74.

[6] E. R. Monsen, New Dietary Reference Intakes proposed to replace the Recommended Dietary Allowances, *Journal of the American Dietetic Association* 96 (1996): 754–755.

[7] Standing Committee on the Scientific Evaluation of Dietary Reference Intakes, Food and Nutrition Board, Institute of Medicine, *Dietary Reference Intakes for Calcium, Phosphorus, Magnesium, Vitamin D, and Fluoride* (Washington, D.C.: National Academy Press, 1997), p. S-5.

[8] M. Y. Hwang, Benefits and dangers of alcohol, *Journal of the American Medical Association* 281 (1999): 104; N. S. Scrimshaw, Nutrition and health from womb to tomb, *Nutrition Today* 31 (1996): 55–67; J. M. McGinnis and W. H. Foege, Actual causes of death in the United States, *Journal of the American Medical Association* 270 (1993): 2207–2212.

[9] R. R. Pate and coauthors, Physical activity and public health: A recommendation from the Centers for Disease Control and Prevention and the American College of Sports Medicine, *Journal of the American Medical Association* 273 (1995): 402–407.

[10] American College of Sports Medicine, The recommended quality and quantity of exercise for developing and maintaining cardiorespiratory and muscular fitness and flexibility in healthy adults, *Medicine and Science in Sports and Exercise* 30 (1998): 975–991.

[11] U.S. Centers for Disease Control and Prevention and American College of Sports Medicine, Summary statement: Workshop on physical activity and public health, *Sports Medicine Bulletin* 28 (1993): 7.

[12] D. A. Leaf, D. L. Parker, and D. Schaad, Changes in VO_{2max}, physical activity, and body fat with chronic exercise: Effects on plasma lipids, *Medicine and Science in Sports and Exercise* 29 (1997): 1152–1159; P. T. Williams, Relationship of distance run per week to coronary heart disease risk factors in 8283 male runners, *Archives of Internal Medicine* 157 (1997): 191–198.

[13] U. M. Kujala and coauthors, Relationship of leisure time, physical activity and mortality: The Finnish twin cohort, *Journal of the American Medical Association* 279 (1998): 440–444; Summary of the Surgeon General's Report

addressing physical activity and health, *Nutrition Reviews* 54 (1996): 280–284; Pate and coauthors, 1995; R. S. Paffenbarger and coauthors, The association of changes in physical-activity level and other lifestyle characteristics with mortality among men, *New England Journal of Medicine* 328 (1993): 538–545; L. Sandvik and coauthors, Physical fitness as a predictor of mortality among healthy, middle-aged Norwegian men, *New England Journal of Medicine* 328 (1993): 533–537.

[14] U.S. Department of Health and Human Services, *Physical Activity and Health: A Report of the Surgeon General Executive Summary* (Washington, D.C.: Government Printing Office, 1996).

[15] S. N. Blair, Physical inactivity and cardiovascular disease risk in women, *Medicine and Science in Sports and Exercise* 28 (1997): 9–10; I. Thune and coauthors, Physical activity and the risk of breast cancer, *New England Journal of Medicine* 336 (1997): 1269–1275; NIH Consensus Development Panel on Physical Activity and Cardiovascular Health, Physical activity and cardiovascular health, *Journal of the American Medical Association* 276 (1996): 241–246; Pate and coauthors, 1995; American Heart Association, Position statement on exercise: Benefits and recommendations for physical activity programs for all Americans, *Circulation* 86 (1992): 340–344.

[16] R. R. Briefel, Nutrition monitoring in the United States, in *Present Knowledge in Nutrition*, 7th ed., ed. E. E. Ziegler and L. J. Filer (Washington, D.C.: International Life Sciences Institute Press, 1996), pp. 517–529.

[17] Ten-year comprehensive plan for the national nutrition monitoring and related research program, *Federal Register*, June 11, 1993.

[18] *Healthy People 2000: National Health Promotion and Disease Prevention Objectives* (Washington, D.C.: U.S. Department of Health and Human Services, 1990).

[19] J. M. McGinnis and P. R. Lee, Healthy People 2000 at mid decade, *Journal of the American Medical Association* 273 (1995): 1123–1129.

[20] P. Kurtzweil, Today's special: Nutrition information, *FDA Consumer*, May/June 1997, pp. 21–25.

241–246; P. T. Williams, High-density lipoprotein cholesterol and other risk factors for coronary heart disease in female runners, *New England Journal of Medicine* 334 (1996): 1298–1303.

[f] G. R. Hunter and coauthors, Fat distribution, physical activity, and cardiovascular risk factors, *Medicine and Science in Sports and Exercise* 29 (1997): 362–369; A. Goulding and coauthors, More exercise, less central fat distribution in women, *Journal of the American Medical Association* 276 (1996): 193–194; A. Tremblay and coauthors, Effect of intensity of physical activity on body fatness and fat distribution, *American Journal of Clinical Nutrition* 51 (1990): 153–157.

[g] G. Perseghin and coauthors, Increased glucose transport-phosphorylation and muscle glycogen synthesis after exercise training in insulin-resistant subjects, *New England*

Journal of Medicine 335 (1996): 1357–1362; S. N. Blair and coauthors, Physical activity, nutrition, and chronic disease, *Medicine and Science in Sports and Exercise* 28 (1996): 335–349.

[h] A. A. Hakim and coauthors, Effects of walking on mortality among nonsmoking retired men, *New England Journal of Medicine* 338 (1998): 94–99.

[i] L. DiPietro, The epidemiology of physical activity and physical function in older people.

Q&A NUTRITION IN PRACTICE

Finding the Truth about Nutrition

Nutrition and health receive so much attention on television, in the popular press, and on the Internet that it is easy to be overwhelmed with conflicting, confusing information. In fact, determining whether nutrition information is accurate can be a challenging task. It is also an important task, because nutrition affects a person both professionally and personally.

A person watches a nutrition report today on television and then reads a conflicting report the next day in the newspaper. Why do nutrition news reports and claims for nutrition products seem to contradict each other so often?

The problem of conflicting messages arises for several reasons:
- Popular media, often faced with tight deadlines and limited time or space to report new information, rush to present the latest "breakthrough" in a headline or a 60-second spot. They can hardly help omitting important facts about the study or studies that the "breakthrough" is based on.
- Despite tremendous advances in the last few decades, scientists still have much to learn about the human body and nutrition. Scientists themselves often disagree on their first tentative interpretations of new research findings, yet these are the very findings that the public hears most about.
- The popular media often broadcast preliminary findings in hopes of grabbing attention and boosting readership or television ratings.
- Commercial promoters turn preliminary findings into advertisements for products or supplements long before the findings have been validated—or disproved. The scientific process requires many experiments or trials to confirm a new finding. Seldom do promoters wait as long as they should to make their claims.
- Promoters are aware that consumers like to try new products or treatments even though they probably will not withstand the tests of time and scientific scrutiny.

So how can a person tell what claims to believe?

The Food and Nutrition Science Alliance (FANSA), whose partners include the American Dietetic Association (ADA), the American Society for Nutritional Sciences (ASNS), and the Institute of Food Technologists (IFT), attempts to help consumers distinguish valid from misleading nutrition information. FANSA has created a list of ten red flags for detecting "junk science" (see Table NP1-1).[1]

Because nutrition misinformation harms the health and economic status of consumers, the ADA works with health care professionals and educators to present sound nutrition information to the public and to actively confront nutrition misinformation.[2] Table NP1-2 offers a list of credible sources of nutrition information.

What about nutrition and health information found on the Internet? How does a person know whether the web sites are reliable?

With hundreds of millions of web sites on the Internet, searching for nutrition and health information can be an overwhelming experience. The Internet offers no guarantee of the accuracy of the information found there, and much of it is pure fiction. Web sites must be evaluated for their accuracy, just like every other source.

To help users find reliable nutrition information on the Internet, Tufts University maintains an online

Table NP1-1
FANSA'S RED FLAGS OF JUNK SCIENCE

1. Recommendations that promise a quick fix.
2. Dire warnings of danger from a single product or regimen.
3. Claims that sound too good to be true.
4. Simplistic conclusions drawn from a complex study.
5. Recommendations based on a single study.
6. Dramatic statements that are refuted by reputable scientific organizations.
7. Lists of "good" and "bad" foods.
8. Recommendations made to help sell a product.
9. Recommendations based on studies published without peer review.
10. Recommendations from studies that ignore differences among individuals or groups

SOURCE: Reprinted with permission from B. Hansen. President's address, 1996: A virtual organization for nutrition in the 21st century, *American Journal of Clinical Nutrition* 64 (1996): 796–799. © American Society for Clinical Nutrition.

rating and review guide called the Nutrition Navigator. The ratings reflect the opinions of a panel of nutrition experts who have scored selected web sites on the basis of their accuracy, depth, and ease of use. In addition to a rating, the Nutrition Navigator provides a review of the web site's content and usefulness.

Navigator.tufts.edu
Tufts University

Also, when using the Internet, pay attention to the last few letters in a web site's name. Generally, an extension of three letters following a dot gives clues about a site's affiliations. For example, a site bearing the letters "gov" is a government site and generally posts valid information. Universities use the extension "edu," which stands for educational institution. The extension "com" stands for commercial, a business web site; and "org" stands for organization. Some credible sites include:

Healthfinder.gov
U.S. Government

www.Quackwatch.com
Stephen Barrett's Quackwatch

eatright.org
American Dietetic Association

www.faseb.org/asns
American Society for Nutritional Sciences

www.ift.org
Institute of Food Technologists

www.ncahf.org
National Council Against Health Fraud

Table NP1-2
CREDIBLE SOURCES OF NUTRITION INFORMATION

Professional health organizations, government health agencies, volunteer health agencies, and consumer groups provide consumers with reliable health and nutrition information. Credible sources of nutrition information include:

- Professional health organizations, especially the American Dietetic Association's National Center for Nutrition and Dietetics (NCND); also the Society for Nutrition Education and the American Medical Association.
- Government health agencies such as the Federal Trade Commission (FTC), the U.S. Department of Health and Human Services (DHHS), the Food and Drug Administration (FDA), and the U.S. Department of Agriculture (USDA).
- Volunteer health agencies such as the American Cancer Society, the American Diabetes Association, and the American Heart Association.
- Reputable consumer groups such as the Better Business Bureau, the Consumers Union, the American Council on Science and Health, and the National Council on Science and Health, and the National Council Against Health Fraud.

Appendix D provides addresses for these and other organizations.

SOURCE: Data from J. M. Ashley and W. T. Jarvis, Position of the American Dietetic Association: Food and nutrition misinformation, *Journal of the American Dietetic Association* 95 (1995): 705–707.

Everyone seems to be giving advice on nutrition. How can a person tell whom to listen to?

Registered dietitians (R.D.'s) and nutrition professionals with advanced degrees (M.S., Ph.D.) are experts (see the glossary on this page). These professionals are probably in the best position to answer a person's nutrition questions. On the other hand, **"nutritionists"** may be experts or quacks, depending on the state where they practice. Some states require people who use this title to meet strict standards. In other states, a "nutritionist" may be any individual who claims a career connection with the nutrition field.

Other purveyors of nutrition information may also lack credentials. A health food store owner may be in the nutrition business simply because it is a lucrative market. The owner may have a background in business or sales and no education in nutrition at all. Such a person is not qualified to provide nutrition information to customers. For accurate nutrition information, seek out a trained professional with a college education in nutrition—an expert in the field of **dietetics.**

What about nurses and other health care professionals?

All members of the health care team share responsibility for helping each client to achieve optimal health, but the registered dietitian is usually the primary nutrition expert. Each of the other team members has a related specialty. Some physicians are specialists in clinical nutrition and are also experts in the field. Other physicians, nurses, and **dietetic technicians (D.T.R.'s)** often assist dietitians in providing nutrition information and may help to administer direct nutrition care. Nurses play central roles in client care management and client relationships. Visiting nurses and home health care nurses may become intimately involved in clients' nutrition care at home, teaching them both theory and cooking techniques. Physical therapists can provide individualized exercise programs related to nutrition—for example, to help control obesity. Social workers may provide practical and emotional support.

What roles might these health professionals play in nutrition care?

Some of the responsibilities of the health care professional might be:
- Helping people understand why nutrition is important to them.

GLOSSARY OF NUTRITION EXPERTS

dietetic technician registered (D.T.R.): a professional who has earned an associate degree or higher; has completed a dietetic technician program approved by the American Dietetic Association (ADA): has passed a national registration exam; and assists in planning, implementing, and evaluating nutritional care.

dietetics: the practical application of nutrition, including the assessment of nutrition status, recommendation of appropriate diets, nutrition education, and the planning and serving of meals.

nutritionist: a person who specializes in the study of nutrition. Some nutritionists are registered dietitians, but others are self-described experts whose training may be minimal or nonexistent. Some states make the term meaningful by allowing it to apply only to people who have master's (M.S.) or doctoral (Ph.D.) degrees from institutions accredited to offer such degrees in nutrition or related fields.

registered dietitian (R.D.): a dietitian who has graduated from a university or college after completing a program of dietetics that has been accredited by the American Dietetic Association (or Dietitians of Canada). The dietitian must serve in an approved internship or coordinated program to practice the necessary skills, pass the association registration examination, and maintain competency through continuing education. Many states require licensing for practicing dietitians. Licensed dietitians (L.D.'s) have met all *state* requirements to offer nutrition advice.

- Answering questions about food and diet.
- Explaining to clients how modified diets work.
- Collecting information about clients that may influence their nutrition status.
- Identifying clients at risk for poor nutrition status (see Chapters 13 and 14) and recommending or taking appropriate action.
- Recognizing when clients need extra help with nutrition problems (in such cases, the problems should be referred to a dietitian or physician).

Health care professionals may routinely perform these nutrition-related tasks:

- Obtaining diet histories.
- Taking weight and height measurements.
- Feeding clients who cannot feed themselves.
- Recording what clients eat or drink.
- Observing clients' responses and reactions to foods.
- Helping clients mark menus.
- Monitoring weight changes.
- Monitoring food and drug interactions.
- Encouraging clients to eat.
- Assisting clients at home in planning their diets and managing their kitchen chores.
- Alerting the physician or dietitian when nutrition problems are identified.
- Charting actions taken and communicating on these matters with other professionals as needed.

Thus, although the dietitian assumes the primary role as the nutrition expert on a health care team, other health care professionals play important roles in administering nutrition care.

Notes

[1] B. Hansen, President's address, 1996: A virtual organization for nutrition in the 21st century, *American Journal of Clinical Nutrition* 64 (1996): 796–799.

[2] J. M. Ashley and W. T. Jarvis, Position of The American Dietetic Association: Food and nutrition misinformation, *Journal of the American Dietetic Association* 95 (1995): 705–707.

CHAPTER 2

CARBOHYDRATES

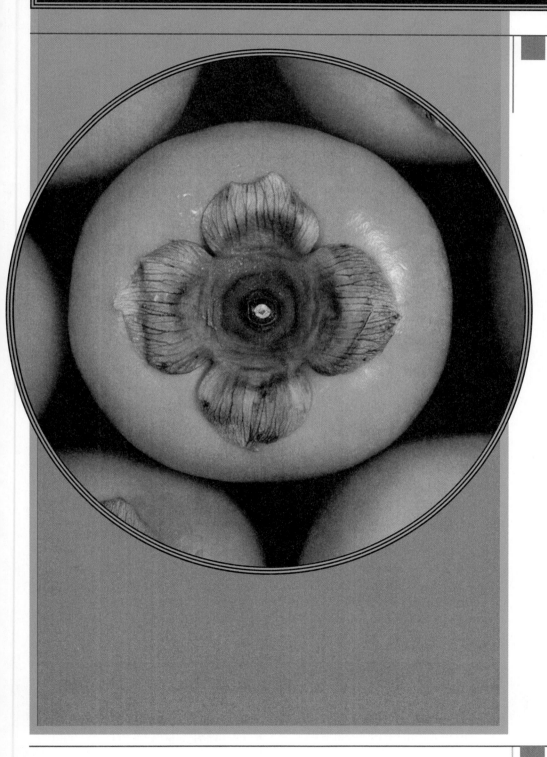

CONTENTS

The Chemist's View
of Carbohydrates

Health Effects of Sugars and
Alternative Sweeteners

Health Effects of Complex
Carbohydrates

Energy Nutrients
in Perspective

Nutrition in Practice:
Nutrition and Dental Health

Grains, vegetables, legumes, fruits, and milk offer ample carbohydrate.

Figure 2-1
CHEMICAL STRUCTURE OF GLUCOSE

On paper, the stucture of glucose has to be drawn flat, but in nature the five carbons and oxygen are roughly in a plane, with the H, OH, and CH₂OH extending out above and below it.

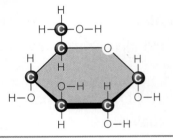

carbohydrates: energy nutrients composed of monosaccharides.
carbo = carbon   hydrate = water

simple carbohydrates (sugars): the monosaccharides (glucose, fructose, and galactose) and the disaccharides (sucrose, lactose, and maltose).

monosaccharide (mon-oh-SACK-uh-ride): a single sugar unit.
mono = one   saccharide = sugar

disaccharide (dye-SACK-uh-ride): a pair of sugar units bonded together.
di = two

complex carbohydrates: long chains of sugars arranged as starch or fiber; also called **polysaccharides.**

glucose (GLOO-kose): a monosaccharide, the sugar common to all disaccharides and polysaccharides; also called blood sugar or dextrose.

Most people would like to feel good all the time. Part of the secret of feeling well is replenishing the body's energy supply with food. That means choosing foods that contain the energy nutrients—carbohydrate and fat, primarily. But which to choose?

Carbohydrate is the preferred energy source for most of the body's functions. The human brain depends on carbohydrate as an energy source. Athletes eat a "high-carb" diet to store as much muscle fuel as possible, and dietary recommendations urge people to eat carbohydrate-rich foods for better health. Many people, however, mistakenly think of carbohydrate-rich foods as "fattening" and avoid them. In truth, people who wish to lose fat and maintain lean tissue can best do so by controlling kcalories, choosing high-carbohydrate, high-fiber foods, and limiting fat-rich foods. All unrefined plant foods—grains, vegetables, legumes, and fruits—provide ample carbohydrate and fiber with little or no fat. Milk is the only animal-derived food that contains significant amounts of carbohydrate.

Carbohydrate shares its fuel-providing responsibility with fat. Fat, however, has disadvantages: it can not be used exclusively as fuel by the brain and central nervous system, and diets high in fat are associated with many chronic diseases. The other energy sources available to the body—protein and alcohol—offer no advantage as fuels. Protein is best left to serve its own diverse functions, as discussed in Chapter 4. Alcohol, of course, has well-known undesirable side effects when used in excess. Alcohol abuse and its relationships with nutrition are the subject of Nutrition in Practice 23.

■■ The Chemist's View of Carbohydrates

Chemists divide the **carbohydrates** into two categories: simple and complex. The **simple carbohydrates (sugars)** are:

- **Monosaccharides** (single sugars).
- **Disaccharides** (double sugars).

The **complex carbohydrates (polysaccharides)** are:

- Starch.
- Glycogen.
- Some fibers.

All of these carbohydrates are composed of the simple sugar **glucose** and other compounds that are much like glucose in composition and structure. Figure 2-1 shows the chemical structure of glucose.

■■ Monosaccharides (Single Sugars)

Three monosaccharides are important in nutrition: glucose, fructose, and galactose. Almost all of the body's cells use glucose as their chief energy source. The body can obtain this glucose from all carbohydrates. In the body, the hormone **insulin,** among others, enables cells to take up glucose from the blood and so helps to keep blood glucose constant (**homeostasis**).

LEARNING LINK

When a person's body fails to produce insulin, or does not respond to insulin appropriately, diabetes mellitus develops. Diabetes, characterized by high blood glucose, is responsive to dietary control, but successful treatment often depends on medical and nutrition therapy. As Chapter 20 explains, nurses and other health care professionals play a vital role in facilitating patient compliance and successful outcome.

Plants also make **fructose,** which is the sweetest of the sugars. It is abundant in fruits, honey, and saps. Glucose and fructose are the most common monosaccharides in nature. The third single sugar, **galactose,** appears in nature only as part of lactose, a disaccharide also known as milk sugar. Only during digestion is galactose freed as a single sugar.

Disaccharides (Double Sugars)

In disaccharides, pairs of single sugars are linked together. Three disaccharides are important in nutrition: maltose, sucrose, and lactose. All three have glucose as one of their single sugars. As Table 2-1 shows, the other monosaccharide is either another glucose (in maltose), or fructose (in sucrose), or galactose (in lactose). The shapes of the sugars in Table 2-1 reflect their chemical structures as drawn on paper.

Table 2-1
THE MAJOR SUGARS

Monosaccharides	Disaccharides
Glucose	Sucrose (glucose + fructose)
Fructose	Lactose (glucose + galactose)
Galactose	Maltose (glucose + glucose)
(found only as part of lactose)	

MALTOSE

The first disaccharide, **maltose,** is a plant sugar that consists of two glucose units. Maltose appears at only one stage in the life of a plant—when the plant is breaking down its stored starch for energy and starting to sprout.

SUCROSE: TABLE SUGAR

Sucrose (table, or white, sugar) is the most familiar of the three disaccharides and is what people mean when they speak of "sugar." This sugar is usually obtained by refining the juice from sugar beets or sugarcane to provide the brown, white, and powdered sugars available in the supermarket, but it occurs naturally in many fruits and vegetables.

When a person eats a food containing sucrose, enzymes in the digestive tract split the sucrose into its glucose and fructose components. Because the body can convert fructose to glucose, one molecule of sucrose can ultimately yield two molecules of glucose.

HONEY VERSUS SUGAR

People often ask: What is the difference between honey and white sugar? Is honey more nutritious? Honey, like white sugar, contains glucose and fructose. The difference is that in white sugar, the glucose and fructose are bonded together in pairs, whereas in honey some of them are paired and some are free single sugars. When you eat either white sugar or honey, though, your body breaks all of the sugars apart into single sugars. It ultimately makes no difference, then, whether you eat single sugars

insulin: a hormone secreted by the pancreas in response to high blood glucose. It promotes cellular glucose uptake and use or storage.

homeostasis: the maintenance of constant internal conditions (such as chemistry, temperature, and blood pressure) by the body's control system.
homeo = the same stasis = staying

fructose: a monosaccharide; sometimes known as fruit sugar. It is abundant in fruits, honey, and saps.
fruct = fruit

galactose: a monosaccharide; part of the disaccharide lactose.

maltose: a disaccharide composed of two glucose units; sometimes known as malt sugar.

sucrose: a disaccharide composed of glucose and fructose; commonly known as table sugar, beet sugar, or cane sugar.
sucro = sugar

Digestion and absorption are the topics of Chapter 5.

Honey, like sugar, contains glucose and fructose.

An orange provides the same sugars as honey or sugar, but the packaging makes a big nutrition difference.

A 12 oz can of cola contains the equivalent of 10 tsp sugar.

lactose: a disaccharide composed of glucose and galactose; commonly known as milk sugar.

lact = milk

starch: a plant polysaccharide composed of glucose and digestible by human beings.

The short chains of glucose units that result from the breakdown of starch are known as dextrins. The word sometimes appears on food labels because dextrins can be used as thickening agents in foods.

linked together, as in white sugar, or the same sugars unlinked, as in honey; they will end up as single sugars in your body. True, honey contains trace amounts of a few vitamins and minerals, but to say that honey is nutritious is misleading.

FRUITS VERSUS SUGAR Some sugar sources are more nutritious than others, though. Consider a fruit such as an orange. The orange provides the same sugars and about the same energy as a tablespoon of sugar or honey, but the packaging makes a big difference in nutrient density. The sugars of the orange are diluted in a large volume of fluid that contains valuable vitamins and minerals, and the flesh and skin of the orange are supported by fibers that also offer health benefits. A tablespoon of honey offers no such bonuses.

COLA BEVERAGES AND SWEETS A cola beverage, containing many teaspoons of sugar, offers no advantages either. Table 2-2 shows sample nutrients supplied by some sugar sources; note the "0s" and "traces" by honey, sugar, and the cola beverage and the substantial numbers by the others. Sucrose is often the principal ingredient of carbonated beverages, candy, cakes, frostings, cookies, and other concentrated sweets, so they are often not nutritious carbohydrate choices.

LACTOSE: MILK SUGAR **Lactose** is the principal carbohydrate of milk. Most human babies are born with the digestive enzymes necessary to split lactose into its two monosaccharide parts, glucose and galactose, so as to absorb it. Breast milk thus provides a simple, easily digested carbohydrate that meets a baby's energy needs; most formulas do, too, because they are made from milk.

LEARNING LINK

After infancy, many people lose the ability to digest lactose efficiently. This condition, known as lactose intolerance, is not the same as milk allergy, which is caused by an immune reaction to the protein in milk. People who are lactose intolerant often avoid milk and milk products completely and thus risk deficiencies of important nutrients such as calcium and vitamin D. Treatment of lactose intolerance centers on dietary strategies to alleviate symptoms without compromising the client's nutritional well-being. Lactose intolerance is discussed further in Chapter 16.

Starch and Glycogen (Energy-Yielding Polysaccharides)

Unlike the sugars, which contain the three monosaccharides—glucose, fructose, and galactose—in different combinations, the polysaccharides, starch and glycogen, are composed almost entirely of glucose. They differ from each other only in the nature of the bonds that link the glucose units together.

STARCH **Starch** is a long, straight or branched chain of hundreds of glucose units linked together. These giant molecules are packed side by side in a rice grain or potato root—as many as a million per cubic inch of food. When a person eats the plant, the body splits the starch into glucose units and uses the glucose for energy.

STARCHY FOODS All starchy foods are plant foods. Grains are the richest food source of starch. In most human societies, people depend on a staple grain for much of their food energy: rice in Asia; wheat in Canada, the

Table 2-2

SAMPLE NUTRIENTS IN SUGARS AND OTHER FOODS

The indicated portion of any of these foods provides approximately 100 kcalories. Notice that for a similar number of kcalories and grams of carbohydrate, milk, legumes, fruits, grains, and vegetables offer more of the other nutrients than do the sugars.

	Size of 100 kcal Portion	Carbohydrate (g)	Protein (g)	Calcium (mg)	Iron (mg)	Vitamin A (µg RE)	Vitamin C (mg)
Foods							
Milk, 1% low-fat	1 c	12	8	300	0.1	144	2
Kidney beans	½ c	20	7	30	1.6	0	2
Apricots	6	24	2	30	1.1	554	22
Bread, whole wheat	1½ slices	20	4	30	1.9	0	0
Broccoli, cooked	2 c	20	12	188	2.2	696	148
Sugars							
Sugar, white	2 tbs	24	0	trace	trace	0	0
Molasses, blackstrap	2½ tbs	28	0	343	12.6	0	0.1
Cola beverage	1 c	26	0	6	trace	0	0
Honey	1½ tbs	26	trace	2	0.2	0	trace

United States, and Europe; corn in much of Central and South America; and millet, rye, barley, and oats elsewhere. A second important source of starch is the legume (bean and pea) family. Legumes include peanuts and "dry" beans such as butter beans, kidney beans, "baked" beans, black-eyed peas (cowpeas), chickpeas (garbanzo beans), and soybeans. Root vegetables (tubers) such as potatoes and yams are a third major source of starch, and in many non-Western societies, they are the primary starch sources.

Grains, legumes, and tubers not only are rich in starch, but also contain abundant dietary fiber, protein, and other nutrients. When nutrition experts advise people to seek out carbohydrate-rich foods to meet most of their energy needs, these are the foods they are recommending.

■■ GLYCOGEN **Glycogen** molecules, which are also made of chains of glucose, are more highly branched than starch molecules. Glycogen stores energy for human beings and animals, just as starch stores energy for plants. Because glycogen does not occur in plants and is found in meats only to a limited extent, it is not important as a nutrient. Nevertheless, glycogen plays an important role in the body as a readily available source of glucose, especially during physical activity.

Athletes, especially endurance athletes such as cyclists and distance runners, know how crucial glycogen stores are to their performance. How much carbohydrate a person eats affects how much glycogen is stored, which in turn influences how much will be used during activity. Thus, diet bears on athletic performance because the more glycogen the athlete stores, the longer the stores will last during physical activity. If the activity is long and hard enough, however, glycogen stores run out almost completely. Endurance athletes are therefore wise to take light carbohydrate snacks or drinks (under 200 kcalories) periodically during activity.

■■ The Fibers

The **fibers** of plants are constituents of plant cell walls. Most fibers are polysaccharides—chains of sugars—just as starch is, but in fibers the sugar units are held together by bonds that human digestive enzymes cannot break. Figure 2-2 (p. 36) shows the difference between starch and the plant fiber cellulose. In addition to cellulose, fibers include the polysaccharides hemicellulose, pectins, gums, and mucilages, as well as the nonpolysaccharide lignins.

glycogen (GLY-co-gen): a polysaccharide composed of glucose, made and stored by liver and muscle tissues of human beings and animals as a storage form of glucose. Glycogen is not a significant food source of carbohydrate and is not counted as one of the complex carbohydrates in foods.

fibers: a general term denoting in plant foods the polysaccharides cellulose, hemicellulose, pectins, gums and mucilages, as well as the nonpolysaccharide lignins, that are not attacked by human digestive enzymes.

Figure 2-2
STARCH AND CELLULOSE MOLECULES COMPARED (SMALL SEGMENTS)

The bonds that link the glucose units together in cellulose are different from the bonds in starch (and glycogen). Human enzymes cannot digest cellulose.

Starch   Cellulose

■■ **FIBERS IN FOODS** Cellulose is the main constituent of plant cell walls, so it is found in all vegetables, fruits, and legumes. Hemicellulose is the main constituent of cereal fibers. Pectins are abundant in vegetables and fruits, especially citrus fruits and apples. The food industry uses pectins to thicken jelly and keep salad dressing from separating. Gums and mucilages have similar structures and are used as additives or stabilizers by the food industry. Lignins are the tough, woody parts of plants; few foods people eat contain much lignin.

■■ **FIBERS IN THE BODY** Although cellulose and other fibers are not broken down by human enzymes, some fibers can be digested by bacteria in the human digestive tract. Bacterial digestion of fibers can generate some absorbable products that can yield energy when metabolized. Food fibers, therefore, can contribute some energy, depending on the extent to which they break down in the body.

■■ **SOLUBLE VERSUS INSOLUBLE FIBERS** Fibers are classified according to several characteristics, including their chemical structure, their digestibility by bacterial enzymes, and their solubility in water. Some fibers are **insoluble,** meaning that they do not dissolve readily in water; other fibers are **soluble—** they do dissolve in water. These distinctions influence the health effects of fibers, which are discussed in a later section.

insoluble fibers: the tough, fibrous structures of fruits, vegetables, and grains; indigestible food components that do not dissolve in water.

soluble fibers: indigestible food components that readily dissolve in water and often impart gummy or gel-like characteristics to foods. An example is pectin from fruit, which is used to thicken jellies.

■■ Health Effects of Sugars and Alternative Sweeteners

Starch-rich and fiber-rich foods such as vegetables, grains, legumes, and fruits should predominate in people's diets; concentrated sweets should account for only 10 percent or less of total kcalories. For most people this means that total carbohydrate intake should increase while sugar intake should decline. The sugars in vegetables, fruits, grains, and milk are acceptable because they are accompanied by many nutrients. In contrast, concentrated sweets contribute many kcalories but relatively few nutrients, and so should be limited. People who want to limit their use of sugar may choose from two sets of alternative sweeteners: sugar alcohols and artificial sweeteners.

■■ Sugars

In the United States, it is estimated that each man, woman, and child consumes over 100 pounds of sugar per year, or a little less than 2 pounds per week.[1] A steady upward trend (see Figure 2-3) in sugar consumption is largely the result of a dramatic increase in consumption of commercially prepared foods and beverages, which contain abundant sugars added to the foods during processing by food manufacturers. In contrast, people are adding less sugar in the kitchen. Recommendations that people reduce their consumption of concentrated sugars to 10 percent or less of total kcalories stem from widely published reports of research performed during the 1970s.[2] These reports implicated

Starch- and fiber-rich foods are the foods to emphasize.

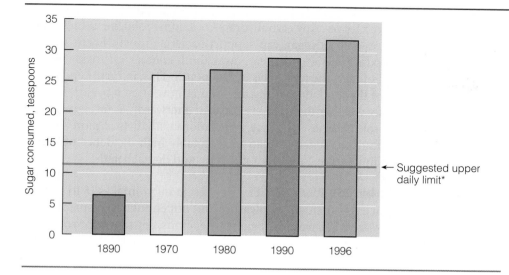

Figure 2-3
DAILY SUGAR CONSUMPTION IN THE UNITED STATES 1890–1996

*Recommended upper limit for a 2200 - kcalorie diet.

SOURCE: U.S. Department of Agriculture, Economic Research Service, *A Dietary Assessment of the U.S. Food Supply: Comparing Per Capita Food Consumption with the Food Guide Pyramid Serving Recommendations* (Washington, D.C.: Government Printing Office 1998), AER#772 p. 25.

sugars (mainly sucrose) as a possible contributing factor in several diseases. Since then, many accusations have been made against the sugar sucrose, but the Food and Drug Administration (FDA) and the National Academy of Sciences have concluded that in the amounts people currently consume, sugar poses no major health risk. Still, some controversy continues. A brief description of accusations pertaining to sugar's effects on health may help clarify the issues. Most commonly, reports accuse sugar of causing obesity, diabetes, hyperactivity and aggressive behavior in children, and dental caries.

⁑ SUGAR AND OBESITY On the first accusation—that sugar causes obesity—the facts are these. Population studies show that obesity rises as sugar consumption increases, but that sugar is not the sole cause. Whenever sugar intakes increase, usually fat and total energy intakes do, too. Simultaneously, physical activity declines. On the other hand, obesity also occurs where sugar intakes are low. In some instances, obese people eat less sugar than thin people do.[3] Fat is more kcalorie dense than sugar and can more easily contribute to obesity. Thus, sugar can contribute to obesity, but does not, by itself, cause obesity.

⁑ SUGAR AND DIABETES On sugar's relation to diabetes, the evidence is conflicting and interesting. In many parts of the world, as sugar consumption has increased, a profound increase in the incidence of one type of **diabetes (type 2)** has occurred. Yet, in other populations, no relationship has been found between diabetes and sugar consumption. Body fatness is one known risk factor: the majority of people with type 2 diabetes are overweight or obese, and evidence shows that weight reduction helps to prevent diabetes or relieve its symptoms. Genetics, age, and physical inactivity are also considered risk factors for type 2 diabetes.[4] Thus, sugar may be a causative factor only when it contributes to obesity.

type 2 diabetes: the more common type of diabetes in which the fat cells resist insulin.

⁑ SUGAR AND BEHAVIOR The accusation that sugar causes hyperactive or aggressive behavior in children stands unproved.[5] Scientific research has failed to demonstrate any consistent effect of sugar on behavior in either normal or hyperactive children.[6] If sugar is related to behavior problems in children, it may be because the sugary foods replace nutrient-dense foods in children's diets, making nutrient deficiencies likely. Many different nutrient deficiencies adversely affect behavior. A lack of nutrients in children's diets, not sugar itself, can in some cases contribute to undesirable behavior.

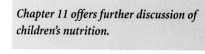

Chapter 11 offers further discussion of children's nutrition.

dental caries: the gradual decay and disintegration of a tooth.

■■ **SUGAR AND DENTAL CARIES** As to whether sugar contributes to **dental caries,** the evidence says yes. Any carbohydrate-containing food, including bread, bananas, or milk, as well as sugar, can support bacterial growth in the mouth. These bacteria produce the acid that eats away tooth enamel. Of major importance is the length of time the food stays in the mouth. This, in turn, depends on the composition of the food, how sticky the food is, how often a person eats the food, and especially whether the teeth are brushed afterward.[7] Total sugar intake still plays a major role in caries incidence; populations with diets of more than 10 percent of kcalories from sugar have an unacceptably high incidence of dental caries.[8] Nutrition in Practice 2 discusses nutrition and dental health.

■■ **RECOMMENDED SUGAR INTAKES** Moderate sugar intakes (5 to 10 percent of total kcalories)—enough for pleasure, but not enough to displace more nutritious foods—are not harmful. Sugar is a delicious, concentrated source of food energy, but it contains no protein, vitamins, or minerals. Eaten in the place of nutrient- and fiber-rich foods, it makes malnutrition likely.

■■ **RECOGNIZING SUGARS** People often fail to recognize sugar in all its forms and so fail to realize how much they consume. To help your clients estimate how much sugar they consume, tell them to treat all of the following concentrated sweets as equivalent to 1 teaspoon of white sugar:

- 1 teaspoon brown sugar, candy, jam, jellies, any corn sweetener, syrup, honey, molasses, or maple sugar.
- 1 tablespoon catsup.
- 1½ ounces carbonated soft drink.

GLOSSARY SUGAR

brown sugar: refined white sugar crystals to which manufacturers have added molasses syrup with natural flavor and color; 91 to 96 percent pure sucrose.

concentrated fruit juice sweetener: a concentrated sugar syrup made from dehydrated, deflavored fruit juice, commonly grape juice; used to sweeten products that can then claim to be "all fruit."

confectioners' sugar: finely powdered sucrose; 99.9 percent pure.

corn sweeteners: corn syrup and sugars derived from corn.

corn syrup: a syrup containing mostly glucose, produced by the action of enzymes on cornstarch. See also *high-fructose corn syrup (HFCS).*

dextrose: an older name for glucose and the name used for glucose in intravenous solutions.

fructose, galactose, glucose: already defined (pp. 32 and 33).

granulated sugar: common table sugar, crystalline sucrose; 99.9 percent pure.

high-fructose corn syrup (HFCS): the predominant sweetener used in processed foods today. HFCS is mostly fructose; glucose makes up the balance.

honey: sugar (mostly sucrose) formed from nectar gathered by bees. An enzyme splits the sucrose into glucose and fructose. Composition and flavor vary, but honey always contains a mixture of sucrose, fructose, and glucose.

invert sugar: a mixture of glucose and fructose formed by splitting sucrose in a chemical process; sold only in liquid form, sweeter than sucrose. Invert sugar is used as an additive to help preserve food freshness and prevent shrinkage.

lactose: already defined (p. 34).

levulose: an older name for fructose.

maltose: already defined (p. 33).

maple sugar: a sugar (mostly sucrose) purified from concentrated sap of the sugar maple tree. Maple sugar is expensive compared with other sweeteners.

molasses: a thick brown syrup, left over from sugarcane juice during sugar refining. Blackstrap molasses contains iron, which comes from the machinery used to process it. This iron is not as well absorbed as the iron in meats and other foods.

raw sugar: the first crop of crystals harvested during sugar processing. Raw sugar cannot be sold in the United States because it contains too much filth (dirt, insect fragments, and the like). Sugar sold domestically as raw sugar has actually gone through about half of the refining steps.

sucrose: already defined (p. 33).

turbinado (ter-bih-NOD-oh) sugar: raw (brown) sugar from which the filth has been washed; legal to sell in the United States.

white sugar: pure sucrose, produced by dissolving, concentrating, and recrystallizing raw sugar.

These portions of sugar all provide about the same number of kcalories. Some are closer to 10 kcalories (for example, 14 kcalories for molasses), while some are over 20 (22 kcalories for honey), so an average figure of 20 kcalories is an acceptable approximation. The glossary on page 38 presents the multitude of names that denote sugar on food labels. The next section discusses sugar substitutes.

Alternative Sweeteners: Sugar Alcohols

The **sugar alcohols** are carbohydrates, but they yield slightly less energy (2 to 3 kcalories per gram) than sucrose (4 kcalories per gram) because they are not absorbed completely.[9] The sugar alcohols are sometimes called **nutritive sweeteners.**

The sugar alcohols occur naturally in fruits; they are also used by manufacturers as a low-energy bulk ingredient in many products. Unlike sucrose, sugar alcohols are fermented to gases in the large intestine by intestinal bacteria. Consequently, side effects such as gas, abdominal discomfort, and diarrhea make the sugar alcohols less attractive than the artificial sweeteners.

The advantage of using sugar alcohols is that they do not contribute to dental caries. Bacteria in the mouth metabolize sugar alcohols much more slowly than sucrose, thereby inhibiting the production of acids that promote caries formation. Sugar alcohols are therefore valuable in chewing gums, breath mints, and other products that people keep in their mouths a while. The FDA allows food labels to carry a health claim (see p. 24 in Chapter 1) about the relationship between sugar alcohols and the nonpromotion of dental caries as long as the FDA criteria for sugar-free status and other criteria are met.[10]

Alternative Sweeteners: Artificial Sweeteners

The **artificial sweeteners** are not carbohydrates. They yield virtually no energy in the amounts typically used and are sometimes called **nonnutritive sweeteners.** Like the sugar alcohols, artificial sweeteners make foods taste sweet without promoting tooth decay. The glossary below offers details on six artificial sweeteners of interest.

SACCHARIN Saccharin has been used for more than 100 years in the United States and is currently used by more than 50 million people, mainly in soft drinks and as the tabletop sweeteners Sweet'n Low and Sweet Thing. Questions about the safety of saccharin arose in 1977 when experiments suggested that it caused bladder tumors in second-generation rats fed high doses.

sugar alcohols: sugarlike compounds; like sugars, they are sweet to taste but yield 2 to 3 kcal per gram, slightly less than sucrose. Examples are maltitol, mannitol, sorbitol, isomalt, lactitol, and xylitol.

nutritive sweeteners: sweeteners that yield energy, including both the sugars and the sugar alcohols.

These chewing gums contain sugar alcohols, which are better than sugar for the teeth, but are not kcalorie-free.

artificial sweeteners: noncarbohydrate, 0- or low-kcalorie synthetic sweetening agents; sometimes called **nonnutritive sweeteners.**

GLOSSARY
ARTIFICIAL SWEETENERS

acesulfame (AY-sul-fame) potassium: a 0-kcalorie artificial sweetener that tastes 200 times as sweet as sucrose; also known as acesulfame-K because K is the chemical symbol for potassium. Acceptable Daily Intake (ADI) = 15 milligrams/kilogram body weight.

alitame (AL-ih-tame): a low-kcalorie artificial sweetener composed of two amino acids (alanine and aspartic acid) that tastes 2000 times as sweet as sucrose; FDA approval pending.

aspartame (ah-SPAR-tame) or ASS-par-tame): a low-kcalorie artificial sweetener composed of two amino acids (phenylalanine and aspartic acid) that tastes 200 times as sweet as sucrose. ADI = 50 milligrams/kilogram body weight.

cyclamate (SIGH-kla-mate): a 0-kcalorie artificial sweetener that tastes 30 times as sweet as sucrose; FDA approval pending in the United States; available in Canada in grocery stores but only as a tabletop sweetener, not as an additive.

saccharin (SAK-ah-ren): a 0-kcalorie artificial sweetener that tastes 500 times as sweet as sucrose; approved in the United States, but available in Canada only in pharmacies and only as a sweetener, not as an additive. ADI = 5 milligrams/kilogram body weight.

sucralose (SUE-kra-lose): a 0-kcalorie artificial sweetener that taste 600 times as sweet as sucrose. ADI = 15 milligrams/kilogram body weight.

The FDA proposed banning saccharin as a result. Public outcry in favor of saccharin was so loud that Congress declared a moratorium on the ban, a moratorium that has been repeatedly extended. In 1991, the FDA withdrew its proposal to ban saccharin, but labels must still carry a consumer warning: "Use of this product may be hazardous to your health. This product contains saccharin, which has been determined to cause cancer in laboratory animals."

Does saccharin cause cancer? The largest population study to date, involving 9000 men and women, showed overall that saccharin use did not raise the risk of cancer. Among certain small groups of the population, however, such as those who both smoked heavily and used saccharin, the risk of bladder cancer was slightly greater. Other studies involving more than 5000 people with bladder cancer showed no association between bladder cancer and saccharin use.[11] Common sense dictates that consuming large amounts of saccharin is probably not safe, but at moderate intake levels, saccharin is currently assumed to be safe for most people. The FDA has set an **Acceptable Daily Intake (ADI)** level for saccharin and for the other artificial sweeteners used in the United States.

Acceptable Daily Intake (ADI): the amount of a sweetener that individuals can safely consume each day over the course of a lifetime without adverse effect. It includes a 100-fold safety factor.

    ▐▐ **ASPARTAME** Aspartame is the active ingredient in NutraSweet, which is used in many commercially prepared foods, and in Equal, a tabletop sweetener. Aspartame was approved by the FDA in 1981 and currently dominates the world market for artificial sweeteners. Aspartame is one of the most studied of all food additives: extensive animal and human studies document its safety. Long-term consumption of aspartame is safe and is not associated with any adverse health effects.[12] Aspartame is approved for use in more than 90 countries.

The FDA's approval of aspartame is based on the assumption that no one will consume more than the ADI of 50 milligrams per kilogram of body weight in a day. This daily intake is indeed a lot: for a 132-pound person, it adds up to 80 packets of Equal or 15 soft drinks sweetened only with aspartame. Most people consume between 2 and 10 milligrams per kilogram of body weight per day. Still, a child who drinks a quart of Kool-Aid sweetened with aspartame on a hot day, and who also has pudding, chewing gum, cereal, and other products sweetened with aspartame, takes in more than the ADI. Although this presents no proven hazard, it seems wise to offer children other foods so as not to exceed the limit.

    ▐▐ **ASPARTAME AND PKU** Although aspartame is considered safe for most people, individuals with the metabolic disorder phenylketonuria (PKU) are an exception. The labels of products that contain aspartame must include information for individuals with PKU. Aspartame contains the amino acid phenylalanine, and people with PKU cannot dispose of it efficiently. Adults with PKU can use some pure aspartame, but children with PKU need to get all their phenylalanine from nutrient-rich foods such as milk and meat (see Nutrition in Practice 6).

    ▐▐ **ACESULFAME POTASSIUM** The FDA approved acesulfame potassium (acesulfame-K) in 1988 after reviewing more than 90 safety studies, conducted over 15 years. Some consumer groups believe that acesulfame-K causes tumors in rats and should not have been approved. The FDA counters that the tumors were not caused by the sweetener, but were typical of those routinely found in rats.

Marketed under the trade names Sunette and Sweet One, acesulfame-K is about as sweet as aspartame. It is used in chewing gum, beverages, instant coffee and tea, gelatins, and puddings, as well as for table use. Unlike aspartame, acesulfame-K holds up well during cooking.

    ▐▐ **SUCRALOSE** Recently approved for use as a sweetener in the United States, sucralose is the only artificial sweetener made from sucrose. Many years of testing have deemed sucralose safe to use and, specifically, not a cause of cancer. Sucralose is not recognized by the body as sugar and therefore passes through unchanged. Sucralose is heat stable and so is useful for cooking and baking; it will soon be showing up in more commercially prepared products and as the tabletop sweetener, Splenda.

OTHER ARTIFICIAL SWEETENERS FDA approval for two other sweeteners—cyclamate and alitame—is still pending. To date, no safety issues have been raised for alitame. Cyclamate, on the other hand, has been battling safety issues for 50 years. Approved by the FDA in 1949, cyclamate was banned in 1970 because of evidence indicating that it caused bladder cancer in rats. In 1985, the National Academy of Sciences concluded that evidence to date indicated that cyclamate did not cause cancer in human beings, but that further studies were warranted. In Canada, cyclamate is restricted to use as a tabletop sweetener on the advice of a physician. In the United States, the FDA is currently reviewing a petition to reapprove the use of cyclamate.

ARTIFICIAL SWEETENERS AND WEIGHT CONTROL Many people eat and drink products sweetened with artificial sweeteners to help them control weight. Does this work? Ironically, a few studies have reported that after consuming such products, people experience heightened feelings of hunger. Despite these reports, most studies find that artificial sweeteners do not heighten feelings of hunger, enhance food intake, or cause weight gain in people. Some studies find, however, that when people reduce their energy intakes by replacing sugar in their diets with artificial sweeteners, they compensate for the reduced energy at later meals.[13] Using artificial sweeteners will not automatically lower energy intake; to successfully control energy intake, a person needs to make informed diet and activity decisions throughout the day.

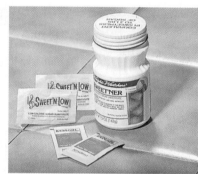

Artificial sweeteners offer the sweet taste of sugar without the kcalories.

LEARNING LINK

Advise clients who choose to use artificial sweeteners to use them in moderation. After all, almost every substance, even water, can be harmful in excess. The American Dietetic Association recommends that artificial sweeteners be used in moderation and only as part of a well-balanced diet.[14]

Health Effects of Complex Carbohydrates

Despite dietary recommendations that people should eat generous servings of complex carbohydrate–rich foods for their health, many people still believe that carbohydrate is the "fattening" component of foods. Gram for gram, carbohydrates contribute fewer kcalories to the body than dietary fat, so a diet of high-carbohydrate foods is likely to be lower in kcalories than a diet of high-fat foods.

The health benefits a person can expect from a diet high in complex carbohydrate foods and low in concentrated sugar are many. Such a diet is almost invariably low in fat, low in food energy, and high in fiber, vitamins, and minerals. All these factors working together can help reduce the risks of obesity, cancer, cardiovascular disease, diabetes, dental caries, gastrointestinal disorders, and malnutrition.

It is difficult to sort out which complex carbohydrates contribute to which health benefits. Starch and fibers almost always occur together in foods (except refined foods), so it is hard to distinguish their effects. Some health effects appear to be especially closely associated with fibers, however, and these are discussed next.

⊞ Fibers

Fiber-rich foods benefit health in many ways (see Table 2-3). Fibers in foods are thought to play a beneficial role in the prevention or management of:

- *Weight control.* Fiber-rich foods contribute little energy and take longer to eat than low-fiber foods, thereby enhancing satiety.[15] High-fiber foods also promote a feeling of fullness as they absorb water. In addition, soluble fibers in a meal slow the movement of food through the upper digestive tract, so a person feels fuller longer. In a study of more than 200 men, researchers assessed body fat and diet composition (energy, carbohydrate, fat, protein, and fiber). Dietary fiber had the strongest correlation with body fat: the men with the highest percentage of body fat consumed significantly less fiber than leaner men.[16] In a different study, researchers found that lean men and women consumed significantly more fiber than obese men and women.[17] A diet high in fiber-rich foods can promote weight loss if those foods displace concentrated fats and sweets.

- *Constipation, hemorrhoids, and diarrhea.* Fibers that both enlarge and soften stools (such as insoluble wheat bran) ease elimination for the rectal muscles, which alleviates or prevents constipation and hemorrhoids.[18] Soluble fibers help to solidify watery stools.

- *Appendicitis.* Some fibers (again, such as wheat bran) help keep the contents of the intestinal tract moving easily. This action helps prevent compaction of the intestinal contents, which could obstruct the appendix and permit bacteria to invade and infect it.

- *Diverticulosis.* Fibers stimulate the muscles of the digestive tract so that they retain their health and tone. This prevents the muscles from becoming weak and the lining of the digestive tract from bulging out in places, as occurs in diverticulosis. In a study of more than 40,000 men, those who ate high-fiber diets had lower risks of diverticulosis than men who ate low-fiber diets did.[19]

- *Colon cancer.* Evidence strongly suggests that high-fiber diets protect against colon cancer.[20] (One recent survey, however, reported no effect.)[21] Populations who consume high-fiber diets have lower rates of colon cancer than similar populations consuming low-fiber diets.[22] Fiber may help prevent colon cancer by diluting, binding, and rapidly removing potentially cancer-causing agents from the colon.

- *Heart disease.* Diets high in fiber, especially cereal fiber, significantly reduce the risk of heart disease.[23] Soluble fibers bind cholesterol compounds and carry them out of the body with the feces, thus lowering the body's cholesterol concentration.[24] High-fiber foods may also lower blood cholesterol indirectly by displacing fatty, cholesterol-raising foods from the diet. Even when dietary fat intake is low, research shows that high intakes of soluble fiber (such as that in apples and oat bran) exert separate and significant blood cholesterol–lowering effects.[25]

- *Diabetes.* Some fibers delay the passage of nutrients from the stomach to the small intestine. This delay slows glucose absorption, thus eliciting a moderate insulin response and a moderate rise in blood glucose. This slow, sustained rise in blood glucose is desirable. The term **glycemic effect** refers to the effect of food on blood glucose; a high glycemic effect reflects fast glucose absorption and a surge in blood glucose, which is undesirable. An extensive long-term study of more than 65,000 women showed that women whose diets had the highest glycemic effect and the lowest cereal fiber content were most vulnerable to type 2 diabetes independent of other dietary constituents or known risk factors.[26]

Chapter 16 describes diverticulosis.

glycemic effect: a measure of the extent to which a food raises the blood glucose concentration and elicits an insulin response, as compared with pure glucose.

Table 2-3

WATER SOLUBILITIES, SOURCES, AND HEALTH EFFECTS OF FIBER

Fiber Type	Major Food Sources	Possible Health Effects
Soluble Gums, mucilages, pectins, psyllium,[a] some hemicellulose	Barley, fruits, legumes, oats, oat bran, rye, seeds, vegetables	These fibers lower blood cholesterol; slow glucose absorption; slow transit of food through upper digestive tract; hold moisture in stools, softening them; are partly fermentable into fragments the body can use.
Insoluble Cellulose, lignin, some hemicellulose	Brown rice, fruits, legumes, seeds, vegetables, wheat bran, whole grains	These fibers soften stools, regulate bowel movements; speed transit of material through small intestine; increase fecal weight and speed fecal passage through colon; reduce colon cancer risk; reduce risks of diverticulosis, hemorrhoids, and appendicitis.

[a]Psyllium, a fiber laxative and a cereal additive, has both soluble and insoluble properties.

▪▪ FIBERS AND HEALTH CLAIMS The FDA authorizes three health claims on food labels concerning fiber. One is for "fiber-containing grain products, fruits, and vegetables and cancer." Another is for "fruits, vegetables, and grain products that contain fiber, and the risk of coronary heart disease," and a third is for "foods that contain oat bran or oatmeal and the risk of coronary heart disease." Chapter 1 describes the criteria foods must meet to bear these health claims.

Even with all these advantages, carbohydrate in the form of raw fiber is not a wonder cure. In some cases it can be detrimental. When too much fiber is consumed, some essential vitamins and minerals may bind to it and be excreted with it without becoming available for the body to use. Also, consuming purified fiber such as cellulose may not confer the same health benefits as consuming cellulose from a food source such as whole grains. Athletes may want to avoid bulky, fiber-rich foods just prior to competition.

▪▪ FIBERS IN FOODS Meals selected according to the Food Guide Pyramid deliver ample fiber. Most people in the United States do not eat this way, though, and their average fiber intakes are about half of the recommended 20 to 35 grams daily.[27] Research shows that only when people meet the grain recommendation and one or both of the vegetable and fruit recommendations (based on the Food Guide Pyramid), do they achieve fiber intakes higher than 20 grams per day.[28] As nurses, you can advise patients to obtain enough fiber by eating ample servings of whole foods, using Figure 2-4 (p. 44) as a guide.

Miscellaneous whole foods also add some fiber: you can get about 1 gram from ½ ounce of nuts, a tablespoon of peanut butter, a large pickle, or 5 olives. A day's meals based on the Daily Food Guide, such as those shown in Figure 2-5 (on p. 46), not only meet carbohydrate recommendations, but provide abundant fiber, too.

LEARNING LINK

Fiber-rich foods can benefit the health of your clients (as well as your own health) in many ways. Fibers help maintain the health of the digestive tract and may protect against or control certain chronic diseases such as diabetes (Chapter 20), heart disease (Chapter 21), and cancer (Chapter 24).

Figure 2-4
FIBER IN SELECTED FOODS

Bread, Cereal, Rice, and Pasta Group

Whole-grain products provide 1 to 2 g of fiber or more per serving:

- ♦ 1 slice whole-wheat or rye bread (1 g).
- ♦ 1 slice pumpernickel bread (2 g).
- ♦ ½ c ready-to-eat 100% bran cereal (10 g).
- ♦ ½ c cooked barley, bulgur, grits, oatmeal (2 to 3 g).

Vegetable Group

Most vegetables contain 2 to 3 g of fiber per serving:

- ♦ 1 c raw bean sprouts.
- ♦ ½ c cooked broccoli, brussels sprouts, cabbage, carrots, cauliflower, collards, corn, eggplant, green beans, green peas, kale, mushrooms, okra, parsnips, potatoes, pumpkin, spinach, sweet potatoes, swiss chard, winter squash.
- ♦ ½ c chopped raw carrots, peppers.

Fruit Group

Fresh, frozen, and dried fruits have about 2 g of fiber per serving:

- ♦ 1 medium apple, banana, kiwi, nectarine, orange, pear.
- ♦ ½ c applesauce, blackberries, blueberries, raspberries, strawberries.
- ♦ Fruit juices contain very little fiber.

Meat and Meat Alternates Group

Many legumes provide about 8 g of fiber per serving:

- ♦ ½ c cooked baked beans, black beans, black-eyed peas, kidney beans, navy beans, pinto beans.

Some legumes provide about 5 g of fiber per serving:

- ♦ ½ c cooked garbanzo beans, great northern beans, lentils, lima beans, split peas.

Most nuts and seeds provide 1 to 3 g of fiber per serving:

- ♦ 1 oz almonds, cashews, hazelnuts, peanuts, pecans, pumpkin seeds, sunflower seeds.

NOTE: Appendix A provides fiber grams for over 2000 foods.

▪▪ Carbohydrates: Food Sources and Recommendations

Grains, vegetables, fruits, and milk are noted for the valuable energy-yielding carbohydrates they contain: starches and dilute sugars. Each class of foods makes its own typical carbohydrate contribution. The Daily Food Guide and the Food Guide Pyramid in Chapter 1 can help you and your clients obtain carbohydrate-rich foods.

▪▪ **BREADS, CEREAL, RICE, AND PASTA**   A serving of most foods in this group—a slice of whole-wheat bread, half an English muffin or bagel, a 6-inch tortilla, or ½ cup of rice, pasta, or cooked cereal—provides abundant carbohydrate as starch.* Some foods in this group, especially baked goods such as biscuits, croissants, muffins, and snack crackers, also contain sugar, fat, or both.

▪▪ **VEGETABLES**   Some vegetables are major contributors of starch in the diet. Just a small white or sweet potato or ½ cup of cooked corn, peas, plantain, or winter squash provides 15 grams of carbohydrate, as much as in a slice of bread, though as a mixture of sugars and starch. A ½-cup portion of carrots, okra, onions, tomatoes, cooked greens, or most other nonstarchy vegetables or a cup of salad greens provides about 5 grams as a mixture of starch and sugars. Each of these foods also contributes a little protein and no fat.

▪▪ **FRUITS**   The size of a typical serving of fruit varies depending on the form of the fruit: ¾ cup of juice; a small banana, apple, or orange; ½-cup of most canned or fresh fruit; or ¼ cup of dried fruit. A typical fruit serving contains an average of about 15 grams of carbohydrate, mostly as sugars, including the fruit sugar fructose. Fruits vary greatly in their water and fiber contents, and therefore their sugar concentrations vary also. With the exception of avocado, which is high in fat, fruits contain insignificant amounts of fat and protein.

▪▪ **MILK, CHEESE, AND YOGURT**   One cup of milk or yogurt or the equivalent (1 cup of buttermilk, ⅓ cup of dry milk powder, or ½ cup of evaporated milk) provides a generous 12 grams of carbohydrate. Cottage cheese provides about 6 grams of carbohydrate per cup, but most other types of cheese contain little if any carbohydrate. The fat content of all milk products varies, an important consideration in choosing among them; Chapter 3 provides the details.

Cream and butter, although dairy products, are not equivalent to milk because they contain little or no carbohydrate and insignificant amounts of the other nutrients important in milk. They are appropriately placed with the fats at the top of the Food Guide Pyramid.

▪▪ **MEAT, POULTRY, FISH, DRY BEANS, EGGS, AND NUTS**   With two exceptions, foods of this group provide almost no carbohydrate to the diet. The exceptions are nuts, which provide a little starch and fiber along with their abundant fat, and dry beans, which are low-fat sources of both starch and fiber. Just a ½-cup serving of beans provides 15 grams of carbohydrate, an amount equal to the richest sources in the Food Guide Pyramid.

▪▪ **RECOMMENDATIONS**   Carbohydrate has no RDA, but there is a recommendation that 55 to 60 percent of total kcalories should come from carbohydrate. The FDA used this guideline in establishing a Daily Value for carbohydrate of 300 grams per day, or 60 percent of kcalories based on 2000 kcalories per day. For people who eat about 1700 kcalories a day, the recommended intake translates to some 900 to 1000 kcalories from carbohydrate per day or, at 4 kcalories per gram, some 225 to 250 grams. Most of this carbohydrate should

---

*Gram values in this section are adapted from the 1995 Exchange System.

be starch, not sugars; thus, many servings of starchy foods are needed to meet this recommendation.

Figure 2-5 offers an example of meals that provide about 1700 kcalories and 225 grams of carbohydrate from nutrient-dense, fiber-rich foods. The

Figure 2-5

A DAY'S MEALS THAT OFFER 1700 KCALORIES AND MEET THE CARBOHYDRATE RECOMMENDATION

Before heading off to classes, a student eats breakfast:
2 shredded wheat biscuits.
1 c 1% low-fat milk.
½ banana (sliced).

Then goes home for a quick lunch:
1 turkey sandwich on whole-wheat bread with mayonnaise and mustard.
1 c vegetable juice (canned).

While studying that afternoon, the student eats a snack:
4 whole-wheat crackers.
1 oz cheddar cheese.
1 apple.

That night, the student makes dinner:

A salad made with:
1 c raw spinach leaves, shredded carrots, and sliced mushrooms.
⅓ c garbanzo beans.
5 lg olives.
1 tbs ranch salad dressing.

A dinner of:
1 c spaghetti with meat sauce.
½ c green beans.
2 tsp butter.
And, for dessert:
1¼ c strawberries (fresh).

Later that evening, the student enjoys a bedtime snack:
3 graham crackers.
1 c 1% low-fat milk.

Total kcal: 1725
53% kcal from carbohydrate
29% kcal from fat
18% kcal from protein

carbohydrate content of a diet can be determined by using a nutrient composition table such as that found in Appendix A, the exchange list system described in Chapter 20, or a computer diet analysis program.

■■ **CARBOHYDRATES ON FOOD LABELS** Food labels list the amount, in grams, of total carbohydrate—including starch, fibers, and sugars—per serving. Fiber grams are also listed separately, as are the grams of sugars. (With this information, consumers can calculate starch grams by subtracting the grams of fibers and sugars from the total carbohydrate.) Sugars reflect both added sugars and those that occur naturally in foods. Total carbohydrate and dietary fiber are also expressed as "% Daily Values" for a person consuming 2000 kcalories; there is no Daily Value for sugars.

■■ Energy Nutrients in Perspective

An uninterrupted flow of energy is so vital to life that other functions are sometimes sacrificed to maintain it. For example, when a child is fed too little food, the food the child does consume will be used for energy to keep the heart and lungs going, while growth comes to a standstill. To go totally without an energy supply, even for a few minutes, is to die. So crucial is this need for energy that all creatures have developed ways of accumulating built-in reserves to protect themselves from being deprived of energy. One major provision against this sort of emergency is glycogen, the storage form of glucose. (The other is body fat, about which the next chapter says more.)

■■ **GLYCOGEN USED FOR ENERGY** When a person does not eat carbohydrate, the person's body rapidly devours its glycogen stores. Stored glycogen can provide glucose whenever the supply runs short, but the liver and muscle cells can store only limited amounts of glycogen. Once this supply is depleted, the body must turn to the other energy nutrients—fat and protein—to meet its energy needs.

■■ **FAT USED FOR ENERGY** Unlike the liver cells, which can store only a limited amount of glycogen, the body's fat cells can store virtually unlimited quantities of fat, so supplies almost never run out. Fat normally is used to meet about half of the body's energy needs, and most tissues can use it as is. The brain and nerves, however, need their energy as glucose, and the body cannot convert fat to glucose. After a long period of glucose deprivation, brain and nerve cells develop the ability to derive about half of their energy from a special form of fat known as **ketones,** but they still require glucose as well. This means that people wanting to lose weight need to eat a certain minimum amount of carbohydrate to meet their energy needs, even when they are limiting their food intakes.

■■ **PROTEIN USED FOR ENERGY** During a fast, when the available glycogen is gone and no carbohydrate is coming in from food, brain and nerve cells demand the glucose they need from the only available source—protein. The body begins to dismantle its own muscles and other lean tissues to generate glucose. Only adequate dietary carbohydrate can prevent this use of protein for energy, and its action in doing so is known as the **protein-sparing effect** of carbohydrate.

Ultimately, after half of the body's protein is used, death occurs. Death from loss of lean body tissues will occur even in an obese person who fasts too long. It should be clear, then, that although carbohydrate is an ideal energy source, fat and sometimes protein are also extremely important in meeting energy demands.

Chapter 7 offers guidelines for weight loss.

ketones (KEY-tones): acidic, fat-related compounds formed from the incomplete breakdown of fat when carbohydrate is not available; technically known as *ketone bodies.*

protein-sparing effect: the effect of carbohydrate in providing energy that allows protein to be used for other purposes.

SELF CHECK

SELF CHECK

1. Carbohydrates are found in virtually all foods **except:**
   a. milks.
   b. meats.
   c. breads.
   d. vegetables.

2. Complex carbohydrates include:
   a. galactose, starch, and glycogen.
   b. starch, glycogen, and fiber.
   c. lactose, maltose, and glycogen.
   d. sucrose, fructose, and glucose.

3. The chief energy source of the body is:
   a. sucrose.
   b. starch.
   c. glucose.
   d. fructose.

4. The primary form of stored glucose in animals is:
   a. glycogen.
   b. cellulose.
   c. starch.
   d. lactose.

5. The polysaccharide that helps form the cell walls of plants is:
   a. cellulose.
   b. starch.
   c. glycogen.
   d. lactose.

6. The two types of alternative sweeteners are:
   a. saccharin and cyclamate.
   b. sugar alcohols and artificial sweeteners.
   c. sorbitol and xylitol.
   d. sucrose and fructose.

7. Which of the following items may denote sugar on food labels?
   a. corn syrup.
   b. aspartame.
   c. xylitol.
   d. cellulose.

8. A diet high in complex carbohydrates is:
   a. most likely low in fat.
   b. most likely low in fiber.
   c. most likely poor in vitamins and minerals.
   d. most likely disease promoting.

9. A fiber-rich diet may help to prevent or control:
   a. some types of cancer.
   b. heart disease.
   c. constipation.
   d. all of the above.

10. The recommended fiber intake is:
    a. 10 to 15 grams per day.
    b. 15 to 25 grams per day.
    c. 20 to 35 grams per day.
    d. 40 to 55 grams per day.

Answers to these questions appear in Appendix H.

CLINICAL APPLICATIONS

Considering the health benefits of carbohydrate-rich foods, especially those that provide starch and fiber, what suggestions would you offer to a client who reports the following:

1. Eats only 3 servings of refined, sugary, breads or cereals each day.

2. Eats one serving of vegetables (usually french fries) each day.

3. Drinks fruit juice once a day, but never eats fruit.

4. Eats cheese at least twice a day, but does not drink milk.

5. Eats large servings of meat at least twice a day.

6. Eats hard candy 2 or 3 times a day.

Notes

[1] U.S. Department of Agriculture, Economic Research Service, *A Dietary Assessment of the U.S. Food Supply: Comparing Per Capita Food Consumption with the Food Guide Pyramid Serving Recommendations* (Washington, D.C.: Government Printing Office, 1998), AER #772.

[2] B. Szepesi, Carbohydrates, in *Present Knowledge in Nutrition,* 7th ed., ed. E. E. Ziegler and L. J. Filer (Washington, D.C.: International Life Sciences Institute Press, 1996), pp. 33–43.

[3] J. O. Hill and A. M. Prentice, Sugar and body weight regulation, *American Journal of Clinical Nutrition* 62 (1995): S264–S274; C. J. Lewis and coauthors, Nutrient intakes and body weights of persons consuming high and moderate levels of added sugars, *Journal of the American Dietetic Association* 92 (1992): 708–713.

[4] A. R. Folsom and coauthors, Increase in fasting insulin and glucose over seven years with increasing weight and inactivity of young adults: The CARDIA study, *American Journal of Epidemiology* 144 (1996): 235–246; G. A. Colditz and coauthors, Weight gain as risk factor for clinical diabetes mellitus in women, *Annals of Internal Medicine* 122 (1995): 481–486.

[5] M. Kinsbourne, Sugar and the hyperactive child, *New England Journal of Medicine* 330 (1994): 355–356; M. L. Wolraich and coauthors, Effects of diets high in sucrose or aspartame on the behavior and cognitive performance of children, *New England Journal of Medicine* 330 (1994): 301–307.

[6] M. L. Wolraich, D. B. Wilson, and J. W. White, The effect of sugar on behavior or cognition in chil-

dren: A meta-analysis, *Journal of the American Medical Association* 274 (1995): 1617–1621.

[7] S. Gibson and S. Williams, Dental caries in preschool children: Associations with social class, toothbrushing habit and consumption of sugars and sugar-containing foods. Further analysis of data from the National Diet and Nutrition Survey of children aged 1.5–4.5 years, *Caries Research* 32 (1999): 101–113.

[8] K. G. Konig and J. M. Navia, Nutritional role of sugars in oral health, *American Journal of Clinical Nutrition* 62 (1995): S275–S283.

[9] S. S. Natah and coauthors, Metabolic response to lactitol and xylitol in healthy men, *American Journal of Clinical Nutrition* 65 (1997): 947–950; J. W. Finley and G. A. Leveille, Macronutrient substitutes, in *Present Knowledge in Nutrition,* 7th ed., ed. E. E. Ziegler and L. J. Filer (Washington, D.C.: International Life Sciences Institute Press, 1996), pp. 581–595.

[10] Food and Drug Administration, Food labeling: Health claims—Sugar alcohols and dental caries, *Federal Register* 61 (August 23, 1996): 43433–43447.

[11] Position of The American Dietetic Association: Use of nutritive and nonnutritive sweeteners, *Journal of the American Dietetic Association* 98 (1998): 580–587.

[12] Position of The American Dietetic Association, 1998.

[13] R. M. Black and G. H. Anderson, Sweeteners, food intake, and selection, in *Appetite and Body Weight Regulation: Sugar, Fat and Macronutrient Substitutes,* ed. J. D. Fernstrom and G. D. Miller (Boca Raton, Fla.: CRC Press, 1994), pp. 125–136.

[14] Position of The American Dietetic Association, 1998.

[15] J. W. Anderson, B. M. Smith, and N. J. Gustafson, Health benefits and practical aspects of high-fiber diets, *American Journal of Clinical Nutrition* 59 (1994): S1242–S1247.

[16] L. H. Nelson and L. A. Tucker, Diet composition related to body fat in a multivariate study of 203 men, *Journal of the American Dietetic Association* 96 (1996): 771–777.

[17] W. C. Miller and coauthors, Dietary fat, sugar, and fiber predict body fat content, *Journal of the American Dietetic Association* 94 (1994): 612–615.

[18] J. H. Cummings and H. N. Englyst, Gastrointestinal effects of food carbohydrate, *American Journal of Clinical Nutrition* 61 (1995): S938–S945.

[19] W. H. Aldoori and coauthors, A prospective study of diet and the risk of symptomatic diverticular disease in men, *American Journal of Clinical Nutrition* 60 (1994): 757–764.

[20] B. S. Reddy, Role of dietary fiber in colon cancer: An overview, *American Journal of Medicine* 106 (1999): S16–S19; D. Kritchevsky, Protective role of wheat bran fiber: Preclinical data, *American Journal of Medicine* 106 (1999): S28–S31; E. Negri and coauthors, Fiber intake and risk of colorectal cancer, *Cancer Epidemiology, Biomarkers, and Prevention* 7 (1998): 667–671.

[21] C. S. Fuchs and coauthors, Dietary fiber and the risk of colorectal cancer and adenoma in women, *New England Journal of Medicine* 340 (1999): 169–176.

[22] Y. Kim and J. B. Mason, Nutrition chemoprevention of gastrointestinal cancers: A critical review, *Nutrition Reviews* 54 (1996): 259–279.

[23] E. B. Rimm and coauthors, Vegetable, fruit, and cereal fiber intake and risk of coronary heart disease among men, *Journal of the American Medical Association* 275 (1996): 447–451.

[24] L. Brown and coauthors, Cholesterol-lowering effects of dietary fiber: A meta-analysis, *American Journal of Clinical Nutrition* 69 (1999): 30–42.

[25] D. J. A. Jenkins and coauthors, Effect on blood lipids of very high intakes of fiber in diets low in saturated fat and cholesterol, *New England Journal of Medicine* 329 (1993): 21–26; J. W. Anderson and coauthors, Prospective, randomized controlled comparison of the effects of low-fat plus high-fiber diets on serum lipid concentrations, *American Journal of Clinical Nutrition* 56 (1992): 887–894.

[26] J. Salmeron and coauthors, Dietary fiber, glycemic load, and risk of noninsulin-dependent diabetes mellitus in women, *Journal of the American Medical Association* 277 (1997): 472–477.

[27] Position of The American Dietetic Association: Health implications of dietary fiber, *Journal of the American Dietetic Association* 97 (1997): 1157–1159.

[28] S. M. Krebs-Smith and coauthors, Characterizing food intake patterns of American adults, *American Journal of Clinical Nutrition* 65 (1997): S1264–S1268.

Q&A NUTRITION IN PRACTICE

Nutrition and Dental Health

Chapter 2 emphasized the health benefits of eating complex carbohydrate–rich foods. Complex carbohydrates may support overall health, but they do not necessarily promote dental health. The carbohydrates people eat and the times they eat them play a major role in the development of **dental caries**—a pervasive health problem throughout the world.

What is dental caries?

Dental caries is an infectious oral disease that develops in the tooth **enamel** (see the accompanying glossary and Figure NP2-1). Caries develops when bacteria that reside in the **plaque** of teeth consume and metabolize carbohydrates, producing

acids that attack the tooth enamel. Thus, at least two main ingredients are required to make dental caries: bacteria and carbohydrates. In addition, factors such as heredity, nutrition status during early tooth development, dental hygiene practices, and fluoride intake influence a person's susceptibility to caries. Poor nutrition during pregnancy, infancy, or early childhood can impair the development of healthy teeth, making caries likely.[1] Table NP2-1 shows the effects of specific nutrient deficiencies on tooth development.

How do carbohydrate-rich foods promote caries development?

The bacteria that promote dental caries thrive on food particles that contain carbohydrate. Both sugar and starch can support bacterial growth. Equally important is the length of time the food stays in the mouth, and this depends on how soon the teeth are brushed after eating and how sticky the food is. The damage a food does relates to both its carbohydrate content and its stickiness. For example, raisins and granola, which adhere to the teeth, cause more caries than a food that is easily rinsed off such as a sugary beverage.

Sugar can be eaten without inviting tooth decay if it is removed from tooth surfaces promptly. Bacterial action is maximal in the first 20 minutes after the first contact. If immediate brushing is not possible, water or

Table NP2-1
NUTRIENT DEFICIENCIES AFFECTING TOOTH DEVELOPMENT

Nutrient Deficiency	Effect on Tooth Development
Protein	Small, irregularly shaped teeth; delayed eruption; high caries susceptibility
Vitamin C	Disturbance of dentin formation
Vitamin A	Disturbance of enamel formation, delayed eruption
Vitamin D	Poor mineralization, pitting, striations
Calcium	Poor mineralization
Phosphorus	Poor mineralization
Magnesium	Enamel underdeveloped
Iron	High caries susceptibility
Zinc	High caries susceptibility
Fluoride	High caries susceptibility

other beverages swished in the mouth after a meal can effectively rinse the teeth. Once-a-day flossing may also effectively control formation of caries, regardless of the carbohydrate content of the diet. Some people may never get caries because they have inherited resistance to them.

Doesn't saliva rinse the mouth and protect the teeth?

Yes, and some foods stimulate more saliva flow than others. Saliva protects against caries formation in several ways. It not only rinses the mouth, but also dilutes the caries-causing acid produced by bacteria, exerts antibacterial action, and provides protective minerals.[2] Foods that elicit saliva flow may therefore defend against caries formation. Not all foods are protective, however; those foods that also contain sugar may promote acid formation. Apples are an example: they stimulate saliva flow, but they also release sugar, so they have both caries-preventing and caries-promoting effects. Clearly, many different factors influence caries development, making it diffi-

Figure NP2-1
A TOOTH

The inner layer of dentin is bonelike material. The outer layer is enamel, which is harder than bone. Caries begins when acid dissolves the enamel that covers the tooth. If it is not repaired, the decay may penetrate the dentin and spread into the pulp of the tooth, causing inflammation and an abscess.

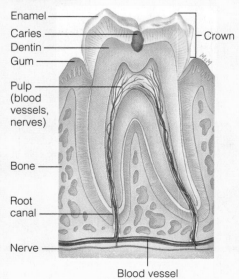

Enamel
Caries
Dentin
Gum
Crown
Pulp (blood vessels, nerves)
Bone
Root canal
Nerve
Blood vessel

GLOSSARY OF DENTAL CARIES TERMS

cariogenic (KARE-ee-oh-JEN-ik): conducive to dental decay.

dental caries (KARE-eez): decay of teeth.

dental plaque (PLACK): a gummy mass of bacteria that grows on teeth and can lead to dental caries and gum disease.

enamel: the hard, white, dense substance that forms a covering for the crown of

cult to predict exactly which foods are **cariogenic.**

www.ada.org
American Dental Association

www.adha.org
American Dental Hygienists' Association

Do any foods prevent caries?

Yes. Some foods stimulate saliva flow and do not contribute to acid formation in the mouth: cheese is an example. Such foods are good choices to eat at the end of a meal. Cheese is a powerful saliva stimulant and does not promote acid formation, so it reduces the cariogenicity of a meal. Furthermore, the high calcium and phosphorus content of cheese supports dental health.

High-fiber foods in general are anticariogenic, especially if their carbohydrate content is low. For example, raw vegetables do not stick to the teeth, and because they require vigorous chewing, they stimulate saliva flow. Cocoa products (including chocolate), coffee, tea, and beer all contain tannin, an acid that prevents caries formation. Table NP2-2 lists some dietary recommendations for controlling dental caries.

Besides foods, what other factors protect against dental caries development?

Research shows that when fluoride is added to the water supply, the children in the community have fewer dental caries than children who drink nonfluoridated water. In fact, water fluoridation is the most effective, least expensive way to provide dental care to everyone. The following recommendations will maximize protection against dental caries:

- Use sugars sparingly; watch for hidden sugars in foods; and use low-sugar or sugar-free products whenever possible.
- Restrict sweets to mealtimes.
- After eating a meal or a between-meal snack, brush and floss or, at least, rinse with water.
- Limit the time that teeth are exposed to sticky foods.
- In any case, brush at least twice daily, and floss at least once daily.
- Visit a dentist regularly; repair damaged teeth.
- Drink fluoridated water; provide infants and children with fluoride supplements when such water is not available.
- Eat a balanced diet composed of a variety of foods that will maintain adequate nutrition status.
- Eat foods that are rich in calcium and phosphorus.
- Eat a variety of firm, fibrous foods that will stimulate saliva flow.

In summary, learning and practicing sound dental hygiene habits, as well as developing eating habits that are consistent with both dental health and nutritional health, will serve a person throughout life.

Table NP2-2

DIETARY RECOMMENDATIONS FOR CONTROLLING DENTAL CARIES

Food Group	Low Cariogenicity (Use When Teeth Cannot Be Brushed Immediately)	High Cariogenicity (Do Not Use unless Followed by Prompt and Thorough Dental Hygiene)
Dairy	Milk, cheese, plain yogurt	Ice cream, ice milk, milk shakes, fruited yogurts, eggnog
Meats/meat alternates	Meat, fish, poultry, eggs, legumes	Peanut butter with added sugar, luncheon meats with added sugar, meats with sugared glazes
Fruits[a]	Fresh, packed in water	Dried (raisins, figs, dates), packed in syrup or juice, jams, jellies, preserves, fruit juices and drinks
Vegetables	Most vegetables	Candied sweet potatoes, glazed carrots
Breads/cereals[b]	Popcorn, toast, hard rolls, pretzels, pizza, bagels	Cookies, sweet rolls, pies, cakes, dry sugared cereals as between-meal snacks, doughnuts, potato chips, granola bars
Other	Sugarless gum, coffee or tea without sugar, nuts	Sugared soft drinks, candy, fudge, caramels, honey, sugars, syrups

[a]Tiny particles of bananas can get lodged between teeth and decompose, increasing risk of caries.
[b]Tiny particles of breads, crackers, and chips can also become lodged in teeth, promoting caries formation.

Notes

[1]D. Fitzsimons and coauthors, Nutritional and oral health guidelines for pregnant women, infants, and children, *Journal of the American Dietetic Association* 98 (1998): 182–189.

[2]K. G. Konig and J. M. Navia, Nutritional role of sugars in oral health, *American Journal of Clinical Nutrition* 62 (1995): S275–S283.

CHAPTER 3

LIPIDS

CONTENTS

▪▪ The Chemist's View of Lipids

▪▪ Fats and Health

▪▪ Fats in Foods

▪▪ Nutrition in Practice:
Vegetarian Diets

Most people know that too much fat in the diet imposes health risks, but they may be surprised to learn that too little does, too. People in the United States, however, are more likely to eat too much fat than too little.

Fat is a member of the class of compounds called **lipids.** The lipids in foods and in the human body include triglycerides (**fats** and **oils**), phospholipids, and sterols.

■■ ENERGY FROM FAT Lipids perform many tasks in the body, but most importantly, they provide energy. A constant flow of energy is so vital to life that, in a pinch, any other function is sacrificed to maintain it. Chapter 2 described one safeguard against such an emergency—the stores of glycogen in the liver that provide glucose to the blood whenever the supply runs short. The body's stores of glycogen are limited, however. In contrast, the body's capacity to store fat for energy is virtually unlimited due to the fat-storing cells of the **adipose tissue.** Unlike most body cells, which can store only limited amounts of fat, the adipose, or fat, cells seem able to expand almost indefinitely. The more fat they store, the larger they grow. Figure 3-1 shows a fat cell. The fat stored in these cells supplies 60 percent of the body's ongoing energy needs during rest. During physical activity or prolonged periods of food deprivation, fat stores may make an even greater energy contribution. In fact, body fat stores are of tremendous importance during physical activity, as long as the activity is not too intense. Fat can be broken down for energy only if oxygen is available. During intense activity, the demand for oxygen becomes too great to permit much use of fat; glycogen becomes the predominant fuel. The person who breathes easily during physical activity allows the muscles to get all the oxygen they need to burn fat.

■■ ROLES OF BODY FAT In addition to supplying energy, fat serves other roles in the body. Natural oils in the skin provide a radiant complexion; in the scalp, they help nourish the hair and make it glossy. The layer of fat

lipids: a family of compounds that includes triglycerides (fats and oils), phospholipids, and sterols.

fats: lipids that are solid at room temperature (70°F or 25°C).

oils: lipids that are liquid at room temperature (70°F or 25°C).

adipose tissue: the body's fat, which consists of masses of fat-storing cells called adipose cells.

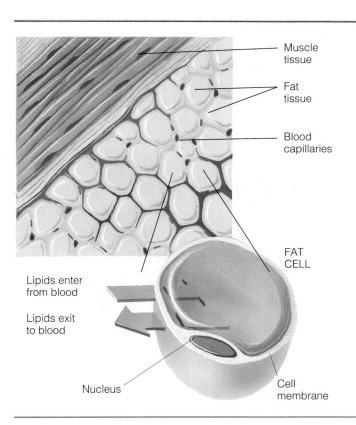

Muscle tissue

Fat tissue

Blood capillaries

FAT CELL

Lipids enter from blood

Lipids exit to blood

Nucleus

Cell membrane

Figure 3-1
A FAT CELL

Within the fat, or adipose, cell, lipid is stored in a droplet. This droplet can greatly enlarge, and the fat cell membrane will grow to accommodate its swollen contents.

Table 3-1
THE FUNCTIONS OF FATS IN THE BODY

Fats in the body:

- Provide energy.
- Insulate the body against temperature extremes.
- Protect the body's vital organs from shock.
- Form the major material of cell membranes.

beneath the skin insulates the body from extremes of temperature. A pad of hard fat beneath each kidney protects it from being jarred and damaged, even during a motorcycle ride on a bumpy road. The soft fat in a woman's breasts protects her mammary glands from heat and cold and cushions them against shock. The fat embedded in muscle tissue shares with muscle glycogen the task of providing energy when the muscles are active. The phospholipids and the sterol cholesterol are cell membrane constituents that help maintain the structure and health of all cells. Table 3-1 summarizes the major functions of fats in the body.

The Chemist's View of Lipids

The diverse and vital functions that lipids play in the body reveal why eating too little fat can be harmful. As mentioned earlier, though, too much fat in the diet seems to be the bigger problem for most people. To understand both the beneficial and the harmful effects that fats exert on the body, a closer look at the structure and function of members of the lipid family is in order.

Triglycerides

When people talk about fat—for example, "I'm too fat" or "That meat is fatty"—they are usually referring to **triglycerides.** Among lipids, triglycerides predominate—both in the diet and in the body. The name *triglyceride* almost explains itself: three **fatty acids** (*tri*) attached to a **glycerol** "backbone." Figure 3-2 shows how three fatty acids combine with glycerol to make a triglyceride.

Fatty Acids

When energy from any energy-yielding nutrient is to be stored as fat, the nutrient is first broken into small fragments. Then the fragments are linked together into chains known as fatty acids. The fatty acids are then packaged, three at a time, with glycerol to make triglycerides.

CHAIN LENGTH AND SATURATION Fatty acids may differ from one another in two ways—in chain length and in degree of saturation. The chain length refers to the number of carbons in a fatty acid. Saturation also refers to

triglycerides (try-GLISS-er-rides): one of the three main classes of lipids; the chief form of fat in foods and the major storage form of fat in the body; composed of glycerol with three fatty acids attached.

tri = three

glyceride = a compound of glycerol

fatty acids: organic compounds composed of a chain of carbon atoms with hydrogens attached and an acid group at one end.

glycerol (GLISS-er-ol): an organic compound, three carbons long, that can form the backbone of triglycerides and phospholipids.

Figure 3-2
TRIGLYCERIDE FORMATION

Glycerol, a small, water-soluble compound, plus three fatty acids, equals a triglyceride.

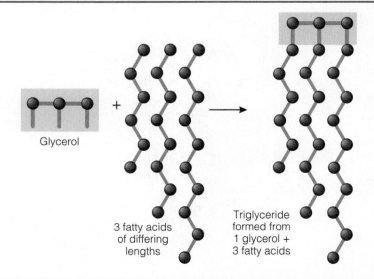

Glycerol

3 fatty acids of differing lengths

Triglyceride formed from 1 glycerol + 3 fatty acids

its chemical structure—specifically, to the number of hydrogens the carbons in the fatty acid are holding. If every available bond is filled to capacity with hydrogen, the chain is called a **saturated fatty acid.** The first zigzag structure in Figure 3-3 represents a saturated fatty acid.

UNSATURATED FATTY ACIDS

In some fatty acids, especially those of plants and fish, hydrogens are missing in the fatty acid chains so the carbons have to form double bonds. The points where the hydrogens are missing are called points of unsaturation, and a chain containing such points is called an **unsaturated fatty acid.** If there is one point of unsaturation, the chain is **monounsaturated.** The second structure in Figure 3-3 is an example.

If there are two or more points of unsaturation, then the fatty acid is polyunsaturated (see the third structure in Figure 3-3). **Polyunsaturated fatty acids** are often abbreviated on food labels as PUFA.

HARD AND SOFT FAT

A triglyceride can contain any combination of fatty acids—long chain or short chain; saturated, monounsaturated, or polyunsaturated. Whether a fat is soft or hard depends on which fatty acids it contains. Fats that contain the shorter-chain or the more unsaturated fatty acids are softer and melt more readily (discussed later in this chapter). Each animal species (including human beings) obeys its own genetic program and makes its own characteristic kinds of triglycerides; but fats in the diet can, within limits, affect the types of triglycerides made in the body. For example, animals raised for food can be fed diets containing softer or harder fats to give their meat or other products softer or harder fat, whichever consumers demand. People, too, incorporate the fatty acids they eat into their own tissues.

ESSENTIAL FATTY ACIDS

The human body can synthesize all the fatty acids it needs from carbohydrate, fat, or protein except for two—**linoleic acid** and **linolenic acid.** Both linoleic acid and linolenic acid are polyunsaturated fatty acids. Because they cannot be made from other substances in the body, they must be obtained from food and are therefore called **essential fatty acids.** Linoleic acid and linolenic acid are found in small amounts in plant and

saturated fatty acid: a fatty acid carrying the maximum possible number of hydrogen atoms (having no points of unsaturation). A saturated fat is a triglyceride that contains three saturated fatty acids.

unsaturated fatty acid: a fatty acid with one or more points of unsaturation where hydrogens are missing (includes monounsaturated and polyunsaturated fatty acids).

monounsaturated fatty acid: a fatty acid that has one point of unsaturation; for example, the oleic acid found in olive oil.

polyunsaturated fatty acid (PUFA): a fatty acid with two or more points of unsaturation. For example, linoleic acid has two such points, and linolenic acid has three. Thus, polyunsaturated *fat* is composed of triglycerides containing a high percentage of PUFA.

linoleic acid, linolenic acid: polyunsaturated fatty acids, essential for human beings.

essential fatty acids: fatty acids that the body requires but cannot make in amounts sufficient to meet its physiological needs.

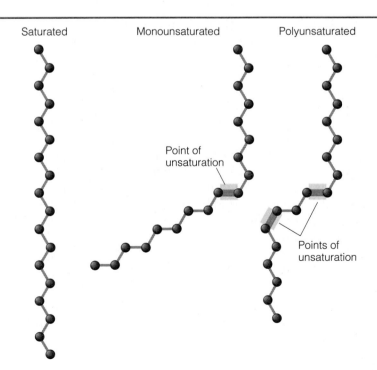

Saturated   Monounsaturated   Polyunsaturated

Point of unsaturation

Points of unsaturation

Figure 3-3
THREE FATTY ACIDS

The more carbon atoms in a fatty acid, the longer its chain. The more hydrogen atoms attached to those carbons, the more saturated the fatty acid.

fish oils, and the body readily stores them, making deficiencies unlikely. From both of these essential fatty acids, the body makes important hormonelike substances that help regulate a wide range of body functions: blood pressure, clot formation, blood lipid concentration, the immune response, the inflammatory response to injury, and many others.[1] These two essential nutrients also serve as structural components of cell membranes.

LINOLEIC ACID: AN OMEGA-6 FATTY ACID Linoleic acid is an **omega-6 fatty acid** found in the seeds of plants and in the oils produced from the seeds. Any diet that contains vegetable oils, seeds, nuts, and whole-grain foods provides enough linoleic acid to meet the body's needs. Researchers have long known and appreciated the importance of the omega-6 fatty acid family.

LINOLENIC ACID AND OTHER OMEGA-3 FATTY ACIDS Linolenic acid belongs to a family of polyunsaturated fatty acids known as **omega-3 fatty acids,** a family that also includes **EPA** and **DHA.** EPA and DHA are found primarily in fish oils. As mentioned, the human body cannot make linolenic acid, but given dietary linolenic acid, it can make EPA and DHA, although the process is slow.

The importance of omega-3 fatty acids was first recognized during the 1980s when research began to unveil impressive roles for EPA and DHA in metabolism and disease prevention. DHA is one of the most abundant structural lipids in the brain, and both EPA and DHA are needed for normal brain development.[2] EPA and DHA are also especially active in the rods and cones of the retina of the eye.[3] Today researchers know that these omega-3 fatty acids are essential for normal growth and development and that they may play an important role in the prevention and treatment of heart disease, hypertension, arthritis, and cancer.[4]

OMEGA-3 SUPPLEMENTS Omega-3 fatty acid supplements are being aggressively marketed as a cure-all for many different diseases without regard for consumer safety. Although some claims are based on research, confirmation that fish oil can prevent or treat heart disease, cancer, or other diseases in individuals or populations is lacking. Experts do agree that adding *fish* to the diet two or three times per week may help to prevent heart disease.[5] The idea that fish oil *supplements* are beneficial and safe in any amount is erroneous, however.

In the first place, the supplements themselves may carry hazards. They may contain toxic amounts of fat-soluble vitamins or pesticide residues. Even when the supplements are not toxic, they may have harmful effects on the body. One potential problem is that the supplement form may not function the same way as the form available directly from foods. Furthermore, omega-3 and omega-6 fatty acids compete for the same slots in the body. Consequently, taking supplements of one can easily induce a deficiency of the other. Still another drawback is that fish oil can cause or aggravate illness by altering blood lipids or blood clotting. Finally, the quantities of omega-3 fatty acids in supplements may vary widely from what the labels say they contain.

EAT DARK-MEAT FISH Questions also remain about the appropriate amount of omega-3 fatty acids for a day's intake. While these questions await answers, the best way for people to increase their intakes of omega-3 fatty acids is to eat several portions of fish each week, particularly the kinds listed in the margin. The darker flesh of fish has the highest fat content, and this is where most omega-3 fatty acids are found.

---

omega: the last letter of the Greek alphabet, used by chemists to refer to the position of the last double bond in a fatty acid.

omega-6 fatty acid: a polyunsaturated fatty acid with its endmost double bond six carbons back from the end of its carbon chain; long recognized as important in nutrition. Linoleic acid is an example.

omega-3 fatty acid: a polyunsaturated fatty acid with its endmost double bond three carbons back from the end of its carbon chain; relatively newly recognized as important in nutrition. Linolenic acid is an example.

EPA, DHA: omega-3 fatty acids made from linolenic acid. The full name for EPA is eicosapentaenoic (EYE-cosa-PENTA-ee-NO-ick) acid. The full name for DHA is docosahexaenoic (DOE-cosa-HEXA-ee-NO-ick) acid.

These fish provide at least 1 g of omega-3 fatty acids, including EPA and DHA, in 100 g of fish (about 3.5 oz):

Anchovy, European.
Bluefish.
Capelin conch.
Herring, Atlantic or Pacific.
Mackerel, Atlantic, chub.
Japanese horse, or king.
Mullet.
Sablefish.
Salmon, all varieties.
Sturgeon, Atlantic or common.
Trout, lake.
Tuna, white albacore or bluefin (not canned light tuna).
Whitefish, lake.

Table 3-2
THE LIPID FAMILY

Triglycerides (fats and oils)
- Glycerol (1 per triglyceride)
- Fatty acids (3 per triglyceride)

  Saturated
  Monounsaturated
  Polyunsaturated
     Omega-6
     Omega-3

Phospholipids (such as the lecithins)

Sterols (such as cholesterol)

Phospholipids

Up to now, this discussion has focused on one of the three classes of lipids, the triglycerides (fats and oils), and their component parts, the fatty acids (see Table 3-2). The other two classes of lipids, the **phospholipids** and sterols, make up only 5 percent of the lipids in the diet, but they are nevertheless worthy of attention. Among the phospholipids, the lecithins are of particular interest.

STRUCTURE OF PHOSPHOLIPIDS
Like the triglycerides, the **lecithins** and some other phospholipids have a backbone of glycerol; they differ from the triglycerides in having only two fatty acids attached to them. In place of the third fatty acid is a phosphate group (a phosphorus-containing acid) and a molecule of **choline** or a similar compound. The fatty acids make phospholipids soluble in fat; the phosphate group enables them to dissolve in water. Such versatility benefits the food industry, which uses phospholipids as **emulsifiers** to mix fats with water in such products as mayonnaise and candy bars.

ROLES OF PHOSPHOLIPIDS
Lecithins and other phospholipids are important constituents of cell membranes. They also act as emulsifiers in the body, helping to keep other fats in solution in the watery blood and body fluids.

PHOSPHOLIPIDS: NOT ESSENTIAL
Lecithins periodically receive attention in the popular press. People may hear that lecithins are a major constituent of cell membranes (true), that the functioning of all cells depends on the integrity of the cell membranes (true), and that consumers must therefore take lecithin supplements (false). The body digests lecithins before it absorbs them, so the lecithins people eat do not reach the body tissues intact. Instead, the lecithins used for building cell membranes are made from scratch by the liver. In other words, the lecithins are not essential nutrients.

Sterols

Sterols are large, complex molecules consisting of interconnected rings of carbon. Cholesterol is the most familiar sterol, but others, such as vitamin D and the sex hormones (for example, testosterone), are important, too.

STEROLS IN FOODS
Both plant and animal-derived foods contain sterols, but only animal-derived foods—meats, eggs, fish, poultry, and dairy products—contain cholesterol. Organ meats, such as liver and kidneys, and eggs are richest in cholesterol; cheeses and meats have less. Shellfish contain many sterols, but much less cholesterol than was thought in the past.

CHOLESTEROL SYNTHESIS
Like the lecithins, cholesterol can be made by the body, so it is not an essential nutrient. Your liver is manufacturing it now,

phospholipids: one of the three main classes of lipids; similar to triglycerides but with choline (or another compound) and a phosphorus-containing acid in place of one of the fatty acids.

lecithins: one type of phospholipid.

choline: a nonessential nutrient that can be made in the body from an amino acid.

emulsifiers: substances that mix with both fat and water and that permanently disperse the fat in the water, forming an emulsion.

sterols: one of the three main classes of lipids; include cholesterol, vitamin D, and the sex hormones (such as testosterone).

as you read, at the rate of perhaps 50,000,000,000,000,000 molecules per second. The raw materials that the liver uses to make cholesterol can all be taken from glucose or saturated fatty acids. In other words, cholesterol can be made from either carbohydrate or fat. Most of the body's cholesterol ends up in the cells, where it performs vital structural and metabolic functions.

■■ CHOLESTEROL'S TWO ROUTES IN THE BODY   After being made, cholesterol either leaves the liver or is transformed there into related compounds such as vitamin D. Cholesterol leaves the liver by two routes:

1. It may be made into **bile,** stored in the gallbladder, and delivered to the intestine.
2. It may travel, via the bloodstream, to all the body's cells.

■■ CHOLESTEROL RECYCLED   The bile that is made from cholesterol in the liver and released into the intestine aids in the digestion and absorption of fat (see Chapter 5). After bile does its job, some is reabsorbed into the body and recycled; the rest is excreted in the feces.

■■ CHOLESTEROL EXCRETED   While bile is in the intestine, some of it may be trapped by certain kinds of dietary fibers or by some medications, which carry it out of the body in the feces. The excretion of bile reduces the total amount of cholesterol remaining in the body.

■■ CHOLESTEROL TRANSPORT   Some cholesterol, packaged with other lipids and protein, leaves the liver via the arteries and is transported to the body tissues by the blood. These packages of lipids and proteins are called lipoproteins. As the lipoproteins travel through the body, tissues can extract lipids from them. Cholesterol's harmful effects in the body occur when it forms deposits in the artery walls. These deposits lead to **atherosclerosis,** a disease that can cause heart attacks and strokes.

■■ Fats and Health

Of all the dietary factors related to chronic diseases prevalent in developed countries, fat is by far the most significant. Excessive intakes of dietary fat contribute to obesity, diabetes, cancer, **cardiovascular disease,** and probably other diseases and disorders as well. Heart disease is the number one killer of adults in the United States. In the interest of good health and disease prevention, the one change that most people should make in their diets is to limit their intakes of total fat. It is especially important to limit saturated fat because the cholesterol that accumulates in arteries is manufactured largely from fragments derived from saturated fat.

■■   Fats and Fatty Acids

Most people realize that elevated blood cholesterol is an important risk factor for heart disease. The higher the blood cholesterol, the greater the risk of heart disease. Most people may not realize, though, that cholesterol in *food* is not the main influential factor in raising *blood* cholesterol. It is total fat, especially saturated fat, that raises blood cholesterol. An extensive review of well-controlled studies on the effects of dietary fats on blood cholesterol reached the following conclusions:

- Saturated fatty acids elevate blood cholesterol and are the main dietary determinants of blood cholesterol levels.*[6]

---

*It should be noted that not all saturated fatty acids have the same cholesterol-raising effect. For example, stearic acid, an 18-carbon fatty acid, does not raise blood cholesterol.

bile: a compound made by the liver from cholesterol and stored in the gallbladder. Bile prepares fat for digestion.

Both the intestine and the liver make lipoproteins. Chapter 5 tells the story of lipid transport.

atherosclerosis (ath-er-oh-scler-OH-sis): a type of artery disease characterized by accumulations of lipid-containing material on the inner walls of the arteries (see Chapter 21).

cardiovascular disease (CVD): a general term for all diseases of the heart and blood vessels (see Chapter 21).

Less influential, but still significant, were the following factors:

- Polyunsaturated fatty acids lower blood cholesterol.
- Monounsaturated fatty acids may lower blood cholesterol slightly or have a neutral effect on it.

Enjoy fish and low-fat foods for good health.

:: SATURATED FAT AND BLOOD CHOLESTEROL In support of these conclusions, researchers found that diets low in saturated fatty acids helped lower blood cholesterol in men.[7] Adding polyunsaturated fatty acids to the diet lowered blood cholesterol further. Probably, the researchers speculated, reducing dietary cholesterol would have reduced blood cholesterol still further. A meta-analysis of more than 30 studies further supports the beneficial effects of reducing dietary saturated fatty acids.[8]

:: OMEGA-3 FATTY ACIDS AND BLOOD CHOLESTEROL The omega-3 fatty acids appear to reduce blood cholesterol. This effect was first noticed when researchers learned that the Inuit peoples of Alaska and Greenland, despite high-energy, high-fat, high-cholesterol diets, enjoyed relative freedom from heart diseases—especially atherosclerosis. Analysis of the foods common in Inuit diets, which derive primarily from marine animals, revealed that they were rich in omega-3 fatty acids, particularly EPA and DHA. Research findings such as these have led the American Heart Association to recommend two to three fish meals a week.

:: OMEGA-3 FATTY ACIDS AND CANCER Omega-3 fatty acids may also help to prevent cancer. High *fat* intakes appear to *promote* cancer, and some animal studies implicate polyunsaturated fatty acids from vegetable oils (mostly omega-6 fatty acids). Available evidence on human beings, however, is controversial and depends on the type of cancer in question.[9] The authors of a review of published studies on the topic concluded that "it seems unlikely that high intakes of vegetable oils raise the risks of breast, colon, or prostate cancer," but they did not rule out the possibility altogether.[10] Much more remains to be learned about the influence of different fatty acids on cancer. In contrast, polyunsaturated fats from fish oils (mostly omega-3 fatty acids) may delay cancer development, slow tumor growth rates, and reduce the size and number of tumors.[11]

:: MONOUNSATURATED FATTY ACIDS AND BLOOD CHOLESTEROL Whether monounsaturated fatty acids exert an independent blood cholesterol–lowering effect or have a neutral effect is unclear. Some research suggests that the cholesterol-lowering effects of monounsaturated fatty acids depend on the amount of cholesterol-elevating saturated fatty acids in the diet.[12] When dietary saturated fatty acids are low, monounsaturated fatty acids seem to have a cholesterol-lowering effect. When dietary saturated fatty acids are high, the cholesterol-lowering effects of monounsaturated fatty acids become less apparent or neutral. In general, compared with saturated fatty acids, monounsaturated fatty acids are cholesterol lowering, although less so than polyunsaturated fatty acids.[13] A benefit to health is seen when monounsaturated fat (such as olive oil or canola oil) replaces saturated fat (such as beef fat) in the diet.[14]

:: RECOMMENDATIONS Dietary guidelines recommend that *total* fat intake should not exceed 30 percent of the day's total energy intake; 20 percent might be ideal. Saturated fats should contribute less than 10 percent; polyunsaturated fats should not exceed 10 percent; monounsaturates should provide the remaining 10 percent or so. Recommendations limiting fat are not intended for infants or children under two years old.

Limiting dietary fat intake to 30 percent or less of total energy requires careful planning at every meal. According to research on U.S. food choices, people typically eat about 34 percent of their energy intakes as fat.[15] Even though this is less than in the recent past, it is more than at the turn of the century, when foods were less highly processed, and it is more than people need. Guidelines for cholesterol intake recommend less than 300 milligrams daily. A later "How to" box offers specific suggestions for putting these guidelines into practice.

LEARNING LINK

Now that you have learned some specifics about dietary fats and health, you can use your knowledge to advise clients who are concerned about the health of their heart and blood vessels. For example, although some clients may already know that the fat they eat, and especially saturated fat, raises blood cholesterol more than the cholesterol they eat, many others will rely on you to clear up their confusion on this matter. Once your clients have a clear understanding of this, you can ask them about the foods they eat and advise them on where to make changes and how to read food labels.

██ Fat Substitutes

As people learn more about the health consequences of high-fat diets, fat substitutes (also called fat replacers) offer hope for the prevention and treatment of heart disease and obesity. Skeptics say that people will use fat replacers the same way they use artificial sweeteners: in *addition* to fats, rather than *instead* of fats. Some research shows that when people use fat substitutes, they eat less fat but more sugar and starch, so their energy intakes remain constant.[16]

Today shoppers can choose from thousands of fat-reduced products on grocery shelves. Fat-free bakery products, cheeses, frozen desserts, and many other products are available that taste rich but offer less than half a gram of fat in a serving. Some of these products use traditional ingredients such as sugar, starch, nonfat milk, soluble fiber, or egg whites in place of fat; others are using specially designed fat replacers.[17]

██ **TYPES OF FAT REPLACERS** Food chemists have been working for decades on ways to reduce the fat in foods. Because the functions of fat in food are both diverse and desirable, it is not an easy ingredient to replace. In general, fat replacers fall into three categories: carbohydrate based (Oatrim and Z-Trim), protein based (Simplesse), and fat based (Salatrim and olestra).[18]

██ **CARBOHYDRATE-BASED REPLACERS: OATRIM AND Z-TRIM** Researchers at the U.S. Department of Agriculture (USDA) developed Oatrim, which is derived from oat fiber. Oatrim can be used as a fat substitute to reduce the energy content of foods such as frozen desserts, salad dressings, soups, and high-fiber baked goods by half. Oatrim is stable when heated, but it cannot be used for frying. Unlike other fat substitutes, Oatrim lowers cholesterol not only by replacing saturated fat, but also by providing fiber.

Z-Trim is made from the seed hulls of oats, peas, soybeans, or rice or the bran from corn or wheat. Z-Trim provides no calories and imparts no flavor to foods. It can be used in baked goods, cheese, and meat products.[19]

██ **PROTEIN-BASED REPLACER: SIMPLESSE** The Food and Drug Administration (FDA) declared Simplesse safe for use in ice cream and frozen desserts in 1990. Simplesse is made from protein—either egg white or milk—

which is processed into mistlike particles similar in consistency to fat. Because the components of Simplesse have long been used in foods, safety studies were not required. This fat substitute mimics the rich taste and texture of fat but cuts the kcalorie content by up to 80 percent. Because proteins coagulate at high temperatures, Simplesse cannot be used for frying or cooking, but it can be used in ice creams, yogurts, salad dressings, mayonnaise, and butter.

■■ FAT-BASED REPLACERS: SALATRIM AND OLESTRA Fat-based replacers can be categorized as modified or synthetic. Modified fat-based replacers are triglycerides that have been altered to contain specific mixtures or arrangements of fatty acids. Salatrim is an example of a modified fat-based replacer. Olestra is the only synthetic fat-based replacer approved so far. Olestra is synthetic because its chemical configuration does not occur in nature.

Olestra, which was formerly known as sucrose polyester, is a synthetic combination of sucrose and fatty acids that looks, feels, and tastes like food fat. Unlike sucrose or fatty acids, though, olestra is indigestible; the body has no way to digest it. Olestra can therefore be substituted for fats without adding kcalories or raising a person's blood lipids.

From some points of view, olestra is the most successful of the fat replacers, for its properties are identical to those of fats and oils when used in frying, cooking, and baking. It can be heated to frying temperatures without breaking down; it performs all of the functions of fat in cakes, pie crusts, and other baked goods; and most remarkably—aside from a slight aftertaste—it tastes like fat. Before jumping for joy, however, the attributes of olestra must be weighed against evidence concerning its safety.

■■ THE SAFETY OF OLESTRA The company that invented olestra has been studying its safety for more than two decades. The results of the studies revealed that olestra causes digestive distress and nutrient losses. The company has addressed each of these problems to the satisfaction of the FDA, which approved olestra for use in snack foods in 1996. Some nutrition experts disagree with the FDA, however, and have come out strongly against olestra's approval. These issues deserve a moment's attention.

Olestra's side effects are rooted in two aspects of its nature. For one thing, because olestra replaces a major food constituent, fat, the substance is consumed in large amounts, measured in many grams per serving. Nonfat potato chips, for example, derive about a third of their weight from olestra. Also, by design, olestra is indigestible. All of the oily olestra eaten in a food passes through the digestive tract and is excreted from the body.

■■ DIGESTIVE PROBLEMS The presence of olestra in the large intestine causes diarrhea, gas, cramping, and an urgent need for defecation in some people. Further, the oil can creep through the feces and leak uncontrollably from the anus. No one yet knows who is most likely to encounter these effects, but some of these symptoms almost always occur when olestra is consumed in large quantities. The FDA decided that these digestive problems were unpleasant but did not constitute a safety problem. When researchers recently provided olestra-containing snacks or ordinary snacks to more than 3100 volunteers, they found no significant increase in digestive distress with olestra.[20]

■■ NUTRIENT LOSSES In the digestive tract, olestra dissolves fat-soluble substances in foods. Consequently, fat-soluble vitamins (vitamins A, D, E, and K) become unavailable for absorption when olestra is present in a meal. Thus, when olestra-containing chips are eaten with other foods, the olestra traps the vitamins those foods contain, robbing the eater of those vitamins. To compensate for this effect, olestra is fortified with vitamins A, D, E, and K. The

FDA ruled that fortification removes the threat of harm from malnutrition that olestra could otherwise cause.

■■ **DO FAT SUBSTITUTES WORK?** Research has not yet shown that fat substitutes promote weight loss or lower blood lipid concentrations. If people use fat replacers literally as replacers for fat in their diets, then perhaps potential health benefits will be realized. Some experts are concerned, however, that people may feel at liberty to eat *more* high-fat foods by rationalizing that they obtain fat "credit" when they eat foods containing fat substitutes. Indeed, this seems to be the case with artificial sweeteners. Although consumption of sugar substitutes has risen since their introduction, sugar consumption has risen as well. Another concern is that people may become so carried away with eating foods containing fat substitutes that they will neglect to eat more nutrient-dense foods such as fresh fruits and vegetables. It seems that with fat substitutes, as with most things in life, moderation is the key to appropriate use.

LEARNING LINK

Fat replacers can be useful for clients who must adhere to diets restricted in total and saturated fat and/or energy (see Chapter 21). Fat replacers offer an effective way to maintain the palatability of energy- or fat-controlled diets, as long as the products are used to *replace* some of the fat or energy that would normally be consumed.

■■ Fats in Foods

Fats are important in foods as well as in the body. Many of the compounds that give foods their flavor and aroma are found in fats and oils. The delicious aromas associated with bacon, ham, and other meats, as well as with onions being sautéed, come from fats. Fats also influence the texture of many foods, enhancing smoothness, creaminess, moistness, or crispness.[21] As mentioned, four vitamins—A, D, E, and K—are soluble in fat. When the fat is removed from a food, many fat-soluble compounds, including these vitamins, are also removed. Table 3-3 summarizes the roles of fats in foods.

Fats are also an important part of most people's ethnic or national cuisines. Each culture has its own favorite food sources of fats and oils. In Canada, canola oil (also known as rapeseed oil) is widely used. In the Mediterranean area, Greeks, Italians, and Spaniards rely heavily on olive oil. Both canola oil and olive oil are rich sources of monounsaturated fatty acids. Asians use the polyunsaturated oil of soybeans. Jewish people traditionally employ chicken fat. Everywhere in North America, butter and margarine are widely used.

These cultural associations along with the very real pleasures fats contribute to meals help explain why fat consumption is so high. Nevertheless, reducing dietary fat is the single most important thing people can do for health; it is the top recommendation of almost every nutrition authority. The first step to reducing dietary fat is finding out where the fat is.

■■ Finding the Fats in Foods

Of the groups in the Food Guide Pyramid, the fats at the tip and the meats (and nuts) one step below always contain fat, but two other food groups—the milk, yogurt, and cheese group and the breads and cereal group—sometimes

---

The FDA requires that olestra-containing foods bear this warning: "This Product Contains Olestra. Olestra may cause abdominal cramping and loose stools. Olestra inhibits the absorption of some vitamins and other nutrients. Vitamins A, D, E, and K have been added."

Table 3-3
THE FUNCTIONS OF FATS IN FOODS

Fats in foods:
- Contribute flavor and aroma.
- Influence the texture, adding creaminess, smoothness, moistness, or crispness.
- Help make foods tender.
- Carry fat-soluble vitamins.

contain fat as well. Vegetables and fruits, if unprocessed, are virtually fat-free, with two notable exceptions: avocados and olives, which are rich in monounsaturated fat. Grains, too, in their natural state contain little or no fat.

■■ **ADDED FATS IN FOODS**   A dollop of dessert topping, a spread of butter on bread, oil or shortening in a recipe, dressing on a salad—all of these are examples of *added* fats. Indeed, all sorts of fats can be added to foods during commercial or home preparation or at the table. The following amounts of these fats contain about 5 grams of pure fat, providing 45 kcalories and negligible protein and carbohydrate:

- 1 teaspoon of oil or shortening.
- 1½ teaspoons of mayonnaise, butter, or margarine.
- 1 tablespoon of regular salad dressing, cream cheese, or heavy cream.
- 1½ tablespoons of sour cream.

These foods provide the majority of added fats to the diet. They are the fats of fried foods and baked goods, sauces and mixed dishes, and dips and spreads.

■■ **HIDDEN FATS IN FOODS**   Other foods that are significant contributors of fat include convenience foods, lunch meats, and other prepared meats. The fat in these foods is sometimes referred to as invisible fat because it does not have the obvious appearance of fat. An ounce of lean meat or low-fat cheese supplies about half its kcalories from fat (28 kcalories from protein and 27 kcalories from fat). An ounce of high-fat meat (such as bologna) or most cheeses supplies 72 percent of its energy from fat (28 kcalories from protein and 72 kcalories from fat). Two tablespoons of peanut butter supply 72 percent of their energy as fat (32 kcalories from protein, 24 kcalories from carbohydrate, and 140 kcalories from fat)! Thus, foods that are usually thought of as protein-rich foods may actually contain more fat energy than protein energy. Note that the values for meat given here are for 1-ounce portions. An average serving of hamburger is usually 3 or 4 ounces. An average dinner steak may be 8 ounces or larger.

■■ **FATS IN THE MILK, YOGURT, AND CHEESE GROUP**   The fat in milk is about 63 percent saturated fat; the cholesterol content is 33 milligrams per cup for whole milk or 4 milligrams for fat-free milk. Thus, choosing fat-free in place of whole milk reduces your intake of cholesterol as well as of saturated fat.

Note that cream and butter do not appear in the milk group. Milk and yogurt are rich in calcium and protein, but cream and butter are not. Cream and butter are fats, as are whipped cream, sour cream, and cream cheese. That is why the food group that includes milk is carefully labeled the "milk, yogurt, and cheese group," not the "dairy group."

■■ **FATS IN THE MEAT, POULTRY, FISH, DRY BEANS AND NUTS, AND EGGS GROUP**
The fats in meats and eggs are about half saturated. The fats in poultry, fish, and nuts are more unsaturated than saturated—a healthier balance. Eating fish instead of meat two or three times a week supports heart health. Fish is not only leaner than most other animal-protein sources, but it is a source of omega-3 fatty acids as well. As noted earlier, research shows that diets rich in fish oils can lower blood cholesterol, just as diets low in fat and saturated fat can.[22]

■■ **FATS IN THE BREAD, CEREAL, RICE, AND PASTA GROUP**   Breads and cereals in their natural state are very low in fat, but fat may be added during processing or cooking. The fat in these foods can be particularly hard to detect, so people must remember which foods stand out as being high in fat. Notable are

Remember, fat is a more concentrated energy source than the other energy nutrients: 1 g carbohydrate or protein = 4 kcal, but 1 g fat = 9 kcal.

Remember that an ounce of meat is not an ounce of protein. An ounce (30 g) of lean meat contains 7 g protein and 3 g fat. The other 20 g are largely water with associated vitamins and minerals.

1 c whole milk:
  8 g fat
  5 g saturated fat
  33 mg cholesterol

1 c reduced-fat milk:
  5 g fat
  3 g saturated fat
  18 mg cholesterol

1 c fat-free milk:
  < 1 g fat
  < 1 g saturated fat
  4 mg cholesterol

LOWER FAT INTAKE—BY FOOD GROUP

Fats can sneak into the diet in every food group. Inspect each group to discover where the fats are, and control them as follows.

MEAT, FISH, AND POULTRY

- Choose fish, poultry, or lean cuts of pork or beef; look for cuts named *round* or *loin* (eye of round, top round, round tip, tenderloin, sirloin, and top loin).
- Trim the fat from pork and beef; remove the skin from poultry.
- Grill, roast, broil, bake, stir-fry, stew, or braise meats. (Don't fry.) When possible, place meat on a rack while cooking so that fat can drain.
- Use lean ground turkey instead of hamburger in recipes.
- Brown ground meats without added fat; then drain off fat.
- Refrigerate meat pan drippings and broth; when the broth solidifies, remove the fat and use the defatted broth in recipes.
- Select tuna packed in water; rinse oil-packed tuna with hot water to remove much of the fat.
- Fill kabob skewers with lots of vegetables and slivers of meat; create main dishes and casseroles by combining a little meat, fish, or poultry with a lot of pasta, rice, or vegetables.
- Make meatless spaghetti sauces and casseroles.
- Eat a meatless meal or two daily (use the milk and other food groups with care as suggested next).

MILK AND CHEESES

- Drink fat-free, low-fat, and reduced-fat milk instead of whole milk.
- Use fat-free, low-fat, and reduced-fat cheeses (such as part-skim ricotta and low-fat mozzarella) instead of regular cheeses.
- Use fat-free yogurt or sour cream instead of regular sour cream.
- Use evaporated fat-free milk instead of cream.
- Enjoy fat-free frozen yogurt, sherbet, or ice milk instead of ice cream.

FRUITS AND VEGETABLES

- Use butter-flavored granules instead of butter or margarine on vegetables.

granola, croissants, biscuits, cornbread, dinner rolls, quick breads, snack and party crackers, muffins, pancakes, and waffles. Packaged breakfast bars often resemble candy bars in their fat and sugar contents.

Cutting Fat Intake and Choosing Unsaturated Fats

Knowing which foods contain the most fat is the first step toward meeting the recommendation to reduce dietary fat in general and saturated fat in particular. As a general rule, a person who eats meat and wishes to reduce both saturated fat and cholesterol intake can eat fewer high-fat meats and dairy foods and fewer eggs, and more poultry (without the skin), fish, and fat-free dairy products. A vegetarian who eats dairy products and eggs can shift to fat-free milk and low-fat cheeses and limit butter and egg intake. Vegetarians who omit animal-derived foods generally eat less saturated fat and consume no cholesterol because plant foods do not contain cholesterol. The accompanying box offers strategies for lowering fat, food group by food group.

- Use fat-free yogurt or fat-free salad dressing instead of sour cream, cheese, mayonnaise, or other sauces on vegetables and in casseroles.
- Select fat-free or low-fat mayonnaise or salad dressings, or use herbs, lemon juice, and spices instead of regular salad dressing.
- Add a little water to thick, bottled salad dressings to dilute the amount of fat each serving provides.
- Eat at least two vegetables (in addition to a salad) with dinner.
- Snack on raw vegetables or fruits instead of high-fat foods like potato chips.
- Enjoy fruit for dessert.

BREADS AND CEREALS

- Use fruit butters or jellies instead of butter or margarine on bread.
- Select breads, cereals, and crackers that are low in fat (for example, bagels instead of croissants).

OTHER FOODS AND COOKING TIPS

- Use a nonstick pan or coat the pan lightly with cooking oil.
- Use egg substitutes in recipes instead of whole eggs or use 2 egg whites in place of each whole egg.
- Use half the margarine, butter, or oil called for in a recipe. (The minimum amount of fat for muffins, quick breads, and biscuits is 1 to 2 tablespoons per cup of flour; for cakes and cookies, 2 tablespoons per cup.)
- Select whipped types of butter, margarine, or cream cheese for use at the table; they contain half the kcalories of the regular types.
- Use wine, lemon juice, or broth instead of butter or margarine when cooking.
- Stir-fry in a small amount of oil; add moisture and flavor with broth, tomato juice, or wine.
- Use variety to enhance enjoyment of the meal: vary colors, textures, and temperatures—hot cooked versus cool raw foods—and use garnishes to complement the food.

▪▪ **FATS AND kCALORIES**   Removing fat from food also removes energy as Figure 3-4 (p. 66) shows. A small pork chop with the fat trimmed to within a half-inch of the lean provides 275 kcalories; with the fat trimmed off completely, it supplies 165 kcalories. A baked potato with butter and sour cream (1 tablespoon each) has 350 kcalories; a plain baked potato has 220 kcalories. The single most effective step you can take to reduce the energy value of a food is to eat it with less fat.

▪▪ **CHOOSING UNSATURATED FATS**   To the extent that a person does eat fats, those to choose are the unsaturated ones. The hardness of fats at room temperature is an indicator: the softer a fat is, the more unsaturated it is. Chicken fat is softer than pork fat, which is softer than beef tallow. Of the three, chicken fat is the most unsaturated, and beef tallow is the most saturated. Unsaturated fats melt more readily. Generally speaking, vegetable and fish oils are rich in polyunsaturates, olive oil and canola oil are rich in monounsaturates, and the harder fats—animal fats—are more saturated (see Figure 3-5 on page 66).

At room temperature, unsaturated fats (such as those found in oil) are usually liquid, whereas saturated fats (such as those found in butter) are solid.

Figure 3-4
FOOD FAT AND kCALORIES

Pork chop with a half-inch of fat (275 kcal and 19 g fat).

Potato with 1 tbs butter and 1 tbs sour cream (350 kcal and 14 g fat).

Whole milk, 1 c (150 kcal and 8 g fat).

Pork chop with fat trimmed off (165 kcal and 8 g fat).

Plain potato (220 kcal and <1 g fat).

Fat-free milk, 1 c (90 kcal and <1 g fat).

Figure 3-5
COMPARISON OF DIETARY FATS

Most fats are mixtures of saturated, monounsaturated, and polyunsaturated fatty acids.

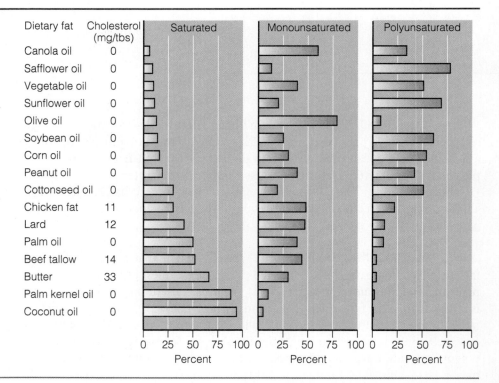

Dietary fat	Cholesterol (mg/tbs)
Canola oil	0
Safflower oil	0
Vegetable oil	0
Sunflower oil	0
Olive oil	0
Soybean oil	0
Corn oil	0
Peanut oil	0
Cottonseed oil	0
Chicken fat	11
Lard	12
Palm oil	0
Beef tallow	14
Butter	33
Palm kernel oil	0
Coconut oil	0

▪▪ DON'T OVERDO FAT RESTRICTION

Although it is very difficult to do, some people actually manage to eat too *little* fat—to their detriment. Among them are young women and men with eating disorders, described in Nutrition in Practice 7. As a practical guideline, it is wise to include the equivalent of at least a teaspoon of fat in every meal.

▪▪ CAUTIONS

If your clients wish to make choices consistent with current recommendations, they need to learn how to read food labels, limit fat in general, and seek out the polyunsaturated and monounsaturated fats in preference to the saturated ones. They must beware, however: vegetable fat or vegetable oil doesn't always mean unsaturated fat. Both coconut oil and palm oil, for example, which are often used in nondairy creamers, are saturated fats, and both raise blood cholesterol.

▪▪ HYDROGENATED FATS

Food producers hydrogenate unsaturated fatty acids to prevent spoilage—the points of unsaturation are vulnerable to attack by oxygen, which makes the oil **rancid. Hydrogenation** also makes unsaturated fatty acids harder—as when corn oil is hydrogenated to make spreadable margarine. Vegetable oils that have been hydrogenated lose their polyunsaturated character and the health benefits that go along with it.

▪▪ *TRANS*-FATTY ACIDS

When polyunsaturated oils are hydrogenated, the fatty acids not only become more saturated, but the hydrogenation process changes the shape of the molecules. In nature, most unsaturated fatty acids are *cis*-fatty acids—meaning that the hydrogens next to the double bonds are on the same side of the carbon chain. When hydrogenated, some of the fatty acids become *trans*-fatty acids—meaning that some of their hydrogens jump to the opposite side of the chain (see Figure 3-6).

Trans-fatty acids have implications for the body's health. Some researchers estimate that *trans*-fatty acids carry a risk to the health of the heart and arteries between that of saturated and unsaturated fats.[23] *Trans*-fatty acids elevate blood cholesterol and thus raise the risk of heart disease and heart attack.[24] Some researchers argue, however, that the relationship between *trans*-fatty acid intake and an increased risk of heart disease remains unproved and that more extensive research is needed.[25]

When news of *trans*-fatty acids' effects on heart health first emerged, some people hastily switched from using margarine back to butter, believing oversimplified reports that margarine provided no heart health advantage over butter. It is true that most margarines and virtually all shortenings are made from mostly hydrogenated fats and therefore contain substantial *trans*-fatty acids—up to 40 percent. Some margarines, however, especially the soft or liquid varieties, are made from unhydrogenated oils. These have long proved to be less likely to elevate blood cholesterol than the saturated fats of butter.

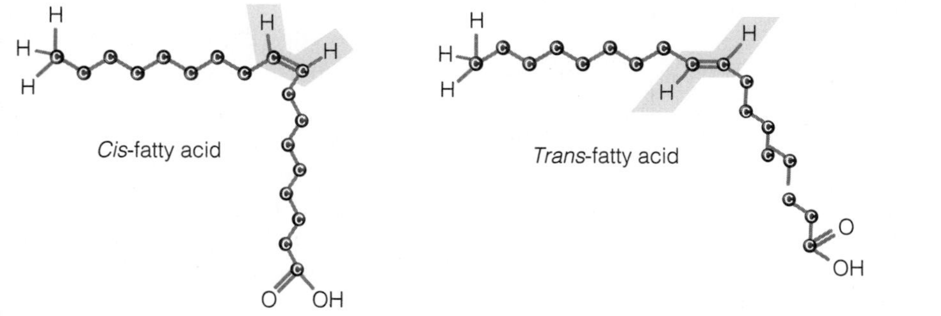

rancid: the term used to describe fats when they have deteriorated, usually by oxidation. Rancid fats often have an "off" odor.

hydrogenation: the process of adding hydrogen to unsaturated fat to make it more solid and resistant to chemical change.

Figure 3-6
CIS- AND *TRANS*-FATTY ACIDS COMPARED

Cis-fatty acid

Trans-fatty acid

These foods are major contributors of trans-fatty acids.

With regard to blood cholesterol, margarine may be just a small contributor. Foods other than margarine contribute far more *trans*-fatty acids to the diet—and more total fat, too.[26] Fast foods, chips, baked goods, and other commercially prepared foods are high in fats containing up to 50 percent *trans*-fatty acids. Overall, consumers are eating more fats containing *trans*-fatty acids than ever before because manufacturers are adding more hydrogenated fats to processed foods.

Trans-fatty acids contribute about 2 to 4 percent of total energy intake, whereas total fat accounts for about 34 percent. Thus, by limiting total fat, one can limit the health risks posed by all types of fat.

Food labels can be misleading in this regard. On "Nutrition Facts" panels, *trans*-fatty acids are counted among the polyunsaturated fats from which they arose, and not with the saturated fats whose health effects they mimic. Further, fast-food chains advertise foods fried in "vegetable oil" when that oil is a hydrogenated type containing abundant *trans*-fatty acids. Some experts are calling for a separate statement of *trans*-fatty acids on food labels and an end to deceptive advertising.[27]

Chapters 2 and 3 have looked briefly at the two major energy fuels in the body—carbohydrate and fat. When used for energy, each has desirable characteristics. The glucose derived from carbohydrate is needed by the brain and nerve tissues and is easily used for energy in other cells. Fat is a particularly useful fuel because the body stores it efficiently and in generous amounts. Chapter 4 looks at protein, a nutrient that can be used as fuel, but whose primary role is to provide machinery for getting things done.

SELF CHECK

1. Three classes of lipids in the body are:
   a. triglycerides, fatty acids, and cholesterol.
   b. triglycerides, phospholipids, and sterols.
   c. fatty acids, phospholipids, and cholesterol.
   d. glycerol, fatty acids, and triglycerides.

2. Fat in the body cannot:
   a. provide energy.
   b. insulate the body against extreme temperatures.
   c. form a part of cell membranes.
   d. make glucose.

3. A fatty acid that has the maximum possible number of hydrogen atoms is known as a(n):
   a. saturated fatty acid.
   b. monounsaturated fatty acid.
   c. PUFA.
   d. essential fatty acid.

4. Essential fatty acids:
   a. are used to make substances that regulate blood pressure, among other functions.
   b. can be made from carbohydrates.
   c. include lecithin and cholesterol.
   d. cannot be found in commonly-eaten foods.

5. To include omega-3 fatty acids in the diet, the American Heart Association recommends eating:
   a. cholesterol-free margarine.
   b. fish oil supplements.
   c. hydrogenated margarine.
   d. two to three fish meals per week.

6. Lecithins and other phospholipids in the body function as:
   a. emulsifiers.
   b. enzymes.
   c. temperature regulators.
   d. shock absorbers.

7. Excess dietary fat contributes to:
   a. heart disease, obesity, diabetes, and cancer.
   b. liver disease, sickle-cell anemia, fatty liver, and ulcerative colitis.
   c. iron-deficiency anemia, Wilson's disease, and fatty liver.
   d. food allergies, hyperglycemia, hepatitis, and ulcerative colitis.

8. Some examples of foods with hidden fats are:
   a. cheese, lettuce, and fruit juices.
   b. peanut butter, cheese, and lunch meats.
   c. fish, rice, and potatoes.
   d. baked potatoes, vegetables, and fruits.

9. Generally speaking, vegetable and fish oils are rich in:
   a. polyunsaturated fat.
   b. saturated fat.
   c. cholesterol.
   d. *trans*-fatty acids.

10. Two ways to lower fat intake are to:
    a. eat no red meat or bread.
    b. fry meat for a shorter time so that it absorbs less fat and substitute sour cream for butter.
    c. look for cuts of meat named round or loin and remove the skin from poultry.
    d. use nuts instead of meat and drink no milk.

Answers to these questions appear in Appendix H.

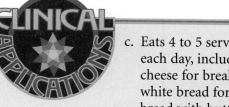

The connection between the overconsumption of fats and chronic diseases (obesity, diabetes, cancer, and cardiovascular disease) emphasize how important it is to be alert to a client's fat intake.

1. What advice would you offer a client who reports the following:

    a. Eats 2 or more 6-ounce servings of meat each day.
    b. Drinks whole milk and eats regular cheddar cheese each day.
    c. Eats 4 to 5 servings of breads and cereals each day, including a bagel with cream cheese for breakfast, a bologna sandwich on white bread for lunch, and biscuits or cornbread with butter to accompany dinner.
    d. Eats one serving of fruit each day and only eats vegetables on occasion.

2. Make a list of low-fat and fat-free foods, beverages, and seasonings that your client can substitute in place of higher fat items.

NUTRITIONTHENET

FOR FURTHER STUDY OF THE
TOPICS IN THIS CHAPTER,
ACCESS THESE WEB SITES.

www.amhrt.org
American Heart Association

cancer.org
American Cancer Society

www.nci.nih.gov
National Cancer Institute

www.ificinfo.health.org/review/
ir-fat.htm
*International Food Information
Council Review: Sorting Out the
Facts on Fat*

www.fda.gov:80/fdac/features/696_
fat.html
*Taking the Fat Out of Food (from
FDA Consumer)*

www.fda.gov:80/opacom/
backgrounders/olestra.html
*FDA Backgrounder on Olestra and
Other Fat Substitutes*

www.nalusda.gov/fnic/Fpyr/
pyramid.html
Food Guide Pyramid

Notes

[1] W. E. Connor, α-Linolenic acid in health and disease. *American Journal of Clinical Nutrition* 69 (1999): 827–828; E. J. Schaefer, Effects of dietary fatty acids on lipoproteins and cardiovascular disease risk: Summary, *American Journal of Clinical Nutrition* 65 (1997): S1655–S1656; C. A. Drevon, Marine oils and their effects, *Nutrition Reviews* 50 (1992): 38–45.

[2] R. Uauy and coauthors, Role of essential fatty acids in the function of the developing nervous system, *Lipids* 31 (1996): S167-S176; M. A. Crawford, The role of essential fatty acids in neural development: Implications of perinatal nutrition, *American Journal of Clinical Nutrition* 57 (1993): S703–S710.

[3] J. Jumpsen and M. T. Clandinin, Lipids and essential fatty acids in brain development, in *Brain Development: Relationship to Dietary Lipid and Lipid Metabolism* (Champaign, Ill.: AOCS Press, 1995), pp. 20–36; Uauy and coauthors, 1996.

[4] N. F. Sheard, Fish consumption and risk of sudden cardiac death, *Nutrition Reviews* 56 (1998): 177–179; A. P. Simopoulos, Epidemiological aspects of omega-3 fatty acids in disease states, in *Handbook of Lipids in Human Nutrition*, ed. G. A. Spiller (New York: CRC Press, 1996), pp. 75–89.

[5] D. S. Siscovick and coauthors, Dietary intake and cell membrane levels of long-chain n-3 polyunsaturated fatty acids and the risk of primary cardiac arrest, *Journal of the American Medical Association* 274 (1995): 1363–1367.

[6] P. M. Kris-Etherton and S. Yu, Individual fatty acid effects on plasma lipids and lipoproteins: Human studies, *American Journal of Clinical Nutrition* 65 (1997): S1628–S1644; R. McPherson and G. A. Spiller, Effects of dietary fatty acids and cholesterol on cardiovascular disease risk factors in man, in *Handbook of Lipids in Human Nutrition*, ed. G. A. Spiller (New York: CRC Press, 1996), pp. 41–49.

[7] A. Nordöy and coauthors, Individual effects of dietary saturated fatty acids and fish oil on plasma lipids and lipoproteins in normal men, *American Journal of Clinical Nutrition* 57 (1993): 634–639.

[8] S. Yu-Poth and coauthors, Effects of the National Cholesterol Education Program's Step I and Step II dietary intervention programs on cardiovascular disease risk factors: A meta-analysis, *American Journal of Clinical Nutrition* 69 (1999): 632–646.

[9] D. P. Rose, Dietary fatty acids and cancer, *American Journal of Clinical Nutrition* 66 (1997): S998–S1003.

[10] P. L. Zock and M. B. Katan, Linoleic acid intake and cancer risk: A review and meta-analysis, *American Journal of Clinical Nutrition* 68 (1998): 142–153.

[11] Simopoulos, 1996; Y. Kim and J. B. Mason, Nutrition chemoprevention of gastrointestinal cancers: A critical review, *Nutrition Reviews* 54 (1996): 259–279.

[12] S. Yu and coauthors, Plasma cholesterol predictive equations demonstrate that stearic acid is neutral and monounsaturated fatty acids are hypocholesteremic, *American Journal of Clinical Nutrition* 61 (1995): 1129–1139.

[13] Kris-Etherton and Yu, 1997.

[14] M. B. Katan, P. L. Zock, and R. P. Mensink, Effects of fats and fatty acids on blood lipids in humans: An overview, *American Journal of Clinical Nutrition* 60 (1994): S1017–S1022.

[15] Federation of American Societies for Experimental Biology, Executive summary from the third report on nutrition monitoring in the United States, *Journal of Nutrition* 126 (1996): S1907–S1936.

[16] J. R. Cotton, J. A. Westrate, and J. E. Blundell, Replacements of dietary fat with sucrose polyester: Effects on energy intake and appetite control in nonobese males, *American Journal of Clinical Nutrition* 63 (1996): 891–896; S. N. Gershoff, Nutrition evaluation of dietary fat substitutes, *Nutrition Reviews* 53 (1995): 305–313; B. J. Rolls and coauthors, Effects of olestra, a noncaloric fat substitute, on daily energy and fat intakes in lean men, *American Journal of Clinical Nutrition* 56 (1992): 84–92.

[17] Position of The American Dietetic Association: Fat replacers, *Journal of the American Dietetic Association* 98 (1998): 463–468.

[18] N. I. Hahn, Replacing fat with food technology, *Journal of the American Dietetic Association* 97 (1997): 15–16.

[19] Hahn, 1997.

[20] R. S. Sandler and coauthors, Gastrointestinal symptoms in 3181 volunteers ingesting snack foods containing olestra or triglycerides: A 6-week randomized, placebo-controlled trial, *Annals of Internal Medicine* 130 (1999): 253–261.

[21] A. Drewnowski, Why do we like fat? *Journal of the American Dietetic Association* 97 (1997): S58–S62.

[22] Nordöy and coauthors, 1993.

[23] ASCN/AIN Task Force on *Trans* Fatty Acids, Position paper on *trans* fatty acids, *American Journal of Clinical Nutrition* 63 (1996): 663–670.

[24] G. J. Nelson, Dietary fat, *trans* fatty acids, and risk of coronary heart disease, *Nutrition Reviews* 56 (1998): 250–252; F. B. Hu and coauthors, Dietary fat intake and the risk of coronary heart disease in women, *New England Journal of Medicine* 337 (1997) : 1491–1499; A. Ascherio and W. C. Willett, Health effects of *trans* fatty acids, *American Journal of Clinical Nutrition* 66 (1997): S1006–S1010.

[25] S. Shapiro, Do *trans* fatty acids increase the risk of coronary artery disease? A critique of the epidemiologic evidence, *American Journal of Clinical Nutrition* 66 (1997): S1011–S1017.

[26] ASCN/AIN Task Force on *Trans* Fatty Acids, 1996.

[27] W. C. Willet and A. Ascherio, Response to the International Life Sciences Institute report on trans fatty acids, *American Journal of Clinical Nutrition* 62 (1995): 524–526; M. B. Katan, European researcher calls for reconsideration of trans fatty acid, *Journal of the American Dietetic Association* 94 (1994): 1097–1098.

Q&A NUTRITION IN PRACTICE

Vegetarian Diets

Eating patterns all along the continuum of dietary choices—from one end, where people eat no foods of animal origin, to the other end, where they eat generous quantities of meat every day—can support or compromise nutritional health. The quality of the diet depends not on whether it consists of all plant foods or centers on meat, but on whether the eater's food choices are based on sound nutrition principles: adequacy of nutrient intakes, balance and variety of foods chosen, appropriate energy intake, and moderation in intakes of substances such as fat, sodium, alcohol, and caffeine that are harmful in excess. People choose to exclude meat and other animal-derived foods from their diets for various reasons—philosophies, health attitudes, or convenience. Some believe that vegetarianism is better for the environment; some, that it is healthier; and some, that it is less costly than the meat-eating alternative. Some just like it better. Whatever the reasons, vegetarians and health professionals who work with them should be aware of the nutrition and health implications of vegetarian diets.

Because vegetarian diets vary in both the types and amounts of animal-derived foods they include, these differences must be considered when evaluating the health status of vegetarians. The glossary on this page defines the various kinds of vegetarian diets.

Vrg.org
The Vegetarian Resource Group

Veg.org/veg
Vegetarian Pages

Are vegetarian diets nutritionally sound?

The American Dietetic Association takes the position that well-planned vegetarian diets offer nutrition and health benefits to adults in general.[1] Research suggests that meat-eating adults who switch to vegetarian diets reduce their risks of heart disease, hypertension, diabetes, some types of cancer, and obesity.[2]

What should be my main concerns when planning a nutritionally sound vegetarian diet?

A vegetarian diet planner faces the same task as other diet planners—obtaining a variety of foods that provide all the needed nutrients within an energy allowance that maintains a healthy body weight. The challenge is to do so using at least one less food group. Since all vegetarians omit meat, and some omit other animal-

GLOSSARY OF VEGETARIAN TERMS

complementary proteins: two or more proteins whose amino acid assortments complement each other in such a way that the essential amino acids missing from one are supplied by the other.

lacto-ovo vegetarians: people who include milk or milk products and eggs, but omit meat, fish, shellfish, and poultry from their diets.

lacto-vegetarians: people who include milk or milk products, but exclude meat, poultry, fish, shellfish, and eggs from their diets.

mutual supplementation: the strategy of combining two protein foods in a meal so that each food provides the essential amino acid(s) lacking in the other.

semivegetarians: people who include some, but not all, groups of animal-derived foods in their diets; they usually exclude meat and may occasionally include poultry, fish, and shellfish; also called *partial vegetarians.*

vegans: people who exclude all animal-derived foods (including meat, poultry, fish, shellfish, eggs, cheese, and milk) from their diets; also called *strict vegetarians* or *total vegetarians.*

derived foods, protein, the nutrient that meat is famous for, merits some discussion here.

Isn't protein a problem in vegetarian diets?

No, protein is not the problem it was once thought to be in vegetarian diets. People who include animal-derived foods such as milk and eggs in their diets need not worry at all about protein deficiency. Even for those who eat only plant-derived foods, protein intakes are usually satisfactory as long as energy intakes are adequate and protein sources are varied.[3] A mixture of proteins from whole grains, legumes, seeds, nuts, and vegetables can provide adequate amounts of all the amino acids.

The idea persists that **vegans** must carefully combine their plant-protein foods in order to obtain the full array of essential amino acids, but new knowledge indicates that this is not necessary.[4] Plant foods can provide more than enough of all the essential amino acids and can sustain people in good health, as long as the diet supplies sufficient energy and does not include too many empty-kcalorie foods. It is true, however, that a meal delivers higher-quality protein when it combines two or more different individual plant-protein sources, each of which supplies amino acids limited in, or missing from, the others. This strategy is called **mutual supplementation.** The protein foods that mutually supplement each other are called **complementary proteins.** Figure NP3-1 (p. 72) shows combinations of plant proteins that provide higher-quality protein than the individual foods alone could supply.

What sorts of food energy intakes do vegetarian diets provide?

Researchers find that vegetarians as a group are closer to a healthy body weight than nonvegetarians. Because

Figure NP3-1
NONMEAT MIXTURES THAT PROVIDE HIGH-QUALITY PROTEIN

Vegetarians who eat no foods from animal sources select foods from two or more of these columns to create high-quality protein combinations:

Grains	Legumes	Seeds and Nuts	Vegetables
Barley	Dried beans	Almonds	Broccoli
Bulgur	(pinto, kidney, navy, etc.)	Cashews	Cabbage
Oats	Dried lentils	Nut butters	Peppers
Pasta	Dried peas	Sesame seeds	Spinach
Rice	Peanuts	Sunflower seeds	Squash
Whole-grain breads	Soy products	Walnuts	

Black beans and rice, a favorite Hispanic combination.

Tofu and stir-fried vegetables with rice, an Asian dish.

obesity impairs health in a number of ways, vegetarians therefore have a health advantage. Vegetarian diets tend to be high in complex carbohydrates and low in fat, characteristics that are consistent with current dietary recommendations aimed at reducing the incidence of obesity and other degenerative diseases in this country.

Not all vegetarians fit the average pattern, though. Obesity does threaten vegetarians who include milk, eggs, and cheese in their diets. They can easily consume both a high-fat diet and excess food energy and so must be careful to select fat-free and low-fat dairy foods and to avoid relying too heavily on these foods in general.

In contrast, people who exclude all animal-derived foods (vegans) may have trouble obtaining *enough* food energy. This is especially true for children and pregnant and lactating women.[5] Vegan diets can fail to provide food energy sufficient to support the growth of a child within a bulk of food small enough for the child to eat.[6] Cereals, legumes, and nuts are the plant foods best suited

to meeting energy needs in a small volume; these foods should be emphasized in a vegan child's diet. Table NP3-1 offers a suggested daily food guide for vegetarians.

Tell me about vitamins and minerals. Does a person eating a vegetarian diet need to take vitamin supplements?

That depends on the kind of vegetarian diet. The **lacto-ovo vegetarian** diet can be complete in all vitamins, but for the vegan, several vitamins may be a problem. One such vitamin is B_{12}. Because vitamin B_{12} occurs only in animal-derived foods, supplements are necessary to prevent deficiency. Women who have adhered to all-plant diets for many years are especially likely to have low vitamin B_{12} stores. Pregnant vegan women, whose needs for vitamin B_{12} are especially high, find it virtually impossible to maintain adequate vitamin B_{12} status without taking supplements or including a reliable food source of the nutrient.

A vitamin B_{12} deficiency can take a long time to develop in adults

because up to four years' worth of the vitamin can be stored in the body. But when the deficiency sets in, it does severe damage to the nervous system (see Chapter 8). In infants, deficiencies set in more rapidly and so threaten their nervous systems early. All vegan mothers must be sure to take the appropriate supplements or to use vitamin B_{12}–fortified products.

What other vitamins do vegans need?

Another vitamin of concern is vitamin D. The milk drinker is protected, provided the milk is fortified with vitamin D, but there is no practical source of vitamin D in plant foods. Regular exposure to the sun will prevent a deficiency, but vegans who are homebound or live in a northern climate or smoggy city probably should take vitamin D supplements. Excesses of vitamin D are toxic, and one should not exceed the recommended daily amount of 5 micrograms.

Riboflavin, another vitamin often obtained from milk, is not a problem for the vegan who eats dark greens frequently in ample servings. The vegan who doesn't consume a lot of greens, however, may not meet riboflavin needs. Nutritional yeast is a rich source of riboflavin for the vegetarian.

So, on a vegan diet, vitamin B_{12}, vitamin D, and riboflavin can be problems if a person is not careful. What about minerals?

For *all* vegetarians, not just the vegan, two minerals may be of concern—iron and zinc. Legumes are an important source of iron in the vegetarian diet. The iron in legumes, however, is not as absorbable as that in meat. In fact, people absorb three times as much iron from a meal that includes meat as from one that does not. Because vitamin C in fruits and vegetables can triple iron absorption from other foods eaten at the same meal, vegetarian meals should be rich in foods offering vitamin C.

Zinc may also be a problem nutrient for vegetarians. It is widespread in plant foods, but its availability may be hindered by the fibers and other binders found in fruits and vegetables. The zinc needs of vegetarians and the effects of mineral binders are subjects of intensive study at the present time. While research continues, vegetarians are advised to eat varied diets that include whole-grain breads well leavened with yeast, which improves the availability of their minerals.

What about calcium for the vegan?

Good thinking. Yes, calcium is of concern. The milk-drinking vegetarian is protected from deficiency, but the vegan must find other sources of calcium. Some good calcium sources are regular and ample servings of stone-ground meal, self-rising flour and meal, legumes, calcium-fortified soy milk, calcium-fortified orange juice, some nuts such as almonds, and certain seeds such as sesame seeds. The choices should be varied because binders in some of these foods may hinder calcium absorption. The vegetarian is urged to use calcium-fortified soy milk in ample quantities regularly. This is especially important for children. Infant formula based on soy is fortified with calcium and can easily be used when cooking foods, even for adults.

Does the vegetarian diet offer any other health advantages?

Yes. Vegetarian protein foods are often higher in fiber, richer in certain vitamins and minerals, and lower in fat than meats. Vegetarians can enjoy a nutritious diet very low in fat provided that they limit other high-fat foods such as butter, cream cheese, sour cream, and nuts. If vegetarians follow the guidelines presented here and plan carefully, they can support their health as well as, or perhaps better than, nonvegetarians.

Abundant evidence supports the idea that vegetarians may actually be healthier than meat eaters. Informed

Table NP3-1
DAILY FOOD GUIDE FOR VEGETARIAN MEAL PLANNING

Food Group	Suggested Daily Servings	Serving Sizes
Breads, cereals, rice, pasta, and other grain products	6 to 11	1 slice bread ½ bun, bagel, or English muffin ½ c cooked cereal, rice, or pasta 1 oz ready-to-eat cereal
Vegetables	3 to 5[a]	½ c cooked or chopped raw vegetables 1 c raw leafy vegetables
Fruits	2 to 4	1 medium-sized piece fresh fruit ¾ c fruit juce ½ c canned, cooked, or chopped raw fruit ¼ c dried fruit
Legumes, nuts, seeds, eggs, and other meat substitutes	2 to 3	½ c cooked legumes ¼ c tofu or tempeh 1 c soy milk 2 tbs peanut butter, nuts, or seeds (these tend to be high in fat, so use sparingly) 1 egg or 2 egg whites
Milk, yogurt, cheese, and other milk products	2 to 3[b]	1 c milk 1 c yogurt 1½ oz cheese

[a]Include 1 cup of dark green vegetables daily to help meet iron requirements.
[b]People who do not use milk or milk products: use soy milk fortified with calcium, vitamin D, and vitamin B_{12}. Other nonmilk calcium-rich food sources are provided in Chapter 9.
SOURCE: Adapted from Position of The American Dietetic Association Vegetarian diets, *Journal of the American Dietetic Association* 97 (1997): 1320.

vegetarians are not only more likely to be at the desired weights for their heights, but to have lower blood cholesterol levels, lower rates of certain kinds of cancer, better digestive function, and more. Even among people who are health conscious, vegetarians generally experience fewer deaths from cardiovascular disease than meat eaters do. Since many vegetarians also abstain from smoking and the consumption of alcohol, dietary practices alone probably do not account for all the aspects of improved health. Clearly, however, they contribute significantly to it.

Notes

[1]Position of The American Dietetic Association: Vegetarian diets, *Journal of the American Dietetic Association* 97 (1997): 1317–1321.

[2]T. J. Key and coauthors, Mortality in vegetarians and nonvegetarians: Detailed findings from a collaborative analysis of 5 prospective studies, *American Journal of Clinical Nutrition* 70 (1999): S516–S524; P. Walter, Effects of vegetarian diets on aging and longevity, *Nutrition Reviews* 55 (1997): S61–S68; T. C. Campbell and C. Junshi, Diet and chronic degenerative diseases: Perspectives from China, *American Journal of Clinical Nutrition* 59 (1994): S1153–S1161.

[3]J. T. Dwyer, Vegetarian eating patterns: Science, values, and food choices—Where do we go from here? *American Journal of Clinical Nutrition* 59 (1994): S1255–S1262.

[4]Position of The American Dietetic Association, 1997.

[5]M. Hebbelinck, P. Clarys, and A. De Malsche, Growth, development, and physical fitness of Flemish vegetarian children, adolescents, and young adults, *American Journal of Clinical Nutrition* 70 (1999): S579–S585.

[6]Hebbelinck, Clarys, and De Malsche, 1999.

CHAPTER 4

PROTEINS AND AMINO ACIDS

CONTENTS

The Chemist's View
of Proteins

Proteins in the Body

Protein and Health

Protein in Foods

Nutrition in Practice:
Supplements and Ergogenic
Aids Athletes Use

People think of proteins as bodybuilding nutrients, the material of strong muscles, and rightly so. No new living tissue can be built without them. Some proteins form structures such as muscle, bone, skin, and other tissues. Other proteins do the cells' work. The energy to fuel that work comes from carbohydrates and fats.

▦ The Chemist's View of Proteins

Proteins are chemical compounds that contain the same atoms as carbohydrates and lipids—carbon (C), hydrogen (H), and oxygen (O)—but proteins are different in that they also contain nitrogen (N) atoms. These nitrogen atoms give the name *amino* (nitrogen containing) to the **amino acids** that form the links in the chains we call proteins.

▦ The Structure of Proteins

Body protein is made up of 20 different amino acids.* All amino acids share a common chemical "backbone," and it is these backbones that are linked together to form proteins. Each amino acid also carries a side group, which varies from one amino acid to another (see Figure 4-1). The side group makes the amino acids differ in size, shape, and electrical charge. The side groups on amino acids are what make proteins so varied in comparison with either carbohydrates or lipids.

▦ PROTEIN CHAINS
The 20 amino acids can be linked end-to-end in a virtually infinite variety of sequences to form proteins. When two amino acids bond together, the resulting structure is known as a **dipeptide.** Three amino acids bonded together form a **tripeptide.** As additional amino acids join the chain, the structure becomes a **polypeptide.** Most proteins are polypeptides 100 to 300 amino acids long.

▦ PROTEIN SHAPES
Polypeptide chains twist into complex shapes. Each amino acid has special characteristics that attract it to, or repel it from, the surrounding fluids and other amino acids. Because of these interactions, polypeptide chains fold and intertwine into intricate coils (see Figure 4-2).

---

*Besides the 20 common amino acids that can be components of proteins, others occur individually (for example, ornithine). Chemists can make still others.

proteins: compounds composed of carbon, hydrogen, oxygen, and nitrogen atoms arranged into strands of amino acids. Some amino acids also contain sulfur atoms.

amino (a-MEEN-oh) acids: building blocks of proteins. Each has a backbone, consisting of an amino group and an acid group attached to a central carbon, and also carries a distinctive side group.
  amino = containing nitrogen

dipeptide (dye-PEP-tide): two amino acids bonded together.
  di = two   peptide = amino acid

tripeptide: three amino acids bonded together.
  tri = three

polypeptide: ten or more amino acids bonded together. An intermediate strand of between four and ten amino acids is an oligopeptide.
  poly = many   oligo = few

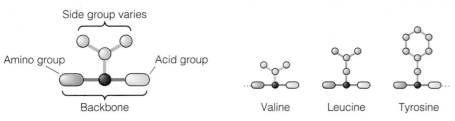

Figure 4-1
AMINO ACID STRUCTURE AND EXAMPLES OF AMINO ACIDS

Side group varies

Amino group

Acid group

Backbone

All amino acids have a "backbone" made of an amino group (which contains nitrogen) and an acid group. The side gorup varies from one amino acid to the next.

Valine   Leucine   Tyrosine

Note that the side group is a unique structure that differentiates one amino acid from another.

Figure 4-2
THE COILING AND FOLDING OF A PROTEIN MOLECULE

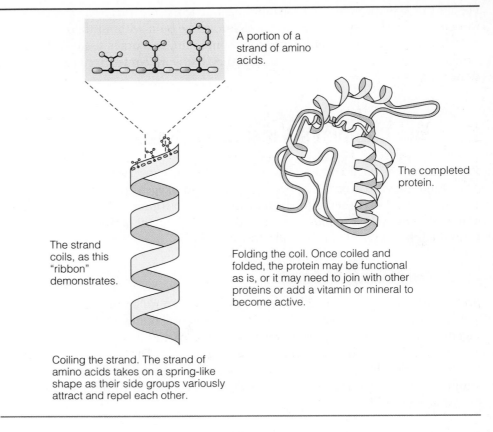

A portion of a strand of amino acids.

The strand coils, as this "ribbon" demonstrates.

Coiling the strand. The strand of amino acids takes on a spring-like shape as their side groups variously attract and repel each other.

Folding the coil. Once coiled and folded, the protein may be functional as is, or it may need to join with other proteins or add a vitamin or mineral to become active.

The completed protein.

---

nonessential amino acids: amino acids that the body can synthesize for itself; also called *dispensable amino acids* because they are dispensable from the diet.

essential amino acids: amino acids that the body cannot synthesize in amounts sufficient to meet physiological need; also called *indispensable amino acids.* Nine amino acids are known to be essential for human adults:
- *Histidine* (HISS-tuh-deen).
- *Isoleucine* (eye-so-LOO-seen).
- *Leucine* (LOO-seen).
- *Lysine* (LYE-seen).
- *Methionine* (meh-THIGH-oh-neen).
- *Phenylalanine* (fen-il-AL-uh-neen).
- *Threonine* (THREE-oh-neen).
- *Tryptophan* (TRIP-toe-fane).
- *Valine* (VAY-leen).

conditionally essential amino acid: an amino acid that is normally nonessential but must be supplied by the diet in special circumstances when the need for it becomes greater than the body's ability to produce it.

The amino acid sequence of a protein determines the specific way the chain will fold.

■■ **PROTEIN FUNCTIONS** The dramatically different shapes of proteins enable them to perform different tasks in the body. Some, such as hemoglobin in the blood (see Figure 4-3), are globular in shape; some are hollow balls that can carry and store materials within them; and some, such as those that form tendons, are more than ten times as long as they are wide, forming stiff, sturdy, rodlike structures.

■■ **Essential Amino Acids**

Proteins in foods do not provide body proteins directly, but rather supply the amino acids from which the body makes its own proteins. The body can make over half of the amino acids for itself; the protein in food does not need to supply these **nonessential amino acids.** But there are other amino acids that the body cannot make at all, and some that it cannot make fast enough to meet its needs. The proteins in foods must supply these nine amino acids to the body; they are therefore called **essential amino acids.**

Sometimes a nonessential amino acid becomes essential. During illness or conditions of trauma, or in other special circumstances such as premature birth, the need for an amino acid that is normally nonessential may become greater than the body's ability to produce it. In such circumstances, that amino acid becomes "conditionally essential." Research suggests that glutamine, normally a nonessential amino acid, may be a **conditionally essential amino acid** for critically ill people (see Chapter 19).[1]

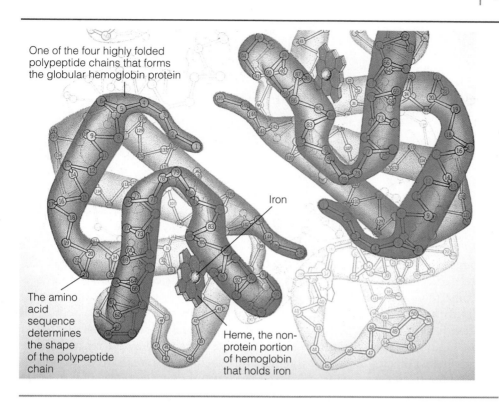

One of the four highly folded polypeptide chains that forms the globular hemoglobin protein

The amino acid sequence determines the shape of the polypeptide chain

Iron

Heme, the non-protein portion of hemoglobin that holds iron

Figure 4-3
A PORTION OF A HEMOGLOBIN MOLECULE

This model represents a portion of a hemoglobin molecule magnified millions of times.

LEARNING LINK

Phenylketonuria (PKU), the most common inborn error of metabolism, provides another example of how a nonessential amino acid becomes conditionally essential. The body normally makes the nonessential amino acid tyrosine from the essential amino acid phenylalanine. Clients with PKU lack the enzyme that converts phenylalanine to tyrosine. Consequently, high concentrations of phenylalanine accumulate and tyrosine becomes deficient. For the client with PKU, then, tyrosine becomes a conditionally essential amino acid that must be supplied by the diet. Inborn errors of metabolism are the topic of Nutrition in Practice 6.

■■ Proteins in the Body

What distinguishes you chemically from any other human being are minute differences in your particular body proteins (enzymes, antibodies, and others). These differences are determined by your proteins' amino acid sequences, which are written into the genes you inherited from your parents and ancestors. The genes direct the making of all the body's proteins.

The human body contains an estimated 10,000 to 50,000 different kinds of proteins. The roles of more than 1000 of these proteins are now known. Only a few of the many roles proteins play are described here, but these should serve to illustrate proteins' versatility, uniqueness, and importance.

■■ ENZYMES **Enzymes** are catalysts that are essential to all life processes. Enzymes in the cells of plants or animals put together the pairs of

enzymes: protein catalysts. A catalyst is a compound that facilitates chemical reactions without itself being changed in the process.

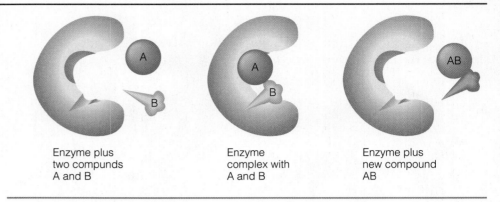

Enzyme plus two compunds A and B

Enzyme complex with A and B

Enzyme plus new compound AB

Figure 4-4
ENZYME ACTION

Enzymes are catalysts: they speed up reactions that would happen anyway, but much more slowly. This enzyme works by positioning two compounds, A and B, so that the reaction between them will be especially likely to take place.

Compounds A and B are attracted to the enzyme's active site and park there for a moment in the exact position that makes the reaction between them most likely to occur. They react by bonding together and leave the enzyme as the new compound, AB.

A single enzyme can facilitate several hundred such synthetic reactions in a second. Other enzymes break compounds apart into two or more products or rearrange the atoms in one compound to make another one.

fluid and electrolyte balance: maintenance of the necessary amounts and types of fluid and minerals in each compartment of the body fluids.

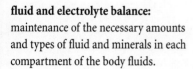

Minerals are helper nutrients. The attraction of protein and mineral particles to water is due to osmotic pressure (see Chapter 9).

edema (eh-DEEM-uh): the swelling of body tissue that occurs when fluid leaks from the blood vessels and accumulates in the interstitial spaces.

sugars that make disaccharides and the long strands of sugars that make starch, cellulose, and glycogen. Enzymes also dismantle these compounds to free their constituent parts and release energy. Enzymes also assemble and disassemble lipids, assemble all other compounds that the body makes, and disassemble all compounds that the body uses for building tissue and other metabolic work. As Figure 4-4 shows, enzymes themselves are not altered by the reactions they facilitate. All enzymes are proteins, and when amino acids have to be put together to make proteins, it is enzymes that put them together, too. In other words, these proteins can even help make other proteins.

The protein story moves in a circle. To follow the circle in nutrition, start with a person eating food proteins. The food proteins are broken down by the body's proteins (digestive enzymes) into amino acids. The amino acids enter the cells of the body, where proteins (enzymes) put the amino acids together in long chains whose sequences are specified by the genes. The chains fold and twist back on themselves to form proteins, and some of these proteins become enzymes themselves. Some of these enzymes break apart compounds; others put compounds together. Day by day, in billions of reactions, these processes repeat themselves, and life goes on. Only living systems can achieve such self-renewal. A toaster cannot produce another toaster; a car cannot fix a broken-down car. Only living creatures and the parts they are composed of— the cells—can duplicate and repair themselves.

FLUID AND ELECTROLYTE BALANCE Proteins help maintain the body's **fluid and electrolyte balance.** As Figure 4-5 shows, the body's fluids are contained in three major body compartments: (1) the spaces inside the blood vessels, (2) the spaces within the cells, and (3) the spaces between the cells (the interstitial spaces outside the blood vessels). Fluids flow back and forth between these compartments, and proteins in the fluids, together with minerals, help to maintain the needed distribution of these fluids.

Proteins are able to help determine the distribution of fluids in living systems for two reasons: first, proteins cannot pass freely across the membranes that separate body compartments, and second, they are attracted to water. A cell that "wants" a certain amount of water in its interior space cannot move the water around directly, but it can manufacture proteins, and these proteins will hold water. Thus, the cell can use proteins to help regulate the distribution of water indirectly. Similarly, the body makes proteins for the blood and the interstitial (intercellular) spaces. These proteins help maintain the fluid volume in those spaces. When too much fluid collects in the interstitial spaces, **edema** results.

Not only is the distribution of the body fluids vital to life, but so is their composition. Special transport proteins in the membranes of cells continuously transfer substances into and out of cells to maintain balance. For example, sodium is concentrated outside the cells, and potassium is concentrated inside. The balance of these two electrolytes is critical to nerve transmission and muscle contraction. Imbalances can cause irregular heartbeats, kidney failure, muscular weakness, and even death.

LEARNING LINK

A major function of the kidneys is to maintain the body's fluid and electrolyte balance and acid-base balance by continuously filtering the blood. The consequences of the nephrotic syndrome, which affects the kidneys and causes sodium retention, edema, and loss of plasma proteins into the urine, can lead to wasting of lean body tissue, and a host of other complications. As Chapter 22 describes, dietary treatment for the nephrotic syndrome involves modifying protein and mineral intakes.

▪▪ ACID-BASE BALANCE Proteins also help maintain the balance between **acids** and **bases** within the body's fluids. Normal body processes continually produce acids and bases, which must be carried by the blood to the kidneys and lungs for excretion. The blood must do this without upsetting its own **acid-base balance.** Blood **pH** is one of the most tightly controlled conditions in the body. If the blood becomes too acidic, vital proteins may undergo **denaturation,** losing their shape and ability to function. A similar situation arises when the balance tips too far toward base. These imbalances are known as **acidosis** and **alkalosis,** respectively, and both can be fatal. Figure 4-6 shows

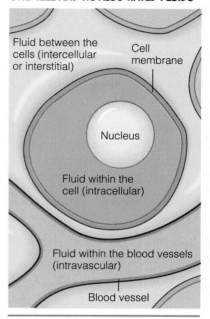

Figure 4-5
ONE CELL AND ITS ASSOCIATED FLUIDS

acids: compounds that release hydrogen ions in a solution.

bases: compounds that accept hydrogen ions in a solution.

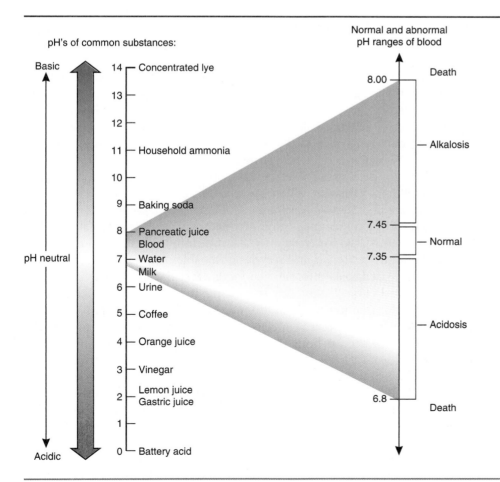

Figure 4-6
THE pH SCALE

A substance's acidity or alkalinity is measured in pH units. Each step down the scale indicates a tenfold increase in the concentration of hydrogen ions. Notice how small the range of normal blood pH is.

acid-base balance: the balance maintained between acid and base concentrations in the blood and body fluids.

pH: the concentration of hydrogen ions. The lower the pH, the stronger the acid. Thus, pH 2 is a strong acid; pH 6 is a weak acid; pH 7 is neutral; and a pH above 7 is alkaline.

denaturation (dee-nay-cher-AY-shun): the change in a protein's shape brought about by heat, acid, or other agents. Past a certain point, denaturation is irreversible.

acidosis (assi-DOE-sis): too much acid in the blood and body fluids.

alkalosis (alka-LOE-sis): too much base in the blood and body fluids.

buffers: compounds that can reversibly combine with hydrogen ions to help keep a solution's acidity or alkalinity constant.

antibodies: large proteins of the blood and body fluids, produced in response to invasion of the body by unfamiliar molecules (mostly proteins). Antibodies inactivate the invaders, called antigens, and so protect the body.

   anti = against

hormones: chemical messengers. Hormones are secreted by a variety of glands in the body in response to altered conditions. Each travels to one or more target tissues or organs and elicits specific responses to restore normal conditions.

Growing children end each day with more bone, blood, muscle, and skin cells than they had at the beginning of the day.

the normal and abnormal pH ranges of body fluids, as well as the pHs of some common substances.

Proteins such as albumin help to prevent acid-base imbalances. In a sense, the proteins protect one another by gathering up extra acid (hydrogen) ions when there are too many in the surrounding medium and by releasing them when there are too few. By accepting and releasing hydrogen ions, proteins act as **buffers,** maintaining the acid-base balance of the blood and body fluids.

■■ ANTIBODIES Other proteins in the blood—the **antibodies**—act against viruses, bacteria, and other disease agents. The antibodies work so efficiently that if a million bacterial cells are injected into the skin of a healthy person, fewer than ten are likely to survive for five hours, which explains why most diseases never have a chance to get started. Without sufficient protein, the body cannot maintain its resistance to disease.

■■ HORMONES The blood also carries messenger molecules known as **hormones,** and *some* hormones are proteins. (Recall that some hormones are sterols, members of the lipid family.) Among the proteins that act as hormones are the thyroid hormones and insulin. Hormones have many profound effects, which will become evident in subsequent chapters.

■■ TRANSPORTERS Some proteins move about in the body fluids, transporting nutrients and other molecules from one organ to another. The protein hemoglobin, which carries oxygen from the lungs to the body's cells, is a prime example. The lipoproteins transport lipids around the body. In addition, special proteins also carry vitamins and minerals.

■■ GROWTH, MAINTENANCE, AND REPAIR The body uses amino acids to build the proteins of all its new tissues. The new tissues may be in an embryo, in a growing child, or in new hair and nails. Proteins also help replace worn-out cells in everyone's body all the time. For example, the millions of cells that line the intestinal tract live for three days; they are constantly being shed and must be replaced. The cells of the skin die and rub off, and new ones grow from underneath. The body uses amino acids to repair damaged tissues, too. The protein collagen serves as the mending material of torn tissue, forming scars to hold the separated parts together.

Both inside and outside the body, then, cells constantly make and break down their proteins. When proteins break down, their component amino acids are liberated to join the general circulation. Some of these amino acids may be promptly recycled into other proteins; others may be stripped of their nitrogen and used for energy. By reusing amino acids to build proteins, however, the body conserves and recycles a valuable commodity.

People need to eat protein-rich foods every day to replace the protein they continuously lose. If the body is growing, it needs more protein than is necessary just for maintenance. Children end each day with more blood cells, more muscle cells, and more skin cells than they had at the beginning of the day. So protein is needed both for routine maintenance (replacement) and growth (addition).

■■ NITROGEN BALANCE If the body maintains the same amount of protein in its tissues from day to day, it is in **nitrogen balance.** If the body adds protein, it is in positive nitrogen balance; if it loses protein, it is in negative nitrogen balance.

Normally, healthy adults are in nitrogen balance; that is, their nitrogen intakes equal their nitrogen outputs. Growing children and pregnant women are in positive nitrogen balance because they are adding new blood, bone, and muscle cells to their bodies. People who are fasting or starving, such as those with anorexia nervosa, and people who suffer from traumas, such as burns (see Chapter 19), are in negative nitrogen balance because their bodies are forced to use protein for energy.

nitrogen balance: the amount of nitrogen consumed (N in) as compared with the amount of nitrogen excreted (N out) in a given period of time.

LEARNING LINK

Chapter 19 shows how the negative nitrogen balance associated with acute medical stress can interfere with the body's ability to maintain its skeletal muscle mass, organ function, immunity, wound healing ability, and nutrient absorption.

Nitrogen equilibrium (zero nitrogen balance): N in = N out.
Positive nitrogen balance: N in > N out.
Negative nitrogen balance: N in < N out.

■■ **ENERGY** Even though amino acids are needed to do the work that only they can perform—build vital proteins—they will be sacrificed to provide energy and glucose if need be. Keeping energy and glucose available are among the body's highest priorities: without energy, cells die; without glucose, the brain and nervous system falter. When glucose or fatty acids are limited, then, cells are forced to use amino acids for energy and glucose. The body does not make a specialized storage form of protein as it does for carbohydrate and fat. Glucose is stored as glycogen in the liver and muscles, fat as triglycerides in the adipose tissue, but body protein is available only as the working and structural components of the tissues. When the need arises, the body dismantles its tissue proteins and uses them for energy. Thus, over time, energy deprivation (starvation) always incurs wasting of lean body tissue as well as fat loss.

The list of protein functions discussed here and summarized in Table 4-1 is by no means exhaustive. Nevertheless, it does give some sense of the immense variety of proteins and their importance in the body.

LEARNING LINK

When working with clients who have anorexia nervosa, you will be reminded firsthand of just how important adequate energy and protein are to health. People with anorexia nervosa are starving themselves to death, and the physical ramifications are similar to those of protein-energy malnutrition (PEM). Dietary treatment includes high-kcalorie, high-protein formulas in addition to regular meals. Nutrition in Practice 7 provides many more details.

Table 4-1
SUMMARY OF FUNCTIONS OF PROTEINS

- *Enzymes.* Proteins facilitate chemical reactions.
- *Fluid and electrolyte balance.* Proteins help to maintain the distribution and composition of various body fluids.
- *Acid-base balance.* Proteins help maintain the acid-base balance of body fluids by acting as buffers.
- *Antibodies.* Proteins act against disease agents to fight diseases.
- *Hormones.* Proteins regulate body processes. Some, but not all, hormones are made of protein.

- *Transportation.* Proteins transport substances such as lipids, minerals, and oxygen around the body.
- *Growth and maintenance.* Proteins form integral parts of most body structures such as skin, tendons, ligaments, membranes, muscles, organs, and bones. As such, they support the growth and repair of body tissues.
- *Energy.* Proteins provide some fuel for the body's energy needs.

▨ Protein and Health

In the short time that scientists have been studying nutrition, no nutrient has been more intensely scrutinized than protein. As you know by now, it is indispensable to life. And it should come as no surprise that protein deficiency can have devastating effects on people's health. But as with the other nutrients, protein in excess can also be harmful; the end of this section discusses the consequences of protein excess.

▨ Protein-Energy Malnutrition

When people are deprived of food and suffer an energy deficit, they degrade their own body protein for energy and indirectly suffer a protein deficiency, as well as an energy deficiency. Because protein and energy deprivation go hand in hand, public health officials have adopted an abbreviation for the overlapping pair: **protein-energy malnutrition (PEM)**. PEM often strikes early in childhood, but it endangers many adults as well. PEM is the most widespread form of malnutrition in the world today. Most of the 33,000 children who die each day are malnourished and suffer from infectious diseases.[2] PEM is prevalent in Africa, Central America, South America, the Middle East, and East Asia, but developed countries including the United States are not immune to it. PEM is common among some population groups in the United States: impoverished Native Americans living on reservations, poor people in inner cities and rural areas, many elderly people, homeless children, and those suffering from the eating disorder anorexia nervosa.[3]

protein-energy malnutrition (PEM): a deficiency of protein and food energy; the world's most widespread malnutrition problem, including both marasmus and kwashiorkor.

mal = bad, poor

LEARNING LINK

As you will see in practice, and learn in later chapters, PEM can be a consequence of many different conditions. PEM has been recognized in people with many chronic diseases such as cancer and AIDS (see Chapter 24) and in those who have severe stresses such as burns or extensive infections (see Chapter 19). The consequences of PEM as a world malnutrition problem are considered here; the problems associated with PEM and illness are described throughout later chapters.

Of all population groups, children are most seriously affected by malnutrition. Children who are thin for their heights may have recently developed PEM, whereas children who are short for their ages may have experienced PEM for extended periods of time. Stunted growth due to PEM is easy to overlook because a small child may look quite normal. Nevertheless, it may be the most common sign of malnutrition in developing countries. Experts estimate that 226 million children have been stunted and are shorter than they should be for their age.[4]

▨ MARASMUS AND KWASHIORKOR—ONE DISEASE OR TWO?

The long-held idea that the two forms of PEM—marasmus and kwashiorkor—are different diseases caused by different nutrient deficiencies has been challenged in recent years. Some researchers now maintain that marasmus and kwashiorkor are two stages of the same disease.[5] Until recently, marasmus was thought to be caused by energy deficiency and kwashiorkor by protein deficiency. In reality, though, marasmus reflects inadequate food intake and therefore inadequate energy, vitamins, and minerals as well as too little protein. Kwashiorkor is usually precipitated by an illness such as measles or other infection. The stress of the illness rapidly depletes energy stores and body protein so that too little protein remains to support optimal body function.[6] Researchers continue to question the exact causes of kwashiorkor, but clearly, an absolute or relative protein deficiency is involved.

▪▪ MARASMUS

Marasmus commonly occurs in children from 6 to 18 months of age in all the overpopulated urban slums of the world. Children in impoverished nations subsist on a weak cereal drink that supplies scant energy and protein of low quality: such food can barely sustain life, much less support growth. Consequently, marasmic children look like little old people—just skin and bones.

Without adequate nutrition, muscles, including the heart muscle, waste and weaken. Because the brain normally grows to almost its full adult size within the first two years of life, marasmus impairs brain development and learning ability. Reduced synthesis of key hormones leads to a metabolism so slow that body temperature drops below normal. There is little or no fat under the skin to insulate against cold. Hospital workers find that the primary need of marasmic children is to be wrapped up and kept warm.

The starving child faces this threat to life by engaging in as little activity as possible—not even crying for food. The body gathers all its forces to meet the crisis, so it cuts down on any expenditure of protein not needed for the heart, lungs, and brain to function. Growth ceases; the child is no larger at age four than at age two. Digestive enzymes are in short supply, the digestive tract lining deteriorates, and absorption fails. The child cannot assimilate what little food is eaten.

Blood proteins, including hemoglobin, are no longer synthesized. Antibodies to fight off invading bacteria are degraded to provide amino acids for other uses, rendering the child vulnerable to infection. Then **dysentery,** an infection of the digestive tract, causes diarrhea, further depleting the body of nutrients. In the marasmic child, once infection has set in, kwashiorkor often follows.[7] The infection that occurs with malnutrition is responsible for two-thirds of the deaths of young children in developing countries.[8]

If caught in time, a child's starvation may be reversed by careful nutrition therapy. The fluid balances are most critical. Diarrhea will have depleted the body's potassium and disturbed other electrolyte balances. Careful correction of fluid and electrolyte imbalances usually raises the blood pressure and strengthens the heart. After the first 24 to 48 hours, protein and energy may be given in small quantities, with intakes gradually increased as tolerated.[9]

▪▪ KWASHIORKOR

Kwashiorkor was originally a Ghanaian word meaning an "evil spirit that infects the first child when the second child is born." If you consider how kwashiorkor often develops, you can easily see how the Ghanaians arrived at this name for the disease. When a mother who has been nursing her first child bears a second child, she weans the first child and puts the second one on the breast. The first child, suddenly switched from nutrient-dense, protein-rich breast milk to a starchy, protein-poor gruel, soon begins to sicken and die. Kwashiorkor typically sets in between the ages of one and three years. As mentioned earlier, some researchers believe that kwashiorkor and marasmus are two stages of the same disease: marasmus represents the body's attempt to adapt to starvation, and kwashiorkor develops when stresses tax nutrient status further and adaptation fails.

Some symptoms of kwashiorkor resemble those of marasmus, but without severe wasting of body fat. Proteins and hormones that previously maintained fluid balance diminish, and fluid leaks into the interstitial spaces. The child's limbs and face become swollen with edema, a distinguishing

marasmus (ma-RAZZ-mus): a disease related to PEM. Marasmus results from severe deprivation, or impaired absorption, of protein, energy, vitamins, and minerals; also called chronic malnutrition because it takes time to develop.

dysentery (DIS-en-terry): an infection of the gastrointestinal tract caused by an amoeba or bacterium that gives rise to severe diarrhea.

dys = bad   entery = intestine

When two variables interact so that each increases the other, synergism (SIN-er-jiz-um) is said to be occurring. Malnutrition and infection are a deadly combination because they work in this way.
  syn = with, together   ergism = work

kwashiorkor (kwash-ee-OR-core or kawsh-ee-or-CORE): a disease related to PEM; also called acute malnutrition because it develops rapidly.

In the photo on the left, the extreme loss of muscle and fat characteristic of marasmus is apparent in the child's matchstick arms and legs. In contrast, the edema and enlarged liver characteristic of kwashiorkor are apparent in the swollen bellies of the children in the photo on the right.

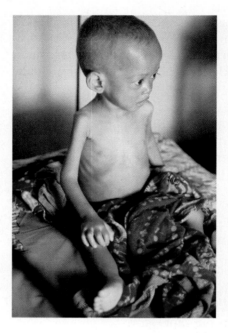
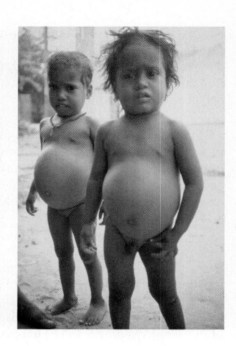

fatty liver: an accumulation of fat in the liver. In PEM, fat accumulates in the liver because no protein is available to form the lipoproteins that normally escort fat molecules in the blood (see Chapter 23).

feature of kwashiorkor; the belly bulges with edema and a **fatty liver** caused by lack of the protein carriers that transport fat out of the liver. The child's hair loses its color; the skin becomes patchy and scaly, sometimes with ulcers and sores that fail to heal.

Experts assure us that we possess the knowledge, technology, and resources to end hunger. Programs that have involved the local people in the process of identifying the problem and devising its solution have met with some success. But until those who have the food, technology, and resources make fighting hunger a priority, the war on hunger will not be won.

Protein Excess

While many of the world's people struggle to obtain enough food and enough protein to survive, in the developed nations protein is so abundant that problems of protein excess are seen. Overconsumption of protein offers no benefits and may pose health risks. For example, protein-rich foods are often high-fat foods that contribute to obesity with its accompanying health risks. Some studies suggest a link between high-meat diets and colon cancer.[10] As in studies of heart disease, however, the effects of protein and fats cannot easily be separated. One study reported an increase in the risk of a type of cancer known as non-Hodgkin's lymphoma with a high intake of animal protein and red meat.[11] The higher a person's intake of protein-rich foods such as meat and milk, the more likely that fruits, vegetables, and grains will be crowded out, making the diet inadequate in other nutrients.

Some research suggests that diets high in protein promote calcium excretion, depleting the bones of their chief mineral. When high protein intakes are accompanied by low calcium intakes, as is typical of many women's diets in the United States, calcium losses incurred by excess protein may compromise bone health.[12] One study of the relationship between protein intake and bone fractures in women reported a higher risk of arm fractures in women with high animal-protein intakes.[13] There are evidently no benefits to be gained by consuming a diet that derives more than 15 percent of its energy from protein. The *Diet and Health* recommendations (see Table C-13 in Appendix C) advise a moderate protein intake—one that falls between the RDA and twice the RDA.

LEARNING LINK

Over the past few decades, many different versions of high-protein, low-carbohydrate diets have come and gone. Remind clients that one reason the diets keep reappearing under different names, with different gimmicks, is that they simply don't work over the long run. After reading about the problems of protein excess, you can also remind clients of the possible health hazards of excess protein.

■■ **PROTEIN RECOMMENDATIONS** The committee that established the RDA states that a generous daily protein allowance for a healthy adult is 0.8 gram per kilogram (2.2 pounds) of appropriate body weight. The protein RDA is adjusted to cover additional needs for building new tissue and so is higher for infants, children, and pregnant and lactating women.

In setting the RDA, the committee assumes that the protein eaten will be of high quality, that it will be consumed together with adequate energy from carbohydrate and fat, and that other nutrients in the diet will be adequate. The committee also assumes that the RDA will be applied only to healthy individuals with no unusual alteration of protein metabolism.

All active people, and especially those who work like athletes, probably need more protein than do sedentary people. Endurance athletes use more protein for fuel than power athletes do, and they retain some, especially in the muscles used for their sport. Power athletes use less protein for fuel, but they still use some and retain much more. Therefore, all athletes in training should attend to protein needs, but should back up the protein with ample carbohydrate. Otherwise, they will burn off as fuel the very protein that they wish to retain in muscle. How much protein, then, should an active person consume? A joint position paper from the American Dietetic Association (ADA) and the Canadian Dietetic Association (CDA) recommends 1.0 to 1.5 grams of protein per kilogram of body weight each day, an amount somewhat higher than the RDA for the general population.[14] Most people in this country receive much more protein than they need, however.

■■ **AMINO ACID SUPPLEMENTS** In view of the high protein intakes of people in the United States and other developed nations, it is surprising that many people feel compelled to take protein and amino acid supplements. Why do people take protein or amino acid supplements? Athletes take them to build muscle. Dieters take them to spare their bodies' protein while losing weight. Popular reports that the amino acid tryptophan could relieve pain and cure depression and insomnia led to widespread public use. More than 1500 people who elected to take tryptophan developed an illness called EMS (short for *eosinophilia-myalgia syndrome*). EMS is characterized by severe muscle and joint pain, limb swelling, an elevated white blood cell count, extremely high fever, and, in at least 38 cases, death.[15] Contaminants in the supplement were determined to be the cause of the disease, and the Food and Drug Administration (FDA) issued a recall of all products containing tryptophan except for specific medical formulas.

A few years ago the FDA asked a panel of scientists from a well-known research group to review the safety of amino acid supplements.[16] The scientists concluded that any use of amino acids as dietary supplements is inappropriate and that some (serine and proline) present a high risk of toxicity. The panel also singled out some groups of people whose growth or altered metabolism makes them especially likely to suffer harm from amino acid supplements:

- All women of childbearing age.
- Pregnant or lactating women.

1989 RDA for protein (adult) = 0.8 g/kg.

To figure your protein need:
1. Find your body weight in pounds (or kilograms).
2. Convert pounds to kilograms, if necessary (pounds divided by 2.2).
3. Multiply kilograms by 0.8 to find total grams of protein recommended.
For example:
1. Weight = 110 lb.
2. 110 lb ÷ 2.2 lb/kg = 50 kg.
3. 50 kg × 0.8 g/kg = 40 g.

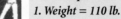

Muscle work builds muscle; protein supplements do not, and athletes do not need them. Protein supplements and other ergogenic aids are discussed in Chapter 4 Nutrition in Practice.

- Infants, children, and adolescents.
- Elderly people.
- People with inborn errors of metabolism that affect their bodies' handling of amino acids.
- Smokers.
- People on low-protein diets.
- People with chronic or acute mental or physical illnesses who take amino acids without medical supervision.

 Also, because weight lifters and bodybuilders may take frequent, massive doses of amino acids, they may suffer harm from the supplements while believing false promises of benefits.

Protein in Foods

To make body protein, a cell must have all the needed amino acids available simultaneously. Therefore, the first important requirement for protein in the diet of a healthy adult is that it should supply at least the 9 essential amino acids and enough nitrogen and energy for the synthesis of the other 11. A protein that fits this description is called a complete protein.

complete protein: a protein containing all the amino acids essential in human nutrition in amounts adequate for human use.

COMPLETE PROTEIN A **complete protein** contains all the essential amino acids in amounts adequate for human use; it may or may not contain all the others. Generally, proteins derived from animal foods (meats, fish, poultry, eggs, and milk) are complete, although gelatin is an exception. Proteins derived from plant foods (legumes, grains, and vegetables) vary more. Some plant proteins are notoriously **incomplete**—for example, corn protein. Others are complete—for example, soy protein. As discussed in Nutrition in Practice 3, the educated vegetarian can design a diet that is adequate in protein by choosing a variety of legumes, whole grains, nuts, and vegetables. Table 4-2 lists the protein contents of foods based on the food groups of the Daily Food Guide in Chapter 1.

incomplete protein: a protein lacking or low in one or more of the essential amino acids.

PROTEIN DIGESTIBILITY AND QUALITY Ideally, a protein should be not only complete, but easily digestible as well so that sufficient numbers of amino acids reach the body's cells to permit them to make the proteins they need. Such a protein is called a **high-quality protein.** One of the finest proteins available by these standards is egg protein. Eggs are a highly valued protein source in developing nations where protein-rich foods are scarce.

high-quality protein: an easily digestible, complete protein.

LIMITING AMINO ACIDS Ideally, dietary protein supplies each amino acid in the amount needed for protein synthesis in the body. If one amino acid is supplied in an amount smaller than is needed, then the total amount of protein that can be synthesized will be limited. The body makes only complete proteins; if one amino acid is missing, the others cannot form a "partial" protein. An essential amino acid supplied in less than the amount needed to support protein synthesis is called a **limiting amino acid.** (By analogy, suppose that a sign maker plans to make 100 identical signs saying "LEFT TURN ONLY." The sign maker needs 200 Ls, 200 Ns, 200 Ts, and 100 of each of the other letters. If only 20 Ls are available, only 10 signs can be made, even if all the other letters are available in unlimited quantities. Suppose further that the sign maker has no place to keep leftover letters—just as the body has no storage place for extra amino acids. If the sign maker doesn't get some additional Ls right away, he will have to throw away all the other letters.)

limiting amino acid: the essential amino acid found in the shortest supply relative to the amounts needed for protein synthesis in the body. It limits the amount of protein the body can make.

Table 4-2
PROTEIN-CONTAINING FOODS

Milk, Cheese, and Yogurt

Each of the following provides about 8 grams of protein:[a]

- 1 c milk, buttermilk, or yogurt (choose reduced-fat, low-fat, or fat-free).
- 1 oz regular cheese (for example, cheddar or swiss).[b]
- ¼ c cottage cheese (choose reduced-fat, low-fat, or fat-free).

Meat, Poultry, Fish, and Alternates

Each of the following provides about 7 grams of protein:

- 1 oz meat, poultry, or fish (choose lean meats to limit fat intake).
- ½ c legumes (navy beans, pinto beans, black beans, lentils, soybeans, and other dried beans and peas).
- 1 egg.[b]
- ½ c tofu (soybean curd).[b]
- 2 tbs peanut butter.[b]
- 1 to 2 oz nuts or seeds.[b]

Breads, Cereals, and Other Grain Products

Each of the following provides about 3 grams of protein:

- 1 slice of bread.
- ½ c cooked rice, pasta, cereals, or other grain foods.

Vegetables

Each of the following provides about 2 grams of protein:

- ½ c cooked vegetables.
- 1 c raw vegetables.

[a]For reference, an adult might need 40 to 100 grams of protein in a day.
[b]These are medium-fat or high-fat choices.

mutual supplementation: the strategy of combining two incomplete protein sources so that the amino acids in one food make up for those lacking in the other food. Such protein combinations are sometimes called *complementary proteins*.

complementary proteins: two or more proteins whose amino acid assortments complement each other in such a way that the essential amino acids missing from one are supplied by the other.

*Reminder: Carbohydrate and fat allow amino acids to be used to build body proteins. This is known as the **protein-sparing** effect of carbohydrate and fat.*

■■ **MUTUAL SUPPLEMENTATION AND COMPLEMENTARY PROTEINS**   If the body does not receive all the essential amino acids it needs, the supply of essential amino acids will dwindle until body organs are compromised. Obtaining enough essential amino acids presents no problem to people who regularly eat complete proteins, such as those of meat, fish, poultry, cheese, eggs, milk, and many soybean products. The proteins of these foods contain ample amounts of all the essential amino acids. An equally sound choice is to eat two incomplete protein foods from plants so that each supplies the amino acids missing in the other. In this strategy, called **mutual supplementation,** the two protein-rich foods are combined to yield **complementary proteins** (see Figure 4-7)—proteins containing all the essential amino acids in amounts sufficient to support health. This concept was illustrated in Figure NP3-1 on p. 72. The two proteins need not even be eaten together, so long as the day's meals supply them both, and the diet provides enough energy and total protein from a variety of sources.

■■ **PROTEIN SPARING**   Dietary protein—no matter how high the quality—will not be used efficiently and will not support growth when energy from carbohydrate and fat is lacking. The body assigns top priority to meeting its energy need and, if necessary, will break down protein to meet this need. After stripping off and excreting the nitrogen from the amino acids, the body will use the remaining carbon skeletons in much the same way it uses those from glucose or fat. A major reason why people must have ample carbohydrate and fat in the diet is to prevent this wasting of protein.

Figure 4-7
AN EXAMPLE OF MUTUAL SUPPLEMENTATION

In general, legumes provide plenty of isoleucine (Ile) and lysine (Lys), but fall short in methionine (Met) and tryptophan (Trp). Grains have the opposite strengths and weaknesses, making them a perfect match for legumes.

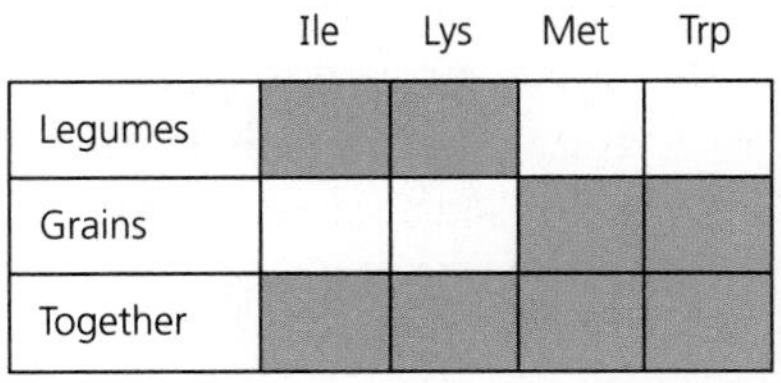

	Ile	Lys	Met	Trp
Legumes	▓	▓		
Grains			▓	▓
Together	▓	▓	▓	▓

SELF CHECK

1. Proteins are chemically different from carbohydrates and fats because they also contain:
   a. iron.
   b. sodium.
   c. nitrogen.
   d. phosphorus.

2. The basic building blocks for protein are:
   a. side groups.
   b. amino acids.
   c. glucose units.
   d. saturated bonds.

3. Enzymes are proteins that, among other things:
   a. defend the body against disease.
   b. regulate fluid and electrolyte balance.
   c. facilitate chemical reactions by changing themselves.
   d. help assemble disaccharides into starch, cellulose, or glycogen.

4. Functions of proteins in the body include:
   a. supplying omega-3 fatty acids for growth, lowering serum cholesterol, and helping with weight contol.
   b. supplying fiber to aid digestion, digesting cellulose, and providing the main fuel source for muscles.

   c. protecting organs against shock, helping the body use carbohydrate efficiently, and providing triglycerides.
   d. supporting growth and maintenance, supplying hormones to regulate body processes, and maintaining fluid and electrolyte balance.

5. The swelling of body tissue caused by the leakage of fluid from the blood vessels into the interstitial spaces is called:
   a. edema.
   b. anemia.
   c. acidosis.
   d. sickle-cell anemia.

6. Major proteins in the blood that protect against bacteria and other disease agents are called:
   a. acids.
   b. buffers.
   c. antigens.
   d. antibodies.

7. Marasmus can be distinguished from kwashiorkor because in marasmus:
   a. only adults are victims.
   b. the cause is usually an infection.
   c. severe wasting of body fat and

   muscle occurs.
   d. the limbs and face swell with edema, and the belly bulges with a fatty liver.

8. The RDA for protein for a healthy adult is _____ gram(s) per kilogram of appropriate or average body weight for height.
   a. 0.5
   b. 0.8
   c. 1.1
   d. 1.4

9. Generally speaking, from which of the following foods are complete proteins derived?
   a. milk, gelatin, and soy
   b. rice, potatoes, and eggs
   c. meats, fish, and poultry
   d. vegetables, grains, and fruits

10. An incomplete protein lacks one or more:
    a. hydrogen bonds.
    b. essential fatty acids.
    c. saturated fatty acids.
    d. essential amino acids.

Answers to these questions appear in Appendix H.

CLINICAL APPLICATIONS

1. Considering the health effects of too little dietary protein, what suggestions would you have for a teenage girl who reports the following information about her food intake:

   a. She never eats any meat or other animal-derived foods because she is a vegan. On a typical day, she eats toast and juice for breakfast, chips, a soft drink, and a piece of fruit for lunch, and a small serving of plain pasta with tomato sauce or steamed vegetables for dinner, along with a glass of water or tea.
   b. She takes amino acid supplements because a friend told her that the only way to get amino acids if she doesn't eat meat is to take them as supplements.

2. Considering the health effects of excess dietary protein, what advice would you have for a college athlete who tells you he wants to bulk up his muscles and reports the following information about his food intake:

   a. He eats large servings of meat (usually red meat) at least twice a day. He drinks whole milk 2 or 3 times a day, and eats eggs and bacon for breakfast almost every day.
   b. He avoids breads, cereals, and pasta in order to save room for protein-rich foods such as meat, milk, and eggs.
   c. He eats a piece of fruit once in a while, but seldom eats vegetables because they are too time-consuming to prepare.

Notes

[1] J. C. Teran, K. D. Mullen, and A. J. McCullough, Glutamine—A conditionally essential amino acid in cirrhosis? *American Journal of Clinical Nutrition* 62 (1995): 897–900.

[2] D. G. Schroeder and R. Martorell, Enhancing child survival by preventing malnutrition, *American Journal of Clinical Nutrition* 65 (1997): 1080–1081.

[3] J. E. Miller and S. Korenman, Nutritional status of poor children in the United States, *Journal of the American Dietetic Association* 95 (1995): 248–250; M. Mowe, T. Bohmer, and E. Kindt, Reduced nutritional status in an elderly population (>70 yrs) is probable before disease and possibly contributes to the development of disease, *American Journal of Clinical Nutrition* 59 (1994): 317–324; J. Dye, Malnourished children in the United States: Caught in the cycle of poverty, *Journal of the American Dietetic Association* 94 (1994): 218; M. L. Taylor and

S. A. Koblinsky, Dietary intake and growth status of young homeless children, *Journal of the American Dietetic Association* 93 (1993): 464–466; J. C. Wolgemuth and coauthors, Wasting malnutrition and inadequate nutrient intakes identified in a multiethnic homeless population, *Journal of the American Dietetic Association* 92 (1992): 834–839; M. Nestle and S. Guttmacher, Hunger in the United States: Rationale, methods, and policy implications of state hunger surveys, *Journal of Nutrition Education* 24 (1992): S18–S22.

[4] The State of the World's Children 1998: A UNICEF Report, Malnutrition: Causes, consequences, and solutions, *Nutrition Reviews* 56 (1998): 115–123.

[5] C. Goplan, The contribution of nutrition research to the control of undernutrition: The Indian experience, *Annual Review of Nutrition* 12 (1992): 1–17.

[6] J. C. Waterlow, Childhood malnutrition in developing nations: Looking back and forward, *Annual Review of Nutrition* 14 (1994): 1–19.

[7] L. Lewinter-Suskind and coauthors, The malnourished child, in *Textbook of Pediatric Nutrition*, 2nd ed., ed. R. M. Suskind and L. Lewinter-Suskind (New York: Raven Press, 1993), pp. 127–140.

[8] P. W. Yoon and coauthors, The effect of malnutrition on the risk of diarrhea and respiratory mortality in children < 2 y of age in Cebu, Philippines, *American Journal of Clinical Nutrition* 65 (1997): 1070–1077.

[9] Lewinter-Suskind and coauthors, 1993.

[10] G. E. Fraser, Associations between diet and cancer, ischemic heart disease, and all-cause

mortality in non-Hispanic white California Seventh-day Adventists, *American Journal of Clinical Nutrition* 70 (1999): S532–S538; L. H. Kushi, E. B. Lenart, and W. C. Willet, Health implications of Mediterranean diets in light of contemporary knowledge: Meat, wine, fats, and oils, *American Journal of Clinical Nutrition* 61 (1995): S1416–S1427.

[11] B. C. H. Chiu and coauthors, Diet and risk of non-Hodgkin lymphoma in older women, *Journal of the American Medical Association* 275 (1996): 1315–1321.

[12] R. P. Heaney, Protein intake and the calcium economy, *Journal of the American Dietetic Association* 93 (1993): 1259–1260.

[13] D. Feskanich and coauthors, Protein consumption and bone fractures in women, *American Journal of Epidemiology* 143 (1996): 472–479.

[14] Position of The American Dietetic Association and The Canadian Dietetic Association: Nutrition for physical fitness and athletic performance for adults, *Journal of the American Dietetic Association* 93 (1993): 691–695.

[15] Food and Drug Administration, Impurities confirmed in dietary supplement 5-hydroxy-L-tryptophan, FDA Talk Paper, August 1998, available from Food and Drug Administration, U.S. Department of Health and Human Services, Public Health Service, 5600 Fishers Lane, Rockville, MD 20857.

[16] S. A. Anderson and D. J. Raiten, eds., *Safety of Amino Acids Used as Supplements* (Bethesda, Md.: Federation of American Societies for Experimental Biology, 1992).

Q&A NUTRITION IN PRACTICE

Supplements and Ergogenic Aids Athletes Use

In a world where body condition and skill are hard won, athletes gravitate to promises that they can easily improve their performance by taking pills, powders, or potions. Chapter 4 discussed some dangers of amino acid supplements, but because protein powders and amino acid supplements are among the most common supplements athletes use, the topic is revisited in this Nutrition in Practice.[1] Athletes often hear well-intended, but unsubstantiated, advice from their coaches and peers recommending that they use certain nutrients, drugs, or procedures to enhance performance. The desire to win is strong, but no amount of wishing can change the fact that an overwhelming majority of supplements marketed for athletes are frauds. If the products have no effect and are harmless, they are only a waste of money; but some products are harmful or actually impair performance, wasting athletic potential as well as money. This Nutrition in Practice looks at some of the so-called magical substances that promise to improve physical performance.

What does ergogenic mean?

Ergogenic means work enhancing or work producing. In connection with athletic performance, ergogenic aids are substances or treatments that purportedly improve athletic performance above and beyond what is possible through training alone. Research findings do not, for the most part, support the claims made for ergogenic aids. A claim that a product is ergogenic should serve as a reminder to consider the source of the claim and ask who may gain from the sale. The glossary defines the commonly used ergogenic aids

discussed in this Nutrition in Practice. Table NP4-1 lists many other substances promoted as ergogenic aids.

Should athletes take protein supplements?

Protein supplements are advocated to improve both strength and endurance, but as discussed in Chapter 4, well-nourished active people and athletes do not need them. Extra protein cannot be forced into the muscles to make them grow. Muscle cells accept nutrients only when they are needed. The cells "decide" what they need based on the messages they receive from the hormones that regulate them and from the demands put upon them. The way to make muscle cells grow, therefore, is to make them work. The only

GLOSSARY OF ERGOGENIC AIDS

anabolic steroids: drugs related to the male sex hormone, testosterone, that stimulate the development of lean body mass.
anabolic = promoting growth
sterols = compounds chemically related to cholesterol

branched-chain amino acids: the amino acids leucine, isoleucine, and valine, which are present in large amounts in skeletal muscle tissue; falsely promoted as fuel for exercising muscles.

caffeine: a natural stimulant found in many common foods and beverages, including coffee, tea, and chocolate; may enhance endurance by stimulating fatty acid release but also causes fluid losses. Overdoses cause headaches, trembling, rapid heart rate, and other undesirable side effects.

carnitine (CAR-ne-teen): a nonessential nutrient made in the body from lysine that helps transport fatty acids across the mitochondrial membrane. Carnitine supposedly "burns" fat and spares glycogen during endurance events, but in reality it does neither.

role for diet in this process is to make protein available, and good diets always do. Although the protein needs of some endurance and strength athletes are higher than those of sedentary people, the additional protein is already present in a well-chosen diet as Chapter 4 described.

Most healthy athletes eating well-balanced diets do not need amino acid supplements either. Advertisers point to research that identifies the **branched-chain amino acids** as the main ones used as fuel by exercising muscles. What the ads leave out is that compared to glucose and fatty acids, branched-chain amino acids provide almost no fuel and that ordinary foods supply them in abundance anyway.

Furthermore, large doses of branched-chain amino acids can raise plasma ammonia concentra-

chromium (CROW-mee-um) picolinate: a trace mineral supplement; falsely promoted as building muscle, enhancing energy, and burning fat. **Picolinate (pick-oh-LYN-ate)** is a derivative of the amino acid tryptophan that seems to enhance chromium absorption.

creatine (KREE-ah-tine): a nitrogen-containing compound that combines with phosphate to form the high-energy compound creatine phosphate (or phosphocreatine) in muscles. Claims that creatine enhances energy use and muscle strength need further confirmation.

DHEA (dehydroepiandrosterone): a hormone made in the adrenal glands that serves as a precursor to the male hormone testosterone; falsely promoted as burning fat, building muscle, and slowing aging. Side effects include acne, aggressiveness, and liver enlargement.

glycogen loading: a regimen of exhausting exercise, followed by eating a high-carbohydrate diet, that enables muscles to temporarily store glycogen beyond their normal capacity; also called *carbohydrate loading* or *glycogen supercompensation.*

tions, which can be toxic to the brain.[2] Branched-chain amino acid supplements are neither effective nor safe and are not recommended.

Some athletes, especially endurance athletes, are taking carnitine supplements. What is carnitine?

Carnitine is a nonessential nutrient made in the body from the amino acid lysine. Endurance athletes believe carnitine will help them burn more fat, thereby sparing glycogen during endurance events. Carnitine is also promoted to bodybuilders as a "fat burner."

In the body, carnitine facilitates the transfer of fatty acids across the mitochondrial membrane. Supplement manufacturers suggest that with more carnitine available, fat oxidation will be enhanced, but this does not seem to be the case. Carnitine supplementation for 7 to 14 days neither raised muscle carnitine concentrations nor influenced fat or carbohydrate oxidation.[3] It did, however, produce diarrhea in half of the men tested. Milk and meat products are good sources of carnitine, and supplements are not needed.

Do athletes need vitamin or mineral supplements?

Some athletes take a vitamin pill right before an event. This practice is pointless, though probably harmless. For one thing, vitamins taken right before competition have no effect; during the event, they are still waiting in the blood and have not yet been assembled into working molecules. Besides, research shows that most bodybuilders' diets provide adequate amounts of vitamins and minerals, and when a nutrient is lacking, the supplements chosen are seldom the ones needed to remedy the deficiencies. If a diet lacks nutrients, it should be modified to provide the needed nutrients. Only if a health professional identifies a clinical nutrient deficiency should nutrient supplements be prescribed—and

Table NP4-1
SUBSTANCES PROMOTED AS ERGOGENIC AIDS

bee pollen: a product consisting of bee saliva, plant nectar, and pollen that confers no benefit on athletes and may cause an allergic reaction in individuals sensitive to it.

blood doping: the process of injecting red blood cells to enhance the blood's oxygen-carrying ability. Risks include dangerous blood clotting, especially in athletes who become dehydrated; infections from non-sterile equipment; transfusion reactions; and dangers of improperly transferred blood. Blood doping is banned in Olympic competitions.

calcium pangamate: a compound once thought to enhance aerobic metabolism, now known to have no such effect.

cell salts: a mineral preparation supposedly prepared from living cells.

coenzyme Q10: a lipid found in cells (mitochondria) shown to improve exercise performance in heart disease patients, but not effective in improving performance of healthy athletes.

DNA and ***RNA (deoxyribonucleic acid*** and ***ribonucleic acid):*** the genetic materials of cells necessary in protein synthesis; falsely promoted as ergogenic aids.

epoetin (eh-poy-EE-tin): a drug derived from the human hormone erythropoietin and marketed under the trade name Epogen; illegally used to increase oxygen capacity.

ginseng: a plant whose extract supposedly boosts energy; side effects to chronic use include nervousness, confusion, and depression.

glycine: a nonessential amino acid; promoted as an ergogenic aid because it is a precursor of the high-energy compound creatine phosphate.

growth hormone releasers: herbs or pills falsely promoted as enhancing athletic performance.

guarana: a reddish berry found in Brazil's Amazon valley that is used as an ingredient in carbonated sodas and taken in powder or tablet form. Guarana is marketed as an ergogenic aid to enhance speed and endurance, an aphrodisiac, a "cardiac tonic," an "intestinal disinfectant," and a smart drug that supposedly improves memory and concentration and wards off senility. Because guarana contains seven times as much caffeine as its relative the coffee bean, there are concerns that high doses can stress the heart and cause panic attacks.

herbal steroids or ***plant sterols:*** lipid extracts of plants, called *ferulic acid, oryzanol, phytosterols,* or *adaptogens;* marketed with false claims that they contain hormones or balance hormonal activity.

inosine: an organic chemical that is falsely said to "activate cells, produce energy, and facilitate exercise," but has been shown actually to reduce the endurance of runners.

octacosanol: an alcohol extracted from wheat germ; often falsely promoted to enhance athletic performance.

phosphate salt: a salt that has been demonstrated to raise the concentration of a metabolically important compound (diphosphoglycerate) in red blood cells and enhance the cells' potential to deliver oxygen to muscle cells. The salts may cause calcium losses from the bones if taken in excess.

royal jelly: a substance produced by worker bees and fed to the queen bees; often falsely promoted as enhancing athletic performance.

sodium bicarbonate: baking soda; an alkaline salt believed to neutralize blood lactic acid and thereby reduce pain and enhance possible workload. Some studies show that sodium bicarbonate in recommended doses can enhance performance of high-intensity exercise (in 1- to 5-minute exercise sessions), but its effects on endurance exercise are unknown. "Soda loading" may cause intestinal bloating and diarrhea.

then only if dietary modification alone cannot remedy the deficiency.

Do vitamin E supplements benefit active people and athletes even if they are not vitamin E deficient?

Many athletes and active people are taking megadoses of vitamin E in hopes of preventing oxidative damage to muscles. Some research suggests vitamin E supplements may protect against exercise-induced oxidative stress and damage.[4] Other research has failed to show a protective effect of vitamin E against oxidative stress during physical activity.[5] Studies examining the long-term effects of vitamin E supplementation are lacking. More research is needed but in the meantime, active people can benefit by eating vitamin E-rich foods such as almonds, wheat germ, and sunflower seeds, as well as antioxidant-rich fruits and vegetables.

Are any of the claims for chromium picolinate true?

Chapter 9 introduces chromium as an essential trace mineral involved in carbohydrate and lipid metabolism. Advertisements in bodybuilding magazines claim that **chromium picolinate,** which is supposed to be more easily absorbed than chromium alone, builds muscle, enhances energy, and burns fat. Such claims derive from one or two initial studies on this mineral. Most studies of chromium picolinate and strength training that have followed, however, show no effects of chromium picolinate supplementation on strength, lean body mass, or body fat.[6] Furthermore, large doses of chromium picolinate may lead to iron deficiency.[7]

Why is creatine so popular?

Interest in—and use of—**creatine** monohydrate supplements to enhance energy production during intense activity has grown dramatically in the last few years.[8] Power athletes such as weight lifters use creatine supplements to enhance stores of the high-energy compound creatine phosphate (or phosphocreatine) in muscles. Theoretically, the more creatine phosphate in muscles, the higher the intensity at which an athlete can train. High-intensity training stimulates the muscles to adapt, which, in turn, improves performance.

The results of some studies suggest creatine supplementation enhances performance of high-density strength activity such as weight lifting.[9] Other research findings, however, conflict with reports that creatine supplements improve strength performance.[10]

Some medical and fitness experts voice concern that, like many performance enhancement supplements before it, creatine is being taken in huge doses (5 to 30 grams per day) before evidence of its value has been ascertained.[11] Even people who eat red meat, which is a creatine-rich food, do not consume near the amounts athletes are taking. Athletes who take megadoses of creatine risk possible long-term side effects such as organ and muscle damage. Despite the uncertainties, creatine supplements are not illegal in international competition.

What is glycogen loading?

The fuel for intense muscular activity is carbohydrate, stored in the muscle as glycogen. Athletes who compete in long-distance endurance events naturally want to have as much energy stored in their muscles as possible. Various techniques were used in the past to trick muscles into storing more glycogen than normal. These **glycogen loading** techniques involved sudden, drastic changes in diet that caused nausea or cramping in some athletes. Other athletes experienced more dangerous effects such as abnormal heart and kidney function.

Exercise physiologists now recommend a modified plan of glycogen loading that confers benefits without such side effects. First, about two or three weeks before competition, the athlete increases activity intensity while eating a normal, high-carbohydrate diet. Then, during the last week before competition, the athlete modifies both physical activity and diet. With respect to activity, the athlete gradually cuts back, resting completely on the day before the event. Meanwhile, with respect to foods, the athlete eats carbohydrate as usual until three days before the competition and then eats a very-high-carbohydrate diet.[12] Endurance athletes who follow this plan can keep going longer than their competitors without ill effects. In a hot climate, extra glycogen confers an additional advantage: as glycogen breaks down, it releases water, which helps to meet the athlete's fluid needs.

Extra glycogen benefits those who must keep going for 90 minutes or longer. The regular, everyday exerciser will not benefit from having larger stores. What that person does need, though, is adequate glycogen from eating a diet high in complex carbohydrates.

Can caffeine improve endurance performance?

Although some research findings support this notion, other findings suggest that **caffeine** has no effect on endurance. If caffeine does enhance endurance, the effect probably occurs because caffeine stimulates fatty acid release, thereby slowing glycogen use. Caffeine is a drug that stimulates the nervous system. The possible benefits must be weighed against caffeine's adverse effects—stomach upset, nervousness, irritability, headache, and diarrhea. Caffeine induces fluid losses that can be potentially hazardous if caffeine-containing fluids are used in place of other fluids by athletes competing in hot environments. The use of caffeine is banned by the International Olympic Committee when it exceeds a dosage equivalent to 5 or 6 cups of coffee in a 2-hour period prior to competition.

Is it safe for athletes to take steroids?

No. **Steroids** have dangerous side effects and are illegal. Technically called androgenic-anabolic steroid drugs, they are derivatives of the male sex hormone testosterone. Testosterone promotes the development of male characteristics (androgenic) and lean body mass (anabolic). Athletes take steroids to stimulate muscle bulking. The American College of Sports Medicine and the American Academy of Pediatrics condemn the use of steroids by athletes, and the International Olympic Committee has banned their use.[13] In support of its position, the committee cites the known toxic side effects and maintains that steroid use is a form of cheating. Competitors who use the drugs put other athletes in the difficult position of either conceding an unfair advantage to abusing competitors or taking steroids and accepting the risk of untoward side effects.

The list of hazards and adverse reactions from steroids continues to grow (see Table NP4-2) amid only a slight decline in use of the drugs. Among the side effects and adverse reactions that steroids produce are cancerous liver tumors that cause the liver to rupture and hemorrhage, impairing its function; testicular

Table NP4-2
ANABOLIC STEROIDS: SIDE EFFECTS AND ADVERSE REACTIONS

Established Side Effects and Adverse Reactions

Acne	Liver disease
Cancer	Liver tumors
Cholesterol increase	Male pattern baldness (in women—irreversible)
Clitoris enlargement	Oily skin (females only)
Death	Peliosis hepatitis (a liver disease)
Edema (water retention in tissue)	Penis enlargement (young boys)
Fetal damage	Priapism (painful, prolonged erections)
Frequent or continuing erections	Prostate enlargement
HDL (which help reduce cholesterol) decrease	Sterility (reversible)
Heart disease	Stunted growth
Hirsutism (hairiness in women—irreversible)	Swelling of feet or lower legs
Increased risk of coronary artery disease (heart attack, stroke)	Testicular atrophy
Jaundice	Yellowing of the eyes or skin

Other Possible Side Effects and Adverse Reactions

Abdominal or stomach pains	Insomnia
Aggressive, combative behavior ("roid rage")	Kidney disease
Anaphylactic shock (from injections)	Kidney stones (from hypercalcemia)
Black, tarry, or light-colored stool	Listlessness
Bone pain	Menstrual irregularities
Breast development (sore or swelling—in men)	Muscle cramps
Chills	Nausea or vomiting
Dark-colored urine	Purple- or red-colored spots on body, inside of mouth or nose
Depression	Rash
Diarrhea	Septic shock (blood poisoning from injections)
Fatigue	Sexual problems
Feeling of abdominal or stomach fullness	Sore tongue
Feeling of discomfort	Unexplained darkening of skin
Fever	Unexplained weight loss
Frequent urge to urinate (mature men)	Unnatural hair growth
Gallstones	Unpleasant breath odor
Headache	Unusual bleeding
High blood pressure	Unusual weight gain
Hives	Urination problems
Hypercalcemia (too much calcium)	Vomiting blood
Impotence	
Increased chance of injury to muscles, tendons, and ligaments, plus longer recovery period from injuries	

SOURCES: K. L. Ropp, No-win situation for athletes, *FDA Consumer,* December 1992, pp. 8–12; National Academy of Sports Medicine policy statement and position paper: Anabolic androgenic steroids, growth hormones, stimulants, ergogenics, and drug use in sports, in B. Goldman and R. Klats, *Death in the Locker Room II: Drugs and Sports* (Chicago: Elite Sports Medicine Publications, 1992), pp. 328–373.

shrinkage in men and masculinization of women; cardiovascular problems (including high blood pressure); and sterility. Athletes using steroids are sure to develop side effects no matter how closely they are being monitored by a trainer or doctor.

The dangers of steroid use cannot be overemphasized. Nurses and other health care professionals are obligated to warn athletes of these dangers. Speak simply and emphatically: the price for the potential competitive edge that steroids confer is dam-

aged health and sometimes life itself. The safest effective way to build muscle has always been through hard training—and always will be.

What is DHEA, and why do some athletes use it?

Some athletes use **DHEA** as an alternative to anabolic steroids. DHEA is a hormone (dehydroepiandrosterone), made in the adrenal glands, that serves as a precursor to the male hormone testosterone. Because DHEA and testosterone concentrations decline

with age, proponents claim DHEA supplements enhance testosterone concentrations, thereby slowing aging and building muscle. Advertisements claim it "burns fat," "builds muscle," and "slows aging," but evidence to support such claims is lacking.

DHEA's short-term side effects include oily skin, acne, body hair growth, liver enlargement, and aggressive behavior.[14] Long-term effects of DHEA use remain to be seen and may take years to become evident. The potential for harm from DHEA supplements is great, and athletes, as well as others, should avoid it. DHEA is banned by the International Olympic Committee and the National Collegiate Athletic Association.

Do any of the substances athletes use to boost performance work?

For the most part, no. Many of these substances have been studied and found to be worthless. Health care professionals can positively influence athletes and others interested in boosting athletic performance by stressing the measures that do help to enhance performance: regular training and sound nutrition.

Notes

[1] E. A. Applegate and L. E. Grivetti, Search for the competitive edge: A history of dietary fads and supplements, *Journal of Nutrition* 127 (1997): S869–S873.

[2] E. Coleman, Branched-chain amino acids and fatigue, *Sports Medicine Digest* 18 (1996): 44.

[3] M. Vukovich, D. L. Costill, and W. J. Fink, Carnitine supplementation: Effect on muscle carnitine and glycogen content during exercise, *Medicine and Science in Sports and Exercise* 26 (1994): 1122–1129.

[4] J. M. McBride and coauthors, Effect of resistance exercise on free radical production, *Medicine and Science in Sports and Exercise* 30 (1998): 67–72; M. Meydani and coauthors, Protective effect of vitamin E on exercise-induced oxidative damage in young and older adults, *American Journal of Physiology* 264 (1993): R992–R998.

[5] J. A. Warren and coauthors, Elevated muscle vitamin E does not attenuate eccentric exercise-induced muscle injury, *Journal of Applied Physiology* 72 (1992): 2168–2175; E. H. Witt and coauthors, Exercise, oxidative damage and effects of antioxidant manipulation, *Journal of Nutrition* 122 (1992): 766–773.

[6] H. C. Lukaski and coauthors, Chromium supplementation and resistance training: Effects on body composition, strength, and trace element status of men, *American Journal of Clinical Nutrition* 63 (1996): 954–965; M. A. Hallmark and coauthors, Effects of chromium and resistance training on muscle strength and body composition, *Medicine and Science in Sports and Exercise* 28 (1995): 139–144.

[7] Lukaski and coauthors, 1996.

[8] Applegate and Grivetti, 1997.

[9] R. B. Kreider and coauthors, Effects of creatine supplementation on body composition, strength, and sprint performance, *Medicine and Science in Sports and Exercise* 30 (1998): 73–82; J. S. Volek and coauthors, Creatine supplementation enhances muscular performance during high-intensity resistance exercise, *Journal of the American Dietetic Association* 97 (1997): 765–770; S. M. Tolar, Creatine is an ergogen for anaerobic exercise, *Nutrition Reviews* 55 (1997): 21–23; C. P. Earnest and coauthors, The effect of creatine monohydrate ingestion on anaerobic power indices, muscular strength, and body composition, *Acta Physiologica Scandinavica* 153 (1995): 207–709.

[10] L. M. Odland and coauthors, Effect of oral creatine supplementation on muscle (PCr) and short-term maximum power output, *Medicine and Science in Sports and Exercise* 29 (1997): 216–219.

[11] M. S. Juhn, J. W. O'Kane, and D. M. Vinci, Oral creatine supplementation in male collegiate athletes: A survey of dosing habits and side effects, *Journal of the American Dietetic Association* 99 (1999): 593–595; T. Noakes, as quoted in M. Gaie, Olympic athletes face heat, other health hurdles, *Journal of the American Medical Association* 276 (1996): 178–180.

[12] M. H. Williams, *Nutrition for Fitness and Sport*, 4th ed. (New York: McGraw-Hill, 1995), pp. 83–120.

[13] D. H. Catlin and T. H. Murray, Performance-enhancing drugs, fair competition and Olympic sport, *Journal of the American Medical Association* 276 (1996): 231–237.

[14] E. Coleman, DHEA—An anabolic aid? *Sports Medicine Digest* 18 (1996): 140–141.

CHAPTER 5

DIGESTION AND ABSORPTION

CONTENTS

■■
Anatomy of the Digestive Tract

■■
The Process of Digestion

■■
The Absorptive System

■■
Transport of Nutrients

■■
The System at Its Best

■■
Nutrition in Practice: Food Safety

The body's ability to transform the foods a person eats into the nutrients that fuel the body's work is quite remarkable. Yet most people probably give little, if any, thought to all the body does with food once it is eaten. This chapter offers the reader the opportunity to learn how the body digests, absorbs, and transports the nutrients in foods and excretes the unwanted substances. The next chapter shows how the body uses the nutrients once they have been absorbed and are traveling in the blood and lymph.

One of the beauties of the digestive tract is that it is selective. Materials that are nutritive for the body are broken down into particles that can be absorbed into the bloodstream. Most of the nonnutritive materials are left undigested and pass out the other end of the digestive tract.

■■ Anatomy of the Digestive Tract

GI tract: the gastrointestinal tract or digestive tract. The principal organs are the stomach and intestines.
gastro = stomach

The **gastrointestinal (GI) tract** is a flexible muscular tube measuring about 15 feet in length from the mouth to the anus.[1] Figure 5-1 (pp. 98–99) traces the path followed by food from one end to the other. The accompanying glossary defines GI anatomy terms. In a sense, the human body surrounds the GI tract. Only when a nutrient or other substance passes through the cells of the digestive tract wall does it actually enter the body.

■■ The Digestive Organs

digestion: the process by which complex food particles are broken down to smaller absorbable particles.

The process of **digestion** begins in the mouth. As you chew, your teeth crush and soften foods, while saliva mixes with the food mass and moistens it for

GLOSSARY GI TERMS

These terms are listed in order from the beginning of the digestive tract to the end.

epiglottis (epp-ee-GLOT-tiss): a cartilage structure in the throat that prevents fluid or food from entering the trachea when a person swallows.
epi = upon (over)
glottis = back of tongue

trachea (TRAKE-ee-uh): the windpipe; the passageway from the mouth and nose to the lungs.

esophagus (e-SOFF-uh-gus): the food pipe; the conduit from the mouth to the stomach.

cardiac sphincter (CARD-ee-ack SFINK-ter): the sphincter muscle at the junction between the esophagus and the stomach.
cardiac = heart

pyloric (pie-LORE-ic) sphincter: the sphincter muscle separating the stomach from the small intestine (also called *pylorus* or *pyloric valve*).
pylorus = gatekeeper

gallbladder: the organ that stores and concentrates bile. When it receives the signal that fat is present in the duodenum, the gallbladder contracts and squirts bile through the bile duct into the duodenum.

pancreas: a gland that secretes enzymes and digestive juices into the duodenum. (This is its exocrine function; it also has the endocrine function of secreting insulin and other hormones into the blood.)

small intestine: a 10-foot length of small-diameter (1-inch) intestine that is the major site of digestion of food and absorption of nutrients.

duodenum (doo-oh-DEEN-um or doo-ODD-ah-num): the top portion of the small intestine (about "12 fingers' breadth" long, in ancient terminology).

jejunum (je-JOON-um): the first two-fifths of the small intestine beyond the duodenum.

ileum (ILL-ee-um): the last segment of the small intestine.

ileocecal (ill-ee-oh-SEEK-ul) valve: the sphincter muscle separating the small and large intestines.

colon or **large intestine:** the last portion of the intestine, which absorbs water. Its main segments are the ascending colon, the transverse colon, the descending colon, and the sigmoid colon.
sigmoid = shaped like the letter S

appendix: a narrow blind sac extending from the beginning of the colon. The appendix stores lymphocytes.

rectum: the muscular terminal part of the GI tract extending from the sigmoid colon to the anus. The rectum stores waste prior to elimination.

anus (AY-nus): the terminal sphincter muscle of the GI tract.

comfortable swallowing. Saliva also helps dissolve the food so that you can taste it; only particles in solution can react with taste buds.

▦ MOUTH TO THE ESOPHAGUS   Once a mouthful of food has been swallowed, it is called a **bolus.** Each bolus first slides across your epiglottis, bypassing the entrance to your lungs. During each swallow, the epiglottis closes off your air passages so that you do not choke.

▦ ESOPHAGUS TO THE STOMACH   Next, the bolus slides down the esophagus, which conducts it through the diaphragm to the stomach. The cardiac **sphincter,** a band of muscle surrounding the esophagus where it enters the stomach, closes behind the bolus so that it cannot slip back. The stomach retains the bolus for a while, adds juices to it (gastric juices are discussed on p. 103), and grinds it into a semiliquid mass called **chyme.** Then bit by bit the stomach releases the chyme through another sphincter, the pyloric sphincter, which opens into the small intestine and then closes behind the chyme.

▦ THE SMALL INTESTINE   At the top of the small intestine, the chyme passes by an opening from the common bile duct, which secretes fluids into the small intestine from two organs outside the GI tract—the gallbladder and the pancreas. The chyme travels down the small intestine through its three segments—the duodenum, the jejunum, and the ileum—a total of about 10 feet of tubing coiled within the abdomen.[2]

▦ THE COLON (LARGE INTESTINE)   Having traveled the length of the small intestine, the chyme passes through another sphincter, the ileocecal valve, into the beginning of the colon (large intestine) in the lower right-hand side of the abdomen. In the colon, the chyme travels up the right-hand side of the abdomen, across the front to the left-hand side, down to the lower left-hand side, and finally below the other folds of the intestines to the back side of the body above the rectum.

▦ THE RECTUM   During chyme's passage to the rectum, the colon withdraws water from it, leaving semisolid waste. The strong muscles of the rectum hold back this waste until it is time to defecate. Then the rectal muscles relax, and the last sphincter in the system, the anus, opens to allow the wastes to pass. Thus, food follows the path shown in the margin.

bolus (BOH-lus): the portion of food swallowed at one time.

sphincter (SFINK-ter): a circular muscle surrounding, and able to close, a body opening.
sphincter = band (binder)

chyme (KIME): the semiliquid mass of partly digested food expelled by the stomach into the duodenum (the top portion of the small intestine).

The path of food through the digestive tract:

- *Mouth.*
- *Esophagus.*
- *Cardiac sphincter (or lower esophageal sphincter).*
- *Stomach.*
- *Pyloric sphincter.*
- *Duodenum (common bile duct enters here), jejunum, ileum.*
- *Ileocecal valve.*
- *Colon.*
- *Rectum.*
- *Anus.*

⌑ LEARNING LINK

As you've just read, each part of the GI tract plays an important role in the digestion of food. Failure of any part to do its job can quickly lead to malnutrition. For example, many elderly people have difficulty chewing because of missing teeth or ill-fitting dentures. Other elderly clients or those with certain types of cancer may have trouble swallowing or keeping food down once it is swallowed. Some clients may have had surgery to remove parts of the GI tract because of cancer or injuries. In all cases, knowledge about the anatomy and physiology of the digestive tract will facilitate your understanding of the appropriate dietary and medical treatment. Later chapters discuss some of the conditions that impair digestive tract function and nutrition status and the diets necessary to their treatment.

▦   The Involuntary Muscles and the Glands

You are usually unaware of all the activity that goes on between the time you swallow and the time you defecate. As is the case with so much else that hap-

Figure 5-1
THE GASTROINTESTINAL TRACT

FIBER	CARBOHYDRATE

Mouth

The mechanical action of the mouth and teeth crushes and tears fiber in food and mixes it with saliva to moisten it for swallowing.

The salivary glands secrete a watery fluid into the mouth to moisten the food. The salivary enzyme amylase begins digestion:

$$\text{Starch} \xrightarrow{\text{amylase}} \text{small polysaccharides, maltose.}$$

Stomach

Fiber is unchanged.

Stomach acid and enzymes start to digest salivary enzymes, halting starch digestion. To a small extent, stomach acid hydrolyzes maltose and sucrose.

Small intestine

Fiber is unchanged.

The pancreas produces enzymes and releases them through the pancreatic duct into the small intestine:

$$\text{Polysaccharides} \xrightarrow[\text{amylase}]{\text{pancreatic}} \text{disaccharides.}$$

Then enzymes on the surfaces of the small intestinal cells break disaccharides into monosaccharides, and the cells absorb them:

$$\text{Maltose} \xrightarrow{\text{maltase}} \text{glucose} + \text{glucose.}$$

$$\text{Sucrose} \xrightarrow{\text{sucrase}} \text{fructose} + \text{glucose.}$$

$$\text{Lactose} \xrightarrow{\text{lactase}} \text{galactose} + \text{glucose.}$$

Colon (large intestine)

Most fiber passes intact through the digestive tract to the colon. Here, bacterial enzymes digest some fiber:

$$\text{Some fiber} \xrightarrow[\text{enzymes}]{\text{bacterial}} \text{fatty acids, gas.}$$

Fiber holds water; regulates bowel activity; and binds cholesterol and some minerals, carrying them out of the body as it is excreted with feces.

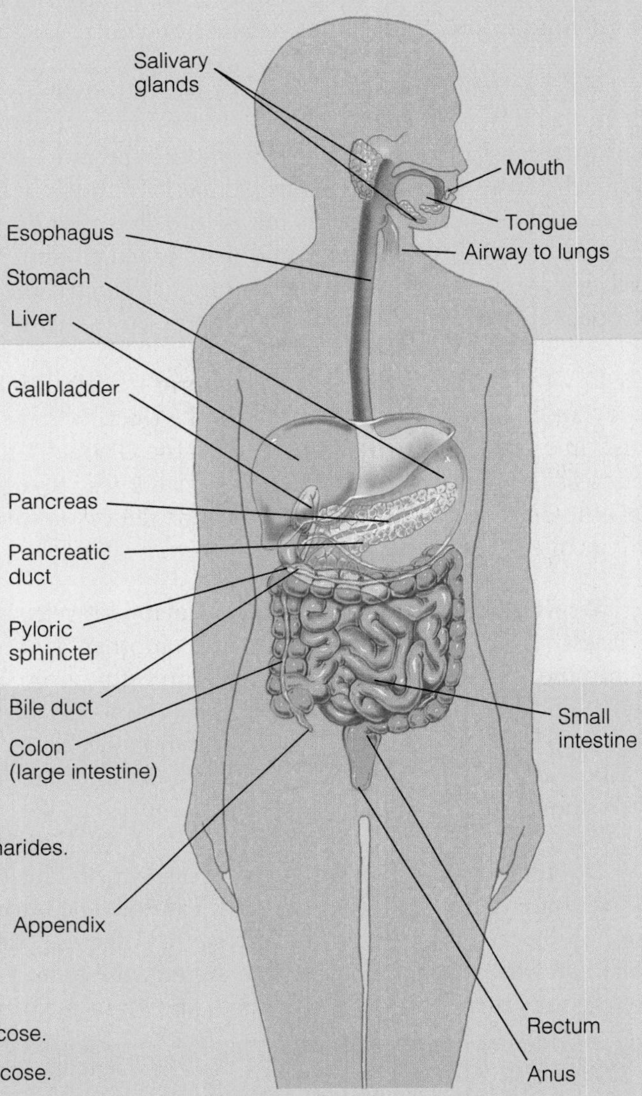

(continued)

Labels: Salivary glands, Mouth, Tongue, Airway to lungs, Esophagus, Stomach, Liver, Gallbladder, Pancreas, Pancreatic duct, Pyloric sphincter, Bile duct, Colon (large intestine), Appendix, Small intestine, Rectum, Anus

pens in the body, the muscles and **glands** of the digestive tract meet internal needs without your having to exert any conscious effort to get the work done.

People consciously chew and swallow, but even in the mouth there are some processes over which you have no control. The salivary glands secrete

Figure 5-1
THE GASTROINTESTINAL TRACT—Continued

FAT	PROTEIN	VITAMINS	MINERALS AND WATER
Mouth Glands in the base of the tongue secrete a fat-digesting enzyme known as lingual lipase. Some hard fats begin to melt as they reach body temperature.	Chewing and crushing moisten protein-rich foods and mix them with saliva to be swallowed.	No action.	The salivary glands add water to disperse and carry food.
Stomach The acid-stable lingual lipase splits one bond of triglycerides to produce diglycerides and fatty acids. The degree of hydrolysis is slight for most fats but may be appreciable for milk fats. The stomach's churning action mixes fat with water and acid. A gastric lipase accesses and hydrolyzes a very small amount of fat.	Stomach acid uncoils protein strands and activates stomach enzymes: Protein $\xrightarrow[\text{HCl}]{\text{pepsin}}$ smaller polypeptides.	Intrinsic factor (see Chapter 8) attaches to vitamin B_{12}.	Stomach acid (HCl) acts on iron to reduce it, making it more absorbable (see Chapter 9). The stomach secretes enough watery fluid to turn a moist, chewed mass of solid food into liquid chyme.
Small intestine Bile flows in from the liver and gallbladder (via the common bile duct): Fat $\xrightarrow{\text{bile}}$ emulsified fat. Pancreatic lipase flows in from the pancreas (via the pancreatic duct): Emulsified fat $\xrightarrow[\text{lipase}]{\text{pancreatic}}$ monoglycerides, glycerol, fatty acids (absorbed).	$\xrightarrow[\text{and intestinal}]{\text{pancreatic}}$ proteases $\xrightarrow[\text{dipeptidases}]{\text{intestinal}}$ amino acids (absorbed) and tripeptidases	Bile emulsifies fat-soluble vitamins and aids in their absorption with other fats. Water-soluble vitamins are absorbed.	The small intestine, pancreas, and liver add enough fluid so that approximately 2 gallons are secreted into the intestine in a day. Many minerals are absorbed. Vitamin D aids in the absorption of calcium.
Colon Some fat and cholesterol, trapped in fiber, exit in feces.		Bacteria produce vitamin K, which is absorbed.	More minerals and most of the water are absorbed.

just enough saliva to moisten each mouthful of food so that it can pass easily down your esophagus.

■■ **GASTRIC MOTILITY** Once you have swallowed, materials are moved through the rest of the GI tract by involuntary muscular contractions. This motion, known as **gastric motility,** consists of two types of movement, peristalsis and segmentation (see Figure 5-2 on p. 100). Peristalsis propels, or pushes; segmentation mixes, with more gradual pushing.

gland: a cell or group of cells that secretes materials for special uses in the body. Glands may be *exocrine glands,* secreting their materials "out" (into the digestive tract or onto the surface of the skin), or *endocrine glands,* secreting their materials "in" (into the blood).

exo = outside   endo = inside
krine = to separate

Figure 5-2
PERISTALSIS AND SEGMENTATION

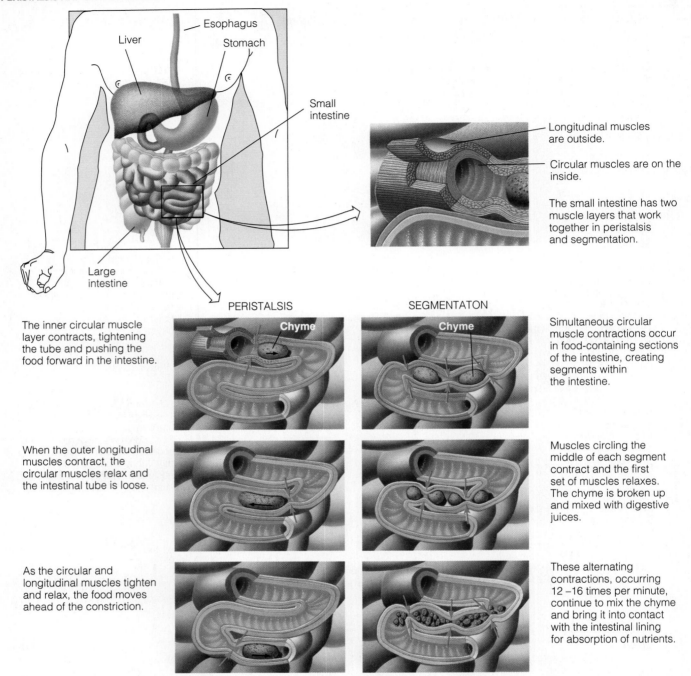

Liver

Esophagus

Stomach

Small intestine

Longitudinal muscles are outside.

Circular muscles are on the inside.

The small intestine has two muscle layers that work together in peristalsis and segmentation.

Large intestine

PERISTALSIS

SEGMENTATON

Chyme

Chyme

The inner circular muscle layer contracts, tightening the tube and pushing the food forward in the intestine.

Simultaneous circular muscle contractions occur in food-containing sections of the intestine, creating segments within the intestine.

When the outer longitudinal muscles contract, the circular muscles relax and the intestinal tube is loose.

Muscles circling the middle of each segment contract and the first set of muscles relaxes. The chyme is broken up and mixed with digestive juices.

As the circular and longitudinal muscles tighten and relax, the food moves ahead of the constriction.

These alternating contractions, occurring 12–16 times per minute, continue to mix the chyme and bring it into contact with the intestinal lining for absorption of nutrients.

gastric motility: spontaneous motion in the digestive tract accomplished by involuntary muscular contractions.

peristalsis (peri-STALL-sis): successive waves of involuntary muscular contractions passing along the walls of the GI tract that push the contents along.

peri = around

stellein = wrap

▪▪ **PERISTALSIS** **Peristalsis** begins when the bolus enters the esophagus. The entire GI tract is ringed with circular muscles that can squeeze it tightly. Surrounding these rings of muscle are longitudinal muscles. When the rings tighten and the long muscles relax, the tube is constricted. When the rings relax and the long muscles tighten, the tube bulges. These actions follow each other continuously and push the intestinal contents along. If you have ever watched a bolus of food pass along the body of a snake, you have a good picture of how these muscles work. The waves of contraction ripple through the GI tract at varying rates and intensities depending on the part of the GI tract and on whether food is present. Peristalsis, aided by the sphincter muscles that surround the GI tract at key places, keeps things moving along.

■■ SEGMENTATION   The intestines not only push but also periodically squeeze their contents as if a string tied around the intestines were being pulled tight. This motion, called **segmentation,** forces the contents back a few inches, mixing them and promoting close contact with the digestive juices and the absorbing cells of the intestinal walls before letting them slowly move along again.

segmentation: a periodic squeezing or partitioning of the intestine by its circular muscles that both mixes and slowly pushes the contents along.

■■ LIQUEFYING PROCESS   Besides forcing the intestinal contents along, the muscles of the GI tract help to liquefy them to chyme so that the digestive enzymes will have access to all their nutrients. The mouth initiates this liquefying process by chewing, adding saliva, and stirring with the tongue to reduce the food to a coarse mash suitable for swallowing. The stomach then further mixes and kneads the food.

■■ STOMACH ACTION   The stomach has the thickest walls and strongest muscles of all the GI tract organs. In addition to the circular and longitudinal muscles, the stomach has a third layer of diagonal muscles that also alternately contract and relax (see Figure 5-3). These three sets of muscles work to force the chyme downward, but the pyloric sphincter usually remains tightly closed so that the stomach's contents are thoroughly mixed and squeezed before being released. Meanwhile, the gastric glands are adding juices. When the chyme is thoroughly liquefied, the pyloric sphincter opens briefly, about three

Figure 5-3
STOMACH MUSCLES

The stomach has three layers of muscles.

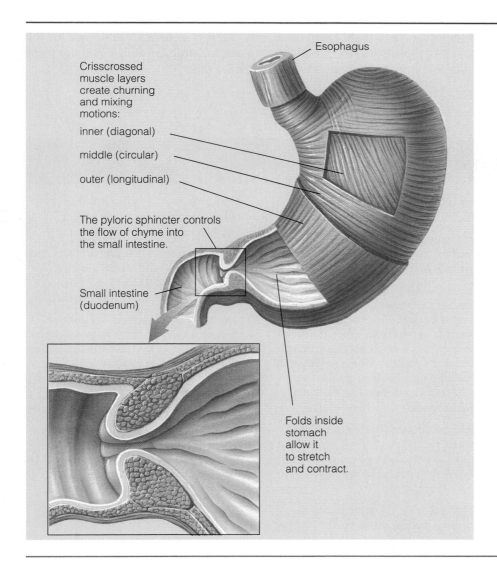

Esophagus

Crisscrossed muscle layers create churning and mixing motions:

inner (diagonal)

middle (circular)

outer (longitudinal)

The pyloric sphincter controls the flow of chyme into the small intestine.

Small intestine (duodenum)

Folds inside stomach allow it to stretch and contract.

times a minute, to allow small portions through. At this point, the intestinal contents no longer resemble food in the least.

▪▪ The Process of Digestion

One person eats nothing but vegetables, fruits, and nuts; another, nothing but meat, milk, and potatoes. How is it that both people wind up with essentially the same body composition? It all comes down to the body rendering food—whatever it is to start with—into the basic units that make up carbohydrate, fat, and protein. The body absorbs these units and builds its tissues from them.

To digest food, five different body organs secrete digestive juices: the salivary glands, the stomach, the small intestine, the liver (via the gallbladder), and the pancreas. Each of the juices has a turn to mix with the food and promote its breakdown to small units that can be absorbed into the body. The glossary (below) defines some of the digestive glands and their juices.

▪▪ Digestion in the Mouth

Digestion of carbohydrate begins in the mouth, where the salivary glands secrete saliva, which contains water, salts, and enzymes (including salivary amylase) that break the bonds in the chains of starch. Saliva also protects the tooth surfaces and linings of the mouth, esophagus, and stomach from attack by molecules that might harm them. The enzymes in the mouth do not, for the most part, affect the fats, proteins, vitamins, minerals, and fiber that are present in the foods people eat (review Figure 5-1).

GLOSSARY
DIGESTIVE GLANDS AND THEIR SECRETIONS

These terms are listed in order from the beginning of the digestive tract to the end.

salivary glands: exocrine glands that secrete saliva into the mouth.

saliva: the secretion of the salivary glands. The principal enzyme is salivary amylase.

amylase (AM-uh-lace): an enzyme that splits amylose (a form of starch). Amylase is a carbohydrase. The ending -*ase* indicates an enzyme; the root tells what it digests. Other examples: protease, lipase.

gastric glands: exocrine glands in the stomach wall that secrete gastric juice into the stomach.
gastro = stomach

gastric juice: the digestive secretion of the gastric glands containing a mixture of water, hydrochloric acid, and enzymes. The principal enzymes are

pepsin (acts on proteins) and lipase (acts on emulsified fats).

hydrochloric acid (HCl): an acid composed of hydrogen and chloride atoms; normally produced by the gastric glands.

mucus (MYOO-cuss): a mucopolysaccharide (a relative of carbohydrate) secreted by cells of the stomach wall that protects the cells from exposure to digestive juices (and other destructive agents). The cellular lining of the stomach wall with its coat of mucus is known as the mucous membrane. (The noun is mucus; the adjective is mucous.)

pepsin: a protein-digesting enzyme (gastric protease) in the stomach. It circulates as a precursor, pepsinogen, and is converted to pepsin by the action of stomach acid.

intestinal juice: the secretion of the intestinal glands; contains enzymes for

the digestion of carbohydrate and protein and a minor enzyme for fat digestion.

bile: an emulsifier that prepares fats and oils for digestion; made by the liver, stored in the gallbladder, and released into the small intestine when needed.

pancreatic (pank-ree-AT-ic) juice: the exocrine secretion of the pancreas, containing enzymes for the digestion of carbohydrate, fat, and protein. Juice flows from the pancreas into the small intestine through the pancreatic duct. The pancreas also has an endocrine function, the secretion of insulin and other hormones.

bicarbonate: an alkaline secretion of the pancreas; part of the pancreatic juice. (Bicarbonate also occurs widely in all cell fluids.)

Digestion in the Stomach

Gastric juice is composed of water, enzymes, and hydrochloric acid. The acid is so strong that it burns the throat if it happens to reflux into the upper esophagus and mouth. The strong acidity of the stomach prevents bacterial growth and kills most bacteria that enter the body with food. You might expect that the stomach's acid would attack the stomach itself, but the cells of the stomach wall secrete mucus, a thick, slimy, white polysaccharide that coats and protects the stomach's lining.

DIGESTIVE ACTIVITIES The major digestive event in the stomach is the initial breakdown of proteins. Other than being crushed and mixed with saliva in the mouth, nothing happens to protein until it comes in contact with the gastric juices in the stomach. There, the acid helps to uncoil (denature) the protein's tangled strands so that the stomach enzymes can attack the bonds. Both the enzyme pepsin and the stomach acid itself act as catalysts in the process. Minor events are the digestion of some fat by a gastric lipase, the digestion of sucrose (to a very small extent) by the stomach acid, and the attachment of a protein carrier to vitamin B_{12}.

The stomach enzymes work most efficiently in the stomach's strong acid, but salivary amylase, which is swallowed with food, does not work in acid this strong. Consequently, the digestion of starch gradually ceases as the acid penetrates the bolus. In fact, salivary amylase becomes just another protein to be digested. The amino acids in amylase end up being absorbed and recycled into other body proteins.

ANTACIDS: THEIR USE AND MISUSE Note that the strong acidity of the stomach is a desirable condition, television commercials for antacids notwithstanding. In a person who eats too quickly or eats too much, the stomach is likely to react with such violence as to cause regurgitation. When this happens, the stomach acid creates a bad taste in the mouth, which the person may interpret as "acid indigestion." Responding to television commercials, an overeater may take antacids to neutralize the stomach acid. But then the stomach will have to secrete more acid to enable the digestive enzymes to do their work. The consumer's stomach ends up with the same amount of acid but has had to work against the antacid to produce it.

Antacids are not designed to relieve the digestive discomfort of the hasty and abusive eater. Their proper use is to correct an abnormal condition, such as that of the person with ulcers, whose stomach or duodenal lining has been attacked by acid. To avoid falling into the same trap as the misguided consumer, remember to eat slowly, chew thoroughly, and perhaps consume less at a sitting.

Digestion in the Small and Large Intestines

By the time food leaves the stomach, digestion of all three energy-yielding nutrients has begun, and the process gains momentum in the small intestine. There, the pancreas, the liver, and the small intestine contribute additional digestive juices through the duct leading into the duodenum. These juices contain digestive enzymes, bicarbonate, and bile.

DIGESTIVE ENZYMES Pancreatic juice contributes enzymes that digest fats, proteins, and carbohydrates. Glands in the intestinal wall also secrete digestive enzymes. (Review the glossary of digestive glands and their secretions on p. 102 for details.)

BICARBONATE The pancreatic juice also contains sodium bicarbonate, which neutralizes the acidic chyme as it enters the small intestine.

Reminder: An emulsifier is a substance that mixes with both fat and water and permanently disperses the fat in the water, forming an emulsion.

Mayonnaise, made from vinegar and oil, would separate as other vinegar-and-oil salad dressings do if food chemists did not blend the vinegar and oil with a third ingredient—an emulsifier. The emulsifier mixes well with the fatty oil and the watery vinegar. In the case of mayonnaise, the emulsifier is lecithin from egg yolks.

intestinal flora: the bacterial inhabitants of the GI tract.

flora = plant growth

GI tract immunity is the topic of Nutrition in Practice 19.

From this point on, the contents of the digestive tract are neutral or slightly alkaline. The enzymes of both the intestine and the pancreas work best in this environment.

BILE Bile is secreted by the liver continuously and is concentrated and stored in the gallbladder. The gallbladder squirts bile into the duodenum whenever fat arrives there. Bile is not an enzyme but an emulsifier that brings fats into suspension in water (see Figure 5-4). After the fats are emulsified, enzymes can work on them, and they can be absorbed. Thanks to all these secretions, all three energy-yielding nutrients are digested in the small intestine.

THE RATE OF DIGESTION The rate of digestion of the energy nutrients depends on the contents of the meal. If the meal is high in simple sugars, digestion proceeds fairly rapidly. If the meal is rich in fat, digestion is slower.

PROTECTIVE FACTORS The intestine contains bacteria that produce a variety of vitamins, including biotin and vitamin K (although bacteria alone cannot meet the need for these vitamins). The GI bacteria also protect people from infections. Provided that the normal **intestinal flora** are thriving, infectious bacteria have a difficult time getting established and launching an attack on the system. In addition, the small intestine and the entire GI tract manufacture and maintain a strong arsenal of defenses against foreign invaders. Several different types of defending cells are present there and confer specific immunity against intestinal diseases.

THE FINAL STAGE The story of how food is broken down into nutrients that can be absorbed is now nearly complete. The three energy-yielding nutrients—carbohydrate, fat, and protein—are disassembled to basic building blocks before they are absorbed. Most of the other nutrients—vitamins, minerals, and water—are absorbed as they are. Undigested residues, such as some fibers, are not absorbed but continue through the digestive tract, providing a semisolid mass that helps stimulate the muscles of the GI tract so that they will remain strong and perform peristalsis efficiently. Fiber also retains water, keeping the stools soft, and carries bile acids, sterols, and fat with it out of the body. Drinking plenty of water in conjunction with eating foods

Figure 5-4
EMULSIFICATION OF FAT BY BILE

Like bile, detergents are emulsifiers and work the same way, which is why they are effective at removing grease spots from clothes. Molecule by molecule, the grease is dissolved out of the spot and suspended in the water, where it can be rinsed away.

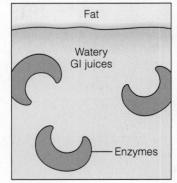 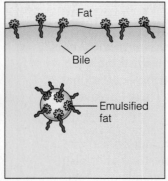 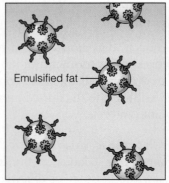 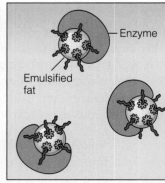

In the stomach, the fat and watery GI juices tend to separate. The enzymes are in the water and can't get at the fat.

When fat enters the small intestine, the gallbladder secretes bile. Bile has an affinity for both fat and water, so it can bring the fat into the water.

Bile's emulsifying action converts large fat globules into small droplets that repel each other.

After emulsification, the enzymes have easy access to the fat droplets.

high in fiber supplies fluid for the fiber to take up. This is the basis for the recommendation to drink water and eat fiber-rich foods to relieve constipation.

The process by which the nutrients are absorbed into the body is discussed in the next section. For the moment, let us assume that the digested nutrients simply disappear from the GI tract as they are ready. Virtually all nutrients are gone by the time the contents of the GI tract reach the end of the small intestine. Little remains but water, a few undissolved salts and body secretions, and undigested materials such as fiber. These enter the large intestine (colon).

In the colon, intestinal bacteria degrade some of the fiber to simpler compounds. The colon itself retrieves from its contents the materials that the body is designed to recycle—water and dissolved salts. The waste that is finally excreted has little or nothing of value left in it. The body has extracted all that it can use from the food.

LEARNING LINK

Chapters 14 and 15 describe consistency-modified diets such as liquid diets, soft diets, and mechanical soft diets. For example, clients who are recently out of surgery are offered only liquids until it is determined that they can tolerate solid foods. Clients who have difficulty chewing and swallowing are offered foods that need little chewing and are easy to swallow. Clients with mouth ulcers may need soft foods that are not salty, spicy, or acidic to minimize irritation. Nurses and other health care professionals monitoring such clients need to be alert to changes in a client's condition that may permit progression to a regular diet with greater variety and enhanced appeal.

The Absorptive System

Within three or four hours after you have eaten a meal, your body must find a way to absorb some two hundred thousand, million amino acid molecules one by one and a comparable number of monosaccharide, monoglyceride, glycerol, fatty acid, vitamin, and mineral molecules as well. The absorptive system is ingeniously designed to accomplish this task.

The Small Intestine

The small intestine is a tube about 10 feet long and about an inch across, yet it provides a surface comparable in area to a tennis court. When nutrient molecules make contact with this surface, they are absorbed and carried off to the liver and other parts of the body.

VILLI AND MICROVILLI How does the intestine manage to provide such a large absorptive surface area? Its inner surface looks smooth, but viewed through a microscope, it turns out to be wrinkled into thousands of folds. Each fold, in turn, is covered with thousands of fingerlike projections called **villi.** The villi are as numerous as the hairs on velvet fabric. A single villus, magnified still more, turns out to be composed of several hundred cells, each covered with microscopic hairs called **microvilli** (see Figure 5-5 on p. 106).

The villi are in constant motion. A thin sheet of muscle lines each villus, enabling it to wave, squirm, and wiggle like the tentacles of a sea anemone.

villi (VILL-ee or VILL-eye): fingerlike projections from the folds of the small intestine. The singular form is **villus.**
   villus = shaggy hair

microvilli (MY-cro-VILL-ee or MY-cro-VILL-eye): tiny, hairlike projections on each cell of every villus that can trap nutrient particles and transport them into the cells. The singular form is **microvillus.**

Figure 5-5
THE SMALL INTESTINAL VILLI

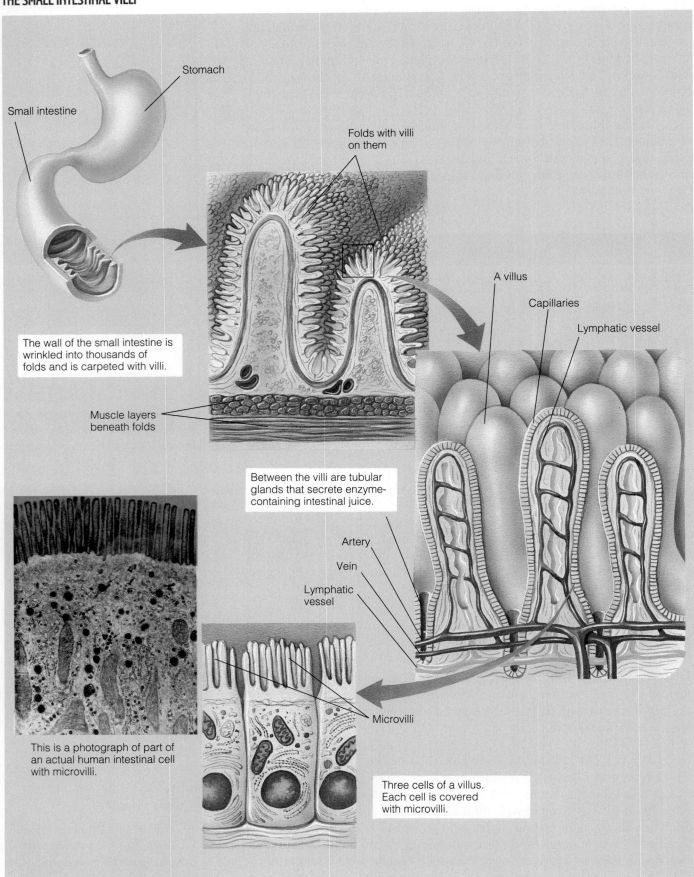

Stomach

Small intestine

Folds with villi on them

The wall of the small intestine is wrinkled into thousands of folds and is carpeted with villi.

Muscle layers beneath folds

A villus

Capillaries

Lymphatic vessel

Between the villi are tubular glands that secrete enzyme-containing intestinal juice.

Artery

Vein

Lymphatic vessel

Microvilli

This is a photograph of part of an actual human intestinal cell with microvilli.

Three cells of a villus. Each cell is covered with microvilli.

Any nutrient molecule small enough to be absorbed is trapped among the microvilli and drawn into a cell beneath them. Some partially digested nutrients are caught in the microvilli, digested further by enzymes there, and then absorbed into the cells.

■■ SPECIALIZATION IN THE INTESTINAL TRACT As you can see, the intestinal tract is beautifully designed to perform its functions. A further refinement of the system is that the cells of successive portions of the tract are specialized to absorb different nutrients. The nutrients that are ready for absorption early are absorbed near the top of the tract; those that take longer to be digested are absorbed further down. The rate at which the nutrients travel through the GI tract is finely adjusted to maximize their availability to the appropriate absorptive segment of the tract when they are ready. The lowly "gut" turns out to be one of the most elegantly designed organ systems in the body.

■■ Release of Absorbed Nutrients

Once a molecule has entered a cell in a villus, the next step is to transmit it to a destination elsewhere in the body by way of the body's two transport systems—the bloodstream and the **lymphatic system.** As Figure 5-5 shows, both systems supply vessels to each villus. Through these vessels, the nutrients leave the cell and enter either the **lymph** or the blood. In either case, the nutrients end up in the blood, at least for a while. The water-soluble nutrients (including the smaller products of fat digestion) are released directly into the bloodstream by way of the capillaries, but the larger fats and the fat-soluble vitamins find access directly into the capillaries impossible because these nutrients are insoluble in water (and blood is mostly water). They require some packaging before they are released.

    The intestinal cells assemble the monoglycerides and long-chain fatty acids into larger triglyceride molecules. These triglycerides, fat-soluble vitamins (when present), and other large lipids (cholesterol and the phospholipids) are then packaged for transport. They cluster together with special proteins to form **chylomicrons,** one kind of **lipoprotein** (lipoproteins are described beginning on p. 109). Finally, the cells release the chylomicrons into the lymphatic system. They can then glide through the lymph spaces until they arrive at a point of entry into the bloodstream near the heart.

lymphatic system: a loosely organized system of vessels and ducts that conveys the products of digestion toward the heart.

lymph (LIMF): the body fluid found in lymphatic vessels. Lymph consists of all the constituents of blood except red blood cells.

chylomicrons (kye-lo-MY-crons): the lipoproteins that transport lipids from the intestinal cells into the body. The cells of the body remove the lipids they need from the chylomicrons, leaving chylomicron remnants to be picked up by the liver cells.

lipoproteins: clusters of lipids associated with proteins that serve as transport vehicles for lipids in the lymph and blood.

LEARNING LINK

Chapter 16 describes irritable bowel syndrome, constipation, and diarrhea—conditions characterized by altered intestinal motility. Because dietary fiber affects GI transit time, fiber-modified diets are frequently used to treat motility disorders. As the previous section described, the rate at which nutrients pass through the digestive tract affects their absorption, so correcting motility problems improves a client's nutrition status.

■■ Transport of Nutrients

Once a nutrient has entered the bloodstream or the lymphatic system, it may be transported to any part of the body and thus become available to any of the cells, from the tips of the toes to the roots of the hair. The circulatory systems are arranged to deliver nutrients wherever they are needed.

The Vascular System

The vascular, or blood circulatory, system is a closed system of vessels through which blood flows continuously in a figure eight, with the heart serving as a pump at the crossover point. On each loop of the figure eight, blood travels a simple route: heart to **arteries** to **capillaries** to **veins** to heart.

The routing of the blood through the digestive system is different, however. The blood is carried to the digestive system (as it is to all organs) by way of an artery, which (as in all organs) branches into capillaries to reach every cell. Blood leaving the digestive system, however, goes by way of a vein, not back to the heart, but to the liver. This vein again branches into capillaries so that every cell of the liver has access to the newly absorbed nutrients that the blood is carrying. Blood leaving the liver then returns to the heart by way of another vein. The route is heart to arteries to capillaries (in intestines) to vein to capillaries (in liver) to vein to heart.

An anatomist studying this system knows there must be a reason for this special arrangement. The liver is located in the circulatory system at the point where it will have the first chance at the materials absorbed from the GI tract. In fact, the liver is the body's major metabolic organ (see Figure 5-6) and must prepare the absorbed nutrients for use by the rest of the body. Furthermore, the liver

artery: a vessel that carries blood away from the heart.

capillary (CAP-ill-ary): a small vessel that branches from an artery. Capillaries connect arteries to veins. Oxygen, nutrients, and waste materials are exchanged across capillary walls.

vein: a vessel that carries blood back to the heart.

The blood arriving at the intestines flows through the mesentery (MEZ-en-terry), a strong, flexible membrane that surrounds and supports the abdominal organs.
mes = middle

Figure 5-6
THE LIVER AND ITS CIRCULATORY SYSTEM

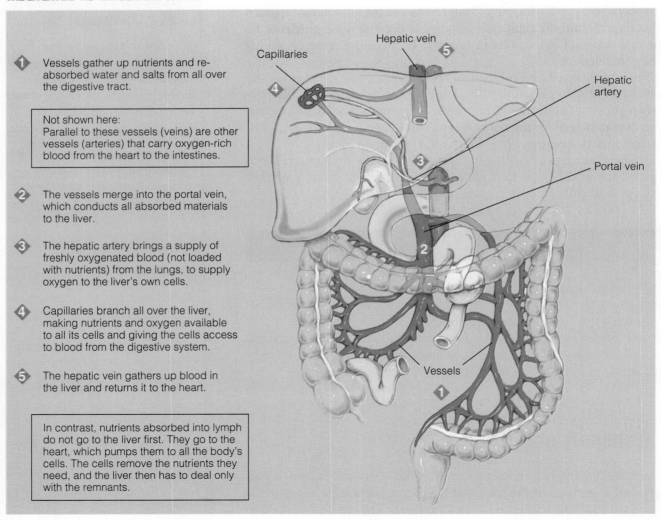

1. Vessels gather up nutrients and re-absorbed water and salts from all over the digestive tract.

   Not shown here:
   Parallel to these vessels (veins) are other vessels (arteries) that carry oxygen-rich blood from the heart to the intestines.

2. The vessels merge into the portal vein, which conducts all absorbed materials to the liver.

3. The hepatic artery brings a supply of freshly oxygenated blood (not loaded with nutrients) from the lungs, to supply oxygen to the liver's own cells.

4. Capillaries branch all over the liver, making nutrients and oxygen available to all its cells and giving the cells access to blood from the digestive system.

5. The hepatic vein gathers up blood in the liver and returns it to the heart.

   In contrast, nutrients absorbed into lymph do not go to the liver first. They go to the heart, which pumps them to all the body's cells. The cells remove the nutrients they need, and the liver then has to deal only with the remnants.

Capillaries

Hepatic vein

Hepatic artery

Portal vein

Vessels

stands as gatekeeper to waylay intruders that might otherwise harm the heart or brain. Chapter 6 offers more information about this noble organ.

LEARNING LINK

As the previous section explained, the liver is the key metabolic organ in the body, performing many diverse functions crucial to overall health and nutrition status. Liver disorders such as hepatitis and cirrhosis can have devastating effects on a client's ability to eat food and metabolize nutrients. Chapter 23 describes liver disorders and emphasizes the importance of nutrition intervention for such disorders.

The Lymphatic System

The lymphatic system is a one-way route for fluids to travel from tissue spaces into the blood. The lymphatic system has no pump; instead, lymph is squeezed from one portion of the body to another like water in a sponge, as muscles contract and create pressure here and there. Ultimately, the lymph collects in a large duct behind the heart. This duct connects to a vein that conducts the lymph into the heart. Thus, some materials from the GI tract enter the lymphatic system before entering the bloodstream.

Transport of Lipids: Lipoproteins

Within the circulatory system, lipids always travel from place to place bundled with protein, that is, as lipoproteins. When physicians measure a person's blood lipid profile, they are interested not only in the types of fat present (such as triglycerides and cholesterol) but also in the types of lipoproteins that carry them.

VLDL, LDL, AND HDL

As mentioned earlier, chylomicrons transport newly absorbed *(diet-derived)* lipids from the intestinal cells to the rest of the body. As chylomicrons circulate through the body, cells remove their lipid contents, so the chylomicrons get smaller and smaller. The liver picks up the chylomicron remnants and assembles new lipoproteins, which are known as **very-low-density lipoproteins (VLDL).** As the body's cells remove triglycerides from the VLDL, the proportion of their contents shifts. As this occurs, VLDL become cholesterol-rich **low-density lipoproteins (LDL).** Lipids returning to the liver for metabolism or excretion from other parts of the body are packaged in lipoproteins known as **high-density lipoproteins (HDL).**

The more lipid in a lipoprotein molecule, the lower the density; the more protein, the higher the density. Both LDL and HDL carry lipids around in the blood, but LDL are larger, lighter, and more lipid filled; HDL are smaller, denser, and packaged with more protein. LDL deliver cholesterol and triglycerides from the liver to the tissues; HDL scavenge excess cholesterol and phospholipids from the tissues and return them to the liver for metabolism or disposal. Figure 5-7 (p. 110) shows the relative sizes and composition of the lipoproteins.

HEALTH IMPLICATIONS OF LDL AND HDL

The distinction between LDL and HDL has implications for the health of the heart and blood vessels. Elevated LDL concentrations in the blood are associated with a high risk of heart disease, and elevated HDL concentrations are associated with a low risk.[3] These associations explain why some people refer to LDL as "bad" cholesterol and HDL as "good" cholesterol. Keep in mind, though, that there is

---

The vein that collects blood from the mesentery and conducts it to capillaries in the liver is the portal vein.
portal = gateway

The vein that collects blood from the liver capillaries and returns it to the heart is the hepatic vein.
hepat = liver

The artery that delivers oxygen-rich blood from the heart to the liver is the hepatic artery.

The duct that conveys lymph toward the heart is the thoracic (thor-ASS-ic) duct. The subclavian vein connects this duct with the right upper chamber of the heart, providing a passageway by which lymph can be returned to the vascular system.

VLDL (very-low-density lipoprotein): the type of lipoprotein made primarily by liver cells to transport lipids to various tissues in the body; composed primarily of triglycerides.

LDL (low-density lipoprotein): the type of lipoprotein derived from VLDL as cells remove triglycerides from them. LDL carry cholesterol and triglycerides from the liver to the cells of the body and are composed primarily of cholesterol.

HDL (high-density lipoprotein): the type of lipoprotein that transports cholesterol back to the liver from peripheral cells; composed primarily of protein.

Lipoproteins and heart disease are discussed in Chapter 21.

Figure 5-7
THE LIPOPROTEINS

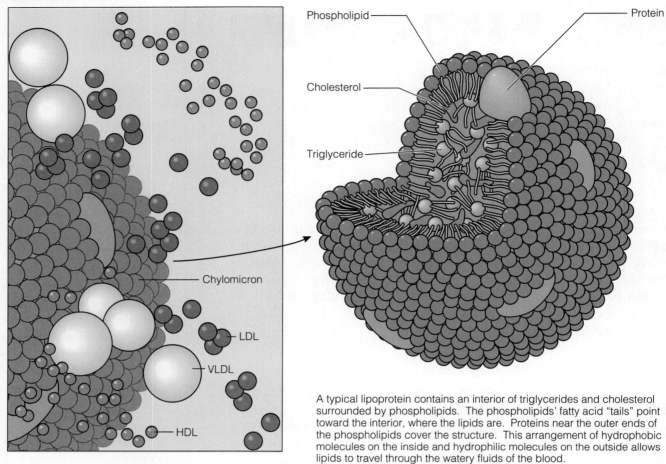

A typical lipoprotein contains an interior of triglycerides and cholesterol surrounded by phospholipids. The phospholipids' fatty acid "tails" point toward the interior, where the lipids are. Proteins near the outer ends of the phospholipids cover the structure. This arrangement of hydrophobic molecules on the inside and hydrophilic molecules on the outside allows lipids to travel through the watery fluids of the blood.

This solar system of lipoproteins shows their relative sizes. Notice how large the fat-filled chylomicron is compared with the others and how the others get progressively smaller as their proportion of fat declines and protein increases.

Chylomicrons contain so little protein and so much triglyceride that they are the lowest in density.

Very-low-density lipoproteins (VLDL) are half triglycerides, accounting for their low density.

Low-density lipoproteins (LDL) are half cholesterol, accounting for their implication in heart disease.

High-density lipoproteins (HDL) are half protein, accounting for their high density.

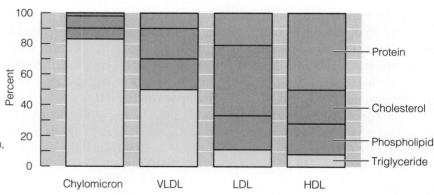

only *one* kind of cholesterol; the differences between LDL and HDL reflect *proportions* of lipids and proteins within them—not the type of cholesterol. Factors that improve the LDL-to-HDL ratio include:

- Weight control (see Chapter 7).
- Polyunsaturated or monounsaturated, instead of saturated, fatty acids in the diet (see Chapter 3).
- Soluble fibers (see Chapter 2).

- Antioxidants (see Chapter 8 and Nutrition in Practice 21).
- Physical activity.

:: The System at Its Best

The GI tract is the first organ in the body to deal with the nutrients that will ultimately maintain the health and nutrition status of the whole body. The intricate architecture of the GI tract makes it sensitive and responsive to conditions in its environment. One condition indispensable to its performance is its own good health. Such lifestyle factors as sleep, physical activity, state of mind, and nutrition affect GI tract health. Adequate sleep allows for repair and maintenance of tissue. Physical activity promotes healthy muscle tone and may protect against cancer of the colon.[4] Mental state profoundly affects digestion and absorption through the activity of nerves and hormones that help regulate these processes. A relaxed, peaceful attitude during a meal enhances digestion and absorption.

SELF CHECK

1. Once food is swallowed, it travels through the digestive tract in this order:
   a. esophagus, stomach, colon, liver.
   b. esophagus, stomach, small intestine, colon.
   c. small intestine, stomach, esophagus, colon.
   d. small intestine, large intestine, stomach, colon.

2. Once chyme travels the length of the small intestine, it passes through the ileocecal valve at the beginning of the:
   a. colon.
   b. stomach.
   c. esophagus.
   d. jejunum.

3. The periodic squeezing or partitioning of the intestine by its circular muscles that both mixes and slowly pushes the contents along is known as:
   a. secretion.
   b. absorption.
   c. peristalsis.
   d. segmentation.

4. An enzyme in saliva begins the digestion of:
   a. starch.
   b. vitamins.

   c. protein.
   d. minerals.

5. Bile is:
   a. an enzyme that splits starch.
   b. an alkaline secretion of the pancreas.
   c. an emulsifier made by the liver that prepares fats and oils for digestion.
   d. a stomach secretion containing water, hydrochloric acid, and the enzymes pepsin and lipase.

6. Which nutrient passes through the large intestine mostly unabsorbed?
   a. fiber
   b. vitamins
   c. minerals
   d. starch

7. The two major nutrient transport systems in the body are:
   a. LDL and HDL.
   b. digestion and absorption.
   c. lipoproteins and chylomicrons.
   d. vascular and lymphatic systems.

8. Within the circulatory system, lipids always travel from place to place bundled with proteins as:
   a. microvilli.
   b. chylomicrons.

   c. lipoproteins.
   d. phospholipids.

9. Elevated LDL concentrations in the blood are associated with:
   a. a high-protein diet.
   b. a low risk of diabetes.
   c. too much physical activity.
   d. a high risk of heart disease.

10. Three factors that improve the LDL-to-HDL ratio include:
    a. polyunsaturated fat, rest, and dietary HDL.
    b. antioxidants, insoluble fibers, and dietary HDL.
    c. saturated fat, antioxidants, and insoluble fibers.
    d. weight control, soluble fibers, and physical activity.

Answers to these questions appear in Appendix H.

CLINICAL APPLICATIONS

1. What suggestions might you have for a client who eats antacids before and after every meal in the belief that the they will prevent or relieve heartburn or acid indigestion?

2. People who experience malabsorption frequently have the most problem digesting fat. Considering the differences in fat, carbohydrate, and protein digestion and absorption, can you offer an explanation?

3. How might you explain the importance of dietary fiber to a client who frequently experiences constipation?

NUTRITION**ON**THE NET

FOR FURTHER STUDY OF THE TOPICS IN THIS CHAPTER, ACCESS THESE WEB SITES.

www.acg.gi.org
American College of Gastroenterology

Notes

[1]W. F. Ganong, *Review of Medical Physiology* (Norwalk, Conn.: Appleton & Lang, 1993), pp. 438–465.

[2]The small intestine in living adults is almost two and a half times shorter than at death, when muscles are relaxed and elongated. Ganong, 1993.

[3]J. M. Dietschy, Theoretical considerations of what regulates low-density-lipoprotein and high-density-lipoprotein cholesterol, *American Journal of Clinical Nutrition* 65 (1997): S1581–S1589; NIH Consensus Conference, Triglyceride, high-density lipoprotein, and coronary heart disease, *Journal of the American Medical Association* 269 (1993): 505–510.

[4]The American Cancer Society 1996 Dietary Guidelines Advisory Committee, *Guidelines on Diet, Nutrition, and Cancer Prevention: Reducing the Risk of Cancer with Healthy Food Choices and Physical Activity,* a booklet.

Q&A NUTRITION IN PRACTICE

Food Safety

Every year **food poisoning** causes some 20 to 80 million cases of GI distress (nausea, vomiting, and diarrhea)—and possibly even more. Many cases of "the flu" may actually be episodes of food poisoning. Each year 9000 people die of food poisoning.[1]

www.fsis.usda.gov
Food Safety and Inspection Service

www.foodsafety.ufl.edu
National Food Safety Database

vm.cfsan.fda.gov/list.html
FDA Center for Food Safety & Applied Nutrition

What is food poisoning?

The term *food poisoning* refers to either food-borne infection or food intoxication. A food-borne infection

GLOSSARY OF FOOD POISONING TERMS

botulism: an often-fatal food poisoning caused by botulin toxin, a toxin produced by bacteria that grow without oxygen.

food poisoning: illness transmitted to human beings through food and water, caused by infectious agents or by toxins that they produce.

pathogens: disease-causing microorganisms.

toxins: poisons. Toxins produced by bacteria come in two varieties: *enterotoxins,* which act in the GI tract, and *neurotoxins,* which act on the nervous system.

traveler's diarrhea: nausea, vomiting, and diarrhea caused by consuming food or water contaminated by any of several organisms, most commonly, *E. coli, Shigella, Campylobacter jejuni,* and *Salmonella.*

is an illness caused by microorganisms, such as *Salmonella* bacteria, that infect people. Food intoxication is caused by **toxins** produced by microorganisms in food or within the digestive tract. In most food-borne illnesses, the symptoms are mild, but for people who are otherwise ill or malnourished, or are very old or young, these relatively mild disturbances can be fatal. If abdominal cramps, headache, vomiting, and diarrhea are the major or only symptoms of your next bout of "flu," chances are excellent that what you really have is food poisoning. The glossary below defines some terms related to food poisoning.

The symptoms of one toxin stand alone as severe and commonly fatal—those of **botulism,** caused by the toxin of a microbe that grows inside improperly canned, home-canned, or vacuum-packed foods or in homemade garlic- or herb-flavored oils stored at room temperature.[2] Botulism danger signs constitute a true medical emergency (see Table NP5-1). Even with medical assistance, survivors can suffer the effects for months, years, or a lifetime. So potent is the botulin toxin that an amount as tiny as a single grain of salt can kill several people within an hour. The botulin toxin is destroyed by heat, so canned foods that have been boiled for ten minutes are generally safe from this threat. Home-canned foods are safe if prepared by following proper canning techniques to the letter.* To prepare herb-flavored oils safely, wash and dry the herbs before adding to the oil. Keep the oil refrigerated, and remove it from the refrigerator only long enough to

*Complete, up-to-date, safe home-canning instructions are included in the USDA's 172-page *Complete Guide to Home Canning* available for purchase from the Superintendent of Documents, Government Printing Office, Washington, DC 20402.

pour out the amount you are going to use immediately. Throw out leftovers at the end of the day.[3] Commercially prepared oils are safer still because they have added acid and processing that cannot be duplicated at home.

How do people get food poisoning?

Transmission of food-borne illness is changing as the food supply changes. In the past, food-borne illness was caused by one person's error in a small setting, such as improperly refrigerated potato salad at a family picnic, and affected only a few. Today, people are eating more foods prepared and packaged by others. Consequently, when a food manufacturer or restaurant chef makes an error, food-borne illness can be epidemic.[4] An estimated 80 percent of reported food-borne illnesses are caused by errors in a commercial setting, such as the improper pasteurization of milk at a large dairy.[5]

In the mid-1990s, when a fast-food restaurant served undercooked burgers tainted with an infectious organism, hundreds of people became ill and at least three people died. This incident and others focused the national spotlight on two important safety issues: disease-causing organisms are commonly found in raw foods, and thorough cooking kills most of these food-borne **pathogens.** These episodes sparked a much-needed overhaul of national food safety programs.

Table NP5-1
WARNING SIGNS OF BOTULISM

- Double vision.
- Weakening muscles.
- Difficulty swallowing.
- Difficulty breathing.
- Slurred speech.

Are foods bought in grocery stores safe?

Canned and packaged foods sold in grocery stores are almost invariably safe, but rare accidents do happen. Batch numbering makes it possible to recall contaminated foods through public announcements via newspapers, television, and radio. In the grocery store, these guidelines can help consumers avoid buying foods that are contaminated:
• Avoid packages with defective seals and wrappers.
• Reject leaking or bulging cans.
• Check safety "buttons" on jars for intact seal.
• Avoid partially frozen foods; those in chest-type freezers should be stored below the frost line.
• Choose packages that have not been damaged, soiled, or punctured.
Improper handling of foods can occur anywhere along the line, from commercial manufacturers to large supermarkets to small restaurants to private homes. Maintaining a safe food supply requires everyone's efforts.

What can people do to protect themselves from food poisoning?

Whether bacteria multiply and cause illness depends, in part, on what happens in the kitchen—whether the kitchen is in a private home, a school cafeteria, a gourmet restaurant, or a canning plant. Foods can provide ideal conditions for bacteria to thrive and produce their toxins. Disease-causing bacteria require:
• Warmth (40° to 140°F).
• Moisture.
• Nutrients.
To prevent bacterial growth, people who prepare foods can do these things: cook foods thoroughly, keep hot foods hot, keep cold foods cold, keep dry foods dry, and keep hands, utensils, and the kitchen clean.

How hot is hot enough?

Cook foods long enough to reach an internal temperature that will kill microbes (see Figure NP5-1). To pre-vent bacterial growth when holding cooked food, keep it at 140°F or higher until it is served. Refrigerate leftover food immediately after serving.

How cold does cold need to be?

Keeping cold foods cold starts when you leave the grocery store. If you are running errands, shop last, so the groceries will not stay in the car too long. (If the ice cream has begun to melt, it has been too long.) Upon arrival home, load foods into the refrigerator or freezer immediately. When serving food cold, allow it to stay at cool room temperature (about 68°F) for no more than two hours. If the room is warm (about 80°F), refrigerate the food after just one hour. Table NP5-2 lists some safe keeping times for foods stored in the refrigerator at 40°F.

Figure NP5-1
FOOD SAFETY TEMPERATURES (FAHRENHEIT)

Bacteria multiply at temperatures between 40° and 140°F. Cook foods to the temperatures on this thermometer and hold them at 140°F or higher.

SOURCE: U.S. Department of Agriculture, 1993.

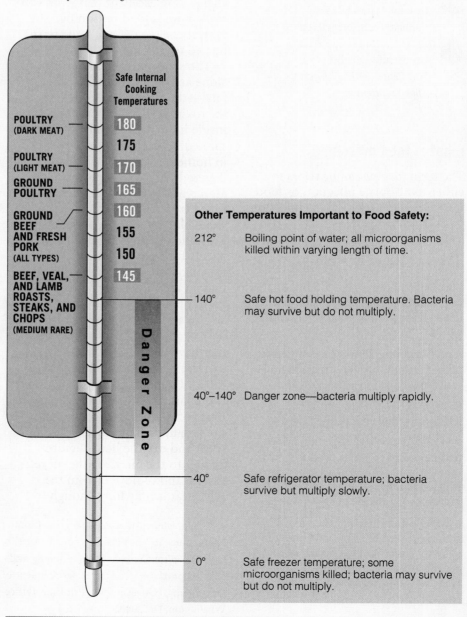

Safe Internal Cooking Temperatures

POULTRY (DARK MEAT) — 180
175
POULTRY (LIGHT MEAT) — 170
GROUND POULTRY — 165
GROUND BEEF AND FRESH PORK (ALL TYPES) — 160
155
150
BEEF, VEAL, AND LAMB ROASTS, STEAKS, AND CHOPS (MEDIUM RARE) — 145

Danger Zone

Other Temperatures Important to Food Safety:

212°   Boiling point of water; all microorganisms killed within varying length of time.

140°   Safe hot food holding temperature. Bacteria may survive but do not multiply.

40°–140° Danger zone—bacteria multiply rapidly.

40°   Safe refrigerator temperature; bacteria survive but multiply slowly.

0°   Safe freezer temperature; some microorganisms killed; bacteria may survive but do not multiply.

Table NP5-2
SAFE REFRIGERATOR STORAGE TIMES (40°F)

1 to 2 Days
Raw ground meats, breakfast or other raw sausages, raw fish or poultry; gravies

3 to 5 Days
Raw steaks, roasts, or chops; cooked meats, vegetables, and mixed dishes; ham slices; mayonnaise salads (chicken, egg, pasta, tuna)

1 Week
Hard-cooked eggs, bacon or hot dogs (opened packages); smoked sausages

2 to 4 Weeks
Raw eggs (in shells); bacon or hot dogs (packages unopened); dry sausages (pepperoni, hard salami); most aged and processed cheeses (swiss, brick)

2 Months
Mayonnaise (opened jar); most dry cheeses (parmesan, romano)

Keeping foods cold applies to defrosting foods before use, too. Bacterial growth begins on thawed portions of food even while the inner core is solidly frozen, so thaw meats or poultry in the refrigerator, not at room temperature. If you must hasten thawing, use cool running water or a microwave oven set to defrost.

Dry foods such as cereals, breads, and powdered mixes will keep for a long time if you protect them against moisture. Store them in cool, dry places, and keep them well sealed against humid air.

What's involved in keeping a kitchen clean enough to protect against food poisoning?

Keeping the kitchen clean includes using freshly washed utensils and laundered towels and washing your hands with hot, soapy water before and after each step of food preparation. If you are ill or have open sores, stay away from food so as not to contaminate it.

To eliminate microbes, you have three choices, each with benefits and drawbacks. One is to destroy the microbes where they reside by washing countertops, cutting boards, and sponges with diluted bleach. The benefit here is that chlorine can kill almost all organisms. The drawback is that chlorine that washes down household drains into the water supply forms chemicals that can harm waterways and fish.

A second option is to use heat. Soapy water heated to 140°F kills most harmful organisms and washes most others away. This method takes effort, though, for you have to use truly scalding water heated well beyond the temperature of the tap. (Most tap water called "hot" is not hotter than 130°F.) The third option is to use clean boards for cutting and washable dishcloths that can be laundered often for wiping. (Save sponges for car washing and other heavy cleaning chores, and keep them away from surfaces that come in contact with raw foods.) For a small initial investment, you can create a truly safe environment in which to prepare your food. These measures all help eliminate microbes and decrease the possibility of cross-contamination, the transfer of microbes from one surface to another.

Some foods are more hospitable to microbial growth than others.

Especially vulnerable are moist foods, nutrient-rich foods, and foods that are chopped or ground.

What precautions need to be taken when preparing meat and poultry?

Because raw meats and poultry require special handling, they now bear labels to instruct consumers on meat safety (see Figure NP5-2). Meats and poultry may contain bacteria, and they provide a moist, nutritious environment that is ideal for microbial growth. Do not leave raw meat or poultry at room temperature for more than a very short time. Chill it thoroughly or cook it promptly, preferably the latter. Remember that every surface and utensil that raw meat has touched is contaminated with microbes and must be washed with hot, soapy water before being used for cooked meat.*

Ground meat is handled more than other kinds of meat and has much more surface exposed to the air for bacteria to land on, so it poses special risks. For safety's sake, cook hamburger and meat loaf thoroughly, using a thermometer to test the internal temperature (review Figure NP5-1).

---

*The USDA's meat and poultry hotline answers questions about meat and poultry safety: 1-800-535-4555.

Figure NP5-2
SAFE HANDLING INSTRUCTIONS FOR MEAT AND POULTRY

Safe Handling Instructions

THIS PRODUCT WAS PREPARED FROM INSPECTED AND PASSED MEAT AND/OR POULTRY. SOME FOOD PRODUCTS MAY CONTAIN BACTERIA THAT CAN CAUSE ILLNESS IF THE PRODUCT IS MISHANDLED OR COOKED IMPROPERLY. FOR YOUR PROTECTION, FOLLOW THESE SAFE HANDLING INSTRUCTIONS.

KEEP REFRIGERATED OR FROZEN. THAW IN REFRIGERATOR OR MICROWAVE.

KEEP RAW MEAT AND POULTRY SEPARATE FROM OTHER FOODS. WASH WORKING SURFACES (INCLUDING CUTTING BOARDS), UTENSILS, AND HANDS AFTER TOUCHING RAW MEAT OR POULTRY.

COOK THOROUGHLY.

KEEP HOT FOODS HOT. REFRIGERATE LEFTOVERS IMMEDIATELY OR DISCARD.

Cook hamburgers to 160°F; color alone does not indicate when a burger is done.

Do not use or even taste a food with an "off" appearance or odor. However, don't trust your senses of smell and sight alone to tell you that foods are safe. Most contamination is not detectable by odor, taste, or appearance. Even hot cooked food, if handled improperly prior to cooking, can cause illness.

How can a person enjoy seafood safely?

For adults and children alike, eating raw or lightly steamed seafood is a risky proposition even when it is prepared as sushi by a master Japanese chef. Eating raw oysters can be dangerous for anyone, but people with liver disease and those with suppressed immune systems are most vulnerable.[6] The microorganisms that lurk in seafood are undetectable even by an expert.*

People who like Japanese sushi know that not all varieties are made from raw fish. Many types are made with cooked seafood and vegetables, avocado, or other delicacies and are

*To speak with an expert about seafood safety, call the FDA seafood hotline: 1-800-FDA-4010.

perfectly safe to enjoy. Also, rumor has it that freezing fish will make it safe to eat raw, but this is only partly true. Freezing fish will kill mature parasitic worms, but only cooking can kill all worm eggs and other microorganisms that can cause illness.

As population density increases along the shores of seafood-harvesting waters, pollution of those waters inevitably invades the seafood living there. To help ensure safe seafood products, the Food and Drug Administration (FDA) requires seafood marketers to adopt food safety practices. Still, unwholesome foods can reach the market. In one season alone, black-market dealers may sell millions of dollars worth of clams and oysters taken illegally from polluted harvesting areas. Experts are unanimous in saying that the risks of eating raw or lightly cooked seafood have become unacceptably high due to environmental contamination.[7] Chemical pollution and microbial contamination can originate in the water or in the boats and warehouses where seafood is cleaned, prepared, and refrigerated. To keep seafood as fresh as possible, people in the industry "keep it cold and keep it clean." Wise consumers eat it cooked.

At some times of the year, seafood may become contaminated with the so-called red tide toxin that occurs during algae blooms. Consumption of seafood contaminated with red tide causes a paralyzing form of food poisoning. The FDA monitors fishing waters for red tide algae and closes waters to fishing whenever it appears.

foodsafety.org/sea.htm
Handling Seafood Safely (from the **National Food Safety Database**)

How can foods be kept safe when away from home?

Picnics are fun and can be safe, too. Choose foods that last without refrigeration, such as fresh fruits and vegetables, breads and crackers,

and canned spreads and cheeses that you can open and use immediately. Aged cheeses, such as cheddar and swiss, do well for an hour or two, but for longer periods, carry them in an ice chest. Mayonnaise resists spoilage because of its acid content, but when mixed with chopped ingredients, such as pasta, meat, or vegetable salads, it spoils quickly. The chopped ingredients offer an extensive surface area for bacteria to invade, and the foods have been in contact with cutting boards, hands, and kitchen utensils that have transmitted bacteria to them. Chill chopped salads well before, during, and after the picnic. Keep mayonnaise itself cold.

What other foods pose special hazards?

Honey has been found to contain dormant bacterial spores that can awaken in the human body to produce the deadly botulin toxin mentioned earlier. Adults are big and strong enough to withstand the doses usually encountered, but infants under one year of age should never be fed honey. (It can also be contaminated with environmental pollutants picked up by the bees.) Honey has been implicated in several cases of sudden infant death.

What should a person do should food-borne illness occur?

Local health departments and the USDA extension service can provide further information about food safety. Should efforts fail and mild food-borne illness develop, drink clear liquids to replace fluids lost through vomiting and diarrhea. If serious food-borne illness is suspected, first call a physician. Then, wrap the remainder of the suspected food and label its container so that it cannot be mistakenly eaten, place it in the refrigerator, and hold it for possible inspection by health authorities.

usda.gov
U.S. Department of Agriculture

How can a person defend against food-borne illness when traveling to foreign countries?

Special food safety concerns arise when people travel. In many parts of the world, food-borne illness is likely to strike tourists even while the local people, eating exactly the same foods prepared the same way, remain healthy. That is because the locals have developed immunity to local disease-causing organisms, but tourists have no such protection. To avoid food-borne illness while traveling:

- Wash your hands often with soap and hot water, especially before handling food or eating.
- Eat only cooked food and canned foods. Eat raw fruits or vegetables only if you have washed them in boiled water and peeled them yourself. Skip salads, raw fish, and shellfish.
- Be aware that water, and ice made from it, may be unsafe, too. Take along disinfecting tablets or an element that boils water in a cup.
- Drink no beverages made with tap water. Drink only treated, boiled, canned, or bottled beverages, and drink them without ice (ice may be made from contaminated water), even if they are not chilled to your liking. Refuse dairy products unless they have been properly pasteurized and refrigerated.
- Do not use the local water, even to brush your teeth, unless you boil or disinfect it first.
- Before you leave on the trip, ask your physician to recommend medicines to take with you in case your efforts to avoid illness fail.

One journalist succinctly sums up these recommendations: "Boil it, cook it, peel it, or forget it."[8] Chances are excellent that if you follow these rules, you will remain well.

What about the safety of drinking water?

Foods are not alone in transmitting food-borne diseases; water is guilty, too. A glass of "water" is more than just H_2O. Infectious organisms found on fresh fruits and vegetables and in raw oysters are transmitted through contaminated water.[9]

Contamination can occur as water travels from the main water supply to homes. Lead or asbestos from old, corroded pipes can contaminate drinking water, as can bacteria and dirt from leaking pipes. People who suspect contamination of their water should have it tested where it flows out, at the tap. Public water systems treat water to remove contaminants that have been detected above acceptable levels. Private well water is usually not treated or cleansed, so people who consume water from private wells are responsible for its safety and should test the water periodically.

epa.gov
Environmental Protection Agency

Some people turn to bottled water as an alternative to tap water. Bottled water is classified as a food, so it is regulated by the FDA and must meet safety standards similar to those set for public water systems. Bottled water must also be processed and labeled according to FDA regulations. Some bottled waters may have minerals or carbonation added. "Carbonated," "seltzer," and "tonic" waters are not considered waters, however, but soft drinks. The FDA requires labels to disclose the sources of bottled waters and to use legally defined descriptive terms.[10]

bottledwater.org
International Bottled Water Association

Safe drinking water is a concern for everyone and must be protected to ensure continued health. To learn more about the water supply in your area, call the local public health agency.*

*For information on safe drinking water in general, call the Environmental Protection Agency's hotline: 1-800-426-4791.

Notes

[1] S. L. Nightingale, From the Food and Drug Administration: National Food Safety Initiative, *Journal of the American Medical Association* 277 (1997): 1664.

[2] C. J. Lackey, Oil, herb, and garlic flavored, http://www.foodsafety.org, site visited on February 5, 1998.

[3] Lackey, 1998.

[4] R. V. Tauxe and J. M. Hughes, International investigations of outbreaks of foodborne disease: Public health responds to the globalization of food, *British Medical Journal* 313 (1996): 1093–1094; T. W. Hennessey and coauthors, A national outbreak of Salmonella enteritidis infections from ice cream, *New England Journal of Medicine* 334 (1996): 1281–1286.

[5] Centers for Disease Control and Prevention, Surveillance for foodborne disease outbreaks, United States, 1988–1992, *Morbidity and Mortality Weekly Report CDC Surveillance Survey* 45 (1996): 1–66.

[6] Raw oyster risk for alcoholics, *FDA Consumer,* May 1996, pp. 23–25.

[7] Seafood safety: Highlights of the Executive Summary of the 1991 Report by the Committee on Evaluation of the Safety of Fishery Products of the Food and Nutrition Board, Institute of Medicine, National Academy of Sciences, *Nutrition Reviews* 49 (1991): 357–363.

[8] R. D. Williams, Boil it, cook it, peel it, or forget it, *FDA Consumer,* September 1991, p. 17.

[9] Outbreaks of Cyclosporiasis—United States, 1997, *Morbidity and Mortality Weekly Report* 46 (1997): 461–462; Vibrio vulnificus infections associated with eating raw oysters—Los Angeles, 1996, *Journal of the American Medical Association* 276 (1996): 937–938; J. W. Besser-Wiek and coauthors, Foodborne outbreak of diarrheal illness associated with Cryptosporidium parvum—Minnesota, 1995, *Morbidity and Mortality Weekly Report* 45 (1996): 783–784.

[10] New bottled water standards, *FDA Consumer,* April 1996, p. 2; V. Lambert, Bottled water: New trends, new rules, *FDA Consumer,* June 1993, pp. 8–11.

CHAPTER 6

METABOLISM AND ENERGY BALANCE

CONTENTS

The Organs and Their
Metabolic Roles

The Body's Energy
Metabolism

The Body's Energy Budget

Nutrition in Practice:
Inborn Errors of Metabolism

Every organ, every tissue, and every cell of the body engages in **metabolism,** the chemical reactions involved in releasing energy, breaking down compounds, making new compounds, and transporting compounds from place to place. Viewed from this perspective, the body is a giant factory that works with astounding efficiency to produce a myriad of products and dispose of a myriad of wastes. All these processes are regulated by hormonal signals that coordinate supply and demand in much the same way as a superb communication system coordinates a smoothly functioning economy.

In disease, metabolic processes always become disturbed, and some diseases are caused by metabolic abnormalities. This chapter provides the normal background against which the disruptions caused by diseases can best be understood.

metabolism: the sum total of all the chemical reactions that go on in living cells.

meta = among

bole = change

The Organs and Their Metabolic Roles

When the body is functioning normally, every organ plays a metabolic role that serves the others. Metabolic reactions also consume or release energy and therefore affect body weight, with consequences for health.

The Principal Organs

Of particular concern to metabolism are the digestive organs, the liver, the pancreas, the circulatory system, and the kidneys. Together, they perform much of the work of breaking down compounds, making new ones, transporting nutrients and oxygen throughout the body, and removing the waste products generated by metabolic processes.

THE DIGESTIVE ORGANS The digestive system has just received attention in Chapter 5. Notable among the digestive system's activities are the physical processes that allow foods to be transported throughout the GI tract; the production of digestive juices and enzymes; the absorption of nutrients; the making of transport proteins to carry lipids and vitamins to other sites in the body; and, in the lower digestive tract, the reabsorption of salts and fluids. The digestive system also possesses the body's most rapidly multiplying cells: when healthy, they replace themselves every few days. As later chapters show, many disorders directly and indirectly affect the GI tract and can lead to failures to ingest, digest, absorb, and metabolize nutrients.

THE LIVER Nutrients absorbed into the bloodstream are conducted to the liver, as described in Chapter 5. The liver is one of the body's most active metabolic factories. It receives nutrients and metabolizes, packages, stores, or ships them out for use by other organs. It metabolizes and stores most vitamins and many minerals. It manufactures bile, which the body uses in emulsifying fat for digestion and absorption. It metabolizes and detoxifies alcohol and drugs, prepares waste products for excretion, and participates in iron recycling and blood cell manufacture. It also makes many proteins necessary for health, including immune factors, transport proteins, and clotting factors. When disorders disrupt liver functions, they can profoundly affect both nutrition and health status, as described in Chapter 23.

THE PANCREAS The pancreas contributes digestive juices to the GI tract, but also has another metabolic function: it produces insulin and other hormones that regulate the body's use of glucose. It is insulin that prompts cells to take up glucose, store fatty acids, and synthesize proteins; insulin also prompts liver cells in particular to store glucose as glycogen. The liver cells can later release glucose back into the blood as needed. Glucose is an indispensable fuel for brain and nerve cells and for red blood cells as well. Its availability is therefore crucial to normal nervous system activity and normal blood chemistry. Abnormalities associated with the digestive functions of the pancreas are described in Chapter 16, and those associated with its hormonal functions are described in Chapter 20.

▪▪ THE HEART AND BLOOD VESSELS

The heart and blood vessels conduct blood with its cargo of nutrients and oxygen to all other body cells and carry wastes and carbon dioxide from them. Diseases of the heart and arteries therefore affect the health of the whole body. Metabolic reactions that affect the heart and blood vessels include, most importantly, the making and transport of lipoproteins, the carriers of cholesterol and other lipids from the liver to the tissues and back again. When lipoproteins deposit cholesterol in the artery walls, coronary heart disease and high blood pressure can result, increasing the risk of disability or death from heart attacks and strokes. Chapter 21 is devoted to these conditions.

▪▪ THE KIDNEYS

The kidneys are also active metabolic organs. Unceasingly, for 24 hours of every day, they filter waste products from the blood to be excreted in the urine and reabsorb needed nutrients, thereby maintaining the blood's delicate chemical balances. The kidneys' cells also produce compounds that help to regulate blood pressure and convert a precursor compound to active vitamin D, thereby helping to maintain the bones. Disorders of the kidneys nearly always involve the heart and the skeleton; kidney disorders are the subject of Chapter 22.

LEARNING LINK

For purposes of teaching and learning, descriptions of individual organs and their functions facilitate understanding, as do the later discussions of the dietary and medical treatments for disorders of each organ. As emphasized throughout this text, however, and as encountered in actual practice, the health and function of one organ invariably rely on the health and function of all of the others. Thus, treatments to relieve the symptoms of one disorder may overlap with or, at times, even conflict with treatments for another. Health care professionals who understand the individual roles of the metabolic organs, as well as the interplay among them, will be prepared to meet the challenges of their profession.

▪▪ Energy for Metabolic Work

The metabolic work that the body's cells do, like all work, requires energy, and food supplies that energy. Food in turn gets its energy from the sun, either directly (in the case of photosynthesizing plants) or indirectly (in the case of animals that eat plants). When chemical reactions in cells release stored energy from energy-yielding nutrients, that energy becomes available to do the cells' work.

▪▪ HEAT ENERGY AND BODY TEMPERATURE

The cells of each organ conduct metabolic activities specific to that organ, but in addition, all cells must maintain themselves, and many must reproduce. To do this, they must have all the essential nutrients available to them: the energy nutrients, the vitamins, and the minerals, as well as water. As cells do their metabolic work, the chemical reactions involved release heat, and this heat keeps the body warm. By regulating the rates at which these metabolic reactions release heat energy, the body maintains its constant normal temperature of about 98.6°F.

■■ ACCELERATED METABOLISM   During medical stress, metabolism speeds up. Fever sometimes develops. An accelerated metabolism signifies that fuels are being burned at a rate more rapid than normal; this may lead to wasting of body organs and loss of weight including loss of vital lean tissue. Chapter 19 describes the metabolic consequences of severe stress and, in particular, of severe infections, deep penetrating wounds, and burns.

As this brief discussion has shown, metabolism occurs all through the body, all the time, and supports normal health. The remainder of this chapter delves into one aspect of metabolism—the metabolism of the energy nutrients and the resulting consequences of underweight and overweight.

The Body's Energy Metabolism

The body manages its energy supply with amazing precision. Consider, for example, that a consistent 1 percent error in energy intake can cause a person to become more than 200 pounds overweight in a lifetime. Yet most people maintain their weight within about a 10- to 20-pound range throughout their lives. How do they do this? How does the body manage excess energy? And how does it manage to survive without food for prolonged periods—as when someone is starving or fasting? The answers to these questions lie in an understanding of metabolism.

Energy metabolism is the sum total of all the chemical reactions that manage energy nutrients in the body. Earlier chapters introduced the energy-yielding nutrients—carbohydrate, fat, and protein—and showed how they are broken down into basic units that are absorbed into the blood. Picking up from there, what becomes of these nutrients? The question is important because it provides insight into proper and improper ways to aid the body in maintaining or losing weight.

energy metabolism: all the reactions by which the body obtains and spends the energy from food or body stores.

Energy metabolism centers on four basic units:

- From carbohydrates: glucose.
- From lipids: glycerol and fatty acids.
- From proteins: amino acids.

Building up Body Compounds

When not needed by the cells for energy, the basic units of the energy-yielding nutrients can be used to build body compounds. The building up of body compounds is known as **anabolism;** this book represents anabolic reactions, wherever possible, by "up" arrows in chemical diagrams (such as those shown in Figure 6-1 on p. 122). Glucose units can be strung together to make glycogen chains. Glycerol and fatty acids can be assembled into triglycerides. Amino acids can be linked together to make proteins. These anabolic reactions, in which simple compounds are put together to form larger, more complex structures, involve doing work and so require energy.

anabolism (an-ABB-o-lism): reactions in which small molecules are put together to build larger ones. Anabolic reactions consume energy.

ana = up

Breaking down Nutrients for Energy

The breaking down of body compounds is known as **catabolism;** catabolic reactions usually release energy and are represented, wherever possible, by

catabolism (ca-TAB-o-lism): reactions in which large molecules are broken down to smaller ones. Catabolic reactions usually release energy.

kata = down

"down" arrows in chemical diagrams (see Figure 6-1). Glycogen can be broken down to glucose, triglycerides to fatty acids and glycerol, and protein to amino acids. When the body needs energy, it breaks any or all of the four basic units—glucose, fatty acids, glycerol, and amino acids—into even smaller units.

glycolysis (gligh-COLL-uh-sis): the metabolic breakdown of glucose to pyruvate.

    glyco = glucose   lysis = breakdown

pyruvate (PIE-roo-vate): pyruvic acid, a 3-carbon compound derived from glucose, glycerol, and certain amino acids in metabolism.

acetyl (ASS-uh-teel or **uh-SEET-ul co-AY) CoA:** a 2-carbon compound (acetate, or acetic acid), to which a molecule of CoA is attached.

CoA (coh-AY): a nickname for a small molecule (coenzyme A) that participates in metabolism.

  ■■ **GLUCOSE BREAKDOWN** Glucose breakdown occurs via a pathway known as **glycolysis.** In glycolysis, glucose is broken down to **pyruvate.*** Most often, pyruvate is then converted to a smaller compound, **acetyl CoA.** In a series of metabolic reactions called the tricarboxylic acid (TCA) cycle, acetyl CoA splits, and its energy is donated to storage compounds, used to do the body's work, or used to produce heat.

    The following sequence is central to an understanding of metabolism:

$$\text{Glucose} \leftrightarrow \text{pyruvate} \rightarrow \text{acetyl CoA} \rightarrow \text{energy.}$$

Notice the two-way arrow between glucose and pyruvate and the one-way arrows after pyruvate. They show that pyruvate can be reconverted to glucose but that acetyl CoA cannot. Any compound that can be converted to pyruvate can be used to make glucose. Any compound that is converted to acetyl CoA cannot be used to make glucose.

  ■■ **FAT BREAKDOWN** Triglycerides (the primary form of fat in the body) cannot be converted to glucose, for the most part, because they consist mostly of fatty acids. Fatty acids are broken down into 2-carbon fragments that combine with CoA to form acetyl CoA. Because fatty acids are broken down to acetyl CoA, they cannot be used to make glucose. The glycerol portion of a triglyceride is interconvertible with pyruvate and can yield glucose, but glycerol represents only about 5 percent of the weight of a triglyceride

---

*The term *pyruvate* means a salt of pyruvic acid. Throughout this book, the ending *–ate* is used interchangeably with *–ic acid;* for our purposes, they mean the same thing.

Figure 6-1
ANABOLIC AND CATABOLIC REACTIONS COMPARED

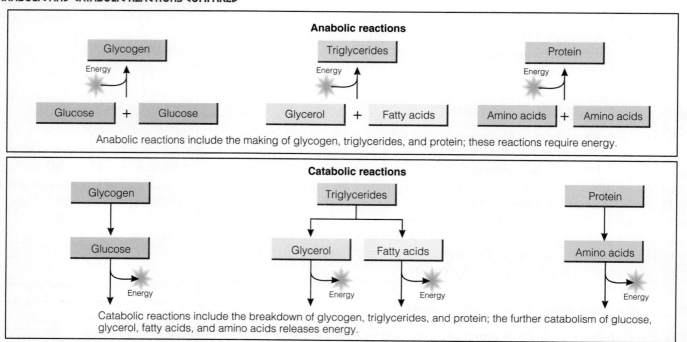

Anabolic reactions include the making of glycogen, triglycerides, and protein; these reactions require energy.

Catabolic reactions include the breakdown of glycogen, triglycerides, and protein; the further catabolism of glucose, glycerol, fatty acids, and amino acids releases energy.

molecule.* Thus, fat is an inefficient source of glucose: about 95 percent of it cannot be converted to glucose at all. Therefore, fat, for the most part, cannot provide energy for the organs (brain and nervous system) that require glucose as fuel. This leaves the task of fueling the nervous system's activities mainly to carbohydrate and protein.

■■ **AMINO ACID BREAKDOWN**   Ideally, amino acids are used to maintain supplies of needed body proteins and will not be used for energy. If they are needed for energy, or if they are consumed in excess, they first undergo **deamination,** a reaction in which they are stripped of their nitrogen. The nitrogen can be used to make other compounds, including the nonessential amino acids, or it can be excreted. With nitrogen removed, about half of the amino acids can be converted to pyruvate and can therefore provide glucose. The other amino acids are converted to acetyl CoA directly or enter the TCA cycle at another point. Thus, protein, unlike fat, is a fairly efficient source of glucose when carbohydrate is not available.

Figure 6-2 (p. 124) depicts the major metabolic pathways involving carbohydrates, fats, and amino acids. Note the central pathway from glucose to pyruvate to acetyl CoA to energy. Also note that all carbohydrates, some amino acids, and the glycerol from fat can be converted to pyruvate and then to glucose. Finally, note that the vast majority of fragments from fat can be used only for energy and not to make glucose. With these understandings, you can follow the events that lead to weight gain and weight loss.

The Body's Energy Budget

The average adult takes in close to a million kcalories a year and expends more than 99 percent of them, maintaining a stable weight for years on end. In other words, the body's energy budget is balanced. Some people, however, eat too much and get fat; others eat too little and get thin. This section examines metabolism from the perspective of the energy budget, looking first at the two forms of unbalanced budget, feasting and fasting, and then at a balanced budget.

The Economics of Feasting

Everyone knows that when people consume more energy than they expend, much of the excess is stored as body fat. Fat can be made from an excess of any energy-yielding nutrient. In addition, excess energy from alcohol is also stored as fat.[1] Fat cells enlarge as they fill with fat, and the body's fat-storing capacity seems to be able to expand indefinitely, as Figure 6-3 (p. 125) shows.

■■ **EXCESS CARBOHYDRATE**   Surplus carbohydrate (glucose) is first stored as glycogen in the liver and muscles, but the glycogen-storing cells have limited capacity. Once glycogen stores are filled, any additional carbohydrate is routed to fat. Thus, excess carbohydrate can contribute to obesity.

■■ **EXCESS FAT**   Surplus dietary fat contributes easily to the body's fat stores. During digestion and metabolism, fat may break down into fragments, such as acetyl CoA, but if the flow of these fragments is rapid enough to meet

deamination: removal of the amino (NH_2) group from a compound such as an amino acid.

The principal nitrogen-excretion product of metabolism is urea (you-REE-uh).

The making of glucose from protein or fat is gluconeogenesis (gloo-co-nee-o-JEN-uh-sis). About 5% of fat (the glycerol portion of triglycerides) and about 50% of protein (the glycogenic amino acids) can be converted to glucose.
  gluco, glyco = glucose
  neo = new
  genesis = making

---

*Figure 3-2 in Chapter 3 showed that a triglyceride consists of glycerol (3 carbons) plus 3 fatty acids (most often 16 to 18 carbons each). Thus, the small glycerol molecule represents only 3 of the 50 or so carbons in the triglyceride.

Figure 6-2
THE CENTRAL PATHWAYS OF ENERGY METABOLISM

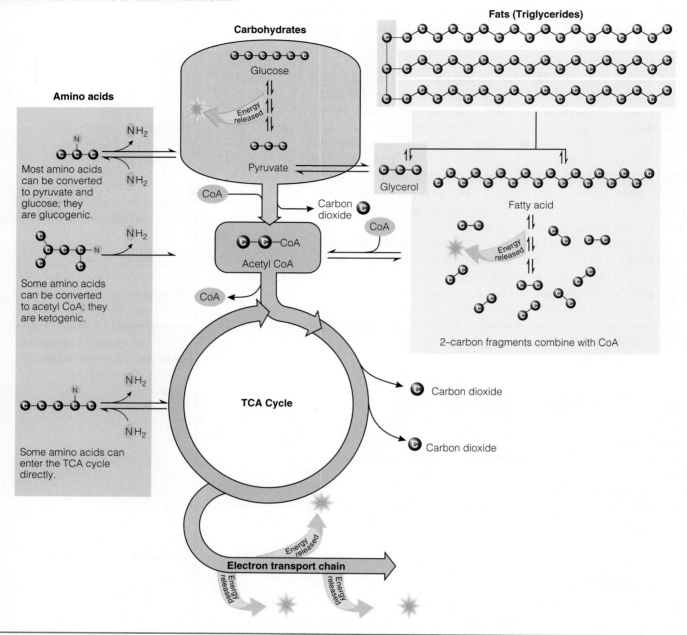

the body's need for energy, any excess fragments that are available will be stored as triglycerides in the fat cells.

EXCESS PROTEIN Surplus protein may encounter the same fate. If not needed to build protein or to meet energy needs, amino acids will lose their nitrogens and be converted through the intermediates, pyruvate and acetyl CoA, to triglycerides. These, too, swell the fat cells and add to body weight. Figure 6-4 shows the metabolic events of feasting.

In summary, the following points deserve repeating:

- Excess energy from carbohydrate, fat, protein, and alcohol will be stored in the body as fat.

Figure 6-3
FAT CELL ENLARGEMENT

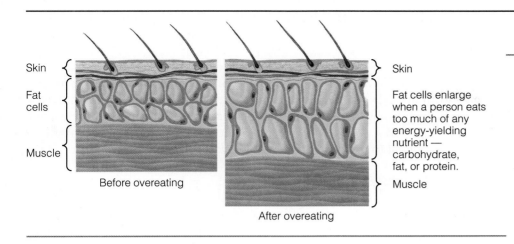

Skin

Fat cells

Muscle

Before overeating

Skin

Fat cells enlarge when a person eats too much of any energy-yielding nutrient — carbohydrate, fat, or protein.

Muscle

After overeating

- Fat from food, as compared with carbohydrate and protein, is especially easy for the body to store in fat tissue.

The Economics of Fasting

The body spends energy all the time. Even when a person is asleep and totally relaxed, the cells of many organs are hard at work. In fact, this cellular work, which maintains all life processes, represents about two-thirds of the total energy a person spends in a day. (The other one-third is the work that a person's muscles do voluntarily during waking hours.)

ENERGY DEFICIT
The body's top priority is to meet the energy needs for this ongoing cellular activity. Its normal way of doing so is by periodic refueling, that is, by eating several times a day. When food is not available, the body uses fuel reserves from its own tissues. If people choose not to eat, we say they are fasting; if they have no choice (as in a famine), we say they are starving. In the body, no metabolic difference exists between fasting and

In fasting, a person voluntarily stops eating food, whereas in starvation the failure to eat is involuntary. The body, however, makes no distinction between the two—metabolically, fasting and starvation are identical.

Figure 6-4
FEASTING

When people overeat, they store energy.

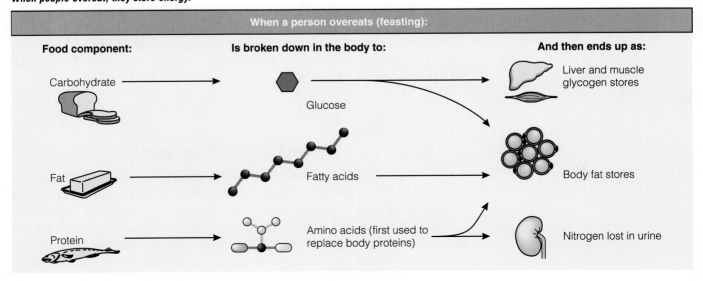

When a person overeats (feasting):		
Food component:	**Is broken down in the body to:**	**And then ends up as:**
Carbohydrate	Glucose	Liver and muscle glycogen stores
Fat	Fatty acids	Body fat stores
Protein	Amino acids (first used to replace body proteins)	Nitrogen lost in urine

Reminder: The liver releases glucose, and the fat cells release fat to fuel the body's cells, but the brain and nervous system always require some glucose.

Reminder: Ketone bodies are acidic, fat-related compounds formed from the incomplete breakdown of fat when carbohydrate is not available. Small amounts of ketone bodies are normally produced during energy metabolism. The abnormal condition known as ketosis results when ketone bodies accumulate in the blood causing the pH of the blood to drop. High blood ketones (ketonemia) sometimes spill into the urine (ketonuria).

Fasting = living on the body's fat and protein.

In fasting, muscle and lean tissues give up protein to supply amino acids for conversion to glucose. This glucose, with ketone bodies produced from fat, fuels the brain's activities.

starving. In either case, the body is forced to switch to a wasting metabolism, drawing on its stores of carbohydrate and fat and, within a day or so, on its vital protein tissues as well.

:: GLYCOGEN USED FIRST As a fast or period of starvation begins, glucose from the liver's stored glycogen and fatty acids from the body's stored fat both flow into cells to fuel their work. Several hours later, most of the glucose is used up—liver glycogen is exhausted. Low blood glucose concentrations serve as a signal to promote further fat breakdown.

:: GLUCOSE NEEDED FOR BRAIN At this point, a few hours into a fast, most of the cells are depending on fatty acids to continue providing fuel. But the nervous system (brain and nerves) and red blood cells cannot use fatty acids; they still need glucose. Even if other energy fuel is available, glucose has to be present to permit the brain's energy-metabolizing machinery to work. Normally, the nervous system consumes about two-thirds of the total glucose used each day—about 400 to 600 kcalories' worth.

:: PROTEIN BREAKDOWN AND KETOSIS Because fat stores cannot provide the glucose needed by the brain and nerves, body protein tissues (such as liver and muscle) always break down to some extent during fasting. In the first few days of a fast, body protein provides about 90 percent of the needed glucose, and glycerol provides about 10 percent. If body protein losses were to continue at this rate, death would ensue within about three weeks. As the fast continues, however, the body finds a way to use fat to provide some fuel to the brain. It adapts by condensing together acetyl CoA fragments derived from fatty acids to produce ketone bodies, which can serve as fuel for some brain cells. Ketone body production rises until, after several weeks of fasting, it is meeting much of the nervous system's energy needs. Still, many areas of the brain rely exclusively on glucose, and body protein continues to be sacrificed to produce it. Figure 6-5 shows the metabolic events that occur during fasting.

:: SLOWED METABOLISM As fasting continues and the body is shifting to partial dependence on ketone bodies for energy, the body simultaneously reduces its energy output (metabolic rate) and conserves both fat and lean tissue. Because of the slowed metabolism, the loss of fat falls to a bare minimum. Thus, although *weight* loss during fasting may be quite dramatic, *fat* loss may actually be less than when at least some food is supplied.

:: HAZARDS OF FASTING The body's adaptations to fasting are sufficient to maintain life for a long period. Mental alertness need not be diminished. Even physical energy may remain unimpaired for a surprisingly long time. Still, fasting is not without its hazards. Among the many changes that take place in the body are:

- Wasting of lean tissues.
- Impairment of disease resistance.
- Lowering of body temperature.
- Disturbances of the body's fluid and electrolyte balances.

For the person who wants to lose weight, fasting is a dangerous way to go. The body's lean tissue continues to be degraded, sometimes amounting to as much as 50 percent of the weight lost. Over the long term, a diet only moderately restricted in energy can actually promote a greater rate of *weight* loss, a faster rate of *fat* loss, and the retention of more lean tissue than a severely restricted fast.[2]

Alterations similar to those in fasting are seen in low-carbohydrate dieting. Renewed food intake, especially of carbohydrate, results in dramatic

Figure 6-5
FASTING

When people are fasting, they draw on stored energy.

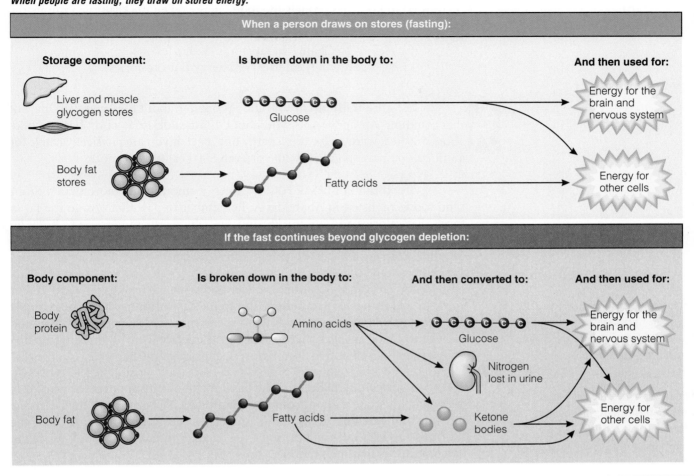

changes in the body's salt and water balance, accounting for most of the wide swings in body weight seen in people on fasts or low-carbohydrate diets.

LEARNING LINK

As you'll read in the Nutrition in Practice that follows Chapter 7, clients with anorexia nervosa become so obsessed with losing weight that they literally starve themselves to the point of endangering their health and, often, their lives. The physical consequences of anorexia nervosa are the same as the hazards of fasting described in the previous section. Anorexia nervosa is a dangerous, life-threatening condition that demands dietary, medical, and psychological intervention.

Energy Balance

If a person's weight is within the appropriate range for height, the person has a balanced energy budget. Food energy intake has equaled energy expenditure, and so, deposits of fat made at one time have been compensated for by withdrawals made at another. In other words, the body uses fat as a savings account for energy. In the case of fat, though, unlike money, more is not better; there is an optimum.

A day's energy balance can be stated like this: change in body fat stores (expressed in kcalories) equals the food energy taken in (kcalories) minus the energy spent on metabolic and other activities (kcalories). More simply:

$$\text{Change in fat stores (kcalories)} =$$
$$\text{energy in (kcalories)} - \text{energy out (kcalories)}.$$

:: ENERGY IN AND ENERGY OUT You know about the "energy in" part of this equation. An apple gives you about 125 kcalories; a candy bar provides about 300 kcalories. As for the "energy out" part, if you are physically active for an hour, you may spend 100, 300, or even 500 kcalories or more.

:: ENERGY VALUES OF FOODS Energy amounts for more than a thousand foods are listed in Appendix A. Remember to pay attention to the fat in foods. Fat calories add up quickly and probably contribute more to body fat stores than do carbohydrate kcalories, and fat contributes little to the feeling of fullness during a meal.[3] Chapter 3 offered strategies for reducing fat intake.

:: ENERGY EXPENDITURES The body spends energy in two major ways: to fuel its **basal metabolism** and to fuel its **voluntary activities.** People can change their voluntary activities to spend more or less energy in a day, and over time they can also change their basal metabolism by building up the body's metabolically active lean tissue, as explained in Chapter 7.

:: BASAL METABOLISM The basal metabolism supports the work that goes on all the time without conscious awareness. The beating of the heart, the inhaling and exhaling of air, the maintenance of body temperature, and the transmission of nerve and hormonal messages to direct these activities are the basal processes that maintain life.

:: BASAL METABOLIC RATE The basal metabolic rate (BMR) is the rate at which the body spends energy for these maintenance activities. This rate varies from person to person and may vary for a single individual with a change in circumstance or physical condition. For example, an infant's metabolic rate relative to body weight is much faster than an adult's to support the infant's extraordinary growth rate. In general, the BMR is fast in people with considerable lean body mass (growing children, physically active people, pregnant women, and males). One way to increase the BMR then is to maximize lean body tissue by participating regularly in endurance and strength-building activities.[4] The BMR is also fast in people who are tall and so have a large surface area for their weight, in people with fever or under stress, in people taking certain medications, and in people with highly active thyroid glands. The BMR is slowed down by loss of lean tissue and depression of thyroid hormone activity due to disease, inactivity, fasting, or malnutrition. Table 6-1 summarizes the factors that speed up and slow down the BMR.

basal metabolism: the energy needed to maintain life when a person is at complete rest after a 12-hour fast. Basal metabolism is normally the largest part of a person's daily energy expenditure.

voluntary activities: the component of a person's daily energy expenditure that involves conscious and deliberate muscular work—walking, lifting, climbing, and other physical activities. Voluntary activities normally require less energy in a day than basal metabolism does.

LEARNING LINK

Later chapters describe several different conditions that either speed up or slow down the basal metabolic rate. Anorexia nervosa, mentioned earlier, slows the BMR because of the loss of lean tissue. In many conditions associated with wasting, such as AIDS, cancer, and severe stress, disease-related metabolic changes keep the BMR high at the expense of lean tissue. For example, in severe stress and trauma, the BMR speeds up to mobilize nutrients needed to repair damage—energy stores are depleted and tissue protein is broken down. In all of these conditions, nutrient needs are high, but loss of appetite is common. Diet therapy aims to restore nutrition status without compromising recovery.

▪▪ **BASAL METABOLIC NEEDS** Basal metabolic needs are surprisingly large. A person whose total energy needs are 2000 kcalories a day spends 1200 to 1400 of them to support basal metabolism. The Clinical Applications feature at the end of this chapter shows how to estimate energy expenditure for basal metabolism.

▪▪ **ENERGY FOR ACTIVITIES** The number of kcalories spent on volun-tary activities depends on three factors: muscle mass, body weight, and activity. The larger the muscle mass required for the activity and the heavier the weight of the body part being moved, the more kcalories are spent. The activity's duration, frequency, and intensity also influence energy costs: the longer, the more frequent, and the more intense the activity, the more kcalories spent per minute. Table 6-2 (p. 130) shows the energy expended on various types of activities. The energy spent on activities, added to the energy required for basal metabolism, equals the total energy spent in a day (see the Clinical Applications feature at the end of this chapter).

▪▪ **ENERGY TO MANAGE FOOD** One component of energy expenditure is not taken into account in the calculations in the Clinical Applications: the energy required for the body to process food. When food is taken into the body, many cells that have been dormant become active. The muscles that move the food through the intestinal tract speed up their rhythmic contrac-tions, and the cells that manufacture and secrete digestive juices begin their tasks. All these and other cells need extra energy as they come alive to partici-pate in the digestion, absorption, and metabolism of food. This stimulation of

Table 6-1
FACTORS THAT AFFECT THE BMR

Factor	Effect on BMR
Age	In youth, the BMR is higher; lean body mass diminishes with age, slowing the BMR.[a]
Height	In tall, thin people, the BMR is higher.[b]
Growth	In children and pregnant women, the BMR is higher.
Body composition	The more lean tissue, the higher the BMR. The more fat tissue, the lower the BMR.[c]
Fever	Fever raises the BMR.[d]
Stresses	Some stresses and certain medications raise the BMR.
Environmental temperature	Both heat and cold raise the BMR.
Fasting/starvation	Fasting/starvation lowers the BMR.[e]
Malnutrition	Malnutrition lowers the BMR.
Hormones	The thyroid hormone thyroxine, for example, is a key BMR regulator; the more thyroxine produced, the higher the BMR.[f]

[a]The BMR begins to decrease in early adulthood (after growth and development cease) at a rate of about 2 percent/decade. A reduction in voluntary activity as well brings the total decline in energy expenditure to 5 percent/decade.
[b]If two people weigh the same, the taller, thinner person will have the faster metabolic rate, reflecting the greater skin surface through which heat is lost by radiation, in proportion to the body's volume.
[c]In general, males tend to have a higher BMR than females due to their greater lean body mass.
[d]Fever raises the BMR by 7 percent for each degree Fahrenheit.
[e]Prolonged starvation reduces the total amount of metabolically active lean tissue in the body, although the decline occurs sooner and to a greater extent than body losses alone can explain. More likely, the neural and hormonal changes that accompany fasting are responsible for changes in the BMR.
[f]The thyroid gland releases hormones that travel to the cells and influence cellular metabolism. Thyroid hormone activity can speed up or slow down the rate of metabolism by as much as 50 percent.

Table 6-2
ESTIMATING DAILY ENERGY EXPENDITURE AT VARIOUS LEVELS OF PHYSICAL ACTIVITY

Level of Intensity	Type of Activity	Activity Factor (× BMR)	Energy Expenditures (kcal/kg/day)
Very light	Seated and standing activities, painting trades, driving, laboratory work, typing, sewing, ironing, cooking, playing cards, playing a musical instrument	1.3 Men 1.3 Women	31 30
Light	Walking on a level surface at 2.5 to 3 mph, garage work, electrical trades, carpentry, restaurant trades, housecleaning, child care, golf, sailing, table tennis	1.6 Men 1.5 Women	38 35
Moderate	Walking 3.5 to 4 mph, weeding and hoeing, carrying a load, cycling, skiing, tennis, dancing	1.7 Men 1.6 Women	41 37
Heavy	Walking with a load uphill, tree felling, heavy manual digging, basketball, climbing, football, soccer	2.1 Men 1.9 Women	50 44
Exceptional	Athletes training in professional or world-class events	2.4 Men 2.2 Women	58 51

SOURCE: Reprinted with permission from *Recommended Dietary Allowances: 10th edition.* Copyright 1989 by the National Academy of Sciences. Courtesy of the National Academy Press, Washington, D.C.

thermic effect of food: an estimate of the energy required to process food (digest, absorb, transport, metabolize, and store ingested nutrients).

cellular activity produces heat and is known as the **thermic effect of food.** The thermic effect of food is generally thought to represent about 10 percent of the total food energy taken in. For purposes of rough estimates, though, the thermic effect of food can be ignored; the 10 percent it may contribute to total energy output is smaller than the probable errors involved in estimating energy input from food or output for activities.

Physical activity spends energy and benefits health in many ways.

SELF CHECK

SELF CHECK

1. The principal organs of metabolism are:

   a. the stomach, heart, lungs, liver, and brain.

   b. the lungs, bloodstream, pancreas, liver, and heart.

   c. the brain, digestive system, heart, stomach, and lungs.

   d. the digestive organs, liver, pancreas, circulatory system, and kidneys.

2. Anabolism is defined as:

   a. the metabolic breakdown of glucose to pyruvate.

   b. removal of nitrogen from a compound such as an amino acid.

   c. reactions in which large molecules are broken down to smaller ones and energy is released.

   d. reactions in which small molecules are put together to build larger ones and energy is consumed.

3. During glycolysis and the tricarboxylic acid (TCA) cycle, glucose is first broken down into ____, then ____, and finally ____ for use in the body.

   a. glycerol, pyruvate, energy

   b. pyruvate, acetyl CoA, energy

   c. acetyl CoA, pyruvate, energy

   d. amino acids, acetyl CoA, energy

4. Fats are catabolized to ____ and ____ for use in the body.

   a. glucose, energy

   b. glycerol, fatty acids

   c. amino acids, energy

   d. glycogen, fatty acids

5. Two functions of protein in the body are:

   a. catabolism to glycogen and synthesis of triglycerides for storage.

   b. catabolism to glycerol and fatty acids for storage and synthesis of new proteins.

   c. maintenance of the body's supply of amino acids and conversion of protein to glucose for energy.

   d. anabolism of glucose to glycogen for storage and conversion of protein to fat for energy.

6. As carbohydrate and fat stores are depleted during fasting or starvation, the body then uses ____ as its fuel source.

   a. alcohol

   b. protein

   c. glucose

   d. triglycerides

7. When carbohydrate is not available to provide energy for the brain, as in starvation, the body produces ketone bodies from:

   a. glucose.

   b. glycerol.

   c. acetyl CoA.

   d. amino acids.

8. Three hazards of fasting are:

   a. water weight loss, decrease in mental alertness, and wasting of lean tissue.

   b. water weight gain, impairment of disease resistance, and lowering of body temperature.

   c. water weight gain, decrease in mental alertness, and impairment of disease resistance.

   d. wasting of lean tissue, impairment of disease resistance, and disturbances of the body's salt and water balance.

9. Two activities that contribute to the basal metabolic rate are:

   a. walking and running.

   b. maintenance of heartbeat and running.

   c. maintenance of body temperature and walking.

   d. maintenance of heartbeat and body temperature.

10. Three factors that affect the body's basal metabolic rate are:

   a. height, weight, and energy intake.

   b. age, body composition, and height.

   c. fever, body composition, and altitude.

   d. weight, fever, and environmental temperature.

Answers to these questions appear in Appendix H.

CLINICAL APPLICATIONS

This Clinical Applications feature shows one way of calculating energy needs. Another way often used in clinical practice is the BEE (basal energy expenditure) method, explained in Chapter 19.

1. *Basal metabolism.* Convert your own or your client's body weight from pounds to kilograms (if necessary). Then multiply by the factor 1.0 kcalorie per kilogram of body weight per hour for men or 0.9 for women.* Then multiply by the 24 hours in a day. For example, suppose your client is a 160-pound man:

   • Change pounds to kilograms: 160 lb ÷ 2.2 lb/kg = 72.7 kg.

   • Multiply weight in kilograms by the BMR factor: 72.7 kg × 1 kcal/kg/hr = 72.7 kcal/hr.

   • Multiply kcalories used in one hour by hours in a day: 72.7 kcal/hr × 24 hr/day = 1744.8 kcal/day.

*Men's metabolic energy needs are assumed to be higher than women's because their hormones induce them to develop more lean tissue than do most women, and lean tissue burns more energy per hour.

Energy for the BMR equals 1745 (rounded) kcalories per day.

2. *Voluntary muscular activity.* To estimate the energy for activities, determine from Table 6-2 (p. 130) the level of intensity that typifies average daily activity. Then multiply the BMR kcalories by the corresponding activity factor. For example, if the 160-pound man engages in mostly light activity, his activity factor would be 1.6. Multiply this factor by his BMR kcalories:

- 1.6 × 1745 kcal/day = 2792 kcal/day.

3. *Total energy needs.* The result, 2792 kcalories/day, expresses his total daily energy needs.

4. Alternatively, total energy expenditure can be estimated in one step based on body weight as shown in the last column of Table 6-2. As an example, for a 160-pound man engaged in mostly light activity:

- 38 kcal/kg/day × 72.7 kg = 2763 kcal/day.

- The difference between 2792 and 2763 is insignificant and acceptable. Either way, the man's energy needs are about 2800 kcalories per day.

NUTRIT**ON**THENET

FOR FURTHER STUDY OF THE TOPICS IN THIS CHAPTER, ACCESS THESE WEB SITES.

www.amhrt.org
American Heart Association

www.kidney.org
National Kidney Foundation

www.asn-online.com
American Society of Nephrology

Notes

[1] P. M. Suter, E. Häsler, and W. Vetter, Effects of alcohol on energy metabolism and body weight regulation: Is alcohol a risk factor for obesity? *Nutrition Reviews* 55 (1997): 157–171; P. M. Suter, Y. Schutz, and E. Jequier, The effect of ethanol on fat storage in healthy subjects, *New England Journal of Medicine* 326 (1992): 983–987.

[2] M. E. Sweeney and coauthors, Severe vs. moderate energy restriction with and without exercise in the treatment of obesity: Efficiency of weight loss, *American Journal of Clinical Nutrition* 57 (1993): 127–134.

[3] A. Golay and E. Bobbioni, The role of dietary fat in obesity, *International Journal of Obesity and Related Metabolic Disorders* 21 (1997): S2–S11.

[4] R. Maughan and K. P. Aulin, Energy costs of physical activity, *World Review of Nutrition and Dietetics* 82 (1997): 18–32; T. J. Horton and C. A. Geissler, Effect of habitual exercise on daily energy expenditure and metabolic rate during standardized activity, *American Journal of Clinical Nutrition* 59 (1994): 13–19.

Q&A NUTRITION IN PRACTICE

Inborn Errors of Metabolism

The discussion in Chapter 6 of normal metabolic processes sets the stage for a closer look at metabolic disorders caused by **genetic** errors in protein synthesis. Each inborn error affects metabolism in a unique way, and some have specific implications for diet.

What is an inborn error of metabolism?

An **inborn error of metabolism** is a genetic error that alters the production of a protein. In many cases, the protein is an enzyme. When the body fails to make an enzyme, makes an enzyme in insufficient amounts, or makes an enzyme with an abnormal structure, body functions that depend on that enzyme cannot proceed. For example, if an enzyme is missing or malfunctioning in the metabolic pathway that converts compound A to compound B, then compound A accumulates and compound B becomes deficient. Both the excess of compound A and the lack of compound B can lead to a variety of problems and, in many cases, can cause death. Furthermore, this imbalance creates excesses and deficiencies in other metabolic pathways that present another array of problems. The accompanying glossary defines related terms.

Do all inborn errors cause severe illness or death?

No, many do not. In some instances, the accumulated compound is not toxic, and the deficient compound is not essential, so individuals experience no problem. They may not even know about the error. In other cases, however, inborn errors have severe consequences, including possible mental retardation. Without prompt diagnosis and treatment, they can be lethal.

What are the treatments for inborn errors?

The primary treatment for many inborn errors is nutrition interven-

GLOSSARY

carrier: an individual who possesses one dominant and one recessive gene for a genetic trait, such as an inborn error of metabolism. When a trait is recessive, a carrier may show no signs of the trait but can pass it on.

dominant gene: a gene that has an observable effect on an organism. If an altered gene is dominant, it has an observable effect even when it is paired with a normal gene; see also *recessive gene*.

galactosemia (ga-LAK-toe-SEE-me-ah): an inborn error of metabolism in which enzymes that normally metabolize galactose to compounds the body can handle are missing and an alternative metabolite accumulates in the tissues, causing damage.

genes: the basic units of hereditary information, made of DNA, that are passed from parent to offspring in the chromosomes. A pair of genes codes for each genetic trait.

inborn error of metabolism: an inherited flaw evident as a metabolic disorder or disease present from birth.

mutation: an alteration in a gene such that an altered protein is produced.
  muta = change

PKU, phenylketonuria (FEN-el-KEY-toe-NEW-ree-ah): an inborn error of metabolism in which phenylalanine, an essential amino acid, cannot be converted to tyrosine. Alternative metabolites of phenylalanine (phenylketones) accumulate in the tissues, causing damage, and overflow into the urine.

recessive gene: a gene that has no observable effect on an organism. If an altered gene is recessive, it has no observable effect as long as it is paired with a normal gene that can produce a normal product. In this case, the normal gene is said to be *dominant*.

tion. With an understanding of the biochemical pathway involved, a clinician can often manipulate the diet to compensate for excesses and inadequacies. Management involves restricting dietary precursors that occur prior to the error in the metabolic pathway, replacing needed products that fail to be produced, or both. The goal of therapy is to:
- Prevent the accumulation of toxic metabolites.
- Replace essential nutrients that are deficient as a result of the defective metabolic pathway.
- Provide a diet that supports normal growth, development, and maintenance.

Meeting these three objectives is a major challenge that was previously unattainable. Knowledge about the body's many biochemical pathways, coupled with current technology for synthesizing formulas with specific nutrient compositions, has greatly enhanced the treatment of inborn errors.

Please provide a specific example.

A classic example that illustrates the principles of treatment is the most common inborn error of metabolism: **phenylketonuria (PKU)**. PKU is one of many inborn errors that affect amino acid metabolism. Other disorders affect not only amino acid metabolism but also carbohydrate, lipid, vitamin, and mineral metabolism.

PKU affects approximately 1 out of every 10,000 newborns in the United States each year. The ability to detect and treat PKU has saved and significantly improved the lives of many people; its example offers hope to those suffering from other inborn errors.

www.wolfenet.com/~kronmal
National PKU News

Figure NP6-1
THE BIOCHEMICAL PATHWAY IN PKU

Normal:

Normally, the amino acid phenylalanine follows two pathways, one in the liver, the other in the kidneys. In the liver, the enzyme phenylalanine hydroxylase adds a hydroxyl group (OH) to produce the amino acid tyrosine. Tyrosine, in turn, produces melanin, the pigmented compound found in skin and brain cells; the neurotransmitters epinephrine and norepinephrine; and the hormone thyroxin. In the kidneys, enzymes convert phenylalanine to by-products that are excreted.

In the liver:

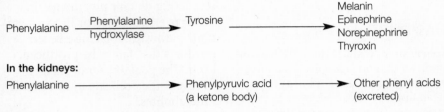

Phenylalanine → (Phenylalanine hydroxylase) → Tyrosine → Melanin / Epinephrine / Norepinephrine / Thyroxin

In the kidneys:

Phenylalanine → Phenylpyruvic acid (a ketone body) → Other phenyl acids (excreted)

In PKU:

Individuals with PKU lack the liver enzyme phenylalanine hydroxylase, impairing conversion of phenylalanine to tyrosine. Phenylalanine accumulates in the liver and blood, reaching the kidneys in abnormally high concentrations. In the kidneys, an aminotransferase enzyme converts phenylalanine to the ketone body phenylpyruvic acid, which spills into the urine—thus the name phenylketonuria.

In the liver:

Phenylalanine (accumulates) → (Phenylalanine hydroxylase (deficient)) → Tyrosine (deficient)

In the kidneys:

Phenylalanine (accumulates) → Phenylpyruvic acid (accumulates) → Other phenyl acids (accumulate)

What causes PKU?

Classic PKU results from a deficiency of an enzyme that converts the essential amino acid phenylalanine to tyrosine (see Figure NP6-1). Without the enzyme, abnormally high concentrations of phenylalanine and other related compounds accumulate and damage the developing nervous system. Simultaneously, the body cannot make tyrosine or other compounds (such as the neurotransmitter epinephrine) that normally derive from tyrosine. Under these conditions, tyrosine becomes an essential amino acid; that is, the body cannot make it, and so the diet must supply it.

How is PKU diagnosed?

PKU is a hidden disease that cannot be seen at birth, yet diagnosis and treatment beginning in the first few days of life can prevent its devastating effects. For these reasons, and because PKU is the most common inborn error of metabolism, all newborns in the United States receive a screening test for PKU.[1] The test must be conducted after the infant has consumed several meals containing protein (usually after 24 hours and before seven days). Before screening became routine, an infant with PKU would suffer the dire consequences of uncorrected high phenylalanine concentrations. At first, the only signs are a skin rash and light skin pigmentation. Between three and six months, signs of developmental delay begin to appear. By one year, irreversible brain damage is clearly evident.

What is the dietary treatment for PKU?

Essentially, the diet restricts phenylalanine and supplements tyrosine to maintain blood concentrations within a safe range. The diet's effectiveness is remarkable; in almost every case, it can prevent the devastating array of symptoms described. As most dietitians can attest, though, the diet is more easily described than designed.

Because phenylalanine is an essential amino acid, the diet cannot exclude it completely. If phenylalanine intake is too low, children suffer bone, skin, and blood disorders; growth and mental retardation; and death. Therefore, the diet must strike a balance, providing enough phenylalanine to support normal growth and health but not enough to cause harm. The problem is not that children with PKU require less phenylalanine than other children do, but that they cannot handle excesses without detrimental effects. To ensure that blood phenylalanine and tyrosine concentrations remain within a safe range, children with PKU receive blood tests periodically and alterations in their diets when necessary. With a controlled phenylalanine intake, children with PKU can lead normal, happy lives.

Special low-phenylalanine formulas are the primary source of energy and protein for children with PKU. Their diets exclude high-protein foods such as meat, fish, poultry, cheese, eggs, milk, nuts, and dried beans and peas. Also excluded are commercial breads and pastries made from regular flour, which has a

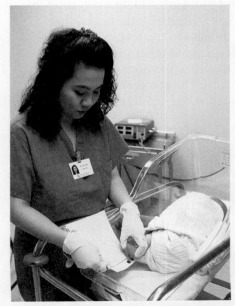

A simple blood test screens newborns for PKU— the most common inborn error of metabolism.

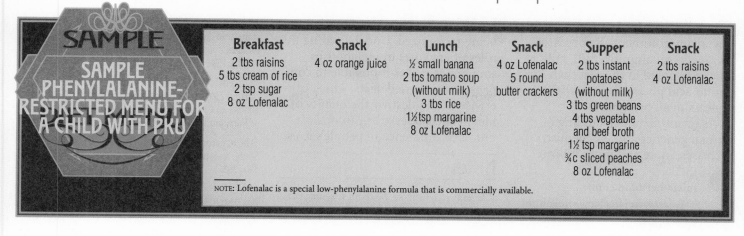

SAMPLE

SAMPLE PHENYLALANINE-RESTRICTED MENU FOR A CHILD WITH PKU

Breakfast	Snack	Lunch	Snack	Supper	Snack
2 tbs raisins	4 oz orange juice	½ small banana	4 oz Lofenalac	2 tbs instant	2 tbs raisins
5 tbs cream of rice		2 tbs tomato soup	5 round	potatoes	4 oz Lofenalac
2 tsp sugar		(without milk)	butter crackers	(without milk)	
8 oz Lofenalac		3 tbs rice		3 tbs green beans	
		1½ tsp margarine		4 tbs vegetable	
		8 oz Lofenalac		and beef broth	
				1½ tsp margarine	
				¾ c sliced peaches	
				8 oz Lofenalac	

NOTE: Lofenalac is a special low-phenylalanine formula that is commercially available.

high phenylalanine content. Basically, the diet allows foods that contain some phenylalanine, such as fruits, vegetables, and cereals, and those that contain none, such as fats, sugars, jellies, and some candies. Clearly, it is impossible to create such a diet using only whole, natural foods, but children who depend primarily on formula for their nourishment risk multiple trace mineral deficiencies.[2] Health care professionals monitor trace mineral status and supplement as needed. The accompanying menu provides a sample phenylalanine-restricted diet for a child with PKU.

Do people with PKU have to stay on the diet for life?

The answer to that question is unclear. During the early years of central nervous system development, the diet is clearly critical to preventing irreversible mental retardation. Less certain is the length of time the nervous system is vulnerable to the PKU defect. Until the late 1970s, researchers assumed that the child with PKU could abandon the special diet after the first few years of life when the central nervous system had completed its development. Unfortunately, however, elevated phenylalanine concentrations in the older child do cause problems such as short attention span, poor short-term memory, and poor eye-to-hand coordination, although the damage is less severe than at an earlier age. A child with PKU who has discontinued the controlled diet may experience problems in school performance,

mood, and behavior. For these reasons, clinicians generally encourage children to continue the low-phenylalanine diet indefinitely.[3] Convincing adolescents to return to the phenylalanine-restricted diet after several years of unrestricted diets requires intense education and reinforcement. Even then efforts are quite often unsuccessful. Reinstitution of a controlled diet, however, does improve blood phenylalanine concentrations, behavior, and IQ scores.

If a woman with PKU becomes pregnant, how does the disorder affect her fetus?

If she has not been following a phenylalanine-restricted diet, her high blood phenylalanine will present a hostile environment to fetal development. The fetus's blood concentrations rise even higher than hers, and she may experience a spontaneous abortion; or her infant may suffer mental retardation, congenital heart disease, and low birthweight. For these reasons, women with elevated phenylalanine concentrations need counseling prior to pregnancy on the problems their condition may create for their children.

Dietary control of maternal PKU does not ensure a successful outcome of pregnancy, but it may protect the fetus, at least in part, if implemented early enough. Women who follow a low-phenylalanine diet from at least one to two months prior to conception and continue it throughout pregnancy are more likely to have

infants with higher birthweights, larger head circumferences, fewer malformations, and higher scores on intelligence tests than the infants of women who begin diet therapy during their pregnancies or not at all.[4] Special formulas that meet the energy, protein, vitamin, and mineral needs of pregnant women with PKU are available.

Are all inborn errors managed the same way as PKU?

In some ways, all are similar, but in other ways each is different. As an example, **galactosemia** is an inborn error of carbohydrate metabolism in which any one of three enzymes that convert galactose to glucose is missing or defective. When infants with galactosemia are given infant formula or breast milk (which contains a galactose unit in each molecule of lactose), they vomit and have diarrhea. The abnormal metabolism causes growth failure, liver enlargement, and other neurological abnormalities that lead to coma and death. Early introduction of a galactose-restricted diet prevents or minimizes most of these symptoms.

Dietary adjustment in galactosemia is simpler than in PKU for two reasons. First, unlike phenylalanine, galactose is not an essential nutrient. The galactosemia diet needs only to exclude galactose, not to provide a calculated dose. Second, galactose occurs primarily in lactose (the sugar in milk), so treatment depends chiefly on the careful elimination of all milk and milk products. This is not to say that the diet is easy to follow; many commercially prepared products contain milk. Still, milk is less widespread in the diet than the amino acid phenylalanine, which appears in all proteins.

As scientific understanding of human genetics and biochemistry increases, more and more inborn errors affecting enzyme function are being recognized. Understanding the roles of enzymes in metabolism sometimes makes it possible to compensate for these defects, which otherwise would destroy the quality of life. In such cases, diet can make a dramatic difference in people's lives.

 miele-herndon.com/ galactosemia/galactosemia.html

Notes

[1]Committee on Genetics, Newborn screening fact sheet, *Pediatrics* 98 (1996): 473–501.

[2]C. Reilly and coauthors, Trace element nutrition status and dietary intake of children with phenylketonuria, *American Journal of Clinical Nutrition* 52 (1990): 159–165.

[3]R. O. Fisch and coauthors, Phenylketonuria: Current dietary treatment practices in the United States and Canada, *Journal of the American College of Nutrition* 16 (1997): 147–151.

[4]The Maternal Phenylketonuria Collaborative Study: A status report, *Nutrition Reviews* 52 (1994): 390–393.

CHAPTER 7

OVERWEIGHT, UNDERWEIGHT, AND WEIGHT CONTROL

CONTENTS

Body Weight and
Body Composition

Causes of Obesity

Risks of Overweight
and Obesity

Aggressive Treatments
of Obesity

Reasonable Strategies
for Weight Loss

Underweight

Nutrition in Practice:
Eating Disorders

Are you pleased with your body weight? If you answered yes, you are a rare individual. Nearly all people in our society think they should weigh more or less (mostly less) than they do. Usually, their primary reason is appearance, but they often perceive, correctly, that their weight is also related to physical health. At the extremes, both overweight and underweight present health risks.

Overweight and underweight both result from unbalanced energy budgets. The simple picture is as follows. Overweight people have consumed more food energy than they have spent and have banked the surplus in their body fat. To reduce body fat, overweight people need to spend more energy than they take in from food. In contrast, underweight people have consumed too little food energy to support their activities and so have depleted their bodies' fat stores and possibly some of their lean tissues as well. To gain weight, they need to take in more food energy than they expend. As you will see, though, the details of the body's weight regulation are quite complex.

This chapter's missions are to examine the problems associated with excessive and deficient body fatness, to present strategies toward solving these problems, and to point out how appropriate body composition, once achieved, can be maintained. The chapter emphasizes overweight because it has been more intensively studied and is a more widespread health problem in the developed countries.

Body Weight and Body Composition

The body's weight reflects its composition—the proportions of its bone, muscle, fat, fluid, and other tissue. All of these body components can vary in quantity and quality: the bones can be dense or porous, the muscles can be well developed or underdeveloped, fat can be abundant or scarce, and so on. By far the most variable tissue, though, is body fat. More than any other component, fat responds to changes in food intake and physical activity, so it is fat that is usually the target of efforts at weight control.

Defining Healthy Body Weight

How much should a person weigh? How can a person know if her weight is appropriate for her height and age? How can a person know if his weight is jeopardizing his health? Such questions seem so simple, yet even the experts cannot agree on the answers.[1] Most often, they try to identify the weights associated with lowest mortality. With this in mind, healthy body weight is defined by three criteria:

- A weight within the suggested range for height, as shown in Table 7-1.
- A fat distribution pattern that is associated with a low risk of illness or death.
- Freedom from all medical conditions that would suggest a need for weight loss.

People who meet all of these criteria may not gain any health advantage by changing their weights. Those who mistakenly think of themselves as overweight even though they meet these criteria for healthy weight may need to revise their self-image. Such people may still want to improve their eating and physical activity habits, but they should do so to reap the rewards of being physically fit, not for the sake of weight loss. Anyone who does not meet all of the above criteria may want to consult with a health care professional, who should carefully consider each criterion in relation to the others. The rest of the chapter examines these three criteria in more detail.

Table 7-1
SUGGESTED WEIGHTS FOR ADULTS, 1995 GUIDELINES

Height[a]	Weight (lb)[a]	
	Midpoint	Range
4'10"	105	91–119
4'11"	109	94–124
5'0"	112	97–128
5'1"	116	101–132
5'2"	120	104–137
5'3"	124	107–141
5'4"	128	111–146
5'5"	132	114–150
5'6"	136	118–155
5'7"	140	121–160
5'8"	144	125–164
5'9"	149	129–169
5'10"	153	132–174
5'11"	157	136–179
6'0"	162	140–184
6'1"	166	144–189
6'2"	171	148–195
6'3"	176	152–200
6'4"	180	156–205
6'5"	185	160–211
6'6"	190	164–216

NOTE: The higher weights in the ranges generally apply to men, who tend to have more muscle and bone; the lower weights more often apply to women, who have less muscle and bone.
[a]Without shoes or clothes.
SOURCE: *Report of the Dietary Guidelines Advisory Committee on the Dietary Guidelines for Americans* (Washington, D.C.: Government Printing Office, 1995).

■■ **WEIGHT FOR HEIGHT**   Scale weight fails to reflect body fatness accurately. Still, health care providers typically compare people's weights with weight-for-height tables. Normally, the assessor uses the midpoint of the weight range for a person of a given height and medium build as a standard. If the person's actual weight is 10 to 20 percent above that, then the person is considered **overweight;** if 20 percent or more above the standard, the person is obese; and if 10 percent below the standard, the person is underweight. Note, however, that weight tables such as Table 7-1 present ranges rather than pinpointing a single ideal weight, a good reminder that there is no one perfect weight that suits everyone.

Standards for desirable weights have steadily increased over the past 35 years. Authorities argue over which weight standards are most appropriate. Some approve, and some disapprove, of the current weight table for not specifying recommendations by sex; the table simply states that higher weights in the ranges generally apply to men and lower weights more often apply to women. Similar controversy focuses on the table not specifying recommendations by age; the guidelines explain that "health risks due to excess weight appear to be the same for older as for younger adults."

■■ **BODY MASS INDEX**   Most health professionals and obesity researchers use a single standard, derived mathematically from the height and weight measures—the **body mass index (BMI):**

$$BMI = \frac{weight\ (kg)}{height\ (m)^2}.$$

A person who takes measurements in pounds and inches can convert them to metric units or can use this modified equation:

$$BMI = \frac{weight\ (lb)}{height\ (in)^2} \times 705.$$

BMI values correlate with disease risks as Table 7-2 shows.[2] Most people with a BMI between 18.5 and 24.9 have few of the health risks typically associated with too-low or too-high body weight. Risks increase as BMI falls below 18.5 or rises above 24.9, reflecting the reality that both underweight and overweight impair health status. Factors such as adverse family medical history, high blood pressure, or tobacco use raise risks independently of BMI. Figure 7-1 (p. 140) presents visual images associated with various BMI values.

BMI values fail to reveal two valuable pieces of information in assessing disease risk. They do not reveal how much of the weight is fat, nor do they

overweight: body weight above some standard of acceptable weight that is usually defined in relation to height (such as the weight-for-height tables).

To convert pounds to kilograms, divide by 2.2.

To convert inches to meters, divide by 39.37.

body mass index (BMI): an index of a person's weight in relation to height; determined by dividing the weight (in kilograms) by the square of the height (in meters).

The inside back cover shows weights for various heights using the BMI to define underweight, healthy weight, overweight, and obesity.

Table 7-2
BMI VALUES AND HEALTH RISKS[a]

BMI	Obesity Class	Risks to Health
<18.5	Underweight	The lower the BMI, the greater the risk
18.5 through 24.9	Normal	Very low risk
25.0 through 29.9	Overweight	Increased risk/high risk[b]
30.0 through 34.9	Class I obesity	High risk/very high risk[b]
35.0 through 39.9	Class II obesity	Very high risk
40.0 or above	Class III obesity	Extreme risk

[a]Risk for type 2 diabetes, hypertension, and cardiovascular disease.
[b]The lower risk applies to men with a waist circumference of 102 centimeters (40 inches) or less and women with a waist circumference of 88 centimeters (35 inches) or less. The higher risk applies to those with a waist circumference above these values.
SOURCE: Data from National Heart, Lung, and Blood Institute Expert Panel, Executive summary of the clinical guidelines on the identification, evaluation, and treatment of overweight and obesity in adults, *Journal of the American Dietetic Association* 98 (1998): 1178–1191.

Figure 7-1
SILHOUETTES AND BMI (ACTUAL BMI SHOWN)

SOURCE: Reprinted from material of the Canadian Dietetic Association.

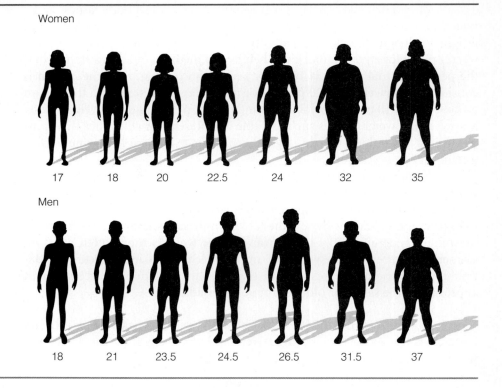

Women

| 17 | 18 | 20 | 22.5 | 24 | 32 | 35 |

Men

| 18 | 21 | 23.5 | 24.5 | 26.5 | 31.5 | 37 |

indicate where the fat is located. For this information, measures of body composition are needed.

Body Composition

For many people, being overweight compared with the standard means that they are over*fat*. This is not the case, though, for athletes with dense bones and well-developed muscles; they may be over*weight* but carry little body fat. Conversely, inactive people may seem to have acceptable weights, but still may carry too much body fat. In addition, the distribution of fat on the body may be even more critical than overfatness alone.

CENTRAL OBESITY Even more than total body fat, fat that collects in the central abdominal area of the body may be especially likely to lead to diabetes, stroke, hypertension, and coronary artery disease. The risk of death from all causes may be higher for those with **central obesity** than for those whose fat accumulates elsewhere in the body.[3] Unlike the fat layers lying just beneath the skin of the abdomen and elsewhere, **intra-abdominal fat,** when mobilized, goes directly to the liver where it is made into cholesterol-carrying low-density lipoprotein (LDL). Fat from elsewhere may arrive in the liver eventually, but it takes a circuitous route that first allows other tissues the chance to pull it from the circulation and metabolize it.

Abdominal fat creates the "apple" profile of central obesity. Fat around the hips and thighs creates more of a "pear" profile. Abdominal fat is common in women past menopause and even more common in men. Even when total body fat is similar, men have more abdominal fat than either premenopausal or postmenopausal women.[4] For those women with abdominal fat, the risks of cardiovascular disease and mortality are increased, just as they are for men.[5] Smokers, too, may carry more of their body fat centrally. A smoker may weigh less than the average nonsmoker, but the smoker's waist circumference may be greater, leading researchers to think that smoking may directly affect body fat distribution.[6] Two other factors that may affect body fat distribution are

central obesity: excess fat on the abdomen and around the trunk of the body.

intra-abdominal fat: fat stored within the abdominal cavity in association with the internal abdominal organs, as opposed to the fat stored directly under the abdominal skin (subcutaneous fat).

intakes of alcohol and physical activity. Alcohol consumption, particularly beer drinking, may favor central obesity.[7] In contrast, regular physical activity seems to prevent abdominal fat accumulation.[8]

:: FATFOLD MEASURES **Fatfold measurements** provide an accurate estimate of total body fat and a fair assessment of the fat's location.[9] About half of the fat in the body lies directly beneath the skin, so the thickness of this subcutaneous fat is assumed to reflect total body fat. Measures taken from central-body sites (around the abdomen) better reflect changes in fatness than those taken from upper sites (arm and back). A skilled assessor can obtain an accurate fatfold measure and then compare the measurement with standards (see Appendix E).

:: WAIST CIRCUMFERENCE **Waist circumference** serves as an indicator of abdominal fatness (see Figure 7-2). Previously, a ratio of waist-to-hip measurements served this purpose, but waist circumference alone has been deemed a valid indicator for both men and women. In general, women with a waist circumference larger than 35 inches and men with a waist circumference larger than 40 inches have a high risk of central obesity–related health problems.[10]

:: How Much Is Too Much Body Fat?

The ideal amount of body fat depends partly on the person. A man with a BMI within the recommended range may have from 10 to 25 percent body fat; a woman, because of her greater quantity of indispensable fat, 18 to 32 percent. For many athletes, a lower-than-average percentage of body fat may be ideal—just enough fat to provide fuel, insulate and protect the body, assist in nerve impulse transmissions, and support normal hormone activity, but not so much as to burden the muscles. For athletes, then, ideal body fat might be 5 to 10 percent for men and 15 to 20 percent for women.

For an Alaskan fisherman, a higher-than-average percentage of body fat is probably beneficial because fat helps prevent heat loss in cold weather. A woman starting a pregnancy needs sufficient body fat to support conception and fetal growth. Below a certain threshold for body fat, individuals may become infertile, develop depression, experience abnormal hunger regulation, or become unable to keep warm. These thresholds differ for each function and for each individual; much remains to be learned about them.

fatfold measure: a clinical estimate of total body fatness in which the thickness of a fold of skin on the back of the arm (over the triceps muscle), below the shoulder blade (subscapular), or in other places is measured with a caliper. (The older, less preferred, term is **skinfold test.**)

waist circumference: a measurement used to assess a person's abdominal fat.

At 6 feet 3 inches tall and 245 pounds, Mike O'Hearn has a BMI greater than 30 and would be considered overweight by most weight-for-height standards. Yet he is clearly not overfat. In fact, his body fat is only 8 percent.

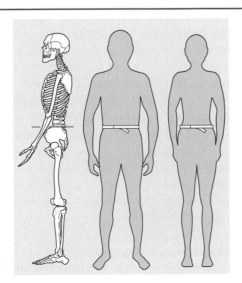

Figure 7-2
MEASURING WAIST CIRCUMFERENCE

Using a nonstretching tape measure, measure the body around the point just above the iliac crest. A healthy waist circumference for men is no larger than 102 centimeters (40 inches); for women, no larger than 88 centimeters (35 inches).

Clearly, the most important criterion of appropriate fatness is health. Researchers find health problems develop when body fat exceeds 22 percent in young men, 25 percent in older men, 32 percent in younger women, and 35 percent in older women; these are the values used to define obesity, and age 40 is the dividing line.

LEARNING LINK

Chapter 13 describes the correct procedures for measuring height and weight, and Appendix E describes measures of body composition. When determining whether a client is overweight or obese, remember that the most important criterion is health. Excess body fat clearly poses greater risks in clients with coronary heart disease, diabetes, or other obesity-related diseases. An overweight client who is physically active and free of disease may need little or no dietary or medical intervention, but may simply need to be assessed on a regular basis.

Causes of Obesity

Henceforth, this chapter will use the term *obesity* to refer to excess body fat. Excess body fat accumulates when people take in more food energy than they spend. Why do they do this? Is it genetic? Metabolic? Psychological? Behavioral? All of these? Most likely, **obesity** has many interrelated causes: many experts in the field speak of several different *obesities*.

obesity: a chronic disease characterized by excessively high body fat in relation to lean body tissue. See p. 141 for body fat percentages that define obesity.

GENETICS AND WEIGHT A person's genetic makeup almost certainly influences the body's tendency to consume or store too much energy or to burn too little.[11] When both parents are obese, the chances that their children will be obese are quite high (up to 80 percent), whereas when neither parent is obese, the chances are relatively small (less than 10 percent). Adoption studies find a similarity in obesity between biological parents and their natural children, but not between adoptive parents and their adopted children.

To determine the relative contributions of genetic and environmental factors to body weight, one group of researchers studied identical and fraternal twins, some of whom were reared together and some apart. Like previous studies, this study found that identical twins were twice as likely to have similar weights as fraternal twins were—even when reared apart. These findings suggest an important role for genetics in determining a person's susceptibility to obesity.[12]

Genetics may influence the way energy is *stored*. When identical twins are given an extra 1000 kcalories a day for 100 days, some pairs gain less than 10 pounds while others gain up to 30 pounds. Within each pair, the amount of weight gain, percentage of body fat, and distribution of fat are similar.

Genetics may also influence how much energy the body *spends*. For example, the differences in basal metabolic rate (BMR) between individuals are greater than can be explained by age, sex, and body composition alone. Similarities within families suggest a genetic influence on BMR. A low metabolic rate is a major risk factor for weight gain.

lipoprotein lipase (LPL): an enzyme mounted on the surface of fat cells (and other cells). It hydrolyzes triglycerides in the blood into fatty acids and glycerol for absorption into the cells. There they are metabolized or reassembled for storage.

LIPOPROTEIN LIPASE Some of the research investigating genetic influence on obesity focuses on the enzyme **lipoprotein lipase (LPL),** which promotes fat storage in fat cells and muscle cells. People with high LPL activity are especially efficient at storing fat. Obese people generally have much more LPL activity in their fat cells than lean people do.

■■ **LEPTIN**   Researchers have discovered a gene in humans called the obesity (*ob*) gene. The obesity gene is expressed in the fat cells and codes for the protein **leptin**.[13] Leptin appears to act primarily on the hypothalamus, suppressing hunger and increasing energy expenditure. As the accompanying photo shows, mice with a defective obesity gene do not produce leptin and can weigh up to three times as much as normal mice. When injected with leptin, the mice lose weight. (Because leptin is a protein, it would be destroyed during digestion if given orally; consequently, it must be given by injection.)

Researchers have identified a genetic deficiency of leptin in human beings as well.[14] An error in the gene that codes for leptin was discovered in two extremely obese children whose blood levels of leptin are barely detectable. Without leptin, their satiety is impaired; the children are constantly hungry and eat considerably more than their siblings or peers.

Most obese people do not have leptin deficiency, however. In fact, in obese people, the more body fat, the more leptin.[15] Researchers speculate that when fat cells are ample, leptin rises in an effort to suppress appetite and inhibit fat storage. Obese people with elevated leptin concentrations may be resistant to its satiating effect.[16] The absence of or resistance to leptin in obesity parallels the scenario of insulin in diabetes: some people have an insulin deficiency (type 1), whereas many others have elevated insulin at first, but are resistant to its glucose-storing effect (type 2).

■■ **FAT CELL DEVELOPMENT**   Another cause of obesity may be the development of excess fat cells during childhood. The amount of fat on a person's body reflects both fat cell *number* and *size*. The number of fat cells increases most rapidly during the growing years of late childhood and early puberty. Fat cell number increases more rapidly in obese children than in lean children, and obese children entering their teen years may already have as many fat cells as do adults of normal weight.

Fat cells can also expand in size. Upon reaching their maximum size, the cells may divide. Thus, obesity develops when a person's fat cells increase in number, in size, or quite often both. With fat loss, the size of the fat cells shrinks, but not their number. For this reason, people with extra fat cells may tend to regain lost weight rapidly. Prevention of obesity, then, is most critical during the growing years when fat cell number is increasing.

■■ **SET-POINT THEORY**   One popular theory of why a person may store too much fat is the **set-point theory.** The set-point theory proposes that body weight, like body temperature, is physiologically regulated. Researchers have noted that most people who lose weight on reducing diets quickly regain all their lost weight. This suggests that somehow the body chooses a weight that it wants to be and defends that weight by regulating eating behaviors and hormonal actions. Research confirms that the body adjusts its metabolism whenever it gains or loses weight—in the direction that returns to the initial body weight: energy expenditure increases with weight gain and decreases with weight loss.[17] These changes in energy expenditure are greater than those predicted based on body composition and help to explain why it is so difficult for an obese person to maintain weight losses. Researchers speculate that an individual's set point for body weight is adjustable, shifting over the life span in response to physiological changes and to genetic, dietary, and other factors.[18]

■■ **ENVIRONMENTAL STIMULI**   To a degree, obesity may be environmentally determined. People may overeat in response to stimuli in their surroundings—

Genes instruct cells to make proteins, and each protein performs a unique function.

leptin: a protein produced by fat cells under the direction of the obesity gene; increases satiety and energy expenditure.
leptos = thin

The mouse on the left is genetically obese—it lacks the gene for producing leptin. The mouse on the right is also genetically obese, but because it receives leptin, it eats less, expends more energy, and is less obese than it would be had it not received the leptin.

*Obesity due to an increase in the **number** of fat cells is hyperplastic obesity. Obesity due to an increase in the **size** of fat cells is hypertrophic obesity.*

set-point theory: the theory that proposes that the body tends to maintain a certain weight by means of its own internal controls.

primarily, the availability of many delectable foods. One food constituent is perceived as especially palatable—fat. Not only does fat deliver twice the kcalories, gram for gram, as protein and carbohydrate, but it also seems to be stored preferentially by the body, and with great efficiency. Of the three energy nutrients, fat stimulates the least energy expenditure after a meal and is least powerful in signaling satiety.[19]

■■ **LEARNED BEHAVIOR** Psychological stimuli also trigger inappropriate eating behaviors in some people. Appropriate eating behavior is a response to **hunger.** Hunger is a drive programmed into people by their heredity. **Appetite,** in contrast, is learned and can lead people to ignore hunger or to overrespond to it. Hunger is physiological, whereas appetite is psychological, and the two do not always coincide.

Food behavior is also intimately connected to deep emotional needs such as the primitive fear of starvation. Yearnings, cravings, and addictions with profound psychological significance can express themselves in people's eating behavior. An emotionally insecure person might eat rather than call a friend and risk rejection. Another person might eat to relieve boredom or to ward off depression.

■■ **PHYSICAL INACTIVITY** The possible causes of obesity mentioned so far all relate to the input side of the energy equation. What about output? People may be obese not because they eat too much, but because they spend too little energy. More than one-third of the overweight population report no physical activity during their leisure time.[20] Obese people observed closely are often seen to eat less than lean people, but they are sometimes so extraordinarily inactive that they still manage to accumulate an energy surplus. Reducing their food intake further would jeopardize health and incur nutrient deficiencies. Physical activity, then, is a necessary component of nutritional health. People must be physically active if they are to eat enough food to deliver all the nutrients needed without unhealthy weight gain.

One hundred years ago, 30 percent of the energy used in farm and factory work came from muscle power; today, only 1 percent does.[21] Modern technology has replaced physical activity at home, at work, and in transportation. Underactivity is probably the single most important contributor to obesity. In turn, television watching may contribute most to physical inactivity.[22]

Watching television contributes to obesity in several ways. First, television viewing requires little energy beyond the resting metabolic rate. Second, it replaces time spent in more vigorous activities. Third, watching television correlates with between-meal snacking, eating the high-kcalorie, high-fat foods most heavily advertised on programs, and influencing family food purchases.

Like all the other "causes" of obesity, inactivity alone fails to explain it fully. Genetics, fat cell development, set point, and overeating all offer possible, but still incomplete, explanations. Most likely, obesity has not one cause, but different causes and combinations of causes in different people. After all, no two people are alike either physically or psychologically. Some causes may be within a person's control, and some may be beyond it. In recent years, the view has been gaining ground that obesity is not simply a matter of undisciplined gluttony. Philosophies of weight control and treatment have been evolving to square with this view.

hunger: the physiological need to eat, experienced as a drive for obtaining food; an unpleasant sensation that demands relief.

appetite: the psychological desire to eat; a learned motivation that is experienced as a pleasant sensation that accompanies the sight, smell, or thought of appealing foods.

Lack of physical activity fosters obesity.

■■ Risks of Overweight and Obesity

Overweight is associated with disease risks. For example, it can precipitate hypertension and bring on strokes.[23] Often weight loss alone can normalize the blood pressure of an overfat person; some people with hypertension can tell you the exact weight at which their blood pressure begins to rise. Weight gain can also precipitate insulin resistance and diabetes in genetically susceptible people.

Despite our nation's preoccupation with body image and weight loss, the incidence of overweight continues to rise dramatically (see Figure 7-3). Approximately one out of three adults, one out of nine teenagers, and one out of seven children in the United States are now overweight.[24] The prevalence of overweight has been increasing and is especially high among women, the poor, and some ethnic groups. If this trend continues, some obesity experts predict that by the year 2230, every adult in the United States will be overweight. Such a dramatic statement sounds preposterous, but it speaks to the reality: obesity is a major public health problem without an apparent solution.[25]

HEALTH RISKS OF OBESITY

The health risks of obesity are so many that it has been declared a disease. In the United States, obesity is second only to tobacco use as the most significant cause of preventable death.[26] Besides diabetes and hypertension already mentioned, other risks threaten obese adults. Among them are high blood lipids, cardiovascular disease, sleep apnea (abnormal ceasing of breathing during sleep), osteoarthritis, abdominal hernias, some cancers, varicose veins, gout, gallbladder disease, respiratory problems (including Pickwickian syndrome, a breathing blockage linked with sudden death), liver malfunction, complications in pregnancy and surgery, flat feet, and even a high accident rate. Each year these obesity-related illnesses cost our nation billions of dollars.[27] The cost in terms of lives is also great. People with lifelong obesity are twice as likely to die prematurely as others.

People want to know exactly how much fat is too fat for health. Ideally, a person has enough fat to meet basic needs but not so much as to incur health risks. Some evidence indicates that being even mildly or moderately overweight aggravates the risk of heart disease.[28] The degree of overweight correlates with the rate of development of cardiovascular disease.[29] Overweight may even have effects years after the excess weight is lost. Even if lean as adults, people who were 20 pounds or more overweight as teenagers may be more likely than others to die of heart disease.[30] On the other hand, some obese people seem to remain healthy and live long despite their body fatness. It may be that genetics determines who among the overweight will be susceptible to diseases and who will stay healthy. Still, the majority of obese people do develop associated health problems.

OTHER RISKS OF OBESITY

Although some obese people seem to escape health problems, few in our society can avoid the social and economic handicaps. People who have been overweight as adolescents are still, seven years later, less likely to be married and more likely to have low household incomes than those who were not overweight earlier.[31] This is especially true for women. In contrast, people with other chronic conditions such as asthma, diabetes, and epilepsy do not differ socially or economically from nonoverweight people.

Our society places enormous value on thinness. Obese people pay more for insurance and for clothing. Psychologically, too, fat people often feel rejected and embarrassed, and this hurts self-esteem.

RISKS OF OBESITY TREATMENT

At any given time, an estimated 30 to 40 percent of all U.S. women (and 20 to 25 percent of all U.S. men) are trying to lose weight, spending up to $40 billion each year to do so. Some of these people do not even need to lose weight. Others need to lose weight, but are not successful; few succeed, and even fewer succeed permanently. In fact, only 5 percent of people who try to lose weight achieve long-term success.

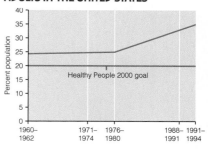

Figure 7-3
PREVALENCE OF OVERWEIGHT AMONG ADULTS IN THE UNITED STATES

A healthy body contains enough lean tissue to support health and the right amount of fat to meet body needs.

Many people assume that every overweight person can achieve slenderness and should pursue that goal. Consider, however, that most overweight people cannot become slender. People vary in their weight tendencies just as they vary in their potentials for height and degrees of health. The question of whether a person should lose weight depends on many factors: the extent of overweight, age, health, and genetics, to name a few. Weight-loss advice, then, does not apply equally to all overweight people. Some people may risk more in the process of losing weight than in remaining overweight. Others may reap significant health benefits with just modest weight loss.

The risks people incur in attempting to lose weight often depend on how they go about it. Weight-loss plans and obesity treatments flourish—some are adequate; many are ineffective and possibly dangerous. Inappropriate ways of treating obesity are listed in the accompanying glossary. The next section addresses aggressive approaches to obesity for those obese people with high risks of medical problems who must lose weight rapidly. The section after that discusses reasonable approaches to overweight for those seeking safe, gradual weight loss.

LEARNING LINK

The clinical chapters later in this text point out how many diseases are linked to obesity and how many of their therapies include weight loss. As nurses, dietitians, or other health care professionals, you will encounter overweight or obese clients who want advice about symptoms or conditions that they may not realize are obesity related. An overweight woman may ask for suggestions about controlling her high blood pressure, for example. An obese man may seek your advice about his sore joints. When addressing these symptoms, be careful not to sound judgmental about the person's weight. Instead, offer the appropriate advice for the client's symptoms, and at the same time, inform the client about the relationship between certain conditions and excess body fatness. Encourage moderate physical activity such as walking or gardening, and offer guidelines for eating nutrient-dense, low-fat foods.

Aggressive Treatments of Obesity

For some obese people, the medical problems caused by their obesity demand treatment approaches that may, themselves, incur some risks. Nevertheless, taking these risks may be deemed necessary because the potential health benefits of weight loss are so great.

Obesity Drugs

Several prescription medications for weight loss have been tried over the years. When used as part of a long-term comprehensive weight-loss program, medications can help obese people lose approximately 10 percent of their weight and maintain that loss for at least a year.[32] Because weight regain commonly occurs with the discontinuation of drug therapy, treatment is long term. And the long-term use of medications poses risks. Medical experts do not yet know whether a person would benefit more from maintaining a 20-pound excess or from taking a medication for a decade to keep the 20 pounds off.

The challenge, then, is to develop an effective medication that can be used over time with minimal side effects and without the potential for abuse.

GLOSSARY

POOR TREATMENT CHOICES FOR OBESITY

diet pills: pills that depress the appetite temporarily; often, physician-prescribed amphetamines (speed). It is generally agreed that these drugs are of little value for weight loss and that their use can cause a dangerous dependency.

diuretic abuse: use of diuretics to promote water excretion by dieters who believe their weight excesses are due to water accumulation. Diuretics promote water loss, not fat loss, and their use can cause dehydration and mineral imbalances.

fad diets: diets based on exaggerated or false theories of weight loss. Such diets are usually inadequate in energy and nutrients. Most fad diets, including the currently popular Zone Diet, advocate essentially the same high-protein, low-carbohydrate diet. Such diets may

offer short-term weight-loss success to some who try them, but they fail to produce long-lasting results for most people. Furthermore, high-protein, low-carbohydrate diets are often high in fat and low in fiber, vitamins, and some minerals. Long-term use of such diets may produce adverse side effects such as nausea, fatigue, constipation, and low blood pressure. Some fad diets are more dangerous to health than obesity itself.

herbal laxatives: laxatives containing senna, aloe, rhubarb root, castor oil, or buckthorn, which are commonly sold as "dieter's tea." Such products cause nausea, vomiting, diarrhea, fainting, and, in some users, possibly death.[a]

---

[a]P. Kurtzweil, Dieter's brews make tea time a dangerous affair, *FDA Consumer*, July/August 1997, pp. 6–11.

herbal products: substances extracted from plants to mimic drugs that suppress appetite. Such substances may have dangerous side effects. St. John's wort, for example, is often prepared in combination with the herbal stimulant ephedrine, extracted from the Chinese plant ma huang. Ephedrine has been implicated in several cases of heart attacks and seizures. (See Nutrition in Practice 24.)

low-carbohydrate diets: diets designed to bring about metabolic responses similar to those of fasting (see Chapter 6, pp. 125–127). Without sufficient carbohydrate, the body cannot use its fat in the normal way, and ketosis results. Many physiological hazards accompany low-carbohydrate diets: high blood cholesterol, mineral imbalances, hypoglycemia, and more.

No such medication currently exists.[33] Two prescription medications, however, are currently in use.

SIBUTRAMINE Sibutramine is an appetite-suppressing drug that works on the brain's neurotransmitters.* Sibutramine enhances satiety and elevates energy expenditure.[34] Sibutramine also raises blood pressure, however, so the Food and Drug Administration (FDA) cautions those with hypertension against using it. Anyone taking sibutramine should monitor blood pressure carefully. As researchers learn more about the molecular chemistry of appetite control, safer appetite-suppressing drugs may be developed.

ORLISTAT The drug orlistat, recently approved by the FDA, takes a different approach to weight loss.† Not an appetite suppressant, orlistat inhibits the production of fat-digesting enzymes in the pancreas and so reduces fat absorption by about 30 percent.[35] As a result of absorbing less fat, people often lose weight.[36] The problem with undigested fat is that it leaves the digestive tract intact, carrying with it fat-soluble vitamins and phytochemicals that would otherwise have been absorbed by the body. Possible side effects of orlistat resemble those of the artificial fat olestra—diarrhea and digestive distress.[37]

Very-Low-kCalorie Diets

Very-low-kcalorie diet (VLCD) plans provide 800 kcalories, at least 1 gram of high-quality protein per kilogram of body weight, little or no fat, and a minimum of 50 grams of carbohydrate (not enough to spare protein). Clients receive an assortment of vitamins and minerals from supplements. Meals consist of a limited number of foods (primarily lean meats, fish, and poultry) each day, a powdered formula available by prescription, or a combination of the two.

---

*Sibutramine's trade name is Merida.

†Orlistat's trade name is XENICAL.

VLCD formulas are designed to be nutritionally adequate, but the body responds to this severe energy restriction as if the person were starving—it conserves energy and prepares to regain weight at the first opportunity. Several changes occur in hormone concentrations, metabolic activities, fluid and electrolyte balances, and organ functions in the effort to meet the challenge of living on a much-less-than-adequate energy intake. For these reasons, a VLCD is appropriate only for short-term use (four months) and under close medical supervision. Common side effects of VLCD include headache, fatigue, dry skin, nausea, and hair loss, as well as others.

Surgery

Surgery as an approach to weight loss is justified in some specific cases of **clinically severe obesity**. Two **gastric partitioning** procedures include **gastric bypass** and **gastric bandings**. Both procedures limit food intake by effectively reducing the size of the stomach. They reduce the size of the outlet from the stomach into the intestine as well, so they delay the passage of food for digestion and absorption.

Early reports of a new procedure for **gastric banding** seem promising.[38] The new technique may be safer than the traditional surgeries because the surgeon makes only two tiny incisions in the abdomen and through them applies a restricting band or pouch around the stomach. So far, patients receiving this type of surgery seem to have less pain, faster recovery, and fewer complications than those having more extensive surgeries.

The long-term safety and effectiveness of gastric surgery depend, in large part, on compliance with dietary instructions (see Chapter 15). Common immediate postsurgical complications include infections, nausea, vomiting, and dehydration; in the long term, vitamin and mineral deficiencies and psychiatric disorders are common. Lifelong medical supervision is necessary for those who choose the surgical route, but in suitable candidates the benefits of weight loss prove worth the risks.

Reasonable Strategies for Weight Loss

The 2000 *Dietary Guidelines* (Table 1-3 on p. 14) suggest that a person should "aim for a healthy weight." The focus is not so much on weight loss as on health gains. Modest weight loss, even when a person is still overweight, can improve control of diabetes and reduce the risks of heart disease by lowering blood pressure and blood cholesterol, especially for those with abdominal fat.

Of course, the same eating and activity habits that improve health often lead to a healthier body weight and composition as well.[39] A loss of 10 to 15 pounds can improve a person's BMI by 2 units, which can significantly improve health—even if the person is still overweight. Successful weight loss, then, is not defined by weight-for-height tables, but by reductions in disease risks.[40] People less concerned with disease risks may prefer to set goals for personal fitness, such as being able to play with children or climb stairs without becoming short of breath.

Whether the goal is health or fitness, weight-loss expectations need to be reasonable. Unreachable targets ensure frustration and failure. If goals are achieved and exceeded, there will be rewards instead of disappointments. The "How to" box offers a way of judging weight-loss diets and programs based on sound nutrition principles.

clinically severe obesity: a BMI of 40 or greater or 100 lb or more overweight for an average adult. A less preferred term used to describe the same condition is *morbid obesity.*

gastric partitioning: a surgical procedure used to treat clinically severe obesity. The operation limits food intake by effectively reducing the size of the stomach and delays gastric emptying by restricting the outlet; also called **gastroplasty.**

gastric bypass: surgery that reroutes food from the stomach to the lower part of the small intestine; creates a chronic, lifelong state of malabsorption by preventing normal digestion and absorption of nutrients.

gastric banding: a surgical means of producing weight loss by restricting stomach size with either a row of surgical staples or a constricting band.

Weight-loss pointers:
- Adopt reasonable expectations about health and weight goals and about how long it will take to achieve them.

RATE SOUND AND UNSOUND WEIGHT-LOSS SCHEMES AND DIETS

Start by giving each diet or program 160 points. Subtract points as instructed, whenever a diet falls short of ideals.

Scoring: 160 = fine.
140–150 = possibly safe with some disadvantage.
120–130 = needs improvement.
110 or below = dangerous to use.

Does the diet or program:

1. Provide a reasonable number of kcalories (not fewer than 1200 kcalories for an average-size person)? If not, give it a minus 10.

2. Provide enough, but not too much, protein (at least the recommended intake or RDA, but not more than twice that much)? If no, minus 10.

3. Provide enough fat for balance but not so much fat as to go against current recommendations (between 20 and 30 percent of kcalories from fat)? If no, minus 10.

4. Provide enough carbohydrate to spare protein and prevent ketosis (100 grams of carbohydrate for the average-size person)? Is it mostly complex carbohydrate (not more than 10 percent of the kcalories as concentrated sugar)? If no to either or both, minus 10.

5. Offer a balanced assortment of vitamins and minerals—that is, foods from all food groups? If it omits a food group (for example, meats), does it provide a suitable substitute? Count five food groups in all: milk/milk products, meat/fish/poultry/eggs/ legumes, fruits, vegetables, and starches/grains. For *each* food group omitted and not adequately substituted for, subtract 10 points.

6. Offer variety, in the sense that different foods can be selected each day? If you'd class it as boring or monotonous, give it a minus 10.

7. Consist of ordinary foods that are available locally (for example, in the main grocery stores) at the prices people normally pay? Or does the dieter have to buy special, expensive, or unusual foods to adhere to the diet? If you would class it as "bizarre" or "requiring special foods," minus 10.

8. Promise dramatic, rapid weight loss (substantially more than 1 percent of total body weight per week)? If yes, minus 10.

9. Encourage permanent, realistic lifestyle changes, including regular physical activity and the behavioral changes needed for weight maintenance? If not, minus 10.

10. Misrepresent salespeople as "counselors" supposedly qualified to give guidance in nutrition and/or general health without a profit motive, or collect large sums of money at the start, or require that clients sign contracts for expensive long-term programs? If so, minus 10.

11. Fail to inform clients about the risks associated with weight loss in general or the specific program being promoted? If so, minus 10.

12. Promote unproven or spurious weight-loss aids such as starch blockers, diuretics, sauna belts, body wraps, passive exercise, ear stapling, acupuncture, electric muscle stimulating (EMS) devices, spirulina, amino acid supplements (e.g., arginine, ornithine), glucomannan, appetite suppressants, "unique" ingredients, and so forth? If so, minus 10.

▦ Diet

No particular eating plan is magical, and no particular food must either be included or avoided. You are the one who will have to live with the plan, so you had better be involved in its planning. Don't think of it as a diet you are going "on"—because then you may be tempted to go "off." The diet is successful only

- Be involved in planning.

- Keep in mind that you will want to maintain your lost weight. Practice needed behaviors as you go.

• Adopt a realistic plan.

• Make the diet adequate by emphasizing nutrient-dense foods.

• Eat small portions of foods at each meal.

• Make grains, legumes, vegetables, and fruits central to your diet plan.

• Select low-fat foods regularly.

• Limit concentrated sweets and alcoholic beverages.

• Drink plenty of water (8 glasses or more a day).

• Learn, practice, and follow a healthful eating plan for the rest of your life.

if the pounds do not return. Think of it as an eating plan that you will adopt for life. It must consist of foods that you like, that are available to you, and that are within your means.

:: A REALISTIC ENERGY INTAKE Choose an energy intake you can live with. You need at least 10 kcalories per pound of current weight each day to lose fat while retaining lean tissue. Nutritional adequacy is difficult for most people to achieve on fewer than 1200 kcalories a day, and most healthy adults should not consume any less than that. You will experience a healthier, more successful weight loss with a small energy deficit that provides an adequate intake than with a large energy deficit that creates feelings of starvation and deprivation, which can lead to an irresistible urge to binge.

:: NUTRITIONAL ADEQUACY Nutritional adequacy should be a high priority. Take a look at the 1200-kcalorie food plan in Table 7-3. Notice that this plan offers the minimum number of servings suggested in the Daily Food Guide (introduced in Chapter 1) and allows a teaspoon of fat at each of three meals. Such an intake would allow most people to lose weight at a satisfactory rate and still meet their nutrient needs with careful food selections. (Women might need an iron supplement.) The other plans provide for higher energy intakes.

:: SMALL PORTIONS Overweight people usually need to learn to eat less food at each meal—one piece of chicken for dinner instead of two, a teaspoon of butter on the vegetables instead of a tablespoon, and one cookie for dessert instead of six. The goal is to eat enough food for energy, nutrients, and pleasure, but not more. This amount should leave a person feeling satisfied—not necessarily full. Keep in mind that even low-fat foods can deliver a lot of kcalories when a person eats large quantities.

:: CARBOHYDRATES, NOT FATS Center meals and snacks on complex carbohydrate foods. Fresh fruits, vegetables, legumes, and whole grains offer abundant vitamins and minerals. They also offer more fiber (which provides bulk and satiety) and far less fat and food energy than refined foods. Researchers compared a diet that restricted fat, but allowed complex carbohydrates to be eaten freely with a more conventional energy-restricted diet, used over six months by obese women.[41] Women in both diet groups lost substantial weight, but those who ate the fat-restricted, complex carbohydrate–rich diet rated it higher in terms of satiety and taste. A person who makes low-fat food selections habitually—even without purposely limiting

Table 7-3
RECOMMENDED NUMBER OF SERVINGS FOR DIFFERENT ENERGY INTAKES

Food Group	Energy Level (kcal)						
	1200	**1500**	**1800**	**2000**	**2200**	**2600**	**3000**
Bread, cereal, rice, and pasta	6	7	8	9	11	13	15
Meat and meat alternates (ounces)	4	5	6	6	6	7	8
Vegetable	3	4	5	5	5	6	6
Fruit	2	3	4	4	4	5	6
Milk and milk products (fat-free)	2	2	2	3	3	3	3
Fat (tsp)	3	5	6	7	8	10	12

NOTE: These patterns follow the Daily Food Guide plan and supply less than 30 percent of kcalories as fat.

energy intake—eats less food, satisfies hunger, and diminishes the desire to eat.[42] How much weight is lost on a low-fat diet depends on the extent of fat reduction and the degree of obesity. A person limiting energy intake would also lose weight, but maintaining that weight loss would be easier following a low-fat diet.[43]

■■ **SUGAR AND ALCOHOL**   A person trying to achieve or maintain a healthy weight needs to pay attention not only to fat, but to sugar and alcohol, too. Using them for pleasure on occasion is compatible with health as long as most daily choices are of nutrient-dense foods.

■■ **ADEQUATE WATER**   Learn to satisfy thirst with water. Water fills the stomach between meals and dilutes the metabolic wastes generated from the breakdown of fat, easing their excretion. Water meets the fluid needs that were formerly met by eating extra food (remember that food provides water).

In summary, adopt an "eating plan for good health" rather than a "diet for weight loss." That way, you will be able to keep the lost weight off.

■■ Physical Activity

Either dieting or physical activity alone can produce some weight loss. Clearly, however, the combination is most effective[44]

■■ **WEIGHT CYCLING**   Those who endeavor to lose weight without physical activity often become trapped in **weight cycling,** the endless repeating rounds of weight loss and regain from "yo-yo" dieting (see Figure 7-4). Over a third of all women report themselves to be perpetual dieters who diet at least once a month.

Fluctuations in body weight appear to increase the risks of chronic diseases and even premature death, independently of obesity itself.[45] Maintaining a stable weight, even if it is overweight, may be less harmful to health than repeated bouts of weight gains and losses. Such concerns should not deter obese people who want to lose weight from trying, but rather should encourage them to commit to lifelong changes (including physical activity) that will maintain weight losses.[46] People who combine diet and physical activity are more likely to lose more fat, retain more muscle, and regain less weight than those who only diet.[47]

■■ **ENERGY EXPENDITURE**   Physical activity makes many contributions to weight loss and maintenance. For one thing, it directly increases energy output by the muscles and cardiovascular system. A 150-pound person walking a brisk 4 miles per hour for 30 minutes spends an extra 185 kcalories on that activity. A football player may spend several thousand extra kcalories on a day of heavy training.

■■ **BMR**   Activity also contributes to energy output in an indirect way—by speeding up basal metabolism.[48] It does this both immediately and over the long term. On any given day, after intense and prolonged exercise, basal metabolism remains elevated for several hours. Over the long term, daily vigorous activity for many weeks gradually shifts body composition toward more lean tissue, which is more active metabolically than fat tissue. The ongoing metabolic rate rises accordingly, and this makes a contribution toward continued weight loss or maintenance.

The raised metabolic rate continues for as long as the person is physically active on a regular basis. The more energy expended in metabolic

Delicious, low-fat, carbohydrate-rich foods such as fresh fruits, vegetables, whole grains, and legumes offer abundant vitamins, minerals, and fiber.

weight cycling: repeated cycles of weight loss and subsequent regain that affect body composition and metabolism. With intermittent dieting, a person rebounds to a higher weight (and a higher body fat content) after each round. The weight-cycling pattern is popularly called the *ratchet effect* or *yo-yo effect* of dieting.

Figure 7-4
THE WEIGHT-CYCLING EFFECT OF REPEATED DIETING

Each round of dieting is followed by a rebound of weight to a higher level than before.

activities, the greater the energy requirement. This means that a person can eat more without gaining weight.

:: **APPETITE CONTROL**   Physical activity also helps to control appetite. People think that exercising will make them hungry, but this is not entirely true.[49] Yes, active people do have healthy appetites, but immediately after a good workout, most people do not feel like eating. They want to shower and may be thirsty, but they are not hungry. The reason is that the body has responded to the stress of activity by mobilizing fuels from storage: glucose and fatty acids are abundant in the blood. At the same time, the body has suppressed its digestive functions. Hard physical work and eating are not compatible.

:: **PSYCHOLOGICAL BENEFITS**   Physical activity helps especially to curb the inappropriate appetite that prompts a person to eat when bored, anxious, or depressed. Weight-control programs encourage people to go out and be active when they're tempted to eat but not really hungry.

Physical activity also helps to reduce stress. Since stress itself is a cue to inappropriate eating behavior for many people, activity can help here, too.

Activity offers still more psychological advantages. The fit person looks and feels healthy, and high self-esteem accompanies these benefits. High self-esteem tends to support a person's resolve to persist in a weight-control effort, rounding out a beneficial cycle.

:: **CHOOSING ACTIVITIES**   What kind of physical activity is best? People seeking to lose weight should choose activities that they enjoy and are willing to do regularly. Nurses and other health care professionals frequently advise people who want to control their body weight and lose fat to engage in activities of low-to-moderate intensity for a long duration, such as an hour-long fast-paced walk. The reasoning behind such advice is that people exercising at low-to-moderate intensity are likely to stick with their activity for longer times and are less likely to injure themselves. In addition, some research suggests that the longer the duration of activity, the greater the contribution fat will make to the fuel mixture, and consequently, the more body fat will be lost—but this conclusion is controversial.[50] Some research refutes the notion that a person "burns more fat during low-intensity activity" and suggests that weight-control benefits are the same from either low- or high-intensity activity.[51] Regardless of the contribution fat makes to the fuel mix, people who engage in regular, vigorous physical activities have less body fat than those who engage in moderately intense activities.[52] Fat use may continue at an accelerated rate for some time after vigorous physical activity has ceased. The conditioned body that is adapted to strenuous and prolonged aerobic activity uses more fat all day long, not just during activity.[53] The bottom line on physical activity and weight and/or fat loss seems to be that total energy expenditure is the main factor, regardless of how a person does it.[54]

In addition to activities such as walking or aerobic dance, there are hundreds of ways to incorporate energy-spending activities into daily routines: take the stairs instead of the elevator, walk to the neighbor's apartment instead of making a phone call, and rake the grass clippings instead of using a bagger. These activities burn only a few kcalories each, but over a year's time they become significant.

:: **SPOT REDUCING**   People sometimes ask about "spot reducing." Unfortunately, no one part of the body gives up fat in preference to another. Fat cells all over the body release fat in response to demand, and the fat is then used by whatever muscles are active. No exercise can remove the fat from any one particular area—and, incidentally, neither can a massage machine that claims to break up fat on trouble spots.

Benefits of physical activity in a weight-control program:

- Short-term increase in energy expenditure (from exercise and from a slight rise in BMR).
- Long-term increase (slight) in BMR.
- Appetite control.
- Stress reduction and control of stress eating.
- Physical, and therefore psychological, well-being.
- High self-esteem.

Regular physical activity helps people achieve and maintain healthy weights.

Physical activity can help with trouble spots in another way, though. Strengthening muscles in a trouble area can help to improve their tone; stretching to gain flexibility can help with posture problems. Thus, cardiorespiratory endurance, strength, and flexibility workouts all have a place in fitness programs.

Behavior and Attitude

Behavior modification once held a key position in weight-loss programs, but its status has diminished.[55] Still, behavior and attitude are important supporting factors in achieving and maintaining appropriate body weight and composition. Changing the behaviors of overeating and underexercising that lead to, and perpetuate, obesity requires time and effort.

BECOMING AWARE OF BEHAVIORS A person who is aware of all the behaviors that create a problem has a head start on developing a solution. First, the person needs to establish a baseline (a record of present eating and physical activity behaviors) against which to measure future progress. It is best to keep a diary (see Figure 7-5) that includes the time and place of meals and snacks, the type and amount of foods eaten, the persons present when food is eaten, and a description of the individual's feelings when eating. The diary should also record physical activities: the kind, the intensity level, the duration, and the person's feelings about them. These entries will help the individual identify possible behaviors to change.

MAKING SMALL CHANGES The "How to" box on p. 154 describes behavioral strategies to support weight control. A particularly attractive feature of these strategies is that they do not involve blaming oneself or putting oneself down—an important element in fostering self-esteem.

MAINTAINING WEIGHT Finally, be aware that it can be hard to maintain weight loss. On arriving at the goal weight after months of self-discipline and new habit formation, the victorious weight loser must not "celebrate" by

behavior modification: the changing of behavior by the manipulation of *antecedents* (cues or environmental factors that trigger behavior), the behavior itself, and *consequences* (the penalties or rewards attached to behavior).

Time	Place	Activity or food eaten	People present	Mood
10:30	School vending machine	6 peanut butter crackers and 12 oz. cola	by myself	starved
12:15	Restaurant	Sub sandwich and 12 oz. cola	friends	relaxed & friendly
3:00	Gym	45 min weight training	work out partner	tired
4:00	Snack bar	Small frozen yogurt	by myself	OK

Figure 7-5
FOOD AND ACTIVITY DIARY

A record of diet and physical activity habits reveals problem areas, the first step toward improving behaviors.

CHANGE BEHAVIORS

Start simply and don't try to master all the behavior changes at once. Attempting too many changes at one time is never successful; a person must set priorities. A person may begin by limiting meals to three a day. When between-meal snacking is no longer a problem, the person may want to change another behavior.

1. To eliminate inappropriate eating cues:
   - Don't buy problem foods (such as ready-to-eat foods).
   - Don't shop when hungry.
   - Don't serve rich sauces and toppings.
   - Let other family members buy, store, and serve their own sweets.
   - Change channels or look away when the television shows food commercials.
   - Shop only from a list and stay away from convenience stores.
   - Carry appropriate snacks from home and avoid vending machines.

2. To suppress the cues that cannot be eliminated:
   - Eat only in one place and in one room.
   - Clear plates directly into the garbage.
   - Create obstacles to the eating of problem foods (for example, make it necessary to unwrap, cook, and serve each one separately).
   - Minimize contact with excessive food (serve individual plates, don't put serving dishes on the table, and leave the table when you have finished eating).
   - Make small portions of food look large (spread food out, serve on small plates).
   - Control states of deprivation (eat regular meals, don't skip meals, avoid getting tired, avoid boredom by keeping cues to fun activities in sight).

3. To strengthen the cues to appropriate eating and physical activity:
   - Eat only at planned times; plan not to eat after a specified time (say, 7:00 or 8:00 P.M.).
   - Encourage others to eat appropriate foods with you.
   - Keep your favorite appropriate foods in the front of the refrigerator.
   - Learn appropriate portion sizes and prepare one portion at a time.
   - Save permitted foods from meals for snacks (and make these your only snacks).
   - Prepare permitted foods attractively.
   - Keep your hiking boots (ski poles, tennis racket) by the door.

4. To engage in desired eating or physical activity behaviors:
   - Slow down (pause several times during a meal, put down utensils between mouthfuls, chew thoroughly before swallowing, swallow before reloading the fork, always use utensils).
   - Leave some food on the plate.
   - Engage in no other activities while eating (such as reading or watching television).
   - Move more (shake a leg, pace, fidget, flex your muscles).
   - Join in and exercise with a group of active people.

5. To arrange or emphasize negative consequences of inappropriate eating:
   - Eat your meals with other people.
   - Ask that others respond neutrally when you deviate from your plan (make no comment). This is a negative consequence because it withholds attention.

6. To arrange or emphasize positive consequences of appropriate behaviors:
   - Update records of food intake, physical activity, and weight change regularly.
   - Arrange for rewards for each unit of behavior change or weight loss.
   - Ask family and friends for reinforcement (praise and encouragement).

resuming old eating habits. Membership in an ongoing weight-control organization and regular, continued physical activity can provide indispensable support for the formerly overweight person who wants to remain trim.

■■ **PERSONAL ATTITUDE** For many people, overeating and being overweight may have become an integral part of their identity. Changing diet and activity behaviors without attention to a person's self-concept invites failure.

Many people overeat to cope with the stresses of life. To break out of that pattern, they must first identify the particular stressors that trigger their urges to overeat. Then, when faced with these situations, they must learn to practice problem-solving skills. When the problems that trigger the urge to overeat are dealt with in alternative ways, people may find that they eat less. The message is that sound emotional health supports the ability to take care of health in all ways—including nutrition, weight control, and fitness.

▟▙ Underweight

For healthy people underweight is far less prevalent than overweight, affecting no more than 10 percent of U.S. adults. Whether the healthy underweight person needs to gain weight is a highly individual matter. People with disorders that lead to undesirable and significant weight loss can quickly develop PEM. People who are healthy at their present weights may stay there; those who are unhealthy might try to gain. Medical advice can help make the distinction.

▟▙ Health Risks of Underweight

The health risks associated with underweight are fewer than those that accompany overweight. Both underweight women and those who have lost a significant amount of weight, however, are more susceptible to osteoporosis. Underweight women may become infertile or may give birth to unhealthy infants. An underweight woman can improve her chances of bearing a healthy infant by gaining weight prior to conception, during pregnancy, or both.

Underweight becomes more hazardous when accompanied by undernutrition. An inadequate supply of nutrients and energy leaves the body underprepared to handle its many metabolic and physical tasks. A person without reserves has a particularly tough battle against medical stresses such as surgery or the wasting diseases of cancer and AIDS. Thus, underweight people are urged to gain lean tissue and body fat (as an energy reserve) and to acquire protective amounts of all the nutrients that can be stored.

▟▙ Strategies for Weight Gain

Some people are unalterably thin by reasons of genetics or early physical influences. Those who wish to gain weight for appearance's sake or to improve athletic performance should be aware that a healthful weight can be achieved only through physical activity, particularly strength training, combined with a high energy intake. Eating many high-kcalorie foods can bring about weight gain, but it will be mostly fat, and this can be as detrimental to health as being slightly underweight. In an athlete, such a weight gain can impair performance. Therefore, in weight gain, as in weight loss, physical activity is an essential component of a sound plan.

▟▙ PHYSICAL ACTIVITY TO BUILD MUSCLES
The person who wants to gain weight should use **weight training** primarily. As activity is increased, energy intake must be increased to support that activity. Eating extra food will then support a gain of both muscle and fat. About 700 to 1000 kcalories a day above normal energy needs is enough to support both the activity and the building of muscle.

▟▙ ENERGY-DENSE FOODS
Energy-dense foods (the very ones eliminated from a successful weight-loss diet) hold the key to weight gain. Pick the

Chapter 13 describes how involuntary weight loss associated with some disorders can hasten the progression of the disease and leave the individual vulnerable to infection and even death from starvation. A "How to" box in Chapter 24 lists strategies for helping clients with wasting disorders gain weight.

Weight-gain pointers:

• **Be physically active and eat to build muscles.**

weight training (also called **resistance training**): the use of free weights or weight machines to provide resistance for developing muscle strength and endurance. A person's own body weight may also be used to provide resistance, as when a person does push-ups, pull-ups, or abdominal crunches.

highest-kcalorie items from each food group—that is, milk shakes instead of fat-free milk, peanut butter instead of lean meat, avocados instead of cucumbers, and blueberry muffins instead of whole-wheat bread. Because fat contains more than twice as many kcalories per teaspoon as sugar does, fat adds kcalories without adding much bulk.

Be aware that health experts recommend a low-fat diet for the general U.S. population because the general population is overweight and at risk for heart disease. Consumption of high-fat foods is not healthy for most people, of course, but may be essential for an underweight individual who needs to gain weight. An underweight person who is physically active and eating a nutritionally adequate diet can afford a few extra kcalories from fat.

■■ THREE MEALS DAILY People wanting to gain weight should eat at least three hearty meals a day. Many people who are underweight have simply been too busy (sometimes for months) to eat enough to gain or maintain weight. Therefore, they need to make meals a priority and plan them in advance. Taking time to prepare and eat each meal can help, as can learning to eat more food within the first 20 minutes of a meal. Another suggestion is to eat meaty appetizers or the main course first and leave the soup or salad until later.

■■ LARGE PORTIONS It is also important for the underweight person to learn to eat more food at each meal: have two sandwiches for lunch instead of one, drink milk from a larger glass, and eat cereal from a larger bowl. Expect to feel full. Most underweight individuals are accustomed to small quantities of food. When they begin eating significantly more, they feel uncomfortable. This is normal and passes over time.

■■ EXTRA SNACKS Since a substantially higher energy intake is needed each day, in addition to eating more food at each meal, it is necessary to eat more frequently. Between-meal snacking offers a solution. For example, a student might make three sandwiches in the morning and eat them between classes in addition to the day's three regular meals.

■■ JUICE AND MILK Beverages provide an easy way to increase energy intake. Consider that 6 cups of cranberry juice add almost 1000 kcalories to the day's intake. kCalories can be added to milk by mixing in powdered milk or packets of instant breakfast.

For people who are underweight due to illness, concentrated liquid formulas are often recommended because a weak person can swallow them easily. A nurse or registered dietitian can recommend high-protein, high-kcalorie formulas or fortified snacks such as candy bars and pudding to help the underweight person maintain or gain weight. Used in addition to regular meals, these supplements can help considerably.

An extreme underweight condition known as anorexia nervosa is sometimes seen in young people who exercise unreasonable self-denial in order to control their weight. They go to such extremes that they become severely undernourished and underweight. The distinguishing feature of a person with anorexia nervosa, as opposed to other thin people, is that the starvation is intentional. Anorexia nervosa is a major eating disorder seen in our society today. Another is bulimia nervosa—compulsive overeating, usually with purging. Eating disorders are the subject of the Nutrition in Practice that follows this chapter.

- Eat energy-dense foods regularly.

- Eat at least three hearty meals a day.

- Eat large portions of foods and expect to feel full.

- Eat snacks between meals.

- Drink plenty of juice and milk.

SELF CHECK

1. The BMI range that correlates with the fewest health risks is:
   a. 16.5 to 20.9.
   b. 18.5 to 24.9.
   c. 25.5 to 30.9.
   d. 30.5 to 34.9.

2. The profile of central obesity is sometimes referred to as a(an):
   a. beer.
   b. pear.
   c. apple.
   d. potato.

3. Two causes of obesity in humans are:
   a. set-point theory and BMI.
   b. genetics and physical inactivity.
   c. genetics and low-carbohydrate diets.
   d. mineral imbalances and fat cell imbalance.

4. The protein produced by the fat cells under the direction of the *ob* gene is called:
   a. leptin.
   b. orlistat.
   c. sibutramine.
   d. lipoprotein lipase.

5. The obesity theory that suggests the body chooses to be at a specific weight is the:
   a. fat cell theory.
   b. enzyme theory.
   c. set-point theory.
   d. external cue theory.

6. Which of the following health risks is **not** associated with being overweight?
   a. hypertension
   b. heart disease
   c. type 1 diabetes
   d. gallbladder disease

7. Which of the following is a reasonable treatment for obesity?
   a. Eat energy-dense foods regularly.
   b. Drink plenty of juice and milk.
   c. Use strength training to build muscles.
   d. Make legumes, grains, vegetables, and fruits central to your diet plan.

8. Physical activity does **not** help a person to:
   a. lose weight.
   b. retain muscle.
   c. maintain weight loss.
   d. lose fat in trouble spots.

9. Suggestions to change behaviors for successful weight control include:
   a. shop only when hungry.
   b. eat in front of the television for distraction.
   c. learn appropriate portion sizes.
   d. eat quickly.

10. Which strategy would **not** help an underweight person to gain weight?
    a. Exercise.
    b. Drink plenty of water.
    c. Eat snacks between meals.
    d. Eat large portions of foods.

Answers to these questions appear in Appendix H.

CLINICAL APPLICATIONS

1. Consider a female client with the following height and weight:
   - Height: 65 in (or 165 cm).
   - Weight: 165 lb (or 75 kg).
   Look up the weight range for a person of your client's height in Table 7-1 on p. 138.
   - Record the entire range: ____ to ____ lb.

   Does your client's weight fall within the suggested range?

   Now calculate your client's BMI using the equations on p. 139.

   - Record your client's BMI: ____.

   Look up the disease risk for a person with your client's BMI value in Table 7-2 on p. 139.
   - Record your client's risk of disease based on the BMI: _____.

2. What information might you want to learn about your client's health habits and family or personal medical history to help you in further determining the client's risk of disease?

3. Check your client's health history. A family or personal medical history of diabetes (type 2), hypertension, or high blood cholesterol signals the need to pay attention to diet and physical activity habits.

NUTRITI**ON**THENET

FOR FURTHER STUDY OF THE
TOPICS IN THIS CHAPTER,
ACCESS THESE WEB SITES.

shapeup.org
Shape Up America!

www.niddk.nih.gov/health/nutrit/
win.htm
Weight-control Information Network

naaso.org
*North American Association for the
Study of Obesity*

Notes

[1] S. M. Garn, Fractionating healthy weight, *American Journal of Clinical Nutrition* 63 (1996): S412–S414.

[2] National Heart, Lung, and Blood Institute Expert Panel, National Institutes of Health, *Clinical Guidelines on the Identification, Evaluation, and Treatment of Overweight and Obesity in Adults* (Washington, D.C.: Government Printing Office, 1998).

[3] A. R. Folsom and coauthors, Body fat distribution and 5-year risk of death in older women, *Journal of the American Medical Association* 269 (1993): 483–487.

[4] S. Lemieux and coauthors, Sex differences in the relation of visceral adipose tissue accumulation to total body fatness, *American Journal of Clinical Nutrition* 58 (1993): 463–467; C. J. Ley, B. Lees, and J. C. Stevenson, Sex- and menopause-associated changes in body-fat distribution, *American Journal of Clinical Nutrition* 55 (1992): 950–954.

[5] M. J. Williams and coauthors, Regional fat distribution in women and risk of cardiovascular disease, *American Journal of Clinical Nutrition* 65 (1997): 855–860.

[6] E. M. Emery and coauthors, A review of the association between abdominal fat distribution, health outcome measures, and modifiable risk factors, *American Journal of Health Promotion,* May/June 1993, pp. 342–353; R. J. Troisi, Cigarette smoking, dietary intake, and physical activity: Effects on body fat distribution—The Normative Aging Study, *American Journal of Clinical Nutrition* 53 (1991): 1104–1111.

[7] B. B. Duncan and coauthors, Association of the waist-to-hip ratio is different with wine than with beer or hard liquor consumption, *American Journal of Epidemiology* 142 (1995): 1034–1038.

[8] G. R. Hunter and coauthors, Fat distribution, physical activity, and cardiovascular risk factors, *Medicine and Science in Sports and Exercise* 29 (1997): 362–369.

[9] C. Orphanidou and coauthors, Accuracy of subcutaneous fat measurement: Comparison of skinfold calipers, ultrasound, and computed tomography, *Journal of the American Dietetic Association* 94 (1994): 855–858.

[10] National Heart, Lung, and Blood Institute Expert Panel, 1998.

[11] L. Perusse and C. Bouchard, Genotype-environment interaction in human obesity, *Nutrition Reviews* 57 (1999): S31–S38.

[12] J. P. Foreyt and W. S. C. Poston II, Diet, genetics, and obesity, *Food Technology* 51 (1997): 70–73; C. Bouchard, Human variation in body mass: Evidence for a role of the genes, *Nutrition Reviews* 55 (1997): S21–S30.

[13] Y. Zhang and coauthors, Positional cloning of the mouse obese gene and its human homologue, *Nature* 372 (1994): 425–431.

[14] C. T. Montague and coauthors, Congenital leptin deficiency is associated with severe early-onset obesity in humans, *Nature* 387 (1997): 903–908.

[15] R. V. Considine, Serum immunoreactive-leptin concentrations in normal-weight and obese humans, *New England Journal of Medicine* 334 (1996): 292–295; S. G. Hassink and coauthors, Serum leptin in children with obesity: Relationship to gender and development, *Pediatrics* 98 (1996): 201–203.

[16] J. Albu and coauthors, Obesity solutions: Report of a meeting, *Nutrition Reviews* 55 (1997): 150–156.

[17] R. L. Leibel, M. Rosenbaum, and J. Hirsch, Changes in energy expenditure resulting from altered body weight, *New England Journal of Medicine* 332 (1995): 621–628; J. M. Kinney, Influence of altered body weight on energy expenditure, *Nutrition Reviews* 53 (1995): 265–268; P. Pasquet and M. Apfelbaum, Recovery of initial body weight and composition after long-term massive overfeeding in men, *American Journal of Clinical Nutrition* 60 (1994): 861–863.

[18] R. E. Keesey and M. D. Hirvonen, Body weight set-points: Determination and adjustment, *Journal of Nutrition* 127 (1997): S1875–S1883.

[19] J. E. Blundell and coauthors, Control of human appetite: Implications for the intake of dietary fat, *Annual Review of Nutrition* 16 (1996): 285–319.

[20] Prevalence of physical inactivity during leisure time among overweight persons—1994, *Morbidity and Mortality Weekly Report* 45 (1996): 185–188.

[21] A. P. Simopoulos, Characteristics of obesity, in *Obesity,* ed. P. Björntorp and B. N. Brodoff (Philadelphia: J. B. Lippincott, 1992), pp. 309–319.

[22] R. E. Anderson and coauthors, Relationship of physical activity and television watching with body weight and level of fatness among children: Results from the Third National Health and Nutrition Survey, *Journal of the American Medical Association* 279 (1998): 938–942.

[23] D. A. McCarron and M. E. Reusser, Body weight and blood pressure regulation, *American Journal of Clinical Nutrition* 63 (1996): S423–S425.

[24] Update: Prevalence of overweight among children, adolescents, and adults—United States, 1988–1994, Morbidity and Mortality Weekly Report 46 (1997): 199–202.

[25] J. Foreyt and K. Goodrick, The ultimate triumph of obesity, *The Lancet* 346 (1995): 134–135.

[26] National Heart, Lung, and Blood Institute Expert Panel, 1998; Albu and coauthors, 1997.

[27] National Task Force on the Prevention and Treatment of Obesity, Long-term pharmocotherapy in the management of obesity, *Journal of the American Medical Association* 276 (1996): 1907–1915.

[28] J. E. Manson and coauthors, Body weight and mortality among women, *New England Journal of Medicine* 333 (1995): 677–685; W. C. Willett and coauthors, Weight, weight change, and coronary heart disease in women, *Journal of the American Medical Association* 273 (1995): 461–465.

[29] W. B. Kannel, R. B. D'Agostino, and J. L. Cobb, Effect of weight on cardiovascular disease, *American Journal of Clinical Nutrition* 63 (1996): S419–S422.

[30] A. Must and coauthors, Long-term morbidity and mortality of overweight adolescents, *New England Journal of Medicine* 327 (1992): 1350–1355; G. A. Bray, Adolescent overweight may be tempting fate, *New England Journal of Medicine* 327 (1992): 1378–1380.

[31] S. L. Gortmaker and coauthors, Social and economic consequences of overweight in adolescence and young adulthood, *New England Journal of Medicine* 329 (1993): 1008–1012.

[32] National Task Force on the Prevention and Treatment of Obesity, 1996.

[33] C. H. Halsted, Is blockade of pancreatic lipase the answer? *American Journal of Clinical Nutrition* 69 (1999): 1059–1060.

[34] D. L. Hansen and coauthors, Thermogenic effects of sibutramine in humans, *American Journal of Clinical Nutrition* 68 (1998): 1180–1186.

[35]L. J. Aronne, Modern medical management of obesity: The role of pharmaceutical intervention, *Journal of the American Dietetic Association* 98 (1998): S23–S26.

[36]M. H. Davidson and coauthors, Weight control and risk factor reduction in obese subjects treated for 2 years with orlistat, *Journal of the American Medical Association* 281 (1999): 235–242.

[37]National Heart, Lung, and Blood Institute Expert Panel, 1998.

[38]R. Weiner, D. Wagner, and H. Bockhorn, Laparoscopic gastric banding for morbid obesity, *Journal of Laparoendoscopic and Advanced Surgical Techniques* 9 (1999): 23–30.

[39]R. P. Abernathy and D. R. Black, Healthy body weights: An alternative perspective, *American Journal of Clinical Nutrition* 63 (1996): S448–S451.

[40]J. G. Meisler and S. St. Jeor, Summary and recommendations for the American Health Foundation's Expert Panel on Healthy Weight, *American Journal of Clinical Nutrition* 63 (1996): S474–S477.

[41]M. Shah and coauthors, Comparison of a low-fat ad libitum complex-carbohydrate diet with a low-energy diet in moderately obese women, *American Journal of Clinical Nutrition* 59 (1994): 980–984.

[42]A. Astrup and coauthors, The role of low-fat diets and fat substitutes in body weight management: What have we learned from clinical studies? *Journal of the American Dietetic Association* 97 (1997): S82–S87; Shah and coauthors, 1994.

[43]S. Toubro and A. Astrup, Ad libitum low-fat, high-carbohydrate diet versus calorie-counting for weight maintenance after major weight loss in obese patients, *British Medical Journal* 314 (1997): 29–34.

[44]S. N. Blair, Diet and activity: The synergistic merger, *Nutrition Today* 30 (1996): 108–112; J. H. Wilmore, Increasing physical activity: Alterations in body mass and composition, *American Journal of Clinical Nutrition* 63 (1996): 254–260.

[45]R. W. Jeffrey, Does weight cycling present a health risk? *American Journal of Clinical Nutrition* 63 (1996): S452–S455.

[46]National Task Force on the Prevention and Treatment of Obesity, Weight cycling, Journal of the American Medical Association 272 (1994): 1196–1202.

[47]A. Geliebter and coauthors, Effects of strength or aerobic training on body composition, resting metabolic rate, and peak oxygen consumption in obese dieting subjects, *American Journal of Clinical Nutrition* 66 (1997): 557–563; B. L. Marks and coauthors, Fat-free mass is maintained in women following a moderate diet and exercise program, *Medicine and Science in Sports and Exercise* 27 (1995): 1243–1251; K. P. G. Kempen, W. H. M. Saris, and K. R. Westerterp, Energy balance during an 8-wk energy-restricted diet with and without exercise in obese women, *American Journal of Clinical Nutrition* 62 (1995): 722–729; R. Ross, H. Pedwell, and J. Rissanen, Effects of energy restriction and exercise on skeletal muscle and adipose tissue in women as measured by magnetic resonance imaging, *American Journal of Clinical Nutrition* 61 (1995): 1179–1185; S. B. Racette and coauthors, Effects of aerobic exercise and dietary carbohydrate on energy expenditure and body composition during weight reduction in obese women, *American Journal of Clinical Nutrition* 61 (1995): 486–494; D. D. Hensrud and coauthors, A prospective study of weight maintenance in obese subjects reduced to normal body weight without weight-loss training, *American Journal of Clinical Nutrition* 60 (1994): 688–694.

[48]H. M. Sjödin and coauthors, The influence of physical activity on BMR, *Medicine and Science in Sports and Exercise* 28 (1996): 85–91.

[49]N. A. King, A. Tremblay, and J. E. Blundell, Effects of exercise on appetite control: Implications for energy balance, *Medicine and Science in Sports and Exercise* 29 (1997): 1076–1089.

[50]P. Arnos, F. Andres, and K. Drowatzky, Fat oxidation and RPE at varied exercise intensities, *Medicine and Science in Sports and Exercise* 25 (1993): S9; F. A. Kulling and coauthors, Identification and evaluation of the exercise intensity which maximizes fat oxidation in young women, *Medicine and Science in Sports and Exercise* 25 (1993): S179.

[51]M. A. Grediagin and coauthors, Exercise intensity does not affect body composition change in untrained, moderately overfat women, *Journal of the American Dietetic Association* 95 (1995): 661–665.

[52]A. Tremblay and coauthors, Effect of intensity of physical activity on body fatness and distribution, American Journal of Clinical Nutrition 51 (1990): 153–157.

[53]T. J. Horton and C. A. Geissler, Effect of habitual exercise on daily energy expenditure and metabolic rate during standardized activity, *American Journal of Clinical Nutrition* 59 (1994): 13–19.

[54]Grediagin and coauthors, 1995.

[55]Albu and coauthors, 1997.

Eating Disorders

An estimated 2 million people in the United States, primarily girls and young women, suffer from the eating disorders **anorexia nervosa** and **bulimia nervosa** (the accompanying glossary defines these and related terms). Many more suffer from **binge eating disorder** or other related conditions that do not meet the strict criteria for anorexia nervosa or bulimia nervosa (described later), but still imperil a person's well-being. Characteristics of disordered eating such as restrained eating, binge eating, purging, fear of fatness, and distortion of body image are common, especially among young middle-class girls.[1] In most other societies, these behaviors and attitudes are much less prevalent.

www.members.aol.com/amanbu
American Anorexia/Bulimia Association

Why do so many young people in our society suffer from eating disorders?

Excessive pressure to be thin is at least partly to blame. Two-thirds of adolescent girls and one-third of adolescent boys are dissatisfied with their body weight.[2] By making thinness the ideal, society pushes people to view a healthy body of normal weight as too fat. Healthy people then take unhealthy actions to lose weight. Severe restriction of food intake may create intense hunger that leads to binges. Research confirms this theory, showing that unhealthy or dangerous diets often precede binge eating in adolescent girls.[3] Energy restriction followed by bingeing can set in motion a pattern of weight cycling, which may make weight loss and maintenance more difficult over time.[4]

People who attempt extreme weight loss are dissatisfied with their bodies to begin with; they may also

be depressed or suffer social anxiety. As weight loss becomes more and more difficult, psychological problems worsen, and the likelihood of developing full-blown **eating disorders** intensifies.

People with anorexia nervosa suffer from an extreme preoccupation with weight loss that seriously endangers their health and even their lives. People with bulimia nervosa engage in episodes of binge eating alternating with periods of severe dieting or self-starvation. Some bulimics also follow binge eating with self-induced vomiting, laxative abuse, or diuretic abuse to "undo the damage."

Are there other groups, besides girls and young women, who are vulnerable to anorexia nervosa and bulimia nervosa?

Yes. Athletes are particularly likely to develop eating disorders.[5] Athletes must often meet stringent weight requirements to compete in their sport. Many athletes report that they engage in behaviors that are typical of people with eating disorders. Female competitors often report being terrified of becoming fat, being obsessed with food, and using laxatives to attempt to control weight. Dancers, jockeys, wrestlers, distance runners, bodybuilders, divers, figure skaters, gymnasts, and others whose body weight and appearance are frequently judged in comparison with an "ideal" are especially prone to develop problems.[6] These athletes may engage in extreme weight-loss practices such as overtraining, prolonged fasting, vomiting, taking diet pills, and using steam baths and saunas to induce sweating.[7]

Men account for about 1 in 20 cases in the general population, but among male athletes and dancers, eating disorders are much more common. Adolescent males normally average about 15 percent of body

weight as fat, but some high school athletes strive to carry only 5 percent or so of their body weight as fat.

Even among athletes, however, women are most vulnerable to developing eating disorders. Many female athletes appear healthy, but in fact may easily develop the three interrelated components of the **female athlete triad**: disordered eating, amenorrhea (the absence of three or more consecutive menstrual cycles), and osteoporosis.[8]

GLOSSARY

anorexia nervosa: an eating disorder characterized by a refusal to maintain a minimally normal body weight, self-starvation to the extreme, and a disturbed perception of body weight and shape; seen (usually) in adolescent girls and young women.
   anorexia = without appetite
   nervosa = of nervous origin

binge eating disorder: an eating disorder whose criteria are similar to those of bulimia nervosa, excluding purging or other compensatory behaviors.

bulimia (byoo-LEEM-ee-uh) nervosa: recurring episodes of binge eating combined with a morbid fear of becoming fat, usually followed by self-induced vomiting or purging.

cathartic: a strong laxative.

cognitive therapy: psychological therapy aimed at changing undesirable behaviors by changing underlying thought processes contributing to these behaviors. In anorexia nervosa, a goal is to replace false beliefs about body weight, eating, and self-worth with health-promoting beliefs.

eating disorder: a disturbance in eating behavior that jeopardizes a person's physical and psychological health.

emetic (em-ETT-ic): an agent that causes vomiting.

female athlete triad: a potentially fatal triad of medical problems: disordered eating, amenorrhea, and osteoporosis.

How does the female athlete triad develop?

Many athletic women engage in self-destructive eating behaviors (disordered eating) because they and their coaches have adopted unsuitable weight standards. An athlete's body must be heavier for height than a nonathlete's body because the athlete's bones and muscles are denser. Weight standards that may be appropriate for others are inappropriate for athletes. Measures such as fatfold measures yield more useful information about body composition.

Many young female athletes severely restrict energy intakes to improve performance, enhance the aesthetic appeal of their performance, or meet the weight guidelines of their specific sports.[9] They fail to realize that the loss of lean tissue that accompanies energy restriction actually impairs their physical performance. Risk factors for the female athlete triad include the following:

- Young age (adolescence).
- Pressure to excel at a chosen physical activity.
- Focus on achieving or maintaining an "ideal" body weight or body fat percentage.
- Participation in endurance sports or competitions that judge performance on aesthetic appeal such as gymnastics, figure skating, or dance.
- Dieting at an early age.
- Unsupervised dieting.

As for amenorrhea, its prevalence among premenopausal women in the United States is about 2 to 5 percent overall, but among female athletes it may be as high as 66 percent.[10] Contrary to previous notions, amenorrhea is not a normal adaptation to strenuous physical training: it is a symptom of something going wrong.[11] Amenorrhea is characterized by low blood estrogen, infertility, and often bone mineral losses.

In general, weight-bearing physical activity, dietary calcium, and the hormone estrogen protect against the bone loss of osteoporosis, but in women with disordered eating and amenorrhea, strenuous activity may impair bone health.[12] Vigorous training combined with low food energy intakes and other life stresses seems to trigger amenorrhea and promote bone loss. Low estrogen leads to diminished bone mass and increased bone fragility. Many amenorrheic athletes have decreased bone density, similar to that of 50- to 60-year-old women, when they should have dense, strong bones. Amenorrheic athletes should be encouraged to consume at least 1500 milligrams of calcium each day, to eat nutrient-dense foods, and to obtain enough food energy to cover the energy expended in physical activity. Future research will focus on the question of hormone replacement therapy for these women.

What can be done to prevent eating disorders in athletes and dancers?

To prevent eating disorders in athletes and dancers, both the performers and their coaches must be educated about links among inappropriate body weight ideals, improper weight-loss techniques, eating disorder development, adequate nutrition, and safe weight-control methods. Coaches and dance instructors should never encourage unhealthy weight loss to qualify for competition or to conform to distorted artistic ideals. Frequent weighings can push young people who are striving to lose weight into a cycle of starving to confront the scale, then bingeing uncontrollably afterward. The erosion of self-esteem that accompanies these events can interfere with the normal psychological development of the adolescent years and set the stage for serious problems later on.

The "How to" box on the next page provides some suggestions to help athletes and dancers protect themselves against developing eating disorders. The next sections describe eating disorders that anyone, athlete or nonathlete, may experience.

What are the characteristics of anorexia nervosa?

Most anorexia nervosa victims are females who come from middle- or upper-class families. Family patterns often include parents who oppose one another's authority and who vacillate between defending and condemning the anorexic child's behavior, confusing the child and disrupting normal parental control.[13]

The person with anorexia nervosa has low self-esteem and is often a perfectionist who works hard to please her parents. She may identify so strongly with her parents' ideals and goals for her that she sometimes feels she has no identity of her own. She earnestly desires to control her own destiny, but she feels controlled by others. When she does not eat, she gains control.

Although families of children with anorexia nervosa often have problems, blame is a useless concept, and parents may suffer deeply from being blamed for their child's illness. In truth, no one knows what causes anorexia nervosa, and it may turn out that the cause is a physical one, and not related to parenting. Rather than judging parents, a more useful tactic is to identify the family's strong points and resources and to prepare them for the job of helping their ill child benefit from treatment.

How does a person know when dieting is going too far?

When a person loses weight to well below the average for her height and is no longer slim, but too slim, and still doesn't stop, she has gone too far. Regardless of how thin she is, she looks in the mirror and sees herself as fat. Central to the diagnosis of anorexia nervosa is a distorted body image that overestimates body fatness. Table NP7-1 (p. 163) shows the criteria that professionals use to diagnose anorexia nervosa. Anorexia nervosa resembles an addiction. The characteristic behavior is obsessive and compulsive. Before drawing conclusions about someone who is

COMBAT EATING DISORDERS

The following guidelines may be useful in combating eating disorders:

- Never restrict food servings to below the numbers suggested for adequacy by the Food Guide Pyramid.
- Eat frequently. People often do not eat frequent meals because of time constraints, but eating can be incorporated into other activities, such as snacking while studying or commuting. The person who eats frequently never gets so hungry as to allow hunger to dictate food choices.
- If not at a healthy weight, establish a reasonable weight goal based on a healthy body composition.
- Allow a reasonable time to achieve the goal. A reasonable loss of excess fat can be achieved at the rate of about 1 percent of body weight per week.
- Establish a weight-maintenance support group with people who share interests.

Specific guidelines for athletes and dancers include:

- Remember that eating disorders impair physical performance. Seek confidential help in obtaining treatment if needed.
- Restrict weight-loss activities to the off-season.
- Focus on proper nutrition as an important facet of your training—as important as proper technique.

chotherapy and supplemental formulas to provide extra nutrients and energy.

High-risk clients may require hospitalization and may need to be force-fed by tube at first to forestall death. This step causes psychological trauma. Medications are commonly prescribed, but to date, they play a limited role in treatment.

Denial runs high among those with anorexia nervosa. Few seek treatment on their own. Almost half of the women who are treated can maintain their body weight within 15 percent of healthy weight; at that weight, many of them begin menstruating again. The other half have poor or fair treatment outcomes, and two-thirds of those treated fight an ongoing mental battle with recurring morbid thoughts about food and body weight.[16] Many relapse into abnormal eating behaviors to some extent. About 5 percent die during treatment, 1 percent by suicide.

extremely thin or who eats very little, remember that diagnosis of anorexia nervosa requires professional assessment.

What is the harm in being very thin?

Anorexia nervosa damages the body much as starvation does. In young people, growth ceases and normal development falters. They lose so much lean tissue that basal metabolic rate slows. Additionally, the heart pumps inefficiently and irregularly, the heart muscle becomes weak and thin, the heart chambers diminish in size, and the blood pressure falls. Electrolytes that help to regulate heartbeat become unbalanced.

Starvation brings other physical consequences as well: impaired immune response, anemia, and a loss of digestive function that worsens malnutrition. Digestive functioning becomes sluggish, the stomach empties slowly, and the lining of the intestinal tract shrinks. The ailing digestive tract fails to sufficiently digest any food the victim may eat. The pancreas slows its production of

digestive enzymes. The person may suffer from diarrhea, further worsening malnutrition.

What kind of treatment helps people with anorexia nervosa?

Treatment of anorexia nervosa requires a multidisciplinary approach that addresses two sets of issues and behaviors: those relating to food and weight, and those involving relationships with oneself and others.[14] Teams of physicians, nurses, psychiatrists, family therapists, and dietitians work together to treat people with anorexia nervosa. Appropriate diet is crucial for normalizing body weight and must be tailored individually to each client's needs.[15] Seldom are clients willing to eat for themselves, but if they are, chances are they can recover without other interventions.

Professionals classify clients based on the risks posed by the degree of malnutrition present. Clients with low risks may benefit from family counseling, **cognitive therapy,** behavior modification, and nutrition guidance; those with greater risks may also need other forms of psy-

How does bulimia nervosa differ from anorexia nervosa?

Bulimia nervosa is distinct from anorexia nervosa and is more prevalent. More men suffer from bulimia

Women with anorexia nervosa see themselves as fat, even when they are dangerously underweight.

nervosa than from anorexia, but bulimia is still more common in women. The secretive nature of bulimic behaviors makes recognition of the problem difficult, but once it is recognized, diagnosis is based on the criteria listed in Table NP7-2.

The typical person with bulimia is well educated, in her early twenties, and close to ideal body weight. She is a high achiever, with a strong feeling of dependence on her parents. She experiences considerable social anxiety and has difficulty establishing personal relationships. She is sometimes depressed and often exhibits impulsive behavior.

Like the person with anorexia nervosa, the person with bulimia spends much time thinking about her body weight and food. Her preoccupation with food manifests itself in secretive binge-eating episodes followed by self-induced vomiting, fasting, or the use of laxatives or diuretics. Such behaviors typically begin in late adolescence after a long series of various unsuccessful weight-reduction diets. People with bulimia commonly follow a pattern of restrictive dieting interspersed with bulimic behaviors and experience weight fluctuations of more than 10 pounds up and down over short periods of time.

Unlike the person with anorexia nervosa, the person with bulimia is aware of the consequences of her behavior, feels that it is abnormal, and is deeply ashamed of it. She feels inadequate and unable to control her eating, so she tends to be passive and to look to men for confirmation of her sense of self-worth. When she is rejected, either in reality or in her imagination, her bulimia becomes worse. If her depression deepens, she may seek solace in drug or alcohol abuse or other addictive behaviors. Many studies show a link between bulimia nervosa and drug and alcohol dependency.[17]

What exactly is binge eating?

Binge eating is unlike normal eating, and the food is not consumed for its nutritional value. The binge eater has a compulsion to eat. A typical binge occurs periodically, is done in secret, usually at night, and lasts an hour or more. A binge frequently follows a period of rigid dieting, so the binge eating is accelerated by hunger. During a binge, the person with bulimia may consume from 1000 to many thousands of kcalories of food. Although the foods consumed during a binge may vary, those typically chosen include sweet, high-kcalorie foods such as cookies, cake, or ice cream.

Table NP7-1
CRITERIA FOR DIAGNOSIS OF ANOREXIA NERVOSA

A person with anorexia nervosa demonstrates the following:
A. Refusal to maintain body weight at or above a minimal normal weight for age and height, e.g., weight loss leading to maintenance of body weight less than 85 percent of that expected; or failure to make expected weight gain during period of growth, leading to body weight less than 85 percent of that expected.
B. Intense fear of gaining weight or becoming fat, even though underweight.
C. Disturbance in the way in which one's body weight or shape is experienced; undue influence of body weight or shape on self-evaluation, or denial of the seriousness of the current low body weight.
D. In females past puberty, amenorrhea, i.e., the absence of at least three consecutive menstrual cycles. (A woman is considered to have amenorrhea if her periods occur only following hormone, e.g., estrogen, administration.)

Two types:
- **Restricting type:** During the episode of anorexia nervosa, the person does not regularly engage in binge eating or purging behavior (i.e., self-induced vomiting or the misuse of laxatives, diuretics, or enemas).
- **Binge eating/purging type:** During the episode of anorexia nervosa, the person regularly engages in binge eating or purging behavior (i.e., self-induced vomiting or the misuse of laxatives, diuretics, or enemas).

SOURCE: Reprinted with permission from the *Diagnostic and Statistical Manual of Mental Disorders*, 4th ed. (Washington, D.C.: American Psychiatric Association, 1994). Copyright 1994 American Psychiatric Association.

Table NP7-2
CRITERIA FOR DIAGNOSIS OF BULIMIA NERVOSA

A person with bulimia nervosa demonstrates the following:
A. Recurrent episodes of binge eating. An episode of binge eating is characterized by both of the following:
1. eating, in a discrete period of time (e.g., within any two-hour period), an amount of food that is definitely larger than most people would eat during a similar period of time and under similar circumstances, and,
2. a sense of lack of control over eating during the episode (e.g., a feeling that one cannot stop eating or control what or how much one is eating).
B. Recurrent inappropriate compensatory behavior in order to prevent weight gain, such as self-induced vomiting; misuse of laxatives, diuretics, enemas, or other medications; fasting; or excessive exercise.
C. Binge eating and inappropriate compensatory behaviors that both occur, on average, at least twice a week for three months.
D. Self-evaluation unduly influenced by body shape and weight.
E. The disturbance does not occur exclusively during episodes of anorexia nervosa.

Two types:
- **Purging type:** The person regularly engages in self-induced vomiting or the misuse of laxatives, diuretics, or enemas.
- **Nonpurging type:** The person uses other inappropriate compensatory behaviors, such as fasting or excessive exercise, but does not regularly engage in self-induced vomiting or the misuse of laxatives, diuretics, or enemas.

SOURCE: Reprinted with permission from the *Diagnostic and Statistical Manual of Mental Disorders*, 4th ed. (Washington, D.C.: American Psychiatric Association, 1994). Copyright 1994 American Psychiatric Association.

HOW TO:

ESTABLISH HEALTHY EATING PATTERNS

The following advice has proved useful for people fighting bulimia nervosa.

- Avoid finger foods; eat foods that require the use of utensils.
- Enhance satiety by eating warm foods.
- Include vegetables, salad, and/or fruit at meals to prolong eating time.
- Choose whole-grain and high-fiber breads and cereals to maximize bulk.
- Eat a well-balanced diet and meals consisting of a variety of foods.
- Use foods that are naturally divided into portions, such as potatoes (rather than rice or pasta); 4- and 8-ounce containers of yogurt, ice cream, or cottage cheese; precut steak or chicken parts; and frozen entrées.
- Include foods containing ample complex carbohydrates (for satiety) and some fat (to slow gastric emptying).
- Eat meals and snacks sitting down.
- Plan meals and snacks, and record plans in a food diary prior to eating.

SOURCE: Reprinted by permission of Lippincott, Williams & Wilkins.

What are the consequences of binge eating?

After a binge, the person may use a **cathartic**—a strong laxative that can injure the lower intestinal tract. Or the person may induce vomiting,

using an **emetic**—a drug intended as first aid for poisoning.

On first glance, purging seems to offer a quick and easy solution to the problems of unwanted kcalories and body weight. Many people perceive such behavior as neutral or even positive, but, in fact, bingeing and purging have serious physical consequences.[18] Fluid and electrolyte imbalances caused by vomiting or diarrhea can lead to metabolic alkalosis, a condition characterized by apathy, confusion, and muscle spasms. Vomiting causes irritation and infection of the pharynx, esophagus, and salivary glands; erosion of the teeth; and dental caries. The esophagus may rupture or tear, as may the stomach. Overuse of emetics depletes potassium concentrations and can lead to death by heart failure.[19]

What is the treatment for bulimia?

As for people with anorexia nervosa, a team approach provides the most effective treatment for people with bulimia. Bulimia is easier to treat than anorexia nervosa in many

For many people with bulimia, guilt, depression, and self-condemnation follow a binge-eating episode.

respects because it seems to be more of a chosen behavior. People with bulimia know that their behavior is abnormal, and many are willing to try to cooperate.

The goal of the dietary plan to treat bulimia is to help clients gain control, establish regular eating patterns, and restore nutritional health. Energy intake should not be severely restricted. The person needs to learn to eat a quantity of nutritious food sufficient to nourish her body and to satisfy hunger (at least 1600 kcalories a day). The accompanying "How to" box offers some ways to begin correcting bulimia nervosa.

Anorexia nervosa and bulimia nervosa are distinct eating disorders, yet they sometimes overlap. Anorexia victims may purge, and victims of both conditions share an overconcern with body weight and the tendency to drastically undereat. The two disorders can also appear in the same person, or one can lead to the other.

At so tender an age as 12 years, beautifully growing, normal-weight female youngsters are already worried that they are too fat. Most are "on diets." Magazines, newspapers, and television all present the message that to be thin is to be beautiful and happy. Anorexia nervosa and bulimia are not a form of rebellion against these unreasonable expectations, but rather the exaggerated acceptance of them. Perhaps a person's best defense against these disorders is to learn to appreciate his or her own uniqueness.

Notes

[1] G. B. Schreiber and coauthors, Weight modification efforts reported by black and white preadolescent girls: National Heart, Lung, and Blood Institute Growth and Health Study, *Pediatrics* 98 (1996): 63–70; L. M. Mellin, C. E. Irwin, and S. Scully, Prevalence of disordered eating in girls: A survey of middle-class children, *Journal of the American Dietetic Association* 92 (1992): 851–853.

[2] D. C. Moore, Body image and eating behavior in adolescents, *Journal of the American College of Nutrition* 12 (1993): 975–980.

[3] G. C. Patton and coauthors, Onset of adolescent eating disorders: Population-based cohort study

over 3 years, *British Journal of Medicine* 318 (1999): 765–768; D. Neumark-Sztainer, R. Butler, and H. Palti, Dieting and binge eating: Which dieters are at risk? *Journal of the American Dietetic Association* 95 (1995): 586–588.

[4]D. Neumark-Sztainer, Excessive weight preoccupation, *Nutrition Today,* March/April 1995, pp. 68–74.

[5]K. K. Yeager and coauthors, The female athlete triad: Disordered eating, amenorrhea, osteoporosis, *Medicine and Science in Sports and Exercise* 25 (1993): 775–777.

[6]A. Yates, Eating disorders in women athletes, *Eating Disorders Review,* July/August 1996, pp. 1–4.

[7]Position stand from the Committee on Sports Medicine and Fitness on promotion of healthy weight-control practices in young athletes, *Pediatrics* 97 (1996): 752–753.

[8]American College of Sports Medicine, Position stand, The female athlete triad, *Medicine and Science in Sports and Exercise* 29 (1997): i–ix.

[9]J. H. Wilson, Nutrition, physical activity, and bone health in women, *Nutrition Research Reviews* 7 (1994): 67–91.

[10]Yeager and coauthors, 1993.

[11]C. L. Otis, American College of Sports Medicine's Ad Hoc Task Force on Women's Issues in Sports Medicine, as quoted in A. A. Skolnick, "Female athlete triad" risk for women, *Journal of the American Medical Association* 270 (1993): 921–923.

[12]N. A. Armsey, Stress injury to bone in the female athlete, *Clinical Sports Medicine* 16 (1997): 197–224.

[13]G. Szmukler and C. Dare, Family therapy of early-onset, short-history anorexia nervosa, in *Family Approaches in Treatment of Eating Disorders,* ed. D. B. Woodside and L. Shekter-Wolfson (Washington, D.C.: American Psychiatric Press, 1991), pp. 25–47.

[14]Position of The American Dietetic Association: Nutrition intervention in the treatment of anorexia nervosa, bulimia nervosa, and binge eating, *Journal of the American Dietetic Association* 94 (1994): 902–907.

[15]A. E. Becker and coauthors, Eating disorders, *New England Journal of Medicine* 340 (1999): 1092–1098.

[16]American Psychiatric Association Workgroup on Eating Disorders, Practice guidelines for eating disorders, I. Disease definition, epidemiology, and nat-

ural history, *American Journal of Psychiatry* 150 (1993): 212–228.

[17]G. T. Wilson, Eating disorders and addictive disorders, in *Eating Disorders and Obesity: A Comprehensive Handbook,* ed. K. D. Brownell and C. G. Fairburn (New York: Guilford Press, 1995), pp. 165–170.

[18]P. W. Meilman, F. A. von Hippel, and M. S. Gaylor, Self-induced vomiting in college women: Its relation to eating, alcohol use, and Greek life, *College Health* 40 (1991): 39–41.

[19]J. E. Mitchell, Medical complications of bulimia nervosa, in *Eating Disorders and Obesity: A Comprehensive Handbook,* ed. K. D. Brownell and C. G. Fairburn (New York: Guilford Press, 1995), pp. 271–275.

CHAPTER 8

THE VITAMINS

CONTENTS

The Fat-Soluble Vitamins

The Water-Soluble Vitamins

Nutrition in Practice:
Vitamin Supplements

Earlier chapters focused primarily on the energy-yielding nutrients—carbohydrate, fat, and protein. This chapter and the next one discuss the nutrients everyone thinks of when nutrition is mentioned—the **vitamins** and minerals.

The vitamins occur in foods in much smaller quantities than do the energy-yielding nutrients, and they themselves contribute no energy to the body. Instead, they serve mostly as facilitators of body processes. They are a powerful group of substances, as their absence attests. Vitamin A deficiency can cause blindness; a lack of niacin can cause mental illness; and a lack of vitamin D can retard growth. The consequences of deficiencies are so dire and the effects of restoring the needed nutrients so dramatic that people spend billions of dollars each year on vitamin supplements to cure many different ailments. Vitamins certainly contribute to sound nutritional health, but supplements do not cure all ills. Actually, a vitamin can cure only the disease caused by a deficiency of that vitamin. The vitamins' roles in supporting optimal health extend far beyond preventing deficiency diseases, however. Emerging evidence points to relationships between low intakes of vitamins and chronic diseases such as cancer and heart disease.

A child once defined vitamins as "what, if you don't eat, you get sick." The description is both insightful and accurate. A more prosaic definition is that vitamins are potent, essential, noncaloric, organic nutrients needed from foods in trace amounts to perform specific functions that promote growth, reproduction, and the maintenance of health and life. Two characteristics distinguish vitamins from energy nutrients:

1. Vitamins do not yield energy when broken down, but assist the enzymes that release energy from carbohydrate, fat, and protein.
2. Vitamins are needed in much smaller amounts than the energy nutrients.

As the individual vitamins were discovered, they were named or given letters, numbers, or both. This led to the confusion that still exists today. This chapter uses the names shown in Table 8-1; alternative names are given in Tables 8-3 and 8-4, which appear later in the chapter.

Vitamins fall naturally into two classes—fat soluble and water soluble. The solubility of a vitamin confers on it many characteristics and determines how it is absorbed, transported, stored, and excreted. This discussion of vitamins begins with the fat-soluble vitamins.

▪▪ The Fat-Soluble Vitamins

The fat-soluble vitamins—A, D, E, and K—usually occur together in the fats and oils of foods, and the body absorbs them in the same way it absorbs lipids. Therefore, any condition that interferes with fat absorption can precipitate a deficiency of the fat-soluble vitamins. Once absorbed, fat-soluble vitamins are stored in the liver and fatty tissues until the body needs them. They are not readily excreted, and unlike most of the water-soluble vitamins, they can build up to toxic concentrations. Excesses of vitamins A, D, and K from supplements can reach toxic levels especially easily.

The capacity to store fat-soluble vitamins affords a person some flexibility in dietary intake. When blood concentrations begin to decline, the body can retrieve the vitamins from storage. Thus, a person need not eat a day's allowance of each fat-soluble vitamin every day, but need only make sure that, over time, average daily intakes approximate recommended intakes. In contrast, water-soluble vitamins must be consumed more regularly because the body does not store them to any great extent.

vitamins: essential, noncaloric, organic nutrients needed in tiny amounts in the diet.

Carbohydrate, fat, and protein are sometimes referred to as macronutrients *because they are measured in foods in gram amounts. Vitamins and minerals are sometimes referred to as* micronutrients *because they are measured in foods in milligram and microgram amounts.*

Table 8-1 **VITAMIN NAMES**
Fat-Soluble Vitamins
Vitamin A
Vitamin D
Vitamin E
Vitamin K
Water-Soluble Vitamins
B vitamins
Thiamin
Riboflavin
Niacin
Pantothenic acid
Biotin
Vitamin B_6
Folate
Vitamin B_{12}
Vitamin C

Vitamin A and Beta-Carotene

vitamin A: a fat-soluble vitamin. Its three chemical forms are *retinol* (the alcohol form), *retinal* (the aldehyde form), and *retinoic acid* (the acid form).

precursor: a compound that can be converted into another compound; with regard to vitamins, compounds that can be converted into active vitamins; also known as **provitamins.**

beta-carotene: a vitamin A precursor made by plants and stored in human fat tissue; an orange pigment.

retinol-binding protein (RBP): the specific protein responsible for transporting retinol.

Measurement of the blood concentration of RBP is a sensitive test of vitamin A status.

cornea (KOR-nee-uh): the hard, transparent membrane covering the outside of the eye.

retina (RET-in-uh): the layer of light-sensitive nerve cells lining the back of the inside of the eye; consists of rods and cones.

night blindness: the slow recovery of vision after exposure to flashes of bright light at night; an early symptom of vitamin A deficiency.

differentiation: the development of specific functions different from those of the original.

Vitamin A has the distinction of being the first fat-soluble vitamin to be recognized. Today, after almost a century of revelations, vitamin A and its plant-derived **precursor, beta-carotene,** are the focus of much research around the world. Much of this intensive research effort has been sparked by accumulating evidence that both vitamin A and beta-carotene may protect against certain types of cancer.

METABOLIC ROLES OF VITAMIN A Vitamin A is a versatile vitamin, playing diverse roles in vision, protein synthesis and cell differentiation (and thereby maintaining the health of body linings and skin), immunity, and reproduction and growth. Three different forms of vitamin A are active in the body: retinol, retinal, and retinoic acid. Each form of vitamin A performs specific tasks. Retinol supports reproduction and is the major transport and storage form of the vitamin; the cells convert retinol to retinal or retinoic acid as needed. Retinal is active in vision, and retinoic acid acts as a hormone, regulating cell differentiation, growth, and embryonic development. A special transport protein, **retinol-binding protein (RBP),** picks up retinol from the liver where it is stored and carries it in the blood.

VITAMIN A IN VISION Vitamin A plays two indispensable roles in the eye. It helps maintain a healthy, crystal-clear outer window, the **cornea;** and it participates in the events of light detection at the **retina.** Figure 8-1 shows vitamin A's site of action inside the eye.

When vitamin A is lacking, the eye has difficulty adapting to changing light levels. At night, after the eye has adapted to darkness, a lag occurs before the eye can see again after a flash of bright light. This lag in the recovery of night vision is known as **night blindness.** Because night blindness is easy to test, it aids in the diagnosis of vitamin A deficiency. Night blindness is only a symptom, however, and may indicate a condition other than vitamin A deficiency.

VITAMIN A IN PROTEIN SYNTHESIS AND CELL DIFFERENTIATION The role that vitamin A plays in vision is undeniably important, but only one-thousandth of the body's vitamin A is in the retina. Much more is in the skin and the linings of organs, where it works at the genetic level to promote protein synthesis and cell **differentiation.** The process of cell differentiation allows each type of cell to mature so that it is capable of performing a specific function.

Figure 8-1
VITAMIN A'S ROLE IN VISION

As light enters the eye, pigments within the cells of the retina absorb the light and generate nerve impulses that travel to the brain. Each pigment contains retinal, the active form of vitamin A.

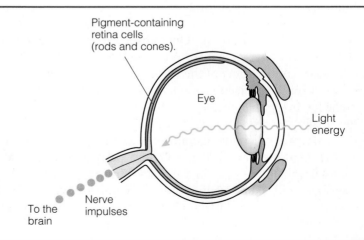

Pigment-containing retina cells (rods and cones).

Eye

Light energy

Nerve impulses

To the brain

All body surfaces, both inside and out, are covered by layers of cells known as **epithelial cells.** The **epithelial tissue** on the outside of the body is, of course, the skin. The epithelial tissues inside the body include the linings of the mouth, stomach, and intestines; the linings of the lungs and the passages leading to them; the lining of the bladder; the linings of the uterus and vagina; and the linings of the eyelids and sinus passageways. The epithelial tissues on the inside of the body must be kept smooth. To ensure that they are, the epithelial cells on their surfaces secrete a smooth, slippery substance (mucus) that coats and protects the tissues from invasive microorganisms and other harmful particles. The **mucous membrane** that lines the stomach also shields its cells from digestion by gastric juices. Vitamin A, by way of its role in cell differentiation, helps to maintain the integrity of the epithelial cells.

:: VITAMIN A IN REPRODUCTION AND GROWTH   Vitamin A also supports reproduction and growth. In men, vitamin A participates in sperm development, and in women, vitamin A promotes normal fetal growth and development.[1]

:: METABOLIC ROLES OF BETA-CAROTENE   For many years scientists believed beta-carotene to be of interest solely as a vitamin A precursor. Eventually, researchers began to recognize beta-carotene as an extremely effective **antioxidant** in the body. Antioxidants are compounds that protect other compounds (such as lipids in cell membranes) from attack by oxygen. Oxygen triggers the formation of compounds known as **free radicals** that can start chain reactions in cell membranes. If left uncontrolled, these chain reactions can damage cell structures and impair cell functions. Oxidative and free-radical damage to cells is suspected of instigating some early stages of cancer and heart disease. Research has identified links between oxidative damage and the development of many other diseases, including age-related blindness, arthritis, cataracts, diabetes, and kidney disease.[2]

Studies of populations suggest that people whose diets are low in beta-carotene have higher incidences of certain types of cancer than those whose diets contain generous amounts of foods rich in beta-carotene.[3] Based on findings that beta-carotene in foods may protect against cancer, researchers designed a study to determine the effects of beta-carotene *supplements* on the incidence of lung cancer among smokers. The researchers expected to see a beneficial effect, but instead found that smokers taking the beta-carotene supplements suffered a *greater* incidence of lung cancer than those taking placebos.[4] Beta-carotene in foods is just one among many nutrients and compounds present in foods. As Table 8-2 shows, many other phytochemicals are also present in foods and may be at least partially responsible for the effect. Until more is known, eating beta-carotene–rich foods, not supplements, is in the best interest of health.

:: VITAMIN A DEFICIENCY   Up to a year's supply of vitamin A can be stored in the body, 90 percent of it in the liver. If a healthy adult were to stop eating vitamin A–rich foods, deficiency symptoms would not begin to appear until after stores were depleted, which would take one to two years. Then, however, the consequences would be profound and severe. Table 8-3, later in this chapter, lists some of them.

In vitamin A deficiency, cell differentiation and maturation are impaired. The epithelial cells flatten and begin to produce **keratin**—the hard, inflexible protein of hair and nails. In the eye, this process leads to drying and hardening of the cornea, which may progress to permanent blindness.

:: BLINDNESS   Vitamin A deficiency is the major cause of childhood blindness in the world, causing more than half a million children to lose their

epithelial (ep-i-THEE-lee-ul) cells: cells on the surface of the skin and mucous membranes.

epithelial tissues: the layers of the body that serve as selective barriers between the body's interior and the environment (examples are the cornea, the skin, the respiratory lining, and the lining of the digestive tract).

mucous membranes: the membranes composed of mucus-secreting cells that line the surfaces of body tissues. (Reminder: *Mucus* is the smooth, slippery substance secreted by these cells.)

antioxidant: a compound that protects other compounds from oxygen by itself reacting with oxygen.

free radical: a highly reactive chemical form that can cause destructive changes in nearby compounds, sometimes setting up a chain reaction.

Nutrition in Practice 21 offers information about free radicals and heart disease.

Reminder: Phytochemicals are nonnutrient compounds found in plant-derived foods that have biological activity in the body.

keratin (KERR-uh-tin): a water-insoluble protein; the normal protein of hair and nails. In vitamin A deficiency, keratin-producing cells may replace mucus-producing cells.

The progressive blindness caused by vitamin A deficiency is called xerophthalmia (zer-off-THAL-mee-uh).
xero = dry
ophthalm = eye

Table 8-2
A SAMPLING OF PHYTOCHEMICALS: CLASSIFICATION, POSSIBLE HEALTH EFFECTS, AND FOOD SOURCES

Name or Class	Possible Effects	Food Sources
Capsaicin	Modulates blood clotting, possibly reducing the risk of fatal clots in heart and artery disease.	Hot peppers
Carotenoids (include beta-carotene, lycopene, and hundreds of related compounds)[a]	May act as antioxidants, possibly reducing risks of cancer and other diseases.	Deeply pigmented fruits and vegetables (apricots, broccoli, cantaloupe, carrots, pumpkin, spinach, sweet potatoes, tomatoes)
Curcumin	May inhibit enzymes that activate carcinogens.	Tumeric, a yellow-colored spice
Flavonoids (include flavones, flavonols, isoflavones, catechin, and others)[b,c]	Many flavonoids may act as antioxidants; scavenge carcinogens; bind to nitrates in the stomach, preventing conversion to nitrosamines; inhibit cell proliferation.	Berries, black tea, celery, citrus fruits, green tea, olives, onions, oregano, purple grapes, purple grape juice, soybeans and soy products, vegetables, whole wheat, wine
Indoles[d]	May trigger production of enzymes that block DNA damage from carcinogens; may inhibit estrogen action.	Broccoli and other cruciferous vegetables (brussels sprouts, cabbage, cauliflower), horseradish, mustard greens
Isothiocyanates (including sulforaphane)	Inhibit enzymes that activate carcinogens; trigger production of enzymes that detoxify carcinogens.	Broccoli and other cruciferous vegetables (brussels sprouts, cabbage, cauliflower), horseradish, mustard greens
Lignans[e]	Block estrogen activity in cells, possibly reducing the risk of cancer of the breast, colon, ovaries, and prostate.	Flaxseed and its oil, whole grains
Monoterpenes (include limonene)	May trigger enzyme production to detoxify carcinogens; inhibit cancer promotion and cell proliferation.	Citrus fruit peels and oils
Organosulfur compounds	May speed production of carcinogen-destroying enzymes; slow production of carcinogen-activating enzymes.	Chives, garlic, leeks, onions
Phenolic acids[c]	May trigger enzyme production to make carcinogens water soluble, facilitating excretion.	Coffee beans, fruits (apples, blueberries, cherries, grapes, oranges, pears, prunes), oats, potatoes, soybeans
Phytic acid	Binds to minerals, preventing free-radical formation, possibly reducing cancer risk.	Whole grains
Phytosterols (genistein and diadzein)	Estrogen inhibition may produce these actions: inhibit cell replication in GI tract; reduce risk of breast, colon, ovarian, prostate, and other estrogen-sensitive cancers; reduce cancer cell survival. Estrogen mimicking may reduce risk of osteoporosis and heart disease.	Soybeans, soy flour, soy milk, tofu, textured vegetable protein, other legume products
Protease inhibitors	May suppress enzyme production in cancer cells, slowing tumor growth; inhibit hormone binding; inhibit malignant changes in cells.	Broccoli sprouts, potatoes, soybeans and other legumes, soy products
Saponins	May interfere with DNA replication, preventing cancer cells from multiplying; stimulate immune response.	Alfalfa sprouts, other sprouts, green vegetables, potatoes, tomatoes
Tannins[c]	May inhibit carcinogen activation and cancer promotion; act as antioxidants.	Black-eyed peas, grapes, lentils, red and white wine, tea

[a]Other carotenoids include alpha-carotene, beta-cryptoxanthin, lutein, and zeaxanthin.
[b]Other flavonoids of interest include ellegic acid and ferulic acid. See also *phytosterols*.
[c]A subset of the larger group *phenolic phytochemicals*.
[d]Indoles include dithiothiones, isothiocyantes, and others.
[e]Lignans act as phytosterols, but their food sources are limited.

An early sign is xerosis (drying of the cornea); the last and most severe stage is keratomalacia (kerr-uh-to-mal-AY-shuh), or total blindness.
malacia = softening, weakening

follicle (FOLL-i-cul): a group of cells in the skin from which a hair grows.

sight every year. More than 100 million children worldwide endure less severe forms of vitamin A deficiency, making them vulnerable to infectious diseases.

■■ INFECTIONS All body surfaces, both inside and out, maintain their integrity with the help of vitamin A. When vitamin A is lacking, cells of the skin harden and flatten, making it dry, rough, scaly, and hard. An accumulation of keratin makes a lump around each hair **follicle** (keratinization).

In the mouth, a vitamin A deficiency results in drying and hardening of the salivary glands, making them susceptible to infection. Secretions of mucus in the stomach and intestines are reduced, hindering normal digestion and absorption of nutrients. Infections of other mucous membranes also become likely.

Vitamin A's role in maintaining the body's defensive barriers may partially explain the relationship between vitamin A deficiency and susceptibility to infection.[5] When children with measles complicated by diarrhea, infections such as pneumonia, or both were given vitamin A supplements, their recovery times and hospital stays were shorter, and their overall survival rates significantly greater, than those of similar children who did not receive vitamin A.[6]

The evidence that vitamin A reduces the severity of measles and measles-related infections and diarrhea has prompted the World Health Organization (WHO) and UNICEF (the United Nations Children's Fund; formerly the United Nations International Children's Emergency Fund) to make control of vitamin A deficiency a major goal in their quest to improve child survival throughout the developing world. The American Academy of Pediatrics has issued a formal statement recommending vitamin A supplementation for certain groups of measles-infected infants and children in the United States.[7]

▪▪ VITAMIN A TOXICITY When the body stores excess vitamin A, toxicity is possible. Normally, toxicity symptoms are likely only when animal-derived foods or supplements are the source of the excess vitamin, for in these sources the vitamin is already in its active form, called **preformed vitamin A.** Plant foods contain the vitamin only as beta-carotene, its inactive precursor form. The precursor does not convert to active vitamin A rapidly enough to cause toxicity.

Overdoses of vitamin A damage the same body systems that exhibit symptoms in vitamin A deficiency (see Table 8-3 later in the chapter). Children are most vulnerable to vitamin A toxicity because, being smaller, they need less than adults, and it is easy to give them too much in pill form. The availability of breakfast cereals, instant meals, fortified milk, and chewable candylike vitamins, each containing 100 percent or more of the recommended daily intake of vitamin A, makes it possible for a well-meaning parent to provide several times the daily allowance of the vitamin to a child within a few hours. Serious toxicity is seen in infants and young children when they are given more than ten times the recommended amount every day for weeks at a time.

Excessive vitamin A poses a **teratogenic** risk.[8] In pregnant women, chronic use of vitamin A supplements providing three to four times the amount recommended for pregnancy can cause malformations of the fetus.[9]

Certain vitamin A relatives are available by prescription as acne treatments. When applied directly to the skin surface, these preparations help relieve the symptoms of acne. Taking massive doses of vitamin A internally will *not* cure acne, however, and may cause the miseries itemized in Table 8-3. Foods are always a better choice than supplements for needed nutrients. The best way to ensure a safe vitamin A intake is to eat generous servings of vitamin A–rich foods. Well-nourished, healthy people do not need vitamin A supplements.

▪▪ BETA-CAROTENE CONVERSION AND TOXICITY When beta-carotene is converted to retinol in the body, losses occur. This is why, rather than expressing the amounts of beta-carotene in foods, nutrition scientists use the RE (retinol equivalent), which expresses the amount of retinol the body actually derives from a plant food after conversion. The body can make one unit of retinol from about six units of beta-carotene.

The accumulation of the hard material keratin around each hair follicle is follicular hyperkeratosis.

preformed vitamin A: vitamin A in its active form.

teratogenic (ter-AT-oh-jen-ik): causing abnormal fetal development and birth defects.

terato = monster
genic = to produce

The units in which vitamin A amounts in foods are expressed are RE (retinol equivalents). A unit used earlier was the IU (international unit).

1 RE = 1 μg retinol.
 = 6 μg beta-carotene.
 = 12 μg of other vitamin A precursor carotenes.

Vitamin A and Beta-Carotene

These foods provide 10 percent or more
of the vitamin A Daily Value (DV = 900 μg RE/day).[a]

RDA for men: 1000 μg RE/day
RDA for women: 800 μg RE/day

Fortified milk: 150 RE[b] (17% DV) per cup

Beef liver: 9124 RE[b] (1014% DV) per
3 oz fried

Sweet potato: 1936 RE[c]
(215% DV) per ½ c mashed

Apricots: 274 RE[c] (30% DV) per 3 fresh
apricots

Spinach: 739 RE[c] (82% DV) per ½ c cooked

Carrots: 1584 RE[c] (176% DV)
per ½ c cooked

[a]The Daily Values are based on a 2000-kcalorie diet.
[b]Preformed vitamin A.
[c]Beta-carotene.

Yellowing of the skin caused by excess carotene in the blood is known as carotenemia (KAR-oh-teh-NEE-me-ah). Carotenemia can be distinguished from jaundice because the mucous membranes lining the eyelids do not turn yellow as they do in jaundice.

As mentioned earlier, beta-carotene from plant foods is not converted to the active form of vitamin A rapidly enough to be hazardous. It has, however, been known to turn people bright yellow if they eat too much. Beta-carotene builds up in the fat just beneath the skin and imparts a yellow cast.

VITAMIN A IN FOODS   In the typical U.S. diet, about half of the vitamin A activity comes from fruits and vegetables that supply the vitamin as its precursor, beta-carotene. The other half of vitamin A activity comes from preformed vitamin A supplied in milk, cheese, butter, and other dairy products; eggs; and meats. Liver is a rich source of preformed vitamin A. Liver offers many nutrients, so eating it periodically may benefit nutritional health, but once every week or so is enough.

Because vitamin A is fat soluble, it is lost when milk is skimmed. Fat-free milk is thus often fortified with vitamin A to compensate. Margarine is also usually fortified so as to provide the same amount of vitamin A as butter. Snapshot 8-1 shows a sampling of the richest food sources of both preformed vitamin A and beta-carotene.

Fast-food meals often lack vitamin A. When fast-food restaurants offer salads with cheese, carrots, and other vitamin A–rich foods, the nutritional quality of their meals greatly improves.

BETA-CAROTENE IN FOODS   Many foods from plants contain beta-carotene, the orange pigment responsible for the bold colors of many fruits and vegetables. Carrots, sweet potatoes, pumpkins, cantaloupe, and apricots are all rich sources, and their bright orange color enhances the eye appeal of the plate. Another colorful group, *dark* green vegetables, such as spinach, other greens, and broccoli, owe their color to both chlorophyll and beta-carotene. The orange and green pigments together give a deep, murky green color to the vegetables. Other colorful vegetables, such as iceberg lettuce, beets, and sweet corn, can fool you into thinking they contain beta-carotene, but these foods derive their color from other pigments and are poor sources of beta-carotene. As for "white" plant foods such as grains and potatoes, they have none. Recommendations to eat *dark* green or *deep* orange vegetables and fruits at least every other day help people to meet their vitamin A needs.

Vitamin D

Vitamin D is different from all the other nutrients in that the body can synthesize it with the help of sunlight. Therefore, in a sense, vitamin D is not an essential nutrient. Given enough sun, people need no vitamin D from foods.

VITAMIN D'S METABOLIC CONVERSIONS The liver manufactures a vitamin D precursor, which migrates to the skin where it is converted to a second precursor with the help of the sun's ultraviolet rays. Next, the liver and then the kidneys alter the second precursor to produce the active vitamin. Vitamin D precursors from plants require the same two conversions by the liver and kidneys to become active. The biologic activity of the active vitamin is 500- to 1000-fold greater than that of its precursor. Diseases that affect either the liver or the kidneys may impair the transformation of precursor vitamin D to active vitamin D and therefore produce symptoms of vitamin D deficiency.

VITAMIN D'S ACTIONS Although known as a vitamin, vitamin D is actually a hormone—a compound manufactured by one organ of the body that has effects on another. The best-known vitamin D target organs are the intestine, the kidneys, and the bones, but scientists have discovered several other vitamin D target tissues, including the brain, the pancreas, the skin, the reproductive organs, and many cancer cells.[10] These discoveries suggest that numerous additional functions for vitamin D may surface, including regulation of the immune system.[11]

VITAMIN D'S ROLES IN BONE Vitamin D is a member of a large, cooperative bone-making and maintenance team composed of nutrients and other compounds, including vitamins A, C, and K; the hormones parathormone and calcitonin; the protein collagen; and the minerals calcium, phosphorus, magnesium, and fluoride. Vitamin D's special role in bone growth is to make calcium and phosphorus available in the blood that bathes the bones. The bones grow denser and stronger as the minerals are deposited from the blood. Vitamin D acts in three ways to maintain blood concentrations of calcium and phosphorus: it stimulates their absorption from the GI tract; it mobilizes calcium and phosphorus from bones into the blood; and it stimulates their retention by the kidneys.

VITAMIN D DEFICIENCY The symptoms of vitamin D deficiency are those of calcium deficiency, as shown in Table 8-3 on pp. 178–179. The bones fail to calcify normally and may grow so weak that they become bent when they have to support the body's weight. A child with **rickets** who is old enough to walk characteristically develops bowed legs, often the most obvious sign of the disease. Worldwide, rickets afflicts a large number of children.

Adult rickets, or **osteomalacia,** occurs most often in women who have low calcium intakes and little exposure to sun and who go through repeated pregnancies and periods of lactation. The bones of the legs may soften to such an extent that a young woman who is tall and straight at 20 may be condemned by repeated pregnancies to become bent, bowlegged, and stooped before she is 30.

Inadequate vitamin D is recognized as a risk factor in **osteoporosis** (reduced bone density). Without sufficient vitamin D, absorption of calcium is limited and bone remodeling is impaired. This combination leads to a loss of bone mass.

VITAMIN D TOXICITY Whereas vitamin D deficiency depresses calcium absorption, blood calcium, and bone mineralization, an excess of

Sunlight promotes vitamin D synthesis in the skin. Exposure to the sun should be moderate, however; excessive exposure may cause skin cancer.

The precursor of vitamin D made in the liver is 7-dehydrocholesterol, which is made from cholesterol. This is one of the body's many "good" uses for cholesterol.

The final, active vitamin is 1-25 dihydroxycholecalciferol, or, more simply, dihydroxy vitamin D.

rickets: the vitamin D–deficiency disease in children.

osteomalacia (os-tee-o-mal-AY-shuh): a bone disease characterized by softening of the bones. Symptoms include bending of the spine and bowing of the legs. The disease occurs most often in adult women.
   osteo = bone
   mal = bad (soft)

osteoporosis: literally, porous bones; reduced density of the bones, also known as *adult bone loss.*

vitamin D does the opposite, as shown in Table 8-3. It enhances calcium absorption, produces high blood calcium, and promotes return of bone calcium into the blood. The excess calcium then tends to precipitate in the soft tissues, forming stones, including kidney stones. Calcification may also harden the blood vessels and is especially dangerous in the major arteries of the heart and lungs, where it can cause death.

Vitamin D in excess is the most toxic of all the vitamins. The amounts of vitamin D in foods available in the United States and Canada are well within safe limits, but supplements containing the vitamin in concentrated form are not. Adults should use caution when taking vitamin D supplements and keep them out of the reach of children.

▓▓ VITAMIN D FROM THE SUN Most of the world's population relies on natural exposure to sunlight to maintain adequate vitamin D nutrition. The sun imposes no risk of vitamin D toxicity. Prolonged exposure to sunlight degrades the vitamin D precursor in the skin, preventing its conversion to the active vitamin. Even lifeguards on southern beaches are safe from vitamin D toxicity from the sun.

Prolonged exposure to sunlight has other undesirable consequences, however, such as premature wrinkling of the skin and risk of skin cancer. These risks may be reduced by using sunscreens. Unfortunately, sunscreens with sun protection factors (SPF) of 8 and above also retard vitamin D synthesis.[12] A strategy to avoid this dilemma is to apply sunscreen after enough time has elapsed to provide sufficient vitamin D. For most people, exposing hands, face, and arms on a clear summer day for 10 minutes, a few times a week, should be sufficient to maintain vitamin D nutrition. Dark-skinned people require longer exposure than light-skinned people, but by three hours, vitamin D synthesis in heavily pigmented skin arrives at the same plateau as in fair skin after 30 minutes.

The ultraviolet rays from tanning lamps and tanning booths may also stimulate vitamin D synthesis, but the hazards outweigh any possible benefits. The Food and Drug Administration (FDA) warns that if the lamps are not properly filtered, people using tanning booths risk burns, damage to the eyes and blood vessels, and skin cancer.[13]

Heavy clouds, smoke, or smog may filter out the ultraviolet rays of the sun that promote vitamin D synthesis. Together with skin pigmentation, smog probably accounts for the development of rickets in dark-skinned people in northern, smoggy cities. For these people, and for those who are unable to go outdoors frequently, dietary vitamin D is especially important.

Vitamin D activity was previously expressed in international units (IU), but as of 1980, it is expressed in micrograms of cholecalciferol, the active form of vitamin D. To convert, use the following factor:
100 IU = 2.5 μg.
400 IU = 10 μg.

▓▓ VITAMIN D IN FOODS Only a few animal foods, notably eggs, liver, butter, some fish, and fortified milk, supply significant amounts of vitamin D. For those who use margarine in place of butter, fortified margarine is a significant source. Infant formulas are fortified with vitamin D in amounts adequate for daily intake. Breast milk is low in vitamin D, so vitamin D supplements are recommended for dark-skinned, breastfed infants who do not have adequate exposure to the sun.[14] These sources, plus any exposure to the sun, provide babies with more than enough of this vitamin.

The fortification of milk with vitamin D is the best guarantee that children will meet their vitamin D needs and underscores the importance of milk in children's diets. Unlike milk, cheese and yogurt are not fortified with vitamin D. Vegans, and especially their children, may have low vitamin D intakes because no fortified plant source except margarine exists. In the United States, breakfast cereals may be fortified with vitamin D, as their labels indicate.[15]

Most adults, especially in sunny regions, need not make special efforts to obtain vitamin D in food. People who are not outdoors much or who live in northern or predominantly cloudy or smoggy areas, however, are advised to make sure their milk is fortified with vitamin D and to drink at least 2 cups a day.

Vitamin E

Like beta-carotene, vitamin E is a fat-soluble antioxidant. It protects other substances from oxidation by being oxidized itself. Table 8-3 summarizes important information about vitamin E.

VITAMIN E AS AN ANTIOXIDANT If there is plenty of vitamin E in the membranes of cells exposed to an oxidant, chances are this vitamin will take the brunt of any oxidative attack, protecting the lipids and other vulnerable components of the membranes. Vitamin E is especially effective in preventing the oxidation of the polyunsaturated fatty acids (PUFA), but it protects all other lipids (for example, vitamin A) as well.

Vitamin E exerts an especially important antioxidant effect in the lungs, where the cells are exposed to high concentrations of oxygen. Vitamin E also protects the lungs from air pollutants that are strong oxidants.

Accumulating evidence suggests that vitamin E also offers protection against heart disease by protecting LDL (low-density lipoproteins) from oxidation.[16] The oxidation of LDL encourages development of atherosclerosis. Vitamin E both from foods and from supplements seems to be effective in defending against heart disease.[17]

VITAMIN E MYTHS While research continues to reveal possible roles for vitamin E, it has also clearly discredited claims that vitamin E improves athletic skill, enhances sexual performance, or cures sexual dysfunction in males. Vitamin E also does not prevent or cure hereditary **muscular dystrophy,** nor does it slow or prevent processes of aging, such as graying of the hair, wrinkling of the skin, or reduced activity of body organs.

VITAMIN E DEFICIENCY When blood concentrations of vitamin E fall below a certain critical level, the red blood cells tend to break open and spill their contents, probably because the PUFA in their membranes oxidize. This classic vitamin E–deficiency symptom, known as **erythrocyte hemolysis,** is seen in premature infants born before the transfer of vitamin E from the mother to the fetus that takes place in the last weeks of pregnancy. Vitamin E treatment corrects erythrocyte hemolysis.

Two other conditions appear to respond to vitamin E therapy. One is a nonmalignant breast disease (**fibrocystic breast disease**), and the other (**intermittent claudication**) is an abnormality of blood flow that causes cramping in the legs.

In human beings, vitamin E deficiency is usually associated with diseases, notably those that cause malabsorption of fat. These include diseases of the liver, gallbladder, and pancreas, as well as various hereditary diseases involving digestion and use of nutrients.

On rare occasions, vitamin E deficiencies develop in people without diseases. Most likely, such deficiencies occur after years of eating diets extremely low in fat; or using fat substitutes, such as diet margarines and salad dressings, as the only sources of fat; or consuming diets composed of highly processed or "convenience" foods. Extensive heating in the processing of foods destroys vitamin E.

VITAMIN E TOXICITY Vitamin E supplement use has risen in recent years as its antioxidant action against disease has been recognized. As a result, signs of toxicity are now known or suspected, although vitamin E toxicity is

Reminder: Oxidation is a type of chemical reaction, so named because oxygen is one of the agents that often brings it about.

muscular dystrophy (DIS-tro-fee): a hereditary disease in which the muscles gradually weaken. Its most debilitating effects arise in the lungs. This disease should not be confused with *nutritional* muscular dystrophy, a vitamin E–deficiency disease of animals characterized by gradual paralysis of the muscles.

erythrocyte (er-REETH-ro-cite) hemolysis (he-MOLL-uh-sis): rupture of the red blood cells, caused by vitamin E deficiency.
  erythro = red
  cyte = cell
  hemo = blood
  lysis = breaking

fibrocystic breast disease: a harmless condition in which the breast develops lumps; sometimes associated with caffeine consumption. In some, it responds to treatment by abstinence from caffeine; in others, it can be treated with vitamin E.
  fibr = fibrous lumps
  cystic = in sacs

intermittent claudication: severe calf pain caused by inadequate blood supply. It occurs when walking and subsides during rest.
  intermittent = at intervals
  claudicare = to limp

Caution: *Other serious conditions can cause lumps in the breast and pain in the legs. Don't self-diagnose; see a physician.*

Vegetable oils, some nuts and seeds such as almonds and sunflower seeds, and wheat germ are vitamin E–rich.

On vitamin bottles, vitamin E activity is often expressed in IU. One IU is the same as 1 mg of the active form of vitamin E.

K stands for the Danish word **koagulation** *(coagulation or "clotting").*

hemorrhagic (hem-o-RAJ-ik) disease: the vitamin K–deficiency disease in which blood fails to clot.

Reminder: The bacterial inhabitants of the digestive tract are known as the intestinal flora.
flora = plant inhabitants

not nearly as common, and its effects are not as serious, as vitamin A or vitamin D toxicity. High doses of vitamin E interfere with the blood-clotting action of vitamin K and enhance the action of anticoagulant medications, leading to hemorrhage. For most individuals, however, ordinary supplemental doses (no more than 800 milligrams daily) taken over a period of months seem to have no adverse health effects.[18]

VITAMIN E IN FOODS Vitamin E is widespread in foods. About 20 percent of the vitamin E in the diet comes from vegetable oils and the products made from them, such as margarine, salad dressings, and shortenings. (Soybean oils and wheat germ oils have especially high concentrations of vitamin E.) Another 20 percent comes from fruits and vegetables. Animal fats, such as butter and milk fat, contain little or no vitamin E. Because vitamin E is readily destroyed by heat processing and oxidation, fresh or lightly processed foods are the best sources of this vitamin.

Vitamin K

Vitamin K has long been known for its role in blood clotting, where its presence can make the difference between life and death. The vitamin also participates in the synthesis of several bone proteins.[19] Research is under way to determine the specific roles of these vitamin K–dependent proteins in bone metabolism and the risk of osteoporosis.[20] Vitamin K may play a role in reducing the risk of hip fracture: in a large study of women, those who ate abundant green vegetables, good sources of vitamin K, suffered hip fractures less often than those with lower intakes.[21]

BLOOD CLOTTING At least 13 different proteins and the mineral calcium are involved in making blood clot. Vitamin K is essential for the activation of seven of these proteins, among them prothrombin, the precursor of the enzyme thrombin (see Figure 8-2).[22] When any of the blood-clotting factors is lacking, **hemorrhagic disease** results. If an artery or vein is cut or broken, bleeding goes unchecked. Of course, this is not to say that hemorrhaging is always caused by a vitamin K deficiency.

INTESTINAL SYNTHESIS Like vitamin D, vitamin K can be obtained from a nonfood source. Bacteria in the intestinal tract synthesize vitamin K that the body can absorb, but people cannot depend on this source alone for their vitamin K.

VITAMIN K DEFICIENCY Vitamin K deficiency is rare, but may occur in two circumstances. First, it may arise in conditions of fat malabsorption.

Figure 8-2
BLOOD-CLOTTING PROCESS

When blood is exposed to air, foreign substances, or secretions from injured tissues, platelets (small, cell-like structures in the blood) release a phospholipid known as thromboplastin. Thromboplastin catalyzes the conversion of the inactive protein prothrombin to the active enzyme thrombin. Thrombin then catalyzes the conversion of the precursor protein fibrinogen to the active protein fibrin that forms the clot.

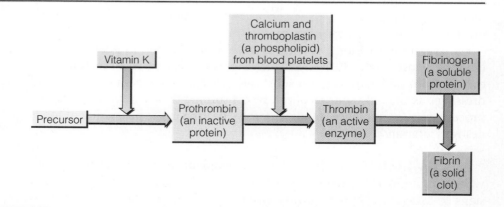

Second, some medications interfere with vitamin K's synthesis and action in the body: antibiotics kill the vitamin K–producing bacteria in the intestine, and anticoagulant medications intentionally interfere with vitamin K metabolism and activity. When vitamin K deficiency does occur, it can be fatal.

VITAMIN K FOR NEWBORNS Newborn infants present a unique case of vitamin K nutrition. An infant is born with a **sterile** digestive tract, and some weeks pass before the vitamin K–producing bacteria become fully established in the infant's intestines. At the same time, plasma prothrombin concentrations are low (this helps prevent blood clotting during the stress of birth, which might otherwise be fatal). A single dose of vitamin K, usually in a water-soluble form, is recommended at birth to prevent hemorrhagic disease in the newborn.

sterile: free of microorganisms, such as bacteria.

VITAMIN K TOXICITY A high intake of vitamin K can reduce the effectiveness of anticoagulant medications used to prevent the blood from clotting. People taking these medications should eat consistent amounts of vitamin K–rich foods from day to day. Vitamin K–toxicity symptoms include red cell hemolysis, **jaundice,** and brain damage (see Table 8-3 on pp. 178–179).

jaundice: yellowing of the skin due to spillover of bile pigments from the liver into the general circulation.

VITAMIN K IN FOODS Many foods contain ample amounts of vitamin K, notably green leafy vegetables, members of the cabbage family, and liver. Milk, meats, eggs, cereal, fruits, and vegetables provide smaller, but still significant, amounts.

LEARNING LINK

The fat-soluble vitamins play many specific roles in the health and function of the eyes, skin, GI tract, bones, teeth, nervous system, and blood; consequently, their deficiencies become apparent in these same areas. Deficiencies of the fat-soluble vitamins are a threat whenever fat absorption is impaired. Later chapters describe disorders of the stomach, intestines, pancreas, and liver, as well as the hereditary disorder cystic fibrosis, that can impair fat absorption and lead to deficiencies of the fat-soluble vitamins. Similarly, disorders of the liver and kidney can interfere with activation of vitamin D, creating a deficiency. In addition, medications used to treat a variety of disorders can alter the absorption, metabolism, and excretion of the fat-soluble vitamins. In all cases, deficiencies should be corrected promptly. Some clients with fat malabsorption can absorb enough of the fat-soluble vitamins by taking a regular supplement. Others may need a water-miscible form of the fat-soluble vitamins to facilitate absorption.

▪▪ The Water-Soluble Vitamins

The B vitamins and vitamin C are the water-soluble vitamins. These vitamins, found in the watery compartments of foods, are distributed into the water-filled compartments of the body. They are easily absorbed into the bloodstream and are just as easily excreted if their blood concentrations rise too high. Thus, the water-soluble vitamins are less likely to reach toxic concentrations in the body than are the fat-soluble vitamins. Foods never deliver excessive amounts of the water-soluble vitamins, but the large doses concentrated in vitamin supplements can reach toxic levels.

Notable food sources of vitamin K include milk, eggs, brussels sprouts, cabbage, and spinach.

Table 8-3

THE FAT-SOLUBLE VITAMINS—A SUMMARY

Vitamin A

Other Names	Deficiency Symptoms	Toxicity Symptoms
Retinol, retinal, retinoic acid; main precursor is beta-carotene	***Blood/Circulatory System***	
	Anemia (small-cell type)[a]	Red blood cell breakage, nosebleeds
Chief Functions in the Body	***Bones/Teeth***	
Vision: health of cornea, epithelial cells, mucous membranes; skin health; bone and tooth growth; reproduction; hormone synthesis and regulation; immunity; cancer protection	Cessation of bone growth, painful joints; impaired enamel formation, cracks in teeth, tendency to decay	Bone pain; growth retardation; increase of pressure inside skull mimicking brain tumor; headaches
	Digestive System	
	Diarrhea, changes in lining	Abdominal cramps and pain, nausea, vomiting, diarrhea, weight loss
Deficiency Disease Name	***Immune System***	
Hypovitaminosis A	Depression; frequent respiratory, digestive, bladder, vaginal, and other infections	Overreactivity
Significant Sources	***Nervous/Muscular Systems***	
Retinol: fortified milk, cheese, cream, butter, fortified margarine, eggs, liver	Night blindness (retinal)	Blurred vision, pain in calves, fatigue, irritability, loss of appetite
Beta-carotene: spinach and other dark leafy greens; broccoli; deep orange fruits (apricots, cantaloupe) and vegetables (squash, carrots, sweet potatoes, pumpkin)	***Skin and Cornea***	
	Keratinization, corneal degeneration leading to blindness,[b] rashes	Dry skin, rashes, loss of hair
	Other	
	Kidney stones, impaired growth	Cessation of menstruation, liver and spleen enlargement

Vitamin D

Other Names	Deficiency Symptoms	Toxicity Symptoms
Calciferol, cholecalciferol, dihydroxy vitamin D; precursor is cholesterol	***Blood/Circulatory System***	
		Raised blood calcium
Chief Functions in the Body	***Bones/Teeth***	
Mineralization of bones (raises blood calcium and phosphorus via absorption from digestive tract, and by withdrawing calcium from bones and stimulating retention by kidneys)	Abnormal growth, misshapen bones (bowing of legs), soft bones, joint pain, malformed teeth	Increased calcium withdrawal
	Nervous/Muscular Systems	
Deficiency Disease Name	Muscle spasms	Excessive thirst, headaches, irritability, loss of appetite, weakness, nausea
Rickets, osteomalacia	***Other***	
Significant Sources		Kidney stones, stones in arteries, death
Self-synthesis with sunlight; fortified milk, margarine, butter, and cereals; eggs, liver, small fish (sardines)		

[a]Small-cell anemia is termed *microcytic anemia;* large-cell type is *macrocytic* or *megaloblastic anemia.*
[b]Corneal degeneration progresses from *keratinization* (hardening) to *xerosis* (drying) to *xerophthalmia* (thickening, opacity, and irreversible blindness).

▓▓ The B Vitamins

Despite advertisements that claim otherwise, the B vitamins do not give people energy. Carbohydrate, fat, and protein—the *energy-yielding* nutrients—supply the fuel for energy. The B vitamins help to burn that fuel, but do not serve as fuel themselves.

▓▓ **COENZYMES** The eight B vitamins were listed in Table 8-1. Each is part of an enzyme helper known as a **coenzyme.** Each B vitamin has other

coenzyme (co-EN-zime): a small molecule that works with an enzyme to promote the enzyme's activity. Many coenzymes have B vitamins as part of their structure.

co = with

Table 8-3
THE FAT-SOLUBLE VITAMINS—A SUMMARY—continued

Vitamin E

Other Names	Deficiency Symptoms	Toxicity Symptoms
Alpha-tocopherol, tocopherol		**Blood/Circulatory System**
Chief Functions in the Body	Red blood cell damage, anemia	Augments the effects of anticlotting medication.
Antioxidant (detoxification of strong oxidants), stabilization of cell membranes, regulation of oxidation reactions, protection of PUFA and vitamin A		**Digestive System**
		General discomfort
		Nervous/Muscular Systems
Deficiency Disease Name	Degeneration, weakness, difficulty walking, leg cramps	(No symptoms reported)
(No name)		**Other**
Significant Sources	Fibrocystic breast disease	(No symptoms reported)
Polyunsaturated plant oils (margarine, salad dressings, shortenings), green and leafy vegetables, wheat germ, whole-grain products, nuts, seeds		

Vitamin K

Other Names	Deficiency Symptoms	Toxicity Symptoms
Phylloquinone, menaquinone, naphthoquinone		**Blood/Circulatory System**
Chief Functions in the Body	Hemorrhaging	Interference with anticlotting medication; vitamin K analogues may cause jaundice
Synthesis of blood-clotting proteins and a protein that binds calcium in the bones		
Deficiency Disease Name		
(No name)		
Significant Sources		
Bacterial synthesis in the digestive tract; liver, green leafy vegetables, cabbage-type vegetables, milk		

important functions in the body as well, but the roles these vitamins play as parts of coenzymes are the best understood. A coenzyme is a small molecule that combines with an enzyme to make it active. With the coenzyme in place, a substance is attracted to the enzyme, and the reaction proceeds instantaneously. Figure 8-3 on p. 180 illustrates coenzyme action.

A coenzyme already mentioned in the last chapter was coenzyme A, or CoA, made from the vitamin pantothenic acid. Various other coenzymes containing thiamin, riboflavin, niacin, or biotin also participate in the release of energy from glucose, amino acids, and fats. A coenzyme containing vitamin B_6 assists enzymes that metabolize amino acids. The making of new cells depends on a folate coenzyme, and the making of this coenzyme depends on vitamin B_{12}.

The eight B vitamins play many specific roles in helping the enzymes perform thousands of different molecular conversions in the body. These vitamins must be present in every cell continuously for the cells to function as they should. As for vitamin C, its primary role, discussed later, is as an antioxidant.

Figure 8-3
COENZYME ACTION

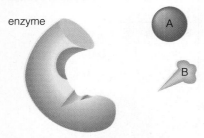

enzyme

Without the coenzyme, compounds A and B don't respond to the enzyme.

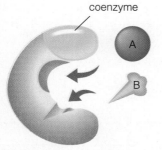

coenzyme

With the coenzyme in place, compounds A and B are attracted to the active site on the enzyme, and they react.

The reaction is completed with the formation of a new product. In this case the product is AB.

beriberi: the thiamin-deficiency disease; characterized by loss of sensation in the hands and feet, muscular weakness, advancing paralysis, and abnormal heart action.

pellagra (pell-AY-gra): the niacin-deficiency disease. Symptoms include the "4 Ds": diarrhea, dermatitis, dementia, and, ultimately, death.

pellis = skin

agra = seizure

refined grain: a product from which the bran, germ, and husk have been removed, leaving only the endosperm.

⠿ B VITAMIN DEFICIENCIES In academic and clinical discussions of the B vitamins, different sets of deficiency symptoms are ascribed to each individual vitamin. Such clear-cut symptoms are found only in laboratory animals that have been fed contrived diets that lack just one nutrient. In reality, a deficiency of any single B vitamin seldom shows up in isolation because people do not eat nutrients one by one; they eat foods containing mixtures of many nutrients. If a major class of foods is missing from the diet, all of the nutrients delivered by those foods will be lacking to various extents.

In only two cases have dietary deficiencies associated with single B vitamins been observed on a large scale in human populations. Diseases have been named for these deficiency states. One of them, **beriberi,** was first observed in Southeast Asia when the custom of polishing rice became widespread. Rice contributed 80 percent of the energy intake of the people in this area, and rice hulls were their principal source of thiamin. When the hulls were removed to make the rice whiter, beriberi spread like wildfire.

The niacin-deficiency disease, **pellagra,** became widespread in the southern United States in the early part of the twentieth century among people who subsisted on a low-protein diet with a staple grain of corn. This diet was unusual in that it supplied neither enough niacin nor enough of its amino acid precursor tryptophan to make the niacin intake adequate.

Even in the cases of beriberi and pellagra, the deficiencies were probably not pure. When foods were provided containing the one vitamin known to be needed, other vitamins that may have been in short supply came as part of the package.

Major deficiency diseases such as pellagra and beriberi no longer occur in the United States and Canada, but more subtle deficiencies of nutrients, including the B vitamins, sometimes are observed. When they do occur, it is usually in people whose food choices are poor because of poverty, ignorance, illness, or poor health habits such as alcohol abuse.

⠿ INTERDEPENDENT SYSTEMS Table 8-4, at the end of this chapter, sums up a few of the better-established facts about B vitamin deficiencies. A look at the table will make another generalization possible. Different body systems depend to different extents on these vitamins. Processes in nerves and in their responding tissues, the muscles, depend heavily on glucose metabolism and hence on thiamin, so paralysis sets in when this vitamin is lacking, but thiamin is important in all cells, not just in nerves and muscles. Similarly, because the red blood cells and GI tract cells divide the most rapidly, two of the first symptoms of a deficiency of folate are a type of anemia and GI deterioration—but again, all systems depend on folate, not just these. The list of symptoms in Table 8-4 is far from complete.

⠿ B VITAMIN ENRICHMENT OF FOODS If the staple food of a region is made from **refined grain,** vitamin B deficiencies are especially likely. One way to protect people from deficiencies is to add nutrients to their staple food, a process known as **fortification** or **enrichment.*** The enrichment of refined breads and cereals has drastically reduced the incidence of iron and B vitamin deficiencies.

The preceding discussion has shown both the great importance of the B vitamins in promoting normal, healthy functioning of all body systems and the severe consequences of deficiency. Now you may want to know how to be sure you and your clients are getting enough of these vital nutrients. The next sections present information on each B vitamin. While reading further, keep in mind that *foods* can provide all the needed nutrients and that supplements are

---

*NOTE: The terms *fortified* and *enriched* may be used interchangeably.

a poor second choice. Some supplements are absurdly costly; but even if they are inexpensive, most people don't need them. Nutrition in Practice 8 discusses uses and choices of supplements in more detail.

Thiamin

All cells use thiamin, which plays a critical role in their energy metabolism. Thiamin also occupies a special site on nerve cell membranes. Consequently, processes in nerves and in their responding tissues, the muscles, depend heavily on thiamin.

THIAMIN NEED

As long as people consume enough food to meet their energy needs—and obtain that energy from nutritious foods—thiamin needs will be met. People who derive a large proportion of their energy from empty-kcalorie items, like sugar or alcohol, however, risk thiamin deficiency, a condition that seems to be reappearing as the population of malnourished and homeless people rises. A person who is fasting or who has adopted a very-low-kcalorie diet needs as much thiamin as when eating enough to meet energy needs.

In developed countries today, abuse of alcohol often leads to a severe form of thiamin deficiency, Wernicke-Korsakoff syndrome.[23] Alcohol contributes energy, but carries almost no nutrients with it and often displaces food. In addition, alcohol impairs absorption of thiamin from the digestive tract and hastens its excretion in the urine, tripling the risk of deficiency. Wernicke-Korsakoff syndrome is characterized by symptoms that are almost indistinguishable from alcohol abuse itself: mental confusion, disorientation, loss of memory, jerky eye movements, and staggering gait. Unlike alcohol toxicity, the syndrome responds quickly to an injection of thiamin, and some experts recommend a precautionary dose for any patients suspected of having the syndrome.[24]

THIAMIN IN FOODS

Thiamin occurs in small quantities in virtually all nutritious foods, but it is concentrated in only a few foods, of which pork is the most commonly eaten. A useful guideline for meeting thiamin needs is to keep empty-kcalorie foods to a minimum and to include ten or more different servings of nutritious foods each day, assuming that each serving will contribute, on the average, about 10 percent of needs. Foods chosen from the bread and cereal group should be either whole grain or enriched. Thiamin is not stored in the body to any great extent, so daily intake is important.

Riboflavin

Like thiamin, riboflavin facilitates energy production in the body. The needs of infants, children, and pregnant women rise rapidly during periods of active growth.

RIBOFLAVIN IN FOODS

Unlike thiamin, riboflavin is not evenly distributed among the food groups. The major contributors of riboflavin to people's diets are milk, milk products, meats, and green vegetables (broccoli, turnip greens, asparagus, and spinach). The riboflavin richness of milk and milk products is a good reason to include these foods in every day's meals. No other commonly eaten food can make such a substantial contribution. People who omit milk and milk products from their diets can substitute generous servings of dark green, leafy vegetables. Among the meats, liver and heart are the richest sources, but all lean meats, as well as eggs, offer some riboflavin.

EFFECTS OF LIGHT

Riboflavin is light sensitive; the ultraviolet rays of the sun or of fluorescent lamps can destroy it. For this reason, milk is sold

fortification: the addition to a food of nutrients that were either not originally present or present in insignificant amounts. Fortification can be used to correct or prevent a widespread nutrient deficiency, to balance the total nutrient profile of a food, or to restore nutrients lost in processing.

enrichment: the addition to a food of nutrients to meet a specified standard. In the case of refined bread or cereal, five nutrients have been added: thiamin, riboflavin, niacin, and folate in amounts approximately equivalent to, or higher than, those originally present and iron in amounts to alleviate the prevalence of iron-deficiency anemia.

Nutritious foods such as pork, legumes, sunflower seeds, and whole-grain breads are valuable sources of thiamin.

Milk and milk products supply much (about 50 percent) of the riboflavin in people's diets, but meats, eggs, green vegetables, and whole-wheat bread are good sources, too.

A food containing 1 mg of niacin and 60 mg tryptophan contains the niacin equivalent of 2 mg, or 2 mg NE.

niacin equivalents (NE): the amount of niacin present in food, including the niacin that can theoretically be made from its precursor tryptophan present in the food.

Niacin-rich foods include meat, fish, poultry, and peanut butter, as well as enriched breads and cereals and a few vegetables.

When a normal dose of a nutrient clears up a deficiency condition, the effect is a physiological one. When a large dose of a nutrient overwhelms a body system and acts like a drug, the effect is a pharmacological one.

The protein avidin in egg whites binds biotin.

in cardboard or opaque plastic containers to protect the riboflavin in the milk from light. In contrast, riboflavin is heat stable, so ordinary cooking does not destroy it.

Niacin

Like thiamin and riboflavin, niacin participates in the energy metabolism of every body cell. Niacin is unique among the B vitamins in that the body can make it from protein. The amino acid tryptophan can be converted to niacin in the body: 60 milligrams of tryptophan yield 1 milligram of niacin. Recommended intakes are therefore stated in **niacin equivalents (NE),** reflecting the body's ability to convert tryptophan to niacin.

NIACIN USED AS A MEDICATION Certain forms of niacin supplements in amounts ten times or more than the dietary recommendation cause "niacin flush," a dilation of the capillaries of the skin with perceptible tingling that, if intense, can be painful. Physicians sometimes use diet and large doses of niacin to lower blood cholesterol to treat atherosclerosis. When used this way, niacin leaves the realm of nutrition to become a pharmacological agent, a drug. As with any medication, self-dosing with niacin is ill advised; large doses may injure the liver, cause ulcers, and produce some symptoms of diabetes.[25]

NIACIN IN FOODS Meat, poultry, and fish contribute about half the niacin equivalents most people consume; enriched breads and cereals contribute about a fourth. Among the vegetables, mushrooms, asparagus, and green leafy vegetables are the richest niacin sources. Niacin is less vulnerable to losses during food preparation and storage than other water-soluble vitamins. Being fairly heat-resistant, niacin can withstand reasonable cooking times, but like other water-soluble vitamins, it will leach into cooking water.

Pantothenic Acid and Biotin

Two other B vitamins—pantothenic acid and biotin—are also important in energy metabolism. Pantothenic acid was first recognized as a substance that stimulates growth. It is a component of a key enzyme that makes possible the release of energy from the energy nutrients. Pantothenic acid is involved in more than 100 different steps in the synthesis of lipids, neurotransmitters, steroid hormones, and hemoglobin. Biotin plays an important role in metabolism as a coenzyme that carries carbon dioxide. This role is critical in the TCA cycle.

PANTOTHENIC ACID AND BIOTIN IN FOODS Both pantothenic acid and biotin are more widespread in foods than the other vitamins discussed so far. There seems to be no danger that people who consume a variety of foods will suffer deficiencies. Claims that pantothenic acid and biotin are needed in pill form to prevent or cure disease conditions are at best unfounded and at worst intentionally misleading.

BIOTIN DEFICIENCY Biotin deficiencies are rare, but have been reported in adults fed artificially by vein without biotin supplementation. Researchers can induce biotin deficiency in animals or human beings by feeding them raw egg whites, which contain a protein that binds biotin and prevents its absorption. Long-term alcohol abuse may also lead to biotin deficiency.[26]

■■ Vitamin B$_6$

Vitamin B$_6$ has been called the "sleeping giant" of vitamins. A surge of research interest in the last decade has not only revealed new knowledge, but has also raised new questions. For example, unlike other water-soluble vitamins, vitamin B$_6$ is stored extensively in muscle tissue.

■■ **METABOLIC ROLES OF VITAMIN B$_6$**   Vitamin B$_6$ has long been known to play roles in protein and amino acid metabolism. In the cells, vitamin B$_6$ helps to convert one kind of amino acid, which the cells have in abundance, to another, which they need in larger amounts. It also aids in the conversion of the amino acid tryptophan to niacin and plays important roles in the synthesis of hemoglobin. Vitamin B$_6$ also assists in releasing stored glucose from glycogen and thus contributes to the regulation of blood glucose. Recent research suggests roles for vitamin B$_6$ in immune function and hormone response.[27] The association between vitamin B$_6$ and immune function is related to the critical role the vitamin plays in protein metabolism. Vitamin B$_6$ deficiency can significantly impair the immune response, perhaps by way of impaired antibody production.[28]

A possible role for vitamin B$_6$ in the treatment of disease is also being investigated. For example, elevated blood levels of the amino acid homocysteine correlate with a high incidence of heart disease.[29] Evidence suggests that low blood concentrations of vitamin B$_6$ and folate are associated with elevated homocysteine concentrations.[30] Researchers conducted a study of more than 80,000 women to determine whether vitamin B$_6$ and folate intake might influence the risk of heart disease.[31] Women with the highest intakes of vitamin B$_6$ from food and supplements combined had lower risks of heart disease than women with low intakes. This was true of folate as well, and the risk was lower still in women with the highest intakes of both vitamins.

■■ **VITAMIN B$_6$ DEFICIENCY**   Besides a weakened immune response, vitamin B$_6$ deficiency is expressed in general symptoms, such as weakness, irritability, and insomnia. Other symptoms include a greasy, flaky dermatitis; anemia; and, in advanced cases, convulsions.

■■ **VITAMIN B$_6$ TOXICITY**   For years it was believed that vitamin B$_6$, like other water-soluble vitamins, could not reach toxic concentrations in the body. Toxic effects of vitamin B$_6$ became known when a physician reported them in women who had been taking more than 2 grams of vitamin B$_6$ daily for two months or more. Most of these women had been attempting to relieve premenstrual syndrome (PMS), the cluster of physical, emotional, and psychological symptoms that some women experience prior to menstruation. The first symptom of toxicity was numb feet; then the women lost sensation in their hands; then they became unable to walk. The women recovered after they discontinued the supplements.

The specific cause or causes of PMS remain undefined, although researchers agree that the hormonal changes of the menstrual cycle must be responsible. Despite a lack of conclusive evidence that vitamin B$_6$ is an effective treatment for PMS, the vitamin and many other unproven remedies remain popular among women suffering with PMS.

■■ **VITAMIN B$_6$ AND PROTEIN INTAKE**   Vitamin B$_6$'s many roles in amino acid metabolism are reflected in dietary needs that are roughly proportional to protein intakes. The RDA for vitamin B$_6$ is more than adequate to handle average protein intakes of 100 grams per day for men and 60 grams per day for women.

Meat, chicken, and fish, as well as some fruits and vegetables, are good sources of vitamin B₆.

neural tube defects: malformations of the brain, spinal cord, or both during embryonic development.

The two main types of neural tube defects are spina bifida (literally "split spine") and anencephaly ("no brain").

Bread products, flour, corn grits, and pasta must be fortified with 1.4 mg per 100 g of food (about ½ c cooked food or 1 slice of bread.

bioavailability: the rate and extent to which a nutrient is absorbed and used.

VITAMIN B₆ IN FOODS

VITAMIN B₆ IN FOODS The richest food sources of vitamin B₆ are protein-rich meat, fish, and poultry. Potatoes, a few other vegetables, and some fruits are good sources, too. Foods lose vitamin B₆ when heated.

Folate

The B vitamin folate is active in cell division. During periods of rapid growth and cell division, such as pregnancy and adolescence, folate needs increase, and deficiency is especially likely. When a deficiency occurs, the replacement of the rapidly dividing cells of the blood and the GI tract falters. Not surprisingly, then, two of the first symptoms of a folate deficiency are a type of anemia and GI tract deterioration (see Table 8-4).

FOLATE, ALCOHOL, AND DRUGS Of all the vitamins, folate appears to be most vulnerable to interactions with alcohol and other drugs. As Nutrition in Practice 23 describes, alcohol-addicted people risk folate deficiency because alcohol impairs folate's absorption and increases its excretion. Furthermore, as people's alcohol intakes rise, their folate intakes decline. Many medications, including aspirin, oral contraceptives, and anticonvulsants, also impair folate status. Smoking exerts a negative effect on folate status as well.

FOLATE AND NEURAL TUBE DEFECTS Research studies confirm the importance of folate in preventing **neural tube defects.**[32] The brain and spinal cord develop from the neural tube, and defects in its orderly formation during the early weeks of pregnancy may result in various central nervous system disorders and death. Folate supplements taken one month before conception and continued throughout the first trimester of pregnancy can prevent neural tube defects.[33] For this reason, the American Academy of Pediatrics and the Public Health Services recommend that all women of childbearing age who are capable of becoming pregnant take 0.4 milligram (400 micrograms) of folate daily.[34]

Although this amount can be met by eating the recommended five servings of fruits and vegetables daily, most women's dietary folate intakes fall short. Furthermore, some research indicates that folate status improves more with supplementation or fortification than with a dietary intake that meets recommendations.[35]

Because most pregnancies are unplanned and neural tube defects arise early in pregnancy before most women realize they are pregnant, the FDA has mandated that refined grain products (flour, cornmeal, pasta, and rice) be fortified to increase average folate intakes. Fortification is expected to prevent half of the 4000 neural tube defects that occur each year. Folate fortification, however, raises safety concerns as well. High doses of folate can complicate the diagnosis of vitamin B₁₂ deficiency, as discussed later. Thus, folate intakes should not exceed 1 milligram per day.[36]

FOLATE IN FOODS As Snapshot 8-2 shows, the best food sources of folate are liver, legumes, beets, and green leafy vegetables (the vitamin's name suggests the word *foliage*). Among the fruits, oranges, orange juice, and cantaloupe are the best sources. With fortification, grain products are good sources of folate, too. The **bioavailability** of added folate is good, and fortification is expected to increase women's average folate intakes by 100 micrograms per day.[37] Heat and oxidation during cooking and storage can destroy up to half of the folate in foods.

Vitamin B₁₂

Vitamin B₁₂ and folate share a special relationship: vitamin B₁₂ assists folate in cell division. Their roles intertwine, but each performs a specific task that the other cannot accomplish.

FOLATE

All of these foods except cantaloupe provide 10 percent or more of the folate Daily Value (DV = 400 µg/day).[a]

RDA for adults: 400 µg/day

Cantaloupe: 15 µg (4% DV) per small wedge (⅙ melon)

Liver: 185 µg (46% DV) per 3 oz fried

Asparagus: 131 µg (33% DV) per ½ c cooked

Spinach: 108 µg (27% DV) per 1 c raw

Pinto beans: 146 µg (37% DV) per ½ c cooked

Beets: 46 µg (12% DV) per ½ c cooked

[a]The Daily Values are based on a 2000-kcalorie diet.

▪▪ VITAMIN B₁₂, FOLATE, AND CELL DIVISION Vitamin B_{12} (in coenzyme form) stands by to accept carbon groups from folate as folate removes them from other compounds. The passing of these carbon groups from folate to vitamin B_{12} regenerates the active form of folate so that it can continue its dismantling tasks. In the absence of vitamin B_{12}, folate is trapped in its inactive, metabolically useless form, unable to do its job. When folate is either trapped due to a vitamin B_{12} deficiency or unavailable due to a deficiency of folate itself, cells that are growing most rapidly, notably the blood cells, are the first to be affected. Thus, a deficiency of either nutrient—vitamin B_{12} or folate—impairs maturation of the blood cells and produces anemia. The anemia is identifiable by microscopic examination of the blood, which reveals many large, immature red blood cells. Either vitamin B_{12} or folate will clear up the anemia.

Large-cell anemia is known as macrocytic or megaloblastic anemia.
macro = large
cyte = cell
mega = large

▪▪ VITAMIN B₁₂ AND THE NERVOUS SYSTEM Although either vitamin will clear up the anemia caused by vitamin B_{12} deficiency, if folate is given when vitamin B_{12} is needed, the result is disastrous, not to the blood but to the nervous system. The reason: vitamin B_{12} also helps maintain nerve fibers. A vitamin B_{12} deficiency can ultimately result in devastating neurological symptoms, undetectable by a blood test. A deceptive folate "cure" of the anemia in vitamin B_{12} deficiency allows the nerve deterioration to progress, leading to paralysis and permanent nerve damage. This interaction between folate and vitamin B_{12} raises safety concerns about the use of folate supplements and fortification of foods.

The way folate masks vitamin B_{12} deficiency underlines a point already made several times: it takes a skilled diagnostician to make a correct diagnosis. A person who self-diagnoses on the basis of a single observed symptom takes a serious risk.

▪▪ VITAMIN B₁₂ ABSORPTION Vitamin B_{12} requires an "**intrinsic factor**"—a compound made inside the body—for absorption from the intestinal tract into the bloodstream. This intrinsic factor is made in the stomach, where it attaches to the vitamin; the complex then passes to the small intestine and is gradually absorbed.

intrinsic: inside the system. Anemia that reflects a vitamin B_{12} deficiency caused by lack of intrinsic factor is known as **pernicious anemia.**

▪▪ LOSS OF INTRINSIC FACTOR In some cases, intrinsic factor production becomes inadequate or ceases altogether—for example, in people who develop chronic inflammation of the lining of the stomach. Some people

The intrinsic factor necessary to prevent pernicious anemia is made in the stomach and aids in the absorption of vitamin B_{12}. The extrinsic factor necessary to prevent pernicious anemia is vitamin B_{12} itself, which must be obtained outside the body from food.

inherit a defective gene for intrinsic factor. Because vitamin B_{12} deficiency in the body may be caused either by a lack of the vitamin in the diet or by the body's inability to absorb the vitamin, a change in diet alone may not correct the deficiency. When absorption failure is the problem, vitamin B_{12} must be supplied by injection.

:: VITAMIN B_{12} IN FOODS A unique characteristic of vitamin B_{12} is that it is found almost exclusively in foods derived from animals. People who eat meat are guaranteed an adequate intake, and lacto-ovo vegetarians (who use milk, cheese, and eggs) are also protected from deficiency. It is a myth, however, that fermented soy products, such as miso (a soybean paste), or sea algae, such as spirulina, provide vitamin B_{12} in its active form. Extensive research shows that the amounts of vitamin B_{12} listed on the labels of these plant products are inaccurate and misleading because the vitamin B_{12} in these products occurs in an inactive, unavailable form. Vegans must take vitamin B_{12} supplements or find other sources of active vitamin B_{12}. Some loss of vitamin B_{12} occurs when foods are heated in microwave ovens.[38]

:: VITAMIN B_{12} DEFICIENCY IN VEGANS Vegans are at special risk for undetected vitamin B_{12} deficiency for two reasons: first, they receive none in their diets, and second, they consume large amounts of folate in the vegetables they eat. Because the body can store 1000 times the amount of vitamin B_{12} used each day, a deficiency may take years to develop in a new vegetarian. When a deficiency does develop, though, it may progress to a dangerous extreme because the deficiency of vitamin B_{12} may be masked by the high folate intake.

LEARNING LINK

The previous discussion of the B vitamins mentions several interactions between individual vitamins and alcohol, cigarette smoke, and medications. Each of the nutrients may interact with substances that can alter the absorption, metabolism, or excretion of the nutrient, but some are more vulnerable than others. Nutrition in Practice 23 discusses alcohol's devastating effects on thiamin and folate, for example. Chapter 13 discusses diet-medication interactions and offers suggestions for nurses when obtaining information about a client's medication use. When medications that interact with nutrients in adverse ways must be used in treatment, the client risks developing a deficiency. Nurses who learn to regularly check for diet-medication interactions will be best prepared to serve their clients.

:: Non-B Vitamins

Other compounds are sometimes inappropriately called B vitamins because, like the true B vitamins, they serve as coenzymes in metabolism. Even if they were essential, however, supplements would be unnecessary because these compounds are abundant in foods.

:: INOSITOL, CHOLINE, AND CARNITINE Among the non-B vitamins are a trio of coenzymes known as inositol, choline, and carnitine. Researchers are exploring the possibility that these substances may be essential. Thus far, only choline has been assigned an Adequate Intake (AI) value.

:: OTHER NON-B VITAMINS Other substances have also been mistaken for essential nutrients. They include para-aminobenzoic acid (PABA), bioflavonoids (vitamin P or hesperidin), and ubiquinone. Other names you may hear are "vitamin B_{15}" (a hoax) and "vitamin B_{17}" (laetrile, a fake cancer-curing drug and not a vitamin by any stretch of the imagination). There is, however, one other water-soluble vitamin of great interest and importance—vitamin C.

Vitamin C

Two hundred years ago, any man who joined the crew of a seagoing ship knew he had only half a chance of returning alive—not because he might be slain by pirates or die in a storm, but because he might contract the dread disease **scurvy.** Then a physician with the British navy found that citrus fruits could cure the disease, and thereafter, all ships were required to carry lime juice for every sailor. (This is why British sailors are still called "limeys" today.) Nearly 200 years later, the antiscurvy factor in citrus fruits was isolated from lemon juice and named **ascorbic acid.** Today, hundreds of millions of vitamin C pills are produced in pharmaceutical laboratories.

scurvy: the vitamin C–deficiency disease.

ascorbic acid: one of the two active forms of vitamin C. Many people refer to vitamin C by this name.
   a = without
   scorbic = having scurvy

METABOLIC ROLES OF VITAMIN C   Vitamin C's action defies a simple, tidy description. It plays many important roles in the body, and its modes of action differ in different situations.

VITAMIN C'S ROLE IN COLLAGEN FORMATION   The best-understood action of vitamin C is its role in helping to form **collagen,** the single most important protein of connective tissue. Collagen serves as the matrix on which bone is formed, the material of scars for wound healing, and an important part of the "glue" that attaches one cell to another. This latter function is especially important in the artery walls, which must expand and contract with each beat of the heart, and in the walls of the capillaries, which are thin and fragile.

collagen: the characteristic protein of connective tissue.
   kolla = glue
   gennan = produc

VITAMIN C AS AN ANTIOXIDANT   Vitamin C is also an important antioxidant. Recall that the antioxidants beta-carotene and vitamin E protect fat-soluble substances from oxidizing agents; vitamin C protects water-soluble substances the same way. Vitamin C's antioxidant action is twofold. First, by being oxidized itself, vitamin C regenerates already-oxidized substances such as iron and copper to their original, active form. Second, in the process, the vitamin removes the damaging oxidizing agent. In the intestines, it protects iron from oxidation and so enhances iron absorption. In the cells and body fluids, it helps to protect other molecules, including the fat-soluble compounds vitamin A, vitamin E, and the polyunsaturated fatty acids.

VITAMIN C IN AMINO ACID METABOLISM   Vitamin C is also involved in the metabolism of several amino acids. Some of these amino acids end up being used to make hormones of great importance in body functioning, among them norepinephrine and thyroxine.

ROLE OF STRESS   During stress, the adrenal glands release large quantities of vitamin C together with the stress hormones epinephrine and norepinephrine. What the vitamin has to do with the stress reaction is unclear, but it is known that stress increases vitamin C needs somewhat.

VITAMIN C AS A POSSIBLE ANTIHISTAMINE   Newspaper headlines touting vitamin C as a cure for colds and cancer have appeared frequently over the years. Some research suggests that vitamin C (2 grams per day for two weeks) may reduce the severity and duration of cold and allergy symptoms by reducing blood histamine concentrations.[39] In other words, vitamin C acts as an

antihistamine. If further research confirms vitamin C's antihistamine effect, its use may permit people to rely less heavily on antihistamine drugs when suffering from cold and allergy symptoms.

▓▓ VITAMIN C'S ROLE IN CANCER PREVENTION AND TREATMENT

The role of vitamin C in the prevention and treatment of cancer is still being studied. A large-scale population study of over 11,000 U.S. adults reported an inverse relationship between all causes of death and vitamin C intake up to a few hundred milligrams.[40] The relationship was stronger for men than for women. In a dozen or so different well-controlled studies, researchers identified individuals with and without cancer and assessed their dietary intakes of vitamin C. They found that people with high vitamin C intakes had lower risks of these cancers than did people with low intakes. The correlation may reflect not just an association with vitamin C, but the broader benefits of a diet rich in fruits and vegetables and low in fat. It does not support the taking of vitamin C supplements to prevent or treat cancer.

▓▓ VITAMIN C DEFICIENCY

When intake of vitamin C is inadequate, the body's vitamin C pool dwindles, and **latent** scurvy appears. The blood vessels show the first deficiency signs. The gums around the teeth begin to bleed easily, and capillaries under the skin break spontaneously, producing pinpoint hemorrhages. Then the symptoms of **overt** scurvy appear. Muscles, including the heart muscle, may degenerate. The skin becomes rough, brown, scaly, and dry. Wounds fail to heal because scar tissue will not form without collagen. Bone rebuilding falters; the ends of the long bones become softened, malformed, and painful; and fractures occur. The teeth may become loose in the jawbone and fall out. Anemia and infections are common. Sudden death is likely, perhaps because of massive bleeding into the joints and body cavities.

It takes only 10 or so milligrams of vitamin C a day to prevent overt scurvy, and not much more than that to cure it. Once diagnosed, scurvy is readily reversible with moderate doses, in the neighborhood of 100 milligrams per day. Such an intake is easily achieved by including vitamin C–rich foods in the diet.

▓▓ VITAMIN C TOXICITY

The easy availability of vitamin C in pill form and the publication of books recommending vitamin C to prevent everything from the common cold to life-threatening cancer have led thousands of people to take large doses of vitamin C. Not surprisingly, instances of vitamin C causing harm have surfaced.

Some of the suspected toxic effects of excess vitamin C have not been confirmed, but others have been seen often enough to warrant concern. Nausea, abdominal cramps, and diarrhea are often reported. Several instances of interference with medical regimens are known. Large amounts of vitamin C excreted in the urine obscure the results of tests used to detect diabetes. People taking anticoagulants may unwittingly counteract the effect of these medicines if they also take massive doses of vitamin C. Large doses of vitamin C can enhance iron absorption too much, resulting in iron overload (see Chapter 9).

People with sickle-cell anemia may be especially vulnerable to excess vitamin C. Those who have a tendency toward **gout,** as well as those who have a genetic abnormality that alters the way they metabolize vitamin C, are more prone to forming stones if they take large doses of vitamin C.

▓▓ WITHDRAWAL REACTION

A person who has taken large doses of vitamin C for a long time may adapt by limiting absorption and destroying and excreting more of the vitamin than usual. If the person then suddenly

latent: the period in the course of a disease when the conditions are present but the symptoms have not begun to appear.

latens = lying hidden

overt: out in the open, full-blown.

ouvrire = to open

As discussed in Chapter 1, the Tolerable Upper Intake Level (UL) is the highest level of daily nutrient intake that is likely to pose no risk of adverse health effects. The UL for vitamin C is 2,000 mg for adults.

The anticoagulants with which vitamin C interferes are warfarin and dicumarol.

gout (GOWT): a metabolic disease in which crystals of uric acid precipitate in the joints.

reduces intake to normal, the accelerated disposal system cannot put on its brakes fast enough to avoid destroying too much of the vitamin. Some case histories have shown that adults who abruptly discontinue megadosing develop scurvy on intakes that would protect other adults. When people stop taking vitamin C in excessive doses, they might be wise to do so gradually.

The temporary withdrawal symptoms experienced by the person who stops overdosing reflect a vitamin C dependency. The body has adjusted to a high intake and so "needs" a high intake until it can readjust.

RECOMMENDED INTAKES OF VITAMIN C   The daily recommendation for vitamin C is 90 milligrams for men and 75 milligrams for women, with an extra 10 milligrams recommended for pregnant women and an additional 45 milligrams for lactating women. The recommendations consider two extremes. At one extreme is the requirement to prevent overt scurvy, or 10 milligrams per day; at the other extreme is the amount at which the body's pool of vitamin C would be full to overflowing, or about 100 milligrams per day.

SPECIAL NEEDS FOR VITAMIN C   As is true of all nutrients, unusual circumstances may raise vitamin C needs. Among the stresses known to do so are infections; burns; surgery; extremely high or low temperatures; toxic doses of heavy metals, such as lead, mercury, and cadmium; and the chronic use of certain medications, including aspirin, barbiturates, and oral contraceptives. Smoking, too, has adverse effects on vitamin C status. Cigarette smoke contains oxidants, which deplete this potent antioxidant.[41] Accordingly, the vitamin C recommendation for people who smoke is an additional 35 milligrams per day above the recommendations for nonsmokers.

SAFE LIMITS FOR VITAMIN C   Few instances warrant the taking of more than 100 to 300 milligrams a day. The risks may not be great for adults who dose themselves with 1 to 2 grams a day, but those taking more than 2 grams, and especially those taking above 8 grams per day, should be aware of the distinct possibility of harm.

VITAMIN C IN FOODS   The inclusion of intelligently selected fruits and vegetables in the daily diet guarantees a generous intake of vitamin C. Even those who wish to ingest amounts well above the recommended intakes can easily meet their goals by eating certain foods (see Snapshot 8-3). Citrus fruits are rightly famous for their vitamin C contents. Certain other fruits and

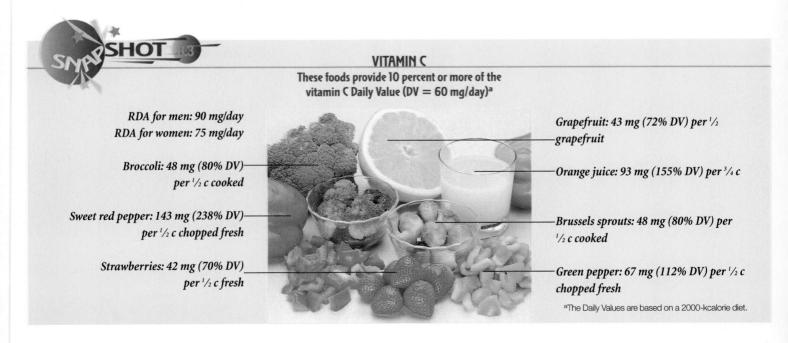

SNAPSHOT 8-3

VITAMIN C
These foods provide 10 percent or more of the vitamin C Daily Value (DV = 60 mg/day)[a]

RDA for men: 90 mg/day
RDA for women: 75 mg/day

Broccoli: 48 mg (80% DV) per ½ c cooked

Sweet red pepper: 143 mg (238% DV) per ½ c chopped fresh

Strawberries: 42 mg (70% DV) per ½ c fresh

Grapefruit: 43 mg (72% DV) per ½ grapefruit

Orange juice: 93 mg (155% DV) per ¾ c

Brussels sprouts: 48 mg (80% DV) per ½ c cooked

Green pepper: 67 mg (112% DV) per ½ c chopped fresh

[a]The Daily Values are based on a 2000-kcalorie diet.

vegetables are also rich sources: cantaloupe, strawberries, broccoli, and brussels sprouts. No animal foods other than organ meats, such as liver and kidneys, contain vitamin C. The humble potato is an important source of vitamin C in Western countries, where potatoes are eaten so frequently that they make substantial vitamin C contributions overall. They provide about 20 percent of all the vitamin C in the average diet. Vitamin C in foods is easily oxidized, so store cut produce and juices in airtight containers.

VITAMIN C AND IRON ABSORPTION Eating foods containing vitamin C at the same meal with foods containing iron can double or triple the absorption of iron from those foods. This strategy is highly recommended for women and children, whose energy intakes are not large enough to guarantee that they will get enough iron from the foods they eat. Table 8-4 summarizes functions, deficiency and toxicity symptoms, and food sources of vitamin C and the other water-soluble vitamins.

In conclusion, a summary of the key points about vitamins—both fat soluble and water soluble—seems in order:

- Vitamins are essential, nonkcaloric, organic nutrients needed in tiny amounts in the diet.
- Vitamins A, D, E, and K are the fat-soluble vitamins. The B vitamins and vitamin C are the water-soluble vitamins.
- The only disease a vitamin will cure is the one caused by a deficiency of that vitamin.
- Vitamins yield no energy for the body when broken down, but they facilitate the release of energy from carbohydrate, fat, and protein.
- Nutritious foods, not supplements, are the best sources of vitamins.

Table 8-4

THE WATER-SOLUBLE VITAMINS—A SUMMARY

Thiamin

Other Names	Deficiency Symptoms	Toxicity Symptoms
Vitamin B$_1$	**Blood/Circulatory System**	
Chief Functions in the Body	Edema, enlarged heart, abnormal heart rhythms, heart failure	(No symptoms reported)
Part of a coenzyme used in energy metabolism, supports normal appetite and nervous system function	**Nervous/Muscular Systems**	
	Degeneration, wasting, weakness, pain, low morale, difficulty walking, loss of reflexes, mental confusion, paralysis	(No symptoms reported)
Deficiency Disease Name		
Beriberi		
Significant Sources		
Occurs in all nutritious foods in moderate amounts; pork, ham, bacon, liver, whole grains, legumes, nuts		

Riboflavin

Other Names	Deficiency Symptoms	Toxicity Symptoms
Vitamin B$_2$	**Mouth, Gums, Tongue**	
Chief Functions in the Body	Cracks at corners of mouth,[a] magenta tongue	(No symptoms reported)
Part of a coenzyme used in energy metabolism, supports normal vision and skin health	**Nervous System and Eyes**	
	Hypersensitivity to light,[b] reddening of cornea	(No symptoms reported)
Deficiency Disease Name	**Other**	
Ariboflavinosis	Skin rash	(No symptoms reported)
Significant Sources		
Milk, yogurt, cottage cheese, meat, leafy green vegetables, whole-grain or enriched breads and cereals		

Niacin

Other Names	Deficiency Symptoms	Toxicity Symptoms
Nicotinic acid, nicotinamide, niacinamide, vitamin B$_3$; precursor is dietary tryptophan	**Digestive System**	
	Diarrhea	Diarrhea, heartburn, nausea, ulcer irritation, vomiting
Chief Functions in the Body	**Mouth, Gums, Tongue**	
Part of a coenzyme used in energy metabolism; supports health of skin, nervous system, and digestive system	Black, smooth tongue[c]	(No symptoms reported)
	Nervous System	
Deficiency Disease Name	Irritability, loss of appetite, weakness, dizziness, mental confusion progressing to psychosis or delirium	Fainting, dizziness
Pellagra		**Skin**
Significant Sources	Flaky skin rash on areas exposed to sun	Painful flush and rash ("niacin flush), sweating
Milk, eggs, meat, poultry, fish, whole-grain and enriched breads and cereals, nuts, and all protein-containing food	**Other**	
		Abnormal liver function, low blood pressure

[a]Cracks at the corners of the mouth are termed *cheilosis* (kee-LOH-sis).
[b]Hypersensitivity to light is *photophobia*.
[c]Smoothness of the tongue is caused by loss of its surface structures and is termed *glossitis* (gloss-EYE-tis).

continued

Table 8-4

THE WATER-SOLUBLE VITAMINS—A SUMMARY—continued

Pantothenic Acid

Other Names	Deficiency Symptoms	Toxicity Symptoms
(None)	Vomiting, intestinal distress	**Digestive System**
Chief Functions in the Body		(No symptoms reported)
	Insomnia, fatigue	**Nervous System**
Part of a coenzyme used in energy metabolism		(No symptoms reported)
Deficiency Disease Name		**Other**
(No name)		Water retention (infrequent)
Significant Sources		
Widespread in foods		

Biotin

Other Names	Deficiency Symptoms	Toxicity Symptoms
(None)	Abnormal heart action	**Blood/Circulatory System**
Chief Functions in the Body		(No symptoms reported)
	Loss of appetite, nausea	**Digestive System**
Part of a coenzyme used in energy metabolism, fat synthesis, amino acid metabolism, and glycogen synthesis		(No symptoms reported)
	Depression, muscle pain, weakness, fatigue	**Nervous/Muscular Systems**
Deficiency Disease		(No symptoms reported)
(No name)	Drying, rash, loss of hair	**Skin**
Significant Sources		(No symptoms reported)
Widespread in foods		

Vitamin B₆

Other Names	Deficiency Symptoms	Toxicity Symptoms
Pyridoxine, pyridoxal, pyridoxamine	Anemia (small-cell type)[d]	**Blood/Circulatory System**
Chief Functions in the Body		Bloating
	Smooth tongue[c]	**Mouth, Gums, Tongue**
Part of a coenzyme used in amino acid and fatty acid metabolism, helps convert tryptophan to niacin, helps make red blood cells		(No symptoms reported)
	Abnormal brain wave pattern, irritability muscle twitching, convulsions	**Nervous/Muscular Systems**
Deficiency Disease Name		Depression, fatigue, impaired memory, irritability, headaches, numbness, damage to nerves, difficulty walking, loss of reflexes, weakness, restlessness
(No name)		
Significant Sources	Irritation of sweat glands, rashes, greasy dermatitis	**Skin**
Green and leafy vegetables, meats, fish, poultry, shellfish, legumes, fruits, whole grains		(No symptoms reported)
		Other
		(No symptoms reported)

Folate

Other Names	Deficiency Symptoms	Toxicity Symptoms
Folic acid, folacin, pteroylglutamic acid	Anemia (large-cell type)[d]	**Blood/Circulatory System**
		(No symptoms reported)

[c]Smoothness of the tongue is caused by loss of its surface structures and is termed *glossitis* (gloss-EYE-tis).
[d]Small-cell anemia is termed *microcytic anemia;* large-cell type is *macrocytic* or *megaloblastic anemia.*

Table 8-4

THE WATER-SOLUBLE VITAMINS—A SUMMARY—continued

Folate (continued)

	Deficiency Symptoms	Toxicity Symptoms
Chief Functions in the Body		***Digestive System***
	Heartburn, diarrhea, constipation	(No symptoms reported)
Part of a coenzyme used in new cell synthesis, helps maintain nerve cells		***Immune System***
	Suppression, frequent infections	(No symptoms reported)
Deficiency Disease		***Mouth, Gums, Tongue***
	Smooth red tongue[c]	(No symptoms reported)
(No name)		***Nervous System***
Significant Sources	Depression, mental confusion, fainting	(No symptoms reported)
		Other
Leafy green vegetables, legumes, seeds, liver		Masks vitamin B_{12} deficiency

Vitamin B_{12}

Other Names	Deficiency Symptoms	Toxicity Symptoms
Cyanocobalamin		***Blood/Circulatory System***
Chief Functions in the Body	Anemia (large-cell type)[d]	(No symptoms reported)
		Mouth, Gums, Tongue
Part of a coenzyme used in new cell synthesis, helps maintain nerve cells	Smooth tongue[c]	(No symptoms reported)
Deficiency Disease		***Nervous/Muscular Systems***
	Fatigue, degeneration progressing to paralysis	(No symptoms reported)
(No name[e])		***Skin***
Significant Sources	Hypersensitivity	(No symptoms reported)
Animal products (meat, fish, poultry, milk, cheese, eggs)		

Vitamin C

Other Names	Deficiency Symptoms	Toxicity Symptoms
Ascorbic acid		***Blood/Circulatory System***
Chief Functions in the Body	Anemia (large-cell type)[d] atherosclerotic plaques, pinpoint hemorrhages	(No symptoms reported)
Helps in collagen synthesis (strengthens blood vessel walls, forms scar tissue, provides matrix for bone growth), thyroxine synthesis, and amino acid metabolism; serves as an antioxidant; strengthens resistance to infection; helps in absorption of iron		***Digestive System***
		Nausea, abdominal cramps, diarrhea, excessive urination
		Immune System
	Suppression, frequent infections	(No symptoms reported)
Deficiency Disease Name		***Mouth, Gums, Tongue***
	Bleeding gums, loosened teeth	(No symptoms reported)
Scurvy		***Muscular/Nervous Systems***
Significant Sources	Muscle degeneration and pain, hysteria, depression	Headache, fatigue, insomnia
Citrus fruits, cabbage-type vegetables, dark green vegetables, cantaloupe, strawberries, peppers, lettuce, tomatoes, potatoes, papayas, mangoes		***Skeletal System***
	Bone fragility, joint pain	(No symptoms reported)
		Skin
	Rough skin, blotchy bruises	Rashes
		Other
	Failure of wounds to heal	Interference with medical tests, aggravation of gout symptoms; deficiency symptoms may appear at first on withdrawal of high doses.

[c]Smoothness of the tongue is caused by loss of its surface structures and is termed *glossitis* (gloss-EYE-tis).
[d]Small-cell anemia is termed *microcytic anemia;* large-cell type is *macrocytic* or *megaloblastic anemia.*
[e]The name *pernicious anemia* refers to the vitamin B_{12} deficiency caused by lack of intrinsic factor, but not to that caused by inadequate dietary intake.

SELF CHECK
SELF CHECK

1. Which of the following vitamins are fat soluble?

   a. vitamins B, C, and E
   b. vitamins B, C, D, and E
   c. vitamins A, C, E, and K
   d. vitamins A, D, E, and K

2. Which of the following describes fat-soluble vitamins?

   a. They include thiamin, vitamin A, and vitamin K.
   b. They cannot be stored to any great extent and so must be consumed daily.
   c. Toxic levels can be reached by consuming citrus fruits and vegetables.
   d. They can be stored in the liver and fatty tissues and can build up toxic concentrations.

3. Night blindness and susceptibility to infection are the result of a deficiency of which vitamin?

   a. niacin
   b. vitamin C
   c. vitamin A
   d. vitamin K

4. Good sources of vitamin D include:

   a. eggs, fortified milk, and sunlight.
   b. citrus fruits, sweet potatoes, and spinach
   c. green leafy vegetables, cabbage, and liver
   d. breast milk, polyunsaturated plant oils, and citrus fruits.

5. Which of the following describes water-soluble vitamins?

   a. They include vitamins D and E.
   b. They are frequently toxic.
   c. They are stored extensively in tissues.
   d. They are easily absorbed and excreted.

6. A coenzyme is:

   a. a fat-soluble vitamin.
   b. an energy-yielding nutrient.
   c. a source of vitamin K.
   d. a molecule that combines with an enzyme to make it active.

7. Good food sources of folate include:

   a. citrus fruits, dairy products, and eggs.
   b. liver, legumes, and leafy green vegetables.
   c. dark green vegetables, corn, and cabbage.
   d. potatoes, broccoli, and whole-wheat bread.

8. Which vitamin is present only in foods of animal origin?

   a. riboflavin
   b. pantothenic acid
   c. vitamin B_{12}
   d. the inactive form of vitamin A

9. Which of the following nutrients is an antioxidant that protects water-soluble substances from oxidizing agents?

   a. beta-carotene
   b. thiamin
   c. vitamin C
   d. vitamin D

10. Eating foods containing vitamin C at the same meal can increase the absorption of which mineral?

    a. iron
    b. calcium
    c. magnesium
    d. folate

Answers to these questions appear in Appendix H.

CLINICAL APPLICATIONS

1. How might a vitamin deficiency weaken a client's resistance to disease?

2. Pull together information from Chapter 1 about the different food groups and the significant sources of vitamins shown in the photos throughout this chapter. Consider which vitamins might be lacking in the diet of a client who reports the following:

   • Dislikes green, leafy vegetables.
   • Never uses milk, milk products, or cheese.
   • Follows a very-low-fat diet.
   • Eats a fruit or vegetable once a day.

   What additional information would help you pinpoint problems with vitamin intake?

NUTRITIONTHENET

FOR FURTHER STUDY OF THE
TOPICS IN THIS CHAPTER,
ACCESS THESE WEB SITES.

vita-men.com *Vita-Men*
bookman.com.au/vitamins/ *Vitamin Update*
cancer.org *American Cancer Society*
www.who.int *World Health Organization*
www.unicef.org *UNICEF*
www.aap.org *American Academy of Pediatrics*
www.nof.org *National Osteoporosis Foundation*
www.fda.gov *Food and Drug Administration*
www.modimes.org *March of Dimes*

Notes

[1] A. C. Ross, Vitamin A and retinoids, in *Modern Nutrition in Health and Disease,* 9th ed., ed. M. E. Shils and coeditors (Baltimore, Md.: Williams & Wilkins, 1999), pp. 305–327; J. A. Olson, Vitamin A, in *Present Knowledge in Nutrition,* 7th ed., ed. E. E. Ziegler and L. J. Filer (Washington, D.C.: International Life Sciences Institute Press, 1996), pp. 109–119.

[2] B. Halliwell, Antioxidants and human disease: A general introduction, *Nutrition Reviews* 55 (1997): 544–549; C. L. Rock, R. A. Jacob, and P. E. Bowen, Update on the biological characteristics of the antioxidant micro-nutrients: Vitamin C, vitamin E, and the carotenoids, *Journal of the American Dietetic Association* 96 (1996): 693–702.

[3] Rock, Jacob, and Bowen, 1996.

[4] K. Smigel, Beta-carotene fails to prevent cancer in two major studies: CARET intervention stopped, *Journal of the National Cancer Institute* 88 (1996): 145; C. S. Omenn and coauthors. Effects of a combination of beta-carotene and

vitamin A on lung cancer and cardiovascular disease, *New England Journal of Medicine* 334 (1996): 1150–1155.

[5] N. S. Scrimshaw and J. P. SanGiovanni, Synergism of nutrition, infection, and immunity, *American Journal of Clinical Nutrition* 66 (1997): S464–S477.

[6] W. W. Fawzi and coauthors, Dietary vitamin A intake and the risk of mortality among children, *American Journal of Clinical Nutrition* 59 (1994): 401–408; W. W. Fawzi and coauthors, Vitamin A supplementation and child mortality: A meta-analysis, *Journal of the American Medical Association* 269 (1993): 898–903.

[7] American Academy of Pediatrics, Committee on Infectious Diseases, Vitamin A treatment of measles, *Pediatrics* 91 (1993): 1014–1015.

[8] K. J. Rothman and coauthors, Teratogenicity of high vitamin intake, *New England Journal of Medicine* 333 (1995): 1369–1373; D. R. Soprano and K. J. Soprano, Retinoids as teratogens, *Annual Review of Nutrition* 15 (1995): 111–132.

[9] Rothman and coauthors, 1995.

[10] A. W. Norman, Vitamin D, in *Present Knowledge in Nutrition,* 7th ed., ed. E. E. Ziegler and L. J. Filer (Washington, D.C.: International Life Sciences Institute Press, 1996), pp. 120–129.

[11] M. F. Holick, Vitamin D, in *Modern Nutrition in Health and Disease,* 9th ed., ed. M. E. Shils and coeditors (Baltimore, Md.: Williams & Wilkins, 1999), pp. 329–345; Norman, 1996.

[12] Holick, 1999.

[13] A. Greely, Dodging the rays, *FDA Consumer,* July/August 1993, pp. 30–33.

[14] American Academy of Pediatrics, *Pediatric Nutrition Handbook,* 4th ed., ed. R. E. Kleinman (Elk Grove Village, Ill.: American Academy of Pediatrics, 1998), pp. 3–20.

[15] C. Lamberg-Allardt and coauthors, Low serum 25-hydroxy vitamin D concentrations and secondary hyperthyroidism in middle-aged white, strict vegetarians, *American Journal of Clinical Nutrition* 58 (1993): 684–689.

[16] J. Regnström and coauthors, Inverse relation between the concentration of low-density lipoprotein vitamin E and severity of coronary artery disease, *American Journal of Clinical Nutrition* 63 (1996): 377–385; K. G. Losonczy, T. B. Harris, and R. J. Havlik, Vitamin E and vitamin C supplements use and risk of all-cause and coronary heart disease mortality in older persons: The Established Populations (or Epidemiologic Studies of the Elderly), *American Journal of Clinical Nutrition* 64 (1996): 190–196; L. H. Kushi and coauthors, Dietary antioxidant

vitamins and death from coronary heart disease in postmenopausal women, *New England Journal of Medicine* 334 (1996): 1156–1162.

[17] Kushi and coauthors, 1996; M. J. Stampfer and coauthors, Vitamin E consumption and the risk of coronary disease in women, *New England Journal of Medicine* 328 (1993): 1444–1449; E. B. Rimm and coauthors, Vitamin E consumption and the risk of coronary disease in men, *New England Journal of Medicine* 328 (1993): 1450–1456.

[18] S. N. Meydani and coauthors, Assessment of the safety of supplementation with different amounts of vitamin E in healthy older adults, *American Journal of Clinical Nutrition* 68 (1998): 311–318; S. N. Meydani and coauthors, Assessment of the safety of high-dose, short-term supplementation with vitamin E in healthy older adults, *American Journal of Clinical Nutrition* 60 (1994): 704–709.

[19] G. Ferland, The vitamin K-dependent proteins: An update, *Nutrition Reviews* 56 (1998): 223–230; P. Dowd and coauthors, The mechanism of action of vitamin K, *Annual Review of Nutrition* 15 (1995): 419–440.

[20] L. J. Sokoll and coauthors, Changes in serum osteocalcin, plasma phylloquinone and urinary g-carboxyglutamic acid in response to altered intakes of dietary phylloquinone in human subjects, *American Journal of Clinical Nutrition* 65 (1997): 779–784; N. C. Brinkley and J. W. Suttie, Vitamin K nutrition and osteoporosis, *Journal of Nutrition* 125 (1995): 1812–1821.

[21] D. Feskanich and coauthors, Vitamin K and hip fractures in women: A prospective study, *American Journal of Clinical Nutrition* 69 (1999): 77–79.

[22] Dowd and coauthors, 1995.

[23] C. G. Harper and coauthors, Prevalence of Wernicke-Korsakoff syndrome in Australia: Has thiamin fortification made a difference? *Medical Journal of Australia* 168 (1998): 542–545.

[24] J. B. Hack and R. S. Hoffman, Thiamine before glucose to prevent Wernicke encephalopathy: Examining the conventional wisdom (letter), *Journal of the American Medical Association* 279 (1998): 583–584.

[25] D. Cervantes-Laurean, N. G. McElvaney, and J. Moss, Niacin, in *Modern Nutrition in Health and Disease,* 9th ed., ed. M. E. Shils and coeditors (Baltimore, Md.: Williams & Wilkins, 1999), pp. 401–411; M. L. Schwartz, Severe reversible hyperglycemia as a consequence of niacin therapy, *Archives of Internal Medicine* 153 (1993): 2050–2052.

[26] D. M. Mock, Biotin, in *Modern Nutrition in Health and Disease,* 9th ed., ed. M. E. Shils and coeditors (Baltimore, Md.: Williams & Wilkins, 1999), pp. 459–466.

[27]J. E. Leklem, Vitamin B-6, in *Present Knowledge in Nutrition,* 7th ed., ed. E. E. Ziegler and L. J. Filer (Washington, D.C.: International Life Sciences Institute Press, 1996), pp. 174–183.

[28]L. C. Rall and S. N. Meydani, Vitamin B_6 and immune competence, *Nutrition Reviews* 51 (1993): 217–225.

[29]E. B. Rimm and coauthors, Folate and vitamin B_6 from diet and supplements in relation to risk of coronary heart disease among women, *Journal of the American Medical Association* 279 (1998): 359–364; J. Selhub and coauthors, Association between plasma homocysteine concentrations and extracranial carotid-artery stenosis, *New England Journal of Medicine* 332 (1995): 286–291; M. J. Stampfer and M. R. Malinow, Can lowering homocysteine levels reduce cardiovascular risk? *New England Journal of Medicine* 332 (1995): 328–329; J. B. Ubbink, Vitamin nutrition status and homocysteine: An atherogenic risk factor, *Nutrition Reviews* 52 (1994): 383–393.

[30]J. V. Woodside and coauthors, Effect of B-group vitamins and antioxidant vitamins on hyperhomocysteinemia: A double-blind, randomized, factorial-design, controlled trial, *American Journal of Clinical Nutrition* 67 (1998): 858–866; Selhub and coauthors, 1995.

[31]Rimm and coauthors, 1998.

[32]C. E. Butterworth, Jr., and A. Bendich, Folic acid and the prevention of birth defects, *Annual Review of Nutrition* 16 (1996): 73–97.

[33]Committee on Genetics, Folic acid for the prevention of neural tube defects, *Pediatrics* 92 (1993): 493–494.

[34]American Academy of Pediatrics, Folic acid for the prevention of neural tube defects, *Pediatrics* 104 (1999): 325–327; Centers for Disease Control and Prevention, Recommendations for use of folic acid to reduce number of spina bifida cases and other neural tube defects, *Journal of the American Medical Association* 269 (1993): 1233, 1236, 1238.

[35]C. J. Cuskelly, H. McNulty, and J. M. Scott, Effect of increasing dietary folate on red-cell folate: Implications for prevention of neural tube defects, *The Lancet* 347 (1996): 657–659.

[36]Folate and neural tube defects: U.S. policy evolves, *Nutrition Reviews* 51 (1993): 358–361.

[37]C. M. Pfeiffer and coauthors, Absorption of folate from fortified cereal-grain products and of supplemental folate consumed with or without food determined by using dual-label-stable isotope protocol, *American Journal of Clinical Nutrition* 66 (1997): 1388–1397.

[38]F. Watanabe and coauthors, Effects of microwave heating on the loss of vitamin B_{12} in foods, *Journal of Agricultural and Food Chemistry* 46 (1998): 206–210.

[39]C. S. Johnston, K. R. Retrum, and J. C. Srilaskshmi, Antihistamine effects and complications of supplemental vitamin C, *Journal of the American Dietetic Association* 92 (1992): 988–989.

[40]J. E. Enstrom, L. E. Kanim, and M. A. Klein, Vitamin C intake and mortality among a sample of the United States population, *Epidemiology* 43 (1992): 194–202.

[41]J. Lykkesfeldt and coauthors, Ascorbic acid and dehydroascorbic acid as biomarkers of oxidative stress caused by smoking, *American Journal of Clinical Nutrition* 66 (1997): 1388–1397.

Q&A NUTRITION IN PRACTICE

Vitamin Supplements

About 50 percent of U.S. adults collectively spend close to $4 billion a year on vitamin supplements.[1] This trend is accelerating as scientists discover more and more links between nutrition and disease prevention. As discussed in the chapter, women of childbearing age may need folate supplements to reduce the risk of neural tube defects. Many recent reports also indicate that the antioxidant nutrients beta-carotene, vitamin C, and vitamin E may be potent protectors against both cancer and heart disease. Before you race out to buy bottles of antioxidant supplements, though, it is important to know that the role of antioxidants in preventing disease remains to be confirmed by hundreds of scientific studies that are currently under way (see Nutrition in Practice 21). Some findings so far are promising, but nevertheless tentative. The main message of this Nutrition in Practice is that most healthy people can get the nutrients they need from foods. Supplements cannot substitute for a healthy diet.

eatright.org/asupple.html

Do foods really contain enough vitamins and minerals to supply all that most people need?

Emphatically, yes, for both healthy adults and children who choose a variety of foods. The Daily Food Guide and the Food Guide Pyramid described in Chapter 1 are the guides to follow to achieve adequate intakes. People who meet their nutrient needs from foods, rather than supplements, have little risk of deficiency or toxicity.

Do some people need supplements?

Yes, some people may suffer marginal nutrient deficiencies due to illness, alcohol or drug addiction, or other conditions that limit food intake. People who may benefit from nutrient supplements in amounts consistent with the RDA include:
- People with nutrient deficiencies.
- People with low energy intakes (less than 1200 kcalories per day), such as habitual dieters.
- People with illnesses that take away the appetite.
- People with illnesses that impair nutrient absorption, such as dis-

eases of the gallbladder, pancreas, and digestive system.

- People taking medications that interfere with nutrient metabolism.
- People who do not use enough milk or milk products (calcium to forestall osteoporosis).
- Elderly people who may have difficulty chewing or swallowing and so do not eat enough food to meet nutrient needs.
- Women who bleed excessively during menstruation (iron supplements).
- Women who are pregnant or lactating (iron and folate).
- Vegans (calcium, iron, zinc, vitamin B_{12}, and vitamin D).
- Newborn infants (a single dose of vitamin K at birth under the direction of a physician).
- People who have infections or injuries or who have undergone major surgery. (The increased metabolic needs associated with these severe stresses are discussed in Chapter 19.)
- Infants, depending on whether they are receiving breast milk or formula and on whether their water contains fluoride (see Chapter 10).

These people are in the minority. For the majority, supplements are not recommended. Nutrients are potentially toxic when taken in large doses, and individual tolerances vary depending on health and age. Whenever a health care professional finds a person's diet inadequate, the right corrective step is to improve the person's food choices and eating patterns, not to begin supplementation.

Why do so many people take supplements?

People frequently take supplements for mistaken reasons, such as "They give me energy" or "They make me strong." Other invalid reasons why people may take supplements include:

- Their feeling of insecurity about the nutrient content of the food supply.
- Their belief that extra vitamins and minerals will help them cope with stress.
- Their belief that supplements can

enhance athletic performance or build lean body tissue without physical work.
- Their desire to prevent, treat, or cure symptoms or diseases ranging from the common cold to cancer.

Ironically, supplement users eat more nutrient-dense diets than nonusers and therefore need supplements less. In addition, little relationship exists between the nutrients people need and the ones they take in supplements. In fact, an argument against supplements is that they may lull people into a false sense of security. A person might eat irresponsibly, thinking, "My supplement will cover my needs."

Do antioxidant supplements prevent cancer and heart disease?

Again, it is better advice to eat a very nutritious diet. Evidence from population studies shows a correlation between low intakes of antioxidant nutrients and a high incidence of disease. Many studies show that low intakes of vegetables and fruits are consistently linked with an increased incidence of cancer.[2]

More than 200 population studies have examined the effects of fruits and vegetables on cancer risk, and the vast majority show that people who eat more of these foods are less likely to develop cancer.[3] Many experts agree that the antioxidant vitamins in these foods are probably the most important protective factors, but they also note that other constituents of fruits and vegetables (see Table 8-2 on p. 170) have not been ruled out as contributing factors.

The way to apply this information is to eat nutritious foods. Before supplementation is recommended as a strategy to prevent cancer or other diseases, researchers must determine the optimal doses to reduce risk and the potential adverse effects of long-term supplementation.

When a person needs a vitamin-mineral supplement, what kind should be used?

Take your health care professional's advice, if it is offered. If you are

selecting a supplement yourself, a single, balanced vitamin-mineral supplement should suffice. Choose the kind that provides all the nutrients in amounts equal to, or very close to, the RDA (remember, you get some nutrients from foods). Avoid individual nutrients, unless prescribed by a health care professional. Avoid preparations that are in excess of the RDA. Avoid preparations that contain items not needed in human nutrition such as inositol.

Can supplement labels help consumers make informed choices?

Yes. To enable consumers to make more informed choices about nutrient supplements, the Food and Drug Administration (FDA), with the encouragement of the American Dietetic Association (ADA), established labeling regulations for supplements.[4] The Dietary Supplement Health and Education Act subjects supplements to the same general labeling requirements that apply to foods. Specifically:

- Nutrition labeling for dietary supplements is now required. The nutrition panel on supplements is called "Supplement Facts" (see Figure NP8-1 on p. 198). The Supplement Facts panel lists the quantity and the percentage of the Daily Value for each nutrient in the supplement. Ingredients that have no Daily Value—for example, sugars and gelatin—appear in a list below the Supplement Facts panel.
- Labels may describe nutrient contents (as "high" or "low") according to specific criteria.
- The FDA authorizes health claims on supplement labels about the relationship between folate and the risk of neural tube defects, calcium and osteoporosis, soluble fiber from whole oats and psyllium husks and heart disease, and sugar alcohols and dental caries.
- Supplement labels are not allowed to include health claims on a number of other nutrient-disease relationships that have been approved for foods.

Figure NP8-1
AN EXAMPLE OF A SUPPLEMENT LABEL

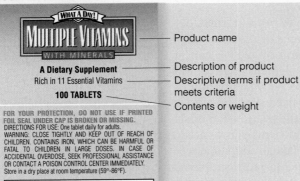

WHAT A DAY!

MULTIPLE VITAMINS

WITH MINERALS

A Dietary Supplement —————— Description of product
Rich in 11 Essential Vitamins —————— Descriptive terms if product meets criteria
100 TABLETS —————— Contents or weight

Product name

FOR YOUR PROTECTION, DO NOT USE IF PRINTED FOIL SEAL UNDER CAP IS BROKEN OR MISSING.
DIRECTIONS FOR USE: One tablet daily for adults.
WARNING: CLOSE TIGHTLY AND KEEP OUT OF REACH OF CHILDREN. CONTAINS IRON, WHICH CAN BE HARMFUL OR FATAL TO CHILDREN IN LARGE DOSES. IN CASE OF ACCIDENTAL OVERDOSE, SEEK PROFESSIONAL ASSISTANCE OR CONTACT A POISON CONTROL CENTER IMMEDIATELY.
Store in a dry place at room temperature (59°-86°F).

Supplement Facts
Serving Size 1 Tablet

The dose

Amount Per Tablet	% Daily Value
Vitamin A 5000 IU (40% Beta Carotene)	100%
Vitamin C 60 mg	100%
Vitamin D 400 IU	100%
Vitamin E 30 IU	100%
Thiamin 1.5 mg	100%
Riboflavin 1.7 mg	100%
Niacin 20 mg	100%
Vitamin B6 2 mg	100%
Folate 400 mg	100%
Vitamin B12 6 mcg	100%
Biotin 30 mcg	10%
Pantothenic Acid 10 mg	100%
Calcium 130 mg	13%
Iron 18 mg	100%
Phosphorus 100 mg	10%
Iodine 150 mcg	100%
Magnesium 100 mg	25%
Zinc 15 mg	100%
Selenium 10 mcg	14%
Copper 2 mg	100%
Manganese 2.5 mg	71%
Chromium 10 mcg	8%
Molybdenum 10 mcg	6%
Chloride 34 mg	1%
Potassium 37.5 mg	1%

The name, quantity per tablet, and "% Daily Value" for all nutrients listed; nutrients without a Daily Value may be listed below.

INGREDIENTS: Dicalcium Phosphate, Magnesium Hydroxide, Microcrystalline Cellulose, Potassium Chloride, Ascorbic Acid, Ferrous Fumarate, Modified Cellulose Gum, Zinc Sulfate, Gelatin, Stearic Acid, Vitamin E Acetate, Hydroxypropyl Methylcellulose, Niacinamide, Calcium Silicate, Citric Acid, Magnesium, Stearate, Calcium Pantothenate, Artificial Colors (FD&C Red No. 40, Titanium Dioxide, FD&C Yellow No. 6 and FD&C Blue No. 2), Selenium Yeast, Manganese Sulfate, Polyethylene Glycol, Cupric Sulfate, Molybdenum Yeast, Chromium Yeast, Vitamin A Acetate, Pyridoxine Hydrochloride, Riboflavin, Sodium Lauryl Sulfate, Thiamin Mononitrate, Beta Carotene, Folic Acid, Polysorbate 80, Vitamin D, Potassium Iodide, Gluten, Biotin.

All ingredients must be listed on the label, but not necessarily in the ingredient list nor in descending order of predominance; ingredients named in the nutrition panel need not be repeated here.

 GUARANTEE
Complete Satisfaction or Your Money Back

Supplements, Inc.
1234 Fifth Avenue
Anywhere, USA

Name and address of manufacturer

- Products may not bear claims to diagnose, treat, cure, or relieve a specific disease.
- Labels may describe the role a nutrient plays in the body, explain how the nutrient performs its function, and indicate that consuming the nutrient is associated with general well-being.
- Labels may claim a substance benefits common complaints such as memory loss or menstrual cramps without proof of effectiveness.

In effect, the Dietary Supplement Health and Education Act resulted in the deregulation of the supplement industry.[5] Unlike food additives or drugs, supplements do not need the FDA's approval before being marketed. Manufacturers alone decide whether their products are safe and effective. Should a problem arise, the burden falls on the FDA to prove that the supplement poses an unreasonable risk and should be removed from the market.

Notes

[1]M. C. Nesheim, Regulation of dietary supplements, *Nutrition Today* 33 (1998): 62–67.

[2]K. A. Steinmetz and J. D. Potter, Vegetables, fruit, and cancer prevention: A review, *Journal of the American Dietetic Association* 96 (1996): 1027–1039.

[3]Steinmetz and Potter, 1996; K. Meister, Fruits, vegetables, and cancer, *Priorities* (a publication of the American Council on Science and Health), Fall/Winter 1993, pp. 27–30.

[4]Commission on Dietary Supplement Labels issues final report, *Journal of the American Dietetic Association* 98 (1998): 270.

[5]Nesheim, 1998.

CHAPTER 9

WATER AND THE MINERALS

CONTENTS

Water and Body Fluids

The Major Minerals

The Trace Minerals

Nutrition in Practice:
Replacing Fluid Losses:
Sports Drinks and
Rehydration Formulas

The body's water cannot be considered separately from the minerals dissolved in it. A person can drink pure water, but in the body, water mingles with minerals to become fluids in which all life processes take place. This chapter begins by discussing the body's fluids and their chief minerals. The focus then shifts to other functions of the minerals.

Water and Body Fluids

Water constitutes about 60 percent of an adult's body weight and a higher percentage of a child's. Every cell in the body is bathed in a fluid of the exact composition that is best for that cell. The body fluids bring to each cell the ingredients it requires and carry away the end products of the life-sustaining reactions that take place within the cell's boundaries. The water in the body fluids:

- Carries nutrients and waste products throughout the body.
- Participates in chemical reactions.
- Serves as the solvent for minerals, vitamins, amino acids, glucose, and many other small molecules.
- Aids in maintaining the body's blood pressure and temperature.
- Acts as a lubricant and cushion around joints.
- Serves as a shock absorber inside the eyes, spinal cord, and amniotic sac surrounding a fetus in the womb.

To support these and other vital functions, the body actively regulates its **water balance.**

Water Balance

The cells themselves regulate the composition and amounts of fluids within and surrounding them. The entire system of cells and fluids remains in a delicate but firmly maintained state of dynamic equilibrium. Imbalances such as **dehydration** and **water intoxication** can occur, but the body quickly restores the balance to normal if it can. The body controls both water intake and water excretion.

WATER INTAKE REGULATION The body can survive for only a few days without water. In healthy people, thirst and satiety govern water intake. Thirst is finely adjusted to ensure a water intake that meets the body's needs. When the blood becomes too concentrated (having lost water but not salt and other dissolved substances), the mouth becomes dry, and the brain center known as the hypothalamus initiates drinking behavior.

Thirst lags behind the lack of water. A water deficiency that develops slowly can switch on drinking behavior in time to prevent serious dehydration, but a deficiency that develops quickly may not. Also, thirst itself does not remedy a water deficiency; a person must pay attention to the thirst signal and take the time to get a drink. The long-distance runner, the gardener in hot weather, and the busy child at play can experience serious dehydration if they fail to drink at regular intervals. With aging, thirst sensations may diminish. Dehydration can threaten elderly people who do not develop the habit of drinking water regularly.

WATER EXCRETION REGULATION Water excretion involves the brain and the kidneys. The cells of the brain's **hypothalamus,** which monitor blood salts, stimulate the **pituitary gland** to release **antidiuretic hormone (ADH)**

Nutrition in Practice 5 addresses consumer concerns about the safety of the water people drink.

water balance: the balance between water intake and water excretion, which keeps the body's water content constant.

dehydration: the loss of water from the body that occurs when water output exceeds water input. The symptoms progress rapidly from thirst, to weakness, to exhaustion and delirium and end in death if not corrected.

water intoxication: the rare condition in which body water contents are too high. The symptoms may include confusion, convulsion, coma, and even death in extreme cases.

hypothalamus (high-poh-THALL-uh-mus): a part of the brain that helps regulate many body balances, including fluid balance.

pituitary (pit-TOO-ih-tary) gland: in the brain, the "king gland" that regulates the operation of many other glands.

ADH (antidiuretic hormone): a hormone released by the pituitary gland in response to high salt concentrations in the blood. The kidneys respond by reabsorbing water.

Water is the most indispensable nutrient of all.

whenever the salts are too concentrated, or the blood volume or blood pressure is too low. ADH stimulates the kidneys to reabsorb water rather than excrete it. Thus, the more water you need, the less you excrete.

If too much water is lost from the body, blood volume and blood pressure fall. Cells in the kidneys respond to the low blood pressure by releasing an enzyme. Through a complex series of events, involving the hormone **aldosterone,** this enzyme also causes the kidneys to retain more water. Again, the effect is that when more water is needed, less is excreted.

:: MINIMUM WATER NEEDED These mechanisms can maintain water balance only if a person drinks enough water. The body must excrete a minimum of about 500 milliliters each day as urine—enough to carry away the waste products generated by a day's metabolic activities. Above this amount, excretion adjusts to balance intake, so the more a person drinks, the more dilute the urine becomes. In addition to urine, some water is lost from the lungs as vapor, some is excreted in feces, and some evaporates from the skin. A person's water losses from all of these routes total about 2½ liters (about 2½ quarts) a day on the average. Table 9-1 shows how fluid intake and output naturally balance out.

:: WATER RECOMMENDATIONS AND SOURCES Water needs vary greatly depending on the foods a person eats, the environmental temperature and humidity, the person's activity level, and other factors. Accordingly, a general water requirement is difficult to establish. Recommendations for adults are expressed in proportion to the amount of energy expended under normal environmental conditions. For the person who expends about 2000 kcalories a day, this works out to 2 to 3 liters, or about 8 to 12 cups. You can tell from the color of the urine whether a person needs more water. Pale yellow urine reflects appropriate dilution.

Plain water best meets people's fluid needs, but milk and fruit juice can also contribute to the day's recommended intake. Alcoholic beverages and those containing caffeine such as coffee, tea, and some sodas are not good water substitutes. Both alcohol and caffeine act as diuretics, causing the body to lose fluids. Many people who drink beer or coffee, then, may end up with a net fluid loss rather than a gain, to the detriment of the body's fluid balance.[1] Foods provide water, too. Most fruits and vegetables contain up to 95 percent water; many meats and cheeses contain at least 50 percent. The energy nutrients in foods also give up water during metabolism.

The enzyme renin (REN-in), released by the kidneys in response to low blood pressure, aids the kidneys in retaining water through the renin-angiotensin mechanism.

aldosterone (al-DOS-ter-own): a hormone secreted by the adrenal glands that stimulates the reabsorption of sodium by the kidneys. Aldosterone also regulates chloride and potassium concentrations.

500 ml = about ½ qt.

Table 9-1
WATER BALANCE

Water Source	Amount (ml)
Liquids	550 to 1500
Foods	700 to 1000
Metabolic water	200 to 300
	1450 to 2800

Water Output	
Kidneys	500 to 1400
Skin	450 to 900
Lungs	350
Feces	150
	1450 to 2800

LEARNING LINK

As you've just read, water balance is crucial to the health and normal functioning of the body. When a person fails to drink enough water to replace obligatory losses, every organ in the body is affected. Some people, such as endurance athletes and infants and children who are ill with infections that cause fever, diarrhea, and vomiting, are especially vulnerable to rapid fluid loss and, thus, dehydration. Sports drinks and rehydration formulas are the topics of the Nutrition in Practice that follows this chapter.

:: Fluid and Electrolyte Balance

When mineral salts dissolve in water, they separate (dissociate) into charged particles known as **ions,** which can conduct electricity. For this reason, a **salt**

ions (EYE-uns): atoms or molecules that have electrical charges.

salt: a compound composed of a positive ion other than H^+ and a negative ion other than OH^-. An example of a salt is potassium chloride (K^+Cl^-).

electrolytes: salts that dissolve in water and dissociate into charged particles called ions.

electrolyte solutions: solutions that can conduct electricity.

The simple statement that water follows salt describes the force that chemists call osmosis.

that partly dissociates in water is known as an **electrolyte.** The body fluids, which contain water and partly dissociated salts, are **electrolyte solutions.**

The body's electrolytes are vital to the life of the cells and must therefore be closely regulated to help maintain the appropriate distribution of body fluids. The major minerals form salts that dissolve in the body fluids; the cells direct where these salts go; and the movement of the salts determines where the fluids flow because water follows salt. Cells use this force to move fluids back and forth across their membranes. Thanks to the electrolytes, water can be held in compartments where it is needed.

Proteins in the cell membranes move ions in or out of the cells. These protein pumps tend to concentrate sodium and chloride outside cells and potassium and other ions inside. By maintaining specific amounts of sodium outside and potassium inside, cells can regulate the exact amounts of water inside and outside their boundaries. Physicians apply the same principle when treating kidney failure; they use an electrolyte solution to draw excess fluid out of the blood (see the discussion of *dialysis* in Nutrition in Practice 22).

Healthy kidneys regulate the body's sodium, as well as its water, with remarkable precision. The intestinal tract absorbs sodium readily, and it travels freely in the blood, but the kidneys excrete unneeded amounts. The kidneys actually filter all of the sodium out of the blood; then, with great precision, they return to the bloodstream the exact amount the body needs to retain. Thus, the body's total electrolytes remain constant, while the urinary electrolytes fluctuate according to what is eaten.

In some cases the body's mechanisms for maintaining fluid and electrolyte balances cannot compensate for a sudden loss of large amounts of fluid and electrolytes. Vomiting, diarrhea, heavy sweating, fever, burns, rapid blood loss, and the like may incur great fluid and electrolyte losses, precipitating an emergency that demands medical intervention.

Acid-Base Balance

The body uses ions not only to help maintain water balance, but also to regulate the acidity (pH) of its fluids. Electrolyte mixtures in the body fluids, as well as proteins, protect the body against changes in acidity by acting as buffers—substances that can accommodate excess acids or bases.

Reminder: Buffers *are compounds that help keep a solution's acidity or alkalinity constant; buffers are capable of neutralizing both acids and bases and thereby maintaining the original concentration of hydrogen ions (pH) in the solution.*

The body's buffer systems serve as a first line of defense against changes in the fluids' acid-base balance. The lungs, skin, GI tract, and kidneys provide other defenses. Of these organ systems, the lungs and kidneys play major roles in maintaining acid-base balance. Disorders of the lungs and kidneys, therefore, impair the body's ability to regulate its acid-base balance.

Not only do all of the major minerals help maintain fluid and electrolyte balance in the body, but they have other important functions as well, as the next sections describe. Table 9-6, at the end of this chapter, offers a summary of the minerals and their functions.

LEARNING LINK

For a person with renal disease or some lung and heart diseases, the physician may order, in addition to other treatments, adjustment of the fluid and electrolyte intake from foods and beverages. Chapter 22 offers much more information on renal disease.

The Major Minerals

Table 9-2 lists the major and trace minerals in the body, and Figure 9-1 shows the amounts found in the body. As you can see, the most prevalent minerals are calcium and phosphorus, the chief minerals of bone. The distinction between the major and the trace minerals does not mean that one group is more important than the other. A deficiency of the few micrograms of iodine needed daily is just as serious as a deficiency of the several hundred milligrams of calcium. The major minerals are so named because they are present, and needed, in larger amounts in the body than the trace minerals.

Although all the major minerals influence the body's fluid balance, sodium, chloride, and potassium are most noted for that role. For this reason, these three minerals are discussed first. Each major mineral also plays other specific roles in the body. Sodium, potassium, calcium, and magnesium are critical to nerve transmission and muscle contractions. Phosphorus and magnesium are involved in energy metabolism. Calcium, phosphorus, and magnesium contribute to the structure of the bones. Sulfur helps determine the shape of proteins.

Sodium

Sodium is the principal electrolyte in the extracellular fluid and the primary regulator of the extracellular fluid volume. When the blood concentration of sodium rises, as when a person eats salted foods, thirst prompts the person to drink water and the body sets in motion mechanisms to attain the appropriate sodium-to-water ratio. Sodium also helps maintain acid-base balance and is essential to muscle contraction and nerve transmission.

SODIUM RECOMMENDATIONS AND FOOD SOURCES

Diets rarely lack sodium. For this reason, recommended intakes have not been set; instead the estimated *minimum* sodium requirement was established. Health recommendations advise a *maximum* intake of *salt*, primarily to help prevent high blood pressure.[2]

Cultures vary in their use of salt. In the United States, men consume an average of 3300 milligrams of sodium (equivalent to about 8 grams of salt) a day. Asian people, whose staple sauces and flavorings are based on soy sauce and monosodium glutamate (MSG), consume the equivalent of about 30 to 40

Table 9-2
THE MAJOR AND TRACE MINERALS

Major Minerals	Trace Minerals
• Calcium	• Arsenic
• Chloride	• Boron
• Magnesium	• Chromium
• Phosphorus	• Cobalt
• Potassium	• Copper
• Sodium	• Fluoride
• Sulfur	• Iodine
	• Iron
	• Manganese
	• Molybdenum
	• Nickel
	• Selenium
	• Silicon
	• Zinc

Estimated minimum *requirement for sodium: 500 mg/day.*
Recommended maximum *intake of salt: 6 g/day (2400 mg sodium).*

5 g salt is about 2 g sodium.
1 g salt = ½ tsp salt.

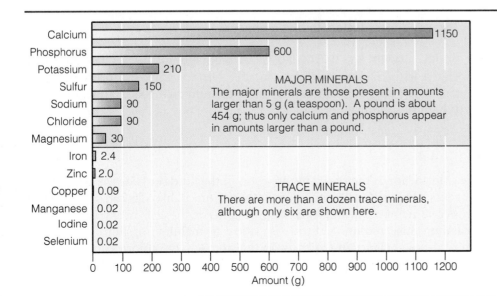

Figure 9-1
THE AMOUNTS OF MINERALS IN A 60-KILOGRAM (132-POUND) HUMAN BODY

MAJOR MINERALS
The major minerals are those present in amounts larger than 5 g (a teaspoon). A pound is about 454 g; thus only calcium and phosphorus appear in amounts larger than a pound.

TRACE MINERALS
There are more than a dozen trace minerals, although only six are shown here.

Calcium 1150
Phosphorus 600
Potassium 210
Sulfur 150
Sodium 90
Chloride 90
Magnesium 30
Iron 2.4
Zinc 2.0
Copper 0.09
Manganese 0.02
Iodine 0.02
Selenium 0.02

Amount (g)

Figure 9-2
WHAT PROCESSING DOES TO THE SODIUM AND POTASSIUM CONTENTS OF FOODS

Note how potassium is lost and sodium is gained as foods become more processed.

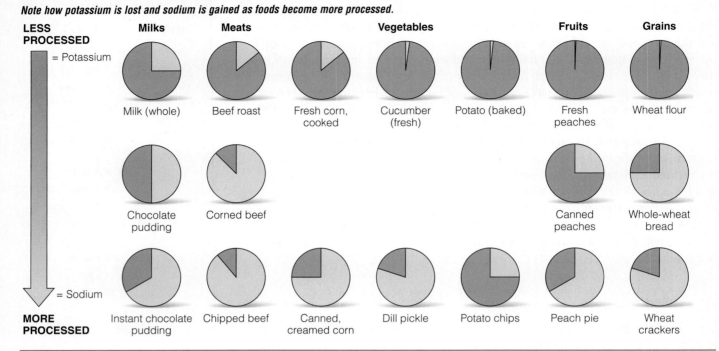

grams of salt per day. In China, Japan, and Korea, high blood pressure is as prevalent, or more so, as in the United States.

Sodium intakes vary widely. People who eat mostly processed foods have the highest sodium intakes, while those who eat mostly whole, unprocessed foods, such as fresh fruits and vegetables, have the lowest intakes. In fact, about three-fourths of the sodium in people's diets comes from salt added to foods by manufacturers. Figure 9-2 shows that processed foods contain not only more sodium but also less potassium than their less processed counterparts.

■■ SODIUM AND BLOOD PRESSURE   Sodium, or the salt that delivers it, contributes to high blood pressure in susceptible people.[3] Susceptibility may increase with potassium, calcium, and magnesium deficiencies. Chapter 21 offers suggestions for avoiding excessive salt intakes and describes the relationships of dietary factors to blood pressure.

■■ SODIUM AND OSTEOPOROSIS   A high sodium intake has also been associated with calcium and bone losses.[4] Dietary advice to prevent osteoporosis might suggest both eating more calcium-rich foods and restricting foods high in sodium.

■■   Chloride

The chloride ion is the major negative ion of the fluids outside the cells, where it occurs primarily in association with sodium. Chloride can move freely across cell membranes and so is also found inside the cells in association with potassium. Like sodium, chloride is critical to maintaining fluid, electrolyte, and acid-base balance in the body. In the stomach, the chloride ion is part of hydrochloric acid, which maintains the strong acidity of the gastric fluids.

Salt is a major food source of chloride, and as with sodium, processed foods are a major contributor of this mineral to people's diets. A chloride rec-

Estimated minimum requirement for chloride: 750 mg/day.

ommendation has not been established, but an estimated minimum requirement has been determined for adults.

Potassium

Potassium is the principal positively charged ion inside the body cells. It plays a major role in maintaining fluid and electrolyte balance and cell integrity. Potassium is also critical to keeping the heartbeat steady. The sudden deaths that occur in severe diarrhea and in children with kwashiorkor are likely due to heart failure caused by potassium loss. Potassium also assists in carbohydrate and protein metabolism.

POTASSIUM DEFICIENCY AND TOXICITY   Potassium deficiency results more often from excessive losses than from deficient intakes. Deficiency arises in abnormal conditions such as diabetic acidosis, dehydration, or prolonged vomiting or diarrhea; potassium deficiency can also result from the regular use of certain medications, including some **diuretics, steroids,** and **cathartics.** One of the earliest symptoms of deficiency is muscle weakness. Inadequate potassium intakes are possible with diets low in fresh fruits and vegetables, but out-and-out deficiencies of potassium are unlikely in healthy people.

In healthy people, potassium toxicity from foods is not a problem because the kidneys excrete excess potassium. Intakes from potassium supplements or the use of potassium-sparing diuretics can cause potassium to reach toxic levels, however, and can cause death.

POTASSIUM RECOMMENDATIONS AND FOOD SOURCES   As with sodium, an estimated minimum requirement for potassium has been determined. Surveys show wide variations in potassium intakes in the United States; people who emphasize fresh fruits and vegetables in their diets have high intakes. Potassium is abundant inside all living cells, both plant and animal, and because cells remain intact until foods are processed, the richest sources of potassium are *fresh* foods of all kinds—especially fruits and vegetables.

POTASSIUM AND BLOOD PRESSURE   Diets low in potassium seem to play an important role in the development of high blood pressure. Research suggests that increasing potassium intakes may both prevent and help to correct hypertension.[5]

LEARNING LINK

Elevated blood pressure, or hypertension, is a major risk factor for coronary heart disease (CHD) and affects millions of adults in the United States. The higher the blood pressure, the greater the risk of heart disease. Chapter 21 discusses other risk factors for hypertension; the relationship of the minerals sodium, chloride, potassium, magnesium, and calcium to hypertension; and diet therapy to reduce blood pressure.

Calcium

Calcium occupies more space in this chapter than does any other major mineral. Other minerals are revisited later in this book, where they play key roles in heart disease and kidney disease. Calcium, though, deserves emphasis here in the normal nutrition part of the book because an adequate intake of calcium early in life helps grow a healthy skeleton and prevent bone disease in later life.

diuretic (dye-yoo-RET-ic): a medication that promotes the excretion of water through the kidneys. Only some diuretics increase the urinary loss of potassium. Others, called potassium-sparing diuretics, result in potassium retention.

steroid (STARE-oid): a medication used to reduce tissue inflammation, to suppress the immune response, or to replace certain steroid hormones in people who cannot synthesize them.

cathartic (ca-THART-ic): a strong laxative.

Estimated minimum requirement for potassium: 2000 mg/day.

Fresh fruits and vegetables provide potassium in abundance.

Figure 9-3
CROSS SECTION OF BONE

The lacy structural elements are trabeculae (tra-BECK-you-lee), which can be drawn on to replenish blood calcium.

cofactor: a mineral element that, like a coenzyme, works with an enzyme to facilitate a chemical reaction.

The regulators are hormones from the thyroid and parathyroid glands, as well as vitamin D. One hormone, parathormone, raises blood calcium. Others, calcitonin and thyrocalcitonin, lower blood calcium by inhibiting release of calcium from bone. The hormonelike vitamin D raises blood calcium by acting at the three sites listed.

calcium rigor: hardness or stiffness of the muscles caused by high blood calcium.

calcium tetany: intermittent spasms of the extremities due to nervous and muscular excitability caused by low blood calcium.

■■ **CALCIUM ROLES IN THE BODY** Calcium owns the distinction of being the most abundant mineral in the body. Ninety-nine percent of the body's calcium is stored in the bones, where it plays two important roles. First, it is an integral part of bone structure. Second, it serves as a calcium bank that is available to the body fluids should a drop in blood calcium occur.

■■ **CALCIUM IN BONE** As bones begin to form, calcium salts form crystals on a matrix of the protein collagen. As the crystals become denser, they give strength and rigidity to the maturing bones. As a result, the long leg bones of children can support their weight by the time they have learned to walk. Many people have the idea that bones are inert, like rocks. Not so. Bones continuously gain and lose minerals in an ongoing process of remodeling. Growing children gain more bone than they lose, and healthy adults maintain a reasonable balance. When withdrawals substantially exceed deposits, however, problems such as osteoporosis develop. Figure 9-3 shows the lacy network of calcium-containing crystals in the bone.

■■ **CALCIUM IN BODY FLUIDS** The 1 percent of the body's calcium that circulates in the fluids as ionized calcium is vital to life. It helps regulate muscle contractions, transmit nerve impulses, clot blood, and secrete hormones, digestive enzymes, and neurotransmitters. Calcium also helps convey signals received at the cell surface to the inside of the cell. Calcium is a **cofactor** for several enzymes as well.

■■ **CALCIUM BALANCE** Blood calcium concentration is tightly controlled. Whenever blood calcium rises too high, a system of hormones and vitamin D promotes its deposit into bone. Whenever blood calcium falls too low, the regulatory system acts in three locations to raise it:

1. The intestine absorbs more calcium.
2. The bones release more calcium.
3. The kidneys excrete less calcium.

Thus, blood calcium rises to normal.

The calcium stored in bone provides a nearly inexhaustible source of calcium for the blood. Even in a calcium deficiency, blood calcium remains normal. Blood calcium changes only in response to abnormal regulatory control, not to diet. Blood calcium above normal causes **calcium rigor:** the muscles contract and cannot relax. Blood calcium below normal causes **calcium tetany**—also characterized by uncontrolled muscle contraction. These conditions are caused by a lack of vitamin D or by abnormal concentrations of the hormones that regulate calcium homeostasis.

Although a chronic *dietary* deficiency of calcium or a chronic deficiency due to poor absorption does not change blood calcium, it does deplete the savings account in the bones. Because this is an important concept, we repeat: it is the bones, not the blood, that are robbed by calcium deficiency.

■■ **CALCIUM AND OSTEOPOROSIS** Bone mass peaks at the time of skeletal maturity (about age 30), and a high peak bone mass is the best protection against later age-related bone loss and fracture. Adequate calcium nutrition during the growing years is essential to achieving optimal peak bone mass.[6] Following menopause, women lose about 15 percent of their bone mass, as do middle-aged and older men. When bone loss has reached such an extreme that bones fracture under even common, everyday stresses, the condition is known as **osteoporosis.** Osteoporosis afflicts as many as 25 million people, mostly women 45 years of age or older.

Both genetic and environmental factors contribute to osteoporosis. Table 9-3 summarizes risk and protective factors for osteoporosis. Osteoporosis is more prevalent in women than men for several reasons. First, women consume only about half as much dietary calcium as men do. Second, at all ages, women's bone mass is lower than men's because women generally have smaller bodies. Finally, bone loss begins earlier in women than in men, and it accelerates after menopause.

In addition to calcium, many other minerals and vitamins, including phosphorus, magnesium, fluoride, and vitamin D, help to form and stabilize the structure of bones. Any or all of these elements are needed to prevent bone loss. The first, most obvious lines of defense, however, are to maintain a life-long adequate intake of calcium and to "exercise it into place." Active bones are denser than sedentary bones.[7] Weight-bearing physical activity, such as walking, running, or dancing, prompts the bones to deposit minerals. It has long been known that when people are confined to bed, both their muscles and their bones lose strength. Muscle strength and bone strength go together: when muscles work, they pull on the bones, and both are stimulated to grow stronger.

CALCIUM AND HYPERTENSION Some evidence suggests that calcium helps prevent hypertension.[8] Studies of populations prone to developing hypertension show that low dietary calcium correlates with a high blood

osteoporosis (oss-tee-oh-pore-OH-sis): literally, porous bones; reduced density of the bones. Also known as *adult bone loss,* it is a condition in which the bones become porous and fragile. The causes of osteoporosis are multiple.

osteo = bone

Table 9-3

RISK AND PROTECTIVE FACTORS THAT CORRELATE WITH OSTEOPOROSIS

Risk Factors	Protective Factors
High Correlation	
Advanced age; postmenopausal	African American
Alcohol abuse	Estrogens, long-term use
Anorexia nervosa	
Caucasian	
Chronic steroid use	
Female sex	
Rheumatoid arthritis	
Surgical removal of ovaries	
Thinness	
Moderate Correlation	
Chronic thyroid hormone use	Having given birth
Cigarette smoking	High body weight
Diabetes (type 1)	High-calcium diet
Early menopause	Regular physical activity
Excessive antacid use	
Low-calcium diet	
Sedentary lifestyle or immobility	
Vitamin D deficiency	
Probably Important But Not Yet Proved	
Caffeine use	Low-sodium diet (later years)
Family history of osteoporosis	
High-fiber diet	
High-protein diet	
High-sodium diet	

SOURCE: Adapted from C. D. Arnaud and S. D. Sanchez, The role of calcium in osteoporosis, *Annual Review of Nutrition* 10 (1990): 397–414.

Table 9-4
RECOMMENDED FLUID MILK INTAKES

Age	Recommended Intake
Children	2 c
Adolescents and young adults	3 c
Adults	2 c
Pregnant and lactating women	3 c
Pregnant and lactating adolescents	4 c
Women past menopause	3 c

Calcium AI:
Adults (19–50 yr): 1000 mg/day.
Adults (51 and older): 1200 mg/day.

Apparently, all fibers in plant foods—cellulose, hemicellulose, pectin, and others—bind calcium to some extent, as do phytate and oxalate. Phytate and oxalate are binders that combine with minerals to form complexes that the body cannot absorb.

pressure.[9] Reports indicate that calcium supplements can sometimes lower blood pressure.[10] Some researchers speculate that calcium's effect on blood pressure is related to its action on the smooth muscle surrounding blood vessels.

■ **CALCIUM RECOMMENDATIONS** As mentioned earlier, blood calcium concentration does not reflect calcium status. Calcium recommendations are therefore based on balance studies, which measure daily intake and excretion. An optimal calcium intake reflects the amount needed to retain the most calcium. The more calcium retained, the greater the bone density (within genetic limits) and, potentially, the lower the risk of osteoporosis. Calcium recommendations during adolescence are set high (1300 milligrams) to help ensure that the skeleton will be strong and dense. Between the ages of 19 and 50, recommendations are lowered to 1000 milligrams a day. For those over 50, recommendations are raised again, to 1200 milligrams, to minimize bone loss. Some authorities advocate calcium recommendations as high as 1500 milligrams per day for women over 50. Many women have intakes well below recommendations.

■ **CALCIUM IN FOODS** Calcium is found almost exclusively in a single food group—milk and milk products. For this reason, dietary recommendations advise daily consumption of reduced-fat, low-fat, or fat-free milk products. A cup of milk offers about 300 milligrams of calcium, so an adult who drinks 2 to 3 cups of milk a day is well on the way to meeting daily calcium needs. Pregnant and lactating adolescents need more (see Table 9-4). The other dairy food that contains comparable amounts of calcium is cheese. One slice of cheese (1 ounce) contains about two-thirds as much calcium as a cup of milk. Cottage cheese, however, contains much less. Snapshot 9-1 shows foods that are rich in calcium, and the "How to" box suggests ways of adding calcium to meals.

Some foods offer large amounts of calcium because of fortification. Calcium-fortified juice, high-calcium milk (milk with extra calcium added), and calcium-fortified cereals are examples.

Among the vegetables, mustard greens, kale, parsley, watercress, and broccoli are good sources of available calcium. Some dark green, leafy vegetables—notably spinach and Swiss chard—appear to be calcium-rich but actually provide very little, if any, calcium to the body. These foods contain binders that prevent calcium absorption.

CALCIUM

All of these foods except the broccoli and the beans provide 10 percent or more of the calcium Daily Value (DV = 1000 mg/day).[a]

AI for adults (19-50 yr): 1000 mg/day
AI for adults (51 and older): 1200 mg/day

Broccoli: 94 mg (9% DV) per ½ c cooked

Sardines: 324 mg (32% DV) per 3 oz

Milk: 300 mg (30% DV) per cup

Pork and beans: 80 mg (8% DV) per ½ c

Cheddar cheese: 306 mg (31% DV) per 1½ oz

Almonds: 100 mg (10% DV) per 2 tbs

[a]The Daily Values are based on a 2000-kcalorie diet.

ADD CALCIUM TO DAILY MEALS

Many people cannot or will not drink enough milk to meet recommendations. To help deliver calcium, try the following suggestions:

1. Powdered fat-free milk is an excellent and inexpensive source of calcium and can be added to many foods (such as baked products and meat loaf) during preparation.
2. Yogurt and kefir (fermented dairy products) are acceptable substitutes for regular milk.
3. Puddings, custards, and baked goods can be prepared using appreciable amounts of milk.
4. Strict vegetarians and people who are either allergic to milk or lactose intolerant can use calcium-rich milk and cheese substitutes such as calcium-fortified soy milk or tofu (bean curd).
5. Small fish, such as canned sardines, and other canned fish prepared with their bones, such as canned salmon, are also rich in calcium.

People may think that taking a calcium supplement is preferable to getting calcium from food, but foods offer important fringe benefits. For example, drinking 2 cups of milk fortified with vitamins A and D will supply substantial amounts of other nutrients. Furthermore, the vitamin D, lactose, fat, and possibly other nutrients in the milk enhance calcium absorption. A calcium supplement supplies only calcium, and in a less absorbable form.

The body is able to regulate its absorption of calcium by altering its production of the calcium-binding protein aided by vitamin D. More of this protein is made if more calcium is needed. Infants and children absorb up to 75 percent of the calcium they ingest, and pregnant women, about 50 percent. Other adults, who are not growing, absorb about 30 percent. Also, calcium seems to be better absorbed if accompanied by an approximately equal amount of phosphorus.

LEARNING LINK

The importance of an adequate calcium intake throughout life to build healthy bones and prevent osteoporosis cannot be overemphasized. Calcium intakes of many adolescents and women fall short of recommendations. Calcium needs during pregnancy and lactation are discussed in Chapter 10; calcium needs during adolescence are discussed in Chapter 11. For people who are lactose intolerant, meeting calcium needs may be challenging, depending on whether they can tolerate milk and milk products at all. Lactose intolerance is included in Chapter 16.

▪▪ Phosphorus

Phosphorus is the second most abundant mineral in the body. About 85 percent of it is found combined with calcium in the crystals of the bones and teeth. As part of one of the body's buffer systems (phosphoric acid), phosphorus is also found in all body cells. Phosphorus is a part of DNA and RNA, the genetic code material present in every cell. Thus, phosphorus is necessary for all growth. Phosphorus also plays many key roles in the transfer of energy that occurs during cellular metabolism. Phosphorus-containing lipids (phospholipids) help transport other lipids in the blood. Phospholipids are also principal components of cell membranes.

Animal protein is the best source of phosphorus because the mineral is so abundant in the cells of animals. Diets that provide adequate energy and protein also supply adequate phosphorus. Dietary deficiencies are unknown. A summary of facts about phosphorus appears in Table 9-6.

⊞ Magnesium

Magnesium barely qualifies as a major mineral. Only about 1 ounce of magnesium is present in the body of a 130-pound person, over half of it in the bones. Most of the rest is in the muscles, heart, liver, and other soft tissues, with only 1 percent in the body fluids. Bone magnesium seems to be a reservoir to ensure that some will be on hand for vital reactions regardless of recent dietary intake.

Magnesium is critical to the operations of hundreds of enzymes. Magnesium acts in all the cells of the soft tissues, where it forms part of the protein-making machinery and is necessary for the release of energy. Magnesium helps relax muscles after contraction and promotes resistance to tooth decay by holding calcium in tooth enamel.

⊞ MAGNESIUM DEFICIENCY
Magnesium deficiency can result from vomiting, diarrhea, alcohol abuse, or protein malnutrition; after surgery in people who have been fed incomplete fluids intravenously for too long; or in people using diuretics. A severe magnesium deficiency causes tetany, an extreme and prolonged contraction of the muscles similar to the calcium tetany described earlier. Magnesium deficiency may also be related to cardiovascular disease and hypertension.[11] Magnesium deficiency is thought to cause the hallucinations commonly experienced during withdrawal from alcohol intoxication.

⊞ MAGNESIUM TOXICITY
Magnesium toxicity is most often reported in older adults who abuse magnesium-containing laxatives, antacids, and other medications. The consequences can be severe: lack of coordination, confusion, coma, and, in extreme cases, death.

⊞ MAGNESIUM INTAKES AND FOOD SOURCES
Dietary intakes of magnesium average about three-quarters of the recommended intake for both men and women in the United States. Dietary intake data, however, do not assess the magnesium contribution of water. In various parts of the country, the water contains both calcium and magnesium and is known as "hard" water. Hard water can contribute significantly to magnesium intakes.

Magnesium-rich food sources include dark green, leafy vegetables; nuts; legumes; whole-grain breads and cereals; seafood; chocolate; and cocoa (see Snapshot 9-2). Magnesium is easily lost from foods during processing, so unprocessed foods are the best choices.

⊞ Sulfur

The body does not use sulfur by itself as a nutrient. Sulfur is included here because it occurs in essential nutrients that the body does use, such as thiamin and certain amino acids. Sulfur is present in all proteins and plays its most important role in shaping strands of protein. The particular shape of a protein enables it to do its specific job, such as enzyme work. Skin, hair, and nails contain some of the body's more rigid proteins, and they have a high sulfur content.

SNAPSHOT

MAGNESIUM

All of these foods except the sunflower seeds provide 10 percent or more of the magnesium Daily Value (DV = 400 mg/day).ᵃ

RDA for men   (19–30 yr): 400 mg/day
(31 and older): 420 mg/day
RDA for women   (19–30 yr): 310 mg/day
(31 and older): 320 mg/day

Oysters: 55 mg (14% DV) per 3 oz steamed

Dried figs: 56 mg (14% DV) per ¼ c

Black-eyed peas: 45 mg (11% DV) per ½ c cooked

Spinach: 78 mg (19% DV) per ½ c cooked

Baked potato: 55 mg (14% DV) per whole small potato

Sunflower seeds (shelled): 21 mg (5% DV) per 2 tbs

ᵃThe Daily Values are based on a 2000-kcalorie diet.

There is no recommended intake for sulfur, and no deficiencies are known. Only a person who lacks protein to the point of severe deficiency will lack the sulfur-containing amino acids.

The Trace Minerals

Figure 9-1, earlier in this chapter, shows how tiny the quantities of trace minerals in the human body are. If you could remove all of them from your body, you would have only a bit of dust, hardly enough to fill a teaspoon. Yet each of the trace minerals performs some vital role for which no substitute will do. A deficiency of any of them can be fatal, and an excess of many can be equally deadly.

Recommendations have been established for the best-known trace minerals—iron, zinc, iodine, selenium, and fluoride. Tentative ranges for safe and adequate daily intakes of others are also published. Still others are recognized as essential nutrients for some animals, but have not been proved to be required for human beings (see Table 9-5 on p. 212). Still others are under study to determine whether they, too, perform indispensable roles in the body.

Iron

Every living cell—both plant and animal—contains iron. Most of the iron in the body is a component of the proteins **hemoglobin** in red blood cells and **myoglobin** in muscle cells. The iron in both hemoglobin and myoglobin helps them carry and hold oxygen and then release it. Hemoglobin in the blood carries oxygen from the lungs to tissues throughout the body. Myoglobin holds oxygen for the muscles to use when they contract. As part of many enzymes, iron is vital to the processes by which cells generate energy. Iron is also needed to make new cells, amino acids, hormones, and neurotransmitters.

The special provisions the body makes for iron's handling show that it is a precious mineral to be tightly hoarded. For example, when a red blood cell dies, the liver saves the iron and returns it to the bone marrow, which uses it to

hemoglobin: the oxygen-carrying protein of the red blood cells.
  hemo = blood
  globin = globular protein

myoglobin: the oxygen-carrying protein of the muscle cells.
  myo = muscle

Table 9-5
TRACE MINERALS

RDA

Iron

Zinc

Iodine

Selenium

Adequate Intake (AI)

Fluoride

Safe and Adequate Daily Dietary Intakes

Copper

Manganese

Chromium

Molybdenum

Known Essential for Animals; Human Requirements under Study

Arsenic

Nickel

Silicon

Boron

Known Essential for Some Animals; No Evidence That Intake by Humans Is Ever Limiting; No Recommendation Necessary

Cobalt

NOTE: The evidence for requirements and essentiality is weak for the trace minerals cadmium, lead, lithium, tin, and vanadium.

transferrin (trans-FERR-in): the body's iron-carrying protein.

The storage proteins are ferritin (FERR-i-tin) and hemosiderin (heem-oh-SID-er-in).

iron deficiency: having depleted iron stores.

iron-deficiency anemia: a blood iron deficiency characterized by small, pale red blood cells; also called **microcytic hypochromic anemia.**

micro = small

cytic = cells

hypo = too little

chrom = color

build new red blood cells. Thus, only tiny amounts of iron are lost, principally in urine, sweat, shed skin, and blood (if bleeding occurs).

Normally, only about 10 to 15 percent of dietary iron is absorbed; but if the body's supply is diminished or if the need increases for any reason (such as pregnancy), absorption increases. The body makes several provisions for absorbing iron. A special protein in the intestinal cells captures iron and holds it in reserve for release into the body as needed; another protein transfers the iron to a special iron carrier in the blood. The blood protein (**transferrin**) carries the iron to tissues throughout the body. When more iron is needed, more of these special proteins are produced so that more than the usual amount of iron can be absorbed and carried. If there is a surplus of iron, special storage proteins in the liver, bone marrow, and other organs store it.

IRON DEFICIENCY Worldwide, iron deficiency is the most common nutrient deficiency, affecting more than one billion people. In developing countries, one-third of the children and women of childbearing age suffer from iron-deficiency anemia.[12] In the United States, iron deficiency is less prevalent, but still affects about 10 percent of toddlers, adolescent girls, and women of childbearing age.[13]

Women are especially prone to iron deficiency during their reproductive years because of blood losses during menstruation. Pregnancy places further iron demands on women. Iron is needed to support the added blood volume, the growth of the fetus, and blood loss during childbirth. Infants (six months or older) and young children receive little iron from their high-milk diets, yet need extra iron to support growth. The rapid growth of adolescence and, for females, the blood losses of menstruation also demand extra iron that a typical teen diet may not provide.

CAUSES OF IRON DEFICIENCY The cause of iron deficiency is usually inadequate intake from ignorance of what foods to choose, from sheer lack of food altogether, or from high consumption of iron-poor foods. In the Western world, high sugar and fat intakes are often responsible for low iron intakes. Blood loss is the primary nonnutritional cause of iron deficiency, especially in poor regions of the world where parasitic infections of the GI tract may lead to blood loss.

TESTS FOR IRON DEFICIENCY The most common tests for iron-deficiency anemia measure the number and size of the red blood cells and the cells' hemoglobin content. Although these tests are easy, quick, and relatively inexpensive, they are late indicators of iron deficiency. Earlier stages of an iron deficiency can be detected by measuring the amount of transferrin in the blood, the amount of iron transferrin is carrying, and the amount of iron in storage.

IRON DEFICIENCY AND ANEMIA The distinction between **iron deficiency** and anemia is important. They often go hand in hand, but people can be anemic without being iron deficient and iron deficient without being anemic. Anemia is a symptom of a wide variety of disorders, some unrelated to nutrition, and some related to nutrients other than iron, such as folate and vitamin B_{12}. (Appendix E lists tests useful in identifying anemia and distinguishing between the major types of nutritional anemias.) In **iron-deficiency anemia,** new red blood cells are smaller and lighter red than normal (see Figure 9-4). The depleted cells cannot carry enough oxygen from the lungs to the tissues, so energy metabolism in all the cells is hindered. The entire body feels the effect.

Anemia is a clinical sign of severe iron deficiency. Other classic symptoms include fatigue, weakness, headaches, apathy, and pallor. A more recently recognized symptom is poor tolerance to cold. One way the body accelerates

Figure 9-4
NORMAL AND ANEMIC BLOOD CELLS

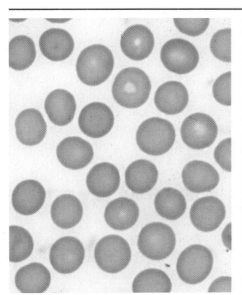

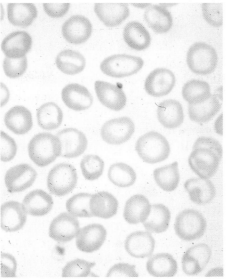

Normal blood cells. Both size and color are normal.

Blood cells in microcytic hypochromic anemia such as that caused by iron deficiency. These cells are small and pale because they contain less hemoglobin.

heat production when the environmental temperature falls involves the neurotransmitter norepinephrine and the thyroid hormones, which speed up the metabolic rate. Iron deficiency impairs temperature regulation in both animals and human beings, probably by interfering with the normal production of these compounds.

Less severe iron deficiency produces symptoms, too. Long before the red blood cells are affected and anemia is diagnosed, a developing iron deficiency affects behavior. Even at slightly lowered iron levels, the complete oxidation of pyruvate is impaired, reducing physical work capacity and productivity. Children deprived of iron become irritable, restless, and unable to pay attention. These symptoms are among the first to appear when the body's iron begins to fall and among the first to disappear when iron status is restored again.

Several mechanisms by which iron deficiency may affect behavior have been proposed. The one most often discussed and researched proposes that even in the earliest stages of iron deficiency, a deficit of iron-dependent neurotransmitter receptors in the brain alters behavior.[14]

■■ **IRON DEFICIENCY IN WOMEN ATHLETES** Physically active young women, especially those who engage in endurance activities such as distance running, are prone to iron deficiency. Iron status may be affected by physical activity in any of several ways. One possibility is that iron lost in sweat contributes to the deficiency.[15] Another possible route of iron loss is red blood cell destruction; blood cells are squashed when body tissues (such as the soles of the feet) make high-impact contact with an unyielding surface such as the ground. In addition, physical activity may cause small blood losses through the digestive tract, at least in some athletes. Habitually low intakes of iron-rich foods, combined with iron losses aggravated by physical activity, may cause iron deficiency in physically active young women.

Iron deficiency impairs physical performance because iron is crucial to the body's handling of oxygen. One consequence of iron-deficiency anemia is impaired oxygen transport. This reduces aerobic work capacity, so the person

One common test for anemia measures the hemoglobin concentration of blood.
- **Norms for adults:**
  Men: ≥13.5 g/100 ml.
  Women: ≥12 g/100 ml.
- **Norms for children:**
  Ages 2–5: ≥11 g/100 ml.
  Ages 6–12: ≥11.5 g/100 ml.
  Note that hemoglobin is measured in grams per 100 ml, but often it is referred to by just the number of grams alone: "hemoglobin, 14."

The effects of iron deficiency on children's behavior is revisited in Chapter 11.

Another common test, the hematocrit, represents the percentage of red blood cells in a whole blood sample.
- **Norms for adults:**
  Men: ≥ 41%.
  Women: ≥ 36%.
- **Norms for children:**
  Ages 2–5: ≥ 34%.
  Ages 6–12: ≥ 35%.

Transferrin can be measured directly or estimated by measuring the total iron-binding capacity (TIBC) and the transferrin saturation.

tires easily. Whether marginal deficiency without anemia impairs physical performance is a point of debate among researchers.[16]

■■ **SPORTS ANEMIA** Early in training, athletes may develop low blood hemoglobin for a while. This condition, sometimes called "sports anemia," probably reflects a normal adaptation to physical training. Aerobic training enlarges the blood volume, and with the added fluid, the red blood cell count per unit of blood drops. True iron deficiency requires treatment with prescribed iron supplements, but the temporary reduced red blood cell count seen early in training corrects itself after a while.

■■ **IRON DEFICIENCY AND PICA** A curious symptom sometimes seen in iron-deficient individuals is an appetite for ice, clay, paste, and other nonnutritious substances. Some people have been known to eat as many as eight trays of ice in a day, for example. This behavior, known as **pica,** has been observed for years, especially in women and children of low-income groups who are deficient in either iron or zinc. After iron is given, pica clears up dramatically within days, long before the red blood cells respond.

■■ **CAUTION ON SELF-DIAGNOSIS** Low hemoglobin may reflect an inadequate iron intake, and if it does, the physician may prescribe iron supplements. However, any nutrient deficiency or disease or agent that interferes with hemoglobin synthesis, disrupts hemoglobin function, or causes a loss of red blood cells can precipitate anemia.

Feeling fatigued, weak, and apathetic is thus a sign that something is wrong, but does not indicate that a person should take iron supplements; it means that the person should consult a physician. In fact, taking iron supplements may be the worst possible thing a person can do, because such supplements can mask a serious medical condition, such as hidden bleeding from cancer or an ulcer. Furthermore, a person can waste precious time in not seeking treatment. Remember, don't self-diagnose.

■■ **IRON OVERLOAD** Normally, the body protects itself against absorbing too much iron by setting up a block in the intestinal cells. The system can be overwhelmed, however, resulting in **iron overload.** Once considered rare, iron overload has emerged as an important disorder of iron metabolism and regulation.[17]

Iron overload is known as **hemochromatosis** and is usually caused by a genetic disorder that enhances iron absorption. Hereditary hemochromatosis is the most common genetic disorder in the United States, affecting about 1.5 million people.[18] Other causes of iron overload include repeated blood transfusions, massive doses of supplementary iron, and other rare metabolic disorders. Long-term overconsumption of iron may cause **hemosiderosis,** a condition characterized by large deposits of the iron-storage protein hemosiderin in the liver and other tissues.

Some of the signs and symptoms of iron overload are similar to those of iron deficiency: apathy, lethargy, and fatigue. Therefore, taking iron supplements before assessing iron status is clearly unwise; hemoglobin tests alone would fail to make the distinction.

Iron overload is characterized by tissue damage, especially in iron-storing organs such as the liver. Infections are likely because bacteria thrive on iron-rich blood. Symptoms are most severe in alcohol abusers because alcohol damages the intestine, further impairing its defenses against absorbing excess iron. Untreated hemochromatosis aggravates the risk of diabetes, liver damage and cancer, heart disease, and arthritis.

Iron overload is more common in men than in women and is twice as prevalent among men as iron deficiency. The fortification of many foods with iron makes it difficult to follow an iron-restricted diet.

The skin of a fair person who is anemic may be noticeably pale, but in all people (including those who are dark skinned), the eye lining, normally pink, will be very pale, even white.

pica (PIE-ka): a craving for nonfood substances; also known as *geophagia (jee-oh-FAY-jee-uh)* when referring to clay-eating behavior.

picus = woodpecker or magpie
geo = earth
phagein = to eat

Binding proteins in the intestinal cells (mucosal ferritin and mucosal transferrin) capture and hold unneeded iron to be shed with the cells, thereby forming a mucosal block to iron absorption.

iron overload: toxicity from excess iron.

hemochromatosis (heem-oh-crome-a-TOE-sis): iron overload characterized by deposits of iron-containing pigment in many tissues, with tissue damage. Hemochromatosis is a hereditary defect in iron metabolism.

hemosiderosis (heem-oh-sid-er-OH-sis): a condition characterized by the deposition of the iron-storage protein hemosiderin in the liver and other tissues.

:: IRON POISONING The rapid ingestion of massive amounts of iron can cause sudden death. The most common cause of accidental poisoning in small children is ingestion of iron supplements or vitamins with iron.[19] The American Academy of Pediatrics has urged the Food and Drug Administration (FDA) to improve the labeling and packaging of iron-containing drugs and supplements. As few as 6 to 12 tablets have caused death in a child.[20] A child suspected of iron poisoning should be rushed to the hospital to have the stomach pumped. Thirty minutes can make a crucial difference.

:: IRON RECOMMENDATIONS The average diet in the United States provides only about 6 to 7 milligrams of iron in every 1000 kcalories. The recommended daily intake for an adult man is 10 milligrams; most men easily eat more than 2000 kcalories, so a man can meet his iron needs without special effort. The recommendation for women during childbearing years, however, is 15 milligrams. Because women have higher iron needs and typically consume fewer than 2000 kcalories per day, they have trouble achieving appropriate iron intakes. On the average, women receive only 10 to 11 milligrams of iron per day. A woman who wants to meet her iron needs from foods must emphasize the most iron-rich foods in every food group.

This chili dinner provides iron and MFP factor from meat, iron from legumes, and vitamin C from tomatoes. The combination of heme iron, nonheme iron, MFP factor, and vitamin C helps to achieve maximum iron absorption.

:: IRON IN FOODS Iron occurs in two forms in foods, one of which is up to ten times more absorbable than the other. The most absorbable form is heme iron, which is bound into the iron-carrying proteins hemoglobin and myoglobin in meats, poultry, and fish. Heme iron contributes a small portion of the iron consumed by most people, but it is absorbed at a fairly constant rate of about 23 percent. The less absorbable form is nonheme iron, found in meats and also in plant foods. People absorb nonheme iron at a lower rate (2 to 20 percent); its absorption depends on several dietary factors and iron stores. Most of the iron people consume is nonheme iron from vegetables, grains, eggs, meat, fish, and poultry. Snapshot 9-3 shows the iron found in usual serving sizes of different foods.

Iron absorption from foods can be maximized through two substances that enhance iron absorption: MFP factor and vitamin C. Meat, fish, and poultry contain a factor (MFP factor) other than heme that promotes the absorption of iron. MFP factor even enhances the absorption of nonheme iron from other foods eaten at the same time. Vitamin C eaten in the same meal also doubles or triples nonheme iron absorption. Additionally, cooking with iron skillets can contribute

Iron RDA:
Men (19 and older): 10 mg/day.
Women (19–50 yr): 15 mg/day.
(>50 yr): 10 mg/day.

About 40% of the iron in meat, fish, and poultry is bound into molecules of heme (HEEM), the iron-holding part of the hemoglobin and myoglobin proteins. This heme iron is much more absorbable than nonheme iron.

SNAPSHOT 9-3

IRON

All of these foods except the tofu and the figs provide 10 percent or more of the iron Daily Value (DV = 18 mg/day).[a]

RDA for men: 10 mg/day
RDA for women: 15 mg/day

Swiss chard: 2.0 mg (11% DV) per ½ c cooked

Clams: 25.2 mg (138% DV) per 3 oz steamed

Navy beans: 2.2 mg (12% DV) per ½ c cooked

Sirloin steak: 2.8 mg (15% DV) per 3 oz cooked

Dried figs: 1.3 mg (7% DV) per ¼ c

Tofu: 1.3 mg (7% DV) per ½ c
[a]The Daily Values are based on a 2000-kcalorie diet.

ADD IRON TO DAILY MEALS

The following set of guidelines can be used for planning an iron-rich diet:

- *Breads and cereals.* Use only whole-grain, enriched, and fortified products (iron is one of the enrichment nutrients).
- *Vegetables.* The dark green, leafy vegetables are good sources of vitamin C and iron. Eat vitamin C–rich vegetables often to enhance absorption of the iron from foods eaten with them.
- *Fruits.* Dried fruits, such as raisins, apricots, peaches, and prunes, are high in iron. Eat vitamin C–rich fruits often with iron-containing foods.
- *Milk and cheese.* Don't overdo foods from the milk group; they are poor sources of iron. But don't omit them either, because they are rich in calcium. Drink fat-free milk to free kcalories to be invested in iron-rich foods.
- *Meats.* Meat, fish, and poultry are excellent iron sources.
- *Meat alternates.* Include legumes frequently. A cup of peas or beans can supply up to 7 milligrams of iron.

iron to the diet. Tea and coffee, on the other hand, interfere with iron absorption. The "How to" box offers suggestions on obtaining adequate iron.

Zinc

Zinc is a versatile trace mineral required as a cofactor by more than 100 enzymes. These zinc-requiring enzymes perform tasks in the eyes, liver, kidneys, muscles, skin, bones, and male reproductive organs. Zinc works with the enzymes that make genetic material; manufacture heme; digest food; metabolize carbohydrate, protein, and fat; liberate vitamin A from storage in the liver; and dispose of damaging free radicals. Zinc also interacts with platelets in blood clotting, affects thyroid hormone function, assists in immune function, and affects behavior and learning performance. Zinc is needed to produce the active form of vitamin A in visual pigments and is essential to wound healing, taste perception, the making of sperm, and fetal development. When zinc deficiency occurs, it impairs all these and other functions.

The body's handling of zinc differs from that of iron, but with some interesting similarities. For example, like iron, extra zinc that enters the body is held within the intestinal cells, and only the amount needed is released into the bloodstream. As with iron, a person's zinc status influences the percentage of zinc the person absorbs from the diet; if more is needed, more is absorbed.

Zinc's main transport vehicle in the blood is the protein albumin. Research suggests that circulating albumin is a main determinant of zinc absorption. This may account for observations that zinc absorption declines in conditions that lower plasma albumin concentrations—for example, pregnancy and malnutrition.

ZINC DEFICIENCY Zinc deficiency in human beings was first reported in the 1960s from studies of growing children and male adolescents in Egypt, Iran, and Turkey. Their diets were typically low in zinc and high in fiber and **phytates** (which impair zinc absorption). The zinc deficiency was marked by dwarfism or severe growth retardation and arrested sexual maturation—symptoms that were responsive to zinc supplementation.

Since that time, zinc deficiency has been recognized elsewhere and is known to affect more than growth. It drastically impairs immune function, causes loss of appetite, and, during pregnancy, may lead to developmental disorders. A detailed list of symptoms of zinc deficiency is presented later in Table

phytates: nonnutrient components of grains, legumes, and seeds. Phytates can bind minerals such as zinc, iron, calcium, and magnesium in insoluble complexes in the intestine, and these complexes are then excreted unused.

9-6. Conditions other than poor diet that contribute to the development of zinc deficiency include loss of blood due to parasitic infections, climates that increase sweat losses, and the practice of clay eating.

Clay eating: see pica, *p. 214.*

Pronounced zinc deficiency is not widespread in developed countries, but deficiencies do occur in the most vulnerable groups of the U.S. population—pregnant women, young children, the elderly, and the poor. Even mild zinc deficiency can result in metabolic changes such as impaired immune response, abnormal taste, and abnormal dark adaptation (zinc is required to produce the active form of vitamin A, retinal, in visual pigments).

Pregnant adolescents are particularly vulnerable because they need zinc for their own growth, as well as for the developing fetus. Vegetarians, especially pregnant vegetarians, who consume large amounts of fiber, phytate, and dairy foods or low levels of protein need to scrutinize their diets for possible zinc deficiency. Research shows that both zinc intake and zinc absorption are reduced in people eating vegetarian diets that include milk and eggs compared with those whose diets include meat, fish, and poultry.[21] The researchers note, however, that despite a greater risk of zinc deficiency in people consuming vegetarian diets, zinc balance can be maintained with the inclusion of zinc-rich whole-grain breads and cereals and legumes.

██ ZINC TOXICITY   Zinc can be toxic if consumed in large enough quantities. A high zinc intake is known to produce copper-deficiency anemia by inducing the intestinal cells to synthesize large amounts of a protein that captures copper in a nonabsorbable form. Accidental consumption of high levels of zinc can cause vomiting, diarrhea, fever, exhaustion, and a host of other symptoms (see Table 9-6, later in the chapter). Large doses can even be fatal.

Zinc RDA:
  Men: 15 mg/day.
  Women: 12 mg/day.

██ ZINC RECOMMENDATIONS AND FOOD SOURCES   The zinc recommendation for men is 15 milligrams per day; for women, 12 milligrams. Zinc intakes of some adults in the United States fall short of recommendations.

Zinc is most abundant in foods high in protein, such as shellfish (especially oysters), meats, and liver. In general, two ordinary servings a day of animal protein provide most of the zinc a healthy person needs. Milk, eggs, and whole-grain products are good sources of zinc if eaten in large quantities. For infants, breast milk is a good source of zinc, which is more efficiently absorbed from human milk than from cow's milk. Commercial infant formulas are fortified with zinc, of course. Snapshot 9-4 shows zinc-rich foods.

ZINC

All of these foods except the beans and peas provide 10 percent or more of the zinc Daily Value (DV = 15 mg/day).[a]

RDA for men: 15 mg/day
RDA for women: 12 mg/day

Yogurt: 2.2 mg (15% DV) per cup

Green peas: 1.0 mg (6% DV) per ½ c

Sirloin steak: 5.5 mg (37% DV) per 3 oz cooked

Black beans: 1.0 mg (6% DV) per ½ c cooked

Oysters: 41.0 mg (273% DV) per 3 oz steamed

Crabmeat: 2.6 mg (17% DV) per 3 oz steamed

[a]The Daily Values are based on a 2000-kcalorie diet.

Zinc supplements are not recommended except for an accurately diagnosed zinc deficiency or when needed for use as a medication to displace other ions in unusual medical circumstances. Normally, it should be possible to obtain enough zinc from the diet.

Selenium

selenium (se-LEEN-ee-um): a trace element.

Selenium is an essential trace mineral that functions as part of an antioxidant enzyme called glutathione peroxidase. Glutathione peroxidase prevents free-radical formation, thus blocking the damaging chain reaction before it begins. Glutathione peroxidase and vitamin E work in concert. If free radicals do form, and a chain reaction starts, vitamin E halts it. Selenium also plays a role in converting thyroid hormone to its active form.[22]

SELENIUM AND CANCER The question of whether selenium protects against the development of some cancers is under investigation. Some research suggests that selenium supplements may reduce the incidence of some types of cancers, but given the potential for harm and the lack of additional evidence, recommendations to take selenium supplements would be premature.[23]

SELENIUM DEFICIENCY Selenium deficiency is associated with heart disease in children and young women living in regions of China where the soil and foods lack selenium. The heart disease is named *Keshan disease* for one of the provinces of China where it was studied.

SELENIUM TOXICITY High doses of selenium are toxic. Selenium toxicity causes vomiting, diarrhea, loss of hair and nails, and lesions of the skin and nervous system.

Selenium RDA:
Adults: 55 μg/day.

SELENIUM RECOMMENDATIONS AND INTAKES Anyone who eats a normal diet composed mostly of unprocessed foods need not worry about meeting selenium recommendations. Selenium is widely distributed in foods such as meats and shellfish and in vegetables and grains grown on selenium-rich soil. Some regions in the United States and Canada produce crops on selenium-poor soil, but people are protected from deficiency because they eat selenium-rich meat and supermarket foods transported from other regions.

Iodine

Iodine occurs in the body in minuscule amounts, but its principal role in human nutrition is well known, and the amount needed is well established. Iodine is an integral part of the thyroid hormones, which regulate body temperature, metabolic rate, reproduction, growth, the making of blood cells, nerve and muscle function, and more.

goiter (GOY-ter): an enlargement of the thyroid gland due to an iodine deficiency, malfunction of the gland, or overconsumption of a thyroid antagonist. Goiter caused by iodine deficiency is *simple goiter*.

IODINE DEFICIENCY When the iodine concentration in the blood is low, the cells of the thyroid gland enlarge in an attempt to trap as many particles of iodine as possible. If the gland enlarges until it is visible, the swelling is called a simple **goiter.** As many as 800 million people border on iodine deficiency, and 200 million people worldwide have goiter. In all but 4 percent of these cases, the cause is iodine deficiency. As for the 4 percent (8 million), those people have goiter because they overconsume plants of the cabbage family and others that contain an antithyroid substance whose effect is not counteracted by dietary iodine.

A thyroid antagonist found in food, which causes **toxic goiter,** *is called a* **goitrogen.**

In addition to causing sluggishness and weight gain, an iodine deficiency may have serious effects on fetal development. Severe thyroid undersecretion during pregnancy causes the extreme and irreversible mental and physical retardation known as **cretinism.** A child with cretinism may have an IQ as low as 20 (100 is normal) and a face and body with many abnormalities. Iodine deficiency is one of the world's most common preventable causes of mental retardation and can be averted if the pregnant woman's deficiency is detected and treated in time.[24]

cretinism (CREE-tin-ism): an iodine-deficiency disease characterized by mental and physical retardation.

▪▪ IODINE TOXICITY Excessive intakes of iodine can enlarge the thyroid gland, just as deficiencies can. In infants, the goiterlike condition can be so severe as to block the airways and cause suffocation.

▪▪ IODINE SOURCES AND INTAKES The ocean is the world's major source of iodine. In coastal areas, seafood, water, and even iodine-containing sea mist are important iodine sources. Further inland, the amount of iodine in the diet is variable and generally reflects the amount present in the soil in which plants are grown or on which animals graze. In the United States and Canada, the use of iodized salt has largely wiped out the iodine deficiency that once was widespread.

The need for iodine is easy to meet by consuming seafood, vegetables grown in iodine-rich soil, and iodized salt. In the United States, you have to read the label to find out whether salt is iodized; in Canada, all table salt is iodized.

Iodine RDA: 150 µg/day.

▪▪ Copper

The body contains about 100 milligrams of copper. About one-fourth is in the muscles, one-fourth is in the liver, brain, and blood, and the rest is in the bones, kidneys, and other tissues. The primary function of copper in the body is to serve as a constituent of enzymes.[25] The copper-containing enzymes have diverse metabolic roles: they catalyze the formation of hemoglobin, help manufacture the protein collagen, assist in the healing of wounds, and help maintain the sheaths around nerve fibers. One of copper's most vital roles is to help cells use iron. Like iron, copper is needed in many reactions related to respiration and energy metabolism.

▪▪ COPPER DEFICIENCY Copper deficiency is rare but not unknown. It has been seen in malnourished children. High intakes of vitamin C and iron interfere with copper absorption and can lead to deficiency.[26]

▪▪ COPPER TOXICITY Some genetic disorders create a copper toxicity. Copper toxicity from foods, however, is unlikely.

▪▪ COPPER RECOMMENDATIONS AND FOOD SOURCES An estimated safe and adequate daily dietary intake has been established for copper. The best food sources of copper include legumes, whole grains, seafood, nuts, and seeds.

Estimated safe and adequate dietary intake for copper: 1.5 to 3.0 mg/day.

▪▪ Manganese

The human body contains a tiny 20 milligrams of manganese, mostly in the bones and glands. Manganese is a cofactor for many enzymes, helping to facilitate dozens of different metabolic processes. Deficiencies of manganese have not been noted in people, but toxicity may be severe. Miners who inhale large quantities of manganese dust on the job over prolonged periods show many symptoms of a brain disease, along with abnormalities in appearance and behavior.

Manganese requirements are low, and plant foods such as nuts, whole grains, and green, leafy vegetables contain significant amounts of this trace mineral. Deficiencies are therefore unlikely.

Estimated safe and adequate dietary intake for manganese: 2.5 to 5.0 mg/day.

■■ Fluoride

Only a trace of fluoride occurs in the human body, but research demonstrates that where diets are high in fluoride during the growing years, crystalline deposits in bones and teeth are larger and more perfectly formed. When bones and teeth become mineralized, first a crystal called hydroxyapatite forms from calcium and phosphorus. Then fluoride replaces the hydroxy portion of hydroxyapatite, forming **fluorapatite,** which makes the bones and teeth more resistant to decay. Once the teeth have erupted, the topical application of fluoride by way of toothpaste or mouth rinse continues to exert a caries-reducing effect.

fluorapatite (floor-APP-uh-tite): the stabilized form of bone and tooth crystal, in which fluoride has replaced the hydroxy portion of hydroxyapatite.

Fluoride AI:
Men: 3.8 mg/day.
Women: 3.1 mg/day.

■■ FLUORIDE DEFICIENCY Where fluoride is lacking in the water supply, the incidence of dental decay is high. Fluoridation of water to raise its fluoride concentration to 1 part per million is recommended as an important public health measure. Those fortunate enough to have had sufficient fluoride during the tooth-forming years of infancy and childhood are protected throughout life from dental decay. Dental problems are of great concern because they can lead to a multitude of other health problems affecting the whole body. Despite fluoride's value, violent disagreement often surrounds the introduction of fluoride to a community. Figure 9-5 shows the extent of fluoridation nationwide.

fluorosis (floor-OH-sis): mottling of the tooth enamel from ingestion of too much fluoride during tooth development.

■■ FLUORIDE SOURCES All normal diets include some fluoride, but drinking water is usually the most significant source. Fish and tea may supply substantial amounts as well.

In some areas, the natural fluoride concentration in water is high, and too much fluoride can damage teeth, causing **fluorosis.** In mild cases, the teeth develop small white specks; in severe cases, the enamel becomes pitted and permanently stained. Fluorosis occurs only during tooth development and cannot be reversed, making its prevention a high priority.

To prevent fluorosis:
• **Monitor the fluoride content of the local water supply.**
• **Supervise toddlers when they brush their teeth and use only a pea-size amount of toothpaste.**
• **Use fluoride supplements only as prescribed by a physician.**

Figure 9-5
FLUORIDATION IN THE UNITED STATES

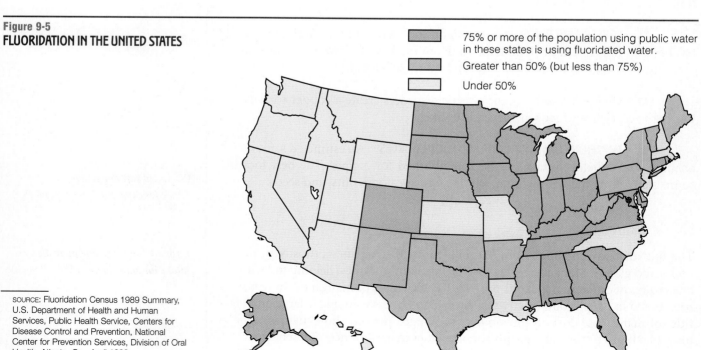

75% or more of the population using public water in these states is using fluoridated water.

Greater than 50% (but less than 75%)

Under 50%

SOURCE: Fluoridation Census 1989 Summary, U.S. Department of Health and Human Services, Public Health Service, Centers for Disease Control and Prevention, National Center for Prevention Services, Division of Oral Health. Atlanta, Ga., April 1993.

Chromium

Chromium is an essential mineral that participates in carbohydrate and lipid metabolism. Chromium enhances the activity of the hormone insulin.[27] Consequently, less insulin is needed to control blood glucose. When chromium is lacking, a diabetes-like condition may develop.

Chromium deficiency is unlikely, given the small amount required and its presence in a variety of foods. The more refined foods people eat, however, the less chromium they obtain from their diets. Unrefined foods such as liver, brewer's yeast, whole grains, nuts, and cheeses are the best sources.

Small organic compounds that enhance insulin's activity are called glucose tolerance factors (GTF). Some glucose tolerance factors contain chromium.

Estimated safe and adequate dietary intake for chromium: 50 to 200 µg/day.

Other Trace Minerals

The trace minerals have been known for decades, but their roles as nutrients are a recent surprise. **Molybdenum** functions as a working part of several metal-containing enzymes, some of which are giant proteins. Deficiencies or toxicities of molybdenum are unknown.

Nickel is recognized as important for the health of many body tissues. Nickel deficiencies harm the liver and other organs. Silicon participates in bone calcification, at least in animals. Tin is necessary for growth in animals and probably in people also. Cobalt is found in the large vitamin B_{12} molecule. The future may reveal that other trace minerals also play key roles. Even arsenic—famous as the death potion in many murder mysteries and known to be a carcinogen—may turn out to be an essential nutrient in tiny quantities.

In summary, the body requires trace minerals in tiny amounts, and they function in similar ways—assisting enzymes all over the body. Eating a diet that consists of a variety of foods is the best way to ensure an adequate intake of these important nutrients. Many dietary factors, including the trace minerals themselves, affect the absorption and availability of these nutrients.

Like the vitamins, the minerals perform a multitude of functions throughout the body. Table 9-6 offers a summary of facts about minerals in the body.

molybdenum (mo-LIB-duh-num): a trace element.

Estimated safe and adequate dietary intake for molybdenum: 75 to 250 µg/day.

Table 9-6
THE MINERALS—A SUMMARY

Mineral Name	Chief Functions In the Body	Deficiency Symptoms	Toxicity Symptoms	Significant Sources
Major Minerals				
Sodium	With chloride and potassium (electrolytes), maintains cells' normal fluid balance and acid-base balance in the body. Also critical to nerve impulse transmission.	Muscle cramps, mental apathy, loss of appetite.	Hypertension.	Salt, soy sauce, processed foods.
Chloride	Part of the hydrochloric acid found in the stomach and necessary for proper digestion.	Growth failure in children; muscle cramps, mental apathy, loss of appetite; can cause death (uncommon).	Normally harmless (the gas chlorine is a poison but evaporates from water); can cause vomiting.	Salt, soy sauce; moderate quantities in whole, unprocessed foods, large amounts in processed foods.
Potassium	Facilitates reactions, including the making of protein; the maintenance of fluid and electrolyte balance; the support of cell integrity; the transmission of nerve impulses; and the contraction of muscles, including the heart.	Deficiency accompanies dehydration; causes muscular weakness, paralysis, and confusion; can cause death.	Causes muscular weakness; triggers vomiting; if given into a vein, can stop the heart.	All whole foods: meats, milk, fruits, vegetables, grains, legumes.
Calcium	The principal mineral of bones and teeth. Also acts in normal muscle contraction and relaxation, nerve functioning, blood clotting, blood pressure, and immune defenses.	Stunted growth in children; adult bone loss (osteoporosis).	Excess calcium is excreted except in hormonal imbalance states (not caused by nutritional deficiency).	Milk and milk products, oysters, small fish (with bones), tofu (bean curd), greens, legumes.
Phosphorus	Important in cells' genetic material, in cell membranes as phospholipids, in energy transfer, and in buffering systems.	Phosphorus deficiency unknown.	Excess phosphorus may cause calcium excretion.	All animal tissues.
Magnesium	Another factor involved in bone mineralization, the building of protein, enzyme action, normal muscular contraction, transmission of nerve impulses, and maintenance of teeth.	Weakness; confusion; depressed pancreatic hormone secretion; if extreme, convulsions, bizarre movements (especially of eyes and face), hallucinations, and difficulty in swallowing. In children, growth failure.[a]	Not known; large doses have been taken in the form of the laxative Epsom salts, without ill effects except diarrhea.	Nuts, legumes, whole grains, dark green vegetables, seafoods, chocolate, cocoa.
Sulfur	A component of certain amino acids; part of the vitamins biotin and thiamin and the hormone insulin; combines with toxic substances to form harmless compounds; stabilizes protein shape by forming sulfur-sulfur bridges.	None known; protein deficiency would occur first.	Would occur only if sulfur amino acids were eaten in excess; this (in animals) depresses growth.	All protein-containing foods.

[a]A still more severe deficiency causes tetany, an extreme, prolonged contraction of the muscles similar to that caused by low blood calcium.

Table 9-6
THE MINERALS—A SUMMARY—continued

Mineral Name	Chief Functions In the Body	Deficiency Symptoms	Toxicity Symptoms	Significant Sources
Trace Minerals				
Iron	Part of the protein hemoglobin, which carries oxygen in the blood; part of the protein myoglobin in muscles, which makes oxygen available for muscle contraction; necessary for the utilization of energy.	Anemia: weakness, pallor, headaches, reduced resistance to infection, inability to concentrate, lowered cold tolerance.	Iron overload: infections, liver injury, possible increased risk of heart attack, acidosis, bloody stools, shock.	Red meats, fish, poultry, shellfish, eggs, legumes, dried fruits.
Zinc	Part of the hormone insulin and many enzymes; involved in making genetic material and proteins, immune reactions, transport of vitamin A, taste perception, wound healing, the making of sperm, and normal fetal development.	Growth failure in children, sexual retardation, loss of taste, poor wound healing.	Fever, nausea, vomiting, diarrhea, muscle incoordination, dizziness, anemia, accelerated atherosclerosis, kidney failure.	Protein-containing foods: meats, fish, shellfish, poultry, grains, vegetables.
Selenium	Part of an enzyme that breaks down reactive chemicals that harm cells; works with vitamin E.	Muscle discomfort, weakness, predisposition to heart disease characterized by cardiac tissue becoming fibrous.	Nausea, abdominal pain, nail and hair changes, nerve damage.	Seafoods, organ meats, other meats, grains and vegetables depending on soil conditions.
Iodine	A component of two thyroid hormones, which help to regulate growth, development, and metabolic rate.	Goiter, cretinism.	Depressed thyroid activity; goiterlike thyroid enlargement.	Iodized salt; seafood; bread; plants grown in most parts of the country and animals fed those plants.
Copper	Necessary for the absorption and use of iron in the formation of hemoglobin; part of several enzymes.	Anemia, bone abnormalities (rare in human beings).	Vomiting, diarrhea.	Meat, drinking water.
Manganese	Facilitator, with enzymes, of many cell processes.	(In experimental animals): poor growth, nervous system disorders, reproductive abnormalities.	Nervous system disorders.	Widely distributed in foods.
Fluoride	An element involved in the formation of bones and teeth; helps to make teeth resistant to decay.	Susceptibility to tooth decay.	Fluorosis (discoloration of teeth), nausea, diarrhea, chest pain, itching, vomiting.	Drinking water (if fluoride containing or fluoridated), tea, seafood.
Chromium	Associated with insulin and required for the release of energy from glucose.	Diabetes-like condition marked by an inability to use glucose normally.	Unknown as a nutrition disorder; occupational exposures damage skin and kidneys.	Meat, unrefined foods, fats, vegetable oils.

SELF CHECK

SELF CHECK

1. Water excretion is governed primarily by the:
   a. liver.
   b. kidneys
   c. brain.
   d. GI tract.

2. People's fluid needs are best met by:
   a. water.
   b. milk.
   c. juice.
   d. sodas.

3. Two situations in which a person may experience fluid and electrolyte imbalances are:
   a. vomiting and burns.
   b. diarrhea and cuts.
   c. broken bones and fever.
   d. heavy sweating and excessive carbohydrate intake.

4. Three-fourths of the sodium in people's diets comes from:
   a. fresh meats.
   b. home-cooked foods.
   c. frozen vegetables and meats.
   d. salt added to food by manufacturers.

5. Which mineral is critical to keeping the heartbeat steady and plays a major role in maintaining fluid and electrolyte balance?
   a. sodium
   b. calcium
   c. potassium
   d. magnesium

6. The two best ways to prevent age-related bone loss and fracture are to:
   a. take calcium supplements and estrogen.
   b. participate in aerobic activity and drink 8 glasses of milk daily.
   c. eat a diet low in fat and salt and refrain from smoking.
   d. maintain a lifelong adequate calcium intake and engage in weight-bearing physical activity.

7. Three good food sources of calcium are:
   a. milk, sardines, and broccoli.
   b. spinach, yogurt, and sardines.
   c. cottage cheese, spinach, and tofu.
   d. swiss chard, mustard greens, and broccoli.

8. Foods high in iron that help prevent or treat anemia include:
   a. green peas and cheese.
   b. dairy foods and fresh fruits.
   c. homemade breads and most fresh vegetables.
   d. meat and dark green, leafy vegetables.

9. Two groups of people who are especially at risk for zinc deficiency are:
   a. Asians and children.
   b. infants and the elderly.
   c. smokers and athletes.
   d. pregnant adolescents and vegetarians.

10. A deficiency of ____ is one of the world's most common preventable causes of mental retardation.
    a. zinc
    b. iodine
    c. selenium
    d. magnesium

Answers to these questions appear in Appendix H.

CLINICAL APPLICATIONS

1. What advice would you give a client who regularly relies on caffeine-containing beverages to quench thirst? Consider the effects of caffeine on both water and mineral balance.

2. Pull together information from Chapter 1 about the different food groups and the significant sources of minerals shown or discussed in this chapter. Consider which minerals might be lacking (or excessive) in the diet of a client who reports the following:

   • Relies on highly processed foods, snack foods, and fast foods as mainstays of the diet.

   • Never uses milk, milk products, or cheese.

   • Dislikes green, leafy vegetables.

   • Never eats meat, fish, poultry, or even meat alternates such as legumes.

   What additional information would help you pinpoint problems with mineral intake?

NUTRITI**ON**THENET

FOR FURTHER STUDY OF THE
TOPICS IN THIS CHAPTER,
ACCESS THESE WEB SITES.

www.nof.org
National Osteoporosis Foundation

www.foodallergy.org
Food Allergy Network

www.aap.org
American Academy of Pediatrics

www.fda.gov
U.S. Food and Drug Administration

www.thyroid.org
American Thyroid Association

Notes

[1] S. M. Kleiner, Water: An essential but overlooked nutrient, *Journal of the American Dietetic Association* 99 (1999): 200–206.

[2] USDA Center for Nutrition Policy and Promotion, Dietary guidance on sodium: Should we take it with a grain of salt? *Nutrition Today* 32 (1997): 250.

[3] A. W. Cowley, Jr., Genetic and nongenetic determinants of salt sensitivity and blood pressure, *American Journal of Clinical Nutrition* 65 (1997): S587–S593.

[4] R. Itoh and Y. Suyama, Sodium excretion in relation to calcium and hydroxyproline excretion in a healthy Japanese population, *American Journal of Clinical Nutrition* 63 (1996): 735–740; A. Devine and coauthors, A longitudinal study of the effect of sodium and calcium intakes on regional bone density in postmenopausal women, *American Journal of Clinical Nutrition* 62 (1995): 740–745.

[5] P. K. Whelton and coauthors, Effects of oral potassium on blood pressure: Meta-analysis of randomized controlled clinical trial, *Journal of the American Medical Association* 277 (1997): 1624–1632.

[6] E. A. Krall and B. Dawson-Hughes, Osteoporosis, in *Modern Nutrition in Health and Disease,* 9th ed., ed. M. E. Shils and coeditors (Baltimore, Md.: Williams & Wilkins, 1999), pp. 1353–1364; R. P. Heaney, Nutrition factors in osteoporosis, *Annual Review of Nutrition* 13 (1993): 287–316.

[7] I. Vuori, Peak bone mass and physical activity: A short review, *Nutrition Reviews* 54 (1996): S11–S14; L. Alekel and coauthors, Contributions of exercise, body composition, and age to bone mineral density in premenopausal women, *Medicine and Science in Sports and Exercise* 27 (1995): 1477–1485.

[8] T. A. Kotchen and J. M. Kotchen, Nutrition, diet, and hypertension, in *Modern Nutrition in Health and Disease,* 9th ed., ed. M. E. Shils and coeditors (Baltimore, Md.: Williams & Wilkins, 1999), pp. 1217–1227; D. A. McCarron, Role of adequate dietary calcium intake in the prevention and management of salt sensitive hypertension, *American Journal of Clinical Nutrition* 65 (1997): S712–S716; Joint National Committee on Detection, Evaluation, and Treatment of High Blood Pressure, The Fifth Report of the Joint National Committee on Detection, Evaluation, and Treatment of High Blood Pressure, *Archives of Internal Medicine* 153 (1993): 154–183.

[9] C. G. Osborne and coauthors, Evidence for the relationship of calcium to blood pressure, *Nutrition Reviews* 54 (1996): 365–381; D. A. McCarron and D. Hatton, Dietary calcium and lower blood pressure: We can all benefit, *Journal of the American Medical Association* 275 (1996): 1128–1129.

[10] H. C. Bucher and coauthors, Effects of dietary calcium supplementation on blood pressure: A meta-analysis of randomized controlled trials, *Journal of the American Medical Association* 275 (1996): 1016–1022.

[11] M. E. Shils, Magnesium, in *Modern Nutrition in Health and Disease,* 9th ed., ed. M. E. Shils and coeditors (Baltimore, Md.: Williams & Wilkins, 1999), pp. 169–189.

[12] C. E. West, Strategies to control nutritional anemia, *American Journal of Clinical Nutrition* 64 (1996): 789–790.

[13] A. C. Looker and coauthors, Prevalence of iron deficiency in the United States, *Journal of the American Medical Association* 277 (1997): 973–976.

[14] E. Pollitt, Iron deficiency and cognitive function, *Annual Review of Nutrition* 13 (1993): 521–537.

[15] M. F. Walter and E. M. Haymes, The effects of heat and exercise on sweat iron loss, *Medicine and Science in Sports and Exercise* 28 (1996): 197–203.

[16] Y. I. Zhu and J. D. Haas, Iron depletion without anemia and physical performance in young women, *American Journal of Clinical Nutrition* 66 (1997): 334–341; J. J. LaManca and E. M. Haymes, Effects of iron repletion on VO_2 max, endurance, and blood lactate in women, *Medicine and Science in Sports and Exercise* 25 (1993): 1386–1392.

[17] J. C. Fleet, Discovery of the hemochromatosis gene will require rethinking the regulation of iron metabolism, *Nutrition Reviews* 54 (1996): 285–292.

[18] Iron overload disorders among Hispanics—San Diego, California, 1995, *Morbidity and Mortality Weekly Report* 45 (1996): 991–993; D. H. G. Crawford and coauthors, Factors influencing disease expression in hemochromatosis, *Annual Review of Nutrition* 16 (1996): 139–160; C. E. McLaren and coauthors, Prevalence of heterozygotes for hemochromatosis in the white population of the United States, *Blood* 84 (1995): 2121–2127.

[19] Pediatricians seek FDA's help in preventing poisoning deaths, *Journal of the American Dietetic Association* 93 (1993): 529.

[20] Keep iron tablets away from children, *FDA Consumer,* May 1993, p. 2.

[21] J. R. Hunt, L. A. Matthys, and L. K. Johnson, Zinc absorption, mineral balance, and blood lipids in women consuming controlled lactovegetarian and omnivorous diets for 8 wk, *American Journal of Clinical Nutrition* 67 (1998): 421–430.

[22] J. R. Arthur, F. Nicol, and G. J. Beckett, Selenium deficiency, thyroid hormone metabolism, and thyroid hormone deiodinases, *American Journal of Clinical Nutrition* 57 (1993): S236–S239.

[23] Letters from V. Herbert, L. H. Kuller, J. S. Parker, and L. C. Clark, Selenium supplementation and cancer rates, *Journal of the American Medical Association* 277 (1997): 880–881; L. C. Clark and coauthors, Effects of selenium supplementation for cancer prevention in patients with carcinoma of the skin—A randomized controlled trial, *Journal of the American Medical Association* 276 (1996): 1984–1985.

[24] G. R. Delong, Effects of nutrition on brain development in humans, *American Journal of Clinical Nutrition* 57 (1993): S286–S290.

[25] R. Uauy, M. Olivares, and M. Gonzalez, Essentiality of copper in humans, *American Journal of Clinical Nutrition* 67 (1998): S952–S959.

[26] B. Lonnerdal, Copper nutrition during infancy and childhood, *American Journal of Clinical Nutrition* 67 (1998): S1046–S1053.

[27] S. Fairweather-Tait and R. F. Hurrell, Bioavailability of minerals and trace elements, *Nutrition Research Reviews* 9 (1996): 295–324.

Replacing Fluid Losses: Sports Drinks and Rehydration Formulas

As Chapter 9 pointed out, water is a crucial nutrient for everyone, but maintaining fluid balance can be of particular concern for those engaged in physical activity and for infants and young children sick with diarrhea. This Nutrition in Practice focuses on sports drinks for athletes and rehydration formulas for children with diarrhea.

Why is fluid balance a special concern for physically active people and athletes?

Maintaining fluid balance is a concern for those engaged in physical activity because the body loses water via sweat. Breathing uses water, too, exhaled as vapor. During physical activity, both routes are significant, and dehydration becomes a threat.

What are the symptoms of dehydration?

Dehydration's first symptom is fatigue: a water loss of even 1 to 2 percent of body weight can reduce a person's capacity to do muscular work.[1] With a water loss of about 7 percent, a person is likely to collapse.[2] Table NP9-1 lists the signs of dehydration.

Why does an active person sweat so much?

Working muscles produce heat. During intense activity, muscle heat production can be 15 to 20 times greater than at rest.[3] The body cools itself by sweating. Each liter of sweat dissipates almost 600 kcalories of heat, preventing a rise in body temperature of almost 10°C. The body routes its blood supply through the capillaries just under the skin, and the skin secretes sweat to evaporate and cool the skin and underlying blood. The blood then flows back to cool the deeper body chambers.

How much fluid do active people lose?

The amount depends on different factors such as the temperature, the intensity of the activity, the duration of the activity, and, of course, the athlete's hydration state prior to activity. Endurance athletes can easily lose 1.5 liters or more of fluid during each hour of activity. To prepare for fluid losses, a person must hydrate before activity. To replace fluid losses, the person must rehydrate during and after activity. (Table NP9-2 presents one schedule of hydration for physical activity.) Even then, in hot weather, the GI tract may not be able to absorb enough water fast enough to keep up with sweat losses, and some degree of dehydration may be inevitable.

What is the best fluid to support physical activity?

For noncompetitive, everyday active people, plain, cool water is recommended, especially in warm weather, for two reasons: it rapidly leaves the digestive tract to enter the tissues where it is needed, and it cools the body from the inside out. For endurance athletes, other beverages may be appropriate. Fluid ingestion during an endurance event has the dual purposes of replenishing water lost through sweating and providing a source of carbohydrate to supplement the body's limited glycogen stores. Carbohydrate depletion brings on fatigue in the athlete, but as already mentioned, fluid loss and the accompanying buildup of body heat can be life-threatening. Thus, the first priority for endurance athletes should be to replace fluids. Many good-tasting sports drinks are marketed for active people.

What do sports drinks have to offer?

First, and most important, of course, sports drinks offer fluids to help offset the losses from physical activity. As discussed earlier, however, plain water can do this, too.

Second, sports drinks supply glucose. A beverage that supplies glucose can be useful during endurance activities lasting 60 minutes or more or during prolonged competitive games that demand repeated intermittent activity.[4] Not just any sweet beverage can meet this need, however, because a carbohydrate concentration greater than 10 percent can delay fluid emptying from the stomach and thereby slow down the delivery of water to the tissues. Most

Table NP9-1
SIGNS OF MILD AND SEVERE DEHYDRATION

Mild (<5% Loss of Body Weight)	Severe (>5% Loss of Body Weight)
Thirst	Pale skin
Sudden weight loss	Bluish lips and fingertips
Rough dry skin	Confusion; disorientation
Dry mouth, throat, body linings	Rapid, shallow breathing
Rapid pulse	Weak, rapid, irregular pulse
Low blood pressure	Thickening of blood
Lack of energy; weakness	Shock; seizures
Impaired kidney function	Coma; death
Reduced quantity of urine; concentrated urine	

sports drinks contain about 7 percent glucose—a safe level for fluid transport.

Third, sports drinks offer sodium and other electrolytes to help replace those lost during physical activity. Sodium in sports drinks also helps to speed fluid absorption from the digestive tract. Most athletes do not need to replace the minerals lost in sweat immediately; a meal eaten within hours of competition replaces these minerals soon enough. Most sports drinks are relatively low in sodium, however, so healthy people who choose to use these beverages run little risk of excessive intake.

In strenuous world-class competitions lasting for many hot, humid days, heavy sweating coupled with drinking large amounts of plain water has been reported to dangerously dilute blood sodium. If an athlete works up a drenching sweat, exceeding 5 to 10 pounds a day (or 3 percent of body weight), for several consecutive days, electrolyte replacement is advised. Athletes who *compete* for longer than six to eight hours each day may especially need to replace sodium.[5]

Equally as effective as commercial sports drinks but much less expensive is a homemade mixture of one-third teaspoon of salt (to provide sodium chloride) and 1 cup of sugar-sweetened fruit juice (to provide potassium and glucose) per quart of water. Avoid electrolyte or salt tablets; they increase potassium losses, irritate the stomach, cause vomiting, and pull water out of the tissues into the digestive tract at first.

Why is fluid balance a concern in infants and young children with diarrhea?

A person of any age with severe diarrhea can quickly develop dehydration and electrolyte imbalances. For infants and young children, these problems can quickly develop even when diarrhea is not severe. Fluid losses from an infant's body become significant because compared with the water in an adult's body, a larger percentage of an infant's body water is located in the interstitial (extracellular) and vascular spaces. Diarrhea—frequent, loose, watery stools—indicates that the chyme has moved too quickly through the intestines for fluid absorption to take place, or that water has been drawn from the cells lining the intestinal tract and added to the food residue. In both cases, the result is the same—extensive fluid and electrolyte losses. If diarrhea continues without treatment, an infant or young child can quickly become dehydrated. The smaller and younger the child, the more life-threatening the condition. Some 300 to 500 infants and young children in the United States die from complications of diarrhea each year.[6]

All children have diarrhea at one time or another, don't they?

Yes, nearly every child suffers from diarrhea at one time or another. Many times an acute case of diarrhea develops and remits in 24 to 48 hours. Well-nourished children with acute diarrhea can usually endure the uncomfortable symptoms without medical treatment. Caregivers can support children during such episodes of diarrhea by eliminating food irritants from the diet, offering clear liquids, and encouraging rest. Fruit juices may aggravate diarrhea and therefore are inappropriate to offer; foods with soluble fiber, such as oatmeal and bananas, may help relieve diarrhea. Children may enjoy such clear liquids as gelatin dessert, carbonated beverages, and Popsicles, but these treats fall short of correcting dehydration. Their low electrolyte content and high **osmolality** make them unsuitable for rehydration therapy.

Nutrient reserves of well-nourished children protect them from the detrimental effects of diarrhea for a short while. The availability of medical treatment and high-quality food in industrialized countries offers children the opportunity to recover, both from fluid and electrolyte losses and from growth losses.

The story is quite different in developing countries where acute diarrhea in a malnourished child threatens life and requires immediate medical attention. In developing countries, more children suffer from malnutrition, and acute diarrhea seriously threatens their tenuous nutritional status. When repeated episodes of diarrhea fall close

Table NP9-2
HYDRATION SCHEDULE FOR PHYSICAL ACTIVITY

When to Drink	Total Amount of Fluid (Consume in 1-cup Servings)
Day of event	Extra fluid throughout the day
2 hours before exercise	About 3 c
10 to 15 minutes before exercise	About 2 c
Every 15 to 30 minutes during exercise	4–8 oz (about 1 qt in 60 to 90 minutes)
After exercise	Replace each pound of body weight lost with 2 c fluid

GLOSSARY

osmolality (OZ-mow-LAL-eh-tee): a measure of the concentration of particles in a solution.

oral rehydration therapy (ORT): the administration of a simple solution of sugar, salt, and water, taken by mouth, to treat dehydration caused by diarrhea. A simple ORT recipe:
1 c boiling water.
2 tsp sugar.
A pinch of salt.

WHO (World Health Organization): an international agency that promotes cooperation for health among nations. WHO carries out programs to control and eradicate disease and strives to improve the quality of human life.

Table NP9-3
ORT FORMULA RECOMMENDED BY THE WORLD HEALTH ORGANIZATION (WHO)

Nutrient	Concentration
Sodium	90 mmol/L[a]
Chloride	80 mmol/L
Potassium	20 mmol/L
Glucose	111 mmol/L
Citrate tribasic or bicarbonate	30 mmol/L

[a]mmol/L = millimoles per liter.

together, these children have little time to recover between bouts. Without full restoration of nutrients, children are progressively less able to defend against future infections.

How can diarrhea be treated to prevent malnutrition?

The potential for accelerated deterioration of nutrition status demands rapid replacement of fluids and nutrients lost through diarrhea. Health care workers around the world treat diarrhea with **oral rehydration therapy (ORT).** (The glossary on p. 227 defines ORT and provides a simple recipe.) The American Academy of Pediatrics recommends ORT as the "preferred treatment of fluid and electrolyte losses caused by diarrhea in children with mild to moderate diarrhea."[7] Traditionally, glucose-based solutions have been the standard treatment for acute diarrhea. These solutions provide an optimal glucose concentration, which favors rapid intestinal absorption of water and sodium. As a result, they effectively reverse dehydration, but may not correct the diarrhea. Rice-based solutions have recently proved more effective in correcting diarrhea and therefore may become the preferred therapy.[8] If diarrhea continues, the child receives water and the

solution alternately. Infants with diarrhea usually tolerate breast milk well, and breastfeeding can alternate with supplements of the solution. Table NP9-3 lists the **World Health Organization (WHO)** standards for ORT formulas; colas, juices, broths, and sports drinks are not appropriate for ORT.

The components of ORT formulas provide needed energy and electrolytes. Except in cases of severe dehydration, ORT can replace the traditional treatment of intravenous (IV) fluid therapy. Intravenous therapy is still most valuable in treating severely dehydrated children; ORT is useful in treating mild-to-moderate cases and following initial IV therapy.

Perhaps the most significant value of ORT is that it is oral—it does not require hospitalization, as IV feedings do. In addition, if the specific WHO-recommended solution is unavailable, caregivers can make other suitable ORT solutions. A mother who lives miles from the nearest pharmacy or clinic and who may not have the resources to purchase medicines can still prepare a solution to refeed her dehydrated infant. A properly prepared rice powder solution facilitates the absorption of electrolytes and water, as well as limiting the duration of diarrhea. The primary role of health care providers then becomes one of educating those who care for children. The people of a community must learn how to prepare a rehydration solution from ingredients available locally. They must learn to measure ingredients carefully and to use sanitary water. Parents will also want to learn how to recognize diarrhea and dehydration symptoms.

The success of ORT in developing countries is central to WHO's effort to counter the dehydration and death commonly associated with diarrhea. In developed countries, however,

physicians are reluctant to adopt ORT, recommending hospitalization and IV fluids instead. Yet treating well-nourished children in developed countries with ORT is as effective as IV therapy and less expensive and invasive.

Dehydration can compromise athletic performance and health. Athletes and young children sick with diarrhea are especially vulnerable to dehydration. Left untreated, dehydration can be life-threatening. Maintaining fluid balance and correcting imbalances early on are in the best interest of health for everyone.

www.who.int
World Health Organization

Notes

[1]C. V. Gisolfi, Fluid balance for optimal performance, *Nutrition Reviews* 54 (1996): S159–S168.

[2]J. E. Greenleaf, Problem: Thirst, drinking behavior, and involuntary dehydration, *Medicine and Science in Sports and Exercise* 24 (1992): 645–656.

[3]Gisolfi, 1996.

[4]X. Shi and C. V. Gisolfi, Fluid and carbohydrate replacement during intermittent exercise, *Sports Medicine* 25 (1998): 157–172.

[5]American College of Sports Medicine, Position stand: Heat and cold illness during distance running, *Medicine and Science in Sports and Exercise* 28 (1996): i–x.

[6]R. D. Williams, Preventing dehydration in children, *FDA Consumer,* July/August 1996, pp. 19–22; P. E. Kilgore and coauthors, Trends of diarrheal disease–associated mortality in US children, 1968–1991, *Journal of the American Medical Association* 274 (1995): 1143–1148.

[7]Provisional Committee on Quality Improvement, Subcommittee on Acute Gastroenteritis, American Academy of Pediatrics, Practice parameter: The management of acute gastroenteritis in young children, *Pediatrics* 97 (1996): 424–433.

[8]E. D. Goldberg and J. R. Saltzman, Rice inhibits intestinal secretions, *Nutrition Reviews* 54 (1996): 36–37.

CHAPTER 10

NUTRITION THROUGH THE LIFE SPAN: PREGNANCY AND INFANCY

CONTENTS

Pregnancy: The Impact
of Nutrition on the Future

Breastfeeding

Nutrition of the Infant

Case Study: Pregnant
Woman with Weight Problem

Nutrition in Practice:
Encouraging Successful
Breastfeeding

The effects of nutrition extend over years. A woman's nutrition prior to and throughout pregnancy and lactation affects not only her own health but also the growth, development, and health of her child, even long after the child has been born. Similarly, sound nutrition is vital to healthy infant development.

Pregnancy: The Impact of Nutrition on the Future

The woman who enters pregnancy with full nutrient stores, sound eating habits, and a healthy body weight has done much to ensure an optimal pregnancy. Then, during the pregnancy itself, if she eats a variety of nutrient-dense foods, her own and her infant's health will benefit further.

Preparing for Pregnancy

Full nutrient stores *before* pregnancy are essential both to conception and to healthy infant development during pregnancy. In the early weeks of pregnancy, before many women are even aware that they are pregnant, significant developmental changes occur that depend on a woman's nutrient stores.

PREPREGNANCY WEIGHT
Appropriate weight prior to pregnancy also benefits pregnancy outcome. Being either underweight or overweight (see p. 237) during pregnancy and childbirth presents medical risks. Underweight women are therefore advised to gain weight before becoming pregnant, and overweight women to lose excess weight. Guidelines for weight gain and loss were offered in Chapter 7.

INFANT BIRTHWEIGHT
Infant birthweight strongly correlates with the mother's prepregnancy weight and is the most potent single predictor of the infant's future health and survival. Compared with normal-weight babies, **low-birthweight** babies are more likely to contract diseases and nearly 40 times more likely to die in the first month of life.[1] Low socioeconomic status may limit a mother's access to medical care and nutritious foods and thus makes low birthweight more likely.[2]

HEALTHY SUPPORT TISSUES
The mother's prepregnancy nutrition is crucial to a healthy pregnancy because it determines whether she will be able to grow healthy support tissues: the **placenta,** the **amniotic sac,** the **umbilical cord,** and the expanding **uterus** (see Figure 10-1). Malnutrition prior to and around conception keeps these tissues from developing fully.[3] If the placenta fails to develop properly, the **fetus** will not receive optimal nourishment.

Nutrient Needs during Pregnancy

Between the moment of conception and the moment of birth, innumerable events determine the course and outcome of fetal development and, ultimately, the health of the newborn infant. With respect to nutrition, each organ needs nutrients most during its own intensive growth period. A nutrient deficiency during one stage of development might affect the heart and, during another stage, the developing limbs.

A woman's nutrient needs during pregnancy and lactation are higher than at any other time in her adult life and are greater for certain nutrients

---

low birthweight (LBW): a birthweight less than 5½ lb (2500 g); indicates probable poor health in the newborn and poor nutrition status of the mother during pregnancy. Normal birthweight for a full-term baby is 6½ to 8½ lb (about 3000 to 4000 g).

Low-birthweight infants are of two different types. Some are premature; they are born early and are of a weight appropriate for gestational age (AGA). Others have suffered growth failure in the uterus; they may or may not be born early, but they are small for gestational age (SGA).

placenta (pla-SEN-tuh): an organ that develops inside the uterus early in pregnancy, in which maternal and fetal blood circulate in close proximity and exchange materials. The fetus receives nutrients and oxygen across the placenta; the mother's blood picks up carbon dioxide and other waste materials to be excreted via her lungs and kidneys.

amniotic (am-nee-OTT-ic) sac: the "bag of waters" in the uterus, in which the fetus floats.

umbilical (um-BIL-ih-cul) cord: the ropelike structure through which the fetus's veins and arteries reach the placenta; the route of nourishment and oxygen into the fetus and the route of waste disposal from the fetus.

uterus (YOO-ter-us): the womb, the muscular organ within which the infant develops before birth.

fetus (FEET-us): the developing infant from eight weeks after conception until its birth.

Figure 10-1
THE PLACENTA

The placenta is composed of spongy tissue in which fetal blood and maternal blood flow side by side, each in its own vessels. The maternal blood transfers oxygen and nutrients to the fetus's blood and picks up fetal wastes to be excreted by the mother. Thus, the placenta facilitates the nutritive, respiratory, and excretory functions that the fetus's digestive system, lungs, and kidneys will provide after birth.

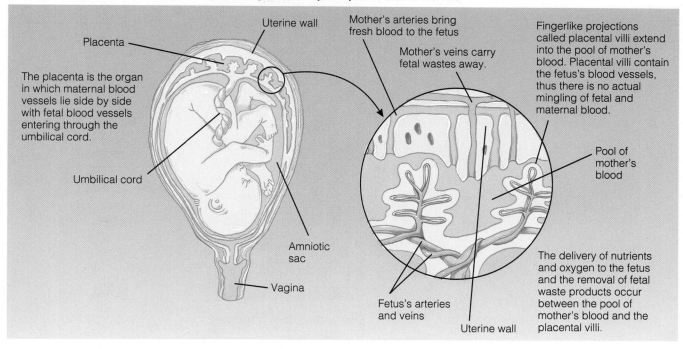

than for others. Figure 10-2 (p. 232) compares the nutrient needs of nonpregnant, pregnant, and lactating women. A study of the figure reveals some of the key needs.

■■ ENERGY A pregnant woman needs extra food energy, but only a little extra—300 kcalories above the allowance for nonpregnant women—and only during the second and third trimesters. A woman can easily obtain 300 kcalories from just one extra serving from each of the five food groups—a slice of bread, a serving of vegetables, an ounce of lean meat, a piece of fruit, and a cup of fat-free milk. Pregnant teenagers, underweight women, or physically active women may require more.

■■ PROTEIN The RDA suggests an added 10 grams of protein per day throughout pregnancy. Many women in the United States exceed the recommended protein intake each day, so they already receive the 10 grams of additional daily protein recommended for pregnancy. In fact, pregnant women in the United States—even those with low incomes who are not participating in food assistance programs—generally receive between 75 and 110 grams of protein a day.

Obtaining enough protein need not pose a problem, even if the diet excludes all foods of animal origin. Pregnant vegetarian women who meet their energy needs by eating ample servings of protein-containing plant foods such as legumes, whole grains, nuts, and seeds meet their protein needs as well. Use of high-protein supplements during pregnancy can be harmful and is discouraged.

■■ CARBOHYDRATE Pregnant women need generous amounts of carbohydrate to spare the protein they eat. If added energy is needed, it is best obtained from carbohydrate.

Pregnancy is often divided into thirds called **trimesters.**

Energy RDA during pregnancy (2nd and 3rd trimesters): + 300 kcal/day.

Protein RDA during pregnancy: + 10 g/day.

Recommended carbohydrate intake: about 50% of energy intake. In a 2000 kcal/day intake, this represents 1000 kcal of carbohydrate, or about 250 g. Four cups of milk will contribute about 50 g carbohydrate. An apple provides 12 g carbohydrate, and a slice of bread provides 12 g, so generous intakes of fruit and grain products are clearly beneficial.

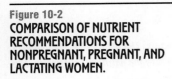

Figure 10-2

COMPARISON OF NUTRIENT RECOMMENDATIONS FOR NONPREGNANT, PREGNANT, AND LACTATING WOMEN.

For actual values, turn to the inside front cover.

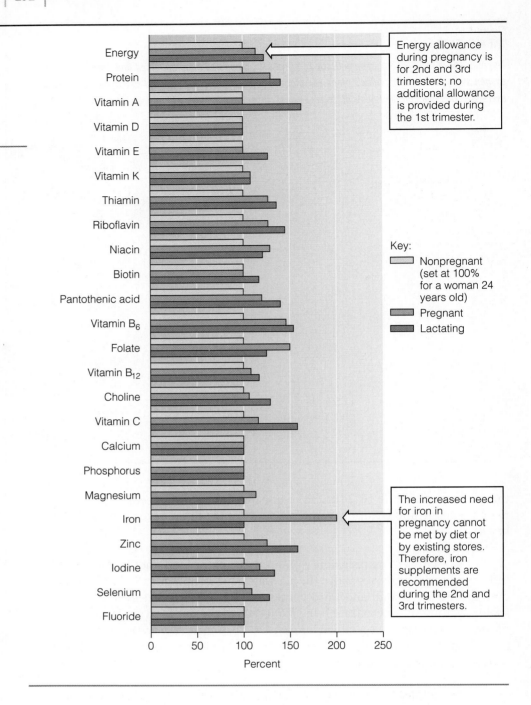

Energy allowance during pregnancy is for 2nd and 3rd trimesters; no additional allowance is provided during the 1st trimester.

Key:
- Nonpregnant (set at 100% for a woman 24 years old)
- Pregnant
- Lactating

The increased need for iron in pregnancy cannot be met by diet or by existing stores. Therefore, iron supplements are recommended during the 2nd and 3rd trimesters.

Nutrients (top to bottom): Energy, Protein, Vitamin A, Vitamin D, Vitamin E, Vitamin K, Thiamin, Riboflavin, Niacin, Biotin, Pantothenic acid, Vitamin B_6, Folate, Vitamin B_{12}, Choline, Vitamin C, Calcium, Phosphorus, Magnesium, Iron, Zinc, Iodine, Selenium, Fluoride

Percent axis: 0, 50, 100, 150, 200, 250

Folate RDA during pregnancy: 600 µg/day.

Vitamin B_{12} RDA during pregnancy: 2.6 µg/day.

VITAMINS The vitamins required for rapid cell proliferation—folate and vitamin B_{12}—are needed in large amounts during pregnancy. New cells are laid down at a tremendous pace as the fetus grows and develops. At the same time, the mother's number of red blood cells rises, so the recommendation for folate increases from 400 to 600 micrograms a day during pregnancy.

As described in Chapter 8, folate plays an important role in preventing neural tube defects. Folate supplements (400 micrograms daily) taken one month before conception and continued throughout the first trimester can prevent neural tube defects.[4] Neural tube defects arise early in pregnancy, however, so the American Academy of Pediatrics, the Public Health Services, the Institute of Medicine, and the March of Dimes, recom-

Table 10-1
RICH FOLATE SOURCES[a]

Natural Folate Sources	Fortified Folate Sources
Liver (3 oz) 185 µg	Multi-Grain Cheerios Plus cereal (1 c) 400 µg[b]
Lentils (½ c) 180 µg	Product 19 cereal (1 c) 400 µg[b]
Chickpeas or pinto beans (½ c) 145 µg	Total cereal (1 c) 400 µg[b]
Asparagus (½ c) 125 µg	Pasta, cooked (1 c) 110 µg
Spinach (1 c raw) 115 µg	Rice, cooked (1 c) 80 µg
Avocado (½ c) 70 µg	Bagel (1 small whole) 50 µg
Orange juice (1 c) 60 µg	Waffles, frozen (2) 40 µg
Beets (½ c) 46 µg	Bread, white (1 slice) 20 µg

[a]Folate amounts for these and 2000 other foods are listed in the Table of Food Composition in Appendix A.
[b]Folate in cereals varies; read the Nutrition Facts panel of the label.

mend that all women capable of becoming pregnant take 400 micrograms of folate daily from supplements or fortified foods or a combination of the two, in addition to eating a variety of foods naturally rich in folate (see Table 10-1).[5]

Supplements and fortified foods offer women a convenient way to ensure sufficient folate regularly and continuously enough to benefit pregnancy.[6] Furthermore, the synthetic form of folate, *folic acid,* in supplements and fortified food is better absorbed than the naturally occurring folate in foods. Thus, folate status improves more with intakes of supplements or fortified foods than with intakes of only natural sources of folate. The foods that naturally contain folate are still important, however, because they are rich sources of other vitamins, minerals, fiber, and the phytochemicals thought to protect against heart disease, cancer, and other diseases.

Generally, even modest amounts of meat, fish, eggs, or milk products together with body stores easily meet the need for vitamin B_{12}. Vegans who exclude all foods of animal origin, however, may need daily supplements to prevent deficiency.

:: SUPPLEMENTS   Dietary improvements are always preferred as the means of correcting nutrient inadequacies, but sometimes nutrient supplementation is necessary. The National Academy of Sciences subcommittee for the publication *Nutrition during Pregnancy* identifies certain women as being nutritionally at risk during pregnancy (see Table 10-2 on p. 234).[7] The subcommittee recommends a multivitamin-mineral supplement containing the nutrient amounts shown in Table 10-3 (p. 234) for these women.

:: VITAMIN D AND CALCIUM FOR BONES   Vitamin D and the minerals involved in building the skeleton—calcium, phosphorus, and magnesium— are in great demand during pregnancy. Insufficient intakes may result in abnormal fetal bone development.

Intestinal absorption of calcium doubles early in pregnancy, when the mother's bones store the mineral. Later, as the fetal bones begin to calcify, there is a dramatic shift of calcium across the placenta. Whether the calcium added to the mother's bones early in pregnancy is withdrawn to build the fetus's bones later is unclear. In the final weeks of pregnancy, more than 300 milligrams a day are transferred to the fetus. Recommendations to ensure an

As noted in Chapters 8 and 9, vitamins and minerals can be toxic in excess. During pregnancy, nutrient excesses (especially of vitamin A) can be potentially harmful. Seek advice from a health care provider before taking supplements.

Calcium AI during pregnancy:
1300 mg/day (14 to 18 yr).
1000 mg/day (19 to 50 yr).

Phosphorus RDA during pregnancy:
1250 mg/day (14 to 18 yr).
700 mg/day (19 to 30 yr).

Magnesium RDA during pregnancy:
400 mg/day (14 to 18 yr)
350 mg/day (19 to 30 yr).
360 mg/day (31 to 50 yr).

Four cups of milk a day will supply 1200 mg calcium. For other food sources of calcium, see Chapter 9.

Table 10-2
WOMEN AT NUTRITIONAL RISK DURING PREGNANCY

- Women who ordinarily consume an inadequate diet due to food faddism or weight-loss dieting, for example
- Women who are lactose intolerant
- Women who are carrying multiple fetuses
- Women who smoke cigarettes or use alcohol or illicit drugs
- Women who are underweight or overweight at conception
- Women who gain insufficient or excessive weight during pregnancy
- Women who lack nutrition knowledge or who have insufficient financial resources to purchase adequate food
- Adolescents

SOURCE: Compiled from information in National Academy of Sciences, Food and Nutrition Board, *Nutrition during Pregnancy* (Washington, D.C.: National Academy Press, 1990).

Fluoride AI during pregnancy:
2.9 mg/day (14 to 18 yr).
3.1 mg/day (19 to 50 yr).

Iron RDA during pregnancy:
30 mg/day.

Table 10-3
NUTRIENT SUPPLEMENTS DURING PREGNANCY[a]

Nutrient	Amount
Folate	300 μg
Vitamin B$_6$	2 mg
Vitamin C	50 mg
Vitamin D	5 μg
Calcium	250 mg
Copper	2 mg
Iron	30 mg
Zinc	15 mg

[a]For pregnant women at nutritional risk (see Table 10-2).
SOURCE: Reprinted with permission from *Nutrition during Pregnancy* © 1990 by the National Academy of Sciences. Courtesy of the National Academy Press, Washington, D.C.

adequate calcium intake during pregnancy are aimed at conserving the mother's bone mass while supplying fetal needs.

For women whose prepregnancy calcium intakes are below recommendations, as most are, increased calcium intakes may be especially important. Milk products offer many advantages over supplements, as emphasized in earlier chapters. Because bones are still actively depositing minerals until about age 25, adequate calcium is especially important for young women. Pregnant women under age 25 who consume less than 600 milligrams of calcium a day need to increase their intakes of milk, cheese, yogurt, and other calcium-rich foods. Alternatively, and less preferably, they may need a daily supplement of 600 milligrams of calcium.

Vegetarians who exclude milk products need calcium-fortified foods such as soy milk. It is worth noting that not all soy milk is fortified with calcium; varieties fortified with calcium and vitamin D are recommended. Calcium-fortified orange juice offers folate and vitamin C as well as calcium.

FLUORIDE Mineralization of the fetus's teeth begins in the fifth month after conception. For this and for bone development, fluoride may be needed. Fluoride crosses the placenta, however, and whether the placenta can defend against excess intakes is questionable. Therefore, fluoride supplements are not recommended for pregnant women who drink fluoridated water. For women who live in communities without fluoridated water, a fluoride supplement may protect fetal teeth.

IRON The body conserves iron especially well during pregnancy: menstruation ceases, and absorption of iron increases up to threefold due to a rise in the blood's iron-absorbing and iron-carrying protein transferrin. Still, iron needs are so high that stores dwindle during pregnancy.

The developing fetus draws on the mother's iron stores to create stores of its own to last through the first four to six months of life. Iron losses also occur with the bleeding that is inevitable at birth.

Few women enter pregnancy with adequate iron stores. Women who enter pregnancy with iron-deficiency anemia have two to three times the normal risk of delivering low-birthweight or preterm infants.[8] For all women not taking supplements containing iron, a daily iron supplement containing

30 milligrams is recommended during the second and third trimesters of pregnancy.[9]

■■ ZINC Zinc is required for DNA and RNA synthesis and thus for protein synthesis. Low blood zinc predicts a low infant birthweight.[10] Zinc is most abundant in foods of high-protein content, such as shellfish, meat, and nuts, but the presence of other trace elements and fiber in foods may adversely affect zinc absorption. For example, iron interferes with the body's absorption and use of zinc, so women taking iron supplements (more than 30 milligrams per day) may also need zinc supplements.

■■ FOOD CHOICES Because food energy needs increase less than nutrient needs, the pregnant woman must select foods of high nutrient density. For most women, appropriate choices include foods such as fat-free milk, fat-free plain yogurt, lean meats, fish, eggs, dark green vegetables, vitamin C–rich fruits, legumes, and whole-grain breads and cereals. Table 10-4 provides a suggested food pattern.

■■ WIC A woman of limited financial means may need help in obtaining needed food and information. At the federal level, the WIC program provides nutrition education and nutritious foods to low-income pregnant women and their children. WIC provides eggs, milk, cereal, juice, cheese, legumes, and peanut butter to infants, children up to age five, and pregnant and breastfeeding women who qualify financially and are at medical or nutritional risk (see also Chapter 14 Nutrition in Practice). For infants given formula, WIC also provides iron-fortified formulas. Studies of the nutritional and health effects of WIC have found that participation in the program benefits both the iron status and the growth and development of infants and children. WIC participation during pregnancy reduces the risks of delivering preterm or low-birthweight infants.[11]

■■ FOOD CRAVINGS AND AVERSIONS Some women develop cravings for, or aversions to, certain foods and beverages during pregnancy. Individual **food cravings** during pregnancy do not seem to reflect real physiological needs. In other words, a woman who craves pickles does not necessarily need salt, nor does a chocolate craving indicate a need for sugar, caffeine, or fat. Similarly, cravings for ice cream are common during pregnancy, but do not signify a calcium deficiency. **Food aversions** and cravings that arise during pregnancy are probably due to hormone-induced changes in taste and sensitivities to smells.

In pregnancy, hemoglobin values of 12 g are not unusual, and 11 g is where the line defining "too low" is often drawn. Appendix E discusses more sensitive measures of iron status.

Food sources of iron:
- *Liver, oysters.*
- *Red meat, fish, other meat.*
- *Dried fruits (raisins, prunes).*
- *Legumes (dried beans, peas, lima beans).*
- *Dark green vegetables.*

Reminder: Vitamin C–rich foods enhance iron absorption from foods.

Zinc RDA during pregnancy: 12 mg/day.

WIC, pronounced WICK, is the acronym for the federal Special Supplemental Food Program for Women, Infants, and Children. The U.S. Department of Agriculture (USDA) funds WIC, and state health departments administer the program.

food craving: a deep longing for a particular food.

food aversion: a strong desire to avoid a particular food.

Cravings for nonfood items such as clay, ice, laundry starch, and cornstarch are known as **pica.**

Table 10-4
DAILY FOOD CHOICES FOR PREGNANT AND LACTATING WOMEN

Food Group	Number of Servings	
	Adults	Pregnant or Lactating Women
Breads/cereals	6 to 11	7 to 11
Vegetables	3 to 5	4 to 5
Fruits	2 to 4	3 to 4
Meat/meat alternates	2 to 3	3
Milk/milk products	2	3 to 4

NOTE: Figure 1-4 in Chapter 1 provides a detailed summary of foods in each group with serving sizes.

LEARNING LINK

LEARNING LINK

As you have just read, energy and nutrient needs during pregnancy are high. Chapters 2, 3, and 4 discussed the importance of the energy nutrients, and Chapters 8 and 9 focused on the major roles of the individual vitamins and minerals. These chapters also emphasized how all the nutrients in the body work together to promote growth and maintain health. Pregnancy provides a vivid example of how crucial each individual nutrient is to the function of other nutrients and to the optimal development of a brand new human being.

Weight Gain

All pregnant women must gain weight: fetal growth and maternal health depend on it. A pregnancy weight gain of 25 to 35 pounds is recommended for women who begin pregnancy at a healthy weight and are carrying a single fetus; for others, see the margin (on p. 237).[12] Some women should strive for gains at the upper ends of the target ranges, notably adolescents who are still growing themselves. Short women (5 feet 2 inches and under) should strive for lower gains. Figure 10-3 shows the components of a weight gain of 30 pounds.

Figure 10-3
COMPONENTS OF WEIGHT GAIN DURING PREGNANCY

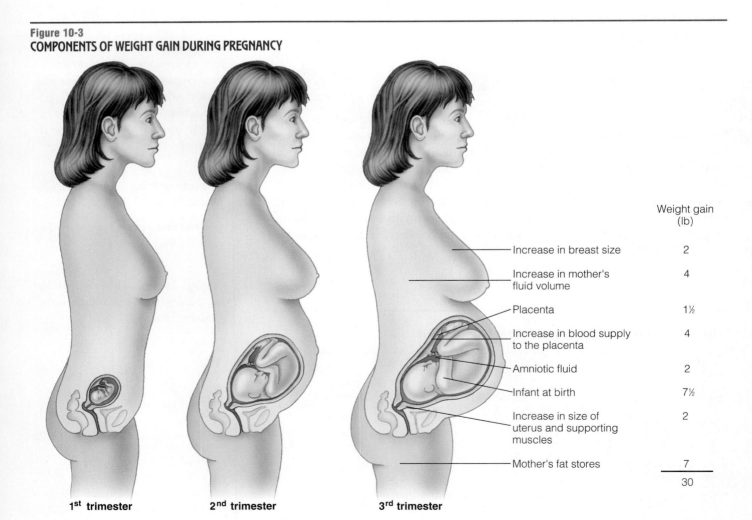

	Weight gain (lb)
Increase in breast size	2
Increase in mother's fluid volume	4
Placenta	1½
Increase in blood supply to the placenta	4
Amniotic fluid	2
Infant at birth	7½
Increase in size of uterus and supporting muscles	2
Mother's fat stores	7
	30

1ˢᵗ trimester   2ⁿᵈ trimester   3ʳᵈ trimester

The ideal pattern of weight gain during pregnancy is about 3½ pounds during the first three months and a pound per week thereafter. Women lose some of the weight gained during pregnancy at delivery and most of the remainder within the following few weeks or months, as blood volume returns to normal and accumulated fluids are lost.

If a woman has gained more than the expected amount of weight early in pregnancy, she should not try to diet in the last weeks. A sudden large weight gain, however, is a danger signal that may indicate the onset of hypertension (see p. 238).

Weight gain during pregnancy, like prepregnancy weight, directly relates to infant birthweight. If the mother does not gain all of the weight recommended, she may give birth to an underweight infant.

Physical Activity

Another lifestyle component that promotes health and well-being is physical activity. The active, physically fit woman experiencing a normal pregnancy can continue her regular activities throughout pregnancy, adjusting the duration and intensity as needed. Her rewards may include reduced stress, better weight control, an easier labor, better posture, less back pain, prevention of gestational diabetes (see the next section), an improved mood and body image, and a faster recovery from childbirth.[13] Potential risks of physical activity during pregnancy include acute conditions such as altered fetal heart rate and hyperthermia and other possible hazards such as premature labor or reduced birthweight.

As is true for everyone, the frequency, duration, and intensity of the activity affect the likelihood of the benefits or risks.[14] Physically fit, low-risk pregnant women can probably tolerate longer activity sessions and more exertion than is currently recommended.[15] Table 10-5 provides some guidelines for physical activity during pregnancy, but every pregnant woman should consult with her own health care provider regarding safe activity guidelines.

Problems in Pregnancy

Just as adequate nutrition and normal weight gain support the health of the mother and growth of the fetus, maternal diseases detract. If discovered early, many diseases can be controlled—another reason why early prenatal care is recommended. Some nutrition measures can help alleviate the most common problems encountered during pregnancy.

Weight-gain recommendations:
- Underweight women: 28 to 40 lb (12.5 to 18.0 kg).
- Normal-weight women: 25 to 35 lb (11.5 to 16.0 kg).
- Overweight women: 15 to 25 lb (7.0 to 11.5 kg).
- Obese women: 15 lb minimum (6.8 kg minimum).
- Twin birth: 35 to 45 lb (16.0 to 20.5 kg).

Weight-for-height categories for prepregnant women are based on body mass index (BMI) measures (see Appendix E):
- Underweight BMI: <19.8.
- Normal BMI: 19.8 to 26.0.
- Overweight BMI: 26.0 to 29.0.
- Obese BMI: >29.0.

A prenatal weight-gain grid (see Appendix E) plots the rate of weight gain during pregnancy.

Table 10-5
PHYSICAL ACTIVITY GUIDELINES FOR PREGNANCY

- Be physically active on a regular basis (at least three times a week), not intermittently.
- Stop exercising if you feel overheated.
- Drink plenty of fluids before, during, and after physical activity.
- Avoid exerting yourself in hot, humid weather; avoid overheating.
- Avoid jarring or jerky motions.
- Avoid any activity that has the potential to cause even mild abdominal trauma.
- Discontinue any activity that causes discomfort.
- Do not exercise while lying on your back after the fourth month.
- Do not allow your heart rate to exceed 150 beats per minute.
- Eat enough to support the energy needs of pregnancy and physical activity.

Pregnant women can enjoy the benefits of physical activity.

gestational diabetes: the detection of abnormal glucose tolerance during pregnancy.

cesarean section: surgical childbirth, in which the infant is delivered through an incision in the woman's abdomen.

preeclampsia: a condition characterized by hypertension, fluid retention, and protein in the urine.

▪▪ **GESTATIONAL DIABETES**   Pregnancy precipitates the onset of diabetes in some women because placental hormones alter the way insulin works or a woman may already have diabetes that is first detected during pregnancy. This condition is known as **gestational diabetes.** In many cases, blood glucose becomes abnormal during pregnancy but returns to normal after the infant is born. Almost one-third of all women with gestational diabetes, however, develop type 2 diabetes later in life, especially if they are overweight. To ensure that the problems of diabetes are dealt with promptly, the American Diabetes Association recommends that most women receive glucose tolerance tests (see Chapter 20) during the sixth month of pregnancy.[16]

Without proper management, gestational diabetes can lead to complications such as a high-birthweight infant and difficult delivery; rates of **cesarean section** delivery are also exceptionally high (almost 30 percent).[17] Children born to women with gestational diabetes are also at risk for obesity and abnormalities of glucose tolerance. An important aspect of nutrition management in gestational diabetes is prevention of excessive weight gain. Chapter 20 provides information about medical nutrition therapy for gestational diabetes. Regular physical activity may help to prevent excessive weight gain, but it must be individualized and medically monitored to ensure safety and effectiveness.[18] Insulin therapy may be required if blood glucose fails to normalize.

▪▪ **HYPERTENSION**   Hypertension complicates pregnancy and affects its outcome in different ways, depending on when the hypertension first develops and on how severe it becomes.[19] Hypertension can be a preexisting chronic condition that develops before a woman becomes pregnant or a transient condition that develops during the pregnancy and subsides after childbirth. In some cases, hypertension that develops during pregnancy warns of the ominous disorder preeclampsia.

▪▪ **PREEXISTING CHRONIC HYPERTENSION**   In addition to the health risks normally imposed by hypertension (heart attack and stroke), high blood pressure increases the risks of a low-birthweight infant or the separation of the placenta from the wall of the uterus before the birth, resulting in stillbirth. Ideally, before a woman with hypertension becomes pregnant, her blood pressure will be under control.

▪▪ **TRANSIENT HYPERTENSION OF PREGNANCY**   Some women first develop hypertension during the second half of pregnancy. Most often, the rise in blood pressure is mild and does not affect the pregnancy adversely. Blood pressure usually returns to normal during the first few weeks after childbirth. This transient hypertension of pregnancy differs from the pregnancy-induced hypertension that accompanies preeclampsia.*

▪▪ **PREECLAMPSIA**   Hypertension may signal the onset of **preeclampsia,** a condition characterized not only by high blood pressure but by protein in the urine and fluid retention (edema). Preeclampsia, which affects less than 10 percent of pregnant women, usually occurs with first pregnancies and almost always after 20 weeks' gestation. Symptoms typically regress within 48 hours of delivery. The edema of preeclampsia is a whole-body edema, distinct from the localized fluid retention women normally experience late in pregnancy.

---

*The Working Group on High Blood Pressure in Pregnancy, convened by the National High Blood Pressure Education Program of the National Heart, Lung, and Blood Institute, has suggested abandoning the term *pregnancy-induced hypertension* because it fails to differentiate between the mild, transient hypertension of pregnancy and the life-threatening hypertension of preeclampsia.

Preeclampsia affects almost all of the mother's organs—the circulatory system, liver, kidneys, and brain. If it progresses, she may experience convulsions; when this occurs, the condition is called **eclampsia.** Maternal mortality during pregnancy is rare in developed countries, but eclampsia is the most common cause. Preeclampsia demands prompt medical attention. Treatment focuses on regulating blood pressure and preventing convulsions.

:: MORNING SICKNESS Unlike the conditions just discussed, the nausea of "morning" (actually, anytime) sickness is usually benign, although distressing to some women. It arises from the hormonal changes taking place early in pregnancy, ranges from mild queasiness to debilitating nausea, and afflicts more than half of all pregnant women. One expert who has worked with women hospitalized with the condition notes that a trigger for morning sickness is what she calls the "radar nose of pregnancy."[20] Many women complain that smells, especially cooking smells, make them sick. Thus, minimizing odors is a key to alleviating morning sickness.

Traditional strategies for alleviating nausea are listed in the margin, but many women benefit most from simply eating the foods they want when they feel like eating. A "How to" box in Chapter 24 provides additional suggestions that may help to combat nausea.

:: HEARTBURN Heartburn, a burning sensation in the lower esophagus near the heart, is common during pregnancy and is also benign. As the growing fetus puts increasing pressure on the woman's stomach, acid may back up and create a burning sensation in her throat.

:: CONSTIPATION As the hormones of pregnancy alter muscle tone and the thriving infant crowds intestinal organs, an expectant mother may complain of constipation, another harmless but annoying condition. A high-fiber diet, physical activity, and a plentiful fluid intake will help relieve this condition. Also, responding promptly to the urge to defecate can help. Laxatives should be used only as prescribed by the physician. Mineral oil should not be used because it robs the body of fat-soluble vitamins.

LEARNING LINK

Maternal diseases such as diabetes and hypertension can compromise pregnancy outcome. Early prenatal care can help detect and control these conditions to ensure healthy pregnancy outcomes. Chapter 20 discusses nutrition, physical activity, and medical treatment for diabetes, and Chapter 21 offers details on treatment for hypertension.

:: Practices to Avoid

A general guideline for the pregnant woman is to eat a normal, healthy diet and practice moderation. A woman's daily choices during pregnancy take on enormous importance. Forewarned, pregnant women can choose to abstain from or avoid potentially harmful practices.

:: SMOKING Smoking adversely affects the pregnant woman's nutrition status, which in turn impairs fetal nutrition. Cigarette smoking increases iron needs and reduces the availability of vitamin B_{12}, vitamin C, folate, and zinc.[21] Smoking also restricts the blood supply to the growing fetus and so limits the delivery of oxygen and nutrients and the removal of wastes. Smoking

---

The normal edema of pregnancy responds to gravity: blood pools in the ankles. The edema of preeclampsia is a generalized edema. The distinction helps with diagnosis.

eclampsia: following preeclampsia, a severe stage in which convulsions occur.

To alleviate the nausea of pregnancy:
- On waking, arise slowly.
- Eat dry toast or crackers before getting out of bed.
- Chew gum or suck hard candies.
- Eat frequent small meals.
- Avoid foods with offensive odors.
- When nauseated, do not drink citrus juice, water, milk, coffee, or tea.
- Take prenatal vitamin and iron supplements on a full stomach or at a time of day when you feel well.
- Avoid shopping for or preparing food when feeling nauseous.

To prevent or relieve heartburn:
- Eat frequent small meals.
- Drink liquids between meals.
- Avoid spicy or greasy foods.
- Sit up while eating.
- Wait an hour after eating before lying down.
- Wait 2 hours after eating before exercising.

also causes these adverse effects: premature births, spontaneous abortions (fetal deaths), and increased risks of infants dying early in life.[22] In addition, sudden infant death syndrome (SIDS) has been positively linked to the mother's cigarette smoking during pregnancy and even to postnatal exposure to smoke in the household.[23] Research also shows that children born to women who smoke during pregnancy may be intellectually impaired.[24] Cigarette smoking is by far the single most important modifiable risk factor responsible for fetal growth retardation in developed countries.[25] The more a mother smokes, the smaller her baby will be. In short, the scientific and medical literature confirms over and over again that smoking during pregnancy is dangerous.

A table of the caffeine amounts in beverages and medicines is provided in Appendix A.

CAFFEINE Caffeine crosses the placenta, and the fetus has a limited ability to metabolize it. So far, no convincing evidence indicates that caffeine causes birth defects in human beings (as it does in animals), but limited evidence suggests that moderate-to-heavy use may lower infant birthweight.[26] One well-designed, carefully controlled study of caffeine's effects on pregnancy showed that moderate caffeine consumption (three cups of coffee daily) is not a risk factor for spontaneous abortion or growth retardation.[27] All things considered, it seems most sensible to limit caffeine consumption to the equivalent of one cup of coffee or two 12-ounce cola beverages a day.

MEDICATIONS Medications taken during pregnancy can cause complications and serious birth defects. For these reasons, women are advised to take medications only if their physicians deem it necessary to protect life and health. Even aspirin can do harm, as revealed by research on more than 3000 pregnant women.[28] Drug labels warn: "As with any drug, if you are pregnant or nursing a baby, seek the advice of a health professional before using this product." For aspirin and ibuprofen, an additional warning immediately follows: "It is especially important not to use aspirin (or ibuprofen) during the last three months of pregnancy unless specifically directed to do so by a doctor, because it may cause problems in the unborn child or excessive bleeding during delivery."

ILLICIT DRUGS The recommendation to avoid drugs during pregnancy includes illicit drugs, of course. Unfortunately, some pregnant women do use illicit drugs. One study of about 30,000 women in California found that 11 percent tested positive for alcohol and illicit drugs.[29] Marijuana or cocaine use during pregnancy adversely affects fetal growth and development.[30] Such drugs of abuse pass easily through the placenta and impair fetal development.[31] Moreover, infants born to drug users face low birthweight, cardiovascular problems, and increased risk of death. If they survive, their cries and behavior at birth are abnormal.[32] They may be hypersensitive or under-aroused; those who test positive for drugs suffer the greatest effects of toxicity and withdrawal.

DIETING Weight-loss dieting, even for short periods, is hazardous during pregnancy. Low-carbohydrate diets or fasts that cause ketosis deprive the growing fetal brain of needed glucose and may impair its development. Energy restriction during pregnancy is dangerous for all women, regardless of their prepregnancy weights.

ALCOHOL Drinking alcohol during pregnancy endangers the fetus. Alcohol can cause irreversible brain damage and mental and physical retardation in the fetus—the abnormalities that define **fetal alcohol syndrome (FAS).**[33] The potential for fetal damage arises when the mother's liver receives

fetal alcohol syndrome (FAS) the cluster of symptoms seen in a person whose mother consumed excess alcohol during her pregnancy; includes mental and physical retardation with facial and other body deformities.

more alcohol than it can detoxify. Alcohol-laden blood then circulates to all parts of the mother's body and freely crosses the placenta to impair fetal development. Alcohol also interferes with placental transport of nutrients to the fetus. FAS is known to occur with as few as 2 drinks a day.

Research using animals shows that one-fifth of the amount of alcohol needed to produce major, outwardly visible defects will surely produce learning impairment in the offspring, a condition known as **fetal alcohol effect (FAE).** Some children show no outward sign of the impairment, but the damage is there on the inside.[34] Others may be short in stature or display subtle facial abnormalities. Most suffer subtle brain damage and perform poorly in school and in social interactions. This early exposure to alcohol may also affect the person's response to both alcohol and other drugs later in life. Even before fertilization, alcohol may damage the ovum or sperm and so lead to abnormalities in the child.

Children born with FAS remain damaged. They may live, but they never fully recover. About 3 in every 1000 children are victims of this preventable damage, making FAS the leading known cause of mental retardation in the world. Moreover, for every baby diagnosed with FAS, three or four with FAE may go undiagnosed until problems develop later in the preschool years. Upon reaching adulthood, such children are ill equipped for employment, relationships, and the other facets of life most adults take for granted.

The Surgeon General has issued a statement that pregnant women should drink absolutely no alcohol. All containers of beer, wine, and distilled liquor now must carry a warning to this effect.

Adolescent Pregnancy

Each year in the United States, about one million adolescent girls between the ages of 12 and 19 become pregnant.[35] Of these, about half choose to continue their pregnancies.[36] Many teenage women, especially the youngest ones, have not had time to store the nutrients needed to support their own rapid growth and development, much less nutrients needed to support pregnancy and the developing fetus. Nutrient shortages place both mother and infant at risk. Pregnant adolescents have more miscarriages, premature births, stillbirths, and low-birthweight infants than do pregnant adult women.[37] Their greatest risk, though, is death of the infant: mothers under 16 bear more babies who die within the first year than do women in any other age group. Clearly, teenage pregnancy is a major public health problem.

Pregnant adolescents suffer many illnesses. The rates of preeclampsia are 50 percent higher in teens than in older women. Other common problems of teen pregnancies are iron-deficiency anemia and prolonged labor.[38]

To support their own and fetal needs, young teenagers (13 to 16 years old) are encouraged to strive for pregnancy weight gains at the upper ends of the ranges recommended for pregnant women.[39] Those who gain between 30 and 35 pounds during pregnancy have lower risks of delivering low-birthweight infants. Adequate nutrition can substantially improve the health of the mother and infant; it is an indispensable component of prenatal care.[40] Pregnant and lactating adolescents can use the Daily Food Guide presented in Table 10-4 (on p. 235), making sure to select at least 4 servings of milk or milk products daily.

Breastfeeding

The American Academy of Pediatrics (AAP) recommends that infants receive breast milk for the first 12 months of life.[41] The American Dietetic Association

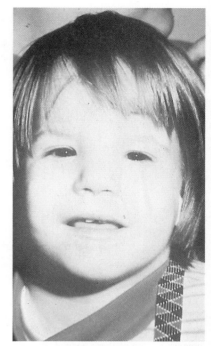

These facial traits are typical of fetal alcohol syndrome, caused by drinking during pregnancy—low nasal bridge, short eyelid opening, underdeveloped groove in center of the upper lip, small midface, short nose, and small head circumference.

fetal alcohol effect (FAE): partial abnormalities from prenatal alcohol exposure that are not sufficient for diagnosis with FAS, but are impairing to the child. Also called *alcohol-related birth defects (ARBD)* or *subclinical FAS.*

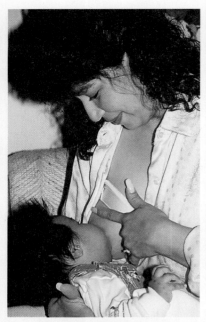

Breastfeeding is a natural extension of pregnancy—the mother's body continues to nourish the infant.

lactation: production and secretion of breast milk for the purpose of nourishing an infant.

(ADA) advocates breastfeeding for the nutritional health it confers on the infant as well as for the physiological, social, economic, and other benefits it offers the mother.[42] Breast milk's unique nutrient composition and protective factors promote optimal infant health and development. The only acceptable alternative to breast milk is iron-fortified formula. Adequate nutrition of the mother supports successful **lactation,** and without it, lactation is likely to falter or fail.

Nutrient Needs during Lactation

By continuing to eat nutrient-dense foods, not restricting energy intake unduly, and drinking fluid at frequent intervals throughout lactation, the mother who chooses to breastfeed her infant will be nutritionally prepared to do so. An inadequate diet does not support the stamina, patience, and self-confidence that nursing an infant demands. Figure 10-2 (on p. 232) shows the differences between a lactating woman's nutrient needs and those of a nonpregnant woman, and Table 10-4 (on p. 235) shows a food pattern that meets those needs.

ENERGY A nursing mother produces about 25 ounces of milk a day, more or less, depending primarily on the infant's demand for milk.[43] To produce milk, a woman needs extra energy—almost 650 kcalories a day above her regular need during the first six months of lactation. To meet this energy need, the woman is advised to eat an extra 500 kcalories of foods each day and let the fat reserves she accumulated during pregnancy provide the rest. Some research suggests that many women may need less energy for milk production than current recommendations.[44] Severe energy restriction (less than 1200 kcalories per day), however, hinders milk production and can compromise the mother's health.[45]

MATERNAL WEIGHT LOSS The period of lactation is the natural time for a woman to lose the extra body fat she accumulated during pregnancy. If she chooses nutrient-dense foods, she will gradually lose weight, even though her energy intake may be greater than normal. In one study, researchers found that breastfeeding or a combination of breastfeeding and formula feeding led to faster loss of body weight during the first month after women delivered their babies than formula feeding alone.[46] After the first month, however, no significant effect of breastfeeding on weight loss was observed. Results of other studies examining the relationship between feeding method and loss of body weight and body fat have been inconsistent.[47] In most studies where breastfeeding duration was three months or longer, researchers found that lactation had a positive influence on weight loss.[48] Most women lose 1 to 2 pounds a month during the first four to six months of lactation; some may lose more, and others may maintain or even gain weight.

SUPPLEMENTS Most lactating women can obtain all the nutrients they need from a well-balanced diet without taking vitamin-mineral supplements. If a woman's diet is inadequate in one or more nutrients, eating foods rich in those nutrients or taking appropriate nutrient supplementation may be advised. Women can produce milk that contains adequate protein, carbohydrate, fat, folate, and most minerals, even when their own supplies are limited.[49] The milk's quality is maintained at the expense of maternal stores. The nutrients in breast milk that are most likely to decline in response to prolonged inadequate maternal intakes are vitamins A, B_6, B_{12}, and D.[50]

WATER Despite previous misconceptions, drinking more fluid does not enable a mother to produce more breast milk. Nevertheless, a lactating woman needs to drink at least 2 quarts of liquids each day to protect herself from dehydration. A convenient way to ensure adequate fluid consumption is to drink a glass of milk, juice, or water each time the infant nurses, as well as with each meal and snack.

:: PARTICULAR FOODS Foods with strong or spicy flavors (such as garlic or onion) may alter the taste of breast milk. A sudden change in the taste of the milk may annoy some infants. Infants who are sensitive to particular foods such as cow's milk protein may become uncomfortable when the mother's diet includes these foods. Most nursing mothers should drink cow's milk, however; the nutrients it provides make a significant contribution to both the infant's and the mother's health.

:: Contraindications to Breastfeeding

Some substances impair maternal milk production or enter the breast milk and interfere with infant development. Some medical conditions prohibit breastfeeding.

:: ALCOHOL In the past, alcohol has been recommended to lactating mothers to ease milk production, despite a lack of scientific support for such recommendations. Alcohol easily enters breast milk and can adversely affect production, volume, composition, and ejection of breast milk, as well as overwhelm an infant's immature alcohol-degrading system.[51] Alcohol concentration in breast milk peaks within one hour after ingestion of even moderate amounts (equivalent to a can of beer). The alcohol may alter the taste of the milk to the disapproval of the nursing infant, who may, in protest, drink less milk than normal. Alcohol consumption by lactating women should be discouraged.

:: CAFFEINE Excessive caffeine consumption during lactation may cause irritability and wakefulness in the breastfed infant. As during pregnancy, caffeine consumption should be moderate.

:: SMOKING Nurses and other health care professionals should actively discourage smoking by lactating women. Research shows that lactating women who smoke produce less milk, and milk with a lower fat content, than mothers who do not smoke.[52] Smoking also exerts numerous harmful effects on both mother and child.[53]

:: MATERNAL ILLNESS If a woman has an ordinary cold, she can continue nursing without worry. If susceptible, the infant will catch it from her anyway and, thanks to the immunological protection of breast milk, may be less susceptible than a formula-fed baby would be. If a woman has a communicable disease such as tuberculosis or hepatitis, which could threaten the infant's health, then mother and baby must be separated. Breastfeeding would be possible only by pumping the mother's breasts several times a day. The Centers for Disease Control recommend that where safe alternatives are available, women who test positive for human immunodeficiency virus (HIV), the virus that causes acquired immune deficiency syndrome (AIDS), should not breastfeed their infants.

In 1998, the United Nations Children's Fund (UNICEF) and the World Health Organization (WHO) revised their guidelines on feeding infants of HIV-positive mothers in developing countries. Previously, in settings where infants who are not breastfed run a particularly high risk of dying from infectious disease and malnutrition, breastfeeding was recommended for all mothers, even those known to be HIV-infected. The current guidelines, however, recommend "avoidance of breastfeeding" as an intervention to prevent mother-to-child transmission.[54] Thus, UNICEF and WHO encourage formula feeding for infants of HIV-infected women even in developing countries.

███ **MEDICATIONS AND ILLICIT DRUGS** Similarly, if a nursing mother must take medication that is secreted in breast milk and that is known to affect the infant, then breastfeeding is contraindicated.[55] Many prescription medications do not reach nursing infants in sufficient quantities to affect them adversely. Some, however, do. As a precaution, a nursing mother should consult with the prescribing physician prior to taking any medication.

Drug addicts, including alcohol abusers, are capable of taking such high doses that their infants can become addicts by way of breast milk. In these cases, too, breastfeeding is contraindicated.

A lactating woman is wise to avoid oral contraceptives until after she has weaned her infant and to use another method of contraception in the meantime. Standard oral contraceptives contain estrogen, which reduces both the volume and the protein content of breast milk.[56]

██ Nutrition of the Infant

Early nutrition affects later development, and early feeding sets the stage for eating habits that will influence nutrition status for a lifetime. Trends change, and experts argue about the fine points, but properly nourishing an infant is relatively simple, overall. Common sense in the selection of infant foods and a nurturing relaxed environment go far to promote an infant's health and well-being.

██ Nutrient Needs during Infancy

An infant grows faster during the first year than ever again, as Figure 10-4 shows. The growth of infants and children directly reflects their nutritional well-being and is an important parameter in assessing their nutrition status. Health care professionals use growth charts to evaluate the growth and development of children from birth to 18 years of age (see Appendix E).

███ **NUTRIENTS TO SUPPORT GROWTH** An infant's birthweight doubles by about four to five months of age, and it triples by the age of one year. (Consider that if an adult, starting at 120 pounds, were to do this, the person's weight would increase to 360 pounds in a single year.) By the end of the first year, the growth rate slows considerably. Between the first and second birthdays, the weight gained amounts to less than 10 pounds.

A newborn infant requires only about 650 kcalories per day, whereas most adults require about 2000 kcalories per day. In comparison to body weight, however, the difference is remarkable. Infants require about 100 kcalories per kilogram of body weight per day; most adults require fewer than 40. A 170-pound adult who tried to eat like an infant would have to ingest over 7000 kcalories a day! Figure 10-5 compares a five-month-old infant's needs per kilogram of body weight with those of an adult male; as you can see, some of the differences are extraordinary. After six months, energy needs increase less rapidly as the growth rate begins to slow, but some of the energy saved by slower growth is spent in increased activity.

███ **WATER** The most important nutrient of all, for infants as for everyone, is the one easiest to forget: water. Conditions that cause fluid loss, such as vomiting, diarrhea, sweating, or obligatory urinary loss without

Figure 10-4
WEIGHT GAIN OF HUMAN INFANTS IN THEIR FIRST FIVE YEARS OF LIFE

In the first year, an infant's birthweight may triple, but over the following several years, the rate of weight gain gradually diminishes.

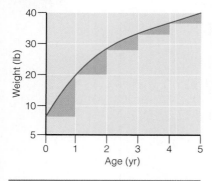

After six months, the energy saved by slower growth is spent in increased activity.

	Infants	Adults
Heart rate (beats/minute)	120 to 140	70 to 80
Respiration rate (breaths/minute)	20 to 40	15 to 20
Energy needs (kcal/body weight)	45/lb (100/kg)	<18/lb (<40/kg)

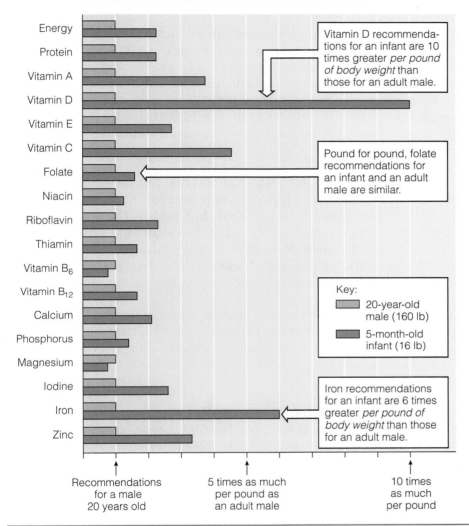

Figure 10-5

NUTRIENT NEEDS OF A FIVE-MONTH-OLD INFANT AND AN ADULT MALE COMPARED ON THE BASIS OF BODY WEIGHT

Infants may be relatively small and inactive, but they use large amounts of energy and nutrients, in proportion to their body size, to keep all their metabolic processes going.

replacement, can rapidly propel an infant into life-threatening dehydration (see Nutrition in Practice 9). In early infancy, breast milk or formula normally provides enough water for a healthy infant to replace water losses from the skin, lungs, feces, and urine.[57] An infant who is exposed to hot weather, has diarrhea, or vomits repeatedly, however, needs supplemental water to prevent dehydration. Infants cannot tell you what they are crying for; remember that they may need water, and let them drink until they quench their thirst.

In developed countries with well-nourished populations, such as the United States and Canada, the dietary practices that influence infants' nutrition status the most center upon which type of milk the infant receives and the age at which solid foods are introduced. The remainder of this discussion is devoted to feeding the infant and identifying the nutrients most often deficient in infant diets.

Figure 10-6
PERCENTAGES OF ENERGY-YIELDING NUTRIENTS IN BREAST MILK AND IN RECOMMENDED ADULT DIETS

The balance of energy-yielding nutrients in human breast milk is ideal for infants and does not resemble the balance recommended for adults.

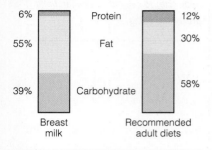

alpha-lactalbumin (lackt-AL-byoo-min): the chief protein in human breast milk, as **casein (CAY-seen)** is the chief protein in cow's milk.

Breast Milk

Breast milk excels as a source of nutrients for the young infant. With the possible exception of vitamin D, breast milk provides all the nutrients a healthy infant needs for the first four to six months of life. It provides many other health benefits as well.

ENERGY NUTRIENTS The energy nutrient balance of breast milk differs dramatically from the balance recommended for adults (see Figure 10-6). Yet for infants, breast milk is the most nearly perfect food, proving that people at different stages of life really do have different nutrient needs.

Tailor-made to meet the nutrient needs of the human infant, breast milk offers its carbohydrate as lactose and its fat as a mixture with a generous proportion of the essential fatty acid linoleic acid. The unique composition of the fat in breast milk, in combination with the fat-digesting enzymes present, contributes to highly efficient fat absorption by the breastfed infant. The protein in breast milk is largely **alpha-lactalbumin,** a protein the human infant can easily digest.

VITAMINS AND MINERALS The vitamin content of breast milk is ample. Even vitamin C, for which cow's milk is a poor source, is supplied generously by the breast milk of a well-nourished mother. The concentration of vitamin D in breast milk, however, is low, and vitamin D deficiency causes impaired bone mineralization in children. Manufacturers fortify cow's milk and infant formulas with vitamin D, and physicians may prescribe vitamin D supplements for dark-skinned breastfed infants who do not receive sufficient exposure to sunlight.

Calcium, phosphorus, and magnesium are present in breast milk in amounts appropriate for the rate of growth expected in a human infant. Breast milk is healthfully low in sodium, and its iron is highly absorbable. Its zinc, too, is absorbed better than zinc from cow's milk, thanks to the presence of a zinc-binding protein.

SUPPLEMENTS FOR INFANTS Pediatricians may prescribe supplements containing vitamin D, iron, and fluoride (after six months of age). Table 10-6 offers a schedule of supplements during infancy.

Table 10-6
SUPPLEMENTS FOR FULL-TERM INFANTS

	Vitamin D[a]	Iron[b]	Fluoride[c]
Breastfed infants:			
Birth to six months of age	√		
Six months to one year	√	√	√
Formula-fed infants:			
Birth to six months of age			
Six months to one year		√	√

[a]Vitamin D supplements are recommended for infants whose mothers are vitamin D deficient and for those who do not receive adequate exposure to sunlight.
[b]Infants four to six months of age need additional iron, preferably in the form of iron-fortified cereal for both breastfed and formula-fed infants and iron-fortified infant formula for formula-fed infants.
[c]The Committee on Nutrition of the American Academy of Pediatrics recommends initiating fluoride supplements at six months of age for breastfed infants, formula-fed infants who receive ready-to-use formulas (these are prepared with water low in fluoride), and those who receive formula mixed with water that contains little or no fluoride (less than 0.3 ppm).
SOURCES: Adapted from American Academy of Pediatrics, Committee on Nutrition, Nutrition and oral health, in *Pediatric Nutrition Handbook,* 4th ed., ed. R. E. Kleinman (Elk Grove Village, Ill.: American Academy of Pediatrics, 1998), pp. 523–529; American Academy of Pediatrics, Committee on Nutrition, Fluoride supplementation for children: Interim policy recommendations, *Pediatrics* 95 (1995): 777.

In addition, as discussed in Chapter 8, the AAP recommends a single dose of vitamin K at birth.[58] In many states, this preventive dose of vitamin K is required by law.

IMMUNOLOGICAL PROTECTION Breast milk offers the infant unsurpassed protection against infection.[59] Protective factors include antiviral agents, antibacterial agents, and other infection inhibitors.

During the first two or three days of lactation, the breasts produce **colostrum,** a premilk substance containing antibodies and white cells from the mother's blood. Colostrum is relatively sterile as it leaves the breast, and the infant cannot contract a bacterial infection from it even if the mother has one. Colostrum contains maternal immune factors that inactivate harmful bacteria within the digestive tract. Later, breast milk also delivers immune factors, although not as many as colostrum. Among them are **bifidus factors** and **lactoferrin.**

Breast milk also contains several enzymes, several hormones (including thyroid hormone and prostaglandins), and lipids, all of which protect the infant against infection. Research suggests that breastfeeding offers better protection against wheezing during the first few months of life than formula feeding does.[60] It seems, too, that breastfed babies are less prone to develop stomach and intestinal disorders during the first few months of life and so experience less vomiting and diarrhea than formula-fed infants do.[61] In fact, research shows that breast milk contains not only antibodies against the most common cause of diarrhea in infants and young children but also another factor that binds to, and inhibits replication of, the infective agent.*[62] Breastfeeding reduces the severity and duration of symptoms associated with this infection. Breastfeeding also protects against other common illnesses of infancy such as middle ear infection and respiratory illness.[63]

Some research shows that children with diabetes (type 1) almost always have antibodies to cow's milk protein in their pancreatic tissues.[64] This finding leads some to suspect that early feeding of cow's milk formula may set the stage for abnormal immune functioning that causes diabetes later on. Other researchers disagree.[65] In any case, breastfeeding prevents early exposure to cow's milk protein. Much remains to be learned about the composition and characteristics of human milk, but clearly it is a very special substance. Nutrition in Practice 10 offers suggestions for successful breastfeeding.

The case study on p. 248 presents a woman who is four months pregnant. Answering the questions offers practice in thinking through some of the issues related to pregnancy and breastfeeding.

Infant Formula

Breastfeeding offers many benefits to both mother and infant, and it should be encouraged whenever possible. The mother who has decided to use formula, however, should be supported in her choice just as the breastfeeding mother should be. She can offer the same closeness, warmth, and stimulation during feedings as the breastfeeding mother can.

Many mothers choose to breastfeed at first but wean their children within the first 1 to 12 months. Before infants reach a year of age, mothers

colostrum (co-LAHS-trum): a milklike secretion from the breast that is rich in protective factors. It is present during the first day or so after delivery, before milk appears.

bifidus (BIFF-id-us, by-FEED-us) factors: factors in colostrum and breast milk that favor the growth of the "friendly" bacterium *Lactobacillus* (lack-toe-ba-SILL-us) *bifidus* in the infant's intestinal tract. This bacterium prevents other, less desirable intestinal inhabitants from flourishing.

lactoferrin (lack-toe-FERR-in): a factor in breast milk that binds iron and keeps it from supporting the growth of the infant's intestinal bacteria.

---

*The most common cause of diarrhea in the United States is rotavirus. More children are hospitalized for rotavirus infection than for any other single cause.

Case Study

PREGNANT WOMAN WITH WEIGHT PROBLEM

Ellen is a 24-year-old woman who is four months pregnant. This is her first pregnancy, and she is eager to learn how to feed herself during pregnancy as well as her infant after birth. She is 5 feet 3 inches tall and currently weighs 150 pounds. Her prepregnancy weight was 148 pounds. Ellen is very concerned about her 2-pound weight gain.

✚ Assess Ellen's weight and determine what her desirable weight would be if she were not pregnant.

✚ Should Ellen be concerned about her 2-pound weight gain? Why or why not?

✚ What advice should you give Ellen about her weight gain during pregnancy?

✚ What other dietary advice would you give her?

✚ Discuss methods of infant feeding with Ellen and describe some of the advantages breastfeeding would offer her.

✚ What are the advantages of formula feeding?

✚ What advice will you give Ellen if she decides to breastfeed?

✚ What information should Ellen have about formula feeding?

Formula preparation:
- **Liquid concentrate (moderately expensive, relatively easy)**—mix with equal part water.
- **Powdered formula (least expensive, lightest for travel)**—read label directions.
- **Ready-to-feed (easiest, most expensive)**—pour directly into clean bottles.

The infant thrives on infant formula offered with affection.

must wean them onto *infant formula,* not onto plain cow's milk of any kind—whole, reduced fat, low fat, or fat-free.

▪▪ **INFANT FORMULA COMPOSITION** Manufacturers can prepare formulas from cow's milk in such a way that they do not differ significantly from human milk in nutrient content. Figure 10-7 illustrates the energy nutrient balance of both. Formulas contain no protective antibodies for human babies, but preventive medical care (vaccinations) and reliable public health measures (clean water) help minimize this disadvantage. The educated mother whose water supply is reliable can prepare safe, sanitary formulas. Lead-contaminated water, however, is a major source of lead poisoning in infants.[66]

▪▪ **INFANT FORMULA STANDARDS** National and international standards have been set for the nutrient contents of infant formulas. U.S. standards are based on AAP recommendations, and the Food and Drug Administration (FDA) mandates quality control procedures to ensure that these standards are met. All standard formulas are therefore nutritionally similar. Small differences in nutrient content are sometimes confusing, but usually not important.

▪▪ **SPECIAL FORMULAS** Standard formulas are inappropriate for some infants (see Figure 10-8 on p. 250). For example, premature babies require special formulas. Many infants allergic to milk protein can drink special formulas based on soy protein. Some infants allergic to cow's milk protein, however, may be allergic to soy protein as well. Soy formulas are lactose-free and so can be used for infants with lactose intolerance as well. For infants with other special needs, many other variations have been formulated.

▪▪ **RISKS OF FORMULA FEEDING** In developing countries and in poor areas of the United States, formula may be unavailable, overdiluted in an attempt to save money, or prepared with contaminated water. Overdilution of formula can cause malnutrition and growth failure. Contaminated formula often causes infections leading to diarrhea, dehydration, and failure to absorb nutrients. Wherever sanitation is poor, breastfeeding should take priority over feeding formula. Breast milk is sterile, and its antibodies enhance an infant's resistance to disease.

■■ **IRON IN FORMULA**   The AAP recommends iron-fortified formula for all formula-fed infants. Low-iron infant formulas have no role in infant feeding. Use of iron-fortified formulas has risen in recent decades and is credited with the decline of iron-deficiency anemia in U.S. infants. Only iron-fortified formula can support normal development in an infant's first year.

■■ **NURSING BOTTLE TOOTH DECAY**   Dentists advise against putting an infant to bed with a bottle. Salivary flow, which normally cleanses the mouth, diminishes as the baby falls asleep. Sucking for long times pushes the jawline out of shape and causes a bucktoothed profile (protruding upper and receding lower teeth). Furthermore, prolonged sucking on a bottle of formula, milk, or juice bathes the upper teeth in a carbohydrate-rich fluid that nourishes decay-producing bacteria. (The tongue covers and protects most of the lower teeth, but they, too, may be affected.) The result is extensive and rapid tooth decay. To prevent **nursing bottle tooth decay,** no child should be put to bed with a bottle as a pacifier.

■■ The Transition to Cow's Milk

During an infant's first six months, formula must supply the nutrients of human milk in similar forms and proportions. Ordinary milk is an inappropriate replacement—primarily because cow's milk provides too little vitamin C and iron and too much sodium and protein. After the first year, the exact formulation of the milk selected is less critical, but milk or a suitable substitute still occupies a place in the diet that no other type of food can fill. Children one to two years of age should not drink reduced-fat, low-fat, or fat-free milk routinely; they need the fat of whole milk. If powdered milk is used, the fat-containing varieties should be chosen.

■■ Introducing First Foods

Changes in the body organs during the first year affect the infant's readiness to accept solid foods. The immature stomach and intestines can digest milk sugar (lactose), but they cannot digest starch until they are several months old. This is one of the many reasons why breast milk and formula are such good foods for a baby; they provide simple, easily digested carbohydrate that supplies energy for the baby's growth and activity.

■■ **THE NEED FOR WATER**   The infant's kidneys are unable to concentrate waste efficiently, so an infant must excrete relatively more water than an adult to carry off a comparable amount of waste. The risk of dehydration is ever present, and it becomes greater once solid foods are introduced. Foods high in protein or electrolytes such as meat and eggs can cause dehydration if offered without water. Water should be offered to infants regularly once they are eating solid food. Water also provides fluid without additional food energy. Many adults would no doubt be healthier had they learned early to quench their thirst with water.

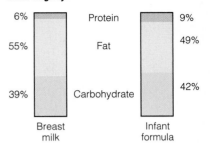

Figure 10-7
PERCENTAGES OF ENERGY-YIELDING NUTRIENTS IN BREAST MILK AND IN INFANT FORMULA

The proportions of energy-yielding nutrients in human breast milk and formula differ slightly.

	Breast milk		Infant formula
Protein	6%		9%
Fat	55%		49%
Carbohydrate	39%		42%

nursing bottle tooth decay: extensive tooth decay due to prolonged tooth contact with formula, milk, fruit juice, or other carbohydrate-rich liquid offered to an infant in a bottle.

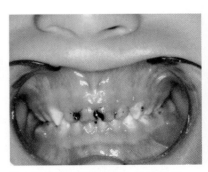

Nursing bottle tooth decay, an extreme example.

Figure 10-8
CHOOSING A FORMULA

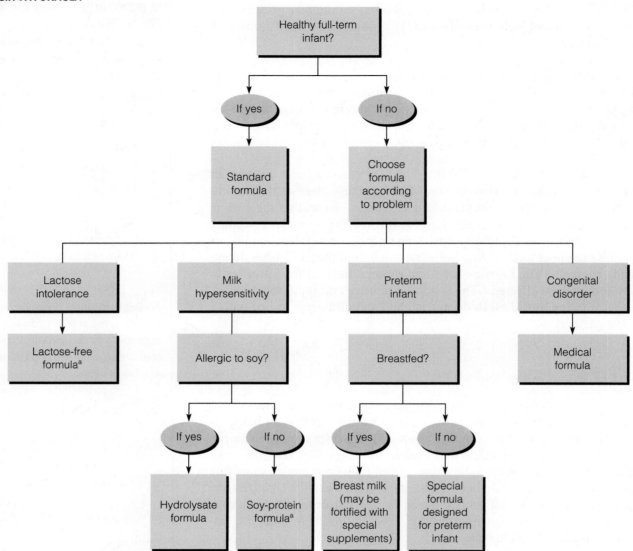

ᵃManufacturers design soy-based formulas for infants with milk sensitivities—whether lactose intolerance or milk allergy. These formulas use corn syrup or sucrose in place of lactose.

■■ **WHEN TO INTRODUCE SOLID FOOD** For an infant receiving formula or breast milk from a healthy, well-nourished mother, additions to the diet are not needed until the infant is four to six months old. Foods may be started gradually beginning sometime between four and six months, depending on the infant's readiness. Indications of readiness for solid foods include:

- The infant can sit with support and control head movements.
- The infant is six months old.

Infants vary, and the program of introducing solid foods depends on the individual infant's developmental readiness, not on any rigid schedule. Table 10-7 presents a suggested sequence for introducing new foods.

The addition of foods to an infant's diet should be governed by three considerations: the infant's nutrient needs, the infant's physical readiness to handle different forms of foods, and the need to detect and control allergic reactions. With respect to nutrient needs, the nutrient needed earliest is iron, then vitamin C.

■ FOODS TO PROVIDE IRON AND VITAMIN C   Iron deficiency is common in young children throughout the world, especially between the ages of six months and three years when they are growing fast and milk, which is a poor source of iron, has a large place in their diets. The iron an infant has stored from before birth typically runs out after the birthweight doubles, long before the end of the first year. This is why cow's milk should not be offered during the first year: it not only displaces breast milk or iron-fortified formula but also causes GI blood loss in many infants. Infants can derive adequate iron first from breast milk or formula with iron, then from iron-fortified cereals, and later from meat or meat alternates such as legumes. Once infants are consuming iron-fortified cereals, parents or caregivers should begin selecting vitamin C–rich foods to go with meals to enhance iron absorption. The best sources of vitamin C are fruits and vegetables.

Vegetables are sometimes introduced to infant diets after fruits, but the opposite order may better favor vegetable acceptance. Once an infant is accustomed to the sweet taste of fruits, the tastes of some vegetables may be less appealing than if they had been offered prior to the fruits.

■ PHYSICAL READINESS FOR SOLID FOODS   The ability to swallow solid food develops at around four to six months, and food offered by spoon helps to develop swallowing ability. Six months is also a good time to introduce the cup. Fruit juices should be diluted with water and served in the cup, not a bottle, once the infant is six months of age or older. Juices should be used moderately, so as not to displace other foods. At eight months to a year, a baby can sit up, can handle finger foods, and begins to teethe. At that time, hard crackers and other hard finger foods may be introduced to promote the development of manual dexterity and control of the jaw muscles. These feedings must occur under the watchful eye of an adult because the infant can also choke on such foods.

Some parents want to feed solids at an earlier age, on the theory that "stuffing the baby" at bedtime promotes sleeping through the night. There is

---

Table 10-7
FIRST FOODS FOR THE INFANT

Breast milk or iron-fortified formula is the only source of nourishment for the first 4 to 6 months. Throughout the first year, the infant's intake of breast milk or iron-fortified formula will gradually decline as solid food intake increases.

Age (mo)	Addition
4 to 6	Iron-fortified rice cereal, followed by other single-grain cereals, mixed with breast milk, formula, or water Pureed vegetables and fruits, one by one (perhaps vegetables before fruits, so the infant will learn to like their less sweet flavors)
6 to 8	Infant breads and crackers Mashed vegetables and fruits, and their juices[a]
8 to 10	Breads and cereals from the table Soft, cooked vegetables and fruit from the table Finely cut meats, fish, chicken, casseroles, cheeses, yogurts, tofu, eggs, and legumes
10 to 12	Continue to introduce a variety of nutritious foods

[a]All baby juices are fortified with vitamin C. Orange juice may cause allergies; apple juice may be a better choice at first. Dilute juices with water and offer in a cup to prevent nursing bottle tooth decay.
SOURCE: Adapted in part from American Academy of Pediatrics, Committee on Nutrition, *Pediatric Nutrition Handbook*, 4th ed., ed. R. E. Kleinman (Elk Grove Village, Ill.: American Academy of Pediatrics, 1998), pp. 43–53.

no proof for this theory, however. On the average, infants start to sleep through the night at about the same age (three to four months) regardless of when solid foods are introduced.

▪▪ **ALLERGY-CAUSING FOODS** New foods should be introduced singly and at intervals spaced to permit detection of allergies. For example, when cereals are introduced, rice cereal is offered first for several days; it causes allergy least often. Wheat cereal is offered last; it is the most common offender. If a cereal causes an allergic reaction (irritability due to skin rash, digestive upset, or respiratory discomfort), its use should be discontinued before going on to the next food.

▪▪ **CHOICE OF INFANT FOODS** Baby foods commercially prepared in the United States and Canada are safe, and except for mixed dinners and heavily sweetened desserts, they generally have high nutrient density. An alternative for the parent who wants the infant to have family foods is to "blenderize" a small portion of the table food (cooked without salt) at each meal.

▪▪ **FOODS TO OMIT** Sweets of any kind (including baby food "desserts") have no place in an infant's diet. The added food energy conveys few if any nutrients to support growth and contributes to obesity. Canned vegetables are also inappropriate for infants; they often contain too much sodium. Honey should never be fed to infants because of the risk of botulism. Infants and even young children have difficulty chewing and swallowing foods such as popcorn, whole grapes, whole beans, hot dog slices, and nuts; consequently, they can easily choke on these foods. Also, an infant's caregiver must be on guard against food poisoning and take precautions against it as described in Nutrition in Practice 5.

▪▪ **FOODS AT ONE YEAR** Whole milk is the best food to supply most of the nutrients the infant needs at one year of age; 2 to 3½ cups a day meet those needs sufficiently. More milk than this displaces iron-rich foods and can lead to the iron-deficiency anemia known as milk anemia. Other foods—meat and meat alternates, iron-fortified cereal, enriched or whole-grain bread, fruits, and vegetables—should be supplied in variety and in amounts sufficient to round out total energy needs. Ideally, a one-year-old will sit at the table, eat many of the same foods everyone else eats, and drink liquids from a cup—not a bottle. The sample menu shows a meal plan that meets a one-year-old's requirements.

Reminder: Milk anemia develops when an excessive milk intake displaces iron-rich foods from the diet.

▪▪ **Looking Ahead**

Probably the most important single measure to undertake during the first year is to encourage eating habits that will support continued normal weight as the child grows. This means introducing a variety of nutritious foods in an inviting way, not forcing the infant to finish the bottle or the baby food jar, avoiding concentrated sweets and empty-kcalorie foods, and encouraging physical activity. Parents should avoid teaching infants to seek food as a reward, to expect food as comfort for unhappiness, or to associate food deprivation with punishment. If infants cry from thirst, give them water, not milk or juice. Infants seem to have no internal "kcalorie counter," and they stop eating when their stomachs feel full. Nutrient-dense, low-kcalorie foods will satisfy as long as they provide bulk.

Normal dental development is also promoted by supplying nutritious foods, avoiding sweets, and discouraging the association of food with reward or comfort. Dental health is the subject of Nutrition in Practice 2.

The AAP recommends against a fat-modified diet during infancy. The available evidence does not warrant dietary manipulation to lower serum cholesterol in infants.

Ideally, a one-year-old eats many of the same foods as the rest of the family.

SAMPLE MENU FOR A ONE-YEAR-OLD DIET MENU	Breakfast	Morning Snack	Lunch	Afternoon Snack	Dinner
	½ c whole milk	½ c yogurt	1 c whole milk	½ c whole milk	1 c whole milk
	½ c cereal	1 to 2 tbs fruit	2 to 3 tbs vegetables	½ slice toast	1 egg or ¼ c tofu
	1 to 2 tbs fruit	Teething crackers	2 tbs chopped meat or well-cooked, mashed legumes	1 tbs peanut butter	½ c potato, rice or pasta 2 to 3 tbs vegetables 2 to 3 tbs fruit

NOTE: Fruit choices should include citrus fruits, melons, and berries, and vegetable choices should include dark green, leafy and deep yellow vegetables.

■■ Mealtimes

The wise parent of a one-year-old offers nutrition and love together. Both promote growth. Children "fed with love" grow more both in weight and in height than children fed the same food in an emotionally negative climate.

The person feeding a one-year-old should be aware that exploring and experimenting are normal and desirable behaviors at this time in a child's life. The child is developing a sense of autonomy that, if allowed to develop, will lay the foundation for later confidence and effectiveness as an individual. The child's impulses, if consistently denied, can turn to shame and self-doubt. In light of the developmental and nutrient needs of one-year-olds, and in the face of their often contrary and willful behavior, a few feeding guidelines may be helpful:

- *Discourage unacceptable behavior (such as standing at the table or throwing food) by removing the child from the table to wait until later to eat.* Be consistent and firm, not punitive. The child will soon learn to sit and eat.
- *Let the child explore and enjoy food.* This may mean the child eats with fingers for a while. Use of the spoon will come in time.
- *Don't force food on children.* Provide children with nutritious foods, and let them choose which ones and how much they will eat. Gradually, they will acquire a taste for different foods. If children refuse milk, provide cheese, cream soups, and yogurt.
- *Limit sweets strictly.* Infants have little room in their 1000-kcalorie daily energy allowance for empty-kcalorie sweets, except occasionally.

These recommendations reflect a spirit of tolerance that serves the best interests of the child emotionally as well as physically. The Nutrition Assessment Checklist helps to identify nutrition-related factors that may help prevent or correct potential problems in pregnant women and infants. Details of nutrition assessment are presented in Chapter 13.

NUTRITION ASSESSMENT CHECKLIST — **FOR PREGNANT WOMEN AND INFANTS**

Health Problems, Signs, and Symptoms

Check the medical record for:

- ☐ Gestational diabetes
- ☐ Hypertension
- ☐ Preeclampsia
- ☐ Neural tube defect in an infant born previously
- ☐ Alcohol or illicit drug abuse
- ☐ Chronic diseases
- ☐ History of previous pregnancies (number, intervals, outcomes, multiple births, and gestational age birthweights)

Note risk factors for complications during pregnancy, including:

- ☐ Cigarette smoking
- ☐ Low socioeconomic status
- ☐ Food faddism
- ☐ Weight-loss dieting
- ☐ Very young or old age
- ☐ Lactose intolerance
- ☐ Significant or prolonged vomiting

Note any complaints of:

- ☐ Morning sickness
- ☐ Heartburn
- ☐ Constipation

Medications

For pregnant women who are using drug therapy for medical conditions, note:

- ☐ Potential for contraindication to breastfeeding

- ☐ GI tract side effects that might reduce food intake or change nutrient needs

Nutrient/Food Intake

For all pregnant women, especially those considered at risk nutritionally, assess the diet for:

- ☐ Total energy
- ☐ Protein
- ☐ Calcium, phosphorus, magnesium, iron, and zinc
- ☐ Folate and vitamin B_{12}
- ☐ Vitamin D

For infants, note:

- ☐ Method of feeding (breastfeeding, formula, or both)
- ☐ Frequency and duration of breastfeeding
- ☐ Amount of infant formula
- ☐ Practice of putting infant to bed with bottle
- ☐ Solid foods the infant is fed, if any
- ☐ Amount of food the infant is fed

Height and Weight

Measure baseline height and weight:

- ☐ Prepregnancy weight
- ☐ Infant birthweight

Reassess weight at each medical checkup and determine whether gains are appropriate. Note:

- ☐ Weight gain during pregnancy
- ☐ Gestational age
- ☐ Weight, length, and head circumference of infants

Laboratory Tests

Monitor the following laboratory tests for pregnant women:

- ☐ Hemoglobin, hematocrit, or other tests of iron status
- ☐ Blood glucose

Monitor the following laboratory tests for infants:

- ☐ Blood glucose of infants born to mothers with gestational diabetes
- ☐ Results of tests for inborn errors

Physical Signs

Blood pressure measurement is a routine measurement in physical exams, but is especially important for pregnant women.

Look for physical signs of:

- ☐ Iron deficiency
- ☐ Edema
- ☐ Protein-energy malnutrition
- ☐ Folate deficiency

SELF CHECK

1. The most important single predictor of an infant's future health and survival is:
   a. the infant's birthweight.
   b. the infant's iron status at birth.
   c. the mother's weight at delivery.
   d. the mother's prepregnancy weight.

2. A mother's prepregnancy nutrition is important to a healthy pregnancy because it determines the development of:
   a. the largest baby possible.
   b. adequate maternal iron stores.
   c. an adequate fat supply for the mother.
   d. healthy support tissues—the placenta, amniotic sac, umbilical cord, and uterus.

3. Two vitamins needed in large amounts during pregnancy for rapid cell proliferation are:
   a. vitamin B_{12} and iron.
   b. calcium and vitamin B6.
   c. folate and vitamin B_{12}.
   d. copper and zinc.

4. For a woman who is at the appropriate weight for height and is carrying a single fetus, the recommended weight gain during pregnancy is:
   a. 40 to 60 pounds.
   b. 25 to 35 pounds.
   c. 10 to 20 pounds.
   d. 20 to 40 pounds.

5. Rewards of physical activity during pregnancy may include:
   a. weight loss.
   b. decreased incidence of pica.
   c. relief from morning sickness.
   d. reduced stress, easier labor, and faster recovery from childbirth.

6. A woman's need for _____ is greater during lactation than during pregnancy.
   a. vitamin D   b. energy
   c. calcium   d. folate

7. Breast milk is recommended for the first 12 months of life because it offers complete nutrition and _____ to the infant.
   a. fluoride
   b. fructose
   c. immunological protection
   d. pica

8. An acceptable substitute for breast milk during the first year is:
   a. low-fat cow's milk.
   b. apple juice.
   c. water.
   d. iron-fortified infant formula.

9. Indications of readiness for solid foods include:
   a. the infant cries a lot.
   b. the infant is two months old.
   c. the infant is not sleeping through the night.
   d. the infant can sit with support and can control head movements.

10. During the first year of life, the most important step to undertake to encourage healthy eating is to:
   a. give food as a reward or for comfort.
   b. give sweets as a reward for eating vegetables.
   c. introduce a variety of nutritious foods in an inviting way.
   d. restrict fat to less than 30 percent of energy intake.

Answers to these questions appear in Appendix H.

CLINICAL APPLICATIONS

1. Consider the different factors in a pregnant woman's history that can affect her nutrition status and the outcome of her pregnancy. Describe what steps you would take to remedy potential problems for the following clients:

   a. A 15-year-old adolescent of low socioeconomic status in her first trimester of pregnancy. She began the pregnancy at a normal, healthy weight, but her weight gain during pregnancy so far has been less than expected. Her favorite beverages are soft drinks; her favorite foods are french fries and boxed macaroni and cheese.

   b. A lactose-intolerant, 22-year-old pregnant woman who has been eating a vegan diet for the past year or so. She began the pregnancy slightly underweight (BMI = 19.5), but her weight gain has been adequate and consistent during the four months of her pregnancy. She complains of feeling tired all the time.

2. What information would you give to a pregnant woman who is considering breastfeeding her infant, but isn't quite sure why she should?

3. The parents of a two-month-old infant have been told by the child's grandparents that they should introduce solid foods to help the baby sleep through the night. What advice would you give the parents?

NUTRITION ON THE NET

FOR FURTHER STUDY OF THE
TOPICS IN THIS CHAPTER,
ACCESS THESE WEB SITES.

acog.com
American College of Obstetricians and Gynecologists

childbirth.org
Childbirth.org

www.modimes.org
March of Dimes

diabetes.org
American Diabetes Association

www.nofas.org
National Organization on Fetal Alcohol Syndrome

www.aap.org
American Academy of Pediatrics

lalecheleague.org
LaLeche League International

www.eatright.org
American Dietetic Association

Notes

[1] National Academy of Sciences, Food and Nutrition Board, *Nutrition during Pregnancy* (Washington, D.C.: National Academy Press, 1990), pp. 176–211.

[2] A. L. Owen and G. M. Owen, Twenty years of WIC: A review of some effects of the program, *Journal of the American Dietetic Association* 97 (1997): 777–782; C. M. Olson, Promoting positive nutritional practices during pregnancy and lactation, *American Journal of Clinical Nutrition* 59 (1994): S525–S531.

[3] W. W. Hay and coauthors, Workshop summary: Fetal growth: Its regulation and disorders, *Pediatrics* 99 (1997): 585–591; Transplacental nutrient transfer and intrauterine growth retardation, *Nutrition Reviews* 50 (1992): 56–57.

[4] American Academy of Pediatrics, Folic acid for the prevention of neural tube defects, *Pediatrics* 104 (1999): 325–327.

[5] American Academy of Pediatrics, Folic acid for the prevention of neural tube defects, *Pediatrics* 104 (1999): 325–327; March of Dimes Resource Center, Folic acid (March of Dimes Birth Defects Foundation, Wilkes-Barre, Pa: 1999); Standing Committee on the Scientific Evaluation of Dietary Reference Intakes, Food and Nutrition Board, Institute of Medicine, *Dietary Reference Intakes for Thiamin, Riboflavin, Niacin, Vitamin B_6, Folate, Vitamin B_{12}, Pantothenic Acid, Biotin, and Choline* (Washington, D.C.: National Academy Press, 1998), pp. 8-1–8-68.

[6] J. E. Brown and coauthors, Predictors of red cell folate level in women attempting pregnancy, *Journal of the American Medical Association* 277 (1997): 548–552.

[7] National Academy of Sciences, 1990, pp. 1–23.

[8] T. O. Scholl and M. L. Hediger, Anemia and iron-deficiency anemia: Compilation of data on pregnancy outcome, *American Journal of Clinical Nutrition* 59 (1994): S492–S501.

[9] National Academy of Sciences, 1990.

[10] J. C. King and C. L. Keen, Zinc, in *Modern Nutrition in Health and Disease,* 9th ed., ed. M. E. Shils and coeditors (Baltimore, Md.: Williams & Wilkins, 1999), pp. 223–239; Y. H. Neggers and coauthors, A positive association between maternal serum zinc concentration and birth weight, *American Journal of Clinical Nutrition* 51 (1990): 678–684.

[11] Owen and Owen, 1997.

[12] National Academy of Sciences, 1990, pp. 63–96.

[13] K. G. Dewey and M. A. McCrory, Effects of dieting and physical activity on pregnancy and lactation, *American Journal of Clinical Nutrition* 59 (1994): S446–S453.

[14] J. M. Pivarnik, Potential effects of maternal physical activity on birth weight: Brief review, *Medicine and Science in Sports and Exercise* 30 (1998): 400–406.

[15] M. C. Hatch and coauthors, Maternal exercise during pregnancy, physical fitness, and fetal growth, *American Journal of Epidemiology* 137 (1993): 1105–1114.

[16] L. C. Tolstoi and J. B. Josimovich, Gestational diabetes mellitus: Etiology and management, *Nutrition Today* 34 (1999): 178–188; The Expert Committee on the Diagnosis and Classification of Diabetes Mellitus, Report of the Expert Committee on the diagnosis and classification of diabetes mellitus, *Diabetes Care* (supplement 1) 21 (1998): 5–19.

[17] C. D. Naylor and coauthors, Cesarean delivery in relation to birth weight and gestational glucose tolerance: Pathophysiology or practice style? *Journal of the American Medical Association* 275 (1996): 1165–1170.

[18] Tolstoi and Josimovich, 1999.

[19] B. M. Sibai, Treatment of hypertension in pregnant women, *New England Journal of Medicine* 335 (1996): 257–265.

[20] N. I. Hahn and M. Erick, Battling morning (noon and night) sickness: New approaches for treating an age-old problem, *Journal of the American Dietetic Association* 94 (1994): 147–148.

[21] National Academy of Sciences, 1990, pp. 390–411.

[22] U.S. Department of Health and Human Services, Public Health Service, *The Health Benefits of Smoking Cessation: A Report of the Surgeon General, 1990* (Washington, D.C.: Government Printing Office, 1990).

[23] American Academy of Pediatrics, Committee on Environmental Health, Environmental tobacco smoke: A hazard to children, *Pediatrics* 99 (1997): 639–642; E. Cutz and coauthors, Maternal smoking and pulmonary neuroendocrine cells in sudden infant death syndrome, *Pediatrics* 98 (1996): 668–672; H. S. Klonoff-Cohen and coauthors, The effect of passive smoking and tobacco exposure through breast milk on sudden infant death syndrome, *Journal of the American Medical Association* 273 (1995): 795–798; E. A. Mitchell and coauthors, Smoking and sudden infant death syndrome, *Pediatrics* 91 (1993): 893–896; K. C. Schoendorf, Relationship of sudden infant death syndrome to maternal smoking during and after pregnancy, *Pediatrics* 90 (1992): 905–908.

[24] D. L. Olds, C. R. Henderson, and R. Tatelbaum, Intellectual impairment in children of women who smoke cigarettes during pregnancy, *Pediatrics* 93 (1994): 221–227.

[25] National Academy of Sciences, 1990, pp. 390–411.

[26] T. S. Hinds and coauthors, The effect of caffeine on pregnancy outcome variables, *Nutrition Reviews* 54 (1996): 203–207; National Academy of Sciences, 1990.

[27] J. L. Mills and coauthors, Moderate caffeine use and the risk of spontaneous abortion and intrauterine growth retardation, *Journal of the American Medical Association* 269 (1993): 593–597.

[28] B. M. Sibai and coauthors, Prevention of preeclampsia with low-dose aspirin in healthy, nulliparous, pregnant women, *New England Journal of Medicine* 329 (1993): 1213–1218.

[29] W. A. Vega and coauthors, Prevalence and magnitude of perinatal substance exposures in California, *New England Journal of Medicine* 329 (1993): 850–854.

[30]F. D. Eyler and coauthors, Birth outcome from a prospective, matched study of prenatal crack/cocaine use I: Interactive and dose effects on health and growth, *Pediatrics* 101 (1998): 229–237; F. D. Eyler and coauthors, Birth outcome from a prospective, matched study of prenatal crack/cocaine use II: Interactive and dose effects on neurobehavioral assessment, *Pediatrics* 101 (1998): 237–241; I. J. Chasnoff and coauthors, Cocaine/polydrug use in pregnancy: Two-year follow-up, *Pediatrics* 89 (1992): 284–289.

[31]D. B. Petitti and C. Coleman, Cocaine and the risk of low birth weight, *American Journal of Public Health* 80 (1990): 25–28; J. J. Volpe, Effect of cocaine use on the fetus, *New England Journal of Medicine* 327 (1992): 399–407.

[32]L. C. Mayes and coauthors, Neurobehavioral profiles of neonates exposed to cocaine prenatally, *Pediatrics* 91 (1993): 778–783; M. J. Corwin and coauthors, Effects of in utero cocaine exposure on newborn acoustical cry characteristics, *Pediatrics* 89 (1992): 1199–1203.

[33]American Academy of Pediatrics, Committee on Substance Abuse and Committee on Children and Disabilities, Fetal alcohol syndrome and fetal alcohol effects, *Pediatrics* 91 (1993): 1004–1006.

[34]S. N. Mattson and coauthors, Heavy prenatal alcohol exposure with or without physical features of fetal alcohol syndrome leads to IQ deficits, *Journal of Pediatrics* 131 (1997): 718–721.

[35]American Academy of Pediatrics, Committee on Adolescence, Adolescent pregnancy—current trends and issues: 1998, *Pediatrics* 103 (1999): 516–520.

[36]Position of The American Dietetic Association: Nutrition care for pregnant adolescents, *Journal of the American Dietetic Association* 94 (1994): 449–450.

[37]M. Story and I. A. Alton, Nutrition issues and adolescent pregnancy, *Nutrition Today* 30 (1995): 142–151.

[38]Position of The American Dietetic Association, 1994; J. L. Beard, Iron deficiency: Assessment during pregnancy and its importance in pregnant adolescents, *American Journal of Clinical Nutrition* 59 (1994): S502–S510.

[39]National Academy of Sciences, 1990, pp. 1–23.

[40]Position of The American Dietetic Association, 1994.

[41]American Academy of Pediatrics, Work Group on Breastfeeding, Breastfeeding and the use of human milk, *Pediatrics* 100 (1997): 1035–1039.

[42]Position of The American Dietetic Association: Promotion of breastfeeding, *Journal of the American Dietetic Association* 97 (1997): 662–666.

[43]K. G. Dewey and coauthors, Maternal versus infant factors related to breast milk and residual milk volume: The DARLING Study, *Pediatrics* 87 (1991): 829–837; National Academy of Sciences, Food and Nutrition Board, *Nutrition during Lactation* (Washington, D.C.: National Academy Press, 1991), pp. 1–19.

[44]J. M. A. van Raaij and coauthors, Energy cost of lactation, and energy balances of well-nourished Dutch lactating women: Reappraisal of the extra energy requirements of lactation, *American Journal of Clinical Nutrition* 53 (1991): 612–619.

[45]Position of The American Dietetic Association, 1997.

[46]F. M. Kramer and coauthors, Breast-feeding reduces maternal lower-body fat, *Journal of the American Dietetic Association* 93 (1993): 429–433.

[47]National Academy of Sciences, 1991, pp. 197–212.

[48]M. J. Heinig and K. G. Dewey, Health effects of breast feeding for mothers: A critical review, *Nutrition Research Reviews* 10 (1997): 35–56.

[49]National Academy of Sciences, 1991, p. 140.

[50]National Academy of Sciences, 1991, p. 140.

[51]J. Liston, Breastfeeding and the use of recreational drugs—Alcohol, caffeine, nicotine, and marijuana, *Breastfeeding Reviews* 2 (1998): 27–30.

[52]Highlights from U.S.D.A. research, *Nutrition Today*, January/February 1993, pp. 4–5.

[53]American Academy of Pediatrics, Committee on Drugs, The transfer of drugs and other chemicals into human milk, *Pediatrics* 93 (1994): 137–150.

[54]Joint United Nations Programme on HIV/AIDS, HIV and infant feeding: Guidelines for decision-makers, Geneva: World Health Organization, 1998, as cited in E. Hormann, Breast-feeding and HIV: What choices does a mother really have? *Nutrition Today* 34 (1999): 189–196.

[55]American Academy of Pediatrics, Committee on Drugs, 1994.

[56]American Academy of Pediatrics, Committee on Drugs, 1994.

[57]American Academy of Pediatrics, Committee on Nutrition, *Pediatric Nutrition Handbook*, 4th ed., ed. R. E. Kleinman (Elk Grove Village, Ill.: American Academy of Pediatrics, 1998), pp. 43–53.

[58]American Academy of Pediatrics, Committee on Nutrition, 1998, pp. 3–20.

[59]A. L. Wright and coauthors, Increasing breast-feeding rates to reduce infant illness at the community level, *Pediatrics* 101 (1998): 837–844.

[60]American Academy of Pediatrics, Work Group on Breastfeeding, 1997.

[61]J. Raisler and coauthors, Breast-feeding and infant illness: A dose-response relationship? *American Journal of Public Health* 89 (1999): 25–30.

[62]D. S. Newburg and coauthors, Role of human-milk lactadherin in protection against symptomatic rotavirus infection, *Lancet* 351 (1998): 1160–1164.

[63]A. H. Cushing and coauthors, Breastfeeding reduces risk of respiratory illness in infants, *American Journal of Epidemiology* 147 (1998): 863–870; D. S. Newburg and J. M. Street, Bioactive materials in human milk, *Nutrition Today* 32 (1997): 191–201.

[64]M. A. Atkinson and T. M. Ellis, Infants' diets and insulin-dependent diabetes: Evaluating the "cows' milk hypothesis" and a role for antibovine serum albumin immunity, *Journal of the American College of Nutrition* 16 (1997): 334–340.

[65]J. M. Norris and coauthors, Lack of association between early exposure to cow's milk protein and b-cell autoimmunity, *Journal of the American Medical Association* 276 (1996): 609–614.

[66]M. W. Shannon and J. W. Graef, Lead intoxication in infancy, *Pediatrics* 89 (1992): 87–90.

Q&A NUTRITION IN PRACTICE

Encouraging Successful Breastfeeding

As discussed in Chapter 10, breastfeeding offers benefits to both mother and infant. The American Academy of Pediatrics (AAP), the American Dietetic Association (ADA), and the U.S. Department of Health and Human Services all advocate breastfeeding as the preferred means of infant feeding.[1] Promotion of breastfeeding is an integral part of the WIC program's nutrition education component.[2] During the late 1980s, breastfeeding was on the decline after reaching a high of about 60 percent in 1984. National efforts to promote breastfeeding seem to be working, at least to some extent. An encouraging trend is emerging, with almost 60 percent of women initiating breastfeeding in 1995.[3] Nevertheless, only about one in five infants is still being breastfed at five to six months of age. Many of the benefits of breastfeeding are enhanced by breastfeeding for at least four months. The AAP recommends that breastfeeding continue for at least a year and thereafter for as long as mutually desired.[4] Increasing the rates of breastfeeding initiation and duration is one of the goals of *Healthy People 2010:*

Increase the proportion of mothers who breastfeed immediately after birth, for the first six months, and preferably, through the infant's first year of life. Increase the proportion of mothers who breastfeed exclusively.[5]

Despite the trend toward increasing breastfeeding, the percentage of women choosing to breastfeed their infants and continuing to do so still falls short of goals.

eatright.org
American Dietetic Association

who.org
World Health Organization

os.dhhs.gov
U.S. Department of Health and Human Services

Why don't more women choose to breastfeed their infants?

Many experts cite two major deterrents: public advertising of infant formula and the medical community's failure to encourage breastfeeding. As an example of the medical lack of encouragement, some hospitals routinely separate mother and infant soon after birth. The child's first feeding then comes from the bottle rather than the breast. Furthermore, many hospitals send new mothers home with free samples of infant formula. The World Health Organization opposes this practice because it sends a misleading message that medical authorities favor infant formula over breast milk for infants. Even in hospitals where women are encouraged to breastfeed and are supported in doing so, little if any assistance is available after hospital discharge when many breastfeeding women still need assistance. Up to half of mothers who initially breastfeed their infants stop within a

GLOSSARY OF BREASTFEEDING TERMS

engorgement: overfilling of the breasts with milk.

letdown reflex: the reflex that forces milk to the front of the breast when the infant begins to nurse.

mastitis: infection of a breast.

rooting reflex: a reflex that causes an infant to turn toward whichever cheek is touched, in search of a nipple.

month—seemingly due to lack of knowledge. Women who receive early and repeated breastfeeding information and support breastfeed their infants longer than other women do.[6] Information and instruction are especially important during the *prenatal* period when most women decide whether to breastfeed or to feed formula. Nurses and other health care professionals can play a crucial role in encouraging successful breastfeeding by offering women adequate, accurate information about breastfeeding that permits them to make informed choices.

If breastfeeding is a natural process, what do mothers need to learn?

Although lactation is an automatic physiological process, breastfeeding requires some learning. This learning is most successful in a supportive environment. It begins with preparatory steps taken before the infant is born.

What are these preparatory steps?

Toward the end of pregnancy and throughout lactation, a woman who intends to breastfeed should stop using soap and lotions on her breasts. The natural secretions of the breasts themselves lubricate the nipple area best. A few weeks before the infant is due, the woman should rub her breasts with a towel or allow them to rub against her outer clothing for a little while each day to toughen the nipples somewhat in preparation for the infant's sucking. A woman who plans to breastfeed should also acquire at least two nursing bras before her infant is born. The bras should provide good support and have drop-flaps so that either breast can be freed for nursing.

How soon after birth should breastfeeding start?

As soon as possible. Immediately after the delivery, for a short period, the infant is intensely alert and intent on sucking. This is the ideal time for the first breastfeeding and facilitates successful lactation.

What does the new mother need to know in order to continue breastfeeding her infant successfully?

She needs to learn how to relax and position herself so that she and the infant will be comfortable and the infant can breathe freely while nursing. She also needs to understand that infants have a **rooting reflex** that makes them turn toward any touch on the face. (The accompanying glossary defines this and other relevant terms.) Consequently, she should touch the infant's cheek to her nipple so that the infant will turn the right way and start to nurse. The mother can then squeeze her areola, the colored ring around the nipple, between two fingers and slip enough of it into the infant's mouth to permit a good hold and strong pumping action (see Figure NP10-1). The nipple must rest well back on the infant's tongue so that the infant's gums will squeeze on the glands that release the milk and swallowing will be effortless. To break the suction, if necessary, the mother can slip a finger between the infant's mouth and her breast.

Does it hurt to have the infant sucking so hard on the breast?

No, because the mother has a **letdown reflex** that forces milk to the front of her breast when the infant begins to nurse, virtually propelling the milk into the infant's mouth. Letdown is necessary for the infant to

Figure NP10-1

INFANT'S GRASP ON MOTHER'S BREAST

The mother squeezes the areola, slipping enough of it into the infant's mouth to promote good pumping action. The infant's lips and gums pump the areola, releasing milk from the mammary glands into the milk ducts that lie beneath the areola.

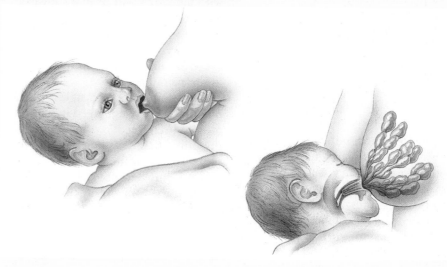

obtain milk easily, and the mother needs to relax for letdown to occur. The mother who assumes a comfortable position in an environment without interruptions will find it easiest to relax.

How long should the infant be allowed to nurse at each feeding?

Although the infant sucks half the milk from the breast within the first 2 minutes, and 80 to 90 percent of it within 4 minutes, sucking on each breast for 10 to 15 minutes is encouraged. The sucking itself, as well as the complete removal of milk from the breast, stimulates the mammary glands to produce milk for the next nursing session. Successive sessions should start on alternate breasts to ensure that each breast is emptied regularly. This pattern maintains the same supply and demand for each breast and thus prevents either breast from overfilling.

Infants should be fed "on demand" and not held to a rigid schedule. The breastfed infant may average 8 to 12 feedings per 24-hour period during the first month or so.

Once the mother's milk supply is well established and the infant's capacity has increased, the intervals between feedings will become longer.

What if a mother wants to skip one or two feedings daily—for example, because she works outside the home?

The mother can express breast milk into a bottle ahead of time, freeze the breast milk, and, when needed, substitute the expressed breast milk for a nursing session. Breast milk can be kept refrigerated for 48 hours or frozen (at a freezer temperature below 0°F) for several months.

The mother can hand express her breast milk or use one of several different breast pumps available. The bicycle-horn type of manual breast pump is not recommended, however, because it is difficult to keep clean. Cylinder-type manual pumps or electric breast pumps are safer and are also more efficient. Alternatively, a mother can substitute formula for the missed feedings and continue to breastfeed at other times.

What about problems associated with breastfeeding such as sore nipples or infection of the breast?

Most problems associated with breastfeeding can be resolved. Many mothers experience sore nipples during the initial days of breastfeeding. Sore nipples need to be treated kindly, but nursing can continue. Improper feeding position is a frequent cause of sore nipples: the mother should make sure the infant is taking the entire nipple and part of the areola onto the tongue. She should nurse on the less sore breast first to get letdown going while the infant is sucking hardest; then she can switch to the sore breast. Between times, she should expose her nipples to light and air to heal them.

Before lactation is well established, when the schedule changes, or when a feeding is missed, the breasts may become full and hard—an uncomfortable condition known as **engorgement.** The infant cannot grasp an engorged nipple and so cannot provide relief by nursing. A gentle massage or warming the breasts with a heating pad or in a shower helps to initiate letdown and to release some of the accumulated milk; then the mother can pump out some of her milk and allow the infant to nurse.

Infection of the breast, known as **mastitis,** is best managed by continuing to breastfeed. By drawing off the milk, the infant helps to relieve pressure in the infected area. The infant is safe because the infection is between the milk-producing glands, not inside them.

Even if everything is going smoothly, the nursing mother should ideally have enough help and support so that she can rest in bed a few hours each day for the first week or so. Successful breastfeeding requires the support of all those who care.

This, plus adequate nutrition, ample fluids, fresh air, and physical activity, will do much to enhance the well-being of mother and infant.

Notes

[1] American Academy of Pediatrics, Work Group on Breastfeeding, Breastfeeding and the use of human milk, *Pediatrics* 100 (1997): 1035–1039; Position of The American Dietetic Association: Promotion of breastfeeding, *Journal of the American Dietetic Association* 97 (1997): 662–666.

[2] A. L. Owen and G. M. Owen, Twenty years of WIC: A review of some effects of the program, *Journal of the American Dietetic Association* 97 (1997): 777–782.

[3] Position of The American Dietetic Association, 1997.

[4] American Academy of Pediatrics, 1997.

[5] *Healthy People 2010: National Health Promotion and Disease Prevention Objectives* (Washington, D.C.: U.S. Department of Health and Human Services, 2000).

[6] C. I. Dungy and coauthors, Effect of discharge samples on duration of breastfeeding, *Pediatrics* 90 (1992): 233–237.

CHAPTER 11

NUTRITION THROUGH THE LIFE SPAN: CHILDHOOD AND ADOLESCENCE

CONTENTS

Early and Middle Childhood

Food Choices and Eating Habits of Children

The Teen Years

Case Study: Boy with Disruptive Behavior

Nutrition in Practice: Childhood Obesity and the Early Development of Chronic Diseases

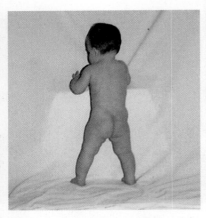

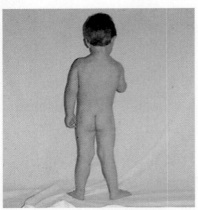

The body shape of a one-year-old (above) changes dramatically by age two (below). The two-year-old has lost much of the baby fat; the muscles (especially in the back, buttocks, and legs) have firmed and strengthened; and the leg bones have lengthened.

Nutrient needs change throughout life, depending on genetics, rates of growth, activity, and many other factors. Nutrient needs also vary from individual to individual, but generalizations are possible and useful. Sound nutrition throughout childhood promotes normal growth and development; facilitates academic and physical performance; and helps prevent obesity, diabetes, heart disease, cancer, and other degenerative diseases in adulthood. As children enter the teen years, a foundation built by years of eating nutritious foods best prepares them to meet the upcoming demands of rapid growth.

☷ Early and Middle Childhood

After the age of one, growth rate slows, but the body continues to change dramatically. At one, infants have just learned to stand and toddle; by two, they walk confidently and are learning to run, jump, and climb. Nutrition and physical activity have helped them prepare for these new accomplishments by adding to the mass and density of their bone and muscle tissue. Thereafter, their bones continue to grow longer and their muscles to gain size and strength, though unevenly and more slowly, until adolescence.

☷ Energy and Nutrient Needs

An infant's appetite declines markedly around the first birthday, consistent with the slowed growth rate. Thereafter, the appetite fluctuates. At times children seem to be insatiable, and at other times they seem to live on air and water. Parents and other caregivers need not worry about this—a child will need and demand much more food during periods of rapid growth than during slow periods. The perfect regulation of appetite in children of normal weight guarantees that their food energy intakes will be right for each stage of growth.[1]

☷ **CHILDREN'S APPETITES** Many people mistakenly believe that they must "make" their children eat the right amounts of food, and children's erratic appetites often reinforce this belief. Research proves otherwise. Researchers studying preschool children for six days found that each child's food energy intake was highly variable from meal to meal, but the total daily energy intake was remarkably constant.[2] The children adjusted their energy intakes at successive meals: if they ate less at one meal, they ate more at the next, and vice versa.

Parents do, however, need to help children choose the right foods, and with overweight children they may need to help more, as described later. Overweight children may not adjust their energy intakes appropriately, but may disregard appetite-regulation signals and eat in response to external cues, such as television commercials.

☷ **ENERGY** Individual children's energy needs vary widely, depending on their growth and physical activity. A one-year-old child needs approximately 1000 kcalories a day; a three-year-old needs slightly more, about 1300 kcalories. By age ten, a child needs about 2000 kcalories a day. Total energy needs increase gradually with age, but energy needs per kilogram of body weight actually decline.

☷ **NUTRIENTS** Steady growth during childhood necessitates a gradual increase in most nutrients. Nutrient recommendations cluster children into age groupings that reflect similarities in growth rate, biological changes, and hormone status (see inside front cover).

Ideally, children accumulate stores of nutrients before adolescence. Then, when they take off on the adolescent growth spurt and their nutrient intakes cannot keep pace with the demands of rapid growth, they can draw on the nutrient stores accumulated earlier. This is especially true of calcium; the denser the bones are in childhood, the better prepared they will be to support adolescent growth and still withstand the inevitable bone losses of later life.[3] Consequently, the way preteen children eat influences their nutritional health during childhood, during their teen years, and for the rest of their lives.

LEARNING LINK

The importance of an adequate calcium intake during the growing years to the health of the bones throughout life cannot be overemphasized. Chapter 9 offered a detailed discussion of calcium's role in protecting against osteoporosis.

:: FOOD PATTERNS FOR CHILDREN   To provide all the needed nutrients, a child's meals and snacks should include a variety of foods from each food group—in amounts suited to the child's appetite and needs. Figure 11-1 (p. 264) offers the Food Guide Pyramid for young children. Notice that this pyramid differs slightly from the one for adults (see Chapter 1) in that a set number of servings, rather than a range, from each food group is suggested. Children two to six years old need at least the specified number of servings from each food group to meet their nutrient needs, but the serving sizes should vary according to age. Generally, children in the four- to six-year age group can eat the serving sizes recommended for adults. Caregivers of children two to three years old should offer servings that are about two-thirds the size of an adult serving. To satisfy the larger appetites of older children and adolescents, caregivers should provide additional servings of the same nutritious foods for the additional energy and nutrients growing children need.

:: CHILDREN'S FOOD CHOICES   Parents and other caregivers can do much to foster the development of healthy eating habits in a child. The challenge is to deliver nutrients in the form of meals and snacks that are nutritious and delicious to children so that they will learn to enjoy a variety of nutritious foods.

Candy, cola, and other concentrated sweets must be limited in children's diets. If such foods are permitted in large quantities, the only possible outcomes are nutrient deficiencies, obesity, or both. Children cannot be trusted to choose nutritious foods on the basis of taste alone; the preference for sweets is innate, and children naturally gravitate to them. In one study, when children were allowed to create meals freely from a variety of foods, they selected foods that provided 25 percent of the kcalories from sugar.[4] Interestingly, when their parents were watching, or even when they were told that their parents would be watching, the children improved their selections. Overweight children, especially, need help in selecting nutrient-dense foods that will meet their nutrient needs within their energy allowances. Underweight children or active, normal-weight children can enjoy higher-kcalorie foods, but these should still be nutritious. Examples are ice cream or pudding in the milk group and whole-grain or enriched pancakes or crackers in the bread group.

Figure 11-1
FOOD GUIDE PYRAMID FOR YOUNG CHILDREN

SOURCE: USDA Center for Nutrition and Policy Promotion, March 1999, Program AID 1649.

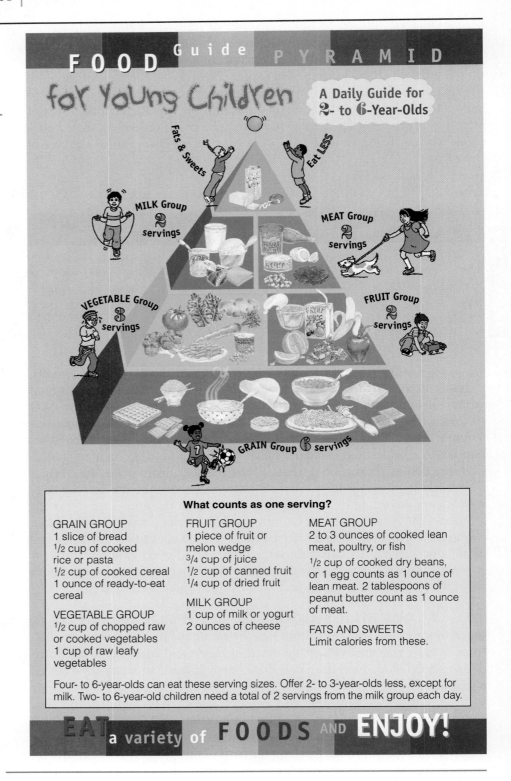

Malnutrition in Children

Most children in the United States and Canada are well nourished. Their average energy intakes are sufficient to support normal growth, and their average nutrient intakes, except for iron, meet or exceed recommendations. Hunger and malnutrition are prevalent among some groups, however. More than 11 million children under 12 years of age in the United States are hungry or are at risk of hunger. A root cause of this hunger is poverty. Many more homeless people are hungry than people who are housed, and children are the fastest-growing segment of the homeless population. Homeless children from large families and from families with single mothers are especially vulnerable to hunger and its adverse consequences.

■■ **EFFECTS OF HUNGER** Both short-term and long-term hunger exert negative effects on behavior and health. Short-term hunger, such as when a child misses a meal, impairs the child's ability to pay attention and to be productive. Hungry children are irritable, apathetic, and uninterested in their environment. Long-term hunger impairs growth and immune defenses. Food assistance programs such as the WIC program and the School Breakfast and National School Lunch Programs are designed to protect against hunger and improve the health of children.

■■ **HUNGER AND SCHOOL PERFORMANCE** Children who eat nutritious breakfasts function better than their peers who do not. A nutritious breakfast that includes cereal is a central feature of a diet that meets the needs of children and supports their healthy growth and development.[5] Young children who participate in the School Breakfast Program improve their scores on achievement tests and are tardy or absent significantly less often than children who qualify for the program but do not participate. Common sense dictates that it is unreasonable to expect anyone to learn and perform work when no fuel has been provided. By the late morning, discomfort from hunger may become distracting even if a child has eaten breakfast. Chronically underfed children suffer all the more.[6]

The problem children face when attempting morning schoolwork on an empty stomach appears to be at least partly due to low blood glucose. The average child up to the age of ten or so needs to eat every four to six hours to maintain a blood glucose concentration high enough to support the activity of the brain and nervous system. The brain is the body's chief glucose consumer, and a child's brain is as big as an adult's is. A child's liver is considerably smaller than an adult's, however—and the liver is the organ responsible for storing glucose (as glycogen) and releasing it into the blood as needed. A child's liver cannot store more than about four hours' worth of glycogen—hence, the need to eat fairly often. Teachers aware of the late-morning slump in their classrooms wisely request that a midmorning snack be provided; it improves classroom performance all the way to lunchtime.

■■ **IRON DEFICIENCY AND BEHAVIOR** In U.S. children and adolescents, as in infants after six months, iron-deficiency anemia is the most prevalent nutrient deficiency. The best-known and most widespread effects of iron deficiency are its impacts on behavior. Most people are familiar with the role of iron in carrying oxygen in the blood. Another important function of iron is transporting oxygen within cells, where it is used to help produce energy. A lack of iron not only causes an energy crisis but also directly affects behavior, mood, attention span, and learning ability. Iron is also involved in the function of many molecules in the brain and nervous system. Much of the research on iron and behavior has focused on the proposal that even in the early stages of iron deficiency an iron-dependent neurotransmitter is altered, and this change, in turn, impairs learning ability and behavior.[7]

Iron deficiency is usually diagnosed by a deficit of iron in the *blood*, after the deficiency has progressed all the way to anemia. A child's *brain*, however, is sensitive to slightly lowered iron concentrations long before the blood deficits appear. Iron's effects are hard to distinguish from the effects of other factors in children's lives, but it is likely that iron deficiency manifests itself in a lowering of "motivation to persist in intellectually challenging tasks," a

Table 11-1
IRON-RICH FOODS CHILDREN LIKE[a]

Breads, cereals, and grains
Canned macaroni (½ c)
Canned spaghetti (½ c)
Cream of wheat (¼ c)
Fortified dry cereals (1 oz)[b]
Noodles, rice, or barley (½ c)
Tortillas (1 flour, 2 corn)
Whole-wheat, enriched, or fortified bread (1 slice)

Vegetables
Baked flavored potato skins (½ skin)
Cooked mushrooms (½ c)
Cooked mung bean sprouts or snow peas (½ c)
Green peas (½ c)
Mixed vegetable juice (1 c)

Fruits
Apple juice (1 c)
Canned plums (3 plums)
Cooked dried apricots (½ c)
Dried peaches (4 halves)
Raisins (1 tbs)

Meats and legumes
Bean dip (¼ c)
Canned pork and beans (⅛ c)
Mild chili or other bean/meat dishes (¼ c)
Meat casseroles (½ c)
Peanut butter and jelly sandwich (½ sandwich)
Lean roast beef or cooked ground beef (1 oz)
Sloppy joes (½ sandwich)

[a]Each serving provides at least 1 milligram iron, or one-tenth of a child's RDA for iron. Vitamin C–rich foods included with these snacks increase iron absorption.
[b]Some fortified breakfast cereals contain more than 10 milligrams iron per half-cup serving (read the labels).

shortening of the attention span, and a reduction of overall intellectual performance. Anemic children perform less well on tests and have more conduct disturbances than their nonanemic classmates. The effects of iron-deficiency anemia are especially detrimental to learning when combined with other nutrient deficiencies.[8]

PREVENTING IRON DEFICIENCY To avert iron-deficiency anemia, children's foods must deliver 10 milligrams of iron or more per day. To achieve this goal, milk intakes must be limited after infancy because milk is a poor source of iron. Children should receive enough milk products to ensure adequate calcium and riboflavin intakes, but no more. That means 2 cups of milk per day up to age 6, increasing to 3 cups per day from age 6 to 12. After age two, if reduced-fat milk is used instead of whole milk, the saved kcalories can be invested in iron-rich foods such as lean meats, fish, poultry, eggs, and legumes. Whole-grain or enriched breads and cereals also contribute iron. Table 11-1 lists iron-rich foods children like.

The iron status of children in the United States is improving. Infant feeding practices such as breastfeeding and use of iron-fortified formulas may be improving iron status in later childhood.[9] Among children from low-income families, the WIC program, which provides supplemental iron-rich foods during infancy and early childhood, appears to be playing a role in improving iron status.

WORLD FOCUS ON IRON DEFICIENCY The prevalence of iron deficiency among children throughout the world has become the focus of major public health organizations. The U.S. Public Health Service lists the reduction of iron deficiency among young children as one of the nation's foremost health priorities. The World Health Organization and a United Nations subcommittee on nutrition have developed a ten-year plan to eliminate iron deficiency.

OTHER NUTRIENT DEFICIENCIES Iron is only one of several dozen nutrients that can be displaced in a diet of nutrient-poor foods. Any of the other nutrients may be lacking as well, and the deficiencies of those nutrients may also cause both behavioral and physical symptoms (see Table 11-2). A child with behavioral symptoms of nutrient deficiencies may be irritable, aggressive, disagreeable, or sad and withdrawn. One might label such a child "hyperactive," "depressed," or "unlikable," but in fact these traits might arise from simple, albeit marginal, malnutrition. Should suspicion of dietary inadequacies be raised, *no matter what other causes may be implicated*, the people responsible for feeding the child should take steps to correct those inadequacies promptly.

Lead Poisoning in Children

The damage caused by malnutrition may be compounded by environmental factors such as lead poisoning. A two-way interaction is typical: lead poisoning can cause an iron deficiency, and an iron deficiency can impair the body's defenses against lead absorption.[10] Adequate calcium may slow lead's absorption or interfere with its toxic effects in the body.[11] Like iron deficiency, mild lead toxicity has nonspecific effects, including diarrhea, irritability, reduced ability of the blood to carry oxygen, and fatigue. The symptoms may be reversible if exposure stops soon enough. With higher levels of lead, the signs become more pronounced, yet pinpointing a cause may still be difficult. Children lose their general cognitive, verbal, and perceptual abilities and develop learning disabilities and behavior problems. Still more severe lead toxicity can cause irreversible nerve damage, paralysis, mental retardation, and death.

Table 11-2
SIGNS OF HEALTH AND MALNUTRITION IN CHILDREN

	Healthy	Malnourished
Hair	Shiny, firm in the scalp	Dull, brittle, dry, loose; falls out
Eyes	Bright, clear pink membranes; adjust easily to darkness	Pale membranes; spots; redness; adjust slowly to darkness
Teeth and gums	No pain or cavities, gums firm, teeth bright	Missing, discolored, decayed teeth; gums bleed easily and are swollen and spongy
Face	Good complexion	Off-color, scaly, flaky, cracked skin
Glands	No lumps	Swollen at front of neck and cheeks
Tongue	Red, bumpy, rough	Sore, smooth, purplish, swollen
Skin	Smooth, firm, good color	Dry, rough, spotty; "sandpaper" feel or sores; lack of fat under skin
Nails	Firm, pink	Spoon-shaped, brittle, ridged
Behavior	Alert, attentive, cheerful	Irritable, apathetic, inattentive, hyperactive
Internal systems	Heart rate, heart rhythm, and blood pressure normal; normal digestive function; reflexes and psychological development normal	Heart rate, heart or blood pressure abnormal; liver and spleen enlarged; abnormal digestion; mental irritability, confusion; burning, tingling of hands and feet; poor balance and coordination
Muscles and bones	Good muscle tone and posture; long bones straight	"Wasted" appearance of muscles; swollen bumps on skull or ends of bones; small bumps on ribs; bowed legs or knock-knees

NOTE: The signs here are consistent with malnutrition but not diagnostic of it.

Paint is the primary source of lead in children's lives.

Lead toxicity is most prevalent among children under age six—as many as 1 million may have blood concentrations high enough to cause mental, behavioral, and other health problems.[12] Table 11-3 lists the symptoms of lead toxicity. Lead aggressively attacks fetuses, infants, and children because the body absorbs lead most efficiently during times of rapid growth. Blood concentrations of lead generally reach a peak in two-year-old children; age two is the typical time of exploring surroundings "hand to mouth" and ingesting lead-tainted dirt and debris. Children's behaviors and activities—putting their hands in their mouths, playing in dirt, and eating nonfood items—favor their chances of exposure to lead. Of all the sources of lead for children, paint remains the most important. In the homes of 3 million young children, leaded surfaces are peeling and deteriorating. Each child in these homes is either already poisoned or in danger of becoming so.

A ban on leaded gasoline and reductions in other uses of lead have dramatically reduced the amount of lead in the environment.[13] The result has been a gratifying decline in children's blood lead concentrations since the 1970s. The decline follows exactly the nation's reduction in use of leaded gasoline, leaded house paint, and lead-soldered food cans. A nationwide lead-monitoring system is now in place, and aggressive community programs are testing and treating children for lead poisoning. Despite such safeguards, the problem of exposure to lead still pervades children's lives (see Figure 11-2 on p. 268). The "How to" box on p. 269 suggests strategies to protect children from lead poisoning.

Table 11-3
SYMPTOMS OF LEAD TOXICITY

- Learning disabilities
- Low IQ
- Behavior problems
- Slow growth
- Iron-deficiency anemia
- Nervous system disorders
- Impaired concentration
- Reduced short-term memory
- Slow reaction time
- Seizures
- Impaired hearing
- Poor coordination

Figure 11-2
SOURCES OF LEAD EXPOSURE

Lead finds its way into the bodies of children when they ingest lead-containing foods, water, dust, or paint chips, or when they breathe lead-laden air.

LEARNING LINK

Chapters 8 and 9 discussed examples of how individual nutrient deficiencies or excesses may enhance or impair the absorption, metabolism, or excretion of other nutrients. Chapter 13 discusses diet-medication interactions. The discussion of lead in this section provides but one example of how contaminants can undermine nutrition status and health through nutrient interactions. Other heavy metal contaminants such as mercury and cadmium behave similarly in the body—they interfere with nutrients that are trying to do their jobs. To protect health, people must defend themselves against contamination by eating nutrient-dense foods and preserving a clean environment.

PROTECT CHILDREN FROM LEAD EXPOSURE

Defensive strategies to protect infants and children from lead exposure include:

- Test children for lead poisoning; effective screening with an appropriate questionnaire is essential to identifying high-risk children and thus treating the devastating effects.[a] About half the pediatricians surveyed report universal screening.[b]
- In contaminated environments, keep small children from putting dirty or old painted objects in their mouths, and make sure children wash their hands before eating. Similarly, keep small children from eating any nonfood items. Lead poisoning has been reported in young children who have eaten pool cue chalk.[c]
- Be aware that other countries do not have the same regulations protecting consumers against lead. Children have been poisoned by eating crayons made in China and drinking fruit juice canned in Mexico.
- Make infant formula from lead-free ingredients. Do not use lead-contaminated water.
- Feed children nutritious meals regularly.

[a]S. J. Schaffer and coauthors, Lead poisoning risk determination in a rural setting, *Pediatrics* 97 (1996): 84–90; M. N. Haan, M. Gerson, and B. A. Zishka, Identification of children at risk for lead poisoning: An evaluation of routine pediatric blood lead screening in an HMO-insured population, *Pediatrics* 97 (1996): 79–83; D. C. Snyder and coauthors, Development of a population-specific risk assessment to predict elevated blood lead levels in Santa Clara County, California, *Pediatrics* 96 (1995): 643–648; American Academy of Pediatrics, Committee on Environmental Health, Lead poisoning: From screening to primary prevention, *Pediatrics* 92 (1993): 176–183.

[b]J. R. Campbell and coauthors, Blood lead screening practices among U.S. pediatricians, *Pediatrics* 97 (1996): 372–377.

[c]Pool cue chalk: A source of environmental lead, *Pediatrics* 97 (1996): 916–917.

Food Allergies

Food allergies are frequently blamed for physical and behavioral abnormalities in children. Food allergies are most common during the first few years of life, but then children typically outgrow them.[14] A true food allergy occurs when a whole food protein or other large molecule enters the body and elicits an immunologic response. (Recall that large molecules of food are normally dismantled in the digestive tract to smaller ones before absorption.) The body's immune system reacts to a food protein or other large molecule as it does to an antigen—by producing antibodies or other defensive agents. A problem that results from exposure to food substances, but does not involve the immune system, is known as **food intolerance.**

ASYMPTOMATIC AND SYMPTOMATIC ALLERGIES
Allergies may have one or two components. They always involve antibodies; they may or may not involve symptoms. A person may produce antibodies without having any symptoms (known as asymptomatic allergy) or may produce antibodies and have symptoms (known as symptomatic allergy). A person who experiences symptoms without producing antibodies, however, does not have an allergy. This means that allergies have to be diagnosed by testing for antibodies.

ALLERGY SYMPTOMS
A symptomatic allergy will exhibit different symptoms depending on the location of the reaction. In the digestive tract, the

food allergies: adverse reactions to foods that involve an immune response; also called *food-hypersensitivity reactions.*

Reminder: **Antigens** *are substances foreign to the body that elicit the formation of antibodies or an inflammation reaction from immune system cells. Food antigens are usually glycoproteins (large proteins with glucose molecules attached).*

Reminder: **Antibodies** *are large proteins that are produced in response to antigens and then inactivate the antigens.*

food intolerance: an adverse response to a food or food additive that does not involve the immune system.

anaphylactic (an-AFF-ill-LAC-tic) shock: a life-threatening whole-body allergic reaction to an offending substance.

allergy may cause nausea or vomiting; in the skin, it may cause rashes; and in the nasal passages and lungs, it may cause inflammation or asthma. A dangerous, generalized, all-systems shock reaction known as **anaphylactic shock** can also occur.

Parents often mistakenly ascribe children's symptoms to food allergies, especially if the symptoms arise after eating. However, stomachaches, headaches, pain, rapid pulse rate, nausea, wheezing, hives, bronchial irritation, coughs, and the like usually have other causes. Only proper testing can distinguish the many possibilities, and such testing is seldom done.

▪▪ **IMMEDIATE AND DELAYED REACTIONS** Allergic reactions to food can occur with different timings, simply classified as immediate and delayed. In both, the antigen interacts immediately with the immune system, but symptoms may appear within minutes or after several (up to 24) hours. Identifying the food that causes an immediate allergic reaction is easy because symptoms correlate closely with the time of eating the food. Identifying the food that caused a delayed reaction is more difficult because the symptoms may not appear until a day after the offending food was eaten; by this time, many other foods will have been eaten, too, complicating the picture.

The foods that most often cause immediate allergic reactions are listed in Table 11-4. Almost 75 percent of reactions are caused by three major foods—eggs, peanuts, or milk. Allergic reactions to single foods are common. Reactions to multiple foods are the exception, not the rule.

▪▪ **OTHER ADVERSE REACTIONS TO FOODS** Adverse reactions to foods that are not true food allergies include:

- A reaction specific to the flavor enhancer monosodium glutamate, or MSG.
- Reactions to chemicals in foods, such as the natural laxative in prunes.
- GI symptoms, such as nausea and indigestion, aggravated by eating any food.
- Enzyme deficiencies, such as lactose intolerance, that cause symptoms superficially indistinguishable from those of food allergy.
- Psychological reactions based on the belief that certain foods cause certain symptoms.

The simple dislike of a food may be a clue to allergy or to any of these reactions.

hyperactivity: inattentive and impulsive behavior that is more frequent and severe than is typical of others a similar age; professionally called **attention deficit hyperactivity disorder (ADHD)**.

▪▪ **FOOD DISLIKES** Parents are advised to watch for signs of food dislikes and take them seriously. Children's food aversions may be the result of nature's efforts to protect them from allergic or other adverse reactions. Test for allergies, and then apply nutrition knowledge conscientiously in deciding how to alter the diet. Don't risk feeding the child an unbalanced diet, which could lead to nutrient deficiencies. Whenever a food is excluded from the diet, care must be taken to include other foods that offer the same nutrients as the omitted food. Remember that children who must avoid certain foods need all their nutrients, just as other children do.

▪▪ **Hyperactivity**

Hyperactivity affects behavior and learning in about 5 percent of young school-aged children. Left untreated, it can interfere with a child's social development and ability to learn. Treatment focuses on relieving the symptoms and controlling the associated problems; there is no cure. Physicians often manage hyperactivity through behavior modification, special educational techniques, psychological counseling, and, in some cases, drug therapy.

Table 11-4
FOODS THAT MOST OFTEN CAUSE ALLERGIES

Eggs	Shellfish
Fish	Soybeans
Milk	Wheat
Peanuts	

SOURCE: Adapted from R. U. Sorensen, M. C. Porch, and L. C. Tu, Food allergy in children, *Textbook of Pediatric Nutrition*, 2nd ed., ed. R. M. Suskind and L. Lewinter-Suskind (New York: Raven Press, 1993), pp. 457–469.

Case Study

BOY WITH DISRUPTIVE BEHAVIOR

Freddie is a six-year-old boy who seldom sits still, often misbehaves, and is frequently sick. Freddie's eating habits are erratic and poor, as is his appetite. He often misses breakfast because he is too tired to get up in time to eat before school. By mid-morning, Freddie is irritable and disruptive in the classroom. At lunchtime he trades the peanut butter and banana sandwich his mother packed in his lunchbox for a piece of cake. After school he hurries home to watch television while he eats his favorite snack—cola and potato chips. At dinnertime Freddie picks at his food because he isn't very hungry. Later on, when it's time for bed, Freddie complains that he's hungry. His parents let him stay up to have a bowl of cereal (the kind with marshmallows) before he finally falls asleep.

✚ What factors in Freddie's daily routine might be contributing to his restless behavior?

✚ Discuss some changes in diet that might improve Freddie's health and disposition.

Parents of hyperactive children sometimes seek help from alternative therapies, including special diets. They mistakenly believe a solution may lie in manipulating the diet—most commonly, by excluding sugar or food additives. Adding carrots or eliminating candy is such a simple solution that many parents eagerly give such diet advice a try. These dietary changes will not solve the problem of true hyperactivity. Studies have consistently found no convincing evidence that sugar causes hyperactivity or worsens behavior.[15] Recommendations to restrict sugar in children's diets to prevent or treat behavior problems are groundless. The accompanying case study offers an opportunity to think about these issues in relation to a specific child.

Children can become excitable, rambunctious, and unruly out of a desire for attention, lack of sleep, overstimulation, watching too much television or playing too many video games, or a lack of physical activity. Such behaviors may suggest that more consistent care is needed. It helps to insist on regular hours of sleep, regular mealtimes, and regular outdoor activity.

■■ Food Choices and Eating Habits of Children

The childhood years are the parents' last chance to influence their children's food choices. Parents are **gatekeepers,** controlling the availability of foods in their children's environments. Gatekeepers who want to promote nutritious choices and healthful habits provide access to nutrient-dense, delicious foods and opportunities for active play at home. Food choices and regular physical activity can not only promote healthy growth but, as mentioned earlier, can also help prevent the degenerative diseases of later life. Many experts agree that early childhood is the time to put into effect practices that, until recently, were recommended only for adults. Childhood obesity and the early development of chronic diseases are the topic of the Nutrition in Practice that follows this chapter.

gatekeeper: with respect to nutrition, a key person who controls other people's access to foods and thereby exerts a profound impact on their nutrition. Examples are the spouse who buys and cooks the food, the parent who feeds the children, and the caregiver in a day-care center.

Mealtimes at Home

Feeding children entails not only providing a variety of nutritious foods but also nurturing the children's self-esteem and well-being. Parents face a number of challenges in preparing meals that both appeal to their children's tastes and provide needed nutrients. Because the interactions between parents and children can set the stage for lifelong attitudes and habits, a child's preferences should be treated with respect, even when nutrient needs must take precedence.

Little children like little tables and little portions.

HONORING CHILDREN'S PREFERENCES Researchers attempting to explain children's food preferences encounter many contradictions. Children say they like colorful foods, yet most often reject green and yellow vegetables while favoring brown peanut butter and white potatoes, apple wedges, and bread. They do like raw vegetables better than cooked ones, though, so it is wise to offer vegetables that are raw or slightly undercooked and crunchy and bright in color. They should be warm, not hot, because a child's mouth is much more sensitive than an adult's. The flavor should be mild (a child has more taste buds), and smooth foods such as mashed potatoes or pea soup should have no lumps (a child wonders, with some disgust, what the lumps might be).

Young children like to eat at little tables and to be served little portions of food. They also love to eat with other children and have been observed to stay at the table longer and eat more food when in the company of their peers. Parents who serve food in a relaxed and casual manner, without anxiety, provide an environment in which a child's negative emotions will be minimized.

AVOIDING POWER STRUGGLES It is not surprising that problems over food often arise during the second or third year, when children begin asserting their independence. Many of these problems stem from the conflict between children's developmental stages and capabilities and parents who, in attempting to do what they think is best for their children, try to control every aspect of eating. Such conflicts can disrupt children's abilities to regulate their own food intakes or to determine their own likes and dislikes. For example, many people share the misconception that children must be persuaded or coerced to try new foods. In fact, the opposite is true. When children are forced to try new foods, even by way of rewards, they are less likely to try those foods again than are children who are left to decide for themselves. The parent is responsible for *what* the child is offered to eat, but the child is responsible for *how much* and even *whether* to eat.[16]

When introducing new foods at the table, parents are advised to offer them one at a time and only in small amounts at first. The more often a food is presented to a young child, the more likely the child will like that food. Whenever possible, the new food should be presented at the beginning of the meal, when the child is hungry, but the child should make the decision to accept or reject it. Parents have their own inclinations and dislikes; so do children. It is best never to make an issue of food acceptance. A power struggle almost invariably establishes a firm pattern of resistance and permanently closes the child's mind.

TELEVISION'S INFLUENCE Watching television adversely affects children's nutritional health. As Chapter 7 pointed out, watching television contributes to obesity. Children who watch a lot of television are likely to become obese. Not only are they inactive, but they often snack on the fattening foods that are advertised.

The average child sees about 10,000 commercials a year, and almost all of them urge viewers to purchase sugarcoated breakfast cereals, candy bars, chips, fast foods, and carbonated beverages. Those foods add sugar, fat, and salt to the diet and displace foods that provide needed nutrients. Many parents and pediatricians believe that food ads aimed at children should be banned because they support corporate profits rather than children's health. Alternatively, parents can teach their children how to evaluate food ads and make healthful choices.

■■ **PREVENTING CHOKING** When feeding children, parents must always be alert to the dangers of choking. A choking child is a silent child—an adult should be present whenever a child is eating. Make sure the child sits when eating; choking is more likely when a child is running or falling. Round foods such as grapes, nuts, hard candies, marshmallows, and hot dog pieces are hard to control in a mouth with few teeth, and they can easily become lodged in the small opening of a child's trachea. Other potentially dangerous foods include tough meat, raw carrots, popcorn, and chips.

■■ **PLAY FIRST** Ideally, each meal is preceded, not followed, by the activity the child looks forward to the most. A number of schools have discovered that children eat a much better lunch if it is served after, rather than before, recess. Otherwise children "hurry up and eat" so that they can go play.

■■ **CHILD PARTICIPATION** Allowing children to help plan and prepare the family's meals provides enjoyable learning experiences and encourages children to eat the foods they have prepared. Vegetables are pretty, especially when fresh, and provide opportunities for children to learn about color, growing things and their seeds, and shapes and textures—all of which are fascinating to young children. Measuring, stirring, decorating, and arranging foods are skills that even a very young child can practice with enjoyment and pride (see Table 11-5).

■■ **SNACKS** Parents may find that their children often snack so much that they are not hungry at mealtimes. Instead of teaching children *not* to snack, teach them *how* to snack. Provide snacks that are as nutritious as the foods served at mealtime. Snacks can even be mealtime foods that are served individually over time, instead of all at once on one plate. When providing snacks to children, think of the food groups and offer such snacks as pieces of cheese, tangerine slices, carrot sticks, and peanut butter on whole-wheat crackers (see Table 11-6 on p. 274). Snacks that are easy to prepare should be readily available to children, especially if they arrive home from school before their parents.

■■ **PREVENTING DENTAL CARIES** Children frequently snack on sticky, sugary foods that stay on the teeth and provide an ideal environment for the growth of bacteria that cause dental caries. Teach children to brush and floss after meals, to brush or rinse after eating snacks, to avoid sticky foods, and to select crisp or fibrous foods frequently.

■■ **SERVING AS ROLE MODELS** In an effort to practice these many tips, parents may overlook perhaps the single most important influence on their children's food habits—themselves. Parents who do not eat carrots should not be surprised when their children refuse to eat carrots. Likewise, parents who dislike the smell of brussels sprouts may not be able to persuade children to try them. Children learn much through imitation. Parents, older siblings, and other caregivers set an irresistible example by sitting with younger children, eating the same foods, and having pleasant conversations during mealtime.[17]

■■ **Nutrition at School**

While parents are doing what they can to establish good eating habits in their children at home, others are preparing and serving foods to their children at

---

Table 11-5
FOOD SKILLS OF PRESCHOOLERS[a]

Age 1–2 years, when large muscles develop, the child:
- uses short-shanked spoon.
- helps feed self.
- lifts and drinks from cup.
- helps scrub, tear, break, or dip foods.

Age 3 years, when medium hand muscles develop, the child:
- spears food with fork.
- feeds self independently.
- helps wrap, pour, mix, shake, or spread foods.
- helps crack nuts with supervision.

Age 4 years, when small finger muscles develop, the child:
- uses all utensils and napkin.
- helps roll, juice, mash, or peel foods.
- cracks egg shells.

Age 5 years, when fine coordination of fingers and hands develops, the child:
- helps measure, grind, cut, and grate.
- uses hand-cranked egg beater with supervision.

[a]These ages are approximate. Healthy, normal children develop at their own pace. SOURCES: Adapted from M. Sigman-Grant, Feeding preschoolers: Balancing nutrition and developmental needs, *Nutrition Today*, July/August 1992, pp. 13–17; A. A. Hertzler, Preschoolers' food handling skills—Motor development, *Journal of Nutrition Education* 21 (1989): 100B–100C.

Table 11-6

HEALTHFUL SNACK IDEAS—THINK FOOD GROUPS, ALONE AND IN COMBINATION

Selecting two or more foods from different food groups adds variety and nutrient balance to snacks. The combinations are endless, so be creative.

Grain Products

Grain products are filling snacks, especially when combined with other foods:

- Cereal with fruit and milk.
- Crackers and cheese.
- Wheat toast with peanut butter.
- Popcorn with grated cheese.
- Oatmeal raisin cookies with milk.

Vegetables

Cut-up fresh, raw vegetables make great snacks alone or in combination with foods from other food groups:

- Celery with peanut butter.
- Broccoli, cauliflower, and carrot sticks with a flavored yogurt or cottage cheese dip.

Fruits

Fruits are delicious snacks and can be eaten alone—fresh, dried, or juiced—or combined with other foods:

- Apples and cheese.
- Bananas and peanut butter.
- Peaches with yogurt.
- Raisins mixed with sunflower seeds or nuts.

Meats and Meat Alternates

Meat and meat alternates add protein to snacks:

- Refried beans with nachos and cheese.
- Tuna on crackers.
- Luncheon meat on wheat bread.

Milk and Milk Products

Milk can be used as a beverage with any snack, and many other milk products, such as yogurt and cheese, can be eaten alone or with other foods as listed above.

The school breakfast must contain at a minimum:
- *One serving of fluid milk.*
- *One serving of fruit or vegetable or full-strength juice.*
- *Two servings of bread or bread alternates; or two servings of meat or meat alternates; or one of each.*

day-care centers and schools. In addition, children begin learning about food and nutrition in the classroom. Meeting the nutrition and education needs of children is critical to supporting their healthy growth and development.[18]

The U.S. government funds several programs to provide nutritious, high-quality meals for children at school. Both the School Breakfast Program and the National School Lunch Program provide meals at a reasonable cost to children from families with the financial means to pay. Meals are available free or at reduced cost to children from low-income families.

■■ SCHOOL BREAKFAST The School Breakfast Program is available in slightly more than half of the nation's schools, and about 5 million children participate in it. Surveys show that the majority of children who eat school breakfasts are from low-income families. As research results continue to emphasize the positive impact breakfast has on school performance and health, campaigns to expand school breakfast programs are under way.

■■ SCHOOL LUNCH More than 25 million children receive lunches through the National School Lunch Program—half of them free or at a reduced price. School lunches are designed to provide at least a third of the

Table 11-7

SCHOOL LUNCH PATTERNS FOR DIFFERENT AGES

Food Group	Preschool (Age)		Grade School through High School (Grade)[a]		
	1 to 2	3 to 4	K to 3	4 to 6	7 to 12
Milk					
1 serving of fluid milk[b]	¾ c	¾ c	1 c	1 c	1 c
Meat or Meat Alternate					
1 serving:					
Lean meat, poultry, or fish	1 oz	1½ oz	1½ oz	2 oz	3 oz
Cheese	1 oz	1½ oz	1½ oz	2 oz	3 oz
Large egg(s)	½	¾	¾	1	1½
Cooked dry beans or peas	¼ c	⅜ c	⅜ c	½ c	¾ c
Peanut butter	2 tbs	3 tbs	3 tbs	4 tbs	6 tbs
Peanuts, soynuts, tree nuts, or seeds[c]	½ oz	¾ oz	¾ oz	1 oz	1½ oz
Vegetable and/or Fruit					
2 or more servings, both to total	½ c	½ c	½ c	¾ c (plus ½ c extra over a week)	¾ c
Bread or Bread Alternate					
Servings[d]	5 per week (minimum ½ per day)	8 per week (minimum 1 per day)	8 per week (minimum 1 per day)	8 per week (minimum 1 per day)	10 per week (minimum 1 per day)

[a]These patterns may be used so long as the meals served meet the *Dietary Guidelines for Americans* and provide one-third of the child's recommendations for nutrients.
[b]Whole milk and unflavored low-fat milk must be offered; flavored milks or fat-free milk may also be offered.
[c]These foods may meet no more than one-half a serving of meat and must be accompanied by other meat or alternate in the meal.
[d]A serving is 1 slice of whole-grain or enriched bread; a whole-grain or enriched biscuit, roll, muffin, or the like; or ½ cup cooked rice, pasta, or other grain.
SOURCE: U.S. Department of Agriculture, 1998.

recommendation for energy, protein, vitamin A, vitamin C, iron, and calcium. They must also include specified numbers of servings from each food group. Table 11-7 shows school lunch patterns for children of different ages.

Parents often rely on school lunches to meet a significant part of their children's nutrient needs on school days. Indeed, students who regularly eat school lunches have higher intakes of energy and nutrients than students who do not. Children do not always like what they are served, however, and school lunch programs must strike a balance between what children want to eat and what will nourish them and guard their health.

Many schoolchildren in the United States have significant risk factors for developing cardiovascular disease.[19] In an effort to help reduce their risk, the U.S. Department of Agriculture (USDA) has ruled that all government-funded meals served at schools must follow the *Dietary Guidelines for Americans*.[20] This admirable decision leaves many schools with a problem, however. Though a school's own cafeteria may serve such meals, private vendors offer unregulated meals, even fast foods, side-by-side with the school lunches. Children receive a mixed message when they are left on their own to choose between the health-supporting school lunch and the high-fat, high-salt, low-nutrient-density foods that their taste buds may prefer.[21]

The American Dietetic Association has set nutrition standards for child-care programs. Among these standards, meal plans should:
- *Be nutritionally adequate.*
- *Involve parents in planning.*
- *Meet the* Dietary Guidelines for Americans.
- *Follow recommended meal patterns while respecting cultural and ethnic differences.*
- *Minimize added fat, sugar, and sodium.*
- *Emphasize fresh fruit, fresh and frozen vegetables, and whole grains.*
- *Respect children's small appetites.*

SOURCE: Position of The American Dietetic Association: Nutrition standards for child-care programs, *Journal of the American Dietetic Association* 99 (1999): 981–988.

■■ The Teen Years

As children pass through **adolescence** on their way to becoming adults, they change in many ways. Their physical changes make their nutrient needs high,

adolescence: the period of growth from the beginning of puberty until full maturity. Timing of adolescence varies from person to person.

and their emotional, intellectual, and social changes make meeting those needs a challenge.

Teenagers make many more choices for themselves than they did as children. They are not fed, they eat; they are not sent out to play, they choose to go. At the same time, social pressures thrust choices at them: whether to drink alcoholic beverages and whether to develop their bodies to meet extreme ideals of slimness or athletic prowess. Their interest in nutrition derives from personal, immediate experiences. They are concerned with how diet can improve their lives now—they engage in crash dieting in order to fit into a new bathing suit, avoid greasy foods in an effort to clear acne, or eat a plate of pasta to prepare for a big sporting event. In presenting information on the nutrition and health of adolescents, this chapter includes these topics of particular interest to teens.

Growth and Development during Adolescence

With the onset of adolescence, the steady growth of childhood speeds up abruptly and dramatically, and the growth patterns of females and males become distinct. Hormones direct the intensity and duration of the adolescent growth spurt, profoundly affecting every organ of the body, including the brain. After two to three years of intense growth and a few more at a slower pace, physically mature adults emerge.

In general, a female's adolescent growth spurt begins at age 10 or 11, and a male's, at 12 or 13. The spurt's duration is about two and a half years. Before **puberty,** the differences between male and female body composition are minimal. During the adolescent spurt, gender differences become apparent in the skeletal system, lean body mass, and fat stores. In males, the lean body mass—muscle and bone—becomes much greater, and in females, fat becomes a larger percentage of the total body weight. On average, males grow 8 inches taller, and females, 6 inches taller. Males gain approximately 45 pounds, and females, about 35 pounds.

Growth charts used for children must be abandoned when the signs of puberty begin to appear. Age in years indicates little about development. One way to monitor teen growth is to compare height and weight with previous measures taken at intervals. Rating scales based on stages of adolescent development are available and widely used to record developmental changes during puberty.

puberty: the period in life in which a person becomes physically capable of reproduction.

Energy and Nutrient Needs

The energy needs of adolescents vary greatly, depending on the current rate of growth, body size, and physical activity. Males' energy needs may be especially high; they experience a more intense growth spurt and, as mentioned, develop more lean body mass than females do. An active boy of 15 may need 4000 kcalories or more a day just to maintain his weight. In general, because females enter their growth spurts earlier and grow less than males, their energy needs peak sooner and decline earlier than those of their male peers. An inactive girl of 15 whose growth is nearly at a standstill may need fewer than 2000 kcalories a day if she is to avoid excessive weight gain. Thus, teenage girls need to pay special attention to being physically active and selecting foods of high nutrient density in order to meet their nutrient needs without exceeding their energy needs.

OBESITY The insidious problem of obesity becomes ever more apparent in adolescence and often continues into adulthood. One in every nine teens is overweight.*[22] The problem is most evident in females, especially those of African American descent. Without intervention, overweight adolescents will face numerous physical and socioeconomic consequences for years to come. The consequences of obesity are so dramatic and our society's attitude toward obese people is so negative that even adolescents of normal weight perceive a need to control their weight. When taken to the extremes, restrictive diets bring dramatic physical consequences of their own, as Nutrition in Practice 7 explains.

:: VITAMINS Recommendations for most vitamins increase during the teen years (see the tables on the inside front cover). Several of the vitamin recommendations for adolescents are similar to those for adults, including the new recommendation for vitamin D. During puberty, both the activation of vitamin D and the absorption of calcium are enhanced, thus supporting the intense skeletal growth of the adolescent years without additional vitamin D.

:: IRON The need for iron during adolescence differs for males and females. Iron needs increase in females as they start to menstruate and in males as their lean body mass develops. Iron intakes often fail to keep pace with increasing needs, especially for females, who typically consume less iron-rich foods such as meat and fewer total kcalories than males. For females, the RDA rises at adolescence and remains high into middle age. For males, the RDA returns to preadolescent values in early adulthood.

:: CALCIUM Adolescence is a crucial time for bone development, and the requirement for calcium reaches its peak during these years.[23] Unfortunately, many adolescents have calcium intakes below the current recommendation of 1300 milligrams per day through age 18.[24] Low calcium intakes during the adolescent growth spurt, especially if paired with physical inactivity, may compromise the development of peak bone mass.[25] In contrast, increasing milk products in the diet to meet calcium recommendations greatly increases bone density.[26] The attainment of maximal bone mass is considered the best protection against age-related bone loss and fractures. Once again, teenage girls are most vulnerable, for their milk—and therefore calcium—intakes begin to decline at the time when their calcium needs are greatest. Furthermore, women experience much greater bone losses than men in later life. In addition to dietary calcium, sports activities during adolescence build strong bones.[27]

:: Food Choices and Health Habits

Adolescents like the freedom to come and go as they choose and eat what they want when they have time. With a multitude of after-school, social, and job activities, they almost inevitably fall into irregular eating habits. At any given time on any given day, a teenager may be skipping a meal, eating a snack, preparing a meal, or consuming food prepared by a parent or restaurant.

---

*For males, obesity is defined as BMI:

    ≥23.0 for 12 to 14 years.

    ≥25.8 for 15 to 17 years.

    ≥26.8 for 18 to 19 years.

For females, obesity is defined as BMI:

    ≥23.4 for 12 to 14 years.

    ≥24.8 for 15 to 17 years.

    ≥25.7 for 18 to 19 years.

Nutritious snacks play an important role in an active teen's diet.

Appendix A provides a table of the caffeine contents of beverages, foods, and medications.

Table 11-8
SOFT DRINK CONSUMPTION BY U.S. ADOLESCENTS

In general, soft drink consumption is inversely associated with the consumption of the nutrients in milk and fruit juices.

This Percentage of Adolescents:	Consumes This Many Soft Drinks Each Day:
22%	More than 26 oz (more than 3¼ c or 2 cans)
28%	13 to 26 oz (1¾ c to 3¼ c or 2 cans)
32%	Up to 13 oz (about 1¾ c or 1 can)
18%	None

SOURCE: Data from L. Harnack, J. Stang, and M. Story, Soft drink consumption among U.S. children and adolescents: Nutritional consequences, *Journal of the American Dietetic Association* 99 (1999): 436–441.

SNACKS Snacks typically provide at least a fourth of the average teenager's daily food energy intake. Most often, favorite snacks are high in fat and low in calcium, iron, vitamin A, vitamin C, and folate.[28] Most adolescents need to eat a greater variety of foods to obtain these nutrients. Table 11-6 on p. 274 shows how to combine foods from different food groups to create healthy snacks. Unfortunately, vending machines rarely offer nutrient-dense options, and nutrition information alone does not convince people to make healthy choices.[29]

BEVERAGES Teenagers frequently drink soft drinks with lunch, supper, and snacks (see Table 11-8). About the only time they select fruit juices is at breakfast. When they drink milk, they are more likely to consume it with a meal (especially breakfast) than as a snack. Soft drinks, when chosen as the primary beverage, may affect the density of the bones because they displace milk from the diet. Because of their greater food intakes, males are more likely to drink enough milk to meet their calcium needs, whereas females typically fall short of calcium recommendations.

Soft drinks present another problem, too, when caffeine intake becomes excessive. Caffeine is a stimulant added during the manufacture of many soft drinks; on the average, caffeine-containing soft drinks deliver between 30 and 55 milligrams of caffeine per 12-ounce can.[30] Caffeine increases the respiration rate, heart rate, blood pressure, and secretion of stress and other hormones. Caffeine seems to be relatively harmless, however, when used in moderate doses (the equivalent of fewer than, say, three 12-ounce cola beverages a day). In greater amounts, it can cause the symptoms associated with anxiety—sweating, tenseness, and inability to concentrate.

EATING AWAY FROM HOME Adolescents eat about one-third of their meals away from home, and their nutritional welfare is enhanced or hindered by the choices they make.[31] A lunch of a hamburger, a chocolate shake, and french fries supplies substantial quantities of many nutrients at a kcalorie cost of about 800, an energy intake many adolescents can afford (see Table 11-9). When they eat this sort of lunch, teens can adjust their breakfast and dinner choices to include fruits and vegetables for vitamin A, vitamin C, folate, and fiber, and lean meats for iron and zinc.

PEER INFLUENCE Adolescents are intensely engaged in day-to-day life with their peers and preparing for their future lives as adults. Adults need to remember that adolescents have the right to make their own decisions—even if those decisions are not in line with the adults' own views. Gatekeepers can set up the environment so that nutritious foods are available and can stand by with reliable nutrition information and advice, but the rest is up to the teenagers. Ultimately, they make the choices.

Problems Adolescents Face

Physical maturity and growing independence present adolescents with new choices to make. The consequences of those choices will influence their nutritional health both today and throughout life. Some teenagers begin using drugs, alcohol, and tobacco; others wisely refrain. Information about the use of these substances is presented here because most people are first exposed to them during adolescence, but it actually applies to people of all ages.

MARIJUANA One out of every three high school students reports having at least tried marijuana.[32] When inhaled by smoking, the active chem-

Table 11-9

SELECTED NUTRIENTS IN A HAMBURGER, LOW-FAT CHOCOLATE SHAKE, AND SMALL SERVING OF FRENCH FRIES

Nutrient	Percentage of RDA for a Male[a]	Percentage of RDA for a Female[a]
Energy	27	37
Protein	47	63
Fat[b]	24	33
Calcium[c]	39	39
Iron	36	29
Zinc	17	22
Vitamin A	0	0
Folate[d]	12	12
Vitamin C[e]	15	18
Sodium[b]	38	38

[a]RDA for a 15- to 18-year-old, moderately active person of average height and weight.
[b]Daily Values used for fat and sodium.
[c]1997 calcium AI value used.
[d]1998 folate RDA.
[e]2000 vitamin C RDA.

icals are rapidly and almost completely absorbed from the lungs.* They then travel in the blood to the various body tissues that metabolize them. The active ingredients from a single marijuana cigarette can linger in the body's fat a month or more before being excreted in the urine.

Marijuana is unique among drugs in that it seems to enhance the enjoyment of eating, especially of sweets, a phenomenon commonly known as "the munchies." Why or how this effect occurs is not known; it may be a social effect induced by suggestibility, or perhaps the drug stimulates appetite. Whatever the reason, prolonged use of the drug does not seem to bring about a weight gain.

■■ **COCAINE** One in 20 high school seniors reports having used cocaine at least once.[33] Cocaine stimulates the nervous system and elicits the stress response—constricted blood vessels, raised blood pressure, widened pupils of the eyes, and increased body temperature. It also drives away feelings of fatigue. Cocaine occasionally causes immediate death—usually by heart attack, stroke, or seizure in an already damaged body system.

Weight loss is common, and cocaine abusers often develop eating disorders. Notably, the craving for cocaine replaces hunger; rats given unlimited cocaine will choose it over food until they starve to death. Thus, unlike marijuana use, cocaine use has major nutritional consequences.

■■ **DRUG ABUSE, IN GENERAL** The effects of other addictive drugs vary in degree but are similar to those caused by cocaine. Drug abusers face the multiple nutrition problems listed in the margin. During withdrawal from drugs, an important part of treatment is to identify and correct these nutrition problems.

■■ **ALCOHOL ABUSE** Sooner or later all teenagers face the decision whether to drink alcohol. The law forbids the sale of alcohol to people under 21, but most adolescents who seek alcohol can obtain it. Four out of five high

Nutrition problems of drug abusers:
- *They buy drugs with money that could be spent on food.*
- *They lose interest in food during "highs."*
- *They use drugs that depress appetite.*
- *Their lifestyle fails to promote good eating habits.*
- *They use intravenous (IV) drugs. They may contract an HIV infection, hepatitis, or other infectious diseases, which increase their nutrient needs. Infections often interfere with appetite and can reduce food intake for many reasons.*
- *Medicines used to treat drug abuse may alter nutrition status.*

---

*The active ingredient of marijuana, which is primarily responsible for its intoxicating effects, is delta-9-tetrahydrocannabinol, or THC.

school students have had at least one alcoholic beverage; about half drink regularly; and one in three students drinks heavily (defined as five or more drinks on at least one occasion in the previous month).[34]

Nutrition in Practice 23 describes how alcohol affects nutrition status. To sum it up, alcohol provides energy but no nutrients, and it can displace nutritious foods from the diet. Alcohol alters nutrient absorption and metabolism, so imbalances develop.

■■ **SMOKING** The prevalence of cigarette smoking among U.S. adolescents is on the rise.[35] Cigarette smoking is a pervasive health problem that causes thousands of people to suffer from cancer and diseases of the cardiovascular, digestive, and respiratory systems. These effects are beyond the scope of nutrition, but smoking cigarettes does influence hunger, body weight, and nutrient status.

Smoking a cigarette eases feelings of hunger. When smokers receive a hunger signal, they can quiet it with cigarettes instead of food. Such behavior ignores body signals and postpones energy and nutrient intake. In rats, nicotine reduces food intake and increases the rate of energy expenditure, causing weight loss.[36]

Indeed, smokers tend to weigh less than nonsmokers and to gain weight when they stop smoking. Weight gain is often a concern for people contemplating giving up cigarettes. They should know that the average person who quits smoking gains less than 10 pounds. Smokers wanting to quit need to prepare for this possibility and adjust their diet and activity habits so as to maintain weight during and after quitting. Smoking cessation programs need to include strategies for weight management.

Nutrient intakes of smokers and nonsmokers differ. Smokers tend to have lower intakes of dietary fiber, vitamin A, beta-carotene, folate, and vitamin C.[37] The association between smoking and low vitamin intake may be noteworthy, considering the altered metabolism of vitamin C in smokers and the protective effect of vitamin A and beta-carotene against some types of cancer.

Research shows that compared to nonsmokers, smokers require an additional 35 milligrams of vitamin C per day to maintain steady body pools. Oxidants in cigarette smoke accelerate vitamin C metabolism and deplete smokers' body stores of this antioxidant; this depletion is even evident to some degree in nonsmokers who are exposed to passive smoke.[38]

Beta-carotene enhances the immune response and protects against some cancer activity.[39] Specifically, the risk of lung cancer is greatest for smokers who have the lowest intakes of carotene. Of course, such evidence should not be misinterpreted. It does not mean that as long as people eat their carrots, they can safely use tobacco. Nor does it mean that beta-carotene supplements would be beneficial. In fact, as mentioned in Chapter 8, some research shows that beta-carotene supplements may have adverse effects in smokers.[40] Smokers were ten times more likely to get lung cancer than nonsmokers. Both smokers and nonsmokers can, however, reduce their cancer risks by eating fruits and vegetables rich in carotene (see Nutrition in Practice 21 for details on antioxidant nutrients and disease prevention).

■■ **SMOKELESS TOBACCO** Nationwide, one in ten high school students reports having used smokeless tobacco products.[41] Like cigarettes, smokeless tobacco use is linked to many health problems, from minor mouth sores to tumors in the nasal cavities, cheeks, gums, and throat. The risk of mouth and throat cancers is even greater than for smoking tobacco. Other drawbacks to tobacco chewing and snuff dipping include bad breath, stained teeth, and blunted senses of smell and taste. Tobacco chewing also damages the gums, tooth surfaces, and jawbones, making it likely that users will lose their teeth in later life.

To review, nutrient needs rise dramatically as children enter the rapid growth phase of the teen years. The busy lifestyles of adolescents add to the challenge of meeting their nutrient needs—especially for iron and calcium. Assessment of nutrition status in healthy children and adolescents can confirm that development is normal or can catch potential problems early. The Nutrition Assessment Checklist highlights problems to look for when working with children and teenagers.

Chapter 13 offers details about nutrition assessment.

NUTRITION ASSESSMENT CHECKLIST — FOR CHILDREN AND ADOLESCENTS

Health Problems, Signs, and Symptoms

Check the medical record for:

- ☐ Food allergies
- ☐ Attention deficit hyperactivity disorder (ADHD)
- ☐ Diabetes or other chronic disorders
- ☐ Eating disorders
- ☐ Lactose intolerance
- ☐ Alcohol, tobacco, or illicit drug abuse
- ☐ Obesity
- ☐ Pregnancy

Medications

For children or adolescents being treated with drug therapy for medical conditions, note:

- ☐ Side effects that might reduce food intake or change nutrient needs
- ☐ Proper administration of medication with respect to food intake

Nutrient/Food Intake

For all children and teens, especially those considered at risk nutritionally, assess the diet for:

- ☐ Total energy
- ☐ Protein
- ☐ Calcium and iron
- ☐ Vitamin A, vitamin C, and folate
- ☐ Fiber

Note the following:

- ☐ Number of days each week a nutritious breakfast is eaten
- ☐ Number of hours the child or teen sleeps each day
- ☐ Number of soft drinks the child or teen drinks each day
- ☐ Number of fast-food meals eaten each day
- ☐ Number and type of snacks eaten each day
- ☐ Type and amount of physical activity
- ☐ Amount of caffeine consumed

Height and Weight

Measure baseline height and weight.

- ☐ Reassess height, weight, and growth patterns at each medical checkup.

- ☐ Note significant obesity or underweight and intervention strategies employed.

Laboratory Tests

Monitor the following laboratory tests for children and adolescents:

- ☐ Hemoglobin, hematocrit, or other tests of iron status
- ☐ Blood glucose for children or teens with diabetes
- ☐ Blood lead concentrations

Physical Signs

Look for physical signs of:

- ☐ Protein energy malnutrition
- ☐ Iron deficiency
- ☐ Vitamin A deficiency
- ☐ Vitamin C deficiency
- ☐ Folate deficiency

SELF CHECK

1. The Food Guide Pyramid for young children differs from the adult Food Guide Pyramid in that:
   a. there are more food groups to choose from.
   b. there are fewer food groups to choose from.
   c. a set number of servings, rather than a range, is suggested.
   d. a range of servings, rather than a set number, is suggested.

2. Children who are hungry may be irritable or apathetic because:
   a. they need vitamin D.
   b. their blood glucose is low.
   c. their blood glucose is high.
   d. they have had too many sweets.

3. Two infant feeding practices that have improved the iron status of older children in the United States are:
   a. feeding infants liver and starches.
   b. breastfeeding and iron-fortified formulas.

c. giving iron supplements and increasing milk intake.

d. feeding solids at an early age and giving iron-fortified milk.

4. Three symptoms of lead toxicity are:

a. diarrhea, irritability, and fatigue.

b. low blood sugar, hair loss, and skin rash.

c. increased heart rate, hyperactivity, and dry skin.

d. bleeding gums, brittle fingernails, and swollen glands.

5. Allergic reactions to foods are most often caused by:

a. corn, rice, or meats.

b. eggs, peanuts, or milk.

c. red meats, milk, or MSG.

d. seafood, dark greens, or lactose.

6. Which of the following is **not** true? Children who watch a lot of television are likely to:

a. become obese.

b. spend less time being physically active.

c. learn healthy eating tips from programs.

d. eat the foods most often advertised on television.

7. When introducing new foods to children:

a. reward children as they try new foods.

b. offer many choices to encourage variety.

c. offer one new food at the end of the meal.

d. offer one new food at the beginning of the meal.

8. During the growth spurt of adolescence:

a. females gain more weight than males.

b. males gain more fat, proportionately, than females.

c. differences in body composition between males and females become apparent.

d. similarities in body composition between males and females become apparent.

9. Two nutrients that are usually lacking in adolescents' diets are:

a. zinc and fat.

b. iron and calcium.

c. protein and thiamin.

d. vitamin A and riboflavin.

10. Smoking increases the need for:

a. iron.   b. folate.

c. vitamin C.   d. vitamin E.

Answers to these questions appear in Appendix H.

CLINICAL APPLICATIONS

1. At two and a half years old, Travis is healthy, though slightly underweight, and headstrong. Travis's mother hovers over him at every meal and insists that he take several bites of every food on his plate, even if he dislikes the food or it is unfamiliar to him. Even though Travis is hungry when he sits down to a meal, his mother's constant urging to get him to take bites quickly quells any interest the child had in eating. Travis simply folds his arms across his chest, closes his mouth tightly, and refuses to eat any more food. After more begging, pleading, and nagging, Travis's mother becomes angry and sends him away from the table. Travis is not allowed to snack between meals because his mother is concerned that snacks will ruin his appetite.

• What factors might be contributing to Travis's refusal to eat?

• Travis's mother is concerned about her son's underweight. What strategies would you suggest to help Travis gain weight?

• What advice would you offer Travis's mother to help her improve mealtimes with her son?

2. Loni is a physically inactive, slightly overweight 15-year-old who started smoking cigarettes at the age of 13 to help her lose weight. She planned to quit smoking as soon as she dropped a few pounds. Smoking reduced her appetite somewhat, but when she did eat, she chose high-fat snack foods, cola drinks, and fast foods such as chicken nuggets and burgers. At 15, Loni is still overweight and is smoking more than a pack of cigarettes a day. She uses her lunch money to fund her cigarette habit, which leaves only a little change for crackers or chips from the vending machine for lunch. She's noticed that her complexion looks dry and off-color, her teeth are less white than they used to be, she's easily fatigued, and she gets sick more often than ever before.

• What would you tell Loni to motivate her to stop smoking?

• What nutrients might be affected by Loni's smoking and how might Loni's diet be adjusted to meet these needs?

• What dietary advice would you suggest to help Loni look and feel healthier?

NUTRITI**ON**THENET

FOR FURTHER STUDY OF THE
TOPICS IN THIS CHAPTER,
ACCESS THESE WEB SITES.

kidshealth.org
Kids Health

www.kidsfood.org
Kids Food Cyber Club

www.ironpanel.org.au
*Australian Iron Status Advisory
Panel*

www.who.int
World Health Organization

phs.os.dhhs.gov/phs/phs.html
Public Health Service

nsc.org/ehc/lead.htm
*Environmental Health Center Lead
Program*

www.foodallergy.org
Food Allergy Network

www.aaaai.org
*American Academy of Allergy,
Asthma and Immunology*

chadd.org
*Children and Adults with
Attentional Disorders*

ncadd.org
*National Council on Alcoholism
and Drug Dependence*

lungusa.org
American Lung Association

Notes

[1] S. Shea and coauthors, Variability and self-regulation of energy intake in young children in their everyday environment, *Pediatrics* 90 (1992): 542–546.

[2] L. L. Birch and coauthors, Effects of a non-energy fat substitute on children's energy and macronutrient intake, *American Journal of Clinical Nutrition* 58 (1993): 326–333; L. L. Birch

and coauthors, The variability of young children's energy intake, *New England Journal of Medicine* 324 (1991): 232–235.

[3] V. Matkovic and J. Z. Ilich, Calcium requirements for growth: Are current recommendations adequate? *Nutrition Reviews* 51 (1993): 171–180.

[4] R. E. Klesges and coauthors, Parental influence on food selection in young children and its relationships to childhood obesity, *American Journal of Clinical Nutrition* 53 (1991): 859–864.

[5] A. F. Suber and coauthors, Dietary sources of nutrients among US children, 1989–1991, *Pediatrics* 102 (1998): 913–923; C. H. Ruxton and T. R. Kirk, Breakfast: A review of associations with measures of dietary intake, physiology, and biochemistry, *British Journal of Nutrition* 78 (1997): 199–214.

[6] J. M. Murphy and coauthors, Relationship between hunger and psychosocial functioning in low-income American children, *Journal of the American Academy of Child and Adolescent Psychiatry* 37 (1998): 163–170.

[7] E. Pollitt, Iron deficiency and cognitive function, *Annual Review of Nutrition* 13 (1993): 521–537.

[8] E. Pollitt, Iron deficiency and educational deficiency, *Nutrition Reviews* 55 (1997): 133–141.

[9] M. C. Holst, Developmental and behavioral effects of iron deficiency anemia in infants, *Nutrition Today* 33 (1998): 27–36.

[10] R. A. Goyer, Toxic and essential metal interactions, *Annual Review of Nutrition* 17 (1997): 37–50; P. Mushak and A. F. Crocetti, Lead and nutrition: Biologic interactions of lead with nutrients, *Nutrition Today* 31 (1996): 12–17.

[11] T. D. Matte, Reducing blood lead levels: Benefits and strategies, *Journal of the American Medical Association* 281 (1999): 2340–2342; Mushak and Crocetti, 1996.

[12] Matte, 1999; D. Farley, Dangers of lead still linger, *FDA Consumer*, January/February 1998, pp. 16–21.

[13] Federal Interagency Forum on Child and Family Statistics, *America's Children: Key National Indicators of Well-Being, 1999* (Washington, D.C.: Government Printing Office, 1999). Yearly updates available from http://www.childstats.gov.

[14] H. A. Sampson and D. D. Metcalfe, Food allergies, *Journal of the American Medical Association* 268 (1992): 2840–2844.

[15] J. W. White and M. Wolraich, Effect of sugar and mental performance, *American Journal of Clinical Nutrition* 62 (1995): S242–S249; M. L. Wolraich and coauthors, Effects of diets high in sucrose or

aspartame on the behavior and cognitive performance of children, *New England Journal of Medicine* 330 (1994): 301–307.

[16] C. Evers, Empower children to develop healthful eating habits, *Journal of the American Dietetic Association* 97 (1997): S116; E. Satter, *How to Get Your Kid to Eat ... But Not Too Much* (Palo Alto, Calif.: Bull Publishing Company, 1987), pp. 13–28.

[17] M. Nahikian-Nelms, Influential factors of caregiver behavior at mealtime: A study of child-care programs, *Journal of the American Dietetic Association* 97 (1997): 505–509.

[18] Position of The American Dietetic Association: Nutrition standards for child-care programs, *Journal of the American Dietetic Association* 99 (1999): 981–988.

[19] J. Anding and coauthors, Blood lipids, cardiovascular fitness, obesity, and blood pressure: The presence of potential coronary heart disease risk factors in adolescents, *Journal of the American Dietetic Association* 96 (1996): 238–242.

[20] K. Schuster, Feds put schools on a lowfat diet, *Food Management*, August 1994, pp. 78–84. Evidence that children grow well on a low-fat diet was provided by L. Van Horn, the principal investigator of the Dietary Intervention Study in Children (DISC), which follows children up to age 18, as cited in Kids grow well on low-fat diet, *Nutrition and the M.D.*, January 1996, pp. 6–7.

[21] The Writing Group for the DISC Collaborative Research Group, Efficacy and safety of lowering dietary intake of fat and cholesterol in children with elevated low-density lipoprotein cholesterol: The Dietary Intervention Study in Children (DISC), *Journal of the American Medical Association* 273 (1995): 1429–1435.

[22] Update: Prevalence of overweight among children, adolescents, and adults—United States, 1988–1994, *Morbidity and Mortality Weekly Report* 46 (1997): 199–202.

[23] A. D. Martin and coauthors, Bone mineral and calcium accretion during puberty, *American Journal of Clinical Nutrition* 66 (1997): 611–615.

[24] S. I. Barr, Associations of social and demographic variables with calcium intakes of high school students, *Journal of the American Dietetic Association* 94 (1994): 260–266, 269.

[25] V. C. Lysen and R. Walker, Osteoporosis risk factors in eighth grade students, *Journal of School Health* 67 (1997): 317–321.

[26] G. M. Chan, K. Hoffman, and M. McMurry, Effects of dairy products on bone and body composition in pubertal girls, *Journal of Pediatrics* 126 (1995): 551–556.

[27] A. M. Fehily and coauthors, Factors affecting bone density in young adults, *American Journal of Clinical Nutrition* 56 (1992): 579–586.

[28] J. G. Dausch and coauthors, Correlates of high-fat/low-nutrient-dense snack consumption among adolescents: Results from two national health surveys, *American Journal of Health Promotion* 10 (1995): 85–88.

[29] S. M. Hoerr and V. A. Louden, Can nutrition information increase sales of healthful vended snacks? *Journal of School Health* 63 (1993): 386–390.

[30] International Food Information Council, Caffeine and health: Clarifying the controversies, *IFIC Review,* May 1993.

[31] B. H. Lin, J. Guthrie, and J. R. Blaylock, *The Diets of America's Children—Influences of Dining Out, Household Characteristics, and Nutrition Knowledge* (Washington, D.C.: U.S. Department of Agriculture, December 1996).

[32] L. Kann and coauthors, Youth risk behavior surveillance—United States, 1993, *Journal of School Health* 65 (1995): 163–171.

[33] Kann and coauthors, 1995.

[34] Kann and coauthors, 1995.

[35] Centers for Disease Control and Prevention, Tobacco use among high school students—United States, 1997, *Journal of the American Medical Association* 279 (1998): 1250.

[36] S. R. Schwid, M. D. Hirvonen, and R. E. Keesey, Nicotine effects on body weight: A regulatory perspective, *American Journal of Clinical Nutrition* 55 (1992): 878–884.

[37] T. A. B. Sanders and coauthors, Essential fatty acids, plasma cholesterol, and fat-soluble vitamins in subjects with age-related maculopathy and matched control subjects, *American Journal of Clinical Nutrition* 57 (1993): 428–433; A. F. Subar, L. C. Harlan, and M. E. Mattson, Food and nutrient intake differences between smokers and non-smokers in the US, *American Journal of Public Health* 80 (1990): 1323–1329.

[38] D. L. Tribble, L. J. Giuliano, and S. P. Fortmann, Reduced plasma ascorbic acid concentrations in nonsmokers regularly exposed to environmental tobacco smoke, *American Journal of Clinical Nutrition* 58 (1993): 886–890.

[39] G. van Poppel, S. Spanhaak, and T. Ockhuizen, Effect of ß-carotene on immunological indexes in healthy male smokers, *American Journal of Clinical Nutrition* 57 (1993): 402–407.

[40] K. Smigel, Beta-carotene fails to prevent cancer in two major studies: CARET intervention stopped, *Journal of the National Cancer Institute* 88 (1996): 145; G. S. Omenn and coauthors, Effects of a combination of beta carotene and vitamin A on lung cancer and cardiovascular disease, *New England Journal of Medicine* 334 (1996): 1150–1155.

[41] Kann and coauthors, 1995.

Q&A NUTRITION IN PRACTICE

Childhood Obesity and the Early Development of Chronic Diseases

Disease of the heart and blood vessels, known as **cardiovascular disease, or CVD,** is the number one killer of adults in the United States and Canada, but CVD begins in childhood (see the accompanying glossary for cardiovascular disease and related terms). Over the past three decades, researchers have been observing how changes in body weight, blood lipids, blood pressure, and individual behaviors correlate with the development of CVD over time—from infancy to childhood through adolescence and into young adulthood. Some major findings have emerged from this research:

- Changes inside the arteries— changes predictive of CVD—are evident in childhood.

GLOSSARY

atherosclerosis (ath-er-oh-scler-OH-sis): a type of artery disease characterized by accumulations of lipid-containing material on the inner walls of the arteries.

> athero = porridge or soft
> scleros = hard
> osis = condition

cardiovascular disease (CVD): a general term for all diseases of the heart and blood vessels. Atherosclerosis is the main cause of CVD. When the arteries that carry blood to the heart muscle become occluded, the heart suffers damage known as **coronary heart disease (CHD).**

> cardio = heart
> vascular = blood vessels

fatty streaks: an accumulation of cholesterol and other lipids along the walls of the arteries.

fibrous plaques: mounds of lipid material, mixed with smooth muscle cells and calcium, which develop in the artery walls in atherosclerosis.

- Obesity in children affects these changes.
- Behaviors that influence the development of obesity and of CVD are learned and begin early in life. These behaviors include overeating, eating high-fat foods, physical inactivity, and cigarette smoking.

This Nutrition in Practice focuses on efforts to prevent childhood obesity and CVD, but the benefits extend to cancer, diabetes, and other chronic diseases as well. The years of childhood are emphasized here, for the earlier in life health-promoting habits become established, the better they will stick.

amhrt.org
American Heart Association

What about genetics? Don't some people inherit the tendency to develop CVD regardless of the lifestyle habits they adopt?

Genetics does not appear to play a *determining* role in CVD; that is, a person is not simply destined at birth to develop CVD.[1] Instead, genetics appears to play a *permissive* role: the potential is inherited, and then CVD will develop, if given a push by poor health choices such as excessive weight gain, poor diet, sedentary lifestyle, and cigarette smoking.

How does CVD develop, and when does its development begin?

Most CVD involves **atherosclerosis**— the accumulation of cholesterol and other blood lipids along the walls of the arteries. Frequently, atherosclerosis alters the flow of blood to the heart and can lead to hypertension and coronary heart disease (CHD), which, in turn, raises the likelihood of a heart attack. When atherosclerosis alters blood flow to the brain, a stroke can result. Infants are born with healthy, smooth, clear arteries, but within the first decade of life, **fatty streaks** may begin to appear. During adolescence, these fatty streaks may begin to turn to **fibrous plaques** (Figure 21-1 in Chapter 21 shows the formation of plaques in atherosclerosis). By early adulthood, the fibrous plaques may begin to calcify and become raised lesions, especially in boys and young men. As the lesions grow more numerous and enlarge, the heart disease rate begins to rise, and the rise becomes dramatic at about age 45 in men and 55 in women. From this point on, arterial damage and blockage progress rapidly, and heart attacks and strokes threaten life. In short, the consequences of atherosclerosis, which become apparent only in adulthood, have their beginnings in the first decades of life.[2]

Atherosclerosis is not inevitable; people can grow old with relatively clear arteries. Early lesions may either progress or regress depending on several factors, many of which reflect lifestyle behaviors. Smoking, for example, is strongly associated with the prevalence of raised lesions, even in young adults.

Parents don't need to worry about their children's blood cholesterol, do they?

Atherosclerotic lesions reflect blood cholesterol: as blood cholesterol increases, more lesions develop. Cholesterol values at birth are similar in all populations; differences emerge in early childhood. In countries where the adults have high blood cholesterol and high rates of CVD, the children also tend to have high blood cholesterol. Conversely, in countries where the adults have low blood cholesterol and low rates of CVD, the children tend to have low blood cholesterol, suggesting that adult heart disease tracks early and that early preventive efforts might reduce the incidence of later CVD.

Such is the case among populations, but individual cholesterol status also becomes established in childhood, as early as one year.[3] The best predictors of a person's blood cholesterol are that person's earlier baseline values: childhood values correlate with values in young adulthood.[4] Quite simply, if you want to know a child's future cholesterol, measure it now. Standard values for cholesterol screening in children and adolescents are listed in Table NP11-1.[5]

Table NP11-1
CHOLESTEROL VALUES FOR CHILDREN AND ADOLESCENTS

Disease Risk	Total Cholesterol (mg/dL)	LDL Cholesterol (mg/dL)
Acceptable	<170	<110
Borderline	170–199	110–129
High	≥200	≥130

NOTE: Adult values appear in Chapter 21. A deciliter (dL) is one-tenth of a liter or 100 milliliters.

Is hypertension a concern for children and adolescents?

Pediatricians routinely monitor blood pressure in children and adolescents.[6] High blood pressure may signal an underlying disease or the early onset of hypertension, which accelerates the development of atherosclerosis. Standard values for hypertension screening in children and adolescents are given in Table NP11-2.[7]

Like atherosclerosis and high blood cholesterol, hypertension may develop in the first decades of life.[8] Children can control their hypertension by participating in regular aerobic activity and by losing weight or maintaining their weight as they grow taller.[9] No evidence suggests a benefit of restricting sodium to lower blood pressure in children and adolescents.[10]

Table NP11-2
HYPERTENSION STANDARDS FOR CHILDREN AND ADOLESCENTS: SYSTOLIC OVER DIASTOLIC PRESSURE (mm Hg)

	6 to 9 yr	10 to 12 yr	13 to 15 yr	16 to 18 yr
High normal	111–121 over 70–77	117–125 over 75–81	124–135 over 77–85	127–141 over 80–91
Significant hypertension	122–129 over 70–85	126–133 over 82–89	136–143 over 86–91	142–149 over 92–97
Severe hypertension	>129 over >85	>133 over >89	>143 over >91	>149 over >97

Is obesity a problem among children today?

Many experts agree that preventing or treating obesity in childhood will reduce the rate of CVD in adulthood. Without intervention, overweight children become overweight adolescents who become overweight adults, and being overweight exacerbates every chronic disease that adults face.[11]

Children are heavier today than they were 20 or so years ago. Since the late 1970s, the prevalence of overweight has almost doubled for children—and more than doubled for adolescents.[12] This pattern is a secular trend—that is, one that cannot be explained by genetics. Diet and lack of physical activity must be responsible.

Is it true that children are eating more food and more fat than ever before?

No. Children's energy intakes have remained relatively stable in recent decades. There has even been a slight decline in fat intake, from 38 to 36 percent of kcalories from fat daily.[13] This slight decline in dietary fat is not enough, however, to have influenced body weight; nor is it enough to meet current dietary recommendations.

Children's dietary fat intakes vary, of course, and some children do eat high-fat diets. Children who prefer high-fat foods tend to consume a relatively large percentage of their energy intake from fat.[14] They also tend to be more overweight than their peers. Particularly noteworthy is the finding that children's fat preferences and consumption correlate with their parents' obesity as well. Such findings confirm the significant roles parents play in teaching children about healthy food choices, providing children with low-fat selections, and serving as role models.

If diet alone is not to blame for the increasing prevalence of obesity among the young, then what is?

Most likely, children have grown more overweight because of their lack of physical activity.[15] An inactive child can become obese even while eating less food than an active child. Today's children are more sedentary and less physically fit than children were 20 years ago.

Watching television accounts for some 24 hours a week of sedentary behavior. Beyond these 24 hours, children spend more sedentary time sitting at computers and playing video games. Both obesity and blood cholesterol correlate with hours of television viewed.[16]

Just as blood cholesterol and obesity track over the years, so does a person's level of physical activity. Researchers studying almost 1000 adolescents found that over half of those who were initially described as inactive remained inactive six years later.[17] Similarly, almost half of those who were physically active remained so. Compared with inactive teens, those who were physically active weighed less, smoked less, ate a diet lower in saturated fats, and had a better blood lipid profile. The message is clear: physical activity offers numerous health benefits, and children who are active today are most likely to be active for years to come.

What else can concerned adults do to help prevent childhood obesity?

In light of all these findings, parents and teachers are encouraged to make major efforts to prevent childhood obesity. Suggestions include the following: encourage children to eat slowly, to pause and enjoy their table companions, and to stop eating when they are full. Teach them how to select low-fat snacks and to serve themselves appropriate portions. Never force children to clean their plates.

Above all, be sensitive in teaching children nutrition principles that can help to prevent obesity. Children can easily get the idea that their worth is tied to their body weight. Some parents fail to realize that society's ideal of slimness can be perilously close to starvation, and that a child encouraged to "diet" cannot obtain the energy and nutrients required for normal growth and development. Even healthy children without eating disorders have been observed to limit their growth through "dieting." Weight gain in truly overweight children can be controlled safely without compromising growth, but should be overseen by a health care professional.

Do pediatricians check cholesterol on a routine visit?

Many children in the United States are not only overweight but also have high blood cholesterol.[18] These children are quite likely to have parents who developed CVD early.[19] For this reason, selective screening is recommended for children and adolescents whose parents or grandparents have CVD; those whose parents have elevated blood cholesterol; and those whose family history is unavailable, especially if other risk factors are evident.[20] Since blood cholesterol in children is a good predictor of adult values, some experts recommend universal screening to identify all children with high blood cholesterol.[21] They note that many children who have high blood cholesterol would be missed under current screening criteria.[22]

Among those children who may have high blood cholesterol, but may not meet screening criteria, are those who are overweight.[23] The incidence of high blood cholesterol in obese children with no other criteria is similar to that of nonobese children with family histories of CVD. In addition to overweight, health care professionals should consider whether children consume a high-fat diet.[24]

Early—but not advanced—atherosclerotic lesions are reversible, making screening and education a high priority. Both those with family histories of CVD and those with multiple risk factors need intervention. Children with the highest risks of developing CVD are sedentary and obese, with high blood pressure and high blood cholesterol. In contrast, children with the lowest risks of heart disease are physically active and of normal weight, with low blood pressure and favorable lipid profiles. Routine pediatric care should identify these known risk factors and provide intervention when needed.

Are adult dietary recommendations appropriate for children?

Regardless of family history, all children over age two should eat a variety of foods and maintain desirable weight. Children should receive 20 to 30 percent of total energy from fat, less than 10 percent from saturated fat, and less than 300 milligrams of cholesterol per day.[25]

Recommendations limiting fat and cholesterol are not intended for infants or children under two years old. Infants and toddlers need a higher percentage of fat to support their rapid growth.

Healthy children over age two can begin the transition to eating according to recommendations by eating fewer high-fat foods, replacing some high-fat foods with low-fat choices, and selecting more fruits and vegetables.[26] All high-fat foods need not be eliminated, though. Healthy meals can still include moderate amounts of a child's favorite foods, even if they are high-fat selections such as french fries and ice cream.[27] Without such additions, diets might be too low in fat, not to mention unappetizing and boring.

Balanced meals need to provide lean meat, poultry, fish, and vegetable sources of protein; fruits and vegetables; whole grains; and low-fat milk products. Such meals can provide enough food energy and nutrients to support growth and maintain blood cholesterol within a healthy range.[28]

Pediatricians warn parents to avoid extremes; they caution that although intentions may be good, excessive food restriction may create nutrient deficiencies and impair growth. Furthermore, parental control over eating may instigate battles and foster attitudes about foods that can lead to inappropriate eating behaviors.

Is there anything else parents or caregivers can do to help children reduce their risks of CVD?

Even though the focus of this text is nutrition, another risk factor for CVD that starts in childhood and carries over into adulthood must also be addressed—cigarette smoking. Each day 5000 children light up for the first time—typically, in grade school. Among high school students, two out of three have tried smoking, and one in seven smokes regularly.[29] Approximately 90 percent of all adult smokers began smoking before the age of 18.[30]

Efforts to teach children about the dangers of smoking need to be aggressive. Children are not likely to consider the long-term health consequences of tobacco use. They are more likely to be struck by the immediate health consequences, such as shortness of breath when playing sports, or social consequences, such as having bad breath. Whatever the context, the message to all children and teens should be clear: don't start smoking. If you've already started, quit.

In conclusion, adult CVD is a major pediatric problem.[31] Without intervention, some 60 million children are destined to suffer its consequences within the next 30 years. Optimal prevention efforts focus on children, especially on those who are overweight. Just as young children receive vaccinations against infectious diseases, they need screening for, and education about, CVD. Many health education programs have been implemented in schools around the country. These programs are most effective when they include education in the classroom, heart-healthy meals in the cafeteria, fitness activities on the playground, and parental involvement at home.

Notes

[1] W. B. Kannel, R. B. D'Agostino, and A. Belanger, Concept of bridging the gap from youth to adulthood—The Framingham Study, an address presented at the Recognition and Prevention of Heart Disease: State of the Art Conference, New Orleans, Louisiana, April 27 and 28, 1994.

[2] H. C. McGill, Childhood nutrition and adult cardiovascular disease, *Nutrition Reviews* 55 (1997): S2–S11.

[3] M. J. T. Kallio and coauthors, Tracking of serum cholesterol and lipoprotein levels from the first year of life, *Pediatrics* 91 (1993): 949–954.

[4] S. Guo and coauthors, Serial analysis of plasma lipids and lipoproteins from individuals 9–21 years of age, *American Journal of Clinical Nutrition* 58 (1993): 61–67.

[5] American Academy of Pediatrics, Committee on Nutrition, Cholesterol in childhood, *Pediatrics* 101 (1998): 141–147.

[6] National High Blood Pressure Education Program Working Group on Hypertension Control in Children and Adolescents, Update on the 1987 Task Force Report on High Blood Pressure in Children and Adolescents: A working group report from the National High Blood Pressure Education Program, *Pediatrics* 98 (1996): 649–658.

[7] American Academy of Pediatrics, Committee on Sports Medicine and Fitness, Athletic participation by children and adolescents who have systemic hypertension, *Pediatrics* 99 (1997): 637–638.

[8] A. R. Sinaiko, Hypertension in children, *New England Journal of Medicine* 335 (1996): 1968–1973.

[9] S. Shea and coauthors, The rate of increase in blood pressure in children 5 years of age is related to changes in aerobic fitness and body mass index, *Pediatrics* 94 (1994): 465–470.

[10] B. Falkner and S. Michel, Blood pressure response to sodium in children and adolescents, *American Journal of Clinical Nutrition* 65 (1997): S618–S621.

[11] S. S. Guo and coauthors, The predictive value of childhood body mass index values for overweight at age 35 y, *American Journal of Clinical Nutrition* 59 (1994): 810–819.

[12] Update: Prevalence of overweight among children, adolescents, and adults—United States, 1988–1994, *Morbidity and Mortality Weekly Report* 46 (1997): 199–202.

[13] T. A. Nicklas and coauthors, Secular trends in dietary intakes and cardiovascular risk factors of 10-year-old children: The Bogalusa Heart Study (1973–1988), *American Journal of Clinical Nutrition* 57 (1993): 930–937.

[14] J. O. Fisher and L. L. Birch, Fat preferences and fat consumption of 3- to 5-year-old children are related to parental obesity, *Journal of the American Dietetic Association* 95 (1995): 759–764.

[15] S. A. Schlicker, S. T. Borra, and C. Regan, The weight and fitness status of United States children, *Nutrition Reviews* 52 (1994): 11–17.

[16] E. Obarzanek and coauthors, Energy intake and physical activity in relation to indexes of body fat: The National Heart, Lung, and Blood Institute Growth and Health Study, *American Journal of Clinical Nutrition* 60 (1994): 15–22.

[17] O. T. Raitakari and coauthors, Effects of persistent physical activity on coronary risk factors in children and young adults: The Cardiovascular Risk in Young Finns Study, *American Journal of Epidemiology* 140 (1994): 195–205.

[18] G. S. Berenson, S. R. Srinivasan, and L. S. Webber, Cardiovascular risk prevention in children: A challenge or a poor idea? *Nutrition, Metabolism and Cardiovascular Diseases* 4 (1994): 46–52.

[19]W. Bao and coauthors, Longitudinal changes in cardiovascular risk from childhood to young adulthood in offspring of parents with coronary artery disease: The Bogalusa Heart Study, *Journal of the American Medical Association* 278 (1997): 1749–1754.

[20]American Academy of Pediatrics, 1998.

[21]L. Van Horn and P. Greenland, Prevention of coronary artery disease is a pediatric problem, *Journal of the American Medical Association* 278 (1997): 1779–1780; Berenson, Srinivasan, and Webber, 1994.

[22]S. J. Wadowski and coauthors, Family history of coronary artery disease and cholesterol: Screening children in disadvantaged inner-city population, *Pediatrics* 93 (1994): 109–113; K. Resnicow and D. Cross, Are parents' self-reported total cholesterol levels useful in identifying children with hyperlipidemia? An examination of current guidelines, *Pediatrics* 92 (1993): 347–354.

[23]M. S. Glassman and S. M. Schwarz, Cholesterol screening in children: Should obesity be a risk factor? *Journal of the American College of Nutrition* 12 (1993): 270–273.

[24]American Academy of Pediatrics, 1998.

[25]American Academy of Pediatrics, 1998.

[26]L. B. Dixon and coauthors, The effect of changes in dietary fat on the food group and nutrient intake of 4- to 10-year old children, *Pediatrics* 100 (1997): 863–872.

[27]E. Satter, A moderate view on fat restriction for young children, *Journal of the American Dietetic Association* 100 (2000): 32–35; M. Sigman-Grant, S. Zimmerman, and P. M. Kris-Etherton, Dietary approaches for reducing fat intake of preschool-age children, *Pediatrics* 91 (1993): 955–960.

[28]Is there a relationship between dietary fat and stature or growth in children three to five years of age? *Pediatrics* 92 (1993): 579–586.

[29]L. Kann and coauthors, Youth risk behavior surveillance—United States, 1993, *Journal of School Health* 65 (1995): 163–171.

[30]Tobacco use and usual source of cigarettes among high school students—United States, 1995, *Journal of the American Medical Association* 276 (1996): 184–185.

[31]Van Horn and Greenland, 1997.

CHAPTER 12

NUTRITION THROUGH THE LIFE SPAN: LATER ADULTHOOD

CONTENTS

Nutrition and Longevity

Nutrition-Related Concerns during Late Adulthood

Energy and Nutrient Needs during Late Adulthood

Food Choices and Eating Habits of Older Adults

Case Study: Elderly Man with a Poor Diet

Nutrition in Practice: Food for Singles

Figure 12-1
THE AGING OF THE U.S. POPULATION

In 1940, 6.8 percent of the population was 65 or older. In 1990, 12.7 percent of us had reached age 65; by 2040, 21.7 percent will have reached age 65; and by 2090, nearly one out of four Americans will be 65 or older. An estimated 25,000 Americans now living are 100 years old or older.

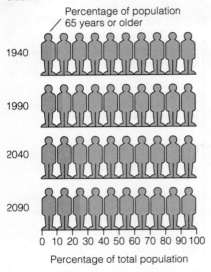

Percentage of total population

life expectancy: the average number of years lived by people in a given society.

life span: the maximum number of years of life attainable by a member of a species.

longevity: long duration of life.

The last two chapters were devoted to stages of the life cycle that require special nutrition attention: pregnancy, lactation, infancy, childhood, and adolescence. Much of the text before that focused on nutrition to support wellness during adulthood. This chapter describes the special nutrition needs of the later adult years.

The most urgent nutrition need of older people, however, is to have made good food choices in the past! All of life's nutrition choices incur health consequences for the better or for the worse. A single day's intakes of nutrients may exert only a minute effect on body organs and their functions, but over years and decades their repeated effects accumulate to have major impacts. This being the case, it is of great importance for everyone, of every age, to pay close attention today to nutrition.

The U.S. population is graying. The majority of citizens are now middle-aged, and the ratio of old people to young is becoming greater, as Figure 12-1 shows. Our society uses the arbitrary age of 65 to define the transition point between middle age and old age, but growing "old" happens day by day, with changes occurring gradually over time. Since 1950 the population of people over 65 has more than doubled. Remarkably, the fastest-growing age group is people over 85 years (see Figure 12-2).[1] The U.S. Bureau of the Census projects that by the year 2040 more than a million Americans will be 100 years old or older.

Life expectancy in the United States is 77 years, up from about 47 years in 1900.[2] Women today live about 6 years longer than men. Advances in medical science are largely responsible for almost doubling the life expectancy during the twentieth century. Improved nutrition and an abundant food supply have also contributed to lengthening life expectancy.[3] The human **life span,** currently estimated at 130 years, is the upper limit of human **longevity,** even given optimal nutrition.[4] With work progressing on unlocking the genetic code, however, the human life span may be extended significantly—or even indefinitely.

The study of the aging process is among the youngest of the scientific disciplines. Not until the twentieth century did human beings achieve a life expectancy worthy of a science devoted to studying it. The idea that nutrition can influence the way the human body ages is particularly appealing, since diet is a factor that people can control and change.

■■ Nutrition and Longevity

What has been learned so far about the effects of nutrition and environment on longevity provides incentive for researchers to keep asking questions about how and why human beings age. Among their questions are:

- To what extent is aging inevitable, and can it be slowed through changes in lifestyle and environment?
- What roles does nutrition play in aging, and what roles can it play in retarding aging?

With respect to the first question, aging is a natural process, programmed into the genes at conception. People can, however, slow the process within the natural limits set by heredity. They can adopt healthy lifestyle habits such as eating nutritious food and engaging in physical activity. With respect to the second question, clearly, good nutrition can retard and ease the aging process in many significant ways.

■■ Slowing the Aging Process

One approach researchers use to search out the secret of long life has been to study older people. Some people are young for their ages, others old for their ages. What makes the difference?

:: HEALTHY HABITS Six lifestyle habits seem to have a profound influence on people's health and therefore on their **physiological age:**[5]

- Sleeping regularly and adequately.
- Eating regular meals, including breakfast.
- Keeping weight under control.
- Engaging in regular physical activity.
- Not smoking.
- Not using alcohol, or using it in moderation.

Over the years, the effects of these lifestyle choices accumulate—that is, those who follow all of the practices are in better health, even if older in **chronological age,** than people who fail to do so. In fact, the physical health of people who adhere to all six practices is comparable to that of people *30 years younger* who follow few or none. Other studies have confirmed that these health habits both extend longevity and support independence in later life.[6] These findings suggest that even though people cannot alter the years of their births, they can alter the probable lengths and quality of their lives. Physical activity seems to be most influential in preventing or slowing the many changes that many people seem to accept as an inevitable consequence of old age. In other words, physical activity and long life seem to go together.[7]

:: PHYSICAL ACTIVITY The many and remarkable benefits of regular physical activity are not limited to the young: older adults who are active weigh less and have greater flexibility, more endurance, and better balance than those who are inactive.[8] They reap additional benefits as well; for example, evening activity helps to eliminate late night trips to the bathroom, moderate endurance activities improve the quality of sleep, and strength training significantly improves mobility and resistance to injury.[9] In fact, regular physical activity is the most powerful predictor of a person's mobility in later years.[10]

Muscle mass and muscle strength tend to decline with aging, making older people vulnerable to falls and immobility. Falls are a major cause of fear, injury, disability, dependence, and even death among older adults. Regular physical activity tones, firms, and strengthens muscles, helping to improve confidence, reduce the risk of falling, and minimize the risk of injury should a fall occur. Strength training, even in frail, elderly people over 85 years of age, has been shown not only to improve muscle strength and mobility but to increase energy expenditure and energy intake, thereby enhancing nutrient intakes.[11] This finding highlights another reason to be physically active: a person spending energy on physical activity can afford to eat more food and with it, more nutrients. People who are committed to an ongoing fitness program have higher energy and nutrient intakes than more sedentary people.[12]

Activities of all kinds are recommended to maintain and promote health. Strength training improves muscle strength, which enhances a person's ability to perform many of life's daily tasks such as climbing stairs and carrying packages.[13] In fact, research shows that muscle strength during midlife (between the ages of 45 and 65) predicts health and disability 25 years later.[14] Among middle-aged, healthy men, those who initially had greater hand grip strength remained stronger and more physically able than their peers who had weaker hand grip strength. Hand grip strength correlates with strength of other muscles and is therefore an indicator of overall strength. The researchers speculate that greater muscle strength during midlife acts as a strength *reserve* later on, protecting older adults from disability even when chronic conditions develop. In short, improving overall strength during early and middle adulthood could potentially lower the risk of later physical disability. Aerobic

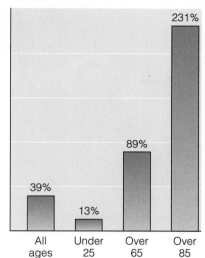

Figure 12-2
U.S. POPULATION GROWTH, 1960 TO 1990

The "oldest old"—those over 85 years— are the fastest-growing age group in the United States. Between 1960 and 1990, the U.S. population grew 39 percent, but the population of those over 85 more than doubled.

physiological age: a person's age as estimated from her or his body's health and probable life expectancy.

chronological age: a person's age in years from his or her date of birth.

Loss of muscle mass, strength, and quality is called sarcopenia.

Strength training promotes strong muscles and bones and healthy appetites.

activity can improve cardiorespiratory endurance and lower blood lipid concentrations.[15] Although aging affects both speed and endurance to some degree, older adults can still train and achieve exceptional performances.

Ideally, physical activity should be part of each day's schedule and should be intense enough to prevent muscle atrophy and to speed up the heartbeat and respiration rate. Healthy older adults who have not been active can ease into a suitable routine. They can start by walking short distances until they can walk at least a mile three times a week, and then they can gradually increase their pace to achieve a 20- to 25-minute mile.

One expert suggests the following physical activity program to maintain good health and function in older adults:[16]

- *Every day.* Spend 60 minutes in some physical activity: gardening, walking, climbing stairs, or simply moving about. This can be for 5 minutes at a time, 12 times a day; 12 minutes at a time, 5 times a day; or any combination of activity to total 60 minutes.
- *Three days a week.* Spend 30 to 45 minutes in vigorous and continuous physical activity, such as swimming, dancing, rowing, or brisk walking.

With persistence, people can achieve great improvements at any age. Training not only benefits physical health but also increases the blood flow to the brain, thereby enhancing both mental ability and mood.[17]

:: RESTRICTION OF kCALORIES Another approach in aging research has been to manipulate animals' diets and look for effects on longevity. These studies have produced some interesting and suggestive findings.[18] For example, rats live longer when their food intake is restricted in the early weeks of their lives, or even when it is restricted after they are mature. Extensive research shows that it is the restriction of food energy rather than restriction of a specific nutrient that exerts the anti-aging action.[19]

Several mechanisms to explain how energy restriction prolongs life in rats have been proposed but not proved. Research suggests that food restriction may extend the life span by delaying age-related diseases and preventing damaging lipid oxidation.[20]

Experiments with food restriction and longevity in rats have *not* suggested any direct applications to human nutrition, though, and the animals given restricted feedings have suffered some distinct disadvantages. For example, the food restriction was so severe that half of the restricted animals died *very* early. The average length of life for the restricted rats was long because the few survivors lived a long time. Extreme starvation to extend life, like any extreme, is not worth the price.

One group of researchers studied the relationship between *moderate* energy restriction (80 percent of usual intake) and aging in human beings.[21] Sixteen middle-aged, nonobese men were studied during ten weeks of energy restriction. Eight matched controls continued their usual diets. The energy-restricted men lost weight (mostly due to loss of fat), their blood pressures dropped significantly, and their HDL cholesterol concentrations rose significantly. Energy restriction had no adverse effects on mental or physical performance. For these men, moderate energy restriction favorably affected disease risk factors such as obesity, blood pressure, and cholesterol.

:: Nutrition and Disease Prevention

Nutrition alone, even if ideal, cannot ensure a long and robust life. Nevertheless, nutrition clearly affects aging and longevity in human beings by way of its role in disease prevention. Among the better-known relationships between nutrition and disease prevention are the following:

- Appropriate energy intake helps prevent *obesity, diabetes,* and related *cardiovascular diseases* such as atherosclerosis and hypertension (Chapters 7, 20, and 21) and may influence the development of some forms of *cancer* (Chapter 24).
- Adequate intakes of essential nutrients prevent *deficiency diseases* such as scurvy, goiter, anemia, and the like (Chapters 8 and 9).
- Variety in food intake, as well as ample intakes of certain vegetables, may be protective against certain types of *cancer* (Chapter 24).
- Moderation in sugar intake helps prevent *dental caries* (Nutrition in Practice 2).
- Appropriate fiber intakes help prevent disorders of the digestive tract such as *constipation, diverticulosis,* and possibly *colon cancer* (Chapters 2, 15, and 16).
- Moderate sodium intake and adequate intakes of potassium, calcium, and other minerals help prevent *hypertension,* at least in people who are genetically predisposed to it (Chapters 9 and 21).
- An adequate calcium intake throughout life helps protect against *osteoporosis* (Chapter 9).

Figure 12-3 illustrates these relationships between diet and degenerative diseases. Note that the figure includes some risk factors such as heredity and age

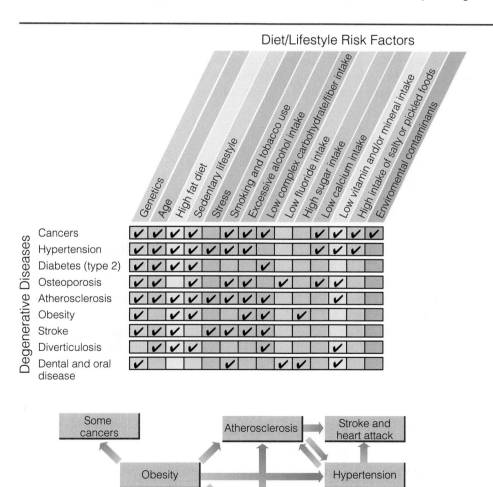

Diet/Lifestyle Risk Factors

Degenerative Diseases:

Disease	Genetics	Age	High fat diet	Sedentary lifestyle	Stress	Smoking and tobacco use	Excessive alcohol intake	Low complex carbohydrate/fiber intake	Low fluoride intake	High sugar intake	Low calcium intake	Low vitamin and/or mineral intake	High intake of salty or pickled foods	Environmental contaminants
Cancers	✔	✔	✔	✔		✔	✔	✔			✔	✔	✔	✔
Hypertension	✔	✔	✔	✔	✔	✔	✔				✔	✔	✔	
Diabetes (type 2)	✔	✔	✔	✔						✔				
Osteoporosis	✔	✔				✔	✔			✔		✔	✔	
Atherosclerosis	✔	✔	✔	✔	✔	✔	✔	✔				✔		
Obesity	✔		✔	✔			✔	✔		✔		✔		
Stroke	✔	✔	✔			✔	✔	✔	✔					
Diverticulosis		✔	✔	✔				✔					✔	
Dental and oral disease	✔					✔			✔	✔		✔		

Some cancers — Obesity — Atherosclerosis — Stroke and heart attack — Hypertension — Diabetes — Gallbladder disease

Figure 12-3
RISK FACTORS AND DEGENERATIVE DISEASES

The chart at the top shows that the same risk factor can affect many chronic diseases. Notice, for example, how many diseases have been linked to a high-fat diet. The chart also shows that a particular disease, such as atherosclerosis, may have several risk factors.

The flow chart at the bottom shows that many of these conditions are themselves risk factors for other chronic diseases. For example, a person with diabetes is likely to develop atherosclerosis and hypertension. These two conditions, in turn, worsen each other. Notice how all these diseases are linked to obesity.

that cannot be modified, but notice how many can. Other, less well-established links between nutrition and disease are being discovered each day. Research that focuses on how life factors affect aging and disease processes is vital to ensuring that more and more people can look forward to long, healthy lives.

LEARNING LINK

To date, scientists who study the aging process have found no specific diet or nutrient supplement that will prolong life—in spite of hundreds of unproven claims to the contrary. One of the most active areas of research along these lines, however, is the connection between antioxidants and disease prevention. Nutrition in Practice 21 offers a discussion of free radicals, antioxidants, and coronary heart disease.

▪▪ Nutrition-Related Concerns during Late Adulthood

Nutrition through the prime years may play a greater role than has been realized in preventing many changes once thought to be inevitable consequences of growing older. The following discussions of cataracts, arthritis, and the aging brain show that nutrition may provide at least some protection against some of the conditions commonly associated with aging.

▪▪ Cataracts

cataracts: thickenings of the eye lenses that impair vision and can lead to blindness.

Cataracts are age-related thickenings in the lenses of the eye that impair vision. If not surgically removed, they ultimately lead to blindness. Cataracts occur even in well-nourished individuals due to ultraviolet light exposure, oxidative damage, injury, viral infections, toxic substances, and genetic disorders. Many cataracts, however, are vaguely called senile cataracts, meaning "caused by aging." In the United States, only about 5 percent of people younger than 50 years have cataracts; by age 65, the percentage jumps to over 50 percent.

Oxidative stress appears to play a significant role in the development of cataracts, and the antioxidant nutrients may help minimize the damage. Studies have reported an inverse relationship between cataracts and dietary intakes of vitamin C, vitamin E, and carotenoids.[22] One study reported that people who had no cataracts took significantly more supplements of vitamins C and E than those who had cataracts.[23]

▪▪ Arthritis

arthritis: inflammation of a joint, usually accompanied by pain, swelling, and structural changes.

osteoarthritis: a painful, chronic disease of the joints caused when the cushioning cartilage in a joint breaks down; joint structure is usually altered, with loss of function; also called *degenerative arthritis*.

The most common type of **arthritis** that disables older people is **osteoarthritis,** a painful swelling of the joints. During movement, the ends of bones are normally protected from wear by cartilage and by small sacs of fluid that lubricate the joint. With age, bones sometimes disintegrate, and the joints become malformed and painful to move. Osteoarthritis afflicts millions of people around the world, especially the elderly. Nutrition quackery to treat arthritis is abundant, but no known food or supplement prevents, relieves, or cures it. Table 12-1 presents some of the many *non*effective dietary treatments for osteoarthritis.

Table 12-1
*Non*effective Dietary Strategies for Arthritis

• Alfalfa tea	• Garlic
• Aloe vera liquid	• Honey
• Amino acid supplements	• Inositol
• Blackstrap molasses	• Kelp
• Burdock root	• Lecithin
• Calcium	• Para-amino benzoic acid (PABA)
• Celery juice	• Raw liver
• Cod liver oil	• Superoxide dismutase (SOD)
• Copper supplements	• Vitamin D
• Dimethyl sulfoxide (DMSO)	• Vitamin megadoses
• Fasting	• Watercress
• Fresh fruit	• Yeast

A known connection between osteoarthritis and nutrition is overweight. Weight loss can help overweight people with osteoarthritis, partly because the joints affected are often weight-bearing joints that are stressed and irritated by having to carry excess poundage. Interestingly, though, weight loss often relieves the worst pain of osteoarthritis in the hands as well, even though they are not weight-bearing joints. Jogging and other weight-bearing activities do not worsen osteoarthritis. In fact, both aerobic activity and weight training offer modest improvements in physical performance and pain relief.[24]

Another type of arthritis, known as **rheumatoid arthritis,** has a possible link to diet through the immune system. In rheumatoid arthritis, the immune system mistakenly attacks the bone coverings as if they were made of foreign tissue. The integrity of the immune system depends on adequate nutrition, and a poor diet may worsen arthritis. It is also possible that in some individuals, certain foods may stimulate the immune system to attack. For example, milk and milk products seem to aggravate arthritis in some people.

Another nutrient linked to rheumatoid arthritis is the omega-3 fatty acid found in fish oil, eicosapentaenoic acid (EPA). Research shows that the same diet recommended for heart health—one low in saturated fat from meats and milk products and high in oils from fish—helps prevent or reduce the inflammation in the joints that makes arthritis so painful. Researchers theorize that EPA probably interferes with the action of prostaglandins, compounds involved in inflammation.

Researchers investigating whether EPA *supplements* benefit sufferers of rheumatoid arthritis have obtained promising results—fewer tender joints and less morning stiffness.[25] The researchers note that, overall, the response to the supplements is modest, but nevertheless consistent.

Another possible link between nutrition and rheumatoid arthritis involves lipid peroxidation. Lipid peroxidation of the membranes within joints causes inflammation and swelling. Vitamin E helps prevent peroxidation, but it has not improved active cases of rheumatoid arthritis. This is not surprising, though, because the vitamin's role in lipid peroxidation is preventive, not restorative.

rheumatoid arthritis: a disease of the immune system involving painful inflammation of the joints and related structures.

The Aging Brain

The brain, like all of the body's organs, responds to both inherited and environmental factors that can enhance or diminish its amazing capacities. One of the challenges researchers face when studying the aging of the brain in human beings is to distinguish among changes caused by normal, age-related, physiological processes; changes caused by diseases; and changes caused by cumulative, extrinsic factors such as diet.

The brain normally changes in some characteristic ways as it ages. For one thing, its blood supply decreases. For another, the number of **neurons,** the brain cells that specialize in transmitting information, diminishes as people age. When the number of nerve cells in one part of the **cerebral cortex** diminishes, hearing and speech are affected. Losses of neurons in other parts of the cortex can impair memory and cognitive function. When the number of neurons in the hindbrain diminishes, balance and posture are affected. Losses of neurons in other parts of the brain affect still other functions.

Clinicians now recognize that much of the cognitive loss and forgetfulness generally attributed to aging is due in part to extrinsic, and therefore controllable, factors such as nutrient deficiencies. In some instances, the degree of cognitive loss is extensive and attributable to a specific disorder such as a brain tumor. In cases such as Alzheimer's disease, deterioration may be genetically determined and will not yield to external approaches.

⁞⁞ ALZHEIMER'S DISEASE In **Alzheimer's disease,** the most prevalent form of **senile dementia,** brain cell death occurs in the areas of the brain that coordinate memory and cognition. Dementia from conditions such as Alzheimer's afflicts 6 to 10 percent of U.S. adults 65 years of age and older—and the rate doubles when milder cases are included.[26] Diagnosis of Alzheimer's depends on its characteristic symptoms: the victim gradually loses memory and reasoning, the ability to communicate, physical capabilities, and eventually life itself.

The causes of Alzheimer's continue to elude researchers, although genetic factors are apparently involved. A newly established connection to a human gene that makes part of a lipoprotein has sparked new hope for prevention of Alzheimer's.*[27] Ultimately, researchers hope to use the genetic findings to develop early detection tests and a cure for this devastating disease.

Treatment involves providing care to clients and support to their families. One medication, Aricept, relieves some of the memory impairment of Alzheimer's and is the most used Alzheimer's medication. Aricept helps only those with mild-to-moderate symptoms of Alzheimer's and does not slow or stop the progression of the disease.[28] Other medications may be used to control depression or behavior problems.

Most people have heard of an association between aluminum and the development of Alzheimer's, although a causal connection seems unlikely. Brain concentrations of aluminum in people with Alzheimer's exceed normal brain concentrations by some 10 to 30 times, but blood and hair aluminum remains normal, indicating that the accumulation is caused by something in the brain itself, not by an overload of aluminum in the body. Thus, the high brain aluminum must be at least partly a result, rather than a cause, of Alzheimer's.

Researchers often find elevated levels of copper, iron, and zinc in the brain tissues of those with Alzheimer's and theorize that these metals may accelerate the progression of the disease, possibly by increasing oxidative stress.[29] Some research casts doubt on the idea that supplements of zinc or other trace minerals can worsen Alzheimer's, but other research lends support to the theory. To err on the side of safety, food sources, not concentrated supplements, of trace minerals may be advisable for people with the disease.[30]

---

*The gene associated with Alzheimer's is apo E4, one of the three apolipoprotein E varieties. A report on the genetic and other aspects of Alzheimer's is available from Alzheimer's Disease Education and Referral Center, P.O. Box 8250, Silver Springs, MD 20907-8250.

neuron: a nerve cell; the structural and functional unit of the nervous system. Neurons initiate and conduct nerve transmissions.

cerebral cortex: the outer surface of the cerebrum.

Alzheimer's disease: a progressive, degenerative disease that attacks the brain and impairs thinking, behavior, and memory.

senile dementia: the loss of brain function beyond the normal loss of physical adeptness and memory that occurs with aging.

Some preliminary, but interesting research raises the possibility of a connection between elevated levels of the amino acid homocysteine and Alzheimer's.[31] As noted in Chapter 8, research shows a correlation between elevated blood levels of homocysteine and atherosclerosis.[32] Researchers are now beginning to recognize an association between atherosclerosis and Alzheimer's, which lends support to the hypothesis that elevated blood levels of homocysteine may play a role in Alzheimer's disease.[33] The possible link between elevated homocysteine and Alzheimer's, in turn, raises the question whether adequate amounts of B vitamins such as vitamin B_6, folate, and vitamin B_{12} may help prevent Alzheimer's. Increased intakes (slightly above the RDA) of vitamin B_6, folate, and vitamin B_{12} help to lower homocysteine.[34] The possibility that these B vitamins may lower the risk of Alzheimer's deserves a closer look.

Both foods and mental challenges nourish the brain.

Maintaining appropriate body weight may be the most important nutrition concern for the person with Alzheimer's. Depression and forgetfulness can lead to poor food intake, and restlessness may increase energy needs. Perhaps the best that a caregiver can do nutritionally for a person with Alzheimer's is to supervise food planning and mealtimes. Providing well-liked and well-balanced meals and snacks in a cheerful atmosphere encourages food consumption. To minimize confusion, offer a few ready-to-eat foods, in bite-size pieces, with seasonings and sauces. To avoid mealtime disruptions, control distractions such as television, children, and the telephone.

■■ **NUTRIENT DEFICIENCIES AND BRAIN FUNCTION**   Poor nutrition in general can affect the brain in other ways. Moderate, long-term nutrient deficiencies may contribute to the loss of memory and cognition that some older adults experience.[35] Table 12-2 summarizes some of the better-known connections between impaired brain function and severe nutrient deficiencies. If long-term, moderate nutrient deficiencies influence the loss of cognitive function that accompanies aging, then the loss may be preventable or at least diminished or delayed through diet.

■■ Energy and Nutrient Needs during Late Adulthood

Knowledge about the nutrient needs and nutrition status of older adults has grown considerably in the last decade or so. The DRI revision that is currently under way includes changes in age groupings (see inside front cover). Older adults are grouped into two age categories—51 to 70 years old, and 71 years and older. After all, the needs of people 50 to 60 years old may be very different from those of people over 80.

Table 12-2
SUMMARY OF NUTRIENT-BRAIN RELATIONSHIPS

Brain Function	Nutrient Deficiency
Short-term memory loss	Folate, vitamin B_{12}, vitamin C
Poor performance in problem-solving tests	Riboflavin, folate, vitamin B_{12}, vitamin C
Dementia	Thiamin, niacin, folate, vitamin B_{12}, zinc
Cognition	Folate, vitamin B_6, vitamin B_{12}, iron
Degeneration of brain tissue	Vitamin B_6

Setting standards for older people is difficult, though, because individual differences become more pronounced as people grow older. One person may tend to omit vegetables from his diet, and by the time he is old, he will have an associated set of nutrition problems. Another may have omitted milk and milk products all her life—her nutrition problems will be different. Also, as people age, they may suffer different chronic diseases and take different medications—both have impacts on nutrient needs. Table 12-3 lists some changes of aging that can affect nutrition. Even before all this, people start out with different genetic predispositions and ways of handling nutrients, and the effects of these become magnified over the years. Researchers have difficulty even defining "healthy aging," a prerequisite to developing recommendations designed to meet the "needs of practically all healthy persons."[36] Still, some generalizations are valid. The next sections give special attention to a few nutrients of concern.

▪▪ Energy and Energy Nutrients

Energy needs decline with advancing age. As a general rule, adult energy needs decline an estimated 5 percent per decade, beginning at about age 30. For one thing, as people age, they usually reduce their physical activity, although they need not do so. For another, lean body mass diminishes, slowing the basal metabolic rate. One researcher contends, however, that if older adults (up to age 75) could avoid changes in body composition (by way of physical activity), they would show little, if any, decrease in basal metabolic rate.[37] Physical activity not only increases energy expenditure but, along with sound nutrition, enhances bone density and supports many body functions as well.[38]

The lower energy expenditures of many older adults require that they eat less food energy to maintain their weights. Accordingly, the energy RDA for adults decreases slightly, beginning at age 51. Energy intakes typically decline in parallel with needs. Still, many older adults are overweight, indicating that their food intakes do not decline enough to compensate for their reduced energy expenditure.

On limited energy allowances, people must select mostly nutrient-dense foods. There is little leeway for sugars, fats, oils, or alcohol. Older adults can follow the Daily Food Guide (see pp. 18–19), making sure to get at least the minimum number of servings from each food group daily.

▪▪ **PROTEIN** The protein needs of older adults appear to be about the same as, or even greater than, those of younger people.[39] Since energy needs

Table 12-3
EXAMPLES OF PHYSICAL CHANGES OF AGING THAT AFFECT NUTRITION

Mouth	Tooth loss, gum disease, and reduced salivary output impede chewing and swallowing. Swallowing disorders and choking may become likely. Discomfort and pain associated with eating may reduce food intake.
Digestive Tract	Intestines lose muscle strength resulting in sluggish motility that leads to constipation (see Chapter 16). Stomach inflammation, abnormal bacterial growth, and greatly reduced acid output impair digestion and absorption. Pain may cause food avoidance or reduced intake.
Hormones	For example, the pancreas secretes less insulin and cells become less responsive, causing abnormal glucose metabolism.
Sensory Organs	Diminished senses of smell and taste can reduce appetite; diminished sight can make food shopping and preparation difficult.
Body Composition	Weight loss and decline in lean body mass lead to lowered energy requirements. May be preventable or reversible through physical activity.
Urinary Tract	Increased frequency of urination may limit fluid intake.

decrease, however, the protein has to be obtained from low-kcalorie sources of high-quality protein, such as lean meats, poultry, fish, and eggs; fat-free and low-fat milk products; legumes; and grains.

CARBOHYDRATE AND FIBER As always, abundant carbohydrate is needed to protect protein from being used as an energy source. Complex carbohydrate foods such as whole grains, vegetables, and fruits are also rich in fiber and essential vitamins and minerals.

Average fiber intakes among older adults fall short of current recommendations. (Fibrous foods require more chewing.) The combination of high-fiber foods and ample water can alleviate constipation—a condition common among older adults and especially among nursing home residents. Physical inactivity and medications also contribute to the high incidence of constipation.

As many as 50 percent of nursing home residents may be malnourished and underweight.[40] For these people, a diet that emphasizes fiber-rich foods such as whole grains, fruits, and vegetables may be too low in concentrated protein and energy. Protein- and energy-dense snacks such as hard-boiled eggs, tuna fish and crackers, peanut butter on graham crackers, and homemade soups are valuable additions to the diets of underweight or malnourished older adults. Enteral formulas served between meals can also add energy and kcalories to a day's intake.

FAT As is true for people of all ages, keeping fat intake to 30 percent of total energy or less continues to be important. Not only are the foods lowest in fat often richest in vitamins, minerals, and phytochemicals, but limiting fat may help retard the development of cancer, atherosclerosis, and other degenerative diseases. For some older adults, though, limiting fat intake too severely may lead to nutrient deficiencies and weight loss—two problems that carry greater health risks in the elderly than overweight.[41]

Water

Dehydration is a risk for older adults, who may not notice or pay attention to their thirst, or who find it difficult and bothersome to get a drink or to get to a bathroom. Older adults who have lost bladder control may be afraid to drink too much water. Despite real fluid needs, older people do not seem to feel thirsty or notice mouth dryness.[42] Many nursing home employees say it is hard to persuade their elderly clients to drink enough water and fruit juices.

Water recommendation for adults: 1 to 1½ oz/kg actual body weight.

Total body water decreases as people age, so even mild stresses such as fever or hot weather can precipitate rapid dehydration in older adults. Dehydrated older adults seem to be more susceptible to urinary tract infections, pneumonia, pressure ulcers, confusion, and disorientation.[43] An intake of 6 to 8 glasses of water a day is recommended. Milk and juices may replace some of this water, but beverages containing alcohol or caffeine cannot because of their diuretic effect.

Vitamins and Minerals

As research reveals more about how specific vitamins and minerals influence disease prevention and how age-related physiological changes affect nutrient metabolism, optimal intakes of vitamins and minerals for different groups of older adults are being defined. This section highlights the vitamins and minerals of greatest concern to older adults. Table 12-4 (on p. 301) summarizes the nutrient concerns of aging.

VITAMIN A Vitamin A stands alone in that it is absorbed and stored more efficiently by the aging GI tract and liver, although its processing

Vitamin A RDA during late adulthood: 1000 µg RE/day (men). 800 µg RE/day (women).

within the body slows slightly. Several studies have reported that healthy older adults have normal concentrations of plasma vitamin A even when their dietary intakes fall below the RDA, suggesting that the current RDA may be too high.[44] The Committee on Dietary Allowances has hesitated to lower recommendations, however, recognizing both the need to prevent vitamin A deficiency and the possibility that the vitamin A precursor beta-carotene might prevent or delay the onset of some age-related diseases.

Vitamin D AI during late adulthood: 10 μg/day.

VITAMIN D Older adults face a greater risk of vitamin D deficiency than younger people do. Only vitamin D–fortified milk provides significant vitamin D, and many older adults drink little or no milk. Consequently, many older adults have vitamin D intakes of less than half of recommendations. Further compromising the vitamin D status of many older people, especially those in nursing homes, is their limited exposure to sunlight. Finally, aging reduces the skin's capacity to make vitamin D and the kidneys' ability to convert it to its active form. To prevent bone loss and to maintain vitamin D status in older people, especially in those who engage in minimal outdoor activity, recommendations for vitamin D were recently raised from 5 to 10 micrograms per day.[45]

Vitamin B_6 RDA during late adulthood: 1.7 mg/day (men). 1.5 mg/day (women).

VITAMIN B_6 Studies on vitamin B_6 reveal that its metabolism is altered with age, resulting in a higher requirement. Many older adults consume far less than the RDA for vitamin B_6.[46] Research suggests that immune response is impaired with vitamin B_6 deficiency. Such findings may have important implications for older adults because the aging process itself seems to be accompanied by a decline in immune function.[47]

Another approach to determining the vitamin B_6 requirements of older adults, as well as the requirements for vitamin B_{12} and folate, focuses on the amino acid homocysteine discussed earlier in connection with Alzheimer's disease. An elevated homocysteine level is recognized as an independent risk factor for heart disease and stroke in the United States.[48] Homocysteine concentrations rise with vitamin B_6, vitamin B_{12}, or folate deficiencies (see Chapter 21).

Vitamin B_{12} RDA during late adulthood: 2.4 μg/day.

VITAMIN B_{12} The DRI committee recommends that adults aged 51 years and older obtain 2.4 micrograms of vitamin B_{12} daily *and* that vitamin B_{12}–fortified foods (such as fortified cereals) or supplements be used to meet much of the DRI recommended intake.[49] The committee's recommendation reflects the finding that between 10 to 30 percent of people older than 50 years lose the ability to produce enough stomach acid to make the protein-bound form of vitamin B_{12} available for absorption. Synthetic vitamin B_{12} is reliably absorbed, however.

One cause of this malabsorption of protein-bound vitamin B_{12} is a condition known as **atrophic gastritis.** The prevalence of atrophic gastritis among those 60 years of age and older is high.

atrophic gastritis: a condition characterized by chronic inflammation of the stomach accompanied by a diminished size and functioning of the mucosa and glands.

FOLATE As is true of vitamin B_6 and vitamin B_{12}, folate intakes of older adults typically fall short of recommendations.[50] The elderly are also more likely to have medical conditions or take medications that can compromise folate status.

Folate RDA during late adulthood: 400 μg/day.

IRON Among the minerals, iron deserves first mention. Iron-deficiency anemia is less common in older adults than in younger people, but it still occurs in some, especially in those with low food energy intakes. Aside from diet, other factors in many older people's lives make iron deficiency likely: chronic blood loss from disease conditions and medicines, and poor iron

Iron RDA during late adulthood: 10 mg/day.

absorption due to reduced secretion of stomach acid and antacid use. Anyone concerned with older people's nutrition should keep these possibilities in mind.

■■ **ZINC**   Zinc intake is commonly low in older people, with many receiving less than half of the recommended amount.[51] In addition, older adults may absorb zinc less efficiently than younger people do. A number of different factors, including medications that older adults commonly use, can impair zinc absorption or enhance its excretion and thus lead to deficiency. Some of the symptoms of zinc deficiency resemble symptoms associated with aging—for example, decline in taste acuity and dermatitis. Whether these symptoms in older adults are attributable to zinc deficiency remains unclear.

Zinc RDA during late adulthood:
15 mg/day (men).
12 mg/day (women).

■■ **CALCIUM**   The appropriate calcium intake for older adults remains controversial. A National Institutes of Health panel has concluded that women over 50 who are not on estrogen replacement therapy and all adults over 65 should receive 1500 milligrams of calcium daily.[52] The current recommendations for calcium for both men and women are higher than previous recommendations, but are still lower than the National Institutes of Health recommendation. As researchers attempt to reach agreement about the calcium requirements of older adults, especially those of women, one thing is clear: the calcium intakes of many people, especially women, in the United States are well below recommendations.

Calcium AI during late adulthood:
1200 mg/day.

■■   **Nutrient Supplements for Older Adults**

People judge for themselves how to manage their nutrition, and some turn to supplements. Advertisers target older people with appeals to take supplements and eat "health" foods, claiming that these products prevent disease and promote longevity. About half of all women over 65 years of age take some type of

Table 12-4
SUMMARY OF NUTRIENT CONCERNS IN AGING

Nutrient	Effect of Aging	Comments
Water	Lack of thirst and decreased total body water make dehydration likely.	Mild dehydration is a common cause of confusion. Difficulty obtaining water or getting to the bathroom may compound the problem.
Energy	Need decreases.	Physical activity moderates the decline.
Fiber	Likelihood of constipation increases with low intakes and changes in the GI tract.	Problems chewing fibrous foods, inadequate water intakes and lack of physical activity, along with some medications, compound the problem.
Protein	Needs stay the same.	Low-fat, high-fiber legumes and grains meet both protein and other nutrient needs.
Vitamin A	Absorption increases.	RDA may be high.
Vitamin D	Increased likelihood of inadequate intake; less likely to go outdoors; skin synthesis declines.	Daily limited sunlight exposure may be of benefit.
Iron	In women, status improves after menopause; deficiencies are linked to chronic blood losses and low stomach acid output.	Adequate stomach acid is required for absorption; antacid or other medicine use may aggravate iron deficiency; vitamin C and meat increase absorption.
Zinc	Intakes may be low and absorption reduced; but needs may also decrease.	Medications interfere with absorption; deficiency may depress appetite and sense of taste.
Calcium	Intakes may be low; osteoporosis common.	Stomach discomfort commonly limits milk intake; calcium substitutes are needed.

Table 12-5
STRATEGIES FOR GROWING OLD GRACEFULLY

- Choose nutrient-dense foods.
- Maintain appropriate body weight.
- Reduce stress by identifying priorities and setting limits.
- For women, consult a physician about planning a strategy to protect against osteoporosis.
- For people who smoke, quit.
- Expect to enjoy sex, and learn new ways of enhancing it.
- Use alcohol only moderately, if at all; use medications only as prescribed.
- Take care to prevent accidents.
- Expect good vision and hearing throughout life; obtain glasses and hearing aids if necessary.
- Be alert to confusion as a disease symptom, and seek diagnosis.
- Control depression through activities and friendships.
- Drink 8 glasses of water every day.
- Practice mental skills. Keep on solving math problems and crossword puzzles, playing cards or other games, reading, writing, imagining, and creating.
- Make financial plans early to ensure security.
- Accept change. Work at recovering from losses; make new friends.
- Cultivate spiritual health. Cherish personal values. Make life meaningful.
- Go outside for sunshine and fresh air as often as possible.
- Be physically active. Walk, run, dance, swim, bike, row, or climb for aerobic activity. Lift weights, do calisthenics, or pursue some other activity to tone, firm, and strengthen muscles. Change activities to suit changing abilities and tastes.
- Be socially active—play bridge, join an exercise group, take a class, teach a class, eat with friends, volunteer time to help others.
- Stay interested in life—pursue a hobby, spend time with grandchildren, take a trip, read, cultivate a garden, or go to the movies.
- Enjoy life.

nutrient supplement, and about one-fifth of older men do. Quite often those who take supplements are not deficient in the nutrients being supplemented.[53]

Elderly people often benefit from a balanced low-dose vitamin and mineral supplement, however. Such supplements supply many of the needed minerals along with the vitamins often lacking in older people's diets without presenting too much of any one nutrient.[54] Many times, those taking such supplements suffer fewer infectious diseases.[55]

Food is still the best source of nutrients for everybody, however. Supplements are just that—supplements to foods, not substitutes for them. For anyone who is motivated to obtain the best possible health, it is never too late to learn to eat well, become physically active, and adopt other lifestyle changes such as quitting smoking, moderating alcohol use, and the like. Table 12-5 offers strategies for growing old gracefully.

The Effects of Medications on Nutrients

As people grow older, the use of medicines—from over-the-counter types such as aspirin and laxatives to prescription medications of all kinds—becomes commonplace and accounts for about 25 percent of all medications sold. Most medications interact with one or more nutrients in several ways, usually resulting in greater-than-normal needs for these nutrients. Chapter 13 discusses diet-medication interactions and describes the many reasons why elderly people are vulnerable to such interactions.

The most common drug that can affect nutrition in older people is alcohol. A recent estimate sets the incidence of alcoholism in people over 60 in our society at 2 to 10 percent. The effects of alcohol on people of all ages are explained in Nutrition in Practice 23.

Food Choices and Eating Habits of Older Adults

To provide any benefit, strategies and interventions to improve a person's nutrition status must be based on knowledge of food preferences and eating patterns. Menus and feeding programs for older adults must take into consideration not only the food likes and dislikes of this diverse group of people but also their living conditions, economic status, and medical conditions. If nutrition intervention is to be successful, it is essential to know what foods people will eat, in what settings they like to eat these foods, and whether they can buy and prepare meals.

Older people are, for the most part, independent, socially sophisticated, mentally lucid, fully participating members of society who report themselves to be happy and healthy. In fact, chronic disabilities among the elderly have declined dramatically over the past decade.[56] Older people spend more money per person on foods to eat at home than other age groups and less money on foods away from home. Manufacturers would be wise to cater to the preferences of older adults by providing good-tasting, nutritious foods in easy-to-open, single-serving packages with labels that are easy to read. Such services enable older adults to maintain their independence; most of them want to take care of themselves and need to feel a sense of control and involvement in their own lives. As discussed earlier, another way older adults can take care of themselves is by remaining or becoming physically active. Physical activity helps preserve one's ability to perform daily tasks and so promotes independence.[57]

Taking time to nourish your body well is a gift you give yourself.

■■ **INDIVIDUAL PREFERENCES** Familiarity, taste, and health beliefs are most influential on older people's food choices. Eating foods that are familiar, especially those that recall family meals and pleasant times, can be comforting. Older adults are choosing low-fat poultry and fish, low-fat milk and milk products, and high-fiber breads and grains, indicating their belief in the importance of diet in supporting good health.[58] Few older adults, however, consume the recommended amounts of milk products.[59]

■■ **MEAL SETTING** The food choices and eating habits of older adults are also affected by the changes in lifestyle that often accompany aging in this society. Whether people live alone, with others, or in institutions affects the way they eat. For example, men living alone are most likely to be poorly nourished. Older adults who live alone do not make poorer food choices than those who live with companions; rather they consume too little food: loneliness is directly related to inadequacies, especially of energy intakes.

■■ **DEPRESSION** Another factor affecting food intake and appetite in older people is depression. Though not an inevitable component of aging, depression is more common with advancing age. Loss of appetite and motivation to cook or even to eat frequently accompanies depression. An overwhelming sense of grief and sadness at the death of a spouse, friend, or family member may leave many people, particularly elderly people, feeling powerless to overcome the depression. The support and companionship of family and friends, especially at mealtimes, can help overcome depression and enhance appetite. The case study (p. 304) presents a man who has several of these problems. Use the suggestions here, and in the Nutrition in Practice that follows

Shared meals can brighten the day and enhance the appetite.

Case Study

ELDERLY MAN WITH A POOR DIET

Mr. Brezenoff is a 75-year-old man who lives alone. He has been losing weight slowly since his wife died a year ago. At 5 feet 8 inches tall, he currently weighs 124 pounds. His previous weight was 150 pounds. In talking with Mr. Brezenoff, you realize that he doesn't even like to talk about food, let alone eat it. "My wife always did the cooking before, and I ate well. Now I just don't feel like eating." You manage to find out that he skips breakfast, has soup and bread for lunch, and sometimes eats a cold-cut sandwich or a frozen dinner for supper. He seldom sees friends or relatives. Mr. Brezenoff has also lost several teeth and doesn't eat any raw fruits or vegetables because he finds them hard to chew. He lives on a meager but adequate income.

✚ Is Mr. Brezenoff's weight within the range of suggested weights for adults his height (see Table 7-1)?

✚ Is his weight loss significant?

✚ What factors are contributing to his poor food intake?

✚ What nutrients are probably deficient in his diet?

✚ Look at Mr. Brezenoff as an individual and suggest ways he can improve his diet and his lifestyle.

✚ What other aspects of Mr. Brezenoff's physical and mental health should you consider in helping him to improve his food intake?

Risk factors for malnutrition in older adults:
- *Disease.*
- *Eating poorly.*
- *Tooth loss, oral pain, or swallowing disorders.*
- *Economic hardship.*
- *Reduced social contact.*
- *Multiple medications.*
- *Involuntary weight loss or gain.*
- *Needs assistance with self-care.*
- *Elderly person older than 80 years.*

this chapter, to help develop solutions. The Nutrition Assessment Checklist helps to pinpoint nutrition-related factors to look for when working with older adults. To *determine* the risk of malnutrition in older clients, health care providers can keep in mind the characteristics listed in the margin. Details of nutrition assessment are presented in Chapter 13.

NUTRITION ASSESSMENT CHECKLIST — FOR OLDER ADULTS

Health Problems, Signs, and Symptoms

Check the medical record for:

- [ ] Chronic diseases (cancer, heart disease, hypertension, diabetes)
- [ ] Dehydration
- [ ] Arthritis
- [ ] Cataracts
- [ ] Alzheimer's disease or other dementia or confusion
- [ ] Alcohol abuse
- [ ] Depression
- [ ] Dental disease or tooth loss
- [ ] Cigarette, cigar, or pipe smoking; use of other tobacco products
- [ ] Swallowing disorders
- [ ] Constipation
- [ ] Inflammation of the stomach (gastritis)

Medications

For older adults being treated with drug therapy for medical conditions, note:

- [ ] Use of multiple medications—prescription and/or over-the-counter medications such as laxatives and pain relievers
- [ ] Side effects that might reduce food intake or change nutrient needs
- [ ] Proper administration of medication with respect to food intake
- [ ] Malnutrition—is the person's nutrition status questionable even before considering side effects of medications that worsen nutrition status?
- [ ] Diminished mental capacity which might interfere with taking correct medications and doses
- [ ] Dehydration (can alter effects of medications)

Nutrient/Food Intake

For all older adults, especially those at risk nutritionally, assess the diet for:

- [ ] Total energy
- [ ] Protein
- [ ] Calcium, iron, and zinc
- [ ] Vitamin B_6, vitamin B_{12}, folate, and vitamin D

Note the following:

- [ ] Number of meals eaten each day
- [ ] Number and ages of people in household
- [ ] Amount of milk consumed each day
- [ ] Type and frequency of outdoor activity
- [ ] Type and frequency of physical activity
- [ ] Financial resources
- [ ] Transportation resources
- [ ] Physical disabilities
- [ ] Mental alertness

Height and Weight

Measure baseline height and weight.

- [ ] Reassess height and weight at each medical checkup.
- [ ] Note significant overweight or underweight, which warrants intervention.
- [ ] Use fatfold measures to reveal altered body composition that may indicate malnutrition and loss of lean tissue.

Laboratory Tests

- [ ] Hemoglobin, hematocrit, or other tests of iron status
- [ ] Serum albumin or other measures of protein status
- [ ] Serum folate
- [ ] Serum B_{12}

Physical Signs

Look for physical signs of:

- [ ] Protein energy malnutrition
- [ ] Iron and zinc deficiency
- [ ] Folate deficiency

SELF CHECK

1. The fastest-growing age group in the United States is:
   a. 21 years of age.
   b. 30 to 45 years of age.
   c. 50 to 70 years of age.
   d. over 85 years of age.

2. Which of the following lifestyle habits can enhance the length and quality of people's lives?
   a. moderate smoking
   b. 6 hours of sleep daily
   c. regular physical activity
   d. moderate alcohol intake

3. Among the better-known relationships between nutrition and disease prevention are:
   a. appropriate fiber intake helps prevent goiter.
   b. moderate sodium intake helps prevent obesity.
   c. moderate sugar intake helps prevent hypertension.
   d. appropriate energy intake helps prevent diabetes and cardiovascular disease.

4. A disease of the immune system that involves painful inflammation of the joints is:
   a. sarcopenia.
   b. osteoarthritis.
   c. senile dementia.
   d. rheumatoid arthritis.

5. Examples of low-kcalorie, high-quality protein foods include:
   a. cottage cheese, sour cream, and eggs.
   b. green and yellow vegetables and citrus fruits.

c. potatoes, rice, pasta, and whole-grain breads.
d. lean meats, poultry, fish, legumes, fat-free milk, and eggs.

6. For malnourished and underweight people, protein- and energy-dense snacks include:

a. fresh fruits and vegetables.
b. yogurt and cottage cheese.
c. whole grains and high-fiber legumes.
d. hard-boiled eggs and peanut butter and crackers.

7. Which of the following does **not** explain why dehydration is a risk for older adults?

a. They do not seem to feel thirsty.
b. Total body water increases with age.

c. They may find it difficult to get a drink.
d. They may have difficulty swallowing liquids.

8. Inadequate milk intake and limited exposure to sunlight contribute to older adults' risk of:

a. vitamin A deficiency.
b. vitamin D deficiency.
c. riboflavin deficiency.
d. vitamin B_6 deficiency.

9. Two risk factors for malnutrition in older adults are:

a. loneliness and multiple medication use.
b. increased energy needs and lack of fiber.
c. decreased mineral absorption and antioxidant intake.

d. high carbohydrate intake and lack of physical activity.

10. Two strategies to improve nutrition status when growing old include:

a. increase vitamin A intake and exercise 30 minutes daily.
b. choose nutrient-dense foods and maintain appropriate weight.
c. avoid high-fiber foods and take a daily vitamin-mineral supplement.
d. eat at least one big meal per day and drink at least 10 glasses of water daily.

Answers to these questions appear in Appendix H.

CLINICAL APPLICATIONS

Ms. Hamilton is an 80-year-old woman in excellent health who lives alone, eats a well-balanced diet, enjoys an active social life, and walks every day. Consider the ways Ms. Hamilton's health and nutrition status might be affected by the following situations:

• Many of Ms. Hamilton's friends pass away or move into extended care facilities.

• Ms. Hamilton falls and breaks her hip.
• Ms. Hamilton begins to feel isolated and depressed.

Describe interventions the nurse can take to help Ms. Hamilton deal with each situation to prevent her from falling into a downward spiral.

NUTRITI**ON**THENET

FOR FURTHER STUDY OF THE
TOPICS IN THIS CHAPTER,
ACCESS THESE WEB SITES.

www.ncoa.org
National Council on Aging

www.cdc.gov/nccdphp/sgr/sgr.htm
*Physical Activity and Health: A
Report of the Surgeon General*

www.ascrs.org
*American Society of Cataract and
Refractive Surgery*

www.nei.nih.gov
National Eye Institute

arthritis.org
Arthritis Foundation

hypercon.com/evolve/oars.htm
Osteoarthritis Research Society

www.rheumatology.org
American College of Rheumatology

www.alz.org
Alzheimer's Association

www.alzheimers.org
*Alzheimer's Disease Education and
Referral Center*

aafp.org/nsi/index.html
Nutrition Screening Initiative

Notes

[1] Position of The American Dietetic Association: Nutrition, aging, and the continuum of care, *Journal of the American Dietetic Association* 96 (1996): 1048–1052.

[2] B. Guyer and coauthors, Annual summary of vital statistics—1998, *Pediatrics* 104 (1999): 1229–1248; Centers for Disease Control and Prevention, National Center for Health Statistics, *Monthly Vital Statistics Report*, October 1996, p. 4.

[3] K. G. Kinsella, Changes in life expectancy 1900–1990, *American Journal of Clinical Nutrition* 55 (1992): S1196–S1202; S. Kobayashi, A scientific basis for the longevity of Japanese in

relation to diet and nutrition, *Nutrition Reviews* 50 (1992): 353–359.

[4] K. G. Manton and E. Stallard, Longevity in the United States: Age and sex-specific evidence on life span limits from mortality patterns 1960–1990, *Journal of Gerontology* 51A (1996): B362–B375.

[5] L. Breslow and N. Breslow, Health practices and disability: Some evidence from Alameda County, *Preventive Medicine* 22 (1993): 86–95.

[6] A. J. Vita and coauthors, Aging, health risks, and cumulative disability, *New England Journal of Medicine* 338 (1998): 1035–1041; A. Z. LaCroix and coauthors, Maintaining mobility in late life: Smoking, alcohol consumption, physical activity, and body mass index, *American Journal of Epidemiology* 137 (1993): 858–869.

[7] U. M. Kujala and coauthors, Relationship of leisure-time, physical activity, and mortality: The Finnish twin cohort, *Journal of the American Medical Association* 279 (1998): 440–444; S. N. Blair and coauthors, Changes in physical fitness and all-cause mortality: A prospective study of healthy and unhealthy men, *Journal of the American Medical Association* 273 (1995): 1093–1098; R. S. Paffenbarger and coauthors, The association of changes in physical-activity level and other lifestyle characteristics with mortality among men, *New England Journal of Medicine* 328 (1993): 538–545.

[8] American College of Sports Medicine, Position stand, Exercise and physical activity for older adults, *Medicine and Science in Sports and Exercise* 30 (1998): 992–1008; L. E. Voorrips and coauthors, The physical condition of elderly women differing in habitual physical activity, *Medicine and Science in Sports and Exercise* 25 (1993): 1152–1157.

[9] A. C. King and coauthors, Moderate-intensity exercise and self-rated quality of sleep in older adults, *Journal of the American Medical Association* 277 (1997): 32–37; W. Evans, Functional and metabolic consequences of sarcopenia, *Journal of Nutrition* 127 (1997): S998–S1003.

[10] LaCroix and coauthors, 1993.

[11] King and coauthors, 1997; M. A. Fiatarone and coauthors, Exercise training and nutritional supplementation for physical frailty in very elderly people, *New England Journal of Medicine* 330 (1994): 1769–1775; W. W. Campbell and coauthors, Increased energy requirements and changes in body composition with resistance training in older adults, *American Journal of Clinical Nutrition* 60 (1994): 167–175.

[12] D. E. Butterworth and coauthors, Exercise training and nutrient intake in elderly women, *Journal of the American Dietetic Association* 93 (1993): 653–657.

[13] W. J. Evans and D. Cyr-Campbell, Nutrition, exercise, and healthy aging, *Journal of the American Dietetic Association* 97 (1997): 632–638.

[14] T. Rantanen and coauthors, Midlife hand grip strength as a predictor of old age disability, *Journal of the American Medical Association* 281 (1999): 558–560.

[15] J. S. Green and S. F. Crouse, The effects of endurance training on functional capacity in the elderly: A meta-analysis, *Medicine and Science in Sports and Exercise* 27 (1995): 920–926.

[16] P. Astrand, Physical activity and fitness, *American Journal of Clinical Nutrition* (supplement) 55 (1992): 1231–1236.

[17] S. M. Avent and D. M. Landers, The effects of exercise on mood in the elderly: A meta-analysis, presented at the annual meeting of the American College of Sports Medicine, Seattle; as yet unpublished, as cited in *Sports Medicine Digest* 21 (1999): 72.

[18] E. J. Masoro, Retardation of aging processes by food restriction: An experimental tool, *American Journal of Clinical Nutrition* (supplement) 55 (1992): 1250–1252.

[19] Masoro, 1992.

[20] R. Weindruch, Caloric restriction and aging, *Scientific American,* January 1997, pp. 46–52; Masoro, 1992.

[21] E. J. M. Velthuis-te Wierik and coauthors, Energy restriction, a useful intervention to retard human ageing? Results of a feasibility study, *European Journal of Clinical Nutrition* 48 (1994): 138–148.

[22] G. E. Bunce, Antioxidant nutrition and cataract in women: A prospective study, *Nutrition Reviews* 51 (1993): 84–86; P. F. Jacques and L. T. Chylack, Epidemiologic evidence of a role for the antioxidant vitamins and carotenoids in cataract prevention, *American Journal of Clinical Nutrition* 53 (1991): S352–S355.

[23] P. F. Jacques and coauthors, Long-term vitamin C supplement use and prevalence of early age-related lens opacities, *American Journal of Clinical Nutrition* 66 (1997): 911–916; J. M. Robertson, A. P. Donner, and J. R. Trevithick, A possible role for vitamins C and E in cataract prevention, *American Journal of Clinical Nutrition* 53 (1991): S346–S351.

[24]W. H. Ettinger and coauthors, A randomized trial comparing aerobic exercise and resistance exercise with a health education program in older adults with knee osteoarthritis: The Fitness Arthritis and Seniors Trial (FAST), *Journal of the American Medical Association* 277 (1997): 25–31.

[25]J. M. Kremer, n-3 fatty acid supplements in rheumatoid arthritis, *American Journal of Clinical Nutrition* 71 (2000): S349–S351.

[26]H. C. Hendrie, Epidemiology of dementia and Alzheimer's disease, *American Journal of Geriatric Psychiatry* 6 (1998): S3–S18.

[27]E. M. Reiman and coauthors, Preclinical evidence of Alzheimer's disease in persons homozygous for the e4 allele for apolipoprotein E, *New England Journal of Medicine* 334 (1996): 752–758; National Institute of Aging, *Progress Report on Alzheimer's Disease*, 1994, NIH publication number 94-3885 (Washington, D.C.: Government Printing Office, 1994).

[28]A. T. Hingley, Alzheimer's: Few clues on the mysteries of memory, *FDA Consumer*, May/June 1998, pp. 27–31.

[29]M. A. Lovely and coauthors, Copper, iron and zinc in Alzheimer's disease senile plaques, *Journal of the Neurological Sciences* 158 (1998): 47-52; C. R. Cornett, W. R. Markesbery, and W. D. Ehmann, Imbalances of trace elements related to oxidative damage in Alzheimer's disease brain, *Neurotoxicology* 19 (1998): 339–345.

[30]M. A. Lovely, C. Xie, and W. R. Markesbery, Protections against amyloid beta peptide toxicity by zinc, *Brain Research* 823 (1999): 88–95; Cornett, Markesbery, and Ehmann, 1998; Lovely and coauthors, 1998; M. P. Cuajungco and G. J. Lees, Zinc metabolism in the brain, *Neurobiology of Disease* 4 (1997): 137–169; F. C. Potocnik and coauthors, Zinc and platelet membrane microviscosity in Alzheimer's disease: The in vivo effect of zinc on platelet membranes and cognition, *South African Medical Journal* 87 (1997): 1116–1119.

[31]J. W. Miller, Homocysteine and Alzheimer's disease, *Nutrition Reviews* 57 (1999): 126–129.

[32]P. M. Ridker and coauthors, Homocysteine and risk of cardiovascular disease among postmenopausal women, *Journal of the American Medical Association* 281 (1999): 1817–1821; K. Robinson and coauthors, Low circulating folate and vitamin B_6 concentrations: Risk factors for stroke, peripheral vascular disease, and coronary artery disease, European COMAC Group, *Circulation* 97 (1998): 437–443; J. Selhub and coauthors, Association between plasma homocysteine concentrations and extracranial carotid-artery stenosis, *New England Journal of Medicine* 332 (1995): 286–291.

[33]A. Hoffman and coauthors, Atherosclerosis, apolipoprotein E, and prevalence of dementia and Alzheimer's disease in the Rotterdam Study, *Lancet* 349 (1997): 151–154, as cited in Miller, 1999.

[34]A. Chait and coauthors, Increased dietary micronutrients decrease serum homocysteine concentrations in patients at high risk of cardiovascular disease, *American Journal of Clinical Nutrition* 70 (1999): 881–887.

[35]A. La Rue and coauthors, Nutritional status and cognitive functioning in a normally aging sample: A 6-y reassessment, *American Journal of Clinical Nutrition* 65 (1997): 20–29.

[36]A. Bendich, Criteria for determining recommended dietary allowances for healthy older adults, *Nutrition Reviews* 53 (1995): S105–S110.

[37]J. V. G. A. Durnin, Energy metabolism in the elderly, in *Nutrition of the Elderly*, ed. H. Munro and G. Schlierf (New York: Vevey/Raven Press, 1992), pp. 51–63.

[38]L. DiPietro, The epidemiology of physical activity and physical function in older people, *Medicine and Science in Sports and Exercise* 28 (1996): 596–660; Durnin, 1992.

[39]W. D. Campbell, Dietary protein requirements of older people: Is the RDA adequate? *Nutrition Today* 31 (1996): 192–197; D. Kritchevsky, Protein requirements of the elderly, in *Nutrition of the Elderly*, ed. H. Munro and G. Schlierf (New York: Vevey/Raven Press, 1992), pp. 109–118.

[40]A. A. Abbase and D. Rudman, Undernutrition in the nursing home: Prevalence, consequences, causes and prevention, *Nutrition Reviews* 52 (1994): 113–122.

[41]P. J. Nestel, Dietary fat for the elderly: What are the issues? in *Nutrition of the Elderly*, ed. H. Munro and G. Schlierf (New York: Vevey/Raven Press, 1992), pp. 119–127.

[42]S. M. Kleiner, Water: An essential but overlooked nutrient, *Journal of the American Dietetic Association* 99 (1999): 200–206.

[43]J. C. Chidester and A. A. Spangler, Fluid intake in the institutionalized elderly, *Journal of the American Dietetic Association* 97 (1997): 23–28; S. A. Gilmore and coauthors, Clinical indicators associated with unintentional weight loss and pressure ulcers in elderly residents of nursing facilities, *Journal of the American Dietetic Association* 95 (1995): 984–992.

[44]R. M. Russell and P. M. Suter, Vitamin requirements of elderly people: An update, *American Journal of Clinical Nutrition* 58 (1993): 4–14.

[45]Standing Committee on the Scientific Evaluation of Dietary Reference Intakes, Food and Nutrition Board, Institute of Medicine, *Dietary Reference Intakes for Calcium, Phosphorus, Magnesium, Vitamin D, and Fluoride* (Washington, D.C.: National Academy Press, 1997).

[46]K. Tucker, Micronutrient status and aging, *Nutrition Reviews* (II) 53 (1995): 9–15.

[47]A. L. de Weck, Immune response and aging, in *Nutrition of the Elderly*, ed. H. Munro and G. Schlierf (New York: Vevey/Raven Press, 1992), pp. 89–97.

[48]Selhub and coauthors, 1995.

[49]C. Ho and coauthors, Practitioners' guide to meeting the vitamin B_{12} Recommended Dietary Allowance for people aged 51 years and older, *Journal of the American Dietetic Association* 99 (1999): 725–727; Standing Committee on the Scientific Evaluation of Dietary Reference Intakes, Food and Nutrition Board, Institute of Medicine, *Dietary Reference Intakes for Thiamin, Riboflavin, Niacin, Vitamin B_6, Folate, Vitamin B_{12}, Pantothenic Acid, Biotin, and Choline* (Washington, D.C.: National Academy Press, 1998), pp. 7-1–7-27.

[50]Tucker, 1995.

[51]R. J. Wood, P. M. Suter, and R. M. Russell, Mineral requirements of elderly people, *American Journal of Clinical Nutrition* 62 (1995): 493–505.

[52]D. V. Porter, Washington update: NIH consensus development conference statement optimal calcium intake, *Nutrition Today*, September/October 1994, pp. 37–40.

[53]H. Payette and K. Gray-Donald, Do vitamin and mineral supplements improve the dietary intake of elderly Canadians? *Canadian Journal of Public Health* 82 (1993): 58–60; W. A. McIntosh and coauthors, The relationship between beliefs about nutrition and dietary practices of the elderly, *Journal of the American Dietetic Association* 90 (1990): 671–675.

[54]R. D. Chandra, Graying of the immune system: Can nutrient supplements improve immunity in the elderly? *Journal of the American Medical Association* 277 (1997): 1398–1399.

[55]M. A. Johnson and K. H. Porter, Micronutrient supplementation and infection in institutionalized elders, *Nutrition Reviews* 55 (1997): 400–404.

[56]K. G. Manton, L. Corder, and E. Stallard, Chronic disability trends in elderly United States population: 1982–1994, *Proceedings of the National Academy of Sciences of the USA* 94 (1997): 2593–2598.

[57]DiPietro, 1996.

[58]Are older Americans making better food choices to meet diet and health recommendations? *Nutrition Reviews* 51 (1993): 20–22.

[59]J. G. Fischer and coauthors, Dairy product intake of the oldest old, *Journal of the American Dietetic Association* 95 (1995): 918–921.

Q&A NUTRITION IN PRACTICE

Food for Singles

Singles of all ages face difficulties in purchasing, storing, and preparing food. Large packages of meat and vegetables are often intended for families of four or more, and even a head of lettuce can spoil before one person can use it all. Many singles live in small dwellings and have little storage space for foods. A limited income presents additional obstacles. The following ideas can help to solve some of these problems.

What advice can help a person on a fixed income acquire groceries?

Once the rent, utilities, and other bills are paid, most people on a fixed income do not have much money for groceries. First, make sure these people know what food assistance programs are available. Such programs are available to older adults who need help obtaining nourishing meals because of financial or other difficulties. Table NP12-1 summarizes food assistance programs for the elderly.

usda.gov/fcs/fs.htm
Food Stamp Program

People who have the means to shop and cook for themselves can cut their food bills just by being wise shoppers. The first decision a person with a tight grocery budget must make is where to shop. Large supermarkets are usually less expensive than convenience stores, but the cost of transportation to the market is a consideration. Once a person decides where to shop, a grocery list that includes specials and coupons will save money and help reduce impulse buying. Specials and coupons are a bargain only when the items featured are those that the shopper needs and uses. Foods that are almost always good buys include rice and fat-free

dry milk, which can be stored on a shelf for months at room temperature; whole wedges of cheese, rather than sliced or shredded cheese; fresh produce in season; cereals that require cooking, instead of ready-to-serve cereals; and dried beans and peas.

All foods, whether featured as specials or not, are bargains only when they are available in quantities that can be used without waste or spoilage. For example, turkey may be on sale and is one of the most economical meats for the nutrients it offers, but only a person with ample storage space for leftovers would benefit from buying a whole turkey.

How can older people and others who live alone buy small quantities when so many foods are packaged for families?

All singles face this problem. Packages of meat and fresh vegeta-

bles often come already wrapped in large servings. Even milk is often available only in gallons or half-gallons and can spoil before one person can use it all. People living alone can try these hints. First, in the dairy case, buy fresh milk in the size best suited for you. If your grocer does not carry pints or half-pints, try a nearby service station or convenience store. Pint-size and even cup-size boxes of heat-treated milk are also available and can be stored unopened on a shelf for up to three months without refrigeration.

Next, among the meats, buy only what you will use. Ask the grocer to break open a package of wrapped meat and rewrap the portion that you need. Alternatively, if you have ample freezer space, you can buy large packages of meat, such as pork chops, ground beef, or chicken, when they are on sale. Then, immediately divide the package into individual

---

Table NP12-1
FOOD ASSISTANCE PROGRAMS FOR OLDER ADULTS

Elderly Nutrition Program

- *Services:* Congregate and home-delivered meals to improve older people's nutrition status. Supportive services include transportation to congregate meal sites; shopping assistance; information and referral; and, to some extent, nutrition counseling and education.
- *Impact:* The Elderly Nutrition Program improves the nutrient content of high-risk older adults' diets and offers socialization and recreation. Many of the nutrition programs around the country go above and beyond federal requirements of congregate and home meals by offering lunch clubs, ethnic meals, and meals for older homeless people. An estimated 25 percent of our nation's elderly poor benefit from these meals.[a]

Food Stamps

- *Services:* Income supplement for low-income households in the form of coupons to purchase food.
- *Impact:* Food stamps serve more as an income supplement for some elderly participants than as a device to improve nutrition status. For other elderly food stamp participants, nutrient intakes are higher than those of nonparticipants with similar incomes.

Meals on Wheels

- *Services:* Direct meal delivery to the homebound elderly, integrated into the meal delivery services provided by the Elderly Nutrition Program.
- *Impact:* Meals on Wheels focuses on filling the need for weekend and holiday meals for homebound elderly people, a service that is limited in the Elderly Nutrition Program.

[a]Federal program nourishes poor elderly, *Journal of the American Medical Association* 278 (1997): 1301.

servings. Wrap them in aluminum foil, not freezer paper: the foil can become the liner for the pan in which you bake or broil the meat, thus saving work over the sink. Don't label these individually; just put them all in a brown bag marked "hamburger" or "chicken thighs" or whatever, along with the date. The bag is easy to locate in the freezer, and you'll know when your supply is running low.

As for meat alternates, buy eggs by the half-dozen—break the carton of a dozen eggs in half. Eggs do keep for long periods, though, if stored in the refrigerator and are such a good source of high-quality protein that you will probably use a dozen before they lose their freshness. Dried beans and peas offer high-quality protein, fiber, and many other nutrients for practically pennies and have a long shelf life.

Among fruits and vegetables, purchase fresh ones individually. Buy only three pieces of each kind of fresh fruit: a ripe one, a semiripe one, and a green one. Eat the first right away, the second soon after, and let the last one ripen on the windowsill. If vegetables are packaged in large quantities, ask the grocer to break open the package so that you can buy what you need. Buy small cans of fruits and vegetables even though they are more expensive per unit. Remember that it is expensive to buy a regular-size can and let the unused portion spoil in the refrigerator. If you have space in your freezer, buy frozen vegetables in large bags rather than in small boxes. You can take out the exact amount you need and close the bag tightly with a rubber band or twist tie. If you return the package quickly to the freezer each time, the vegetables will stay fresh for a long time.

Finally, breads and cereals usually must be purchased in large quantities. When you buy bread, take out the amount you will use in a few days and store the rest in the freezer. Cereal grains (rice, barley, oatmeal) and pastas, like the dried beans mentioned earlier, generally have a long shelf life if you keep them sealed in jars.

What are some suggestions for when a person just has to buy more food than he or she can use?

One suggestion is to make mixtures of leftovers that are already on hand. A thick stew prepared from leftover green beans, carrots, cauliflower, broccoli, and any meat with added onion, pepper, celery, and potatoes makes a complete and balanced meal—except for milk, but then powdered milk can be added to the stew.

Another suggestion is to set aside a shelf in the kitchen for rows of glass jars containing staple items that a person usually cannot buy in single-serving quantities—rice, tapioca, lentils and other dry beans, flour, cornmeal, fat-free dry milk, pasta, cereal, or coconut, to name only a few possibilities. Freeze each filled jar for one night first to kill any insect eggs that might be present. The jars will then keep bugs out of the food indefinitely. The jars make an attractive display and are reminders of different choices on hand to vary menus. Cut the directions-for-use labels from the packages and tape them on the jars.

Creative chefs think of various ways to use foods when only large amounts are available. For example, a head of cauliflower can be divided into thirds. Cook one-third and eat it as a hot vegetable. Put the other two-thirds into a vinegar and oil marinade for use as an appetizer or in a salad. Keep half a package of frozen vegetables with other vegetables to be used in soup or stew.

Also, when one person cooks a lot of food, that person can invite someone to share it. Before long, the guest will probably invite the host or hostess back, and both people will get to enjoy a meal they wouldn't have thought to cook for themselves.

What suggestions can help a person who doesn't drink milk get enough calcium?

Try using fat-free dry milk. It is the greatest convenience food there is.

Dry milk can be used in just about everything: hamburgers, gravies, soups, casseroles, sauces, and beverages, such as iced coffee. The taste is negligible, but 5 heaping tablespoons are the equivalent of a cup of fresh milk.

People can increase their calcium intakes by making soup stock from pork and chicken bones soaked in vinegar. The bones release their calcium into the acid medium, and the vinegar boils off when the stock is boiled. One tablespoon of such stock may contain over 100 milligrams of calcium. Something can then be cooked in this stock every day: vegetables, rice, and stews. And, of course, this stock can be used as a soup base.

Sometimes single people, especially older people, don't feel like cooking a meal and may not get the nutrients they need.

People need to take advantage of times when they do feel like cooking and cook several meals at a time. For example, boil three potatoes with skins. Eat one hot with margarine and chives. When the others have cooled, use one to make a potato-cheese casserole ready to be put into the oven for the next evening's meal. Then slice the third one into a covered bowl and pour the juice from pickles over it. The pickled potato will keep several days in the refrigerator and can be used later in a salad.

Depending on freezer space, make double or triple portions of a dish that takes time to prepare: a casserole, vegetable pie, or meat loaf. The little aluminum trays from frozen foods can be saved and used to freeze the extra servings. Be sure to date the packages and use the oldest first. The work will seem worthwhile when several meals are prepared at once.

An occasional frozen dinner can also be useful—although expensive—if it makes the difference between a person's eating and not eating. Many such dinners that are now available are low in kcalories and nutritious. Adding a fresh salad, a whole-wheat roll, and a glass of milk can make a nice meal.

But encourage people to socialize, too. Sometimes, the loneliness that results from a single person's isolation can impair the person's appetite and motivation to cook appetizing meals.

One more suggestion for those who are alone at mealtime is this: make it a special occasion. One way to do this is to set the table with a tablecloth, a napkin, a full set of utensils, and fresh flowers. Set a pot of stew or homemade soup with vegetables and fresh herbs on low heat to cook, and make a salad. Get comfortable in a stuffed chair, and enjoy a book or some soothing music until the rich aroma of your simmering dinner beckons. After serving your plate, light a candle, dim the lights, savor the food, and relish some of the best company you will ever have—your own.

CHAPTER 13

NUTRITION AND NURSING ASSESSMENTS

CONTENTS

Nutrition in Health Care

Health History

Physical Examinations

Case Study: Risk Factors
for Malnutrition in a
Computer Scientist
following a Car Accident

Nutrition in Practice:
Nutrition and Mental Health

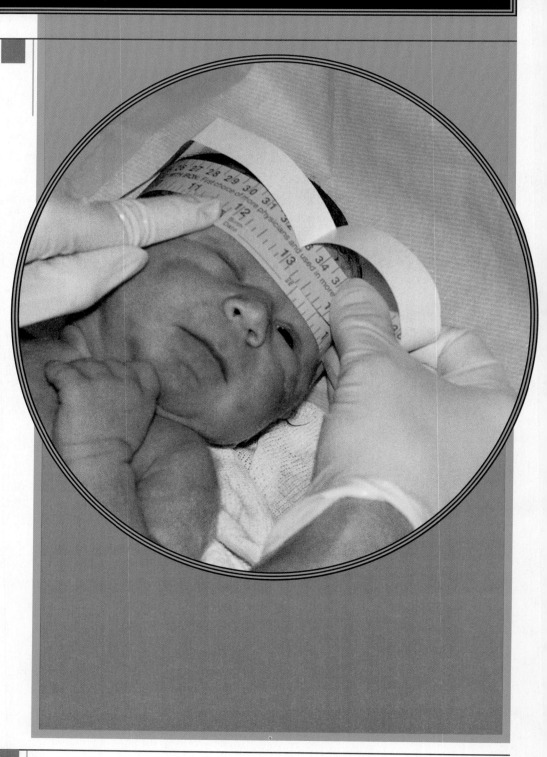

The first twelve chapters of this book have shown how nutrition supports a healthy body and an alert mind throughout life. Turning now to clinical nutrition, the remaining chapters of this book describe how changes in health affect nutrient needs. This chapter explains nurses' responsibilities for nutrition care and shows how nurses can use data from nursing assessments to recognize malnutrition. The next chapter shows how the remaining steps of the **nursing process** can be used to address the nutrition needs of clients.

Nutrition in Health Care

Clients requiring health care services have a variety of physical, emotional, intellectual, social, and spiritual needs that both define their health and affect their response to therapies. Nurses, with their medical expertise and their frequent and intimate contact with clients, have a unique opportunity to identify these needs and address them in an **individualized** and **holistic** manner. Such an approach is particularly useful in meeting clients' nutrition needs because a variety of personal factors influence food choices (see Chapter 1).

For the busy nurse who has many responsibilities, it may sometimes be easy to put a client's nutrition needs on the back burner. After all, the effects of providing nutrition care are not always as apparent as the immediate and visible effects of some medical treatments. Sometimes, however, the consequences of good nutrition can be quite dramatic. Correcting protein-energy malnutrition (PEM), for example, is a case in point. The person with PEM is likely to have multiple nutrient deficiencies, and without adequate energy, protein, vitamins, and minerals, the body is unable to maintain immune defenses, mend broken bones, heal wounds, utilize medications, and support organ function. The compromised person may then fall victim to additional complications, and a downward spiral is set in motion (see Figure 13-1). Yet PEM often goes unrecognized, particularly in the elderly, and even if it is recognized, steps may not be taken to correct the problem or prevent its progression.[1]

Nurses who routinely think "nutrition" and stay alert to the signs of PEM can make a remarkable difference in helping clients maintain or improve health and quality of life. As an added bonus, attention to nutrition can help save time and minimize health care costs by preventing, forestalling, and efficiently treating medical conditions and complications.

Responsibility for Nutrition Care

Each member of the health care team has a unique role in nutrition care. By working together, they can ensure that all nutrition problems are quickly addressed.

PHYSICIANS Physicians hold the ultimate responsibility for meeting all the client's medical needs, including nutrition. The physician prescribes the client's diet and writes the **diet order** in the medical record. The physician may also write orders that relate to nutrition care, including orders for comprehensive nutrition assessments, diet counseling, and evaluation of food intakes. Physicians, in turn, rely on registered dietitians and nurses to alert them to nutrition problems, suggest strategies for correcting these problems, and provide requested nutrition services.

DIETITIANS Not all facilities employ dietitians, and even in those that do, the dietitian cannot see every client. Most often, the dietitian will see only those clients with written orders for special diets or other nutrition services. In these cases, the dietitian may conduct complete **nutrition assessments** and provide **medical nutrition therapy.**

Figure 13-1
ILLNESS, MALNUTRITION, AND IMMUNITY

Regardless of where a person enters the spiral, the effects of illness, malnutrition, and impaired immunity can interact to compromise recovery and worsen malnutrition.

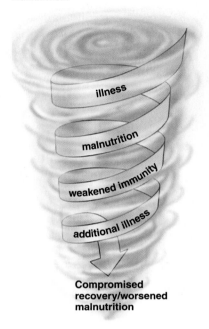

illness

malnutrition

weakened immunity

additional illness

Compromised recovery/worsened malnutrition

nursing process: a systematic approach to identifying and addressing the nursing needs of clients, including nutrition. The nursing process consists of five steps:
1. Assessment.
2. Diagnosis.
3. Outcome identification and planning.
4. Implementation.
5. Evaluation.

individualized: to treat a person as a unique being rather than as one of many others.

holistic (hoe-LIS-tik): to treat a person as a whole being rather than as a sum of many parts.

diet order: a physician's written statement in the medical record of the client's diet prescription.

nutrition assessment: the evaluation of the many factors that influence or reflect a client's nutrient and nutrition education needs.

medical nutrition therapy: provision of a client's nutrient and nutrition education needs based on a complete nutrition assessment.

In this book the term diet-medication interactions *includes interactions between medications and foods, individual nutrients, and nonnutrient food components.*

nutrition status: the state of the body's nutritional health.

■■ **NURSES** If nutrition problems are overlooked when a client seeks health care, the physician's written orders may not reflect the need for a dietitian's nutrition services. Consequently, nutrition problems may not be resolved in a timely manner—or at all. For clients with PEM or those who may readily develop PEM, the consequences can be substantial. The best chance of preventing such a situation lies with the nurse. Indeed, one of the most important nutrition-related responsibilities for the nurse is to identify clients who need nutrition services and communicate that information to the physician or dietitian.

Another important nutrition-related task for the nurse is to be alert to potential diet-medication interactions, which can alter a drug's effectiveness or affect a client's nutrition status. The pharmacist serves as an invaluable resource for providing information about diet-medication interactions, which are described later in this chapter.

Dietitians recognize the unique relationship that nurses have with clients and rely on nurses to communicate pertinent information. Nurses also provide direct nutrition services, such as encouraging clients to eat, measuring height and weight, recording information about food intake, and answering questions about special diets. In facilities that do not employ dietitians, the nurse often assumes the primary responsibility for nutrition care.

■■ **OTHER HEALTH CARE PROFESSIONALS** Other health care professionals may also assist with nutrition care. Dietetic technicians directly assist dietitians. Nursing assistants, home health care aides, social workers, pharmacists, physical therapists, and occupational therapists can be instrumental in alerting dietitians or nurses to nutrition problems and sharing relevant information about clients' health status, personal histories, or concerns. The greater the input of all members of the health care team, the greater the likelihood that nutrition care will be practical and successful.

■■ **Illness and Nutrition Status**

As the earlier chapters of this book have shown, poor nutrition can lead to health problems. Excessive energy intakes, for example, contribute to the development of many chronic diseases including heart disease, hypertension, and diabetes. Inadequate energy can also lead to illness by depressing the immune system. Poor health, in turn, often leads to nutrition problems. People who are most likely to develop nutrition problems, especially PEM, include:

- Those with compromised **nutrition status** prior to developing a health problem.
- Those with multiple disorders or chronic diseases, because such conditions significantly tax the body's nutrient stores and treatment often requires the use of multiple medications over long periods of time.
- Those with markedly elevated metabolic rates, because hypermetabolism dramatically increases the body's demands for nutrients.
- Those with altered organ function, especially dysfunction of the liver, kidneys, or heart, because damaged organs interfere with the body's metabolism of nutrients, excretion of waste products, and distribution of nutrients and medications to the cells.
- Those who are very young, especially premature infants, and those who are elderly, because these groups are more likely to have marginal or poor nutrition status, multiple or chronic illnesses, and altered organ function.

■■ **HEALTH PROBLEMS** Health problems can lead to malnutrition by impairing a person's ability to eat, interfering with the digestion and absorp-

Table 13-1

EXAMPLES OF HEALTH CONDITIONS ASSOCIATED WITH A HIGH RISK OF PROTEIN-ENERGY MALNUTRITION[a] (PEM)

Acquired immune deficiency syndrome (AIDS)

Alcoholism

Anorexia nervosa

Burns, second or third degree covering more than 20% of body surface area

Cancer, some types

Chronic obstructive pulmonary disease, later stages

Chronic pancreatitis

Congestive heart failure, later stages

Crohn's disease

Cystic fibrosis

Depression, especially in the elderly

Dysphagia

Feeding disabilities

Infections

Kidney failure, later stages

Liver failure, later stages

Malabsorption

Septicemia

Surgical resection of the mouth, esophagus, and small intestine

Trauma

[a]The conditions listed here frequently lead to PEM. Many other conditions can lead to malnutrition, especially if nutrition-related problems are not addressed and corrected early in the course of the problem.

tion of nutrients, raising nutrient needs, and altering the ways nutrients are metabolized and excreted. Table 13-1 lists some health conditions that can seriously affect nutrition status.

TREATMENTS Not only health problems, but their treatments as well can interfere with the body's use of nutrients. The use of medications and their interactions with diet are described in the next section.

In addition to medications, other treatments can also affect nutrition status. Preparation for diagnostic tests and surgeries may require that a person not eat. For the person who needs many such tests and procedures, food intake may be very poor at a time when nutrient needs may be especially high due to illness. Surgeries, especially those of the GI tract, can limit food intake or interfere with the digestion and absorption of nutrients. The highly restrictive diets used in the treatment of some disorders limit a person's food choices and can become monotonous—interfering with the client's motivation to eat.

IMMOBILITY AND PRESSURE SORES Bed rest (a treatment for some conditions) and immobility (a consequence of some medical conditions) further compromise nutrition status. Without physical activity, the muscles and bones begin to lose nitrogen and calcium, respectively, which can further contribute to malnutrition.

Immobility, poor food intake, and PEM are all associated with the development of **pressure sores.** Pressure sores can form wherever there is constant pressure on the skin. The elderly and people who are unable to respond to pain or change body positions are most likely to develop pressure sores. Pressure sores can be extremely painful and can easily become infected, raising the body's need for nutrients.

pressure sores: the breakdown of skin and underlying tissues due to constant pressure and lack of oxygen to the affected area; also called **decubitus (dee-CUE-bih-tis) ulcers** or **bedsores.**

■■ **INDIRECT EFFECTS** In addition to the direct physical effects of medical conditions on nutrition status, the costs of health care can seriously affect a person's economic status and lifestyle choices, limiting the ability to obtain enough food and the space and equipment necessary to store and prepare the food. With chronic disorders that demand substantial commitment of time for management, such as cystic fibrosis, diabetes, and chronic renal failure, nutrition problems are quite likely.

LEARNING LINK

Earlier chapters have described the ways that excessive energy and nutrient intakes can harm health, and later chapters describe how weight reduction can assist in the treatment of a variety of health problems. As a review of Table 13-1 shows, however, many medical conditions frequently lead to PEM, and even people who are overweight can develop PEM during the course of an illness.

As you read through Chapters 15 through 24, you will be learning more about medical conditions, symptoms, and treatments and the ways they impair nutrition status. In addition, Nutrition Assessment Checklists at the ends of the chapters highlight assessment findings relevant to a particular group of disorders. For now, remember that nutrition risk is most often associated with reduced food intake, impaired digestion and excretion of nutrients, and altered metabolism.

■■   **Diet-Medication Interactions**

Diet-medication interactions can range from mild to severe. Medications can alter nutrient needs, and foods and food components (both nutrients and nonnutrients) can alter a medication's effectiveness or a person's tolerance of a medication. Adverse diet-medication interactions are most likely to occur if medications are taken over long periods, if several medications are taken, if liver or renal function is poor, if nutrition status is poor or deteriorating, or if nutrient needs are high. Several of these conditions may coexist, as is often true for the elderly. Consider that elderly people are more likely to have:

- Chronic diseases that require the use of multiple medications over long periods of time.
- Altered organ function.
- Malnutrition due to chronic diseases, changes in organ function and mental and emotional health, and financial difficulties that limit access to an adequate diet.
- Problems taking medications as prescribed, either due to changes in mental function or due to physical or financial difficulties that affect their ability to acquire and pay for medications.

With so many factors at work to increase the risk of diet-medication interactions in the elderly, it is not surprising that studies of institutionalized elderly people suggest that multiple medication use may significantly affect the nutrition status of this population.[2]

As the population expands and people live to older ages, the number of people on medications is growing at a rapid pace. Many new medications reach the market each year, and the use of potent medications with an extended duration of action is also on the rise. All of these factors underscore the need for nurses and other health care professionals to understand the

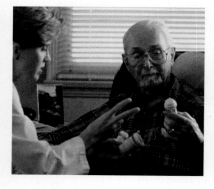

mechanisms of diet-medication interactions, identify them when they occur, and prevent them whenever possible. The "How to" box on p. 319 provides practical suggestions for managing diet-medication interactions.

Foods and food components can alter the absorption, metabolism, and excretion of medications. Likewise, medications can alter food intake and the absorption, metabolism, and excretion of nutrients. Although this discussion focuses on medical drugs, both prescription drugs and nonprescription over-the-counter (OTC) drugs, keep in mind that illicit drugs, nutrient supplements, and herbal preparations (see Nutrition in Practice 24) can also interact with diet. Table 13-2 lists the general classes of medications notable for their interactions with foods and food components. The following sections describe these interactions, and Table 13-3 on p. 318 summarizes this information and provides specific examples.

■■ FOOD INTAKE Medications can reduce food intake by suppressing the appetite, altering taste sensations or the sense of smell, drying the mouth, or leading to mouth lesions that make eating difficult. Amphetamines used to treat attention deficit hyperactivity disorder provide an example: they may effectively improve concentration and behavior, but they also suppress appetite, alter taste perceptions, dry the mouth, and cause nausea. Sedatives and medications that cause excessive drowsiness may make a person too tired to eat. Conversely, some medications stimulate the appetite and lead to undesirable weight gain. An example is thioridazine hydrochloride (Mellaril), an antipsychotic agent.

■■ ABSORPTION Foods frequently affect medication absorption. Foods reduce the absorption of one antihypertensive drug, captopril, and improve the absorption of another, propranolol. In some cases, foods delay, but do not reduce, a medication's absorption. An aspirin taken on an empty stomach works faster than when it is given with food, but because aspirin can cause nausea, taking it with food can reduce the nausea. Individual nutrients and nonnutrients in foods can also affect drug absorption. Among the more common substances in foods that can bind with medications and reduce their absorption are minerals, **phytates,** and **oxalates.**

Laxatives provide an example of how medications can interfere with nutrient absorption. Some laxatives move foods rapidly through the intestine, reducing the time available for nutrient absorption. Other laxatives reduce nutrient absorption for other reasons. For example, fat-soluble vitamins (notably, vitamin D) dissolve in and are excreted along with mineral oil, an indigestible oil that is sometimes used as a laxative. Calcium, too, is excreted.

phytate: a nonnutrient found in the husks of grains, legumes, and seeds.

oxalate: a nonnutrient found in significant amounts in rhubarb, spinach, beets, nuts, chocolate, tea, wheat bran, and strawberries.

Table 13-2

CLASSES OF MEDICATIONS THAT CAN AFFECT NUTRITION STATUS

• Amphetamines and other stimulants	• Antineoplastics
• Analgesics	• Antiulcer agents
• Antacids	• Antiviral agents
• Antibiotics	• Catabolic steroids
• Anticonvulsants	• Diuretics
• Antidepressants	• Hormonal agents
• Antidiabetic agents	• Immunosuppressive agents
• Antidiarrheals	• Laxatives
• Antifungal agents	• Oral contraceptives
• Anithyperlipemics	• Vitamin and other nutrient preparations
• Antihypertensives	

Table 13-3

MECHANISMS AND EXAMPLES OF DIET-MEDICATION INTERACTIONS

Medications Can Alter Food Intake by:

- Altering the appetite (amphetamines suppress the appetite).
- Interfering with taste or smell (methotrexate changes taste perceptions).
- Inducing nausea or vomiting (digitalis can do both).
- Changing the oral environment (phenobarbital can cause dry mouth).
- Causing sores or inflammation of the mouth (methotrexate can cause painful mouth ulcers).

Medications Can Alter Nutrient Absorption by:	**Foods Can Alter Medication Absorption by:**
• Changing the acidity of the digestive tract (antacids can interfere with iron absorption).	• Changing the acidity of the digestive tract (candy can change the acidity, thereby causing slow-acting asthma medication to dissolve too quickly).
• Altering motility of the digestive tract (laxatives speed motility, causing the malabsorption of many nutrients).	• Stimulating secretion of digestive juices (griseofulvin is absorbed better when taken with foods that stimulate the release of digestive enzymes).
• Inactivating enzyme systems (neomycin may reduce lipase activity).	• Altering rate of absorption (aspirin is absorbed more slowly when taken with food).
• Damaging mucosal cells (chemotherapy can damage mucosal cells).	• Binding to drugs (calcium binds to tetracycline, limiting drug absorption).
• Binding to nutrients (some antacids bind phosphorus).	• Competing for absorption sites in the intestines (dietary amino acids interefere with levodopa absorption this way).

Medications and Nutrients Can Interact and Alter Metabolism by:

- Acting as structural analogs (as anticoagulants and vitamin K do).
- Competing with each other for metabolic enzyme systems (as phenobarbital and folate do).
- Altering enzyme activity and contributing pharmacologically active substances (as monoamine oxidase inhibitors and tyramine do).

Medications Can Alter Nutrient Excretion by:	**Foods Can Alter Medication Excretion by:**
• Altering reabsorption in the kidneys (some diuretics increase the excretion of sodium and potassium).	• Changing the acidity of the urine (vitamin C can alter urinary pH and limit the excretion of aspirin).
• Displacing nutrients from their plasma protein carriers (aspirin displaces folate).	

An added danger with all laxatives is that a person who uses them daily for a long time may find that the intestines can no longer function without them. The more often laxatives are used, the more likely that nutrient deficiencies will develop.

Nutrients and medications can interact and reduce the absorption of both. A classic example is the interaction between the antibiotic tetracycline and the minerals calcium and iron. When either of these minerals and tetracycline are taken at the same time, the mineral binds to the drug and both are excreted. To circumvent this problem, clients are instructed to take tetracycline on an empty stomach at least one hour before or two hours after meals or after using milk, milk products, calcium-containing antacids, and iron supplements.

:: METABOLISM Nutrients and nonnutrients can significantly affect a medication's metabolism. Some medications, for example, resemble vitamins in structure, and it is this property that makes them effective. Vitamin K and the anticlotting medication warfarin (Coumadin) provide an example. Because warfarin resembles vitamin K, warfarin interferes with the synthesis of clotting factors that require vitamin K. The prescribed warfarin dose depends, in part, on how much vitamin K is in the diet. If a person's vitamin K intake changes, as may happen when people change their diets in the summer to include more leafy green vegetables, then the physician has to alter the medication dose. Another example is methotrexate, which resembles folate in

MANAGE DIET- MEDICATION INTERACTIONS

With hundreds of known diet-medication interactions and many more to be identified in the future, how can the nurse deal with all the information? Although this box specifically addresses diet-medication interactions, nurses also need to remain alert for drug-drug interactions, side effects unrelated to nutrition, effects of medications on diagnostic tests, and other nursing considerations. The process is difficult at first, but becomes easier with clinical experience. At first, looking up and learning about the medications you encounter in clinical practice is an inescapable task. To do so, you must purchase a current drug guide (many are written specifically for nurses) or have access to such a guide in the facility where you work. In hospitals or extended care facilities where nurses administer medications to clients, the pharmacy may supply a list of precautions that apply to the medications clients receive.

Begin with a list of the medications prescribed for a particular client. As you study each medication, make a note of:

- The appropriate method of administering the medication (twice daily or at bedtime, for example).
- How the medications should be given with respect to foods or specific nutrients (give on an empty stomach, give with food, do not give with milk, give iron supplements at least 2 hours apart from the medication dose, for example).
- How the medication should be given with respect to other medications.
- Side effects that can affect food intake (nausea, vomiting, or sedation, for example), nutrient needs (hypokalemia or hyperglycemia, for example), or dietary recommendations (constipation or flatlence, for example).

The nurse, pharmacist, or dietitian should review all precautions related to medications with the client, explain signs of potential nutrient deficiencies or nutrition-related problems, and advise clients of actions to take if problems arise.

Clients, particularly those who must take multiple medications, may need help figuring out when to take each medication to avoid medication-medication or diet-medication interactions. A later section of this chapter describes ways you can uncover information about a client's usual eating habits. You will need this information for medications that must be administered with regard to food intake or specific dietary components. If you need more information about a medication or potential interaction, remember to ask the pharmacist.

For medications that can lead to nutrient deficiencies or alter nutrient needs, remain alert for signs of nutrient imbalances, especially when:

- Imbalances are commonly noted with the use of medication.
- The adverse effect persists over time.
- The client is in a high-risk group (see p. 314).

Remember to alert the dietitian if you suspect a problem or feel that the client can benefit from nutrition counseling.

structure (see Figure 13-2 on p. 320). People who take methotrexate to treat certain cancers or rheumatoid arthritis can develop severe folate deficiencies.

Aspirin can also alter folate metabolism but in a different way. Aspirin competes with folate for its protein carrier, thus hindering the body's use of the vitamin. When aspirin is used over long periods of time, health care

Figure 13-2
FOLATE AND METHOTREXATE

By competing for the enzyme that activates folate, methotrexate prevents cancer cells from obtaining the folate they need to multiply. In the process, normal cells are also deprived of the folate they need.

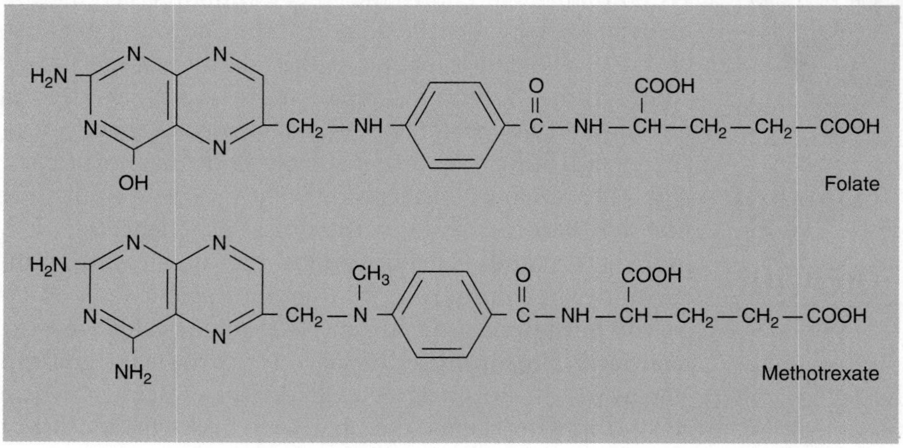

professionals should ensure that either the diet or supplements are supplying sufficient folate to meet the added needs.

Tyramine, a substance found in some foods, and monoamine oxidase (MAO) inhibitors, medications prescribed to treat certain forms of severe depression, provide an example of a potentially fatal diet-medication interaction. MAO inhibitors block the action of the enzyme in the brain that normally inactivates tyramine. When people who take MAO inhibitors consume large amounts of tyramine, tyramine remains active and stimulates the release of the neurotransmitter norepinephrine. Severe headaches and hypertension can result, and if blood pressure rises high enough, it can be fatal. For this reason, people taking MAO inhibitors are advised to restrict their intakes of foods rich in tyramine (see Table 13-4).

In some cases, nutrients must be available to maximize a medication's effectiveness. The medication aldendronate sodium (Fosamax), used to increase bone mass and prevent osteoporosis in postmenopausal women, for example, depends on an adequate supply of vitamin D and calcium, either from the diet or from supplements, for maximum effectiveness.

Among the notable foods and food components that can affect drug metabolism are grapefruit juice (but not other citrus juices), caffeine, and nat-

Table 13-4
FOODS RESTRICTED IN A TYRAMINE-CONTROLLED DIET

Beverages	Red wines including chianti, sherry[a]
Cheeses	Aged cheeses, American, camembert, cheddar, gouda, gruyère, mozzarella, parmesan, provolone, romano, roquefort, stilton[b]
Meats	Liver; dried, salted, smoked, or pickled fish; sausage; pepperoni; salami; dried meats
Vegetables	Fava beans; Italian broad beans; sauerkraut; snow peas; fermented pickles and olives
Other	Brewer's yeast;[c] all aged and fermented products; soy sauce in large amounts; cheese-filled breads, crackers, and desserts; salad dressings containing cheese

NOTE: The tyramine contents of foods vary from product to product depending on the methods used to prepare, process, and store the food. In some cases, as little as 1 ounce of cheese can cause a severe hypertensive reaction in people taking monoamine oxidase inhibitors. In general, the following foods contain small enough amounts of tyramine that they can be consumed in small quantities: ripe avocado, banana, yogurt, sour cream, acidophilus milk, buttermilk, raspberries, and peanuts.
[a]Most wine and domestic beer can be consumed in small quantities.
[b]Unfermented cheeses, such as ricotta, cottage cheese, and cream cheese, are allowed.
[c]Products made with baker's yeast are allowed.

ural licorice. An area currently generating a great deal of interest and research is the interaction between grapefruit juice and a variety of medications. Grapefruit juice has been shown to raise blood levels of several medications, including some calcium channel blockers used to treat hypertension, the cholesterol-lowering drugs lovastatin and simvastatin (and possibly other cholesterol-lowering drugs in the same family), the antihistamine terfenadine, the hormone ethinyl estradiol, the sedative medazolam, the antianxiety agent buspirone, the anticonvulsant carbamazepine, the antiviral agent saquinavir, and the immunosuppressant cyclosporine.[3]

Caffeine, which acts as a central nervous system stimulant, diuretic, and muscle relaxant, can potentiate the actions of some medications. Natural licorice can complicate drug therapy using diuretics and antihypertensive agents because it promotes sodium retention and potassium excretion. Most licorice sold in the United States, however, is not natural licorice, but rather a flavored substitute that does not interact with medications.

▪▪ EXCRETION Urinary acidity affects the reabsorption of medications from the kidneys back into the blood. An acidic urine limits the excretion of acidic drugs like aspirin. Consequently, large doses of vitamin C taken along with aspirin increase the urine's acidity, and aspirin remains in the blood longer.

Medications can also alter urinary excretion of nutrients. For example, some diuretics accelerate the excretion of calcium, potassium, magnesium, and zinc.

▪▪ OTHER INGREDIENTS IN MEDICATIONS Besides the active ingredients, medications may contain other substances such as sugar, sorbitol, lactose, sodium, and caffeine. For most people who use medications on occasion and in small amounts, such ingredients pose no problems. When medications are taken regularly or in large doses, however, people with specific problems may need to be aware of these additional ingredients and their effects.

Many liquid preparations contain sugar or sorbitol to make them taste better. For people who must regulate their intakes of carbohydrates, such as people with diabetes, the amount of sugar in medications must be considered. Large doses of liquids containing sorbitol may result in diarrhea. The lactose added as a filler to some medications may cause problems for people who cannot digest lactose or those who cannot metabolize galactose.

Reminder: Lactose is composed of two simple sugars, glucose and galactose.

Antibiotics and antacids often contain sodium. People who take Alka-Seltzer may not realize that a single two-tablet dose may exceed their safe sodium intakes for a whole day. Medications given by vein provide water and frequently provide sodium, potassium, and other electrolytes, or dextrose (the name for glucose in intravenous solutions). The contributions these nutrients make must be considered when a client's diet is restricted in any of these nutrients. Interactions between medications, tube feedings, and intravenous feedings can also occur and will be described in Chapters 17 and 18.

LEARNING LINK

As you gain clinical experience, you will feel more comfortable about looking for diet-medication interactions. To help ease the transition from learning to experience, boxes in Chapters 15 through 24 describe diet-medication interactions that apply to specific medical conditions. Table E-1 in Appendix E shows where you can find information about different classes of medications.

▓ Identifying Risk for Malnutrition

With an understanding of the potential effects of illnesses and their treatments on nutrition status, how can nurses identify nutrition problems? The first step of the nursing process, assessment, provides the information nurses need; all that is required is to think nutrition when reviewing it (see Table 13-5). Nurses routinely complete both a health history and a physical examination when clients are first admitted to health care facilities and regularly reassess clients throughout their course of care. Ongoing assessments evaluate how well a care plan is working and identify new problems that may arise. Assessments evaluate a client's physical, emotional, intellectual, social, and spiritual health to identify potential and current health problems, factors that contribute to these problems, and lifestyle factors, resources, and motivation for addressing health problems.

▓ **NUTRITION SCREENING** Although nurses can use assessment information to identify actual and potential nutrition problems in any health care setting, formal procedures for **nutrition screening** are most likely to be used routinely in hospitals and extended care facilities. The Joint Commission on Accreditation of Healthcare Organizations (JCAHO) recommends that nutrition screenings be completed shortly after admission, ideally within 24 hours, and at regular intervals thereafter. A person may be admitted to a health care facility in good nutrition status, but nutrient stores may decline as a consequence of the client's medical condition.

nutrition screening: a tool for quickly identifying clients most likely to be at risk for malnutrition so that they can receive complete nutrition assessments.

Table 13-5
INFORMATION RELEVANT TO NUTRITION STATUS GATHERED DURING NURSING ASSESSMENTS

Area of Assessment	What It Can Reveal
Health • Current health problem(s) • Client's and family's health history • Previous surgeries • Medication history • Alternative therapies	Information about nutrient needs, nutrition status, or the need for dietary measures to prevent health problems.
Personal Information • Age • Gender • Cultural/ethnic group • Occupation • Role in family • Self-perception/self-concept	Factors that affect nutrient needs (age, gender, food choices) as well as attitudes and beliefs about seeking health care and following recommendations.
Review of body systems • Hypertension • Mobility status • Visual impairment	Factors that may affect nutrient needs or the ability to prepare and eat foods.
Activities of daily living • Type of diet • Usual weight • Appetite and changes in appetite • Use of alcohol, caffeine, and tobacco • Fluid intake • Urinary and bowel elimination patterns • Sleep pattern • Activity pattern • Ability to perform self-care	Typical diet, appetite, activity, and elimination patterns that can suggest the need for dietary changes, risk for malnutrition (appetite), or the ability to prepare and eat foods.
Physical examination • Height • Weight • Physical signs of malnutrition • Laboratory tests	Problems with growth and development, risk for PEM, obesity, and potential nutrient deficiencies or excesses.

NUTRITION RISK The exact criteria that determine nutrition risk vary from facility to facility and are based, in part, on the client population. The screening criteria for women in a maternal and infant clinic, for example, differ from those used in a hospital that serves a geriatric population. Each facility's nutrition screening tool will include those factors most likely to identify risk for malnutrition among the clients it serves. The following sections highlight factors that indicate risk for malnutrition.

Health History

Health histories identify current and potential health problems, medication use, personal information that affects health needs, **activities of daily living,** and a **review of body systems.** As Figure 13-3 shows, various components of the health history are interrelated, and all must be considered to capture a complete picture of a person's health status.

Health Problems and Treatments

Clients with some health problems (see Table 13-1 on p. 315) are considered to be at high risk for PEM based solely on these findings. At facilities with formal nutrition screening procedures, the dietitian would then conduct a complete nutrition assessment. Besides identifying risk for PEM, the nurse's review of clients' health problems and treatment can uncover other nutrition-related information including:

- The need for diet modifications.
- Physical disabilities that may interfere with a person's ability to prepare and eat adequate amounts of food.
- Disorders that require treatments or lifestyle modifications that demand a great deal of time, motivation, or financial resources.

activities of daily living: a description of the client's lifestyle and ability to manage self-care. Activities of daily living include patterns of food intake, urinary and bowel elimination, activity and exercise, and rest and sleep.

review of body systems: a brief review of clinical signs or symptoms associated with each organ system.

Figure 13-3

INTERRELATIONSHIPS AMONG HEALTH, MEDICATIONS, NUTRITION, AND PERSONAL FACTORS

Attention to the interrelationships among the components of a client's history helps health care professionals provide high-quality, cost-effective, and realistic care. Here the dark lines indicate direct relationships, and the light lines indicate indirect relationships.

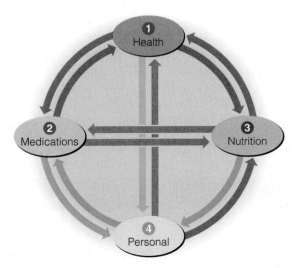

1 Health status can alter nutrient requirements, alter nutrition status, and directly affect the need for medications. Health care costs and the effects of health status on quality of life can alter lifestyle and seriously affect economic status.

2 Medications can alter nutrient requirements and affect nutrition status. Medications can have positive effects, negative effects, or both positive and negative effects on health. The costs of medications and their side effects can affect economic status, lifestyle, and quality of life.

3 Nutrition status can affect a client's health status, which, in turn, can affect the need for medications. Nutrition status and food intake can also alter the way medications are digested, absorbed, metabolized, and excreted. Although many personal factors are independent of nutrition status, poor nutrition and health status can affect a person's quality of life and have an impact on lifestyle. The costs of some nutrition therapies as well as a person's quality of life can interfere with a person's ability to maintain an optimal economic status.

4 Personal factors can have an impact on nutrient needs, food availability, and food choices. Personal factors can affect health status by altering nutrition status. Personal factors can also increase the predisposition to some diseases, such as hypertension and diabetes, and can affect people's ability to pay for health care, as well as their attitudes and beliefs about seeking health care and/or following health care recommendations.

- Potential health problems that modified diets might help to prevent or delay.
- Potential diet-medication interactions.

Health assessments include a record of all medications the client takes (prescription and over-the-counter), the dose, and the number of times the medication is taken daily. Remember to ask clients if they use any type of nutrient supplements or alternative therapies, and note the types and amounts of these preparations as well. Once medical conditions, symptoms, and treatments with nutrition implications are noted, assessment of other portions of the health history can be used to confirm suspected problems or uncover problems that may not be apparent from the medical diagnoses.

 ∷ PERSONAL FACTORS Personal factors identified in nursing assessments often include age, gender, number of people in the household, education level, occupation, income, ethnic background, cultural orientation, religious affiliation, perception of health status, sources of psychological stress, coping skills, and self-concept. Personal factors help direct the assessment and nursing process. For example, age and gender help pinpoint nutrient needs and cue the nurse to look for nutrition problems common to specific age groups. For another example, the nurse who discovers that a client typically leaves health care decisions in the hands of a significant other is reminded to include the significant other in the assessment and remainder of the nursing process.

For all age groups, an evaluation of intellectual, spiritual, social, psychological, and financial factors helps identify beliefs about health and health care, influences on food choices, resources for dealing with health and nutrition problems, and constraints that may interfere with prescribed interventions. When personal factors suggest potential nutrition problems, the nurse uses other assessment data to confirm whether a problem exists. For example, if a client's personal information suggests limited financial resources to purchase adequate amounts of foods, the nurse remains alert to the possibility of malnutrition throughout the remainder of the assessment.

∷ ACTIVITIES OF DAILY LIVING This portion of the health history includes a description of the client's patterns of food intake, urinary and bowel elimination, activities and exercise, and sleep and rest. Nurses review food intake patterns to help identify nutrient inadequacies, excessive food intake, and eating habits that might contribute to health problems. Nurses can also use food intake data to help clients plan medication schedules (see p. 319). The next section explains how to collect and review food intake information.

Reduced food intake accompanies many health problems and frequently contributes to PEM. As a general rule, clients who eat less than half of their recommended energy intakes for five or more days risk developing PEM. For clients who report reduced food intakes, assessment of weight and rate of weight loss (described later in this chapter), as well as symptoms, clinical findings, and personal factors that contribute to reduced food intake, assists in identifying appropriate interventions.

LEARNING LINK

Recall that part of the nursing assessment evaluates the client's physical activity and exercise patterns. With this information in hand, you can estimate a client's energy needs by using Table 6-2 in Chapter 6. As for pregnant women, infants, children, teens, and adults, Chapters 10 through 12 described their changing energy needs. Table 19-2 in Chapter 19 shows another formula for estimating energy needs that is frequently used in hospitals.

In clinical practice, however, the nurse may not have the time or the information on hand to estimate energy needs and instead relies on body weight to determine if the client's intake is adequate or excessive. If an adult is maintaining a desirable body weight over time, then that person is already consuming an appropriate energy intake, for example. A later section of this chapter describes how to evaluate weight changes.

With respect to bowel elimination patterns, a history of diarrhea or constipation alerts the nurse to check for dietary patterns that may contribute to these problems. Activity and exercise patterns provide information about the client's level of physical fitness and energy needs as well as the client's ability to walk, stand, prepare foods, and eat without assistance.

Methods for Gathering Food Intake Data

Simple approaches for gathering information about clients' food intakes include the 24-hour recall, the usual intake method, and the food frequency questionnaire. To provide meaningful results, the nurse needs to inquire about portion sizes, beverage consumption, and how foods and beverages are prepared.

24-HOUR RECALL OR USUAL INTAKE

To get a general picture of food intake and eating habits, the 24-hour recall asks the client to recount everything eaten or drunk during the previous day. The usual intake method is similar, but asks the client to recount everything eaten in a typical day and at what times meals and snacks are routinely eaten.

To obtain data about a person's 24-hour or usual intake, an inquiry might begin with "What is the first thing you ate or drank yesterday (or, you usually eat or drink) during the day?" Similar questions follow until the data are complete. When using the 24-hour recall, always ask clients if their intake that day was usual. If not, find out how it varies from the usual intake. By carefully asking questions to verify information clients provide, usual eating habits can be identified. For example, one person may always eat an afternoon snack, another may never eat breakfast, and still another may avoid milk.

Clients may forget to mention the beverages they drink and the condiments, sweeteners, and spices they add to foods, unless specifically prompted to do so. Clients may state they ate 2 slices of toast, for example, and may not think to add that they spread 4 pats of butter and 2 tablespoons of jelly on the toast.

FOOD FREQUENCY QUESTIONNAIRE

Food frequency questionnaires used for nursing assessments typically ask clients how their daily food intake compares to the Daily Food Guide. Clients may be asked how many servings of breads, cereals, or grain products; vegetables; fruits; meat, poultry, fish, and alternatives; milk, cheese, and yogurt; and fats, oils, and sweets they eat in a typical day. Clients are often asked specific questions to determine their fat intakes; for example, "How often do you eat red meat in a typical week?" or "What type of milk do you drink?"

ANALYSIS OF FOOD INTAKE INFORMATION

Although 24-hour recalls, usual intakes, and food frequency questionnaires provide a great deal of useful information, none of the methods provides a degree of accuracy that allows the nurse to determine specific nutrient imbalances. Clients' recollections of their intakes, the amounts of foods consumed, and the ways foods are prepared are estimates at best. Instead, the nurse can use the information to determine if a more detailed analysis of food intake by the dietitian is warranted.

The nurse looks for dietary patterns by comparing the client's intake to the Daily Food Guide and dietary guidelines. Are all food groups included in appropriate amounts? If not, which nutrients may be excessive or deficient (see Figure 1-4 in Chapter 1)? Are food choices varied, or is the diet monotonous? Does the diet provide adequate fiber or excessive amounts of refined sugars? Are the total amount of fat and the types of fat (including saturated, monounsaturated, and polyunsaturated fats and cholesterol) appropriate? Is caffeine or alcohol consumption excessive?

In assessing the client's diet, remember to keep the client's health problems in mind. Consider a client who reports problems with sleeping, for example. Such a problem should alert the nurse to look for sources of caffeine in the diet and the times of day the client uses such products. For a client with lactose intolerance, the nurse looks for nonmilk sources of calcium in the diet.

For clients in hospitals and extended care facilities who are eating poorly or losing too much weight, it may be important to document what and how much they are eating. Nurses play an important role in completing kcalorie counts, a method for determining clients' food intakes that requires cooperation between the nursing service and the dietary department.

kCALORIE COUNTS In facilities that serve meals, client's food intake can be directly observed and analyzed. Nurses or nursing assistants (or, in some cases, the client) write down all the foods and beverages as well as the amounts of each the client receives. After the client finishes the food, beverage, or meal, the amount that is left is also recorded. Dietitians or dietary technicians retrieve the records, deduce what has been eaten, and use the information to estimate nutrient intake.

FOOD RECORDS When the nurse or dietitian needs more information about a client's eating habits and food intake, a food record maintained over several days provides valuable information. Food records are particularly useful for monitoring a client's response to and compliance with medical nutrition therapy and ascertaining a client's tolerance for certain foods. The client writes down all foods and beverages consumed, the times foods are eaten, the amounts consumed, and methods of preparation. Often the person is asked to record other information as well, depending on the purpose of the food record. If the record is to be used to help a person change eating behaviors and lose weight, it might also include information about the person's mood, the occasion (party, holiday, family meal), behaviors associated with eating food (watching TV, driving in the car while on the way to work, sitting at the table with the family), and physical activity. Figure 7-5 in Chapter 7 provided an example of this type of record. When the purpose of the food record is to establish blood glucose control (see Chapter 20), records include details of medication administration, physical activity, illness, and the results of blood glucose monitoring. When the purpose of the record is to establish food tolerances (such as the amount of lactose a person can handle), food records also include symptoms associated with eating (for example, cramps, diarrhea, nausea, or hives).

Food records help pinpoint problem food patterns so that solutions can be implemented. Unfortunately, they require a great deal of time to complete, and clients must be highly motivated to keep them accurately. Another drawback is that clients may either consciously or unconsciously change their eating behaviors while keeping the records. If clients understand their diets but are not following them, the clients may record what they believe they should be eating, rather than what they are actually eating.[4]

The health history contains abundant information to help nurses identify possible malnutrition or nutrition-related problems. Physical examinations can help verify this information and provide additional clues to nutrition status.

■■ Physical Examinations

In addition to health histories, nursing assessments include physical examinations. The portions of the physical examination that are routinely collected and have implications for nutrition status include measures of growth and development, physical findings, and biochemical and diagnostic tests. Table 13-6 lists these and other physical measurements and tests that provide information about nutrition status and indicates what each measure reflects.

Measurements of physical characteristics of the body, such as height and weight, are also called **anthropometric (an-throw-poe-MEH-trick)** *measurements.*
> *anthropos = human*
> *metric = measure*

■■ Measures of Growth and Development

Nurses evaluate growth and development in children by measuring weight, length or height, and head circumference and plotting the data on standard growth charts (see Appendix E). For adults, nurses evaluate body weight in relation to a person's height. Weight is measured for all children and adults. Height (or length) and weight are routinely measured in all health care settings; therefore, learning to identify the best equipment and use the correct measurement techniques is a skill that will be very valuable in nursing practice. The "How to" box on p. 328 shows how to accurately measure weight.

Recumbent length measurements can be significantly greater than standing height measurements, so using the correct measurement is important.[5] Length is the correct measurement for children under three when the growth chart from birth to 36 months is used to plot growth data; height is the correct measurement for children who are two or older when growth charts for children over two years are used to plot growth data. Length may be measured in adults and children who are unable to stand erect for medical or physical reasons. Because length measurements are significantly greater than height measurements, however, weight data must be interpreted carefully. The "How to" box on p. 329 describes how to measure length and height.

Nurses measure head circumference in infants and young children to confirm that brain development is proceeding normally. To measure head circumference, the nurse places a nonstretchable tape so that it encircles the

Table 13-6
BODY MEASUREMENTS USEFUL IN ASSESSING NUTRITION STATUS

Measurement	What It Reflects
Abdominal girth measurement	Abdominal fluid retention and abdominal organ size
Hand grip strength	Ability of muscles to perform work (protein status)
Height-weight	Overnutrition and undernutrition; growth in children
%IBW, %UBW,[a] recent weight change	Overnutrition and undernutrition
Head circumference	Brain growth and development in infants and children under age two
Fatfold	Subcutaneous and total body fat
Midarm muscle circumference	Skeletal muscle mass (protein status)
Size of reaction to skin test	Immune function (protein status)
Waist circumference	Fat distribution

[a]%IBW = percent ideal body weight; %UBW = percent usual body weight.

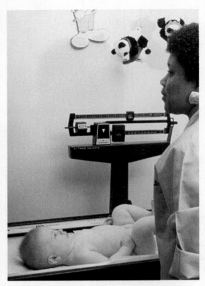

Infants are measured on scales that allow them to sit or lie down while they are being weighed.

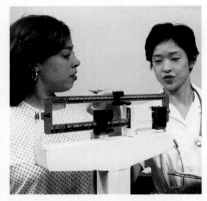

Beam balance scales provide accurate weight measurements for older children and adults.

HOW TO:

MEASURE WEIGHT

To improve the accuracy of weight measurements, keep in mind these tips:

- Always measure—never ask! Self-reported weights are often inaccurate and are frequently lower than measured weights.
- Use the right equipment. Beam balance or electronic scales that have been calibrated and checked for accuracy at regular intervals provide the most reliable weight measurements. Infant scales are equipped with platforms that allow infants to sit or lie down while being weighed. Once children can stand erect, regular scales provide accurate weight measurements. Special scales and hospital beds with built-in scales assist in weighing clients who are unable to stand on regular scales.
- Follow standard procedures. For children under two or three (depending on the growth chart used), weigh the child without clothing or diapers. For all others, try to use the same scale and take weight measurements at about the same time of the day (preferably before breakfast), in about the same amount of clothing (without shoes), and after the person has voided.
- Record the measurement immediately either to the nearest ¼ pound or 0.1 kilogram.

largest part of the head: just above the eyebrows, just above the point where the ears attach, and around the occipital prominence at the back of the head. Immediately record the measurement to the nearest ⅛ inch or 0.5 centimeter.

ANALYSIS OF MEASURES IN INFANTS AND CHILDREN Single measurements are of limited value; measurements taken periodically allow a more meaningful assessment of growth patterns. Although individual growth patterns vary, a child's height and weight generally fall within about the same percentile, and both remain roughly at about the same percentile throughout childhood.

In addition to nutrition, genetic factors influence height; therefore, the nurse considers the heights of family members in analyzing height measurements. Height below the 5th percentile or a sudden drop in a previously steady growth pattern suggests growth retardation, an important sign of malnutrition. A weight percentile significantly lower than the height percentile—for example, weight in the 25th percentile and height in the 95th percentile—suggests that the child is underweight. Conversely, a weight percentile significantly higher than the height percentile—for example, weight in the 95th percentile and height in the 25th percentile—suggests that the child is overweight.

Head circumference percentiles reflect brain growth, which occurs rapidly before birth and during early infancy and is nearly complete by age seven. Malnutrition before birth and during childhood can impair brain development. In addition to nutrition factors, certain disorders and genetic variation can also influence head circumference.

ANALYSIS OF WEIGHT MEASURES FOR ADULTS Health care professionals typically compare weights with weight-for-height standards. (See Chapter 7 for a discussion of healthy body weight and weight standards.) Each facility or practitioner decides which standard to use based on a review of the literature and clinical judgment. Some commonly used standards for weight assessment include:

- The body mass index (BMI) and tables based on the BMI observed to be consistent with health (see p. 139 and the inside back cover).

MEASURE LENGTH AND HEIGHT

Tips for measuring length and height include:

- Always measure—never ask! Self-reported heights are often greater than measured heights. If height is not measured, document that the height is self-reported.
- Measure length for infants and young children using a measuring board with a fixed headboard and movable footboard. It often takes two people to measure length. One person gently holds the infant's head against the headboard; the other straightens the infant's legs and moves the footboard to the bottom of the infant's feet.

Using a measuring tape is a less exacting way to measure length in infants and is the method used to measure length for others who cannot stand erect. To use this method, straighten out the infant's or person's body, make a mark at the top of the head, make another mark at the bottom of the heel, and then measure the distance between the two marks.

- Measure height against a wall to which a nonstretchable measuring tape or stick has been fixed. Ask the person to stand erect without shoes and with heels together. The person's line of sight should be horizontal, with the heels, buttocks, shoulders, and head touching the wall. Place a ruler or other flat, stiff object on the top of the head at a right angle to the wall and carefully note the height measurement. Although less accurate, the measuring rod of a scale can also be used to measure height. To use this method, follow the same general procedure, asking the person to face away from the scale. Take extra care to ensure that the client is standing erect and that the line of sight is horizontal.
- Immediately record length and measurements to the nearest ¼ inch or 0.5 centimeter.

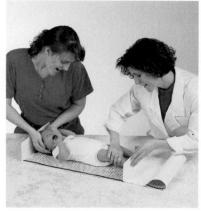

Lying still with legs straight for a length measurement can be a trying experience for an infant.

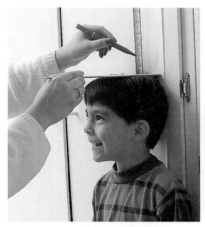

Standing erect allows for an accurate height measurement.

- The Suggested Weights for Adults table (see p. 138).
- A quick method of estimating desirable body weight shown in Table 13-7.

The standard chosen can be used to compare a person's actual weight with desirable weight, a figure commonly known as the percent ideal

Reminder: Weight-for-height tables suggest a weight range, rather than pinpointing one weight—a helpful reminder that desirable weights are estimates at best.

Table 13-7
QUICK ESTIMATE OF DESIRABLE BODY WEIGHT

Men

For 5 ft, consider 106 lb a reasonable weight.

For each inch over 5 ft, add 6 lb.

Subtact 6 lb for each inch under 5 ft.

Add 10% for a large-framed individual; subtract 10% for a small-framed individual.

Example: A man 5 ft 8 in tall (medium frame) would start at 106 lb, add 48, and arrive at a reasonable weight of 154 lb.

Women

For 5 ft, consider 100 lb a reasonable weight.

For each inch over 5 ft, add 5 lb.

Subtract 5 lb for each inch under 5 ft.

Add 10% for a large-framed individual; subtract 10% for a small-framed individual.

Example: A woman 5 ft 6 in tall (medium frame) would start at 100 lb, add 30, and arrive at a reasonable weight of 130 lb.

ESTIMATE %IBW AND %UBW

To estimate %IBW, compare the client's current weight with the desirable body weight from standard weight-for-height tables:

$$\%IBW = \frac{\text{actual body weight}}{\text{desirable (ideal) weight}} \times 100.$$

For example, suppose you wish to calculate %IBW for a man who is 5 feet 8 inches tall and weighs 123 pounds. For desirable (ideal) weight, use the midpoint of the upper half of the weight range in Table 7-1 on p. 138. In this example, the desirable weight is 154 pounds.

$$\%IBW = \frac{123\ lb}{154\ lb} \times 100 = 80\%.$$

The man in this example is at 80 percent of his ideal body weight. A look at Table 13-8 indicates that at 80 percent IBW he is mildly underweight.

This man has lost 15 pounds in the last month. To calculate his %UBW, use this formula:

$$\%UBW = \frac{\text{actual weight}}{\text{usual weight}} \times 100.$$

Calculate the usual weight (138 pounds) by adding the weight loss (15 pounds) to the current body weight (123 pounds).

$$\%UBW = \frac{123\ lb}{138\ lb} \times 100 = 89\%.$$

The man is at 89 percent of his usual body weight. A look at Table 13-8 reveals that a person at 89 percent UBW is mildly underweight. Based on %UBW, the degree of underweight is less severe than the %IBW implied, because this man has consistently weighed less than standard weight. Nevertheless, his recent rate of weight change is very significant: he has lost almost 4 pounds per week.

Although the term ideal body weight *is a misnomer, it is the term most likely to be used in health care settings, and so it is used here.*

body weight (%IBW). Actual weight can also be compared to usual body weight to generate the percent usual body weight (%UBW), which considers what weight is normal for a particular individual. The %UBW is particularly useful for evaluating the degree of nutrition risk associated with acute illness. For example, in cases where an overweight individual becomes acutely ill and is rapidly losing weight, the health care professional relying on the %IBW may inadvertently overlook significant weight loss. The "How to" box above and Table 13-8 below show how to calculate and evaluate %IBW and %UBW.

Table 13-8
WEIGHT AS AN INDICATOR OF NUTRITION STATUS

%IBW	%UBW	Nutrition Status
>120	—	Obese
110–120	—	Overweight
90–109	—	Adequate
80–89	85–95	Mildly underweight
70–79	75–84	Moderately underweight
<70	<75	Severely underweight

In assessing weight loss, consider not only the amount of loss, but also the rate. A person who loses less than 5 percent UBW over a six-month period is at minimal risk for poor nutrition status. A loss of 5 to 10 percent UBW over six months is significant, and a loss of greater than 10 percent UBW is highly significant.[6]

Keep in mind that a person's state of hydration can significantly influence body weight measurements. Disorders or medications that cause fluid retention can mask significant weight loss. Conversely, the person who is dehydrated may have a body weight that is deceptively low.

■■ **WEIGHT GAIN DURING PREGNANCY** One of the anthropometric measures most predictive of an infant's birthweight is the mother's amount and pattern of weight gain during pregnancy. Chapter 10 describes normal weight gains related to pregnancy, and Appendix E provides an example of a chart used to monitor weight gain during pregnancy. Patterns of weight gain that deviate from these require further investigation.

LEARNING LINK

Chapter 7 mentioned that an important component of body weight is the amount and distribution of body fat. Nursing assessments do not routinely include body composition measurements such as triceps fatfold (a measure of body fat), the waist circumference (see p. 141), and bioelectrical impedance (a measure used to estimate lean body mass), although dietitians may include these measurements as part of a complete nutrition assessment. Chapter 7 and Appendix E provide more information about these measurements. Underwater weighing, isotope studies, X rays, ultrasonography, and computerized axial topography (CAT scans) have also been used to determine body composition. The expense and practical application of these tests limit their clinical use, although they are useful research tools.

■■ Physical Signs of Malnutrition

Some physical signs with nutrition implications are obvious. The nurse can simply look at some people and determine that they are overweight, underweight, pale, lethargic, confused, or unable to feed themselves, to give a few examples. Other signs, such as those associated with nutrient deficiencies or toxicities, may require closer examination. Signs of malnutrition appear most rapidly in parts of the body where cell replacement occurs at a high rate, such as the hair, skin, and digestive tract (including the mouth and tongue). Table 13-9 on p. 332 lists physical signs of PEM and of vitamin and mineral malnutrition in general. The summary tables in Chapters 8 and 9 list physical signs of specific vitamin and mineral malnutrition.

■■ **PHYSICAL SIGNS OF FLUID IMBALANCE** Among the most useful physical signs of nutrition status are those that reflect dehydration and fluid retention (see Table 13-10 on p. 332). Many illnesses upset fluid balances, and attention to physical signs of dehydration or fluid retention can help interpret the results of weight and laboratory tests (described later).

■■ **LIMITATIONS OF PHYSICAL SIGNS** Identifying and interpreting physical signs of malnutrition requires knowledge, skill, and clinical judgment.

Physical signs of malnutrition are evident in parts of the body where cells are replaced at a rapid rate.

Table 13-9
PHYSICAL SIGNS OF NUTRIENT DEFICIENCIES

Body System	Acceptable Findings	Malnutrition Findings	What the Findings May Reflect
Hair	Shiny, firm in the scalp	Dull, brittle, dry, loose; falls out	PEM
Eyes	Bright, clear pink membranes; adjust easily to light	Pale membranes; spots; redness; adjust slowly to darkness	Vitamin A, the B vitamins, zinc and iron status
Teeth and gums	No pain or caries, gums firm, teeth bright	Missing, discolored, decayed teeth; gums bleed easily and are swollen and spongy	Mineral and vitamin C status
Face	Clear complexion without dryness or scaliness	Off-color, scaly, flaky, cracked skin	PEM, vitamin A, and iron status
Glands	No lumps	Swollen at front of neck	PEM and iodine status
Tongue	Red, bumpy, rough	Sore, smooth, purplish, swollen	B vitamin status
Skin	Smooth, firm, good color; measurable reaction to skin tests[a]	Dry, rough, spotty; "sandpaper" feel or sores; lack of fat under skin; little or no reaction to skin tests	PEM, essential fatty acid deficiency, vitamin A, the B vitamins, and vitamin C status
Nails	Firm, pink	Spoon-shaped, brittle, ridged, pale	Iron status
Internal systems	Regular heart rhythm, heart rate, and blood pressure; no impairment of digestive function, reflexes, or mental status	Abnormal heart rate, heart rhythm, or blood pressure; enlarged liver, spleen; abnormal digestion; burning, tingling of hands, feet; loss of balance, coordination; mental confusion, irritability, fatigue	PEM and mineral status
Muscles and bones	Muscle tone; posture, long bone development appropriate for age	"Wasted" appearance of muscles; poor grip strength; swollen bumps on skull or ends of bones; small bumps on ribs; bowed legs or knock-knees	PEM, mineral and vitamin D status

[a]Skin tests measure immune function and are influenced by nutrition status as well as many other factors. To perform a skin test, the nurse injects organisms that produce a reaction in most people just under the skin. Raised, hardened areas of skin at the injection site should be visible in 24 to 48 hours.

Table 13-10
PHYSICAL SIGNS OF DEHYDRATION AND FLUID RETENTION

Dehydration
- Sunken eyes
- Hollow cheekbones
- Dry mucous membranes
- Loss of skin turgor (elasticity)[a]
- Weak cry[b]
- Depression of the anterior fontanel[b]
- Deep, gasping respirations
- Weak, rapid pulse
- Thirst
- Reduced urinary output
- Weight loss

Fluid Retention
- Edema
- Ascites (abdominal fluid retention)
- Elevated blood pressure
- Increased urinary output
- Weight gain

[a]May not be a useful parameter in the elderly.
[b]Findings specific to infants.

Many physical signs are nonspecific; they can reflect any of several nutrient deficiencies as well as conditions unrelated to nutrition.[7] For example, cracked lips may be caused by sunburn, windburn, dehydration, or any of several B vitamin deficiencies. Diet intake patterns can provide further support for a suspected deficiency. Weight measurements can confirm that a person is underweight and quantify the degree of underweight. Laboratory tests can help verify a person's state of hydration or vitamin-mineral status.

■■ Biochemical Analysis

Biochemical analyses or laboratory tests can provide information about protein-energy balance, vitamin and mineral status, fluid status, body composition, organ function, and metabolic status. They can also help determine if nutrition therapy is appropriate or if a person is complying with a special diet. Some tests useful in assessing nutrition status and response to diet therapy are described in this section. Other tests, such as serum glucose, cholesterol, and tests that define fluid and electrolyte balance, acid-base balance, and organ function, help pinpoint disease-related problems with nutrition implications. Table 13-11 lists some biochemical tests with nutrition implications. Tests important in specific disorders will be discussed in the appropriate chapters.

To interpret laboratory test results that fall outside the normal range, all the factors that influence the test result must be considered. For laboratory tests that suggest malnutrition, the person's state of health and hydration can influence test results independently of nutrition factors. With dehydration, lab results may be deceptively high; with fluid retention, lab results may be deceptively low.

Table 13-11

EXAMPLES OF ROUTINE HOSPITAL LABORATORY TESTS WITH NUTRITION IMPLICATIONS

Test	Uses
Hematology	
Hemoglobin (Hg)	To detect anemia and determine state of hydration.
Hematocrit (Hct)	To detect anemia and determine state of hydration.
White blood cells (WBC)	To detect infection and determine total lymphocyte count.
Mean corpuscular volume (MCV)	To detect anemia and determine its causes.
Mean corpuscular hemoglobin (MCH)	To detect anemia and determine its causes.
Mean corpuscular hemoglobin concentration (MCHC)	To detect anemia and determine its causes.
Blood Chemistry	
Proteins	
Total protein[a]	To detect PEM and various nutrient imbalances.
Albumin	To detect PEM and determine state of hydration.
Transferrin	To detect PEM and iron status, and monitor response to feeding.
Prealbumin	To detect PEM and monitor response to feeding.
Electrolytes	
Sodium	To check state of hydration.
Potassium	To monitor acid-base balance and renal function and detect imbalances.
Chloride	To monitor acid-base balance and detect GI losses of chloride (from vomiting or nasogastric suctioning).
Carbon dioxide	To monitor acid-base balance.
Other	
Glucose	To detect diabetes mellitus, pancreatic tumors, and hypoglycemia and monitor glucose intolerance.
Glycated hemoglobin	To monitor blood glucose control over the past 2–3 months.
Blood urea nitrogen	To monitor renal function and determine state of hydration.
Calcium	To detect hormonal imbalances, certain malignancies, and calcium imbalances.
Phosphorus	To detect imbalances and PEM and monitor renal function and response to feeding.
Magnesium	To monitor renal function and response to feeding and detect PEM.
Cholesterol	To assess risk of heart disease and possibility of obstructive jaundice.
Uric acid	To detect gout and determine state of hydration.
Serum creatinine	To monitor renal function and determine state of hydration.
Serum enzymes	
Creatinine phosphokinase (CPK)	To monitor heart function and muscle damage.
Lactic dehydrogenase (LDH)	To monitor heart and renal function.
Alanine transaminase (ALT, formerly SGPT)	To monitor heart and liver function.
Aspartate transaminase (AST, formerly SGOT)	To monitor heart and liver function.
Alkaline phosphatase	To monitor liver function.
Serum amylase	To monitor pancreatic function.
Serum lipase	To monitor pancreatic function.

NOTE: This table presents a partial listing of the major uses of certain commonly performed lab tests that have implications for nutrition.
[a]More than half of the total protein is albumin.

The body's way of handling nutrients also affects lab tests. The low blood concentration of a nutrient, for example, may reflect a primary deficiency of that nutrient, but it may also be secondary to a deficiency of one or several other nutrients involved in the transport or metabolism of that nutrient. Nutrient concentrations in the blood and urine sometimes reflect recent intakes rather than long-term intakes. Thus, blood concentrations of a nutrient may be normal, even when tissue levels are deficient.

The serum is the watery portion of the blood that remains after removal of the cells and clot-forming material; plasma is the fluid that remains when unclotted blood is centrifuged. Usually, serum and plasma concentrations are similar, but plasma samples are more likely to clog mechanical blood analyzers, so serum samples are preferred.

It is beyond the scope of this chapter to describe all lab tests used to assess nutrition status, define organ function, and develop and monitor care plans. Instead, the emphasis is on lab tests used to evaluate protein status—serum proteins and the total lymphocyte count. Appendix E provides tables of lab tests useful in detecting various vitamin and mineral deficiencies, including those associated with nutrition-related anemias. Appendix E also provides information about nitrogen balance studies, which are sometimes used to evaluate protein status and needs.

■■ **BIOCHEMICAL TESTS OF PROTEIN STATUS** Along with other assessment techniques, serum proteins such as albumin, transferrin, transthyretin, and retinol-binding protein can help evaluate a person's current protein status and response to nutrition therapy. The larger the body's pool of a serum protein, and the slower the metabolism of that protein, the longer it takes for the protein to be affected by diet. Serum proteins reflect both protein intake (the availability of amino acids) and the distribution of proteins. During some severe stresses, such as extensive burn wounds, serum proteins may shift from the blood to extracellular fluid, for example. Thus, serum proteins will be low, even though the body's pool may not be low. Furthermore, serum proteins are synthesized in the liver, so serum levels can also reflect liver function. As Table 13-12 shows, a variety of medical conditions and treatments can alter levels of serum proteins and the total lymphocyte count independently of nutrition status.

■■ **ALBUMIN** Albumin is the most abundant serum protein, and its lab tests are inexpensive and readily available. Serum albumin concentrations are altered by many medical conditions, however, and serum albumin is slow to reflect changes in nutrition status because it is plentiful in the body and breaks down slowly. In people with chronic PEM, serum albumin levels remain normal despite a depletion of body proteins; the levels fall only after prolonged malnutrition. Likewise, albumin concentrations increase slowly with appropriate nutrition support, so albumin is not a sensitive indicator of response to nutrition therapy.

■■ **TRANSFERRIN** The serum protein transferrin transports iron, and its concentrations reflect both protein-energy and iron status. As with albumin, markedly reduced transferrin levels indicate severe PEM. Transferrin breaks down in the body more rapidly than albumin, but is still relatively slow to

Blood and urine samples provide clues for assessing nutrition status.

Table 13-12
MEDICAL CONDITIONS AND TREATMENTS AFFECTING SERUM PROTEIN LEVELS AND THE TOTAL LYMPHOCYTE COUNT

Lab Test	Facts That Influence Values
Albumin	Chronic PEM, metabolic stress, liver disease, kidney disease (nephrotic syndrome), and eclampsia may lower serum levels.
Tranferrin	Chronic PEM, metabolic stress, liver disease, kidney disease (nephrotic syndrome), and the use of some antibiotics lower serum levels; pregnancy, iron deficiency, and use of oral contraceptives raise serum levels.
Transthyretin	PEM, metabolic stress, hemodialysis, and hypothyroidism lower serum levels; kidney disease and the use of corticosteroids may raise serum levels.
Retinol-binding protein	PEM, metabolic stress, liver disease, hyperthyroidism, vitamin A deficiency, and cystic fibrosis may lower serum levels; kidney disease may raise serum levels.
Total lymphocyte count	PEM, metabolic stress, and the use of chemotherapy, immunosuppressants, and corticosteroids lower levels; infections raise levels.

respond to changes in protein intake. Thus, it may not be a sensitive indicator of response to nutrition therapy.[8] Furthermore, interpreting transferrin levels as a measure protein-energy status is difficult when iron deficiency is present. Transferrin rises as iron deficiency grows worse and falls as iron status improves.

■■ TRANSTHYRETIN AND RETINOL-BINDING PROTEIN Transthyretin and retinol-binding protein levels decrease rapidly during PEM and respond quickly to changes in protein intake. Thus, lab tests of transthyretin and retinol-binding protein are more sensitive tests of protein status, but they are also more expensive than serum albumin, which is routinely available for hospitalized clients. Therefore, tests of transthyretin and retinol-binding protein are often reserved for clients with disorders that markedly change metabolic rates and can rapidly and profoundly affect nutrition status.

■■ OTHER SERUM PROTEINS Other serum proteins that may be useful in assessing the response to nutrition therapy include insulin-like growth factor (IGF-1) and fibronectin. These indicators of protein status are less practical and more expensive to measure than the other serum proteins and are not commonly used.

■■ TOTAL LYMPHOCYTE COUNT PEM compromises the immune system, reducing the number of white blood cells (lymphocytes), which are important in resisting and fighting infections. Two laboratory test results, the number of white blood cells and the percentage of lymphocytes, are used to derive the total lymphocyte count. The total lymphocyte count is inexpensive and easy to obtain, but its value in nutrition assessment is limited because so many variables can affect the body's number of white blood cells and the percentage of lymphocytes. Table 13-13 provides standards for evaluating serum proteins and the total lymphocyte count.

One of the most important nutrition-related tasks for nurses is to identify clients who are likely to develop malnutrition. Fortunately, identifying such clients does not require the nurse to collect more information, but rather to "think nutrition" when reviewing it.

■■ Follow-Up Care

When nurses identify clients with potential or actual nutrition problems, those clients are referred to the dietary department for further evaluation. All clients

Transthyretin is also known as prealbumin; *or* thyroxin-binding prealbumin.

Insulin-like growth factor (IGF-1) is also known as somatomedin-C.

Total lymphocyte count (mm³) = white blood cells (mm³) × percent lymphocytes.

Table 13-13
RELATIONSHIP AMONG DEGREE OF UNDERNUTRITION, SERUM PROTEINS, AND THE TOTAL LYMPHOCYTE COUNT

Indicator	Degree of Depletion			
	Normal	**Mild**	**Moderate**	**Severe**
Albumin (g/100 ml)	≥3.5	2.8–3.4	2.1–2.7	<2.1
Transferrin (mg/100 ml)	>200	150–200	100–149	<100
Transthyretin (mg/100 ml)	16–30	10–15	5–9	<5
Retinol-binding protein[a] (mg/100 ml)	2.6–7.6	—	—	—
Total lymphocyte count (mm³)	2500	<1500	<1200	<800

NOTE: To convert albumin (g/100 ml) to standard international units (g/L), multiply by 100. To convert transferrin (mg/100 ml) to standard international units (g/L), multiply by 0.01.
[a]Levels less than normal suggest compromised protein status. The actual degree of depletion (mild, moderate, and severe) has not been defined.

should be reevaluated throughout the course of care because nutrition status may change. In addition, nurses can help prevent nutrition problems by taking these steps:

- Check the client's tray to see if food is being eaten.
- If the client is to receive no food or is unable to eat, find out how long it has been since the client has eaten. Ask if the client is expected to be able to eat soon.
- If the client is unable to eat, determine whether adequate nutrients are being delivered by tube (Chapter 17) or by vein (Chapter 18).

Communicate problems to the dietitian or physician, record problems in the medical record, and follow up to make sure the problem is being addressed.

Nursing assessments can provide a wealth of information about a client's nutrition status and risk for malnutrition. The case study that follows provides an example of how nursing assessment data can be used to identify risk for malnutrition. The next chapter shows how assessment data are translated into care plans.

Case Study

RISK FACTORS FOR MALNUTRITION IN A COMPUTER SCIENTIST FOLLOWING A CAR ACCIDENT

Ms. Green, a 38-year-old computer scientist, was admitted to the hospital for emergency surgery to repair a broken hip following a car accident. Other injuries included a wound over her left eye and bruises on her arm. She is in stable condition following surgery. The nurse completes a nursing assessment of Ms. Green and notes the following information with nutrition implications:

- Ms. Green has been in good health over the past ten years, although she has experienced a gradual weight gain (height, 5 feet, 7 inches; current weight, 150 pounds). Her current energy and nutrient needs are high as a consequence of her medical condition.
- Prior to hospitalization, Ms. Green was taking one multivitamin-mineral supplement daily and no other medications or herbal preparations. Morphine sulfate is now being administered for pain although Ms. Green seems alert.
- Ms. Green lives alone and states that she is usually too busy to prepare meals, so she often eats out.
- A review of Ms. Green's usual eating habits reveals that she skips breakfast and enjoys foods from all food groups although she seldom eats the recommended servings of fruits and vegetables.
- Ms. Green has been on a diet in the three weeks prior to her accident and had lost 10 pounds.
- Ms. Green appears overweight and pale.
- Available information from the lab report shows a mildly depleted serum albumin (3g/100 ml) and a moderately depleted total lymphocyte count (1000 mm^3).

✚ What factors from the health history and physical examination suggest possible nutrient imbalances?

✚ Estimate Ms. Green's desirable body weight and calculate her percent ideal body weight and percent usual body weight. Consider Ms. Green's recent weight change. What is a safe rate of weight loss (see p. 331)? How does Ms. Green's weight loss compare with the safe rate? Does her recent weight change increase her risk for malnutrition?

✚ What do Ms. Green's lab values suggest with respect to her protein status?

✚ What factors besides protein should be considered when interpreting Ms. Green's lab test results?

SELF CHECK

1. The nurse has contacted the dietitian about a client who has been eating very poorly for about three days and continues to lose weight. Besides identifying nutrition risk and alerting the dietitian, which of the following would be a nutrition-related task for the nurse?

   a. conducting a complete nutrition assessment
   b. encouraging the client to eat when meals are served
   c. providing diet instructions
   d. allowing the dietitian to assume total responsibility for the client's nutrition care

2. A nurse completes a nursing assessment of an elderly woman recently admitted to the hospital. The nurse recognizes that all of the following factors place the client at high risk for malnutrition **except:**

   a. the client has a history of congestive heart failure.
   b. the client takes several prescription medications.
   c. the client lives with her husband in a middle-income neighborhood.
   d. the client's organ function is altered as a consequence of aging.

3. Mr. Salpingo has experienced a loss of appetite, difficulty swallowing, and a reduced secretion of saliva. The nurse recognizes that Mr. Salpingo may be at risk of malnutrition due to:

   a. altered digestion and absorption.
   b. altered metabolism.
   c. reduced food intake
   d. altered excretion of nutrients.

4. Adverse diet-medication interactions are more likely to occur if:

   a. nutrient requirements are very high.
   b. one drug is taken exclusively.
   c. nutrition status is adequate.
   d. medications are used for short periods of time.

5. A client is taking the anticoagulant warfarin. The nurse should periodically evaluate the client's intake of:

   a. folate.
   b. vitamin K.
   c. vitamin D.
   d. calcium.

6. For a client taking the antibiotic tetracycline, which foods and medications should not be taken at the same time as the tetracycline?

   a. milk, calcium-containing antacids, iron supplements
   b. milk, calcium-containing antacids, folate supplements
   c. grapefruit juice, calcium-containing antacids, iron supplements
   d. grapefruit juice, milk, and calcium-containing antacids

7. The nurse uses information from food intake patterns to:

   a. uncover specific nutrient deficiencies.
   b. analyze nutrient needs.
   c. evaluate food preparation methods.
   d. determine general eating habits.

8. Both height and weight measurements:

   a. are affected by fluid status.
   b. cannot be taken on people who are bedridden.
   c. are routinely used measurements.
   d. require equipment that is not readily available in most health care facilities.

9. The %IBW of a person who weighs 185 pounds and has a desirable body weight of 150 pounds is:

   a. 150 percent.
   b. 123 percent.
   c. 50 percent.
   d. 23 percent

10. A client has just begun to eat after days without significant amounts of food. Which of the following laboratory tests would the nurse expect to respond most quickly to changes in energy and protein intakes?

    a. albumin
    b. transferrin
    c. retinol-binding protein
    d. total lymphocyte count

Answers to these questions appear in Appendix H.

CLINICAL APPLICATIONS

1. Describe the possible nutrition implications of these findings from a client's health history: age 73, lives alone, recently lost spouse, uses a walker, no natural teeth or dentures, history of hypertension and diabetes, several medications prescribed.

2. Drawing upon the discussion of laxatives (and especially mineral oil) on p. 317, discuss why elderly people are at particular risk for both developing diet-medication interactions and suffering serious consequences. Consider the reasons and implications for the following statements to guide your thinking: elderly people are more likely to use laxatives, elderly people are more likely to have chronic diseases and take many medications, elderly people are more likely to be malnourished, and elderly people are more likely to develop osteoporosis.

3. Calculate the percent ideal body weight and percent usual body weight for a man who is 5 feet 11 inches tall and has a current weight of 160 pounds and a usual body weight of 180 pounds. State which standards you chose for determining desirable body weight. What additional information will be important for you to find out about this man's weight loss?

Notes

[1] D. H. Sullivan, S. Sun, and R. C. Walls, Protein-energy undernutrition among elderly hospitalized patients, *Journal of the American Medical Association* 281 (1999): 2013–2019.

[2] C. W. Lewis, E. A. Frogillo, and D. A. Roe, Drug-nutrient interactions in long-term care facilities, *Journal of the American Dietetic Association* 95 (1995): 309–315; R. N. Varma, Risk for drug-induced malnutrition is unchecked in elderly patients in nursing homes, *Journal of the American Dietetic Association* 94 (1994): 192–194.

[3] T. Kantola, K. T. Kivisto, and P. J. Neuvonen, Grapefruit juice greatly increases serum concentrations of lovastatin and lovastatin acid, *Clinical Pharmacology and Therapeutics* 63 (1998): 397–402; J. J. Lilja and coauthors, Grapefruit juice–simvastatin interaction: Effect on serum concentrations of simvastatin, simvastatin acid, and HMG-CoA reductase inhibitors, *Clinical Pharmacology and Therapeutics* 64 (1998): 477–483; J. J. Lilja and coauthors, Grapefruit juice substantially increases plasma concentration of buspirone, *Clinical Pharmacology and Therapeutics* 64 (1998): 655–660; S. K. Garg and coauthors, Effect of grapefruit juice on carbamazepine bioavailibility in patients with epilepsy,

Clinical Pharmacology and Therapeutics 64 (1998): 286–288; E. B. Feldman, How grapefruit juice potentiates drug bioavailability, *Nutrition Reviews* 55 (1997): 398–400.

[4] D. J. Mela and J. I. Aaron, Honest but invalid: What subjects say about recording their food intakes, *Journal of the American Dietetic Association* 97 (1997): 791–793.

[5] A. Trudell, Nutrition-clinical assessment of the pediatric patient, *Support Line,* August 1996, pp. 5–8.

[6] J. Hillhouse, Reliability of commonly used anthropometrics in adult hospitalized patients, *Support Line,* August 1996, pp. 9–13.

[7] S. G. Morrisson, Clinical nutrition physical examination, *Support Line,* August 1996, pp. 1–4.

[8] M. Russell, Serum proteins and nitrogen balance: Evaluating response to nutrition support, *Support Line,* February 1995, pp 3–8.

Q&A NUTRITION IN PRACTICE

NUTRITION AND MENTAL HEALTH

Mental health and nutritional health go together. Mentally healthy people have the capacity to feed themselves well. Well-nourished people experience none of the nutrient deficiencies that might contribute to poor mental health. By the same token, when either type of health is impaired, both may be affected. People with mental and emotional problems often have poor diets; and people who are malnourished are often in poor mental health. Nurses who recognize the relationships between nutrition and mental health are well-prepared to offer effective care.

How can emotional and mental health affect what a person eats?

To understand the connections between nutrition and mental health, consider what ordinary anxiety does to your own eating habits. Do you lose your appetite? Overeat? Eat "junk" foods instead of balanced meals? Your reaction depends on your personality type. Temporary anxiety may have little effect on nutrition, but if such stresses become chronic, the resulting changes in eating habits can lead to underweight,

overweight, or nutrient deficiencies and imbalances. Mental health problems that are difficult to overcome, such as depression, have profound effects on nutrition status.

In what ways can depression affect nutrition status?

The feelings of worthlessness and hopelessness that accompany depression can leave a person feeling drained of energy. Quite often people with depression lose interest in caring for themselves and in participating in usual activities such as eating, socializing, or pursuing hobbies. When people fail to care for themselves and isolate themselves from

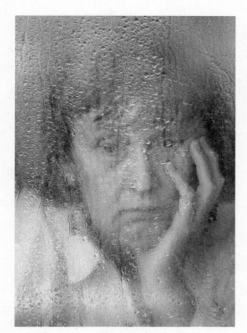

Depression and loneliness can profoundly affect nutrition status.

pleasurable activities and friendships, depression deepens and becomes a self-aggravating condition.

Not surprisingly, many conditions increase the risk of both malnutrition and depression. A serious illness such as cancer, for example, with its accompanying pain, loss of physical independence, and economic hardships can trigger both depression and malnutrition.

Although depression and its consequences affect people of all ages, depression among the elderly is pervasive and is a common cause of weight loss among those in nursing homes.[1] Depression in the elderly often results from loneliness associated with social isolation and the loss of loved ones or a sense of purpose. Many authorities believe that loneliness is particularly relevant to depression and malnutrition among the elderly.

In what ways does loneliness lead to nutrition problems?

For human beings, eating is as much a social and psychological event as a biological one. Without companionship, appetite diminishes. Some 6 million adults over age 65 live alone. Their most pressing need seems to be for companionship; food takes second place. Social interaction is important to mental health, and elderly people of all classes in our society, both the financially secure and the poverty-stricken, tend to become socially isolated. Jack Weinberg, professor of psychiatry at the University of Illinois, wrote perceptively of this problem:

> In our efforts to provide the aged with a proper diet, we often fail to perceive it is not what the older person eats but with whom that will be the deciding factor in proper care for him. The oft-repeated complaint of the older patient that he has little incentive to prepare food for only himself is not merely a statement of fact but also a rebuke to the questioner for failing to perceive his isolation and aloneness and to realize that food . . . for one's self lacks the condiment of another's presence which can transform the simplest fare to the ceremonial act with all its shared meaning.[2]

A sad spiral can set in when a lonely person begins to neglect to eat well. Malnutrition worsens the apathy felt due to loneliness. Then the person has even less energy with which to secure nourishment. Watch for this spiral in all people, and especially in elderly people who live alone and in those who have recently lost a spouse or other loved one and are grieving.

Can something be done to prevent such a spiral?

Health care professionals must be alert to signs of depression in a person of any age and be aware of its potential influence on nutrition status. For the elderly, health care professionals may regard depression as a normal consequence of aging and overlook the potential for addressing it.[3] When depression is recognized early, the person can receive appropriate treatment before health and nutrition status markedly deteriorate. Table NP13-1 lists some nursing

Table NP13-1
EXAMPLES OF NURSING DIAGNOSES THAT MAY BE RELATED TO DEPRESSION AND LONELINESS[a]
• Body image disturbance
• Caregiver role strain
• Chronic low self-esteem
• Chronic pain
• Decisional conflict
• Diversional activity deficient
• Hopelessness
• Impaired social interaction
• Ineffective individual coping
• Powerlessness
• Risk for loneliness
• Self-esteem disturbance
• Sleep pattern disturbance
• Social isolation
• Spiritual distress
• Unilateral neglect

[a]North American Nursing Diagnosis Association–approved nursing diagnoses.

diagnoses related to depression and loneliness that should alert the nurse to look for nutrition problems. Nurses, counselors, and social workers can help elderly clients work through depression and find solutions to their loneliness. Nurses and dietitians can help clients understand how depression affects food intake and health and how eating a well-balanced diet can prevent health and nutrition problems. Meal plans that emphasize foods that are easy to prepare and eat can help some clients sustain their nutrient needs when they lack the motivation to eat.

In what ways can poor diets affect mental health?

Nutrient deficiencies can impair mental health. Protein-energy malnutrition in pregnant women, for example, can cause mental retardation in their children. Children malnourished early in life show behavioral and social deficits as well as physical retardation. Also, children who are neglected early in life show a greater tendency to suffer from severe malnutrition than children who receive love and attention.

MENTAL ILLNESS GLOSSARY

Alzheimer's (ALTZ-high-merz) disease: a degenerative disease of the brain involving progressive memory loss and major structural changes in the brain's nerve cells.

delusions (dee-LOO-shuns: inappropriate beliefs not consistent with the individual's own knowledge and experience.

dementia (dee-MEN-she-ah): irreversible loss of mental function.

mood disorders: mental illness characterized by episodes of severe depression or excessive excitement (mania) or both.

paranoia (PAR-ah-NOY-ah): mental illness characterized by delusions of persecution.

schizophrenia (SKITz-oh-FREN-ee-ah): mental illness characterized by an altered concept of reality and, in some cases, delusions and hallucinations.

Whenever you see a malnourished child, ask yourself if the child requires emotional as well as physical intervention. And whenever you see mental or emotional illness, look to the child's nutrition, too.

Individual nutrient deficiencies can also affect mental health. Table 12-2 on p. 297 lists some brain functions affected by nutrient deficiencies. People with B vitamin deficiencies often exhibit symptoms ranging from confusion, apathy, fatigue, memory deficits, and irritability to delirium and psychoses. Severe niacin deficiencies, for example, can lead to **dementia.**[4] (The glossary above defines dementia and other terms related to mental illness.) Deficiencies of folate and vitamin B_{12} are also associated with depression and dementia.[5] Folate shows a particularly strong correlation with depression. Depression is a common manifestation of folate deficiency, and conversely, people diagnosed with depressive disorders frequently have low serum or red blood cell folate levels.[6] In the latter case, it is not clear whether folate deficiency leads to depression or depression leads to low folate levels by altering food intake. Both folate and vitamin B_{12} deficiencies are associated with dementia and memory impairment.[7]

Is it possible that malnutrition might be responsible for the senility seen in some elderly people?

Sometimes the confusion caused by a vitamin or mineral deficiency is incorrectly diagnosed as senility. An elderly person may even be wrongly confined to a nursing home. The story is told of a woman who exhibited the classic signs of senility—mental confusion, inability to make decisions, and forgetting to perform important tasks, such as turning off a stove burner. The woman's family decided to move her into a nursing home. While she was waiting for a place there, her family took her into their own home. After several weeks of eating good meals and enjoying social stimulation, the woman became her old self again and was able to return to her home. This story has been repeated with many variations and serves to remind us to think about loneliness and nutrition before concluding that a person is senile and needs institutional care. What harm could there be in first trying good, balanced meals served with tender, loving care?

What nutrition considerations apply to people who are truly mentally ill?

Nutrition in psychiatric care is a specialty all its own because there are so many connections. Mental illnesses characterized by depression, illogical thinking, dementia, **paranoia, delusions,** and inappropriate eating habits can alter food intake and thus interfere with nutrition status. Such disorders include **schizophrenia, Alzheimer's disease, mood disorders,** substance abuse, and eating disorders

People with mental disorders characterized by depression risk poor nutrition status for reasons already described. People with mental illnesses characterized by illogical thinking or dementia may have little interest in food or may be unable to make appropriate food choices. Those who are paranoid may believe that foods are being used to poison them. People suffering from delusions may attribute magical powers to certain foods and insist on eating only those foods. Medications used in the treatment of mental illnesses can also interact with and alter nutrition status.

What are the most common mental disorders with nutrition implications?

Among the most common mental disorders with considerable effects on nutrition status are alcoholism and eating disorders. The relationship between alcoholism and nutrition is treated in Nutrition in Practice 23, and eating disorders are discussed in Nutrition in Practice 7.

Nutrition affects the brain and the mind, and the brain and the mind affect the way people eat. All are interrelated, and the wise health care professional will keep these interrelationships in focus.

Notes

[1] G. J. Kennedy, The geriatric syndrome of late-life depression, *Psychiatric Services* 46 (1995): 43–48.

[2] J. Weinberg, Psychological implications of the nutritional needs of the elderly, *Journal of the American Dietetic Association* 60 (1972): 293–296.

[3] C. Ryan and M. E. Shea, Recognizing depression in older adults: The role of the dietitian, *Journal of the American Dietetic Association* 96 (1996): 1042–1044.

[4] J. E. Morley, Nutritional modulation of behavior and immunocompetence, *Nutrition Reviews* (supplement 2) 52 (1994): 6–8.

[5] T. Bottiglieri, Folate, vitamin B_{12}, and neuropsychiatric disorders, *Nutrition Reviews* 54 (1996): 382–390.

[6] J. E. Alpert and M. Fava, Nutrition and depression: The role of folate, *Nutrition Reviews* 55 (1997): 145–149.

[7] Bottiglieri, 1996.

CHAPTER 14

NUTRITION INTERVENTION

CONTENTS

Care Plans

Special Diets

Medical Records

Case Study: Computer
Scientist with Car
Accident Injuries

Nutrition in Practice:
Community Nutrition

This chapter describes how the nurse can use the remaining steps in the nursing process—diagnosis, outcome identification and planning, implementation, and evaluation—to address nutrition-related problems identified through the nursing assessment. Assuring that a person's nutrient needs are met is a key part of this process, so the chapter also describes medical nutrition therapy and looks at how clients' nutrition needs are communicated among health care team members.

Care Plans

The goals of medical nutrition therapy are:
- *To meet the client's nutrient needs.*
- *To meet the client's needs for nutrition education.*

Once the nurse has collected assessment information, the next steps are to identify a nutrition-related nursing diagnosis, when appropriate, and develop a plan of action. The plan sets goals and expected outcomes that aim to resolve the client's immediate and long-term nutrition problems. It also describes the interventions necessary to achieve the goals.

Nutrition-Related Nursing Diagnoses

Although numerous nursing diagnoses have nutrition-related implications, the North American Nursing Diagnosis Association (NANDA) has developed the following diagnoses that specifically relate to nutrition:

- *Altered nutrition: More than body requirements.* This nursing diagnosis describes the nutrition problems of people who are overweight or obese.
- *Altered nutrition: Less than body requirements.* This nursing diagnosis describes the nutrition problems of people with protein-energy malnutrition (PEM) and those who are likely to develop PEM. These people may be unable to eat enough food to meet their body's metabolic demands or may lose weight despite a seemingly adequate nutrient intake.
- *Altered nutrition: Risk for more than body requirements.* This nursing diagnosis describes the nutrition problems of people who are likely to become overweight or obese. The people may have a family history of obesity or habits that may lead to weight gain such as eating large meals at the end of the day or eating in response to anxiety.

Although this text describes many disorders requiring attention to nutrition, many more disorders have nutrition implications. When the nurse encounters such a disorder, nursing diagnoses alert the nurse to potential nutrition problems. Several neurological conditions, for example, lead to problems with swallowing (see Chapter 15). Regardless of the medical diagnosis, a nursing diagnosis of impaired swallowing alerts the nurse to the need for nutrition intervention. Table 14-1 lists examples of nursing diagnoses that cue the nurse to think nutrition.

LEARNING LINK

Chapter 7 described many behaviors associated with overweight and obesity. Identifying which factors apply to a particular client is important in determining which interventions might be helpful in correcting the problem. The chapters that follow will provide many examples of ways in which reduced food intake, inadequate digestion and absorption, or altered metabolism can result in weight loss. As with overweight and obesity, it is important to understand the specific causes of weight loss in order to determine effective strategies for dealing with it.

Table 14-1

EXAMPLES OF NURSING DIAGNOSES THAT MAY BE RELATED TO RISK FOR MALNUTRITION[a]

- Altered nutrition: More than body requirements
- Altered nutrition: Less than body requirements
- Altered nutrition: Risk for more than body requirements
- Altered growth and development
- Altered oral mucous membrane
- Chronic confusion
- Chronic pain
- Diarrhea
- Feeding self-care deficit
- Impaired memory
- Impaired physical mobility
- Impaired skin integrity
- Impaired swallowing
- Ineffective breastfeeding
- Ineffective infant feeding pattern

NOTE: Table NP13-1 in Nutrition in Practice 13 lists examples of nursing diagnoses that may be related to depression and loneliness—all of these should alert the nurse to look for malnutrition.

[a]North American Nursing Diagnosis Association–approved nursing diagnoses.

▪▪ Outcome Identification and Planning

The client and the client's caregivers are central to the planning process. To establish realistic and achievable goals, the plan must take into account the client's food habits, lifestyle, and personal factors identified in the nursing assessment. The care plan should also agree with the plans of other members of the health care team to ensure the most consistent and cost-effective care. If a dietitian is working with the client, for example, then the nursing care plan and nutrition care plan should be coordinated. In some cases, plans for addressing nutrition problems are incorporated into a total medical care plan developed by the health care team. Such plans, called **clinical pathways, critical pathways,** or **care maps,** are charts or tables that outline a coordinated plan of care for a specific medical diagnosis, treatment, or procedure.

▪▪ **GOALS** When establishing goals, the nurse states them in terms of measurable outcomes, such as target ranges for body weight or blood glucose levels, because the success of therapy can more readily be determined if goals are measurable. Consider, for example, a problem of diarrhea. If a medication is causing the diarrhea, an appropriate strategy might be to prevent dehydration by encouraging the client to consume ample fluids and electrolytes while the physician tries substituting other medications. The client should be informed of this plan. Measurable goals might include a target range for serum electrolytes, urinary output, and blood pressure, as well as reduced frequency of bowel elimination and the production of stools of normal volume and consistency. (Chapter 16 describes dietary strategies for treating diarrhea.) If the diarrhea results from lactose intolerance, the goals may be the same, but the problem-solving strategy is to temporarily eliminate milk and milk products.

▪▪ **NUTRITION EDUCATION** In addition to attending to a client's nutrient needs, care plans also consider the client's needs for nutrition education. For the client who is experiencing diarrhea as a result of lactose intolerance, nutrition education is a key part of the plan. The plan will include

clinical pathways, critical pathways, or **care maps:** charts or tables that outline a plan of care for a specific diagnosis, treatment, or procedure, with a goal of providing the best outcome at the lowest cost. The plan, developed by the health care team after a careful study of each facility's unique client population, is regularly reassessed and improved.

counseling sessions to teach the client how to maintain a nutritionally adequate diet that limits the intake of milk and milk products. The nurse or dietitian might measure goals for nutrition education by ensuring that the client can verbally identify foods containing milk or milk products, plan a day's menus appropriate for the restrictions, and find sources of milk on food labels.

Data gathered during an assessment help the nurse or dietitian to determine the best way to present nutrition information, the amount of information the client will be able to handle, and the level of the client's interest in nutrition and health. Keep in mind that the nutrition education plan must be flexible to accommodate the client's goals, understanding of the information presented, and motivation to practice the suggestions offered.

■■ INSTRUCTIONS REGARDING DIET-MEDICATION INTERACTIONS For people on medications, remember to address potential diet-medication interactions as a part of nutrition education. Nurses, dietitians, or pharmacists should provide clients with advice on how to take medications in relation to food or specific nutrients, signs of potential nutrient deficiencies or nutrition-related problems, and actions to take if problems arise.

■■ IMPLEMENTATION OF THE CARE PLAN Once a care plan is developed, the next step is to implement it by providing both the appropriate diet and education. In an in-patient health care facility (such as a hospital, nursing home, or in-patient mental health care facility), the diet part appears deceptively simple: appropriate foods are delivered to clients. Behind the scenes, however, the dietitian carefully plans the diet for each client, the dietitian or dietetic technician checks the clients' menus, and the foodservice department assures that the appropriate foods are prepared and delivered. Nutrition in Practice 16 describes how foodservice systems operate. Equally important, once delivered, the meals must be eaten by the clients.

Nurses, thanks to their frequent daily contact with clients, can offer important support in the educational aspect of client care. Clients often think of questions long after the dietitian has left, and they most often ask the nurse. The nurse who is confident of the answers should provide them. If not sure of the answers, the nurse should express honest uncertainty and reassure the client that the information will be provided. The nurse should then inform the dietitian that the client needs a follow-up visit.

■■ ONGOING EVALUATION While the planned strategies are being implemented, the nurse keeps track of how they are working. If, for example, a client on a weight-reduction diet fails to lose weight, a change may be needed. Is the client eating too much? Is the client too inactive? Perhaps the client should keep a food and activity record to help identify problems with the weight-loss plan.

If a client's situation changes, so may nutrition status and nutrient needs. For example, when a pregnant woman delivers her baby, she will need instructions on how to feed her infant. She will also need information on how to revise her diet to support lactation (if she is breastfeeding) or to return to a healthy weight. The accompanying case study affords an opportunity to apply some of the planning principles introduced here.

A care plan may be ideal, but may still fall short of meeting goals if a client is unable or unwilling to comply with it. If the client is unable to comply, reassessing communication techniques may help. Perhaps the level of instruction needs to be simplified or cultural differences addressed. If the client is unwilling, despite the best efforts of health care professionals, little can be done except to try later when the client may be more receptive.

Case Study

COMPUTER SCIENTIST WITH CAR ACCIDENT INJURIES

To practice using assessment data to develop a care plan, answer the questions that follow drawing upon the information presented here as well as in the case study about Ms. Green on p. 336. The nurse reviews the information from a follow-up assessment of Ms. Green and determines that Ms. Green's current medical status indicates a need for sufficient kcalories and protein to minimize weight loss and the depletion of serum proteins. She is on a regular diet. The nurse checks Ms. Green's trays frequently and notes that she is eating well. The dietitian has been contacted, and once Ms. Green's medical condition has stabilized and she has recovered from her injuries, the dietitian will confer with Ms. Green's physician and recommend a safe weight-loss diet.

✚ Contrast Ms. Green's current nutrient needs with her long-term needs. How should the care plan reflect these changing needs? Describe methods the nurse can use to determine if the nutrition care plan is effective while Ms. Green is in the hospital.

✚ Assuming that Ms. Green is now fully recovered from her injuries and using Table 14-1 as a guide, which nursing diagnosis applies to Ms. Green? What factors in Ms. Green's history need to be considered in developing a realistic weight-loss plan? Other than adjusting kcalories, what other changes in eating habits might the nurse suggest to Ms. Green?

✚ Once Ms. Green is discharged from the hospital, it is unlikely that she will see the dietitian who provided the diet instructions again. Consider ways that follow-up care might be arranged.

■■ Special Diets

An essential part of any care plan that includes nutrition is to provide the appropriate amounts of energy, protein, carbohydrate, fat, vitamins, major minerals, trace elements, and water in whatever form best meets the client's needs. For example, a person who cannot chew needs soft foods; a person who is in a coma may need to have a formula delivered by tube into the GI tract; and a person with heart disease needs a low-fat diet.

■■ Standard and Modified Diets

Standard or **regular** diets include all foods and provide all the nutrients in amounts appropriate for healthy people. **Modified diets** are used when standard diets fail to meet the specific needs of clients. Modifying the standard diet is much like tailoring a suit. A tailored suit is the same suit after alterations—only it fits better. In the case of a modified diet, the tailoring may involve changing the consistency; adjusting the amounts of energy, individual nutrients, or fluid; altering the number of meals; or including or eliminating certain foods. Nutrition in Practice 16 shows how a hospital menu can be modified for different diets. The chapters that follow provide many examples of modified diets, and the next section introduces liquid and soft diets, which are routinely used in facilities that serve meals.

It is helpful to think about modified diets in terms of the nutrition-related problems they relieve rather than in terms of specific disorders. Two people with the same disorder may need two different diets. Conversely, two people with different disorders may benefit from the same diet. Consider two people with cancer: one may need a diet that will help control nausea; the other may need a high-kcalorie diet. Now consider a pregnant woman with nausea. She may benefit from the same recommendations as those for the first person with cancer.

■■ DIET ORDERS As Chapter 13 described, the physician prescribes the client's diet and writes the diet order in the medical record. The physician

standard (or **regular diets**): diets that include all foods and meet the nutrient needs of a healthy person.

modified (or **therapeutic diets**): regular diets that are adjusted to meet special nutrient needs. Such diets can be adjusted in consistency, level of energy and nutrients, amount of fluid, or number of meals, or by the inclusion or elimination of certain foods.

often relies on the dietitian or nurse to suggest a diet prescription or make recommendations when changes or clarity in the diet order appear warranted.

To avoid confusion, physicians should provide clear and precise diet orders. For example, a "low-sodium diet" order should specify the amount of sodium; otherwise, "low-sodium" could be interpreted to mean any amount from 500 to 4000 milligrams. For uncomplicated diets, such as low-sodium diets, the foodservice department often sends a preselected diet to the client until orders are clarified, and the level of restriction is recorded in the medical record. For more complicated diets, such as renal diets, meals will not be sent until the order is clarified.

Occasionally, diet orders may be inappropriate. For example, the physician may describe an obese individual as "well-nourished" and order a regular diet. Unless the nurse identifies the problem, the client will receive an inappropriate diet and no nutrition advice. As another example, a physician may order that a client should receive no food or fluids after midnight for a lab test to be conducted in the morning. The doctor assumes that the order will be discontinued after the test, but it may not be. The client may miss several meals before someone notices the error. These examples illustrate situations where communication between health care professionals can make a difference in client care. Whenever you notice inappropriate diet orders, contact the dietitian or alert the physician.

▪▪ **DIET MANUALS** The exact foods excluded from or included on a specific modified diet, and even the name given to the diet, may differ among health care facilities, generally in minor ways. These variations reflect different schools of thought regarding diet; institutional **diet manuals** are consulted as a standard practice.

In large facilities, the staff of dietitians may compile a diet manual, subject to approval by the hospital administrator, several physicians, and representatives of the nursing service. Small facilities may adopt the diet manual of another hospital or an organization such as the state dietetic association. The diet manual describes the foods allowed and not allowed on each diet, outlines the rationale and indications for use of each diet, provides information on the nutritional adequacy of the diets, and offers sample menus. The dietary department uses the manual to design menus for each diet. Diet manuals are usually available on each nursing floor, and the nurse should consult the manual whenever questions arise regarding diets.

Routine Progressive Diets

In health care facilities that serve meals, a common diet order reads, "progress diet from clear liquids to a regular diet as tolerated." Such a **diet progression** is typical for clients after uncomplicated surgeries, for clients suffering from nausea or diarrhea, and for clients who have not been eating foods orally and are beginning to eat again. Progressive diets frequently include clear-liquid diets, full liquid diets, and soft, low-fiber, or low-residue diets.

▪▪ **CLEAR-LIQUID DIETS** **Clear liquids** are often the first foods offered to clients on progressive diets. A clear-liquid diet consists of foods that are liquid and transparent at body temperature, such as gelatin, tea, and broth. Clear liquids provide minimal stimulation to the GI tract, and they are easily and almost completely absorbed; they help determine if the digestive system is working well enough to handle more complex foods. Table 14-2 lists the foods allowed on clear-liquid diets, and the sample menu shows an example of a day's meals. Once the person tolerates clear liquids, the diet may progress to full liquids.

An order to give a client nothing orally (including food, beverages, and medications) reads NPO, *the abbreviation for* non per os, *which means "nothing by mouth."* PO *stands for* per os, *which means "by mouth" or "orally."*

diet manual: a book that describes the foods allowed and restricted on a diet, outlines the rationale and indications for use of each diet, and provides sample menus.

progressive diet: a diet that changes as a client's tolerances permit.

clear liquids: foods that are liquid and transparent at body temperature.

Nurses consult manuals to clarify what foods are included on or excluded from different diets.

Table 14-2
FOODS INCLUDED ON LIQUID DIETS

Clear-Liquid Diets	Full-Liquid Diets
Bouillon	All clear liquids
Broth, clear	Butter
Carbonated beverages	Commercially prepared liquid formulas (all)
Coffee, regular and decaffeinated	Cooked cereals, strained
Commercially prepared clear-liquid formulas	Cream
Fruit drinks	Custard
Fruit ices	Ice cream, plain
Fruit juices, strained	Instant breakfast drinks
Gelatin	Margarine
Hard candy	Milk, all types
Honey	Pudding
Lemonade	Sherbet
Popsicles	Soups, strained vegetable, meat, or cream
Salt	Sour cream
Salt substitutes	Vegetable juices, strained
Sugar	Vegetable purees, diluted in cream soups
Sugar substitutes	Yogurt
Tea, regular and decaffeinated	

■■ **FULL-LIQUID DIETS** A full-liquid diet includes both clear and opaque liquid foods and semiliquid foods (see Table 14-2 and the sample menu on p. 348). Full-liquid diets may be used as a person progresses from clear liquids to regular foods or for people who are unable to chew or swallow regular foods for medical reasons or who are too ill to eat. Like clear-liquid diets, unsupplemented, standard full-liquid diets are deficient in many nutrients and should be used only temporarily. Supplementing the diet with a formula (described next) can help provide nutrients and prevent boredom. Another option is to skip the full-liquid diet altogether and advance the diet to very small servings of soft foods.

Note that many foods allowed on full-liquid diets contain lactose. The same conditions for which liquid diets are prescribed often result in temporary lactose intolerance. When potential problems with lactose intolerance are suspected, lactose-containing liquids cannot be provided.

■■ **CAUTIOUS USE OF LIQUID DIETS** Liquids are offered in small amounts at first to make sure the person can tolerate them. Many clients willingly accept liquid diets because they are too ill to eat. When left on the diet for any length

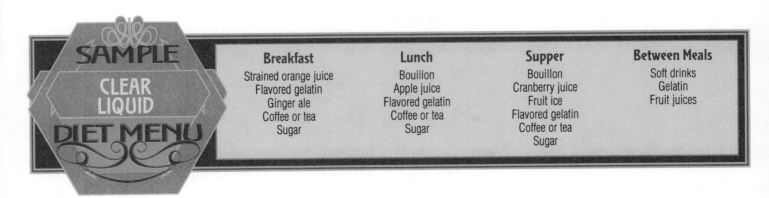

SAMPLE CLEAR LIQUID DIET MENU

Breakfast	Lunch	Supper	Between Meals
Strained orange juice	Bouillon	Bouillon	Soft drinks
Flavored gelatin	Apple juice	Cranberry juice	Gelatin
Ginger ale	Flavored gelatin	Fruit ice	Fruit juices
Coffee or tea	Coffee or tea	Flavored gelatin	
Sugar	Sugar	Coffee or tea	
		Sugar	

Breakfast	Lunch	Supper	Between Meals
Orange juice	Apricot nectar	Apple juice	Milk shakes (made
Strained oatmeal	Yogurt, plain	Creamed soup	with plain ice cream),
Milk	Pudding	Custard	ice cream, eggnog,
Sugar	Milk	Milk	pudding, custard,
Margarine	Coffee or tea	Coffee or tea	or gelatin
	Sugar	Sugar	

of time, however, people may understandably find these foods unappetizing and boring. Even more disturbing, liquid diets are deficient in energy and most nutrients, especially in relation to the high energy and nutrient needs often imposed by illness. As mentioned, liquid diets are temporary diets; to meet the energy and nutrient needs of clients who must stay on liquid diets for more than a few days, formulas are recommended. (Chapter 17 describes formula diets in detail.) Formulas are of a known nutrient composition, they can be given in quantities that meet all nutrient needs, they can be selected to meet a variety of medical needs and to reduce the likelihood of GI problems, and almost all are lactose-free.

▪▪ INTRODUCTION OF SOLID FOODS Once a person is tolerating liquids, the next step is solid foods. The exact diet used as the first solid diet depends on the client's medical condition. Some clients begin receiving regular foods, but sometimes clients better tolerate the shift from liquid diets to solid diets if they first receive soft, low-fiber, or low-residue diets.

▪▪ SOFT DIETS Soft diets provide soft but solid foods (such as tender meats and soft fresh fruits) that are lightly seasoned and moderate in fiber. Generally, soft foods are offered in frequent small meals. As Table 14-3 shows, soft diets include foods that are easy to chew, digest, and absorb. Soft diets limit high-fiber foods, nuts, foods with seeds, and coconut. By limiting highly seasoned foods that might irritate the GI tract and high-fiber foods that are more likely to produce gas, soft diets minimize the risk of indigestion, nausea, distention, cramping, and other GI upsets. In addition to their use in a progressive diet, soft diets also benefit people with temporary indigestion or nausea. Clients tolerating soft diets are then progressed to regular diets that provide the full spectrum of nutrients and include all foods.

▪▪ LOW-FIBER AND LOW-RESIDUE DIETS Low-fiber and low-residue diets serve to reduce the total fecal volume (**residue**). Low-fiber/low-residue foods

residue: the total amount of material in the colon; includes dietary fiber and undigested food, intestinal secretions, bacterial cell bodies, and cells shed from the intestinal mucosa.

Table 14-3
FOODS INCLUDED ON SOFT DIETS

Meat and Meat Alternates	Tender, moist meats, fish, or poultry; mild cheeses, creamy peanut butter, and eggs
Milk and Milk Products	Milk, milk products, yogurt without seeds or nuts
Fruits and Vegetables	Cooked, canned, or soft fresh fruits such as melons; fruit juice; soft-cooked vegetables except those likely to produce gas (see Table 16-1 in Chapter 16)
Grains	Refined white or light rye bread, rolls, or crackers; cooked or ready-to-eat cereals without nuts or seeds
Other	Mild condiments; salt; sugar; mildly seasoned broths or soups; all nonalcoholic beverages; desserts without nuts, seeds, or coconut

Table 14-4
LOW-FIBER FOODS

Meat	Tender meat, poultry, seafood, and eggs
Breads and Cereals	Refined breads, cereals, rice, and pasta
Fruits	Cooked or canned peeled apples, apricots, peaches, pineapple, plums; bananas, cherries, cranberry sauce, mandarin oranges, pomegranates, tangerines; fruit juice
Vegetables	Cooked bean sprouts, cabbage, onions, summer squash; celery, endive, lettuce, tomato paste, tomato puree, vegetable juice, water chestnuts, watercress
Other	Avocado, nuts

are least likely to obstruct an intestinal tract that is narrowed by inflammation or scarring or in which GI motility is slow. Thus, a low-fiber or low-residue diet may be prescribed following surgery of the intestinal tract. Low-fiber diets restrict high-fiber foods and tough meats. Low-residue diets limit these foods and also limit milk and milk products. Table 14-4 lists foods low in fiber. The sample menu below shows a day's meals for a low-fiber/low-residue diet.

■■ **ADVANCING THE DIET**   The nurse is often responsible for monitoring the client's tolerance and readiness to advance through the stages of a progressive diet. Indigestion, nausea, vomiting, diarrhea, cramping, or other GI upsets indicate intolerance. At each progressive step, a client may be intolerant to particular foods (orange juice or milk, for example), rather than to the diet itself. In such a case, the offending food is withheld temporarily.

■■ **HELPING THE CLIENT EAT**   As Chapter 13 described, many medical conditions and treatments can depress the appetite. Nurses and their assistants play a central role in helping clients to eat. The accompanying "How to" box on p. 351 offers suggestions for improving clients' food intakes.

Providing appropriate nutrients in an appropriate form is an important component of medical nutrition therapy. The nurse is the person most likely to observe how well a client understands the diet modification and tolerates the foods provided. The medical record, described next, provides a vehicle for the nurse to communicate these observations to the rest of the health care team.

Soft, low-fiber, and low-residue diets are similar, and in facilities that serve meals, menus for these three diets are often the same. All three diets restrict high-fiber foods. Soft diets limit highly seasoned foods, low-residue diets limit milk and milk products, and both low-fiber and low-residue diets limit serving sizes of allowed fruits and vegetables.

SAMPLE LOW-FIBER/LOW-RESIDUE DIET MENU

Breakfast	Lunch	Supper	Snack
Orange juice	Baked fish	Roast beef	Applesauce
Soft-cooked egg	White rice	Mashed potatoes	Vanilla wafers
Puffed rice cereal	Green beans	Cooked carrots	
White bread toast	Small banana	Canned peaches	
Coffee or tea	Roll	Roll	
Milk for cereal	Margarine	Margarine	
Creamer	Coffee or tea	Coffee or tea	
	Sugar	Sugar	
	Creamer	Creamer	

The medical record serves as a tool for communicating information about a client's health and responses to treatments.

▚ Medical Records

Maintaining strong professional communication networks benefits both health care professionals and their clients. Conversely, miscommunication between professionals can result in inappropriate or ineffective therapy with serious consequences for clients' health. Professionals have many opportunities to discuss client concerns and can record these concerns in the medical record.

Medical records are legal documents that record a client's health history; the assessment, diagnosis, and prognosis of medical problems; the measures being taken to treat those problems; and the results of tests and therapy. Writing in the medical records allows health care professionals to document the actions taken to comply with physicians' orders, the client's responses to those actions, and recommendations. This information helps determine if medical orders are being followed and directs future care.

▚ Types of Medical Records

Medical records can be organized in many ways. In today's cost-conscious health care environment, time constraints compel health care professionals to minimize the time spent in documenting medical care.[1] Traditional problem-oriented medical records (POMR), which focus on a client's medical problems and the strategies being used to address those problems, are gradually being replaced by outcome- or goal-oriented medical records, which focus on observable medical goals. A problem-oriented medical record for a person with diabetes mellitus, for example, would list diabetes mellitus as a problem and then define the strategies used to control the diabetes, such as a 1600-kcalorie diet, oral hypoglycemic agents, and the like. A goal-oriented medical record for the same client would list the client's goal weight and target blood glucose measurements, and the health care team would document how these values have responded to therapy.

▚ The Medical Record and Nutrition Care

Learn how to effectively use the medical record in the facility where you work. Regardless of the approach used, be sure the client's medical record includes important nutrition-related information. Examples of important information include:

- Documentation of nutrition screening.
- Assessment data relevant to nutrition status.
- Recommended medical nutrition therapy, including goals.
- Acceptance and tolerance of current diet.
- Problems with food intake.
- Documentation of diet counseling, including the client's response.
- Any planned follow-up or referral to another person or agency.

Effective nutrition care addresses the unique needs of the individual, framing nutrition needs in the context of the person's personal and medical needs. This chapter has described techniques for building effective nutrition care plans. The remaining chapters describe how nutrition care can support recovery when health problems arise.

HOW TO:

HELP CLIENTS IMPROVE THEIR FOOD INTAKES

1. Empathize. If the person is frightened, angry, or confused, show that you care and are there to help. Imagine feeling too sick to move or too tired to sit up. Show that you understand how difficult eating may be.

2. Motivate. Be sure the client understands how important nutrition is to recovery.

3. Help clients select foods they like and mark menus appropriately. Call the dietitian if the client needs extra help. When appropriate and permissible, let a friend or family member bring in favorite foods from outside the hospital. This may be especially helpful for clients with strong ethnic, religious, or personal food preferences.

4. Solve eating problems. Encourage clients who feel full after a short time to eat the most nutritious foods first and save liquids until after meals. For clients who are weak or tired, suggest foods that require little effort to eat. Eating a roast beef sandwich, for example, requires less effort than cutting and eating a steak; drinking soup is easier than eating it with a spoon. For clients who either fill up quickly when eating or are weak or tired, smaller meals combined with snacks, such as a sandwich at bedtime and milk shakes or instant breakfast drinks between meals, can improve intake considerably.

5. Help clients prepare for meals. Encourage clients to wash their hands and faces and to brush their teeth or rinse their mouths before eating. Help them get comfortable, either in bed or in a chair. Adjust the extension table to a comfortable distance and height, and make sure it is clean. A clean, odor-free room also helps. Take these steps before the tray arrives, so the meal can be served promptly and at the right temperature.

6. Check for accuracy and appearance. When the food cart arrives, check the client's tray. Confirm that the client is receiving the right diet, that the foods on the tray are the ones the client marked on the menu, and that the foods look appealing. Order a new tray if foods are not appropriate.

7. Help with eating. Help clients who need assistance in opening containers or cutting foods and those who are unable to feed themselves.

8. Take a positive attitude toward the hospital's food. Never say something like "I couldn't eat this stuff either." Instead, say, "The foodservice department really tries to make foods appetizing. I'm sure we can find a solution."

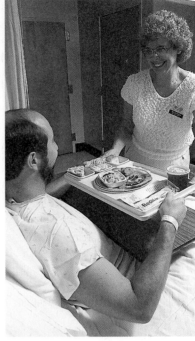

People enjoy eating when they feel comfortable and cared for.

SELF CHECK

1. The nurse analyzing information from a comprehensive assessment recognizes that a client's poor appetite and weight of less than 80% IBW place him at risk for malnutrition. An appropriate nutrition-related nursing diagnosis for this client would be:

   a. altered nutrition: more than body requirements.

   b. altered nutrition: less than body requirements.

   c. altered nutrition: risk for more than body requirements.

   d. feeding self-care deficit.

2. Which of the following describes a measurable, realistic, and attainable goal for the client in self-check 1:

   a. The client will eat a high-kcalorie, high-protein snack before bedtime.

   b. The client will gain at least 2 pounds a month.

   c. The client will eat at least 1500 kcalories more than his current intake each day.

   d. The client will learn the kcalorie contents of at least ten foods a day.

3. The most important factor(s) affecting how the nurse **presents** nutrition information to a client is the:

   a. client's nutrient needs and nutrition status.

   b. client's ability level and motivation.

   c. client's health history.

   d. medical record.

4. A nurse instructs a client on a 1200-kcalorie weight-reduction diet and exercise program and expects her to lose about 1 pound a week. The next step for the nurse is to:

a. assume the plan is working.
b. formulate a new nursing diagnosis.
c. find a new strategy for meeting weight-loss goals.
d. plan a follow-up meeting with the client to see how well the plan is working.

5. The following statements describe diet orders **except:**

a. the physician prescribes diet orders.
b. diet orders should be written clearly and precisely.
c. diet orders should never be questioned.
d. the physician may write an order that reads, "progress diet from clear liquids to a regular diet as tolerated."

6. A nurse notices a food on a client's tray and is not sure if that food is allowed on the client's diet. An appropriate action for the nurse to take would be to:

a. check the care plan.
b. check the diet order.
c. check the diet manual.
d. check the medical record.

7. The nurse checking the tray of a client on a clear-liquid diet would expect to see all of the following foods **except:**

a. cola-flavored soft drink.
b. beef bouillon.
c. strained orange juice.
d. cream of potato soup.

8. Which of the following statements describes both clear- and full-liquid diets?

a. They can be used for long periods of time.
b. They are deficient in nutrients and are temporary diets.
c. They are contraindicated for people who are unable to chew or swallow regular foods.
d. They are contraindicated for people experiencing gastrointestinal disturbances.

9. A client's diet order following surgery reads, "progress diet from clear liquids to a regular diet as tolerated." What signs and symptoms should the nurse look for before advancing the diet?

a. nausea, vomiting, diarrhea, and cramping
b. headache or fever
c. dehydration or overhydration
d. pain and confusion

10. A soft diet for a client with indigestion would include the following advice **except:**

a. limit foods that produce gas.
b. limit highly seasoned foods.
c. use foods that are chopped or pureed.
d. limit high-fiber foods.

Answers to these questions appear in Appendix H.

CLINICAL APPLICATIONS

1. A client who recently suffered a heart attack is on a special diet. The client tells you that the dietitian has talked to him about his diet and that he is totally confused. He confides that diet is the last thing on his mind right now. What actions should you take?

2. An initial nutrition screening of an elderly client admitted to the hospital for surgery revealed that the client was not at risk for poor nutrition status. The client was not on a modified diet and was not referred for nutrition assessment, and so has not seen a dietitian. Following surgery, ongoing assessment revealed that the client developed several complications, and recovery has been slower than expected. The nurse working with the client notices that the client has eaten only minimal amounts of food for several days. Describe several steps that can be taken to uncover and address problems the client might be having with food.

3. Clients respond differently to treatments, and usually there is more than one treatment option. Consider the following scenario. The physician orders a routine progressive diet for a client who has not eaten an oral diet for two weeks. The client had problems tolerating a clear-liquid diet at first, but by the end of two days was ready to advance to a full-liquid diet. (Note: This is a longer-than-usual amount of time for a client to adjust to a clear-liquid diet.) The physician suspects that the client may experience temporary lactose intolerance and have even more problems tolerating a full-liquid diet. Describe the potential benefits and drawbacks of each of these options:

• Advance the client to an unsupplemented, lactose-free full-liquid diet for a day, and monitor the client's response.

• Provide a lactose-free full-liquid diet supplemented with a formula, and monitor the client's response.

• Skip the full-liquid step in the diet progression, and offer lactose-free soft foods in limited amounts.

NUTRITION**ON**THENET

FOR FURTHER STUDY OF THE
TOPICS IN THIS CHAPTER,
ACCESS THESE WEB SITES.

www.nanda.org
*North American Nursing Diagnosis
Association*

www.dietitian.com/
medtherapy.html
Ask the Dietitian

Note

[1]C. J. Klein, J. B. Bosworth, and C. E. Wiles,
Physicians prefer goal-oriented note format more
than three to one over other outcome-focused
documentation, *Journal of the American Dietetic
Association* 97 (1997): 1306–1310.

Q&A NUTRITION IN PRACTICE

COMMUNITY NUTRITION

As Chapters 13 and 14 have shown, the nursing process can be used as a tool for meeting the highly individualized nutrition needs of each client. For the most part, however, clients do not receive such individualized attention until they develop a health problem. Community nutrition programs, on the other hand, must attempt to meet the general nutrition needs of large numbers of people. Their goals are to provide nutrition education and to prevent the energy and nutrient imbalances that might impair health.

How are the community's nutrition needs identified?

The approach to solving community nutrition problems is similar to the nursing process. As Chapter 1 described, the process begins with nutrition surveys—a community approach to assess the eating habits, nutritional health, and nutrition knowledge of the community in general. From these surveys, scientists and health care professionals determine which nutrition problems require further action, either in the form of additional research or services aimed at correcting problems.

The *Healthy People 2010* objectives (see p. 22) represent the collection and analysis of numerous nutrition surveys and research information.

What are some examples of community nutrition education programs?

Efforts to provide sound nutrition education for all citizens include the Daily Food Guide, the Food Guide Pyramid, and nutrition labeling, to name a few. An educational effort aimed at helping people make small but positive changes in their eating habits is the National 5 a Day Better Health Program. Rather than focusing on all food groups, the objective of this program is to encourage people to eat at least 5 servings of fruits and vegetables a day. Few people in the United States eat the recommended servings of fruits and vegetables, and concentrating on "5 a day" is a simple step. By adding fruits and vegetables, the person obtains additional fiber, vitamin C, and beta-carotene.

Many government agencies (federal, state, and local) and health care organizations make avail-

able easy-to-understand and colorfully illustrated educational materials aimed at educating large numbers of people about nutrition and food safety. Among the federal agencies that provide nutrition education materials are the National Institutes of Health (NIH), the U.S. Department of Agriculture (USDA), and the Food and Drug Administration (FDA). Examples of health care organizations that provide nutrition education materials are the American Dietetic Association and the American Heart Association. Appendix D lists these and many other nutrition resources.

Community-based efforts to feed citizens include food pantries that provide groceries.

In what ways can nurses use these educational materials?

Nurses can use materials from these agencies in any health care setting to show clients how to apply nutrition information to their everyday lives. A helpful way to offer suggestions might be to conduct a usual intake with the client and review the results. First praise the client's positive habits. If the client usually drinks at least 2 to 3 servings of milk, for example, point out just how important that is. Then offer suggestions for changes that require minimal adjustments to the client's routine. If the client always drinks whole milk, recommend using reduced-fat milk instead. Once the client becomes adjusted to the change, he may be willing to try an even lower fat milk. Use the Food Guide Pyramid to point out other foods low in fat and high in nutrient density and encourage their use (see Figure 1-4 on pp. 18–19). Another way to help clients use the pyramid is to help them plan menus for a day or two.

What community programs aim to prevent nutrient deficiencies?

An important component of community nutrition efforts are the food assistance programs that aim to meet the nutrient needs of large groups of people directly. Efforts to prevent hunger and nutrient deficiencies at the community level often target low-income groups. The lack of financial resources to purchase adequate amounts of nutritious foods greatly affects the nutrition status of a community. Other factors that contribute to hunger and poor nutrition include abuse of alcohol and other drugs, physical and mental illness, lack of awareness of available food assistance programs, and the reluctance of people, particularly the elderly, to accept what they perceive as "welfare" or "charity." At present, several food assistance programs are in effect. They include the Food Stamp Program and programs directed at the needs of children and the elderly.

What is the Food Stamp Program, and how does it work?

The Food Stamp Program, administered by the USDA, is the largest of the federal food assistance programs, both in amount of money spent and in number of people participating. Over 27 million people in the United States receive food stamps at a budget cost of over $22 billion per year.[1]

The USDA issues food stamp coupons through state social services or welfare agencies to households. The number of stamps a household receives depends on its size and income. Recipients may use the coupons to purchase food and food-bearing plants and seeds, but not to buy tobacco, cleaning items, alcohol, or other nonfood items.

Which food assistance programs serve children?

Food assistance programs aimed specifically at meeting the nutrient needs of children include the School Lunch and Breakfast Programs and the Child Care Food Program. The School Lunch and Breakfast Programs, also administered by the USDA, enable schools to provide low-income students with meals at no cost while charging other students somewhat less than the full costs of their meals. More than 26 million children receive lunches through the School Lunch Program—half of them free or at a reduced price. School lunches are designed to provide at least a third of the recommended intake for each of many nutrients.

The School Breakfast Program is available in slightly more than half of the nation's schools, and about 5 million children participate in it. The school breakfast must provide at least a fourth of the recommended intake for each of many nutrients. The Child Care Food Program operates similarly and provides funds to organized child-care programs.

The USDA also administers the Special Supplemental Food Program for Women, Infants, and Children (WIC), which provides health care referrals, nutrition education, and food packages or vouchers for specific foods to low-income pregnant women and their young children. WIC provides eggs, milk, cereal, juice, cheese, legumes, peanut butter, and infant formula to infants, children up to age five, and pregnant and breastfeeding women who qualify financially and have a high risk of medical or nutrition problems. Prenatal WIC participation can effectively reduce infant mortality, low birthweight, and newborn medical costs.[2] For every dollar spent on WIC, an estimated three dollars in medical costs are saved. In 1992, participation in WIC reduced first-year medical expenses for infants by $1.19 billion.[3]

What food assistance programs meet the needs of the elderly?

Federal food assistance programs are also in place to meet the needs of the elderly. The federal Elderly Nutrition Program is intended to improve older people's nutrition status and enable them to avoid health problems, continue living in communities of their own choice, and stay out of institutions. Its specific goals are to provide low-cost, nutritious meals; opportunities for social interaction; homemaker education and shopping assistance; counseling and referral to social services; and transportation.

The Elderly Nutrition Program provides for both congregate meal programs and the delivery of meals to those who are homebound. The home-delivery program, known as Meals on Wheels, ensures a nutritious meal, but its recipients miss out on the social benefit of the congregate meal sites (see Nutrition in Practice 13); every effort is made to persuade older people to come to the shared meals, if possible. These programs provide at least one meal a day that meets a third of the recommended nutrient intakes for this age group; they must operate five or more days a week. Many programs voluntarily offer additional services: provisions for modified diets, food pantries, ethnic meals, and delivery of meals to the homeless.

Do any organizations besides the government offer food assistance to the community?

To supplement federal programs, private efforts have sprung up in many communities, where concerned citizens work through local agencies and churches to provide food assistance. Community-based food pantries provide groceries, and soup kitchens serve prepared meals. The meals often average 1000 kcalories each, but most homeless people receive fewer than one and a half meals a day, so many are inadequately nourished.

Why do nurses need to know about food assistance programs?

Public health nurses are generally well acquainted with food assistance programs, and often many of their clients receive such assistance. Nurses not involved in a community health program may assume that with the many food assistance programs available, people who need assistance are getting it. That is not always the case. For example, of the estimated 2 million homeless people in the United States who are eligible for food assistance, only 15 percent of single adults and 50 percent of families receive food stamps. Regardless of the setting in which you see clients, encourage those who may be having financial problems to talk with a social worker who can assess their eligibility for food assistance programs. Remember to approach the subject in a nonjudgmental and tactful manner—the client may feel uncomfortable about seeking assis-

tance. Be proactive—don't assume that someone else is already addressing the problem.

Nurses can serve as valuable members of community groups seeking to provide food assistance. Table NP14-1 shows how individuals can assist in local hunger relief efforts; it presents a 14-step program for developing a hunger-free community.

Notes

1. U.S. Department of Commerce, *Statistical Abstract of the United States, 1994* (Washington, D.C.: Bureau of the Census, 1994), p. 386.

2. P. A. Buescher and coauthors, Prenatal WIC participation can reduce low birth weight and newborn medical costs: A cost-benefit analysis of WIC participation in North Carolina, *Journal of the American Dietetic Association* 93 (1993): 163–166.

3. S. Avruch and A. P. Cackley, Savings achieved by giving WIC benefits to women prenatally, *Public Health Reports*, 110 (1995): 27–34.

Table NP14-1
FOURTEEN WAYS COMMUNITIES CAN ADDRESS THEIR HUNGER PROBLEMS

1. Establish a community-based emergency food delivery network.
2. Assess community hunger problems and evaluate community services. Create strategies for responding to unmet needs.
3. Establish a group of individuals, including low-income participants, to develop and implement policies and programs to combat hunger and the threat of hunger; monitor responsiveness of existing services; and address underlying causes of hunger.
4. Participate in federally assisted nutrition programs that are easily accessible to targeted populations.
5. Integrate public and private resources, including local businesses, to relieve hunger.
6. Establish an education program that addresses the food needs of the community and the need for increased local citizen participation in activities to alleviate hunger.
7. Provide information and referral services for accessing both public and private programs and services.
8. Support programs to provide transportation and assistance in food shopping, where needed.
9. Identify high-risk populations and target services to meet their needs.
10. Provide adequate transportation and distribution of food from all resources.
11. Coordinate foodservices with parks and recreation programs and other community-based outlets to which residents of the area have easy access.
12. Improve public transportation to human service agencies and food resources.
13. Establish nutrition education programs for low-income citizens to enhance their food purchasing and preparation skills and make them aware of the connections between diet and health.
14. Establish a program for collecting and distributing nutritious foods—either agricultural commodities in farmers' fields or prepared foods that would have been wasted.

SOURCE: House Select Committee on Hunger, legislation introduced by Tony P. Hall, excerpted in *Seeds*, Sprouts edition, January 1992, p. 3 with permission, © SEEDS Magazine, P.O. Box 6170; Waco, TX 76706. For more guidance on developing a hunger-free community, write: Hunger Free, House Select Committee on Hunger, 505 Ford House Office Building, Washington, DC 20515.

CHAPTER 15

NUTRITION AND
UPPER GI TRACT DISORDERS

CONTENTS

■■ Disorders of the Mouth
and Esophagus

■■ Disorders of the Stomach

■■ Case Study: Accountant with
Reflux Esophagitis

■■ Nutrition in Practice:
Helping People with
Feeding Disabilities

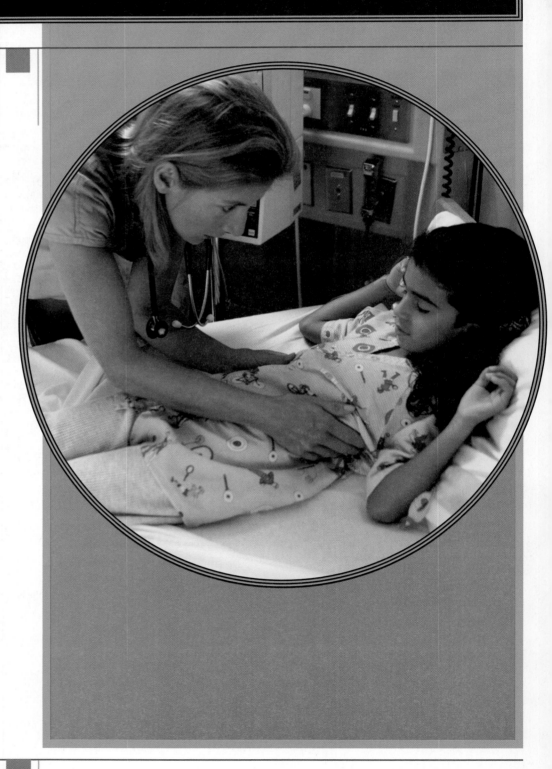

This chapter marks your introduction to specific diseases and how they affect nutrient needs and nutrition status. Your study begins with the upper GI tract where nutrients from foods enter the body. A healthy upper GI tract allows the body to ingest foods and begin digesting them, but these functions can be impaired by physical and emotional disorders.

Disorders of the Mouth and Esophagus

The upper GI tract consists of the mouth, throat, esophagus, and stomach. These organs accomplish feats that people rarely think about: chewing, swallowing, and the first steps of digestion. When things go wrong, however, health and nutrition status can suffer.

Difficulties Chewing

Many conditions can temporarily or permanently interfere with chewing. They range from heavy sedation and pain to injuries, mouth ulcers, and missing teeth or ill-fitting dentures (see Table 15-1). If problems with chewing persist, people may eat too little, lose too much weight, and suffer the consequences of deteriorating nutrition status. Modifying the diet in texture helps ease the task of chewing.

The process of chewing is also called **mastication**.

The person without teeth is said to be **edentulous**.

DIET OPTIONS People's tolerances of food consistencies vary greatly. Following surgery to repair a broken jaw, for example, a person may be able to consume liquids only. With time, however, the person may progress to regular foods that are modified in texture to make them easy to chew and swallow—a **mechanical soft diet**. Unlike the soft diet described in Chapter 14, mechanical soft diets include all foods and seasonings. Nurses and dietitians work together with the client, family, and caregivers to identify which foods the client can safely handle; such foods may be naturally soft foods such as fish or bananas, tender cooked whole foods, ground or chopped foods, or pureed foods. The goal is to provide a wide variety of foods that are as similar as possible to those of a regular diet. Such a strategy enhances appetite and minimizes the likelihood of nutrient deficiencies. Therefore, soft natural foods, tender-cooked foods, and chopped foods are provided whenever possible; only when the person cannot chew or swallow adequate amounts of these foods are pureed foods provided.

mechanical soft diet: a diet that excludes only those foods that a person cannot chew; also called a **dental soft diet** or **edentulous diet**.

The person recovering from a broken jaw, described above, may progress from liquids to pureed foods, then to ground or chopped foods, and then to regular foods. In some cases, however, a person may need to follow a mechanical soft diet permanently. Moist, soft-textured foods, such as casseroles and foods prepared with sauces and gravies, often work well for clients who can handle some solid foods. Drinking liquids along with meals makes it easier to chew and swallow.

CAUTIOUS USE OF PUREED FOODS Pureed foods can meet all nutrient needs, but the monotony of

Table 15-1
CONDITIONS THAT MAY INTERFERE WITH CHEWING AND SWALLOWING

Achalasia	Ill-fitting dentures
Alzheimer's disease	Missing teeth
Broken jaw	Multiple sclerosis
Cancer	Myasthenia gravis
Candida albicans (thrush) infection	Oral herpes infection
Chemotherapy	Parkinson's disease
Congenital defects of upper GI tract	Periodontal disease
Dental caries	Radiation therapy of the head and neck
Dryness of mouth	Sensitivity of mouth to hot or cold
Dysphagia	Strokes
Guillain-Barré syndrome	Surgery of the mouth, head, or neck
Head injury	Ulcers of mouth, gums, or esophagus
HIV Infection	

Can you tell that the foods in this photo are actually pureed foods shaped with commercial thickeners?

IMPROVE ACCEPTANCE OF PUREED FOODS

Take a moment to think about a meal of pureed foods. A typical dinner of baked chicken, mashed potatoes, and green beans is blenderized to white mush, more white mush, and a green blob. The foods may taste great, but on seeing the plate, the person may be reluctant to try the first bite. To stimulate the appetite, use creative techniques such as these for preparing and serving food:

- Encourage clients and their caregivers to prepare a variety of favorite foods and blenderize them to a tolerable consistency. The smell of favorite foods and the thought of consuming them stimulate the appetite.
- Consider color when planning meals. The meal of baked chicken, mashed potatoes, and green beans, described above, can be made more appealing by substituting mashed sweet potatoes for the white potatoes. Arranging foods attractively on a plate with appropriate garnishes also adds color and eye appeal.
- Serve foods at the right temperature and puree them so that they are smooth and thick—not watery and thin. Commercially available thickeners add shape, texture, and even nutrients to food.
- Experiment with seasonings and spices to enliven food flavors, excluding only those that the person cannot tolerate for personal or medical reasons. Seasoning food to accommodate personal tastes adds flavor, allows for variety, and improves the appetite.
- Supplement the diet with nutritious liquids such as milk, instant breakfast drinks, or liquid formulas (see Chapter 17).

Efforts to improve the acceptance of pureed foods can go a long way toward helping people to eat and maintain or improve their weights.[a] When efforts to improve a client's intake of pureed foods are unsuccessful, feeding the person by tube becomes an option (see Chapter 17).

---

[a]D. Cassens, E. Johnson, and S. Keelan, Enhancing taste, texture, appearance, and presentation of pureed food improved resident quality of life and weight status, *Nutrition Reviews* (supplement) 11 (1996): 51–54.

eating food of the same consistency for every meal, day after day, can create a psychological block to eating, and malnutrition can follow. The "How to" box above offers suggestions for clients and caregivers that can improve the acceptance of pureed foods.

Foodservice departments in nursing homes, rehabilitation centers, and other long-term care facilities that provide clients with pureed foods over extended periods of time face a difficult challenge. Fortunately, a variety of commercial products are available to thicken and shape pureed foods to give them an appetizing appearance. Other factors may also interfere with a person's desire to eat, however. Clients in long-term care facilities lack the comforts of home and may be overwhelmed by their medical conditions, isolation, and loss of control. Addressing these emotional needs is critical to stimulating the appetite.

mouth ulcers: lesions or sores in the lining of the mouth. Certain drugs, radiation therapy, and some disorders, such as oral herpes virus infections, can cause mouth ulcers.

tonsillectomy: (tawn-sill-ECK-tah-me): surgical removal of the tonsils.

:: MOUTH AND THROAT PAIN The mechanical soft diet for people with **mouth ulcers** or inflammation of the lips or throat (following a **tonsillectomy,** for example) provides moist, soft-textured foods and eliminates dry, spicy, salty, or acidic foods (such as citrus fruits and tomatoes) that may be painful to eat. In addition, nuts or seeds in foods (such as pecans or poppy seeds in breads) and sticky foods (like peanut butter) can cause discomfort. Heat may intensify pain, too. Foods are served no warmer than room temperature, although many clients prefer cold foods and beverages.

■■ **REDUCED FLOW OF SALIVA** People with mouth dryness due to a reduced flow of saliva often prefer moist, soft foods. Clients can moisten foods with sauces and gravies. Salty foods and snacks dry the mouth and should be avoided. Encourage these clients to practice good oral hygiene; when salivary flow is reduced, the mouth is poorly defended against dental caries. Clients can increase salivary secretions by sucking on sugarless candy, chewing sugarless gum, eating lemon or lime ices, or using drugs that stimulate the flow of saliva.

■■ Difficulties Swallowing

Problems in swallowing, known as **dysphagia,** can arise from many causes, including aging, nervous system diseases, injuries, developmental disabilities, and strokes. A person with dysphagia may be unable to initiate swallowing, to chew foods and mix them with saliva, or to push foods to the back of the throat and into the esophagus. Some forms of dysphagia specifically interfere with the muscular contractions of the esophagus.

> **dysphagia (dis-FAY-gee-ah):** difficulty in swallowing.
> dys = bad
> phagein = to eat

THE NURSING DIAGNOSIS **altered oral mucous membranes** should serve as a reminder to help clients adjust their food selections to ease chewing and swallowing and minimize pain.

> *Dysphagia that occurs when the cardiac sphincter fails to relax and allow swallowed food to enter the stomach is known as* achalasia (ACK-ah-LA-zee-ah).

■■ **SUBTLE AND DANGEROUS** Dysphagia sometimes goes undiagnosed, especially when caused by aging, because symptoms are not always obvious. All people "catch food in their throat" at one time or another, so a person who experiences this feeling more frequently may consider the condition normal and ignore it. Eventually, the person may eat less and develop nutrient deficiencies.

When undetected, dysphagia can be dangerous for other reasons as well. Unlike healthy people, people with dysphagia may fail to cough in response to the presence of food in the trachea. Food particles may then pass unnoticed into the lungs (**aspiration**), allowing bacteria to multiply. Such "silent" aspiration, which has been observed in people following strokes, carries with it the risk of pneumonia and death.

> **aspiration:** the drawing in of food, gastric secretions, or liquid into the lungs.

THE NURSING DIAGNOSES related to dysphagia are **impaired swallowing** and **risk of aspiration.**

■■ **SIGNS OF DYSPHAGIA** Health care professionals should be alert to subtle symptoms of dysphagia including an unexplained decline in food intake or repeated bouts of pneumonia. Other symptoms include pain on swallowing, weight loss, fear of eating certain foods or any food at all, a feeling that food is sticking in the throat, a tendency to hold food in the mouth rather than swallowing it, coughing or choking during meals, frequent throat clearing, drooling, or a change in voice quality. Depending on the cause of dysphagia, the voice may sound hoarse, nasal, or "wet." Diagnosis is based on extensive testing that may include cranial nerve assessment, X rays, fluoroscopy, and measurements of esophageal sphincter pressure and esophageal peristalsis.

■■ **DIET FOR DYSPHAGIA** The mechanical soft diet for dysphagia leaves little room for error or experimentation. Speech pathologists, physicians, nurses, and dietitians work together to assess the client's swallowing abilities and design an individualized diet. Many clients with dysphagia tolerate mildly spiced and moderately sweet foods that are served at room temperature. Sticky foods and small pieces of food or foods that break into small pieces when eaten (such as rice or dry chopped meats) can be difficult to handle. Foods with more than one texture, such as yogurt with fruit or dry cereal with milk, are also difficult for many clients to handle. Clients often experience the most difficulty swallowing true liquids. Thickened liquids or smooth solids, such as milk shakes or puddings, are good choices; these foods flow slowly enough to allow time to coordinate swallowing movements. Commercial thickeners, pudding, yogurt, or baby cereal can be used to thicken liquids.

Appropriate body positioning during meals limits the risk of choking. Ensure that the client's body is upright at a 90-degree angle with hips flexed, feet flat on the floor, and the head tilted slightly forward.

With time, swallowing function may improve. The health care team continuously monitors the person and expands the diet to include additional food as tolerated. Ideally, the diet is progressed to a regular diet, although this is not always possible. Tube feedings (discussed in Chapter 17), delivered directly into the intestine to minimize the risk of aspiration, may improve the nutrition status of clients who are unable to eat enough to meet their nutrient needs or those whose swallowing function deteriorates.

▪▪ Disorders of the Stomach

Once swallowed, food travels down the esophagus into the stomach. The stomach retains the bolus for a while, adding acids, fluids, and enzymes to the mixture before releasing its contents into the intestine. Disorders of the stomach range from occasional bouts of indigestion to serious conditions that require surgical resections.

▪▪ Indigestion and Reflux Esophagitis

Indigestion, or **dyspepsia,** is a vague term used to describe **epigastric** pain. Emotional tension, medical conditions, medications, eating too much, eating too fast, and chewing inadequately can all cause indigestion. Signs and symptoms associated with indigestion include fullness, early satiety, belching, hiccups, **heartburn,** and regurgitation of the stomach's acid fluids into the mouth (acid indigestion). The "How to" box provides suggestions for preventing indigestion.

When acid indigestion becomes a chronic problem, the highly acidic gastric fluids irritate the esophagus and cause a painful inflammation called **reflux esophagitis.** Heartburn, chest pain, chronic cough, hoarseness, and sore throat are symptoms of reflux esophagitis. Severe inflammation may narrow the inner diameter of the esophagus, and dysphagia may result. Severe and chronic irritation may also lead to **esophageal ulcers** and bleeding, and **Barrett's esophagus,** a condition associated with cancer of the esophagus.[1]

Aging, certain medications, and the use of feeding tubes that pass from the nose to the stomach increase the risk of reflux esophagitis. Reflux esophagitis frequently occurs in people with asthma, peptic ulcers (described later), and **hiatal hernias.**

▪▪ HIATAL HERNIA

The esophagus joins the stomach at the cardiac sphincter. Normally, this sphincter sits within the hiatus of the diaphragm and is reinforced by it. The esophagus lies completely above, and the stomach completely below, the diaphragm. Sometimes, however, the diaphragm weakens, and a portion of the stomach slides up through it, an abnormality called a hiatal hernia. Once a hiatal hernia forms, the cardiac sphincter, which holds foods in the stomach, no longer is reinforced by the surrounding diaphragm,

dyspepsia: vague abdominal pain; also called indigestion.

epigastric: the region of the body just above the stomach.

epi = above

gastric = stomach

heartburn: a burning sensation felt behind the sternum caused by the presence of gastric juices in the esophagus; also called **pyrosis (pie-ROE-SIS).**

reflux esophagitis: (eh-sof-ah-JYE-tis): the backflow or regurgitation of gastric contents from the stomach into the esophagus, causing inflammation of the esophagus; also called **gastroesophageal reflux (GER).**

The narrowing of the inner diameter of the esophagus from inflammation and scarring is called an **esophageal stricture.**

esophageal ulcers: lesions or sores in the lining of the esophagus.

Barrett's esophagus: changes in the cells of the esophagus associated with chronic gastroesophageal reflux that raise the risk of cancer of the esophagus.

hiatal hernia: a protrusion of a portion of the stomach through the esophageal hiatus of the diaphragm.

HOW TO:

PREVENT INDIGESTION

To prevent indigestion, advise clients to take these steps:

- Eat at regular times and in a relaxed atmosphere.
- Eat slowly and chew food thoroughly.
- Eat small meals.
- Avoid gas-forming foods. Remember that milk may be one of these foods if lactose intolerance is present. (There is more on lactose intolerance in the next chapter.)
- Avoid excessive amounts of fats, alcohol, and caffeine-containing foods and beverages.
- Avoid any food known to cause indigestion.
- Keep a food record, noting when indigestion occurs in relationship to foods eaten.

and it becomes easy for acidic gastric juices to reflux into the esophagus. Because reflux occurs more readily in people with hiatal hernias, they frequently suffer from heartburn and esophagitis. Figure 15-1 shows the normal relationship of the upper GI tract to the diaphragm and the changes that occur with reflux and with a hernia.

DIETARY PREVENTION AND TREATMENT Treatment for active reflux esophagitis aims to prevent reflux and reduce inflammation of the esophagus

Figure 15-1

THE UPPER GI TRACT, ACID REFLUX, AND HIATAL HERNIA

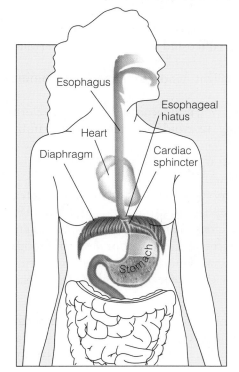
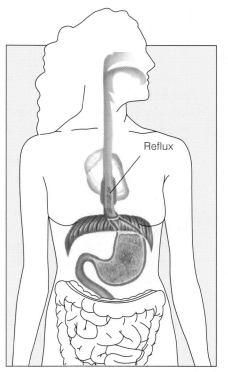
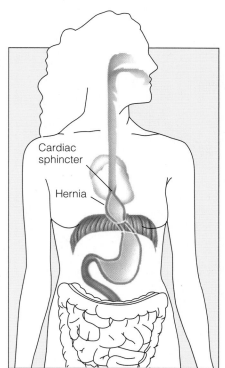

The stomach normally lies below the diaphragm, and the esophagus passes through the esophageal hiatus. The cardiac sphincter prevents reflux of stomach contents.

Whenever the pressure in the stomach exceeds the pressure in the esophagus, as can occur with overeating and overdrinking, the chance of reflux increases. The resulting "heartburn" is so-named because it is felt in the area of the heart.

Acid reflux often occurs as a consequence of a hiatal hernia. The most common type, a sliding hiatal hernia, results when part of the stomach, with the cardiac sphincter, slip through the diaphragm.

Table 15-2
SUBSTANCES THAT RELAX THE CARDIAC SPHINCTER

Alcohol

Anticholinergic agents

Calcium channel blockers

Chocolate

Cigarette smoking

Diazepam

Garlic

High-fat foods

Meperidine

Onions

Peppermint and spearmint oils

Theophylline

and its associated pain by reducing gastric acidity, reducing pressure in the stomach, and eliminating foods, substances, or activities that weaken cardiac sphincter pressure (see Table 15-2). The "How to" box offers specific suggestions.

■■ **DRUG THERAPY**   Turn on the television set and quite likely you will see advertisements touting antacids and a new generation of medications that are highly effective in both preventing and relieving indigestion. The ads imply that rather than eat sensibly and listen to your body's signals about the kinds of foods you are eating or the manner in which you are eating them, you can eat anything, anyway and then simply take medication to avoid the discomforts of indigestion. All medications carry risks, however, and clients should be encouraged to prevent simple indigestion without medication. For clients with reflux esophagitis, however, physicians often prescribe antacids and antisecretory agents. Antacids neutralize gastric acidity, and antisecretory agents suppress or inhibit gastric acid secretion. The Diet-Medication Interactions box on p. 371 includes a discussion of the medications described here. If medical

HOW TO:

PREVENT AND TREAT REFLUX ESOPHAGITIS

To prevent and treat reflux esophagitis and its associated discomfort, recommend that clients:

- Eat small meals and drink liquids one hour before or one hour after meals to avoid distending the stomach. A distended stomach exerts pressure on the cardiac sphincter.
- Relax during mealtimes, eat foods slowly, and chew them thoroughly to avoid swallowing air and distending the stomach.
- Limit foods that weaken cardiac sphincter pressure or increase gastric acid secretion, including fat, alcohol, caffeine, decaffeinated coffee and tea, chocolate, spearmint, and peppermint.
- Lose weight, if necessary. Overweight tends to increase abdominal pressure.
- Refrain from lying down, bending over, and wearing tight-fitting clothing or belts, particularly after eating, to avoid increasing pressure in the stomach.
- Elevate the head of the bed by 4 to 6 inches. Keeping the chest higher than the stomach helps prevent reflux.
- Refrain from smoking cigarettes. Smoking relaxes the cardiac sphincter.
- During periods of active reflux esophagitis, avoid foods and beverages that irritate the esophagus, such as citrus fruits and juices, tomatoes and tomato-based products, pepper, spices, and very hot or very cold foods, according to individual tolerances. If a client cannot tolerate citrus fruits and juices, ensure that the diet provides adequate vitamin C from other foods or supplements.

Individual tolerances of types and amounts of foods and spices vary markedly especially during periods of active esophagitis. Health care professionals can help clients pinpoint food intolerances by advising them to keep a record of the types and amounts of foods and beverages consumed, the time of consumption, GI symptoms, and time of occurrence. Assessment of the record by a dietitian or health care professional provides the basis for determining the types and amounts of food and food components that the client can handle without discomfort.

management fails, however, surgery may be indicated. The accompanying case study provides questions that review the nutrition needs of a client with a hiatal hernia and reflux esophagitis.

Nausea and Vomiting

Nausea is the feeling that one is about to vomit. Nausea is a symptom of many medical conditions, medications, and treatments and can be triggered by emotional stress, certain odors, motions, and even disturbing sights.

Simple vomiting is certainly unpleasant and wearying, but is no cause for alarm. Prolonged vomiting, however, requires medical attention. With vomiting, foods and medications fail to reach the intestine and are unavailable to the body. Large amounts of fluids and electrolytes are also expelled, raising the risk of dehydration and nutrient imbalances. Infants and children can quickly become seriously dehydrated as a consequence of vomiting, because their bodies contain a greater percentage of water than an adult body.

TREATMENT OF NAUSEA AND VOMITING   For treatment of nausea and vomiting related to a medical condition, treatment of that condition often alleviates the problem. Nausea and vomiting associated with medications are sometimes temporary and often resolve after a few doses. If not, a physician may change the medication, if possible. The suggestions in the "How to" box can help relieve nausea. Physicians may also prescribe antinauseants and antiemetics (see the Diet-Medication Interactions box on p. 371).

Most antiulcer agents suppress or inhibit gastric acid secretion. Those that do are also classed as antisecretory agents or anti-GERD (GastroEsophageal Reflux Disease).

Case Study

ACCOUNTANT WITH REFLUX ESOPHAGITIS

Mrs. Scarlatti, a 49-year-old accountant, recently underwent a complete physical examination. She told her physician that she had been feeling fairly well until she began experiencing heartburn, which has become more frequent and more painful. The heartburn usually occurs after she has eaten a large meal, particularly when she lies down after eating. After inspecting the interior of her esophagus, stomach, and duodenum with a long tube equipped with an optical device called a gastroscope, the physician diagnosed a sliding hiatal hernia and reflux esophagitis.

Mrs. Scarlatti's past health history shows no indication of significant health problems. During her last physical, the physician did advise her to stop smoking cigarettes and to lose 20 pounds, which she has yet to do. During a nursing assessment, Mrs. Scarlatti mentioned that her life is pretty hectic because it is the middle of the tax season. She usually does not have time for breakfast, eats lunch hurriedly while continuing to work, and eats a large dinner around 8:00 p.m. She generally drinks one or two alcoholic beverages before going to sleep. Her current height and weight are 5 feet 6 inches and 170 pounds.

✚ Explain to Mrs. Scarlatti what a sliding hiatal hernia is, and describe how the hernia leads to reflux and heartburn.

✚ From the brief history provided, list the factors and behaviors that increase Mrs. Scarlatti's chances of experiencing gastric reflux and the pain of heartburn. What recommendations can you make to help her change these behaviors?

✚ What drugs might the physician prescribe and why?

HOW TO: MINIMIZE NAUSEA

The following suggestions can help clients alleviate nausea:

- Relax before eating and avoid overeating. Breathe deeply and slowly if you begin to experience nausea. Eat small meals and save liquids for between meals to prevent distention of the stomach.
- Drink liquids between meals, especially when vomiting is a problem. Although individual tolerances vary, many clients tolerate cold or carbonated liquids and juices best.
- Identify individual food intolerances and aromas that precipitate nausea; avoid these foods and aromas, if possible. High-fat foods and highly spiced foods are often poorly tolerated. Cold or room-temperature foods may be better tolerated than hot foods, especially if food aromas trigger nausea. Ask others to prepare foods, if possible.
- Eat carbohydrate-rich, low-fat foods before getting out of bed if nausea occurs soon after you wake up. Crackers or bread can be kept at bedside.
- If nausea is a problem at specific times of the day, avoid eating or drinking immediately before, after, or during those times.
- Try drinking ginger or peppermint teas, which provide relief for some people.
- Relax after eating, but don't lie down. Get fresh air after meals and wear nonconstrictive clothing.

Physicians may also prescribe antinauseants or antiemetics to help control nausea and vomiting. Counsel clients to take these medications at least 30 minutes before eating to ensure effectiveness during mealtimes.

■■ DIETARY INTERVENTIONS FOR VOMITING If efforts to control nausea fail, foods are withheld for a few hours. If tolerated, clear liquids can replace fluids and electrolytes. If oral liquids cannot be tolerated, intravenous (IV) fluids may be given until vomiting resolves. Providing all nutrients by vein (Chapter 18) may be indicated in cases of severe and prolonged vomiting, especially for clients who are malnourished.

■■ Gastritis

gastritis: inflammation of the stomach lining.

Gastritis is a common disorder in which the mucosal lining of the stomach becomes inflamed and painful. The person with gastritis may complain of anorexia, nausea, vomiting, and epigastric pain. Although gastritis generally resolves with treatment, unresolved gastritis can lead to ulcers (described in the next section), hemorrhage, shock, obstruction, perforation, and an increased risk of gastric cancer.

The bacterial infection associated with acute and chronic gastritis is caused by Helicobacter pylori.

■■ ACUTE GASTRITIS Acute gastritis most often follows the repeated use of aspirin or other medications that irritate the gastric mucosa. Alcohol abuse, food irritants, food allergies, food poisoning, radiation therapy, metabolic stress, and bacterial infections can also cause gastritis.

For the person with gastritis who cannot eat because of nausea or vomiting, foods are generally withheld for a day or two. The diet then progresses from liquids to a highly individualized diet that aims to eliminate foods that stimulate gastric acid secretion, irritate the gastric mucosa, or cause problems for the individual. For lack of a better name, such a diet is widely referred to as a liberal bland diet (see Table 15-3), because it replaced a highly restrictive bland diet, which was invalidated by research. Antacids, antisecretory agents, and antibiotics (amoxicillin and metronidazole) may also be prescribed. The Diet-Medication Interactions box on p. 371 discusses these medications.

CHRONIC GASTRITIS Chronic gastritis may be associated with aging, peptic ulcer disease (described in the next section), and conditions that cause the chronic reflux of basic fluids from the duodenum into the stomach (gastric reflux). Many people with chronic gastritis have **pernicious anemia,** and, indeed, pernicious anemia may lead to the diagnosis. In some cases the cause of chronic gastritis is unknown.

The symptoms of chronic gastritis are generally milder than those of acute gastritis, but they persist over time. As gastritis progresses, the gastric cells gradually shrink, and gastric secretions decline.

A liberal bland diet may help to relieve GI symptoms in some cases. When intrinsic factor is lacking, vitamin B_{12} is given by injection to bypass the need for absorption.

Ulcers

Ulcers can develop both inside and outside the body, but the term *ulcer* used alone generally refers to a *peptic ulcer*—an erosion of the top layer of cells from the GI tract lining. This erosion leaves the underlying layers of cells exposed to gastric juices. When the gastric juices reach the capillaries, the ulcer bleeds, and when they reach the nerves, they cause pain.

CAUSES OF ULCERS Three major causes of ulcers have been identified: *Helicobacter pylori* infections, the use of certain anti-inflammatory medications, and disorders that cause excessive gastric acid secretion. One such disorder, the **Zollinger-Ellison syndrome,** results from a tumor of the pancreas that produces gastrin, which, in turn, stimulates the production of gastric acid. Severe peptic ulcer disease follows.

A recent study of over 45,000 men suggests that dietary factors may aid in the prevention of ulcers.[2] High intakes of vitamin A, fruits and vegetables, and dietary fiber were associated with a lower incidence of duodenal ulcers during the six-year study period. Further research is necessary to determine if other closely related nutrition factors may actually account for the protective effects observed.

TREATMENT FOR ULCERS Treatment aims at relieving pain, healing the ulcer, and minimizing the likelihood of recurrence. Drug therapy plays the primary role in treatment. Antibiotics are used to treat bacterial infections. Antiulcer agents include antisecretory agents to suppress or inhibit gastric acid secretion and protect the stomach and duodenal lining from acid erosion. As for diet, advise the client to follow a liberal bland diet, relax during mealtimes, eat slowly, chew foods thoroughly, and avoid overating.

Gastric Surgery

Several surgical procedures affect the functions of the stomach. During a **gastrectomy,** the surgeon removes either a portion or all of the stomach. Figure 15-2 illustrates three common gastrectomy procedures. Another type of gastric surgery, **pyloroplasty,** enlarges the pyloric sphincter (the sphincter at the junction of the stomach and small intestine) so that the basic intestinal

---

Table 15-3
THE LIBERAL BLAND DIET

A bland diet includes all foods except:

- Those an individual identifies as irritating to the GI tract.
- Alcohol.
- Caffeine and caffeine-containing beverages (including cola beverages, cocoa, coffee, and tea).
- Decaffeinated coffee and tea.
- Pepper and spicy foods except as tolerated.

NOTE: The liberal bland diet shown here has replaced the earlier "traditional bland diet," which was highly restrictive and invalidated by research.

---

pernicious anemia: anemia caused by a lack of intrinsic factor and the consequent malabsorption of vitamin B_{12}.

In atrophic gastritis, all layers of the stomach's mucosal cells are inflamed. Early diagnosis and treatment help limit damage to the gastric cells and prevent complications, including malnutrition.

ulcer: an open sore or lesion. A **peptic ulcer** is an erosion of the top layer of cells from the mucosa of the lower esophagus, the stomach, or the small intestine. Ulcers may also develop in the mouth, upper esophagus, and other parts of the intestine, or on the skin.

The medications associated with ulcers are nonsteroidal anti-inflammatory agents such as ibuprofen.

Zollinger-Ellison syndrome: marked hypersecretion of gastric acid and consequent peptic ulcers caused by a tumor of the pancreas.

gastrectomy (gas-TREK-tah-mee): surgery to remove all (total gastrectomy) or part (subtotal or partial gastrectomy) of the stomach.

pyloroplasty (pie-LOOR-o-PLAS-tee): surgery that enlarges the pyloric sphincter.

vagotomy (vay-GOT-o-mee): surgery that severs the nerves that stimulate gastric acid secretion.

gastric partitioning: surgery for severe obesity that limits the functional size of the stomach.

dumping syndrome: the symptoms that result from the rapid entry of undigested food into the jejunum: sweating, weakness, and diarrhea shortly after eating and hypoglycemia later.

fluids reflux into the stomach and neutralize gastric acidity. During a **vagotomy,** the surgeon severs the nerves that stimulate gastric acid production. A vagotomy may accompany either a gastrectomy or a pyloroplasty in some cases. In **gastric partitioning,** a treatment for severe obesity, the stomach remains intact, but all or a portion of it is bypassed.

▪▪ **DUMPING SYNDROME** When the portion of the stomach containing the pyloric sphincter has been removed, bypassed, or disrupted, one problem that may occur is the **dumping syndrome** (see Figure 15-3 on p. 367). A typical scenario goes something like this: Mrs. Clark had a gastric resection about a week ago and has just begun to eat solid foods. She swallows the food and about 15 minutes later begins to feel weak and dizzy. She looks pale, her heart beats rapidly, and she breaks out in a sweat. Shortly thereafter, she develops diarrhea. What causes this sequence of events?

Without a functional pyloric sphincter, Mrs. Clark's stomach has lost control over the rate at which food enters the intestine. Now food gets "dumped" rapidly into the jejunum. (The duodenum is short, and even if it has not been bypassed surgically, food passes quickly through it into the jejunum.) As the mass of food is digested, the intestinal contents rapidly become concentrated (hypertonic). Fluids from the body move into the intestinal lumen to dilute the concentration. Consequently, the volume of circulating blood falls rapidly, causing weakness, dizziness, and a rapid heartbeat. The large volume of hypertonic fluid and unabsorbed material in the jejunum causes pain and hyperperistalsis, and diarrhea results.

Two to three hours later, Mrs. Clark experiences many of the same symptoms again: dizziness, fainting, nausea, and sweating. This time the cause is different. Mrs. Clark's intestines efficiently absorbed so much glucose from the meal that her blood glucose rose quickly. Her pancreas responded by overproducing insulin, which made the blood glucose fall quickly. Now low blood glucose (hypoglycemia) is causing the symptoms.

Not all people experience the diarrhea of the dumping syndrome following gastric procedures. Even fewer develop hypoglycemia. Many people who initially experience the dumping syndrome gradually adapt to a fairly regular diet. Nevertheless, a postgastrectomy diet serves as a preventive measure following the immediate postsurgical period and benefits clients with prolonged or severe problems.

▪▪ **POSTSURGICAL CARE** In the immediate postsurgical period, the client receives no foods or fluids by mouth. Health care professionals monitor fluid and electrolyte balances carefully, and the physician corrects imbalances promptly. After several days, liquids and then solids are gradually introduced in small amounts offered frequently.

Figure 15-2
TYPICAL GASTRIC SURGERY RESECTIONS

In a gastric resection, part or all of the stomach is surgically removed. The dashed lines show the removed section.

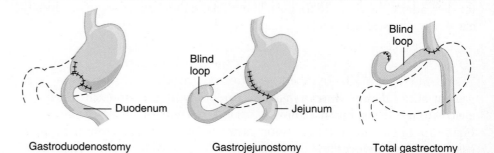

Gastroduodenostomy   Gastrojejunostomy   Total gastrectomy

Figure 15-3
DUMPING SYNDROME

When partially digested food rapidly enters the jejunum, it is quickly digested and creates a concentrated mass. Fluid from the intestinal capillaries enters the jejunum, diminishing blood volume and stimulating peristalsis. The result: low blood pressure and diarrhea.

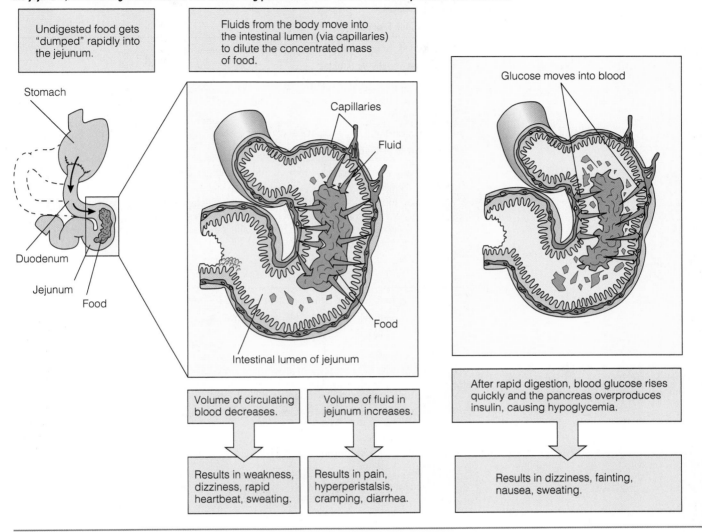

Undigested food gets "dumped" rapidly into the jejunum.

Fluids from the body move into the intestinal lumen (via capillaries) to dilute the concentrated mass of food.

Stomach

Duodenum

Jejunum

Food

Capillaries

Fluid

Intestinal lumen of jejunum

Food

Glucose moves into blood

Volume of circulating blood decreases.

Volume of fluid in jejunum increases.

After rapid digestion, blood glucose rises quickly and the pancreas overproduces insulin, causing hypoglycemia.

Results in weakness, dizziness, rapid heartbeat, sweating.

Results in pain, hyperperistalsis, cramping, diarrhea.

Results in dizziness, fainting, nausea, sweating.

■■ **THE POSTGASTRECTOMY DIET** Postgastrectomy diets limit the total amount of carbohydrate as well as the amount of simple sugars (see Table 15-4 on p. 368). To provide energy and slow the passage of foods through the stomach, the diet emphasizes foods high in protein and provides additional energy from fat

Diet advice to offer the client along with the postgastrectomy diet is provided in the "How to" box on p. 369. Dietitians carefully tailor the postgastrectomy diet to meet the person's tolerances. Initially, the dietitian visits the client after each meal to check the person's tolerance for food. With time, the symptoms of the dumping syndrome often resolve or improve. Gradually, most people begin to tolerate limited amounts of concentrated sweets, larger quantities of food, and some liquids with meals. Sometimes adding pectin and guar gum (types of dietary fibers) to the diet can help control dumping syndrome. If dietary and medical management of the dumping syndrome fail to adequately resolve the problem, additional surgery may be necessary.

■■ **WEIGHT LOSS AND NUTRIENT DEFICIENCIES** After gastric surgery, many people experience significant weight loss and develop nutrient deficiencies.

Table 15-4

POSTGASTRECTOMY DIET[a]

Meat and Meat Alternatives
Any type allowed.
Milk and Milk Products
Withheld initially and then gradually introduced as tolerated.
Grains and Starchy Vegetables
Allowed (up to 5 servings per day): Plain breads, crackers, rolls, unsweetened cereal, rice, pasta, corn, lima beans, parsnips, peas, white potatoes, sweet potatoes, pumpkin, yams, winter squash.
Excluded: Sweetened cereal; cereal containing dates, raisins, or brown sugar.
Nonstarchy Vegetables
Allowed (unlimited): Cabbage, celery, Chinese cabbage, cucumbers, lettuce, parsley, radishes, watercress.
Allowed (up to two ½ c servings per day as individual tolerances permit): Asparagus, bean sprouts, beets, broccoli, brussels sprouts, carrots, cauliflower, eggplant, green pepper, greens, mushrooms, okra, onions, rhubarb, sauerkraut, string beans, summer squash, tomatoes, turnips, zucchini.
Excluded: Vegetables prepared with sugar or creamed.
Fruits
Allowed (up to 3 servings per day): Unsweetened fruits and fruit juices.
Excluded: Sweetened fruits and fruit juices, dates, raisins.
Fats
Any type allowed.
Beverages
Allowed: Coffee, tea, artificially sweetened drinks.
Excluded: Alcohol; sweetened milk, beverages, and fruit drinks; cocoa.
Other
Excluded: Cakes, cookies, ice cream, sherbet, honey, jam, jelly, syrup, and sugar.

[a]Clients with dumping syndrome who are unable to tolerate a sufficient variety or volume of foods over long periods of time often require nutrient supplements.

Other potential problems associated with gastric surgery, including early satiety, postsurgical pain, the desire to prevent the symptoms of the dumping syndrome, reflux esophagitis, and, sometimes, dysphagia, can limit food intake.

In addition to the malabsorption that occurs whenever foods pass rapidly through the intestine, fat malabsorption (see Chapter 16) can occur whenever clients have undergone surgery that bypasses the duodenum. When food bypasses the duodenum, the trigger for the release of hormones that mediate the secretion of enzymes and bile to aid fat digestion and absorption is also bypassed, resulting in fat malabsorption. Fat malabsorption can also occur as a consequence of reduced gastric acid secretion, which can lead to bacterial overgrowth (described in detail in Chapter 16) in the stomach or upper small intestine. Bacterial overgrowth contributes to the malabsorption not only of fat but also of fat-soluble vitamins (especially vitamin D), as well as folate, vitamin B_{12}, and calcium.

■■ **ANEMIA** Iron-deficiency anemia is another common problem following gastric surgery, although it may take several years to develop. When iron's exposure to gastric acid is limited, less iron is converted to its absorbable form. The rapid transit of food through the duodenum (where 50 percent of

ADJUST MEALS TO PREVENT THE DUMPING SYNDROME

The person on a postgastrectomy diet benefits from the following suggestions:

- Avoid concentrated sweets (sugar, cookies, cakes, pies, or soft drinks) because the body digests and absorbs these carbohydrates rapidly and breaks them down into many particles that draw fluids from the body into the intestine.
- Eat frequent small meals to fit the reduced storage capacity of the stomach.
- Drink liquids in small amounts about 45 minutes before or after meals, not with them. This precaution prevents overloading the stomach's reduced storage capacity and slows the transit of food from the stomach to the intestine.
- Lie down immediately after eating to help slow the transit of food to the intestine. Clients who experience esophageal reflux, however, should not lie down after eating.
- Be aware that lactose intolerance (described in Chapter 16) may develop and add to the problem of diarrhea and abdominal pain. Enzyme-treated milk and milk products should also be avoided because the enzymes break down lactose to glucose and galactose—simple sugars that further concentrate the intestinal contents and promote dumping. Discontinue all types of milk and milk products until recovery is under way. Then try them gradually and in small amounts.

iron absorption normally occurs), inadequate iron intake, and blood loss also contribute to the problem. An iron supplement helps correct the deficiencies.

Inadequate intake and malabsorption can also lead to anemia caused by folate deficiency and, less often, by vitamin B_{12} deficiency. Although one might expect vitamin B_{12} deficiencies to be common after gastric surgery because intrinsic factor production could be impaired, surgeons often avert this problem by leaving intact a small part of the stomach that produces intrinsic factor. To correct deficiencies, clients receive supplements.

■■ BONE DISEASE People who experience fat malabsorption following gastric surgery also malabsorb vitamin D and calcium. After many years, a significant number of people who have undergone gastrectomies develop a bone disease similar to osteomalacia.[3] Vitamin D and calcium supplements are frequently provided, although the bone disease is often resistant to treatment.

■■ GASTRIC PARTITIONING Unlike other gastric surgeries, weight loss is a goal following gastric partitioning. Most clients lose weight for about 12 to 18 weeks after surgery and lose 50 to 60 percent of their excess body weight.[4]

The safety and effectiveness of gastric partitioning depend, in large part, on compliance with medical nutrition therapy, which aims to prevent nutrient deficiencies and promote eating and lifestyle habits that will help the client lose weight and maintain a desirable weight loss. Poor dietary habits may prevent weight loss, rupture staples, or obstruct the small opening left for food to pass into the lower stomach. Other postsurgical complications include infections, nausea, vomiting, dehydration, dumping syndrome, esophageal reflux, and, as a result of all this and more, depression.

■■ DIET FOLLOWING GASTRIC PARTITIONING After surgery, clients typically follow a mechanical soft diet and are cautioned to chew foods thoroughly to reduce the risk of dislodging surgical staples and to prevent foods from obstructing the opening into the lower stomach. After about three months, clients can generally begin to eat solid foods. Regardless of the food consistency, clients tolerate only small amounts of food (about a cup of food at a time); overeating or overdrinking can cause nausea, reflux, and vomiting. Although the reasons are unclear, clients undergoing gastric bypass surgery frequently report that foods taste sweeter to them than before surgery; others develop an aversion to red meat.[5]

FOOD SELECTIONS Clients must understand that their selections of foods and beverages influence the extent of their weight loss. If they drink high-kcalorie liquids or eat high-kcalorie foods all the time, even if only in small quantities, they will not lose weight. Their selections also affect the nutritional quality of the diet. Nutrient deficiencies, particularly of vitamin B_{12}, iron, and vitamin D, are common. Careful planning, diligent compliance with the prescribed diet, and vitamin-mineral supplements help ensure the adequacy of a diet that strictly limits energy intake.

Indigestion, reflux esophagitis, nausea, and vomiting are frequent GI problems that can lead to weight loss and malnutrition. When treatment for a medical condition requires gastric surgery, the impact on nutrition status can be severe. The accompanying Nutrition Assessment Checklist highlights nutrition-related problems to watch for when working with clients with disorders of the upper GI tract, and the Diet-Medication Interactions box discusses the medications mentioned in this chapter.

NUTRITION ASSESSMENT CHECKLIST — FOR PEOPLE WITH UPPER GI TRACT DISORDERS

Health Problems, Signs, and Symptoms

Does the client's health history reveal medical conditions or treatments that:

☐ Interfere with chewing and swallowing?
☐ Result in indigestion, nausea, or vomiting?

Does the client have a medical diagnosis of:

☐ Hiatal hernia or reflux esophagitis?
☐ Gastritis?
☐ Pernicious anemia?
☐ Peptic ulcers?

For a client who has undergone gastric surgery, check for the following complications:

☐ Dumping syndrome
☐ Malabsorption
☐ Anemia
☐ Bone disease

Medications

Check the medication and dosing schedule for:

☐ Medications that may cause indigestion, nausea, and vomiting. Note that many medications can cause nausea, especially the first few doses. Remember to give medications with food, when possible, to help alleviate nausea.
☐ Clients receiving tetracycline for *Helicobacter pylori* infections. Do not give tetracycline one hour before or two hours after giving milk and milk products or antacids. Do not give tetracycline two hours before or three hours after giving iron supplements.
☐ Clients receiving antisecretory agents and iron supplements. Give the antisecretory agent two hours before or after the iron supplement.

Nutrient/Food Intake

Confer with clients and the dietitian to help pinpoint food intolerances for clients with:

☐ Indigestion/nausea
☐ Active reflux esophagitis, gastritis, or peptic ulcers
☐ Dumping syndrome

For clients on long-term mechanical soft diets, regularly note:

☐ Appetite
☐ Variety of foods being offered and eaten
☐ Consistencies the client can handle

Height and Weight

Measure baseline height and weight. Address weight loss early to prevent malnutrition for clients with:

☐ Any condition requiring mechanical soft diets for long periods of time
☐ Dysphagia
☐ Indigestion and nausea of long duration
☐ Dumping syndrome/malabsorption

Laboratory Tests

Check laboratory tests for signs of dehydration for clients with:

☐ Persistent vomiting

☐ Dumping syndrome

Check laboratory tests for nutrition-related anemias (see Appendix E) for people with:

☐ Gastritis

☐ Gastric surgeries

☐ Conditions that require long-term use of antisecretory agents

Physical Signs

Look for physical signs of:

☐ Dehydration (especially for people with persistent vomiting or dumping syndrome)

☐ Iron deficiency (especially for people who have had gastric surgery or those on antisecretory agents)

☐ Vitamin B_{12} deficiency (especially for people with chronic gastritis or those who have undergone gastric partitioning)

DIET-MEDICATION INTERACTIONS

Antacids

Aluminum-containing antacids contribute aluminum to the diet and may cause constipation or lead to phosphorus deficiency. Long-term or inappropriate use can lead to aluminum toxicity.

Calcium-containing antacids contribute calcium to the diet and may cause constipation. Concurrent use with vitamin D supplements or foods containing large amounts of vitamin D can lead to elevated blood calcium levels (hypercalcemia).

Magnesium-containing antacids contribute magnesium to the diet and may cause constipation; long-term use may lead to magnesium toxicity.

Sodium bicarbonate, an antacid that contributes sodium to the diet, can alter serum electrolyte levels and raise the pH of the blood. Sodium bicarbonate should not be taken with milk—hypercalcemia can result.

Antibiotics (for *Helicobacter pylori* infections)

When *amoxicillin* is given without regard to food, nausea and diarrhea are common side effects. Encourage clients to take with food to reduce nausea.

Metronidazole may cause taste alterations (metallic taste), and no alcohol should be used during treatment and for 24 hours afterward. Alcohol can react with metronidazole and result in a disulfram-like reaction with symptoms that include nausea, vomiting, headache, cramps, and flushing of the skin.

The classic interactions of nutrients and *tetracycline* were mentioned in Chapter 13. Calcium, magnesium, zinc, aluminum, antacids that contain any of these nutrients, and vitamin-mineral supplements should not be given from one hour before to two hours after a tetracycline dose. Iron supplements should not be given from two hours before to three hours after a tetracycline dose. Tetracycline can also cause nausea and diarrhea.

Antiemetics, antinauseants

The nutrition-related side effects of *antiemetics* vary. Two antiemetics, *dronabinol* (a derivative of marijuana) and *prochlorperazine*, stimulate the appetite and can result in weight gain. Prochlorperazine also increases the urinary excretion of riboflavin. Prochlorperazine, *granisetron,* and *odansteron* may cause constipation.

Metoclopramide can cause lethargy but only rarely causes GI side effects.

Antisecretory agents

Antisecretory agents may interfere with iron absorption. When iron supplements are necessary, they should be given two hours before or after taking these medications. People taking some antisecretory agents should avoid the use of the herb pennyroyal; antisecretory agents may increase the formation of toxic metabolites from pennyroyal.

SELF CHECK

1. The nurse working with a client on a mechanical soft diet recognizes that:
   a. only pureed food should be given to minimize the risk of dysphagia.
   b. the client can have any food that can be safely chewed and swallowed.
   c. highly seasoned foods are always restricted.
   d. such diets cannot be planned to meet total nutrient needs.

2. For a person with dysphagia:
   a. the diet requires a great deal of experimentation to uncover which foods the person can tolerate.
   b. the diet is based on a health care team's assessment of the person's swallowing abilities and close monitoring of the person's ability to handle different foods.
   c. meals should be eaten in bed with the head tilted back.
   d. coughing during meals indicates that the person is able to clear the throat and is not at risk for aspiration.

3. Reflux esophagitis is:
   a. an inflammation of the esophagus caused by the backflow of acidic gastric fluids from the stomach.
   b. a protuberance of a portion of the stomach above the cardiac sphincter.
   c. an erosion of the lining of the esophagus.
   d. an obstruction of the lower esophagus that results in dysphagia.

4. The nurse working with a client who has a hiatal hernia and reflux esophagitis recognizes that the client understands her diet when she says:
   a. "I need to eat three meals a day and drink liquids with my meals."
   b. "I need to eat foods slowly, relax during mealtimes, and lie down after meals."
   c. "I need to drink more citrus juices and eat more tomato-based products."
   d. "I need to limit my intake of fat, alcohol, caffeine, and decaffeinated coffee and tea."

5. For the client with persistent vomiting, the major nutrition-related concern(s) are:
   a. dehydration and malnutrition.
   b. reflux esophagitis.
   c. emotional distress.
   d. peptic ulcers.

6. Liberal bland diets help clients during times when gastritis or ulcers are active because they eliminate foods that:
   a. cause dysphagia.
   b. significantly increase gastric acid production.
   c. are difficult to chew and swallow.
   d. contain simple sugars.

7. Nutrition concerns most commonly associated with chronic gastritis include:
   a. malnutrition and pernicious anemia.
   b. iron-deficiency anemia and protein malabsorption.
   c. dumping syndrome and pernicious anemia.
   d. dehydration and electrolyte imbalances.

8. A bland diet and antisecretory agents are recommended for a client with peptic ulcers. The client complies with the suggestions in Table 15-3, but continues to complain of indigestion and gastric pain. The most appropriate advice would be to:
   a. try a mechanical soft diet.
   b. eat large meals.
   c. keep a record of food intake, gastric symptoms, and when symptoms occur.
   d. drink milk with each meal.

9. Which of the following snacks would be an appropriate choice for a client on a postgastrectomy diet?
   a. milk shake
   b. saltine crackers with peanut butter
   c. cookies
   d. eggnog

10. The nurse assessing a client who underwent a gastrectomy three years ago should be alert to signs of:
    a. problems chewing food.
    b. vitamin C deficiency.
    c. iron-deficiency anemia.
    d. individual intolerances.

Answers to these questions appear in Appendix H.

CLINICAL APPLICATIONS

1. People on mechanical soft diets differ in the kinds of foods they can handle, in the lengths of time that diet modifications remain necessary, and in the help they need from health care professionals. Think about the difference between working with a person who has had no teeth for years and a person who recently had mouth surgery and is just beginning to eat again. Describe some nutrition-related concerns you might have for the person who has been following a mechanical soft diet for years. How would these concerns differ for a person who needs the mechanical soft diet only temporarily? Contrast the amounts of time a nurse or dietitian might need to spend with the two types of clients.

2. Many of the diets described in this chapter are highly individualized. A particular food may cause discomfort for one person and have no effect on another. Describe practical ways to keep track of food intolerances.

3. Review the chapter to find disorders that often occur as a consequence of aging. Referring to Chapter 12, describe the effects of aging on the upper GI tract and relate these changes to the disorders you find.

NUTRIT**ON**THENET

FOR FURTHER STUDY OF THE TOPICS IN THIS CHAPTER, ACCESS THESE WEB SITES.

www.dysphagia.com
Dysphagia Resource Center

www.gerd.com
GERD Information Resource Center

www.gutfeelings.com
Gut Feelings

www.helico.com
The Helicobacter Foundation

www.acg.gi.org
American College of Gastroenterology

www.hoptechno.com/book35.htm
Peptic Ulcer (U.S. Department of Health and Human Services)

Notes

[1] J. Lagergren and coauthors, Symptomatic gastroesophageal reflux as a risk factor for esophageal adenocarcinoma, *New England Journal of Medicine* 340 (1999): 825–831.

[2] W. H. Aldoori and coauthors, Prospective study of diet and the risk of duodenal ulcer in men, *American Journal of Epidemiology* 145 (1997): 42–50.

[3] J. Grant, G. Chapman, and M. K. Russell, Malabsorption associated with surgical procedures and treatments, *Nutrition in Clinical Practice* 11 (1996): 43–52.

[4] F. P. Kennedy, Medical management of patients following bariatric surgery, presented at the Eight Annual Advances and Controversies in Clinical Nutrition, sponsored by the Mayo Clinic Foundation, Dallas, Texas, April 5–7, 1998.

[5] J. C. Burge and coauthors, Changes in patients' taste acuity after Roux-en-Y gastric bypass for clinically severe obesity, *Journal of the American Dietetic Association* 95 (1995): 666–670.

Q&A NUTRITION IN PRACTICE

Helping People with Feeding Disabilities

Chapter 15 described problems encountered when people have difficulties with chewing and swallowing. This Nutrition in Practice discusses a broader problem faced by thousands of people who must cope with disabilities that interfere with the process of eating, including some that interfere with chewing and swallowing. These obstacles can arise at any time in a person's life and from any number of conditions. An infant may be born with a physical impairment such as cleft palate; an adolescent may suffer nerve damage from injuries sustained in a car accident; a middle-aged adult may lose motor control following a stroke; an older adult may struggle with the pain of arthritis or the mental deterioration of dementia. Table NP15-1 lists some of the conditions that may lead to feeding problems.

In what ways can disabilities impair eating and nutrition status?

To get food from the table to the stomach requires an amazing number of individual coordinated motions. Consider an infant just beginning to learn to feed himself. At first, the infant cannot sit upright and finds it impossible to hold a spoon. Every single little action—from picking up the utensil, to biting and chewing, to swallowing—requires coordinated movements. Any injury or disability that interferes with these movements can lead to feeding problems.

Disabilities may have nutrition-related consequences beyond their effects on the process of eating. For example, a person who has problems with sight, or who cannot drive or walk or carry groceries, or who cannot plan meals and think through what to buy, has a disability that affects eating. Disabilities of any type can cause people to have trouble maintaining adequate nutrition status.[1] Their number one nutrition problem is inadequate food intake, which leads to malnutrition, underweight, and, in children, poor growth.[2] Children and adolescents with severe disabilities are especially vulnerable to these problems.

A disability can also affect energy needs. For example, a disability may make it difficult for a person to engage in enough physical activity to support a healthy appetite or, conversely, to burn energy. Then the person's nutrient intake may either fail to meet nutrient needs or result in excessive weight gain. As another example, a person who has lost a limb to amputation has altered energy needs. Energy needs are reduced in proportion to the weight and metabolism represented by the missing limb, but may be increased if extra energy is necessary to do ordinary activities, such as walking on crutches. As still another example, people with involuntary motor activity may have exceedingly high energy needs.

In addition to the nutrition risks imposed by the disability itself, many conditions that lead to feeding problems require medications, which may also affect nutrition status. Furthermore, people who have feeding disabilities often encounter emotional and social problems. For example, children fail to receive the social training that mealtimes provide, and older people miss the social stimulation that goes with eating in the company of others.

How can health care professionals promote independent eating for people with disabilities?

The evaluation and treatment of feeding problems ideally involve the joint efforts of several health care professionals, including a dietitian, a psychologist, an occupational therapist, a physical therapist, a speech-language pathologist, a dentist, and one or several nurses. Table NP15-2

Table NP15-1
CONDITIONS THAT MAY LEAD TO FEEDING PROBLEMS

The following conditions may lead to feeding problems by interfering with a person's ability to suck, bite, chew, swallow, or coordinate hand-to-mouth movements:

- Accidents
- Amputations
- Arthritis
- Birth defects
- Cerebral palsy
- Cleft palate
- Down's syndrome
- Head injuries
- Huntington's chorea
- Hydrocephalia

- Language, visual, or hearing impairment
- Microcephalia
- Multiple sclerosis
- Muscle weakness
- Muscular dystrophy
- Neuromotor dysfunction
- Parkinson's disease
- Polio
- Spinal cord injuries
- Stroke

Adaptive feeding equipment can help clients with disabilities gain independence.

Table NP15-2
FEEDING EVALUATION

Oral Ability

Sucking

Oral prehension

Swallowing

Breathing and swallowing coordination

Drooling

Lips

Tongue size, thrust, mobility

Biting

Munching

Chewing

Drinking

Response to input

Dental Health

Structural malformation

Impaired oral-motor ability

Delayed weaning

Use of cariogenic foods or medications

Occlusion

Teeth

Caries

Gingiva

Oral hygiene

Palate

Pain on exam

Hypersensitivity

Teething stage

Body Position

General tone and movement

Reflex activity

Head control

Sitting balance

Placement of feet

Usual feeding position

Hand Use

Palmar grasp

Pincer grasp

Opposition finger/thumb

Hand-to-mouth control

Developmental Feeding

Breast _____ bottle _____ weaned _____

Baby food _____ junior food _____

mashed table food _____ minced foods _____

Cut table foods _____ regular table foods _____

Closes hands in on bottle

Hand to mouth/sucks on fingers

Teething biscuit, holds and brings to mouth

Finger feeds

Opposes lips to rim of cup

Attempts to grasp spoon

Grasps spoon

Dips spoon in dish

Brings spoon to mouth

Holds bottle and drinks independently

Grasps cup

Raises cup to mouth

Lifts cup, drinks, and replaces

Scoops well with spoon

Feeds independently with spoon

Drinks with straw

Spears with fork

Spreads with knife

Feeding Environment

Time of feedings (note number and length of feedings)

Atmosphere of feedings (tense, pleasant, unpleasant)

Person responsible for feeding

Parental and/or caregiver's attitude toward feeding

Past successful and unsuccessful methods

Positive and negative reinforcing behaviors

Behavioral problems

Diet History, Including:

Total fluid intake

Types of foods consumed

Method of feeding

Texture modifications

Food aversions, intolerances, or allergies

Use of food as rewards

Medications

Type

Dosage

Time given

Bowel Concerns

Regular

Constipation

Diarrhea

SOURCES: Adapted from J. J. Cafferky, Nutrition assessment of children with developmental disabilities: Special considerations, *Support Line* (a newsletter of Dietitians in Nutrition Support), June 1993; R. B. Howard, Nutritional support of the developmentally disabled child, *Textbook of Pediatric Nutrition*, ed. R. M. Suskind (New York: Raven Press, 1981), pp. 577–582. By permission.

shows the multiple observations health care professionals use to assess feeding skills and nutrition needs. Because each case is different, health care professionals with a thorough understanding of the many variables that affect eating can best provide effective intervention.

What roles do the various team members play in dealing with feeding problems?

The physician diagnoses the client's medical problems, prescribes treatments and medications, and coordinates the health care team. The nurse identifies a plan to address the client's health needs and educates the client and caregivers about the plan. The dietitian assesses the

client's nutrition status, plans a diet, monitors medications for possible diet-medication interactions, and provides nutrition counseling.[3] A speech-language pathologist most often evaluates a client's ability to chew and swallow and trains the client to use lips, tongue, and throat for eating and speaking. The occupational therapist may work with the client to evaluate the need for special feeding devices and then show the client how to use appropriate devices. A dentist may evaluate the client's dental health and provide instructions for maintaining oral hygiene. The health care team faces many challenges in solving feeding problems, but the personal satisfaction of seeing someone gain control of eating can be very rewarding.

How does the process of teaching someone to eat independently begin?

Direct observation of the client during mealtimes allows the health care team to teach the appropriate feeding techniques, monitor the client's or caregiver's understanding of the techniques, and evaluate how well the care plan is working. Table NP15-3 provides examples of strategies health care professionals can use to tackle different feeding and nutrition-related problems. For example, a child with a feeding problem may be overly sensitive to oral stimulation. In this case, the helper uses strategies to desensitize the client to oral stimulation. The helper may start by gently stroking

Table NP15-3
FEEDING AND NUTRITION-RELATED PROBLEMS AND SUGGESTED INTERVENTIONS

Inability to Suck
- Use cold substances around lips to stimulate sucking.
- Use a cloth soaked with water for child to suck.
- Try different types of nipples.
- As child's ability begins to improve, change to nipple with a smaller hole.

Inability to Chew
- Place a small amount of food between back teeth and move jaw up and down. A mirror may help demonstrate and point out various body parts.
- Place foods such as peanut butter on lips and encourage client to wash lips with tongue. Gradually change from pureed foods to solid foods (sprinkle crackers in soup, etc.).

Inability to Swallow
- Close jaw and lips of client together (swallowing is easiest when the mouth is closed).
- Stroke throat upward under chin.
- Offer next bite of food only after client swallows.
- Demonstrate—let client feel *you* swallow.

Inability to Grasp
- Allow client to finger feed.
- Guide client in exploring mouth.
- Cut food into small pieces.
- Place your hand over client's hand and help client to grasp spoon.
- Use adaptive equipment.
- Make sure bowl is stabilized (suction, tape).
- Use plates with high straight sides, or build higher edge using aluminum foil.

Poor Hand-Mouth Coordination
- Pour sand, etc.
- Exercise with ball.
- Exercise with push-pull objects.
- Study body parts with client, if appropriate.

Impaired Vision
- Place meats and vegetables consistently in same areas of plate.

Overweight
- Cut down snacks and high-kcalorie foods.
- Refrain from rewarding with food.
- Increase exercise and leisure-time activities.

Underweight
- Increase number of meals per day.
- Include high-kcalorie foods, especially liquid supplements.
- Encourage proper exercise.

Lack of Nutrition Education
- Work with families.
- Stress the importance of proper nutrition for *all* family members.
- Teach proper feeding environment (good eating habits, eating positions).
- Provide nutrition instruction materials.

SOURCE: Adapted with permission from S. Calvert and F. Davies, Nutrition of children with handicapping conditions, *Dietetic Currents* 4 (1977): 13–17.

the client's face with a hand, washcloth, or soft pliable toy. (This can be done playfully, making a game of it.) When the child can tolerate touch on less-sensitive areas of the face such as the forehead, cheeks, and lips, then the helper may begin to slowly and firmly rub the gums, palate, and tongue. With time, the client may be better able to tolerate the presence of food in the mouth.

What special feeding devices are available for helping people to eat?

Figure NP15-1 shows a few of many special feeding devices and describes

Figure NP15-1

EXAMPLES OF ADAPTIVE FEEDING DEVICES

Utensils

Rocker knife

Roller knife

People with only one arm or hand may have difficulty cutting foods and may appreciate using a *rocker knife* or a *roller knife*.

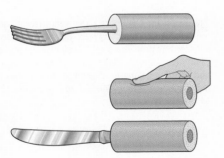

People with a limited range of motion can feed themselves better when they use *flatware with built-up handles*.

People with extreme muscle weakness may be able to eat with a *utensil holder*.

For people with tremors, spasticity, and uneven jerky movements, *weighted utensils* can aid the feeding process.

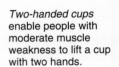

Battery-powered feeding machines enable people with severe limitations to eat with less assistance from others.

Plates

People who have limited dexterity and difficulty maneuvering food find *scoop dishes* or *food guards* useful.

People with uncontrolled or excessive movements might move dishes around while eating and may benefit from using *unbreakable dishes with suction cups*.

Cups

People with limited neck motion can use a *cutout plastic cup*.

Two-handed cups enable people with moderate muscle weakness to lift a cup with two hands.

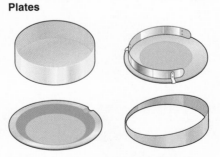

People with uncontrolled or excessive movements might prefer to drink liquids from a *covered cup* or glass with a *slotted opening* or *spout*.

A soft, flexible long plastic straw may also ease the task of drinking.

their uses. These devices can make a remarkable difference in a person's ability to eat independently. For a person who cannot grasp an ordinary fork, for example, a special fork may be the key to future health.

What happens if the person with a feeding disability is unable to learn to eat enough food after training?

Sometimes, despite the best efforts of all involved, the client still can't eat enough food by mouth or may experience complications such as frequent choking or vomiting. Delivering feedings through tubes placed directly in the stomach (or intestine) can improve nutrition status and ease problems with choking or vomiting.[4]

With so many things to learn and practice, how do clients and caregivers cope?

The time and patience required to learn and handle the many tasks required to care for a person with disabilities often result in a great deal of frustration and distress for both the client and the caregiver. A child with cerebral palsy, for example, may take ten times longer to eat than a child without disabilities. Some mothers have reported spending up

to seven hours a day feeding these children. Besides feeding, the caregiver must often cope with many other tasks that also require a great deal of time. Caregivers may feel overwhelmed with responsibility and have little time to care for themselves and other family members. Psychologists can offer counseling to clients or caregivers to help them adjust; all members of the health care team can offer emotional support and practical suggestions to ease caregivers' responsibilities and frustrations.

www.eatright.org/adap0297b.html
American Dietetic Association Position Paper on Nutrition in Comprehensive Program Planning for Persons with Developmental Disabilities

www.wvu.edu/~uacdd/nutrition/nutrition.htm
University Affiliated Center for Developmental Disabilities

Successful therapy for people with feeding disabilities requires the involvement of many health care professionals and depends on the accurate identification of impaired feeding skills and appropriate interventions. Ideally, with training people with disabilities attain total independence—they are able to prepare, serve, and eat nutritionally adequate food daily without help. In some cases, this goal can be attained with the help of caregivers. The combined efforts of the health care team can support both clients and caregivers in enhancing quality of life and in achieving independence to the greatest degree possible.

Notes

[1]Position of The American Dietetic Association: Nutrition services for children with special needs, *Journal of the American Dietetic Association* 95 (1995): 800–812; Position of The American Dietetic Association: Nutrition in comprehensive program planning for persons with developmental disabilities, *Journal of the American Dietetic Association* 92 (1992): 613–615.

[2]R. D. Stevenson, Feeding and nutrition in children with developmental disabilities, *Pediatric Annals* 24 (1995): 225–260.

[3]H. H. Cloud, Expanding roles for dietitians working with persons with developmental disabilities, *Journal of the American Dietetic Association* 97 (1997): 129–130.

[4] R. Tawfik and coauthors, Caregivers' perceptions following gastrostomy in severely disabled children with feeding problems, *Developmental Medicine and Child Neurology* 39 (1997): 746–751; G. Björnestam and coauthors, Long-term home jejunostomy feeding of young children, *Nutrition in Clinical Practice* 14 (1999): 247–249.

CHAPTER 16

NUTRITION AND LOWER GI TRACT DISORDERS

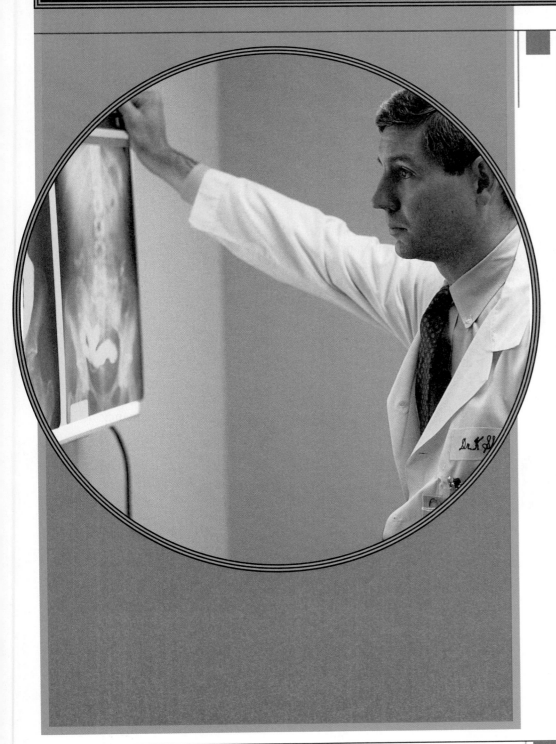

CONTENTS

Common Problems of the Lower Intestine

Fat Malabsorption

Disorders of the Large Intestine

Case Study: College Student with Crohn's Disease

Nutrition in Practice: Food and Foodservice in the Hospital

As Chapter 5 noted, the intestine is "the" organ of digestion. Some health problems and treatments affect the rate at which food passes through the intestine; others affect the absorption of nutrients.

Common Problems of the Lower Intestine

Several common GI tract disorders—constipation, diarrhea, irritable bowel syndrome, and lactose intolerance—are characterized by changes in the rate at which food passes through the intestine. In each case, medical nutrition therapy may help alleviate the discomforts clients experience.

Constipation

constipation: the condition of having infrequent or difficult bowel movements.

Constipation describes a symptom, not a disease. Each person's GI tract has its own cycle of waste elimination, which depends on its owner's physical makeup, the type of food eaten, when it was eaten, and when the person makes time to defecate. For some people, bowel movements occur daily; for others, several days may pass between movements. Only when people pass stools that are difficult or painful to expel or when they experience a reduced frequency of bowel movements from their typical pattern are they constipated. Abdominal discomfort, headaches, backaches, and the passing of gas sometimes accompany constipation.

CAUSES OF CONSTIPATION

Constipation may be associated with fluid and electrolyte imbalances, hormonal imbalances, certain diseases of the GI tract, chronic laxative abuse, stress, lack of physical activity, and a variety of medications, including narcotic analgesics, anticholinergics (which reduce GI tract motility), aluminum-containing antacids, and some antihypertensives. In addition, herbal preparations, including chondrotin sulfate and glucosamine, can lead to constipation.[1] Expectant mothers frequently experience constipation as the growing fetus crowds intestinal organs and hormonal changes alter intestinal muscle tone. Constipation is a problem for many elderly people, particularly those in extended care facilities. Some people, especially elderly people, are bothered by problems with constipation and take laxatives to treat it, although objective measurements of stool consistency and frequency fail to confirm constipation.[2]

TREATMENTS FOR CONSTIPATION

When constipation occurs as a result of an underlying medical condition, treatment of that disorder may alleviate the problem. Medical nutrition therapy includes following a high-fiber diet, drinking plenty of fluids, eating prunes, drinking prune juice, and engaging in regular physical activity. Ideally, laxatives, especially bulk-forming agents (**hydrophilic colloids**), are prescribed only when other measures fail to relieve constipation (see the Diet-Medication Interactions box on pp. 398–399).

hydrophilic colloids: laxatives composed of fibers that work like dietary fibers. They attract water in the intestine to form a bulky stool, which then stimulates peristalsis. Metamucil and Fiberall are examples.

THE NURSING DIAGNOSIS constipation or **perceived constipation** cues the nurse to check the diet pattern for fiber and fluid intake, the diet-medication interactions associated with laxative use, and the level of physical activity.

HIGH-FIBER DIETS

As Chapter 2 described, high-fiber foods include whole-grain breads and cereals, fruits, vegetables, and legumes. High-fiber diets help maintain a healthy GI tract by adding volume and weight to the stool, normalizing the transit of undigested materials through the intestine, and minimizing pressure within the colon.

■■ **INTESTINAL GAS** An uncomfortable side effect of high-fiber diets is excessive gas production, which can cause a bloated feeling, abdominal pain, and **flatus.** Undigested and unabsorbed dietary fibers may pass into the colon where bacteria may metabolize them and produce gas in the process. Thus, many high-fiber foods, including legumes, some grains, fruits, and vegetables, are also gas-forming foods (see Table 16-1). Note that in addition to fibers, any undigested and unabsorbed food can cause intestinal gas. Thus, high-fat foods, fructose, and sugar alcohols (sorbitol, mannitol, xylitol, and maltitol) taken in large amounts may be incompletely absorbed and cause gas for some people. Unabsorbed lactose can cause gas for people with lactose intolerance (described later in this chapter). The "How to" box on the next page offers suggestions for helping clients adjust their fiber intakes so that they will experience minimal discomfort from gas. Some clients take antiflatulents to help control intestinal gas production.

LEARNING LINK

Many factors contribute to the high incidence of constipation, as well as a high rate of laxative use, in the elderly. With aging, the intestinal wall loses strength and elasticity, and gastrointestinal motility slows. In addition, compared to others, the elderly person is:

- Less likely to experience thirst in response to dehydration.
- Less likely to be physically active, especially if living in an extended care facility.
- More likely to use medications that can contribute to constipation.

Furthermore, adding fiber to the diet may be more difficult for many elderly people. Changing food habits that have been established for many years is not easy, and for those with missing teeth or poorly fitted dentures, chewing high-fiber foods may be difficult. Thus, the elderly person may rely on laxatives to alleviate constipation. When working with elderly clients with constipation, help them find high-fiber foods they can chew and enjoy, emphasize drinking water and other fluids at regular times throughout the day, and encourage physical activity.

■■ **Diarrhea and Dehydration**

Diarrhea occurs either when fluids are drawn from the GI tract lining and added to the food residue or when the intestinal contents move so quickly through the GI tract that fluids are not absorbed. In both cases, the result is frequent, watery bowel movements.

Diarrhea is not a disease, but a symptom of many medical conditions and a complication of some medical treatments, including many medications. It can be acute, lasting less than two weeks, or chronic, lasting longer. Mild diarrhea that remits in 24 to 48 hours is seldom a cause for concern unless the person is already dehydrated. A person with severe, persistent diarrhea may rapidly become dehydrated, lose weight, and develop multiple nutrient deficiencies. A child or infant can lose proportionately more fluid and weight than an adult and can develop severe dehydration and malnutrition in a short time. Table 13-10 in Chapter 13 listed physical findings associated with dehydration.

THE NURSING DIAGNOSES diarrhea alerts the nurse to the potential for fluid and electrolyte imbalances and nutrient deficiencies.

Table 16-1
FOODS OR SUBSTANCES THAT MAY PRODUCE GAS

Apples	High-fat meats
Asparagus	Honey
Beer	Legumes (dried beans and peas)
Bran	
Broccoli	Mannitol
Brussels sprouts	Milk and milk products
Cabbage	
Carbonated beverages	Nuts
	Onions
Cauliflower	Peppers, green
Corn	Prunes
Cream sauces	Radishes
Cucumbers	Raisins
Fried foods	Sorbitol
Fructose	Soybeans
Gravy	Wheat

flatus (FLAY-tuss): gas in the intestinal tract or the expelling of gas from the intestinal tract, especially through the anus.

diarrhea: the frequent passage of watery bowel movements.

Beans, broccoli, cabbage, and onions produce gas in many people. People troubled by gas need to determine which foods bother them and then eat those foods in moderation.

During the first few weeks on a high-fiber diet, a person may feel bloated, pass gas frequently, or experience heartburn. These suggestions can help:

ADD FIBER TO THE DIET

- *Go slow.* Add high-fiber foods gradually and in small portions at first. Increase portion sizes and add foods as tolerance improves.
- *Add fluids.* Fiber attracts water as it moves through the intestine.
- *Experiment.* Try small servings of various fiber-containing foods at first and adopt those that are most pleasing.
- *Mix high-fiber foods with other foods.* Sprinkle bran flakes, wheat germ, or raisins on salads or applesauce. Add bran or mashed legumes to meat loaf. Add legumes and other high-fiber vegetables to soups and salads.

Secretory diarrhea results from an acclerated movement of fluids and electrolytes from the intestinal capillaries into the lumen of the intestine.

Osmotic diarrhea results from an increase in the osmolality of the intestinal contents due to unabsorbed nutrients or drugs. The diarrhea associated with the dumping syndrome (see Chapter 15) is an example.

Severe, chronic diarrhea that does not respond to treatment is intractable diarrhea.

CAUSES OF DIARRHEA Acute diarrhea that occurs abruptly in a healthy person frequently results from viral, bacterial, or protozoal infections or as a side effect of medications or herbal remedies. It can also occur in a person who begins to eat foods or begins a tube feeding after a period of fasting, starvation, or intravenous nutrition. Infants frequently develop diarrhea when given formulas their immature GI tracts cannot tolerate or when they are ill. When used in large quantities, food additives such as sorbitol and olestra may cause diarrhea in some people.

Chronic diarrhea can occur as a result of disorders that alter GI tract motility. Such disorders include irritable bowel syndrome, lactose intolerance, and some infections, including those caused by some parasites and human immunodeficiency virus (HIV).

TREATMENTS FOR DIARRHEA The treatments for diarrhea require resolving the primary medical condition. If a food is responsible, that food must be omitted from the diet. If a medication is responsible, a different medication or different form of medication (injectable versus oral, for example) may alleviate the problem. Infections are treated with appropriate anti-infective agents. Antidiarrheals slow GI motility and are often recommended along with other therapies to treat diarrhea (see the Diet-Medication Interactions box on pp. 389–399).

FLUID AND ELECTROLYTE REPLACEMENT Until the diarrhea is resolved, treatment includes replacement of lost fluids and electrolytes to prevent dehydration. For mild cases of diarrhea, fluids and electrolytes can be replaced using fruit juices, sports drinks, caffeine-free carbonated beverages, tea, and broth with crackers. In mild-to-moderate cases of diarrhea, oral rehydration formulas—simple solutions of water, salts, and sugar—provide needed fluids and electrolytes (see Nutrition Practice 9). In severe cases, it may become necessary to stop placing demands on the GI tract by withholding all foods and beverages until the diarrhea remits, usually in a day or two. During bowel rest, intravenously administered fluids and electrolytes replace losses.

MEDICAL NUTRITION THERAPY Often people with diarrhea can tolerate solid foods, although they may benefit from temporarily avoiding highly seasoned foods, foods high in fat, foods that cause gas, lactose-containing foods, caffeinated beverages, and any food that aggravates the diarrhea. If eating aggravates diarrhea, clients may be advised to drink only clear liquids (including oral rehydration formulas). Once the diarrhea remits, the diet gradually advances to full liquids, to soft foods, and then to regular foods as tolerated. The diet temporarily excludes lactose and caffeine. Frequent small meals are easiest to tolerate at first. Permanent dietary changes may be necessary for diarrhea caused by food sensitivities or allergies. For some people, adding fiber to the diet may reduce diarrhea because some fibers absorb water and help regulate transit time through the intestine.

Irritable Bowel Syndrome

Irritable bowel syndrome is a common motility disorder characterized by chronic abdominal pain associated with diarrhea, constipation, or alternating episodes of both.[3] The person may also experience bloating, flatulence, the passage of mucus with stools, indigestion, and nausea. Symptoms frequently occur shortly after a person eats and resolve temporarily following a bowel movement.

In the United States, irritable bowel syndrome affects about 5 million people annually and occurs more frequently in women.[4] Episodes of active symptoms often decrease with age.

CAUSES OF IRRITABLE BOWEL SYNDROME
The causes of irritable bowel syndrome remain elusive, but stress and anxiety are believed to be contributing factors. Other contributing factors may include abnormal GI tract motility and hypersensitivity to intestinal distention. Foods and food components, including gas-forming foods (review Table 16-1), high-fat foods, fat substitutes, lactose, fructose, sorbitol, caffeine, and alcohol, may aggravate the symptoms of irritable bowel syndrome, but are not believed to cause it. Medications and herbal remedies can also aggravate diarrhea or lead to constipation.

TREATMENTS FOR IRRITABLE BOWEL SYNDROME
Medical therapy for irritable bowel syndrome frequently includes stress management along with medical nutrition therapy. Drug therapy, often reserved for cases that do not respond to stress management and medical nutrition therapy, may include antidepressants, anticholinergics, antidiarrheals, and laxatives (see Diet-Medication Interactions box on pp. 398–399).

MEDICAL NUTRITION THERAPY
To help minimize abdominal distention, improve digestion, and reduce the chance of swallowing air with foods (which can contribute to abdominal distention and gas in the intestine), clients with irritable bowel syndrome are advised to avoid eating too much or too fast and to chew foods thoroughly. Clients respond uniquely to specific dietary interventions for irritable bowel syndrome, and no one diet is suitable for everybody. Food records that include food intake, fluid intake, stool consistencies, other GI symptoms, and stress levels can help identify individual food intolerances.

Many clients benefit from a high-fiber, fat-restricted diet. The dietitian often adjusts the types of fiber (soluble and insoluble) depending on whether the client experiences primarily constipation or diarrhea and on whether gas is a problem. Some clients tolerate the fiber from hydrophilic colloids better than dietary fibers. Highly seasoned foods, lactose, fructose, sorbitol, caffeine, and alcohol are restricted to the extent that they cause discomfort for the individual.

Lactose Intolerance

Unlike diarrhea and constipation, which can occur as a consequence of many health-related problems, the cause of lactose intolerance is clear. It results from

irritable bowel syndrome: an intestinal disorder of unknown cause. Symptoms include abdominal discomfort and cramping, diarrhea, constipation, or alternating diarrhea and constipation.

Many people with irritable bowel syndrome benefit from eating high-fiber foods.

a deficiency of lactase, the enzyme that splits the disaccaharide lactose into the two monosaccharides glucose and galactose.

CAUSES OF LACTASE DEFICIENCY In rare cases, a person is born with a lactase deficiency; more often, lactase activity gradually diminishes with age. Lactose intolerance is prevalent among people in certain ethnic groups including those of Mediterranean origin, African Americans, Asians, Jews, Mexicans, and Native Americans. Permanent or temporary lactase deficiencies can develop as a consequence of any disorder or condition that damages the delicate intestinal microvilli including malnutrition, radiation therapy, and many of the conditions described later in this chapter.

Vigorous advertising campaigns to promote products for lactose intolerance have led many people to believe they have the condition, when, in fact, they do not.[5] Some believe they are intolerant to even the smallest amounts of lactose. These people may unnecessarily eliminate milk and milk products from their diets and inadvertently develop calcium and vitamin D deficiencies in the process.

SYMPTOMS OF LACTOSE INTOLERANCE Undigested lactose causes the intestinal contents to become hypertonic, which results in cramps, distention, and diarrhea. Bacteria in the intestine metabolize the undigested lactose to irritating acids and gases, which further contribute to cramping, diarrhea, and flatulence. The severity of the symptoms varies among individuals.[6]

TREATMENTS FOR LACTOSE INTOLERANCE Medical nutrition therapy provides relief from the symptoms of lactose intolerance. Lactose-restricted diets are highly individualized and most often limit, but do not exclude, milk and milk products. Dietitians advise clients to test their tolerance for lactose by gradually increasing consumption of lactose-containing foods to the point that precipitates symptoms of lactose intolerance. Many lactose-intolerant individuals can tolerate 1 to 2 cups of milk a day without significant symptoms, provided that the milk is taken with food and intake is divided throughout the day.[7] Some tolerate chocolate milk better than plain milk. Often people can eat cheeses, particularly aged cheeses.[8] Most tolerate yogurt well, although they may need to experiment with different brands to find the one that works best.

People can also add a lactase enzyme preparation to milk before they drink it or take enzyme tablets whenever they eat lactose-containing foods. Lactose-free milk and milk products that have been treated with lactase are also available. Products to aid lactose digestion are often unnecessary, however, because most people can usually tolerate a fair amount of lactose.

People with temporary lactose intolerance are advised to temporarily restrict all milk and milk products. Then lactose-containing foods are gradually reintroduced in small amounts.

PREVENTING CALCIUM AND VITAMIN D DEFICIENCIES People who restrict milk and milk products risk calcium and vitamin D deficiencies. To help prevent such deficiencies, encourage people to include milk and milk products in their diets to the extent that they can tolerate these foods, with or without the use of enzyme preparations. People who fail to receive adequate amounts of calcium from milk and milk products should be encouraged to eat other food sources of calcium, including calcium-fortified juices and cereals, broccoli, mustard greens, kale, and sardines. For people who cannot get enough calcium from foods, calcium supplements may be indicated, particularly for children, adolescents, and pregnant, lactating, or postmenopausal women. Vitamin D

Reminder: The "How to" box on p. 209 offers suggestions for adding calcium to meals.

deficiency is not a problem if the person gets regular exposure to sunlight; otherwise a supplement may be necessary.

Fat Malabsorption

Disorders involving the stomach, pancreas, intestine, and liver can all lead to fat malabsorption. Chapter 15 described how gastric surgery can lead to fat malabsorption. The next few sections show how disorders of the pancreas and intestine can interfere with the absorption of nutrients. Chapter 23 covers liver disorders, which can alter the production or secretion of bile, a necessary step in the absorption of fat.

Fat malabsorption can rapidly lead to malnutrition. The loss of fat in the stools means that valuable food energy, essential fatty acids, fat-soluble vitamins, and some minerals are lost as well (see Figure 16-1). Protein may also be malabsorbed, although usually to a lesser extent than fat.

THE NURSING DIAGNOSES altered nutrition: less than body requirements, fluid volume deficit, and **risk for fluid volume deficit** frequently apply to people who experience fat malabsorption.

The body excretes unabsorbed fat in the stools, causing the type of diarrhea known as **steatorrhea.** In addition to nutrient losses, malabsorption syndromes and their treatments may lead to malnutrition by reducing nutrient intakes and raising nutrient needs, all of which can profoundly threaten nutrition status (see Table 16-2 on p. 386).

steatorrhea (STEE-ah-toe-REE-ah): fatty diarrhea characterized by loose, foamy, foul-smelling stools.

ESSENTIAL FATTY ACID DEFICIENCY Among the fats lost in steatorrhea are the essential fatty acids. Studies show that many people with severe fat malabsorption develop essential fatty acid deficiencies.[9]

Figure 16-1
THE CONSEQUENCES OF FAT MALABSORPTION

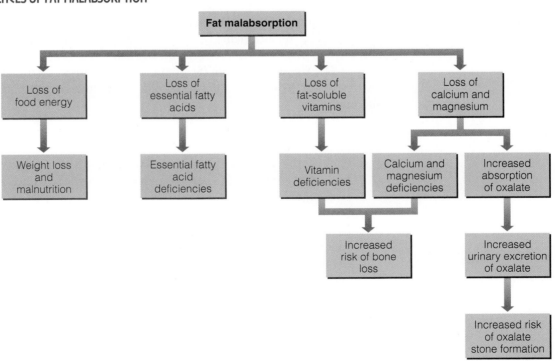

Table 16-2
POSSIBLE CAUSES OF MALNUTRITION IN MALABSORPTION SYNDROMES

Reduced Nutrient Intake	Excessive Nutrient Losses	Raised Nutrient Needs
Abdominal pain	Blood loss	High basal energy expenditure
Anorexia	Diarrhea	Infection
Bowel rest	Fistulas	Medications
Emotional stress	General malabsorption	Surgery
Food intolerance	Intestinal losses of serum proteins	
Indigestion	Medications	
Medications	Steatorrhea	
Nausea	Vomiting	
Obstructions		

soaps: chemical compounds formed between a basic mineral (such as calcium) and unabsorbed fatty acids. Soaps give steatorrhea its foamy appearance.

Reminder: Oxalate is a nonnutrient found in significant amounts in rhubarb, spinach, beets, nuts, chocolate, tea, wheat bran, and strawberries.

■■ **VITAMIN AND MINERAL MALABSORPTION** Fat-soluble vitamins are excreted in the stools along with unabsorbed fat. Some minerals, including calcium and magnesium, form **soaps** with unabsorbed fatty acids and are not available for absorption. Vitamin D losses further aggravate calcium malabsorption. Bone loss is a possible consequence.

■■ **OXALATE STONES** The binding of calcium to fatty acids can cause another problem. **Oxalate,** which is present in some foods, normally binds with some of the calcium in the gut and is excreted with it. But when fatty acids bind with calcium, the oxalate remains unbound. The intestine absorbs the unbound oxalate, but the body cannot metabolize it, and so excretes it in the urine. High urinary oxalate favors the formation of kidney stones (see Chapter 22).

■■ **Medical Nutrition Therapy**

To treat malabsorption successfully, the underlying disorder must be diagnosed and treated. High-kcalorie, high-protein diets are prescribed to meet energy and protein needs. Dietary fat is often modified in type or amount; otherwise enzyme replacements (described later) are given to aid absorption.

■■ **FAT-RESTRICTED DIETS** Table 16-3 provides instructions for a fat-restricted diet. Whenever fat is restricted, the following diet-planning principles apply:

- Fat is restricted only as much as is necessary to prevent steatorrhea because the person needs food energy.
- Part of the fat may be given as medium-chain triglycerides (MCT) because these fats are easier to digest and absorb than long-chain triglycerides (LCT).
- Foods are provided in frequent small meals because fat is best tolerated in small amounts.
- Fat-soluble vitamins are provided in a water-miscible form when fat malabsorption is severe. Water-miscible, fat-soluble vitamins mix readily with water and can be absorbed without fat.
- To reduce the risk of oxalate stones, clients with fat malabsorption may be advised to limit foods high in oxalate, including spinach, rhubarb, beets, nuts, chocolate, tea, wheat bran, and strawberries.

The "How to" box on p. 388 offers suggestions for helping people accept fat-restricted diets and includes tips for using MCT.

Table 16-3
FAT-RESTRICTED DIET (35 GRAMS)

Use:

1. Fat-free milk, fat-free cheeses, fat-free yogurt, sherbet, and fruit ices.

2. Low-fat egg substitutes and up to three regular eggs per week.

3. Up to 6 oz of lean meats and poultry without skin daily.

4. Up to 3 servings of fat daily. One serving is any one of the following:

   1 tsp butter, margarine, shortening, oil, or mayonnaise
   1 tbs reduced-fat butter, margarine, or mayonaise
   1 strip crisp bacon
   1 tbs salad dressing or 2 tbs reduced-fat salad dressing
   ⅛ avocado
   2 tbs cream (half and half)
   10 small nuts
   8 large black olives or 10 large green olives
   If fat is used to cook or season food, it must be taken from this allowance.

5. All vegetables prepared without fat.

6. All fruits prepared without fat.

7. Plain white or whole-grain bread; fat-free cereals, pasta, rice, noodles, and macaroni.

8. Clear soups.

9. Angel food cake and fruit whips made with gelatin, sugar, and egg-white meringues.

10. Jelly, jam, honey, gumdrops, jelly beans, and marshmallows.

Do not use:

1. Whole milk, chocolate milk, whole-milk cheeses, and ice cream.

2. Pastries, cakes, pies, sweet rolls, breads, or vegetables made with fat.

3. More than one egg a day, fried or fatty meats (sausage, luncheon meats, spareribs, frankfurters), duck, goose, or tuna packed in oil (unless well drained).

4. More than 3 servings of fat.

5. Desserts, candy, or anything made with chocolate, nuts, or foods not allowed.

6. Creamed soups made with whole milk.

Suggestions:

1. To make the diet still lower in fat, reduce the fat and meat (and egg) servings.

2. To raise the fat content, give additional fat or meat servings.

3. To improve acceptance of the diet, check the fat content of a well-liked food and allow that food if possible. Use the exchange system fat list for alternate suggestions for fat servings (see Appendix C).

NOTE: A "How to" box in Chapter 3 provides additional tips for lowering fat in the diet.

Most naturally occurring fats are LCT. LCT contain fatty acids with at least 14 carbon atoms, and they require lipase and bile for digestion. MCT contain fatty acids with 8 to 12 carbon atoms, and they require minimal lipase and no bile for digestion. MCT supply almost as many kcalories as regular fats but do not provide essential fatty acids, so some LCT are required.

■■ **ENZYME REPLACEMENT THERAPY** **Enzyme replacements** are used when malabsorption occurs due to chronic and severe damage to the pancreas. Because a severely damaged pancreas also fails to secrete enough bicarbonate to provide an optimal pH for enzymes to function, people with pancreatic insufficiency often take antisecretory agents to limit gastric acid production. At best, enzyme replacements taken with meals lessen the malabsorption of fat and protein, but may not fully correct it.

pancreatic enzyme replacements: extracts of pork or beef pancreatic enzymes that are taken as supplements to aid digestion.

The next few sections describe disorders that can lead to fat malabsorption. Because the causes of malabsorption vary, medical nutrition therapy varies somewhat as well.

IMPROVE ACCEPTANCE OF FAT-RESTRICTED DIETS

Fat-restricted diets can be difficult to follow. Fats give flavors, aromas, and textures to foods—characteristics that people may miss.[a] Unlike some diets that can be introduced gradually, the fat-restricted diet for malabsorption must be implemented immediately without giving the person time to adapt to the changes. These suggestions may help:

- Provide clients with tips for making foods palatable while lowering fat intake, such as those found in the box on p. 64.
- Remind clients that new fat-free and low-fat products appear on market shelves daily, and most people find these products very acceptable. Caution clients to avoid products containing fat substitutes (such as olestra), however. While healthy people can use fat substitutes in appropriate amounts, people with digestive problems and fat-soluble vitamin malabsorption may aggravate their conditions.

People who use MCT need additional advice:

- Advise clients to add MCT to the diet gradually. Nausea, vomiting, diarrhea, abdominal pain, and distention can result from using too much MCT all at once.
- Recommend that clients improve the palatability of MCT oil by substituting it for regular oil in salad dressing and for cooking and baking and by adding it to beverages, desserts, and other dishes.
- Warn clients that MCT products are expensive, and explain that these products can be purchased at pharmacies and are sometimes covered by medical insurance.

---

[a]A. Drewnowski, Why do we like fat? *Journal of the American Dietetic Association* (supplement) 97 (1997): 58–62.

Pancreatitis

Normally, the pancreas stores digestive enzymes in an inactive form to protect itself from digestion. In pancreatitis, however, digestive enzymes are activated within the pancreas and begin to damage the organ itself. The blood picks up some of these enzymes; thus, serum amylase and lipase rise and serve as indicators of acute pancreatitis.

ACUTE PANCREATITIS Pancreatitis most often develops as a consequence of gallstones or alcoholism; sometimes, though, the reasons are unclear because a variety of medical conditions and some medications can also precipitate pancreatitis. In most cases, an episode of pancreatitis subsides in about a week, but in other cases, life-threatening complications including shock, renal failure (Chapter 22), respiratory failure (Chapter 21), pancreatic hemorrhages, **fistulas,** and **abscesses** may develop.

MEDICAL NUTRITION THERAPY FOR ACUTE PANCREATITIS Initially, food is withheld, fluids and electrolytes are provided intravenously, and a tube is inserted into the stomach to suction gastric secretions and help relieve pain and distention. Oral intake begins when abdominal pain subsides and serum amylase returns to normal or near-normal levels. The diet progresses from liquids to a fat-restricted diet and then a regular diet as tolerated. Tube feeding and intravenous nutrition may be necessary in severe cases or when complications arise.

fistula (FIS-too-lah): an abnormal opening formed between two organs or between an internal organ and the skin.

abscess: an accumulation of pus, caused by local infection, that contains live microorganisms and immune system cells. Treatment involves draining the abscess.

CHRONIC PANCREATITIS

When severe pancreatitis or repeated episodes of pancreatitis permanently damage the pancreas, absorption, especially of fat, becomes permanently impaired. Chronic pancreatitis is most commonly associated with alcoholism.

Abdominal pain is often severe and unrelenting, vomiting is frequent, and severe weight loss is common. Sometimes pancreatitis also damages the cells that produce the glucose-regulating hormones insulin and glucagon.

MEDICAL NUTRITION THERAPY FOR CHRONIC PANCREATITIS

Medical nutrition therapy aims to maintain optimal nutrition status, reduce steatorrhea (if present), and avoid subsequent attacks of acute pancreatitis. Clients are strongly advised to avoid alcohol. Enzyme replacements taken with foods that are moderately restricted in fat help the person digest and absorb protein and fat while minimizing steatorrhea. Clients who develop glucose intolerance, diabetes, or hypoglycemia require additional dietary adjustments (see Chapter 20).

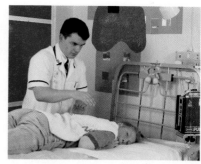

Chest physical therapy (postural drainage) promotes maximal lung function for people with cystic fibrosis.

LEARNING LINK

Alcoholism and alcohol abuse stress many organs and result in numerous complications with implications for nutrition status. The person who develops chronic pancreatitis as a consequence of excessive alcohol intake may have multiple nutrient deficiencies and alterations in metabolism that require additional dietary adjustments (see Chapter and Nutrition in Practice 23).

Cystic Fibrosis

People with **cystic fibrosis,** the most common fatal genetic disorder of the white population, produce secretions of thick, sticky mucus that may seriously impair the function of many organs, most notably the lungs and pancreas.[10] Just a few decades ago, few infants born with cystic fibrosis survived to adulthood. Today, thanks to early detection and advances in medical therapy and nutrition care, the outlook is much brighter, with some surviving into their forties and even fifties.

CONSEQUENCES OF CYSTIC FIBROSIS

Cystic fibrosis has three major consequences with nutrition implications: chronic lung disease, malabsorption, and the loss of electrolytes in the sweat. People with cystic fibrosis produce secretions of thick, sticky mucus that clog the lungs' airways and the ducts of the pancreas and liver. Labored breathing and thick, stagnant mucus in the bronchial tubes provide an ideal environment for bacteria to multiply. Lung infections occur frequently and are the usual cause of death. Damage to the pancreas and bile duct interferes with the secretion of digestive enzymes, pancreatic juices, pancreatic hormones, and bile. Malabsorption of many nutrients, glucose intolerance, and diabetes are frequent complications.

MEDICAL NUTRITION THERAPY FOR CYSTIC FIBROSIS

Although the person with cystic fibrosis may have a healthy appetite, meeting the high energy and nutrient needs imposed by labored breathing, repeated infection, and the loss of nutrients through malabsorption presents a considerable challenge. With such high-energy needs, fat restrictions are inappropriate. An unrestricted high-kcalorie, high-protein diet carefully tailored to individual tolerances supports nutritional health, while enzyme replacements help control steatorrhea and relieve abdominal pain. Multivitamin and fat-soluble vitamin

cystic fibrosis: a hereditary disorder characterized by the production of thick mucus that affects many organs, including the pancreas, lungs, liver, heart, gallbladder, and small intestine.

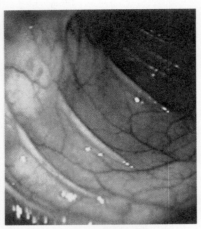

The normal colon has a smooth, shiny surface with a visible pattern of fine blood vessels.

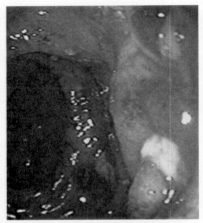

In ulcerative colitis, the colon appears inflamed and reddened, and ulcers are visible.

inflammatory bowel diseases (IBD): diseases characterized by inflammation of the bowel.

Crohn's disease: inflammation and ulceration along the length of the GI tract, often with granulomas.

ulcerative colitis (ko-LYE-tis): inflammation and ulceration of the colon.

granuloma (gran-you-LOH-mah): a tumor or growth that contains foreign organisms surrounded by immune system cells and covered with a fibrous coat.

peritonitis: infection and inflammation of the membrane lining the abdominal cavity caused by leakage of infectious organisms through a perforation (hole) in an abdominal organ.

supplements are routinely recommended. The liberal use of table salt is encouraged to replace losses of electrolytes in the sweat.

Breast milk, standard infant formulas, and hydrolyzed infant formulas can all meet the nutrient needs of infants with cystic fibrosis, provided that enzyme replacements are given as well. The breastfed infant is also given additional table salt mixed with water to replace electrolytes lost through sweat.

For all people with cystic fibrosis, every effort is made to maintain an appropriate weight for height. If weight falls between 85 and 90 percent of desirable weight for height, diets should also include high-kcalorie snacks and formula supplements to meet energy needs.[11] A client whose weight falls below 75 percent of desirable weight for height may benefit from home nutrition programs that include oral diets during waking hours and tube feedings at night.

Inflammatory Bowel Diseases

The two most prevalent **inflamatory bowel diseases** are **Crohn's disease** and **ulcerative colitis.** In Crohn's disease, cracklike ulcers and sometimes **granulomas** accompany intestinal inflammation. Crohn's disease most often affects the ileum and colon, but can affect the entire GI tract. The person often experiences intermittent fatigue, abdominal pain, diarrhea, and weight loss.

Ulcerative colitis develops only in the large intestine. During active episodes, ulcerative colitis causes an almost continuous diarrhea with malabsorption and great losses of fluids and electrolytes. Bloody diarrhea, cramping, abdominal pain, anorexia, and weight loss are clinical manifestations of the disorder. Anemia may develop as a consequence of bleeding and malabsorption.

COMPLICATIONS ASSOCIATED WITH INFLAMMATORY BOWEL DISEASES As inflammatory bowel disease progresses, fibrous tissue forms in the intestine, reducing its absorptive capacity, narrowing the intestinal lumen, and sometimes creating an obstruction. Localized infections can develop. The intestine may also rupture, which can lead to **peritonitis.**

Fistulas may develop if an inflamed loop of intestine sticks to another loop of intestine, to another organ, or to the skin and gradually erodes. If a fistula forms between the stomach or the upper portion of the small intestine and the colon, ingested food is shunted directly into the colon, and malabsorption worsens. Bacteria from the colon can then invade the stomach or upper small intestine, contributing to further malabsorption, increasing the risk of serious infections, and causing severe inflammation, nausea, and vomiting. (Bacterial overgrowth is discussed later in this chapter.) If a fistula forms between the small intestine and the skin, significant malabsorption, dehydration, and sepsis can follow.[12] The potential nutrition-related side effects of medications used in the treatment of inflammatory bowel diseases, which may include analgesics, antidiarrheals, anti-inflammatory agents (prednisone), and sulfasalazine, are described in the Diet-Medication Interactions box on pp. 398–399. Surgery may be necessary to remove a diseased or obstructed portion of the intestine or to repair a fistula.

MEDICAL NUTRITION THERAPY FOR INFLAMMATORY BOWEL DISEASES Restoring and maintaining nutrition status can be a challenging task. Growth failure in children, protein-energy malnutrition (PEM), and deficiencies of calcium, magnesium, zinc, iron, vitamin B_{12}, folate, vitamin C, and fat-soluble vitamins are commonly reported. Low serum albumin and multiple nutrient deficiencies threaten immune function and may reduce the effectiveness of drug therapy.

For people with active Crohn's disease who have intestinal obstructions or fistulas, foods and fluids may be withheld temporarily while fluids and electrolytes are replaced intravenously. In some cases, feeding tubes can be placed so as to bypass a fistula or partial obstruction.

For people with active ulcerative colitis, no dietary interventions seem to lessen disease activity. People with severe abdominal pain and diarrhea need complete bowel rest.

Intravenous nutrition provides nutrients when oral or tube feedings significantly aggravate pain and diarrhea; when the bowel is obstructed; when complete bowel rest might help a fistula to close; or when oral or tube feedings cannot meet nutrient requirements. Once oral intake is possible, the diet gradually progresses to a high-kcalorie, high-protein diet. Fat may be restricted for people with fat malabsorption. Fish oils have been reported to help prevent recurrences of active Crohn's disease and, to a lesser extent, ulcerative colitis, but further research is needed to clarify such a potential benefit.[13]

Low-fiber diets (see Chapter 14) may be recommended for people with partial intestinal obstructions or those likely to develop obstructions. The diet is adjusted for individual intolerances to specific foods or food components, such as lactose intolerance. The diet prescription frequently includes vitamin-mineral supplements. The case study below presents a person with Crohn's disease.

▪▪ Bacterial Overgrowth

The colon normally houses a bacterial population, but the small intestine is protected from bacterial overgrowth by gastric acid, which kills bacteria, and peristalsis, which flushes microorganisms through the small intestine before they can multiply. Conditions that disrupt these protective mechanisms can result in bacterial overgrowth in the stomach and small intestine. Bacteria in the small intestine partly dismantle bile salts, which are essential for fat digestion and absorption. Fat malabsorption occurs as a result. Figure 16-2 on p. 392 repeats the figure from Chapter 5 that shows how bile prepares fat for digestion, this time illustrating how bacteria interfere with that process. The bacteria also compete with the body for vitamin B_{12} and folate, limiting the available supply and leading to vitamin B_{12} and folate deficiencies.

Case Study

COLLEGE STUDENT WITH CROHN'S DISEASE

Lilinoe, a 19-year-old college student, was admitted to the hospital to manage an active episode of Crohn's disease. When Lilinoe was first diagnosed with Crohn's disease, she was 18 years old, weighed 120 pounds, and was 5 feet 7 inches tall. Her normal weight had been 130 pounds until the disease symptoms began to appear. Since then, she has been hospitalized several times for recurrent attacks of Crohn's disease. As anticipated from her weight history, Lilinoe's nutrition assessment shows that she is suffering from PEM. In talking with Lilinoe, you discover that she is very particular about the foods she eats and often simply does not eat. Her physician has ordered an easy-to-absorb formula diet to be fed to Lilinoe by tube.

✚ Review Lilinoe's weight history. What is her desirable weight? What measures could have been taken to help Lilinoe avoid weight loss?

✚ What possible benefits might an easy-to-absorb formula offer Lilinoe? Why might the physician prefer a tube feeding rather than oral feedings? How might Lilinoe be fed if she develops an obstruction or fistula?

✚ Consider Lilinoe's long-term dietary management. What type of diet should she eventually follow? What goals should be set for weight gain? Why is it important to reassess her nutrition status regularly?

✚ Given Lilinoe's age and stage of development, what emotional concerns might she be experiencing? What interventions might be planned to enhance her emotional health and well-being?

Figure 16-2
BACTERIAL OVERGROWTH AND STEATORRHEA

Normal bile action

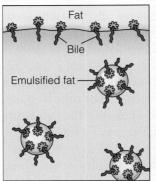

Bile's emulsifying action converts large fat globules into small droplets that repel each other.

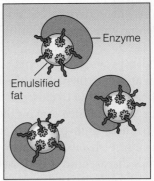

After emulsification, the enzymes have easy access to the fat droplets.

Impaired bile action

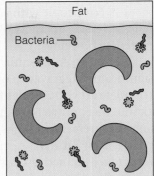

Bacterial overgrowth damages the bile, so that it is ineffective in fat digestion and absorption. The result is fat malabsorption and steatorrhea.

blind loop syndrome: the problems of fat malabsorption and vitamin B_{12} and folate deficiencies that result from the overgrowth of bacteria in a bypassed segment of the intestine.

▪▪ CAUSES OF BACTERIAL OVERGROWTH In some types of gastric surgery, a portion of the small intestine is bypassed, causing stasis in that portion of the intestine and allowing bacteria to flourish (see Figure 15-2 on p. 366). The bypassed portion is called a blind loop, and the symptoms associated with bacterial overgrowth are called the **blind loop syndrome.**

Other conditions that can lead to bacterial overgrowth by significantly reducing gastric acid secretions include chronic gastritis (see p. 365), medications (antisecretory agents), and HIV infections (see Chapter 24). Small bowel obstructions and nerve dysfunction associated with diabetes (Chapter 20) can lead to bacterial overgrowth by altering peristalsis.

▪▪ MEDICAL NUTRITION THERAPY FOR BACTERIAL OVERGROWTH Clients receive fat-restricted diets, vitamin B_{12} injections, and oral folate supplements to control steatorrhea and prevent vitamin B_{12} and folate deficiencies. Antibiotics are given to kill bacteria.

▪▪ Short-Bowel Syndrome

The treatment of inflammatory bowel diseases, cancer of the intestine, intestinal obstructions, fistulas, diverticulitis, or impaired blood supply to the intestine may include surgery to remove a portion of the small intestine. The length, location, and health of the remaining intestine determine the degree to which nutrient absorption is affected. Figure 16-3 reviews nutrient absorption in the GI tract and describes how absorption is affected by surgical resection. When the absorptive surface of the small intestine is significantly reduced, **short-bowel syndrome**—characterized by malabsorption, weight loss, muscle wasting, bone disease, hypocalcemia, hypomagnesemia, and anemia—can result.

short-bowel or short-gut syndrome: severe malabsorption that may occur when the absorptive surface of the small bowel is reduced, resulting in diarrhea, weight loss, bone disease, hypocalcemia, hypomagnesemia, and anemia.

▪▪ ADAPTATION After an intestinal resection, a remarkable adaptive response occurs in the portion that remains: it gets longer, thicker, and wider, and it either absorbs nutrients more efficiently or begins to absorb nutrients it did not absorb before. The presence of nutrients in the remaining gut appears to stimulate this adaptation—a good reason to use the GI tract (rather than the veins) to deliver nutrients as early as possible.[14] Specific dietary constituents, such as the amino acid glutamine, short-chain fatty acids, fiber, and growth hormone, may also aid in this adaptation. Generally, up to 50 percent of the intestine can be resected without serious nutrition consequences. Remarkably, even

Figure 16-3

NUTRIENT ABSORPTION AND CONSEQUENCES OF INTESTINAL SURGERIES

About 90 to 95 percent of nutrient absorption takes place in the first half of the small intestine. After a resection, nutrient absorption may be reduced.

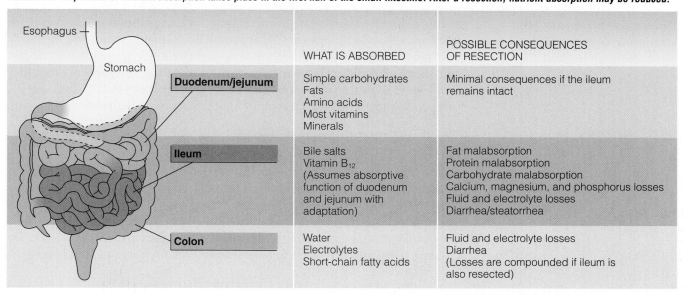

	WHAT IS ABSORBED	POSSIBLE CONSEQUENCES OF RESECTION
Duodenum/jejunum	Simple carbohydrates Fats Amino acids Most vitamins Minerals	Minimal consequences if the ileum remains intact
Ileum	Bile salts Vitamin B$_{12}$ (Assumes absorptive function of duodenum and jejunum with adaptation)	Fat malabsorption Protein malabsorption Carbohydrate malabsorption Calcium, magnesium, and phosphorus losses Fluid and electrolyte losses Diarrhea/steatorrhea
Colon	Water Electrolytes Short-chain fatty acids	Fluid and electrolyte losses Diarrhea (Losses are compounded if ileum is also resected)

resections of up to 80 percent may be well tolerated, provided that the terminal ileum, the ileocecal valve, and the colon remain intact. When the ileum has been resected, however, the absorption of fat, protein, carbohydrate, fat-soluble vitamins, vitamin B$_{12}$, calcium, and magnesium can be impaired. The ileum is also where bile salts are normally reabsorbed. Without bile salt reabsorption, the body's pool of bile salts diminishes, and fat malabsorption worsens.

■■ **MEDICAL NUTRITION THERAPY FOR SHORT-BOWEL SYNDROME** Immediately after surgery, intravenous fluids serve to maintain fluid and electrolyte balances. For resections of less than 50 percent of the intestine, oral nutrition begins after a few days. For more extensive resections, some nutrients are provided through the GI tract (usually by tube feeding) in limited amounts as early as possible to stimulate adaptation, while intravenous nutrition meets nutrient needs.

Once oral diets are possible, clients whose colons remain intact benefit from diets high in complex carbohydrates, restricted in fat, and low in oxalate.[15] Bacteria in the colon can metabolize unabsorbed carbohydrate to short-chain fatty acids, which can then be absorbed and used for energy. People whose colons do not remain intact following surgery often have difficulty absorbing both carbohydrate and fat. In such cases, intravenous nutrition is often needed permanently to supply all or part of nutrient needs. Dietitians work closely with clients to help them identify individual food intolerances and to offer suggestions to help control the rate at which foods pass through the intestinal tract.[16]

Clearly, the disorders that lead to fat malabsorption can seriously impair nutrition status. Celiac disease, described next, can also lead to fat malabsorption, but because it arises from a different cause, a fat-restricted diet does not play a role in its treatment.

■■ **Celiac Disease**

Celiac disease is a hereditary disorder with an incidence that appears to be increasing and may be as high as 1 in 250 people.[17] In celiac disease, the intestinal

celiac (SEE-lee-ack) disease: a sensitivity to a part of the protein gluten that causes flattening of the intestinal villi and malabsorption; also called **gluten-sensitive enteropathy (EN-ter-OP-ah-thee)** or **celiac sprue.**

gluten (GLUE-ten): a protein found in wheat. **Gliadin (GLIGH-ah-din)** is the fraction of gluten that causes the toxic effects in celiac disease. Corresponding protein fractions in barley, rye, and possibly oats also have these effects.

mucosal cells become sensitive to certain fractions of proteins found in some grains, including wheat, rye, barley, and possibly oats. **Gluten** is a protein found in wheat, and **gliadin** is the fraction of gluten that causes sensitivity in celiac disease. Other grains have corresponding protein fractions that can cause sensitivity in celiac disease. These protein fractions act as toxic substances, damaging the intestinal villi and leading to malabsorption. Clinical manifestations of the disease may range from simple iron and folate deficiencies to severe weight loss, fatigue, and anemia.[18] Reduced bone mineral density and bone diseases are common.[19] People with unexplainable bone disease may actually have celiac disease that has gone undiagnosed because GI symptoms are either mild or absent.[20]

■■ EFFECTS ON NUTRITION STATUS Although the severity of malabsorption varies, people with celiac disease may malabsorb many nutrients, notably, fat, protein, carbohydrate, vitamin K, folate, vitamin B_{12}, iron, and calcium. Consequently, people with celiac disease may experience steatorrhea, diarrhea, weight loss, and malnutrition. Lactose intolerance is common. Anemia may occur as a result of iron, folate, or vitamin B_{12} deficiency. A vitamin K deficiency may precipitate clotting abnormalities, and the person may bleed easily. Calcium deficiency can result in bone diseases, tetany, fractures, and bone pain.

■■ MEDICAL NUTRITION THERAPY FOR CELIAC DISEASE Lifelong adherence to a gluten-free diet serves as the primary treatment for celiac disease. Once the person follows a gluten-free diet for a few weeks, the intestinal changes reverse almost completely. With early diagnosis and treatment, children and adolescents with celiac disease who adhere to a gluten-free diet can achieve normal bone mass and avoid bone diseases.[21] In adults, bone demineralization is not always reversible.[22] Lactose intolerance may be permanent.

■■ GLUTEN-FREE DIETS The treatment for celiac disease sounds deceptively simple: eliminate gluten. Actually, the diet eliminates not only wheat (because it contains gluten), but also traditionally eliminates rye, barley, and oats because protein fractions from these grains may also damage the intestinal cells. Wheat, rye, barley, and oats are common components of many foods (see Table 16-4 on p. 395), and many foods contain hidden sources of these grains, some of which are not listed on food labels.

Acceptable substitutes for wheat flour in recipes include tapioca and soybean, arrowroot, buckwheat, and potato flours. Low-gluten wheat starch flour is also available, but because this product contains some gluten, it is not an acceptable flour substitute for people with celiac disease.

Although gluten-free diets traditionally eliminate oats, studies suggest that moderate amounts of oats may be consumed without adverse effects.[23] Oats may be contaminated with wheat, however, and many practitioners believe that more evidence is needed before the inclusion of oats on gluten-free diets can be recommended.

This international symbol identifies gluten-free foods.

■■ Disorders of the Large Intestine

This section describes two disorders of the large intestine with nutrition implications. In diverticular disease, diet plays a role in both prevention and treatment. Diet also plays a role in controlling the complications associated with colostomies and ileostomies.

■■ Diverticular Disease of the Colon

diverticula (dye-ver-TIC-you-la): sacs or pouches that develop in the weakened areas of the intestinal wall (like bulges in an inner tube where the tire wall is weak).

Sometimes pouches of the intestinal wall (called **diverticula**) bulge out through the muscles surrounding the large intestine, often at points where blood vessels

Table 16-4
GLUTEN-FREE DIET

Meat and Meat Alternates

Any allowed except those that are breaded, prepared with bread crumbs, or creamed.

Milk and Milk Products

Any allowed if client is not intolerant to lactose except milk mixed with Ovaltine, commercial chocolate milk with a cereal additive, milk beverages flavored with malt, pudding thickened with wheat flour, or ice cream or sherbet containing gluten stabilizers.

Fruits and Vegetables

Any allowed except those that are breaded, prepared with bread crumbs, or creamed.

Starches and Grains

Allowed: Bread, cereal, or dessert products made from cornmeal, soybean flour, rice flour, or potato flour; tapioca; cornmeal, popcorn, and hominy; rice, cream of rice, puffed rice, and rice flakes; potato chips.

Not allowed:[a] Bread, cereal, or dessert products made from wheat, rye, barley, and oats; wheat starch; commercially prepared mixes for biscuits, cornbread, muffins, pancakes, cakes, cookies, or waffles; bran; pasta, macaroni, and noodles; malt; pretzels; wheat germ; doughnuts; ice cream cones; matzo.

Other

Not allowed: Beer; ale; certain whiskeys (Canadian rye); alcohol-based extracts; cereal beverages (Postum); root beer; commercial salad dressings that contain gluten stabilizers; distilled white vinegar; soy sauce; soups containing any ingredients not allowed (such as barley or noodles); products made with hydrolyzed vegetable protein.

---

[a]Note that many special products made with allowed ingredients are available. Gluten-free pastas and macaroni, for example, are acceptable substitutes for regular pastas and macaroni.

Figure 16-4
DIVERTICULA

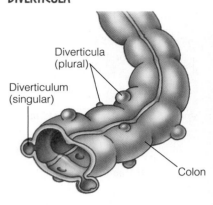

Diverticula (plural)

Diverticulum (singular)

Colon

enter the muscles (see Figure 16-4). Evidence suggests that diverticula form when strong intestinal contractions pinch off segments of the intestine; pressure then builds in the segments and forces parts of the membrane to balloon outward through the muscle layer. Diverticular disease often occurs with aging.

■■ **DIVERTICULOSIS AND DIVERTICULITIS** People with **diverticulosis** are frequently symptom-free and unaware of the disorder. In some people, however, fecal material gets trapped in the diverticula. A localized area of inflammation and infection develops—a condition called **diverticulitis.** People with diverticulitis may suffer from abdominal pain and distention, alternating episodes of diarrhea and constipation, indigestion, flatus, and fever. Occasionally, a diverticulum ruptures, causing a localized or sometimes life-threatening infection (peritonitis). If the diverticula become inflamed repeatedly, the intestinal wall can form scar tissue and thicken (fibrosis), narrowing the intestinal lumen and creating an obstruction. An inflamed bowel segment can also stick to other pelvic organs, forming a fistula.

THE NURSING DIAGNOSES constipation, diarrhea, risk for infection, and **pain** that may be associated with diverticular diseases have nutrition-related implications.

■■ **MEDICAL NUTRITION THERAPY** People with diverticulosis are advised to adopt high-fiber diets to reduce pressure in the colon and stimulate peristalsis and to avoid foods with seeds, such as okra and strawberries, because the seeds may get trapped in the diverticula and cause irritation. During periods of active diverticulitis, however, fiber is temporarily restricted. A study of over 45,000 people suggests that high-fiber, low-fat diets are associated with a lower incidence of symptomatic diverticular disease.[24]

diverticulosis: the condition of having diverticula.

diverticulitis: the condition of having infected diverticula.

Colostomies and Ileostomies

Treatment for some medical conditions and surgeries affecting the large intestine necessitate that the feces be diverted from all or portions of the colon. A **colostomy** is a surgical procedure that creates an opening (**stoma**) from the colon through the abdominal wall to the surface of the skin to allow for defecation when feces cannot pass through the colon and anus. A pouch placed over the stoma collects the feces. Permanent colostomies are performed when intestinal obstructions, lesions, or tumors necessitate the removal of all or a portion of the large intestine. Temporary colostomies allow an injured or inflamed colon time to heal. In an **ileostomy,** the stoma is created from the ileum, and the entire colon is bypassed. In an alternative to an ileostomy, the ileal pouch/anal anastomosis, the surgeon removes the diseased colon and rectal tissue and connects the ileum to the anus. A temporary ileostomy allows time for the tissue to heal; thereafter, defecation occurs through the anus, rather than a stoma. Figure 16-5 shows examples of a colostomy and an ileostomy. In this section the word *ostomy* is used to refer to both colostomies and ileostomies.

The consistency of the stools following colostomies and ileostomies varies depending on both the length and the portion of the resected bowel. In general, the greater the length of colon left intact, the more formed the stool. Thus, ileostomies, which bypass the entire colon, frequently result in watery stools.

MEDICAL NUTRITION THERAPY Once solid foods are permitted following surgery, people who have undergone colostomies or ileostomies often receive low-fiber, bland diets to prevent obstructing the ostomy, promote healing, and prevent GI upsets. Encourage people to judiciously include other foods as soon as possible, however. Foods are added one at a time and in small amounts so that their effects can be assessed. If the added food presents problems, the person can try it again in a few weeks or months.

colostomy (ko-LOSS-toe-me): surgery that creates an opening from a portion of the colon through the abdominal wall and out through the skin.

stoma (STOH-ma): a surgically formed opening.

stoma = window

ileostomy (ILL-ee-OSS-toe-me): surgery that creates a stoma from the ileum through the abdominal wall and out through the skin.

Figure 16-5
COLOSTOMY AND ILEOSTOMY

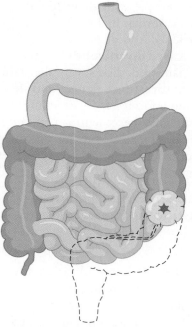

In a colostomy, the rectum and anus are removed, and the stoma is formed from the remaining colon.

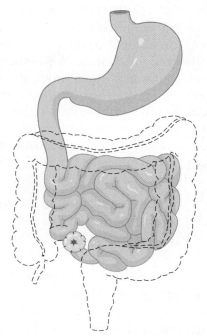

In an ileostomy, the entire colon, rectum, and anus are removed, and the stoma is formed from the ileum.

THE NURSING DIAGNOSES altered nutrition: less than body requirements, body image disturbance, diarrhea, fluid volume deficit, risk for fluid volume deficit, impaired skin integrity, and **pain** that may apply to people who have undergone colosotomies and ileostomies have nutrition-related implications.

■■ **PREVENTING OBSTRUCTIONS**   Some foods are more likely than others to be incompletely digested and obstruct the stoma. These include stringy foods such as celery, spinach, and bean sprouts; foods with tough skins such as dried fruits, raw apples, and corn; foods with seeds such as okra; mushrooms; and nuts. Some of these foods may be used if the client cuts the food into small pieces and chews them thoroughly. An undigested mushroom, for example, may act as a plug and obstruct a stoma, but if it is cut into small pieces, it may be tolerated.

■■ **ENCOURAGING FLUIDS**   People with ostomies need extra fluids because they are absorbing less fluid from the large intestine. They may tend to restrict their fluid intakes, however, for fear of aggravating diarrhea. Explain that drinking fluids helps prevent constipation and dehydration, and that excess fluid taken above and beyond the amount lost through the stoma will be absorbed and excreted by the kidneys; fluids will not aggravate diarrhea.

■■ **CONTROLLING DIARRHEA**   People with ostomies may benefit from foods that thicken the stool and help control diarrhea. These foods include applesauce, bananas, cheeses, creamy peanut butter, and starchy foods such as breads, rice, and potatoes. Foods that may aggravate diarrhea include apple, grape, and prune juices; highly seasoned foods; and caffeine. The foods mentioned here are suggestions only; what works for the individual is determined by trial and error.

■■ **REDUCING GAS AND ODORS**   People with ostomies are often concerned about gas and odors associated with foods. Gas-forming foods in general were listed in Table 16-1, but certain foods in particular seem to cause gas and odors for people with ostomies: asparagus, beans, beer, broccoli, brussels sprouts, cabbage, carbonated beverages, cauliflower, eggs, fish, garlic, and onions. Foods thought to reduce odors include buttermilk, cranberry juice, parsley, and yogurt.

This chapter has shown how disorders of the intestine can have effects that range from mild discomfort to severe malnutrition. Clients often need emotional support to handle the many lifestyle changes that conditions such as cystic fibrosis, inflammatory bowel diseases, and the need for colostomies and ileostomies incur. The accompanying Nutrition Assessment Checklist highlights nutrition-related findings important to consider for clients with disorders of the lower GI tract. The Diet-Medication Interactions box provides examples of interactions of medications used to treat disorders of the lower GI tract.

NUTRITION ASSESSMENT CHECKLIST — FOR PEOPLE WITH LOWER GI TRACT DISORDERS

Health Problems, Signs, and Symptoms

Does the client have a medical diagnosis of:

- ☐ Irritable bowel syndrome?
- ☐ Lactose intolerance?
- ☐ Pancreatitis?
- ☐ Cystic fibrosis?
- ☐ Crohn's disease?
- ☐ Ulcerative colitis?
- ☐ Celiac disease?
- ☐ Diverticular disease?

Has the client had surgery that includes:

- ☐ Intestinal resection?
- ☐ Colostomy?
- ☐ Ileostomy?

Does the client have the following symptoms or complications:

- ☐ Constipation
- ☐ Diarrhea/dehydration
- ☐ Lactose intolerance
- ☐ Malabsorption
- ☐ Essential fatty acid deficiency
- ☐ Bone disease
- ☐ Oxalate stones
- ☐ Infection
- ☐ Fistulas
- ☐ Obstructions

☐ Bacterial overgrowth of the intestine

Medications

Check for medications or herbal remedies that may:

- ☐ Cause constipation or diarrhea
- ☐ Interfere with food intake including those that cause nausea, vomiting, cramps, dry mouth, or drowsiness
- ☐ Alter nutrient needs

Nutrient/Food Intake

Note the following conditions and contact the dietitian if you suspect a problem with:

- ☐ Poor appetite or limited food intake
- ☐ Food intolerances (including amount of fat for those with malabsorption)
- ☐ Inadequate fiber/fluid intake for those with constipation
- ☐ Inadequate calcium intake for those with lactose intolerance
- ☐ Fluid intake

Height and Weight

Measure baseline height and weight. Address weight loss early to prevent malnutrition for clients with:

- ☐ Severe or persistent diarrhea
- ☐ Malabsorption

Laboratory Tests

Check laboratory tests for signs of dehydration for clients with:

- ☐ Severe or persistent diarrhea
- ☐ Malabsorption
- ☐ Intestinal resections

Check laboratory tests for signs of nutrition-related anemias and nutrient deficiencies (see Appendix E) for clients with:

- ☐ Severe or persistent diarrhea
- ☐ Malabsorption

Physical Signs

Look for physical signs (see Table 13-9 on p. 332) of:

- ☐ Dehydration (especially for people with severe or persistent diarrhea or malabsorption)
- ☐ PEM
- ☐ Essential fatty acid deficiencies (especially for people with fat malabsorption)
- ☐ Folate and vitamin B_{12} deficiencies (especially for people with bacterial overgrowth)
- ☐ Mineral deficiencies (especially for people with severe and persistent diarrhea, lactose intolerance, or fat malabsorption)

DIET-MEDICATION INTERACTIONS

Analgesics

All *analgesics* can cause nausea, vomiting, and GI upsets. When oral intake is permitted, giving the medication along with food can minimize these side effects. Narcotic analgesics can lead to constipation; these medications can also lead to lethargy, which can contribute to reduced food intake.

Antianxiety agents

Antianxiety agents can cause drowsiness, which can interfere with food intake. *Alprazolam* and *chlordiazepoxide* can stimulate the appetite and lead to weight gain. The herb kava should not be used together with alprazolam. *Diazepam, lorazepam,* and *oxazepam* can cause constipation, diarrhea, or dry mouth. *Meprobamate* can cause drowsiness. Caution clients on antianxiety agents to avoid alcohol.

Antidepressants

Antidepressants frequently cause mouth dryness, nausea, vomiting, and constipation. *Nefazodone* should be administered without food or with a consistent amount of food to ensure proper drug availability.

Antidiarrheals

Antidiarrheals seldom result in significant nutrition-related side effects, with the exception of opium and paragoric, which can cause nausea and vomiting, constipation, and lethargy. Encourage clients taking *psyllium* to drink plenty of fluids.

Antiflatulents

Antiflatulents seldom result in significant nutrition-related side effects.

Antiemetics

Antiemetics frequently lead to drowsiness. *Prochlorperazine* and *thiethyperazine maleate* can also lead to mouth dryness and constipation.

Anti-inflammatory agents (for inflammatory bowel diseases)

Prednisone can stimulate the appetite and lead to fluid retention—both can lead to weight gain. Long-term use also leads to negative nitrogen and calcium balances and osteoporosis. Clients taking prednisone should be encouraged to eat high-protein, high-calcium, high-potassium diets. Clients who are not losing fluids through diarrhea and malabsorption may need to limit sodium.

Anti-infective agents

Numerous *anti-infective agents* with a variety of nutrition-related interactions may be used in the treatment of diarrhea, pulmonary infections (cystic fibrosis), bacterial overgrowth of the intestine, and infections that arise in clients recovering from surgery of the intestine. Consult a pharmacist or drug guide for specific interactions.

Laxatives

Common side effects of *laxatives* include nausea and cramps. *Mineral oils* can reduce the absorption of fat-soluble vitamins. Clients on laxatives are encouraged to eat high-fiber foods and drink generous amounts of fluids.

Pancreatic enzyme replacements

Enzyme replacements sometimes cause nausea and may interfere with the body's response to oral iron supplements. Enzyme replacements must be taken with meals and snacks. Enteric-coated forms should not be crushed or chewed. Capsules containing enteric-coated microspheres (tiny spheres of medication) may be sprinkled on soft foods (such as applesauce), provided they can be swallowed without chewing and are followed by a glass of water or juice.

Uncategorized drugs (for inflammatory bowel diseases)

Sulfasalazine may cause nausea, vomiting, and diarrhea. The drug should be taken with food at regular intervals during the day. Encourage the client to drink fluids. *Infliximab* (used in the treatment of Crohn's disease) may cause nausea, abdominal pain, fever, and fatigue.

SELF CHECK

1. The nurse caring for an elderly client with constipation encourages the client to eat a:

   a. low-fat diet rich in potassium.
   b. low-fiber diet rich in calcium.
   c. gluten-free diet rich in protein.
   d. high-fiber diet and drink plenty of fluids.

2. The nurse recognizes that for people with permanent lactose intolerance, lactose-restricted diets:

   a. strictly limit lactose from all sources.
   b. require the use of lactase-containing digestive aids.
   c. include lactose-containing foods according to individual tolerances.
   d. require careful review of the client's intake of folate and vitamin B_{12}.

3. Nutrition problems associated with fat malabsorption syndromes include all of the following **except:**

   a. bone diseases.
   b. oxalate kidney stones.
   c. weight loss and malnutrition.
   d. essential amino acid deficiencies.

4. The nurse counsels the client with chronic pancreatitis to:

   a. eat high-fat foods.
   b. follow a very-low-fat diet.
   c. moderately restrict alcohol.
   d. follow a moderate fat-restricted diet and use enzyme replacements to improve fat absorption.

5. Dietary recommendations for people with cystic fibrosis include:

a. fluid restrictions.
b. a fat-restricted diet.
c. limited use of table salt.
d. a high-kcalorie, high-protein diet.

6. The nurse working with a client with an inflammatory bowel disease recognizes that all of the following can affect nutrient needs **except:**

a. fistulas.
b. medications.
c. fat malabsorption.
d. dumping syndrome.

7. Common nutrition problems associated with bacterial overgrowth in the stomach and small intestine include:

a. permanent lactose intolerance.
b. sensitivity to gluten and gliaden.
c. increased absorption of bile salts and constipation.
d. fat malabsorption and folate and vitamin B_{12} deficiencies.

8. The nurse recognizes that a client understands the basics of a gluten-free diet when the client states, "I must avoid all products containing:

a. wheat, corn, and oats."
b. barley, soybeans, and corn."
c. wheat, barley, and rye."
d. tapioca, buckwheat, and oats."

9. Diets for people who have undergone extensive intestinal resections but whose colons remain intact should be:

a. gluten-free.
b. low in simple sugars and high in protein and fat.
c. low in carbohydrate to prevent lactose intolerance.
d. high in complex carbohydrate and restricted in fat.

10. Long-term management of diverticular disease includes a:

a. low-fiber, lactose-free diet.
b. high-fiber, lactose-free diet.
c. low-fiber diet that omits foods with seeds.
d. high-fiber diet that omits foods with seeds.

Answers to these questions appear in Appendix H.

CLINICAL APPLICATIONS

1. Treatments for diseases with the same consequences often differ. With this in mind, describe the similarities and differences in the treatment of malabsorption for people with chronic pancreatitis and people with cystic fibrosis. Contrast these treatments with the treatments of malabsorption for people with celiac disease and people with lactose intolerance.

2. Using Table 16-3 on p. 387 as a guide, plan a day's menu for a diet providing 35 grams of fat. Take care to make the meals both palatable and nutritious. How can this menu be improved using the suggestions in the box on p. 388?

3. As stated in this chapter, treatment of celiac disease is deceptively simple—eliminate gluten. Take a trip to the grocery store and randomly select ten of your favorite snack and convenience foods. Check the labels of these products and see if they are allowed on gluten-restricted diets. (As you complete this part of the assignment, keep in mind that the labels may not list all offending ingredients.) Find acceptable substitutes for the products that are not allowed. No doubt, this will be a challenging assignment.

NUTRITI**ON**THENET

FOR FURTHER STUDY OF THE
TOPICS IN THIS CHAPTER,
ACCESS THESE WEB SITES.

www.niddk.nih.gov/health/digest/
pubs/gas/gas.htm
Gas in the Digestive Tract
(from the National Digestive
Diseases Clearinghouse)

www.acg.gi.org
American College of
Gastroenterology

www.niddk.nih.gov/health/digest/
pubs/irrbowel/irrbowel.htm
Irritable Bowel Syndrome
(from the National Digestive
Diseases Clearinghouse)

www.ccfa.org
Crohn's and Colitis Foundation of
America

www.niddk.nih.gov/health/digest/
pubs/divert/divert.htm
Diverticulosis and Diverticulitis

www.uoa.org
United Ostomy Association

Notes

[1] J. E. Suneson, Irritable bowel syndrome: A practical approach to medical nutrition therapy, *Support Line,* February 1999, pp. 11–15.

[2] D. Harari and coauthors, Bowel habit relation to age and gender: Findings from the National Health Interview Survey and clinical implications, *Archives of Internal Medicine* 156 (1996): 315–320; C. S. Probert and coauthors, Evidence for the ambiguity of the term constipation: The role of irritable bowel syndrome, *Gut* 35 (1994): 1455–1458.

[3] L. R. Schiller, Irritable bowel syndrome: One physician's perspective, *Support Line,* October 1998, pp. 3–8.

[4] Suneson, 1999.

[5] F. L. Suarez, D. A. Savaiano, and M. D. Levitt, A comparison of symptoms after the consumption of milk or lactose-hydrolyzed milk by people with self-reported lactose intolerance, *New England Journal of Medicine* 333 (1997): 1–4; F. L. Suarez and coauthors, Tolerance to the daily ingestion of two cups of milk by individuals claiming lactose intolerance, *American Journal of Clinical Nutrition* 65 (1997): 1502–1506.

[6] M. Lee and S. D. Kransinski, Human adult-onset lactase decline: An update, *Nutrition Reviews* 56 (1998): 1–8.

[7] Suarez and coauthors, 1997.

[8] S. R. Hertzler and coauthors, How much lactose is low lactose? *Journal of the American Dietetic Association* 96 (1996): 243–246.

[9] P. B. Jeppesen and coauthors, Essential fatty acid deficiency in patients with severe fat malabsorption, *American Journal of Clinical Nutrition* 65 (1997): 837–843.

[10] C. E. Beck and coauthors, Improvement in the nutritional and pulmonary profiles of cystic fibrosis patients undergoing bilateral sequential lung and heart-lung transplantation, *Nutrition in Clinical Practice* 12 (1997): 45–53.

[11] A.S.P.E.N. Board of Directors, Practice guidelines: Cystic fibrosis, *Journal of Parenteral and Enteral Nutrition* (supplement) 17 (1993): 44.

[12] S. Fukuchi and coauthors, Nutrition support of patients with enterocutaneous fistulas, *Nutrition in Clinical Practice* 13 (1998): 59–65.

[13] A. Belluzi and coauthors, Effect of an enteric-coated fish-oil preparation on relapses in Crohn's disease, *New England Journal of Medicine* 334 (1996): 1557–1560; Y. Kim, Can fish oil supplements maintain Crohn's disease in remission? *Nutrition Reviews* 54 (1996): 248–257.

[14] K. A. Tappenden, Nutritional regulation of intestinal adaptation, *Support Line,* August 1998, pp. 18–23.

[15] T. A. Byrne and coauthors, A new treatment option for patients with short-bowel syndrome: Bowel rehabilitation with growth hormone, glutamine, and a modified diet, *Support Line,* February 1996, pp. 1–7.

[16] T. C. Lykins and J. Stockwell, Comprehensive modified diet simplifies nutrition management of adults with short-bowel syndrome, *Journal of the American Dietetic Association* 98 (1998): 309–315.

[17] T. Not and coauthors, Celiac disease risk in the USA: High prevalence of antiendomysium antibodies in healthy blood donors, *Scandinavian Journal of Gastroenterology* 33 (1998): 494–498.

[18] J. R. Saltzman and B. D. Clifford, Identification of the triggers of celiac sprue, *Nutrition Reviews* 52 (1994): 317–319.

[19] S. Mora and coauthors, Reversal of low bone density with a gluten-free diet in children and adolescents with celiac disease, *American Journal of Clinical Nutrition* 67 (1998): 477–481; G. R. Corazza and coauthors, Propeptide of type I procollagen is predicative of posttreatment bone mass gain in adult celiac disease, *Gastroenterology* 113 (1997): 67–71.

[20] J. L. Shaker and coauthors, Hypocalcemia and skeletal disease as presenting features of celiac diseases, *Archives of Internal Medicine* 157 (1997): 1013–1016.

[21] Mora and coauthors, 1998.

[22] Corazza and coauthors, 1997.

[23] T. Thompson, Do oats belong in a gluten-free diet? *Journal of the American Dietetic Association* 97 (1997): 1033–1037.

[24] W. H. Aldoori and coauthors, A prospective study of diet and risk of symptomatic diverticular disease in men, *American Journal of Clinical Nutrition* 60 (1994): 757–764.

Q&A NUTRITION IN PRACTICE

Food and Foodservice in the Hospital

When medical conditions require treatment in a hospital or long-term care facility, foods are prepared for and provided to the clients. Many clients lose their appetites as a result of their medical condition or their emotional distress. Even when appetites are healthy, comments about hospital foods—both positive and negative—are common. The hospital's foodservice department, under the direction of an administrative dietitian or foodservice manager, faces a challenge in planning, preparing, and delivering meals designed to accommodate dozens of special diets and food preferences. Although this discussion focuses on hospitals, much of the information applies to any health care facility that serves food to large groups of people, including nursing homes, assisted living centers, rehabilitation centers, and residential mental health care facilities. When people in hospitals eat poorly, they can make up for nutrient deficits by eating well when they return home. Residents of a long-term care facility, however, do not have this option. For this reason, foodservice departments in long-term care facilities must make even greater efforts to ensure that their clients receive nutritious and appetizing foods.

Why are remarks about a person's experience with hospital food so common?

In the hospital, eating offers clients familiarity in an otherwise strange environment. Most people generally look forward to eating, and for many people in the hospital, a healthy appetite signals a return to health. Eating may become even more enjoyable than usual. It is also one of the few hospital experiences where clients have a choice. Consider that clients usually cannot choose when

they will receive tests, how many times blood will be drawn, what nurse will care for them, or what time they will have surgery. But they usually can select their meals, and they can use those meals to exercise control or express their feelings: they can eat or refuse to eat!

When a client complains about hospital food, the complaint may have little to do with the food itself, but instead may serve as a way to vent fear, frustration, anger, and physical pain. Clients need opportunities to express their feelings, and often a problem can be resolved simply by listening and providing emotional support.[1]

Don't clients sometimes have real complaints about food?

Yes, of course. Many complaints stem from hospital foods not being prepared in the way the client prefers—a considerable problem when the client must eat three meals a day for several days in the hospital. Unfortunately, too, hot foods may not be hot and cold foods may not be cold by the time they arrive in the client's room. In addition, the client receives meals at specified times regardless of hunger and often must eat in bed without companionship, which can be more of

a chore than a pleasurable experience. Many disorders, medications, and treatments can dramatically alter taste perceptions and lead to complaints about food.[2] Meals may also be unwelcome if the person is in pain.

What can be done to help alleviate these problems?

The majority of people in the hospital will eat adequate amounts of food, even if they complain about it. Their intakes may decrease slightly, but the deficit will be easy to correct once they are at home eating familiar foods. Chapter 14 provides suggestions for helping people to eat. For people in pain, administering pain medications so that they will be effective during mealtimes can be helpful. For people with altered taste perceptions, the dietitian can work closely with them to uncover the tastes and food preferences that they can tolerate and enjoy.

In some cases, the foodservice department must be contacted to solve food-related problems. In one study, researchers found that clients expressed the most satisfaction with meals that were served attractively and tasted good and with cold foods that were served at the correct temperature.[3] Some problems can be

Foodservice departments prepare foods to accommodate dozens of special diets and hundreds of food preferences.

handled directly by the person caring for the client. For example, the nurse providing a tray to the client can make sure foods and utensils are arranged attractively before they are served.

Who is responsible for the operations of the foodservice department?

The responsibility of budgeting, planning, preparing, and serving appropriate meals rests with either a chief administrative dietitian or a foodservice manager. In some facilities, foodservice companies from outside the hospital perform these duties.

Clinical dietitians work directly with clients to assess their nutrition status, plan appropriate diets, and provide nutrition education. In some facilities, dietetic technicians assist dietitians in both administrative and clinical responsibilities. Other dietary employees include clerks, porters, and other assistants. Keep in mind that many dietary employees do not have formal education in nutrition, and their ability to interpret diet orders and provide accurate information is limited.

How does the foodservice department know what foods to serve each client?

Most hospitals provide menus from which clients can select their meals. A client who must follow a modified diet receives menus that include only foods specified in the hospital's diet manual for that particular diet. By allowing a choice, this system helps to ensure that clients receive foods they enjoy and will eat. An added advantage for people on special diets is that they become familiar with their diets by marking appropriate menus.

Although procedures vary somewhat between facilities, generally dietary employees deliver menus to each client's room early in the day and pick them up again later in the day. Each menu shows the client's name and room number, as well as the name of the meal, the type of diet, and the day the menu will be served. Generally, the client makes selections for the next day or for the next few days to give the foodservice department time to collect the menus and estimate the amount and type of food to prepare. Menus are usually color coded by diet, which helps foodservice employees put the right types of foods on food trays. Color-coding also helps the person delivering the tray to quickly determine if the right diet has been delivered. Figure NP16-1 on pp. 404–405 shows lunch menus for several different diet menus and explains how each menu might be used.

If clients are making their own selections, how can they be receiving foods they don't like?

Several problems with menus can occur. Clients may not receive foods they enjoy if they fail to mark the menus or inadvertently make the wrong food selections. Consider these possible scenarios:

- A client may have trouble seeing, reading, understanding, or physically marking menus.
- Clients may not understand that their selections will be for the next (or another) day.
- Clients may be out of their rooms (for tests, procedures, or physical activity) or asleep when the menus arrive; when the clients return or wake up, they may not see the menus or may have missed the menu pickup time.
- Clients may be too ill or too disinterested in food to make menu selections.

If menus are not marked or if a menu is lost, the client receives meals preselected by the foodservice department.

Occasional problems with menu selections can usually be corrected simply by explaining the menu system to clients or taking extra time to help them mark menus. If clients continue to complain about food selections, contact the dietetic technician or dietitian.

What happens to the menus once they are collected?

Once food selections have been made and menus collected, menus are often checked by a member of the foodservice staff (usually, a dietetic technician or dietitian) to make sure that selections are appropriate. Completed menus can provide valuable clues about a person's usual eating habits or understanding of a modified diet. In checking menus, for example, the technician may notice that one person on a regular diet is selecting very little or that another is selecting too many foods. In another case, the technician may notice that a person on a kcalorie-restricted diet is not selecting the appropriate number of servings from each food group. Such problems suggest the need for further instruction.

Do all facilities offer selective menus?

Not all. Facilities that do not offer a selective menu serve a standard house diet, adjusting the menu for individual food preferences. For example, clients can request simple changes, such as the substitution of one vegetable for another.

How do foodservice departments prepare foods to meet the needs of a variety of diets?

The logistics of preparing food tailored to each modified diet can be overwhelming. For this reason, foodservice departments use systems designed to limit costs and minimize errors. Foods prepared for regular and soft/bland/low-fiber/low-residue diets are prepared with some fat and salt, because these dietary components are not restricted on such diets. Note that the other diet menus shown in Figure NP16-1 provide a number of low-fat (LF) or low-sodium (LS) or low-sodium, low-fat (LSLF) foods.

If the foodservice department were to prepare a food (baked chicken, for example) for each different diet, it would have to prepare regular baked chicken, low-fat baked chicken, low-sodium baked chicken, and low-sodium, low-fat baked chicken. Using the system illustrated in the menus of Figure NP16-1, only two types of baked chicken need to

Figure NP16-1
SAMPLE LUNCH MENUS

Lunch

REGULAR   **SUNDAY**
Meats
Baked chicken❤   Fried fish
Hamburger on bun with chips
(with lettuce and tomato)
Starchy Vegetables
Cornbread dressing   Parsleyed potatoes❤
Vegetables
Baby carrots❤   Stewed tomatoes
Soup/Salad   **Dressings**
Coleslaw   French
Clam chowder   Thousand Island
Gelatin   Italian
Tossed salad❤   Diet Thousand Island❤
Desserts
Apple pie   Butterscotch pudding
Fresh fruit❤
Breads
Dinner roll   Bran bread❤
White bread   Crackers
Wheat bread
Beverages & Condiments
Coffee   Sugar
Decaf. coffee   Sugar substitute
Hot tea   Herb seasoning
Decaf. hot tea   Creamer
Iced tea   Lemon
Whole milk   Mustard
Buttermilk   Mayonnaise
2% milk   Catsup
Fat-free milk❤   Margarine
Chocolate milk

PLEASE DO NOT LEAVE MENU ON THE TRAY
NAME _____ ROOM _____

Lunch

SOFT/BLAND/LOW RESIDUE   **SUNDAY**
Meats
Baked chicken   Baked fish(cod)
Hamburger on bun
Starchy Vegetables
Rice   Boiled potatoes
Vegetables
Baby carrots   Green beans
Soup/Salad/Juice   **Dressings**
Gelatin   Mayonnaise
Lemonade   Catsup
Tomato soup
Desserts
Apple pie   Pears
Breads
Dinner roll   Crackers
White bread
Beverages & Condiments
Decaf. coffee   Sugar
Decaf. hot tea   Sugar substitute
Decaf. iced tea   Creamer
Hot chocolate   Lemon
Whole milk   Margarine
2% milk
Buttermilk
Fat-free milk

NO PEPPER
PLEASE DO NOT LEAVE MENU ON THE TRAY
NAME _____ ROOM _____

Lunch

KCALORIE RESTRICTED, DIABETIC
1200 CALORIES   **SUNDAY**
LF = Low Fat   LSLF = Low Sodium, Low Fat
Meat Exchange (Select _1_)
LSLF Baked chicken (2oz)   LSLF Baked fish (2oz)
LSLF Hamburger on bun (with lettuce and
tomato, 2oz meat) omit 2 starches
Starch Exchange (Select _1_)
Clam chowder (1 c)   LF Dinner roll (1)
LSLF Rice (⅓c)   White bread (1 slice)
LSLF Boiled potatoes (½c)   Wheat bread (1 slice)
Angel food cake   Bran bread (1 slice)
(1" slice)   Crackers (6)
Vegetable Exchange (Select _2_)
LSLF Baby carrots   LSLF Green beans
(½ c)   (½ c)
Fruit Exchange (Select _1_)
Diet pears (½ c)   Fresh fruit
Milk Exchange (Select _1_)
Whole milk (1 c) omit 2 fats   Buttermilk (1 c)
2% Milk (1 c) omit 1 fat   Fat-free milk (1 c)
Fat Exchange (Select _1_)
Margarine (1 tsp)   Creamer (1 = ½ fat)
Diet mayonnaise (½ oz)
kCalorie-free Foods
Coffee   LSLF Coleslaw (½ c)
Decaf. coffee   Tossed salad (1 c)
Hot tea   Diet gelatin (½ c)
Decaf. hot tea   Diet French
Iced tea   Diet Thousand Island
Sugar substitute   Diet Italian
Lemon   Mustard
Herb seasoning   Diet catsup

PLEASE DO NOT LEAVE MENU ON THE TRAY
NAME _____ ROOM _____

People on regular diets select the foods of their choice. The regular menu may also be used for high-kcalorie, high-protein diets. The menu items marked with a heart, guide people in selecting foods that are lower in fat, cholesterol, sodium, and caffeine or higher in fiber than other menu choices.

Foods for soft/bland/low-residue diets are similar to those for regular diets. Foods from the regular menu that are not appropriate have been eliminated from the menu, and substitutions have been made. For clients on bland diets, decaffeinated coffee and tea would be crossed off the menu.

For kcalorie-restricted and diabetic diets, the number of exchanges allowed is written on the menu beforehand. (This example uses a 1200-kcalorie diet.) Note that the meat exchange is written in 2-ounce portions so that 1 serving = 2 exchanges. (Chapter 20 describes the exchange system.)

be prepared—one with some fat and salt, and the other without fat or salt.

Keep in mind that baked chicken is only one of many menu items in a day, and you can see why preparing individual foods for each diet is not feasible. Instead, clients can add allowed ingredients to food. For example, the person on a low-salt diet could add margarine to a serving of vegetables, the person on a low-fat diet could add salt to a serving of rice.

How does the foodservice delivery system work?

Sometimes foods are prepared in a main kitchen, assembled on trays, and heated in areas close to the clients' rooms. In other cases, foods are delivered directly from the main kitchen. In either case, foodservice personnel deliver food carts directly to the nursing unit; then nursing or foodservice personnel take a tray to each client. Efficient delivery of foods to the nursing unit and then to the client helps ensure that clients receive foods at the appropriate temperature.

Once the client is finished eating, the tray is returned to the food cart. Foodservice personnel pick up the carts and return them to the foodservice department.

How is this information useful to the nurse?

Nurses are most likely to hear clients' complaints about food or problems with foodservice. Help clients—and save needless aggravation and time—by learning about the foodservice system in the health care facility where you work. Better yet, ask to spend a few hours or a day working with different foodservice employees to see firsthand how the department operates and what problems they encounter. If that is not possible, learn the facility's procedures for ordering diets, making diet changes, reporting problems with a client's tray, and making special requests. Remember that requests are not simply made by one individual to another. Often many people are involved in processing a single request, and the number of requests made during any one meal can be

Figure NP16-1
SAMPLE LUNCH MENUS—continued

Lunch

LOW-FAT/LOW CHOLESTEROL/
CARDIAC   SUNDAY
LF = Low Fat   LSLF = Low Sodium, Low Fat
Meats
LSLF Baked chicken   LSLF Baked fish (cod)
LSLF Hamburger on bun
(with lettuce and tomato)
Starchy Vegetables
LSLF Rice   LSLF Boiled potatoes
Vegetables
LSLF Baby carrots   LSLF Green beans
Soup/Salad/Juice   **Dressings**
LSLF Coleslaw   Diet French
Gelatin   Diet Thousand Island
Tomato soup   Diet Italian
LS Chicken broth
Tossed salad
Desserts
Pears   Angel food cake
       Fresh fruit
Breads
LF Dinner roll   Bran bread
White bread   Crackers
Wheat bread   LS Crackers
Beverages & Condiments
Coffee   Creamer
Decaf. coffee   Sugar
Hot tea   Sugar substitute
Decaf. hot tea   Herb seasoning
Iced tea   Lemon
Buttermilk   Margarine
Fat-free milk   Mustard
       Diet mayonnaise
       Catsup

PLEASE DO NOT LEAVE MENU ON THE TRAY

NAME _____ ROOM _____

Lunch

LOW SODIUM   SUNDAY
LF = Low Fat   LSLF = Low Sodium, Low Fat
Meats
LSLF Baked chicken   LSLF Baked fish (cod)
LSLF Hamburger on bun
(with lettuce and tomato)
Starchy Vegetables
LSLF Rice   LSLF Boiled potatoes
Vegetables
LSLF Baby carrots   LSLF Green beans
Soup/Salad/Juice   **Dressings**
LSLF Coleslaw   Diet French
LS Chicken broth   Diet Thousand Island
Apple juice   Diet Italian
Tossed salad
Desserts
Angel food cake   Pears
       Fresh fruit
Breads
Dinner roll   Bran bread
White bread   LS Crackers
Wheat bread
Beverages & Condiments
Coffee   Sugar
Decaf. coffee   Sugar substitute
Hot tea   Creamer
Decaf. hot tea   Lemon
Iced tea   Herb seasoning
Whole milk   Margarine
2% Milk   Diet mustard
Fat-free milk   Diet mayonnaise
       Diet catsup
NO SALT

PLEASE DO NOT LEAVE MENU ON TRAY

NAME _____ ROOM _____

Lunch

RENAL   SUNDAY
LF = Low Fat   LSLF = Low Sodium, Low Fat
Meats (2 oz)
LSLF Baked chicken   LSLF Baked fish
LSLF Hamburger on bun (with lettuce)
Starchy Vegetables
LSLF Rice   LSLF Dialyzed potatoes
Vegetables
LSLF Baby carrots   LSLF Green beans
Soup/Salad/Juice   **Dressings**
Lemonade   Diet French
LSLF Coleslaw   Diet Thousand Island
Tossed salad   Diet Italian
(no tomato)
Desserts
Pears   Apple pie
Breads
Dinner roll   Bran bread
White bread   LS Crackers
Wheat bread
Beverages & Condiments
Coffee   Sugar
Decaf. coffee   Sugar substitute
Hot tea   Creamer
Decaf. hot tea   Lemon
Iced tea   Margarine
       Diet mustard
       Mayonnaise
NO SALT

PLEASE DO NOT LEAVE MENU ON THE TRAY

NAME _____ ROOM _____

People on low-fat, low-cholesterol diets who also need kcalorie restriction receive a kcalorie-restricted menu to control portion sizes and number of servings. Both menus provide low-fat, low-cholesterol foods. Foods not appropriate for a low-fat, low-cholesterol diet, such as whole milk, would be crossed off the menu beforehand.

Low-sodium menus are similar to those provided for low-fat, low-cholesterol diets, but they eliminate high-sodium foods, such as tomato soup. The person on a low-sodium, low-fat, low-cholesterol diet selects foods from a low-fat menu with high-sodium foods crossed off beforehand. If the person is also on a low-kcalorie diet, foods would be selected from a low-kcalorie menu with high-sodium foods crossed off the menu beforehand.

Renal diets must be highly individualized, and the person checking the menu has to carefully consider the client's selections and make appropriate changes when necessary.

considerable. Translating requests (for example, preparing another tray) takes time, and delays are unavoidable. The best strategy is prevention—make sure that clients mark menus and that requests are called in as early as possible to allow the foodservice department the time to process the request.

One of the most important things to know about a facility's foodservice system is the time of meal assembly so that you can call in requests well in advance. Once tray assembly begins, dietary employees are extremely busy, and requests will be difficult to process.

With so many people and steps involved in foodservice, and so many clients with individual dietary needs and food preferences, it is easy to see many opportunities for problems to arise. An understanding of how the foodservice department operates will enable you to tackle problems efficiently and avoid needless frustration.

Notes

[1] M. Bélanger and L. Dubé, The emotional experience of hospitalization. Its moderation and its role in patient satisfaction with foodservice, *Journal of the American Dietetic Association* 96 (1996): 354–360.

[2] M. A. Hess, Taste: The neglected nutritional factor, *Journal of the American Dietetic Association* (supplement 2) 97 (1997): 205–207.

[3] P. A. O'Hara and coauthors, Taste, temperature, and presentation predict satisfaction with foodservices in a Canadian continuing-care hospital, *Journal of the American Dietetic Association* 97 (1997): 401–405.

CHAPTER 17

ENTERAL FORMULAS

CONTENTS

■■ Enteral Formulas:
What Are They?

■■ Enteral Formulas:
Who Needs What?

■■ Tube Feedings:
How Are They Given?

■■ From Tube Feedings
to Table Foods

■■ Case Study: Graphics
Designer Requiring
Enteral Nutrition

■■ Nutrition in Practice:
Nutrition and Cost-
Conscious Health Care

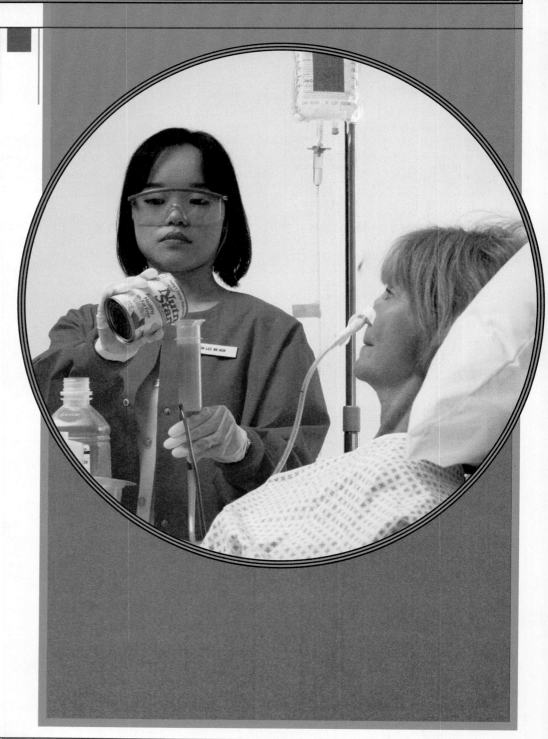

To meet nutrient needs, a person must be able to eat, digest, and absorb nutrients in the amounts necessary to satisfy metabolic demands. Most people, whether healthy or ill, can meet these needs with conventional foods. As Chapters 15 and 16 have shown, however, illnesses may interfere with eating, digestion, and absorption to such a degree that conventional foods fail to deliver necessary nutrients. If a poor appetite is the primary nutrition problem, **enteral formulas** given orally can help clients meet nutrient needs, if the clients can drink them in sufficient amounts.

For clients who cannot eat or drink enough to meet nutrient needs, however, it may be necessary to deliver nutrients by tube or by vein. Formulas provided either orally or by tube are enteral feedings, the subject of this chapter. Enteral feedings are possible whenever a client can digest and absorb nutrients via the GI tract. Otherwise, feedings are given by vein as parenteral feedings, the subject of the next chapter. (Figure 18-1 in the next chapter summarizes some of the factors involved in deciding the most appropriate way to feed a client.)

THE NURSING DIAGNOSES altered nutrition: less than body requirements is often appropriate for the person who needs an enteral formula.

> **enteral formulas:** liquid diets intended for oral use or for tube feedings.
> enteron = intestine

▪▪ Enteral Formulas: What Are They?

The number of enteral formulas on the market is staggering (some examples are listed in Appendix G). Most formulas are available in ready-to-use form or in powdered form. They are designed to meet a variety of medical and nutrition needs and can be either used alone or given along with other foods. The health care team, dietitian, or physician identifies each client's nutrient needs through a careful medical and nutrition assessment and then selects a formula that meets these needs.

Whenever formula is the primary source of nutrients, **complete formulas** are necessary. Such is the case when a client is on a tube feeding or an oral liquid diet for more than a few days. Complete formulas, when given in appropriate amounts, supply all the nutrients a client needs. Complete formulas can also be (and often are) used in smaller quantities to supplement table foods.

> **complete formulas:** formulas designed to supply all needed nutrients when provided in sufficient volume.

▪▪ Types of Formulas

Formulas are classified in many ways, but for purposes of this book, it is reasonable to think of two major kinds categorized by the type of protein they supply. Standard formulas contain complete proteins, whereas hydrolyzed formulas contain fragments of proteins, including free amino acids, dipeptides, and tripeptides.

▪▪ STANDARD FORMULAS   **Standard formulas** are appropriate for people who are able to digest and absorb nutrients without difficulty. Most standard formulas contain one or a combination of **protein isolates** (purified proteins). A few formulas, called *blenderized formulas,* derive their protein primarily from pureed meat (a mixture of whole proteins).

▪▪ HYDROLYZED FORMULAS   To simplify the body's digestive work, a complete protein can be hydrolyzed—that is, partially broken down to yield small peptides. Alternatively, a formula can be made from free amino acids. For simplicity, this text calls both types hydrolyzed. **Hydrolyzed formulas** are

> **standard formula:** a liquid diet that contains complete molecules of protein; also called **intact** or **polymeric formula.**
>
> **protein isolate:** a protein that has been separated from a food. Examples include casein from milk and albumin from egg.
>
> **hydrolyzed formula:** a liquid diet that contains broken-down fragments of protein, such as amino acids and short peptide chains; also called **monomeric formula.**

often low in fat or provide some fat from medium-chain triglycerides (MCT) to ease digestion and absorption. People who cannot digest nutrients well may benefit from hydrolyzed formulas.

■■ **MODULAR FORMULAS** Unlike complete formulas, a few formulas, called **modules,** provide essentially a single nutrient (protein, carbohydrate, or fat). In addition to commercial modular formulas, intravenous nutrients and even table foods (vegetable oil or corn syrup, for example) can serve as modules. Modules can then be added to enteral formulas to alter nutrient composition (for example, to add kcalories or protein). Modules can also be combined with other modules and liquid vitamin and mineral preparations to construct individualized formulas for clients with unique nutrient needs. Designing, preparing, and delivering such a formula is a challenge that requires in-depth nutrition knowledge and skills.

■■ **Distinguishing Characteristics**

Formulas differ not only in the form of protein they contain but also in the amount of energy they provide and the source of energy nutrients. Although formulas differ, most fit into general categories and can be used interchangeably. For example, standard formulas provide similar amounts of energy and nutrients and require the same degree of digestive function. They may derive their nutrients from different sources, but the sources are all of similar quality. Thus, the physician or dietitian often has many choices in selecting a formula for an individual client.

■■ **NUTRIENT DENSITY** Standard formulas provide about 1.0 kcalorie per milliliter. Nutrient-dense formulas provide 1.2 to 2.0 kcalories per milliliter and meet energy and nutrient needs in a smaller volume. Thus, nutrient-dense formulas often benefit clients with high nutrient needs or those requiring fluid restrictions. Formulas also vary in the percentage of energy provided from protein, fat, and carbohydrate.

■■ **RESIDUE AND FIBER** The positive health effects of dietary fibers suggest that fiber-enriched formulas would be the best choice for most people. Why, then, do many standard formulas have a low-to-moderate residue content? The answer is that formulas are most often used for relatively short periods of time and low-to-moderate residue formulas are least likely to cause gas and abdominal distention. Consequently, they are often well tolerated by many who need them: people with GI tract disorders (inflammatory bowel diseases or partial obstructions, for example), those who have undergone surgeries of the GI tract, or those beginning enteral nutrition after periods of GI tract disuse.

Fiber-enriched formulas are available for people who need them, however. People who depend on tube feedings for long periods of time, those with constipation, and some people with short-bowel syndrome (see Chapter 16) are likely to benefit from fiber-enriched formulas. One type of dietary fiber, pectin, may be effective in controlling some forms of diarrhea in people receiving enteral formulas.[1]

■■ **OSMOLALITY** **Osmolality** is a measure of the concentration of molecular and ionic particles in a solution. A formula that approximates the osmolality of the blood serum (about 300 milliosmoles per kilogram) is referred to as an **isotonic formula.** A **hypertonic formula** has a higher osmolality than serum.

Most people tolerate both isotonic and hypertonic formulas without difficulty. When hypertonic formulas are delivered directly into the intestine,

---

modules: formulas or foods that provide primarily a single nutrient and are designed to be added to other formulas or foods to alter nutrient composition. They can also be combined together to create a highly individualized formula.

Caution: Although intravenous nutrients can be used enterally, the reverse is not true. Enteral formulas cannot be delivered by vein without serious consequences.

For practical purposes, 1 ml (milliliter) is equivalent to 1 cc (cubic centimeter).

Since hydroloyzed formulas are almost completely absorbed, they leave little residue in the intestine.

osmolality (OZ-mow-LAL-eh-tee): (a measure of the concentration of particles in a solution, expressed as the number of milliosmoles (mOsm) per kilogram.

isotonic formula: a formula with an osmolality similar to that of blood serum (300 mOsm/kg).
  iso = the same
  ton = tension

hypertonic formula: a formula with an osmolality greater than that of blood serum.
  hyper = greater, more

however, the hyperosmolar load can result in diarrhea, a situation analogous to the dumping syndrome (see Chapter 15). To prevent this problem, hypertonic formulas delivered into the intestine are initially delivered at a slow, even rate and gradually increased as tolerated. In some cases, the formula may be diluted at first, the strength gradually increased, and then the rate increased.

Like table foods, enteral formulas can meet a variety of dietary needs. With that in mind, it is important to identify which people benefit from enteral formulas.

Enteral Formulas: Who Needs What?

All people with functional GI tracts who cannot get the nutrients they need from table foods can potentially benefit from enteral formulas. They could also get nutrients from parenteral nutrition, but enteral nutrition is preferable whenever it is possible. Compared to parenteral nutrition, enteral nutrition helps maintain normal gut function better, causes fewer complications, and is less costly. Enteral feedings help stimulate intestinal adaptation following intestinal resections and long periods of GI tract disuse.

Similarly, oral feedings are preferred to tube feedings whenever a person can drink enough of the formula. In so doing, clients avoid the stress of having a feeding tube inserted and nurses save valuable time. Compared to feeding by tube, orally provided formulas are less costly and less likely to result in complications.

Oral Feedings

Formulas provided orally can meet all nutrient needs for people who can tolerate only liquids and those who need hydrolyzed formulas. If people can drink enough of the formula, they can avoid being fed by tube.

More often, formulas provided orally supplement a conventional diet. Some people can eat table foods, but not in the quantities they need to meet nutrient needs. Enteral formulas provide a reliable source of nutrients and work particularly well for adding energy and protein to the diet. Psychologically, liquids seem less filling than foods, and they are easier for debilitated, weak, or tired clients to handle.

When a client uses an oral formula, taste becomes an important consideration. Allowing clients to sample different products and flavors and select the ones they like best helps ensure acceptance.[2]

Tube Feedings

Tube feedings are simply complete formulas delivered through a tube into the stomach or intestine. An individual who has a functional GI tract but is unable to eat enough food (or the appropriate type) by mouth to meet nutrient needs may need a tube feeding. Candidates for tube feedings include:

- People with physical problems that seriously interfere with chewing and swallowing.
- People with no appetite for an extended time.
- People with a partial obstruction, fistula, or altered motility in the upper GI tract.
- People in a coma.
- People with high nutrient requirements.
- People who are unable to ingest a hydrolyzed formula orally.

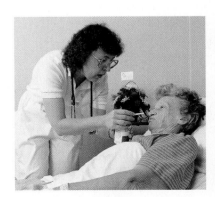

Formula supplements often help meet nutrient needs when clients cannot eat enough conventional foods.

Figure 17-1
FEEDING TUBE PLACEMENT SITES

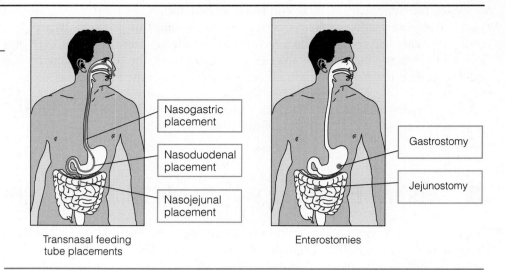

Transnasal feeding tube placements

Enterostomies

The final location of the feeding tube determines the type of feeding. If a feeding tube is passed through a gastrostomy into the duodenum or jejunum, the feeding is intestinal rather than gastric.

■■ **FEEDING TUBE PLACEMENT** Feeding tubes are inserted into different locations along the GI tract depending on the client's medical problems and the estimated length of time that the feeding will be required. Figure 17-1 shows various feeding tube placement sites, and the accompanying glossary describes these sites.

When clients are expected to be on tube feedings for less than four weeks, feeding tubes are frequently inserted through the nose and passed into the stomach or intestine. For infants, feeding tubes are often inserted through the mouth to the stomach before each feeding and removed immediately after the feeding to allow the infant to breathe easily and reduce the risk of aspiration. When a client will be on a tube feeding for longer than four weeks, or when a feeding tube cannot be passed through the nose, esophagus, or stomach due to an obstruction or for other medical reasons, an opening can be made into the stomach or jejunum. Tube enterostomies can be made either

GLOSSARY

FEEDING TUBE PLACEMENT SITES

These terms are listed in order from the nose to lower organs of the digestive system.

transnasal: through the nose. A **transnasal feeding tube** is one that is inserted through the nose.

   naso = nose

nasogastric (NG): from the nose to the stomach.

nasoenteric: from the nose to the stomach or intestine. *Nasoenteric feedings* include nasogastric, nasoduodenal, and nasojejunal feedings. Most clinicians use nasoenteric to refer to nasoduodenal and nasojejunal feedings only.

nasoduodenal (ND): from the nose to the duodenum.

nasojejunal (NJ): from the nose to the jejunum.

orogastric: from the mouth to the stomach. This method is often used to feed infants because they breathe through their noses, and tubes inserted through the nose can hinder the infant's breathing. The tube is inserted before, and removed after, each feeding.

enterostomy (EN-ter-OSS-toe-mee): a gastric or jejunal opening made surgically or under local anesthesia through which a feeding tube can be passed.

gastrostomy (gas-TROSS-toe-mee): an opening in the stomach made surgically or under local anesthesia through which a feeding tube can be passed. The technique for creating a gastrostomy

under local anesthesia is called **percutaneous endoscopic gastrostomy,** or **PEG** for short. When the feeding tube is guided from such an opening into the jejunum, the procedure is called **percutaneous endoscopic jejunostomy (PEJ),** a misnomer because the enterostomy site is in the stomach.

jejunostomy (JEE-ju-NOSS-toe-mee): an opening in the jejunum made surgically or under local anesthesia through which a feeding tube can be passed. The technique for creating a jejunostomy under local anesthesia is called a **direct endoscopic jejunostomy (DEJ).** Note: Some clinicians also refer to this procedure as a PEJ, which is a more accurate use of the term than the more common use described above.

Table 17-1
COMPARISON OF FEEDING TUBE SITES[a]

Insertion Method and Feeding Site	Advantages	Disadvantages
Transnasal	Does not require surgery or incisions for placement.	Easy to remove by disoriented clients; long-term use may irritate the nasal passages, throat, and esophagus.
Nasogastric	Easiest to insert and confirm placement; feedings can often be given intermittently and without an infusion pump.	Highest risk of aspiration in compromised clients.
Nasoduodenal and nasojejunal	Lower risk of aspiration in compromised clients; allow for enteral nutrition earlier than gastric feedings following severe stress; may allow for enteral feeding when partial obstructions, fistulas, or other medical conditions prevent gastric feeding.	More difficult to insert and confirm placement; feedings require an infusion pump for administration; may take longer to reach nutrition goals.
Tube enterostomies	Allow gastroesophageal sphincter to remain closed, lowering the risk of aspiration; more comfortable than transnasal insertion for long-term use.	May require general anesthesia for insertion; require incisions; greater risk of complications from the insertion procedure; greater risk of infection; may cause skin irritation around the insertion site.
Gastrostomy	Feedings can often be given intermittently and without a pump; easier to insert than a jejunostomy.	Moderate risk of aspiration in high-risk clients.
Jejunostomy	Lowest risk of aspiration; allows for enteral nutrition earlier following severe stress; may allow for enteral feeding when partial obstructions, fistulas, or medical conditions prevent gastric feeding.	Most difficult to insert; feedings require an infusion pump for administration; may take longer to reach nutrition goals.

[a]Relative to each tube feeding site. The actual advantages and disadvantages of different insertion procedures depend on the person's medical condition.

surgically or nonsurgically using local anesthesia. Table 17-1 compares some of the features of various tube feeding sites. The "How to" box on p. 412 suggests ways to reduce anxiety for clients beginning a tube feeding.

■■ **FEEDING TUBE CHARACTERISTICS** Feeding tubes are soft and flexible and come in a variety of diameters and lengths. Many have special characteristics that make them desirable for specific purposes. For example, tubes with double lumens allow for gastric decompression and intestinal feedings at the same time.

The particular feeding tube to use depends on the client's age and size, medical condition, how the tube will be placed (transnasally or through an enterostomy), and how far the final placement will be from the insertion site (from the nose to the stomach or intestine, for example). Once the appropriate length is selected, the smallest tube through which the formula will flow without clogging the tube is selected. Unclogging a tube is a difficult procedure that interrupts the feeding schedule, and inserting a new tube causes stress and anxiety for the client and raises the cost of the feeding.

■■ **FORMULA SELECTION** To select an appropriate formula requires a logical approach. Figure 17-2 on p. 413 shows some of the considerations involved. In a nutshell, the formula that meets the client's medical and nutrient needs with the lowest risk of complications and at the lowest cost is the best choice. If no formula can be found that meets the client's needs, then modules can be used to create an appropriate formula.

Selecting a formula that meets the client's nutrient needs is paramount. Nutrient requirements are estimated based on a careful assessment of the client's age, nutrition status, medical condition, ability to digest and absorb nutrients, and metabolic rate. Standard formulas are appropriate for the vast majority of clients. Clients with a functional, but impaired, GI tract may benefit from hydrolyzed formulas. Besides the client's ability to digest and absorb

The outer diameter of a feeding tube is measured using the French scale, where each unit is about one-third of a millimeter. Thus, the outer diameter of a 10 French feeding tube is a little over 3 mm. The inner diameter varies depending on the thickness of the material used to construct the tube.

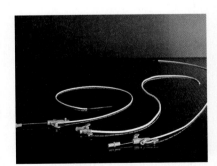

Feeding tubes come in many lengths and diameters. The thin wires protruding from the end of the feeding tube are stylets, which stiffen the tube to ease insertion and are discarded thereafter. The Y connector (shown here in yellow) provides a port for administering medications and water without disrupting the feeding.

HELP CLIENTS COPE WITH TUBE FEEDINGS

The thought of being "force-fed" is frightening to many people. One person may envision a large feeding tube and fear that the procedure will be extremely painful. Another may have heard about tube feedings only from the popular press and associate them with irreversible comas. All clients benefit when they understand the insertion procedure, the expected duration of the tube feeding, and the strategic role that nutrition plays in recovery from disease. These pointers can help nurses prepare clients for transnasal tube feedings:

- Allow clients to see and touch the feeding tube. Seeing firsthand that the tube is soft and narrow (only about half the diameter of a pencil) often alleviates anxiety. Show clients how the feeding apparatus is attached to the feeding tube, and explain how the feeding will work. Use dolls or stuffed toys to demonstrate tube insertion and feeding procedures to a young child.
- Explain that the client remains fully alert during the procedure and helps pass the tube by swallowing. A numbing solution sprayed on the back of the throat minimizes discomfort and prevents gagging during the procedure.
- Tell the client that once the tube has been inserted, most people become accustomed to its presence within a few hours. In most cases, the client can easily swallow foods and liquids with the tube in place. If permitted, favorite foods or beverages can still be enjoyed.
- Assure the client that the tube feeding will be temporary, if such assurance is appropriate.

Although a tube feeding may be frightening for some, for others, it is a relief. People who understand that they should eat, but cannot do so, may be relieved to receive sound nutrition without any effort. As they feel better and begin to eat again, the volume of the tube feeding can often be reduced and then discontinued when oral intake is adequate.

Some people feel a loss of control over their lives; others feel self-conscious about how the feeding tube looks or awkward about moving around with the equipment. A few simple measures can help:

- Involve older children, teens, and adults in the decision-making and care process whenever possible. Clients can help arrange daily feeding schedules, and some can also perform many of the feeding procedures themselves.
- Show clients how to manipulate the feeding equipment so that they can get out of bed and move around.
- Recommend that clients walk around and socialize with others, if permissible.
- Recommend that clients maintain contact with friends and keep busy with hobbies and activities they enjoy. This measure is especially important for children, teens, and those on long-term feedings.
- For infants and children, keep the developmental age of the child in mind and work with parents to ensure that appropriate feeding skills are mastered (see Nutrition in Practice 15). For infants, providing a pacifier during feedings helps maintain the associations between sucking, swallowing, eating, and fullness. When possible, some of the tube feeding formula may be provided by bottle or by spoon to further develop skills.

The more complex the procedure that a health care professional is responsible for, the easier it becomes to focus on the procedure and forget about the client's emotions. No matter how many technicalities you have to keep in mind, remember to stay focused on the person receiving your care.

Figure 17-2
SELECTING A FORMULA

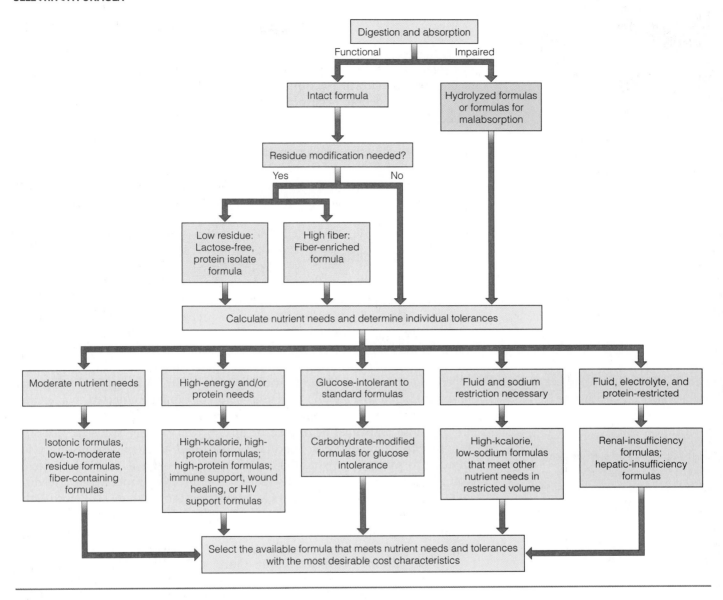

nutrients, other nutrition-related factors that affect the selection of a formula include:

- *The client's energy, protein, fluid, and nutrient needs.* High nutrient needs must be met in the volume of formula the client can tolerate. If nutrients (including water) must be restricted, the selected formula can deliver only the prescribed amount of nutrients in a day's volume.
- *The need for residue or fiber modifications.* The choice of formulas is narrowed when a person needs a low-residue or a high-fiber diet.
- *Individual tolerances (food allergies and sensitivities).* Most formulas are lactose-free because temporary and permanent lactose intolerances are common problems for people who need enteral formulas.

In addition, health care facilities cannot stock all formulas, so formula selection is limited by availability. In the final analysis, the dietitian or physician can make only an educated guess in selecting the best formula for an individual. The health care team monitors each person's nutrition status and responses to the formula to help ensure that individual needs are being met.

Tube Feedings: How Are They Given?

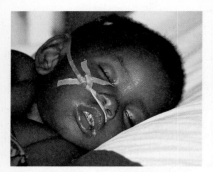

The safe delivery of a tube feeding helps clients meet their medical and nutrition goals with minimal discomfort.

Once a feeding route has been selected and the feeding tube inserted, attention turns to delivering the formula. Just as the actual foods provided on special diets may vary from one health care facility to the next, the exact procedures for safely providing tube feedings to clients vary as well. The procedures presented in the following sections are suggested guidelines.

Many people beginning a tube feeding are seriously ill or malnourished. Using the safest methods of preparing and administering the formula helps minimize the risk of complications, which can delay or prevent the attainment of medical and nutrition goals. Of all health care professionals, nurses hold the greatest responsibility for ensuring that formulas are safely administered. Although the following sections specifically address the preparation and administration of tube feedings in health care institutions, the principles apply to people who receive tube feedings at home as well.

Safe Handling

open system: a system where formulas are decanted from their original packaging and then added to a feeding container before they can be administered through a feeding tube.

closed system: a system where formulas come prepackaged and ready to be attached to a feeding tube for administration.

People who are ill or malnourished may have suppressed immune systems that make them vulnerable to infection from food-borne illness. To prevent contamination, all personnel involved in preparing or delivering formulas should handle them only in clean environments, using clean equipment and clean hands.

Formulas are available in both **open systems** and **closed systems**. Open systems require that a formula be removed from its original packaging and transferred to a feeding container before it can be connected to a feeding tube and delivered to the client. Formulas that come in standard cans and bottles or those that must be diluted or mixed from powder are examples. The foodservice department or pharmacy most often assumes responsibility for diluting formulas or mixing them from powder. Formulas in closed systems come prepackaged in containers that can be directly connected to a feeding tube. In the case of either open or closed systems, formulas are labeled with the client's name, room, date, and time of preparation (if necessary) and then sent to the nursing station.

AT THE NURSING STATION Once a formula reaches the nursing station, the nursing staff assumes responsibility for its safe handling. The following steps reduce the likelihood of formula contamination when an open feeding system is used:

- Before opening a can of formula, carefully clean the can opener and the lid. If you do not use the entire can at one feeding, label the can with the time it was opened.
- Store and refrigerate the unused portion of formula promptly.
- Discard unlabeled or improperly labeled containers and all opened containers of formula not used within 24 hours.

AT BEDSIDE Prevent the risk of bacterial infections by following these procedures:

- Hang no more than an 8-hour supply of formula administered through an open system or no more than a 24-hour supply of formula administered through a closed system.
- Never add fresh formula to formula that has been hanging.
- Change the feeding container and the tubing attached to it (except the feeding tube itself) every 24 hours.

Closed feeding systems save nursing time and may significantly reduce the risk of bacterial contamination.

Initiating and Progressing a Tube Feeding

Two of the most serious complications associated with tube feedings are the inadvertent placement of a transnasal tube into the respiratory tract and the aspiration of formula from the stomach into the lungs. In both cases, potentially fatal complications can follow. To minimize the risk of such complications, most facilities use X rays to verify the position of the feeding tube before a feeding is initiated and confirm the tube's position several times a day. Because lying flat increases the risk of aspiration, the client's upper body is elevated to at least a 30- to 45-degree angle during the feeding and for 30 minutes after the feeding whenever possible.[3]

THE NURSING DIAGNOSIS **risk for aspiration** applies to people on tube feedings, especially those receiving gastric feedings.

FORMULA DELIVERY TECHNIQUES A day's volume of formula can be given in relatively large amounts at intervals or in relatively small amounts continuously throughout the day. A client may start on a continuous feeding and gradually be converted to an intermittent feeding. Each method has specific uses, advantages, and disadvantages.

INTERMITTENT FEEDINGS **Intermittent feedings** are best tolerated when they are delivered into the stomach (not intestine) and no more than 250 to 400 milliliters is given over 30 minutes or more using the gravity drip method or an infusion pump. The larger the volume of formula needed to meet nutrient needs, the more frequently feedings are delivered. (People who have very high nutrient needs benefit from either high-nutrient-density formulas or continuous feedings.) Rapid delivery (in 10 minutes or less) of a large volume of formula (250 to 300 milliliters)—called **bolus feeding**—often leads to complaints of abdominal discomfort, nausea, fullness, and cramping. This makes sense. After all, people do not gobble down a meal in just a few minutes, especially when they are not feeling well.

Intermittent feedings may be difficult for a client to tolerate, and because there is a greater volume of formula in the stomach at one time, the risk of aspiration is higher than with continuous feedings. Intermittent feedings work well for clients able to tolerate them, however. Often clients gradually adapt to larger volumes of formula given over shorter periods of time. Such feedings mimic the usual pattern of eating and allow the client freedom of movement between meals. They also require less time, making them less costly and easier for people to use at home.

CONTINUOUS FEEDINGS **Continuous feedings** are delivered slowly and in constant amounts over a period of 8 to 24 hours. Such feedings benefit people who have received no food through the GI tract for a long time, those who are hypermetabolic, and those receiving intestinal feedings. Although continuous feedings are often well tolerated and are associated with a relatively low risk of aspiration, they require infusion pumps to help ensure accurate and constant flow rates; the pump makes the feedings more costly to provide and limits the client's freedom of movement. The "How to" box on p. 416 explains several ways to plan tube feeding schedules.

FORMULA VOLUME AND STRENGTH Recommended formula administration schedules vary among institutions; the ones provided here are

To assist in the early identification of aspiration, food coloring is sometimes added to formulas so that the formula can be easily distinguished from pulmonary secretions.

intermittent feeding: the delivery of about 250 to 400 ml of formula over 30 minutes or more.

bolus feeding: the delivery of about 250 to 300 ml of formula over 10 minutes or less.

continuous feeding: slow delivery of formula in constant amounts over an 8- to 24-hour period.

Caution: The bright lights, beeps, and controls of an infusion pump may make playing with the pump irresistible to young children. Keep the infusion pump at a safe distance to prevent a child from toppling the pump or IV pole, possibly injuring the child or damaging the equipment.

PLAN A TUBE FEEDING SCHEDULE

After selecting a formula that meets the client's medical and nutrient needs, the planner, usually a dietitian, determines the volume of formula per day that meets those needs. Consider a client who needs 2000 milliliters of formula per day. If the client is to receive the formula intermittently six times a day, he needs about 330 milliliters of formula at each feeding (2000 ml ÷ 6 feedings = 333 ml/feeding). Alternatively, if he is to receive the same volume of formula eight times a day, then he needs 250 milliliters (or about one can of ready-to-feed formula) at each feeding (2000 ml ÷ 8 feedings = 250 ml/feeding). He will probably tolerate this volume of formula best if it is given to him over 30 minutes or more at each feeding. If the client is to receive the formula continuously over 24 hours, he needs about 85 milliliters of formula each hour (2000 ml ÷ 24 hr = 83 ml/hr).

examples. Almost all people can receive undiluted formula (either isotonic or hypertonic) at the start of a feeding. Formulas are given slowly at first and the volume gradually increased. In rare cases, formulas may need to be diluted initially, the strength increased gradually, and then the rate advanced.

Intermittent feedings may start at about 100 to 150 milliliters at each feeding and then be increased by 50 to 100 milliliters each day until the goal volume is reached.[4] Continuous feedings may start at about 25 to 50 milliliters per hour. If the person tolerates the formula, the rate of feeding can be increased by about 25 milliliters per hour every 4 to 12 hours (see the margin note), depending on the location of the feeding tube (gastric or intestinal), the person's medical condition, and the nutrient density of the formula. For both intermittent and continuous feedings, if the new rate is not tolerated, back up and proceed more slowly, giving the person more time to adapt. If clients on an intermittent feeding cannot tolerate the feeding, they should be switched to a continuous feeding schedule.

> *As an example of a volume progression for a tube feeding, start the feeding at 50 ml/hr at full strength and then progress as follows:*
> • *After 6 hours: 75 ml/hr.*
> • *After 12 hours: 100 ml/hr.*
> • *After 18 hours: 125 ml/hr.*

■■ SUPPLEMENTAL WATER In addition to the formula itself, water provided through the feeding tube helps to hydrate the client and to prevent clogged feeding tubes. As a guideline, adults require about 2000 milliliters (approximately 2 quarts) of water daily. In kidney, liver, and heart diseases, water may need to be restricted. Clients with fever, excessive sweating, severe vomiting, diarrhea, fistula drainage, high-output ostomies, blood loss, or open wounds require supplemental water.

> *The following guideline is one of several used to estimate fluid requirements for adults and children:*
> • *Allow 100 ml/kg for each of the first 10 kg of body weight.*
> • *Allow 50 ml/kg for each of the next 10 kg of body weight.*
> • *Allow 20 ml/kg for each kg of body weight above 20 kg.*

THE NURSING DIAGNOSIS fluid volume excess or **fluid volume deficit** describes clients who are receiving too much or too little water to meet needs.

Attention to indicators of body water balance can help determine how much additional water an individual needs. In alert adults, thirst is a good indicator of water needs; a person complaining of thirst generally needs water. In the elderly, however, thirst may be slow to develop in response to dehydration. Tables 13-10 and 13-11 on pp. 332 and 333, respectively, listed physical and laboratory indices of dehydration.

> *Formulas themselves contain considerable amounts of water and meet a substantial portion of clients' water needs. Standard formulas contain about 850 ml of water per liter of formula. Higher-kcalorie formulas contain less water: formulas that contain 1.5 kcal/ml or 2.0 kcal/ml provide about 775 and 600 ml of water per liter of formula, respectively.*

THE NURSING DIAGNOSES risk for fluid volume deficit is indicated for people on tube feedings, especially for infants, the elderly, and people who are unconscious.

Clients often receive water by syringe or gravity bag at room (for gastric feedings) or body (for intestinal feedings) temperature.[5] Automatic flush pumps that deliver a selected volume of water each hour are also available.[6] For clients on continuous feedings, the feeding tube is generally flushed with water every 4 hours. For clients on intermittent feedings, the feeding tube is generally flushed with water after each feeding.

■■ GASTRIC RESIDUALS   Nurses measure gastric residuals to ensure that the stomach is emptying properly and to prevent nausea, vomiting, and possible aspiration of formula into the lungs. The gastric residual is the volume of formula that remains in the stomach from a previous feeding. It is measured by gently withdrawing the gastric contents through the feeding tube using a syringe. The gastric residual is measured before each feeding for intermittent feedings and every 4 hours for continuous feedings. The gastric residual that is considered excessive varies among facilities, but ranges from about 75 to 150 milliliters. If the residual is excessive, the feeding is held for an hour or two, and then the residual is rechecked. If excessive residuals persist, the physician may withhold the feeding, reduce the rate of administration, or begin drug therapy to stimulate gastric emptying.

■■   **Delivering Medications through Feeding Tubes**

Clients receiving tube feedings are also likely to be receiving numerous medications. Often these medicines are delivered through feeding tubes, and in some cases, complications can occur.

Keep in mind that enteral formulas can interact with medications in the same ways that foods can. The vitamin K in a formula can interact with the drug warfarin (see Chapter 13) just as the vitamin K in foods can. The health care team must consider the effects of drug therapy on nutrient requirements, the effects of the formula on medication absorption, the effects of medications on the physical properties of formulas, and the prevention of complications.

■■ MEDICATION FORMS   A medication may come in any of several forms including tablets, liquids, injectables, and intravenous. People on tube feedings have functional GI tracts, and oral medications are less costly and easier to deliver than injectable or intravenous forms. Thus, clinicians often prefer to use oral medications for clients on tube feedings. The following guidelines may help prevent medication-medication interactions, medication-formula interactions, or clogged feeding tubes when medications are delivered through feeding tubes:

- Give medications by mouth instead of by tube whenever possible.
- Use the liquid form of medications whenever available. Dilute hypertonic or thick liquid medications with water before administering them through the feeding tube.
- Use injectable or intravenous forms of medications if they are not available in liquid form.
- Ideally, tablets should not be crushed and administered through feeding tubes. If using tablets is unavoidable, crush the tablets to a fine powder and mix with water before administering them.
- Never crush time-released tablets or enteric-coated medications. In these cases, another medication form must be given.
- Unless the compatibility of multiple medications is known, do not mix medications together or mix medications with the formula hanging in the feeding container. Instead, give each medication individually using the separate port whenever available.
- Flush the feeding tube with warm (body temperature) water before and after administering each medication.

Like hypertonic formulas, hypertonic liquid medications can cause diarrhea. Examples of hypertonic medications include potassium chloride elixir (a potassium supplement), liquid multivitamin preparations, Tagamet liquid, and theophylline elixir.

Table 17-2

SELECTED MEDICATIONS THAT ARE INCOMPATIBLE WITH SOME FORMULAS

Aluminum hydroxide	MCT oil
Chlorpromazine concentrate	Mellaril concentrate
Cibalith-S syrup	Mellaril oral solution
Cimetidine	Paregoric elixir
Dimetane elixir	Potassium chloride
Dimetapp elixir	Reglan syrup
Feosol elixir	Riopan
Fleet's phosphosoda	Robitussin expectorant
Gevrabon liquid	Sudafed syrup
Klorvess syrup	Thorazine concentrate
Mandelamine Forte suspension	Zinc sulfate capsules

NOTE: These substances may be compatible with some formulas and not others.
SOURCES: P. E. Burns, L. McCall, and R. Wirsching. Physical compatibility of enteral formulas with various common medications, *Journal of the American Dietetic Association* 88 (1988): 1094–1096; A. J. Cutle, E. Altman, and L. Lenkel, Compatibility of enteral products with commonly employed drug additives, *Journal of Parenteral and Enteral Nutrition* 7 (1983): 186–191; Z. M. Pronsky, *Food-Medication Interactions*, 11th ed. (Pottstown, Pa.: Food-Medication Interactions, 2000).

- Avoid medications known to be incompatible with formulas. Bulk-forming agents, for example, can clog the feeding tube. Other medications that may be incompatible with some formulas are listed in Table 17-2.

■■ **GI SIDE EFFECTS** Medications can trigger GI side effects, including nausea, vomiting, and diarrhea, that are also frequent complications of tube feedings. Thus, medications should be investigated as possible causes of such complications. Medications are a frequent culprit of diarrhea associated with tube feedings. In one study, the most common cause of diarrhea (responsible for about half the cases) in clients receiving tube feeding was medications containing sorbitol.[7]

■■ **ADDITIONAL CONSIDERATIONS** The location of the feeding tube (whether gastric or intestinal) is also relevant when administering medications. Medications designed to dissolve in the mouth should not be given through a feeding tube. Medications that depend on the stomach's acidic environment for absorption may be poorly absorbed if delivered directly into the duodenum or jejunum. Similarly, a medication that is optimally absorbed in the duodenum may be poorly absorbed in the jejunum. In such cases, oral, intravenous, or injectable forms of the medication should be provided.

In some cases, formulas can alter medication absorption. One example is phenytoin, a medication used to control seizures. Absorption of phenytoin may be markedly reduced when a person is on continuous tube feedings. Although opinions of the best way to handle this problem vary, some clinicians suggest that feedings should be stopped for 2 hours before and 2 hours after giving phenytoin.[8] For clients requiring continuous feedings, the rate of delivery is increased during the times the feeding is given to ensure that nutrient needs are met.

Some clients have a specific type of jejunostomy, called a needle catheter jejunostomy, *through which phenytoin cannot be delivered. In such cases, phenytoin is given intravenously.*

■■ **Addressing Tube Feeding Complications**

Table 17-3 summarizes complications associated with tube feedings and shows that many problems can be prevented or corrected by selecting the formula and feeding route wisely, preparing the formula correctly, and delivering it

Table 17-3
CAUSES AND PREVENTION OR CORRECTION OF TUBE FEEDING COMPLICATIONS

Complications	Possible Causes	Preventive/Corrective Measures
Aspiration of formula	Compromised gastroesophageal sphincter, delayed gastric emptying	Use nasoenteric, gastrostomy, or jejunostomy feedings in high-risk clients; check tube placement; elevate head of bed during and 45 minutes after feeding; check gastric residuals.
Clogged feeding tube	Formula too thick for tube	Select appropriate tube size; flush tubing with water before and after giving formula; use infusion pump to deliver thick formulas; remedies reported to help unclog feeding tubes include cola, cranberry juice, meat tenderizer, and pancreatic enzymes.
	Medications delivered through feeding tube	Use oral, liquid, or injectable medications whenever possible; dilute thick or sticky liquid medications with water before administering; crush tablets to a fine powder and mix with water; flush tubing with water before and after medications are given; give medications individually; do not add medications to the feeding container.
Constipation	Low-fiber formula	Provide additional fluids; use high-fiber formula.
	Lack of exercise	Encourage walking and other activities, if appropriate.
Dehydration and electrolyte imbalance	Excessive diarrhea	See items under *Diarrhea*.
	Inadequate fluid intake	Provide additional fluid.
	Carbohydrate intolerance	Use continuous drip administration of formula; monitor blood glucose; select a formula with a lower amount or different type of carbohydrate.
	Excessive protein intake	Monitor blood electrolyte levels; reduce protein intake.
Diarrhea, cramps, abdominal distention	Bacterial contamination	Use fresh formula every 24 hours; store opened or mixed formula in a refrigerator; rinse feeding bag and tubing before adding fresh formula; change feeding apparatus every 24 hours; prepare formula with clean hands using clean equipment in a clean environment.
	Lactose intolerance	Use lactose-free formula in lactose-intolerant and high-risk clients.
	Hypertonic formula	Use small volume of formula and increase volume gradually.
	Rapid formula administration	Slow administration rate or use continuous drip feedings.
	Malnutrition/low serum albumin	Use small volume of dilute formula and increase volume and concentration gradually.
Hyperglycemia	Diabetes, hypermetabolism, drug therapy	Check blood glucose; slow administration rate; provide adequate fluids; select a formula with a lower amount or different type of carbohydrate.
Nausea and vomiting	Obstruction	Discontinue tube feeding.
	Delayed gastric emptying	Check gastric residual; slow administration rate, use continuous drip feedings, or discontinue tube feeding.
	Intolerance to concentration or volume of formula	Use small volume of formula and increase volume and concentration gradually; use continuous drip feedings.
	Psychological reaction to tube feeding	Address client's concerns.
Skin irritation at enterostomy site	Leakage of GI secretions and friction caused by the tube	Keep site clean; inspect area for redness, tenderness, and drainage; use protective skin cream.

NOTE: Many of the complications presented here can be caused by the client's primary disorder or drug therapy rather than the tube feeding itself. In such a case, the corrective measure would include treatment of the disorder or a change in drug therapy. Additionally, other corrective measures that require a physician's order are not shown here.

appropriately. Attention to the person's primary medical condition and medications is important as well. Table 17-4 on p. 420 provides a sample monitoring schedule that helps ensure early detection of problems that may be encountered when clients are fed by tube.

■■ **TYPES OF COMPLICATIONS** Failure to estimate nutrient needs correctly or to ensure that the selected formula meets these needs limits the client's ability to achieve or maintain adequate nutrition status. Mechanical problems, such as a clogged feeding tube, a malfunctioning feeding pump, or a tube that becomes dislodged from its appropriate location, can interrupt the feeding schedule. Other complications related to the formula or its administration can

Table 17-4
SUGGESTED GUIDELINES FOR MONITORING CLIENTS ON TUBE FEEDINGS

Before starting a new feeding:	Complete a nutrition assessment.
	Check tube placement.
Before each intermittent feeding:	Check client position.
	Check gastric residual.
	Check tube placement.
After each intermittent feeding:	Flush feeding tube with water
Every half hour:	Check gravity drip rate, when applicable.
Every hour:	Check pump drip rate, when applicable.
Every 4 hours:	Check vital signs, including blood pressure, temperature, pulse, and respiration.
Every 6 hours:	Check blood glucose; monitoring blood glucose can be discontinued after 48 hours if test results are consistently negative in a nondiabetic client.
Every 4 to 6 hours of continuous feeding:	Check client position
	Check gastric residual.
	Flush feeding tube with water.
Every 8 hours:	Check intake and output.
	Check specific gravity of urine.
	Check tube placement.
	Chart client's total intake of, acceptance of, and tolerance to tube feeding.
Every day:	Weigh client.
	Change feeding container and attached tubing.
	Clean feeding equipment.
Every 7 to 10 days:	Reassess nutrition status.
As needed:	Observe client for any undesirable responses to tube feeding; for example, delayed gastric emptying, nausea, vomiting, or diarrhea.
	Check nitrogen balance.
	Check laboratory data.
	Chart significant details.

result in nausea, vomiting, diarrhea, cramps, constipation, delayed gastric emptying, abdominal distention, and aspiration.

Metabolic complications such as dehydration, electrolyte imbalance, and elevated blood glucose can also occur. Complications further stress the client and make it more difficult to achieve medical and nutrition goals.

■■ **WHAT TO CHART** Chapter 13 emphasized the importance of the medical record as a legal document and communication tool. Before reimbursing for the cost of tube feedings, many types of insurance (including Medicare) and managed care organizations require that the physician document the client's need for a tube feeding as well as justify the use of an infusion pump or a special formula, when appropriate. Medicare does not cover the cost of tube feeding unless the physician believes the condition necessitating the tube feeding will last for at least three months. Furthermore, Medicare does not cover the cost of tube feeding for clients with functioning GI tracts who are unable to eat due to a lack of appetite.

The health care team should routinely document the following information for clients on tube feedings:

- The nutrition goals for the client.
- The condition necessitating the tube feeding.

- The placement—both where (gastric or intestinal) and how (transnasal or enterostomy)—of the feeding tube and the type and size of the feeding tube.
- The formula selected to meet nutrition goals and its nutrient composition.
- The recommended administration schedule (concentration and rate) and method of delivery (intermittent or continuous, gravity drip or infusion pump).
- The client's education regarding the nutrition goals and the tube feeding procedure.
- The client's physical and emotional responses to the tube insertion and tube feeding procedure.
- The client's tolerances to the formula and administration schedule, complications (if any), and corrective actions recommended.
- All substances delivered through the feeding tube including the formula, additional water, medications, and any substances used to unclog the feeding tube (see Table 17-3).
- Confirmation that measures for monitoring the client have been taken.
- The reason why a tube feeding was interrupted or could not be delivered, if necessary.

The medical record serves as both a legal record and an invaluable source of information for investigating problems so that corrective measures can be taken promptly.

■■ From Tube Feedings to Table Foods

Once the medical condition that required a tube feeding resolves, the client can gradually shift to an oral diet as the volume of formula is tapered off. The client should be meeting about two-thirds of the estimated nutrient

Case Study

GRAPHICS DESIGNER REQUIRING ENTERAL NUTRITION

Mrs. Innis is a 24-year-old graphics designer who suffered multiple fractures when she fell from a cliff while hiking. She has been in the hospital for seven days and has no appetite. Mrs. Innis has lost 8 pounds over the course of her hospitalization. Due to the nature of her injuries, Mrs. Innis is in traction and is immobile, although the head of her bed can be elevated to 45 degrees. From the history, the dietitian determined that Mrs. Innis's nutrition status was adequate prior to hospitalization. The health care team agrees that a nasoduodenal tube feeding should be instituted before nutrition status deteriorates further. The standard formula selected for the feeding is lactose-free, and Mrs. Innis's nutrient requirements can be met with 2200 milliliters of the formula per day.

✚ What steps can the nurse take to prepare Mrs. Innis for tube feeding? Why might nasoduodenal placement of the feeding tube be preferred to nasogastric placement?

✚ The physician's orders specify that the feeding should be given continuously over 18 hours. Develop a tube feeding schedule.

✚ What parameters should be monitored to ensure that Mrs. Innis's fluid needs are being met? How can additional fluids be given? Describe precautions that should be taken if Mrs. Innis is to receive medications through the feeding tube.

✚ After three days of feeding, Mrs. Innis develops diarrhea. Look at Table 17-3 on p. 419 to determine the possible causes. What measures can be taken to correct the various causes of diarrhea?

✚ What tube feeding information should be charted in Mrs. Innis's medical record? When Mrs. Innis is ready to eat table foods again, what steps will the health care team take?

needs by mouth before the tube feeding is discontinued.[9] In many cases, the person can begin to drink the same formula that is being delivered by tube. As clients begin to take more food or formula orally, they can receive less formula by tube. Some people cannot make the transition to oral intake for medical reasons and go home on tube feedings. Chapter 18 discusses how home tube feedings work. The case study on p. 421 helps you to consider the many factors involved in tube feedings, and the Nutrition Assessment Checklist reviews key points for assessing nutrition status in people receiving tube feedings.

Tube feeding is a practical solution to feeding a person who is unable to receive adequate nourishment by mouth. A person without a functional GI tract, however, cannot benefit from a tube feeding. In such a case, intravenous nutrition (the subject of the next chapter) can be a lifesaving treatment option.

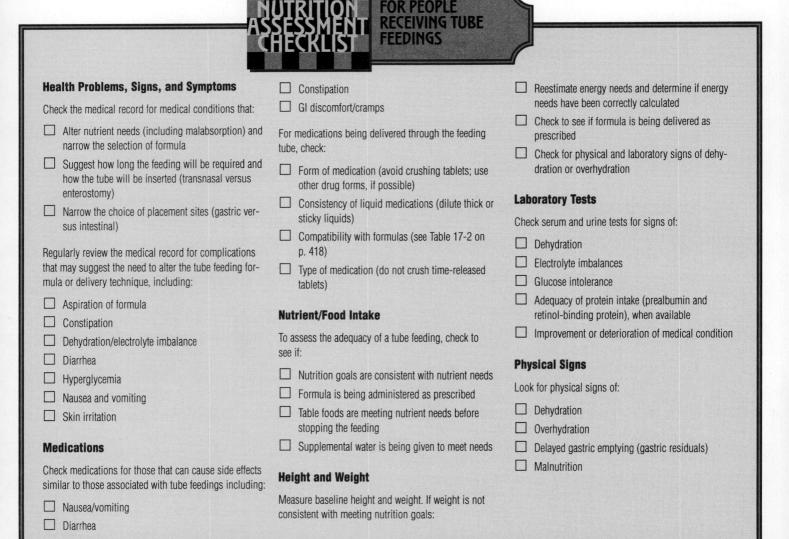

NUTRITION ASSESSMENT CHECKLIST — FOR PEOPLE RECEIVING TUBE FEEDINGS

Health Problems, Signs, and Symptoms

Check the medical record for medical conditions that:

☐ Alter nutrient needs (including malabsorption) and narrow the selection of formula

☐ Suggest how long the feeding will be required and how the tube will be inserted (transnasal versus enterostomy)

☐ Narrow the choice of placement sites (gastric versus intestinal)

Regularly review the medical record for complications that may suggest the need to alter the tube feeding formula or delivery technique, including:

☐ Aspiration of formula

☐ Constipation

☐ Dehydration/electrolyte imbalance

☐ Diarrhea

☐ Hyperglycemia

☐ Nausea and vomiting

☐ Skin irritation

Medications

Check medications for those that can cause side effects similar to those associated with tube feedings including:

☐ Nausea/vomiting

☐ Diarrhea

☐ Constipation

☐ GI discomfort/cramps

For medications being delivered through the feeding tube, check:

☐ Form of medication (avoid crushing tablets; use other drug forms, if possible)

☐ Consistency of liquid medications (dilute thick or sticky liquids)

☐ Compatibility with formulas (see Table 17-2 on p. 418)

☐ Type of medication (do not crush time-released tablets)

Nutrient/Food Intake

To assess the adequacy of a tube feeding, check to see if:

☐ Nutrition goals are consistent with nutrient needs

☐ Formula is being administered as prescribed

☐ Table foods are meeting nutrient needs before stopping the feeding

☐ Supplemental water is being given to meet needs

Height and Weight

Measure baseline height and weight. If weight is not consistent with meeting nutrition goals:

☐ Reestimate energy needs and determine if energy needs have been correctly calculated

☐ Check to see if formula is being delivered as prescribed

☐ Check for physical and laboratory signs of dehydration or overhydration

Laboratory Tests

Check serum and urine tests for signs of:

☐ Dehydration

☐ Electrolyte imbalances

☐ Glucose intolerance

☐ Adequacy of protein intake (prealbumin and retinol-binding protein), when available

☐ Improvement or deterioration of medical condition

Physical Signs

Look for physical signs of:

☐ Dehydration

☐ Overhydration

☐ Delayed gastric emptying (gastric residuals)

☐ Malnutrition

SELF CHECK

1. Which of the following statements is correct?

   a. Hydrolyzed formulas are made from pureed meats.

   b. Standard formulas contain whole proteins or protein isolates.

   c. Standard formulas contain free amino acids or small peptide chains.

   d. Hydrolyzed formulas may contain protein isolates or whole proteins.

2. When a client cannot meet nutrient needs from table foods:

   a. parenteral nutrition is preferred to tube feedings.

   b. tube feedings are preferred to parenteral nutrition.

   c. parenteral nutrition is preferred to enteral formulas provided orally.

   d. enteral formulas provided by tube are preferred to formulas provided orally.

3. The nurse recognizes that for a client with a high risk of aspiration who is expected to eat table foods in about a month, the most appropriate placement of the feeding tube would be:

   a. nasogastric.

   b. nasoenteric.

   c. gastrostomy.

   d. jejunostomy.

4. The nurse recognizes that in selecting an appropriate enteral formula for a client, the primary consideration is the:

   a. formula's osmolality.

   b. client's nutrient needs.

   c. availability of infusion pumps.

   d. formula's cost and availability.

5. What step can the nurse take to prevent bacterial contamination of tube feeding formulas?

   a. Deliver the formula continuously.

   b. Do not change the feeding bag and attached tubing.

   c. Discard all opened containers of formula not used in 24 hours.

   d. Add formula to the feeding container before it empties completely.

6. Compared to intermittent feedings, continuous feedings:

   a. require a pump for infusion.

   b. allow greater freedom of movement.

   c. are more like normal patterns of eating.

   d. are associated with more GI side effects.

7. A client needs 1800 milliliters of formula a day. If the client is to receive the formula intermittently every 4 hours, he will need _____ milliliters of formula at each feeding.

   a. 225

   b. 300

   c. 400

   d. 425

8. The term that describes the volume of formula that remains in the stomach from a previous feeding is:

   a. residue.

   b. osmolar load.

   c. gastric residual.

   d. intermittent feeding.

9. The nurse using the feeding tube to deliver medications recognizes that:

   a. medications generally do not result in GI complaints.

   b. medications can be added directly to the feeding container.

   c. thick or sticky liquid medications and crushed tablets can clog feeding tubes.

   d. enteral formulas do not interact with medications in the same way that foods do.

10. Tube feedings can gradually be discontinued when:

   a. discharge planning begins.

   b. the client experiences hunger.

   c. the medical condition resolves.

   d. the client is able to eat foods or drink formula in sufficient amounts.

Answers to these questions appear in Appendix H.

CLINICAL APPLICATIONS

1. Complex procedures, such as those necessary to insert feeding tubes and deliver tube feedings, require attention to many technical details, making it easy to focus on the procedure rather than the client. Imagine that you need a transnasal tube feeding. How might you react to the news that you need the feeding and to the insertion procedure? What things would you miss most about eating table foods? Think about ways a nurse might help a client deal with these feelings.

2. Review Chapters 15 and 16 and note symptoms or disorders that may require the use of tube feedings. For each, consider when and why a tube feeding might be appropriate and which conditions, if any, might require hydrolyzed formulas.

NUTRITION THE NET

FOR FURTHER STUDY OF THE TOPICS IN THIS CHAPTER, ACCESS THESE WEB SITES.

www.clinnutr.org
American Society for Parenteral and Enteral Nutrition

magi.com/~cpena
Canadian Parenteral-Enteral Nutrition Association

www.espen.org
European Society of Parenteral and Enteral Nutrition

Notes

[1] E. A. Emergy and coauthors, Banana flakes control diarrhea in enterally fed patients, *Nutrition in Clinical Practice* 12 (1997): 72–75.

[2] A. Skipper, C. Bohac, and M. B. Gregoire, Knowing brand name affects patient preferences for enteral supplements, *Journal of the American Dietetic Association* 99 (1999): 91–92.

[3] E. H. Elpern, Pulmonary aspiration in hospitalized adults, *Nutrition in Clinical Practice* 12 (1997): 286–289.

[4] M. H. DeLegge and B. M. Rhodes, Continuous versus intermittent feedings: Slow and steady or fast and furious? *Support Line,* October 1998, pp. 11–15.

[5] J. Lipp, L. M. Lord, and L. H. Scholer, Fluid management in enteral nutrition, *Nutrition in Clinical Practice* 14 (1999): 232–237.

[6] M. D. Coleman and K. A. Tappenden, Enteral feeding equipment: From the 15th century to the millennium, *Support Line,* October 1999, pp. 22–27.

[7] M. S. Williams and coauthors, Diarrhea management in enterally fed patients, *Nutrition in Clinical Practice* 13 (1998): 225–229.

[8] J. Hatton and B. Magnuson, How to minimize interaction between phenytoin and enteral nutrition: Two approaches, *Nutrition in Clinical Practice* 11 (1996): 28–31.

[9] T. C. Lykins, Nutrition support clinical pathways, *Nutrition in Clinical Practice* 11 (1996): 16–20.

 NUTRITION IN PRACTICE

Nutrition and Cost-Conscious Health Care

Decades of medical research have resulted in an explosion of knowledge and technologies to diagnose and treat diseases. This astounding progress, however, has come at a tremendous financial cost. The United States spends more money on health care than any other nation, yet the high cost of medical care has put it out of the reach of many citizens. Without attention to cost containment, the health status of the nation is threatened; sophisticated medical services do little good if people cannot use them. The medical community, which once embodied the idealistic approach of sparing no cost when it came to health care, has embraced the reality that cost is an element of quality.[1] To discuss all the ramifications of cost containment for health care delivery is beyond the scope

of this discussion. Instead, this discussion highlights major trends that affect medical services, including nutrition services, and some ways nutrition can help reduce health care costs.

 www.namcp.com
National Association of Managed Care Physicians

 www.nahc.org
Homecare Online

 www.hcaa-homecare.com
Home Care Association of America

 www.cdnhomecare.on.ca
Canadian Home Care Association

What steps can be taken to reduce health care costs?

Among other things, efforts to control health care costs aim to eliminate

duplication of services, reduce the number of hours spent in client care, maximize the use of health care professionals, and limit access to unnecessary services. The challenge is to cut costs without compromising the quality of care.

Efforts to control costs have triggered a major change in health care delivery: the transformation from the **indemnity,** or fee-for-service, system to the **managed care** system (see the accompanying glossary). The traditional fee-for-service health insurance plan allows clients to use the physicians and health care facilities of their choice and then reimburses the clients for a percentage of the cost of the services covered by the plan. Managed care organizations, which include **health maintenance organizations (HMOs)** and **preferred provider organizations (PPOs),** strive to provide high-quality, low-cost care by controlling the access to and cost of services. Each client selects a primary care physician, who then determines what services the client needs. Managed care is moving toward a **capitated** system; that is, a system where the health care provider agrees to supply all the services a client needs for a set monthly fee. Because the providers assume the financial responsibility if they fail to perform within a predetermined budget, they have a strong incentive to control costs. Managed care organizations, which were rare just 15 years ago, are increasingly replacing traditional insurance plans. Although managed care organizations have been effective at controlling costs, many consumers and health care professionals worry that quality of care has been compromised.

How do health care professionals measure quality of care?

In cost-conscious health care, quality of care is most often defined in terms of outcomes.[2] Examples of desirable outcomes include the person's ability to function independently (the more care a person needs, the more expenses incurred), reduced number or length of hospital stays, prevention of diseases and complications, and extended survival time. Services are cost-justified when they have a reasonable chance of

GLOSSARY OF TERMS RELATED TO HEALTH CARE INSURANCE

capitation: prepayment of a set fee per client in exchange for medical services.

health maintenance organization (HMO): a form of managed care that limits the subscriber's choice of health care professionals to those affiliated with the organization and controls access to services by directing care through a primary care physician.

indemnity insurance: traditional fee-for-service insurance.

managed care: a health care delivery system that aims to provide cost-effective health care by coordinating services and limiting access to services.

preferred provider organization (PPO): a form of managed care that encourages subscribers to select health care providers from a group that has contracted with the organization to provide services at lower costs.

improving a client's outcome. Services that produce the best outcomes at the lowest costs are the most cost-effective. The more expensive the procedure or test, the more critical justifying its cost becomes. Thus, it may not be cost-effective to perform a nutrition assessment on every client admitted to a hospital, whereas nutrition screening is cost-effective. For another example, providing enteral formulas orally to

clients who can drink them is more cost-effective than providing the formula by tube.

How can nutrition affect the quality of care?

A review of some of the desirable outcomes mentioned earlier reveals many ways that nutrition can improve quality of care and reduce health care costs. Adequate nutrition can significantly improve a client's ability to function independently, prevent or forestall diseases, limit complications, and maintain quality of life. All of these factors can help prevent the need for medical treatments, hospitalization, or lengthy hospital stays. Both timely identification of nutrition problems and appropriate interventions are essential to improving outcomes and cutting costs. Malnutrition, a persistent and common problem in hospitalized clients, is associated with significantly longer hospital stays, higher costs, and the need for home health care.[3] Based on one analysis of data from several sources, researchers estimate that early attention to malnutrition in hospitalized clients can reduce the length of stay and may result in cost savings of approximately $8300 per hospital bed per year.[4]

Medical nutrition therapy can also have a significant impact on the costs of common chronic diseases such as diabetes, hypertension, cardiovascular diseases, and cancer. To ensure that nutrition is addressed in standards that guide the care of clients, health care professionals must document the cost savings of nutrition intervention.

What types of care standards and treatments include nutrition?

Health care professionals and the organizations that represent them are working together to define outcome-oriented standards of practice and measures of care. The Joint Commission on Accreditation of Healthcare Organizations (JCAHO), an agency that oversees

Cost-cutting measures have redefined the responsibilities of health care professionals as well as the settings in which they work.

the accreditation of health care facilities, requires compliance with nutrition care standards to identify, address, and monitor each client's nutrition needs.[5] The standards promote coordination and communication among disciplines in an effort to improve quality of care in a cost-efficient manner.[6]

Why is it important to promote coordination and communication among health care professionals?

One of the positive trends spurred by cost-cutting measures is the replacement of traditional health care delivery, which is discipline- or department-specific, with a multidisciplinary or team approach. By working together, health care professionals can coordinate their services, eliminate unnecessary services, and solve problems in a timely and efficient manner. Traditionally, for example, nurses performed nursing functions and left the responsibility for the nutrition care of clients to the dietitian. When a client was admitted to the hospital, the nurse would perform a nursing assessment, and the dietitian, a nutrition assessment. Yet much of the information from these assessments overlaps. As Chapter 13 showed, an alert nurse can use the nursing assessment to determine the client's nutrition risk and make appropriate referrals to the dietitian. The dietitian, in turn, can rely on much of the information the nurse collects during the nursing assessment to complete a nutrition assessment. Nutrition support teams that include several health care disciplines can ensure quality nutrition care while reducing health care costs.

What is a nutrition support team?

Nutrition support teams often include a physician, nurse, dietitian, and pharmacist. The team develops standards and protocols for nutrition screening, nutrition assessment, tube feedings, and parenteral nutrition; identifies and monitors the care of people who need nutrients from enteral formulas or parenteral nutrition; and serves as consultants to the rest of the institution's staff. Team members identify and solve problems regarding clients, procedures, equipment, and other issues appropriate to their specialty areas. The team also analyzes new products and reviews current research. Figure NP17-1 shows the responsibilities of individual team members as well as those shared by the team. All of these functions serve to improve quality of care and reduce health care costs.

In what ways can the nutrition support team provide high-quality, cost-effective care?

Consider this example. During team rounds, the dietitian mentions a recurring problem with the collection of urine for nitrogen balance studies, which has invalidated several test results, wasted time (and money), and eliminated a source of information that could improve client care. The nurse recalls that many of the nurses are having difficulty with the procedure and suggests an in-service session on the proper techniques for nitrogen balance studies. The nurse agrees to schedule and conduct the session with the help of the dietitian. In an alternative scenario, the pharmacist (reacting to the dietitian's comments) relays information about a recent article describing a cost-effective and efficient procedure for conducting nitrogen balance studies. The team members agree to review the article and decide at the next meeting whether to test the new procedure.

Figure NP17-1
THE NUTRITION SUPPORT TEAM

The physician
- Diagnoses medical problems
- Performs medical procedures
- Coordinates and prescribes therapy
- Directs and supervises team
- Approves guidelines and protocols
- Consults with other physicians

The nurse
- Assesses nursing needs
- Performs direct client care
- Explains medical procedures and treatment plans
- Instructs clients regarding medical care
- Acts as a liaison between team and nursing staff
- Coordinates discharge plans

All team members
- Review current research
- Analyze new products
- Develop guidelines
- Provide in-service training
- Monitor clients
- Correct problems
- Educate clients
- Evaluate the outcome of the care provided and cost savings
- Promote the appropriate use of nutrition support
- Improve communications among team members and between the team and other health care professionals

The dietitian
- Assesses nutrition status
- Determines clients' nutrient needs
- Recommends appropriate diet therapy
- Reevaluates clients regularly
- Instructs clients about their diets
- Acts as a liaison between the team and the dietary department

The pharmacist
- Recommends appropriate drug therapy
- Identifies medication-medication and diet-medication interactions
- Identifies medication-related complications
- Educates clients about their medications
- Acts as a liaison between the team and the pharmacy

Clients benefit from many eyes noting problems and many brains searching for solutions.

In a similar manner, the team relies on the expertise of its members to develop protocols for cost-effective, high-quality care. One example might be the development of a clinical pathway—a plan that focuses on a desirable outcome for a specific diagnosis, procedure, or treatment. For the nutrition support team, examples might include clinical pathways for enteral and parenteral nutrition. Such a plan defines a time frame for each intervention with a goal of providing the best outcome at the lowest cost. Once the plan is in place, each team member assists in studying unexpected outcomes or variances from the expected course of treatment and fine-tuning the plan as necessary to achieve the best outcomes.

Specific examples of cost savings generated by nutrition support teams include:

• Preventing overuse of parenteral nutrition. In one hospital that implemented procedures to reduce the inappropriate use of parenteral nutrition, quality nutrition care could be achieved with an estimated cost savings of over $225,000 for one year.[7] In another study, reducing the inappropriate use of parenteral nutrition was estimated to generate a potential cost savings of $500,000 per year.[8]

• Selecting cost-effective enteral and parenteral formulas and supplies. In one hospital seeking to reduce formula waste and the incidence of clogged feeding tubes and encourage appropriate formula administration techniques, clinicians studied the costs and benefits of replacing open feeding systems with closed feeding systems and intravenous pumps with a new type of pump (automatic flush pump). The potential cost savings generated by reducing the number of clogged feeding tubes was estimated at $43,350 per month.[9] In addition, the cost savings due to the reduced time nurses spent manually flushing feeding tubes was estimated at $40,150 per year.

These are but a few of a growing number of examples that suggest nutrition support teams can improve the quality of care and generate cost savings at the same time.

Besides the increased use of teams, what other changes have occurred due to cost-cutting measures?

Another change is an increased use of skilled assistants in health care. A dietetic technician assisting a clinical dietitian, for example, may collect the data for a nutrition assessment, and the dietitian may analyze the data. Because the dietitian's time is saved for tasks that require a dietitian's unique skills, the dietitian can manage more clients, and money is saved.

As more skilled assistants are added, communication and coordination become even more crucial. In the case of nutrition care, all persons involved must understand their roles and perform their functions effectively to ensure quality of care.

Another major change affecting health care professionals has been a great reduction in the length of hospital stays and the growth of home health care. Early discharges lower hospital costs. As a result, hospital staffs have been reduced, and there is less time to identify nutrition problems. Clients may be discharged before they have regained health, and they may require additional medical and nutrition care in outpatient settings and at home. For this reason, it has become more important that clients have access to nutrition services in other settings.

How do nutrition services fit into the health care systems outside the hospital?

In most managed care settings, the primary care physician treats medical nutrition therapy as specialty care that requires referral. Medical nutrition therapy is a relatively inexpensive service, and when appropriately used, it can help prevent, forestall, and treat diseases and their complications. Thus, it is a potentially cost-

effective service. The authors of a review article estimate a potential savings of billions of health care dollars from an increased public awareness of the roles of nutrition in preventing several common disorders (type 2 diabetes, hypertension, coronary heart disease, stroke, cancer, osteoporosis, and depression) and consequent changes in eating habits.[10] In representing nutrition professionals, the American Dietetic Association takes the position that managed care organizations and integrated health delivery systems should provide medical nutrition therapy as an essential component of health care and that it should be provided by qualified nutrition professionals.[11]

Estimates suggest that more than 8 million Americans currently receive home care to meet both long- and short-term medical needs.[12] Nurses most often provide nutrition services that are part of a home care treatment plan, although the number of dietitians in home health care is expected to grow.[13]

Health care professionals are challenged to ensure that home care agencies and managed care organizations understand the cost-effectiveness of nutrition services in helping clients maintain health and quality of life. Achieving this goal requires that concerned health care professionals conduct research to document the ways in which timely and appropriate nutrition services improve clients' health outcomes in a cost-efficient manner. Nutrition advocates must take a proactive stand in ensuring that nutrition services are included in all settings.

The full implications of cost containment and its impact on the nation's health remain to be seen. In the words of one clinician, "Our American society is in the midst of the most far-reaching and profoundly disturbing uncontrolled study in the history of health care. We are experiencing major changes in the way we practice and pay for health care with very little evidence that these changes will achieve the desired outcomes."[14] In the midst of these changes, health care professionals have often been pushed to provide better outcomes with less personnel and time. Little wonder, then, that the changes are often met with resistance. Yet significant improvements in quality of care continue to be realized.[15] Health care professionals who react positively to the inevitable changes have the opportunity to make a significant difference in ensuring the success of the health care system for the future.

Notes

[1]A. Bothe, Consensus: We should not lower quality to cut costs, *Nutrition in Clinical Practice* (supplement) 10 (1995): 1–7.

[2]D. A. August, Outcomes research, nutrition care, and the provider-payer relationship, *Nutrition in Clinical Practice* (supplement) 13 (1998): S8–S11.

[3]C. S. Chima and coauthors, Relationship of nutritional status to length of stay, hospital costs, and discharge status of patients hospitalized in the medicine service, *Journal of the American Dietetic Association* 97 (1997): 975–978.

[4]H. N. Tucker and S. G. Miguel, Cost containment through nutrition intervention, *Nutrition Reviews* 54 (1996): 111–121.

[5]D. Dougherty and coauthors, Nutrition care given new importance in JCAHO standards, *Nutrition in Clinical Practice* 10 (1995): 26–31.

[6]Bothe, 1995.

[7]N. M. Pace and coauthors, Performance model anchors successful nutrition support protocol, *Nutrition in Clinical Practice* 12 (1997): 274–279.

[8]E. B. Trujillo and coauthors, Metabolic and monetary costs of avoidable parenteral nutrition use, *Journal of Parenteral and Enteral Nutrition* 23 (1999): 109–113.

[9]M. P. Petnicki, Cost savings and improved patient care with use of a flush enteral feeding pump, *Nutrition in Clinical Practice* (supplement) 13 (1998): S39–S41.

[10]As cited in D. J. Rodriquez, Managed care: Key concepts and nutritional implications, *Support Line*, October 1999, pp. 8–15.

[11]Position of The American Dietetic Association: Nutrition services in managed care, *Journal of the American Dietetic Association* 96 (1996): 391–395.

[12]National Association of Home Care web page (www.nahc.org), site visited on November 3, 1999.

[13]P. S. Anthony, The business of home care, *Support Line*, October 1999, pp. 16–21; M. S. Schiller, M. B. Arensberg, and B. Kantor, Administrators' perceptions of nutrition services in home health care agencies, *Journal of the American Dietetic Association* 98 (1998): 56–61.

[14]J. R. Wesley, Managing the future of nutrition support, *Journal of Parenteral and Enteral Nutrition* 20 (1996): 391–395.

[15]P. J. Schneider, A. Bothe, and M. Bisognago, Improving the nutrition support process: Assuring that more patients receive optimal nutrition support, *Nutrition in Clinical Practice* 14 (1999): 221–226.

CHAPTER 18

PARENTERAL NUTRITION

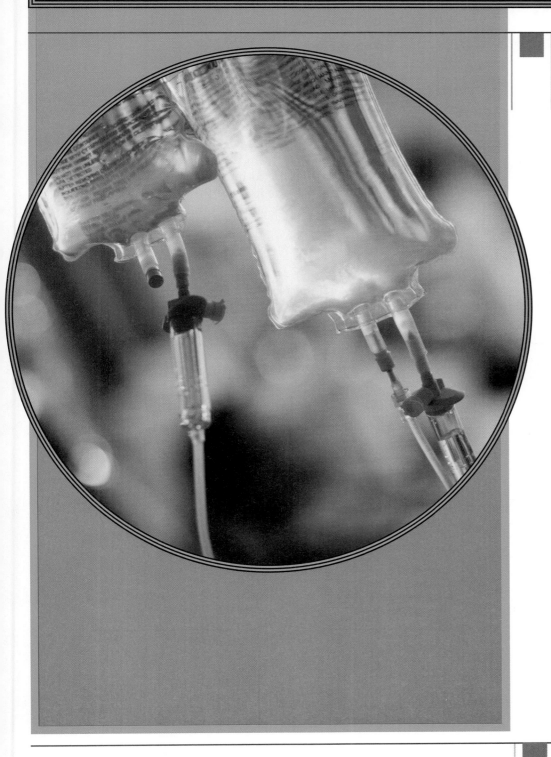

CONTENTS

Intravenous Solutions:
What Are They?

Intravenous Nutrition:
Who Needs What?

Intravenous Solutions:
How Are They Given?

From Parenteral
to Enteral Feedings

Specialized Nutrition
Support at Home

Case Study: Mail
Carrier Requiring
Parenteral Nutrition

Nutrition in Practice:
Ethical Issues in
Nutrition Care

The science of medical nutrition as we know it today was shaped tremendously by the demonstration in 1968 that nutrient needs could be met by vein.[1] Practitioners now had the means to feed people who otherwise might have died from malnutrition. With time, clinicians learned more and more about the solutions and delivery techniques to provide **parenteral nutrition.** They also discovered that although parenteral nutrition is a lifesaving procedure, it is also very costly and is associated with serious complications including liver dysfunction, progressive kidney problems, bone disorders, and many nutrient deficiencies. These findings prompted a renewed appreciation for the GI tract, and clinicians today subscribe to the adage "If the GI tract works, use it." Only when people cannot meet all of their nutrient needs using the enteral route should they receive total parenteral (or **intravenous**) nutrition support. Figure 18-1 summarizes the decision-making process in selecting the most appropriate feeding method.

parenteral nutrition: delivery of nutrient solutions directly into a vein, bypassing the GI tract.

> para = outside
> enteron = intestine

intravenous (IV): through a vein.
> intra = within
> vena = vein

■■ Intravenous Solutions: What Are They?

As is true of all medical nutrition therapy, the decision to use intravenous (IV) nutrition, the method of delivery, and the type and amount of nutrients to provide are based on a thorough assessment of the client's medical condition and nutrient needs. Infusion of IV nutrients immediately changes blood levels of fluids, electrolytes, and other nutrients and, therefore, requires vigilant attention to the individual's responses and adjustments to the formula, if appropriate.

Reminder: Formulas intended for enteral use cannot be delivered intravenously.

■■ Intravenous Nutrients

A variety of nutrients can be administered by vein. The IV solutions may contain any or all of the essential nutrients: water, amino acids, carbohydrate, fat,

Figure 18-1
SELECTING A FEEDING METHOD

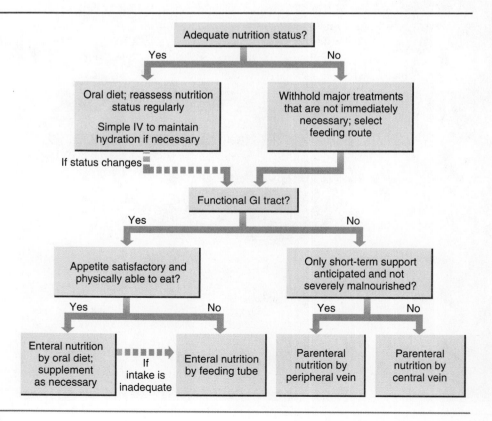

The labels of IV solutions use several abbreviations, including:

- **D:** dextrose.
- **W:** water.
- **NS:** normal saline (0.9 percent sodium chloride solution).

Interpret these abbreviations as follows:

- **D_5W or D5:** Read as 5 percent dextrose in water (the subscript following the *D* tells you the percentage of dextrose in the solution) or 5 percent dextrose.
- **$D_{10}W$ or D10:** Read as 10 percent dextrose in water or 10 percent dextrose.
- **D_5½ NS:** Read as 5 percent dextrose in a ½ normal saline solution (0.45 percent sodium chloride).

vitamins, and minerals. The "How to" box explains how to read IV solution abbreviations.

AMINO ACIDS By the time protein from food reaches the bloodstream, it has been broken down into amino acids. Thus, IV solutions supply amino acids rather than whole proteins. Standard IV solutions contain both essential and nonessential amino acids to meet the body's protein needs. Special solutions may contain only essential amino acids or large amounts of certain amino acids and small amounts of others to meet protein needs for specific medical conditions. Products designed for liver failure, for example, may contain more branched-chain amino acids and fewer aromatic amino acids (see Chapter 23).

Sometimes a nonessential amino acid may be omitted from standard parenteral solutions because it does not mix well or is unstable in the solution. For example, glutamine, which may be a conditionally essential amino acid for some clients, is not stable in IV solutions. Providing glutamine as a dipeptide solves the instability problem, and studies suggest that short-chain peptides can be digested to free amino acids by enzymes bound to cell membranes.

> *Reminder: A conditionally essential amino acid is one that is normally nonessential but must be supplied by the diet when the body's need for it exceeds the body's ability to produce it. Glutamine may be a conditionally essential amino acid following intestinal resections or during recovery from severe stress.*

CARBOHYDRATE Standard IV solutions provide carbohydrate as dextrose (the name used for glucose in IV solutions). The form of dextrose used in IV solutions (**dextrose monohydrate**) contains some water—a form more suitable for IV solutions. Intravenous dextrose provides 3.4 kcalories per gram, whereas glucose provides 4.

dextrose monohydrate: a form of glucose that contains a molecule of water and is stable in IV solutions. IV dextrose solutions provide 3.4 kcal/g.

LIPID Lipid emulsions deliver fat in IV solutions. Lipid emulsions are provided either daily or periodically (once or twice a week). If provided daily, IV fat serves as a concentrated source of energy; if offered less often, it serves primarily as a source of essential fatty acids. A 500-milliliter bottle of a 10 percent fat emulsion provides 550 kcalories (1.1 kcalories per milliliter). The same volume of a 20 percent fat emulsion provides 1000 kcalories (2.0 kcalories per milliliter).

Lipid emulsions are contraindicated for some newborns with markedly elevated **bilirubin** levels, people with some types of hyperlipidemia, people with severe liver disease, and those with severe egg allergies. Cautious use of IV fats is recommended for people with atherosclerosis, moderate liver disease, blood coagulation disorders, pancreatitis, and some lung problems. After long-term administration, brown pigments may accumulate in certain liver cells, but

bilirubin: a pigment in the bile whose blood concentration may rise as a result of some medical conditions, especially those that affect the liver.

Hyperlipidemia and atherosclerosis are discussed in Chapter 21, and fatty liver is described in Chapter 23.

these pigments disappear after parenteral therapy is discontinued; their effects on liver function are unknown. Prolonged IV lipid use may lead to fatty liver and reduce the number of blood platelets and white blood cells.

:: MICRONUTRIENTS Vitamins, electrolytes (minerals), and trace elements may all be used in IV solutions. Currently available IV multivitamin formulations for adults meet the recommendations of the Nutrition Advisory Group of the American Medical Association. These recommendations do not include vitamin K (and vitamin K is not added to IV multivitamin formulations) due to potential interactions with anticoagulant medications.[2] Instead, adults receive an appropriate dose of vitamin K in one weekly dose. Pediatric multivitamin solutions, on the other hand, contain vitamin K.

Failure to include vitamins in parenteral nutrition solutions can result in serious nutrient deficiencies and even death. During 1988, for example, a national shortage of parenteral multivitamin formulations led to the omission of vitamins in parenteral solutions and resulted in severe thiamin deficiencies and three deaths.[3]

Some electrolytes (particularly, calcium and phosphorus) can precipitate with other IV solution components, potentially resulting in life-threatening complications. As an indication of the seriousness of this problem, the Food and Drug Administration warned health care professionals that a precipitate of calcium phosphate might have been responsible for two deaths and at least two cases of respiratory distress in one institution. Pharmacists skilled in formulating IV solutions follow specific mixing procedures to minimize the risk of precipitation and protect the stability of IV solutions.

:: OTHER ADDITIVES Intravenous medications are sometimes added directly to the IV solution or infused into IV tubing through a separate port. Common examples include heparin, insulin, cimetidine, ranitidine, and famotidine. Providing medications along with parenteral nutrition solutions saves time and avoids the need for a separate infusion site. Interactions between medications and IV solutions, however, can and do occur. Only medications proved to be physically compatible with, and biochemically stable in, IV solutions are safe to administer together with the solution. When medications are added directly to the IV solution, and the total volume of solution is not infused, then the client does not receive the full dose of medication. Conversely, if the addition of a medication to an IV solution is not noted on the medication record, the physician may inadvertently reorder the medication, and the client may suffer potentially severe consequences.

:: Intravenous Solutions

Intravenous solutions can be compounded in many ways. The client's medical condition and nutrient needs determine what goes into an IV solution. The concentrations of energy nutrients and electrolytes are often expressed as percentages (see the "How to" box), although experts recommend that physicians' orders for IV solutions and labels for their containers state the formula's composition in grams per day and grams per liter.[4]

:: SIMPLE INTRAVENOUS SOLUTIONS Simple IV solutions typically contain 5 percent dextrose and normal saline. Other electrolytes or salts may be added as needed. Often 3 liters of the solution are provided daily and deliver about 150 grams of dextrose or about 500 kcalories per day. Although simple IV solutions provide minimal amounts of energy, they fall far short of meeting energy and nutrient needs. The volume of simple IV solutions required to meet energy needs exceeds the volume of fluid the body can safely handle.

CALCULATE ENERGY AND ENERGY NUTRIENT CONTENT OF IV SOLUTIONS EXPRESSED AS PERCENTAGES[a]

The percentage of a substance in an IV solution tells you how many grams of that substance are present in 100 milliliters. For example, a 5 percent dextrose solution contains 5 grams of dextrose per 100 milliliters. A 0.9 percent normal saline solution contains 0.9 gram sodium per 100 milliliters.

Suppose a person is receiving 1800 milliliters of a complete nutrient IV solution containing 20 percent dextrose, 5 percent amino acids, and 1 percent lipid. First, determine the number of grams of each energy nutrient:

- For dextrose:

$$\frac{20 \text{ g dextrose}}{100 \text{ ml}} = \frac{X \text{ g dextrose}}{1800 \text{ ml}}.$$

$$\frac{20 \text{ g} \times 1800 \text{ ml}}{100 \text{ ml}} = 360 \text{ g dextrose}.$$

- For amino acids:

$$\frac{5 \text{ g amino acids}}{100 \text{ ml}} = \frac{X \text{ g amino acids}}{1800 \text{ ml}}.$$

$$\frac{5 \text{ g} \times 1800 \text{ ml}}{100 \text{ ml}} = 90 \text{ g amino acids}.$$

- For lipid:

$$\frac{1 \text{ g lipid}}{100 \text{ ml}} = \frac{X \text{ g lipid}}{1800 \text{ ml}}.$$

$$\frac{1 \text{ g lipid} \times 1800 \text{ ml}}{100 \text{ ml}} = 18 \text{ g lipid}.$$

Next, calculate the total kcalories in 1800 milliliters of the solution by multiplying the kcalories per gram and then adding the totals:

360 g dextrose × 3.4 kcal/g = 1224 kcal
90 g amino acids × 4.0 kcal/g = 360 kcal
18 g lipid × 9.0 kcal/g[b] = 162 kcal
Total = 1746 kcal.

[a]The preferred method for ordering energy nutrients in IV solutions and labeling their containers is to state the amount of each energy nutrient in grams per day and grams per liter. National Advisory Group on Standards of Practice Guidelines for Parenteral Nutrition, *Journal of Parenteral and Enteral Nutrition* 22 (1998): 49–66.

[b]The number of kcalories per gram of IV lipid is actually more than 9.0 kcalories per gram, but 9.0 kcalories per gram is an acceptable estimate.

■■ **COMPLETE NUTRIENT SOLUTIONS** Complete nutrient solutions provide all of the essential nutrients including amino acids, dextrose, fatty acids, vitamins, minerals, and trace elements. Many facilities use standard forms to assist physicians in writing orders for complete nutrient solutions. Solutions intended to be delivered through the **peripheral veins** typically contain from 5 to 10 percent dextrose (like simple IVs) and 3 to 5 percent amino acids and provide about 1500 to 2000 kcalories per day; lipid emulsions contribute more than half of the total kcalories.

THE NURSING DIAGNOSIS altered nutrition: less than body requirements often applies to the person who needs complete nutrient solutions.

peripheral veins: the small-diameter veins that carry blood from the arms and legs.

The dextrose concentration of solutions infused through peripheral veins should not exceed 10 percent, and the osmolarity of the solution should not exceed 600 milliosmoles per liter (mOsm/L).

central veins: the large-diameter veins located close to the heart.

Complete nutrient solutions intended to be delivered through the **central veins** meet energy needs primarily from concentrated dextrose solutions. Recommendations suggest that these solutions provide 70 to 85 percent of the nonprotein energy from dextrose and 15 to 30 percent from fat.[5] If lipids are not used as an energy source, however, essential fatty acid requirements may be met by giving IV lipids periodically (two to three times per week).

■■ Intravenous Nutrition: Who Needs What?

The different types of IV solutions deliver nutrients for different purposes. The composition of the solution and the delivery method depend on the person's immediate medical and nutrient needs, nutrition status, and anticipated length of time on IV nutrition support. When enteral feeding is not possible and complete nutrient solutions are required, they should be initiated before nutrition status becomes severely compromised. It is much easier to maintain nutrition status than to try to replenish lost nutrient stores.

IV catheter: a thin tube inserted into a vein. Additional tubing connects the IV solution to the catheter.

■■ **SIMPLE INTRAVENOUS INFUSIONS** People with medical conditions that disrupt fluid and electrolyte or acid-base balances may need simple IV solutions to help them maintain or restore their body's normal homeostasis (Chapters 15 and 16 provided many examples). Simple IV solutions are delivered through peripheral **IV catheters** inserted into the small-diameter peripheral veins, such as those in the forearm and back of the hand. For people who are not expected to be able to eat for more than a few days, especially those who are malnourished or have high nutrient requirements, simple IV solutions are inappropriate, however.

THE NURSING DIAGNOSES fluid volume deficit and **risk for fluid volume deficit** often apply to people who need simple IVs.

peripheral parenteral nutrition (PPN): the provision of an IV solution that meets nutrient needs delivered through the peripheral veins.

■■ **PERIPHERAL PARENTERAL NUTRITION (PPN)** For some people, nutrient needs can be met using the peripheral veins. **Peripheral parenteral nutrition (PPN)** relies on IV lipid emulsions to provide a concentrated source of kcalories in a form that is isotonic and less irritating to the blood than concentrated dextrose solutions, which irritate the peripheral veins and cause them to collapse.

The relatively large volume of fluid necessary to deliver a limited number of kcalories and the need for accessible and strong peripheral veins limit the use of PPN. Peripheral parenteral nutrition best suits people with normal renal function who need only short-term nutrition support (about 7 to 14 days), people who need additional nutrients temporarily to supplement an oral diet or tube feeding, and those in whom inserting an IV catheter into a central vein is medically unsound. People who need fluid restrictions, people who have high-energy requirements, and people for whom lipid emulsions are contraindicated are not candidates for PPN. In addition, the need for strong peripheral veins to provide PPN solutions limits the use of PPN for people whose veins are weak or collapse easily.

central total parenteral nutrition (central TPN): the provision of an IV solution that meets nutrient needs delivered through a central vein.

■■ **TOTAL PARENTERAL NUTRITION (TPN) BY CENTRAL VEIN** Another method used to meet nutrient needs by vein is **central total parenteral nutrition,** or **central TPN** for short. In central TPN, the tip of a central venous catheter is either placed directly into a large-diameter central vein or threaded into a central vein through a peripheral vein (see Figure 18-2). The advantage of central TPN is that it allows the infusion of concentrated IV solutions so that energy and nutrient needs can be met in small volumes. The central veins

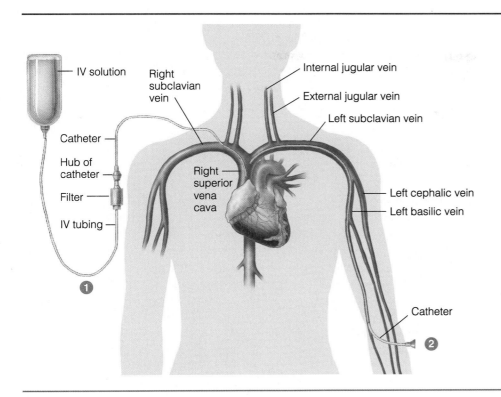

Figure 18-2
CENTRAL TPN

❶ *Traditionally, central TPN catheters enter the circulation at the right subclavian vein and are threaded into the superior vena cava with the tip of the catheter lying close to the heart. Sometimes catheters are threaded into the superior vena cava from the left subclavian vein, the internal jugular vein, or the external jugular vein.*

❷ *Peripherally inserted central catheters usually enter the circulation at the basilic or cephalic vein and are guided up toward the heart so that the catheter tip rests in a central vein, usually the superior vena cava.*

lie close to the heart, where a large volume of blood rapidly dilutes the TPN solution. Consider that even under resting conditions the heart pumps over a gallon of blood out of its chambers and into the arteries each minute. By the time the TPN solution reaches the peripheral veins, it is no longer concentrated enough to irritate the blood vessels.

Central TPN is indicated whenever parenteral nutrition will be required for long periods of time, when nutrient requirements are high, or when people are severely malnourished (see Table 18-1 on p. 436). Compared to catheters inserted directly into a central vein, **peripherally inserted central catheters** are associated with fewer insertion-related and infection-related complications, and they are less costly. Although access to a strong peripheral vein is still required, once inserted, a peripherally inserted central catheter generally remains usable longer than a standard peripheral IV catheter. Thus, when people need TPN to meet nutrient needs for weeks or months or when they need nutrition support but have few strong veins remaining, a peripherally inserted central catheter is an option.[6]

peripherally inserted central catheter (PICC): a catheter inserted into a peripheral vein and advanced into a central vein.

Intravenous Solutions: How Are They Given?

Intravenous nutrition is like tube feedings in that careful attention to formula selection, preparation, and delivery helps support nutrition status and minimize the risks of complications. To prevent bacterial contamination and ensure the stability of IV solutions, they should be shielded from light and refrigerated. As Table 18-2 on p. 437 shows, many of the risks associated with IV nutrition are more serious than those associated with enteral nutrition.

Insertion and Care of Intravenous Catheters

Insertion of peripheral catheters for PPN is the same as for simple IV solutions. Skilled nurses can place peripherally inserted catheters for simple IVs,

Table 18-1
POSSIBLE INDICATIONS FOR CENTRAL TPN

Acquired immune deficiency syndrome (AIDS)

Bone marrow transplants

Extensive small bowel resections

GI tract obstructions

High-output enterocutaneous fistulas

Hypermetabolic disorders, or major surgery, when it is anticipated that the GI tract will be unusable for more than 2 weeks

Intractable diarrhea

Intractable vomiting

Low birthweight with necrotizing enterocolitis (severe GI inflammatory disease) or bronchopulmonary dysplasia (chronic lung disease)

Radiation enteritis (inflammation of intestine caused by radiation)

Severe acute pancreatitis

Severe malnutrition if surgical or intensive medical intervention is necessary

Severe nausea and vomiting associated with pregnancy (hyperemesis gravidarum) when they last more than 14 days

When it is anticipated that adequate enteral nutrition cannot be established within 14 days of hospitalization

NOTE: Parenteral nutrition is indicated only when enteral nutrition is contraindicated. If short-term parenteral nutrition support is anticipated (less than 14 days), PPN might be preferred.
SOURCE: Adapted from A.S.P.E.N. Board of Directors, Guidelines for the use of parenteral and enteral nutrition in adult and pediatric patients, *Journal of Parenteral and Enteral Nutrition* (supplement) 17 (1993): 1–49.

PPN, or TPN, but a catheter for direct central vein access requires surgical insertion by a qualified physician. The client may be awake for the procedure, but is given a local anesthetic. Unnecessary apprehension can be avoided by explaining the procedure to the client.

The difficulty of maintaining the integrity of peripheral veins frequently interferes with the ability to sustain PPN. Veins may become inflamed and sometimes infected. Often the infusion catheter must be removed and reinserted at a new site; consequently, long-term feedings are difficult and rarely indicated. As described earlier, peripherally inserted central catheters are less irritating to the veins and can often remain in place longer than PPN catheters.

Infections can develop at the catheter site in both PPN and central TPN. Compared with peripherally inserted catheters (for either PPN or central TPN), though, catheters inserted directly into a central vein present a greater risk of introducing disease-causing microorganisms into the bloodstream because the catheter location is so near the heart. Nurses inspect the catheter site regularly and use aseptic techniques when changing the dressing covering a catheter insertion site.

THE NURSING DIAGNOSIS risk for infection applies to people receiving PPN or central TPN.

Administration of Complete Nutrient Solutions

Just as a tube feeding is started slowly to give the GI tract time to adapt to the formula, a complete nutrient solution is started slowly to allow time for the body to adapt to the glucose concentration and osmolality of the solution. An infusion pump ensures an accurate and steady delivery rate. Rapid changes in the infusion rate, especially for central TPN, can cause severe hyperglycemia or hypoglycemia, which can lead to coma, convulsions, and even death, so all changes must be made gradually and cautiously and infusion pumps must be

Reminder: The presence of disease-causing organisms in the blood is sepsis—a life-threatening complication that may occur as a consequence of parenteral nutrition.

To prevent hyperglycemia, the final volume of a PPN or central TPN solution should provide no more than 5 to 7 g carbohydrate per kilogram body weight per day.

Reminder: Keeping IV poles and infusion pumps at a safe distance from young children helps prevent potential injuries to the child from toppling the pole or pump and potential damage to the equipment.

Table 18-2
COMPLICATIONS ASSOCIATED WITH CENTRAL TPN

Catheter- or Care-Related Complications

Air leaking into catheter, obstructing blood flow (air embolism)

Air or gas in the chest (pneumothorax)

Blood clot (thrombosis)

Blood in the chest (hemothorax)

Catheter inadvertently placed in subclavian artery (arterial puncture)

Catheter occlusion

Catheter tip broken off, obstructing blood flow (catheter embolism)

Fluid in the chest (hydrothorax)

Hole or tear in heart made by catheter tip (myocardial perforation)

Improperly positioned catheter tip

Infection

Inflammation of vein (phlebitis)

Infusion pump malfunctions

Sepsis

Metabolic or Nutrition-Related Complications

Acid-base imbalances

Bone demineralization

Coma from excessive glucose load (hyperosmolar, hyperglycemic, nonketotic coma)

Dehydration

Electrolyte imbalances

Elevated blood glucose (hyperglycemia)

Elevated liver enzymes

Essential fatty acid deficiency

Fatty liver

Fluid overload

High blood ammonia levels (hyperammonemia)

Low blood glucose (hypoglycemia)

Trace element deficiencies

Vitamin and mineral deficiencies

---

checked regularly for malfunctions. Problems are more likely to occur in people with organ dysfunction or in infants with immature organ systems.

THE NURSING DIAGNOSES fluid volume excess, fluid volume deficit, and **risk for fluid volume deficit** may apply to people on TPN solutions.

Electrolytes and blood glucose are monitored vigilantly initially and periodically thereafter. If tests indicate electrolyte imbalances or unacceptably high blood glucose, the causes are investigated and treated. Although specific protocols vary, Table 18-3 on p. 438 provides an example of guidelines for monitoring clients on PPN and central TPN.

When the administration of solution gets behind or ahead of schedule, the drip rate should be adjusted to the correct hourly infusion rate, but no attempt should be made to speed up or slow down the rate to meet the originally ordered volume. When a person is being taken off TPN, the infusion rate of the solution must be tapered off gradually to prevent hypoglycemia. PPN can be discontinued without tapering.

■■ **INTRAVENOUS LIPID INFUSION** Traditionally, IV fat emulsions are infused separately from the base solution containing amino acids, dextrose,

Total nutrient admixtures contain dextrose, amino acids, lipids, and micronutrients. The IV lipids give the admixture its milky white color.

Table 18-3

SUGGESTED GUIDELINES FOR MONITORING CLIENTS RECEIVING PARENTERAL NUTRITION

Before starting TPN:	Complete nutrition assessment.
	Record weight.
	Confirm placement of catheter tip by X ray.
	Check blood glucose, electrolytes, chemical profile, and complete blood count.
Every 4 to 6 hours:	Check blood glucose.[a]
	Monitor vital signs.
	Check pump infusion rate.
Daily:	Monitor weight changes.[b]
	Record intake and output.
	Inspect catheter site for signs of inflammation and infection.
Weekly:	Reassess nutrition status.
	Monitor serum proteins, ammonia, enzymes, triglycerides, cholesterol, and bleeding indices.
	Check the complete blood count.
As needed:	Monitor serum transferrin, electrolytes, calcium, magnesium, phosphorus, blood urea nitrogen, and creatinine.

[a]If blood glucose levels are stable, they can be checked once daily after the first week.
[b]After the first week, weight can usually be measured less often (about twice a week).

and micronutrients. People sometimes experience adverse reactions to IV lipid emulsions, particularly when the IV lipids are given in large amounts or administered too rapidly. Immediate reactions may include fever, warmth, chills, backache, chest pain, allergic reactions, palpitations, rapid breathing, wheezing, cyanosis, nausea, and an unpleasant taste in the mouth. To guard against adverse reactions, the client receives only small amounts of lipid emulsion over the first 15 to 30 minutes. After that time, the rate can be increased.

When IV lipid emulsions are used as an energy source, they are often added directly to the base solution and infused along with it. The use of total nutrient admixtures for clients in the hospital as well as at home has grown dramatically. Total nutrient admixtures must be compounded carefully, refrigerated prior to use, and mixed gently before they are infused.

> **Total nutrient admixtures** *are also called* **three-in-one (3-in-1) admixtures** *or* **all-in-one admixtures.**

cyclic parenteral nutrition: the continuous administration of central TPN solution for 8 to 12 hours a day with time periods when no nutrients are infused during the rest of the day.

■■ **CYCLIC INFUSION** A person on **cyclic parenteral nutrition** receives a TPN solution at a constant rate for 8 to 12 hours a day with breaks in the infusion during the rest of the day. The infusion can be given during the night to allow freedom for routine daytime activities and, therefore, is suited to long-term TPN, especially for clients who receive TPN at home. When a person receives a TPN solution continuously, insulin levels stay high. As a result, the person cannot mobilize fat stores for energy or for essential fatty acids; eventually, fat may be deposited in the liver. Cyclic TPN reverses these problems. Additionally, fewer kcalories seem to be effective in maintaining nitrogen balance, probably because the person uses body fat for energy. Some people, however, cannot tolerate the delivery of a day's volume of solution over a short period of time.

■■ From Parenteral to Enteral Feedings

Although some clients will need parenteral nutrition for the rest of their lives, most often the medical problem causing the need for IV nutrition resolves, and the client can gradually shift to an enteral diet. The transition requires careful planning, especially when the client has been receiving no nutrients enterally or the feeding has continued for a long time. During long periods of disuse, the intestinal villi shrink and lose some of their function. Reintroducing nutrients to the GI tract at the appropriate rate and volume will stimulate the progressive restoration of the villi's normal structure and function and prevent malabsorption and GI discomforts.

■■ **TRANSITIONAL FEEDINGS** The transition from IV feeding to an enteral diet can be accomplished in different ways and often involves a combination of feeding methods. One way is to start an oral diet while the person is still on IV nutrition. The diet is often progressive, beginning with liquids provided in small amounts. If the person cannot eat enough food to meet at least 50 percent of daily nutrient needs within a few days, and intake does not seem to be improving, a tube feeding may be considered. Whether a person is given a tube feeding or provided an oral diet, the volume of the IV solution is reduced as the volume of enteral feeding is increased. The person who cannot tolerate enteral feedings can still rely on TPN to meet nutrient needs. Parenteral nutrition can be discontinued when at least some 70 to 75 percent of estimated energy needs are being met by oral intake, tube feeding, or a combination of the two.[7] Chapter 17 described the transition from tube feedings to table foods.

■■ **PSYCHOLOGICAL EFFECTS** Returning to oral intake after having been fed either intravenously or by tube can have a variety of psychological effects. Some people may be extremely eager to eat again, and food can be an important morale booster. Others may be apprehensive about eating, particularly if they have had extensive GI problems. Appetite may be slow to return for some. In such circumstances, all members of the health care team can support the successful reintroduction of food. Recognize the person's concerns, and provide reassurances that help will be available throughout the process. The many decisions surrounding the provision of TPN require careful consideration. The case study below presents an example for review.

Case Study

MAIL CARRIER REQUIRING PARENTERAL NUTRITION

Mr. Rossi, a 37-year-old mail carrier, has been admitted to the hospital for Crohn's disease (see Chapter 16). He has been steadily losing weight and appears emaciated. A thorough examination indicates that Mr. Rossi needs surgery as soon as possible to remove a portion of his small intestine. In the meantime, he cannot be placed on enteral feedings. The nutrition assessment reveals protein-energy malnutrition.

Mr. Rossi is placed on central TPN before surgery. He progresses well, gains weight, and undergoes surgery one week after admission.

✚ What factors in Mr. Rossi's history indicate the need for central TPN? How would you explain the need for TPN to Mr. Rossi?

✚ Describe the components of a typical TPN solution. Calculate the energy content of 1 liter, a solution providing 17 grams dextrose, 4 grams amino acids, and 2 grams lipid.

✚ Consider some of the physiological and psychological problems Mr. Rossi might face when enteral nutrition is reintroduced. How will the health care team know when it will be safe to take Mr. Rossi off central TPN from a nutrition standpoint? Describe the different ways the transition from TPN to enteral feedings can be accomplished.

▪▪ Specialized Nutrition Support at Home

Occasionally, a client must continue to receive nutrition support (tube feedings or parenteral nutrition) after the primary medical condition has stabilized. In such a case, home nutrition support may be an option.

Since the first report of a person sent home successfully on TPN in 1969,[8] the use of home nutrition support has expanded rapidly. According to recent estimates, about 40,000 people receive parenteral nutrition at home, and more than 150,000 receive enteral nutrition at home.[9] As the number of people benefiting from home nutrition programs continues to grow, health care professionals who work with these programs are gaining valuable experience and improving the quality of home nutrition support. Medical supply companies provide the equipment, formulas, and service necessary to support home nutrition care.

▪▪ The Basics of Home Programs

As with tube feedings and TPN in the hospital, the main objective of home nutrition support is to maintain or achieve adequate nutrition status. Nutrition support at home, however, has an added dimension: it provides a sense of independence in familiar surroundings. If you have ever been in the hospital, or taken a long trip for that matter, you probably remember the comfort you experienced when you returned to your own bed, knew where things were, and could get things when you needed them.

▪▪ CANDIDATES FOR HOME NUTRITION SUPPORT   The nutrition support team most frequently determines whether a client is a candidate for a home nutrition program. In addition to medical considerations, the candidate for home nutrition support and those who care for that person must have rational, stable personalities so they can successfully handle the problems that arise. They must be capable of learning the necessary techniques and of dealing with complications. They must be willing to comply with recommendations. They must also have adequate financial resources and access to the equipment, supplies, and professional support that are integral components of a successful home program.

▪▪ ROLES OF HEALTH CARE PROFESSIONALS   The health care team not only assesses the client's medical and nutrition needs, but also evaluates the client's home environment so that the team can plan practical feeding and care schedules and find practical solutions to the unique problems the client will face at home. Health care teams who work with home nutrition programs must also understand the regulations that govern payment for home nutrition programs by both government and private insurance. Team members use this knowledge to document the need for services to ensure maximum reimbursement for the client.

Once a home nutrition program is initiated, the physician directs the program and monitors the client's care. Nurses train clients in the appropriate techniques; dietitians monitor nutrition status; pharmacists coordinate the delivery of formulas, medications, and supplies with the home care company; and social workers provide insurance assistance. All team members are expected to answer questions and provide emotional support. Home visits by a nurse, and sometimes the dietitian, help ensure that the client is able to comply with the demands of the home care program. A qualified nurse, dietitian, or physician must be available to answer questions and handle problems as they arise.

:: COST CONSIDERATIONS One recent analysis found that the average cost of maintaining a home parenteral nutrition program was $70,000 per year with a range of $15,000 to $169,000 per year.[10] Costs for home enteral nutrition programs are lower (average $18,000 per year with a range of $5000 to $50,000 per year), but can still be quite substantial. Insurance and Medicare often pay for a portion of the costs, but people on home programs are also more likely than others to have high medical costs for managing their diseases as well. The economic impact of home nutrition programs is often the primary concern for people on these programs.

To compare the expense of receiving nutrients from parenteral or enteral solutions with table foods, consider that the average monthly "grocery bill" for parenteral and enteral nutrition is $5800 and $1500, respectively, per person.

:: ADJUSTMENTS FOR CLIENTS Although home nutrition programs can extend and improve the quality of life, clients may struggle with the many lifestyle adjustments the programs entail. In addition to the economic impact of the therapy, clients may find the demands of the schedule to be frustrating, inconvenient, and monotonous. Among physical complaints, clients cite frequent urination (due to the high volume of fluids), disturbed sleep (often as a consequence of waking up to go to the bathroom), and weakness.[11] Lifestyle complaints focus on sleep, travel, exercise, and social life.[12] Nevertheless, most clients report that their lives improve with home nutrition therapy.[13] Many people who require nutrition support at home can eat some foods by mouth, and some resume activities, such as going to work, driving, and playing sports.

THE NURSING DIAGNOSES that may apply to clients (and their caregivers) who receive parenteral or enteral nutrition at home include **impaired social interaction, social isolation, risk for loneliness, caregiver role strain, impaired physical mobility, fatigue, activity intolerance,** and **sleep pattern disturbance.**

:: How Home Enteral and Parenteral Nutrition Programs Work

People on home enteral nutrition programs commonly have cancer or neurological disorders that interfere with swallowing. People on home TPN often have acquired immune deficiency syndrome (AIDS), cancer, Crohn's disease, or other intestinal disorders. Sometimes home TPN is relatively short term, lasting for weeks or months as opposed to lifelong.

:: HOME ENTERAL NUTRITION Both transnasal tubes and tube enterostomies provide access to the GI tract for home care tube feedings (see Chapter 17). When possible, intermittent feeding schedules are arranged so that clients are free to move around between meals. Some people are able to meet some of their nutrient needs by eating during the day and using tube feedings at night. Clients on continuous feedings who must use pumps can obtain small, lightweight pumps that are easily concealed in carrying cases that allow freedom of movement.

:: HOME TPN Different types of home TPN programs are currently in use. Ideally, clients are given as much responsibility for their own care as they can handle. For example, a client who is capable of changing the catheter dressings is trained to do so. Typically, caregivers also learn the techniques so that they can assist as needed.

Special catheters designed for long-term use are often inserted for home TPN. Many times these catheters are tunneled under the skin so that the catheter exit site lies in the abdominal area where the client can access and care for it easily. The day's volume of TPN solution is frequently delivered within 8 to 12 hours using an infusion pump. The client infuses the solution while

Portable pumps and convenient carrying cases allow clients who require nutrition support at home to move about freely.

sleeping or at some other convenient time and thus is free to move about unencumbered for much of the day. For those who need continuous feedings, a lightweight carrying case that holds a small pump and IV bags enables the client to move around freely.

Unquestionably, tube feedings and parenteral nutrition provide a lifesaving feeding alternative for people who cannot eat traditional diets either in the hospital or at home. The Nutrition Assessment Checklist reviews areas of concern for people on parenteral nutrition support.

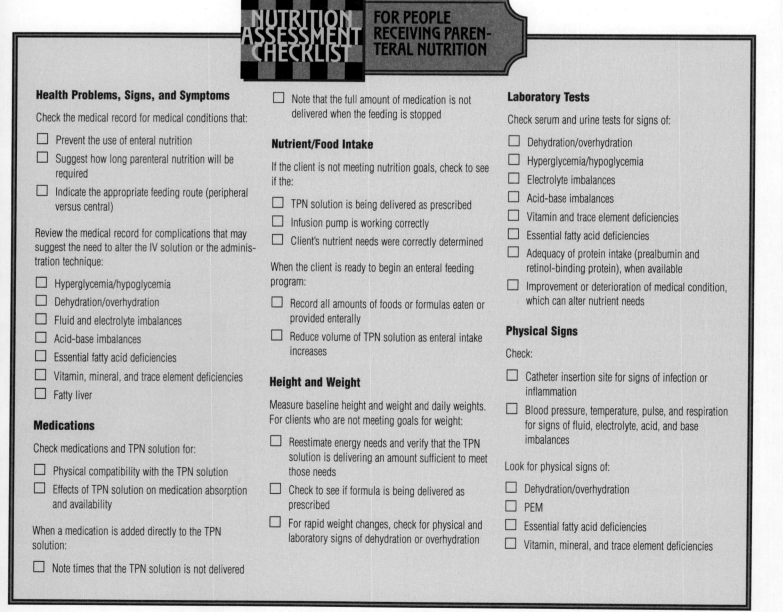

NUTRITION ASSESSMENT CHECKLIST

FOR PEOPLE RECEIVING PARENTERAL NUTRITION

Health Problems, Signs, and Symptoms

Check the medical record for medical conditions that:

- ☐ Prevent the use of enteral nutrition
- ☐ Suggest how long parenteral nutrition will be required
- ☐ Indicate the appropriate feeding route (peripheral versus central)

Review the medical record for complications that may suggest the need to alter the IV solution or the administration technique:

- ☐ Hyperglycemia/hypoglycemia
- ☐ Dehydration/overhydration
- ☐ Fluid and electrolyte imbalances
- ☐ Acid-base imbalances
- ☐ Essential fatty acid deficiencies
- ☐ Vitamin, mineral, and trace element deficiencies
- ☐ Fatty liver

Medications

Check medications and TPN solution for:

- ☐ Physical compatibility with the TPN solution
- ☐ Effects of TPN solution on medication absorption and availability

When a medication is added directly to the TPN solution:

- ☐ Note times that the TPN solution is not delivered

- ☐ Note that the full amount of medication is not delivered when the feeding is stopped

Nutrient/Food Intake

If the client is not meeting nutrition goals, check to see if the:

- ☐ TPN solution is being delivered as prescribed
- ☐ Infusion pump is working correctly
- ☐ Client's nutrient needs were correctly determined

When the client is ready to begin an enteral feeding program:

- ☐ Record all amounts of foods or formulas eaten or provided enterally
- ☐ Reduce volume of TPN solution as enteral intake increases

Height and Weight

Measure baseline height and weight and daily weights. For clients who are not meeting goals for weight:

- ☐ Reestimate energy needs and verify that the TPN solution is delivering an amount sufficient to meet those needs
- ☐ Check to see if formula is being delivered as prescribed
- ☐ For rapid weight changes, check for physical and laboratory signs of dehydration or overhydration

Laboratory Tests

Check serum and urine tests for signs of:

- ☐ Dehydration/overhydration
- ☐ Hyperglycemia/hypoglycemia
- ☐ Electrolyte imbalances
- ☐ Acid-base imbalances
- ☐ Vitamin and trace element deficiencies
- ☐ Essential fatty acid deficiencies
- ☐ Adequacy of protein intake (prealbumin and retinol-binding protein), when available
- ☐ Improvement or deterioration of medical condition, which can alter nutrient needs

Physical Signs

Check:

- ☐ Catheter insertion site for signs of infection or inflammation
- ☐ Blood pressure, temperature, pulse, and respiration for signs of fluid, electrolyte, acid, and base imbalances

Look for physical signs of:

- ☐ Dehydration/overhydration
- ☐ PEM
- ☐ Essential fatty acid deficiencies
- ☐ Vitamin, mineral, and trace element deficiencies

SELF CHECK

1. A product that cannot be delivered intravenously is:
   a. dextrose.
   b. amino acids.
   c. lipid emulsions.
   d. hydrolyzed enteral formula.

2. The nurse recognizes that a simple IV solution is most appropriate for people who:
   a. are malnourished.
   b. have high nutrient needs.
   c. cannot eat for long periods of time.
   d. are well nourished and are expected to eat in a few days.

3. The nurse working with a client receiving central TPN recognizes that the IV lipid emulsions provided two or three times a week serve primarily as a source of:
   a. vitamin C.
   b. essential fatty acids.
   c. fat-soluble vitamins.
   d. concentrated energy.

4. Compared to solutions delivered by peripheral vein, solutions delivered by central vein provide:
   a. more fat.
   b. more dextrose.
   c. a lower osmolality.
   d. more vitamins and minerals.

5. Among the following, the people who are **least** likely to benefit from peripheral parenteral nutrition (PPN) are those:
   a. who need long-term IV nutrition support.
   b. on oral or tube feedings who need additional nutrients temporarily.
   c. with normal renal function who need short-term IV support.
   d. in whom inserting an IV catheter into a central vein would be difficult.

6. The person who is a good candidate for central TPN rather than PPN:
   a. is well-nourished.
   b. does not have high nutrient requirements.
   c. needs long-term parenteral nutrition support.
   d. has strong peripheral veins and moderate nutrient needs.

7. The solutions infused through peripherally inserted central catheters are the same as those infused for:
   a. PPN.
   b. TPN through central veins.
   c. simple IVs.
   d. enteral feedings.

8. Which type of IV insertion procedure is associated with the highest risk of sepsis?
   a. central TPN catheter
   b. peripheral catheter for PPN
   c. peripheral catheter for simple IVs
   d. peripherally inserted central catheter

9. A gradual transition from an IV feeding to an oral diet is primarily designed to:
   a. improve the appetite.
   b. prevent hypoglycemia.
   c. prevent apprehension about eating.
   d. ensure that nutrient needs will continue to be met.

10. The health care team evaluating a client's ability to manage a home enteral or parenteral nutrition program would be least concerned with:
    a. the client's financial resources.
    b. the client's psychological status.
    c. the exact composition of the enteral formula or IV solution.
    d. the client and caregivers' abilities to learn the necessary techniques and follow medical instructions.

Answers to these questions appear in Appendix H.

CLINICAL APPLICATIONS

1. One liter of a TPN solution contains 500 milliliters of 50 percent dextrose and 500 milliliters of 5 percent amino acids. Determine the daily kcalorie and protein intakes of a person who receives 2 liters of such a TPN solution. Calculate the average daily energy intake if the person also receives 500 milliliters of a 20 percent fat emulsion three times a week.

2. Consider what it must be like to be on a home TPN program with no foods allowed by mouth. What would be the advantages of being at home instead of in the hospital? What would be the disadvantages? Think of how you would manage feedings. How would you feel about the time, costs, and commitment required to maintain this therapy? If you were not permitted to take any foods by mouth, how would you feel about not being able to eat after a long time? How would you handle holidays and special occasions that often center on food?

NUTRITION ON THE NET

FOR FURTHER STUDY OF THE
TOPICS IN THIS CHAPTER,
ACCESS THESE WEB SITES.

www.clinnutr.org
American Society for Parenteral and Enteral Nutrition

magi.com/~cpena
Canadian Parenteral-Enteral Nutrition Association

www.espen.org
European Society of Parenteral and Enteral Nutrition

www.wizvax.net/oleyfdn
The Oley Foundation

Notes

[1] D. W. Wilmore and S. J. Dudrick, Growth and development of an infant receiving all nutrients exclusively by way of the vein, *Journal of the American Medical Association* 203 (1968): 860–864.

[2] National Advisory Group on Standards of Practice Guidelines for Parenteral Nutrition, *Journal of Parenteral and Enteral Nutrition* 22 (1998): 49–66.

[3] As cited in National Advisory Group on Standards of Practice Guidelines for Parenteral Nutrition, 1998.

[4] National Advisory Group on Standards of Practice Guidelines for Parenteral Nutrition, 1998.

[5] National Advisory Group on Standards of Practice Guidelines for Parenteral Nutrition, 1998.

[6] J. Z. Rogers, K. McKee, and E. McDermott, Peripherally inserted central venous catheters, *Support Line*, October 1995, pp. 6–9; S. C. Loughran and M. Borzatta, Peripherally inserted central catheters: A report of 2506 catheter days, *Journal of Parenteral and Enteral Nutrition* 19 (1995): 133–136.

[7] T. C. Clark, Nutrition support clinical pathways, *Nutrition in Clinical Practice* 11 (1996): 16–20; The 1995 A.S.P.E.N. standards for nutrition support: Hospitalized patients, *Nutrition in Clinical Practice* 10 (1995): 206–207.

[8] M. E. Shils and coauthors, Long-term parenteral nutrition through an external arteriovenous shunt, *New England Journal of Medicine* 283 (1970): 314–343.

[9] P. Reddy and M. Malone, Cost and outcome analysis of home parenteral and enteral nutrition, *Journal of Parenteral and Enteral Nutrition* 22 (1998): 302–310.

[10] Reddy and Malone, 1998.

[11] M. F. Winkler, Nutrition support from hospital to home, Presented at the 82nd Annual Meeting of the American Dietetic Association, October 21, 1999.

[12] Reddy and Malone, 1998.

[13] M. Malone, Effect of home nutrition support on patient's lifestyle (abstract), *Journal of Parenteral and Enteral Nutrition* (supplement) 19 (1995): 23.

Q&A NUTRITION IN PRACTICE

Ethical Issues in Nutrition Care

Enteral and parenteral feedings meet nutrient needs and support recovery at times when table foods cannot. In some cases, such support can improve the quality of a declining life. In other cases, however, nutrition support can prolong life by merely delaying death, and the remaining life may be of low quality. Thus, the very availability of special nutrition support forces health care professionals and society to face **ethical** issues. (Terms relating to ethical issues appear in the glossary on p. 445.)

www.aslme.org
American Society of Law, Medicine, and Ethics

www.asbh.org
American Society for Bioethics and Humanities

www.choices.org
Choice in Dying

www.ama—assn.org
The American Medical Association

www.mindspring.com/~scottr/will.html
Provides links related to living wills.

Is it ever morally and legally appropriate to withhold or withdraw nutrition support?

In attempting to answer a question such as this one, ethics experts weigh such factors as:
- The client's right to make decisions concerning his or her own well-being.
- The potential benefits of the treatment versus the potential risk the treatment poses to the client's health or quality of life.
- The client's right to be fully informed of a treatment's risks and benefits in a fair and honest manner.
- When the client cannot speak for herself or himself, the ability of the

GLOSSARY

advance directive: the means by which competent adults record their preferences for future medical interventions. A living will and durable power of attorney are types of advance directives.

artificial feeding: parenteral and enteral nutrition; feeding by a route other than the normal ingestion of food.

comatose: in a state of deep unconsciousness from which the person cannot be aroused.

competent: having sufficient mental ability to understand a treatment, weigh its risks and benefits, and comprehend the consequences of refusing or accepting the treatment.

durable power of attorney: a legal document in which one competent adult authorizes another competent adult to make decisions for her or him in the event of incapacitation. The phrase "durable power" means that the agent's authority survives the client's incompetence; "attorney" refers to an attorney-in-fact (not an attorney-at-law).

ethical: in accordance with moral principles or professional standards. Socrates described *ethics* as "how we ought to live."

legal: established by law.

living will: a document signed by a competent adult that specifically states whether the person wishes aggressive treatment in the event of terminal illness or irreversible coma from which the person is not expected to recover.

persistent vegetative state: exhibiting motor reflexes but without the ability to regain cognitive behavior, communicate, or interact purposefully with the environment.

terminal illness: a progressive, irreversible disease that will lead to death in the near future.

clients truly comprehend the complexity and potential finality of their decisions, and it may be equally as difficult to determine whether a treatment will ultimately pose more risks than provide benefits. In reality, the answers often lie tangled in personal values, charged emotions, and legal conflict.

Isn't there a moral obligation to give nutrition support whenever it will prolong life?

Not necessarily. Health care professionals readily recommend whatever form of nutrition is necessary to support clients who have any reasonable chance of recovering from a disease. Clearly, health care professionals cannot rightfully withhold nutrition support because of poor judgment or negligence. The decision of whether to feed a client becomes less clear, however, when clients are **terminally ill** or in **persistent vegetative states,** or when they (or their caregivers) simply refuse specialized nutrition support because they feel the quality of their lives is poor.[1] Do we (as a society) allow them such choices? Are health care professionals morally and legally obligated to comply with, or to deny, such requests? Furthermore, who determines when clients are **competent** to speak for themselves? If a client is **comatose** or incompetent to make such a decision, who, if anyone, should be allowed to make such life-and-death decisions? These

unanswered questions are but a few that have evolved along with advances in medical technology and nutrition support; a discipline known as "medical ethics" has developed out of the need to discuss and solve problems such as these.

Who decides what will be done in cases where there are ethical dilemmas?

Most often the client (when competent) and the client's family work out their own answers based on extensive consultation with the physician and the health care team. Health care professionals must provide decision makers with the information they need to fully understand the disease, its treatments, the potential benefits of treatments, and the risks of treatments in order for them to make a truly informed choice. When the client or client's family makes a decision that the health care team or the facility housing the client believes may make the team or facility liable for malpractice, the conflict of interests may give rise to court cases. Then the court is charged with defining the problems and solving them.

How have courts resolved nutrition support issues?

One of the most widely publicized cases involving nutrition support

caregivers or family to promote the client's well-being without selfish intent.
• The client's right to expect that health care professionals and caregivers will honor his or her stated wishes about health care.
When sifting through ethical dilemmas, answers rarely come easily. It may be difficult to determine whether

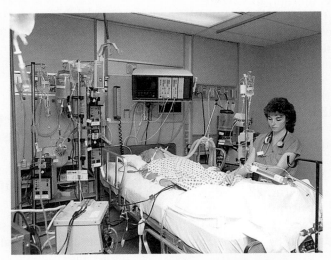

When is it morally and legally appropriate to withhold or withdraw special nutrition support?

issues was that of Nancy Cruzan. Nancy Cruzan was a young woman who suffered permanent and irreversible brain damage after a car crash in 1983. For eight years, she was in a persistent vegetative state—awake but unaware. Her physicians and parents held no hope for her recovery, yet given food and water, she might have lived for another 30 years. Her parents requested permission to discontinue tube feeding, but the Missouri Supreme Court rejected their request in 1987. The court held that Cruzan never definitively stated her "right to die" wishes and that her parents had no **legal** right to make such a request for her. The court stated that preserving life, no matter what its quality, takes precedence over all other considerations.

Cruzan's parents appealed the decision, and in 1989 the U.S. Supreme Court agreed to hear their arguments. The Cruzans tried to convince the Supreme Court that their once independent and vivacious daughter would not want to live in a vegetative state. No one questioned that Cruzan's parents knew their daughter's wishes better than anyone and had the highest and most loving motives. The question for the Court was whether families (or anyone) can make life-and-death decisions on behalf of incompetent persons. The Supreme Court recognized that competent adults have the right to stop life-sustaining treatment, but said that for incompetent adults, these treatments cannot be withdrawn without "clear and convincing" evidence that the incompetent person would refuse treatment.[2] The Court, therefore, upheld the Missouri Supreme Court's decision. It took still another round of court battles before additional evidence convinced the Missouri Supreme Court to allow Nancy Cruzan's tube feeding to be discontinued. Her feeding tube was removed, and she died from dehydration two weeks later.

Are such cases unusual?

Unfortunately, no. Some 30,000 other families of people who live in a persistent vegetative state face the same dilemma. The High Court's decision had widespread implications for these people and the health professionals who take care of them. Ethical concerns are more likely to arise as the population of elderly clients grows and new medical advances capable of sustaining life are developed. End-of-life decisions touch all of us because they influence the extent to which our society views life-sustaining treatment as optional not only for our clients, but for our families and ourselves.

What is the prevailing attitude about a person's right to make decisions about life-sustaining treatments?

The emerging ethical, medical, and legal consensus seems to support the view that individual rights outweigh those of the state. Competent individuals have a legal right to refuse medical treatment—including nourishment and hydration—even when medical experts consider that treatment necessary to sustain life. Most people believe that it is acceptable to withhold nourishment and hydration when a competent person desires to forgo such treatment, when an incompetent person has given **advance directives** (described later) about it, or when a person is in a permanently unconscious state.[3] In other words, even when treatment is lifesaving and its refusal may bring an earlier death, clients' rights remain paramount.

How can people protect their rights to retain or refuse treatment in the event that they are unable to make a decision when life-sustaining treatment is necessary?

Health care professionals should routinely encourage their competent clients to use advance directives to express ahead of time their preferences regarding medical treatments, including **artificial feedings,** should terminal illness, coma, or incompetency develop.[4] Each client's preferences should be noted in the medical records.

Clients should expect that physicians and facilities will comply with their preferences and not merely tolerate them grudgingly. Health care professionals should reassure clients that their decisions to refuse life-sustaining treatments, including nutrition support, will not mean that other care will be withheld. The client who makes such a choice still deserves meticulous physical care, pain management, and compassionate emotional support.[5] Any health care professional who is unwilling to abide by the client's stated preferences should arrange for continuing care by another equally qualified professional and then withdraw from that client's care.

One form of an advance directive is a **living will.** Another is the **durable power of attorney.** All states have statutes governing the use of advance directives; health care professionals should be aware of the regulations of the states where they work.[6] Some states' laws specifically address the conditions under which nutrition support can be withheld or withdrawn.

What is a living will?

A living will allows a competent adult to express clear directions regarding medical treatment in the event that the person is unable to make the necessary decisions at that time. The living will may specify that no extraordinary treatments should be administered, or alternatively, it may declare that every effort should be made to maintain life. People who prefer that nourishment and hydration be continued or discontinued need to write this specification into their living wills to ensure that their wishes are known.

What is a durable power of attorney?

A durable power of attorney allows a competent adult to designate another competent adult (usually a relative or close friend) as an agent to make health care decisions in the event of incapacitation. In essence, it says, "I give this person the right to make health care decisions on my behalf should I become unable to make them."

When people give another person a durable power of attorney, they should appoint someone who understands their health care preferences, is familiar with the contents of their living will (when available), and can be trusted to make decisions that reflect those preferences. Before serious medical conditions arise, the living will and hypothetical scenarios should be discussed with family members and the person who holds the durable power of attorney to ensure that they understand what medical care is desired. A copy of any advance directive should be a part of the client's medical record. The person who plans ahead for future care in the case of a terminal illness or irreversible state of unconsciousness relieves others of the guilt and some of the anxiety of having to make decisions. Imagine the anguish a family member may go through in directing the health care team to stop nutrition support, knowing that doing so will hasten death. That decision is a little easier if the family member knows it is what the individual would want, or if a legal document takes the decision out of the family's hands altogether.

The questions raised here have no easy answers, yet decisions must be made. Each case must be carefully considered individually. Most hospitals and extended care facilities have established ethics committees to deal with issues such as those presented here. Health care team members should ensure that their disciplines are represented on such committees and become familiar with their profession's ethics policies and guidelines.

Notes

[1] R. DeChicco, A. Trew, and D. L. Seidner, What to do when the patient refuses a feeding tube, *Nutrition in Clinical Practice* 12 (1997): 228–230.

[2] *Cruzan v. Director, Missouri Department of Health,* 497 U.S. 261, 110 S.Ct. 2841, 111 L.Ed.2d 224 (1990).

[3] R. Burck, Feeding, withdrawing, and withholding: Ethical perspectives, *Nutrition in Clinical Practice* 11 (1996): 243–253.

[4] *Advance Directives: The Role of Health Care Professionals* (Columbus, Ohio: Ross Products Division, Abbott Laboratories, 1996).

[5] V. M. Herrmann, Ethics in nutrition support, presented at the Eighth Annual Advances and Controversies in Clinical Nutrition, Dallas, Texas, April 5–7, 1998.

[6] B. Dorner and coauthors, The "to feed or not to feed" dilemma, *Journal of the American Dietetic Association* (supplement 2) 97 (1997): 172–176.

NUTRITION IN SEVERE STRESS

CONTENTS

The Body's Responses
to Stress

Nutrition Support
following Acute Stress

Case Study: Journalist
with a Third-Degree Burn

Nutrition in Practice:
Intestinal Immunity
and Stress

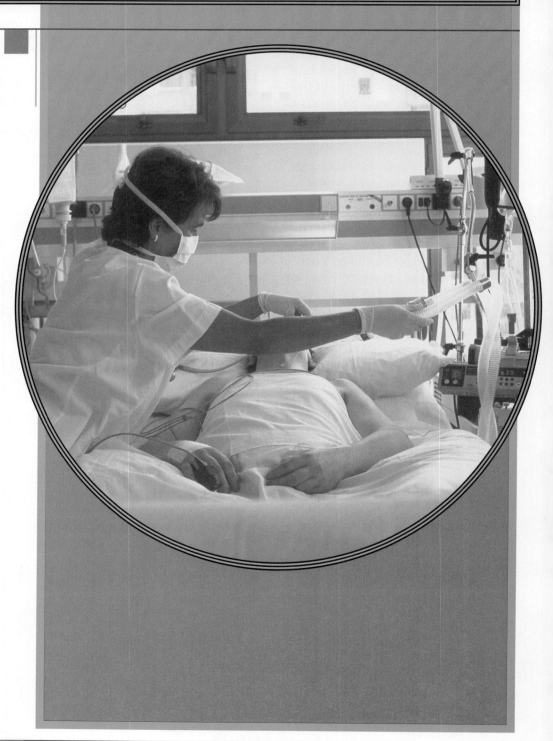

Stress describes a state in which the body's internal balance (homeostasis) is disrupted by a threat to the person's well-being. Some stresses fall within the body's normal and healthy functioning and are known as physiological stresses. Outside these limits, stresses imposed by illness, infection, or tissue injury are pathological stresses. All illnesses threaten the body and impair nutrition status to some extent, but the burden of acute stresses that occur without warning and involve extensive infection or tissue damage is especially great. When confronted with such a severe stress, the body takes action to reestablish its balance. These actions constitute an adaptation that is sustained for as long as is necessary or until the body becomes so exhausted as to cause death.

The Body's Responses to Stress

In response to acute stress, the body alters blood flow to the site of injury or infection to prevent or halt the spread of invading organisms. The metabolic rate speeds up (hypermetabolism) to mobilize nutrients into glucose and amino acid pools. From these pools the body can synthesize the special factors it needs to limit and repair damage and to regain homeostasis. Immune system factors and hormones mediate the response, which ultimately affects many body systems.

Researchers are working to elucidate the body's complex and interrelated responses to inflammation and infection. Much of that research focuses on **cytokines**—proteins that direct changes in the cardiovascular and nervous systems and stimulate the production of the many cells necessary to attack and destroy foreign organisms.[1] The accompanying glossary defines terms related to acute stress.

> **THE NURSING DIAGNOSES** that have nutrition implications and may apply to people with critical illnesses include **altered nutrition: less than body requirements, risk of infection, risk for altered body temperature, altered tissue perfusion, fluid volume excess, fluid volume deficit, risk for fluid volume deficit, impaired gas exchange, inability to sustain spontaneous ventilation, impaired tissue integrity,** and **impaired skin integrity.**

IMMUNE SYSTEM RESPONSES The body's natural system of defense against pathogens—the immune system—enables the body to fight off infectious

Although all illnesses threaten the body's internal balances, the healthy body adapts to minor stresses quickly and efficiently.

stress: the state in which a body's internal balance (homeostasis) is upset by a threat to a person's physical well-being (stressor).

Acute stresses occur suddenly and are severe, and with appropriate therapy, they resolve. The terms acute stress *and* severe stress *are used in this chapter to refer to pathological stresses that rapidly and markedly raise the body's metabolic rate and significantly upset its normal internal balance. Examples of stressors that lead to acute stress are infection, surgery, burns, fractures, and deep, penetrating wounds (gunshot wounds, fistulas, or surgical incisions). These stressors may lead to tissue injury or tissue death (necrosis), inflammation, and shock.*

GLOSSARY
OF STRESS-RELATED TERMS

abscess: an accumulation of pus that contains live microorganisms and immune system cells. Treatment involves draining the abscess.

counterregulatory hormones: hormones such as glucagon, cortisol, and catecholamines that oppose insulin's actions and promote catabolism.

cytokines (SIGH-toe-kynes): proteins that help regulate inflammatory responses. Some cause anorexia, fever, and discomfort.

granuloma: a grainlike growth that contains live microorganisms and immune system cells enclosed within a fibrous coat.

inflammatory response: the changes orchestrated by the immune system when tissues are injured by such forces as blows, wounds, foreign bodies (chemicals, microorganisms), heat, cold, electricity, or radiation.

sepsis: the presence of infectious microorganisms in the bloodstream.

shock: a sudden drop in blood volume that disrupts the supply of oxygen to the tissues and organs and the return of blood to the heart. Shock is a critical event that must be corrected immediately.

systemic inflammatory response syndrome (SIRS): the complex of symptoms (see the text) that occur as a result of immune and inflammatory factors in response to tissue damage. In severe cases, SIRS may progress to multiple organ failure.

agents. The immune system defends the body so alertly and silently that most healthy people are unaware that thousands of microbes attack them every day. Occasionally, though, an infection succeeds in making a person ill, and the immune system must then mount a vigorous counterattack. Serious infections greatly tax the body, and if the counterattack fails, death follows.

Tissues damaged by trauma, heat, chemicals, or infectious agents render the body vulnerable to invading organisms. The body's response to tissue damage—the **inflammatory response**—repairs the tissue damage and inactivates and removes foreign particles. As part of the response, blood flow to the injured area is rerouted to seal it off from the rest of the body while repair is under way. Meanwhile immune factors disarm any invaders that may have gained entry to the site. Together, these responses protect the body from the spread of infection.

If the immune factors fail to disarm the invaders but succeed in preventing an infection from spreading, an **abscess** or **granuloma** may form. If local defenses are overwhelmed and the invaders gain entry into the bloodstream, **sepsis** develops. Microbes in the bloodstream trigger the production of additional immune system factors. The invading organisms, surrounded by immune system cells, may attach to and damage blood vessel walls, which further stimulates the inflammatory responses. The resulting changes in circulation can disrupt blood flow to organs and may eventually lead to organ failure and **shock.**

■■ HORMONAL RESPONSES The hormonal changes characteristic of the immediate stress response shift the balance between insulin, which promotes the storage of carbohydrate and lipid and the synthesis of protein, and the **counterregulatory hormones,** which promote the breakdown of glycogen, the mobilization of fatty acids from lipids, and the synthesis of glucose from protein (see Table 19-1). Although insulin levels rise, the rise in counterregulatory hormones is greater still. These changes result in negative nitrogen balance: the body breaks down and uses more protein than it receives and syn-

---

Table 19-1
METABOLIC EFFECTS OF HORMONAL CHANGES DURING SEVERE STRESS

Hormone	Alteration	Metabolic Effect
Catecholamines	Increase	Glucagon release increases.
		Insulin-to-glucagon ratio decreases.[a]
		Glycogen breakdown increases.
		Glucose production from amino acids increases.
		Mobilization of free fatty acids increases.
Cortisol	Increases	Mobilization of free fatty acids increases.
		Glucose production from amino acids increases.
Glucagon	Increases	Insulin-to-glucagon ratio decreases.[a]
		Glucose production from amino acids increases.
		Glycogen breakdown increases.
		Storage of glucose, amino acids, and fatty acids decreases.
Insulin	Increases	Breakdown of body fat inhibited; blood glucose levels rise despite the increase.
Antidiuretic hormone	Increases	Retention of water increases.
Aldosterone	Increases	Retention of sodium increases.
		Excretion of potassium increases.

NOTE: These changes are part of the immediate stress response. As adaptation occurs and recovery is in progress, hormone levels gradually return to normal.
[a]The net effect of a decrease in the ratio of insulin to glucagon is that catabolism predominates.

thesizes. The protein loss comes primarily from skeletal muscle, connective tissue, and the gut; protein synthesis in the liver and the tissues that produce immune cells actually increases. Thus, skeletal muscle, connective tissue, and the gut are sacrificed to supply the glucose and amino acids necessary to respond to stress.[2]

As a result of the hormonal changes, the metabolic rate and blood glucose level rise. High insulin levels suppress the mobilization of essential fatty acids from body stores, and clinical signs of essential fatty acid deficiencies may develop. Figure 19-1 illustrates the differences between the body's responses to simple fasting and to metabolic stress to show why acute stresses rapidly deplete nutrient stores.

Other hormonal changes promote the retention of water and sodium and the excretion of potassium. Hypermetabolism generally peaks at about 3 to 4 days and subsides in about 7 to 10 days. With recovery, hormone levels gradually return to normal.

■■ **CLINICAL SIGNS AND SYMPTOMS** The inflammatory and immune responses to stressors result in redness, swelling, heat, and pain at the injury site. Body temperature, heart rate, and respiratory rate increase and anorexia develops. The person often experiences malaise—a feeling of discomfort and uneasiness. The symptoms associated with the inflammatory response are collectively called the **systemic inflammatory response syndrome (SIRS)**. Laboratory tests reveal elevated blood glucose (hyperglycemia), elevated blood urea nitrogen (from protein catabolism), negative nitrogen balance, increased retention of fluid and sodium, and increased excretion of potassium. Blood concentrations of albumin, iron, and zinc fall; white blood cell counts are often markedly elevated.

The acute phase of the stress response occurs immediately following an injury or infection and is characterized by hypermetabolism. During the adaptive phase, the body begins to recover from stress. If adaptation succeeds, recovery follows. If adaptation fails, exhaustion leads to death.

Cytokines are believed to contribute to the metabolic and GI tract changes as well as to the anorexia, fever, and malaise that accompany stress.

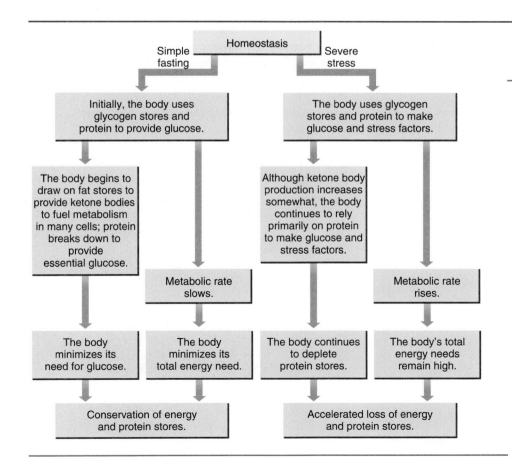

Figure 19-1
METABOLIC RESPONSES TO FASTING AND STRESS

■■ **GI TRACT RESPONSES** During stress, blood flow to the GI tract diminishes, and GI motility slows. Altered blood flow may hinder the stomach's ability to protect itself from gastric acid. Consequently, ulcers may form in the stomach or small intestine. The cells of the intestinal tract gradually shrink and may lose some of their absorptive and immune functions as gut proteins are sacrificed to support the body's defenses.

■■ **EFFECTS ON NUTRITION STATUS** For a well-nourished person with a short-term stress, the duration of negative nitrogen balance (loss of protein) is acceptable because sufficient protein remains to support defense systems and maintain vital functions. Acute stresses that markedly elevate the metabolic rate over longer periods of time, however, rapidly deplete energy reserves and break down protein tissues. Consequently, they may lead to **acute malnutrition** in a previously healthy person or worsen preexisting protein-energy malnutrition (PEM). If the person is malnourished or the stress is extreme or of long duration, however, the loss of lean body mass and the consequent depletion of vital proteins compromise the function of the immune system, heart, lungs, kidneys, and GI tract. Once organ systems begin to fail, recovery becomes severely compromised. Table 19-2 lists the consequences of stress that can lead to malnutrition.

■■ **EFFECTS ON MEDICATIONS** Stresses, and the malnutrition that may accompany them, can alter the body's use of medications and worsen health and nutrition status. Many medications are transported in the blood bound to blood proteins such as albumin, and low blood albumin is a symptom of both stress and PEM. Without sufficient carriers, medications may be slow to reach their sites of action. Once medications do arrive at their target cells, the lack of carriers may delay the medications' transport to the liver and kidneys, where they are often detoxified and excreted. Thus, medications may take a long time to work and then may remain active for longer than intended. In addition, acutely stressed people often receive multiple medications, which complicates decisions about the correct medication dosage and raises the risk of diet-medication interactions. Furthermore, when medications are given in an oral form, the effects of both stress and malnutrition on GI tract function can alter the absorption of both medications and nutrients. The box on p. 453 describes diet-medication interactions commonly used in the treatment of acute stresses.

■■ Nutrition Support following Acute Stress

Following acute stress, the immediate concerns of the health care team are to restore blood flow and oxygen transport and to prevent or treat infection. Possible measures include providing intravenous (IV) solutions to correct

acute malnutrition: PEM that develops rapidly due to a sudden and dramatic demand for nutrients. The person suffering from acute malnutrition may be of normal weight or may be overweight, but serum protein levels are low. **Chronic malnutrition,** on the other hand, represents an insufficient intake of energy and protein over long periods of time and is characterized by underweight, depleted fat stores, and normal serum protein levels. When the chronically malnourished person suffers an acute stress, or when acute malnutrition extends over long periods of time, the person suffers the consequences of both forms, and recovery is far more difficult.

Table 19-2
POSSIBLE CAUSES OF MALNUTRITION DURING ACUTE STRESS

Reduced Nutrient Intake	Excessive Nutrient Losses	Raised Nutrient Needs
Anorexia	Blood loss	Fever
Emotional stress	Immobility	Hypermetabolism
Immobility	Malabsorption	Infection
Location of injury	Medications	Medications
Malaise	Urinary losses	Pressure sores
Medications	Vomiting	
Nausea/vomiting		

DIET-MEDICATION INTERACTIONS

- **Analgesics:** See box on p. 398.
- **Antianxiety agents:** See box on p. 398.
- **Antidiarrheals:** See box on p. 399.
- **Antiemetics/antinauseants:** See box on p. 371.
- **Anti-infective agents:** Numerous anti-infective agents with a variety of nutrition-related interactions may be used to treat infections related to acute stresses. Consult a pharmacist or drug guide for specific interactions.
- **Anti-inflammatory agents:** See box on p. 399.

- **Sedatives:** Sedation itself may lead to reduced food intake. In addition, sedatives may cause nausea. Of note is the sedative *propofol,* which is a lipid-based sedative that provides 1.1 kcalorie per milliliter and can provide a significant amount of energy from fat. When propofol is provided to acutely stressed clients, enteral and parenteral formulas must be adjusted to avoid overfeeding.

fluid and electrolyte imbalances, giving transfusions, removing dead tissues, draining abscesses, and administering antibiotics. Nutrition support following stress helps to minimize nutrient losses, preserve organ function, and maintain immune defenses—all of which aid the client's recovery. Once hypermetabolism subsides, nutrition support aims to restore nutrient deficits and promote positive energy and nitrogen balance.

The removal of dead tissues resulting from burns and other wounds, called **debridement** *(dee-BREED-ment), speeds healing and helps prevent infection.*

Nutrient Needs

Feeding the acutely stressed individual is a formidable challenge. Nourishment must be introduced cautiously and adjustments must be made frequently throughout the course of the stress as needed. For the person with severely depleted nutrient stores, both underfeeding and overfeeding, as well as introducing nutrients too rapidly, can result in serious complications that compromise recovery.

> **THE NURSING DIAGNOSIS altered nutrition: risk for less than body requirements** frequently applies to acute stress, but **altered nutrition: risk for more than body requirements** should also be considered.

Nurses who care for people with acute stresses have many responsibilities, of which nutrition support is only one. The sections that follow may seem a bit overwhelming, but as you read, keep in mind that nurses are not responsible for the diet prescription. Nurses who understand the "whats, whys, and hows" of the nutrition plan, however, will feel confident that the extra steps they take to feed such clients are medically necessary and well worth the effort.

The physiologic and metabolic complications associated with reintroducing adequate energy and nutrients too rapidly in people with depleted nutrient stores are collectively called the **refeeding syndrome.** *These complications can include malabsorption, cardiac insufficiency, respiratory distress, congestive heart failure, convulsions, coma, and possibly death.*

LEARNING LINK

Because the course of critical illness may involve failure of one or more organs, nutrient needs must often be adjusted to an even greater extent than described here. Remember to keep the considerations of critical illness in mind when you read about acute heart failure and acute respiratory failure in Chapter 21, acute renal failure in Chapter 22, and liver failure in Chapter 23.

FLUIDS AND ELECTROLYTES To restore circulation and prevent dehydration and electrolyte imbalances, the medical team must act quickly to stabilize the body's blood volume and fluid and electrolyte balances. This step is critical, for without adequate blood volume, oxygen, nutrients, and medications cannot be delivered to the cells, organ systems cannot function properly, and waste products cannot be eliminated from the body. Conversely, providing too much fluid stresses the heart, which must pump a greater blood volume.

The physician determines fluid needs based on clinical measures such as blood pressure, heart rate, respiratory rate, urinary output, level of consciousness, and body temperature. Serum electrolytes are closely monitored, and adjustments are made as necessary.

With catabolism, blood levels of electrolytes normally concentrated in the intracellular fluids (potassium, phosphorus, and magnesium) rise. Once catabolism subsides, these electrolytes along with glucose and amino acids move back into the cells to begin rebuilding tissues. Without careful attention to replacement, blood levels of these electrolytes can plummet, resulting in life-threatening complications.

ENERGY Supplying too little energy further compromises nutrition status, raises the risk of infections, and interferes with wound healing and lung function.[3] Supplying too much energy further elevates the metabolic rate, which, in turn, increases oxygen use and carbon dioxide production. Then the heart and lungs, already working hard as a consequence of stress, must work even harder to keep the body's gases in balance. If these vital organs are weakened by preexisting malnutrition, the stress is even greater.

Indirect calorimetry is widely believed to provide the best estimate of energy needs during acute stress, especially when the stress is particularly severe.[4] Using indirect calorimetry requires special equipment (metabolic cart) and skilled clinicians to measure and interpret the results, which limits its use in many facilities.[5] Instead, clinicians often rely on estimates such as the Harris-Benedict equation to estimate basal energy needs (see Table 19-3). Additional kcalories are added to the basal energy estimate to meet the demands of stress. For most stresses, energy needs range from 100 to 120 percent of the basal energy expenditure, a range that is not significantly greater than a typical diet provides. Head and severe burn injuries, however, raise energy needs to a greater extent than other stresses.

indirect calorimetry: an indirect estimate of resting energy needs made by measuring the amount of carbon dioxide expired (carbon dioxide production) and the amount of oxygen inspired (oxygen consumption).

Many clinicians estimate energy needs in critically ill people based on body weight and provide 25 to 30 kcal/kg body weight.

Table 19-3
THE HARRIS-BENEDICT EQUATION FOR ESTIMATING BASAL ENERGY EXPENDITURE (BEE)[a]

Women:

$$BEE = 655 + (9.6 \times wt^b \text{ in kg}) + (1.7 \times ht \text{ in cm}) - (4.7 \times age \text{ in yr})$$

Men:

$$BEE = 66 + (13.7 \times wt^b \text{ in kg}) + (5 \times ht \text{ in cm}) - (6.8 \times age \text{ in yr})$$

Add to BEE for stress:

　0–20% for most stresses

　30–50% for head injuries

　20–100% for severe burns

NOTE: wt = weight; ht = height; yr = years; kg = kilograms; cm = centimeters.
[a]Basal metabolic rate (BMR, described on p. 128) and BEE express the same thing: basal energy need. The equation for BMR is traditionally used in physiology and fitness laboratories; that for BEE, in hospitals. The two equations yield slightly different results, each suitable for the purposes intended. All results are approximations; all require judgment in their application.
[b]Use actual body weight, not ideal body weight.
SOURCES: M. M. McMahon, Nutrition support of hospitalized patients, presented at the Eighth Annual Advances and Controversies in Clinical Nutrition, Mayo Clinic Foundation, April 5–7, 1998; D. J. Rodriguez, Nutrition in major burn patients: State of the art, *Support Line,* August 1995, pp. 1–8; E. Weeks and M. Elia, Observations on the patterns of 24-hour energy expenditure changes in body composition and gastric emptying in head-injured patients receiving nasogastric tube feeding, *Journal of Parenteral and Enteral Nutrition* 20 (1996): 31–37.

HOW TO:

ESTIMATE ENERGY AND PROTEIN NEEDS FOLLOWING SEVERE STRESS

Bernadette is a 39-year-old female, who is 5 feet 3 inches tall and weighs 130 pounds. She recently underwent extensive surgery. Her energy needs can be estimated using the Harris-Benedict equation as follows:

Weight in kilograms = 130 lb ÷ 2.2 kg/lb = 59 kg.
Height in centimeters = 63 in × 2.54 cm/in = 160 cm.
BEE = 655 + (9.6 × wt in kg) + (1.7 × ht in cm) − (4.7 × age in yr).
655 + (9.6 × 59 kg) + (1.7 × 160 cm) − (4.7 × 39)
= 655 + 566 + 272 − 183 = 1310 kcal.

Next add 20% × BEE for surgery:

1310 kcal × 20% = 262 kcal.
1310 + 262 = 1572 kcal.

Bernadette needs about 1572 kcalories to meet her BEE and additional energy needs due to surgery; clinicians monitor weight changes to determine if actual needs are higher or lower. Her energy needs will change as stress resolves.

Protein needs for Bernadette can be estimated at 1.0 to 1.5 grams of protein per kilogram of body weight per day. Use her weight of 59 kilograms to make the calculation:

59 kg × 1.0 g/kg = 59 g protein.
59 kg × 1.5 g/kg = 89 g protein.

Bernadette needs an estimated 59 to 89 grams of protein daily. Some clinicians subtract the kcalories provided by protein from the total energy needs and then provide the remaining calories as carbohydrate and fat.[a] Others meet the energy needs with carbohydrate and fat only and do not consider the kcalories contributed by protein. In either case, clinicians can monitor serum proteins (see Chapter 13) or use nitrogen balance studies to determine whether the estimate is meeting actual protein needs.

[a]J. M. Miles and J. A. Klein, Should protein be included in calorie calculations for a TPN prescription? *Nutrition in Clinical Practice* 11 (1996): 204–206.

The effects of obesity on the heart, lungs, and immune function, as well as the insulin resistance typical of obesity, are additional concerns for the obese person who suffers a severe stress. Although it is no less important that timely and appropriate nutrition support be given to the person who is obese, researchers report that obese people may benefit by providing energy at about half of their estimated needs, provided that protein needs are met.[6]

▪▪ PROTEIN AND AMINO ACIDS Stressed people with normal kidney and liver function generally need between 1.0 and 1.5 grams of protein per kilogram of body weight per day. People with severe burns may need more. Only after the hypermetabolic stage of severe stress subsides can negative nitrogen balance be fully corrected. The "How to" box above shows an example of how to estimate energy and protein needs.

An amino acid receiving wide attention in relation to stress is glutamine. Glutamine is a nonessential amino acid in healthy people, but the demands for glutamine during stress may exceed the body's ability to synthesize sufficient quantities; thus, glutamine may become a conditionally essential amino acid. Glutamine provides fuel for the intestinal cells and plays important roles in maintaining immune function and promoting wound healing.[7] Glutamine supplementation may reduce the incidence of infection in acutely stressed people and in people undergoing bone marrow transplants. [8]

■■ **CARBOHYDRATES AND LIPIDS** Nonprotein energy sources spare protein, and the amount of energy provided as carbohydrate or lipid has ramifications for stressed people. Carbohydrates provide a readily usable source of energy, but the body can metabolize only a fixed amount (about 500 grams per day) of glucose during stress. Excess glucose serves no benefit, but contributes to hyperglycemia, which can adversely affect fluid and electrolyte balances and may also increase the risk of infections in stressed people.[9]

Lipids provide energy and essential fatty acids, but given in excess, they can also tax metabolic functions and hamper immune responses. Clinicians often supply nonprotein kcalories through a mixture of about 70 percent carbohydrates and 30 percent lipids.[10]

■■ **MICRONUTRIENTS** The vitamin and mineral needs of people in stress are highly variable, and specific requirements are unknown. The needs for many B vitamins increase when energy and protein intakes increase. Some micronutrients act as cofactors in the many metabolic reactions that are occurring, so their stores may dwindle quickly. Other micronutrients play specific roles in healing wounds and mending broken bones. Levels of antioxidant nutrients fall, and supplementing these nutrients may prove to be beneficial. [11]

During stress, blood iron may fall as iron moves from the blood into the liver for storage. Iron is not given to compensate, however, because the shift robs invading organisms of the iron they need to grow, and thus helps prevent the spread of infection.

■■ **GROWTH HORMONE AND INSULIN-LIKE GROWTH FACTOR** Researchers have begun to study nondietary factors that might improve nitrogen balance and lessen the impact of severe stress. Two such factors, growth hormone and insulin-like growth factor-1 (IGF-1), also stimulate the growth of intestinal cells and may play an important role in protecting the intestine during stress (see Nutrition in Practice 19).

■■ **Delivery of Nutrients following Stress**

Selecting the appropriate amounts and types of nutrients to help people recover from stress is only part of medical nutrition therapy. Just as important is supplying nutrients in a form that best serves the body's ability to recover.

■■ **ORAL DIETS** Well-nourished clients with mild stresses who are expected to eat within 7 to 10 days following stress receive simple IV solutions to maintain fluid and electrolyte balances and provide minimal kcalories. Once GI tract motility returns, they begin an oral diet that often progresses from clear liquids to soft foods and then to regular foods as tolerated (see Chapter 14). Many people have great difficulty eating enough food to meet nutrient needs following an acute stress. Use the suggestions in the "How to" box in Chapter 14 on p. 351 to help improve food intake.

■■ **TUBE FEEDINGS AND PARENTERAL NUTRITION** People with severe malnutrition, those who are not expected to be able to eat within 7 to 10 days, or those who have high nutrient needs benefit from tube feedings or parenteral nutrition. To prevent abdominal distention, nausea, vomiting, and the possible aspiration of foods or formula into the lungs, oral or gastric feedings have to wait until gastric motility is restored. Peristalsis returns more quickly to the small intestine than to the stomach, however, and feeding formula directly into the small intestine through a tube is not only possible, but provides advantages over parenteral nutrition.[12] Early feeding (initiated within 48 hours following stress) stimulates intestinal blood flow, function, and adaptation and may minimize hypermetabolism and help prevent bacterial translocation (see Nutrition in Practice 19). Most significantly, early enteral feeding improves recovery following stress by reducing septic complications.[13] Early enteral feedings are not possible in all cases, however, particularly when blood flow to the intestine is disrupted either because of the person's medical condition or as a consequence of medications.[14] Additionally, some clients may need both enteral feedings and parenteral nutrition to meet nutrient needs. Once oral feeding is possible, tube feedings or parenteral nutrition is gradually discontinued (see Chapters 17 and 18).

:: STRESS FORMULAS Clinicians eager to improve a client's outcome often rely on formulas designed to meet nutrient needs during stress. Many such formulas are of high nutrient density. Some contain extra vitamins A and C, zinc, and other nutrients designed to promote wound healing. Formulas designed to improve immune function often contain added glutamine, arginine, nucleotides, and omega-3 fatty acids. Although such formulas appear to be beneficial for specific situations, further research is necessary to determine if their impact on recovery justifies their use. The accompanying case study of a client with burns tests your knowledge of nutrition and severe stress, and the Nutrition Assessment Checklist summarizes information necessary to monitor the nutrition care of stressed clients.

Stresses marked by extensive tissue damage and hypermetabolism can place tremendous demands on the body. Then the body uses its arsenal of defenses to survive and regain health. Recovery depends, in part, on the body's receiving the energy and nutrients required to mount a defense, repair damaged tissues, and replenish nutrient reserves.

Arginine may help minimize negative nitrogen balance, improve wound healing, and stimulate the immune system.

Nucleotides *are nitrogen-containing components of RNA and DNA that are important to the structure and function of intestinal cells and may help improve immune function.*

Case Study

JOURNALIST WITH A THIRD-DEGREE BURN

Mr. Sampson, a 48-year-old journalist, has been admitted to the emergency room. He suffered a severe burn covering over 40 percent of his body when he was trapped in a burning building. His wife told the nurse that Mr. Sampson's height is 6 feet and that he weighs about 175 pounds. The physician ordered lab work, including serum proteins; the results are not back yet.

+ Identify Mr. Sampson's immediate postinjury needs. What measures might be taken to meet those needs?

+ Considering Mr. Sampson's condition, what problems might the health care team encounter in getting information from him about his preburn nutrition status?

+ Calculate Mr. Sampson's energy and protein needs to support burn healing (use 2 × the BEE for energy and 2 to 3 grams of protein per kilogram of body weight). What other nutrients might be of concern?

+ Describe the possible benefits of early enteral nutrition to Mr. Sampson. How might these benefits be particularly important following a severe burn injury?

NUTRITION ASSESSMENT CHECKLIST — FOR PEOPLE EXPERIENCING ACUTE STRESS

Health Problems, Signs, and Symptoms

Check the medical record regularly to determine:

☐ Type of stress
☐ Severity of stress
☐ Stage of stress
☐ Route of feeding (oral/tube feeding/parenteral)
☐ If any organ system is compromised

Review the medical record for complications that may be related to underfeeding or overfeeding:

☐ Dehydration/fluid overload
☐ Hyperglycemia
☐ Electrolyte imbalances
☐ Acid-base imbalances

Medications

Record all medications and note:

☐ Signs that medication dosage may be inappropriate
☐ Signs of nutrient deficiencies
☐ kCalories provided from propofol, if appropriate

For people on medications who are able to eat an oral diet:

☐ Note any medications that may result in reduced food intake or nutrient imbalances
☐ Provide pain medications at times when they will be effective during meals, when appropriate
☐ Provide antinauseants or antiemetics at times when they will be effective during meals, when appropriate

Nutrient/Food Intake

If the client is not meeting nutrition goals:

☐ Calculate nutrient intake from enteral formulas or parenteral solutions
☐ Determine the contribution of nutrients from oral foods and the sedative propofol, if appropriate
☐ Investigate causes, including incorrect determination of nutrient needs or need to alter prescription based on client's unique responses

For clients eating an oral diet:

☐ Note the problems client is having with food
☐ Consider interventions to correct problems with eating and take action

Height and Weight

Measure baseline height and weight and daily weights. Note that body weight changes erratically in acutely ill clients due to the large amounts of fluids required for resuscitation. Once clients are in the adaptive stage of stress, if their weights are not meeting goals:

☐ Reestimate energy needs
☐ Check to see if client is receiving the diet (or formula) that has been prescribed
☐ Consider the need to change the energy prescription to meet weight goals
☐ Consider kcalories contributed from propofol, if appropriate

Laboratory Tests

Laboratory tests that often change as a consequence of stress itself and require careful interpretation as indicators of nutrition status during acute stress include:

☐ Albumin
☐ Transferrin
☐ Prealbumin
☐ Total lymphocyte count (white blood cell counts are often elevated)

Check laboratory tests for signs of:

☐ Dehydration/fluid overload
☐ Hypertriglyceridemia
☐ Electrolyte imbalances
☐ Acid-base imbalances
☐ Vitamin and trace element deficiencies
☐ Essential fatty acid deficiencies
☐ Response to protein intake (nitrogen balance studies, prealbumin, and retinol-binding protein), when available
☐ Organ dysfunction or return to normal organ function

Physical Signs

Regularly assess vital signs including:

☐ Blood pressure
☐ Pulse
☐ Temperature
☐ Respiration

Look for physical signs of:

☐ PEM (muscle mass and strength; body fat)
☐ Essential fatty acid deficiencies
☐ Dehydration/fluid overload
☐ Nutrient deficiencies and excesses

SELF CHECK

1. The acute stresses described in this chapter are characterized by:
   a. chronic malnutrition.
   b. tissue damage and hypermetabolism.
   c. reduced protein synthesis in the liver.
   d. hormonal changes that protect skeletal muscle.

2. Cytokines are proteins that:
   a. repair damaged tissues.
   b. destroy microorganisms.
   c. direct the inflammatory response to stress.
   d. oppose insulin's action and result in the catabolism of protein in skeletal muscle, connective tissue, and the gut.

3. The nurse recognizes that the complication most likely to lead to multiple organ failure following an acute severe stress is:
   a. sepsis.
   b. catabolism of protein.

c. essential fatty acid deficiency.

d. low serum albumin and transferrin.

4. Which of the following metabolic changes accompany acute stress?

a. Proteins are broken down in the liver.

b. Protein synthesis in the liver decreases.

c. The body conserves protein as it does in simple fasting.

d. Proteins are broken down in skeletal muscle, connective tissue, and the gut.

5. Which of the following statements with respect to nutrition and severe stress is true?

a. A previously well-nourished person can develop acute malnutrition if the stress is extreme or prolonged.

b. A person with chronic malnutrition who suffers an acute stress does not require immediate attention to energy and nutrient needs.

c. A person with either acute or chronic malnutrition has the energy reserves and protein needed to respond successfully to stress.

d. A person with chronic malnutrition who suffers an acute stress generally does not need tube feedings or TPN unless he or she will be unable to eat for 7 to 10 days.

6. The nurse may use all of the following parameters to assess a person's fluid status during stress **except:**

a. iron status.

b. urinary output.

c. blood pressure.

d. body temperature.

7. Which of the following statements correctly describes an appropriate diet for acute stress?

a. The diet is always high in kcalories.

b. The diet has little effect on blood glucose levels.

c. The amounts of carbohydrates, lipids, and proteins that supply energy make little difference.

d. Although the diet must supply adequate energy, energy needs are generally not high, except for some people with extensive burns or head injuries.

8. Depending on the type of stress, the amount of protein a stressed person who weighs 150 pounds needs can range from about _____ grams of protein per day.

a. 70 to 100

b. 70 to 200

c. 130 to 200

d. 130 to 300

9. The use of oral diets in the immediate poststress period is not possible because severe stress:

a. slows gastric motility.

b. decreases the appetite.

c. alters plasma amino acid levels.

d. results in increased blood flow to the GI tract.

10. The major reason early enteral nutrition following a severe stress must be introduced by tube is that:

a. appetite is depressed.

b. hydrolyzed diets are necessary.

c. gastric feeding is not medically possible.

d. oral diets cannot meet energy and nutrient needs.

Answers to these questions appear in Appendix H.

CLINICAL APPLICATIONS

1. Returning to Bernadette from the "How to" box on p. 455 and assuming that she can tolerate an intact enteral formula, find at least three formulas in Appendix G that the health care team might select if she needs a tube feeding. Determine the volume of each formula that would be needed to meet Bernadette's energy and protein needs. Would this volume also meet the recommendations for vitamins and minerals?

2. Bennie is a well-nourished seven-year-old who develops the flu and has a fever of 101°F for two days. Describe how this stress could temporarily affect his nutrition status. How would your concerns differ if Bennie were a seven-year-old hospitalized for injuries suffered in a car accident, who develops the flu and a fever and is unable to eat for several days?

3. Susan is a 28-year-old woman admitted to the hospital following a car accident in which she broke several bones, ruptured a portion of her small intestine, and suffered a severe burn. She has been in the hospital for several weeks and is now eating table foods. Aside from the nutrient demands imposed by the stresses she withstood, describe how the following factors can impair her nutrition status:

• Susan's injuries are painful.

• Susan's medications cause extreme drowsiness.

• Susan is depressed.

• Susan is often out of her room for X rays and other diagnostic tests when her menus and food trays arrive.

• Susan's food intake is often restricted for diagnostic tests she will be receiving.

How might these problems be resolved to improve Susan's ability to eat?

4. Reviewing the information about the metabolic changes directed by hormones in response to acute stress (pp. 450–451), suggest a reason why a previously well-nourished person who develops acute malnutrition as a consequence of stress might have normal body fat stores.

NUTRITON**THE**NET

FOR FURTHER STUDY OF THE
TOPICS IN THIS CHAPTER,
ACCESS THESE WEB SITES.

www.aacn.org
*American Association of Critical
Care Nurses*

www.execulink.com/~caccn
*Canadian Association of Critical
Care Nurses*

www.ameriburn.org
American Burn Association

www.burnrehab.com
Alberta Burn Rehabilitation Society

Notes

[1]H. R. Chang and B. Bistrian, The role of cytokines in the catabolic consequences of infection and injury, *Journal of Parenteral and Enteral Nutrition* 22 (1998): 156–166.

[2]L. L. Moldawer, Cytokines and the cachexia response to acute inflammation, *Support Line*, April 1996, pp. 1–6.

[3]S. A. McClave and coauthors, Are patients fed appropriately according to their caloric requirements? *Journal of Parenteral and Enteral Nutrition* 22 (1998): 375–381.

[4]S. A. McClave and D. A. Spain, Indirect calorimetry should be used, *Nutrition in Clinical Practice* 13 (1998): 143–145.

[5]C. S. Ireton-Jones and J. D. Jones, Should predictive equations or indirect calorimetry be used to design nutrition support regimens? *Nutrition in Clinical Practice* 13 (1998): 141–143.

[6]P. S. Choban, J. C. Burge, and L. Flancbaum, Nutrition support of obese hospitalized patients, *Nutrition in Clinical Practice* 12 (1997): 149–154; J. C. Burge and coauthors, Efficacy of hypocaloric total parenteral nutrition in hospitalized obese patients: A prospective, double-blind, randomized trial, *Journal of Parenteral and Enteral Nutrition* 18 (1994): 203–207.

[7]H. Saito, S. Furukawa, and T. Matsuda, Glutamine as an immunoenhancing nutrient, *Journal of Parenteral Nutrition* (supplement) 23 (1999): S59–S61.

[8]A. P. J. Houdijk, R. J. Nijveldt, and P. A. M. van Leeuwen, Glutamine-enriched enteral feeding in trauma patients: Reduced infectious morbidity is not related to changes in endocrine and metabolic responses, *Journal of Parenteral and Enteral Nutrition* (supplement) 23 (1999): S52–S58; A. P. J. Houdijk and coauthors, Randomized trial of glutamine-enriched enteral nutrition on infectious morbidity in patients with multiple trauma, *Lancet* 352 (1998): 772–776; P. R. Schloerb and M. Amare, TPN with glutamine in bone marrow transplantation and other clinical applications (a randomized double blind study), *Journal of Parenteral and Enteral Nutrition* 17 (1993): 407–413; T. R. Ziegler and coauthors, Clinical and metabolic efficacy of glutamine-supplemented parenteral nutrition after bone marrow transplantation, *Annals of Internal Medicine* 116 (1992): 821–828.

[9]M. M. McMahon, Nutrition support of hospitalized patients, presented at the Eighth Annual Advances and Controversies in Clinical Nutrition, Mayo Clinic Foundation, April 5–7, 1998; J. J. Pomposelli and coauthors, Early postoperative glucose control predicts nosocomial infection rate in diabetic patients, *Journal of Parenteral and Enteral Nutrition* 22 (1998): 77–81.

[10]McMahon, 1998.

[11]K. M. Oldham and P. E. Bowen, Oxidative stress in critical care: Is antioxidant supplementation beneficial? *Journal of the American Dietetic Association* 98 (1998): 1001–1008.

[12]W. W. Souba, Nutritional support, *New England Journal of Medicine* 336 (1997): 41–48; B. Beier, E. A. Bergman, and M. J. Morrissey, Factors related to the use of early postoperative enteral feedings in thoracic and abdominal surgery patients in the United States, *Journal of the American Dietetic Association* 97 (1997): 293–295; S. Trice, G. Melnik, and C. P. Page, Complications or costs of early postoperative parenteral versus enteral nutrition in trauma patients, *Nutrition in Clinical Practice* 12 (1997): 114–119.

[13]K. A. Kudsk, Immunologic support: Enteral vs parenteral feeding, in *Enteral Nutrition Support*, Report of the First Ross Conference on Enteral Devices, Ross Laboratories, 1996, pp. 70–74.

[14]G. Minard, Early enteral feeding—Is it safe for every patient? *Nutrition in Clinical Practice* 13 (1998): 79–80.

Q&A NUTRITION IN PRACTICE

Intestinal Immunity and Stress

The vital roles of the intestine in the digestion and absorption of nutrients are widely appreciated, but the significance of the GI tract in preventing infections has only recently received the attention it deserves. An emerging area of study focuses on factors that maintain the structure and functions of the GI tract. In late 1999, experts from all over the world gathered at the 1st International Conference on Intestinal Growth Factors and Nutrients in Health and Disease.[1] This Nutrition in Practice describes GI tract immunity and reviews some of the dietary and growth factors that help protect the structure and function of the intestine to protect this immunity.

Controversies abound in this area, partly because much of the research has been conducted on animals, and animals respond differently to stress and the lack of enteral nutrition than people do. From what is known, however, it seems clear that preserving GI tract immune function has far-reaching effects on health and nutrition status.

What role does the GI tract play in immunity?

A protective coating of mucus lines the entire GI tract. The mucus contains antimicrobial chemicals and enzymes to destroy foreign bodies and forms a slippery coat that prevents them from attaching to the lining of the GI tract. To reach the intestine, invaders must also avoid destruction by the highly acidic contents of the stomach. Invaders that enter the intestine directly (through breaks in tissue) or avoid destruction by mucus secretions or gastric acidity encounter formidable obstacles in the intestinal tract.

Recall from Chapter 5 that the surface of the intestine is lined with fingerlike projections called villi. Healthy villi are crowded close together, forming a physical barrier that hinders passage between them. Consequently, substances can pass from the intestine to the inside of the body only by crossing the cells' membranes, and the cells are remarkably efficient at keeping foreign materials out.

Interspersed among the villi are mucus-secreting cells and lymph tissues that house immune cells to fend off invaders (see Figure NP19-1) and cytokines to induce immune

Figure NP19-1
IMMUNE CELLS OF THE INTESTINAL VILLI

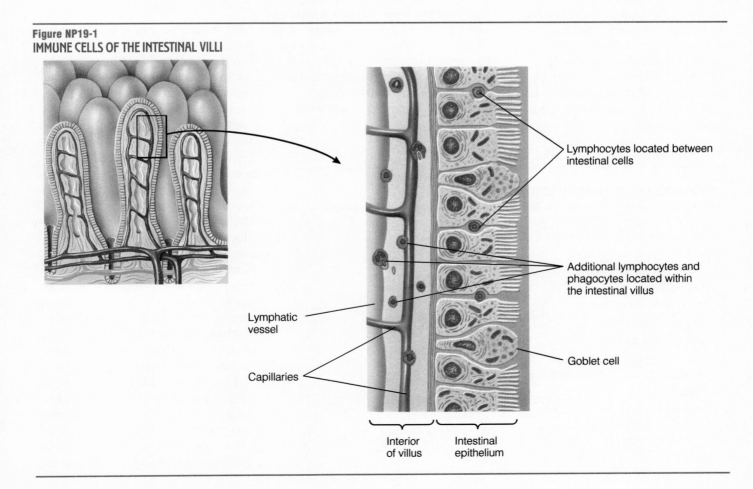

Lymphocytes located between intestinal cells

Additional lymphocytes and phagocytes located within the intestinal villus

Goblet cell

Lymphatic vessel

Capillaries

Interior of villus

Intestinal epithelium

responses.[2] To appreciate the vast importance of the intestinal lymph tissue in fighting foreign invaders, consider that its surface area is more than 200 times greater than the surface area of the skin and that 70 to 80 percent of the body's immunologic-secreting cells are located within the intestine.[3]

The large intestine also supports a bacterial population that inhibits the growth of harmful bacteria by competing with them for nutrients and space. The normal bacterial flora also produce short-chain fatty acids that prevent harmful microbes from sticking to the intestinal surface.

How can these functions be disrupted by stress?

Conditions such as stress and malnutrition divert necessary nutrients from the intestinal tract, thus limiting the ability of the intestinal cells to maintain their structure. An impaired structure compromises the GI tract's barrier function, normal bacterial flora, and immune cell function. Consequently, infectious agents may cross the intestinal barrier and enter the body—a process called translocation (see Table NP19-1).[4] Translocation of infectious agents in small quantities may stimulate the immune system, but extensive translocation may cause

serious infection and even death. Research suggests that translocation may be an important factor in the development of sepsis and multiple organ failure that may accompany severe stress.[5]

How is translocation associated with multiple organ failure?

The characteristics of multiple organ failure are remarkably similar in most cases, even though the original stresses may differ, suggesting that common factors may be involved. For one thing, multiple organ failure typically does not develop until days or weeks after the initial stress. People who develop multiple organ failure generally develop sepsis initially, and the first organs to fail are the lungs, followed by the liver and kidneys. Some of the infectious agents associated with sepsis and multiple organ failure arise from the intestinal tract.[6] Clinical evidence to support a role for intestinal translocation as a primary factor, or even one of the factors, leading to sepsis and multiple organ failure is lacking, and the theory remains unproved. In clinical practice, however, measures to support intestinal integrity and maintain immune function, including early enteral nutrition (see Chapter 19), are widely accepted and used.

In what ways can early enteral nutrition protect GI tract integrity and immune function?

Animal studies indicate that when nutrients are not present in the GI tract, intestinal cells atrophy and gradually become dysfunctional, even when nutrients are being delivered by vein. Most clinicians believe that early enteral nutrition following stress is superior to parenteral nutrition in maintaining the structure and functions of the intestinal tract, although this has not been proved in humans.[7] A primary objective of early enteral feeding is to preserve intestinal immunity, but it is not clear whether preserving GI tract integrity can prevent extensive translocation. In addition to early enteral nutrition, researchers are looking for specific dietary constituents that might protect the health of intestinal cells and GI immunity. Among these dietary constituents, glutamine is the most widely studied.

What is glutamine?

Glutamine is an amino acid abundant both in food and in the blood. When healthy people fail to get enough glutamine from their diets, they synthesize the glutamine they need from branched-chain amino acids. Glutamine provides fuel for the intestinal cells as well as for some immune cells. After the intestinal cells have metabolized glutamine, the liver uses the end products, alanine and ammonia, to make glucose and urea, respectively. Thus, glutamine is important in gluconeogenesis (the synthesis of glucose) and as a source of nitrogen. Glutamine also serves in the replication of rapidly dividing cells, including those of the intestine and immune system. These cells use glutamine to synthesize nucleotides (the basic units of RNA and DNA) as well as to synthesize nonessential amino acids.

During stress, the body's demands for glutamine may exceed the body's ability to synthesize it, and glutamine may become a conditionally essential amino acid. Intravenous feeding solutions traditionally do not contain glu-

Table NP19-1
CONDITIONS THAT INCREASE THE LIKELIHOOD OF TRANSLOCATION

Altered structure and function of GI tract barrier:

    Prolonged fasting or lack of enteral nutrients

    Injury to the GI tract

    Inflammatory responses

    Malnutrition

Changes in bacterial flora:

    Lack of enteral nutrients

    Decreased GI tract motility

    Use of broad-spectrum antibiotics

Compromised function of immune factors:

    Malnutrition

    Hypermetabolism

SOURCE: Adapted from M. T. DeMeo, The role of enteral nutrition in maintaining the structural and functional integrity of the gastrointestinal tract, in *Enteral Nutrition Support,* Report of the First Ross Conference on Enteral Devices, Ross Laboratories, 1996, pp. 4–8.

tamine, and many enteral formulas contain only small amounts. Limited evidence suggests that adding glutamine to standard TPN solutions can help preserve intestinal structure.[8] Other research suggests that glutamine-supplemented TPN can prevent atrophy of lymph tissue in the intestine.[9] As Chapter 19 described, studies have shown a benefit to glutamine supplementation in reducing the incidence of infection in acutely stressed people and people undergoing bone marrow transplants.

Are there other dietary factors that can help maintain the GI tract?

Dietary fiber also appears to help maintain GI tract integrity. Dietary fibers escape digestion in the small intestine, but in the colon, intestinal bacteria digest fibers and produce short-chain fatty acids. Short-chain fatty acids provide fuel for the cells of the colon, stimulate intestinal cell growth, enhance intestinal blood flow, bolster secretion of pancreatic enzymes, and promote sodium and water absorption in the colon. Short-chain fatty acids may also exert an anti-inflammatory effect in the colon, which may prove beneficial in the treatment of ulcerative colitis (see Chapter 16).[10]

What are growth factors, and how do they affect the intestinal cells?

As the name implies, growth factors stimulate the growth of tissues. The two that have been most widely studied are growth hormone and insulin-like growth factor-1 (IGF-1). Growth hormone appears to mediate its effects on intestinal cells through the regulation of IGF-1, which appears to directly stimulate intestinal cell growth following stress or long periods of GI tract disuse.[11] Human studies suggest that glutamine may exert some of its positive effects on the intestinal cells by raising growth hormone levels. Whether the effects of growth factors include protection against translocation remains to be determined.

Evidence is mounting to support the theory that a breach in the intestinal barrier can be a significant source of infection in acute stress. Clearly, studies of the possible role of early enteral nutrition and glutamine supplementation in severe stress are promising. Although many practitioners are currently using these strategies to improve their clients' recoveries, further studies are needed to confirm the benefits. More work must be done to determine if a breakdown of the intestinal barrier is truly a major route of infection during stress. Carefully controlled studies in human beings are lacking, and little work has been done to determine under which conditions, and in what amounts, glutamine, short-chain fatty acids, and growth factors might be safe and effective. Hopes are high, however. The prospects of supporting the GI immune system and preventing further decline in stressed people are inviting, and the potential benefits may prove to be lifesaving.

Notes

[1]D. W. Wilmore, Introduction, *Journal of Parenteral and Enteral Nutrition* (supplement) 23 (1999): S1–S2.

[2]I. Takahashi and H. Kiyono, Gut as the largest immunologic tissue, *Journal of Parenteral and Enteral Nutrition* (supplement) 23 (1999): S7–S12.

[3]Takahashi and Kiyono, 1999: P. Brandzaeg and coauthors, Immunobiology and immunopathology of human gut mucosa, *Gastroenterology* 97 (1989): 1562–1584.

[4]M. T. DeMeo, The role of enteral nutrition in maintaining the structural and functional integrity of the gastrointestinal tract, in *Enteral Nutrition Support*, Report of the First Ross Conference on Enteral Devices, Ross Laboratories, 1996, pp. 4–8.

[5]DeMeo, 1996; E. V. Shronts, Enteral versus parenteral nutrition: A clinical review, *Support Line*, June 1996, pp. 10–13; K. A. Kudsk, Clinical applications of enteral nutrition, *Nutrition in Clinical Practice* 9 (1994): 165–171.

[6]Shronts, 1996.

[7]T. O. Lipman, Grains or veins: Is enteral nutrition really better than parenteral nutrition? A look at the evidence, *Journal of Parenteral and Enteral Nutrition* 22 (1998): 167–182.

[8]R. R. J. Van Der Hulst and coauthors, Glutamine and the preservation of gut integrity, *Lancet* 222 (1993): 243–255.

[9]J. Li and coauthors, Glycyl-L-glutamine-enriched total parenteral nutrition maintains small intestine gut-associated lymphoid tissue and upper respiratory tract immunity, *Journal of Parenteral and Enteral Nutrition* 22 (1998): 31–36.

[10]A. Andoh, T. Bamba, and M. Sasaki, Physiological and anti-inflammatory roles of dietary fiber and butyrate in intestinal functions, *Journal of Parenteral and Enteral Nutrition* 23 (1999): S70–S73.

[11]T. Inaba and coauthors, Effects of growth hormone and insulin-like growth factor 1 (IGF-1) treatments on the nitrogen metabolism and hepatic IFG-1-messenger RNA expression in postoperative parenterally fed rats, *Journal of Parenteral and Enteral Nutrition* 20 (1996): 325–331.

CHAPTER 20

NUTRITION AND DIABETES MELLITUS

CONTENTS

What Is Diabetes Mellitus?

Treatment of Diabetes Mellitus

Diabetes Management throughout Life

Case Study: Child with Type 1 Diabetes

Nutrition in Practice: Mastering Diabetes Control

As Chapter 19 showed, stresses that disturb the body's metabolic activities can have severe and even fatal consequences. Acute stresses alter the body's normal balance between insulin and the counterregulatory hormones, which raises blood glucose and alters energy metabolism. The balance between insulin and the counterregulatory hormones is also disrupted in **diabetes mellitus**—this time because the body's production of insulin falters. Unlike acute stresses, however, diabetes mellitus is a chronic disorder for which there is no cure.

What Is Diabetes Mellitus?

Diabetes mellitus describes a group of metabolic disorders characterized by elevated blood glucose and altered energy metabolism and caused by defective insulin secretion, defective insulin action, or a combination of the two. There are two major types of diabetes, and their distinguishing features are summarized in Table 20-1. A later section describes the special case of gestational diabetes. Some other types of diabetes can occur as a consequence of genetic disorders, diseases of the exocrine pancreas, hormonal imbalances, drugs or chemicals, certain infections, and immune system disorders.

diabetes (DYE-uh-BEET-eez) mellitus (MELL-ih-tus or **mell-EYE-tus):** a group of metabolic disorders of glucose regulation and utilization.

    diabetes = passing through (the body)
    mellitus = honey-sweet (sugar)

Reminder: Insulin is the hormone that, among other roles, enables many cells to take up glucose from the blood and store energy fuels. Insulin is produced in the beta cells of the islets of Langerhans—a specific type of endocrine cell in the pancreas. The counterregulatory hormones, which include glucagon, epinephrine, norepinephrine, cortisol, and growth hormone, oppose insulin's actions.

LEARNING LINK

Recall from Chapter 16 that diabetes can arise from the destruction of pancreatic cells as a consequence of chronic pancreatitis and cystic fibrosis. Chapter 19 described how glucose rises as a response to severe infections. Some people who develop certain infections may develop diabetes permanently as a consequence.

Types of Diabetes

Over 10 million people in the United States have been diagnosed with diabetes, and estimates suggest that almost 5.5 million more may have diabetes that goes

Table 20-1
FEATURES OF TYPE 1 AND TYPE 2 DIABETES

	Type 1	Type 2
Other names	IDDM[a]	NIDDM[a]
	Juvenile-onset diabetes	Adult-onset diabetes
	Ketosis-prone diabetes	Ketosis-resistant diabetes
	Brittle diabetes	Lipoplethoric diabetes
		Stable diabetes
Age of onset	<20 (mean age, 12)	>40
Associated conditions	Viral infection, heredity	Obesity, heredity, aging
Insulin required?	Yes	Sometimes
Cell response to insulin	Normal	Resistant
Symptoms	Relatively severe	Relatively moderate
Prevalence in diabetic population	5 to 10%	90 to 95%

[a]The names IDDM (insulin-dependent diabetes mellitus) and NIDDM (noninsulin-dependent diabetes mellitus) frequently appear in references and generally describe type 1 and type 2 diabetes, respectively.

undiagnosed.[1] Undiagnosed diabetes is especially dangerous because its damaging effects (described later) may begin to develop years before the symptoms appear.[2] The symptoms of diabetes include increased urination, hunger, and thirst; unexplained weight loss; fatigue; and irritability.

■■ **TYPE 1 DIABETES** In **type 1 diabetes,** the less common of the two major types of diabetes (about 5 to 10 percent of diagnosed cases), the pancreas cannot synthesize insulin. Blood glucose rises abnormally high, but cannot enter the cells that need it for energy. Without insulin, the body's energy metabolism is dramatically altered with such severe consequences that people with type 1 diabetes cannot survive unless they inject insulin regularly. In many cases of type 1 diabetes, the individual appears to inherit a defect in which immune cells mistakenly attack and destroy insulin-producing pancreatic cells. People with siblings or parents with type 1 diabetes risk developing the disease themselves. Studies are currently under way to determine whether type 1 diabetes can be prevented.[3]

■■ **TYPE 2 DIABETES** The predominant type of diabetes mellitus (90 to 95 percent of cases), and the type most likely to go undiagnosed, is **type 2 diabetes.** The pancreas produces insulin, and the cells respond to it, but with less sensitivity. As blood glucose rises, the pancreas makes more insulin, and blood insulin rises to abnormally high levels (hyperinsulinemia). During this period of impaired glucose tolerance, the body is able to maintain blood glucose within a fairly normal range but at a cost. The chronic demand for insulin gradually exhausts the beta cells of the pancreas, and finally insulin production falters as the disease progresses.

Although the exact causes are unknown, impaired glucose tolerance and type 2 diabetes appear to be associated with obesity, especially abdominal fat, physical inactivity, and aging. Obesity aggravates **insulin resistance:** as body fat increases, body tissues become less and less able to respond to insulin. People most likely to develop type 2 diabetes include:

- Those who are obese.
- Those who have immediate family members with diabetes.
- Members of high-risk ethnic populations (African Americans, Asian and Pacific Islanders, Hispanic Americans, and Native Americans).
- Those who are over age 45.
- Women who have given birth to babies weighing over 9 pounds or who have been diagnosed with diabetes while pregnant. (Gestational diabetes is described in a later section.)

A nationwide multicenter study (the Diabetes Prevention Program) is currently under way to determine whether early intervention for people with impaired glucose tolerance can prevent type 2 diabetes.[4]

THE NURSING DIAGNOSES altered nutrition: more than body requirements and **altered nutrition: risk for more than body requirements** frequently apply to people with type 2 diabetes.

■■ **DIAGNOSIS OF DIABETES** The primary criteria used to diagnose diabetes include:

- Random blood glucose samples (one taken without regard to food intake) that exceed 200 milligrams per 100 milliliters in a person with symptoms of diabetes.
- A blood glucose of 126 milligrams per 100 milliliters or greater in a person who has been fasting for at least 8 hours.

---

type 1 diabetes: the less common type of diabetes in which the person produces no insulin at all.

Immune system disorders in which the body destroys its own cells or tissues are called autoimmune disorders.

type 2 diabetes: the more common type of diabetes that develops gradually and is associated with insulin resistance.

People with impaired glucose tolerance *may have fasting glucose levels higher than normal but not high enough to diagnose diabetes; others may have normal blood glucose levels most of the time, but when given a large amount of glucose, their blood glucose rises too high.*

The Expert Committee on the Diagnosis and Classification of Diabetes Mellitus defines obesity as greater than 120% desirable body weight or a body mass index of 27 or higher.

insulin resistance: the condition in which the cells fail to respond to insulin as they do in healthy people.

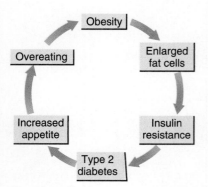

- A blood glucose level of 200 milligrams per 100 milliliters or greater at any time during a glucose tolerance test—a test that measures a person's fasting blood glucose and then retests it several times after a specific amount of glucose is given orally.

Any one of these tests can suggest diabetes, but a positive result must be confirmed by performing a second test (using any of the three criteria) on a subsequent day.

Acute Complications of Diabetes

Figure 20-1 presents an overview of the metabolic changes and acute complications that occur in uncontrolled diabetes. The metabolic consequences of type 1 diabetes are more immediate and severe than those of type 2 diabetes because in type 1 diabetes no glucose enters the cells. Disruption of energy metabolism and exposure of the tissues to high glucose concentrations result in both acute and chronic complications. The glossary on p. 468 defines diabetes-related symptoms and complications.

HYPERGLYCEMIA, DEHYDRATION, AND GLYCOSURIA With insufficient or ineffective insulin, blood glucose rises and hyperglycemia results. High blood glucose creates an osmotic effect, drawing water from the tissues into the blood. The high blood glucose concentration eventually exceeds the kidneys' ability to reabsorb glucose (the **renal threshold**), and the excess glucose is excreted in the urine along with fluids and electrolytes. This series of events explains two symptoms of uncontrolled diabetes: intense thirst and increased urination.

Symptoms of hyperglycemia:
- **Intense thirst and, sometimes, hunger.**
- **Increased urination.**
- **Blurred vision.**
- **Fatigue.**
- **Acetone breath.**
- **Labored breathing.**

renal threshold: the point at which a blood constituent that is normally reabsorbed by the kidneys reaches a level so high the kidneys cannot reabsorb it.

The renal threshold for glucose is generally reached when blood glucose rises above 180 mg/100 ml.

Figure 20-1

METABOLIC CONSEQUENCES AND ACUTE CLINICAL MANIFESTATIONS OF UNTREATED TYPE 1 AND TYPE 2 DIABETES

As you can see, when glucose cannot enter the cells, a cascade of metabolic changes follows. In type 2 diabetes, some glucose enters the cells. Because the cells are not "starved" for glucose, the body does not shift into the metabolism of fasting.

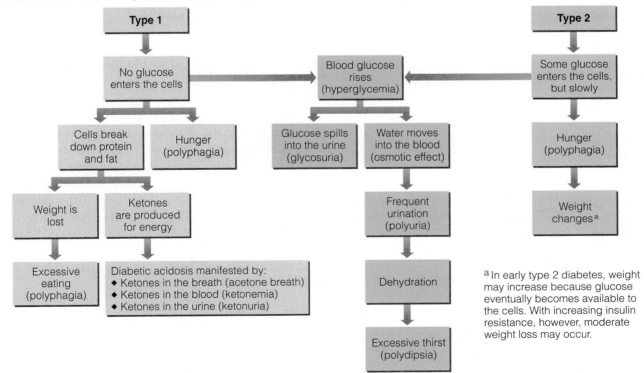

a In early type 2 diabetes, weight may increase because glucose eventually becomes available to the cells. With increasing insulin resistance, however, moderate weight loss may occur.

GLOSSARY
OF DIABETES-RELATED SYMPTOMS AND COMPLICATIONS

acetone breath: a distinctive fruity odor that can be detected on the breath of a person who is experiencing ketosis.

diabetic coma: unconsciousness precipitated by hyperglycemia, dehydration, ketosis, and acidosis in people with diabetes.

gangrene: death of tissue due to a deficient blood supply and/or infection.

gastroparesis: delayed gastric emptying.

glycosuria (GLY-ko-SUE-ree-ah) or glucosuria (GLUE-ko-SUE-ree-ah): glucose in the urine, which generally occurs when blood glucose exceeds 180 mg/100 ml.

hyperglycemia: elevated blood glucose. Normal fasting blood glucose is less than 110 mg/100 ml. Fasting blood glucose between 110 and 125 mg/100 ml suggests impaired glucose tolerance; values of 126 mg/100 ml or higher suggest diabetes.

hyperosmolar hyperglycemic nonketotic coma: coma that occurs in uncontrolled type 2 diabetes precipitated by the presence of hypertonic blood and dehydration.

hypoglycemia: low blood glucose.

ketonemia: ketones in the blood.

ketonuria: ketones in the urine.

macroangiopathies: disorders of the large blood vessels.

microangiopathies: disorders of the capillaries.

nephropathy: a disorder of the kidneys.

neuropathy: a disorder of the nerves.

polydipsia (POLL-ee-DIP-see-ah): excessive thirst.

polyphagia (POLL-ee-FAY-gee-ah): excessive eating.

polyuria (POLL-ee-YOU-ree-ah): excessive urine production.

retinopathy: a disorder of the retina.

THE NURSING DIAGNOSIS fluid volume deficit or **risk for fluid volume deficit** applies to people with uncontrolled diabetes.

A person with diabetes may develop hyperglycemia in response to the timing or amount of carbohydrate eaten, the improper use of medications, or a rise in counterregulatory hormone levels. Thus, hyperglycemia can even be present when the person with diabetes arises in the morning, as a consequence of an overnight fast (**dawn phenomenon**); when blood glucose levels are high and the person engages in strenuous physical activity; when a person uses too much insulin (causing **rebound hyperglycemia**); or when the person develops an illness or infection. Even a minor illness such as a cold or flu can cause blood glucose to rise dramatically. A major concern is the prevention of dehydration and vomiting.

dawn phenomenon: early morning hyperglycemia that develops in response to elevated levels of counterregulatory hormones that act to raise blood glucose after an overnight fast. Without adequate insulin, the glucose cannot enter the cells and remains in the blood.

rebound hyperglycemia: hyperglycemia resulting from excessive secretion of counterregulatory hormones in response to excessive insulin and consequent low blood glucose levels; also called the **Somogyi (so-MOHG-yee) effect.**

Reminder: Ketone bodies are produced by the incomplete breakdown of fat when glucose is not available to the cells.

KETOSIS AND COMA In undiagnosed type 1 diabetes, the lack of insulin continuously deprives the cells of the energy fuels they need, and the body responds by mobilizing protein and fat for energy. The liver responds to the mobilization of fatty acids by making ketone bodies, which accumulate in the blood and urine, resulting in acidosis. A similar sequence of events can happen in type 2 diabetes, most often as a consequence of an illness or infection. (Recall that insulin resistance accompanies stress or infection.) A fruity odor on the breath of a person with undiagnosed or poorly controlled diabetes (**acetone breath**) reflects the presence of the ketone acetone. If the combination of hyperglycemia, dehydration, and acidosis becomes severe enough, a potentially fatal coma may follow. Severe hyperglycemia and ketosis in the person with diabetes constitute a medical emergency that is treated in the hospital by carefully administering insulin and correcting fluid, electrolyte, and acid-base imbalances using intravenous (IV) fluids.

NONKETOTIC COMA Another kind of coma can occur as a consequence of extremely high blood glucose and dehydration without ketosis. This problem is more common in elderly people with type 2 diabetes because they may not recognize thirst and drink enough water to compensate for high blood glucose levels. This kind of coma is called hyperosmolar hyperglycemic nonketotic coma.

WEIGHT LOSS With the loss of glucose and ketone bodies (both energy sources) in the urine, combined with protein breakdown, serious weight loss may follow, especially in type 1 diabetes. People with uncontrolled or poorly controlled type 1 diabetes are likely to be thin despite eating apparently adequate amounts of foods. Although people with type 2 diabetes are often overweight, they may experience a gradual weight loss as a consequence of increasing insulin resistance.

THE **NURSING DIAGNOSIS** altered nutrition: less than body requirements frequently applies to the person with newly diagnosed type 1 diabetes.

HYPOGLYCEMIA Hypoglycemia is a consequence, not of untreated diabetes, but rather of inappropriate management. It can result from too much insulin or glucose-lowering medications, strenuous physical activity, skipped or delayed meals, inadequate food intake, vomiting, or severe diarrhea. The mental confusion and shakiness that are symptoms of hypoglycemia (see the margin) may make it difficult for the person to recognize the problem and take corrective actions. Furthermore, the warning signs become less noticeable over time, and people with longstanding diabetes risk severe hypoglycemia.

LEARNING LINK

Hypoglycemia can occur for reasons unrelated to diabetes. In some otherwise healthy people with hypoglycemia, blood glucose levels fall too low as the body shifts from a fed to a fasting state. Chapter 15 described the hypoglycemia that may occur in people who have undergone gastric surgery. In this case, the rapid absorption of glucose following a meal, rapidly raises blood glucose levels and causes too much insulin to be released by the pancreas. Ultimately, blood glucose falls too low as a consequence.

Chronic Complications of Diabetes

Chronic hyperglycemia damages the structures of the blood vessels and nerves (see the glossary). Circulation becomes poor and nerve function falters. Infections are more likely to occur due to poor circulation coupled with glucose-rich blood and urine. Infections may go undetected due to impaired nerve function; gangrene may follow. People with diabetes must pay special attention to hygiene and keep alert for early signs of infection.

THE **NURSING DIAGNOSIS** risk for infection applies to people with diabetes.

Elevated blood glucose may have damaging effects on blood vessels and nerves even before a diagnosis of diabetes is made. The combination of insulin resistance, secretion of more and more insulin to maintain blood glucose levels, obesity, hypertension, elevated LDL and triglycerides, and lowered HDL is frequently observed in people with both type 2 diabetes and cardiovascular disease and is called **syndrome X.**

CARDIOVASCULAR DISEASES Atherosclerosis (see Chapter 21) tends to develop early, progress rapidly, and be more advanced at the time of diagnosis in people with diabetes.[5] More than 80 percent of people with diabetes die as a consequence of cardiovascular diseases, especially heart attacks. If

---

Symptoms of hypoglycemia:
- **Hunger.**
- **Headache.**
- **Sweating.**
- **Shakiness.**
- **Nervousness.**
- **Confusion.**
- **Disorientation.**
- **Slurred speech.**

Notice that many of the symptoms of hypoglycemia are those of alcohol intoxication. If the true problem goes unrecognized, the person may die. To prevent such a tragic mistake, advise every person with diabetes to wear identification in the form of a bracelet or necklace.

Hypoglycemia in a person who uses insulin is also called an insulin reaction or insulin shock.

syndrome X: the combination of insulin resistance, hyperinsulinemia, obesity, hypertension, elevated LDL and triglycerides, and reduced HDL that is frequently associated with type 2 diabetes and cardiovascular disease; also called **insulin-resistance syndrome** and **metabolic syndrome.**

nerve function is also impaired, the person may suffer a heart attack and not even realize it.

SMALL BLOOD VESSEL DISORDERS Disorders of the smallest blood vessels—the capillaries—may also develop and lead to loss of kidney function and retinal degeneration and loss of vision. About 85 percent of people with diabetes have nephropathy, retinopathy, or both. Consequently, diabetes is a leading cause of both kidney failure and blindness.

NEUROPATHY Nerve tissues may also deteriorate, resulting in neuropathy. At first, the person may experience a painful prickling sensation, often in the arms and legs, which progresses until the person loses sensations in the hands and feet. Injuries to these areas may go unnoticed, and infections can progress rapidly. If tissues die as a consequence, amputation of the affected limb (usually the toes, feet, or legs) may be necessary. Neuropathy can also delay gastric emptying. When the stomach empties slowly after a meal (gastroparesis), the person may experience a premature feeling of fullness, nausea, vomiting, weight loss, and poor blood glucose control due to irregular nutrient absorption.[6]

Treatment of Diabetes Mellitus

A diagnosis of diabetes can be devastating. The parents of a young child with type 1 diabetes may feel overwhelmed, angry, anxious, and even guilty. An adolescent may feel that it is the end of the world. An adult with type 2 diabetes may fear possible complications and resent having to readjust lifestyles and adopt a new diet. To control blood glucose successfully, the person must master the complex task of coordinating diet, physical activity, and medications. On the bright side, such mastery helps a person live a full and active life and significantly reduces the risk of chronic complications. This chapter's Nutrition in Practice describes how clients and health care professionals work together to help clients gain maximum control over diabetes.

TREATMENT GOALS The goals of both medical and nutrition therapy for diabetes are to maintain blood glucose within a fairly normal range, achieve optimal blood lipid levels (see Chapter 21), control blood pressure (Chapter 21), support health and well-being, and prevent and treat complications. The most important of these goals is blood glucose control. A major multicenter clinical trial (the Diabetes Control and Complications Trial, or DCCT) clearly showed that tightly controlling blood glucose reduces the risks of onset and progression of nephropathy, retinopathy, and neuropathy by about 50 percent. Although this study specifically addressed the management of diabetes for people who use insulin, most clinicians believe that the results apply to people with type 2 diabetes as well.

TREATMENT PLANS Because many factors influence blood glucose and clients must often change many lifestyle behaviors to control diabetes and its complications, accurate assessment and monitoring of a person's diet, physical activity, medications, and health status are vital in helping clients set acceptable goals for managing their diseases. To promote success, the health care team plans and adjusts therapy for the client's medical needs, motivational level, educational ability, and lifestyle. Successful diabetes education takes time and must be flexible to accommodate changing needs.

◼◼ Medical Nutrition Therapy for Diabetes

The diet for diabetes parallels a healthy diet for all people in both amounts and types of nutrients. Attention to all energy nutrients is important: controlling carbohydrate prevents hyper- and hypoglycemia; controlling protein can help prevent loss of kidney function; controlling fat helps prevent cardiovascular complications. In addition, the diet plan must be coordinated with medications and physical activities. The most appropriate professional to design and implement medical nutrition therapy for diabetes is the dietitian.

◼◼ **ENERGY** Medical nutrition therapy for diabetes first focuses on providing food energy in the amount necessary to achieve or maintain a healthy and realistic body weight and to support growth in children and pregnant women. The diet planner calculates the person's average daily energy intake to determine an appropriate energy level, monitors height and weight measures periodically, and adjusts the diet as necessary.

People with type 2 diabetes frequently benefit from weight-loss diets. Even moderate weight loss (10 to 20 pounds) can help reverse insulin resistance, improve the blood lipid profile, and reduce blood pressure. A diet plan that moderately restricts energy intake (250 to 500 kcalories less than the current average intake) provides for a realistic and gradual weight loss.

◼◼ **PROTEIN** Protein provides about 10 to 20 percent of the total kcalories in the diet for diabetes. Providing adequate, but not excessive, protein may help delay the onset or progression of kidney disease (see Chapter 22). At the first sign of kidney disease, people with diabetes may need to restrict protein to 0.8 gram per kilogram of body weight (the same as the RDA).[7] If kidney disease progresses, protein may need to be restricted further.

After considering the energy provided by protein, the remaining kcalories are distributed between carbohydrate and fat. The next sections describe some of the considerations that apply.

An early sign of impending kidney disease is the excretion of small amounts of albumin in the urine or microalbuminemia.

◼◼ **CARBOHYDRATE** The person with diabetes needs to have glucose available throughout the day, but not so much at any one time that blood glucose levels rise too high. Of all energy nutrients, carbohydrates have the greatest effect on blood glucose. Once eaten, carbohydrates raise blood glucose in about an hour. Some protein from a meal may also be converted to glucose, but very slowly and in small amounts. Fat is not converted to glucose in significant amounts.

For people with diabetes, consistent timing and composition of meals and snacks from day to day improve glucose control. The amount of carbohydrate to include depends on the diet plan; a typical plan provides about 45 to 60 percent of the total kcalories from carbohydrate.

A person who engages in regular physical activity, takes prescribed medication regularly, and eats about the same amount of carbohydrate at about the same time each day is likely to have a safe amount of glucose available to the body when it is needed. Too much carbohydrate at a meal or snack can cause hyperglycemia. Skipping meals or eating too little can lead to hypoglycemia. An evening snack is especially important because it helps sustain the person's blood glucose through the night. In addition to the problems associated with hypoglycemia described earlier, a person who frequently experiences hypoglycemia and necessarily uses carbohydrate to treat it may experience excessive weight gain.

The diet for diabetes emphasizes a consistent intake of carbohydrate from well-balanced meals spaced evenly throughout the day.

◼◼ **COMPLEX CARBOHYDRATES VERSUS SIMPLE SUGARS** Encourage clients with diabetes to select foods rich in complex carbohydrates: whole-grain breads and cereals, legumes, fruits, and vegetables. In addition to carbohydrates, these foods provide fiber, vitamins, and minerals and offer many health benefits (see Chapter 2).

Traditionally, concentrated sweets were strictly excluded from the diet for diabetes, but now they are restricted only to the same extent as they are for all people. The total amount of carbohydrate is of greater concern in diabetes than the type of carbohydrate. The person with diabetes can consume concentrated sweets and alternative nutritive sweeteners as a limited part of a healthy diet, as long as they are counted as part of the carbohydrate allowance. Artificial sweeteners, and products made from them, contain minimal kcalories and can be used in place of sugar.

FAT People with diabetes who have acceptable blood lipid concentrations benefit from a fat intake consistent with the *Dietary Guidelines for Americans* (30 percent or less of kcalories from fat, less than 10 percent of kcalories from saturated fat, and less than 300 milligrams of cholesterol). People with diabetes who have elevated LDL may need to restrict saturated fat intake to 7 percent or less of total kcalories and cholesterol to less than 200 milligrams daily (see Chapter 21). To lower fat and cholesterol intakes, clients can use low-fat and fat-free milk and milk products and lean meats, among other strategies (see the "How to" box on p. 64).

For people with elevated triglycerides and VLDL, another diet approach may be useful. These clients may be able to control their blood glucose and improve their blood lipid levels with a moderately higher fat and lower carbohydrate intake, provided that additional fat kcalories come from monounsaturated fat.[8]

All people with diabetes who use reduced-fat products must use them cautiously.[9] They may overeat in the belief that they are saving kcalories by eating reduced-fat products. Additionally, carbohydrate often replaces the fat in these products. This carbohydrate must be considered as part of the carbohydrate allowance in the diet plan.

SODIUM People with diabetes frequently develop hypertension, and, for some, limiting sodium can help reduce blood pressure. Recommendations for appropriate sodium intakes vary from 2400 to 3000 milligrams of sodium per day for all people, including those with diabetes. People with diabetes and hypertension may benefit from restricting sodium to 2400 milligrams or less per day.[10] Chapter 21 provides more information about sodium-restricted diets.

ALCOHOL The person whose blood glucose is well controlled can usually include some alcoholic beverages with the consent of the physician. Alcohol can cause hypoglycemia in any person, and people with diabetes who take medications to lower blood glucose are particularly likely to develop this complication. People who take insulin are advised to drink only moderate amounts (no more than two drinks per day), with meals, and in addition to the usual meal plan. To protect against hypoglycemia, no foods should be omitted. (Remember, too, that the person with hypoglycemia may appear to be intoxicated and alcohol use can add confusion to a potentially dangerous situation.)

People with a history of alcohol abuse, pancreatitis, abnormal blood lipids, or neuropathy and women who are pregnant are strongly cautioned to avoid alcohol completely. Alcohol use is also discouraged for people who are overweight; if it is used, alcohol should be substituted for fat exchanges. Drinks that contain simple sugars (mixers, sweet wines, and liqueurs) are best avoided. If they are used, the person must count their carbohydrate contents as part of the daily carbohydrate allowance. The combination of alcohol and some **oral antidiabetic agents** may cause flushing of the skin and a rapid heartbeat (disulfiram-like reaction).

Reminder: Authorities recommend an intake of 20 to 35 g of dietary fiber a day, although an even higher intake may be desirable.

Reminder: Nutritive sweeteners are carbohydrates and include sorbitol, mannitol, and xylitol.

People with very high triglyceride levels (≥1000 mg/100 ml) need to restrict all types of dietary fat to less than 10% of kcalories.

The fat substitute olestra does not contribute kcalories or carbohydrate.

One alcoholic drink is defined as 1½ oz of distilled liquor, 12 oz of beer, or 5 oz of wine.

Note: Light beer contains the same amount of alcohol as regular beer, with half the carbohydrate.

For people who need to lose weight, 1 drink = 2 fat exchanges.

oral antidiabetic agents: medications taken by mouth to lower blood glucose levels in people with type 2 diabetes.

▪▪ MICRONUTRIENTS Unless the person with diabetes has a vitamin, mineral, or trace element deficiency, micronutrient needs for people with diabetes do not differ from those of the healthy population. People who take some types of diuretics (see Chapter 21), however, may need to take potassium supplements.

▪▪ MISSED MEALS AND ILLNESSES When people with diabetes are ill, their blood glucose levels often rise. During illness, clients may be advised to increase their doses of medication, reduce carbohydrate intakes somewhat, or a combination of both. However, they need to have some carbohydrate to forestall hypoglycemia; so do people who must miss a meal for any reason. If appetite is poor, people can use juice, flavored gelatin, soft drinks, or frozen juice bars to meet their carbohydrate needs.

Hospitals employ different procedures for clients with diabetes who miss a meal. One procedure provides at least half the prescribed carbohydrate and kcalories within three hours of the missed meal. If this cannot be done, the physician may change the insulin schedule, give IV dextrose, and/or change the diet prescription to include more simple carbohydrates.

▪▪ TREATING HYPOGLYCEMIA If a client does develop hypoglycemia for any reason, judicious treatment prevents overcorrection and subsequent hyperglycemia. As soon as the symptoms are observed, the person needs 10 to 15 grams of carbohydrate. People who take oral agents that interfere with the digestion of sucrose and complex carbohydrates need to take glucose to treat hypoglycemia; for all others, any carbohydrate source that is readily available and easy to eat is a good choice. Advise clients to avoid foods that also contain fat because fat slows the absorption of carbohydrate. Blood glucose is then checked within 15 to 20 minutes to see if it has risen to an acceptable level. If not, an additional 10 to 15 grams of carbohydrate are given, and blood glucose is rechecked. The procedure continues until blood glucose returns to an acceptable range. Advise clients to carry some convenient source of carbohydrates with them at all times, so they can act immediately when hypoglycemic symptoms occur. People prone to **nocturnal hypoglycemia** may be advised to wake up during the night and test their blood glucose (described later in the chapter), eat a snack at bedtime, participate in strenuous activity earlier in the day, or reduce the insulin dose following evening physical activity.

If hypoglycemia becomes severe, the person may be disoriented, unable to recognize a hypoglycemic reaction, and unable to swallow safely. In such cases, the person needs to receive IV glucose or the hormone glucagon or both to counteract the insulin reaction. Without treatment, the person may lapse into shock and die.

▪▪ ENTERAL AND PARENTERAL FORMULAS The indications for enteral and parenteral nutrition for people with diabetes are the same as for other people (see Chapters 17 and 18). When people with diabetes require enteral or parenteral formulas, however, adjustments may need to be made for the large amounts of carbohydrates the formulas contain. Often health care professionals provide or adjust insulin doses to cover the higher carbohydrate loads. If insulin fails to control hyperglycemia, people on parenteral nutrition may need to receive less energy from dextrose and more from IV lipid emulsions. People who cannot tolerate the carbohydrate in standard enteral formulas may benefit from specially designed formulas that contain less total carbohydrate (see Appendix G).

▪▪ Meal-Planning Strategies

No single approach to medical nutrition therapy meets everyone's needs, and dietitians use several approaches to help clients achieve blood glucose control.

Easy-to-eat sources of carbohydrate (10 to 15 g per serving):
- 2 to 3 tsp honey.
- 4 to 5 hard candies (such as Lifesavers).
- 5 to 6 large jelly beans.
- 4 to 6 oz sweetened soft drink.
- 4 oz orange or other fruit juice.
- 1 tbs icing from a can or tube.
- Glucose gel or tablets (check label for amount).

nocturnal hypoglycemia: hypoglycemia that occurs while a person is sleeping.

Some diet strategies teach clients to use food guides or simple menus to plan diets. Traditionally, however, diet planners use the exchange system to help people with diabetes plan their diets.

Appendix C includes the complete U.S. exchange system, and Appendix B includes the Canadian system.

▪▪ EXCHANGE LISTS The exchange system sorts foods into three main groups by their proportions of carbohydrate, fat, and protein. The foods in these three groups—the carbohydrate group, the fat group, and the meat and meat substitutes group (protein)—are then organized into several exchange lists.

The carbohydrate group includes these exchange lists:

- Starch (cereals, grains, pasta, breads, crackers, snacks, starchy vegetables, and dried beans, peas, and lentils).
- Fruit.
- Milk and some milk products (fat-free and very low fat, low fat, and whole).*
- Other carbohydrates (desserts and snacks with added sugars and fats).
- Vegetables.

The fat group includes typical fats such as butter, margarine, oil, and salad dressing as well as high-fat foods such as nuts, olives, bacon, avocados, coconut, and cream cheese. The meat and meat substitutes group includes high-protein foods. Table 20-2 below shows the energy and energy nutrients found in each exchange list and Figure 20-2 on pp. 475–476 shows examples of foods and portion sizes for each group of foods.

By strictly defining portion sizes, all of the foods in a given exchange list provide approximately the same amounts of energy (kcalories) and energy nutrients (carbohydrate, fat, and protein). Any food on a list can then be exchanged, or traded, for any other food on that same list without affecting a plan's energy balance. The "How to" box on pp. 477–480 shows the basics of how to plan a diet for diabetes using exchange lists.

---

*The exchange lists may also refer to fat-free products as nonfat.

Table 20-2
THE EXCHANGE LISTS: ENERGY NUTRIENTS PER SERVING

Group/Lists	Carbohydrate(g)	Protein (g)	Fat (g)	Energy (kcal)
Carbohydrate Group				
Starch	15	3	1 or less	80
Fruit	15	—	—	60
Milk				
Fat-free and very low fat	12	8	0–3	90
Low-fat	12	8	5	120
Whole	12	8	8	150
Other carbohydrates	15	varies	varies	varies
Vegetable	5	2	—	25
Meat and Meat Substitutes Group				
Meat				
Very lean	—	7	0–1	35
Lean	—	7	3	55
Medium-fat	—	7	5	75
High-fat	—	7	8	100
Fat Group				
Fat	—	—	5	45

Figure 20-2

THE EXCHANGE SYSTEM: EXAMPLES OF FOODS, PORTION SIZES, AND ENERGY-NUTRITION CONTRIBUTIONS

The Carbohydrate Group

Starch

1 starch exchange is like:
1 slice bread.
¾ c ready-to-eat cereal.
½ c cooked pasta, rice noodles, or bulgar.
⅓ c cooked rice.
½ c cooked legumes.[a]
½ c corn, peas, or yams.
1 small (3 oz) potato.
½ bagel, English muffin, or bun.
1 tortilla, waffle, roll, taco, or matzoh.
(1 starch = 15 g carbohydrate, 3 g protein, 0–1 g fat, and 80 kcal.)

Vegetables

1 vegetable exchange is like:
½ c cooked carrots, greens, green beans, brussels sprouts, beets, broccoli, cauliflower, or spinach.
1 c raw carrots, radishes, or salad greens.
1 lg tomato.
(1 vegetable = 5 g carbohydrate, 2 g protein, and 25 kcal.)

Fruits

1 fruit exchange is like:
1 small banana, nectarine, apple, or orange.
½ large grapefruit, pear, or papaya.
½ c orange, apple, or grapefruit juice.
17 small grapes.
⅓ cantaloupe (or 1 c cubes).
2 tbs raisins.
1½ dried figs.
3 dates.
1½ carambola (star fruit).
(1 fruit = 15 g carbohydrate and 60 kcal.)

Milks (fat-free and very low fat)

1 fat-free milk exchange is like:
1 c fat-free or 1% milk.
¾ c fat-free yogurt, plain.
1 c fat-free or low-fat buttermilk.
½ c evaporated fat-free milk.
⅓ c dry fat-free milk.
(1 fat-free milk = 12 g carbohydrate, 8 g protein, 0–3 g fat, and 90 kcal.)

[a]Count as 1 very lean meat exchange *plus* 1 starch exchange.
[b]The higher the fat content of milk and milk products, the greater the amount of saturated fat and cholesterol.

Milks (low-fat)[b]

1 reduced-fat milk exchange is like:
1 c 2% milk.
¾ c low-fat yogurt, plain.
(1 reduced-fat milk = 12 g carbohydrate, 8 g protein, = g fat, and 120 kcal.)

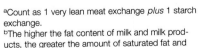

Milks (whole)

1 whole-milk exchange is like:
1 c whole milk.
½ c evaporated whole milk.
(1 whole milk = 12 g carbohydrate, 8 g protein, 8 g fat, and 150 kcal.)

Other carbohydrates

1 other carbohydrates exchange is like:
2 small cookies.
1 small brownie or cake.
5 vanilla wafers.
1 granola bar.
½ c ice cream.
(1 other carbohydrate = 15 g carbohydrate and may be exchanged for 1 starch, 1 fruit, or 1 milk. Because many items on this list contain added sugar and fat, their fat and kcalorie values vary, and their portion sizes are small.)

Continued

Figure 20-2

THE EXCHANGE SYSTEM: EXAMPLES OF FOODS, PORTION SIZES, AND ENERGY-NUTRITION CONTRIBUTIONS—continued

The Meat and Meat Substitutes Group (Protein)

Meat and substitutes (very lean)
1 very lean meat exchange is like:
1 oz chicken (white meat, no skin).
1 oz cod, flounder, or trout.
1 oz tuna (canned in water).
1 oz clams, crab, lobster, scallops, shrimp, or imitation seafood.
1 oz fat-free cheese.
½ c cooked legumes.ᶜ
¼ c fat-free or low-fat cottage cheese.
2 egg whites (or ¼ c egg substitute).
(1 very lean meat = 7 g protein, 0–1 g fat, and 35 kcal).

Meats and substitutes (lean)
1 lean meat exchange is like:
1 oz beef or pork tenderloin.
1 oz chicken (dark meat, no skin).
1 oz herring or salmon.
1 oz tuna (canned in oil, drained).
1 oz low-fat cheese or luncheon meats.
(1 lean meat = 7 g protein, 3 g fat, and 55 kcal.)

Meats and substitutes (medium-fat)
1 medium-fat meat exchange is like:
1 oz ground beef.
1 oz pork chop.
1 egg (high in cholesterol, limit 3 per week).
¼ c ricotta.
4 oz tofu.
(1 medium-fat meat = 7 g protein, 5 g fat, and 75 kcal.)

Meats and substitutes (high-fat)ᵈ
1 high-fat meat exchange is like:
1 oz pork sausage.
1 oz luncheon meat (such as bologna).
1 oz regular cheese (such as cheddar or swiss).
1 slice bacon
1 small hot dog (turkey or chicken).ᵉ
2 tbs peanut butter.ᶠ
(1 high-fat meat = 7 g protein, 8 g fat, and 100 kcal.)

The Fat Group

Fats
1 fat exchange is like:
1 tsp butter.ᵍ
1 tsp margarine or mayonnaise (1 tbs reduced fat).
1 tsp any oil.
1 tbs salad dressing (2 tbs reduced fat).
8 large black olives.
10 large peanuts.
⅛ medium avocado.
1 slice bacon.ᵍ
2 tbs shredded coconut.ᵍ
1 tbs cream cheese (2 tbs reduced fat).ᵍ
(1 fat = 5 g fat and 45 kcal.)

ᶜCount as 1 very lean meat exchange *plus* 1 starch exchange
ᵈThese foods are high in saturated fat, cholesterol, and kcalories.
ᵉA beef or pork hot dog counts as 1 high-fat meat exchange *plus* 1 fat exchange.
ᶠCount as 1 high-fat meat exchange *plus* 1 fat exchange.
ᵍButter, bacon, coconut, and cream cheese contain saturated fats.

PLAN A DIET FOR DIABETES USING EXCHANGE LISTS

The diet planner (usually a dietitian) carefully considers the client's lifestyle and medical goals in planning a diet for diabetes. Using exchange lists to plan diets takes time at first, but with practice, the planning process becomes routine. This box describes a simplified diet plan.

The first step is to assess each individual to determine what weight is reasonable and how many kcalories will be necessary to achieve or maintain that body weight. For this example, we will use a man who is 6 feet tall and is comfortable with the weight of 178 pounds that he has maintained throughout his adult life. From an assessment of food intake, the dietitian estimates that the man has maintained his weight on about 2900 kcalories per day with 25 percent of kcalories from protein, 45 percent from carbohydrate, and 30 percent from fat.

1. The first step is to determine the grams of protein, carbohydrate, and fat recommended for a diet for diabetes.
   - 10 to 20% of the kcalories from protein.
   - 45 to 60% of the kcalories from carbohydrate.
   - 30% or less of the kcalories from fat.

   For 2900 kcalories, this division of nutrients translates into grams as follows:
   - Protein:

$$10\% \times 2900 \text{ kcal} = 290 \text{ kcal.}$$
$$290 \text{ kcal} \div 4 \text{ kcal/g} = 73 \text{ g.}$$

$$20\% \times 2900 \text{ kcal} = 580 \text{ kcal.}$$
$$580 \text{ kcal} \div 4 \text{ kcal/g} = 145 \text{ g.}$$

Thus, the man needs between 73 and 145 grams of protein.
   - Carbohydrate:

$$45\% \times 2900 \text{ kcal} = 1305 \text{ kcal.}$$
$$1305 \text{ kcal} \div 4 \text{ kcal/g} = 326 \text{ g.}$$

$$60\% \times 2900 \text{ kcal} = 1740 \text{ kcal.}$$
$$1740 \text{ kcal} \div 4 \text{ kcal/g} = 435 \text{ g.}$$

Thus, the man needs between 326 and 435 grams of carbohydrate.
   - Fat:

$$30\% \times 2900 \text{ kcal} = 870 \text{ kcal.}$$
$$870 \text{ kcal} \div 9 \text{ kcal/g} = 97 \text{ g.}$$

Thus, the man needs about 97 grams of fat or less.

2. The dietitian recognizes that the client will need to make dietary changes to conform to a healthy eating plan. To minimize the changes the client must make, the dietitian decides to plan the diet to include 20 percent protein or 580 kcalories. Thus, 80 percent or 2320 kcalories remain for carbohydrate and fat. After reviewing information about the man's blood lipids, which are within acceptable limits, the dietitian plans the diet to keep fat at the current level of 30 percent (870 kcalories). This means that 50 percent of the kcalories (1450 kcalories) remain for carbohydrate.

$$2900 \text{ total kcal} - 580 \text{ protein kcal} - 870 \text{ fat kcal} =$$
$$1450 \text{ carbohydrate kcal.}$$

Continued

PLAN A DIET FOR DIABETES USING EXCHANGE LISTS

3. To translate the kcalories from fat and carbohydrate to grams:

$$870 \text{ fat kcal} \div 9 \text{ kcal/g} = 96 \text{ g.}^*$$
$$1450 \text{ carbohydrate kcal} \div 4 \text{ kcal/g} = 363 \text{ g.}^*$$

Thus, the diet will provide 2900 kcalories with a distribution of 20 percent protein (145 grams or 580 kcalories), 50 percent carbohydrate (363 grams or 1450 kcalories), and 30 percent fat (96 grams or 870 kcalories).

4. Now it is time to translate the diet prescription into a meal plan. Table 20-2 on p. 474 shows the grams of carbohydrate, protein, and fat and the energy value in each serving on an exchange list. Using this table and the client's food intake record as a guide, the dietitian first plans servings of foods that contain carbohydrate, then protein, and finally fat, trying to match foods as closely as possible to the client's usual food intake. This process takes practice and requires some adjusting based on trial and error. Most often, the final result does not fit the meal plan exactly, but comes close. Table 20-3 shows how the dietitian might plan a day's exchanges for the man in this example. Note that the plan falls within the guidelines of the Daily Food Guide on pp. 18–19. The plan uses fat-free milk and lean meat exchanges for calculations. Lower-fat foods are encouraged. If the client occasionally chooses to use another type of milk or meat, the number of fat servings must be adjusted accordingly. For example, if the client eats 4 ounces of a high-fat meat (32 grams of fat) instead of lean meat (12 grams of fat), he must then use four fewer fat exchanges during the day (20 grams of fat). The plan shown in Table 20-3 does not include the "other carbohydrates" list; starches and other foods (described later) can be substituted for foods on this list.

*To limit fat, round down the actual 96.6 g to 96. Round up the 362.5 g carbohydrate to 363.

Table 20-3
A DAY'S EXCHANGES FOR A SAMPLE 2900-kCALORIE DIET

Exchange Group/List	Number of Exchanges	Carbohydrate (g)	Protein (g)	Fat[a] (g)
Carbohydrate group[b]				
Starch	13	195	39	0
Fruit	7	105	—	—
Milk	3	36	24	0
Vegetable	6	30	12	—
Meat and meat substitutes group				
Lean	10	—	70	30
Fat group	13	—	—	65
Total grams		366	145	95
Total kcalories		1464	580	855
% kcalories		50.5	20	29.5

[a]To ease calculation, exchanges from the carbohydrate groups are assumed to have 0 grams fat. If the client uses a fat-containing exchange, the fat can be deducted from the daily fat allowance.
[b]Foods from the "other carbohydrates" list can be substituted for a starch, fruit, or milk list exchange. Any fat in the selected food is then deducted from the daily fat allowance.

PLAN A DIET FOR DIABETES USING EXCHANGE LISTS

5. Distribute foods into meals that fit the client's usual eating patterns. Table 20-4 shows how the day's exchanges might be divided for the man in this example. With this information in hand, the dietitian and client can begin to fill in the plan with real foods to create a sample menu such as the one shown in Table 20-5 below. The client is reminded to eat about the same amount of carbohydrate at about the same time each day.

6. Teach clients how to tailor the diet to meet their own preferences. For example, foods from the starch, fruit, milk, and other carbohydrate lists contain similar amounts of energy and carbohydrate and can be

Table 20-4
DIVIDING A DAY'S EXCHANGES BETWEEN MEALS AND SNACKS

Exchange Group/List	Number of Exchanges[a]	Breakfast	Lunch	Midafternoon Snack	Supper	Bedtime Snack
Carbohydrate group						
Starch	13	3	3	2	3	2
Fruit	7	2	1	1	1	2
Milk	3	1	1		1	
Vegetable	6		3		3	
Meat and meat substitutes group						
Lean	10		3	1	4	2
Fat group	13	3	3	2	3	2

[a]From Table 20-3.

Table 20-5
TRANSLATING A DAY'S EXCHANGES INTO A DAY'S MEALS

Breakfast

½ c bran cereal (1 starch)
1 c fat-free milk (1 milk)
1 bagel (2 oz) (2 starch)
2 tsp margarine (2 fat)
1 banana (2 fruit)
Coffee

Lunch

1 cup pasta (2 starch) served with:
  ½ c spaghetti sauce (1 starch, 1 fat)
  2 oz meatballs (medium fat) (2 meat, 2 fat)
  2 tbs grated parmesan cheese
  (1 meat, ½ fat)
1½ c broccoli (3 vegetables)
17 small grapes (1 fruit)
1 c fat-free milk (1 milk)

Midafternoon Snack

1 fat-free granola bar (2 starch)
2 tbs peanut butter (1 meat, 1½ fat) spread
over slices of 1 small apple (1 fruit)

Supper

4 oz grilled salmon (4 meat)
⅔ c rice (2 starch)
1 whole-wheat roll (1 starch)
1 c cooked carrots (2 vegetables)
1 tsp margarine (1 fat)
1 c mixed green salad (1 vegetable)
Salad dressing made with 2 tsp olive oil
  (2 fat) and wine vinegar
1 c diced cantaloupe (1 fruit)
1 c fat-free milk (1 milk)

Bedtime Snack

½ c lowfat cottage cheese (2 meat)
1 c fruit cocktail (2 fruit)
1 small frosted cupcake (2 starch, 1 fat)
1 c fat-free milk (1 milk)

NOTE: Compared to Table 20-4, this menu provides one fewer fat exchange at breakfast and one-half fat exchange more at lunch.

Continued

PLAN A DIET FOR DIABETES USING EXCHANGE LISTS

substituted for one another from time to time. Regular substitution is discouraged, however, because each list makes unique contributions to other nutrient needs. A client who regularly substitutes fruit for milk, for example, may not be getting enough calcium. The client who regularly substitutes milk for a starch or fruit may not be getting enough fiber.[a]

7. Three servings of free foods can be included as long as they are spread throughout the day. Free foods contain up to 20 kcalories and 5 grams of carbohydrate per serving. A serving of food that contains a carbohydrate-based fat substitute that provides 5 grams or less carbohydrate counts as a free food.

[a]M. Wheeler, M. J. Franz, and P. Barrier, Helpful hints: Using the 1995 exchange lists for meal planning, *Diabetes Spectrum* 8 (1995): 325–326.

The exchange system provides a valuable tool for helping clients recognize sources and types of fat. By including items like bacon and avocados on the fat list, the exchange system alerts users to foods that are unexpectedly high in fat. The fat list also shows which fats are monounsaturated, polyunsaturated, and saturated to help clients incorporate specific strategies for fat-restricted diets. The exchange system encourages users to think of fat-free milk as milk and of whole milk as milk with added fat; and to think of very lean meats as meats and of lean, medium-fat, and high-fat meats as meats with added fat. To that end, foods on the milk and meat lists are separated into categories based on their fat contents. Meat exchanges that are particularly high in cholesterol are noted. The starch list also specifies grain products that contain added fat.

CARBOHYDRATE COUNTING Another widely used diet strategy is called carbohydrate counting.[11] It teaches clients to focus mainly on the carbohydrate contents of foods. Clients learn to eat consistent amounts of carbohydrates at meals and snacks. They must have the motivation to monitor their blood glucose and keep records of their glucose levels (see Nutrition in Practice 20), the time and amount of carbohydrate they eat at each meal, and the time and amount of insulin or oral antidiabetic agents they use. They must also be able to perform the mathematical operations necessary to calculate their carbohydrate intakes. The "How to" box on p. 481 provides more information about carbohydrate counting. Regardless of the diet strategy, all clients receive instructions on planning well-balanced and healthy meals, eating consistent amounts of foods at regular times, and maintaining a desirable weight.

Physical Activity

The client with diabetes should be carefully evaluated to determine a safe type and amount of physical activity. For people with type 2 diabetes, a regular program of physical activity improves blood glucose control, contributes to weight loss, improves blood lipid levels, and lowers blood pressure. Although physical activity has not been shown to aid blood glucose control in people with type 1 diabetes, its value lies in its benefits to the cardiovascular system. People who use insulin, however, need to take special precautions when exercising.

PHYSICAL ACTIVITY AND BLOOD GLUCOSE LEVELS People who need insulin should check their blood glucose before and after engaging in physical

Physical activity plays an important role in the management of diabetes.

HOW TO:

HELP CLIENTS COUNT CARBO-HYDRATES

1. Begin by calculating how much carbohydrate the person usually eats at each meal and snack. For an example, assume that the client is a woman who maintains her weight and controls her blood glucose levels on an intake of 1700 kcalories. Use exchanges to determine her usual carbohydrate intake. In this example, the client and dietitian determined that her usual eating pattern includes:

- Breakfast: 2 starch, 1 fruit, 1 milk.
- Lunch: 2 starch, 1 fruit, 2 vegetables.
- Midafternoon snack: 1 fruit, 1 milk.
- Supper: 2 starch, 1 fruit, 2 vegetables.
- Bedtime snack: 1 starch, 1 fruit.

2. Next, use Table 20-2 to determine the grams of carbohydrate in each meal. Breakfast, for example, contains:

$$2 \text{ starch} = 2 \times 15 \text{ g carbohydrate} = 30 \text{ g}$$
$$1 \text{ fruit} = 1 \times 15 \text{ g carbohydrate} = 15 \text{ g}$$
$$1 \text{ milk} = 1 \times 12 \text{ g carbohydrate} = 12 \text{ g}$$
$$\text{Total} = 57 \text{ g carbohydrate.}$$

Using the same method to calculate the carbohydrate from other meals and snacks, the carbohydrate contents are as follows:

- Lunch: 55 g.
- Midafternoon snack: 27 g.
- Supper: 55 g.
- Bedtime snack: 30 g.

3. Teach the client to include these same amounts of carbohydrate consistently at each meal and snack. Clients can use exchanges, food labels, and food composition tables to determine the carbohydrate contents of the foods they eat.

4. Encourage clients to carefully weigh and measure all foods initially. Once clients are familiar with portion sizes, they can weigh and measure foods less often, but should still check portion sizes occasionally.

5. Remind clients that they still need to be aware of the total amount of food they eat as well as the type and amount of fat they are eating.

Clients who use insulin learn how much insulin they need to cover their carbohydrate intakes. Then, if they change their carbohydrate intakes on occasion, they can adjust their insulin dose accordingly.

---

activity. Physical activity should not be undertaken if blood glucose is too low (less than 100 milligrams per 100 milliliters) or too high (greater than 300 milligrams per 100 milliliters). If blood glucose is too low, hypoglycemia can quickly develop. If blood glucose is too high, exercising can cause blood glucose levels to rise even higher.

:: PHYSICAL ACTIVITY AND FOOD INTAKE All persons, especially those with diabetes, need to make sure they are adequately hydrated before and during physical activity by drinking liquids throughout the day. The person who needs insulin may also need to eat before, during, and after vigorous physical activity. Especially important is carbohydrate, which is readily available from fruits, fruit juices, yogurt, crackers, and other starches. The amount of carbohydrate needed depends on the type of activity, its duration, the client's individual responses, and the results of blood glucose tests.

Drug Therapy for Diabetes

All people with type 1 diabetes need insulin to control blood glucose and utilize energy. People with type 2 diabetes can sometimes control their blood glucose without medications by using a combination of diet and physical activity. When these measures fail, one or a combination of oral antidiabetic agents may be prescribed. Oral antidiabetic agents do not replace diet and physical activity in diabetes management; advise clients to continue all therapies as instructed. Oral antidiabetic agents have a maximum dose; if blood glucose cannot be controlled at the maximum dose, the physician may prescribe insulin or insulin analogs, alone or in combination with oral antidiabetic agents. The Diet-Medication Interactions box below includes the agents described here. Medications used to treat cardiovascular and renal complications that may accompany diabetes are described in Chapters 21 and 22, respectively.

ORAL ANTIDIABETIC AGENTS The number of oral agents available to treat people with type 2 diabetes has grown markedly in recent years. Some oral agents work by stimulating the release of insulin from the beta cells, some work by reducing insulin resistance and depressing the liver's production of glucose without raising insulin levels, and still others reduce the rate of complex carbohydrate and sucrose digestion in the intestine and slow the rate of glucose absorption. People using oral agents that slow the rate of carbohydrate absorption must use glucose to treat episodes of hypoglycemia.

> *Sulfonylureas and replaglinide are also called* hypoglycemic agents *because they stimulate insulin release and can cause hypoglycemia. Sulfonylureas are the type of antidiabetic agent that reacts with alcohol and can cause a disulfiram-like reaction (flushing of the skin and a rapid heartbeat). The thioglitazones (rosiglitazone and piolitazone) work primarily by lessening peripheral insulin resistance. The alpha-glucosidase inhibitors, acarbose and meglitol, reduce the rate of digestion of complex carbohydrate and some sugars.*

DIET-MEDICATION INTERACTIONS

Alpha-glucosidase inhibitors (acarbose and meglitol)

Alpha-glucosidase inhibitors are taken at the start of each meal. Clients using these agents must use glucose to treat hypoglycemia. Nutrition-related side effects include abdominal pain, gas, and diarrhea.

Insulin

The timing of *insulin* administration varies, depending on the action of the insulin prescribed. Insulin may cause hypoglycemia. Ask clients who take insulin about their use of the following herbs, which may affect blood glucose levels: basil, bay, bee pollen, burdock, garlic dust, ginseng, and sage.

Lispro (insulin analog)

Lispro is taken 5 to 10 minutes before meals. Lispro lowers the risk of hypoglycemia compared to insulin.

Metformin

Metformin is taken once or twice a day before meals (breakfast or breakfast and dinner). GI side effects are uncommon, but metformin may cause diarrhea or leave a metallic taste in the mouth.

Repaglinide

Repaglinide is taken before meals. Repaglinide can cause hypoglycemia and diarrhea.

Sulfonyureas (chlorpropamide, glipzide, glyburide, glimepiride)

Sulfonyureas are generally taken one or two times a day, before meals. Disulfiram-like reactions can occur when large amounts of alcohol are taken (especially with *chlorpropamide*). These medications can also cause hypoglycemia.

Thioglitazones (rosiglitazone, piolitazone)

Thioglitazones are taken once or twice a day before meals. Nutrition-related side effects are uncommon. One medication of this type, *troglitazone*, was recently taken off the market due to concerns that its use might be a cause of liver failure.

⠿ INSULIN AND INSULIN ANALOGS For people who need insulin, commercial insulin is available in different forms that act with different timings so that it can be delivered in a manner that mimics the body's normal insulin actions as closely as possible. As Figure 20-3 shows, insulins can be either rapid acting (regular), intermediate acting (NPH and lente), or long acting (ultralente). Insulin analog (lispro) is a rapid-acting insulin whose amino acid composition has been modified so that it works faster and has a shorter duration of action.[12] As a result, lispro reduces after-meal hyperglycemia to a greater extent than regular insulin and is also associated with a lower risk of hypoglycemia between meals and during the night.

⠿ INSULIN DELIVERY People who need insulin must inject it or use pumps to deliver the insulin they need. Insulin is a protein, so if it is taken orally, digestive enzymes reduce it to amino acids. Because the pancreas of a person who needs insulin cannot synthesize insulin or cannot synthesize enough of it from amino acids, it would not be available to the body. The person who chooses injections often receives **multiple daily injections**—a mixture of two or more types of insulin three or more times daily. Single injections are seldom effective.

External pumps, about the size of a beeper, hold enough insulin to meet needs for two or three days. From the pump, insulin enters the body through tubing and a needle inserted into the abdominal area. Personal preferences, motivational level, and financial considerations guide clients in deciding which delivery system works best for them.

Some people experience a temporary remission from diabetes after their initial treatment with insulin—a time referred to as the "honeymoon phase." Perhaps with insulin treatment and relief from hyperglycemia, the insulin-producing cells of the pancreas become able to function again—but only temporarily.

⠿ INSULIN AND FOOD INTAKE Normally, the body secretes a constant, baseline amount of insulin at all times and more after blood glucose rises following a meal. The person with type 1 diabetes often receives NPH (intermediate-acting) insulin to meet baseline needs and regular (rapid-acting) insulin or insulin analogs to process energy nutrients after meals. People with type 2 diabetes may be treated with insulin alone, or they may use it in combination with oral antidiabetic agents. Often a single injection of NPH (intermediate-acting) insulin is given at bedtime. Insulin analogs may also be used in the treatment of type 2 diabetes.

Injections are one option for delivering insulin to people with diabetes.

Although the availability of insulin has been lifesaving, insulin therapy cannot achieve the same degree of blood glucose control as a body that produces its own insulin.

multiple daily injections: delivery of different types of insulin by injection three or more times daily.

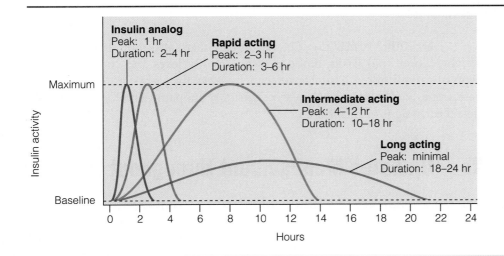

Insulin analog
Peak: 1 hr
Duration: 2–4 hr

Rapid acting
Peak: 2–3 hr
Duration: 3–6 hr

Intermediate acting
Peak: 4–12 hr
Duration: 10–18 hr

Long acting
Peak: minimal
Duration: 18–24 hr

Insulin activity — Maximum — Baseline

Hours — 0 2 4 6 8 10 12 14 16 18 20 22 24

Figure 20-3
ACTIONS OF INSULIN TYPES

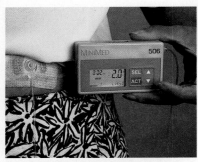

For people who use insulin pumps, insulin passes through tubing that enters the body through the abdomen as shown.

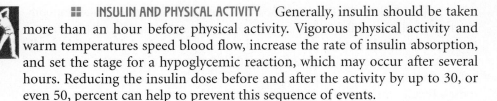

INSULIN AND PHYSICAL ACTIVITY   Generally, insulin should be taken more than an hour before physical activity. Vigorous physical activity and warm temperatures speed blood flow, increase the rate of insulin absorption, and set the stage for a hypoglycemic reaction, which may occur after several hours. Reducing the insulin dose before and after the activity by up to 30, or even 50, percent can help to prevent this sequence of events.

PANCREAS TRANSPLANTS   For people with type 1 diabetes who encounter serious problems managing their diseases with insulin, pancreas transplants have been successful in providing functional, insulin-producing beta cells. Most often, a pancreas transplant is combined with a kidney transplant because the person often has significant kidney problems as well. A combination pancreas-kidney transplant can eliminate the need for insulin and for dialysis (see Chapter 22) and greatly enhance the person's quality of life.

Monitoring Diabetes Management

Clients and the health care team depend on several tests to see how well they are managing the diabetes. Clients learn to monitor their blood glucose levels at home, most often using blood glucose meters. At medical appointments, the health care team reviews and records the client's test results and looks for patterns that suggest the need for adjustments in the treatment plan. Even clients who carefully follow their diets, administer their medications correctly, and follow a regular program of physical activity sometimes develop hyperglycemia and hypoglycemia. Table 20-6 summarizes strategies for adjusting treatment plans to correct problems with hyperglycemia and hypoglycemia. Nutrition in Practice 20 provides information about how the health care team works with each client to develop an individualized treatment plan.

Glycated hemoglobin (GHb) is also called glycosylated hemoglobin. Acceptable GHb values vary because different laboratories use different methods for measuring GHb. In the Diabetes Control and Complications Trial, groups receiving intensive treatment achieved an average GHb of 7.2% (normal <6%).

GLYCATED HEMOGLOBIN   In addition to checking blood glucose records, physicians monitor blood glucose control by evaluating the percentage of glycated hemoglobin (GHb). As blood glucose rises, glucose attaches to amino acids on hemoglobin molecules and remains there until the red blood cells that carry the hemoglobin die (about 120 days). Glycated hemoglobin reflects diabetes control over the past two to three months, rather than just prior to the test.

URINARY KETONES   Health care professionals often recommend that clients with consistently high blood glucose also monitor ketones in the urine, especially during illness. As described earlier, high blood glucose may predispose the person to ketosis and coma. Such a course is especially likely during illness.

OTHER MEASURES   The health care team also monitors the client's weight, blood lipid levels, blood pressure, and reflexes and checks for early signs of complications. Urine tests can help detect the early signs of kidney disease, eye exams help detect the early signs of retinopathy, and foot exams help detect early signs of infection.

Blood glucose monitoring helps people with diabetes maintain blood glucose in a safe range.

Diabetes Management throughout Life

The overall approach to diabetes management remains the same throughout life, but special problems may become apparent at different ages. Health care professionals who are aware of these problems are best prepared to provide practical and useful advice.

Table 20-6

STRATEGIES FOR MANAGING HYPERGLYCEMIA AND HYPOGLYCEMIA

Hyperglycemia	Possible Solutions[a]
Before breakfast	• Adjust dose of intermediate-acting insulin at bedtime.[b]
Before lunch	• Adjust morning dose of rapid-acting insulin.[b]
	• Reduce amount of carbohydrate at breakfast.
	• Reduce or omit midmorning snack.
	• Change time of breakfast or midmorning snack.
	• Add physical activity after breakfast.
Before dinner	• Adjust afternoon dose of rapid-acting insulin.[b]
	• Reduce carbohydrate at lunch.
	• Reduce or omit midafternoon snack.
	• Change time of lunch or midafternoon snack.
	• Add physical activity between lunch and dinner.
At bedtime	• Adjust insulin dose before dinner.[b]
	• Reduce amount of carbohydrate at dinner.
	• Reduce or omit evening snack.
	• Add physical activity after dinner.

Hypoglycemia	Possible Solutions[a]
Before breakfast	• Adjust dose of intermediate- or long-acting insulin at bedtime.[b]
	• Add carbohydrate at evening snack.
	• Avoid strenuous activity late in the day.
Before lunch	• Adjust morning dose of rapid-acting insulin.[b]
	• Add carbohydrate at breakfast.
	• Add a morning snack.
	• Change time of breakfast, lunch, or morning snack.
	• Adjust physical activity schedule.
Before dinner	• Adjust afternoon dose of rapid-acting insulin.[b]
	• Add carbohydrate at lunch.
	• Add an afternoon snack.
	• Change time of lunch, dinner, or afternoon snack.
	• Adjust physical activity schedule.
At bedtime	• Adjust insulin dose before dinner.[b]
	• Add carbohydrate at dinner.
	• Add an evening snack.
	• Change time of dinner or evening snack.

[a]Skilled health care professionals gather additional data to find the best solution for problems with blood glucose control. Is the problem an isolated occurrence or a pattern? Has food intake changed? If yes, why? Has the activity level changed? Has illness been a problem? Whenever altering the diet or physical activity plan is difficult for the client, insulin is adjusted to correct problems, if possible.
[b]Insulin doses can be adjusted in amount or timing or both.

▪▪ Diabetes Management in Childhood

Like those of all children, the energy and nutrient needs of children with diabetes keep changing throughout the growing years. Children's appetites and activities vary widely from day to day. A child may eat like a horse one day and like a mouse the next. A teen may spend hours walking around the mall one day and spend the next day watching TV. Growth and activity influence the needs for food and insulin, and management must adjust to meet those needs.

     Teenagers often have intense difficulty accepting a diagnosis of diabetes. At a time when they are striving to develop their identity with a group and to be as similar to their peers as possible, they are faced with an

Because a child's activities vary from day to day, food and insulin needs may also change.

unwelcome diagnosis and new rules they are expected to follow; their response may be denial and refusal to cooperate. Adolescents especially need to know that they can manage their disease themselves—that it won't turn them into dependent children.

■■ **MEAL PLANS**   To support growth and development, children with diabetes need flexible, balanced meals and snacks that offer wide varieties of foods from each of the food groups. Dietitians often teach children and caregivers carbohydrate counting to provide flexibility from day to day. Concentrated sweets are allowed within the context of a healthy diet. Caregivers need not force children to finish meals, but should encourage them not to skip meals either, because hypoglycemia can result, particularly if the child uses insulin. Meals are best taken at about the same times each day, and children with diabetes can eat the same foods as the rest of the family.

■■ **FAMILY LIFESTYLES**   Successful diet management incorporates prescribed meals into existing family lifestyles and eating patterns. For children who take insulin, a typical eating plan may include three meals with two to three snacks a day. Snacks before bedtime help prevent hypoglycemia while the child is sleeping, especially if the child engages in strenuous activity late in the day. Caregivers should vary meals and snacks to prevent boredom, provide enough to share with friends, and avoid identifying foods as "good" or "bad." Such connotations create unrealistic expectations or fears and invite the development of manipulative eating behaviors. The case study below presents a child with type 1 diabetes.

■■   **Diabetes Management in Pregnancy**

All women who become pregnant face new challenges that include altering their diets to meet the demands of the growing fetus and addressing compli-

Case Study

CHILD WITH TYPE 1 DIABETES

One year ago, Yusuf, a 12-year-old boy, was diagnosed with type 1 diabetes. The initial diagnosis was made after Yusuf's parents became concerned when he began to lose weight, urinate excessively, and complain of thirst. Aware of a family history of diabetes, the parents quickly sought medical help. Since that time, Yusuf's diabetes has been well controlled. Recently, however, Yusuf was admitted to the emergency room, complaining of nausea, vomiting, and intense thirst. He had a fever, and his blood glucose records from the previous day showed that his blood glucose was high throughout the day. The physician observed that Yusuf was confused and breathing with difficulty and also noted the smell of acetone on his breath. Urine tests were positive for glycosuria and ketonuria, and Yusuf's blood glucose was 400 milligrams per 100 milliliters. The diagnosis was hyperglycemia, ketosis, and acidosis.

✚ Describe the metabolic events that led to the symptoms associated with diabetes (before diagnosis), as well as those associated with ketosis and acidosis. Were Yusuf's physical symptoms and laboratory tests consistent with this diagnosis? How can you distinguish between diabetic ketoacidosis and hypoglycemia?

✚ When Yusuf recovers, what advice can you offer him to prevent future incidents of ketosis and acidosis? Assume that Yusuf had instructions for a diet for diabetes. What dietary modifications might help him if he becomes ill again?

✚ Think about and discuss the influence of Yusuf's age on his outlook and ability to cope with diabetes. What problems does his age pose? Consider some ways you might help him deal with these problems. Regarding his future, describe the possible role of diet in preventing the chronic complications of diabetes.

cations that occur as a consequence of pregnancy. Pregnancy in all women elevates blood insulin and alters insulin resistance. Blood insulin begins to rise soon after conception, and the cells respond by storing energy nutrients to provide for the developing fetus. Later in pregnancy, insulin remains high, but the cells become insulin resistant. Levels of hormones that act antagonistically to insulin rise. This hormonal shift signals the body to stop storing energy fuels and to begin allowing the fetus to rapidly take up energy nutrients. Because normal pregnancy stresses the glucose regulatory system in these ways, women with diabetes should expect control to become more difficult during pregnancy.

:: HEALTH RISKS ASSOCIATED WITH DIABETES DURING PREGNANCY Women with either type 1 or type 2 diabetes who are contemplating pregnancy should know that uncontrolled diabetes in early pregnancy raises the risk of spontaneous abortions.[13] Infants exposed to high blood glucose and ketones risk birth defects. High blood glucose levels also "overfeed" the growing fetus, resulting in a large infant that is difficult to deliver. The extra glucose also means that the fetus must make extra insulin to handle the load, which may lead to severe hypoglycemia in the infant after birth. Women with diabetes who are contemplating pregnancy need to receive preconception care, which aims to achieve the best possible blood glucose control before pregnancy, and continued care to maintain control during pregnancy. Women in the Diabetes Control and Complications Trial who tightly controlled their blood glucose levels and became pregnant experienced rates of spontaneous abortion and birth defects similar to those experienced in pregnant women without diabetes.[14]

:: GESTATIONAL DIABETES Women who have never had diabetes or never knew that they had it may be diagnosed with diabetes during pregnancy (gestational diabetes). Gestational diabetes is the most common medical complication of pregnancy.[15] The American Diabetes Association recommends that women be screened for diabetes between 24 and 28 weeks of gestation, with the exception of women who meet all of these criteria:[16]

- Less than 25 years of age.
- Normal body weight.
- No first-degree relatives with diabetes.
- Not of a high-risk ethnic origin including Hispanic, Native American, Pacific Islander, and African American.

Women with gestational diabetes have a high risk of requiring a cesarean delivery or experiencing pregnancy-induced hypertension. Infants may be large and may experience complications after birth including severe hypoglycemia.

:: BLOOD GLUCOSE MONITORING Obstetricians recommend blood glucose monitoring for all pregnant women with any type of diabetes. Establishing blood glucose control is important to the health of both mother and infant.

:: MEDICAL NUTRITION THERAPY For pregnant women with diabetes, individualized diets tailored to meet the added nutrient demands of pregnancy and carefully coordinated with insulin therapy (when necessary) are central to therapy. The diet plan aims to provide adequate but not excessive kcalories to support appropriate weight gain (see the weight-gain recommendations in the margin on p. 237). Carbohydrate may be moderately restricted (40 to 45 percent of total kcalories) to help blood glucose levels from rising too high after

The hormones that oppose the action of insulin during late pregnancy are placental lactogen, cortisol, prolactin, *and* progesterone.

The term that describes infants with large bodies at birth is macrosomia.

Reminder: High blood pressure that develops during pregnancy is known as pregnancy-induced hypertension *and may signal the onset of other complications (see Chapter 10).*

Oral antidiabetic agents are not currently recommended to treat diabetes during pregnancy.

meals. Limiting carbohydrate at breakfast helps maintain morning blood glucose levels in an acceptable range until counterregulatory hormone levels diminish. Frequent small meals help assure an ongoing supply of glucose without inducing hyperglycemia. A bedtime snack is recommended to prevent nocturnal hypoglycemia and ketosis in the mother and to provide fuel to the developing fetus.

■■ PREVENTIVE MEASURES AFTER PREGNANCY For most women with gestational diabetes, glucose tolerance returns to normal after pregnancy, but some women with gestational diabetes and their offspring develop diabetes (usually type 2 diabetes) later in life, especially if they are overweight.[17] For this reason, health care professionals closely monitor women who have experienced gestational diabetes and their offspring and recommend that they seek medical attention if they develop symptoms suggestive of diabetes. Education centers on strategies to achieve and maintain a healthy weight and to develop a regular program of physical activity.

■■ Diabetes Management in Later Life

The elderly also face special problems in dealing with diabetes. They have greater risks of hyperglycemia and hypoglycemia because of reduced appetite, altered thirst regulation, altered organ function, depression or mental deterioration, multiple medication use, and other medical conditions that complicate diabetes control.

■■ BLOOD GLUCOSE CONTROL Many elderly people have type 2 diabetes, and with aging, their insulin resistance may progress until they can no longer maintain glucose levels within acceptable ranges with diet and oral drugs. The prospect of daily insulin injections and the need for additional blood glucose monitoring may be overwhelming to an elderly client. Those who have also suffered a loss of vision as a consequence of aging or diabetes may have difficulty drawing correct insulin doses, giving self-injections, or monitoring blood glucose. The inability to perform these necessary tasks may make it impossible for the person to live independently.

■■ FINANCIAL AND SOCIAL CONSIDERATIONS The elderly may also lack the financial resources or social support necessary to help them cope with their diabetes. Without the financial resources to purchase medications and medical supplies and to visit the physician regularly, health and nutrition status may deteriorate. Lack of social interaction and consequent depression can further impair health and nutrition status. Caring health care professionals address these problems and help elderly clients find solutions.

Diabetes is a common disorder associated with many acute and chronic complications. To minimize the risk of complications, people with diabetes learn to carefully control their blood glucose levels. Successful management requires support from the diabetes care team—physicians, nurses, dietitians, and counselors. All members of the team coordinate instructions with each other so that their clients receive consistent advice. People with diabetes and those who work with them should realize that the learning process takes time. After their initial introduction to the world of diabetes, clients benefit from frequent follow-up visits to address problems, expand their knowledge, and promote independence. The Nutrition Assessment Checklist highlights areas of concern for people with diabetes.

NUTRITION ASSESSMENT CHECKLIST — FOR PEOPLE WITH DIABETES

Health Problems, Signs, and Symptoms

Check the medical record to determine:

- ☐ Type of diabetes
- ☐ Duration of diabetes
- ☐ Acute and chronic complications
- ☐ Other medical conditions, including pregnancy, that alter nutrient needs

Medications

For clients with preexisting diabetes who use antidiabetic agents, insulin, or both, note:

- ☐ Type(s) of antidiabetic agent or insulin
- ☐ Administration schedule

Check for other medications and note diet-medication interactions, including:

- ☐ Antilipemics (to lower blood lipids, see Chapter 21)
- ☐ Antihypertensive agents (to reduce blood pressure, see Chapter 21)
- ☐ Diuretics (to reduce blood pressure, see Chapter 21)

Nutrient/Food Intake

To devise an acceptable meal plan and coordinate medications, obtain:

- ☐ An accurate and thorough record of food intake and usual eating habits
- ☐ An account of usual physical activities

At medical checkups, reassess the client's ability to:

- ☐ Maintain an appropriate energy intake
- ☐ Maintain a consistent intake of carbohydrate
- ☐ Adjust the diet for missed meals due to illness
- ☐ Use appropriate amounts and types of foods to treat hypoglycemia

Height and Weight

Take accurate baseline height and weight measurement as a basis for:

- ☐ An appropriate energy intake
- ☐ Initial insulin therapy

Reassess height and weight for children and weight for adults periodically to ensure that the meal plan provides an appropriate energy intake.

Laboratory Tests

Check the following tests to monitor the success of diabetes therapy:

- ☐ Results of home blood glucose monitoring
- ☐ Glycated hemoglobin
- ☐ Blood lipids
- ☐ Microalbuminemia, when available

Physical Signs

Look for signs of:

- ☐ Dehydration, especially in the elderly
- ☐ Nutrient deficiencies and excesses

SELF CHECK

1. Which of the following is characteristic of type 1 diabetes?
   a. The pancreas makes no insulin.
   b. It frequently goes undiagnosed.
   c. It is the predominant form of diabetes.
   d. Insulin secretion is ineffective in preventing hyperglycemia.

2. Which of the following describes type 2 diabetes?
   a. Immune factors play a role.
   b. The pancreas makes no insulin.
   c. Ketosis and acidosis are common complications.
   d. Chronic complications may have begun to develop before it is diagnosed.

3. The chronic complications associated with diabetes result from:
   a. alterations in kidney function.
   b. weight gain and hypertension.
   c. damage to blood vessels and nerves.
   d. infections that deplete nutrient reserves.

4. The nurse working with a client with diabetes emphasizes that the diet should provide:
   a. a very low intake of fat.
   b. more protein than regular diets.
   c. a restricted intake of simple sugars and concentrated sweets.
   d. a consistent carbohydrate intake from day to day and at each meal and snack.

5. Which of the following is true regarding the use of alcohol in a diet for diabetes?
   a. A serving of alcohol should always be exchanged for two fat exchanges.
   b. Alcohol can cause hypoglycemia in all people, including those with diabetes.
   c. Even people with well-controlled blood glucose levels should refrain from all alcohol use.
   d. Some types of insulin can cause flushing of the skin and a rapid heartbeat in people with type 2 diabetes.

6. The meal-planning strategy that is most effective for the person with diabetes is:
   a. carbohydrate counting.
   b. the exchange list system.
   c. food guides and sample menus.
   d. the one that best helps the client control blood glucose levels.

7. Which of the following best describes insulin therapy in a

person receiving both NPH insulin and an insulin analog?

a. Since NPH insulin is of long duration, there is no need for the insulin analog.

b. NPH insulin covers basal insulin needs while the insulin analog covers the carbohydrate from meals.

c. NPH insulin covers the carbohydrate from meals while the insulin analog covers basal insulin needs.

d. Since glucose is not available from food between meals, insulin analog alone would cover the person's insulin needs.

8. Which of the following describes the treatment plan for a person with type 1 diabetes who practices intensive therapy?

a. The person cannot use a pump to deliver insulin.

b. The person uses one type of insulin one or two times daily.

c. The person must monitor blood glucose several times a day.

d. After learning to control blood glucose levels, the person can quit monitoring blood glucose levels.

9. Sudden hyperglycemia in a person who has consistently maintained good blood glucose control can be precipitated by:

a. infections or illnesses.

b. chronic alcohol ingestion.

c. undertreatment of hypoglycemia.

d. conditions that lower levels of counterregulatory hormones.

10. Women with pregnancies complicated by diabetes:

a. often need less carbohydrate at breakfast.

b. generally benefit from larger meals and a snack at bedtime.

c. need more carbohydrate than women with diabetes who are not pregnant.

d. need more kcalories than healthy women to support the pregnancy.

Answers to these questions appear in Appendix H.

CLINICAL APPLICATIONS

1. Using the "How to" box on pp. 477–480, plan a diet using the exchange lists for a sedentary woman with type 1 diabetes who is 5 feet 9 inches tall and weighs 160 pounds. Assume that the distribution of kcalories will be 55 percent from carbohydrate, 20 percent from protein, and 25 percent from fat. Develop a sample menu.

2. An important part of learning is being able to apply knowledge and guidelines to real-life situations. Using Table 20-6 on p. 485 as a guide, think about the possible remedies for either hyperglycemia or hypoglycemia. Describe at least one situation when it might be preferable to alter the insulin dose and one situation when it might be preferable to alter the carbohydrate intake.

3. Take a trip to a pharmacy and price these items: blood glucose meter, test strips appropriate for the glucose meter you select, lancets, insulin, and syringes. Determine the approximate cost of insulin for a person who uses 14 units of regular insulin and 26 units of NPH insulin in three injections daily. Then estimate the cost of testing blood glucose four times daily. Consider how an external pump might affect the total cost of managing diabetes. How does the need for a well-balanced diet influence the cost of diabetes care?

Notes

[1]American Diabetes Association web site, www.diabetes.org, visited on November 15, 1999.

[2]The Expert Committee on the Diagnosis and Classification of Diabetes Mellitus, Report of the Expert Committee on the diagnosis and classification of diabetes mellitus, *Diabetes Care* (supplement 1) 21 (1998): 5–19.

[3]E. A. Simone, D. R. Wegmann, and G. S. Eisenbarth, Immunologic "vaccination" for the prevention of autoimmune diabetes (type 1a), *Diabetes Care* (supplement 2) 22 (1999): B7–B15.

[4]A. I. Adler, The Diabetes Prevention Program, *Diabetes Care* 22 (1999): 543–544; W. Y. Fujimoto, A national multicenter study to learn whether type II diabetes can be prevented: The Diabetes Prevention Program, *Clinical Diabetes* 15 (1997): 13–15.

[5]B. Janand-Delenne and coauthors, Silent myocardial ischemia in patients with diabetes, *Diabetes Care* 22 (1999): 1396–1400.

[6]V. Valentine, J. A. Barone, and J. V. C. Hill, Gastropathy in patients with diabetes: Current concepts and treatment recommendations, *Diabetes Spectrum* 11 (1998): 248–253.

[7]American Diabetes Association, *Nutrition Recommendations and Principles for People with Diabetes Mellitus* (American Diabetes Association, 1998).

[8]A. Garg, High-monounsaturated fat diets for patients with diabetes mellitus: A meta-analysis, *American Journal of Clinical Nutrition* (supplement) 67 (1998): S577–S582; J. P. Parnett and A. Garg, Medical nutrition therapy for patients with diabetes mellitus: Role of dietary fats, *On the Cutting Edge: Diabetes Care and Education,* Fall 1997, pp. 5–6; A. M. Coulston, Monounsaturated fats for people with diabetes, *On the Cutting Edge: Diabetes Care and Education,* Fall 1997, pp. 14–16; A. Garg and coauthors, Effects of varying carbohydrate content of diet in patients with non-insulin-dependent diabetes mellitus, *Journal of the American Medical Association* 271 (1994): 1421–1428.

[9]American Diabetes Association, Role of fat replacers in diabetes medical nutrition therapy, *Diabetes Care* (supplement 1) 21 (1998): S64–S65.

[10]American Diabetes Association, *Nutrition Recommendations and Principles for People with Diabetes Mellitus,* 1998.

[11]S. J. Gillespie, K. Kulkarni, and A. E. Daly, Using carbohydrate counting in diabetes clinical practice, *Journal of the American Dietetic Association* 98 (1998): 897–905.

[12]American Diabetes Association, Lispro: A new fast-acting insulin option, *Diabetes Spectrum* 9 (1996): 253.

[13]American Diabetes Association, Position statement: Preconception care of women with diabetes, *Diabetes Care* (supplement 1) 21 (1998): 56–59.

[14]The Diabetes Control and Complications Trial Research Group: Pregnancy outcomes in the Diabetes Control and Complications Trial, *American Journal of Obstetrics and Gynecology* 174 (1996): 1343–1353.

[15]D. B. Carr and S. Gabbe, Gestational diabetes: Detection, management, and implications, *Clinical Diabetes* 16 (1998): 4–11.

[16]The Expert Committee on the Diagnosis and Classification of Diabetes Mellitus, 1998; American Diabetes Association web site, www.diabetes.org, visited on November 16, 1999.

[17]American Diabetes Association, Position statement: Gestational diabetes mellitus, *Diabetes Care* (supplement 1) 22 (1999): S60–S61.

Q&A NUTRITION IN PRACTICE

Mastering Diabetes Control

A healthy person goes about daily activities with little thought to how the body will react to everyday routines or disruptions to those routines. If you usually eat breakfast at 8:00 A.M., you may choose to sleep in and not eat breakfast on weekends without a second thought. If your friend asks you to play tennis and the match interferes with dinner, you simply eat later. If you get hungry during the day, you eat a snack. If you're not hungry at your usual dinner hour, you wait and eat later. For people with diabetes, even such simple variations in a daily schedule require thought and adjustment. They need to learn facts, master techniques, and develop new attitudes and behaviors that will provide for a healthy life. With assistance from the health care team, clients with the motivation can learn to manage their diabetes, rather than allow their disease to control their lives.

www.diabetes.org
American Diabetes Association

www.joslin.org
Joslin Diabetes Center

www.aadenet.org
American Association of Diabetes Educators

How can health care professionals help clients take control of their disorders?

Health care professionals who simply "prescribe" remedies and then expect their clients to comply fail to consider the impact that lifestyle changes have on a person's quality of life. Clients can easily be overwhelmed, and their motivation and compliance may be poor. Instead, an education approach that recognizes clients as central in making health care decisions and determining health care interventions that they can live with has a better chance for success.

Together with the client, the diabetes management team includes physicians, nurses, dietitians, and counselors with training in diabetes management.[1] The diabetes management team provides clients with the information, options for care, and skills they need to make decisions about their treatment plans. The Diabetes Control and Complications Trial identified a team approach as an important strategy in the management of diabetes. Clients in the study spent more than 80 percent of their time during follow-up visits with nonphysician team members; about 60 percent of their time was spent with nurses.[2]

How can health care professionals leave important health decisions up to clients without endangering their health?

Health care professionals cannot knowingly encourage clients to prac-

For a person with diabetes, even simple changes in routine require planning and adjustments.

tice behaviors that are medically harmful. Nor can they force clients to follow advice. Some aspects of disease management are critical for survival. Other aspects may be beneficial but less critical. Still other aspects may be ideal but less pressing when considered in the total context of a treatment plan.

A person with type 1 diabetes, for example, must take insulin or face death, but the person's goals, motivations, finances, and abilities help determine if traditional therapy or intensive therapy is more appropriate. One client with diabetes may choose traditional therapy even though it means less than the best blood glucose control, while another may be eager to learn intensive therapy. Health care professionals explain the ramifications of treatment choices and may continue to encourage clients to consider different options throughout the course of care, but ultimately they respect the individual's right to make health care decisions without judgment.

What is the difference between traditional and intensive therapy?

People who practice traditional therapy follow the same meal plan from day to day and take the same dose of insulin at the same time each day, usually in the morning and evening. Although traditional therapy generally prevents dangerously high or low blood glucose levels, most often blood glucose stays well above normal levels. People who practice intensive therapy, on the other hand, learn to adjust their insulin doses to cover their food intakes and tightly control their blood glucose levels. The person practicing intensive therapy uses either multiple injections or pumps to supply a low dose of insulin at all times and extra insulin to cover meals.

What must clients understand in order to practice intensive therapy?

All newly diagnosed clients and their families need diabetes management training and counseling, and the learning process takes time. Even highly motivated clients who listen attentively need several counseling sessions to learn the basics of diabetes management. Diabetes education is a continuous process that is a routine part of diabetes management. People who are contemplating intensive therapy must already have a thorough understanding of their diabetes and have the knowledge and skills to take that knowledge a step further. In order to practice intensive therapy, clients also need to learn about:

- *Carbohydrate counting.* Clients need to know how to figure out how much carbohydrate they eat at meals and snacks (see Chapter 20).
- *Medication adjustments.* Clients need to know how to adjust their medications when they eat more or less than the usual amount of carbohydrate.
- *Self-monitoring.* Clients need to learn how to use blood glucose meters, draw blood samples, record and interpret results, and bring glucose levels within a desirable range.
- *Self-management.* Clients need to learn how to keep an accurate record of blood glucose test results, food intake, physical activity, illnesses, and medication administration so that they can understand how their bodies respond to diabetes and how they can gain control over their disease. Clients need the support of the diabetes management team to help them design a diabetes management plan, cope with their fears, stay motivated when they feel overwhelmed, gain confidence in their abilities to manage the disorder, and live a high-quality life.

What can health care professionals do to help clients design management plans that meet their needs?

Regardless of whether a client chooses traditional or intensive therapy, health care professionals facilitate the learning process by working out a highly individualized plan that carefully considers the client's current lifestyle, motivation, goals, and ability to grasp new concepts and make lifestyle changes. The plan must be flexible to accommodate changing needs; a person who is highly motivated at one point, for example, may become totally discouraged at another time and need to restructure goals temporarily. At each step, the health care team balances medical needs with the individual's goals and motivation.

As Chapter 20 describes, a complete assessment serves as the first step in formulating a treatment plan. The more detailed the assessment, the more closely the plan can be tailored to the individual's existing lifestyle and the better the chances for success.[3] Using assessment data, the health care team sets long-term medical goals; to enable the team and the client to measure the success of therapy, the goals are stated in terms of measurable outcomes such as target ranges for glycated hemoglobin, blood lipids, and body weight. The team also sets short-term goals. For example, the team determines acceptable fasting and after-meal blood glucose goals for each client. Because maintaining near-normal blood glucose levels is always a goal, clients learn to self-monitor their blood glucose levels at home.

How is this task accomplished?

To monitor blood glucose, the client pricks a finger to get a blood sample and transfers the blood to a disposable strip. Most often clients rely on strips that work with computerized meters that measure the glucose concentration from the blood sample. At first, the person performs blood tests at least seven times a day: before each meal, two hours after each meal, and at bedtime. As people learn how their blood glucose responds to food intake and physical activity, they can test less often—about three or four times daily.

Motivated clients who wish to tightly control their blood glucose levels keep accurate records of food intake, physical activity, blood glucose measurements, medication administration, and illnesses. Using the records, the health care team can help clients learn about the unique ways their blood glucose changes in response to foods. From this information, the health care team calculates the carbohydrate-to-insulin ratio—that is, the number of units of insulin a person needs per gram of carbohydrate. The lower the ratio, the more insulin a person must use to cover carbohydrate intake. Knowing their carbohydrate-to-insulin ratio can help clients adjust their insulin doses when they eat more or less than they usually do.

Throughout the educational process, clients learn to use their records to look for patterns that suggest they are consistently having problems with blood glucose control at a particular time or times of the day. They learn what can cause their blood glucose to fluctuate and what strategies they can use to handle the problems (review Table 20-6 on p. 485).

How do emotional issues enter into diabetes management?

In addition to their diabetes-related tasks, clients have many other responsibilities related to work, school, family, and community. It should come as little surprise that even when clients know what to do and why they should do it, they may be unable to carry out the plan at times. Many people with diabetes report feeling overwhelmed and frustrated by the multitude of self-care demands.[4] In the words of one diabetes educator: "No one but another person who has diabetes can fully appreciate the demands of diabetes. It is 24 hours a day, 365 days a year (except on leap year when you get an extra day of diabetes). It involves all manner of imposition and deprivation. And even when you do everything right, there are no guarantees."[5]

People with diabetes often feel that health care professionals, family, and friends "blame" them for having diabetes-related problems and complications.[6] They may feel guilt and remorse because their efforts at diabetes control were not good enough to prevent complications. Health care professionals are wise to remain nonjudgmental and to recognize that diabetes control must be balanced with quality of life. Clients also need to know that they may experience complications even when they are doing everything possible. Insisting on perfection can only lead to frustration.

Parents and other family members also face the challenge of living with diabetes. The intensity of the situation can either reinforce or disrupt family unity. Parents may resent the demands of caring for a child with a chronic illness and then may experience guilt for having those feelings. They may feel anxious and be reluctant to allow their child to follow the diabetes care plan without their constant assistance. Especially in the case of an older child, they may press their care and control on a child who needs to develop autonomy and self-care. Parents may also become emotionally upset when they see their child feeling anxious, depressed, or withdrawn. Parents need time to work through these feelings. They might want to attend meetings for parents of children with diabetes. Such meetings offer opportunities to share feelings, ideas, and frustrations with others in similar situations. Sometimes just knowing that you're not alone helps.

What other resources are available?

Many hospitals and medical centers offer clients educational programs designed to expand their knowledge and promote independence. Some programs encourage children to bring friends, which makes the experience more comfortable and fun. Some programs are designed specifically for parents, grandparents, and other caregivers.

Children can combine education and summer vacation at camps designed especially for children with diabetes. These camps offer the chance to learn more about diabetes while "living" the lifestyle with companions under supervision. Children trade snack ideas, try new recipes, and help prepare meals. Older children assist younger ones, and all benefit.

The results of the Diabetes Control and Complications Trial clearly show that tightly managing diabetes can dramatically reduce the complications associated with it. Helping clients with diabetes make the necessary adjustments is a continuous process that balances medical and individual needs.

Notes

[1] M. Bayless and C. Martin, The team approach to intensive diabetes management, *Diabetes Spectrum* 11 (1998): 33–37.

[2] Bayless and Martin, 1998.

[3] S. M. Strowig, Improved methods of therapy in diabetes, *Diabetes Spectrum* 11 (1998): 16–17.

[4] W. H. Polonsky, Listening to our patients' concerns: Understanding and addressing diabetes-specific emotional distress, *Diabetes Spectrum* 9 (1996): 8–11; R. R. Rubin, Life's work they have not chosen, *Diabetes Spectrum* 8 (1995): 308.

[5] Rubin, 1995; K. F. McFarland, The power of words, *Diabetes Spectrum* 8 (1995): 308.

[6] McFarland, 1995.

CHAPTER 21

NUTRITION AND DISORDERS OF THE HEART, BLOOD VESSELS, AND LUNGS

CONTENTS

Atherosclerosis
and Hypertension

Prevention and
Treatment of CHD

Heart Failure and Strokes

Disorders of the Lungs

Case Study: History
Professor with
Cardiovascular Disease

Nutrition in Practice:
Free Radicals,
Antioxidants, and CHD

cardiovascular diseases (CVD): diseases of the heart and blood vessels.

The heart receives oxygen-rich blood from the lungs and delivers this oxygen (as well as nutrients) to the body's tissues by way of the blood vessels. Any disorder of the heart or blood vessels disrupts this delivery; without an adequate supply of fluid, oxygen, and nutrients, the body's ability to perform its metabolic activities falters. **Cardiovascular diseases (CVD)** are the leading cause of death in the United States and around the world. Men are more likely to develop heart disease and to do so at an earlier age than women. However, the outdated notion that heart disease is a "man's disease" has given way to the grim reality that CVD currently account for more deaths in women than the next 16 causes of death combined.[1] This chapter first examines the two most common CVD: atherosclerosis and hypertension. Then it describes the major consequences of these conditions—heart failure and strokes. It continues with a discussion of disorders of the lungs.

Atherosclerosis and Hypertension

Most CVD involve both atherosclerosis and hypertension. When either is a problem, the likelihood that the other will develop is great. Therefore, they often occur together and each raises the risk of complications further. Thus, treatment strategies often aim to prevent or treat both atherosclerosis and hypertension. The accompanying glossary defines terms related to CVD.

Atherosclerosis

Atherosclerosis usually begins with the accumulation of plaques along the inner arterial walls, especially at branch points (see Figure 21-1). These plaques obstruct blood flow—both directly and indirectly—through the arteries, which damages tissues and raises blood pressure.

DIRECT EFFECTS ON BLOOD FLOW Plaques develop as mounds of lipids (mostly cholesterol) and become covered with fibrous connective tissue.

GLOSSARY
OF TERMS RELATED TO CARDIOVASCULAR DISEASES (CVD)

aneurysm (AN-you-ris-um): a ballooning out of a portion of a blood vessel (usually an artery) due to weakness of the vessel's wall.

angina: a painful feeling of tightness or pressure in the area in and around the heart and often radiating to the back, neck, and arms; caused by a lack of oxygen to an area of heart muscle.

atherosclerosis: the form of CVD characterized by the buildup of plaques along the inner walls of the arteries, which narrow the lumen of the artery and restrict blood flow to the tissue supplied by the artery.

cerebral thrombosis: a blood clot that blocks the flow of blood through an artery that feeds the brain.

coronary heart disease (CHD): heart damage that results from an inadequate blood supply to the heart; also called **coronary artery disease (CAD).**

coronary thrombosis: a blood clot that blocks the flow of blood through an artery that feeds the heart.

embolism: the obstruction of a blood vessel by an **embolus (EM-boh-luss),** or traveling clot, causing sudden tissue death.
  embol = to plug

hypertension: higher-than-normal blood pressure.

plaques (PLACKS): mounds of lipid material (mostly cholesterol) with some macrophages (a type of white blood

cell) covered with fibrous connective tissue and embedded in artery walls. With time, plaques may enlarge and harden as the fibrous coat thickens and calcium is deposited in them.
  placken = patch or plate

platelets: tiny, disc-shaped bodies in the blood that play an important role in clot formation.

thrombosis: the formation of a **thrombus,** or blood clot.
  thrombo = clot

Figure 21-1
THE FORMATION OF PLAQUES IN ATHEROSCLEROSIS

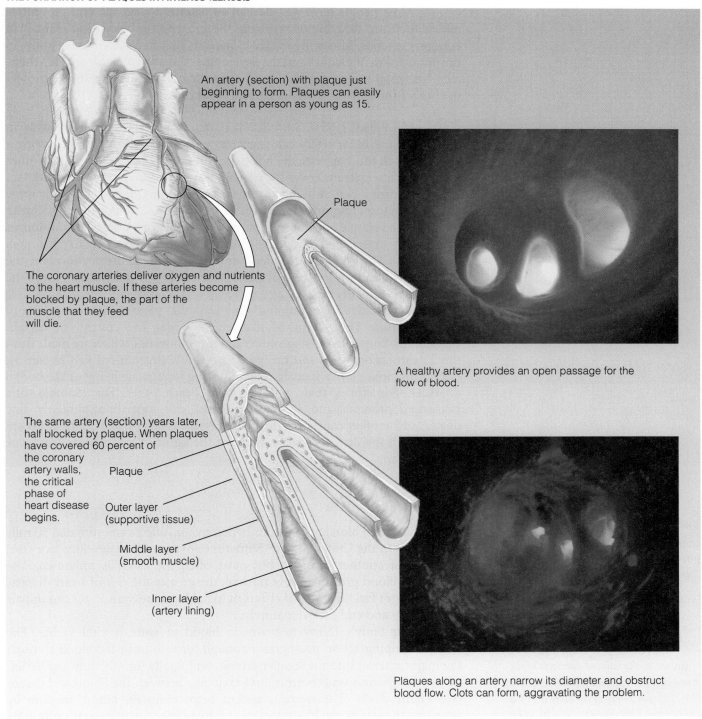

An artery (section) with plaque just beginning to form. Plaques can easily appear in a person as young as 15.

The coronary arteries deliver oxygen and nutrients to the heart muscle. If these arteries become blocked by plaque, the part of the muscle that they feed will die.

Plaque

A healthy artery provides an open passage for the flow of blood.

The same artery (section) years later, half blocked by plaque. When plaques have covered 60 percent of the coronary artery walls, the critical phase of heart disease begins.

Plaque

Outer layer (supportive tissue)

Middle layer (smooth muscle)

Inner layer (artery lining)

Plaques along an artery narrow its diameter and obstruct blood flow. Clots can form, aggravating the problem.

They gradually enlarge and harden as the fibrous coat thickens and minerals are deposited in them. As plaques grow, they directly obstruct blood flow. Although the plaques of atherosclerosis may obstruct blood flow through any blood vessel, the coronary arteries are most often affected (coronary artery disease). When coronary artery disease damages the heart muscle (coronary heart disease, or CHD), the person may experience pain and pressure in and around the area of the heart (angina). If blood flow to the heart is cut off, that area of the heart muscle dies, and a heart attack results. When blood flow to the brain is obstructed, a transient ischemic attack or stroke may result.

Most people have well-developed plaques by age 30, and preventive efforts focus on delaying or reversing the progression of existing plaques in coronary arteries. Mounting evidence suggests that free radicals play a role in the progression of plaques and the development of atherosclerosis.[2] This chapter's Nutrition in Practice explores how antioxidant nutrients might protect the body from free-radical damage and heart disease.

▪▪ INDIRECT EFFECTS ON BLOOD FLOW A sudden spasm or surge in blood pressure in an artery can tear away part of the fibrous coat covering a plaque. When this happens, the body responds to the tear as it would to other tissue injuries. Platelets cover the damaged area, and together with other factors, they form a clot. A blood clot may stick to a plaque and gradually grow large enough to close off a blood vessel (thrombosis). A clot may also break free from the plaque and travel through the circulatory system until it lodges in a small artery and suddenly shuts off the blood flow to that area (embolism). The gradual or sudden loss of blood flow to the portion of tissue supplied by the clotted artery robs the tissue of oxygen and nutrients, and the tissue may eventually die.

Researchers are examining other ways that platelets may contribute to atherosclerosis. The action of platelets is under the control of certain eicosanoids, known as prostaglandins and thromboxanes, which are made from the 20-carbon omega-6 and omega-3 fatty acids (introduced in Chapter 3). Each eicosanoid plays a specific role in helping to regulate many of the body's activities. Sometimes their actions oppose each other. For example, one eicosanoid prevents, and another promotes, clot formation; similarly, one dilates, and another constricts, the blood vessels. When omega-3 fatty acids are abundant in the diet, they make more of the kinds of eicosanoids that favor heart health.

▪▪ Hypertension

Chronic elevated blood pressure, or hypertension, affects an estimated 50 million people in the United States.[3] Sometimes hypertension develops as a consequence of another disorder, but most often the cause is unknown. The higher the blood pressure above normal, the greater the risk of heart disease. People cannot feel the physical effects of high blood pressure, but it can impair life's quality and end life prematurely.

The body's ability to maintain blood pressure is vital to life. The heart's pumping action must create enough force to push the blood through the major arteries into the smaller arteries and finally into the tiny capillaries, whose thin porous walls permit fluid exchange between the blood and tissues (see Figure 21-2). The nervous system helps maintain blood pressure by adjusting the size of the blood vessels and by influencing the heart's pumping action. The kidneys also help regulate blood pressure by setting in motion mechanisms that change the blood volume. The narrower the blood vessels or the greater the volume of blood in the circulatory system, the harder the heart must pump (and the more pressure the heart must create) to feed the tissues.

▪▪ CONSEQUENCES OF HYPERTENSION Strain on the heart's pump, the left ventricle, can enlarge and weaken it until finally it fails (heart failure is described later). Constant high pressure damages the arterial walls and may cause them to balloon out and burst (aneurysm). Aneurysms that go undetected can lead to massive bleeding and death, particularly when a large vessel such as the aorta is affected. In the small arteries of the brain, an aneurysm may

Optimal resting blood pressure for adults averages about 120 over 80 millimeters of mercury (mm Hg). Blood pressure of 140 over 90 mm Hg or higher increases the risks of heart attacks and strokes (see Table 21-1 on p. 501).

The resistance to the flow of blood caused by the reduced diameter of the smallest arteries and capillaries at the periphery of the body is called peripheral resistance.

Figure 21-2

BLOOD PRESSURE AND FLUID EXCHANGE

At the same time the heart pushes blood into an artery, the small-diameter arteries and capillaries at its other end resist the blood's flow (peripheral resistance). Both actions contribute to the pressure inside the artery. Another determining factor is the volume of fluid in the circulatory system, which depends in turn on the number of dissolved particles in that fluid.

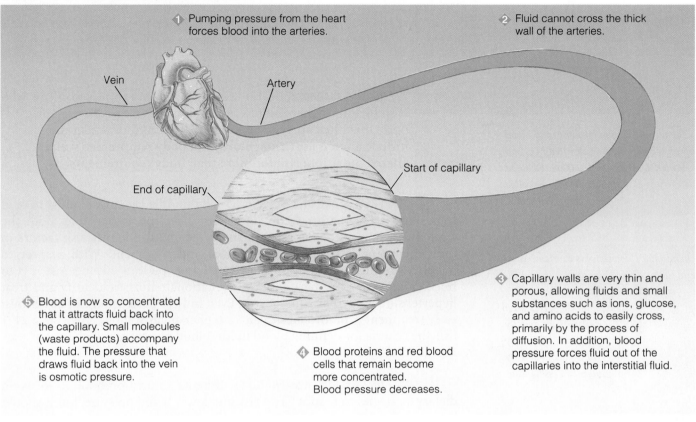

1 Pumping pressure from the heart forces blood into the arteries.

2 Fluid cannot cross the thick wall of the arteries.

Vein

Artery

Start of capillary

End of capillary

3 Capillary walls are very thin and porous, allowing fluids and small substances such as ions, glucose, and amino acids to easily cross, primarily by the process of diffusion. In addition, blood pressure forces fluid out of the capillaries into the interstitial fluid.

5 Blood is now so concentrated that it attracts fluid back into the capillary. Small molecules (waste products) accompany the fluid. The pressure that draws fluid back into the vein is osmotic pressure.

4 Blood proteins and red blood cells that remain become more concentrated. Blood pressure decreases.

lead to stroke, and in the eye, it may lead to blindness. Likewise the kidneys may be damaged when the heart is unable to adequately pump blood through them.

▓▓ INTERRELATIONSHIPS BETWEEN HYPERTENSION AND ATHEROSCLEROSIS
Hypertension injures the arterial walls, and plaques and clots readily form at damage points. Thus, hypertension sets the stage for atherosclerosis or makes it worse. Once plaques or clots develop, they may reduce the diameter of an artery and raise blood pressure even higher. Hypertension and atherosclerosis are mutually aggravating conditions, and both contribute to acute and chronic heart failure and strokes.

▓▓ Risk Factors for Coronary Heart Disease (CHD)

Efforts to reduce the incidence of cardiovascular diseases focus on preventing CHD. The margin lists the major risk factors for CHD as identified by the American Heart Association.[4] Note that one of the risk factors is hypertension because efforts to control hypertension also reduce the risk of CHD. Many risk factors are interrelated—lack of physical activity, overweight, hypertension, elevated blood cholesterol, and type 2 diabetes frequently occur together, for example. Other factors, including emotional stress, may also contribute to the risk of CHD.

Risk factors for CHD that cannot be modified:
- **Increasing age.**
- **Male sex.**
- **Heredity (including race). Children of parents with heart disease are at higher risk, as are African Americans.**

Risk factors for CHD that can be modified:
- **Smoking.**
- **Elevated blood cholesterol levels[a] (LDL).**
- **Hypertension.[a]**
- **Physical inactivity.**
- **Obesity and overweight.[a]**
- **Diabetes mellitus.[a]**

[a]*Risk factors that have relationships with diet.*

Recall from Chapter 20 that obesity contributes to insulin resistance, and insulin resistance is associated with the development of type 2 diabetes, hypertension, and elevated LDL and triglyceride levels (a combination of symptoms known as syndrome X). These many interrelationships may explain why 80 percent of people with diabetes die from cardiovascular diseases.

The technical term for abnormal blood lipids is dyslipidemia; *elevated LDL and low HDL are examples. Elevated blood lipids may also be called* hyperlipidemia.

Reminder: Cholesterol is carried in several lipoproteins, chief among them LDL and HDL (see Chapter 3 for details). Remember them this way:
• HDL = High-density lipoprotein = Healthy
• LDL = Low-density lipoprotein = Less healthy.

Reminder: Triglycerides are carried in VLDL (very-low-density lipoproteins).

THE NURSING DIAGNOSIS health-seeking behaviors applies to clients who wish to reduce their risk of CHD and hypertension. Nursing diagnoses that often apply to people at risk for cardiovascular diseases include **altered nutrition: more than body requirements** and **altered nutrition: risk for more than body requirements.**

PREVALENCE OF RISK FACTORS By middle age, most adults have at least one risk factor for CHD, and many have more than one. Both the United States and Canada recommend screening to identify risk factors in individuals and offer preventive advice and treatment. With respect to hypertension, the single most effective step people can take against it is to find out whether they have it. A major national effort to identify and treat hypertension is currently under way. Even mild hypertension can be serious; early treatment promotes health and a higher-quality, longer life. Table 21-1 lists the criteria for defining blood lipids, blood pressure, and obesity in relation to CHD risk.

DIET-RELATED RISK FACTORS It befits a nutrition book to focus on dietary strategies to reduce CHD risk and lower blood pressure, but it should be noted that these strategies may not reduce the risk of heart attack and stroke as successfully as quitting smoking and taking prescribed medicines. Dietary changes and physical activity can be effective, however, in modifying several major risk factors: blood cholesterol levels, hypertension, obesity and overweight (Chapter 7), and diabetes mellitus (Chapter 20).

Prevention and Treatment of CHD

Treatment plans for preventing and treating CHD, including medical nutrition therapy, aim to normalize blood lipids and blood pressure, improve modifiable risk factors, and prevent complications. The blood cholesterol linked to CHD risk is LDL cholesterol, which contributes to the accumulation of plaque on the artery walls. HDL also carry cholesterol, but it is believed that they carry cholesterol away from the arteries and back to the liver; elevated HDL may be *protective* against heart disease. Like LDL, elevated triglycerides are also linked to CHD; however, when triglycerides are elevated, HDL are generally low, and researchers have not determined whether elevated triglycerides alone are a risk factor for CHD.[5]

TREATMENT STRATEGIES Lifestyle changes (including quitting smoking, improving diet, and increasing physical activity) are the first step in reducing the risk of CHD. If these lifestyle changes fail to normalize blood lipids or blood pressure, medications may be necessary (described later). For people with diabetes, controlling blood glucose is another important strategy.

Screening people for hypertension is a first step toward early detection and prevention of complications.

Table 21-1
STANDARDS FOR CHD RISK FACTORS

LDL Cholesterol[a]	Total Cholesterol[b]
<130 mg/dL = desirable	<200 mg/dL = desirable
130–159 mg/dL = borderline high	200–239 mg/dL = borderline high
>160 mg/dL = high	>240 mg/dL = high

HDL Cholesterol	Triglycerides (Fasting)[d]
HDL ≤35 mg/dL indicates risk[c]	<200 mg/dL = desirable
LDL-to-HDL ratio:	200–400 mg/dL = borderline high
Men: >5.0 indicates risk	400–1000 mg/dL = high
Women: >4.5 indicates risk	>1000 mg/dL = very high

Blood Pressure[e]

	Systolic pressure		Diastolic pressure
Optimal	<120	and	<80
Normal	<130	and	<85
High-normal	130–139	or	85–89
Hypertension			
Mild (stage 1)	140–159	or	90–99
Moderate (stage 2)	160–179	or	100–109
Severe (stage 3)	>180	or	>110

Weight (Body Mass Index)

Overweight:	BMI 25–29.9
Obese:	BMI ≥30

[a]For people with existing CHD, desirable LDL cholesterol values are lower (≤100 mg/dL).
[b]To convert cholesterol (mg/dL) to standard international units (mmol/L), multiply by 0.02586. For cholesterol values for children and adolescents, see Table NP11-1 on p. 285.
[c]This HDL value may be too low for women; no alternative value has yet been proposed. NIH Consensus Development Panel on Triglyceride, High-Density Lipoprotein, and Coronary Heart Disease, Triglyceride, high-density lipoprotein, and coronary heart disease, *Journal of the American Medical Association* 269 (1993): 505–510.
[d]High triglycerides alone normally do not indicate *direct* risk, but may reflect lipoprotein abnormalities associated with CHD. The risk of CHD increases as triglyceride levels increase in people with other risk factors. High triglycerides also occur in conditions such as kidney disease and diabetes, which suggest a high CHD risk.
[e]For blood pressure standards for children, see Table NP11-2 on p. 285.
SOURCES: Blood lipid standards adapted from The Expert Panel, Summary of the second report of the National Cholesterol Education Program (NCEP) Expert Panel on Detection, Evaluation, and Treatment of High Blood Cholesterol in Adults (Adult Treatment Panel II), *Journal of the American Medical Association* 269 (1993): 3015–3023; hypertension standards adapted from the Sixth Report of the Joint National Committee on Prevention, Detection, Evaluation, and Treatment of High Blood Pressure, National High Blood Pressure Education Program, National Heart, Lung, and Blood Institute, National Institutes of Health, November 1997, p. 11.

For people with advanced CHD, therapy may also include surgery to restore blood flow to the affected organ.

■■ **PHYSICAL ACTIVITY** Physical activity deserves attention in all preventive and treatment programs for heart disease. Physical activity can help promote weight loss and loss of body fat, strengthen the cardiovascular system, lower LDL, raise HDL, reduce hypertension, and improve glucose tolerance in people with type 2 diabetes.[6] Physical activity can also improve mental health. Frequent and sustained aerobic, endurance-type activities such as brisk walking, undertaken faithfully for 30 minutes or more as a daily or every-other-day routine, may be most effective.

■■ **SMOKING** Smoking stresses the cardiovascular system by depriving the heart of oxygen and raising the blood pressure. It also damages platelets, making blood clot formation more likely. Toxins in cigarette smoke damage the blood vessels, setting the stage for plaque formation.

Regular aerobic exercise can help to defend against heart disease by strengthening the heart muscle, promoting weight loss, reducing blood pressure, and improving blood lipid and blood glucose levels.

▐▐ Medical Nutrition Therapy

As Table 21-2 shows, the goals of medical nutrition therapy for preventing and treating CHD focus on reducing LDL cholesterol. A panel of experts recommends a two-step plan, shown in Table 21-3.[7] If blood lipids do not improve on the Step I diet, therapy moves on to Step II. People with existing CHD or those who have already been following a fat-modified diet prior to seeking treatment start on the Step II diet.

Diet strategies to reduce hypertension have typically included recommendations to control weight, reduce salt intake, and limit alcohol. Although health authorities continue to recommend these modifications, they also recommend a diet that provides adequate amounts of potassium, calcium, and magnesium and that limits saturated fats and cholesterol. Results of the Dietary Approaches to Stop Hypertension (DASH) trial show that a diet rich in fruits, vegetables, and low-fat dairy products combined with a reduced total fat and saturated fat intake can lower blood pressure to a significant degree.[8] Although the study did not measure the effects of the DASH diet on blood lipids, the diet is consistent with the recommendations of the Step I and Step II diets. Table 21-4 shows the daily servings of foods on the DASH diet.

LEARNING LINK

Dietitians often use the exchange system described in Chapter 20 and presented in Appendixes B and C to help clients determine which types of fats are in foods. The exchange system can be useful, but it has to be adapted somewhat. On the DASH diet, for example, a serving of meat is 3 ounces whereas in the exchange system, a serving of meat is 1 ounce. Therefore, 1 serving of meat on the DASH diet is equivalent to 3 servings of meat in the exchange system.

▐▐ **ENERGY** Dietary plans to reduce LDL cholesterol, triglycerides, and hypertension include attention to weight reduction or weight maintenance (see Chapter 7). Obesity, especially abdominal obesity, is associated with high blood lipids, hypertension, and diabetes. Weight loss improves LDL cholesterol and triglyceride levels as well as insulin resistance in people with type 2 diabetes. Weight loss is also one of the most effective and long-lasting treatments for hypertension. Even a modest weight loss of 10 pounds may significantly lower blood pressure.

Table 21-2
DIETARY TREATMENT GUIDELINES BASED ON LDL CHOLESTEROL

Risk Factor Status	Cholesterol Values When Diet Therapy Should Begin	Goals for Cholesterol Values with Diet Therapy
• Without existing CHD and with one risk factor	≥160 mg/dL[a]	<160 mg/dL
• Without existing CHD and with two or more risk factors	≥130 mg/dL[b]	<130 mg/dL
• With existing CHD	>100 mg/dL	≤100 mg/dL

[a]Note that ≥160 mg/dL indicates high risk.
[b]Note that ≥130 mg/dL indicates borderline-high risk.
SOURCE: Adapted from The Expert Panel, Summary of the second report of the National Cholesterol Education Program (NCEP) Expert Panel on Detection, Evaluation, and Treatment of High Blood Cholesterol in Adults (Adult Treatment Panel II), *Journal of the American Medical Association* 269 (1993): 3015–3023.

Table 21-3
DIET STRATEGIES TO REDUCE LDL CHOLESTEROL

	Step I	Step II
Energy	Adequate to achieve or maintain desirable weight	Adequate to achieve or maintain desirable weight
Total fat[a]	≤30%	≤30%
Saturated fat[a]	<10%	<7%
Polyunsaturated fat[a]	<10%	<10%
Monounsaturated fat[a]	5–15%	5–15%
Cholesterol	<300 mg/day	<200 mg/day

[a]All fats except cholesterol are expressed as percentages of total food energy.
SOURCE: Adapted from E. J. Schaefer, New recommendations for the diagnosis and treatment of plasma lipid abnormalities, *Nutrition Reviews* 51 (1993): 246–252.

■■ TOTAL FAT AND SATURATED FAT The Step I diet recommends a total fat intake of less than 30 percent of daily kcalories, with saturated fat no more than one-third of that and dietary cholesterol less than 300 milligrams a day. Such a diet restricts obvious sources of fat, saturated fat, and cholesterol and can be accomplished by following the practical suggestions for reducing fat outlined in Chapter 3 on pp. 64–65.

The Step II diet further reduces saturated fat to 7 percent of daily kcalories and cholesterol to less than 200 milligrams per day. For persons requiring a Step II diet, the help of a registered dietitian can ensure that saturated fat and cholesterol are reduced without sacrificing the nutritional quality of the diet.

Compared to the typical American diet, the DASH diet provides more fruits, vegetables, and grains and less meat, poultry, fish, fat, and salt. The "How to" box on pp. 504–505 suggests ways to help clients implement the DASH diet.

■■ OTHER FATS In addition to the expert panel's dietary recommendations for lowering LDL cholesterol, other fat-related strategies for reducing CHD risk are under investigation. Research suggests that a diet that provides

Notice that the low-fat diet described here controls both the amount of fat and the amount of saturated, monounsaturated, and polyunsaturated fat. By comparison, the low-fat diet described in Chapter 16's discussion of fat malabsorption controls the amount of fat and sometimes includes fat from medium-chain triglycerides (MCT).

Table 21-4
THE DIETARY APPROACHES TO STOP HYPERTENSION (DASH) DIET

Food Group	Servings per Day[a]	Serving Sizes[b]
Grains	7–8	1 slice bread or 1/2 c dry cereal, rice, or pasta
Vegetables	4–5	1 c raw leafy vegetables, ½ c cooked vegetables, 6 oz vegetable juice
Fruits	4–5	1 medium fruit; ½ c fresh, frozen, or canned fruit; ¼ c dried fruit, 6 oz fruit juice
Milk, fat-free or low-fat	2–3	1 c fat-free or low-fat milk or yogurt, 1½ oz fat-free or low-fat cheese
Meats, poultry, fish	2 or less	3 oz cooked
Nuts, seeds, legumes	½–1[c]	⅓ c or 1½ oz nuts, 2 tbs or ½ oz seeds, ½ c cooked legumes
Fats and oils	2–3	1 tsp soft margarine or vegetable oil, 1 tbs reduced-fat margarine, 1 tbs reduced-fat mayonnaise, 2 tbs light salad dressing

[a]Servings for a person consuming a 2000-kcalorie diet.
[b]In general, serving sizes are consistent with those used in the Food Guide Pyramid.
[c]Four to five servings per week. Selecting legumes more often than nuts and seeds can significantly reduce the total amount of fat in the diet.
SOURCE: Adapted from L. J. Appel and coauthors, A clinical trial of the effects of dietary patterns on blood pressure, *New England Journal of Medicine* 336 (1997): 1117–1124.

HELP CLIENTS IMPLEMENT THE DASH DIET

For most people, following the DASH diet translates into major changes in the foods they eat. The diet may be easier to adopt if changes are made gradually. These suggestions may help clients adapt:

- Gradually increase servings of fruits, vegetables, and grains to reduce gas and bloating that may accompany a higher-than-normal fiber intake. Remember to drink plenty of water as well.
- For those who eat only one or two servings of fruits or vegetables a day, start by including at least two servings of fruits and/or vegetables at each meal (6 servings). Then add one fruit and one vegetable as a snack (total 8 servings).
- Cut back on servings of meat, poultry, and fish by eating one-third to one-half the amount usually eaten at each meal. The goal is to include no more than 6 ounces of meat, poultry, and fish a day.
- For those who usually don't drink milk, or only have one serving, start by adding a glass of fat-free or very-low-fat (1%) milk at one meal. Then add a glass at another meal. A third glass of milk can be used as a snack or taken with a third meal.
- Reduce salt intake by gradually cutting down on the salt added to food and used in cooking. The goal is to reduce salt intake to about 1 teaspoon a day.

The box on pp. 64–65 offered many tips for reducing fat and the type of fat in the diet. To ease diet compliance, suggest that clients include:

- Whole-grain breads and cereals instead of refined breads and cereals.
- Fresh, frozen, dried, or canned fruit packed in its own juice.
- Fresh, frozen, or canned vegetables that contain no added salt.
- Fat-free or 1% fat milk and low-fat and fat-free yogurt.
- Casseroles, pasta meals, and stir-fry dishes to limit meat and increase servings of vegetables and grains.

Continued

up to 45 percent of total kcalories from fat can effectively lower LDL and triglycerides, provided that the additional fat comes from monounsaturated sources. As an added benefit, the diet does not appear to lower HDL, as low-fat diets sometimes do.

Fish oils, rich in omega-3 fatty acids, have received much attention with respect to CHD risk. A large and carefully conducted study failed to show a correlation between fish consumption and CHD risk.[9] The authors noted, however, that very few men in the study group ate no fish at all, and eating fish one or two times per week may confer the same benefits as eating fish five to six times per week. Two servings of fish can easily be included on the DASH diet or the Step I and Step II diets. Although eating one or two fish meals per week may be beneficial and is certainly safe, there is not enough evidence to recommend the use of fish oil supplements, which carry their own risks (see p. 56).[10]

The Nurses' Health Study, a study of over 80,000 women, found that reducing total fat was not as important in reducing CHD risk in women as replacing saturated fats with unsaturated fats and reducing the intake of *trans*-fatty acids.[11] Many margarines and processed foods contain *trans*-fatty acids (see Chapter 3), which tend to raise total cholesterol and LDL and lower HDL. Margarines in tub or liquid form may contain fewer *trans*-fatty acids than stick margarines; margarines that contain no *trans*-fatty acids are also available. It is difficult to determine the *trans*-fatty acid content of processed foods because

HELP CLIENTS IMPLEMENT THE DASH DIET —CONTINUED

- Two or more vegetarian meals per week.
- Sodium-free spices such as basil, bay leaves, curry, garlic, ginger, lemon, mint, oregano, pepper, rosemary, and thyme to season foods.
- Low-salt or salt-free products.

Offer clients these suggestions for snacks:

- Frozen low-fat or fat-free yogurt.
- Fruit.
- Raw vegetables.
- Unsalted pretzels.
- Unsalted nuts (watch serving size) with raisins.
- Plain popcorn without butter, margarine, or salt.

Recommend that clients avoid these high-salt foods:

- Pickles, olives, and sauerkraut.
- Cured or smoked meats, such as beef jerky, bologna, corned or chipped beef, frankfurters, ham, luncheon meats, salt pork, and sausage. Many of these foods are also high in fat.
- Salty or smoked fish, such as anchovies, caviar, salted or dried cod, herring, sardines, and smoked salmon.
- Salted snack foods like potato chips, pretzels, popcorn, nuts, and crackers. Many of these foods are also high in fat.
- Bouillon cubes, seasoned salts, and soy, steak, Worcestershire, and barbecue sauces.
- Cheeses, especially processed cheeses. Many cheeses are also high in fat.
- Salted, canned, and instant soups.
- Prepared horseradish, catsup, and mustard.

Clients with lactose intolerance unable to tolerate the recommended servings of dairy foods can use the digestive aids described in Chapter 16.

---

food labels do not provide this information. A solution to this problem is to limit processed foods to the greatest degree possible.

■■ **COMPLEX CARBOHYDRATES** The DASH diet derives much of its energy from foods rich in complex carbohydrates. Including fiber-rich foods may have additional cholesterol-lowering benefits. When used together with a diet low in both fat and saturated fat, the fiber found in oats, barley, and pectin-rich fruits and vegetables may help reduce blood lipids further.[12]

A new product marketed as an aid to lower cholesterol is margarine made with plant sterols. Plant sterols are found in complex carbohydrates and products derived from them. Plant sterols resemble cholesterol in structure, reduce the absorption of dietary cholesterol, and lower blood cholesterol levels.[13] Even vegetarians, however, may find it difficult to consume enough plant sterols from foods to lower blood cholesterol; thus, the margarines serve to supplement plant sterols in the diet.

High-fiber diets tend to be rich in minerals (described later), which may help reduce blood pressure. High-fiber diets also tend to be good sources of folate and other B vitamins, which may help lower blood homocysteine levels. Studies have shown a positive association between elevated blood homocysteine and the risk of CHD.[14] How homocysteine increases the risk of CHD is unknown, but it may damage the artery walls or promote thrombosis. Although folate and the other B vitamins can reduce homocysteine levels, the

People with elevated triglycerides should avoid simple sugars, which often cause triglyceride levels to rise.

question remains whether such a reduction also lowers CHD risk.[15] For many people, selecting at least five servings of folate-rich fruits and vegetables daily should be sufficient to normalize homocysteine concentrations.

:: ANTIOXIDANT NUTRIENTS A discussion of dietary factors to reduce CHD risk must include the antioxidant nutrients—a hot topic in both the popular press and scientific journals. Evidence continues to mount that antioxidant nutrients, particularly vitamin E, may help protect against CHD by preventing the damaging effects of free radicals. This Chapter's Nutrition in Practice provides more information about this topic.

:: MINERALS Minerals appear to play important roles in blood pressure regulation. High intakes of sodium and low intakes of potassium, calcium, and magnesium have been associated with hypertension for some time, but data from studies have been inconclusive and difficult to interpret. Perhaps the effect of each mineral is too small to detect, but the combined effect of all minerals together results in a significant effect.[16] The DASH trial did not try to sort out how individual nutrients affect blood pressure, but rather how dietary patterns affect it. For people following the DASH diet at energy levels of about 2000 kcalories, the diet provided about 4.5 grams of potassium, 1250 milligrams of calcium, and 500 milligrams of magnesium—approximately two to three times the amounts in typical American diets. Although the DASH diet limits sodium to about 3000 milligrams per day, hypertension experts recommend a slightly lower sodium intake (2400 milligrams or about a teaspoon of salt per day).[17]

:: ALCOHOL Moderate consumption of alcohol (no more than two drinks a day) may reduce the risk of CHD by raising HDL cholesterol and preventing clot formation.[18] Beneficial effects of alcohol are more apparent in people over age 50, in those with other risk factors, and in those with high LDL.[19] These findings pose a dilemma for health care professionals who are well aware of the potentially damaging effects of alcohol. Furthermore, alcohol can raise triglycerides, and people with elevated triglycerides are advised to restrict alcohol. Most authorities do not advise clients who do not drink alcohol to begin to do so to reduce their risk of CHD. If clients do drink alcohol, clinicians stress that moderation is key.

:: Drug Therapy

When used together with diet therapy and a physical activity program, drug therapy can effectively lower blood lipids. Lipid-lowering medications carry health risks, however, and are costly. Therefore, physicians as a rule do not prescribe medications to treat hyperlipidemia until after a six-month trial of intensive diet therapy and physical activity has proved unsuccessful in lowering blood lipid concentrations. For people with very high LDL cholesterol (greater than 220 milligrams per deciliter), a shorter diet trial may be considered. Besides lipid-lowering medications, medications used in the treatment of cardiovascular diseases may include antihypertensives and diuretics to reduce blood pressure, aspirin and anticoagulants to prevent clot formation, and nitroglycerin to alleviate angina. (Clients with type 2 diabetes may also be taking oral antidiabetic agents.) All of these medications are associated with significant risks and nutrition-related side effects (see the Diet Medication Interactions box), a problem compounded by the fact that drug therapy often includes multiple medications and continues for many years or even life. Elderly clients especially risk diet-medication interactions.

DIET-MEDICATION INTERACTIONS

Anticoagulants (including aspirin)

Aspirin can be given with food to reduce nausea and GI distress. Long-term use of aspirin may lead to folate and vitamin C deficiencies. *Ticlopidine* is best absorbed along with foods. Ticlopidine may cause nausea, GI distress, and diarrhea. For people taking *warfarin,* a consistent amount of vitamin K must be provided from day to day to assure the medication's effectiveness. Warfarin may also cause diarrhea. The herbs dong quai, ginko, cayenne, and feverfew and garlic can affect clotting times and must be used cautiously by people who are also taking anticoagulants.

Antihypertensives

All clients on *antihypertensives* should avoid natural licorice and limit alcohol. People on *ACE inhibitors* should avoid salt substitutes containing potassium and use potassium supplements (if prescribed) cautiously because these agents can raise blood potassium levels. One ACE inhibitor, *fosinopril,* should not be given along with calcium or magnesium supplements or calcium- or magnesium-containing antacids because calcium and magnesium reduce the drug's absorption. *Alpha-adrenergic blockers* can lead to weight gain and fatigue. *Beta-blockers,* which also act as antianginals, should be given with foods to reduce nausea and GI distress. Grapefruit juice should not be given along with *calcium channel blockers* because it can potentiate the effects of the drugs. *Clonidine* may dry the mouth and lead to constipation and drowsiness.

Antilipemics

Cholestyramine can cause nausea, GI distress, and constipation and can lead to fat-soluble vitamin deficiencies because it reduces their absorption. *Colestipol* is less likely than cholestyramine to cause nausea and lead to fat-soluble vitamin deficiencies, but it can lead to constipation. *Gemfibrozil* can lead to nausea and GI distress. The *statins (lovastatin, pravastatin, simvastin)* should not be given with grapefruit juice. People taking statins should limit alcohol use.

Cardiac glycosides

Digitoxin, digoxin, and *digitalis* can lead to anorexia and nausea. Low blood potassium and elevated blood calcium levels can lead to drug toxicity, and clients are advised to eat a high-potassium diet. If they use calcium and vitamin D supplements, they must be used cautiously. Magnesium-containing antacids or supplements given along with *cardiac glycosides* can reduce drug absorption. The herb hawthorn can potentiate the effects of cardiac glycosides.

Diuretics

Diuretics can cause potassium imbalances and lead to muscle weakness, unexplained numbness and tingling sensations, irregular heartbeats, and cardiac arrest. *Thiazide* and *loop diuretics* increase the urinary excretion of potassium, and clients on these diuretics are encouraged to include rich sources of potassium daily and may be prescribed potassium supplements. When taken internally, aloe vera can potentiate the loss of potassium in people taking potassium-wasting diuretics. Potassium-sparing diuretics (*amiloride, spironolactone,* and *triamterene*), on the other hand, lead to potassium retention. Clients on potassium-sparing diuretics should avoid excessive potassium intakes, potassium supplements, and salt substitutes that contain potassium. Sometimes a combination of both types of diuretics is used to avoid potassium imbalances.

Although research is lacking, some people use herbal and other remedies to protect the cardiovascular system. Some of these remedies include garlic, hawthorn, and black and green teas. Garlic is a relatively safe remedy that, when used along with dietary changes, may modestly lower serum cholesterol levels. Garlic may also reduce blood pressure, but the change is usually mild and of short duration. Hawthorn is also a relatively safe herb that may potentiate the action of antihypertensive agents and possibly lower the dose of medication necessary to reduce blood pressure. Black and green teas are sources of one type of antioxidant that may play a role in CHD prevention. The Diet-Medication Interactions box mentions herbs used for other reasons that may interact with medications prescribed for cardiovascular diseases.

▓ Heart Failure and Strokes

When atherosclerosis and hypertension run their courses, the consequences can be fatal. Most often, these conditions lead to heart attacks, chronic heart failure, and strokes.

▓ Heart Attacks

The heart receives nutrients and oxygen, not from inside its chambers, but from arteries on its surface (review Figure 21-1 on p. 497). As mentioned earlier, a **heart attack,** or myocardial infarction (MI), occurs when the supply of blood to the heart muscle is blocked. Treatment aims to relieve pain, stabilize the heart rhythm, and reduce the heart's workload.

heart attack: sudden tissue death caused by blockages of vessels that feed the heart muscle; also called **myocardial infarction (MI), cardiac arrest,** or **acute heart failure.**

myo = muscle
cardial = heart
infarct = tissue death

▓ IMMEDIATE CARE Like an accident victim, a heart attack victim is in shock at first, and diet therapy cannot begin until shock resolves. After several hours of observation, the person can usually begin to eat again, as outlined in the "How to" box. Medical nutrition therapy aims to reduce the work of the heart and therefore restricts total energy, the amount of food or drink at each feeding, sodium, and caffeine. Because nausea is a common problem following an MI, liquids are offered first. Low-sodium foods prevent fluid retention, and soft foods can help prevent nausea and abdominal distention, which can push the diaphragm up toward the heart and stress the heart muscle. Temperature extremes can stimulate nerves that slow the heart rate, so foods are offered at moderate temperatures. Caffeine stimulates the metabolic rate and is usually restricted.

THE NURSING DIAGNOSES with nutrition implications that apply to clients who have had a heart attack include **constipation, altered tissue perfusion, fluid volume excess, decreased cardiac output,** and **fatigue.**

▓ LONG-TERM DIET THERAPY After the person is out of immediate danger (in about five to ten days), the diet is tailored to meet individual needs and to deal with conditions such as hyperlipidemia, hypertension, obesity, and diabetes. Whatever else may be necessary, a diet to further reduce the risk of CHD and hypertension is appropriate. Such a diet can be planned to provide three meals a day, but people who still have chest pain after a heart attack may

HOW TO: ▽

MANAGE DIETS AFTER A MYOCARDIAL INFARCTION (MI)

- Offer nothing by mouth until shock resolves.
- After several hours, give a 1000- to 1200-kcalorie diet that progresses from low-sodium liquids to low-sodium soft foods of moderate temperature in frequent small feedings.
- After five to ten days, adjust the diet to meet individual needs, generally to three meals a day.
- Although somewhat controversial, many practitioners restrict caffeine completely during the first few days after an MI and generally recommend a moderate restriction (no more than 3 cups of a caffeine-containing beverage per day) thereafter.

continue to benefit from eating frequent small meals. Advise clients to eat slowly and to avoid strenuous physical activity before and after meals.

■■ **LIFESTYLE CHANGES**   Often people who have experienced an MI are eager to apply strategies to reduce their risks of further CHD, and nurses can use the opportunity to offer sound and useful counsel. Continue to offer support and encouragement for as long as possible. Clients may readily return to their old habits when their symptoms disappear if they have not fully incorporated new healthful behaviors into their lives.

■■ Chronic Heart Failure (CHF)

While heart attacks represent acute heart failure, heart failure can also occur gradually. Coronary heart disease, hypertension, and obesity are common causes. A person may develop chronic heart failure following a heart attack or a severe stress. The conditions that lead to **chronic, or congestive, heart failure (CHF)** cause the heart muscle to work unusually hard. As a result, the heart muscle enlarges and gradually weakens as it strains to supply adequate blood to the tissues.

■■ **CONSEQUENCES OF CHF**   As heart failure progresses, reduced blood flow impairs the function of all organs. Reduced blood flow to the kidneys triggers the retention of fluid, further stressing the heart and compounding the stagnation of fluid in the body. Peripheral, pulmonary, and hepatic edema may develop as the person becomes increasingly "congested" with excess fluids. Pulmonary edema increases the likelihood of respiratory infections, which can further stress the heart and lungs.

■■ **CHF AND NUTRITION STATUS**   As CHF progresses, energy needs increase because organ systems, particularly the heart and lungs, must work extra hard to maintain their functions. At the same time, the disrupted blood flow limits the supply of nutrients and oxygen to the organs and tissues. Repeated respiratory infections can further tax nutrition status. People in the later stages of CHF are often unable to eat enough to meet energy and protein demands; oral intake may be limited due to anorexia, altered taste sensitivity, intolerance to food odors, physical exhaustion, the diet used for treatment (described later), and medications. Weight loss in people with CHF may go unnoticed until it has progressed considerably, because edema masks their underweight condition. Thus, severe protein-energy malnutrition (PEM) is a frequent consequence, particularly in the later stages of the disease. Malnutrition can further contribute to the weakness of heart muscle and lungs and the development of respiratory infections.

THE NURSING DIAGNOSES that often apply to people with congestive heart failure include **altered nutrition: less than body requirements, altered tissue perfusion, fluid volume excess, decreased cardiac output, ineffective airway clearance, ineffective breathing pattern,** and **risk for infection.**

■■ **TREATMENT OF CHF**   Drug therapy for CHF includes diuretics to reduce the fluid volume and cardiac glycosides to increase the strength of heart muscle contractions. People taking thiazide or loop diuretics and cardiac glycosides are likely to develop potassium deficiency and may be prescribed a potassium supplement as well. Stool softeners may be prescribed, particularly for elderly clients who frequently experience constipation, because straining to empty the bowels can stress the heart. Initially, bed rest helps reduce the heart's

chronic or **congestive heart failure (CHF):** a syndrome in which the heart can no longer adequately pump blood through the circulatory system.

Enlargement of the heart is called **cardiomegaly (CAR-dee-oh-MEG-ah-lee).**
   cardio = heart
   mega = large

Chronic PEM that develops as a consequence of heart failure is called **cardiac cachexia (ka-KEKS-ee-ah).** *Research suggests that cytokines play a role in the development of PEM in the late stages of CHF.*
   kakos = bad
   hexia = condition

workload. Once recovery is under way, the person must rest frequently and avoid overexertion.

 MEDICAL NUTRITION THERAPY The person newly diagnosed with CHF who is overweight benefits from a safe weight-loss program. Medical nutrition therapy for the wasting associated with later stages of CHF aims to preserve or restore nutrition status and reduce the work of the heart. Providing adequate energy is vital, but providing too much energy increases the body's metabolic rate, stressing the heart. Likewise, giving too much fluid and sodium expands the body's fluid volume, which also taxes the heart.

A diet to help normalize blood lipids and blood pressure is also appropriate. Sodium may sometimes be restricted to 2 grams a day. Dietary fiber is carefully adjusted: the goal is to provide some fiber to prevent constipation, but to avoid amounts and types of fibers that produce gas and abdominal distention. When clients are unable to eat adequate amounts of table foods, formulas of high-nutrient density, which provide energy and protein with less fluid, may help clients meet their nutrient needs.

Strokes

Temporary interference with blood flow to the brain may result in a **transient ischemic attack (TIA),** a condition that causes changes in mental status that may last for a few minutes or a few hours. People who experience a TIA may or may not develop a total blockage of blood flow to a portion of the brain that results in a **stroke,** or cerebral vascular accident (CVA). Most strokes occur as a consequence of atherosclerosis, hypertension, or a combination of the two.

 COMPLICATIONS AFFECTING FOOD INTAKE For some stroke victims, recovery is unremarkable; others may require months of rehabilitative therapy. Victims may suffer from temporary or permanent problems that interfere with the ability to communicate, and sometimes they develop dysphagia (see Chapter 15). The inability to communicate effectively makes it difficult for clients to tell health care professionals about foods they can or would like to eat or about problems they may be having with swallowing foods.

Following a stroke, some people develop physical problems that interfere with their ability to prepare foods or with the physical process of eating. Such problems can include an inability to grasp utensils or coordinate movements that bring foods or liquids from the table to the mouth. Nutrition in Practice 15 described ways to handle such problems.

THE NURSING DIAGNOSES that may apply to people following a stroke include **altered cerebral tissue perfusion, ineffective airway clearance, risk for aspiration, impaired physical mobility,** and **feeding self-care deficit.**

 MEDICAL NUTRITION THERAPY Clients with significant problems with chewing and swallowing may require tube feedings until they have recovered from the stroke and a speech pathologist determines which foods they can safely chew and swallow. Long-term diet therapy for stroke victims depends on the underlying medical condition but often includes a diet to help lower blood lipids and blood pressure. Food energy may need to be limited due to inactivity. A client who must relearn how to walk or use other muscle groups can best do so if not overweight. Underweight can also hinder physical rehabilitation, another reason to ensure that the person does not become malnourished. The accompanying case study presents a client with CHD. Carefully consider the questions posed to review the information presented in this chapter.

---

transient ischemic attack (TIA): a temporary reduction in blood flow to the brain that causes temporary symptoms that depend on the part of the brain that is affected. Some common symptoms include light-headedness, visual disturbances, paralysis, staggering, numbness, or dysphagia.

stroke: an event in which the blood flow to a part of the brain is cut off; also called a **cerebral vascular accident (CVA).**

cerebro = brain
vascular = blood vessels

Reminder: Dysphagia *refers to an inability to coordinate swallowing movements appropriately.*

Case Study

HISTORY PROFESSOR WITH CARDIOVASCULAR DISEASE

Mr. Jablonski, a 48-year-old history professor, has a blood lipid profile that includes elevated LDL cholesterol. He is 5 feet 7 inches tall and weighs 200 pounds. Mr. Jablonski has a family history of CHD. His diet history shows excessive intakes of food energy, cholesterol, total fat, saturated fat, and salt. He smokes a pack of cigarettes a day, and his lifestyle leaves him little time for physical activity. Mr. Jablonski also has hypertension, for which diuretics have been prescribed. He frequently forgets to take his pills, though, and his blood pressure is often quite high.

➕ Name the risk factors for CHD and hypertension in Mr. Jablonski's history. Which of them can he control? Which can be helped by diet? What complications might you expect if his condition goes untreated?

➕ What type of diet, if any, would you recommend to treat Mr. Jablonski's high LDL cholesterol? Explain the rationale for each diet change. How will his current diet change? Prepare a day's menus for Mr. Jablonski. What type of diet, if any, would you recommend for Mr. Jablonski's hypertension? What suggestions might you offer to help him make the necessary diet changes? Are these diet changes consistent with those you would recommend for high LDL cholesterol?

➕ What laboratory and clinical tests would you expect to see monitored regularly? Why?

➕ Name at least three ways in which Mr. Jablonski could benefit from losing weight. How does physical activity fit into a weight-loss plan? Describe other ways Mr. Jablonski might benefit from a physical activity program.

➕ Discuss nutrition considerations if Mr. Jablonski should suffer a heart attack or stroke. Describe the relationships of these disorders to elevated blood lipids and hypertension.

▪▪ Disorders of the Lungs

The heart and lungs work together to ensure the delivery of oxygen to the cells of the body and eliminate the carbon dioxide generated by metabolic processes. Lung disorders can be acute or chronic; both can affect nutrition status.

▪▪ Acute Respiratory Failure

Recall from Nutrition in Practice 19 that the lungs are often the first organs to fail in people following a severe stress or in those with sepsis. With hypermetabolism, the body avidly consumes oxygen and produces carbon dioxide, and the lungs must work hard to keep the body's gases in balance. At the same time, the inflammatory responses to stress increase the permeability of the lungs' **alveoli,** fluids accumulate in the lungs' interstitial spaces, and pulmonary edema may develop. If the problem becomes severe enough, the lungs lose their elasticity and ability to exchange gases, resulting in **respiratory failure.** Inflammatory responses can also lead to acute respiratory failure in people with severe allergic reactions or acute pancreatitis.

Acute respiratory failure can also occur when an embolus lodges in the lungs or when toxic substances such as gastric contents (aspiration pneumonia), smoke, or some inhaled toxic drug directly damage the lung tissue. Premature infants may develop acute respiratory failure because their lungs have not fully developed.

▪▪ **CONSEQUENCES OF ACUTE RESPIRATORY FAILURE** To compensate for altered gas exchange, the person breathes faster, the heart rate increases, breathing

alveoli (al-VEE-oh-lie): air sacs in the lungs. One sac is an *alveolus.*

respiratory failure: failure of the lungs to exchange gases. In **acute respiratory failure,** the lungs fail over a short period of time.

The cluster of symptoms associated with acute respiratory failure is called the **acute respiratory distress syndrome (ARDS).**

becomes labored, and the person becomes restless. Lack of oxygen in the blood causes the person's skin to become pale and develop a bluish tint. If respiratory failure progresses to a severe stage, mental status diminishes, and the person may lapse into a coma. The weakened lungs may fail to push oxygenated blood to the heart, which may also fail. The combination of poor respiration, poor circulation, and poor local nutrition makes respiratory infections more likely. Clients who recover from respiratory failure, however, often regain normal lung function.

> *Lack of oxygen in the blood is called hypoxemia (hie-pox-EE-meh-uh), and the bluish discoloration of the skin caused by a lack of oxygen is called cyanosis (sigh-ah-NO-sis).*

TREATMENT OF ACUTE RESPIRATORY FAILURE   Treatment focuses on correcting the underlying disorder and preventing the progression of acute respiratory failure. The person often requires the use of a **mechanical ventilator.** Intravenous solutions are used to correct fluid and electrolyte and acid-base imbalances. Diuretics help mobilize fluids.

mechanical ventilator: a machine that "breathes" for the person who can't. In normal respiration, the lungs expand, which draws air into the lungs. With mechanical ventilation, air is forced into the lungs at regular intervals using pressure.

MEDICAL NUTRITION THERAPY   Medical nutrition therapy aims to provide enough energy and protein to support lung function and prevent infections without overtaxing the compromised respiratory system. Fluid and sodium restrictions may be necessary to ease pulmonary edema. Overfeeding generates excess carbon dioxide, so total kcalories must be carefully controlled. Glucose metabolism generates more carbon dioxide than does fat metabolism, but this effect is of lesser importance when total energy needs are not exceeded.[20] Actual energy needs are highly variable. Ideally, clinicians use indirect calorimetry to determine the ratio of carbon dioxide produced to oxygen consumed; then this ratio can be used to estimate total energy needs and the appropriate mix of carbohydrate and fat to meet those needs. Alternatively, clinicians use estimates such as the Harris-Benedict equation to predict energy needs (see Table 19-3 on p. 454) and prevent overfeeding.

People recovering from respiratory failure who are being weaned from mechanical ventilators also have high energy and protein needs. The person's lungs were weak before mechanical ventilation began, and they become weaker with disuse. As mechanical ventilation decreases, the lungs must do more work—a stressful and energy-consuming process.

TUBE FEEDINGS AND PARENTERAL NUTRITION   Often people with acute respiratory failure are unable to meet their nutrient needs with an oral diet. Tube feedings are preferred to parenteral nutrition, and intestinal feedings are preferred to gastric feedings for people in respiratory failure because they reduce the risk of aspiration. Special nutrient-dense pulmonary formulas that provide less carbohydrate and more fat are available, but other high-nutrient formulas can also meet nutrient needs.[21]

Chronic Obstructive Pulmonary Disease (COPD)

chronic obstructive pulmonary disease (COPD): one of several disorders, including emphysema and bronchitis, that interfere with respiration.

emphysema (EM-feh-SEE-mah): a type of COPD in which the lungs lose their elasticity and the victim has difficulty breathing; often occurs along with bronchitis.

bronchitis (bron-KYE-tis): inflammation of the lungs' air passages.

Chronic obstructive pulmonary disease (COPD) is a term that describes several conditions characterized by persistent obstruction of airflow through the lungs. Unlike acute respiratory failure, which may occur suddenly, lung failure resulting from COPD is gradual. The two major types of COPD are **emphysema** and chronic **bronchitis.** COPD ranks as the fourth leading cause of death in the United States.[22] Smoking is a primary risk factor for COPD, and most people with COPD are smokers. Other risk factors include exposure to environmental pollution (including exposure of nonsmokers to cigarette smoke) and, possibly, repeated respiratory tract infections.

Experts believe that COPD may be largely preventable by encouraging people not to smoke and controlling environmental pollution. Although COPD has no cure, people with COPD who smoke can help preserve lung function if they quit smoking before the disorder has progressed too far.

■■ **CONSEQUENCES OF COPD**   Regardless of the type of COPD, the lungs gradually lose their functional surface area and strength, making it difficult to deliver oxygen to the blood and to remove carbon dioxide from it. As lung function becomes increasingly compromised, pulmonary infections, respiratory failure, and heart failure can follow.

THE NURSING DIAGNOSES that apply to people with COPD include **altered nutrition: less than body requirements, fatigue, impaired gas exchange, ineffective airway clearance, ineffective breathing pattern,** and **risk for aspiration.**

Health care professionals urge clients not to smoke to reduce the incidence of COPD or preserve lung function in those who already have the disorder.

■■ **COPD AND NUTRITION STATUS**   People with COPD frequently experience weight loss, PEM, and infections and account for many cases of malnutrition in hospitals. The extent of malnutrition is associated with the severity of the pulmonary disease. Body weight is a predictor of survival in people with COPD, and low body weight is associated with reduced muscle mass, respiratory function, and immune competence.[23] Weight loss, which may be rapid and dramatic, may occur for many reasons, including the following:

- Anorexia and poor food intake. These may be caused by depression and anxiety; chronic mouth breathing, which can alter tastes for foods; difficulty breathing while preparing and eating food; and gastric discomfort from swallowing air while eating.[24]
- High-energy expenditures associated with labored breathing.
- Medications, including anti-inflammatory agents, diuretics, and antibiotics, which alter nutrient requirements and compromise nutrition status.
- Use of oxygen masks, because people cannot eat while using them.
- Repeated infections, which raise nutrient needs and deplete nutrient stores. Poor nutrition status, in turn, opens the way for further infection, a vicious cycle.

Cytokines are believed to play a role in the development of PEM associated with COPD.

■■ **TREATMENT OF COPD**   The main treatment for COPD is to stop smoking. People with COPD are encouraged to receive vaccinations to prevent influenza and pneumonia because infections significantly impair lung function. Likewise, clients are often encouraged to take antibiotics at the first sign of bacterial infection. Medications for COPD frequently include bronchodilators to limit respiratory muscle spasms and corticosteroids to reduce inflammation of lung tissue. People with severe COPD may need long-term oxygen therapy to help relieve symptoms.

■■ **MEDICAL NUTRITION THERAPY**   Like people with CHF, people newly diagnosed with COPD or those with mild cases who are overweight benefit from safe weight-loss programs. Excessive body weight increases the work of the lungs in maintaining respiration. As the disease progresses, unintentional weight loss is more likely to occur. At this stage, maintaining or repleting nutrient stores helps to maintain lung function and prevent lung infections. Malnourished clients replete nutrient stores best when refed gradually, with a goal of providing a high-kcalorie, high-protein diet without stressing the lungs. Overfeeding people with COPD and PEM, however, can be as harmful as underfeeding them. Overfeeding and excessive carbohydrate intakes produce high carbon dioxide levels, which the already stressed lungs must work hard to expel.

Meeting energy and protein needs from easy-to-eat foods provided in frequent small meals works best. When appetite is poor, clients may benefit from enteral formulas provided either orally or by tube. Pulmonary formulas that provide more kcalories from fat and fewer from carbohydrate than standard formulas are available, but there is little evidence to suggest that pulmonary formulas are superior to standard enteral formulas for managing COPD.[25]

Chronic diseases of the heart and lungs gradually lead to loss of organ function and can eventually progress to acute events that are fatal. Health care professionals seek to identify people at risk for heart and lung disorders and encourage clients to modify behaviors known to increase that risk. The Nutrition Assessment Checklist helps to identify nutrition-related factors that may help to prevent or treat people with heart and lung disorders.

NUTRITION ASSESSMENT CHECKLIST — DISORDERS OF THE HEART, BLOOD VESSELS, AND LUNGS

☐ Health Problems, Signs, and Symptoms

Check the medical record for a medical diagnosis of:

☐ Coronary heart disease
☐ Hypertension
☐ Congestive heart failure
☐ Acute respiratory failure
☐ COPD

Review the medical record for complications related to CVD and COPD:

☐ Heart attacks
☐ Stroke
☐ Weight loss related to congestive heart failure or COPD

Note risk factors for CHD related to diet, including:

☐ Elevated LDL cholesterol
☐ Obesity or overweight
☐ Diabetes

Medications

For people who are using drug therapy for cardiovascular and respiratory diseases, note:

☐ Side effects that may alter food intake
☐ Potential for folate and vitamin C deficiencies for people taking aspirin
☐ Consistency of vitamin K intake from day to day for people taking warfarin
☐ Consistency of use of dong quai, ginko, cayenne, feverfew, and garlic for people taking anticoagulants
☐ Intake of potassium (including potassium from salt substitutes) for people taking ACE inhibitors, cardiac glycosides, and diuretics
☐ Intake of fat-soluble vitamins for people taking cholestyramine
☐ Use of alcohol and licorice for people taking antihypertensive agents

☐ Use of grapefruit juice for people taking calcium channel blockers and statins
☐ Use of calcium- and magnesium-containing antacids and supplements for people taking fosinopril and cardiac glycosides
☐ Use of hawthorn for people taking cardiac glycosides

Nutrient/Food Intake

For all clients, especially those with CHD or hypertension, or those with risk factors for CHD or hypertension, assess the diet for:

☐ Total energy
☐ Total fat and sources of saturated fat, monounsaturated fat, polyunsaturated fat, and cholesterol
☐ *Trans*-fatty acids
☐ Fiber
☐ Salt, potassium, calcium, and magnesium
☐ Folate and B vitamins
☐ Alcohol

Considerations for clients with complications of the heart or lungs include:

☐ Energy and nutrient intake for clients in the later stages of congestive heart failure or COPD or those with acute respiratory failure
☐ Physical disabilities that interfere with clients' abilities to prepare and eat foods and/or dysphagia following a stroke

Height and Weight

Measure baseline height and weight and reassess weight at each medical checkup.

Note whether clients are meeting weight goals, including:

☐ Weight loss for clients who are overweight, especially those who have or are at risk for CHD, hypertension, congestive heart failure, or COPD
☐ Weight maintenance for people in the later stages of congestive heart failure or COPD

Remember that weight may be deceptively high in people who are retaining fluids, especially those in the later stages of :

☐ Congestive heart failure
☐ COPD

Laboratory Tests

Monitor the following laboratory tests for people with cardiovascular diseases or those at risk for CHD and hypertension:

☐ Blood lipids
☐ Blood glucose for people with diabetes
☐ Blood potassium for people taking diuretics and/or cardiac glycosides
☐ Indicators of fluid retention for people with heart or respiratory failure

Physical Signs

Blood pressure measurement is routine in physical exams, but is especially important for people:

☐ With cardiovascular diseases or following a heart attack or stroke
☐ At risk for CHD or hypertension

Look for physical signs of:

☐ Folate and vitamin C deficiencies in people taking aspirin
☐ Potassium imbalances (muscle weakness, numbness and tingling sensations, irregular heartbeats) in people taking diuretics or cardiac glycosides
☐ Fat-soluble vitamin deficiencies in people taking cholestyramine
☐ Fluid overload in people in the later stages of congestive heart failure, acute respiratory failure, and COPD

SELF CHECK

SELF CHECK

1. Modifiable risk factors for CHD include:

   a. increasing age and stress.
   b. heredity, smoking, and salt intake.
   c. obesity, diabetes, and elevated LDL.
   d. hypertension and low dietary cholesterol.

2. Which of the following statements is true?

   a. Low blood pressure reduces life expectancy.
   b. The wider the diameter of the arteries, the greater the blood pressure.
   c. Hypertension aggravates atherosclerosis, but atherosclerosis does not aggravate hypertension.
   d. Obesity, central obesity, and insulin resistance are associated with the development of hypertension.

3. The nutrition-related strategy that appears to be most effective and long lasting in lowering blood pressure is:

   a. weight control.
   b. total fat restriction.
   c. saturated fat restriction.
   d. potassium supplementation.

4. A health care professional working with a client with elevated LDL recommends which of the following diet modifications?

   a. low total fat, low saturated fat, and low fiber
   b. low total fat, low saturated fat, and low cholesterol
   c. low total fat, low saturated fat, and high cholesterol
   d. low total fat, low monounsaturated fat, and high carbohydrate

5. The Step II diet differs from the Step I diet in its recommendations for:

   a. weight loss.
   b. complex carbohydrates.
   c. saturated fat and cholesterol.
   d. total fat, monounsaturated fat, and cholesterol.

6. Diet-related components that may be protective against CHD include:

   a. protein and vitamin K.
   b. simple sugars and zinc.
   c. saturated fats and *trans*-fatty acids.
   d. fiber, B vitamins, and antioxidants.

7. Which of the following statements about alcohol and cardiovascular diseases might a health care professional relay to a client who wishes to start using alcohol to lower the risks of CHD?

   a. Alcohol lowers blood pressure.
   b. Alcohol intake should not exceed more than four drinks a day.
   c. A reduced risk of CHD is observed at all levels of alcohol intake.
   d. Risks associated with alcohol use may outweigh its benefits in protecting against CHD.

8. For people receiving drug therapy for either hyperlipidemia or hypertension:

   a. physical activity programs must be restricted.
   b. the risk of diet-medication interactions is high.
   c. diet restrictions and physical activity programs are no longer effective.
   d. the risk of potassium imbalances is great for people taking anticoagulants.

9. Diet strategies to help a person recover from a myocardial infarction include:

   a. high-fiber foods to prevent constipation.
   b. foods of moderate temperatures to avoid slowing the heart rate.
   c. three larger meals to allow the heart more time to rest between meals.
   d. high-kcalorie, high-protein, low-sodium liquids at first to speed repair of the heart and prevent fluid retention.

10. Factors that frequently contribute to wasting in people with either congestive heart failure or COPD include:

    a. anorexia, medications, and respiratory infections.
    b. dehydration, fat malabsorption, and lactose intolerance.
    c. essential fatty acid deficiencies and potassium imbalances.
    d. vitamin K deficiencies, dysphagia, and low-fiber, low-potassium diets.

Answers to these questions appear in Appendix H.

CLINICAL APPLICATIONS

1. Consider the risk factors for CHD. Describe possible interrelationships among the factors. For example, a female over age 55 is also at higher risk of diabetes; a person with diabetes is more likely to have hypertension.

2. Look at the DASH diet shown in Table 21-4. As the chapter described, the DASH diet is consistent with the Step I and Step II diets, and it also provides plenty of fiber. Suggest ways that the person following the DASH diet might accomplish these additional dietary modifications:

   • Eat a higher percentage of the total fat from monounsaturated sources.
   • Reduce intake of *trans*-fatty acids.
   • Include omega-3 fatty acids.

NUTRIT**ION**THE**NET**

FOR FURTHER STUDY OF THE
TOPICS IN THIS CHAPTER,
ACCESS THESE WEB SITES.

www.americanheart.org
American Heart Association

www.hsf.ca
Heart and Stroke Foundation of Canada

www.nhlbi.nih.gov
National Heart, Lung, and Blood Institute

www.lungusa.org
American Lung Association

www.lung.ca
Canadian Lung Association

[1]American Heart Association, About women, heart disease, and stroke, www.amhrt.org, site visited on November 19, 1999.

[2]B. Halliwell, Antioxidants and human disease, A general introduction, *Nutrition Reviews* (supplement 2) 55 (1997): 44–52.

[3]As cited in American Heart Association, Cardiovascular diseases, www.amhrt.org, site visited on November 22, 1999.

[4]American Heart Association, Risk factors and coronary heart disease: AHA scientific position, www.amhrt.org, site visited on November 19, 1999.

[5]S. M. Grundy and coauthors, Primary prevention of coronary heart disease: Guidance from Framingham, *Circulation* 97 (1998): 1876–1887.

[6]G. F. Fletcher and coauthors, Statement on exercise: Benefits and recommendations for physical activity programs for all Americans, *Circulation* 94 (1996): 857–862.

[7]The Expert Panel, Summary of the second report of the National Cholesterol Education Program (NCEP) Expert Panel on Detection, Evaluation, and Treatment of High Blood Pressure in Adults (Adult Treatment Panel II), *Journal of the American Medical Association* 269 (1993): 3015–3023.

[8]L. J. Appel and coauthors, A clinical trial of the effects of dietary patterns on blood pressure, *New England Journal of Medicine* 336 (1997): 1117–1124.

[9]A. Ascherio and coauthors, Marine *n*-3 fatty acids, fish intake, and the risk of coronary heart disease among men, *New England Journal of Medicine* 332 (1995): 977–982.

[10]N. J. Stone, Fish consumption, fish oil, lipids, and coronary heart disease, *Circulation* 94 (1996): 2337–2340.

[11]F. B. Hu and coauthors, Dietary fat intake and the risk of coronary heart disease in women, *New England Journal of Medicine* 21 (1997): 1491–1499.

[12]L. Van Horn, Fiber, lipids, and coronary heart disease, *Circulation* 95 (1997): 2701–2704.

[13]H. Gylling and coauthors, Reduction of serum cholesterol in postmenopausal women with previous myocardial infarction and cholesterol malabsorption induced by dietary sitostanol ester margarine: Women and dietary sitostanol, *Circulation* 96 (1997): 4226–4231; B. V. Howard and coauthors, Phytochemicals and cardiovascular disease: A statement for healthcare professionals from the American Heart Association, *Circulation* 95 (1997): 2591–2593.

[14]E. Arnesen and coauthors, Serum total homocysteine and coronary heart disease, *International Journal of Epidemiology* 24 (1995): 704–709; M. J. Stampfer and coauthors, A prospective study of plasma homocyst(e)ine and risk of myocardial infarction in US physicians, *Journal of the American Medical Association* 268 (1992): 877–880.

[15]M. J. Stampfer and E. B. Rimm, Folate and cardiovascular disease, *Journal of the American Medical Association* 275 (1995): 1929–1930.

[16]M. B. Zemel, Dietary pattern and hypertension: The DASH study, *Nutrition Reviews* 55 (1997): 303–308.

[17]National Institutes of Health, National Heart, Lung, and Blood Institute, The DASH diet, NIH Publication no. 98–4082, September 1998.

[18]P. R. Ridker and coauthors, Association of moderate alcohol consumption and plasma concentrations of endogenous tissue–type plasminogen activator, *Journal of the American Medical Association* 272 (1994): 929–933; J. M. Gaziano and coauthors, Moderate alcohol intake, increased levels of high-density lipoproteins and its subfractions, and decreased risk of myocardial infarction, *New England Journal of Medicine* 329 (1993): 1829–1834.

[19]H. O. Hein, P. Suadicani, and F. Gyntelberg, Alcohol consumption, serum low density lipoprotein cholesterol concentration and risk of ischaemic heart disease: Six year follow up in the Copenhagen male study, *British Journal of Medicine* 312 (1996): 731–736; C. S. Fuchs and coauthors, Alcohol consumption and mortality among women, *New England Journal of Medicine* 332 (1995): 1245–1250.

[20]A. M. Malone, Is a pulmonary enteral formula warranted for patients with pulmonary dysfunction? *Nutrition in Clinical Practice* 12 (1997): 168–171.

[21]Malone, 1997.

[22]T. L. Petty and G. G. Weinmann, Building a national strategy for the prevention and management of research in chronic obstructive pulmonary disease, *Journal of the American Medical Association* 277 (1997): 246–253.

[23]K. Gray-Donald and coauthors, Nutritional status and mortality in chronic obstructive pulmonary disease, *American Journal of Respiratory and Critical Care Medicine* 153 (1996): 961–966.

[24]K. M. Chapman and L. Winter, COPD: Using nutrition to prevent respiratory function decline, *Geriatrics* 51 (1996): 37–42.

[25]Malone, 1997.

Q&A NUTRITION IN PRACTICE

Free Radicals, Antioxidants, and CHD

Health-conscious consumers have eagerly been swept up in the advertising campaign for antioxidants, which has made claims of preventing cardiovascular diseases, cancer, and aging. Each bit of fresh news about antioxidants convinces more people to use antioxidant supplements, yet few consumers truly understand what antioxidants are or how they work. The purpose of this Nutrition in Practice is to explain how antioxidants work and how one notable antioxidant, vitamin E, might function in preventing CHD. Such a discussion necessarily includes an explanation of free-radical formation—the process that causes the damage that the antioxidant nutrients protect against.

 www.americanheart.org
American Heart Association

 www.hsf.ca
Heart and Stroke Foundation of Canada

 www.nhlbi.nih.gov
National Heart, Lung, and Blood Institute

What are free radicals?

Many types of free radicals exist, but the ones that are most important in the human body derive from oxygen. Oxygen is basic to life, and yet, ironically, it can also be toxic.[1] Most of the oxygen we breathe in is used by the body's cells to produce energy. A small amount of oxygen, however, is not used to produce energy, but instead reacts with electrons generated during energy production. This oxygen forms free radicals—molecules with unpaired electrons. A molecule with an unpaired electron is chemically unstable and highly reactive; a free radical quickly reacts with a stable molecule to capture an electron and return to a stable state. In the process, the stable molecule loses an electron and becomes a free radical itself, and a chain reaction is set in motion that damages other molecules.

In addition to being generated during metabolism, free radicals can also form in the body from exposure to sunlight, ozone, radiation, tobacco smoke, and environmental pollution. Sometimes the immune system intentionally generates free radicals as a means of killing or inactivating bacteria and viruses. With free radicals being generated by these various sources, virtually all of the body's cells are under constant attack from free radicals.

How do the cells manage to survive such an environment?

The body would not survive without defense mechanisms to keep free-radical formation in check. Enzymes that convert free radicals to less toxic substances provide one line of defense. Many of these enzymes depend on nutrient cofactors, including copper, zinc, manganese, and selenium. In addition to the action of enzymes, the body relies on nutrients, such as beta-carotene, vitamin C, and vitamin E, and a variety of nonnutrients to act as antioxidants.

How can free radicals harm the body if the body's defenses are countering their effects?

The defense systems keep free-radical formation in check, but free-radical damage is not completely prevented. The term *oxidative stress* describes the serious imbalance between free-radical formation and antioxidant defenses when the defenses are overwhelmed.

Sometimes cells under oxidative stress adapt, and mild oxidative stress may actually make cells more resistant to oxidative damage. But if the oxidative stress is severe or prolonged, damage is self-aggravating. Free radicals can damage lipid and protein molecules in cell membranes and even DNA.[2] As a result, cells may be injured or lose their function, and mutations can arise in DNA. Tissue damage activates the immune system (see Chapter 19), which then generates more free radicals. In addition, when cells break down, they release metals (mainly iron and copper) that may catalyze chemical reactions that generate free radicals. When this occurs, the damage becomes progressive. Aging is associated with both an increased production of free radicals and the accumulation of unrepaired damage from free radicals. Thus, free-radical formation has been implicated as a factor in aging and in the development of chronic diseases including CHD and cancer, inflammatory diseases (such as Crohn's disease and arthritis), and degenerative diseases (such as Alzheimer's and Parkinson's diseases).

Vegetable oils and margarines are the most commonly eaten sources of vitamin E—the antioxidant most closely related to a reduced risk of CHD.

How exactly do free radicals relate to CHD?

As Chapter 21 describes, high LDL cholesterol is a major risk factor for CHD. Data suggest that the oxidation of LDL by free radicals promotes the adhesion of certain immune system cells (macrophages) to the artery walls and exposes receptors that allow the macrophages to fill with oxidized LDL.[3] The immune system cells undergo changes and eventually become the fatty materials characteristic of plaques. Oxidized LDL also promote clot formation and damage the blood vessel walls, which allows the entry of other plaque-forming factors.

Is there any way to prevent oxidative stress?

This is perhaps the million dollar question. With free radicals implicated in so many diseases, researchers are busy trying to find an answer to this question. Not surprisingly, many investigations focus on antioxidant nutrients. With respect to cardiovascular disease, vitamin E has shown the strongest correlation. Epidemiological studies suggest a negative correlation between vitamin E status and death rates from heart disease.[4] High death rates from heart disease correlate more strongly with low vitamin E concentrations than with either cholesterol or blood pressure. Although the evidence is suggestive of a protective effect of vitamin E in heart disease, it is by no means conclusive; other factors or a combination of factors may be responsible for the findings.

In the body, vitamin E is carried along with cholesterol in LDL. Some studies suggest that vitamin E may slow the progression of plaques that have already formed in the arteries or may lower the risk of developing a heart attack in people with existing CHD.[5] To date there has been no carefully designed study to compare the effects of vitamin E and a placebo in altering complications or death rates from CHD.[6]

Is there any harm in taking vitamin E to prevent CHD until more is known?

It is certainly safe to eat a diet rich in natural sources of vitamin E (vegetable oils, polyunsaturated margarines, some nuts, and wheat germ). However, the amount of vitamin E that appears to be protective against CHD is difficult to consume from a typical diet, particularly from a diet that obtains less than 30 percent of its energy from fat. Therefore, vitamin E supplements may be necessary to exert a protective effect on CHD risk. When provided in large amounts, vitamin E may have pharmacological, rather than nutritional, effects. The impact of taking a single antioxidant for many years is unknown. Such a practice could possibly interfere with normal physiological processes, perhaps immune defenses.[7] Until further studies verify whether vitamin E can protect against heart disease and determine what dose is effective and safe, most authorities do not recommend vitamin E supplements.

Another safe strategy is to include a wide variety of fruits and vegetables—rich sources of folate, other B vitamins, phytochemicals, and fiber. Folate and other B vitamins can lower homocysteine levels; elevated homocysteine stimulates the oxidation of LDL. Although studies have not confirmed that lowering homocysteine levels by eating adequate amounts of folate and other B vitamins reduces CHD risk, including fruits and vegetables clearly presents no health risks.

Much remains to be discovered about cardiovascular diseases and the many factors that might prevent or delay their development. Health care professionals have the responsibility for staying abreast of new developments so that they can help clients balance fact and fiction and thus protect their health.

Notes

[1] B. Halliwell, Antioxidants and human disease: A general introduction, *Nutrition Reviews* (supplement 2) 55 (1997): 44–52.

[2] V. M. Sardesai, Role of antioxidants in health maintenance, *Nutrition in Clinical Practice* 10 (1995): 19–25.

[3] B. Frei, Reactive oxygen species and antioxidant vitamins: Mechanisms of action, *American Journal of Medicine* (supplement 3A) 97 (1994): 5–13.

[4] K. F. Gey and coauthors, Increased risk of cardiovascular disease at suboptimal plasma concentrations of essential antioxidants: An epidemiological update with special attention to carotene and vitamin C, *American Journal of Clinical Nutrition* 57 (1993): 333–336; K. F. Gey and coauthors, Inverse correlation between plasma vitamin E and mortality from ischemic heart disease in cross-cultural epidemiology, *American Journal of Clinical Nutrition* 53 (1991): 326–334.

[5] N. G. Stephens and coauthors, Randomised controlled trial of vitamin E in patients with coronary disease: Cambridge Heart Antioxidant Study (CHAOS), *Lancet* 347 (1996): 781–786; H. N. Hodis and coauthors, Serial coronary angiographic evidence that antioxidant vitamin intake reduces progression of coronary artery atherosclerosis, *Journal of the American Medical Association* 273 (1995): 1849–1854.

[6] P. O. Kwiterovich, The effect of dietary fat, antioxidants, and pro-oxidants on blood lipids, lipoproteins, and atherosclerosis, *Journal of the American Dietetic Association* (supplement) 97 (1997): 31–41.

[7] Sardesai, 1995.

CHAPTER 22

NUTRITION AND RENAL DISEASES

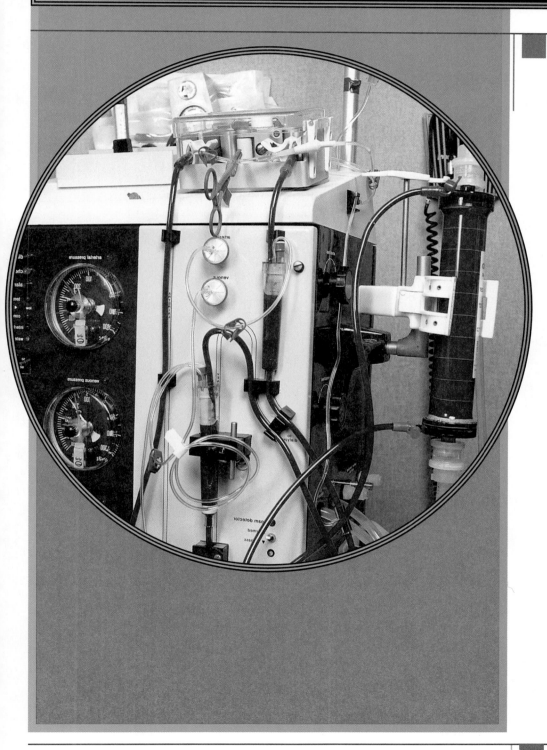

CONTENTS

Kidney Stones

The Nephrotic Syndrome

Acute Renal Failure

Chronic Renal Failure

Case Study: Store Manager with Acute Renal Failure

Nutrition in Practice: Dialysis and Nutrition

glomerulus (glow-MARE-you-lus): a cup-shaped membrane enclosing a tuft of capillaries within a nephron. (The plural is *glomeruli*.)

nephrons (NEF-rons): the working units of the kidneys. Each nephron consists of a glomerulus and a tubule.

tubule: a tubelike structure that surrounds the glomerulus and descends through the nephron. A pressure gradient between the glomerular capillaries and the tubule returns needed materials to the blood and moves wastes into the tubule to be excreted in the urine.

The kidneys certainly illustrate the adage that good things come in small packages. Each bean-shaped kidney is only about the size of a fist, yet the kidneys carry much of the responsibility for maintaining the body's chemical balance—and that's only one of their roles. Figure 22-1 shows the location of the kidneys and the structure of a **nephron**—one of the kidneys' functional units. The **glomerulus** serves as a gate through which fluids and other blood components enter the nephron and pass through the **tubule** to be cleansed and returned to the body. Healthy kidneys, by continuously filtering the blood, maintain the body's fluid and electrolyte and acid-base balances and eliminate metabolic waste products. They also secrete an enzyme, **renin,** that helps regulate blood pressure and produce the hormone **erythropoietin,** which stimulates red blood cell production (see glossary on p. 526). Additionally, the kidneys convert vitamin D to its most active form, and they play an important role in maintaining healthy bone tissue.

Figure 22-1

THE URINARY TRACT, A KIDNEY, AND A NEPHRON

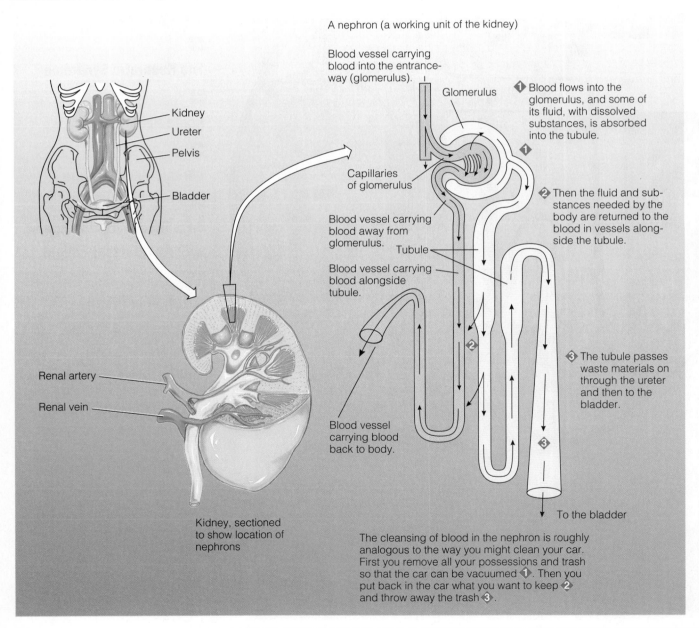

A nephron (a working unit of the kidney)

Blood vessel carrying blood into the entrance-way (glomerulus).

Glomerulus

❶ Blood flows into the glomerulus, and some of its fluid, with dissolved substances, is absorbed into the tubule.

Capillaries of glomerulus

Blood vessel carrying blood away from glomerulus.

❷ Then the fluid and substances needed by the body are returned to the blood in vessels alongside the tubule.

Tubule

Blood vessel carrying blood alongside tubule.

❸ The tubule passes waste materials on through the ureter and then to the bladder.

Kidney

Ureter

Pelvis

Bladder

Renal artery

Renal vein

Blood vessel carrying blood back to body.

To the bladder

Kidney, sectioned to show location of nephrons

The cleansing of blood in the nephron is roughly analogous to the way you might clean your car. First you remove all your possessions and trash so that the car can be vacuumed ❶. Then you put back in the car what you want to keep ❷ and throw away the trash ❸.

Kidney Stones

Kidney stones are the most common disorder that affects the kidneys and urinary tract. About 500,000 people in the United States develop kidney stones each year; most of them are men over 20 years old. Kidney stones are often quite painful, although sometimes they produce no symptoms. Stones often recur, but recurrences may be preventable.

Kidney stones develop when stone constituents become concentrated in the urine and form crystals that grow. The stone constituents vary, but more than three-fourths of them contain calcium as calcium oxalate, calcium phosphate, or a combination of calcium, oxalate, and phosphate. Stones that contain calcium oxalate or a combination of calcium oxalate and phosphate are by far the most common.

Some people with calcium stones excrete normal amounts of calcium in the urine, and others excrete excess calcium (**hypercalciuria).** The reason people with normal amounts of calcium in their urine form calcium stones is unclear, although they may lack factors that normally inhibit stone formation. People with hypercalciuria are either more efficient at absorbing calcium from the intestine or more wasteful in their excretion of calcium than most people.

Less commonly, stones are composed of uric acid, the amino acid cystine, or magnesium ammonium phosphate (known as struvite). Struvite stones are associated with recurring urinary tract infections and are more common in women. Table 22-1 lists some disorders associated with kidney stone formation.

LEARNING LINK

Earlier chapters described two conditions associated with kidney stones. Chapter 13 discussed how immobilization leads to negative calcium balance, which raises the amount of calcium excreted in the urine, thus favoring the formation of stones. Chapter 16 described how fat malabsorption syndromes sometimes lead to kidney stones by raising urinary oxalate levels.

Consequences of Kidney Stones

In most cases, kidney stones pose no serious medical problems, especially when they are few and small. Small stones (less than one-fifth of an inch in diameter) may readily pass through the ureters and out of the body via the urine with minimal treatment (review Figure 22-1).

THE NURSING DIAGNOSES risk for infection, altered urinary elimination, and **pain** may apply to people with kidney stones.

RENAL COLIC
Large stones cannot pass easily through the ureters. When a large stone enters a ureter, it produces a sharp, stabbing pain, called **renal colic.** Typically, the pain starts suddenly in the back and intensifies as the stone follows the ureter's course down the abdomen toward the groin. The intense pain is often accompanied by nausea and vomiting. When the stone reaches the bladder, the pain subsides abruptly.

URINARY TRACT COMPLICATIONS
Symptoms associated with kidney stones may include frequent urination, urgency of urination, **dysuria,** and **hematuria.** Large stones that fail to pass through the ureter may cause a urinary tract obstruction or infection and serious bleeding.

Kidney stones are also called **renal calculi.** *Stones that have passed from the kidney to the ureter are technically urethral stones, but most people use the term* **kidney stone** *to describe stones anywhere in the urinary tract.*
renal = kidney

hypercalciuria (HIGH-per-kal-see-YOU-ree-ah): excessive urinary excretion of calcium.

renal colic: the severe pain that accompanies the movement of a kidney stone through the ureter to the bladder.

dysuria (dis-YOU-ree-ah): painful or difficult urination.

hematuria (HE-mah-TOO-ree-ah): blood in the urine.

Table 22-1
CONDITIONS ASSOCIATED WITH KIDNEY STONES

Cystinuria
Fat malabsorption
Glucocorticoid excess
Gout
Hyperparathyroidism
Hyperthyroidism
Immobilization (prolonged bed rest or paralysis)
Malignancies (some types)
Osteoporosis
Paget's disease
Recurrent urinary tract infections
Renal tubular acidosis
Vitamin D toxicity

Prevention and Treatment of Kidney Stones

Clients with a small kidney stone may pass the stone more readily if they drink plenty of fluids (more than 3 liters a day). They may also need anti-infective agents if an infection is present. People who excrete too much calcium sometimes need thiazide diuretics to help reduce the amount of calcium in the urine. People who absorb too much calcium may benefit from medications (cellulose sodium phosphate) that prevent the absorption of calcium in the intestine (see the Diet Medication Interactions box on p. 536). Uric acid stones are often treated with allopurinol, a medication that reduces urinary uric acid concentrations. Large stones that block the flow of urine or cause an infection require removal either surgically or, more commonly by using shock waves to break the stone into pieces small enough to pass through the urinary tract.

Preventing urinary tract infections is an important strategy for preventing struvite stones. Studies suggest that cranberry juice may help prevent urinary tract infections in women by preventing bacteria from adhering to the inner lining of the urinary tract.[1]

Treatment of the underlying medical condition is necessary to help prevent recurrences of kidney stones. Dietary measures vary according to the composition of the stone, but all advice to prevent stones includes this recommendation: increase fluid intake to dilute the urine. People who have had kidney stones need to drink enough fluid (mostly water) to produce a urine volume of at least 2 liters per day. Generally, this requires a total intake of about 3 to 4 liters of fluid throughout the day. People who are physically active or live in warm climates may need additional fluids. People with fevers, diarrhea, or vomiting also need additional fluids until these conditions resolve. Medical nutrition therapy for specific types of kidney stones is described next.

CALCIUM AND OXALATE STONES Medical nutrition therapy for people with hypercalciuria includes the recommendation to avoid excessive calcium intakes. Instead, calcium is modestly limited to the recommended adequate intake appropriate for age and sex. It is important that calcium intakes not fall too low, however. Those with hypercalciuria who follow a low-calcium diet generally excrete more calcium than they ingest, indicating that they are losing calcium from their bones. Calcium restriction is not appropriate in the treatment of calcium stones unrelated to hypercalciuria, including stones related to fat malabsorption (see Chapter 16).

All people with calcium oxalate stones are advised to limit their intakes of foods high in oxalate—spinach, rhubarb, beets, nuts, chocolate, tea, wheat bran, and strawberries. **Hyperoxaluria** increases the likelihood of calcium oxalate stone formation more than hypercalciuria does. Because megadoses of vitamin C over long times raise urinary oxalate concentrations, people at risk for oxalate stones are advised to avoid vitamin C supplements.

Results of the Nurses Health Study (see Chapter 21) suggest that, for women, a high intake of calcium from foods may actually lower the risk of developing kidney stones. Using calcium supplements, however, especially if they are taken without meals or with foods low in oxalate, may raise the risk of kidney stones.[2] Whether it is the calcium from foods or some other constituent in the calcium-containing food that provides protection remains unknown.

In addition to calcium and oxalate, attention to salt intake may also be important in preventing calcium oxalate stones. Excess salt increases urinary calcium excretion in all people, but causes a proportionately greater amount of calcium to be excreted in people with hypercalciuria.[3] For people taking thiazide diuretics, moderate salt restriction helps limit the urinary excretion of both potassium and calcium. Further research is necessary to determine whether moderate salt restriction can prevent calcium stones from forming.

hyperoxaluria (HIGH-per-OX-all-YOU-ree-ah): excessive urinary excretion of oxalate.

Reminder: When calcium and oxalate are eaten together, they bind together and both are excreted, which may benefit the person who forms calcium oxalate stones.

URIC ACID STONES Uric acid stones are frequently associated with gout, a metabolic disorder characterized by elevated levels of uric acid in the blood and urine. Uric acid stones form when the urine becomes persistently acid, contains excessive uric acid, or both. Purine-restricted diets are commonly prescribed to prevent uric acid stones. A purine-restricted diet limits red meats, particularly organ meats, anchovies, sardines, and meat extracts. The benefits of such a diet are unproven, but avoiding excessive protein may be useful, and foods high in purines are also high in protein. Alcohol intake is also limited.

CYSTINE STONES Cystine stones form when an inherited disorder of amino acid metabolism (cystinuria) causes the abnormal excretion of cystine in the urine. Because the body makes cystine from the amino acid methionine, a diet restricted in methionine-rich foods benefits the person with cystine stones. Medications to reduce urinary acidity may also be beneficial.

Although kidney stones do not alter kidney function directly, they arise in the kidneys and can lead to complications that affect kidney function. The conditions described in the remainder of this chapter are characterized by changes in renal function.

The Nephrotic Syndrome

The nephrotic syndrome is not a disease, but rather a distinct cluster of symptoms caused by dysfunction of the glomerular capillaries. The symptoms include the loss of plasma proteins into the urine, low serum albumin, edema, and elevated blood lipids. Damage to the kidneys from diabetic nephropathy, infections, blood clots in the renal veins, and some drugs and toxins can precipitate the nephrotic syndrome. The nephrotic syndrome is sometimes an early sign of renal failure, especially in people with diabetes (see Chapter 20). In other cases, treatment of the underlying condition can correct the disorder.

Consequences of the Nephrotic Syndrome

The consequences of the nephrotic syndrome include protein-energy malnutrition (PEM), anemia, infection, and accelerated atherosclerosis. Many of the consequences of this disorder are the same as those of malnutrition, which is not surprising because both alter protein status. If the nephrotic syndrome progresses to renal failure, the person develops uremia and other manifestations, as a later section describes.

> **THE NURSING DIAGNOSES altered nutrition: less than body requirements, risk for infection, altered tissue perfusion,** and **fluid volume excess** may apply to people with the nephrotic syndrome.

LOSS OF BLOOD PROTEINS As plasma proteins escape through the urine, blood proteins plummet. Figure 22-2 on p. 524 shows the consequences of protein losses.

EDEMA While plasma proteins are lost as a consequence of the nephrotic syndrome, sodium is reabsorbed by the nephrons in greater amounts than normal.[4] Sodium retention results in edema.

ALTERED BLOOD LIPIDS Elevated cholesterol, triglycerides, LDL, and VLDL and low HDL are characteristic of the nephrotic syndrome. Blood

Figure 22-2
CONSEQUENCES OF URINARY PROTEIN LOSSES IN THE NEPHROTIC SYNDROME

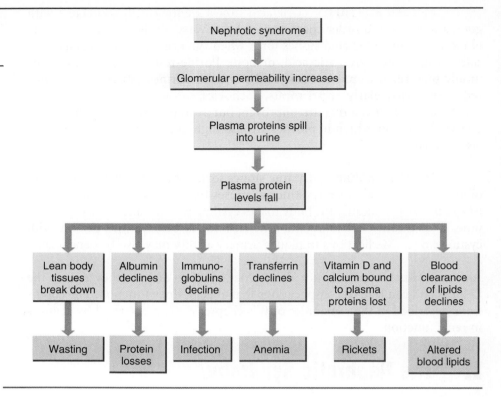

coagulation disorders are also common, and people with the nephrotic syndrome frequently develop blood clots. All of these factors combine to raise the risk of cardiovascular disease and stroke in people with the nephrotic syndrome.

▦ Treatment of the Nephrotic Syndrome

Medical treatment of the nephrotic syndrome first requires treatment of the underlying disorder and then drug and medical nutrition therapy to resolve symptoms. Medications may include anti-infective agents, anticoagulants, antihypertensives, anti-inflammatory agents, antilipemics, and diuretics (see the Diet-Medication Interactions box on p. 536). Diet is central to preventing PEM and alleviating edema.

▦ **ENERGY** A diet adequate in energy (about 35 kcalories per kilogram of body weight per day) sustains weight and spares protein. Weight loss or infections signal the need for additional kcalories. People who are obese may benefit from lowering their energy intakes to help control blood lipid levels.

▦ **PROTEIN** In the past, high-protein diets (about 120 grams per day) were often prescribed for people with the nephrotic syndrome, stemming from the belief that high dietary protein would compensate for protein losses. High-protein diets accelerate protein synthesis, but they also incur greater urinary protein losses.[5] Furthermore, extra dietary protein may accelerate deterioration of renal function. For these reasons, protein is provided at about the RDA (0.8 to 1.0 gram per kilogram of body weight per day).

▦ **FAT** A diet low in fat, saturated fat, and cholesterol, such as the plans described in Chapter 21, can help control the elevated blood lipids associated with the nephrotic syndrome. Often, however, people with the nephrotic syndrome are unable to control blood lipids adequately using diet alone, and physicians prescribe antilipemic agents as well.

SODIUM Because a person with the nephrotic syndrome avidly retains sodium, sodium is generally restricted. Although the level of sodium restriction varies depending on the individual's response to diuretics, the diet generally provides about 2 grams of sodium (see Table 22-2). Clients placed on thiazide or loop diuretics should be encouraged to select foods rich in potassium (see p. 205).

Acute Renal Failure

The kidneys may fail suddenly or deteriorate gradually. In acute **renal failure,** the nephrons suddenly lose function and are unable to maintain homeostasis. The degree of renal dysfunction varies from mild to severe. With prompt treatment, acute renal failure may be reversible in some cases, but in others, the damage is permanent.

Table 22-2
TWO-GRAM SODIUM-RESTRICTED DIET

Foods Restricted	Number of Servings Daily	Serving Size	Sodium per Serving (mg)
Regular breads and cooked cereals	4	1 slice bread or ½ c cereal	125
Fresh, frozen, or canned vegetables without salt: artichokes; beets; carrots; celery; beet, collard, dandelion, mustard, and turnip greens; kale; swiss chard; white turnips; low-sodium vegetable juice	Avoid excessive use	½ cup	50
Canned or frozen vegetables with salt; frozen corn, lima beans, mixed vegetables, and peas	2	½ cup	250
Regular fat-free, whole, and evaporated milk and milk products	2	8 oz	120
Fresh and fresh frozen meats, poultry, and freshwater fish; low-sodium canned meats and fish, peanut butter, cheese; unsalted soybeans, textured vegetable protein, and cottage cheese	8	1 oz	25
Eggs	1	1	70
Regular butter and margarine	6	1 tsp	50

Foods Allowed

1. Low-sodium breads, bread products, and cereals; bread products made without salt and with low-sodium baking powder; puffed rice and wheat and shredded wheat cereals; rice; pasta.
2. Fresh, unsalted frozen, and low-sodium canned vegetables (except those listed above); low-sodium tomato juice.
3. All fruits and fruit juices.
4. Unsalted butter, margarine, nuts, and gravy; low-sodium salad dressings and mayonnaise; shortening.
5. Low-sodium catsup, mustard, and tabasco sauce.
6. Soups, casseroles, and recipes made with allowed foods and food ingredients.

Foods Not Allowed (Unless Calculated into the Diet)

1. Table, celery, garlic, and onion salts; reduced-sodium salts; regular catsup, mustard, and tabasco sauce; monosodium glutamate; Worcestershire, barbeque, and soy sauces; baking powder and soda.
2. Instant and quick-cooking hot cereals; commercial bread products made from self-rising flour or cornmeal, salt, baking powder, or baking soda; salted snack foods such as potato chips, corn chips, tortilla chips, popcorn, and pretzels.
3. Sauerkraut, pickles, and salted vegetable juices.
4. Maraschino cherries; crystallized or glazed fruits, and dried fruits with sodium sulfite added.
5. Buttermilk, chocolate milk, instant milk mixes, regular cheeses, and prepared pudding mixes; commercial ice cream, sherbet, and frozen desserts.
6. Cured, canned, salted, or smoked meats, poultry, and fish such as bacon, luncheon meats, corned beef, kosher meats, and canned tuna and salmon; imitation fish products; salted textured vegetable protein; regular peanut butter; salted nuts.
7. Salt pork and bacon fat; commercial salad dressings and mayonnaise; olives; regular gravy.
8. Regular canned soups and bouillon.

Reduced blood flow to the kidneys is a prerenal cause of acute renal failure. A urinary tract obstruction, which can be a consequence of a kidney stone, is a postrenal cause of acute renal failure. Damage to the kidneys' cells is an intrarenal cause of acute renal failure.

pre = before
post = after
intra = within

Acute renal failure frequently develops when blood flow to the kidneys suddenly declines, often as a result of a severe stress such as heart failure, shock, or severe blood loss after surgery or trauma. In these cases, acute renal failure is imposed on catabolic illness, and mortality rates are high. When acute renal failure develops for other reasons, such as urinary tract obstructions, recovery rates are much better. Infections, toxins, and some drugs can directly damage the kidneys' cells and lead to acute renal failure.

Consequences of Acute Renal Failure

As renal function falters, the composition of the blood and urine changes. The rate at which the kidneys form **filtrate,** the **glomerular filtration rate (GFR),** declines, upsetting fluid and electrolyte and acid-base balances. A sudden and precipitous drop in the GFR and urine output characterizes acute renal failure. As the nephrons fail, the body's principal nitrogen-containing metabolic waste products—blood urea nitrogen (BUN), creatinine, and uric acid—accumulate in the blood (**uremia**). In a person who is catabolic, nitrogen balance is negative and large amounts of nitrogen are excreted. It is easy to see how uremia and its clinical manifestations—the **uremic syndrome**—can quickly develop. (The accompanying glossary defines terms related to renal function.)

> **THE NURSING DIAGNOSES altered nutrition: less than body requirements, risk for infection, altered tissue perfusion, fluid volume excess, fluid volume deficit, decreased cardiac output, altered oral mucous membrane** (see uremic syndrome in glossary), and **anxiety** may apply to people with acute renal failure.

ELECTROLYTE LEVELS RISE As blood electrolyte levels rise, serious consequences can follow. In a person who is catabolic, electrolyte levels can rise rapidly as the body's cells use protein for energy and release potassium,

---

GLOSSARY
OF TERMS RELATED TO RENAL FUNCTION

continuous renal replacement therapy (CRRT): a slow and continuous type of dialysis used in the treatment of acute renal failure.

dialysis (dye-AL-ih-sis): removal of waste from the blood through a semipermeable membrane using the principles of simple diffusion and osmosis. The two main types of dialysis are hemodialysis and peritoneal dialysis (see Nutrition in Practice 22).

erythropoietin (eh-RITH-row-POY-eh-tin): a hormone secreted by the kidneys in response to oxygen depletion or anemia that stimulates the bone marrow to produce red blood cells.
erthro = red
poi = to make

glomerular filtration rate (GFR): the rate at which the kidneys form filtrate. Normally, the GFR is about 120 ml/min.

Falling GFR signifies deteriorating kidney function. When the GFR falls drastically, to about 10 ml/min, dialysis is usually initiated.

renal failure: failure of the kidneys to maintain normal function to a degree that requires dialysis or kidney transplantation for survival.

renal filtrate: the fluid that passes from the blood through the capillary walls of the glomeruli, eventually forming urine.

renal insufficiency: reduced renal function but not to the degree that requires dialysis or kidney transplantation.

renin (REN-in): an enzyme secreted by the kidneys in response to a reduced blood flow that triggers the release of the hormone aldosterone from the adrenal glands. Aldosterone, in turn, signals the kidneys to retain sodium and fluid.

uremia (you-REE-me-ah): abnormal accumulation of nitrogen-containing substances, especially urea, in the blood; also called **azotemia (AZE-oh-TEE-me-ah).** Normal BUN (blood urea nitrogen) levels are 10 to 20 mg/dL. A BUN of 50 to 150 mg/dL indicates serious impairment of renal function. BUN may rise to as high as 150 to 250 mg/dL in end-stage renal disease.

uremic syndrome: the many symptoms that may accompany the buildup of nitrogen-containing waste products in the blood including fatigue; diminished mental alertness; agitation; muscle twitches and cramps; anorexia, nausea, vomiting; inflammation of the mucous membranes of the mouth; unpleasant tastes in the mouth; dry, scaly, itchy skin; skin hemorrhages; gastritis and GI bleeding; and diarrhea.

phosphorus, and magnesium in the process. Elevated potassium (hyperkalemia) is of particular concern because potassium imbalances can alter the heart rate and lead to heart failure.

BLOOD VOLUME CHANGES In the early stages of acute renal failure, blood pressure may rise dangerously high as fluids accumulate. Later in the course of acute renal failure, the kidneys cannot conserve water, and the person begins to excrete large amounts of fluid and electrolytes. If recovery occurs, kidney function gradually normalizes.

Treatment of Acute Renal Failure

The primary goal in acute renal failure is to treat the underlying cause in order to prevent permanent or further damage to the kidneys. For example, if severe blood loss is the problem, a blood transfusion may be needed to restore blood volume. A combination of medical nutrition therapy, drug therapy, and **dialysis** may be undertaken to restore fluid and electrolyte balances and to minimize blood concentrations of toxic waste products. (Nutrition in Practice 22 describes dialysis and **continuous renal replacement therapy,** a type of dialysis used specifically in the treatment of acute renal failure, and discusses how nutrient needs are affected by various dialysis procedures.)

ENERGY If the person in acute renal failure receives too little energy, body proteins break down, raising blood urea and potassium even higher and taxing the kidneys further. Thus, wasting and malnutrition complicate recovery. The person's energy needs depend on the rate of catabolism, so indirect calorimetry provides the best estimate of energy needs.[6] When indirect calorimetry is not available, clinicians use formulas to estimate energy needs as described in Chapter 19 (see p. 454).

PROTEIN Although protein contributes nitrogen and can tax the kidneys, protein is necessary to prevent complications such as impaired wound healing, infections, muscle wasting, and negative nitrogen balance—complications that could prove fatal. Accordingly, clinicians often provide a liberal protein intake, even if it means the person will need dialysis. Actual protein needs of people with acute renal failure depend on the degree of renal function, the metabolic rate, and nutrition status. People who are not on dialysis receive about 0.6 to 1.0 gram of protein per kilogram of body weight per day. Those on dialysis may have a more liberal protein intake because nitrogenous waste products are removed during the procedure.[7]

FLUIDS Fluid balances are carefully restored in clients who are either overhydrated or dehydrated. Thereafter, health care professionals determine fluid needs by measuring urine output and then adding about 500 milliliters for water lost through the skin, lungs, and perspiration. More fluids are needed if the person is vomiting; has diarrhea, a high fever, or a high-output fistula; or is otherwise losing fluids. In the oliguric stage, the person needs small amounts of fluids. In the diuretic stage, urine volume may increase significantly, and large amounts of fluids may have to be provided.

ELECTROLYTES Sodium may be restricted in the oliguric phase, but this may change as the person enters the diuretic phase. Generally, potassium, phosphorus, and magnesium are also restricted; lab values must be monitored carefully, however, especially in those who are catabolic. Once the catabolic period of stress subsides, potassium, phosphorus, and magnesium may shift back into the intracellular fluid, and blood levels can plummet.

The early phase of acute renal failure, when urine volume is reduced, is the **oliguric phase.** *The phase characterized by large fluid and electrolyte losses in the urine is the* **diuretic phase.** *The gradual return of renal function marks the* **recovery phase.**

■■ **ENTERAL AND PARENTERAL NUTRITION** Because clients with acute renal failure are often severely stressed, they frequently receive their nutrients from tube feedings or TPN. Special enteral and parenteral formulas for renal failure meet energy and protein needs in small volumes. Compared with standard enteral formulas, renal formulas have less protein, fewer electrolytes, and more kcalories per milliliter (see Appendix G). TPN formulas are compounded with mixtures of both nonessential and essential amino acids at lower concentrations and dextrose at higher concentrations than standard TPN solutions. Electrolytes are added in appropriate amounts.

The carbohydrate-dense enteral and parenteral formulas used for acute renal failure may exacerbate the hyperglycemia and insulin resistance that frequently accompany both severe stress and renal failure. Fat can be used to add kcalories, reduce the need for carbohydrate, and limit the glucose load. Insulin may be provided to lower blood glucose.

■■ **DRUG THERAPY** In the oliguric phase of acute renal failure, diuretics may be used to mobilize fluids. Medications called exchange resins may be used to treat hyperkalemia. These medications, provided by mouth or through an enema, cause an exchange of sodium for potassium in the colon, and the potassium is then excreted in the stool.

As mentioned, insulin may be provided to help lower blood glucose. Insulin also lowers blood potassium and phosphorus in two ways. First, as insulin moves glucose into the cells, potassium and phosphorus follow. Second, as an anabolic hormone, insulin minimizes tissue breakdown, and consequently, the cells retain potassium and phosphorus. The Diet-Medication Interactions box on p. 536 provides more information about nutrition and medications used in the treatment of renal diseases. The accompanying case study helps direct your thoughts toward the needs of a client with acute renal failure.

■■ Chronic Renal Failure

Unlike acute renal failure, in which the GFR drops suddenly and sharply, in chronic renal failure the GFR gradually deteriorates, and renal function is irre-

Case Study

STORE MANAGER WITH ACUTE RENAL FAILURE

Mrs. Calley is a 35-year-old store manager admitted to the hospital's intensive care unit. She was first seen in the emergency room after she sustained multiple and severe injuries in an auto accident. She had lost so much blood she almost died before reaching the hospital. Her injuries include a fractured leg, broken ribs, a collapsed lung, and internal bleeding. Following emergency surgery to stop the internal bleeding and repair injuries, she developed acute renal failure. Mrs. Calley is 5 feet 3 inches tall and weighs 125 pounds.

Mrs. Calley has a urine volume of less than 50 ml/day and a BUN of 75 mg/dL. A test of GFR could not be performed due to the low volume of urine she was excreting.

✚ Describe the most probable reason why Mrs. Calley developed acute renal failure. What other problems can cause acute renal failure? Describe the phases of acute renal failure.

✚ What are Mrs. Calley's dietary needs in the early phase? What waste products and electrolytes are of greatest concern? Why? What factors do you have to keep in mind in determining Mrs. Calley's energy and protein needs during acute renal failure? How will these needs change if dialysis is begun?

✚ How will Mrs. Calley's nutrient needs change as she progresses to the second stage of acute renal failure? Why? As you read through the discussion of chronic renal failure, consider how Mrs. Calley's needs would change if her renal failure became chronic.

versibly altered. Chronic renal failure develops from conditions that progressively and permanently damage the kidneys, notably certain infections, renal artery obstructions, diabetic nephropathy (Chapter 20), hypertension (Chapter 21), atherosclerosis (Chapter 21), and congestive heart failure (Chapter 21). Recall that some people who develop the nephrotic syndrome eventually develop chronic renal failure. In a few cases, chronic renal failure results from a disorder that rapidly causes irreversible kidney damage, as may occur following acute renal failure.

▪▪ Consequences of Chronic Renal Failure

In the early stages of chronic renal failure, the body compensates for the loss of some nephron function by enlarging the remaining functional nephrons. The hypertrophied nephrons work so efficiently that the GFR may fall to 75 percent of its normal rate before the symptoms of renal failure appear.

The capacity of the kidneys to function despite loss of some nephrons is referred to as **renal reserve.**

The body eventually exhausts the overworked nephrons, and renal function deteriorates further. (This effort is similar to that of the pancreatic beta cells in type 2 diabetes, which at first produce more and more insulin in response to high blood glucose and later become exhausted and unable to produce adequate insulin.) In **end-stage renal disease,** the GFR drops below 20 percent of normal.

end-stage renal disease: the severe stage of renal failure in which dialysis or a kidney transplant is necessary to sustain life.

THE NURSING DIAGNOSES altered nutrition: less than body requirements, risk for infection, altered tissue perfusion, fluid volume excess, impaired tissue integrity, altered oral mucous membrane, altered sexuality patterns, altered growth and development, and **altered thought processes** may apply to the person with chronic renal failure.

▪▪ **UREMIC SYNDROME** As renal function deteriorates, nitrogen-containing waste products accumulate in the blood, and the uremic syndrome (see p. 526) develops. In the later stages, urea (which can be excreted through the sweat) may crystallize on the skin, a symptom known as **uremic frost.** Nausea, vomiting, diarrhea, gastritis, and GI bleeding frequently accompany the uremic syndrome.

Reminder: The nitrogen-containing waste products that accumulate in renal failure include blood urea nitrogen (BUN), creatinine, and uric acid.

uremic frost: the appearance of urea crystals on the skin.

▪▪ **BLOOD CHEMISTRY CHANGES** In addition to nitrogen-containing waste products, the body retains excess fluids, electrolytes, and acids produced during metabolism and normally excreted in the urine. As a consequence, the person develops edema and acidosis, which stress the cardiovascular and pulmonary systems. Elevated blood potassium can trigger irregular heartbeats that further stress the heart and can cause it to fail. Elevated phosphorus upsets the body's balance of phosphorus and calcium, often leading to bone disease.

Reminder: Acidosis is the condition of having too much acid in the blood.

Hormonal balances are also upset. Altered hormone levels together with the retention of waste products lead to hypertension, hyperglycemia, elevated blood lipids (especially triglycerides), low serum albumin, anemia, and altered bone metabolism.

Bone disorders resulting from calcium and phosphorus imbalances in renal disease are called **renal osteodystrophies (OS-tee-oh-DIS-tro-fees).**

▪▪ **OTHER COMPLICATIONS** Accelerated atherosclerosis and cardiovascular diseases frequently accompany renal failure, and they can aggravate hypertension and raise the risk of congestive heart failure, heart attacks, and pulmonary edema. Anemia may result from depressed erythropoietin synthesis by the damaged kidneys, restrictive diets, nausea and vomiting, GI blood losses, and blood losses through dialysis and from frequent blood testing.

▪▪ **GROWTH FAILURE AND WASTING** Both children and adults with chronic renal disease frequently develop wasting and PEM. Nutrition status

Table 22-3
POSSIBLE CAUSES OF WASTING IN RENAL FAILURE

Reduced Nutrient Intake	Excessive Nutrient Losses	Raised Nutrient Needs
Anorexia	Dialysis	Hormonal alterations
Fatigue	Diarrhea	Infection
Medications	GI bleeding	Medications
Nausea	Medications	
Pain	Numerous blood tests	
Restrictive diet	Poor absorption	
Taste alterations	Vomiting	

becomes more difficult to maintain as renal disease progresses. Table 22-3 summarizes the causes of wasting associated with renal failure. Children with renal disease need nutrition intervention before the end of puberty if they are to make up growth deficits. Adults with renal disease can maintain or restore their nutrition status, avoid complications, and improve quality of life by attending to diet as well.

■■ Treatment of Chronic Renal Failure

Treatment of chronic renal failure includes diet, medications, dialysis, and, sometimes, kidney transplants. Medications may include antiemetics, antihypertensives, antilipemics, antisecretory agents, cardiac glycosides, diuretics, insulin, and phosphate binders.

Table 22-4 summarizes nutrient needs for chronic renal failure and shows how hemodialysis and peritoneal dialysis affect them. The complexity of the renal diet, as well as its critical role in the treatment of renal disease, underscores the need for a specialist, a renal dietitian, to educate clients and provide

Continuous renal replacement therapy is not used in the treatment of chronic renal failure.

Table 22-4
NUTRIENT NEEDS IN CHRONIC RENAL FAILURE

Nutrients[a]	Renal Insufficiency (Predialysis)	Hemodialysis	Peritoneal Dialysis
Energy (kcal/kg)	35–40	30–35	25–35
Protein (g/kg)	0.6–0.8	1.2–1.4	1.2–1.5
Fluid (ml)	Typically not restricted	500–750 plus daily urine output, or 1000 if anuric	≥2000
Sodium (g)	2–4	2–3	2–4
Potassium (g/kg)	Typically not restricted	3–4	Typically not restricted
Phosphorus (mg/g protein)	10–12[b]	12–15[b]	12–15[b]
Supplements			
Calcium (mg)	1000–1500	1000–1500	1000–1500
Folate (mg)	1	1	1
Vitamin B$_6$ (mg)	5	10	10
Vitamin D	As appropriate	As appropriate	As appropriate

NOTE: The actual amounts of these nutrients in the diet must be highly individualized based on each person's responses. For example, calcium supplementation may be as high as 3000 milligrams per day.
[a]Besides the specific nutrients listed, all others should meet recommended amounts.
[b]The extent of phosphorus restriction depends on serum phosphorus. The goal is to maintain serum phosphorus between 4.5 and 6.0 milligrams per deciliter. Often, phosphate binders are useful for this purpose.
SOURCES: Adapted from Meeting the challenge of the renal diet: A preview of the "National Renal Diet" educational series, *Journal of the American Dietetic Association* 93 (1993): 637–639; J. A. Beto, Which diet for which renal failure: Making sense of the options, *Journal of the American Dietetic Association* 95 (1995): 898–903.

diet plans. Nurses do not need to know the specifics of medical nutrition therapy, but they must understand the general concepts in order to communicate effectively with clients.

▪▪ ENERGY All people with renal failure need adequate energy (see Table 22-4) to maintain a desirable weight and prevent protein catabolism. For children, an intake of at least 100 kcalories per kilogram of body weight is desirable, but 80 kcalories per kilogram of body weight is considered reasonable. At the minimum, energy intake should meet the RDA.[8]

▪▪ PROTEIN Providing the right amount of protein to the person in renal failure is like walking a tightrope. Given too little protein, the person develops malnutrition. Given too much protein, blood urea (the toxic waste product of protein metabolism) rises. For people with renal insufficiency, restricting protein may help protect the remaining nephrons, although human studies have not clearly demonstrated a beneficial effect.[9] Clinicians generally recommend a protein-restricted diet that derives about two-thirds of its protein from sources such as eggs, milk, meat, poultry, and fish. The remaining protein is derived from plant sources. A diet that includes protein from both animal and plant sources may be best because it combines the high quality of animal proteins with the low–saturated fat, low-cholesterol nature of plant proteins.

As renal failure progresses, some clinicians prescribe very-low-protein diets supplemented with essential amino acids or essential amino acid precursors (keto acids). Once the client begins dialysis, protein restrictions can be relaxed somewhat because dialysis removes nitrogenous waste products and incurs protein losses (see this Chapter's Nutrition in Practice).

> To evaluate dialysis and guide diet therapy, renal teams may use a mathematical model that takes into account the kidneys' ability to clear urea and the person's protein catabolic rate. The technique is called **urea kinetic modeling**.

▪▪ LIPID AND CARBOHYDRATE The ideal renal diet restricts total fat, saturated fat, and cholesterol to help control elevated blood lipid levels. A diet rich in complex carbohydrates helps minimize elevated blood glucose and triglycerides. People on peritoneal dialysis (see Nutrition in Practice 22) may need to further restrict their intake of total carbohydrate, and especially simple carbohydrates, because they absorb a considerable amount of glucose as a consequence of the dialysis procedure.

▪▪ SODIUM AND FLUIDS As renal failure progresses, the person excretes less urine and cannot handle normal amounts of sodium and fluids. At this point, limiting sodium and fluids helps to prevent hypertension, edema, and heart failure. Individual needs for sodium and fluids are determined by carefully monitoring each person's weight, blood pressure, urine output, and blood electrolyte levels. A rapid rise in body weight and blood pressure suggests that the person is retaining sodium and fluid; conversely, a rapid decline in body weight and blood pressure (a desirable outcome of dialysis) indicates fluid loss.

Fluids are not restricted in renal insufficiency until urine output decreases. For the person who is neither dehydrated nor overhydrated, daily fluid needs amount to the daily urine output plus 500 to 750 milliliters to provide for insensible water losses. Once a person is on dialysis, sodium and fluid intakes are controlled to allow a weight gain of about 2 pounds (of fluid) between dialysis treatments, although larger weight gains are common.[10] The margin lists foods considered part of the fluid allowance.

> Foods and other substances that are considered part of the fluid allowance on fluid-restricted diets include:
> - Cream.
> - Frozen yogurt.
> - Fruit ice.
> - Gelatin.
> - Ice.
> - Ice cream.
> - Ice milk.
> - Liquid medications.
> - Popsicles.
> - Sherbet.
> - Soup.

▪▪ POTASSIUM Most people with renal insufficiency and those on peritoneal dialysis can handle typical intakes of potassium. People on hemodialysis, however, may develop hyperkalemia between dialysis treatments. In such cases, potassium may be moderately restricted to about 2 to 3 grams per day. Remember, however, that individual needs vary. People taking thiazide and loop diuretics may need to adjust their potassium intakes accordingly.

> Caution: People with renal failure who must restrict potassium should avoid salt substitutes that contain potassium and products made from such salt substitutes.

PHOSPHORUS, CALCIUM, AND VITAMIN D

Controlling blood phosphorus and calcium levels in renal failure may help slow the progression of the disease and help prevent bone diseases. As renal function declines, blood phosphorus levels rise and active vitamin D levels fall. The low amount of active vitamin D limits calcium absorption, and the combination of high blood phosphorus and limited availability of calcium leads to the loss of calcium from the bones and consequent bone disorders.

Dietary phosphorus restrictions help control rising blood phosphorus. Fortunately, when a client follows a protein-restricted diet, phosphorus is restricted as well. In addition, the person must limit foods high in phosphorus.

Physicians may also prescribe medications that bind phosphorus in the GI tract and make it unavailable for absorption. Calcium and aluminum salts and, most recently, a binder that contains neither calcium nor aluminum can be used to bind phosphorus. The use of aluminum-containing phosphate binders is discouraged because aluminum toxicity is a problem for many people on dialysis.

In addition to restricting phosphorus, most people with renal disease need calcium supplements. Supplemental vitamin D in its active form can help maintain blood calcium and prevent bone disease. Some people, however, develop hypercalcemia. The renal team monitors serum calcium closely to prevent both low and high blood calcium and prescribes calcium and vitamin D supplements accordingly. Calcium-containing phosphate binders provide some calcium, but the absorption of calcium from these products varies widely. For people who tend to develop hypercalcemia, the newer phosphate binders that do not contain calcium may be useful.

OTHER VITAMINS

People with renal failure frequently develop folate and vitamin B_6 deficiencies because of the restrictive diet, loss of vitamins during dialysis, drug therapy, and altered metabolism. Medical nutrition therapy often includes folate and vitamin B_6 in generous amounts, along with the recommended amounts of the other water-soluble vitamins.[11] Adequate amounts of folate, vitamin B_6, and vitamin B_{12} may also help to lower serum homocysteine levels (see Chapter 21), which may help to protect against cardiovascular disease. Furthermore, these vitamins are important in preventing anemia, a common complication of renal disease.

Because people with renal failure have high blood oxalate levels, intakes of vitamin C from both the diet and supplements are often limited to less than 100 milligrams per day.[12] When oxalate levels are elevated, oxalate crystals can become deposited in soft tissues and result in complications, including kidney stones and heart attacks. Supplementation of fat-soluble vitamins other than vitamin D is usually not necessary.

TRACE MINERALS

The administration of human erythropoietin along with the B vitamins and iron needed to synthesize hemoglobin is effective in treating iron-deficiency anemia, once a common and persistent problem in people with chronic renal disease. Clients should be cautioned to avoid iron supplements that also contain vitamin C.

People on dialysis frequently complain of anorexia, altered taste perceptions (dysgeusia), and a decreased sexual drive, symptoms typical of zinc deficiency. Clients who have these symptoms may need zinc supplements if their serum zinc levels are inadequate. Recall that children have particularly high needs for zinc, and zinc deficiencies can contribute to growth retardation.

ENTERAL AND PARENTERAL NUTRITION

Enteral formulas and parenteral nutrition can provide nutrients to people with chronic renal failure who are unable to eat adequate amounts of foods. Formulas designed for use in renal failure are described on p. 528.

Foods high in phosphorus include milk and milk products, eggs, cheese, peanut butter, bran cereal, sardines, and legumes.

The active form of supplemental vitamin D, or calcitriol, can be given orally or intravenously.

Caution: People with renal failure should avoid aluminum- and magnesium-containing antacids and laxatives or enemas containing magnesium; aluminum toxicity and serious hypermagnesemia can result.

■■ **DIET PLANNING** The ideal renal diet presents a challenge: provide adequate energy, but restrict protein, fat, and sometimes simple carbohydrate. Complex carbohydrates, which might appear to be the ideal energy source, are often also rich sources of fiber, but because fiber absorbs fluids, which are often restricted, complex carbohydrates must be used cautiously. Diet planners must accept that under such circumstances, no diet is truly ideal. They must recognize that the need to adjust protein and electrolytes outweighs the need to restrict fat and simple carbohydrate.

To help meet energy needs, clients include as many complex carbohydrates as their diet plans allow. They supplement their meals with formulas high in kcalories but restricted in protein and electrolytes and use foods such as sugars (hard candy and jelly) and fats (margarine and oil) freely. The person with elevated blood lipids is advised to restrict fat and modify the type of fat to whatever extent is possible. The person with diabetes or hyperglycemia is advised to eat a consistent carbohydrate intake at regular intervals and to adjust insulin to cover carbohydrate intake.

To help individuals on renal diets find foods they will accept and enjoy, food lists similar to the exchange system are available. Whereas the exchange system for diabetes groups foods by their energy, carbohydrate, protein, and fat contents, renal food lists group foods by their energy, protein, sodium, potassium, and phosphorus contents. The "How to" box provides suggestions to ease the task of complying with a renal diet.

HELP CLIENTS COMPLY WITH A RENAL DIET

The following suggestions can assist clients in complying with the renal diet:

1. To keep track of fluid intake:

   - Fill a container with an amount of water equal to your total fluid allowance. Each time you use a liquid food or beverage, discard an equivalent amount of water from the container. The amount remaining in the container will show you how much fluid you have left for the day.
   - Be sure to save enough fluid to take medications.

2. To help control thirst:

   - Chew gum or suck hard candy.
   - Freeze fluids so they take longer to consume.
   - Add lemon juice to water to make it more refreshing.
   - Gargle with refrigerated mouthwash.

3. To prevent the diet from becoming monotonous:

   - Experiment with new combinations of allowed foods.
   - Use favorite foods whenever possible.
   - Substitute nondairy products for regular dairy products. Nondairy products are lower in protein, phosphorus, and potassium than regular dairy foods, and they can substitute for milk and add energy to the diet.
   - Add zest to foods by seasoning with garlic, onion, chili, curry powder, oregano, pepper, or lemon juice.
   - Consult a dietitian when you want to eat restricted foods. Many restricted foods can be used occasionally and in small amounts if the diet is carefully adjusted.

∷ DIET COMPLIANCE The challenges dietitians face in designing renal diets pale in comparison with those encountered by clients and caregivers who must follow a complicated medical plan of which diet is only a part. To help clients understand their medical plans and support their efforts, the members of the health care team must communicate effectively with their clients and, just as important, listen to them. Within a tangled web of grossly altered metabolic processes, dialysis lines, and toxic waste products is a person, often a frightened or discouraged one. All members of the health care team need to understand the plan if they are to offer the most effective support.

A nutrition education approach that emphasizes self-management appears to be a useful strategy for helping clients comply with a renal diet.[13] This approach guides clients in identifying goals, selecting strategies to meet goals, and evaluating progress.

∷ Kidney Transplants

A preferable alternative to dialysis in end-stage renal disease is a kidney transplant. Kidney transplants can successfully restore kidney function and promote normal growth. For this reason, transplants are particularly desirable in children. Given a choice, many would prefer transplants, but suitable kidney donors cannot always be found. Of the more than 35,000 people awaiting transplants, only about 11,000 will receive them.[14]

∷ IMMUNOSUPPRESSIVE DRUG THERAPY After receiving a new kidney, the person must take very large doses of immunosuppressive drugs to prevent rejection (see the Diet-Medication Interactions box on p. 536). Muscular weakness, GI bleeding, protein catabolism, carbohydrate intolerance, sodium retention, fluid retention, hypertension, weight gain, and a characteristic puffy-faced appearance commonly accompany immunosuppressive therapy. Infections and increased susceptibility to malignant tumors are also common. Diuretics are frequently prescribed to promote the excretion of fluid and sodium and to prevent hypertension.

∷ DIET INTERVENTIONS Once recovery is under way, the degree of renal function guides diet therapy. Typical post-transplant diet modifications appear in Table 22-5. Protein is provided in amounts adequate to prevent the protein catabolism that immunosuppressants may incur, but not so much as to tax renal function. Because blood lipids are frequently elevated, clients are advised to follow a fat-modified diet. Sodium restriction helps prevent fluid retention and hypertension. Depending on the type of diuretic prescribed and the person's responses, potassium intake may have to be adjusted.

A person with a kidney transplant may reject the new kidney either temporarily or permanently. During these times, dialysis must be reinstituted, and the person must return to the prescribed diet for renal failure. Clients may find this regression difficult to accept and should be prepared for the possibility before it occurs.

For people with kidney stones, diet strategies may be useful in treating existing stones and preventing recurrences. People with the nephrotic syndrome or renal failure may develop malnutrition as a consequence of the disorder and hasten its progression. Use the Nutrition Assessment Checklist to review assessment findings relevant to people with disorders of the kidneys and urinary tract.

Table 22-5
DIETARY GUIDELINES FOLLOWING A KIDNEY TRANSPLANT

Energy Adequate to achieve or maintain desirable body weight.

Protein 1 gram per kilogram of body weight. (Adjust based on renal function tests.)

Fat ≤30 percent of total kcalories; ≤10% saturated fat; ≤300 milligrams cholesterol.

Sodium 3 to 4 grams per day.

Potassium Adjust according to individual needs.

SOURCE: Adapted from J. A. Beto, Which diet for which renal failure: Making sense of the options, *Journal of the American Dietetic Association* 95 (1995): 898–903.

NUTRITION ASSESSMENT CHECKLIST FOR PEOPLE WITH RENAL AND URINARY TRACT DISORDERS

Health Problems, Signs, and Symptoms

Check the medical record to determine:

☐ Type of kidney stone

☐ Degree of renal function

☐ Cause of nephrotic syndrome or renal failure

☐ Type of dialysis, if appropriate

☐ If client has received a kidney transplant

Review medical record for complications that may alter nutrient needs:

☐ PEM

☐ Severe stress

☐ Infection

☐ Anemia

☐ Diabetes mellitus

☐ Hyperlipidemia

☐ Hypertension

☐ Heart failure (acute or chronic)

Medications

For clients with kidney stones, note diet-medication interactions for medications taken by the client, including:

☐ Anti-infective agents

☐ Diuretics (see Chapter 21)

☐ Exchange resins (cellulose sodium phosphate)

☐ Antigout agents

Note that clients with nephrotic syndrome, renal insufficiency, or renal failure and those who have had a kidney transplant risk medication-related malnutrition for many reasons, including:

☐ Long-term use of medications

☐ Multiple medication use with many of the medications having significant effects on nutrition status

☐ Altered renal function, which compounds nutrition risks

☐ Preexisting malnutrition due to the disorder itself and complications of the disorder

☐ Reduced food intake, altered digestion and absorption, altered metabolism, and the altered excretion of nutrients due to the medications as well as the disorders themselves

For all clients with kidney diseases, note:

☐ Use of over-the-counter medications and supplements that may contain electrolytes that must be controlled

☐ Use of herbs and other remedies, which can have a significant impact on clients who suffer from malnutrition and renal insufficiency or renal failure and who use multiple medications

Nutrient/Food Intake

For people with kidney stones or a past history of kidney stones:

☐ Stress the importance of drinking plenty of fluids regularly throughout the day.

☐ Assess intake of calcium, oxalate, salt, and protein as appropriate for type of stone.

For people who wish to try cranberry juice cocktail to prevent urinary tract infections, explain that:

☐ Studies are limited and they were conducted on older women; but, at worst, cranberry juice cocktail is not harmful.

☐ The amount of cranberry juice cocktail used by study participants was 10 ounces per day.

☐ A 10-ounce serving of regular cranberry juice cocktail has about 180 kcalories, while the same serving of low-kcalorie cranberry juice has about 60 kcalories.

For clients with the nephrotic syndrome, renal insufficiency, or renal failure, or those who have undergone kidney transplants, regularly assess intakes of:

☐ Energy

☐ Protein

☐ Sodium

☐ Potassium

In addition, for clients with renal insufficiency or renal failure, assess intakes of:

☐ Fluid

☐ Phosphorus

☐ Calcium

☐ Vitamins

☐ Minerals

Height and Weight

Take accurate baseline height and weight measurements. Keep in mind that:

☐ Fluid retention in people with the nephrotic syndrome or renal failure can mask malnutrition.

☐ For people on dialysis, the weight measured immediately following a dialysis treatment, called the "dry weight," most accurately reflects the person's true weight.

☐ Rapid weight gain between dialysis treatments often reflects fluid retention. For clients who regularly have problems with fluid retention, review fluid intake to ensure that the client understands and is complying with diet recommendations.

☐ Weight loss is expected and intentional following a dialysis treatment.

Laboratory Tests

Note that serum protein levels are often low in people with the nephrotic syndrome or renal failure. Review the following laboratory test results to assess degree of renal function and response to treatments:

☐ Glomerular filtration rate (GFR)

☐ Creatinine

☐ Blood urea nitrogen (BUN)

☐ Electrolytes

Check laboratory test results for complications associated with renal disease including:

☐ Anemia

☐ Hyperglycemia

☐ Hyperlipidemia

☐ Hyperparathyroidism (bone diseases)

Physical Signs

For clients with the nephrotic syndrome and renal insufficiency or renal failure, look for physical signs of:

☐ Dehydration and fluid retention

☐ Iron deficiency

☐ Uremia

☐ Bone diseases

☐ Hyperkalemia

☐ Zinc deficiencies

DIET-MEDICATION INTERACTIONS

Anticoagulants

For *anticoagulants,* see p. 507.

Antiemetics

For *antiemetics,* see p. 507.

Antigout

Allopurinol and *colchicine* should both be given with meals. Nutrition-related side effects of allopurinol are uncommon, but colchicine can cause nausea, vomiting, GI distress, and diarrhea. Colchicine reduces the absorption of vitamin B_{12}, and the person may need to take vitamin B_{12} supplements.

Antihypertensives

For *antihypertensives,* see p. 507.

Anti-infective agents

Numerous *anti-infective agents* may be used in the treatment of kidney stones, nephrotic syndrome, and acute or chronic renal failure. Consult a current drug guide for specific interactions.

Anti-inflammatory agents

For *anti-inflammatory agents,* see p. 399.

Antilipemics

For *antilipemics,* see p. 507.

Antisecretory agents

For *antisecretory agents,* see p. 371.

Cardiac glycosides

For *cardiac glycosides,* see p. 507.

Diuretics

For *diuretics,* see p. 507.

Exchange resins

Cellulose sodium phosphate (prescribed for people with hypercalciuria) can cause altered tastes, GI distress, and diarrhea and can lead to low blood levels of magnesium. Clients should take the medication with meals and should also take a magnesium supplement at least one hour before or after taking the resin. *Sodium polystyrene* (prescribed to reduce blood potassium levels) can cause anorexia, nausea, vomiting, and constipation and low blood levels of potassium (intentional) and calcium. The powder can be mixed with sorbitol-containing syrup to combat constipation. Calcium-containing antacids or calcium supplements should not be taken within several hours of taking sodium polystyrene.

Immunosuppressants

Immunosuppressants have multiple effects on many organ systems, may be toxic to the kidneys and liver, and can lead to many problems including nausea, vomiting, and a high risk of infections. *Cyclosporine* can elevate blood potassium, and it should not be given with grapefruit or grapefruit juice (unless prescribed). The person taking cyclosporine should not use potassium supplements or salt substitutes containing potassium. *Azathioprine* can also lead to pancreatitis. Lymphocyte immune globulin and muromonab-cd3 can lead to pulmonary edema. See also *anti-inflammatory agents* on p. 399.

Phosphate binders

For *calcium acetate, calcium carbonate, calcium citrate, aluminum carbonate,* and *aluminum hydroxide,* see calcium and aluminum-containing antacids on p. 371. *Sevelamer hydrochloride* does not contain calcium or aluminum. It is given with meals and can cause nausea, flatulence, GI distress, diarrhea, and, less frequently, constipation.

SELF CHECK
SELF CHECK

1. Which of the following is **not** a function of the kidneys?
   a. activation of vitamin K
   b. maintenance of acid-base balance
   c. elimination of metabolic waste products
   d. maintenance of fluid and electrolyte balance

2. Treatment for all kidney stones includes:
   a. a protein-restricted diet.
   b. a methionine-restricted diet.
   c. calcium intake that meets, but does not exceed the DRI.
   d. fluid intake to maintain a urine volume of at least 2 liters a day.

3. People with calcium oxalate stones may benefit from diets that restrict:
   a. oxalate and sodium.
   b. calcium and potassium.
   c. protein and methionine.
   d. calcium and phosphorus.

4. A person with the nephrotic syndrome is prone to infections due to losses of:
   a. albumin.
   b. transferrin.
   c. lean body mass.
   d. immunoglobulins.

5. Diet recommendations for the nephrotic syndrome include:
   a. no diet restrictions.
   b. sodium restrictions.
   c. protein intakes that are less than the RDA.
   d. protein intakes that are 1.5 to 2.0 times the RDA.

6. The electrolytes that may rise rapidly in people with acute renal failure who are catabolic include:
   a. sodium, phosphorus, and calcium.
   b. sodium, potassium, and phosphorus.
   c. potassium, phosphorus, and magnesium.
   d. potassium, phosphorus, and calcium.

7. The health care team estimates fluid requirements for clients with acute renal failure by adding _____ milliliters to the amount of urine output.
   a. 100 to 150
   b. 300 to 450
   c. 500 to 750
   d. 750 to 1000

8. Complications commonly associated with chronic renal failure may include:
   a. growth failure and renal colic.
   b. nausea, vomiting, and reflux esophagitis.
   c. anemia, edema, and potassium deficiencies.
   d. anemia, bone disease, cardiovascular disease, and malnutrition.

9. Which of the following nutrients may be unintentionally restricted when a person follows a renal diet?
   a. fluid
   b. calcium
   c. potassium
   d. phosphorus

10. The nurse recognizes that compared to the person with renal failure who is not on dialysis, the diet of a person on dialysis is:
    a. lower in protein.
    b. higher in protein.
    c. lower in potassium and phosphorus.
    d. higher in potassium and phosphorus.

Answers to these questions appear in Appendix H.

CLINICAL APPLICATIONS

1. Consider that a person with chronic renal failure may need multiple medications to control disease progression and treat symptoms and complications. For people with diabetes and hyperlipidemia who develop renal failure, medications might include insulin, antihypertensives, diuretics, antilipemics, antiemetics (for nausea), antisecretory agents, and phosphate binders. Review the nutrition-related side effects of these medications. Describe the ways that medications can make it harder for people to maintain nutrition status.

2. Using the "How to" box on p. 533 as a guide, give suggestions for helping people adjust to different aspects of their renal diets. Can you think of any other suggestions?

NUTRITI**ON**THENET

FOR FURTHER STUDY OF THE
TOPICS IN THIS CHAPTER,
ACCESS THESE WEB SITES.

www.kidney.org
National Kidney Foundation

www.kidney.ca
Kidney Foundation of Canada

www.niddk.nih.gov
*National Institute of Diabetes and
Digestive and Kidney Diseases*

www.renalnet.org
*Renalnet Kidney Information
Clearinghouse*

www.aakp.org
*American Association of Kidney
Patients*

www.ohf.org
*Oxalosis and Hyperoxaluria
Foundation*

www.afud.org
*American Foundation for Urologic
Disease*

Notes

[1] S. Ahuja, B. Kaack, and J. Roberts, Loss of fimbrial adhesion with the addition of Vaccinum macorcarpon to the growth medium of P-fimbriated Escheria coli, *Journal of Urology* 159 (1998): 559–562; J. Avorn and coauthors, Reduction of bacteriuria and pyuria after ingestion of cranberry juice, *Journal of the American Medical Association* 271 (1994): 751–754.

[2] G. C. Curhan and coauthors, Comparison of dietary calcium with supplemental calcium and other nutrients as factors affecting the risk for kidney stones in women, *Annals of Internal Medicine* 126 (1997): 497–504.

[3] W. J. Burtis and coauthors, Dietary hypercalciuria in patients with calcium oxalate stones, *American Journal of Clinical Nutrition* 60 (1994): 424–429.

[4] S. R. Orth and E. Ritz, The nephrotic syndrome, *New England Journal of Medicine* 338 (1998): 1202–1211.

[5] R. Rodrigo and M. Pino, Proteinuria and albumin homeostasis in the nephrotic syndrome: Effect of dietary protein intake, *Nutrition Reviews* 54 (1996): 337–347.

[6] D. J. Rodriguez and W. M. Sandoval, Nutrition support in acute renal failure patients: Current perspectives, *Support Line*, December 1997, pp. 3–7.

[7] J. D. Kopple, The nutrition management of the patient with acute renal failure, *Journal of Parenteral and Enteral Nutrition* 20 (1996): 3–12.

[8] A.S.P.E.N. Board of Directors, Practice guidelines: Kidney failure—Pediatric, *Journal of Parenteral and Enteral Nutrition* (supplement) 17 (1993): 43.

[9] The Modification of Diet in Renal Disease Study Group, The effects of dietary protein restriction and blood pressure control on the progression of chronic renal disease, *New England Journal of Medicine* 330 (1994): 877–884.

[10] J. A. Beto, Which diet for which renal failure: Making sense of the options, *Journal of the American Dietetic Association* 95 (1995): 898–903.

[11] R. Makoff and H. Gonick, Renal failure and concomitant derangement of micronutrient metabolism, *Nutrition in Clinical Practice* 14 (1999): 238–246.

[12] Beto, 1995.

[13] B. P. Gillis and coauthors, Nutrition intervention program of the Modification of Diet in Renal Disease Study: A self-management approach, *Journal of the American Dietetic Association* 95 (1995): 1288–1294.

[14] National Kidney Foundation, About kidney disease, www.kidney.org, site visited November 30, 1999.

Q&A NUTRITION IN PRACTICE

Dialysis and Nutrition

Although there is no perfect substitute for one's own kidneys, **dialysis** offers a life-sustaining treatment option for people with end-stage renal disease. Dialysis can serve as a permanent treatment for kidney failure or as a temporary measure to sustain life until a suitable kidney donor can be found. Dialysis also benefits the person with acute renal failure who needs immediate help in restoring normal blood balances. All health care professionals need to understand that renal diseases alter nutrient needs and that dialysis affects those needs. Those who routinely work with clients with renal diseases, however, need to understand more about dialysis procedures. This Nutrition in Practice describes the different types of dialysis procedures and explains why different procedures affect nutrient needs in different ways.

www.kdf.org.sg
Kidney Dialysis Foundation

What is dialysis?

Dialysis is a procedure that removes excess fluids and wastes from the blood by employing the principles of simple **diffusion** and **osmosis** across a semipermeable membrane (see the accompanying glossary). For **hemodialysis** and **peritoneal dialysis,** a solution similar in composition to normal blood plasma, called the **dialysate,** is placed on one side of the semipermeable membrane; the person's blood flows by on the other side. A semipermeable membrane is like a filter; different types of membranes have pores of different sizes—the size of the pores determines which molecules will pass through the membrane and which will not. Small molecules like urea and electrolytes cross the membrane freely; large molecules like proteins are less likely to cross the membrane.

How do the dialysate and semipermeable membrane work together to remove wastes?

Wastes are removed from the blood by adjusting the composition of the dialysate. When the concentration of electrolytes in the dialysate is low, for example, the electrolytes (which are small molecules) cross the membrane and diffuse out of the blood. The dialysate can also add needed components back into the blood. For a person with acidosis, for example,

During hemodialysis, shown here, blood passes through a dialyzer where wastes are extracted. The cleansed blood is then returned to the body.

GLOSSARY OF TERMS RELATED TO DIALYSIS

continuous renal replacement therapy (CRRT): a method for slowly and continuously removing fluids and wastes from the blood during acute renal failure that relies on filtration, filtration plus replacement fluids, dialysis, or a combination of methods.

dialysate (dye-AL-ih-sate): a solution used during dialysis to draw wastes and fluids from the blood.

dialysis (dye-AL-ih-sis): removal of wastes and fluids from the blood using the principles of simple diffusion and osmosis through a semipermeable membrane.

dialyzer (dye-ah-LYES-er): the machine used for hemodialysis; also called an *artificial kidney.*

diffusion: movement of solutes from an area of high concentration to one of low concentration.

hemodialysis: removal of fluids and wastes from the blood by passing it through a dialyzer.

hemofiltration: removal of fluids and wastes from the blood using ultrafiltration and fluid replacement.

osmosis: movement of water from an area of low solute concentration to one of high solute concentration.

peritoneal dialysis: removal of wastes and fluids from the body using the peritoneal membrane as a semipermeable membrane.

ultrafiltration: a method of removing fluids and small to medium-sized molecules from the blood by using pressure to transfer the blood across a semipermeable membrane.

bases such as bicarbonate can be added to the dialysate. The base moves by diffusion into the person's blood to ease acidosis.

Excess fluids are removed from the blood by osmosis. Water moves from a less concentrated solution to a more concentrated solution. Sodium moves into the dialysate along with water. Dialysis can filter excess substances only from the blood. Thus, electrolytes like potassium and phosphorus that reside mainly in the intracellular fluid are more difficult to control with dialysis.[1]

What is the difference between hemodialysis and peritoneal dialysis?

In hemodialysis, the blood enters a machine called a **dialyzer** or artificial kidney. The dialyzer houses a series of tubes made from synthetic semipermeable membranes. Blood is pumped out of the person's body and into the dialyzer where it flows between the tubing that carries the dialysate. Wastes are extracted and blood is returned to the body.

In peritoneal dialysis, the peritoneal membrane (the membrane that covers the abdominal membranes) serves as the semipermeable membrane. The dialysate is infused directly through a tube into the peritoneal space—the space within the person's abdomen that overlays the intestine. Fluid is kept in the abdomen for about 30 minutes and then is drained from the abdomen through a tube by gravity. The procedure is repeated with fresh fluids. One complete exchange may take about 45 minutes, and the complete treatment may require about six exchanges.

Are there any other kinds of dialysis?

For people in acute renal failure, another procedure, called **continuous renal replacement therapy (CRRT),** allows removal of fluids or wastes.[2] CRRT is usually reserved for people with acute renal insufficiency or renal failure who cannot tolerate hemodialysis or peritoneal dialysis for medical reasons.

Several different methods can be used to perform CRRT, depending on the individual's needs. If the purpose is to remove fluids only, no dialysate is used. Instead blood is circulated through a filter where pressure gradients force all blood components small enough to pass through the pores of a semipermeable membrane out of the blood (**ultrafiltration**). The final composition of the filtrate is similar to that of the blood but without plasma proteins and blood cells, which are too large to pass through the membrane. Thus, excess electrolytes and waste products are not removed, but the fluid volume is reduced. In some cases, part of the fluid removed during ultrafiltration is replaced with intravenous fluids that contain only those components needed by the client (**hemofiltration**). If blood containing electrolytes and wastes is removed and some of the fluid is replaced with a very dilute dextrose solution, for example, the concentration of electrolytes and wastes in the blood and the amount of fluids is eventually lowered. The filter can be adjusted to perform dialysis as well as filtration, so slow, continuous dialysis is also possible, and the methods can be combined so that ultrafiltration and dialysis can take place. Figure NP22-1 illustrates the differences between diffusion, osmosis, and ultrafiltration.

How do the different forms of dialysis alter nutrient needs?

Compared to people with renal failure who are not on dialysis, people on dialysis receive diets higher in protein. In part this is because dialysis can remove the nitrogen-containing waste products of protein metabolism. In addition, some amino acids and smaller proteins may be lost in the dialysate. In hemodialysis, protein losses average about 5 to 8 grams per treatment. Protein losses from peritoneal dialysis are slightly higher (about 10 to 20 grams per day) because more blood proteins pass into the dialysate through the peritoneal membrane than through the synthetic membrane used for hemodialysis. Clients with acute renal failure often receive TPN solutions during the time that they are undergoing CRRT. People undergoing CRRT retain about 90 percent of infused amino acids.[3]

For people on dialysis, dextrose (glucose) is added to the dialysate. Glucose helps draw fluid into the dialysate through osmosis, but the person undergoing dialysis can also absorb some of it. The amount of glucose absorbed during hemodialysis is insignificant. In peritoneal dialysis, however, glucose is slowly absorbed across the peritoneal membrane; this makes the dialysate less efficient at drawing fluid out of the blood and also means that the person may absorb significant amounts of glucose. The glucose may provide as much as 600 to 800 kcalories per day. For people on CRRT, the absorption

Figure NP22-1
DIFFUSION, OSMOSIS, AND ULTRAFILTRATION

Diffusion

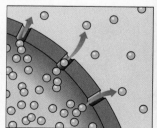

Small molecules (electrolytes and waste products) move from an area of high concentration to an area of low concentration by diffusion.

Osmosis

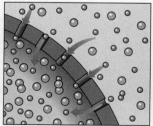

Water moves from an area of low concentration to an area of high concentration by osmosis. In other words, water moves from an area that has fewer particles dissolved in it to an area with more particles dissolved in it.

Ultrafiltration

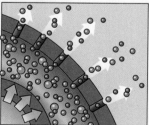

Pressure squeezes water and small molecules through the pores of a semipermeable membrane during ultrafiltration.

of glucose is also considerable and can range from about 35 to 45 percent to as high as 60 to 70 percent, depending on the type of CRRT and many other factors.[4] Because people undergoing CRRT are often catabolic, they frequently experience hyperglycemia; the carbohydrate provided must be carefully adjusted to avoid serious problems with hyperglycemia.

Another difference between hemodialysis and peritoneal dialysis with respect to nutrition concerns potassium. In hemodialysis, the dialysate contains potassium; because a relatively large volume of blood is pumped through the dialyzer, potassium levels can drop rapidly, and hypokalemia can interfere with heart function. In peritoneal dialysis and CRRT, the exchange of potassium between the blood and the dialysate is much slower, and blood potassium does not change rapidly. In peritoneal dialysis and CRRT, potassium may be added in varying amounts or left out of the dialysate altogether, depending on the person's blood potassium levels.

Since peritoneal dialysis can remove more potassium, the person on peritoneal dialysis generally does not have to restrict dietary sources of potassium. The person on hemodialysis, however, does need to moderately restrict potassium between dialysis treatments.

Dialysis and CRRT help remove wastes and fluids normally removed by functional kidneys. In so doing, dialysis and CRRT reduce the symptoms of uremia, hypertension, and edema, as well as the risk of heart failure. The hormonal functions of the kidneys, however, are not restored.

Notes

[1] D. I. Charney, Medical treatment in renal disease: Basic concepts in dialysis, *Support Line,* February 1998, pp. 3–7.

[2] D. J. Rodriguez and W. M. Sandoval, Nutrition support in acute renal failure patients: Current perspectives, *Support Line,* December 1997, pp. 3–7.

[3] D. C. Kaufman and coauthors, Adjustment of nutrition support with continuous hemodiafiltration in a critically ill patient, *Nutrition in Clinical Practice* 14 (1999): 120–123; Rodriguez and Sandoval, 1997.

[4] As cited in Kaufman and coauthors, 1999.

CHAPTER 23

NUTRITION AND LIVER DISORDERS

CONTENTS

Fatty Liver and Hepatitis

Cirrhosis

Liver Transplants

Case Study: Carpenter
with Cirrhosis

Nutrition in Practice:
Nutrition and Alcohol Abuse

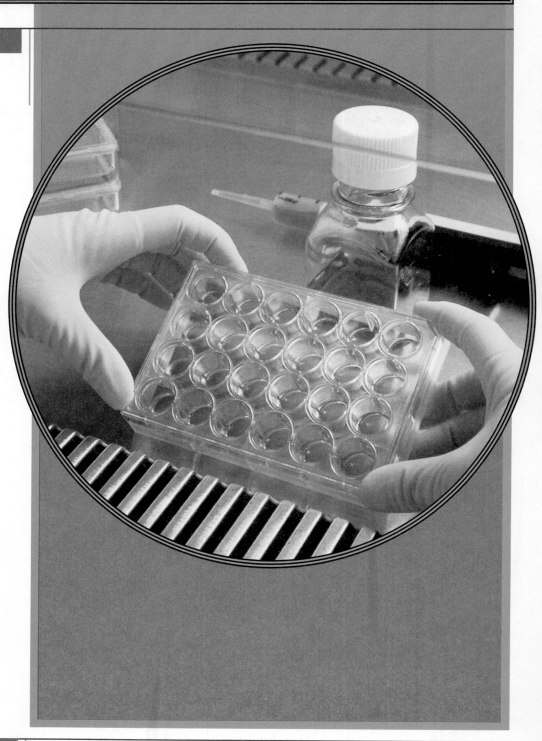

The liver is the metabolic crossroads of the body, and its health is crucial to every body function. The liver receives nutrients and then metabolizes, packages, stores, or ships them out for use by other organs. It plays a pivotal role in controlling blood glucose levels. It manufactures bile, which the body uses to prepare fat for digestion and absorption. It synthesizes albumin and other proteins. The liver also detoxifies drugs, packages excess nitrogen in urea so that it can be excreted by the kidneys, and participates in iron recycling, the manufacture of red blood cells, and blood clotting. No wonder that disorders of the liver can profoundly affect both nutrition and general health status. The liver conditions described in this chapter are not diseases, but rather forms of liver damage that can occur as a consequence of many disorders.

▪▪ Fatty Liver and Hepatitis

Fatty liver and hepatitis are two of the more common signs of liver dysfunction. Both enlarge the liver. Dietary factors may play a role in the development of both disorders, although both may also arise from causes unrelated to diet.

▪▪ Fatty Liver

Fatty liver is a clinical finding common to many conditions. Fatty liver may develop from damage to liver cells or from altered nutrient availability to the liver. Most commonly, fatty liver develops from the liver's exposure to toxic substances such as alcohol (see Nutrition in Practice 23), from excessive weight gain, or as a consequence of diabetes mellitus. Fatty liver can also develop as a consequence of inadequate intake of protein (as in protein-energy malnutrition, or PEM), an infection or malignant disease, drug therapy (such as corticosteroids or tetracycline), long-term TPN, or small bowel bypass surgery. Fatty liver does not arise from eating too much fat alone; excess total kcalories (from either fat or carbohydrate) and high blood glucose levels (as found in poorly controlled diabetes and in some people on TPN) appear to have a greater association with fatty liver.

> **fatty liver:** an accumulation of triglycerides in the liver resulting from many conditions, including exposure to excessive alcohol, excessive weight gain, and diabetes mellitus; also called **hepatic steatosis, steatohepatitis,** and **fatty infiltration of the liver.**

THE NURSING DIAGNOSES **altered nutrition: more than body requirements** and **altered nutrition: less than body requirements** may be related to fatty liver.

The exact reasons why fats (triglycerides) accumulate in the liver are unknown, but the liver may either synthesize too much fat, oxidize too little, take up too much from the blood, release too little back to the blood, or make a combination of these errors. In severe cases, liver weight may increase from a typical weight of about 3 pounds to as much as 11 pounds, with triglycerides increasing from 5 percent to as much as 40 percent of liver weight. Laboratory findings associated with fatty liver may include elevated serum transaminases (ALT and AST), alkaline phosphatase, and bilirubin.

> *The two transaminase enzymes that may be elevated in liver disease are* **alanine transaminase (ALT)** *and* **aspartate transaminase (AST).**

▪▪ **CONSEQUENCES OF FATTY LIVER** Fatty liver alone usually causes no harm. Fatty liver associated with TPN, for example, may resolve as the feeding continues, when the feeding is changed to a cyclic infusion, or when TPN is discontinued. In other cases, however, the liver's accumulation of fat suggests the presence of an underlying disorder that can progress to permanent liver damage, if left untreated.

▪▪ **TREATMENT OF FATTY LIVER** The appropriate therapy for fatty liver focuses on eliminating the cause and reversing its effects. Fatty liver related to obesity requires weight reduction. Fatty liver related to diabetes mellitus

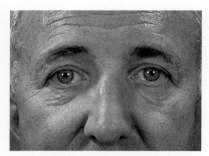

Jaundice is easy to see in the whites of the eyes.

requires control of blood glucose levels. Fatty liver caused by malnutrition requires a gradual introduction of a high-kcalorie, high-protein diet adequate in all other nutrients. Fatty liver due to alcohol abuse requires abstinence from alcohol and an adequate diet to replenish nutrient stores. Fatty liver caused by drug therapy requires alternative medications or other therapies.

Hepatitis

In hepatitis, damage to liver cells causes inflammation and enlargement of the liver. Most often the damage occurs as a consequence of a viral infection. Of the three most common viruses known to cause hepatitis (A, B, C), hepatitis A virus (HAV) is the highly contagious form that can be spread through contaminated foods and water and between family members, although it usually resolves and does not become chronic. Hepatitis B virus (HBV) and hepatitis C virus (HCV) can lead to serious, permanent liver damage. An estimated 1.2 million Americans are believed to be carriers of HBV, and another 3.5 million are carriers of HCV. Other, less common viruses can also cause hepatitis, and hepatitis can occur as a consequence of damage to liver cells by chronic and excessive alcohol ingestion, certain medications and illicit drugs, or other chemical toxins.

jaundice (JON-dis): a characteristic yellowing of the skin, whites of the eyes, mucous membranes, and body fluids resulting from the accumulation of bile pigments (bilirubin) in the blood.

SYMPTOMS OF HEPATITIS   Initially, the symptoms may be mild and hepatitis may go undiagnosed. In some cases, the person may develop flu-like symptoms including fatigue, anorexia, nausea, vomiting, diarrhea or constipation, fever, and pain in the area of the liver. If hepatitis progresses to a severe stage, yellow bile pigments accumulate in the inflamed liver and spill into the blood, causing **jaundice** and producing dark urine. Serum transaminase levels (AST and ALT) rise, and the liver becomes enlarged and tender.

THE NURSING DIAGNOSES altered nutrition: less than body requirements and **fatigue** may apply to people with hepatitis.

CONSEQUENCES OF HEPATITIS   The origin and type of hepatitis, the extent of liver damage, and the person's response to treatment all determine how seriously the disease will affect health. In many cases, liver cells gradually regenerate and liver function recovers. Recovery from HAV generally occurs within six months. Approximately 10 percent of people with HBV and 85 percent of those with HCV develop chronic hepatitis, which increases the risk of cirrhosis (described next), liver cancer, and liver failure. Far less frequently, severe hepatitis progresses rapidly and leads to acute liver failure and death.

People with chronic hepatitis and cirrhosis who rely on complementary therapies may use milk thistle to protect the liver. Although its benefits have not been proven, some studies support a protective effect.

MEDICAL NUTRITION THERAPY   Liver cells need nutrients to help them recover from hepatitis. Recovery also aims to reduce further insult to the liver cells; thus, the person must abstain from alcohol and avoid taking any unnecessary medications. The person with hepatitis who is in good nutrition status receives a regular, well-balanced diet. The malnourished person with hepatitis receives a high-kcalorie, high-protein diet to replenish nutrient stores. For the person with mild anorexia, frequent small meals, enteral formula supplements, or both may be helpful. For the person with persistent vomiting, parenteral nutrition is an alternative.

Cirrhosis

cirrhosis (sih-ROW-sis): an advanced form of liver disease in which scar tissue replaces liver cells that have permanently lost their function.

Cirrhosis is a serious consequence of chronic liver disease. Liver cells are damaged to the point that they cannot regenerate. Scar tissue (fibrosis) forms

GLOSSARY
OF TERMS RELATED TO CIRRHOSIS

biliary cirrhosis: severe liver damage from prolonged inflammation or obstruction of the bile ducts. This form can lead to chronic pancreatitis and fat malabsorption.

cardiac cirrhosis: severe liver damage associated with the later stages of congestive heart failure (see Chapter 21).

glycogen storage disease: one of several inherited disorders that can lead to severe liver damage. People with glycogen storage diseases lack one of the enzymes that allow them to utilize glycogen stores efficiently, and glycogen accumulates in the liver and other organs.

hemochromatosis (HE-moe-CROW-mah-toe-sis): an inherited disorder that can lead to severe liver damage. People with hemochromatosis absorb too much iron from the intestine and store too much iron in the liver.

idopathic (id-ee-oh-PATH-ic) cirrhosis: severe liver damage for which the cause cannot be identified.

Laennec's (LAY-eh-necks) cirrhosis: severe liver damage related to alcohol abuse.

postnecrotic cirrhosis: severe liver damage that develops as a complication of chronic hepatitis.

Wilson's disease: an inherited disorder that can lead to severe liver damage. People with Wilson's disease absorb too much copper from the intestine, and they have too little of the protein that transports copper from the liver to the sites where it is needed.

within the liver, altering the structure of the liver cells and interrupting the blood flow through it (see the photos in this chapter's Nutrition in Practice).

Chronic alcohol abuse is the most common cause of cirrhosis in the United States, although not all people with cirrhosis are alcohol abusers, and not all alcohol abusers develop cirrhosis. Other causes of cirrhosis include some infections (including chronic hepatitis), prolonged obstruction or diseases of the bile ducts, severe heart disease, severe reactions to medications, and exposure to toxic chemicals. People with inherited disorders that cause them to accumulate either iron or copper in their livers can also develop cirrhosis, as can those with inherited disorders that alter the liver's ability to use glucose. The accompanying glossary defines terms related to cirrhosis.

Consequences of Cirrhosis

Unlike healthy liver tissue, which is soft and flexible, scar tissue is unyielding, a difference that leads to major consequences. Many people with cirrhosis have no symptoms in the early stages. As the disorder progresses, symptoms may include nausea, vomiting, weight loss, liver enlargement, edema, jaundice, mental disturbances, and itching from the accumulation of bile pigments under the skin. The liver progressively loses function, and in the end stages, fails. Table 23-1 on p. 546 presents standards for laboratory tests often used to diagnose and monitor the extent of liver disease. Figure 23-1 shows many of the consequences of cirrhosis described in the sections that follow.

PORTAL HYPERTENSION
The portal vein and the hepatic artery carry 11½ quarts of blood every minute to the miles of intermeshed capillaries and arterioles within the liver. This huge volume of blood cannot pulse easily through the scarred tissue of a cirrhotic liver. Consequently, blood flow to and through the liver decreases, blood backs up, and pressure in the portal vein rises sharply, causing **portal hypertension.**

THE NURSING DIAGNOSES altered nutrition: less than body requirements, fluid volume excess, and **altered thought processes** apply to people with the end stages of cirrhosis.

The liver sometimes fails suddenly (fulminant liver failure) as a consequence of acute stresses or some types of hepatitis. Ingestion of chaparral tea, a complementary therapy promoted as an antioxidant and pain reliever, has led to liver failure requiring transplantation. Another herb, pennyroyal (used to treat coughs and upset stomachs), may also be toxic to liver cells.

Reminder: The portal vein carries nutrients from the GI tract to the liver. Normally, almost all the blood that flows from the intestine passes through the liver. The hepatic vein returns blood from the liver to the heart. The hepatic artery delivers oxygen-rich blood from the heart back to the liver.

portal hypertension: elevated blood pressure in the portal vein caused by obstructed blood flow through the liver.

Table 23-1

STANDARDS FOR LABORATORY TESTS USED TO DIAGNOSE AND MONITOR LIVER DISEASE

Test	Normal Values	Values in Liver Disease
Albumin	3.5–5.0 g/100 ml	Decreased
Alkaline phosphatase	Varies[a]	Normal or elevated
ALT (formerly SGPT)[b]	Varies[a]	Elevated
Ammonia	<50 µg/100 ml	Elevated
AST (formerly SGOT)[b]	Varies[a]	Elevated
Bilirubin (direct)	0.1–0.3 mg/100 ml	Elevated
Prothrombin time	10–13 seconds	Prolonged

NOTE: To convert albumin (g/100 ml) to standard international (SI) units, multiply by 10; to convert ammonia (µg/100 ml) to SI units, multiply by 0.5872; to convert bilirubin (mg/100 ml) to SI units (µmol/L), multiply by 17.10.
[a]Reference ranges vary depending on the test used. Consult laboratory report for normal ranges.
[b]ALT = alanine transaminase; SGPT = serum glutamic pyruvic transaminase; AST = aspartate transaminase; SGOT = serum glutamic oxaloacetic transaminase.

collaterals: small branches of a blood vessel that develop to divert blood flow away from the obstructed liver; also called **shunts.**

varices (VAIR-ih-seez): blood vessels that have become twisted and distended.

■■ **COLLATERALS AND ESOPHAGEAL VARICES**   With normal blood flow through the portal vein obstructed, pressure forces some of the blood from the portal vein to detour through the smaller vessels around the liver. These **collaterals,** or **shunts,** can develop throughout the GI tract, but often occur in the area of the esophagus. As pressure builds, the collaterals become enlarged and twisted, forming **varices.** Esophageal varices bulge into the lumen of the esophagus, and they can rupture and lead to massive bleeding that can be fatal.

Figure 23-1
THE CONSEQUENCES OF CIRRHOSIS

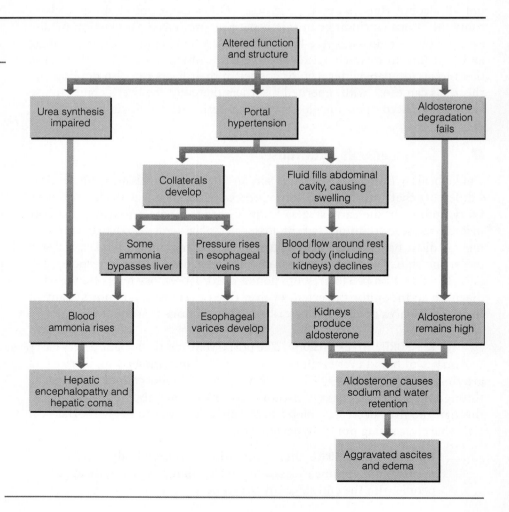

In addition, blood from ruptured varices can travel to the intestine, where it can serve as a source of ammonia (described later).

■■ **ASCITES AND EDEMA**   The rising pressure in the portal vein forces plasma out of the liver's blood vessels into the abdominal cavity, causing the abdomen to swell. This accumulation of fluid, or edema, in the abdominal cavity is called **ascites.** At the same time, the diseased liver is unable to synthesize adequate amounts of the protein albumin, which normally creates a pressure that draws fluids from tissues back into the blood. As fluid accumulates in the abdomen, less blood reaches the kidneys, a sign the body interprets as fluid depletion. In response, the body makes more aldosterone and antidiuretic hormone, hormones that expand the body's fluid volume by triggering retention of sodium and water in the kidneys. As a result of sodium and water retention, ascites worsens, and edema may spread to other parts of the body (peripheral edema). To make matters worse, the diseased liver cannot dispose of aldosterone as it usually does, so aldosterone levels remain high. Thus, ascites is a self-aggravating condition. Ascites frequently causes early satiety and nausea and also raises the basal metabolic rate.

■■ **ELEVATED BLOOD AMMONIA LEVELS**   Ammonia is a normal but toxic product of protein metabolism. The healthy liver detoxifies ammonia by removing it from circulation and converting it to urea. As liver disease progresses, ammonia-laden blood bypasses the liver by way of the collaterals, the liver may be unable to detoxify the ammonia it does retrieve, and blood ammonia rises. Elevated blood ammonia disrupts central nervous system function.

■■ **HEPATIC ENCEPHALOPATHY AND HEPATIC COMA**   Liver diseases can alter mental function (**hepatic encephalopathy**) and lead to **hepatic coma**—a dangerous complication that develops in some people when liver function deteriorates significantly. The causes of encephalopathy and hepatic coma remain elusive, although elevated blood ammonia appears to play a role. Events that can precipitate a coma include GI bleeding, infection, constipation, and the use of sedatives. Elevated blood ammonia is associated with encephalopathy and hepatic coma, but the degree of ammonia elevation does not correlate with the severity of symptoms. A possible explanation for this poor correlation is that *blood* ammonia does not always parallel *brain* ammonia concentration. When brain ammonia is high, the body produces greater quantities of two substances (glutamine and ketoglutarate), and the degree of their elevation tends to correlate with the severity of symptoms.

Blood amino acid patterns also change in the later stages of liver disease; aromatic amino acid levels rise, and branched-chain amino acid levels fall. These alterations may add ammonia to the blood and may contribute to encephalopathy and hepatic coma in some people.

The person with hepatic encephalopathy may suffer from irritability, short-term memory loss, and an inability to concentrate. The person with impending hepatic coma exhibits changes in judgment, personality, or mood. The person may be unable to draw even a simple shape, such as a star. A sweet, musty, or pungent odor may develop on the breath. **Flapping tremor** may also develop in the precoma state. Just before passing into a coma, the person becomes very difficult to arouse.

■■ **INSULIN RESISTANCE**   Although the reasons remain unclear, insulin resistance, hyperinsulinemia, and hyperglycemia develop in many people with chronic liver disease. In some cases, the pancreas is unable to produce enough insulin to overcome the insulin resistance, and diabetes mellitus develops.

ascites (ah-SIGH-teez): a type of edema characterized by the accumulation of fluid in the abdominal cavity.

An elevated blood ammonia level is called hyperammonemia (HYE-per-AM-moe-KNEE-me-ah). *Normal blood ammonia levels are less than 50 μg/100 ml.*

hepatic encephalopathy (en-SEF-ah-LOP-ah-thee): mental changes associated with liver disease that may include irritability, short-term memory loss, and an inability to concentrate.
  encephalo = brain
  pathy = disease

hepatic coma: a state of unconsciousness that results from severe liver disease.

Reminder: Phenylalanine and tyrosine are aromatic amino acids; leucine, lysine, and valine are branched-chain amino acids.

The odor that may develop in people with impending hepatic coma is called fetor hepaticus (FEE-tor he-PAT-eh-cuss).

flapping tremor: uncontrolled movement of the muscle group that causes the outstretched arm and hand to flap like a wing; occurs in disorders that cause encephalopathy; also called **asterixis (AS-ter-ICK-sis).**

Table 23-2
POSSIBLE CAUSES OF WASTING IN CIRRHOSIS

Reduced Nutrient Intake	Excessive Nutrient Losses	Raised Nutrient Needs
Abdominal pain	Blood loss through GI bleeding	Ascites (raises metabolic rate)
Anorexia		Infections
Early satiety	Diarrhea	Malnutrition
Esophageal varices	Malabsorption	Medications
Medications	Medications	
Nausea	Steatorrhea	
Restrictive diet	Vomiting	
Vomiting		

Reminder: Alcoholism is a common cause of chronic pancreatitis.

The combined symptoms of liver and kidney failure that occur as a consequence of severe liver damage are called hepatorenal (HEP-ah-toe-REE-nal) syndrome.

For people with hemochromatosis, treatment may include surgically opening a vein to withdraw blood (phlebotomy), which reduces the body's stores of iron by stimulating red blood cell production.

WASTING As liver deterioration progresses, malnutrition and wasting become evident. Table 23-2 summarizes possible causes of wasting in people with cirrhosis, which can include anorexia and reduced food intake, altered metabolism, and malabsorption. Fat malabsorption can occur if the liver disease interferes with bile salt production, if the flow of bile from the liver is obstructed, or if the person develops pancreatic insufficiency.

OTHER CONSEQUENCES When liver function deteriorates significantly, the alterations in fluid and electrolyte balance as well as the buildup of waste products in the blood can precipitate kidney failure. Bone disorders that are generally unresponsive to vitamin D and calcium supplementation can also become a problem.

Treatment of Cirrhosis

Treatment of cirrhosis aims to preserve remaining organ function, reverse the damage that has occurred to whatever extent is possible, and prevent and control complications. Treatment depends, in part, on addressing the underlying medical condition leading to cirrhosis. When cirrhosis has progressed to a severe stage, liver transplantation becomes a treatment option.

DRUG THERAPY One goal of drug therapy in cirrhosis is to control blood ammonia levels. About two-thirds of the body's ammonia comes from the intestine where bacteria make ammonia from undigested dietary proteins as well as from protein from shed intestinal cells and GI tract bleeding. Digestive enzymes also produce ammonia in the process of dismantling dietary proteins. Thus, drug therapy often includes antibiotics to limit the growth of intestinal bacteria and laxatives to speed intestinal transit time.

LEARNING LINK

As described on p. 547, blood ammonia levels play a role in the development of hepatic coma, and many conditions that can precipitate hepatic coma lead directly to elevated blood ammonia levels. GI bleeding provides the GI tract bacteria with protein from which ammonia is produced, and constipation and the use of sedatives (which lead to constipation) allow the bacteria in the intestine more time to make ammonia. Infections and any acute stress result in the breakdown of protein, thereby producing more ammonia to be detoxified. Recall also that during acute stress, the liver is called upon to synthesize the many factors necessary to mount a stress response, and such a burden can easily overwhelm a failing liver.

In addition to these medications, antihypertensives may be used to control portal hypertension, and diuretics may be used to reduce fluid retention and prevent ascites and edema.[1] Oral antidiabetic agents may be used to control hyperglycemia. Cholestyramine, an antilipemic, may be used to alleviate the excessive itching that often accompanies liver disease. Megestrol acetate and dronabinol may help improve the appetite and stimulate weight gain.[2] Interferon and ribavirin may be used to improve immune responses in people with cirrhosis caused by viral hepatitis. People with Wilson's disease may be treated with penicillamine, which binds with copper and allows it to be excreted in the urine. The diet-medication interactions of these drugs are described in the box below.

Medical Nutrition Therapy

Medical nutrition therapy for people with cirrhosis carefully considers each client's needs for energy, protein, sodium, and fluid (see Table 23-3 on page 550). Individual needs vary considerably and depend on the complications present in each case. In all cases, clients with cirrhosis must abstain completely from alcohol to protect the liver from further injury.

ENERGY The diseased liver needs energy to protect its functioning cells and to prevent protein catabolism. People with ascites, malabsorption, or infections have high energy needs. For people with ascites, energy needs are based on the person's desirable or estimated dry weight (weight without ascites) to avoid overestimating energy needs.[3]

DIET-MEDICATION INTERACTIONS

Antidiabetic agents

For *antidiabetic agents*, see p. 482.

Antihypertensives

For *antihypertensives*, see p. 507.

Anti-infectives

The antibiotic frequently used to control blood ammonia levels in liver disease is *neomycin*. Because neomycin kills bacteria in the intestine, the person may need a vitamin K supplement. The person may be prone to other intestinal infections that ordinarily are suppressed by normal bacterial flora. Neomycin can also be toxic to the kidneys. The antiviral agent *ribavirin* may be used in the treatment of chronic hepatitis infections. Nutrition-related side effects of ribavirin include anemia and fatigue.

Antilipemics

For *antilipemics*, see p. 507.

Appetite stimulates

Megestrol acetate and *dronabinol* (a derivative of marijuana) stimulate the appetite and result in weight gain. GI side effects are uncommon, although some people develop nausea, vomiting, and diarrhea. Dronabinol produces euphoria in some people and can reach toxic levels in people with liver dysfunction.

Diuretics

Spironolactone is a potassium-sparing diuretic frequently used to treat ascites. People taking spironolactone should avoid foods high in potassium, potassium supplements, and salt substitutes that contain potassium. See also p. 507.

Immunosuppressants

Immunosuppressants are used for people who undergo liver transplants. For *cyclosporine* and a general discussion of immunosuppressants, see p. 536. Another

continued

immunosuppressant sometimes used to prevent tissue rejection following a liver transplant is *tacrolimus*. Tacrolimus can be given with food to reduce nausea, vomiting, and GI upsets. Other nutrition-related side effects include anorexia, diarrhea, constipation, anemia, hyperglycemia, potassium imbalances, low blood magnesium levels, and ascites.

Interferon

Interferon can lead to nausea, vomiting, weight loss, fever, fatigue, and depression.

Laxatives

The laxative commonly used in the treatment of liver disease is *lactulose*. Lactulose can cause belching, cramps, and diarrhea. Encourage clients to drink fluids to replace fluid lost through diarrhea.

Penicillamine

Penicillamine is used in the treatment of Wilson's disease. Foods decrease the absorption of penicillamine, and it should be taken on an empty stomach. Nutrition-related side effects may include altered taste perceptions, anorexia, nausea, vomiting, diarrhea, vitamin B_6 deficiency, iron deficiency, and loss of protein in the urine. The person should avoid foods and supplements that contain copper. Iron supplements should be given at least 2 hours before or after penicillamine is given.

Table 23-3
DIET GUIDELINES FOR CIRRHOSIS

Energy
- Without ascites, malnutrition, or infection: 120% of basal energy expenditure (BEE); see Table 19-3 on p. 454.
- With ascites, malnutrition, or infection: 150–175% of BEE.

Protein
- Not sensitive to protein: 1.0 to 1.5 g protein/kg body weight/day.
- Protein sensitive: Start with 0.5 to 0.7 g protein/kg body weight/day; increase gradually to level of tolerance with a goal of at least 1.0 g protein/kg body weight/day and up to 1.5 g protein/kg body weight/day if the person is malnourished.
- Very sensitive to protein (unable to tolerate at least 0.8 g protein/kg body weight/day): Consider vegetable-based diets or enteral supplements enriched with branched-chain amino acids and restricted in aromatic amino acids.

Carbohydrate
- Not restricted.
- For people with insulin resistance or diabetes, provide up to 50 to 60% of kcalories from carbohydrates (mainly complex carbohydrates) with a consistent carbohydrate intake from day to day and at each meal and snack.

Fat
- Not restricted unless fat malabsorption is present.
- With fat malabsorption: Restrict fat only as necessary to control steatorrhea (see Chapter 16); use medium-chain triglycerides (MCT) to increase kcalories as necessary.

Sodium
- Restrict only as necessary to control ascites, but not less than 2 g sodium/day in most cases.

Fluid
- Restrict to 1.0 to 1.5 liters per day for people with ascites who also have low blood sodium levels.

Vitamins and minerals
- Ensure adequate intake from diet or supplements based on individual needs.

SOURCE: Adapted from J. Hasse and coauthors, Nutrition therapy for end-stage liver disease: A practical approach, *Support Line*, August 1997, pp. 8–15.

■■ **PROTEIN** For people with cirrhosis who do not show signs of impending coma, protein restrictions are unnecessary, and a high-protein diet (1.0 to 1.5 grams of protein per kilogram of body weight per day) is appropriate.[4] For people who develop signs of impending coma, the cause is investigated, and the underlying condition is treated. If a person falls into a coma, all protein is withheld temporarily until the cause is determined and corrective measures are taken.

In a small number of people (7 to 9 percent), dietary protein can precipitate hepatic encephalopathy.[5] For people who are protein sensitive, the diet plan is low in protein at first, and the level is gradually increased. The final amount of protein that is optimal depends on the individual's nutrition status and tolerance to protein. For a person who is malnourished, up to 1.5 grams of protein per kilogram of body weight is the final goal. The actual level of protein the person tolerates may be less than this amount, however. Sometimes medications can be adjusted to control symptoms, allowing a higher protein intake.

In some cases, the person who is protein sensitive can tolerate only a little protein (0.50 to 0.75 gram of protein per kilogram of body weight per day). Although research is limited and controversial, many clinicians recommend special enteral (or parenteral, when needed) formulas that are low in aromatic amino acids and high in branched-chain amino acids for these clients (see Appendix G).[6] Some studies suggest that vegetable proteins may be better tolerated than meat proteins, perhaps because vegetables contain fewer ammonia-forming constituents and aromatic amino acids and more branched-chain amino acids than meats do. In addition, the fiber from plant foods speeds intestinal transit time, reducing the amount of time for ammonia absorption from the intestine.

The foods shown here provide about 60 grams of protein, the entire day's protein allowance for some people with liver disease.

■■ **CARBOHYDRATE AND FAT** In a diet for liver disease, carbohydrate generally provides 50 to 60 percent of energy needs. A high-carbohydrate (50 grams) snack at bedtime may help prevent excessive breakdown of fat and protein during an overnight fast.[7] People with hyperglycemia should follow the guidelines for diets in diabetes; that is, eat mostly complex carbohydrates and eat them at consistent times throughout the day.

Because fat helps make foods appetizing and delivers energy efficiently, fat plays an important role in the diet of a person with cirrhosis. Fat is restricted only if the person develops fat malabsorption. In these cases, medium-chain triglycerides (MCT) can provide additional kcalories.

■■ **SODIUM AND FLUID** For people with ascites, the diet restricts sodium. Fluids are restricted only when people have low blood sodium levels—a sign that the person is overloaded with fluid. In these cases, fluids may need to be restricted to about 1.0 to 1.5 liters per day. In liver disease, the total amount of sodium in the body is excessive, even though blood concentrations may sometimes be low. Thus, adding sodium to the diet is inappropriate. To assess changes in fluid balance, nurses monitor weight and measure abdominal girth. Nurses measure abdominal girth by placing a tape measurer around the back and over the person's umbilicus. Rapid weight gain and an increasing abdominal girth indicate fluid retention; sudden weight loss and decreasing abdominal girth indicate successful fluid excretion.

Low blood sodium is hyponatremia (HIGH-poe-nay-TREE-mee-ah).

The lower the sodium intake, the more quickly ascites resolves. Very-low-sodium diets are unpalatable to many people, however, so most clinicians recommend a diet that allows from 2 to 4 grams of sodium. (Table 22-2 on p. 525 presents a 2-gram sodium-restricted diet plan.)

■■ **VITAMINS** The liver's central role in the metabolism and storage of vitamins and minerals, combined with coexisting conditions (such as malabsorption, alcoholism, and malnutrition), explains why nutrient deficiencies commonly occur in people with liver diseases. Virtually all people with advanced liver disease require supplements. Physicians determine which nutrients to supplement by monitoring serum levels and checking for clinical signs of deficiencies.

The B vitamins serve as cofactors for the liver's many metabolic reactions and repair work; deficiencies of thiamin, vitamin B_6, riboflavin, and folate are common. Fat-soluble vitamins may be malabsorbed if steatorrhea develops. If the diseased liver fails to synthesize adequate amounts of retinol-binding protein, body tissues may not receive the vitamin A they need. Vitamin D deficiencies may develop from the inability of the liver to perform its roles in vitamin D metabolism. Although the bone diseases associated with liver disease are generally not responsive to vitamin D supplements, it is important to ensure adequate vitamin D intakes to prevent further problems. Vitamin K deficiencies can prolong the time it takes for blood to clot, a dangerous complication that increases the risk of massive bleeding from the GI tract. (Recall that liver damage itself can interfere with blood coagulation.)

One laboratory test that evaluates the time it takes for blood to clot is called the prothrombin time. *Both vitamin K deficiency and liver disease can prolong the prothrombin time.*

LEARNING LINK

As Chapter 8 described, the healthy liver manufactures a vitamin D precursor that migrates to the skin where ultraviolet rays activate it to a second precursor. The second precursor or vitamin D from foods requires activation by both the liver and the kidneys. Thus, either liver or kidney disease can produce symptoms of vitamin D deficiency.

⠿ MINERALS   People with cirrhosis can develop calcium deficiencies from three causes: steatorrhea, low serum albumin (albumin, which carries calcium in the blood, is manufactured in the liver), and impaired vitamin D metabolism. Thus, it is important that the diet or supplements provide adequate calcium to prevent deficiencies. Magnesium deficiencies are also common. Zinc stores may also be depleted as a consequence of liver damage, and limited research suggests that zinc deficiencies may be implicated in the development of hepatic encephalopathy.[8]

⠿ ENTERAL AND PARENTERAL NUTRITION   People with cirrhosis face many difficulties in eating enough food to maintain nutrition status. The "How to" box offers tips on how to help people with cirrhosis cope with the demands of their diets. If the person with cirrhosis cannot take enough food or formula by mouth, tube feedings or TPN is indicated. As mentioned earlier, enteral and parenteral formulas designed for liver disease provide fewer aromatic and more branched-chain amino acids than standard formulas (see Appendix G). Both enteral and parenteral nutrition have been used successfully in people with cirrhosis.

People with bleeding esophageal varices will be unable to consume food by mouth and are often given simple intravenous solutions to maintain fluid and electrolyte balance. Parenteral nutrition should be considered if the person is malnourished or the health care team anticipates that oral intake will not resume for an extended time. The case study on p. 554 asks you to use clinical knowledge and judgment in answering questions about cirrhosis.

HOW TO:
HELP THE PERSON WITH CIRRHOSIS EAT ENOUGH FOOD

People with cirrhosis often have difficulty eating enough food to prevent malnutrition and its consequences. Ascites and gastrointestinal symptoms such as nausea and vomiting can interfere with food intake. The person with encephalopathy may be confused about what to eat or have little interest in eating. People with sodium restrictions may have difficulty adjusting to a low-salt diet. To facilitate diet compliance:

- Individualize the diet plan based on each person's symptoms and responses. The diet should restrict protein, fat, sodium, or fluid only when such restrictions are warranted.
- Use the suggestions for improving food intake in the "How to" box in Chapter 23.
- Advise clients to eat frequent small meals and to eat a high-carbohydrate bedtime snack. Eating frequent small meals can help with nausea and can also improve glucose tolerance. For people who are protein sensitive, eating small amounts of protein frequently throughout the day improves tolerance.
- Recommend the suggestions in the box on p. 533 for people who must restrict fluids and proteins.
- Point out that there is probably no food that cannot be incorporated into the diet on special occasions and in limited amounts. Advise clients to talk with the dietitian about how to change their meal plans so that favorite foods can be incorporated from time to time.

The low-sodium diet used in the treatment of cirrhosis may be more restrictive than the low-salt diet recommended for hypertension. Table 22-2 on p. 525 describes a 2-gram sodium-restricted diet. Some suggestions that can help clients adhere to their sodium restrictions include:

- Suggest that clients replace the salt they use for cooking and seasoning with herbs or spices like basil, bay leaves, curry, garlic, ginger, lemon, mint, oregano, rosemary, and thyme.
- Suggest that clients experiment with low-sodium products to find the ones they like. (People on potassium-sparing diuretics should be cautioned to avoid salt substitutes that replace sodium with potassium.)
- Advise clients to check food labels to learn the sodium content of the foods they eat. They may be able to find similar products with lower sodium contents.

Continue to offer support and encouragement to the client with cirrhosis. Severe weight loss is less likely to occur if nutrition intervention is provided before problems progress too far.

Liver Transplants

If liver failure progresses to a severe and irreversible state, a liver transplant may be an option. A transplant candidate must often wait for a liver donor before surgery is possible. Wise health care professionals use this time to identify and correct nutrient imbalances whenever possible.

NUTRITION BEFORE TRANSPLANTS In severe liver failure, malnutrition has often progressed for some time. Clinicians report malnutrition in over 70 percent of liver transplant recipients and note that malnutrition increases

Case Study

CARPENTER WITH CIRRHOSIS

Mr. Sloan, a 48-year-old carpenter, has been hospitalized many times. He recognizes his problem with alcohol abuse and has entered alcohol rehabilitation programs several times over the last few years. Nevertheless, he is still drinking. Mr. Sloan was recently admitted to the hospital, and a diagnosis of alcoholic cirrhosis has been confirmed. At 5 feet 7 inches tall, Mr. Sloan, who once weighed 150 pounds, now weighs 120 pounds. He is jaundiced and looks thin, although his abdomen is distended with ascites. He has advanced liver disease and is showing signs of impending hepatic coma. Laboratory findings include elevated AST, ALT, alkaline phosphatase, and blood ammonia. Compare these findings with Table 23-1 to determine if they are consistent with liver disease.

✚ Can you explain to Mr. Sloan what cirrhosis is and what its consequences are? From the limited information available, what can you determine about Mr. Sloan's nutrition status? What medical problem makes it difficult to interpret Mr. Sloan's actual weight? How can his weight measurements help determine if his condition is improving? Calculate Mr. Sloan's energy needs. What factors influence protein needs for a person with impending hepatic coma? What signs suggest that a person is in a precoma state?

✚ Why is Mr. Sloan's abdomen distended? Explain the development of ascites in liver disease and how diet is adjusted.

✚ Would you expect Mr. Sloan's blood ammonia levels to be high? Why or why not?

✚ Describe portal hypertension, jaundice, and esophageal varices. How would Mr. Sloan's diet be changed if he were found to have bleeding esophageal varices?

the risk of complications following a liver transplant.[9] The person equipped with adequate nutrient stores faces the transplant better prepared to fight infections, heal wounds, and mount a stress response.

❚❚ NUTRITION FOLLOWING TRANSPLANTS Following liver transplantation, liver function determines nutrient needs. All people are hypermetabolic after surgery, and energy needs are high. Immunosuppressant drugs given to prevent tissue rejection can contribute to nutrient imbalances by causing nausea, vomiting, diarrhea, and mouth sores. The nutrition support team uses indirect calorimetry to estimate energy needs and carefully monitors clinical and laboratory data to determine specific nutrient needs.

Although TPN has been the traditional source of nutrients in the post-transplant period, researchers report equal success with intestinal tube feedings.[10] Early enteral nutrition may reduce the incidence of infection, a particularly important consideration for people with suppressed immune systems.

Halting or delaying the progression of liver diseases depends in large part on attention to nutrition and nutrition assessment parameters. The accompanying Nutrition Assessment Checklist reviews important points to keep in mind when assessing the nutrition status of people with liver diseases.

NUTRITION ASSESSMENT CHECKLIST

FOR PEOPLE WITH LIVER DISORDERS

Health Problems, Signs, and Symptoms

Check the medical record to determine:

☐ Type of liver disease

☐ Cause of liver disease

☐ If client has received a liver transplant

Review the medical record for complications that may alter medical nutrition therapy including:

☐ Malnutrition

☐ Esophageal varices

☐ Ascites

☐ Hepatic encephalopathy

☐ Hepatic coma

☐ Insulin resistance/diabetes mellitus

☐ Heart failure

☐ Pancreatitis

☐ Malabsorption

☐ Renal failure

Medications

For clients with liver disease, note that risk of diet-medication interactions is very high because many medications are metabolized in the liver. The risk of interactions is intensified for clients with:

☐ Ascites (medications may take a long time to reach the liver)

☐ Renal failure (medications are often metabolized further in the kidneys and excreted in the urine)

☐ Malnutrition

☐ Multiple medication prescriptions and long-term medication use

Nutrient/Food Intake

For clients with fatty liver, pay special attention to the client's intake of:

☐ Energy, if client is overweight or malnourished, has diabetes, or is receiving TPN

☐ Carbohydrate, if client has diabetes or is receiving TPN

☐ Alcohol

For clients with hepatitis and cirrhosis, make a note of:

☐ Appetite

☐ Adequacy of energy and nutrient intake

☐ Alcohol use

For clients with protein-sensitive encephalopathy:

☐ Ensure that total energy intake is adequate.

☐ Work with the dietitian to find the level of protein restriction that is appropriate for the individual.

☐ Base energy needs on desirable or estimated dry weight to avoid overfeeding.

Height and Weight

Take baseline height and weight measurements and monitor weight regularly. For clients with ascites and edema:

☐ Use weight to monitor the degree of fluid retention.

☐ Remember that the client may be malnourished even though weight may be deceptively high.

Laboratory Tests

Note that albumin and serum proteins are often reduced in people with liver disease and are difficult to interpret as an indicator of nutrition status. Review the following laboratory test results to assess liver function (see Table 23-1):

☐ Albumin

☐ Alkaline phosphatase

☐ ALT

☐ Ammonia

☐ AST

☐ Bilirubin

☐ Prothrombin time

Check laboratory test results for complications associated with liver failure including:

☐ Anemia

☐ Fluid retention

☐ Hyperglycemia

☐ Renal function tests

Physical Signs

Look for physical signs of:

☐ Fluid retention (ascites and edema)

☐ PEM (muscle wasting and unintentional weight loss)

☐ B vitamin deficiencies

☐ Fat-soluble vitamin deficiencies

☐ Calcium deficiency

☐ Magnesium deficiency

☐ Potassium imbalances

☐ Zinc deficiency

SELF CHECK

1. Which of the following diet strategies would be most appropriate for helping to reverse fatty liver associated with diabetes mellitus?
   a. low-protein diet
   b. fat-restricted diet
   c. fluid- and sodium-restricted diet
   d. energy to achieve or maintain a desirable weight with a consistent intake of carbohydrate

2. Which of the following statements about hepatitis is true?
   a. Chronic hepatitis can progress to cirrhosis.
   b. Regardless of the type of hepatitis, symptoms are severe.
   c. People with hepatitis always need high-kcalorie, high-protein diets.
   d. HCV infections are often mild and can be spread through contaminated foods and water.

3. The consequences of cirrhosis are primarily due to:
   a. chronic malnutrition.
   b. chronic alcohol abuse.
   c. liver cell damage and altered hepatic blood flow.

d. elevated blood ammonia, amino acid, and sodium levels.

4. Esophageal varices are a dangerous complication of liver disease primarily because they:

a. interfere with food intake.
b. can lead to massive bleeding.
c. divert blood flow from the GI tract.
d. cause portal hypertension and collateral development.

5. The nurse recognizes that the condition(s) in liver disease most likely to lead to ascites is (are):

a. portal hypertension.
b. rising blood ammonia levels.
c. elevated serum albumin levels.
d. insulin resistance and diabetes mellitus.

6. For the person with cirrhosis, the nurse recognizes irritability, short-term memory loss, and an inability to concentrate as possible signs of:

a. coma.
b. encephalopathy.

c. hyperammonemia.
d. hepatorenal syndrome.

7. Medical nutrition therapy for people with cirrhosis includes diets that are:

a. restricted in fat.
b. high in kcalories and protein.
c. based on symptoms and responses to treatment.
d. restricted in protein, carbohydrate, sodium, and fluid.

8. Which of the following statements regarding diet and cirrhosis is false?

a. People with ascites need to restrict sodium.
b. People with steatorrhea may need to restrict fat.
c. People with ascites and low blood sodium levels need to restrict sodium and fluid.
d. People who are protein sensitive always need to restrict protein to 0.50 to 0.75 gram of protein per kilogram of body weight per day.

9. With respect to vitamins and minerals, people with cirrhosis:

a. may develop calcium deficiencies.
b. seldom require nutrient supplements.
c. frequently develop vitamin C deficiencies.
d. may develop clotting abnormalities associated with vitamin A deficiency.

10. For the person undergoing a liver transplant:

a. immunosuppressant drugs seldom alter nutrient needs.
b. enteral nutrition is contraindicated in the post-transplant period.
c. attention to nutrition before a transplant improves chances of recovery.
d. provided adequate nutrition is given following a transplant, nutrition status before surgery has little impact on recovery.

Answers to these questions appear in Appendix H.

CLINICAL APPLICATIONS

1. Think about the problems a person might encounter when following an 1800-kcalorie diet that is restricted to 40 grams of protein. On such a diet, just one scrambled egg, 3 ounces of meat, a cup of milk, and two slices of bread could use up the total protein allowance. Using Appendix A for reference, calculate the kcalories these foods provide. Then consider what other foods to add to the diet to meet energy needs without adding protein.

   Now compare the result with the Daily Food Guide on pp. 18–19. Which food groups have you offered in the recommended quantities? Which food groups are in short supply on such a diet? Which nutrients might be low? How might fats and sugars be useful in such a diet?

2. Chapter 19 described acute stresses, and Nutrition and Practice 19 discussed how the combination of hypermetabolism and infection can lead to multiple organ failure. The liver, kidneys, and respiratory system may all fail as a consequence of multiple organ failure. On a sheet of paper make three columns, one for stress, one for liver failure, and one for kidney failure. Under each column, list the energy, protein, fat, fluid, and other nutrient considerations for each disorder individually. Which nutrient modifications are common to all disorders? Do some of the modifications necessary for one disorder conflict with those for another? Consider why providing adequate nutrients, but not providing too much, is important for supporting the function of each organ.

3. Think about the relationship of ammonia to urea (see p. 547), and describe why you would expect blood ammonia levels to rise in liver disease and blood urea nitrogen levels to rise in renal disease.

NUTRITION**THE**NET

FOR FURTHER STUDY OF THE
TOPICS IN THIS CHAPTER,
ACCESS THESE WEB SITES.

www.liverfoundation.org   *American Liver Foundation*
www.liver.ca   *Canadian Liver Foundation*
www.livertx.org   *Children's Liver Alliance*
www.scn.org/health/hepatitis/   index.htm   *Hepatitis Education Project*
www.hepfi.org   *Hepatitis Foundation International*
www2.hepb.org   *Hepatitis B Foundation*
www.centerspan.org   *Center Span Transplant News Network*

Notes

[1] J. S. Crippin, Medical management of end-stage liver disease: A bridge to transplantation, *Support Line,* August 1997, pp. 3–7.

[2] C. Gurk-Turner, Management of the metabolic complications of liver disease: An overview of commonly used pharmacologic agents, *Support Line,* August 1997, pp. 17–19.

[3] J. Hasse and coauthors, Nutrition therapy for end-stage liver disease: A practical approach, *Support Line,* August 1997, pp. 8–15.

[4] C. Corish, Nutrition and liver disease, *Nutrition Reviews* 55 (1997): 17–20.

[5] As cited in Hasse and coauthors, 1997.

[6] A. Fabri and coauthors, Overview of randomized clinical trials of oral branched-chain amino acid treatment in chronic hepatic encephalopathy, *Journal of Parenteral and Enteral Nutrition* 20 (1996): 159–164.

[7] W. Chang and coauthors, Effects of extra-carbohydrate supplementation in the late evening on energy expenditure and substrate oxidation in patients with liver cirrhosis, *Journal of Parenteral and Enteral Nutrition* 21 (1997): 96–99.

[8] Hasse and coauthors, 1997.

[9] J. Pikul and coauthors, Degree of preoperative malnutrition is predictive of postoperative morbidity and mortality in liver transplant recipients, *Transplantation* 57 (1994): 469–472; J. Hasse, Nutrition and transplantation, *Nutrition in Clinical Practice* 8 (1993): 3–4.

[10] J. M. Hasse, Early enteral nutrition support in patients undergoing liver transplantation, *Journal of Parenteral and Enteral Nutrition* 19 (1995): 437–443; C. Wicks and coauthors, Comparison of enteral feeding and total parenteral nutrition after liver transplantation, *Lancet* 344 (1994): 469–472.

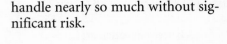

NUTRITION IN PRACTICE

Nutrition and Alcohol Abuse

As Chapter 23 described, alcoholism is a frequent cause of liver disorders. When used in excess, alcohol has a toxic effect on many organ systems, and its impacts on nutrition are so profound that they deserve special attention here.

Like all drugs, alcohol—properly termed ethanol, the active ingredient of alcoholic beverages—offers both benefits and hazards. Wine, beer, and other fermented beverages have been associated with pleasure and relaxation for more than 5000 years. People have always known that these beverages affected their moods, sensations, and behavior. Taken in moderation, alcohol can relax people, reduce their inhibitions, encourage social interactions, and possibly reduce the risk of coronary heart disease (see Chapter 21). Taken in excess, alcohol can be devastatingly destructive. The key to unlocking the benefits of alcohol without unleashing its toxic effects lies with the concept of moderation.

www.niaaa.nih.gov
National Institute on Alcohol Abuse and Alcoholism

www.ncadd.org
National Council on Alcoholism and Drug Dependence

www.health.org/aboutn.htm
National Clearinghouse for Alcohol and Drug Information

What constitutes moderate alcohol use?

People's tolerances to alcohol differ, and it is impossible to name an exact amount of alcohol that is appropriate for everyone. The amount a person can drink safely is highly individual and depends on genetics, health, sex, body composition, age, and family history. To provide people with a guideline, authorities have attempted to set limits that are acceptable for most healthy adults. This amount is supposed to be enough to elevate mood with a minimum risk of long-term harm to health. An accepted definition of moderation is not more than two drinks a day for the average-sized man and not more than one drink a day for the average-sized woman or for any person age 65 and over. Women and elderly people have less total body water than men, and therefore, the same amount of alcohol reaches higher alcohol concentrations. A drink is any alcoholic beverage that delivers ½ ounce of pure ethanol:

- 4 to 5 ounces of wine.
- 12 ounces of wine cooler.
- 12 ounces of regular or light beer.
- 1½ ounces of hard liquor (80 proof whiskey, scotch, brandy, rum, gin, or vodka).

Doubtless, some people could consume slightly more; others could not handle nearly so much without significant risk.

What happens to alcohol in the body?

Unlike the energy nutrients, which undergo digestion before they can be absorbed, 95 percent of the alcohol consumed is absorbed directly in the stomach or the small intestine. Once inside the bloodstream, alcohol easily travels to the body cells. To rid the body of alcohol, the liver must metabolize it. The liver performs this function at a fixed rate—that is, enzymes in the liver can metabolize only so much alcohol at a time, regardless of how much alcohol is present.

Within minutes of ingestion, alcohol reaches the brain where it quickly depresses inhibitions and judgments and affects motor skills and reaction times. If drinking continues, the breathing rate and heart rate slow. Death can follow if alcohol depresses breathing too low. Such an occurrence is uncommon, however. More likely, the body

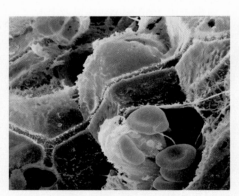

This micrograph of a normal liver shows several hepatocytes (reddish-brown color), bile ducts (green), red blood cells (bright red), and a macrophage (yellow).

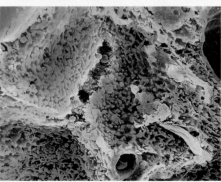

In a liver damaged by cirrhosis, the hepatocytes lose their structure and flexibility, altering the flow of blood and bile through the liver and ultimately interfering with the liver's ability to carry out its functions.

reacts to heavy drinking by vomiting to expel additional alcohol before it can be absorbed, or people lapse into comas before they can drink a fatal dose.

What are the long-term effects of alcohol on the body?

The easy access that alcohol has into the cells explains why so many organ systems can be affected by excessive alcohol intakes. The organ systems most often affected include the liver, the GI tract (including the pancreas), the heart, and the central nervous system. (The effects of alcohol on the developing fetus are described in Chapter 10.) Chronic, excessive alcohol use can lead to fatty liver, alcoholic hepatitis, and cirrhosis (see Chapter 23). In the GI tract, chronic alcohol abuse can slow motility and cause inflammation along the length of the GI tract (see Chapters 15 and 16). Reflux esophagitis, nausea, gastritis, ulcers, and chronic pancreatitis are common in people who drink too much alcohol. Alcohol consumption also increases the risks of cancer of the mouth, throat, and esophagus.[1] Heavy drinkers experience a high incidence of hypertension, heart attacks, arrhythmias, and deterioration of the heart muscle (see Chapter 21). Even moderate alcohol use alters brain function, and excessive alcohol use shrinks and destroys brain cells.

Other effects of alcohol on the body include sexual dysfunction and an increased susceptibility to lung infections and lung cancer. Women who drink alcohol may have a greater risk of breast cancer than women who do not drink.[2]

The more a person drinks, the greater the risks of long-term health consequences. Yet alcohol remains the most widely abused drug in the world. *Alcohol abuse* refers to patterns of drinking that result in health problems, social problems, or both. *Alcohol addiction,* often called alcoholism or alcohol dependence, refers to a disease that is characterized by abnormal alcohol-seeking behavior that leads to impaired

control over drinking.* Alcohol abusers and alcohol-addicted individuals experience many of the same harmful effects of alcohol consumption; the distinguishing characteristics of alcohol addiction are physical dependence on alcohol and an impaired ability to control alcohol intake. Alcohol abuse and addiction exert a heavy toll on the health of the 18 million people in the United States who meet the criteria for these conditions. The effects of alcohol on nutrition and metabolism—both directly and as a consequence of alcohol-related diseases—are significant.

_____

*Alcohol addiction means the same thing as alcohol dependence. This book uses the term *addiction* because it is more self-explanatory.

Every alcohol abuser and alcohol-addicted person should be considered at risk for poor nutrition status.

What are the direct effects of alcohol on nutrition status?

Alcohol is rich in energy (7 kcalories per gram), but like pure fat or sugar kcalories, the kcalories are empty, providing energy without nutrients. People who use moderate amounts of alcohol regularly may gain weight as a result of the extra kcalories they consume. For heavy drinkers, however, weight loss is more likely. Alcohol produces euphoria, which depresses appetite, and the more alcohol people drink, the less likely that they will eat enough food to obtain adequate nourishment. Table NP23-1 shows

Table NP23-1

kCALORIES IN ALCOHOLIC BEVERAGES AND MIXERS

Beverage	Amount (oz)	Energy (kcal)
Beer		
Regular	12	150
Light	12	78–131
Nonalcoholic	12	32–82
Distilled liquor (gin, rum, vodka, whiskey)		
80 proof	1½	100
86 proof	1½	105
90 proof	1½	110
Liqueurs		
Coffee liqueur	1½	175
Coffee and cream Liqueur	1½	155
Crème de menthe	1½	185
Mixers		
Club soda	12	0
Cola	12	150
Cranberry juice Cocktail	8	145
Diet drinks	12	2
Ginger ale	12	125
Grapefruit juice	8	95
Orange juice	8	110
Tomato or vegetable juice	8	45
Tonic	12	124
Wine		
Dessert	3½	160
Nonalcoholic	8	14
Red	3½	75
Rosé	3½	75
White	3½	70
Wine cooler	12	150

the kcalories in typical alcoholic beverages and mixers.

Nutrient deficiencies are an almost inevitable result of alcohol abuse, not only because the person who drinks heavily eats less and obtains fewer nutrients from food, but also because alcohol interferes with the way the body processes nutrients. Some researchers believe that energy utilization from large doses of alcohol is impaired because the alcohol is diverted into a metabolic pathway that generates less useful energy. Once alcohol has damaged the liver, GI tract, or cardiovascular system, many additional factors begin to impair a person's nutrition status, as previous chapters have described.

In what ways does alcohol affect how the body processes nutrients?

Alcohol produces major changes in the way the body absorbs and metabolizes nutrients. As Chapter 20

described, alcohol can induce hypoglycemia in any person. Hypoglycemia is most likely to develop in people who take a drink after they have been fasting for several hours. By-products of alcohol metabolism inhibit the synthesis of glucose from amino acids. When there is no food to supply glucose directly, hypoglycemia may develop.

Alcohol also interferes with the availability and activation of virtually every vitamin and many minerals. For example, as Figure NP23-1 shows, alcohol impairs absorption of thiamin. Furthermore, a liver damaged by alcohol may be unable to activate thiamin for optimal use by the body. Severe thiamin deficiencies cause major brain dysfunction characterized by disordered thinking, feeling, and memory and disturbances of motor coordination (Wernike-Korsakoff syndrome).

Alcohol abuse causes a folate deficiency that is self-aggravating. When alcohol is present, the body behaves as if it were actively trying

to expel folate from its sites of action and storage. The liver, which normally stores folate to meet the body's needs, instead releases folate into the blood. As blood folate rises, the kidneys react as if the body had an excess of folate and begin to excrete the perceived "excess." To make matters worse, alcohol also interferes with the action of what little folate is left, inhibiting the production of new cells, especially the rapidly dividing cells of the intestine and the blood. Damage to the intestine from both folate deficiency and alcohol toxicity impairs the intestine's ability to continuously release and retrieve folate as it normally does, and it also fails to absorb any folate that may trickle in from food as well. The combination of poor folate status and alcohol consumption can lead to anemia and has been implicated in promoting colorectal cancer.[3]

Alcohol abuse alters metabolism of many other vitamins, including vitamins B_6, B_{12}, and A. One of the products of alcohol metabolism dislodges vitamin B_6 from its protective binding protein so that it is destroyed. Alcohol inhibits vitamin B_{12} absorption, both directly and indirectly, by suppressing the secretion of the intrinsic factor that facilitates the vitamin's absorption from the intestine into the bloodstream. Alcohol does not impair absorption of vitamin A, but even moderate alcohol use has been shown to deplete liver stores of the vitamin.

Alcohol also promotes water excretion by the kidneys. Important minerals, such as zinc, magnesium, and potassium, are lost with the water. In short, alcohol directly and profoundly affects nutrition status.

Can an adequate diet protect the body from the harmful effects of alcohol?

No. Eating well and even taking supplements of protein, vitamins, and minerals do not protect the drinker. Even just a couple of drinks set in motion the destructive processes

Figure NP23-1
ALCOHOL'S EFFECT ON THIAMIN ABSORPTION

In the presence of alcohol, intestinal cells fail to absorb thiamin, except at very high concentrations. Similar effects occur for other vitamins.

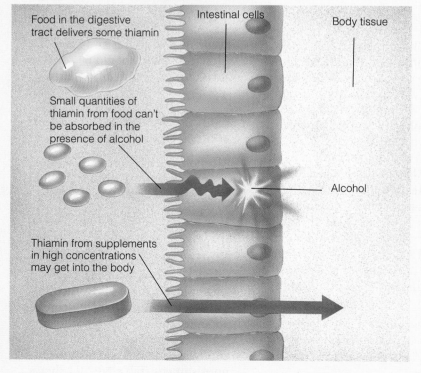

Food in the digestive tract delivers some thiamin

Intestinal cells

Body tissue

Small quantities of thiamin from food can't be absorbed in the presence of alcohol

Alcohol

Thiamin from supplements in high concentrations may get into the body

described, but if the drinking has been moderate, the next day's abstinence can repair the damage. Someplace between total abstinence and the extreme of alcoholism, there may be alcohol intakes moderate enough not to harm health, but the more a person drinks, the closer to a dangerous extreme that person comes.

Notes

[1] M. J. Thun and coauthors, Alcohol consumption and mortality among middle-aged and elderly U.S. adults, *New England Journal of Medicine* 337 (1997): 1705–1714; D. M. Winn, Diet and nutrition in the etiology of oral cancer, *American Journal of Clinical Nutrition* 61 (1995): S437–S445; C. S. Lieber, Herman Award Lecture, 1993: A personal perspective on alcohol, nutrition, and the liver, *American Journal of Clinical Nutrition* 58 (1993): 430–442.

[2] S. A. Smith-Warner and coauthors, Alcohol and breast cancer in women: A pooled analysis of cohort studies, *Journal of the American Medical Association* 279 (1998): 535–540.

[3] A. E. Rogers, Methyl donors in the diet and responses to chemical carcinogens, *American Journal of Clinical Nutrition* (supplement) 61 (1995): S659–S665; Folate, alcohol, methionine, and colon cancer risk: Is there a unifying theme? *Nutrition Reviews* 52 (1994): 18–20.

CHAPTER 24

NUTRITION, CANCER, AND HIV INFECTION

CONTENTS

Cancer

HIV Infection

Case Study: Travel Agent
with HIV Infection

Nutrition in Practice:
Alternative Therapies

Although cancer and HIV (human immunodeficiency virus) infections are distinct disorders, from a nutrition standpoint, they share many similarities. Both disorders can impair many organ systems that affect nutrient needs, and both can lead to severe wasting. In addition, both require medical nutrition therapy that is highly individualized based on the symptoms manifested and the organ systems involved.

People with cancer take comfort from the support of others and from the knowledge that medical science is waging an unrelenting battle to find better treatments and possible cures.

Cancer

Although the thought of cancer often strikes fear in people, the prognosis for most people with cancer today is far brighter than in the past. New techniques for detecting **cancers** early and innovative therapies to treat cancers offer hope and encouragement.

Cancer is not a single disorder; instead, there are many different cancers. They have different characteristics, occur in different locations in the body, take different courses, and require different treatments. Whereas an isolated, nonspreading type of skin cancer may be removed in a physician's office with no observable effect on nutrition status, advanced cancers (especially those of the GI tract, pancreas, and liver) can seriously impair nutrition status.

cancers: diseases that result from the unchecked growth of cells.

How Cancers Develop

The genes in a healthy body work together to regulate cell division and ensure that each new cell is a replica of the parent cell. In this way, the healthy body grows, replacing dead cells and repairing damaged ones. Cancers develop from mutations in the genes that normally regulate cell division. The mutations silence the genes that ordinarily monitor replicating DNA for chemical errors. The affected cells seemingly have no brakes to halt cell division. As the abnormal mass of cells, called a malignant **tumor,** grows, blood vessels form to supply the tumor with the nutrients it needs to support its growth. Eventually, the tumor invades healthy tissue and may spread. In leukemia (cancer of the blood-forming cells of the bone marrow), the cancer cells do not form a tumor, but rather circulate with the blood through other tissues where they can accumulate. Clinicians describe cancers by their type, size, and extent, specifically noting if the tumor has spread to surrounding lymph nodes or to distant sites in the body. Figure 24-1 on p. 564 illustrates tumor development.

tumor: a new growth of tissue forming an abnormal mass with no function; also called a **neoplasm (NEE-oh-plazm).** Tumors that multiply out of control, threaten health, and require treatment are **malignant (ma-LIG-nant).** Tumors that stop growing without intervention or can be removed surgically and most often pose no threat to health are **benign (bee-NINE).**

malignus = of bad kind
benign = mild

GENETIC FACTORS Some cancers appear to have a genetic component. A person with a family history of colon cancer, for example, has a greater risk of colon cancer than a person without such a genetic predisposition. This does not mean, however, that the person *will* develop cancer, only that the risk is greater.

IMMUNE FACTORS A healthy immune system recognizes foreign cells and destroys them. Researchers theorize that an ineffective immune system may not recognize tumor cells as foreign, thus allowing tumor growth. Aging affects immune function, and the incidence of cancer increases with age. Medications that suppress the immune system (immunosuppressants) and viral infections (including HIV infection) and other disorders that severely tax the immune system may increase the risk of cancer.

ENVIRONMENTAL FACTORS Among environmental factors, exposure to radiation and sun, water and air pollution, and smoking are known to cause cancer. As Table 24-1 on p. 565 shows, dietary constituents are also associated with an increased risk of certain cancers. Some dietary factors may initiate

Cancers are classified by the tissues or cells from which they develop:

- **Adenomas (ADD-eh-NO-mahz)** arise from glandular tissues.
- **Carcinomas (KAR-see-NO-mahz)** arise from epithelial tissues.
- **Gliomas (gly-OH-mahz)** arise from the glial cells of the central nervous system.
- **Leukemias (loo-KEY-mee-ahz)** arise from the blood-forming cells of the bone marrow.
- **Lymphomas (lim-FOE-mahz)** arise from lymph tissue.
- **Melanomas (MEL-ah-NO-mahz)** arise from pigmented skin cells.
- **Sarcomas (sar-KO-mahz)** arise from muscle, bone, or connective tissue.

Figure 24-1
TUMOR FORMATION

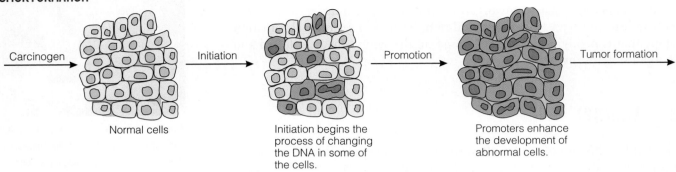

Carcinogen → Initiation → Promotion → Tumor formation

Normal cells

Initiation begins the process of changing the DNA in some of the cells.

Promoters enhance the development of abnormal cells.

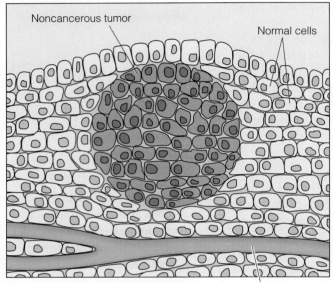

Noncancerous tumor

Normal cells

Blood vessel

A noncancerous (benign) tumor usually grows within a self-contained capsule. It does not invade nearby tissue, nor does it spread.

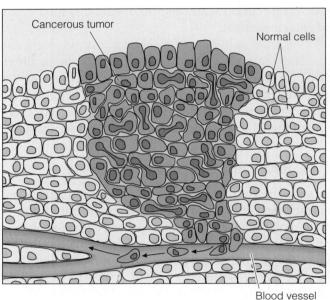

Cancerous tumor

Normal cells

Blood vessel

A cancerous (malignant) tumor usually grows out of control and may spread to other parts of the body through the blood or lymph systems.

A cancer that spreads from one part of the body to another is said to metastasize *(meh-TAS-tah-size).*

carcinogenic (CAR-sin-oh-JEN-ick):
producing cancer. A substance that produces or enhances the development of cancer is a **carcinogen (car-SIN-oh-jen).**
  carcin = cancer
  genesis = gives rise to

Factors that cause mutations that give rise to cancer, such as radiation and carcinogens, are called initiators.

cancer development, others may promote cancer development once it has started, and still others may protect against the development of cancer.

■■ **DIETARY FACTORS—CANCER INITIATORS** We do not know to what extent diet contributes to cancer development, although some experts estimate that diet may be linked to a third or more of all cases. Consequently, many people think that certain foods are **carcinogenic,** especially those that contain additives or pesticides. As Nutrition in Practice 5 explained, however, our food supply is one of the safest in the world. Additives that have been approved for use in foods are not carcinogenic. Some pesticides are carcinogenic at high doses, but not at the concentrations allowed on fruits and vegetables.

The incidence of cancers, especially stomach cancers, is high in parts of the world where people eat a lot of heavily pickled or salt-cured foods that produce carcinogenic nitrosamines. Most commercial manufacturers in the United States use different preservative methods, and all are carefully controlled to minimize carcinogenic contamination. Alcohol has also been associated with a high incidence of some cancers, especially of the mouth, esophagus, and liver (see Nutrition in Practice 23). Breast cancer has also been associated with alcohol. Beverages such as beer and scotch may contain damaging nitrosamines as well as alcohol.[1] The amounts of these compounds in

Table 24-1
FACTORS ASSOCIATED WITH SPECIFIC CANCERS

Cancer Sites	High Incidence Associated with:	Protective Effect Associated with:
Pancreatic cancer	*Possibly* meat, cholesterol	Fruits and vegetables, *possibly* vitamin C and fiber
Esophageal cancer	Alcohol, tobacco, and especially combined use; preserved foods (such as pickles); high intakes of vitamin A supplements	Fruits and vegetables
Stomach cancer	Salt-preserved foods (such as dried, salted fish); *possibly* grilling and barbecuing	Fresh fruits and vegetables; safe food handling methods; vitamin C
Colorectal cancer	Fat (particularly saturated fat), meat, and alcohol (especially beer)	Vegetables; physical activity; *possibly* calcium; omega-3 fatty acids
Liver cancer	Infection with hepatitis B or C; aflatoxins; alcohol; iron overload	
Lung cancer	Cigarette smoking primarily; *possibly* alcohol, fat (notably saturated fat), and cholesterol	Fruits and vegetables, especially green and yellow ones; *possibly* carotenes, vitamin C, and selenium; physical activity
Breast cancer	Obesity; alcohol; *possibly* meat and dietary fat	Fruits and vegetables, especially green and yellow ones; *possibly* carotenes, fiber, and physical activity
Ovarian cancer	No dietary risk factors have been established; inversely correlated with oral contraceptive use	Fruits and vegetables, especially green and yellow ones
Cervical cancer	Folate deficiency; cigarette smoking	*Possibly* fruits, vegetables, carotenes, and vitamin C
Endometrial cancer	Obesity, estrogen therapy, hypertension, and diabetes (type 2)	*Possibly* fruits and vegetables
Bladder cancer	Cigarette smoking; *possibly* artificial sweeteners, and alcohol	Fruits and vegetables, especially green and yellow ones
Prostate cancer	High fat intake, especially saturated fats from meats	Fruits and vegetables, especially green and yellow ones

NOTE: Findings based on epidemiological studies.
SOURCES: C. Marwick, Global review of diet and cancer links available, *Journal of the American Medical Association* 278 (1997): 1650–1651; *Diet and Health: Implications for Reducing Chronic Disease Risk* (Washington, D.C.: National Academy Press, 1989), pp. 594–600; J. H. Weisburger, Nutritional approach to cancer prevention with emphasis on vitamins, antioxidants, and carotenoids, *American Journal of Clinical Nutrition* (supplement) 53 (1991): 226–237; R. G. Ziegler, Vegetables, fruits, and carotenoids and the risk of cancer, *American Journal of Clinical Nutrition* (supplement) 53 (1991): 251–259; Potential mechanisms for food-related carcinogens and anti-carcinogens: A scientific status summary by the Institute of Food Technologists' Expert Panel on Food Safety and Nutrition, *Food Technology* 47 (1993): 105–118.

alcoholic beverages currently on the market are not considered harmful—assuming consumption in moderate amounts.

▪▪ DIETARY FACTORS—CANCER PROMOTERS Unlike carcinogens, which initiate cancers, some dietary components may accelerate cancers that have already begun to develop. Studies suggest that certain dietary fats eaten in excess may promote cancer, in part by contributing to obesity. More specifically, linoleic acid, the omega-6 fatty acid of vegetable oils, has been implicated in enhancing cancer development in some animals; in contrast, omega-3 fatty acids from fish oils appear to delay cancer development.[2]

▪▪ DIETARY FACTORS—ANTIPROMOTERS It seems apparent that foods may also contain antipromoters. Almost without exception, epidemiological studies find a link between eating plenty of fruits and vegetables and a low incidence of cancers. The fiber in fruits and vegetables helps to protect against some cancers by speeding up the transit time of all materials through the GI tract so that the colon walls are not exposed to cancer-causing substances for long. In addition to fiber, fruits and vegetables contain both nutrients and nonnutrients (phytochemicals) that protect against cancer. The nutrients beta-carotene, vitamin C, vitamin E, and some phytochemicals act as antioxidants and may help to prevent tissue damage caused by oxygen-derived free radicals that may give rise to cancer. Other phytochemicals common to many

Factors that favor the development of cancers once they have begun are called **promoters.**

Factors that oppose the development of cancer are called **antipromoters.**

vegetables, especially those of the cabbage family (cruciferous vegetables), can activate enzymes that are capable of destroying carcinogens.

■■ **DIETARY RECOMMENDATIONS FOR REDUCING CANCER RISK**   Current dietary recommendations for cancer prevention include the following guidelines:

- Control weight and prevent obesity.
- Reduce consumption of total fat to 30 percent or less of total energy intake and saturated fat to 10 percent or less of total energy.
- Increase fiber intake to 20 to 35 grams per day.
- Include a variety of fresh vegetables and fruits (including deep yellow and dark green cruciferous vegetables) in the daily diet.
- Consume salt-cured, salt-pickled, nitrite-cured, and smoked foods in moderation.
- Consume alcoholic beverages in moderation, if at all.

One additional recommendation is in order: vary food choices. This last suggestion is based on an important concept that applies specifically to the prevention of cancer initiation—dilution. Eating a variety of foods dilutes the negative qualities of any one food. For example, it is safe to eat some salt-cured meats, but not at every meal. Combine such foods with a variety of others so that any carcinogen that may be present will be diluted in the total diet.

THE NURSING DIAGNOSIS health seeking behaviors applies to people who wish to reduce their risk of cancer by modifying their diets.

■■   **Consequences of Cancer**

Once cancer develops, its consequences depend on its location, severity, and treatment. Cancer of the lungs, for example, will have different consequences than cancer of the kidneys. In the early stages, some cancers produce no symptoms, and the person may be unaware of any threat to health. Efforts to detect cancers early aim to identify cancers before they have progressed too far.

■■   **WASTING ASSOCIATED WITH CANCER**   Loss of appetite, weight loss, depletion of lean body mass and serum proteins, and debilitation typify the **cancer cachexia syndrome,** which occurs in as many as 80 percent of people with cancer before they die.[3] Weight loss is often evident at the time cancer is diagnosed, and malnutrition may be the ultimate cause of death in many cases.

THE NURSING DIAGNOSES altered nutrition: less than body requirements, risk for infection, risk for injury, risk for impaired skin integrity, fatigue, and **risk for activity intolerance** apply to some people with cancer cachexia.

Many factors appear to play a role in the wasting associated with cancer, although the exact causes are unknown. Cytokines, proteins produced as the immune system wages a battle against the tumor cells, appear to play an important role.[4] Whatever the cause, the combination of poor appetite, accelerated and altered metabolism, and the diversion of nutrients to support tumor growth simultaneously reduces the supply of energy and nutrients and raises the demand for them.

People who develop cachexia swiftly fall into a downward spiral. Once weight loss and wasting have been set in motion, the debilitation and general poor health that follow make it even more difficult for the person to eat. Figure 24-2 summarizes some of the factors that contribute to the cancer cachexia syndrome.

cancer cachexia (ka-KEKS-ee-ah) syndrome: a syndrome that frequently accompanies many types of cancer and is characterized by anorexia, loss of lean body mass, depletion of serum proteins, and debilitation.

Reminder: Cytokines are proteins secreted as part of the body's immune responses.

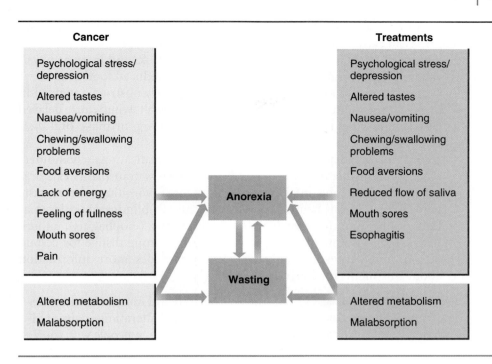

Figure 24-2
THE CANCER CACHEXIA SYNDROME: CONTRIBUTING FACTORS

Anorexia and wasting contribute to each other. Cancer and its treatments make both problems worse.

Cancer

Psychological stress/depression

Altered tastes

Nausea/vomiting

Chewing/swallowing problems

Food aversions

Lack of energy

Feeling of fullness

Mouth sores

Pain

Altered metabolism

Malabsorption

Treatments

Psychological stress/depression

Altered tastes

Nausea/vomiting

Chewing/swallowing problems

Food aversions

Reduced flow of saliva

Mouth sores

Esophagitis

Altered metabolism

Malabsorption

Anorexia

Wasting

LEARNING LINK

When you began your study of clinical nutrition in Chapter 13, you learned that identifying protein-energy malnutrition (PEM) is a vital role that nurses play in the nutrition care of clients. Perhaps now that you have read about the many devastating effects that different disorders can wreak on nutrition status, a reminder will have more meaning. Wasting depletes the body's proteins, whether it occurs as a consequence of acute stresses or chronic disorders like cancer, HIV infections, malabsorption syndromes, chronic heart failure, chronic obstructive pulmonary diseases, renal failure, or liver failure. Studies have shown that once lean body mass is significantly depleted, death will follow regardless of the cause of the protein loss.[5] Without adequate protein, the body is poorly equipped to maintain immune defenses, support organ function, absorb nutrients, mend damaged tissues, and utilize medications. The same disorders that lead to wasting require many nursing interventions—don't let the client's nutrition needs slip through the cracks!

■■ **ANOREXIA AND REDUCED FOOD INTAKE** Anorexia is a major contributor to wasting associated with cancer. Some factors that can contribute to anorexia or otherwise reduce food intake in the person with cancer include:

- *Chronic nausea and early satiety.* People with cancer frequently experience nausea and a premature feeling of fullness after eating small amounts of food.
- *Fatigue.* People with cancer often tire easily and lack energy to prepare and eat meals. Once wasting is evident, these tasks become even more difficult for the person to handle.
- *Pain.* People in pain may have little interest in eating, particularly if eating makes the pain worse.
- *Psychological stress.* The very diagnosis of cancer can cause so much distress that eating becomes unimportant. The person's fear and anxiety

about the medical, personal, and financial concerns created by the diagnosis can compound stress. Once wasting is evident, people may become depressed by their inability to perform routine tasks and by their physical appearance, and depression can also lead to reduced food intake.

- *Obstructions.* A tumor may partially or completely obstruct any portion of the GI tract and interfere with chewing and swallowing; cause delayed gastric emptying, early satiety, nausea, or vomiting; or make oral diets impossible.

- *Cancer therapy.* Therapy for cancer including medications, chemotherapy, radiation therapy, surgery, and bone marrow transplants can dramatically affect food intake by causing nausea, vomiting, altered taste perceptions, diminished taste sensitivity (mouth blindness), inflammation of the mouth (stomatitis) and esophagus (esophagitis), mouth ulcers, reduced flow of saliva, food aversions (strong dislike for certain foods), and depression. (A later section provides more information about treatments for cancer.)

> **THE NURSING DIAGNOSES** hyperthermia, altered oral mucous membrane, altered protection, fatigue, sensory/perceptual alterations, pain, anxiety, and fear increase the risk of poor food intake in people with cancer.

■■ **METABOLIC ALTERATIONS AND NUTRIENT LOSSES** In cancer, metabolic pathways use nutrients in inefficient ways that demand more energy and waste vital protein tissues. These metabolic alterations explain why some people with cancer fail to regain lean body mass even when they are receiving adequate energy and nutrients. Some people with cancer are hypermetabolic and have high nutrient needs; others may become hypermetabolic as a consequence of surgery or infection. Chemotherapy can interfere with normal metabolic pathways and create nutrient deficiencies as well.

Like anorexia, nutrient losses can develop as a consequence of either cancer itself or the treatment for it. Causes of nutrient losses include inadequate digestion, malabsorption, vomiting, and **radiation enteritis.** Cancers of the pancreas, liver, or small intestine frequently lead to malabsorption. Radiation therapy to the small intestine can cause radiation enteritis, which can result in malabsorption, chronic blood loss, fluid and electrolyte imbalances, and, sometimes, intestinal obstructions and fistulas. Intestinal function may return after radiation therapy ends, but for some, the changes are permanent. Severe diarrhea and malabsorption, with fluid losses often exceeding 10 liters a day, can occur in people who undergo bone marrow transplants and then reject the transplanted tissue.

> **THE NURSING DIAGNOSES** diarrhea, fluid volume deficit, and risk for fluid volume deficit may apply to people with cancer. Many other nursing diagnoses with nutrition implications may apply to people with specific types of cancer.

■■ **Treatments for Cancer**

The primary treatments for cancer—**radiation therapy, chemotherapy,** surgery, or any combination of the three—aim to annihilate cancer cells, relieve pain, alleviate symptoms, and prevent tumor growth. Table 24-2 summarizes the nutrition-related side effects of radiation therapy and chemotherapy, and Table 24-3 shows how various cancer surgeries can affect nutrition status. The use of **bone marrow transplants** to treat certain cancers has increased markedly.[6] To prepare a person for a bone marrow transplant, high doses of chemotherapy and sometimes whole-body radiation are used to eradicate

---

Cancer-induced causes of nutrient losses can include:
- **Inadequate digestion.**
- **Malabsorption.**
- **Vomiting.**
- **Diarrhea.**

radiation enteritis: inflammation and scarring of the intestinal cells caused by exposure to radiation.

radiation therapy: the use of radiation to arrest or destroy cancer cells.

chemotherapy: the use of drugs to arrest or destroy cancer cells. Drugs used for chemotherapy are called **chemotherapeutic** or **antineoplastic agents.**

bone marrow transplant: the replacement of diseased bone marrow in a recipient with healthy bone marrow from a donor; sometimes used as a treatment for breast cancer, leukemia, lymphomas, and certain blood disorders.

Table 24-2

POSSIBLE CAUSES OF WASTING ASSOCIATED WITH RADIATION AND CHEMOTHERAPY

	Reduced Nutrient Intake	Accelerated Nutrient Losses	Altered Metabolism
Radiation	Anorexia	Chronic blood loss from intestine and bladder	Secondary effects of malnutrition or infection
	Damage to teeth and jaws	Diarrhea	
	Esophagitis	Fistula formation	
	Mouth ulcers	Intestinal obstructions	
	Nausea	Malabsorption	
	Reduced salivary secretions	Radiation enteritis	
	Taste alterations	Vomiting	
	Thick salivary secretions		
	Vomiting		
Chemotherapy	Abdominal pain	Diarrhea	Fluid and electrolyte imbalances
	Anorexia	Intestinal ulcers	Hyperglycemia
	Mouth ulcers	Malabsorption	Interference with vitamins or other metabolites
	Nausea	Vomiting	Negative nitrogen and calcium balance
	Taste alterations		Secondary effects of malnutrition or infection
	Vomiting		

cancer cells. Thus, the nutrition-related side effects of these treatments apply to bone marrow transplants as well. In addition, immunosuppressants, given to help prevent **tissue rejection** following a bone marrow transplant, have multiple effects on nutrition status. The Diet-Medication Interactions box on p. 579 includes interactions from medications used in the treatment of cancer.

tissue rejection: destruction of healthy donor cells by the recipient's immune system, which recognizes the donor cells as foreign; also called **graft-versus-host disease (GVHD).**

Table 24-3

POSSIBLE EFFECTS OF SURGERIES FOR CANCERS ON NUTRITION STATUS

Head and Neck Resection

Difficulty in chewing/swallowing	Inability to chew/swallow

Esophageal Resection

Diarrhea	Reduced gastric motility
Fistula formation	Steatorrhea (fat malabsorption)
Reduced gastric acid secretion	Stenosis (constriction)

Gastric Resection

Dumping syndrome	Lack of gastric acid
General malabsorption	Vitamin B_{12} malabsorption
Hypoglycemia	

Intestinal Resection

Blind loop syndrome	Malabsorption
Diarrhea	Hyperoxaluria
Fluid and electrolyte imbalance	Steatorrhea

Pancreatic Resection

Diabetes mellitus	Malabsorption

Megestrol acetate is believed to exert its effects on appetite by altering the synthesis and release of certain cytokines.

▪▪ MEDICATIONS TO COMBAT ANOREXIA AND WASTING To help people with advanced cancers combat anorexia, medications that stimulate the appetite have gained wide use. One of the most promising medications, megestrol acetate, stimulates the appetite and promotes weight gain. Dronabinol (a medication containing the principal psychoactive ingredient in marijuana) works as both an appetite stimulant and an antiemetic, although side effects including euphoria and confusion limit its use in some cases.

Even when diets supply seemingly adequate amounts of energy and nutrients, reversing wasting is often ineffective. People may be able to regain weight, but the weight gain is often in the form of fat rather than lean body mass. Under investigation are other medications, notably anabolic steroids (see the box on p. 579), which may help to restore lean body mass.

▪▪ OTHER MEDICATIONS Depending on the organ systems affected by cancer and by the side effects of treatments, many medications may be used in treatment. In addition to antineoplastic agents (chemotherapy), medications commonly used to treat symptoms of cancer include antinausea agents, antidiarrheals, analgesics, and sedatives.

▪▪ ALTERNATIVE THERAPIES People who feel they are making little progress in their fight against cancer or who think conventional medicine offers little hope of recovery may turn to alternative therapies, the subject of this chapter's Nutrition in Practice. Still others may choose to use alternative therapies along with traditional medical treatments.

▪▪ Medical Nutrition Therapy

Although nutrition cannot change the ultimate outcome of cancer, attention to nutrition can bolster the immune system and help people maintain their strength and nutrition status while undergoing stressful treatments and coping with their diseases. Compared to malnourished people with cancer, well-nourished people with cancer enjoy a better quality of life—that is, they feel better, function better, are stronger and more active, and eat more.[7] Malnourished people with cancer may also have a poorer response to cancer treatments and a shorter survival time than people in good nutrition status.[8] At a minimum, meeting nutrient needs eliminates the additional stresses imposed by malnutrition.

▪▪ EARLY NUTRITION INTERVENTION For people with cancer, early nutrition intervention takes a high priority. The initial nutrition assessment evaluates the individual's current nutrition status and establishes baseline parameters from which to monitor changes. Early intervention helps to prepare the person for the stresses ahead and helps detect and correct deficiencies before the task becomes monumental.

▪▪ ORAL DIETS For people with cancer, medical nutrition therapy must account for the organ systems involved and the type and severity of the cancer (see Table 24-4). In addition, a thorough nutrition assessment uncovers specific symptoms that each person is experiencing, and the interventions address these symptoms. The "How to" box on pp. 572–574 outlines some strategies for improving oral intake for people who are having problems eating and maintaining weight. For these people, the diet often aims to supply about 150 percent of the basal energy expenditure (see Table 19-3 on p. 454) and 1.5 to 2.0 grams of protein per kilogram of body weight per day. The sug-

Table 24-4

DIETARY CONSIDERATIONS FOR SPECIFIC CANCERS

Cancer Sites	Dietary Considerations
Brain	Physical feeding disabilities (see Nutrition in Practice 15); chewing and swallowing problems (see Chapter 15).
Head/neck	Chewing and swallowing problems.
Mouth/esophagus	Chewing and swallowing problems; if obstructed, tube feeding below the obstruction may be necessary.
Stomach	Nausea, vomiting; if obstructed, tube feeding below the obstruction or TPN may be necessary; if resection is performed, a postgastrectomy diet (see Chapter 15) may be needed; nutrient deficiencies due to bacterial overgrowth (Chapter 16) may occur.
Intestine	If obstructed, tube feeding or TPN may be necessary; resections or inflammation may cause multiple nutrition problems (see Chapter 16); fat- and lactose-restricted diet may be useful.
Liver	Protein-, sodium-, and fluid-restricted diet may be necessary (see Chapter 23).
Pancreas	Fat-restricted diet and enzyme replacements may be necessary (see Chapter 16); diabetic diet may be necessary if insulin production is affected (see Chapter 20).
Kidneys	Protein-, electrolyte-, and fluid-controlled diet may be necessary (see Chapter 22).

NOTE: The considerations listed here are specific to the type of cancer; they do not include other nutrition-related concerns, such as anorexia, nausea, and vomiting.

gestions are numerous and detailed, reflecting both the complexity of the problems and the importance of offering specific suggestions to deal with specific problems. Clients who are unable to eat adequate amounts of table foods often benefit from nutrient-dense enteral formula supplements. Limited research suggests that formulas supplemented with fish oils may help moderate weight loss in people with cancer.[9]

Not all clients with cancer lose weight. Women diagnosed with breast cancer often gain, rather than lose weight, and this weight gain can be distressing. Weight gain may continue for some time after diagnosis. By encouraging physical activity, regularly assessing weight and energy intakes, and recommending strategies to correct problems early, nurses and dietitians can help clients avoid unnecessary weight gain.[10]

THE NURSING DIAGNOSES altered nutrition: more than body requirements and altered nutrition: risk for more than body requirements are appropriate for some people with cancer.

■■ TUBE FEEDINGS AND TPN In general, a tube feeding or TPN is not routinely recommended for an adequately nourished or mildly malnourished person with cancer who is unable to eat. Special nutrition support is indicated when anorexia persists or when a person is severely malnourished and is about to undergo aggressive cancer therapy. Each case is decided individually, and the use of special nutrition support is more likely when the person's chances of recovery or of significant response to treatment are good, or when the type of cancer is associated with a high risk of death from malnutrition.[11] People requiring head and neck resections, for example, may need long-term tube feedings and may need to continue tube feedings at home. People with severe radiation enteritis may require home TPN.

HELP CLIENTS HANDLE FOOD-RELATED PROBLEMS

For people with cancer or HIV infections, many problems can interfere with eating. It is important to find out what problems clients are having with food. The following list offers possible solutions to improve food intake based on answers to the question "What problems are you having with eating or food?" For all clients, explain how eating can help them feel better. Not all of the suggestions will work for each client; encourage clients to experiment and find the ones that work best.

I just don't have an appetite.
- Eat small meals and snacks at regular times each day.
- Eat the most food at the time of day when you feel the best.
- Use nutrient-dense foods for meals and snacks. (Suggestions are provided later.)
- Eat nutrient-dense foods first.
- Indulge in favorite foods throughout the day.
- Avoid drinking large amounts of liquids with meals.
- Eat in a pleasant and relaxed environment.
- Listen to your favorite music or enjoy a program on TV while you eat.
- Eat with family and friends.
- Serve foods attractively.
- Take a walk before you eat.

I am too tired to fix meals and eat.
- Let friends and family members prepare food for you.
- Use foods that are easy to prepare and eat like sandwiches, frozen dinners, meals from take-out restaurants, instant breakfast drinks, liquid formulas, and supplements in candy bar and pudding form.
- Eat nutrient-dense foods first.

Foods just don't taste right.
- Brush your teeth or use a mouthwash before you eat.
- Add sauces and seasonings to meats.
- Eat meats cold or at room temperature.
- Use eggs, fish, poultry, and dairy products instead of meats.
- Try new foods and experiment with herbs and spices.
- Use plastic, rather than metal, eating utensils.
- Ask your doctor about zinc supplements. If you have a deficiency, your tastes may change.

I am nauseated a lot of the time, and sometimes I throw up.
- If you feel nauseated in the morning, eat a few dry saltine crackers before you get out of bed.
- If you take antinausea medications, take the medication so that it will work during meals.
- Eat frequent small meals.
- Eat dry salty foods like saltine crackers, baked pretzels, and plain salted popcorn.
- Chew foods thoroughly and eat at a relaxed pace.
- Try breathing slowly and deeply when you begin to experience nausea.
- Avoid spicy and high-fat foods.
- Avoid foods with odors that make you nauseated. It may help to avoid foods with strong odors or to ask others to prepare meals, if possible.
- Save most liquids for after meals. Clear liquids or popsicles after meals help prevent dehydration.

HELP CLIENTS HANDLE FOOD-RELATED PROBLEMS

- Sip cold, carbonated beverages that do not contain caffeine when you begin to experience nausea.
- Don't lie down after eating.
- Rest after meals and loosen tight clothing. Getting fresh air after eating can also help.
- If you experience vomiting, use clear liquids like broths, carbonated beverages, juices, jello, or popsicles to replace fluids and electrolytes.
- Try using ginger or peppermint teas, which provide relief for some people.
- If you become nauseated from chemotherapy treatments, avoid eating for at least 2 hours before treatments.

Some of the foods I really used to like, I can't stand anymore.

- Save your favorite foods for times when you are not feeling nauseated or sick to your stomach.
- Maintain a food-free "window" of an hour or so before and after you have treatments or take medications that cause nausea or vomiting.

I am having problems chewing and swallowing food.

- Experiment with food consistencies to find the ones you can handle best. Thin liquids, true solids, and sticky foods (like peanut butter) are often difficult to swallow.
- Add sauces and gravies to dry foods.
- Drink fluids with meals to ease chewing and swallowing.
- Try using a straw to drink liquids.
- Tilt your head forward and backward to see if you can swallow easier with the head positioned differently.

I have sores in my mouth and they hurt when I eat.

- Use cold or frozen foods; they are often soothing.
- Try soft, soothing foods like ice cream, milk shakes, bananas, applesauce, mashed potatoes, cottage cheese, and macaroni and cheese.
- Avoid foods that irritate mouth sores like citrus fruits and juices, tomatoes and tomato-based products, spicy foods, foods that are very salty, foods with seeds (like poppy seeds and sesame seeds) that can be trapped in the sore, and coarse foods like raw vegetables and toast.
- Ask your doctor about using a local anesthetic solution like lidocaine before eating to reduce pain.
- Use a straw for drinking liquids, to bypass the sores.

My mouth is really dry.

- Rinse your mouth with warm salt water or mouthwash frequently, and drink liquids between meals.
- Ask your doctor or pharmacist about medications that can help with dry mouth.
- Use sour candy or gum to stimulate the flow of saliva.
- Make sure you brush your teeth and floss regularly to prevent cavities and oral infections.

I am having trouble with diarrhea.

- Drink plenty of fluids. Salty broths and soups, diluted fruit juices, and sports drinks are good choices. For severe diarrhea, try commercially prepared oral rehydration formulas (see Nutrition in Practice 9).
- Eat foods with soluble fiber such as bananas, peeled apples, applesauce, peeled pears, oranges, peeled white or sweet potatoes, oatmeal, barley, millet, couscous, noodles, rice, white bread toast, and saltine crackers;

Continued

HELP CLIENTS HANDLE FOOD-RELATED PROBLEMS

temporarily limit foods with insoluble fibers like whole-wheat bread, bran, raw vegetables, corn, peas, nuts and seeds, and skins of fruit.

- Temporarily avoid foods that cause gas like dried beans and peas, broccoli, cucumbers, and onions.
- Take lactase enzyme replacements when you use dairy products because you may also experience lactose intolerance while you are having diarrhea. You may be able to tolerate low-fat yogurt.
- Avoid high-fat foods and foods made with lots of sugar.
- Avoid caffeine.
- Eat smaller meals more often.
- Check with your doctor about using digestive enzyme replacements if you have diarrhea for a long time.

I am having trouble with constipation.
- Drink plenty of fluids. Try warm fluids, especially in the morning.
- Eat whole-grain breads and cereals, nuts, fresh fruits, prunes, prune juice, and raw vegetables. Avoid refined carbohydrates like white bread, white rice, and pasta.
- Exercise regularly.

I need to gain weight, but my blood lipids are elevated.
- To gain weight, you will need to eat more fat, but the type of fat you use is important. Use unhydrogenated margarine in tubs (some are available that contain no *trans*-fatty acids), but don't use the low-kcalorie types.
- Use more monounsaturated fats like olive, canola, and peanut oils for baking and frying and in salad dressings and dips. Snack on avocados, nuts, and peanut butter. Make guacamole dip and spread it on vegetables and low-fat crackers. Spread peanut butter on celery, cucumbers, fruit, and low-fat crackers.
- To increase protein in your diet without adding too much of the fat you need to avoid, eat larger-than-usual servings of chicken and fish, especially salmon. Add chicken, fish, or low-fat and fat-free cheeses to sauces, soups, casseroles, and vegetables. Use instant breakfast drinks made with low-fat milk, or use commercially available supplements that contain no more than 30 percent fat. Add low-fat milk or milk powder to meat loaves, casseroles, soups, puddings, and cereals. Use low-fat yogurt.
- Use plenty of dried fruits. Add nuts and dried fruits to desserts, cereals, and salads.

I need help figuring out how to eat more energy and protein. (Note: All of the suggestions directly above apply here as well, but for some people the need for energy and protein outweighs the need to restrict the type of fat.)
- Use whole milk and regular yogurt instead of the low-fat or fat-free varieties.
- Use plenty of butter, margarine, mayonnaise, cream cheese, oil, and salad dressings on breads, sandwiches, potatoes, vegetables, salads, pasta, and rice.
- Use yogurt, sour cream, or a sour cream dip with vegetables.
- Add whipping cream to desserts and hot chocolate, or use it to lighten coffee.
- Use cream instead of milk with cereal.

Every malnourished person with cancer or HIV infection (described later in the chapter) who cannot eat an adequate diet orally is a potential candidate for a tube feeding or TPN. This chapter describes uses of tube feedings or TPN as they are applied to people with cancer and HIV infections who have a chance of recovery (for cancer) or a reasonable life expectancy. When incurable cancer or HIV infection has reached its final stages, however, the person, caregivers, and the health care team should consider some important ethical issues before a tube feeding or parenteral nutrition support is undertaken (see Nutrition in Practice 18). What is the prognosis if the wasting can be corrected or reversed? Will the person survive longer? Will quality of life improve? Will treatments have a better chance of success? Does the client understand that nutrition support will not change the ultimate outcome of the disease? Does the client understand the costs of nutrition support? Does the client want aggressive nutrition support? Special nutrition support should be undertaken only if the client clearly understands the benefits and risks and chooses it.

■■ NUTRITION SUPPORT AND BONE MARROW TRANSPLANTS The person undergoing a bone marrow transplant routinely receives TPN before and after the transplant because the GI tract is severely compromised by the preparatory procedure. Researchers have found that adding glutamine to the TPN solution results in fewer infections and shorter hospital stays for people undergoing bone marrow transplants.[12]

When GI tract function returns, the person begins to receive foods orally, and TPN is gradually tapered off. After a bone marrow transplant, nutrition problems can be severe and debilitating, especially for people who reject the transplant and develop GI complications. Early oral feedings often start with lactose-free, low-residue, low-fat liquids to maximize absorption and minimize nausea, vomiting, and fat malabsorption. Gradually, solid foods are introduced. For about three months after the transplant, the diet excludes most fresh fruits and vegetables, undercooked meats, poultry and eggs, and ground meats to minimize the risk of food-borne infections. Thereafter, clients are advised to follow safe food-handling practices to minimize the risks of food-borne illnesses (see Nutrition in Practice in Chapter 5). Fiber, lactose, and fat are gradually added to the diet as individual tolerances allow. Because the transplant recipient also receives immunosuppressants, which often incur negative nitrogen and calcium balances, the final goal is to provide a high-kcalorie, high-protein, high-calcium diet. In addition, physicians often prescribe calcium and vitamin D supplements. Individuals with persistent diarrhea are encouraged to eat high-potassium foods.

■■ HIV Infection

For many years, the devastating effects of infection by the **human immunodeficiency virus (HIV),** the infection that eventually causes **acquired immune deficiency syndrome (AIDS),** seemed unstoppable. Although the disease still has no cure, remarkable progress has been made in extending the life expectancy of people with HIV infections.

■■ Consequences of HIV Infection

HIV is transmitted from one person to another by direct contact with contaminated body fluids, most often through sexual intercourse, through

human immunodeficiency virus (HIV): a virus that progressively hampers the function of the immune system and leaves its host defenseless against other infections and cancer and eventually causes AIDS. The most common HIV is HIV-1—the virus described in this chapter.

acquired immune deficiency syndrome (AIDS): the severe complications associated with the end stages of HIV infection. A person with an HIV infection is diagnosed with AIDS when that person develops certain infections or severe wasting (AIDS-defining illnesses).

contaminated needles or blood products, or from mother to infant during pregnancy or lactation. The virus attacks the immune system and leaves its victims defenseless against **opportunistic illnesses**—illnesses that would cause few, if any, symptoms in a person with a healthy immune system. HIV infections progress in stages. The virus gradually destroys cells with a specific protein called CD4+ on their surfaces. Among the cells most affected are the **CD4+ lymphocytes,** essential components of the immune system. At first, the number of CD4+ lymphocytes declines gradually. As the infection continues, depletion of CD4+ lymphocytes progressively impairs immune function; in the final stages, fatal complications develop.

With improved treatments for HIV infection, the progression of the disorder has been slowed dramatically. On average, it takes about ten years for an HIV infection to progress to AIDS (the final stage of infection). Clinicians monitor the progress of HIV infections by measuring concentrations of CD4+ lymphocytes and circulating virus (viral load), as well as by monitoring clinical symptoms.

■■ SYMPTOMS AND OPPORTUNISTIC ILLNESSES In the early stages of HIV infection, the individual remains symptom-free. As the infection progresses, early symptoms may include fatigue, skin rashes, fevers, diarrhea, joint pain, night sweats, weight loss, oral lesions and infections, and other opportunistic illnesses that are not life-threatening. In the final stages, the person develops frequent and eventually fatal complications, such as severe weight loss; recurrent bacterial pneumonia; serious infections of the central nervous system, GI tract, and skin; cancers; and severe diarrhea. Infection with hepatitis C, which often leads to chronic hepatitis, is common in people with HIV infections, especially those who use injected drugs or those who acquired the HIV infection from contaminated blood products. In the years since improved treatment for HIV infections has become available, the incidence of many opportunistic illnesses and severe malnutrition has fallen significantly.[13]

■■ BODY COMPOSITION ALTERATIONS AND WASTING Prior to improved treatments, severe wasting and debilitation were common manifestations of HIV infections. Although involuntary weight loss and wasting remain a problem for many people with AIDS, people on aggressive therapies (described later) who are not experiencing complications often gain weight.[14] For some people, the weight gain is composed primarily of fat, and the distribution of fat is altered.[15] Compounding the problem is that, with aging, all people lose lean body mass and gain fat. As people live with HIV infections for significantly longer periods of time, both HIV infection and aging favor the deposition of body fat and the loss of lean body mass. A person who develops a "pot belly" as a consequence of aging most often also accumulates fat in the face, arms, and legs as well. Many people with HIV infections, however, accumulate fat around the abdominal area (central fat) and lose fat from the face, arms, and legs. Fat deposits not only under the skin in the abdominal area, but also around the organs located there. Breast enlargement, thick necks, fat at the top of the back, and multiple benign growths composed of fat have also been observed. Central body fat, whether as a consequence of aging or of HIV infection, is associated with hyperlipidemia, impaired glucose tolerance, and increased risks of cardiovascular disease and diabetes.

People with HIV infections often lose weight when they are experiencing complications that alter their appetites or interfere with the absorption or metabolism of nutrients; they regain weight when they are feeling better. Like the wasting that accompanies advanced cancers, wasting and debilitation frequently accompany AIDS. Much as in cancer, the wasting associated with

---

opportunistic illnesses: illnesses that would not normally occur or would cause only minor problems in the healthy population, but can cause great harm when the immune system is compromised.

CD4+ lymphocyte: a type of white blood cell that has a specific protein on its surface and is a necessary component of the immune system.

The cluster of symptoms and disorders that sometimes occurs before AIDS develops is called AIDS-related complex (ARC).

Wasting, an AIDS-defining illness, is the involuntary loss of greater than 10% of body weight accompanied by chronic diarrhea, lethargy, and/or fever lasting longer than 30 days.

The redistribution of fat that can occur in HIV infections as well as in other disorders is called lipodystrophy (LIP-oh-DISS-tro-fee).

The accumulation of abdominal fat associated with HIV infections is sometimes called protease paunch.

The accumulation of fat at the top of the back is sometimes called a buffalo hump.

Benign tumors composed of fat are called lipomas (lih-POE-mah).

AIDS has many causes: anorexia and inadequate food intake, altered metabolism, excessive nutrient losses, and diet-medication interactions.

THE NURSING DIAGNOSES altered nutrition: less than body requirements, risk for infection, hyperthermia, diarrhea, fluid volume deficit, risk for fluid volume deficit, impaired tissue integrity, impaired skin integrity, altered protection, activity intolerance, and **fatigue** may occur as a consequence of HIV infection.

Vitamin, mineral, and trace element deficiencies are also common in people with HIV infections. Many deficiencies have been documented, even in people who are consuming recommended amounts of these nutrients.

:: ANOREXIA AND REDUCED NUTRIENT INTAKE People with HIV infections frequently suffer from anorexia and inadequate nutrient intakes, particularly as the disease progresses. HIV-related causes of anorexia and reduced food intake include:

- *Psychological stress and pain.* As in cancer, fear, depression, and anxiety over the HIV diagnosis, the prognosis, and the medical, personal, and financial problems that lie ahead, as well as the pain associated with complications of the disorder, can destroy the appetite.
- *Oral infections.* Infections and fever cause anorexia. In addition, oral infections associated with HIV infection cause further problems. **Thrush,** a common oral infection associated with HIV infection, can alter taste sensitivity, reduce the flow of saliva, and cause pain on swallowing. Oral infections caused by **herpes virus** can cause painful mouth ulcers that interfere with chewing and swallowing.
- *Respiratory infections.* Pneumonia and tuberculosis, frequent complications associated with HIV infection, cause fever and pain that contribute to anorexia. The person who must use an oxygen mask to improve breathing may find it difficult to eat.
- *GI tract complications and altered organ function.* In addition to oral infections, people with HIV infection may experience belching, reflux esophagitis, gastritis, and heartburn that may cause nausea and interfere with eating. Intestinal complications, including infections, diarrhea, and constipation, and hepatitis can contribute to reduced food intake and food aversions.
- *Fatigue, lethargy, and dementia.* Anemia and fatigue are common complications of HIV infections, even in the early stages. In the later stages of HIV, lethargy and dementia frequently occur and may interfere with food intake. The individual may not care or even remember to eat.
- *Cancer.* As previously described, cancer often leads to anorexia. **Kaposi's sarcoma,** a cancer associated with HIV infection, can cause lesions and obstructions in the esophagus that make eating painful.
- *Medical treatments.* Medications used to treat HIV infection, associated infections, and cancer often cause anorexia, taste alterations, nausea, and vomiting and reduce food intake. Food aversions may also arise.

THE NURSING DIAGNOSES hopelessness, powerlessness, pain, chronic pain, anxiety, fear, altered oral mucous membranes, hyperthermia, diarrhea, constipation, fatigue, activity intolerance, risk for activity intolerance, sensory/perceptual alterations, and **altered thought processes** should alert the nurse to look for problems with food intake.

:: METABOLIC ALTERATIONS AND NUTRIENT LOSSES The metabolic alterations associated with repeated infections and cancer contribute to wasting in people with HIV infections. In addition to anorexia and altered metabolism, nutrient losses can occur as a consequence of diarrhea and malabsorption,

thrush: a fungal infection of the mouth and esophagus caused by *Candida albicans* that coats the tongue with a milky film, alters taste sensations, and causes pain on chewing and swallowing. The technical term for this infection is **candidiasis.**

herpes virus: a virus that can lead to mouth lesions and may also affect the lower GI tract, causing diarrhea.

Kaposi's (cap-OH-seez) sarcoma: a type of cancer rare in the general population but common in people with HIV infection.

which often occur late in the course of HIV infection. An estimated 50 to 90 percent of people with AIDS experience significant diarrhea and malabsorption. Both the structure and the function of the intestinal cells are altered as a consequence of HIV infection and other GI tract infections. The prolonged use of antibiotics to treat infections and antisecretory agents and antacids to relieve nausea can lead to bacterial overgrowth in the upper small intestine, further contributing to malabsorption. Food, water, and enteral formulas may serve as a source of infectious agents, and people with advanced HIV infections are highly susceptible to food-borne illness. In most cases, diarrhea is recurrent, and typical losses average less than 1 liter of diarrheal fluids daily. Diarrhea caused by parasites, however, may be severe and unresponsive to medications—the person may lose 10 or more liters of diarrheal fluids daily. Treatments common among people with HIV infections, especially anti-infective agents, chemotherapy, and radiation therapy, can accelerate nutrient losses due to vomiting, diarrhea, and malabsorption. Megadoses of vitamin C and other home remedies that some people with HIV infections use may also cause diarrhea. Once malnutrition is under way, it, too, contributes to malabsorption.

The combination of diarrhea and malabsorption associated with HIV infection is called HIV enteropathy (EN-ter-OP-a-thee).

▪▪ Treatments for HIV Infection

Treatments for HIV infection focus on slowing the course of the infection, controlling symptoms, and alleviating pain. Drug therapy often includes a combination of medications that disrupt HIV at different stages of replication. The combined drug therapy called highly active antiretroviral therapy (HAART) has made a remarkable difference in the treatment of HIV infection. In 1997, for the first time since 1990, AIDS was no longer among the top ten causes of death in the United States, falling from eighth to fourteenth place.[16] Antiviral agents can have many interactions with diet as the box on p. 579 describes.

The medications used to treat HIV infections are antiviral agents. Antiviral agents include nonnucleoside reverse transcriptase inhibitors, nucleoside reverse transcriptase inhibitors, and protease inhibitors.

▪▪ MEDICATIONS TO COMBAT ANOREXIA AND WASTING   The medications megestrol acetate and dronabinol (described on p. 570) serve to stimulate the appetite and help people with HIV infections to gain weight. More recently, human growth hormone has been approved to treat HIV-related wasting. Growth hormone helps restore lean body mass. Testosterone may also be beneficial in HIV-related wasting and is currently being studied.[17] Reduced testosterone levels are associated with loss of lean body mass, and levels often decrease as an HIV infection progresses. When testosterone levels are low, administering testosterone increases lean body mass and benefits quality of life. Other studies have shown that the combination of megestrol acetate and testosterone can improve appetite and increase weight and lean body mass.

Reminder: Growth hormone and testosterone are called anabolic agents.

LEARNING LINK

Medications to prevent wasting, specifically the loss of lean body mass, have most often been studied in people with AIDS, perhaps because wasting has been a hallmark of the disorder. Now that people have a chance to live longer, supporting quality of life takes on added significance. The same medications (growth hormone and testosterone) are currently under investigation for their potential value in treating acute stresses and cancer and other chronic disorders associated with wasting including alcoholic hepatitis, chronic obstructive pulmonary disease, and end-stage renal disease.[18]

DIET-MEDICATION INTERACTIONS

Anabolic agents

Testosterone, testosterone derivatives (*oxandrolone, nandrolone,* and *oxymetholone*), and *growth hormone* may be used to promote weight gain, specifically a gain of lean body mass. These medications can be taken without regard to food, and nutrition-related side effects are uncommon.

Antidiabetic agents

For *antidiabetic agents,* see p. 482.

Antidiarrheals

For *antidiarrheals,* see p. 399.

Anti-infectives

Many medications may be used to treat the infections associated with both cancer and HIV infections (consult a drug guide for specific examples). The antiviral agents specifically used to treat HIV infections are described below.

Antilipemics

For *antilipemics,* see p. 507.

Antinauseants

For *antinauseants,* see p. 371.

Antineoplastics

About 80 antineoplastic agents are in use today. Many of these agents can cause nausea and vomiting. Most often nausea and vomiting begin a few hours after a treatment and resolve shortly thereafter. In some cases, the problems may last for a few days. The suggestions in the "How to" box on pp. 573–575 can help alleviate these problems. Antinauseants may also be prescribed. Mouth sores, taste alterations, fatigue, anemia, diarrhea, and constipation are also associated with many agents. *Megestrol acetate* is used as both an antineoplastic agent and an appetite stimulant. Megestrol acetate and some antineoplastic agents that are hormones (including testosterone) lead to weight gain. Fluid and sodium retention and edema can also occur, although these side effects are not common.

Antisecretory agents

For *antisecretory agents,* see p. 371.

Antivirals

The nonnucleoside reverse transcriptase inhibitors include *delavirdine* and *nevirapine,* which must be taken 1 hour apart from antacids and without regard to food. Neither is associated with significant nutrition-related side effects. The nucleoside reverse transcriptase inhibitors include *lamivudine (3TC), zidovudine (AZT), stavudine, zalcitabine (ddC),* and *didanosine (ddl).* Lamivudine, zidovudine, and stavudine can be taken without regard to food; zidovudine can cause nausea and vomiting. Clients who use zalcitabine and didanosine should avoid aluminum- and magnesium-containing antacids. Didanosine can lead to nausea and vomiting and should be taken 1 hour before or 2 hours after meals. The protease inhibitors include *indinavir, saquinavir, ritonavir,* and *nelfinavir.* Indinavir should be taken 1 hour before or 2 hours after meals; however, people who experience nausea and abdominal pain can eat a light meal (less than 300 kcalories, 6 grams of protein, and 3 grams of fat). Indinavir can lead to kidney stones, so clients taking this medication are encouraged to drink liquids (6 liters a day). Saquinavir should be taken within 2 hours of a meal. Ritonavir and nelfinavir are better absorbed along with food and should be taken with meals. Ritonavir can cause nausea, vomiting, diarrhea, taste alterations, and abdominal pain. The most common side effect of nelfinavir is mild-to-moderate diarrhea, although nausea and abdominal pain may also occur.

Appetite stimulants

For *appetite stimulants,* see p. 549.

Immunosuppressants

For *immunosuppressants,* see p. 536.

Laxatives

For *laxatives,* see p. 399.

▪▪ **OTHER MEDICATIONS** Other medications are used to control infections and complications that frequently occur in people with HIV infections. Anti-infective agents, antinauseants, oral antidiabetic agents, antilipemics, and antidiarrheals are examples.

Nutrition provides an edge in maintaining quality of life and encouraging independence.

Reminder: Monounsaturated fat is found in olive, canola, and peanut oils; avocados; nuts; and peanut butter. Omega-3 fatty acids are found primarily in fish oils.

Medical Nutrition Therapy

Nutrition assessment and counseling should begin as soon as a diagnosis of HIV infection has been made. The initial assessment provides baseline data from which to monitor progress throughout the course of the disease. For people with HIV infections on aggressive combination drug therapies, assessment should include an evaluation of body composition. Clinics that treat clients with HIV infections may use bioelectrical impedance analysis to estimate how much muscle and lean tissue, fat, and water a person's body contains (see Appendix E). Although the technique does have drawbacks—it can't tell where fat is, for example—it provides a simple and convenient analysis. Measurements repeated at intervals allow clinicians to make adjustments to diet and drug therapies.

ORAL DIETS For people on aggressive combination drug therapies, diet advice often mimics that for clients with cardiovascular diseases: achieve or maintain a desirable weight, reduce fat to less than 30 percent of total kcalories, reduce saturated fat, limit *trans*-fatty acids, and replace saturated fats with monounsaturated fats and omega-3 fatty acids. People with elevated triglycerides benefit from limiting simple sugars. People with glucose intolerance or diabetes benefit from eating a consistent intake of carbohydrate throughout the day and using mostly complex carbohydrates. A combination of aerobic activity and resistance training may also play an important role in preventing central obesity.[19] As Chapter 20 described, physical activity can also help reduce weight and lower insulin resistance.

For clients who are losing weight, a high-kcalorie (150 percent of basal energy expenditure), high-protein (1.5 to 2.0 grams per kilogram of body weight) oral diet helps halt weight loss and restore weight. Enteral formulas and supplements in the form of candy bars and pudding can help boost nutrient intake. The "How to" box on pp. 572–574 offers other suggestions for improving the appetite and adding kcalories and protein to the diet.

All people with HIV infections experience anorexia from time to time. Almost all experience bouts of diarrhea and constipation as well. It is important to find out what problems the person is having and make suggestions based on the symptoms the person experiences.

LEARNING LINK

The sheer number of issues that affect people with HIV infection and the many ways of addressing these issues are staggering. If you have access to a computer, you can gain valuable insights into what it must be like to deal with the disorder by visiting www.thebody.com. Of particular interest are the forums on a variety of problems (including wasting), where people with HIV describe problems they are encountering and ask for advice from experts.

VITAMINS AND MINERALS Vitamin and mineral needs for people with HIV are highly variable, and little information is available concerning specific needs. Anemia, fatigue, and neuropathy, common clinical findings in people with HIV infections, are also associated with B vitamin and iron deficiencies. A vitamin and mineral supplement that contains at least 100 percent of recommended intakes is frequently prescribed to ensure an adequate intake. Iron may be an exception. Some people with HIV may have elevated blood levels of iron, and excess iron may be harmful to the liver and interfere with immune function. For people with HIV infections who are not menstruating or pregnant, the physician should determine if iron supplementation is appropriate. Some clinicians recommend a variety of nutrient supplements, including antioxidant nutrients, and supplemental nonessential amino acids, including glutamine, cysteine, and carnitine, although research to support the use of these supplements is lacking.

▪▪ FOOD SAFETY To prevent infection from food-borne microorganisms, clients are provided with instructions for the safe handling and preparation of foods (see Nutrition in Practice in Chapter 5). Water can also be a source of food-borne illnesses for people with HIV infections. Using water that has been boiled for one minute for cooking and making ice cubes helps prevent problems. Filtered water and bottled water are safer than tap water. (Caution: Some bottled waters contain tap water.)

▪▪ TUBE FEEDINGS AND TPN People with HIV infections need aggressive nutrition support whenever they are unable to consume oral diets that are sufficient to prevent nutrition complications and unintentional weight loss. Tube feedings are preferred whenever the GI tract is functional. Tube feedings can be given at night to supplement oral diets during the day. Preventing bacterial contamination of the formula is particularly important because of the susceptibility of HIV-infected individuals to GI infections. TPN is generally reserved for people with HIV infections who are unable to tolerate enteral nutrition, but need to maintain their nutrition status while undergoing a therapy that is expected to improve their condition. For people with severe HIV-related malabsorption, orally administered hydrolyzed formulas containing medium-chain triglycerides and supplemented with glutamine may be as effective as TPN in reversing weight loss and wasting.[20]

People with GI tract obstructions, severe vomiting, or GI infections affecting the entire small intestine may benefit from TPN. As with enteral formulas, the concern for infection related to TPN is magnified in people with HIV infection. The accompanying case study discusses nutrition concerns of a person with HIV infection. The Nutrition Assessment Checklist applies to people with cancer and HIV infections or AIDS.

Case Study

TRAVEL AGENT WITH HIV INFECTION

Three years ago, Mr. Sands, a 34-year-old travel agent, sought medical help when he began feeling run-down and developed a painful white coating over his mouth and tongue. The presence of thrush and anemia alerted Mr. Sands's physician to the possibility of an HIV infection. When Mr. Sands tested positive for an HIV infection, he and his family and friends were devastated by the news, but those closest to him have remained supportive. Mr. Sands has gained weight and developed central obesity and hyperlipidemia during the three years since he began combination drug therapy. Mr. Sands is 6 feet tall and currently weighs 190 pounds. He occasionally develops diarrhea and sometimes anorexia.

✚ What is Mr. Sands's percent ideal body weight (%IBW)? Describe an appropriate diet for Mr. Sands.

✚ What suggestions can you give Mr. Sands for the times he has trouble with diarrhea or anorexia? What factors can lead to diarrhea or contribute to anorexia in people with HIV infections?

✚ Describe how HIV infection can lead to wasting as the disease progresses to the later stages.

NUTRITION ASSESSMENT CHECKLIST FOR PEOPLE WITH CANCER OR HIV INFECTIONS

Health Problems, Signs, and Symptoms

Check the medical record to determine:

- ☐ Type of cancer
- ☐ Stage of HIV infection

Review the medical record for complications that may alter medical nutrition therapy including:

- ☐ Malnutrition/wasting
- ☐ Altered organ function
- ☐ Anorexia
- ☐ Nausea/vomiting
- ☐ Mouth ulcers
- ☐ Taste alterations
- ☐ Dry mouth
- ☐ Diarrhea/malabsorption
- ☐ Constipation

Medications

For clients with cancer or HIV infections:

- ☐ Make a note of all medications the client is taking and remain alert for possible diet-medication interactions.
- ☐ Give antinauseants or pain medications at times when they will be most effective during meals.
- ☐ Ask about use of alternative therapies, including herbal preparations and megadoses of vitamins.

For clients with cancer who require chemotherapy:

- ☐ Recommend strategies to prevent food aversions (see the box on pp. 572–574).

- ☐ Offer suggestions for handling complications associated with medications.

For clients on antivirals for HIV infections:

- ☐ Note that some antivirals can be taken without regard to foods, some are better absorbed when taken with foods, and still others must be taken on an empty stomach.
- ☐ Help clients work out an acceptable medication schedule that considers the client's lifestyle and the timing of the medication dose with food intake.
- ☐ Offer suggestions for handling complications associated with medications.

Nutrient/Food Intake

For clients with poor food intakes and weight loss:

- ☐ Determine the reason(s) for reduced food intake.
- ☐ Offer suggestions for improving intake based on specific problems the client is having.
- ☐ Provide interventions before weight loss progresses too far.

For clients with HIV infections who experience weight gain and hyperlipidemia and/or glucose intolerance:

- ☐ Assess diet for total energy; total fat and types and amounts of specific fats; and total carbohydrate, fiber, and simple sugars.
- ☐ Recommend a low-fat, low–saturated fat, energy-controlled diet for hyperlipidemia.
- ☐ Recommend a low-fat, low–saturated fat, energy-controlled diet for glucose intolerance or diabetes with a consistent carbohydrate intake from day to day.

Height and Weight

Take baseline height and weight measurements, monitor weight regularly, and make dietary adjustments promptly, if necessary. Baseline and periodic body composition measurements are important for people with HIV infections who take combination antiviral medications.

Laboratory Tests

Note that albumin and serum proteins may be reduced for people with cancer or HIV infections, especially those who are experiencing wasting. Check laboratory tests for indications of:

- ☐ Changes in organ function
- ☐ Anemia
- ☐ Dehydration

For people with HIV infections, the progression of the HIV infection is evaluated by checking:

- ☐ CD4+ counts
- ☐ Viral loads

Physical Signs

Look for physical signs of:

- ☐ Wasting and PEM
- ☐ Fluid status (especially for those with fever, vomiting, or diarrhea)
- ☐ Mouth ulcers

This chapter brings to a close your introduction to normal and clinical nutrition. Congratulations! You have received an abundance of information since you first began your study. The normal nutrition chapters of this text provided you with current recommendations to promote optimal health. You learned how the body transforms foods into nutrients and how those nutrients support the body's well-being. The clinical chapters addressed you as a future health care professional, concerned with the well-being of others during times of illness. We hope this text has served you well and that you will remember, when selecting food for yourself or when making recommendations for others, to honor the body.

SELF CHECK
SELF CHECK

1. Energy and protein needs for people with wasting due to cancer or HIV infections are often about _____ of the basal energy expenditure (BEE) and _____ grams of protein per kilogram of body weight per day.

   a. 100 percent; 1.5 to 2.0
   b. 120 percent; 1.0 to 1.2
   c. 150 percent; 1.5 to 2.0
   d. 200 percent; 1.0 to 1.2

2. Which of the following statements describes wasting associated with cancer and HIV infections?

   a. Anorexia is a major factor in the development of wasting.
   b. Wasting always accompanies cancer and HIV infection.
   c. Altered metabolism plays no role in the development of wasting.
   d. Unlike wasting associated with acute stresses, cytokines do not contribute to wasting in cancer and HIV infections.

3. Practical advice for a person with cancer or HIV infection who has trouble preparing and eating foods due to fatigue might include:

   a. prepare dinner for friends.
   b. prepare fresh vegetables every day.
   c. make homemade ice cream for snacks.
   d. keep premixed breakfast drinks or enteral formulas in the refrigerator for snacks.

4. Oral diets after bone marrow transplants:

   a. limit calcium.
   b. test tolerance for lactose.
   c. restrict high-protein foods.
   d. avoid raw fruits and vegetables.

5. Mouth sores in people with HIV infections are most frequently due to:

   a. dehydration.
   b. oral infections.
   c. malabsorption.
   d. food-borne illnesses.

6. A tube feeding or TPN is most likely to benefit people with cancer or HIV infection if:

   a. they do not wish to prolong their lives.
   b. there are no further treatments available to them.
   c. they have been told they have no other alternatives.
   d. malnutrition may have an undesirable effect on their ability to receive additional treatments.

7. Which of the following statements is true with respect to the diarrhea that often occurs as a consequence of HIV infection?

   a. Megadoses of vitamin C can help resolve it.
   b. HIV and secondary infections can play roles in its development.
   c. Fibers from wheat bran and raw vegetables are useful in controlling it.
   d. Diarrrhea can always be corrected with appropriate nutrition therapy and fluid replacement.

8. Which people are most prone to infections arising from foods, enteral formulas, and TPN?

   a. people who have undergone radiation therapy
   b. people with oral infections or cancer of the GI tract
   c. people who have undergone chemotherapy or surgery
   d. people with HIV infections or those who have undergone bone marrow transplants

9. The changes in body fat seen in many people receiving aggressive treatments for HIV infections include:

   a. increased central and peripheral fat.
   b. decreased central and peripheral fat.
   c. increased central and decreased peripheral fat.
   d. decreased central and increased peripheral fat.

10. Which of the following statements is true?

   a. People with HIV infections may gain weight.
   b. People with cancer never have to worry about gaining weight.
   c. People with cancer never have to worry about losing weight.
   d. People with HIV infections never have to worry about losing weight.

Answers to these questions appear in Appendix H.

CLINICAL APPLICATIONS

1. Many disorders can lead to wasting. For some of these disorders, such as fat malabsorption (Chapter 16), diet is a cornerstone of treatment. For others, such as congestive heart failure, chronic obstructive pulmonary disease, cancer, and HIV infection, nutrition plays a supportive role. What determines whether nutrition plays a major or a supportive role in the treatment of a disorder? Review the effects of PEM on pp. 82–84 and p. 313. Carefully consider how severe malnutrition can further debilitate people with wasting due to cancer or HIV infection.

2. The suggestions for handling food-related problems in the "How to" box on pp. 572–574 appear simple enough, but many of the suggestions may be difficult to implement in some cases. What suggestions

to control nausea and vomiting contradict suggestions to add kcalories and protein? What other contradictions can you find? How might a health care provider deal with such contradictions?

3. Consider problems associated with nutrition in a 36-year-old woman with a malignant brain tumor affecting her ability to move the right side of her body (including the tongue) and to speak coherently. She has an expected length of survival of six months and is taking a pain medication that makes

her nauseated and sleepy. What would be a realistic goal of nutrition support? If she is right-handed, how can her impairment interfere with eating? What suggestions might you have for overcoming this problem? How might nutrition be affected by her problems with communication? Describe ways that the medications she is taking can affect her nutrition status. Would tube feedings or TPN be appropriate for this woman? Why or why not?

NUTRITION ON THE NET

FOR FURTHER STUDY OF THE TOPICS IN THIS CHAPTER, ACCESS THESE WEB SITES.

www.cancer.org
The American Cancer Society

www.aicr.org
American Institute for Cancer Research

www.thebody.com
The Body: A Multimedia AIDS and HIV Information Resource

www.aegis.com
AIDS Education Global Information System

Notes

[1] H. Hwang, J. Dwyer, and R. M. Russel, Diet *Heliobacter pylori* infection, food preservation and gastric cancer risk: Are there new roles for preventative factors? *Nutrition Reviews* 52 (1994): 75–83.

[2] E. A. de Deckere, Possible beneficial effect of fish and fish n-3 polyunsaturated fatty acids in breast and colorectal cancer, *European Journal of Cancer Prevention* 8 (1999): 213–221; M. W. Pariza, Animal studies: Summary, gaps, and future research, *American Journal of Clinical Nutrition* 66 (1997): 1539–1540.

[3] As cited in A. M. Herrington, J. D. Herrington, and C. A. Church, Pharmacologic options for the treatment of cancer, *Nutrition in Clinical Practice* 12 (1997): 101–113.

[4] M. Puccio and L. Nathanson, The cancer cachexia syndrome, *Seminars in Oncology* 24 (1997): 277–278.

[5] D. P. Kotler and coauthors, Magnitude of body cell mass depletion and timing of death from wasting in AIDS, *American Journal of Clinical Nutrition* 50 (1989): 444–447.

[6] T. Duell and coauthors, Health and functional status of long-term survivors of bone marrow transplantation, *Annals of Internal Medicine* 126 (1997): 182–184.

[7] P. O'Gorman, D. C. McMillan, C. S. McArdle, Impact of weight loss, appetite, and the inflammatory response on quality of life in gastrointestinal cancer patients, *Nutrition and Cancer* 32 (1998): 76–80.

[8] S. Mercadante, Parenteral versus enteral nutrition in cancer patients: Indications and practice, *Supportive Care in Cancer* 6 (1998): 85–93.

[9] M. D. Barber and coauthors, Fish oil–enriched nutritional supplement attenuates progression of the acute-phase response in weight-losing patients with advanced cancer, *Journal of Nutrition* 129 (1999): 1120–1125; J. M. Daly and coauthors, Enteral nutrition with supplemental arginine, RNA, and omega-3 fatty acids in patients after operation: Immunologic, metabolic, and clinical outcome, *Surgery* 112 (1992): 56–67.

[10] C. L. Rock, Factors associated with weight gain in women after diagnosis of breast cancer, *Journal of the American Dietetic Association* 99 (1999): 1212–1218, 1221.

[11] M. Marian, Cancer cachexia: Prevalence, mechanisms, and interventions, *Support Line*, April 1998, pp. 3–12.

[12] P. R. Schloerb and M. Amare, Total parenteral nutrition with glutamine in bone marrow transplantation and other clinical applications (randomized, double-blind study), *Journal of Parenteral and Enteral Nutrition* 17 (1993): 407–413; T. R. Ziegler and coauthors, Clinical and metabolic efficacy of glutamine-supplemented parenteral nutrition after bone marrow transplantation, *Annals of Internal Medicine* 116 (1992): 821–828.

[13] A. Mocroft and coauthors, Changes in AIDS-defining illnesses in a London clinic, 1987–1998, *Journal of Acquired Immune Deficiency Syndrome* 21 (1999): 401–407; Centers for Disease Control and Prevention, Surveillance for AIDS-defining opportunistic illness, 1992–1997, *Morbidity and Mortality Weekly Report*, April 19, 1999 (accessed at www.thebody.com/cdc/oi/oi.html on December 9, 1999).

[14] E. Gomez, Wasting and body changes: What do we know so far? *PWA Newsline*, September 1998, pp. 34–36.

[15] A. Carr and coauthors, Diagnosis, prediction, and natural course of HIV-1 protease-inhibitor-associated lipodystrophy, hyperlipidemia, and diabetes mellitus: A cohort study, *Lancet* 353 (1999): 2093–2099; M. Mann and coauthors, Unusual distributions of body fat in AIDS patients: A review of adverse events reported to the Food and Drug Administration, *AIDS Patient Care Standards* 13 (1999): 287–295.

[16] J. Henkel, Attacking AIDS with a "cocktail" therapy, www.thebody.com/fda/cocktail.html, site visited on December 9, 1999.

[17] Centers for Disease Control and Prevention, The use of testosterone in AIDS wasting syndrome, *AIDS Clinical Care* 11 (1999): 25; J. C. Loss, The use of anabolic agents in HIV disease, *Support Line*, June 1999, pp. 23–28.

[18] E. Silverman, Anabolic agents: Do they have roles in treatment of acute and chronic disease states? *Support Line*, June 1999, pp. 19–22; D. C. Gore, Justification for anabolic agents in critical illness, *Support Line*, June 1999, pp. 4–6.

[19] R. Roubenoff and coauthors, A pilot study of exercise training to reduce trunk fat in adults with HIV-associated fat redistribution, *AIDS* 13 (1999): 1373–1375.

[20] D. P. Kotler, L. Fogelman, and A. R. Tierney, Comparison of total parenteral nutrition and an oral, semielemental diet on body composition, physical function, and nutrition-related costs of patients with malabsorption due to acquired immunodeficiency syndrome, *Journal of Parenteral and Enteral Nutrition* 22 (1998): 120–126.

Q&A NUTRITION IN PRACTICE

Alternative Therapies

Alternative therapies have become increasingly popular in recent years. Sales of herbal products in the United States amounted to $3.24 billion in 1997.[1] Some states license alternative therapy practitioners in areas such as **chiropractic, acupuncture, homeopathy,** and **naturopathy.**[2] Although many mainstream health care professionals remain skeptical about alternative therapies, many consumers have become skeptical of, and feel overwhelmed by, the costly and high-tech diagnostic tests and treatments that conventional medicine offers. They want to try a simpler approach to health care, and they want to take more responsibility for maintaining their own health and finding cures for their own diseases, especially when traditional medical therapies prove ineffective or offer little hope for marked improvement.

Can you give some examples of alternative therapies?

Table NP24-1 lists selected fields of alternative medicine, and the glossary on p. 586 defines terms. Notice that most alternative medicines fall outside the field of nutrition, but

Digoxin, a medication commonly prescribed for abnormal heart rhythms, derives from the foxglove plant.

nutrition itself can be an alternative therapy. An herb may be a food in some cultures and may also contain nutrients. Conversely, a food such as garlic may also be considered to be a supplement or alternative therapy.

Furthermore, many alternative therapies include specific dietary adjustments.[3] Most of the dietary recommendations presented throughout this text are based on scientific evidence and do not fall into the alternative category; strategies that are still experimental, however, do. For example, alternative therapists may recommend megadoses of antioxidant supplements or **macrobiotic diets** to help prevent chronic diseases, whereas most registered dietitians would advise people to eat at least five servings of vegetables and fruits daily instead.

How do people use alternative therapies?

People may use alternative therapies to foster good health or to prevent and treat diseases and symptoms ranging from anxiety and headaches to cancer and HIV infection. Many people use alternative therapies in addition to, rather than in place of, conventional therapies. Thus, **complementary,** or **integrative, therapy** combines conventional medical therapy with alternative therapy; the term *complementary therapy* is often used interchangeably with *alternative therapy.* Most people seek alternative therapies for health promotion or to treat minor ailments or side effects of medications. People with cancer or HIV infections, for example, may rely on conventional therapies to treat their diseases, but may use alternative therapies to treat the side effects of these conventional therapies, including nausea, vomiting, diarrhea, and constipation. The use of alternative therapies may help people feel better, access is easy, and sometimes no other options seem to

work. In cases where symptoms are chronic and subjective, such as pain and fatigue, the chances of finding relief are often as good with an alternative therapy as they are with a placebo, standard medical intervention, or even nonintervention.

Why are many mainstream health care professionals skeptical about alternative therapies?

Unlike conventional medical therapies, alternative therapies have not

Table NP24-1
FIELDS OF ALTERNATIVE MEDICINE AND SELECTED EXAMPLES

Mind-body interventions
  Biofeedback
  Faith healing
  Hypnotherapy
  Imagery
  Meditation
Bioelectromagnetic applications in medicine
  Electroacupuncture
  Microwave resonance therapy
Alternative systems of medical practice
  Acupuncture
  Ayurveda
  Homeopathic medicine
  Naturopathic medicine
Manual healing methods
  Biofield therapeutics
  Chiropractic
  Massage therapy
Pharmacological and biological treatments
  Cartilage therapy
  Chelation therapy
  Ozone therapy
Herbal medicine
Diet and nutrition in the prevention and treatment of chronic disease
  Macrobiotic diets
  Orthomolecular medicine

SOURCE: Alternative Medicine: Expanding Medical Horizons, A report to the National Institutes of Health and Alternative Medical Systems and Practices in the United States (Washington, D.C.: Government Printing Office, 1992).

GLOSSARY OF ALTERNATIVE MEDICINE TERMS

acupuncture (AK-you-PUNK-cher): a technique that involves piercing the skin with long thin needles at specific anatomical points to relieve pain or illness. Acupuncture sometimes uses heat, pressure, friction, suction, or electromagnetic energy to stimulate the points.

alternative therapies: approaches to medical diagnosis and treatment that are not fully accepted by the established medical community. As such, they are not widely taught at U.S. medical schools or practiced in U.S. hospitals; also called *adjunctive, unconventional,* or *unorthodox therapies.*

aroma therapy: a technique that uses oil extracts from plants and flowers (usually applied by massage or baths) to enhance physical, psychological, and spiritual health.

ayurveda (EYE-your-VAY-dah): a traditional Hindu system of improving health by using herbs, diet, meditation, massage, and yoga to stimulate the body to make its own natural drugs.

bioelectromagnetic medical applications: the use of electrical energy, magnetic energy, or both to stimulate bone repair, wound healing, and tissue regeneration.

biofeedback: the use of special devices to convey information about heart rate, blood pressure, skin temperature, muscle relaxation, and the like to enable a person to learn how to consciously control these medically important functions.

biofield therapeutics: a manual healing method that directs a healing force from an outside source (commonly God or another supernatural being) through the practitioner and into the client's body; commonly known as "laying on of hands."

cartilage therapy: the use of cleaned and powdered connective tissue, such as collagen, to improve health.

chelation therapy: the use of ethylene diamine tetraacetic acid (EDTA) to bind with metallic ions, thus healing the body by removing toxic metals.

chiropractic (KYE-roe-PRAK-tik): a manual healing method of manipulating vertebrae to relieve musculoskeletal pain suspected of causing problems with internal organs.

complementary or **integrative therapy:** therapy that incorporates both traditional medical and alternative treatments.

faith healing: healing by invoking divine intervention without the use of medical, surgical, or other traditional therapy.

herbal medicine: the use of plants to treat disease or improve health; also known as *botanical medicine* or *phytotherapy.*

homeopathic (home-ee-OP-ah-thick) medicine: a practice based on the theory that "like cures like," that is, that substances that cause symptoms in healthy people can cure those symptoms when given in very dilute amounts.

homeo = like
pathos = suffering

hypnotherapy: a technique that uses hypnosis and the power of suggestion to improve health behaviors, relieve pain, and heal.

imagery: a technique that guides clients to achieve a desired physical, emotional, or spiritual state by visualizing themselves in that state.

iridology: the study of changes in the iris of the eye and their relationships to disease.

macrobiotic diet: a diet consisting of brown rice, miso soup, sea vegetables, and other traditional Japanese foods.

massage therapy: a healing method in which the therapist manually kneads muscles to reduce tension, increase blood circulation, improve joint mobility, and promote healing of injuries.

meditation: a self-directed technique of relaxing the body and calming the mind.

naturopathic medicine: a system that integrates traditional medicine with botanical medicine, clinical nutrition, homeopathy, acupuncture, East Asian medicine, hydrotherapy, and manipulative therapy.

orthomolecular medicine: the use of large doses of vitamins to treat chronic disease.

ozone therapy: the use of ozone gas to enhance the body's immune system.

been widely taught in medical schools, are not practiced in hospitals, and are usually not reimbursed by medical insurance companies.[4]

Although some alternative therapies have been proved to be safe and effective, most information on alternative therapies comes from folklore, tradition, testimonial accounts, and poorly conducted clinical trials. In short, many practitioners are unfamiliar with alternative therapies and their appropriate uses and doses, and scientific evidence proving their safety and effectiveness is lacking. Some say alternative therapies simply do not work; others argue that the established medical community has not given these therapies a fair trial.

One thing is clear—people are using alternative therapies regardless of whether mainstream medical practitioners approve. One of the most disturbing facts about alternative therapies is that most people use them without telling their physician or other health care professional. Furthermore, most physicians don't ask.[5] Such a lack of communication increases the risk that an alternative therapy may interact with conventional treatments or cause deterioration in a client's health condition. As an example, consider a person with an HIV infection who uses megadoses of vitamin C and develops diarrhea from it. As Chapter 24 described, the person with an HIV infection can experience diarrhea due to many causes. If the health care team is unaware that the person is using megadoses of vitamin C, many tests may be conducted before the cause of diarrhea is found—if a cause is found at all.

Are alternative therapies safe?

Many alternative therapies can be safe if they are used by the right people for the right reason and in the right amounts. The herbal remedy aloe may be appropriately applied topically to soothe sunburn, for example, or taken internally for its laxative effect. Injecting aloe as a treatment for cancer, however, is not only unproved, but its use has resulted in serious complications and several deaths.[6] Defining the proper dose of an herbal remedy is also a challenge because research has yet to define which remedies are effective and how much is most beneficial.

Ideally, an alternative therapy provides benefits with little or no risk. The use of ginseng as an adjunct in the management of type 2 diabetes (see Chapter 20), for example, correlates with improved measures of blood glucose control without adverse side effects.[7] Though still in the experimental stage, such findings, if replicated, hold promise that this alternative therapy may one day become accepted medical practice.

Some alternative therapies are innocuous, providing little or no benefit for little or no risk. Sipping a cup of warm tea with a pleasant aroma, for example, won't cure heart disease, but it may improve the person's mood and help relieve tension. Given no physical hazard and little financial risk, such therapies are acceptable. In contrast, other products and procedures are downright dangerous, posing great risks while providing no benefits. One example is the folk practice of geophagia (eating earth or clay), which can cause GI impaction and impair iron absorption. Clearly, such therapies are too harmful to be used. Figure NP24-1 shows how the benefit-risk relationship can help clients and health care professionals determine whether an alternative therapy is safe.

Alternative therapies can be dangerous if people delay or discontinue conventional treatments that have been proved effective for remedies that tout "miracle cures." The consequences can be severe and irreversible. Furthermore, some alternative therapies, particularly those that include megadoses of vitamins and herbal remedies, can interact with medications, and some can have potentially serious consequences. Garlic and ginger, for example, can magnify the anticlotting effects of the medication warfarin; high doses of vitamin K can reduce warfarin's effectiveness.

nccam.nih.gov
National Center for Complementary and Alternative Medicine

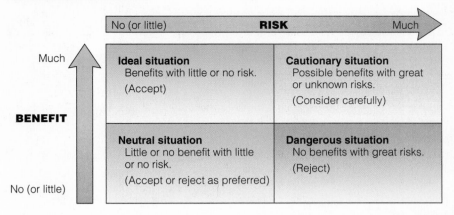

Figure NP24-1
RISK-BENEFIT RELATIONSHIPS

	No (or little) ——— **RISK** ——— Much
Much ↑ **BENEFIT**	**Ideal situation** Benefits with little or no risk. (Accept)	**Cautionary situation** Possible benefits with great or unknown risks. (Consider carefully)
No (or little)	**Neutral situation** Little or no benefit with little or no risk. (Accept or reject as preferred)	**Dangerous situation** No benefits with great risks. (Reject)

www.cancer.org
American Cancer Society

www.thebody.com
The Body: A Multimedia AIDS and HIV Information Resource

www.herbalgram.org
American Botanical Council

www.fda.gov/fdahomepage.html
FDA Center for Food Safety and Applied Nutrition

www.herbs.org
Herbal Research Foundation

What other risks do herbal remedies carry?

Before medications became widely available, a myriad of herbs and plants were used to cure aches and ills with varying degrees of success. Upon scientific scrutiny, dozens of these folk remedies reveal their secrets. For example, myrrh, a plant resin used as a painkiller in ancient times, does indeed have an analgesic effect.[8] The herb valerian, which has long been used as a tranquilizer, contains oils that have a sedative effect. Senna leaves, brewed as a laxative tea, produce compounds that act as a potent cathartic drug. The compounds that plants make are so beneficial that today they contribute to more than half of our modern medicines. Table NP24-2 lists selected herbs, their common uses, and risks.

Thus, herbs are drugs, even though they are often less potent than manufactured drugs. In the United States, however, herbs are classified not as drugs, but rather as dietary supplements. Under the Dietary Supplement Health and Education Act, herbs can be marketed without prior approval from the Food and Drug Administration (FDA), and unlike foods and food additives, which require the manufacturer to provide proof of safety, the FDA has the burden of proving that an herbal product is not safe.[9] Yet people who use supplements and the health care professionals who counsel them are faced with an almost countless array of herbs, which include a multitude of brands, strengths, and formulations of the same product.[10]

Some factors that can affect an individual's risk of adverse effects from herbal supplements include the person's age, genetic makeup, nutrition, state of health, and concurrent use of medications, as well as how much of the supplement the person takes and how long it is used. Some issues that relate to the safety of herbal preparations include:

• *True identification of herbs.* Most mint teas are safe, for instance, but some varieties contain the highly

Table NP24-2
SELECTED HERBS, THEIR COMMON USES, AND RISKS

Common Name	Claims and Uses	Risks[a]
Aloe (gel)	Promote wound healing (topical use), alleviate constipation (oral use)	Generally considered safe
Chamomile (flowers)	Relieve indigestion	Generally considered safe
Chaparral (leaves and twigs)	Slow aging, "cleanse" blood, heal wounds	Acute, toxic hepatitis
Comfrey (leafy plant)	Soothe nerves	Liver disease
Echinacea (roots)	Alleviate symptoms of colds, flus, and infections; promote wound healing; boost immunity	Generally considered safe
Ephedra (stems) or ma huang	Promote weight loss	Rapid heart rate, tremors, seizures, insomnia, headaches
Feverfew (leaves)	Prevent migraine headaches	Generally considered safe; may cause mouth irritation, swelling, ulcers, and GI distress
Garlic (bulbs)	Lower blood lipids and blood pressure	Generally considered safe; may cause garlic breath, body odor, gas, and GI distress; inhibits blood clotting
Ginkgo (tree leaves)	Improve memory, relieve vertigo	Generally considered safe; may cause headache, GI distress, and dizziness
Ginseng (roots)	Boost immunity, increase endurance	Generally considered safe; may cause insomnia and high blood pressure
Goldenseal (roots)	Relieve indigestion, treat urinary infections	Generally considered safe
Jin bu huan (Chinese herbal product)	Relieve pain	Slows heart rate, breathing, and central nervous system functioning; hepatitis
Laetrile (apricot pits)	Treat cancer	Cyanide poisoning
Pennyroyal	Treats cough and stomach upsets	Liver toxicity and alters blood clotting
Saw palmetto (ripe fruits)	Relieve symptoms of enlarged prostate; diuretic; enhance sexual vigor; enlarge mammary glands	Generally considered safe
St. John's wort (leaves and tops)	Relieve depression and anxiety	Generally considered safe; may cause fatigue and GI distress
Valerian (roots)	Calm nerves, improve sleep	Generally considered safe
Yohimbe (tree bark)	Enhance "male performance"	High blood pressure, fatigue, kidney failure, seizures

[a]Allergies are always a possible risk.
SOURCES: V. E. Tyler, *The Honest Herbal: A Sensible Guide to the Use of Herbs and Related Remedies* (New York: Pharmaceutical Products Press, 1993); New safety measures are proposed for dietary supplements containing ephedrine alkaloids, *Journal of the American Medical Association* 278 (1997): 15; C. L. Bartels and S. J. Miller, Herbal and related remedies, *Nutrition in Clinical Practice* 12 (1998): 5–19; N. Spaulding-Albright, A review of some herbal and related products commonly used in cancer patients, *Journal of the American Dietetic Association* 97 (1997):S208–S215.

toxic pennyroyal oil. Mistakenly used to treat colic, mint tea laden with pennyroyal has been blamed for the liver and neurological injuries of at least two infants, one of whom died.[11] In another case, neurological and cardiovascular complications developed in three children, and hepatitis was reported in three adults after ingesting the Chinese herb jin bu huan.[12] Chemical analysis of the herbal preparation revealed that its active ingredient was actually from another herb.

- *Mislabeling of active ingredients.* Chemical analysis of the Chinese herbal preparation tung shueh, which has been linked to acute renal failure, revealed the presence of two synthetically manufactured drugs, mefenamic acid (a nonsteroidal anti-inflammatory agent) and diazepam (an antianxiety agent), that were not listed as active ingredients on the label.[13]
- *Purity of herbal preparations.* Potentially toxic quantities of arsenic and mercury have been detected in traditional Chinese herbal balls used to treat fever, rheumatism, and cataracts.[14]
- *Adverse reactions and toxicity levels of herbs.* As is true of all drugs, herbs may produce undesirable reactions. Herbal preparations containing ephedrine, commonly known as ma huang and used to promote weight loss, act as strong central nervous system stimulants, causing rapid heart rate, nervousness, headaches, insomnia, and even death.
- *Safe dosages of herbs.* Herbs that are effective contain active ingredients that need to be administered in proper doses. Each of these active ingredients has a different potency, time of onset, duration of activity, and consequent effects, making the plant itself too unpredictable to be useful. Consider that labels of over-the-counter bronchodilators containing ephedrine at doses of from 12.5 to 25.0 milligrams instruct users to take one dose two or three times per day; yet chemical analysis of one herbal preparation containing ephedrine revealed that each tablet contained 45 milligrams of ephedrine and 20 milligrams of caffeine (another stimulant), and the label instructed the user to take five tablets per dose.[15]

Not only are herbal preparations not strictly regulated, but their labels may carry unsubstantiated health claims as long as the following disclaimer also appears: "Has not been evaluated by the Food and Drug Administration."

What can health care professionals do to help?

First of all, it is important that health care professionals maintain open communications with their clients about alternative therapies. Most herbal remedies are safe, even though some may not be effective. Health care professionals need to help clients sift through the available information and determine if the potential benefits outweigh the risks.

Health care professionals increase the likelihood of open communications by adopting a nonjudgmental, culturally sensitive, respectful, and open-minded attitude toward clients. Working together with their clients, health care professionals can develop medical plans that include both conventional and alternative therapies.

For a client who is considering an alternative therapy, discuss what is known about the therapy's safety and effectiveness. (If you are thinking, "I don't know anything about herbs or some other therapy," you are not alone. Most of us simply cannot avoid doing some research before giving clients information.) If you feel unable to help the client sufficiently, licensed, alternative practitioners can be consulted.[16]

Recommend that the client try only one alternative therapy at a time, so the effectiveness of the measure can be evaluated. One clinician recommends that the client keep a symptom record, logging the severity and frequency of symptoms as well as factors that make the symptom worse or better.[17]

Alternative therapies come in a variety of shapes and sizes. Both their benefits and their risks may be either small, none, or great. Accept the beneficial, or even neutral, practices with an open mind and reject only those practices known to cause harm. Making healthful choices requires knowing what all the choices are.

Notes

[1] As cited in C. L. Bartels and S. J. Miller, Herbal and related remedies, *Nutrition in Clinical Practice* 12 (1998): 5–19.

[2] D. Shattuck, Complementary medicine: Finding a balance, *Journal of the American Dietetic Association* 97 (1997): 1367–1369.

[3] K. K. Hamilton, An overview of herbal and nutritional integrative medicine: A registered dietitian's perspective, *Support Line,* August 1998, pp. 5–9.

[4] National Center for Complementary and Alternative Medicine, www.nccam.nih.gov/nccam/what-is-cam/faq/shtml, site visited on December 16, 1999.

[5] D. M. Eisenberg, Advising patients who seek alternative medical therapies, *Annals of Internal Medicine* 127 (1997): 61–69.

[6] American Cancer Society, Aloe, www.cancer.org/alt_therapy/index.html, site visited on December 16, 1999.

[7] S. A. Sotanbiemi, E. Haapakoski, and A. Rautio, Ginseng therapy in noninsulin-dependent diabetic patients: Effects on psychophysical performance, glucose homeostasis, serum lipids, serum aminoterminalpropeptide concentration, and body weight, *Diabetes Care* 10 (1995): 1373–1375.

[8] P. Lipkin, An ancient salve dampens pain, *Science News* 149 (1996): 20.

[9] Bartels and Miller, 1998.

[10] R. Chang, Quality and standards in dietary supplements, *Support Line,* August 1998, pp. 3–4.

[11] J. A. Bakerink and coauthors, Multiple organ failure after ingestion of pennyroyal oil from herbal tea in two infants, *Pediatrics* 98 (1996): 944–947.

[12] R. S. Horowitz and coauthors, The clinical spectrum of jin bu huan toxicity, *Archives of Internal Medicine* 156 (1996): 899–903.

[13] A. B. Abt and coauthors, Chinese herbal medicine induced acute renal failure, *Archives of Internal Medicine* 155 (1995): 211–212.

[14] E. O. Espinoza, M. J. Mann, and B. Bleasdell, Arsenic and mercury in traditional Chinese herbal balls, *New England Journal of Medicine* 333 (1995): 803–804.

[15] Bartels and Miller, 1998.

[16] L. Hanna, Complementary and alternative medicine, www.sfaf.org/treatment/bet/b35/b35cam.html, site visited on December 16, 1999.

[17] Eisenberg, 1997.

APPENDIXES

CONTENTS

A
Table of Food Composition

B
Canada: Recommendations,
Choice System, and Labels

C
United States: Recommendations
and Exchanges
World Health Organization:
Recommendations

D
Nutrition Resources

E
Nutrition Assessment:
Supplemental Information

F
Aids to Calculation

G
Enteral Formulas

H
Answers to Self Check Questions

A Table of Food Composition

CONTENTS

Table of Food Composition

This edition of the table of food composition contains more complete values for several nutrients than any comparable table.[a] These include dietary fiber; saturated, monounsaturated, and polyunsaturated fat; vitamin B_6; vitamin E; folate; magnesium; and zinc. The table includes a wide variety of foods from all food groups and is updated yearly to reflect current food patterns. For example, this edition includes many new nonfat items; several new ethnic items such as adzuki beans, tahitian taro, and gai choy chinese mustard; and a new selection of vegetarian foods.

Sources of Data

To achieve a complete and reliable listing of nutrients for all the foods, over 1200 sources of information are researched. Government sources are the primary base for all data for most foods. In addition to USDA data (from Release 12 and surveys), provisional USDA information—both published and unpublished—is included.

Even with all the government sources available, however, some nutrient values are still missing; and as the USDA updates various data, it sometimes reports conflicting values for the same items. To fill in the missing values and resolve discrepancies, other reliable sources of information are used. These sources include journal articles, food composition tables from Canada and England, information from other nutrient data banks and publications, unpublished scientific data, and manufacturers' data.

The data for brand foods are listed as provided by the food manufacturers and the food chain restaurants. This information changes often because recipes and formulations are modified to meet consumer preferences, and the data are usually limited to those nutrients required for food labels. To provide more complete information, values for several nutrients are often estimated based on known values for major ingredients.

Accuracy

The energy and nutrients in recipes and combination foods vary widely, depending on the ingredients. The amounts of various fatty acids and cholesterol are influenced by the type of fat used (the specific type of oil, vegetable shortening, butter, margarine, etc.).

Estimates of nutrient amounts for foods and nutrients include all possible adjustments in the interest of accuracy. When multiple values are reported for a nutrient, the numbers are averaged and weighted with consideration of the original number of analyses in the separate sources. Whenever water percentages are available, estimates of nutrient amounts are adjusted for water content. When no water is given, water percentage is assumed to be that shown in the table. Whenever a reported weight appeared inconsistent, many kitchen tests were made, and the average weight of the typical product was given as tested.

When estimates of nutrient amounts in cooked foods are derived from reported amounts in raw foods, published retention factors are applied. Data for combination foods are modified to include newer data for major ingredients.

Considerable effort has been made to report the most accurate data available. The table is revised annually, and the authors welcome any suggestions or comments for future editions.

Average Values

It is important to know that many different nutrient values can be reported for foods, even by reliable sources. Many factors influence the amounts of nutrients in foods, including the mineral content of the soil, the method of processing, genetics, the diet of the animal or the fertilizer of the plant, the season of the year, methods of analysis, the difference in moisture content of the samples analyzed, the length and method of storage, and methods of cooking the food. The mineral content of water also varies according to the source.

Although each nutrient is presented as a single number, each number is actually an average of a range of data. More detailed reports from the USDA, for example, indicate the number of samples and the standard deviation of the data. One can also find different reported values for foods as older data are replaced with newer data from more recent analyses using newer analytical techniques. Therefore, nutrient data should be viewed and used only as a guide, a close approximation of nutrient content.

Dietary Fiber

There can be many different reported values for dietary fiber in foods because information depends on the type of analytical technique used. The fiber data in this table are primarily from the USDA/ARS Human Nutrition Information

---

[a]This food composition table has been prepared for West-Wadsworth Publishing Company and is copyrighted by ESHA Research in Salem, Oregon—the developer and publisher of the Food Processor®, Genesis® R&D, and the Computer Chef® nutrition software systems. The major sources for the data are from the USDA, supplemented by more than 1200 additional sources of information. Because the list of references is so extensive, it is not provided here, but is available from the publisher.

Service in Hyattsville, Maryland; Composition of Foods by Southgate and Paul (England); and many journal articles.

Vitamin A

Vitamin A is reported in retinol equivalents (RE). The amount of vitamin A can vary by the season of the year and the maturity of the plant. Reported values in both dairy products and plants are higher in summer and early fall than in winter. The values reported here represent year-round averages. The organ meats of all animal products (liver especially) contain large amounts of vitamin A, which vary widely, depending on the background of the animal. The vitamin is also present in very small amounts in regular meat and is often reported as a trace.

Vitamin E

Vitamin E is actually a combination of various forms of this nutrient, and the measure of alpha-tocopherol equivalents (α-TE) summarizes the activity of the various types of tocopherols and tocotrienols into one measure.

Fats

Total fats, as well as the breakdown of total fats to saturated, monounsaturated, and polyunsaturated fats, are listed in the table. The fatty acids seldom add up to the total due to rounding and to other fatty acid components that are not included in these basic categories, such as trans-fatty acids and glycerol. Trans-fatty acids can comprise a large share of the total fat in margarine and shortening (hydrogenated oils) and in any foods that include them as ingredients.

Enrichment-Fortification

The mandatory enrichment values for foods are presented as appropriate, including the new values for folate in grain products.

Niacin

Niacin values are for preformed niacin and do not include additional niacin that may form in the body from the conversion of tryptophan.

Using the Table

The items in this table have been organized into several categories, which are listed at the head of each right-hand page. As the key shows, each group has been color-coded to make it easier to find individual items.

In an effort to conserve space, the following abbreviations have been used in the food descriptions and nutrient breakdowns:

- diam = diameter
- ea = each
- enr = enriched
- f/ = from
- frzn = frozen
- g = grams
- liq = liquid
- pce = piece
- pkg = package
- w/ = with
- w/o = without
- t = trace
- 0 = zero (no nutrient value)
- blank space = information not available

Table A–1

Food Composition

(Computer code number is for West Diet Analysis program)   (For purposes of calculations, use "0" for t, <1, <.1, <.01, etc.)

Computer Code Number	Food Description	Measure	Wt (g)	H₂O (%)	Ener (cal)	Prot (g)	Carb (g)	Dietary Fiber (g)	Fat (g)	Fat Breakdown (g) Sat	Mono	Poly
	BEVERAGES											
	Alcoholic:											
	Beer:											
1	Regular (12 fl oz)	1½ c	356	92	146	1	13	1	0	0	0	0
2	Light (12 fl oz)	1½ c	354	95	99	1	5	0	0	0	0	0
1506	Nonalcoholic (12 fl oz)	1½ c	360	98	32	1	5	0	0	0	0	0
	Gin, rum, vodka, whiskey:											
3	80 proof	1½ fl oz	42	67	97	0	0	0	0	0	0	0
4	86 proof	1½ fl oz	42	64	105	0	<1	0	0	0	0	0
5	90 proof	1½ fl oz	42	62	110	0	0	0	0	0	0	0
	Liqueur:											
1359	Coffee liqueur, 53 proof	1½ fl oz	52	31	175	<1	24	0	<1	.1	t	.1
1360	Coffee & cream liqueur, 34 proof	1½ fl oz	47	46	154	1	10	0	7	4.5	2.1	.3
1361	Crème de menthe, 72 proof	1½ fl oz	50	28	186	0	21	0	<1	t	t	.1
	Wine, 4 fl oz:											
6	Dessert, sweet	½ c	118	72	181	<1	14	0	0	0	0	0
7	Red	½ c	118	88	85	<1	2	0	0	0	0	0
8	Rosé	½ c	118	89	84	<1	2	0	0	0	0	0
9	White medium	½ c	118	90	80	<1	1	0	0	0	0	0
1592	Nonalcoholic	1 c	232	98	14	1	3	0	0	0	0	0
1593	Nonalcoholic light	1 c	232	98	14	1	3	0	0	0	0	0
1409	Wine cooler, bottle (12 fl oz)	1½ c	340	90	169	<1	20	<1	<1	t	t	t
1595	Wine cooler, cup	1 c	227	90	113	<1	13	<1	<1	t	t	t
	Carbonated:											
10	Club soda (12 fl oz)	1½ c	355	100	0	0	0	0	0	0	0	0
11	Cola beverage (12 fl oz)	1½ c	372	89	153	0	39	0	0	0	0	0
12	Diet cola w/aspartame (12 fl oz)	1½ c	355	100	4	<1	<1	0	0	0	0	0
13	Diet soda pop w/saccharin (12 fl oz)	1½ c	355	100	0	0	<1	0	0	0	0	0
14	Ginger ale (12 fl oz)	1½ c	366	91	124	0	32	0	0	0	0	0
15	Grape soda (12 fl oz)	1½ c	372	89	160	0	42	0	0	0	0	0
16	Lemon-lime (12 fl oz)	1½ c	368	89	147	0	38	0	0	0	0	0
17	Orange (12 fl oz)	1½ c	372	88	179	0	46	0	0	0	0	0
18	Pepper-type soda (12 fl oz)	1½ c	368	89	151	0	38	0	<1	.1	0	0
19	Root beer (12 fl oz)	1½ c	370	89	152	0	39	0	0	0	0	0
20	Coffee, brewed	1 c	237	99	5	<1	1	0	<1	t	0	t
21	Coffee, prepared from instant	1 c	238	99	5	<1	1	0	<1	t	0	t
	Fruit drinks, noncarbonated:											
22	Fruit punch drink, canned	1 c	248	88	117	0	29	<1	<1	t	t	t
1358	Gatorade	1 c	241	93	60	0	15	0	0	0	0	0
23	Grape drink, canned	1 c	250	87	125	<1	32	<1	0	0	0	0
1304	Koolade sweetened with sugar	1 c	262	90	97	0	25	0	<1	t	t	t
1356	Koolade sweetened with nutrasweet	1 c	240	95	43	0	11	0	0	0	Mono	0
26	Lemonade,frzn concentrate (6-oz can)	¾ c	219	52	396	1	103	1	<1	.1	t	.1
27	Lemonade, from concentrate	1 c	248	89	99	<1	26	<1	<1	t	t	t
28	Limeade, frzn concentrate (6-oz can)	¾ c	218	50	408	<1	108	1	<1	t	t	.1
29	Limeade, from concentrate	1 c	247	89	101	0	27	<1	<1	t	t	t
24	Pineapple grapefruit, canned	1 c	250	88	118	<1	29	<1	<1	t	t	.1
25	Pineapple orange, canned	1 c	250	87	125	3	29	<1	0	0	0	0
	Fruit and vegetable juices: see Fruit and Vegetable sections											
	Ultra Slim Fast, ready to drink, can:											
30411	Chocolate Royale	1 ea	350	84	220	10	38	5	3	1	1	.5
30415	French Vanilla	1 ea	350	84	220	10	38	5	3	1	1.5	.5
30413	Strawberries n' cream	1 ea	350	83	220	10	42	5	3	1	1.5	.5
1357	Water, bottled: Perrier (6½ fl oz)	1 ea	192	100	0	0	0	0	0	0	0	0
1594	Water, bottled: Tonic water	1½ c	366	91	124	0	32	0	0	0	0	0
	Tea:											
30	Brewed, regular	1 c	237	100	2	0	1	0	<1	t	0	t
1662	Brewed, herbal	1 c	237	100	2	0	<1	0	<1	t	t	t
32	From instant, sweetened	1 c	259	91	88	<1	22	0	<1	t	t	t
31	From instant, unsweetened	1 c	237	100	2	<1	<1	0	0	0	0	0

Chol (mg)	Calc (mg)	Iron (mg)	Magn (mg)	Pota (mg)	Sodi (mg)	Zinc (mg)	VT-A (RE)	Thia (mg)	VT-E (a-TE)	Ribo (mg)	Niac (mg)	V-B6 (mg)	Fola (µg)	VT-C (mg)
0	18	.11	21	89	18	.07	0	.02	0	.09	1.61	.18	21	0
0	18	.14	18	64	11	.11	0	.03	0	.11	1.39	.12	14	0
0	25	.04	32	90	18	.04	0	.02	0	.09	1.63	.18	22	0
0	0	.02	0	1	<1	.02	0	<.01	0	<.01	<.01	0	0	0
0	0	.02	0	1	<1	.02	0	<.01	0	<.01	<.01	0	0	0
0	0	.02	0	1	<1	.02	0	<.01	0	<.01	<.01	0	0	0
0	1	.03	2	16	4	.02	0	<.01	0	.01	.07	0	0	0
7	8	.06	1	15	43	.07	20	0	.12	.03	.04	.01	0	0
0	0	.03	0	0	2	.02	0	0	0	0	<.01	0	0	0
0	9	.28	11	109	11	.08	0	.02	0	.02	.25	0	<1	0
0	9	.51	15	132	6	.11	0	.01	0	.03	.1	.04	2	0
0	9	.45	12	117	6	.07	0	<.01	0	.02	.09	.03	1	0
0	11	.38	12	94	6	.08	0	<.01	0	.01	.08	.02	<1	0
0	21	.93	23	204	16	.19	0	0	0	.02	.23	.05	2	0
0	21	.93	23	204	16	.19	0	0	0	.02	.23	.05	2	0
0	19	.92	18	152	29	.2	<1	.02	.02	.02	.15	.04	4	6
0	13	.61	12	102	19	.13	<1	.01	.02	.02	.1	.03	3	4
0	18	.04	4	7	75	.35	0	0	0	0	0	0	0	0
0	11	.11	4	4	15	.04	0	0	0	0	0	0	0	0
0	14	.11	4	0	21	.28	0	.02	0	.08	0	0	0	0
0	14	.14	4	7	57	.18	0	0	0	0	0	0	0	0
0	11	.66	4	4	26	.18	0	0	0	0	0	0	0	0
0	11	.3	4	4	56	.26	0	0	0	0	0	0	0	0
0	7	.26	4	4	40	.18	0	0	0	0	.05	0	0	0
0	19	.22	4	7	45	.37	0	0	0	0	0	0	0	0
0	11	.15	0	4	37	.15	0	0	0	0	0	0	0	0
0	18	.18	4	4	48	.26	0	0	0	0	0	0	0	0
0	5	.12	12	128	5	.05	0	0	0	0	.53	0	<1	0
0	7	.12	10	86	7	.07	0	0	0	<.01	.67	0	0	0
0	20	.52	5	62	55	.3	3	0	0	.06	.05	0	3	73
0	0	.12	2	26	96	.05	0	.01	0	0	0	0	0	0
0	7	.25	10	87	2	.07	<1	.02	0	.02	.25	.05	2	40
0	42	.13	3	3	37	.08	0	0	0	<.01	<.01	0	<1	31
0	17	.65	5	50	50	.26	2	.02	0	.05	.05	0	5	77
0	15	1.58	11	147	9	.17	22	.06	0	.21	.16	.05	22	39
0	7	.4	5	37	7	.1	5	.01	0	.05	.04	.01	5	10
0	11	.22	9	129	0	.09	0	.02	0	.02	.22	0	9	26
0	7	.07	2	32	5	.05	0	<.01	0	<.01	.05	0	2	7
0	17	.77	15	153	35	.15	9	.07	0	.04	.67	.1	26	115
0	12	.67	15	115	7	.15	13	.07	0	.05	.52	.12	27	56
5	400	2.7	140	530	220	2.24	525	.52	7	.59	7	.7	120	21
5	400	2.8	140	450	460	2.1	525	.52	7	.59	7	.7	120	21
5	400	2.7	140	450	460	2.24	525	.52	7	.59	7	.7	120	21
0	27	0	0	0	2	0	0	0	0	0	0	0	0	0
0	4	.04	0	0	15	.37	0	0	0	0	0	0	0	0
0	0	.05	7	88	7	.05	0	0	0	.03	0	0	12	0
0	5	.19	2	21	2	.09	0	.02	0	.01	0	0	1	0
0	5	.05	5	49	8	.08	0	0	0	.05	.09	<.01	10	0
0	5	.05	5	47	7	.07	0	0	0	<.01	.09	<.01	1	0

Table A–1

Food Composition   (Computer code number is for West Diet Analysis program)   (For purposes of calculations, use "0" for t, <1, <.1, <.01, etc.)

Computer Code Number	Food Description	Measure	Wt (g)	H₂O (%)	Ener (cal)	Prot (g)	Carb (g)	Dietary Fiber (g)	Fat (g)	Fat Breakdown (g)		
										Sat	Mono	Poly
DAIRY												
	Butter: see Fats and Oils, #158,159,160											
	Cheese, natural:											
33	Blue	1 oz	28	42	99	6	1	0	8	5.2	2.2	.2
34	Brick	1 oz	28	41	104	6	1	0	8	5.3	2.4	.2
35	Brie	1 oz	28	48	93	6	<1	0	8	4.9	2.2	.2
36	Camembert	1 oz	28	52	84	6	<1	0	7	4.3	2	.2
37	Cheddar:	1 oz	28	37	113	7	<1	0	9	5.9	2.6	.3
38	1" cube	1 ea	17	37	68	4	<1	0	6	3.6	1.6	.2
39	Shredded	1 c	113	37	455	28	1	0	37	24	10.6	1.1
1406	Low fat, low sodium	1 oz	28	65	48	7	1	0	2	1.2	.6	.1
	Cottage:											
984	Low sodium, low fat	1 c	225	83	162	28	6	0	2	1.4	.7	.1
40	Creamed, large curd	1 c	225	79	232	28	6	0	10	6.4	2.9	.3
41	Creamed, small curd	1 c	210	79	216	26	6	0	9	6	2.7	.3
42	With fruit	1 c	226	72	280	22	30	0	8	4.9	2.2	.2
43	Low fat 2%	1 c	226	79	203	31	8	0	4	2.8	1.2	.1
44	Low fat 1%	1 c	226	82	164	28	6	0	2	1.5	.7	.1
46	Cream	1 tbs	15	54	52	1	<1	0	5	3.3	1.5	.2
983	low fat	1 tbs	15	64	35	2	1	0	3	1.7	.9	.1
47	Edam	1 oz	28	42	100	7	<1	0	8	4.9	2.3	.2
48	Feta	1 oz	28	55	74	4	1	0	6	4.2	1.3	.2
49	Gouda	1 oz	28	41	100	7	1	0	8	4.9	2.2	.2
50	Gruyère	1 oz	28	33	116	8	<1	0	9	5.3	2.8	.5
51	Gorgonzola	1 oz	28	43	97	6	1	0	8	5		
1676	Limburger	1 oz	28	48	92	6	<1	0	8	4.7	2.4	.1
53	Monterey Jack	1 oz	28	41	104	7	<1	0	8	5.3	2.4	.3
54	Mozzarella, whole milk	1 oz	28	54	79	5	1	0	6	3.7	1.8	.2
55	Mozzarella, part-skim milk, low moisture	1 oz	28	49	78	8	1	0	5	3	1.4	.1
56	Muenster	1 oz	28	42	103	7	<1	0	8	5.3	2.4	.2
2422	Neufchatel	1 oz	28	62	73	3	1	0	7	4.1	1.9	.2
1399	Nonfat cheese (Kraft Singles)	1 oz	28	61	44	6	4	0	0	0	0	0
59	Parmesan, grated:	1 oz	28	18	128	12	1	0	8	5.5	2.4	.2
57	Cup, not pressed down	1 c	100	18	456	42	4	0	30	19.7	8.7	.7
58	Tablespoon	1 tbs	6	18	27	2	<1	0	2	1.2	.5	t
60	Provolone	1 oz	28	41	98	7	1	0	7	4.9	2.1	.2
61	Ricotta, whole milk	1 c	246	72	428	28	7	0	32	20.4	8.9	.9
62	Ricotta, part-skim milk	1 c	246	74	339	28	13	0	19	12.1	5.7	.6
63	Romano	1 oz	28	31	108	9	1	0	8	4.8	2.2	.2
64	Swiss	1 oz	28	37	105	8	1	0	8	5	2	.3
976	low fat	1 oz	28	60	50	8	1	0	1	.9	.4	t
	Pasteurized processed cheese products:											
65	American	1 oz	28	39	105	6	<1	0	9	5.5	2.5	.3
66	Swiss	1 oz	28	42	93	7	1	0	7	4.5	2	.2
67	American cheese food, jar	½ c	57	43	187	11	4	0	14	9	4.1	.4
68	American cheese spread	1 tbs	15	48	43	2	1	0	3	2	.9	.1
982	Velveeta cheese spread, low fat, low sodium, slice	1 pce	34	62	61	9	1	0	2	1.5	.7	.1
	Cream, sweet:											
69	Half & half (cream & milk)	1 c	242	81	315	7	10	0	28	17.3	8	1
70	Tablespoon	1 tbs	15	81	19	<1	1	0	2	1.1	.5	.1
71	Light, coffee or table:	1 c	240	74	468	6	9	0	46	28.8	13.4	1.7
72	Tablespoon	1 tbs	15	74	29	<1	1	0	3	1.8	.8	.1
73	Light whipping cream, liquid:	1 c	239	63	698	5	7	0	74	46.1	21.7	2.1
74	Tablespoon	1 tbs	15	63	44	<1	<1	0	5	2.9	1.4	.1
75	Heavy whipping cream, liquid:	1 c	238	58	821	5	7	0	88	54.7	25.5	3.3
76	Tablespoon	1 tbs	15	58	52	<1	<1	0	6	3.4	1.6	.2
77	Whipped cream, pressurized:	1 c	60	61	154	2	7	0	13	8.3	3.8	.5
78	Tablespoon	1 tbs	4	61	10	<1	<1	0	1	.6	.3	t
79	Cream, sour, cultured:	1 c	230	71	492	7	10	0	48	29.9	13.9	1.8
80	Tablespoon	1 tbs	14	71	30	<1	1	0	3	1.8	.8	.1

Chol (mg)	Calc (mg)	Iron (mg)	Magn (mg)	Pota (mg)	Sodi (mg)	Zinc (mg)	VT-A (RE)	Thia (mg)	VT-E (a-TE)	Ribo (mg)	Niac (mg)	V-B6 (mg)	Fola (μg)	VT-C (mg)
21	148	.09	6	72	391	.74	64	.01	.18	.11	.29	.05	10	0
26	189	.12	7	38	157	.73	85	<.01	.14	.1	.03	.02	6	0
28	51	.14	6	43	176	.67	51	.02	.18	.15	.11	.07	18	0
20	109	.09	6	52	236	.67	71	.01	.18	.14	.18	.06	17	0
29	202	.19	8	28	174	.87	85	.01	.1	.1	.02	.02	5	0
18	123	.12	5	17	106	.53	51	<.01	.06	.06	.01	.01	3	0
119	815	.77	31	111	702	3.51	342	.03	.41	.42	.09	.08	21	0
6	197	.2	8	31	6	.86	17	.01	.05	.01	.02	.02	5	0
9	137	.31	11	194	29	.85	25	.04	.25	.36	.29	.16	27	0
33	135	.31	12	190	911	.83	108	.05	.27	.37	.28	.15	27	0
31	126	.29	11	177	851	.78	101	.04	.26	.34	.26	.14	26	0
25	108	.25	9	151	915	.65	81	.04	.21	.29	.23	.12	22	0
19	155	.36	14	217	918	.95	45	.05	.13	.42	.32	.17	30	0
10	138	.32	12	193	918	.86	25	.05	.25	.37	.29	.15	28	0
16	12	.18	1	18	44	.08	66	<.01	.14	.03	.01	.01	2	0
8	17	.25	1	25	44	.11	33	<.01	.07	.04	.02	.01	3	0
25	205	.12	8	53	270	1.05	71	.01	.21	.11	.02	.02	5	0
25	138	.18	5	17	312	.81	36	.04	.01	.24	.28	.12	9	0
32	196	.07	8	34	229	1.09	49	.01	.1	.09	.02	.02	6	0
31	283	.05	10	23	94	1.09	84	.02	.1	.08	.03	.02	3	0
30	170	.18			280		43							
25	139	.04	6	36	224	.59	88	.02	.18	.14	.04	.02	16	0
25	209	.2	8	23	150	.84	71	<.01	.09	.11	.03	.02	5	0
22	145	.05	5	19	104	.62	67	<.01	.1	.07	.02	.02	2	0
15	205	.07	7	26	148	.88	53	.01	.13	.1	.03	.02	3	0
27	201	.11	8	37	176	.79	88	<.01	.13	.09	.03	.02	3	0
21	21	.08	2	32	112	.15	74	<.01	.26	.05	.03	.01	3	0
4	221	0		81	427		126		0	.1				0
22	385	.27	14	30	521	.89	48	.01	.22	.11	.09	.03	2	0
79	1375	.95	51	107	1861	3.19	173	.04	.8	.39	.31	.1	8	0
5	82	.06	3	6	112	.19	10	<.01	.05	.02	.02	.01	<1	0
19	212	.15	8	39	245	.9	74	<.01	.1	.09	.04	.02	3	0
124	509	.93	28	258	207	2.85	330	.03	.86	.48	.26	.11	30	0
76	669	1.08	36	308	308	3.3	278	.05	.53	.45	.19	.05	32	0
29	298	.22	11	24	336	.72	39	.01	.2	.1	.02	.02	2	0
26	269	.05	10	31	73	1.09	71	.01	.14	.1	.03	.02	2	0
10	269	.05	10	31	73	1.09	18	.01	.05	.1	.02	.02	2	0
26	172	.11	6	45	400	.84	81	.01	.13	.1	.02	.02	2	0
24	216	.17	8	60	384	1.01	64	<.01	.19	.08	.01	.01	2	0
36	327	.48	17	159	678	1.7	125	.02	.4	.25	.08	.08	4	0
8	84	.05	4	36	202	.39	28	.01	.11	.06	.02	.02	1	0
12	233	.15	8	61	2	1.13	22	.01	.17	.13	.03	.03	3	0
89	254	.17	25	315	98	1.23	259	.08	.27	.36	.19	.09	6	2
6	16	.01	2	19	6	.08	16	<.01	.02	.02	.01	.01	<1	<1
159	231	.1	21	293	95	.65	437	.08	.36	.35	.14	.08	6	2
10	14	.01	1	18	6	.04	27	<.01	.02	.02	.01	<.01	<1	<1
265	166	.07	17	231	82	.6	705	.06	1.43	.3	.1	.07	9	1
17	10	<.01	1	14	5	.04	44	<.01	.09	.02	.01	<.01	1	<1
326	154	.07	17	179	89	.55	1001	.05	1.5	.26	.09	.06	9	1
21	10	<.01	1	11	6	.03	63	<.01	.09	.02	.01	<.01	1	<1
46	61	.03	6	88	78	.22	124	.02	.36	.04	.04	.02	2	0
3	4	<.01		6	5	.01	8	<.01	.02	<.01	<.01	<.01	<1	0
102	267	.14	26	331	123	.62	449	.08	1.3	.34	.15	.04	25	2
6	16	.01	2	20	7	.04	27	<.01	.08	.02	.01	<.01	2	<1

Table A-1

Food Composition   (Computer code number is for West Diet Analysis program)   (For purposes of calculations, use "0" for t, <1, <.1, <.01, etc.)

Computer Code Number	Food Description	Measure	Wt (g)	H₂O (%)	Ener (cal)	Prot (g)	Carb (g)	Dietary Fiber (g)	Fat (g)	Fat Breakdown (g) Sat	Mono	Poly
	DAIRY—Continued											
	Cream products—imitation and part dairy:											
81	Coffee whitener, frozen or liquid	1 tbs	15	77	20	<1	2	0	1	1.4	t	0
82	Coffee whitener, powdered	1 tsp	2	2	11	<1	1	0	1	.6	t	0
83	Dessert topping, frozen, nondairy:	1 c	75	50	239	1	17	0	19	16.4	1.2	.4
84	Tablespoon	1 tbs	5	50	16	<1	1	0	1	1.1	.1	t
85	Dessert topping, mix with whole milk:	1 c	80	67	151	3	13	0	10	8.6	.7	.2
86	Tablespoon	1 tbs	5	67	9	<1	1	0	1	.5	t	t
88	Dessert topping, pressurized	1 c	70	60	185	1	11	0	16	13.3	1.3	.2
87	Tablespoon	1 tbs	4	60	11	<1	1	0	1	.8	.1	t
91	Sour cream, imitation:	1 c	230	71	478	6	15	0	45	40.9	1.3	.1
92	Tablespoon	1 tbs	14	71	29	<1	1	0	3	2.5	.1	t
89	Sour dressing, part dairy:	1 c	235	75	418	8	11	0	39	31.3	4.6	1.1
90	Tablespoon	1 tbs	15	75	27	1	1	0	2	2	.3	.1
	Milk, fluid:											
93	Whole milk	1 c	244	88	150	8	11	0	8	5.1	2.7	.3
94	2% reduced-fat milk	1 c	244	89	121	8	12	0	5	2.9	1.3	.2
95	2% milk solids added	1 c	245	89	125	9	12	0	5	2.9	1.4	.2
96	1% lowfat milk	1 c	244	90	102	8	12	0	3	1.6	.7	.1
97	1% milk solids added	1 c	245	90	104	9	12	0	2	1.5	.7	.1
98	Nonfat milk, vitamin A added	1 c	245	91	85	8	12	0	<1	.3	.1	t
99	Nonfat milk solids added	1 c	245	90	90	9	12	0	1	.4	.2	t
100	Buttermilk, skim	1 c	245	90	99	8	12	0	2	1.3	.6	.1
	Milk, canned:											
101	Sweetened condensed	1 c	306	27	982	24	166	0	27	16.8	7.4	1
103	Evaporated, nonfat	1 c	256	79	199	19	29	0	1	.3	.2	t
	Milk, dried:											
104	Buttermilk, sweet	1 c	120	3	464	41	59	0	7	4.3	2	.3
105	Instant, nonfat, vit A added (makes 1 qt)	1 ea	91	4	326	32	47	0	1	.4	.2	t
106	Instant nonfat, vit A added	1 c	68	4	243	24	35	0	<1	.3	.1	t
107	Goat milk	1 c	244	87	168	9	11	0	10	6.5	2.7	.4
108	Kefir	1 c	233	88	149	8	11	0	8			
	Milk beverages and powdered mixes:											
	Chocolate:											
109	Whole	1 c	250	82	209	8	26	2	8	5.3	2.5	.3
110	2% fat	1 c	250	84	179	8	26	1	5	3.1	1.5	.2
111	1% fat	1 c	250	84	158	9	26	1	2	1.5	.7	.1
	Chocolate-flavored beverages:											
112	Powder containing nonfat dry milk:	1 oz	28	1	101	3	22	<1	1	.7	.4	t
113	Prepared with water	1 c	275	86	138	4	30	3	2	.9	.5	t
114	Powder without nonfat dry milk:	1 oz	28	1	98	1	25	2	1	.5	.3	t
115	Prepared with whole milk	1 c	266	81	226	9	31	1	9	5.5	2.6	.3
116	Eggnog, commercial	1 c	254	74	343	10	34	0	19	11.3	5.7	.9
974	Eggnog, 2% reduced-fat	1 c	254	85	189	12	17	0	8	3.7	2.7	.7
1027	Instant Breakfast, envelope,powder only:	1 ea	37	7	131	7	24	<1	1	.3	.1	t
1028	Prepared with whole milk	1 c	281	77	280	15	36	<1	9	5.4	2.5	.3
1029	Prepared with 2% milk	1 c	281	78	252	15	36	<1	5	3.3	1.5	.2
1283	Prepared with 1% milk	1 c	281	79	233	15	36	<1	3	1.9	.9	.1
1284	Prepared with nonfat milk	1 c	282	80	216	16	36	<1	1	.7	.3	t
117	Malted milk, chocolate, powder:	3 tsp	21	1	79	1	18	<1	1	.5	.2	.1
118	Prepared with whole milk	1 c	265	81	228	9	30	<1	9	5.5	2.6	.4
1661	Ovaltine with whole milk	1 c	265	81	225	9	29	<1	9	5.5	2.5	.4
119	Malted mix powder, natural:	3 tsp	21	2	87	2	16	<1	2	.9	.4	.3
120	Prepared with whole milk	1 c	265	81	236	10	27	<1	10	6	2.8	.6
121	Milk shakes, chocolate	1 c	166	71	211	6	34	1	6	3.8	1.8	.2
122	Milk shakes, vanilla	1 c	166	75	184	6	30	1	5	3.1	1.4	.2
	Milk desserts:											
134	Custard, baked	1 c	282	79	296	14	30	0	13	6.6	4.3	1
1548	Low-fat frozen dessert bars	1 ea	81	72	88	2	19	0	1	.2	.1	.4
	Ice cream, vanilla (about 10% fat):											
123	Hardened: ½ gallon	1 ea	1064	61	2138	37	251	0	117	72.4	33.8	4.4
124	Cup	1 c	132	61	265	5	31	0	14	9	4.2	.5
126	Soft serve	1 c	172	60	370	7	38	0	22	12.9	6	.8

Chol (mg)	Calc (mg)	Iron (mg)	Magn (mg)	Pota (mg)	Sodi (mg)	Zinc (mg)	VT-A (RE)	Thia (mg)	VT-E (a-TE)	Ribo (mg)	Niac (mg)	V-B6 (mg)	Fola (µg)	VT-C (mg)
0	1	<.01		29	12	<.01	1	0	.24	0	0	0	0	0
0		.02		16	4	.01	<1	0	<.01	<.01	0	0	0	0
0	5	.09	1	14	19	.02	64	0	.14	0	0	0	0	0
0		.01		1	1	<.01	4	0	.01	0	0	0	0	0
8	72	.03	8	121	53	.22	39	.02	.11	.09	.05	.02	3	1
<1	5	<.01		8	3	.01	2	<.01	.01	.01	<.01	<.01	<1	<1
0	4	.01	1	13	43	.01	33	0	.12	0	0	0	0	0
0		<.01		1	2	0	2	0	.01	0	0	0	0	0
0	6	.9	15	370	235	2.71	0	0	.34	0	0	0	0	0
0		.05	1	22	14	.16	0	0	.02	0	0	0	0	0
13	266	.07	23	381	113	.87	5	.09	.29	.38	.17	.04	28	2
1	17	<.01	1	24	7	.06	<1	.01	.02	.02	.01	<.01	2	<1
33	290	.12	33	371	120	.93	76	.09	.24	.39	.2	.1	12	2
18	298	.12	33	376	122	.95	139	.09	.17	.4	.21	.1	12	2
18	314	.12	35	397	128	.98	140	.1	.17	.42	.22	.11	13	2
10	300	.12	34	381	123	.95	144	.09	.1	.41	.21	.1	12	2
10	314	.12	35	397	128	.98	145	.1	.1	.42	.22	.11	13	2
4	301	.1	28	407	126	.98	149	.09	.1	.34	.22	.1	13	2
5	316	.12	35	419	130	1	149	.1	.1	.43	.22	.11	13	2
9	284	.12	27	370	257	1.03	20	.08	.15	.38	.14	.08	12	2
104	869	.58	79	1135	389	2.88	248	.27	.65	1.27	.64	.16	34	8
9	742	.74	69	850	294	2.3	300	.11	.01	.79	.44	.14	22	3
83	1420	.36	132	1910	620	4.82	65	.47	.48	1.9	1.05	.41	57	7
17	1119	.28	106	1551	500	4.01	646	.38	.02	1.58	.81	.31	45	5
12	836	.21	80	1159	373	3	483	.28	.01	1.18	.61	.23	34	4
28	327	.12	34	498	122	.73	137	.12	.22	.34	.68	.11	1	3
		.3	33	373	107									
30	280	.6	32	418	149	1.03	72	.09	.23	.4	.31	.1	12	2
17	285	.6	33	423	151	1.03	143	.09	.13	.41	.31	.1	12	2
7	288	.6	33	425	152	1.03	148	.09	.06	.41	.32	.1	12	2
1	91	.33	23	199	141	.41	1	.03	.04	.16	.16	.03	0	<1
3	129	.47	33	270	198	.6	1	.04	.06	.21	.22	.04	0	1
0	10	.88	27	165	59	.43	1	.01	.11	.04	.14	<.01	2	<1
32	301	.8	53	497	165	1.28	77	.1	.21	.43	.32	.1	12	2
149	330	.51	47	419	138	1.17	203	.09	.58	.48	.27	.13	2	4
194	269	.71	32	367	155	1.26	197	.11	1.01	.55	.21	.15	30	2
4	105	4.74	84	350	142	3.16	554	.31	5.31	.07	5.27	.42	105	28
38	396	4.86	117	719	262	4.09	630	.41	5.51	.47	5.46	.52	118	31
23	401	4.86	118	726	264	4.12	693	.41	5.41	.48	5.46	.53	118	31
14	406	4.86	118	731	266	4.12	698	.41	5.36	.48	5.47	.53	118	31
9	407	4.83	112	755	268	4.14	703	.4	5.3	.42	5.47	.52	118	31
1	13	.48	15	130	53	.17	4	.04	.08	.04	.42	.03	4	<1
34	305	.61	48	498	172	1.09	79	.13	.26	.44	.62	.13	16	3
34	384	3.76	53	620	244	1.17	901	.73	.32	1.26	10.9	1.02	32	34
4	63	.15	19	159	104	.21	18	.11	.08	.19	1.1	.09	10	1
37	355	.26	53	530	223	1.14	95	.2	.32	.59	1.31	.19	22	3
22	188	.51	28	332	161	.68	38	.1	.11	.41	.27	.08	6	1
18	203	.15	20	289	136	.6	53	.07	.1	.3	.31	.09	5	1
245	316	.85	39	431	217	1.49	169	.09	.68	.64	.24	.14	28	1
1	82	.07	10	111	47	.26	38	.03	.07	.11	.06	.03	3	1
468	1361	.96	149	2117	851	7.34	1244	.44	0	2.55	1.23	.51	53	6
58	169	.12	18	263	106	.91	154	.05	0	.32	.15	.06	7	1
157	225	.36	21	304	105	.89	265	.08	.64	.31	.16	.08	15	1

A

Table A–1

Food Composition

(Computer code number is for West Diet Analysis program)   (For purposes of calculations, use "0" for t, <1, <.1, <.01, etc.)

Computer Code Number	Food Description	Measure	Wt (g)	H₂O (%)	Ener (cal)	Prot (g)	Carb (g)	Dietary Fiber (g)	Fat (g)	Fat Breakdown (g)		
										Sat	Mono	Poly
	DAIRY—Continued											
	Ice cream, rich vanilla (16% fat):											
127	Hardened: ½ gallon	1 ea	1188	60	2554	49	264	0	154	88.9	41.5	5.5
128	Cup	1 c	148	57	357	5	33	0	24	14.8	6.9	.9
1724	Ben & Jerry's	½ c	108		230	4	21	0	17	10		
	Ice milk, vanilla (about 4% fat):											
129	Hardened: ½ gallon	1 ea	1048	68	1456	40	238	0	45	27.7	12.9	1.7
130	Cup	1 c	132	68	183	5	30	0	6	3.5	1.6	.2
131	Soft serve (about 3% fat)	1 c	176	70	222	9	38	0	5	2.9	1.3	.2
	Pudding, canned (5 oz can = .55 cup):											
135	Chocolate	1 ea	142	69	189	4	32	1	6	1	2.4	2
136	Tapioca	1 ea	142	74	169	3	27	<1	5	.9	2.3	1.9
137	Vanilla	1 ea	142	71	185	3	31	<1	5	.8	2.2	1.9
	Puddings, dry mix with whole milk:											
138	Chocolate, instant	1 c	294	74	326	9	55	3	9	5.4	2.7	.5
139	Chocolate, regular, cooked	1 c	284	74	315	9	51	3	10	5.9	2.8	.4
140	Rice, cooked	1 c	288	72	351	9	60	<1	8	5.1	2.4	.3
141	Tapioca, cooked	1 c	282	74	321	8	55	0	8	5.1	2.3	.3
142	Vanilla, instant	1 c	284	73	324	8	56	0	8	4.9	2.4	.4
143	Vanilla, regular, cooked	1 c	280	75	311	8	52	0	8	5.1	2.4	.4
	Sherbet (2% fat):											
132	½ gallon	1 ea	1542	66	2127	17	469	8	31	17.9	8.3	1.2
133	Cup	1 c	198	66	273	2	60	1	4	2.3	1.1	.2
144	Soy milk	1 c	245	93	81	7	4	3	5	.7	1	2.6
2301	Soy milk, fortified, fat free	1 c	240	88	110	6	22	1	0	0	0	0
	Yogurt, frozen, low-fat											
1584	Cup	1 c	144	65	229	6	35	0	8	4.9	2.3	.3
1512	Scoop	1 ea	79	74	78	4	15	0	<1	.1	t	t
	Yogurt, lowfat:											
1172	Fruit added with low-calorie sweetener	1 c	241	86	122	12	19	1	<1	.2	.1	t
145	Fruit added	1 c	245	74	250	11	47	<1	3	1.7	.7	.1
146	Plain	1 c	245	85	155	13	17	0	4	2.4	1	.1
147	Vanilla or coffee flavor	1 c	245	79	209	13	34	0	3	2	.8	.1
148	Yogurt, made with nonfat milk	1 c	245	85	137	15	19	0	<1	.3	.1	t
149	Yogurt, made with whole milk	1 c	245	88	150	9	11	0	8	5.1	2.2	.2
	EGGS											
	Raw, large:											
150	Whole, without shell	1 ea	50	75	74	6	1	0	5	1.5	1.9	.7
151	White	1 ea	33	88	16	3	<1	0	0	0	0	0
152	Yolk	1 ea	17	49	61	3	<1	0	5	1.6	2	.7
	Cooked:											
153	Fried in margarine	1 ea	46	69	91	6	1	0	7	1.9	2.8	1.3
154	Hard-cooked, shell removed	1 ea	50	75	77	6	1	0	5	1.6	2	.7
155	Hard-cooked, chopped	1 c	136	75	211	17	2	0	14	4.4	5.5	1.9
156	Poached, no added salt	1 ea	50	75	74	6	1	0	5	1.5	1.9	.7
157	Scrambled with milk & margarine	1 ea	61	73	101	7	1	0	7	2.2	2.9	1.3
1681	Egg substitute, liquid:	½ c	126	83	106	16	1	0	4	.8	1.1	2
1254	Egg Beaters, Fleischmann's	½ c	122		60	12	2	0	0	0	0	0
1262	Egg substitute, liquid, prepared	½ c	105	80	100	14	1	0	4	.8	1.1	1.9
	FATS and OILS											
158	Butter: Stick	½ c	114	16	817	1	<1	0	92	57.7	27.4	3.4
159	Tablespoon:	1 tbs	14	16	100	<1	<1	0	11	7.1	3.4	.4
8025	Unsalted	1 tbs	14	18	100	<1	<1	0	11	7.1	3.4	.4
160	Pat (about 1 tsp)	1 ea	5	16	36	<1	<1	0	4	2.5	1.2	.2
1682	Whipped	1 tsp	3	16	21	<1	<1	0	2	1.5	.7	.1
	Fats, cooking:											
1363	Bacon fat	1 tbs	14		125	0	0	0	14	6.3	5.9	1.1
1362	Beef fat/tallow	1 c	205	0	1849	0	0	0	205	103	87.3	8.2
1364	Chicken fat	1 c	205		1845	0	0	0	205	61.1	91.6	42.8
161	Vegetable shortening:	1 c	205	0	1812	0	0	0	205	52.1	91.2	53.5
162	Tablespoon	1 tbs	13	0	115	0	0	0	13	3.3	5.8	3.4

PAGE KEY: A–4 = Beverages   A–6 = Dairy   A–10 = Eggs   A–10 = Fat/Oil   A–14 = Fruit   A–20 = Bakery   A–28 = Grain   A–32 = Fish   A–34 = Meats   A–38 = Poultry   A–40 = Sausage   A–42 = Mixed/Fast   A–46 = Nuts/Seeds   A–50 = Sweets   A–52 = Vegetables/Legumes   A–62 = Vegetarian Foods   A–64 = Misc   A–66 = Soups/Sauces   A–68 = Fast   A–84 = Convenience   A–88 = Baby foods

Chol (mg)	Calc (mg)	Iron (mg)	Magn (mg)	Pota (mg)	Sodi (mg)	Zinc (mg)	VT-A (RE)	Thia (mg)	VT-E (a-TE)	Ribo (mg)	Niac (mg)	V-B6 (mg)	Fola (μg)	VT-C (mg)
1081	1556	2.49	143	2102	725	6.18	1829	.58	4.4	2.16	1.13	.57	107	9
90	173	.07	16	235	83	.59	272	.06	0	.24	.12	.06	7	1
95	150	.36			55		214		0					0
147	1456	1.05	157	2211	891	4.61	493	.61	0	2.78	.94	.68	63	8
18	183	.13	20	279	112	.58	62	.08	0	.35	.12	.09	8	1
21	276	.11	25	389	123	.93	51	.09	0	.35	.21	.08	11	2
4	128	.72	30	256	183	.6	16	.04	.18	.22	.49	.04	4	3
1	119	.33	11	148	168	.38	0	.03	.13	.14	.44	.14	4	1
10	125	.18	11	160	192	.35	9	.03	.18	.2	.36	.02	0	0
32	300	.85	53	488	835	1.23	62	.1	.18	.41	.28	.11	12	3
34	315	1.02	43	463	293	1.28	74	.09	.17	.49	.29	.1	11	2
32	297	1.09	37	372	314	1.09	58	.22	.17	.4	1.28	.1	11	2
34	293	.17	34	372	341	.96	76	.08	.23	.4	.21	.11	11	2
31	287	.2	34	364	812	.94	71	.09	.18	.39	.21	.1	11	2
34	300	.14	36	381	448	.98	76	.08	.17	.4	.21	.09	11	2
77	833	2.16	123	1480	709	7.4	216	.39	.88	1.05	1.48	.52	77	66
10	107	.28	16	190	91	.95	28	.05	.11	.13	.19	.07	10	9
0	10	1.42	47	345	29	.56	7	.39	.02	.17	.36	.1	4	0
0	400	1.44		20	60		0	.07		.1	3			0
3	206	.43	20	304	125	.6	82	.05	.07	.32	.41	.11	9	1
1	137	.07	13	175	53	.67	1	.03	<.01	.16	.08	.04	8	1
3	369	.61	41	550	139	1.83	6	.1	.17	.45	.5	.11	32	26
10	372	.17	36	478	143	1.81	27	.09	.07	.44	.23	.1	23	2
15	448	.2	43	573	172	2.18	39	.11	.1	.52	.28	.12	27	2
12	419	.17	40	537	161	2.03	32	.1	.08	.49	.26	.11	26	2
4	488	.22	47	625	187	2.38	5	.12	.01	.57	.3	.13	30	2
31	296	.12	28	380	114	1.45	73	.07	.22	.35	.18	.08	18	1
213	24	.72	5	60	63	.55	95	.03	.52	.25	.04	.07	23	0
0	2	.01	4	47	54	<.01	0	<.01	0	.15	.03	<.01	1	0
218	23	.6	2	16	7	.53	99	.03	.54	.11	<.01	.07	25	0
211	25	.72	5	61	162	.55	114	.03	.75	.24	.03	.07	17	0
212	25	.59	5	63	62	.52	84	.03	.52	.26	.03	.06	22	0
577	68	1.62	14	171	169	1.43	228	.09	1.43	.7	.09	.16	60	0
212	24	.72	5	60	140	.55	95	.02	.52	.21	.03	.06	17	0
215	43	.73	7	84	171	.61	119	.03	.8	.27	.05	.07	18	<1
1	67	2.65	11	416	223	1.64	272	.14	.61	.38	.14	<.01	19	0
0	80	2.16			170	200		80		.59				0
1	63	2.51	10	394	211	1.55	258	.11	.58	.34	.12	<.01	13	0
250	27	.18	2	30	942	.06	860	.01	1.8	.04	.05	<.01	3	0
31	3	.02		4	116	.01	106	<.01	.22	<.01	.01	0	<1	0
31	3	.02		4	2	.01	106	<.01	.22	<.01	.01	0	<1	0
11	1	.01		1	41	<.01	38	0	.08	<.01	<.01	0	<1	0
7	1	<.01		1	25	<.01	23	0	.05	<.01	<.01	0	<1	0
14		0			76	<.01	0	0	.31	0	0	0	0	0
223	0	0	0		<1	0	0	0	3.08	0	0	0	0	0
174	0	0	0	0	0	0	0	0	5.54	0	0	0	0	0
0	0	0	0	0	0	0	0	0	17	0	0	0	0	0
0	0	0	0	0	0	0	0	0	1.08	0	0	0	0	0

Table A-1

Food Composition   (Computer code number is for West Diet Analysis program)   (For purposes of calculations, use "0" for t, <1, <.1, <.01, etc.)

Computer Code Number	Food Description	Measure	Wt (g)	H₂O (%)	Ener (cal)	Prot (g)	Carb (g)	Dietary Fiber (g)	Fat (g)	Fat Breakdown (g) Sat	Mono	Poly
	FATS and OILS—Continued											
163	Lard:	1 c	205	0	1849	0	0	0	205	81.1	87	28.3
164	Tablespoon	1 tbs	13	0	117	0	0	0	13	5.1	5.5	1.8
	Margarine:											
165	Imitation (about 40% fat), soft:	1 c	232	58	800	1	1	0	90	17.9	36.4	32
166	Tablespoon	1 tbs	14	58	48	<1	<1	0	5	1.1	2.2	1.9
167	Regular, hard (about 80% fat):	½ c	114	16	820	1	1	0	92	18	40.8	29
168	Tablespoon	1 tbs	14	16	101	<1	<1	0	11	2.2	5	3.6
169	Pat	1 ea	5	16	36	<1	<1	0	4	.8	1.8	1.3
170	Regular, soft (about 80% fat):	1 c	227	16	1625	2	1	0	183	31.3	64.7	78.5
171	Tablespoon	1 tbs	14	16	100	<1	<1	0	11	1.9	4	4.8
2056	Saffola, unsalted	1 tbs	14	20	100	0	0	0	11	2	3	4.5
2057	Saffola, reduced fat	1 tbs	14	37	60	0	0	0	8	1.3	2.7	4.4
172	Spread (about 60% fat), hard:	1 c	227	37	1225	1	0	0	138	32	59	41.1
173	Tablespoon	1 tbs	14	37	76	<1	0	0	9	2	3.6	2.5
174	Pat	1 ea	5	37	27	<1	0	0	3	.7	1.2	1
175	Spread (about 60% fat), soft:	1 c	227	37	1225	1	0	0	138	29.3	71.5	31.3
176	Tablespoon	1 tbs	14	37	76	<1	0	0	9	1.8	4.4	1.9
2160	Touch of Butter (47% fat)	1 tbs	14		60	0	0	0	7	1.5	3.1	1.5
	Oils:											
1585	Canola:	1 c	218	0	1927	0	0	0	218	15.5	128	64.5
1586	Tablespoon	1 tbs	14	0	124	0	0	0	14	1	8.2	4.1
177	Corn:	1 c	218	0	1927	0	0	0	218	29.4	54.1	131
178	Tablespoon	1 tbs	14	0	124	0	0	0	14	1.9	3.5	8.4
179	Olive:	1 c	216	0	1909	0	0	0	216	29.4	159	21.3
180	Tablespoon	1 tbs	14	0	124	0	0	0	14	1.9	10.3	1.4
1683	Olive, extra virgin	1 tbs	14		126	0	0	0	14	2	10.8	1.3
181	Peanut:	1 c	216	0	1909	0	0	0	216	40	99.8	71.3
182	Tablespoon	1 tbs	14	0	124	0	0	0	14	2.6	6.5	4.6
183	Safflower:	1 c	218	0	1927	0	0	0	218	19.8	26.4	162
184	Tablespoon	1 tbs	14	0	124	0	0	0	14	1.3	1.7	10.4
185	Soybean:	1 c	218	0	1927	0	0	0	218	32	50.8	126
186	Tablespoon	1 tbs	14	0	124	0	0	0	14	2.1	3.3	8.1
187	Soybean/cottonseed:	1 c	218	0	1927	0	0	0	218	39.5	64.3	105
188	Tablespoon	1 tbs	14	0	124	0	0	0	14	2.5	4.1	6.7
189	Sunflower:	1 c	218	0	1927	0	0	0	218	25.3	42.5	143
190	Tablespoon	1 tbs	14	0	124	0	0	0	14	1.6	2.7	9.2
	Salad dressings/sandwich spreads:											
191	Blue cheese, regular	1 tbs	15	32	76	1	1	0	8	1.5	1.8	4.2
1040	Low calorie	1 tbs	15	79	15	1	<1	<1	1	.2	.5	.4
1684	Caesar's	1 tbs	12	36	56	1	<1	<1	5	1	3.7	.5
192	French, regular	1 tbs	16	38	69	<1	3	0	7	1.5	1.3	3.5
193	Low calorie	1 tbs	16	71	21	<1	3	0	1	.1	.2	.5
194	Italian, regular	1 tbs	15	40	70	<1	2	0	7	1	1.7	4.2
195	Low calorie	1 tbs	15	84	16	<1	1	<1	1	.2	.3	.9
	Kraft, Deliciously Right:											
2150	1000 Island	1 tbs	16		35	0	4	0	2	.5		
2153	Bacon & tomato	1 tbs	16		31	1	2	0	3	.5		
2154	Cucumber ranch	1 tbs	16		31	0	1	0	3	.5		
2151	French	1 tbs	16		25	0	3	0	1	.2		
2152	Ranch	1 tbs	16		52	0	3	0	5	.8		
199	Mayo type, regular	1 tbs	15	40	58	<1	4	0	5	.7	1.3	2.7
1030	Low calorie	1 tbs	14	54	36	<1	3	0	3	.4	.7	1.4
	Mayonnaise:											
197	Imitation, low calorie	1 tbs	15	63	35	<1	2	0	3	.5	.7	1.6
196	Regular (soybean)	1 tbs	14	17	100	<1	<1	0	11	1.7	3.2	5.8
1488	Regular, low calorie, low sodium	1 tbs	14	63	32	<1	2	0	3	.5	.6	1.4
1493	Regular, low calorie	1 tbs	15	63	35	<1	2	0	3	.5	.7	1.6
198	Ranch, regular	1 tbs	15	39	80	0	<1	0	8	1.2		
2251	Low calorie	1 tbs	14	69	30	<1	1	0	2	.5		
1685	Russian	1 tbs	15	34	74	<1	2	0	8	1.1	1.8	4.4
1502	Salad dressing, low calorie, oil free	1 tbs	15	88	4	<1	1	<1	<1	0	0	0

Chol (mg)	Calc (mg)	Iron (mg)	Magn (mg)	Pota (mg)	Sodi (mg)	Zinc (mg)	VT-A (RE)	Thia (mg)	VT-E (a-TE)	Ribo (mg)	Niac (mg)	V-B6 (mg)	Fola (µg)	VT-C (mg)
195		0			<1	.23	0	0	2.46	0	0	0	0	0
12		0			<1	.01	0	0	.16	0	0	0	0	0
0	41	0	4	59	2227	0	1853	.01	5.41	.05	.03	.01	2	<1
0	2	0		4	134	0	112	<.01	.33	<.01	<.01	<.01	<1	<1
0	34	.07	3	48	1075	0	911	.01	14.6	.04	.03	.01	1	<1
0	4	.01		6	132	0	112	<.01	1.79	<.01	<.01	<.01	<1	<1
0	1	<.01		2	47	0	40	<.01	.64	<.01	<.01	0	<1	<1
0	60	0	5	86	2447	0	1813	.02	27.2	.07	.04	.02	2	<1
0	4	0		5	151	0	112	<.01	1.68	<.01	<.01	<.01	<1	<1
	0	0			0		51							0
	0	0			115		51							0
0	47	0	4	68	2256	0	1813	.02	11.4	.06	.04	.01	2	<1
0	3	0		4	139	0	112	<.01	.7	<.01	<.01	<.01	<1	<1
0	1	0		1	50	0	40	0	.25	<.01	<.01	0	<1	<1
0	47	0	4	68	2256	0	1813	.02	20.5	.06	.04	.01	2	<1
0	3	0		4	139	0	112	<.01	1.26	<.01	<.01	<.01	<1	<1
0	0	0		0	110		100		1.27					
0	0	0	0	0	0	0	0	0	45.8	0	0	0	0	0
0	0	0	0	0	0	0	0	0	2.94	0	0	0	0	0
0	0	0	0	0	0	0	0	0	46	0	0	0	0	0
0	0	0	0	0	0	0	0	0	2.95	0	0	0	0	0
0		.82		0	<1	.13	0	0	26.8	0	0	0	0	0
0		.05		0	<1	.01	0	0	1.74	0	0	0	0	0
									1.74					
0		.06			<1	.02	0	0	27.9	0	0	0	0	0
0		<.01			<1	<.01	0	0	1.81	0	0	0	0	0
0	0	0	0	0	0	0	0	0	94	0	0	0	0	0
0	0	0	0	0	0	0	0	0	6.03	0	0	0	0	0
0		.04		0	0	0	0	0	39.7	0	0	0	0	0
0		<.01		0	0	0	0	0	2.55	0	0	0	0	0
0	0	0	0	0	0	0	0	0	61.5	0	0	0	0	0
0	0	0	0	0	0	0	0	0	3.95	0	0	0	0	0
0	0	0	0	0	0	0	0	0	110	0	0	0	0	0
0	0	0	0	0	0	0	0	0	7.08	0	0	0	0	0
3	12	.03	0	6	164	.04	10	<.01	1.4	.01	.01	.01	1	<1
<1	13	.07	1	1	180	.04	<1	<.01	.13	.01	.01	<.01	<1	<1
12	23	.2	3	21	207	.13	7	<.01	.72	.02	.5	.01	2	1
0	2	.06	0	13	219	.01	21	<.01	1.35	<.01	0	<.01	1	0
0	2	.06	0	13	126	.03	21	0	.19	0	0	0	0	0
0	1	.03		2	118	.02	4	<.01	1.56	<.01	0	<.01	1	0
1		.03	0	2	118	.02	0	0	.22	0	0	0	0	0
2	0	0		27	160		0		.19					0
1	0	0		21	155		0		.75					0
0	0	0		10	232		0		.73					0
0	0	0		7	130		50		.42					0
0	0	0		5	165		0		1.31					0
4	2	.03		1	107	.03	13	<.01	.6	<.01	<.01	<.01	1	0
4	2	.03		1	99	.02	9	<.01	.6	<.01	0	<.01	1	0
4		0		1	75	.02	0	0	.96	0	0	0	0	0
8	3	.07		5	79	.02	12	0	1.65	0	<.01	.08	1	0
3	0	0	0	1	15	.01	1	0	.53	<.01	0	0	<1	0
4		0		1	75	.02	0	0	.96	0	0	0	0	0
5	0	0			105		0							0
5	5	0			120		0		.7					0
3	3	.09		24	130	.06	31	.01	1.53	.01	.09	<.01	2	1
0	1	.04	2	7	256	<.01	<1	0	0	0	<.01	<.01	<1	<1

Table A–1

Food Composition (Computer code number is for West Diet Analysis program) (For purposes of calculations, use "0" for t, <1, <.1, <.01, etc.)

Computer Code Number	Food Description	Measure	Wt (g)	H₂O (%)	Ener (cal)	Prot (g)	Carb (g)	Dietary Fiber (g)	Fat (g)	Fat Breakdown (g) Sat	Mono	Poly
	FATS and OILS—Continued											
	Salad dressing, no cholesterol											
1605	Miracle Whip	1 tbs	15	57	48	0	2	0	4	1.1	1.1	2.1
203	Salad dressing, from recipe, cooked	1 tbs	16	69	25	1	2	0	2	.5	.6	.3
200	Tartar sauce, regular	1 tbs	14	34	74	<1	1	<1	8	1.5	2.6	4.1
1503	Low calorie	1 tbs	14	63	31	<1	2	<1	2	.4	.6	1.3
201	Thousand island, regular	1 tbs	16	46	60	<1	2	0	6	1	1.3	3.2
202	Low calorie	1 tbs	15	69	24	<1	2	<1	2	.2	.4	.9
204	Vinegar and oil	1 tbs	16	47	72	0	<1	0	8	1.5	2.4	3.9
	Wishbone:											
2180	Creamy Italian, lite	1 tbs	15		26	<1	2		2	.4	.9	.7
2166	Italian, lite	1 tbs	16	90	6	0	1		<1	0	.2	.1
8427	Ranch, lite	1 tbs	15	56	50	0	2	0	4	.7		
	FRUITS and FRUIT JUICES											
	Apples:											
	Fresh, raw, with peel:											
205	2¾" diam (about 3 per lb w/cores)	1 ea	138	84	81	<1	21	4	<1	.1	t	.1
206	3¼" diam (about 2 per lb w/cores)	1 ea	212	84	125	<1	32	6	1	.1	t	.2
207	Raw, peeled slices	1 c	110	84	63	<1	16	2	<1	.1	t	.1
208	Dried, sulfured	10 ea	64	32	156	1	42	6	<1	t	t	.1
209	Apple juice, bottled or canned	1 c	248	88	117	<1	29	<1	<1	t	t	.1
210	Applesauce, sweetened	1 c	255	80	194	<1	51	3	<1	.1	t	.1
211	Applesauce, unsweetened	1 c	244	88	105	<1	28	3	<1	t	t	t
	Apricots:											
212	Raw, w/o pits (about 12 per lb w/pits)	3 ea	105	86	50	1	12	3	<1	t	.2	.1
	Canned (fruit and liquid):											
213	Heavy syrup	1 c	240	78	199	1	52	4	<1	t	.1	t
214	Halves	3 ea	120	78	100	1	26	2	<1	t	t	t
215	Juice pack	1 c	244	87	117	2	30	4	<1	t	t	t
216	Halves	3 ea	108	87	52	1	13	2	<1	t	t	t
217	Dried, halves	10 ea	35	31	83	1	22	3	<1	t	.1	t
218	Dried, cooked, unsweetened, w/liquid	1 c	250	76	213	3	55	8	<1	t	.2	.1
219	Apricot nectar, canned	1 c	251	85	141	1	36	2	<1	t	.1	t
	Avocados, raw, edible part only:											
220	California	1 ea	173	73	306	4	12	8	30	4.5	19.5	3.5
221	Florida	1 ea	304	80	340	5	27	16	27	5.3	14.8	4.5
222	Mashed, fresh, average	1 c	230	74	370	5	17	11	35	5.6	22.1	4.5
	Bananas, raw, without peel:											
223	Whole, 8¾" long (175g w/peel)	1 ea	118	74	109	1	28	3	1	.2	t	.1
224	Slices	1 c	150	74	138	2	35	4	1	.3	.1	.1
1285	Bananas, dehydrated slices	½ c	50	3	173	2	44	4	1	.3	.1	.2
225	Blackberries, raw	1 c	144	86	75	1	18	8	1	t	.1	.3
	Blueberries:											
226	Fresh	1 c	145	85	81	1	20	4	1	t	.1	.2
227	Frozen, sweetened	10 oz	284	77	230	1	62	6	<1	t	.1	.2
228	Frozen, thawed	1 c	230	77	186	1	51	5	<1	t	t	.1
	Cherries:											
229	Sour, red pitted, canned water pack	1 c	244	90	88	2	22	3	<1	.1	.1	.1
230	Sweet, red pitted, raw	10 ea	68	81	49	1	11	2	1	.1	.2	.2
231	Cranberry juice cocktail, vitamin C added	1 c	253	85	144	0	36	<1	<1	t	t	.1
1411	Cranberry juice, low calorie	1 c	237	95	45	0	11	<1	0	0	0	0
232	Cranberry-apple juice, vitamin C added	1 c	245	83	164	<1	42	<1	0	0	0	0
233	Cranberry sauce, canned, strained	1 c	277	61	418	1	108	3	<1	t	.1	.2
234	Dates, whole, without pits	10 ea	83	22	228	2	61	6	<1	.2	.1	t
235	Dates, chopped	1 c	178	22	490	4	131	13	1	.3	.3	.1
236	Figs, dried	10 ea	190	28	485	6	124	18	2	.4	.5	1.1
	Fruit cocktail, canned, fruit and liq:											
237	Heavy syrup pack	1 c	248	80	181	1	47	2	<1	t	t	.1
238	Juice pack	1 c	237	87	109	1	28	2	<1	t	t	t
	Grapefruit:											
	Raw 3¾" diam (half w/rind = 241g)											
239	Pink/red, half fruit, edible part	1 ea	123	91	37	1	9	2	<1	t	t	t
240	White, half fruit, edible part	1 ea	118	90	39	1	10	1	<1	t	t	t
241	Canned sections with light syrup	1 c	254	84	152	1	39	1	<1	t	t	.1

PAGE KEY: A–4 = Beverages A–6 = Dairy A–10 = Eggs A–10 = Fat/Oil A–14 = Fruit A–20 = Bakery A–28 = Grain A–32 = Fish A–34 = Meats A–38 = Poultry A–40 = Sausage A–42 = Mixed/Fast A–46 = Nuts/Seeds A–50 = Sweets A–52 = Vegetables/Legumes A–62 = Vegetarian Foods A–64 = Misc A–66 = Soups/Sauces A–68 = Fast A–84 = Convenience A–88 = Baby foods

A–15

Chol (mg)	Calc (mg)	Iron (mg)	Magn (mg)	Pota (mg)	Sodi (mg)	Zinc (mg)	VT-A (RE)	Thia (mg)	VT-E (a-TE)	Ribo (mg)	Niac (mg)	V-B6 (mg)	Fola (μg)	VT-C (mg)
0	0	<.01	0	0	102	0	2	0	.64	0	0	0	0	0
9	13	.08	1	19	117	.06	20	.01	.3	.02	.04	<.01	1	<1
7	3	.13		11	99	.02	9	<.01	2.24	<.01	0	<.01	1	<1
3	2	.09		5	83	.02	2	0	.83	<.01	.01	<.01	<1	<1
4	2	.1		18	112	.02	15	<.01	.18	<.01	<.01	<.01	1	0
2	2	.09		17	150	.02	14	<.01	.18	<.01	0	<.01	1	0
0	0	0	0	1	<1	0	0	0	1.41	0	0	0	0	0
<1	0			148			0	0	.56	0	0			0
0	1	0		255			2	0	.24	0	0			<1
2	0	0		120			0							0
0	10	.25	7	159	0	.05	7	.02	.44	.02	.11	.07	4	8
0	15	.38	11	244	0	.08	11	.04	.68	.03	.16	.1	6	12
0	4	.08	3	124	0	.04	4	.02	.09	.01	.1	.05	<1	4
0	9	.9	10	288	56	.13	4	0	.35	.1	.59	.08	0	2
0	17	.92	7	295	7	.07	<1	.05	.02	.04	.25	.07	<1	2
0	10	.89	8	156	8	.1	3	.03	.03	.07	.48	.07	2	4
0	7	.29	7	183	5	.07	7	.03	.02	.06	.46	.06	1	3
0	15	.57	8	311	1	.27	274	.03	.93	.04	.63	.06	9	10
0	22	.72	17	336	10	.26	295	.05	2.14	.05	.9	.13	4	7
0	11	.36	8	168	5	.13	148	.02	1.07	.03	.45	.06	2	4
0	29	.73	24	403	10	.27	412	.04	2.17	.05	.84	.13	4	12
0	13	.32	11	178	4	.12	183	.02	.96	.02	.37	.06	2	5
0	16	1.65	16	482	3	.26	253	<.01	.52	.05	1.05	.05	4	1
0	40	4.18	42	1222	7	.65	590	.01	1.25	.07	2.36	.28	0	4
0	18	.95	13	286	8	.23	331	.02	.2	.03	.65	.05	3	2
0	19	2.04	71	1096	21	.73	106	.19	2.32	.21	3.32	.48	113	14
0	33	1.61	103	1483	15	1.28	185	.33	2.37	.37	5.84	.85	162	24
0	25	2.35	90	1377	23	.97	140	.25	3.08	.28	4.42	.64	142	18
0	7	.37	34	467	1	.19	9	.05	.32	.12	.64	.68	22	11
0	9	.46	43	594	1	.24	12	.07	.4	.15	.81	.87	29	14
0	11	.57	54	746	1	.3	15	.09	0	.12	1.4	.22	7	3
0	46	.82	29	282	0	.39	23	.04	1.02	.06	.58	.08	49	30
0	9	.25	7	129	9	.16	14	.07	1.45	.07	.52	.05	9	19
0	17	1.11	6	170	3	.17	11	.06	2.02	.15	.72	.17	19	3
0	14	.9	5	138	2	.14	9	.05	1.63	.12	.58	.14	15	2
0	27	3.34	15	239	17	.17	183	.04	.32	.1	.43	.11	19	5
0	10	.26	7	152	0	.04	14	.03	.09	.04	.27	.02	3	5
0	8	.38	5	45	5	.18	1	.02	0	.02	.09	.05	<1	90
0	21	.09	5	52	7	.05	1	.02	0	.02	.08	.04	<1	76
0	17	.15	5	66	5	.1	1	.01	0	.05	.15	.05	<1	78
0	11	.61	8	72	80	.14	6	.04	.28	.06	.28	.04	3	6
0	27	.95	29	541	2	.24	4	.07	.08	.08	1.83	.16	10	0
0	57	2.05	62	1160	5	.52	9	.16	.18	.18	3.92	.34	22	0
0	274	4.24	112	1352	21	.97	25	.13	9.5	.17	1.32	.43	14	2
0	15	.72	12	218	15	.2	50	.04	.72	.05	.93	.12	6	5
0	19	.5	17	225	9	.21	73	.03	.47	.04	.95	.12	6	6
0	13	.15	10	159	0	.09	32	.04	.31	.02	.23	.05	15	47
0	14	.07	11	175	0	.08	1	.04	.29	.02	.32	.05	12	39
0	36	1.02	25	328	5	.2	0	.1	.63	.05	.62	.05	22	54

Table A–1

Food Composition   (Computer code number is for West Diet Analysis program)   (For purposes of calculations, use "0" for t, <1, <.1, <.01, etc.)

Computer Code Number	Food Description	Measure	Wt (g)	H₂O (%)	Ener (cal)	Prot (g)	Carb (g)	Dietary Fiber (g)	Fat (g)	Fat Breakdown (g) Sat	Mono	Poly
	FRUITS and FRUIT JUICES											
	Grapefruit juice:											
242	Fresh, white, raw	1 c	247	90	96	1	23	<1	<1	t	t	.1
243	Canned, unsweetened	1 c	247	90	94	1	22	<1	<1	t	t	.1
244	Sweetened	1 c	250	87	115	1	28	<1	<1	t	t	.1
	Frozen concentrate, unsweetened:											
245	Undiluted, 6-fl-oz can	¾ c	207	62	302	4	72	1	1	.1	.1	.2
246	Diluted with 3 cans water	1 c	247	89	101	1	24	<1	<1	t	t	.1
	Grapes, raw European (adherent skin):											
247	Thompson seedless	10 ea	50	81	35	<1	9	<1	<1	.1	t	.1
248	Tokay/Emperor, seeded types	10 ea	50	81	35	<1	9	<1	<1	.1	t	.1
	Grape juice:											
249	Bottled or canned	1 c	253	84	154	1	38	<1	<1	.1	t	.1
	Frozen concentrate, sweetened:											
250	Undiluted, 6-fl-oz can, vit C added	¾ c	216	54	387	1	96	1	1	.2	t	.2
251	Diluted with 3 cans water, vit C added	1 c	250	87	128	<1	32	<1	<1	.1	t	.1
1410	Low calorie	1 c	253	84	154	1	38	<1	<1	.1	t	.1
252	Kiwi fruit, raw, peeled (88 g with peel)	1 ea	76	83	46	1	11	3	<1	t	t	.2
253	Lemons, raw, without peel and seeds (about 4 per lb whole)	1 ea	58	89	17	1	5	2	<1	t	t	.1
	Lemon juice:											
254	Fresh:	1 c	244	91	61	1	21	1	0	0	0	0
255	Tablespoon	1 tbs	15	91	4	<1	1	<1	0	0	0	0
256	Canned or bottled, unsweetened:	1 c	244	92	51	1	16	1	1	.1	t	.2
257	Tablespoon	1 tbs	15	92	3	<1	1	<1	<1	t	t	t
258	Frozen, single strength, unsweetened:	1 c	244	92	54	1	16	1	1	.1	t	.2
2298	Tablespoon	1 tbs	15	92	3	<1	1	<1	<1	t	t	t
	Lime juice:											
260	Fresh:	1 c	246	90	66	1	22	1	<1	t	t	.1
261	Tablespoon	1 tbs	15	90	4	<1	1	<1	<1	t	t	t
262	Canned or bottled, unsweetened	1 c	246	92	52	1	16	1	1	.1	.1	.2
263	Mangos, raw, edible part (300 g w/skin & seeds)	1 ea	207	82	135	1	35	4	1	.1	.2	.1
	Melons, raw, without rind and contents:											
264	Cantaloupe, 5" diam (2⅓ lb whole with refuse), orange flesh	½ ea	276	90	97	2	23	2	1	.2	t	.3
265	Honeydew, 6½" diam (5¼ lb whole with refuse), slice = ⅟₁₀ melon	1 pce	160	90	56	1	15	1	<1	t	t	.1
266	Nectarines, raw, w/o pits, 2¼" diam	1 ea	136	86	67	1	16	2	1	.1	.2	.3
	Oranges, raw:											
267	Whole w/o peel and seeds, 2⅝" diam (180 g with peel and seeds)	1 ea	131	87	62	1	15	3	<1	t	t	t
268	Sections, without membranes	1 c	180	87	85	2	21	4	<1	t	t	t
	Orange juice:											
269	Fresh, all varieties	1 c	248	88	112	2	26	<1	<1	.1	.1	.1
270	Canned, unsweetened	1 c	249	89	105	1	24	<1	<1	t	.1	.1
271	Chilled	1 c	249	88	110	2	25	<1	1	.1	.1	.2
	Frozen concentrate:											
272	Undiluted (6-oz can)	¾ c	213	58	339	5	81	2	<1	.1	.1	.1
273	Diluted w/3 parts water by volume	1 c	249	88	112	2	27	<1	<1	t	t	t
1345	Orange juice, from dry crystals	1 c	248	88	114	0	29	0	0	0	0	0
274	Orange and grapefruit juice, canned	1 c	247	89	106	1	25	<1	<1	t	t	t
	Papayas, raw:											
275	½" slices	1 c	140	89	55	1	14	3	<1	.1	.1	t
276	Whole, 3½" diam by 5⅛" w/o seeds and skin (1 lb w/refuse)	1 ea	304	89	119	2	30	5	<1	.1	.1	.1
1031	Papaya nectar, canned	1 c	250	85	143	<1	36	1	<1	.1	.1	.1
	Peaches:											
277	Raw, whole, 2½" diam, peeled, pitted (about 4 per lb whole)	1 ea	98	88	42	1	11	2	<1	t	t	t
278	Raw, sliced	1 c	170	88	73	1	19	3	<1	t	.1	.1
	Canned, fruit and liquid:											
279	Heavy syrup pack:	1 c	262	79	194	1	52	3	<1	t	.1	.1
280	Half	1 ea	98	79	72	<1	19	1	<1	t	t	t

PAGE KEY: A–4 = Beverages A–6 = Dairy A–10 = Eggs A–10 = Fat/Oil A–14 = Fruit A–20 = Bakery A–28 = Grain A–32 = Fish A–34 = Meats A–38 = Poultry A–40 = Sausage A–42 = Mixed/Fast A–46 = Nuts/Seeds A–50 = Sweets A–52 = Vegetables/Legumes A–62 = Vegetarian Foods A–64 = Misc A–66 = Soups/Sauces A–68 = Fast A–84 = Convenience A–88 = Baby foods

A–17

A

Chol (mg)	Calc (mg)	Iron (mg)	Magn (mg)	Pota (mg)	Sodi (mg)	Zinc (mg)	VT-A (RE)	Thia (mg)	VT-E (a-TE)	Ribo (mg)	Niac (mg)	V-B6 (mg)	Fola (µg)	VT-C (mg)
0	22	.49	30	400	2	.12	2	.1	.12	.05	.49	.11	25	94
0	17	.49	25	378	2	.22	2	.1	.12	.05	.57	.05	26	72
0	20	.9	25	405	5	.15	0	.1	.12	.06	.8	.05	26	67
0	56	1.01	79	1001	6	.37	6	.3	.37	.16	1.6	.32	26	248
0	20	.35	27	336	2	.12	2	.1	.12	.05	.54	.11	9	83
0	5	.13	3	92	1	.02	3	.05	.35	.03	.15	.05	2	5
0	5	.13	3	92	1	.02	3	.05	.35	.03	.15	.05	2	5
0	23	.61	25	334	8	.13	3	.07	0	.09	.66	.16	7	<1
0	28	.78	32	160	15	.28	6	.11	.38	.2	.93	.32	9	179
0	10	.25	10	52	5	.1	2	.04	.12	.06	.31	.1	3	60
0	23	.61	25	334	8	.13	3	.07	0	.09	.66	.16	7	<1
0	20	.31	23	252	4	.13	14	.01	.85	.04	.38	.07	29	74
0	15	.35	5	80	1	.03	2	.02	.14	.01	.06	.05	6	31
0	17	.07	15	303	2	.12	5	.07	.22	.02	.24	.12	31	112
0	1	<.01	1	19	<1	.01	<1	<.01	.01	<.01	.01	.01	2	7
0	27	.32	19	249	51	.15	5	.1	.22	.02	.48	.1	25	60
0	2	.02	1	15	3	.01	<1	.01	.01	<.01	.03	.01	2	4
0	19	.29	19	217	2	.12	2	.14	.22	.03	.33	.15	23	77
0	1	.02	1	13	<1	.01	<1	.01	.01	<.01	.02	.01	1	5
0	22	.07	15	268	2	.15	2	.05	.22	.02	.25	.11	20	72
0	1	<.01	1	16	<1	.01	<1	<.01	.01	<.01	.01	.01	1	4
0	29	.57	17	185	39	.15	5	.08	.17	.01	.4	.07	19	16
0	21	.27	19	323	4	.08	805	.12	2.32	.12	1.21	.28	29	57
0	30	.58	30	853	25	.44	889	.1	.41	.06	1.58	.32	47	116
0	10	.11	11	434	16	.11	6	.12	.24	.03	.96	.09	10	40
0	7	.2	11	288	0	.12	101	.02	1.21	.06	1.35	.03	5	7
0	52	.13	13	237	0	.09	27	.11	.31	.05	.37	.08	40	70
0	72	.18	18	326	0	.13	38	.16	.43	.07	.51	.11	54	96
0	27	.5	27	496	2	.12	50	.22	.22	.07	.99	.1	75	124
0	20	1.1	27	436	5	.17	45	.15	.22	.07	.78	.22	45	86
0	25	.42	27	473	2	.1	20	.28	.47	.05	.7	.13	45	82
0	68	.75	72	1435	6	.38	60	.6	.68	.14	1.53	.33	330	294
0	22	.25	25	473	2	.12	20	.2	.47	.04	.5	.11	109	97
0	62	.2	2	50	12	.1	551	<.01	0	.04	0	0	143	121
0	20	1.14	25	390	7	.17	30	.14	.17	.07	.83	.06	35	72
0	34	.14	14	360	4	.1	39	.04	1.57	.04	.47	.03	53	86
0	73	.3	30	781	9	.21	85	.08	3.4	.1	1.03	.06	116	188
0	25	.85	7	77	12	.37	27	.01	.05	.01	.37	.02	5	7
0	5	.11	7	193	0	.14	53	.02	.69	.04	.97	.02	3	6
0	8	.19	12	335	0	.24	92	.03	1.19	.07	1.68	.03	6	11
0	8	.71	13	241	16	.24	86	.03	2.33	.06	1.61	.05	8	7
0	3	.26	5	90	6	.09	32	.01	.87	.02	.6	.02	3	3

Table A–1
Food Composition

(Computer code number is for West Diet Analysis program)   (For purposes of calculations, use "0" for t, <1, <.1, <.01, etc.)

Computer Code Number	Food Description	Measure	Wt (g)	H₂O (%)	Ener (cal)	Prot (g)	Carb (g)	Dietary Fiber (g)	Fat (g)	Fat Breakdown (g) Sat	Mono	Poly
	FRUITS and FRUIT JUICES—Continued											
281	Juice pack:	1 c	248	87	109	2	29	3	<1	t	t	t
282	Half	1 ea	98	87	43	1	11	1	<1	t	t	t
283	Dried, uncooked	10 ea	130	32	311	5	80	11	1	.1	.4	.5
284	Dried, cooked, fruit and liquid	1 c	258	78	199	3	51	7	1	.1	.2	.3
	Frozen, slice, sweetened:											
285	10-oz package,vitamin C added	1 ea	284	75	267	2	68	5	<1	t	.1	.2
286	Cup, thawed measure, vitamin C added	1 c	250	75	235	2	60	4	<1	t	.1	.2
1032	Peach nectar, canned	1 c	249	86	134	1	35	1	<1	t	t	t
	Pears:											
	Fresh, with skin, cored:											
287	Bartlett, 2½" diam (about 2½ per lb)	1 ea	166	84	98	1	25	4	1	t	.1	.2
288	Bosc, 2⅛" diam (about 3 per lb)	1 ea	139	84	82	1	21	3	1	t	.1	.1
289	D'Anjou, 3" diam (about 2 per lb)	1 ea	209	84	123	1	32	5	1	t	.2	.2
	Canned, fruit and liquid:											
290	Heavy syrup pack:	1 c	266	80	197	1	51	4	<1	t	.1	.1
291	Half	1 ea	76	80	56	<1	15	1	<1	t	t	t
292	Juice pack:	1 c	248	86	124	1	32	4	<1	t	t	t
293	Half	1 ea	76	86	38	<1	10	1	<1	t	t	t
294	Dried halves	10 ea	175	27	459	3	122	13	1	.1	.2	.3
1033	Pear nectar, canned	1 c	250	84	150	<1	39	1	<1	t	t	t
	Pineapple:											
295	Fresh chunks, diced	1 c	155	86	76	1	19	2	1	t	.1	.2
	Canned, fruit and liquid:											
	Heavy syrup pack:											
296	Crushed, chunks, tidbits	½ c	127	79	99	<1	26	1	<1	t	t	.1
297	Slices	1 ea	49	79	38	<1	10	<1	<1	t	t	t
298	Juice pack, crushed, chunks, tidbits	1 c	250	83	150	1	39	2	<1	t	t	.1
299	Juice pack, slices	1 ea	47	83	28	<1	7	<1	<1	t	t	t
300	Pineapple juice, canned, unsweetened	1 c	250	85	140	1	34	<1	<1	t	t	.1
	Plantains, yellow flesh, without peel:											
301	Raw slices (whole=179 g w/o peel)	1 c	148	65	181	2	47	3	1	.2	t	.1
302	Cooked, boiled, sliced	1 c	154	67	179	1	48	4	<1	.1	t	.1
	Plums:											
303	Fresh, medium, 2⅛" diam	1 ea	66	85	36	1	9	1	<1	t	.3	.1
304	Fresh, small, 1½" diam	1 ea	28	85	15	<1	4	<1	<1	t	.1	t
	Canned, purple, with liquid:											
305	Heavy syrup pack:	1 c	258	76	230	1	60	3	<1	t	.2	.1
306	Plums	3 ea	138	76	123	<1	32	1	<1	t	.1	t
307	Juice pack:	1 c	252	84	146	1	38	3	<1	t	t	t
308	Plums	3 ea	138	84	80	1	21	1	<1	t	t	t
1698	Pomegranate, fresh	1 ea	154	81	105	1	26	1	<1	.1	.1	.1
	Prunes, dried, pitted:											
309	Uncooked (10 = 97 g w/pits, 84 g w/o pits)	10 ea	84	32	201	2	53	6	<1	t	.3	.1
310	Cooked, unsweetened, fruit & liq (250 g w/pits)	1 c	248	70	265	3	70	16	1	t	.4	.1
311	Prune juice, bottled or canned	1 c	256	81	182	2	45	3	<1	t	.1	t
	Raisins, seedless:											
312	Cup, not pressed down	1 c	145	15	435	5	115	6	1	.2	t	.2
313	One packet, ½ oz	½ oz	14	15	42	<1	11	1	<1	t	t	t
	Raspberries:											
314	Fresh	1 c	123	87	60	1	14	8	1	t	.1	.4
315	Frozen, sweetened	10 oz	284	73	293	2	74	12	<1	t	t	.3
316	Cup, thawed measure	1 c	250	73	258	2	65	11	<1	t	t	.2
317	Rhubarb, cooked, added sugar	1 c	240	68	278	1	75	5	<1	t	t	.1
	Strawberries:											
318	Fresh, whole, capped	1 c	144	92	43	1	10	3	1	t	.1	.3
	Frozen, sliced, sweetened:											
319	10-oz container	10 oz	284	73	273	2	74	5	<1	t	.1	.2
320	Cup, thawed measure	1 c	255	73	245	1	66	5	<1	t	t	.2
	Tangerines, without peel and seeds:											
321	Fresh (2⅜" whole) 116 g w/refuse	1 ea	84	88	37	1	9	2	<1	t	t	t
322	Canned, light syrup, fruit and liquid	1 c	252	83	154	1	41	2	<1	t	t	t

PAGE KEY: A–4 = Beverages A–6 = Dairy A–10 = Eggs A–10 = Fat/Oil A–14 = Fruit A–20 = Bakery A–28 = Grain A–32 = Fish A–34 = Meats A–38 = Poultry A–40 = Sausage A–42 = Mixed/Fast A–46 = Nuts/Seeds A–50 = Sweets A–52 = Vegetables/Legumes A–62 = Vegetarian Foods A–64 = Misc A–66 = Soups/Sauces A–68 = Fast A–84 = Convenience A–88 = Baby foods

A–19

A

Chol (mg)	Calc (mg)	Iron (mg)	Magn (mg)	Pota (mg)	Sodi (mg)	Zinc (mg)	VT-A (RE)	Thia (mg)	VT-E (a-TE)	Ribo (mg)	Niac (mg)	V-B6 (mg)	Fola (μg)	VT-C (mg)
0	15	.67	17	317	10	.27	94	.02	3.72	.04	1.44	.05	8	9
0	6	.26	7	125	4	.11	37	.01	1.47	.02	.57	.02	3	4
0	36	5.28	55	1294	9	.74	281	<.01	0	.28	5.69	.09	<1	6
0	23	3.38	33	826	5	.46	52	.01	0	.05	3.92	.1	<1	10
0	9	1.05	14	369	17	.14	79	.04	2.53	.1	1.85	.05	9	268
0	7	.92	12	325	15	.12	70	.03	2.23	.09	1.63	.04	8	236
0	12	.47	10	100	17	.2	65	.01	.2	.03	.72	.02	3	13
0	18	.41	10	208	0	.2	3	.03	.83	.07	.17	.03	12	7
0	15	.35	8	174	0	.17	3	.03	.69	.06	.14	.02	10	6
0	23	.52	12	261	0	.25	4	.04	1.05	.08	.21	.04	15	8
0	13	.58	11	173	13	.21	0	.03	1.33	.06	.64	.04	3	3
0	4	.17	3	49	4	.06	0	.01	.38	.02	.18	.01	1	1
0	22	.72	17	238	10	.22	2	.03	1.24	.03	.5	.03	3	4
0	7	.22	5	73	3	.07	1	.01	.38	.01	.15	.01	1	1
0	59	3.68	58	933	10	.68	1	.01	0	.25	2.4	.13	0	12
0	12	.65	7	32	10	.17	<1	<.01	.25	.03	.32	.03	3	3
0	11	.57	22	175	2	.12	3	.14	.15	.06	.65	.13	16	24
0	18	.48	20	132	1	.15	1	.11	.13	.03	.36	.09	6	9
0	7	.19	8	51	<1	.06	<1	.04	.05	.01	.14	.04	2	4
0	35	.7	35	305	2	.25	10	.24	.25	.05	.71	.18	12	24
0	7	.13	7	57	<1	.05	2	.04	.05	.01	.13	.03	2	4
0	42	.65	32	335	2	.27	1	.14	.05	.05	.64	.24	58	27
0	4	.89	55	739	6	.21	167	.08	.4	.08	1.02	.44	33	27
0	3	.89	49	716	8	.2	140	.07	.22	.08	1.16	.37	40	17
0	3	.07	5	114	0	.07	21	.03	.4	.06	.33	.05	1	6
0	1	.03	2	48	0	.03	9	.01	.17	.03	.14	.02	1	3
0	23	2.17	13	235	49	.18	67	.04	1.81	.1	.75	.07	6	1
0	12	1.16	7	126	26	.1	36	.02	.97	.05	.4	.04	3	1
0	25	.86	20	388	3	.28	255	.06	1.76	.15	1.19	.07	7	7
0	14	.47	11	213	1	.15	139	.03	.97	.08	.65	.04	4	4
0	5	.46	5	399	5	.18	0	.05	.85	.05	.46	.16	9	9
0	43	2.08	38	626	3	.44	167	.07	1.22	.14	1.65	.22	3	3
0	57	2.75	50	828	5	.59	77	.06	<.01	.25	1.79	.54	<1	7
0	31	3.02	36	707	10	.54	1	.04	.03	.18	2.01	.56	1	10
0	71	3.02	48	1088	17	.39	1	.23	1.02	.13	1.19	.36	5	5
0	7	.29	5	105	2	.04	<1	.02	.1	.01	.11	.03	<1	<1
0	27	.7	22	187	0	.57	16	.04	.55	.11	1.11	.07	32	31
0	43	1.85	37	324	3	.51	17	.05	1.28	.13	.65	.1	74	47
0	37	1.63	32	285	2	.45	15	.05	1.13	.11	.57	.08	65	41
0	348	.5	29	230	2	.19	17	.04	.48	.05	.48	.05	13	8
0	20	.55	14	239	1	.19	4	.03	.2	.09	.33	.08	25	82
0	31	1.68	20	278	9	.17	6	.04	.4	.14	1.14	.08	42	118
0	28	1.5	18	250	8	.15	5	.04	.36	.13	1.02	.08	38	106
0	12	.08	10	132	1	.2	77	.09	.2	.02	.13	.06	17	26
0	18	.93	20	197	15	.6	212	.13	.86	.11	1.12	.11	12	50

Table A–1

Food Composition   (Computer code number is for West Diet Analysis program)   (For purposes of calculations, use "0" for t, <1, <.1, <.01, etc.)

Computer Code Number	Food Description	Measure	Wt (g)	H₂O (%)	Ener (cal)	Prot (g)	Carb (g)	Dietary Fiber (g)	Fat (g)	Fat Breakdown (g) Sat	Mono	Poly
	FRUITS and FRUIT JUICES—Continued											
323	Tangerine juice, canned, sweetened	1 c	249	87	125	1	30	<1	<1	t	t	.1
	Watermelon, raw, without rind and seeds:											
324	Piece, ¹⁄₁₆ wedge	1 pce	286	91	91	2	20	1	1	.1	.3	.4
325	Diced	1 c	152	91	49	1	11	1	1	.1	.2	.2
	BAKED GOODS: BREADS, CAKES, COOKIES, CRACKERS, PIES											
326	Bagels, plain, enriched, 3½" diam.	1 ea	71	33	195	7	38	2	1	.2	.1	.5
1663	Bagel, oat bran	1 ea	71	33	181	8	38	3	1	.1	.2	.3
	Biscuits:											
327	From home recipe	1 ea	60	29	212	4	27	1	10	2.6	4.2	2.5
328	From mix	1 ea	57	29	191	4	28	1	7	1.6	2.4	2.5
329	From refrigerated dough	1 ea	74	27	276	4	34	1	13	8.7	3.4	.5
330	Bread crumbs, dry, grated (see # 364, 365 for soft crumbs)	1 c	108	6	427	13	78	3	6	1.4	2.3	1.7
2087	Bread sticks, brown & serve	1 ea	57	34	150	7	28	1	1	.5	.5	.5
	Breads:											
331	Boston brown, canned, 3¼" slice	1 pce	45	47	88	2	19	2	1	.1	.1	.3
332	Cracked wheat (¼ cracked-wheat & ¾ enr wheat flour): 1-lb loaf	1 ea	454	36	1180	39	225	25	18	4.2	8.6	3.1
333	Slice (18 per loaf)	1 pce	25	36	65	2	12	1	1	.2	.5	.2
334	Slice, toasted	1 pce	23	30	65	2	12	1	1	.2	.5	.2
335	French/Vienna, enriched: 1-lb loaf	1 ea	454	34	1243	40	236	14	14	2.9	5.5	3.1
337	Slice, 4¾ x 4 x ½"	1 pce	25	34	68	2	13	1	1	.2	.3	.2
336	French, slice, 5 x 2½"	1 pce	25	34	68	2	13	1	1	.2	.3	.2
	French toast: see Mixed Dishes, and Fast Foods, #691											
2083	Honey wheatberry	1 pce	38	38	100	3	18	2	1	0	.5	
338	Italian, enriched: 1-lb loaf	1 ea	454	36	1230	40	227	12	16	3.9	3.7	6.3
339	Slice, 4½ x 3¼ x ¾"	1 pce	30	36	81	3	15	1	1	.3	.2	.4
340	Mixed grain, enriched: 1-lb loaf	1 ea	454	38	1135	45	211	29	17	3.7	6.9	4.2
341	Slice (18 per loaf)	1 pce	26	38	65	3	12	2	1	.2	.4	.2
342	Slice, toasted	1 pce	24	32	65	3	12	2	1	.2	.4	.2
343	Oatmeal, enriched: 1-lb loaf	1 ea	454	37	1221	38	220	18	20	3.2	7.2	7.7
344	Slice (18 per loaf)	1 pce	27	37	73	2	13	1	1	.2	.4	.5
345	Slice, toasted	1 pce	25	31	73	2	13	1	1	.2	.4	.5
346	Pita pocket bread, enr, 6½" round	1 ea	60	32	165	5	33	1	1	.1	.1	.3
	Pumpernickel (⅔ rye & ⅓ enr wheat flr):											
347	1-lb loaf	1 ea	454	38	1135	40	216	29	14	2	4.2	5.6
348	Slice, 5 x 4 x ⅜"	1 pce	26	38	65	2	12	2	1	.1	.2	.3
349	Slice, toasted	1 pce	29	32	80	3	15	2	1	.1	.3	.4
350	Raisin, enriched: 1-lb loaf	1 ea	454	34	1243	36	237	19	20	4.9	10.5	3.1
351	Slice (18 per loaf)	1 pce	26	34	71	2	14	1	1	.3	.6	.2
352	Slice, toasted	1 pce	24	28	71	2	14	1	1	.3	.6	.2
353	Rye, light (⅓ rye & ⅔ enr wheat flr): 1-lb loaf	1 ea	454	37	1175	39	219	26	15	2.9	6	3.6
354	Slice, 4¾ x 3¾ x ⁷⁄₁₆"	1 pce	32	37	83	3	15	2	1	.2	.4	.3
355	Slice, toasted	1 pce	24	31	68	2	13	2	1	.2	.3	.2
356	Wheat (enr wheat & whole-wheat flour): 1-lb loaf	1 ea	454	37	1160	43	213	25	19	3.9	7.3	4.5
357	Slice (18 per loaf)	1 pce	25	37	65	2	12	1	1	.2	.4	.2
358	Slice, toasted	1 pce	23	32	65	2	12	1	1	.2	.4	.2
359	White, enriched: 1-lb loaf	1 ea	454	35	1293	36	225	9	26	5.4	5.9	12.6
360	Slice	1 pce	42	35	120	3	21	1	2	.5	.5	1.2
361	Slice, toasted	1 pce	38	29	119	3	21	1	2	.5	.5	1.2
366	Whole-wheat: 1-lb loaf	1 ea	454	38	1116	44	209	31	19	4.2	7.6	4.5
367	Slice (16 per loaf)	1 pce	28	38	69	3	13	2	1	.3	.5	.3
368	Slice, toasted	1 pce	25	30	69	3	13	2	1	.3	.5	.3
	Bread stuffing, prepared from mix:											
369	Dry type	1 c	200	65	356	6	43	6	17	3.5	7.6	5.2
370	Moist type, with egg and margarine	1 c	232	65	390	9	51	5	17	3.4	7.4	4.9

PAGE KEY: A–4 = Beverages A–6 = Dairy A–10 = Eggs A–10 = Fat/Oil A–14 = Fruit A–20 = Bakery A–28 = Grain A–32 = Fish A–34 = Meats A–38 = Poultry A–40 = Sausage A–42 = Mixed/Fast A–46 = Nuts/Seeds A–50 = Sweets A–52 = Vegetables/Legumes A–62 = Vegetarian Foods A–64 = Misc A–66 = Soups/Sauces A–68 = Fast A–84 = Convenience A–88 = Baby foods

A–21

A

Chol (mg)	Calc (mg)	Iron (mg)	Magn (mg)	Pota (mg)	Sodi (mg)	Zinc (mg)	VT-A (RE)	Thia (mg)	VT-E (a-TE)	Ribo (mg)	Niac (mg)	V-B6 (mg)	Fola (µg)	VT-C (mg)
0	45	.5	20	443	2	.07	105	.15	.22	.05	.25	.08	11	55
0	23	.49	31	332	6	.2	106	.23	.43	.06	.57	.41	6	27
0	12	.26	17	176	3	.11	56	.12	.23	.03	.3	.22	3	15
0	52	2.53	21	72	379	.62	0	.38	.02	.22	3.24	.04	62	0
0	9	2.19	40	145	360	1.48	<1	.23	.17	.24	2.1	.14	57	<1
2	141	1.74	11	73	348	.32	14	.21	1.45	.19	1.77	.02	37	<1
2	105	1.17	14	107	544	.35	15	.2	.23	.2	1.72	.04	3	<1
5	89	1.64	9	87	584	.29	24	.27	.44	.18	1.63	.03	6	0
0	245	6.61	50	239	931	1.32	<1	.83	.95	.47	7.4	.11	118	0
0	60	2.7			290		0	.22		.1	1.6			0
<1	31	.94	28	143	284	.22	5	.01	.13	.05	.5	.04	5	0
														0
0	195	12.8	236	804	2442	5.63	0	1.63	2.56	1.09	16.7	1.38	277	
0	11	.7	13	44	135	.31	0	.09	.14	.06	.92	.08	15	0
0	11	.7	13	44	135	.31	0	.07	.14	.05	.83	.07	7	0
0	341	11.5	123	513	2764	3.95	0	2.36	1.07	1.49	21.6	.19	431	0
0	19	.63	7	28	152	.22	0	.13	.06	.08	1.19	.01	24	0
0	19	.63	7	28	152	.22	0	.13	.06	.08	1.19	.01	24	0
0	20	.72			200		0	.12	.24	.07	.8			0
0	354	13.3	123	499	2651	3.9	0	2.15	1.26	1.33	19.9	.22	431	0
0	23	.88	8	33	175	.26	0	.14	.08	.09	1.31	.01	28	0
0	413	15.8	241	926	2210	5.77	0	1.85	2.79	1.55	19.8	1.51	363	1
0	24	.9	14	53	127	.33	0	.11	.16	.09	1.14	.09	21	<1
0	24	.9	14	53	127	.33	0	.08	.16	.08	1.02	.08	16	<1
0	300	12.3	168	645	2719	4.63	9	1.81	1.56	1.09	14.3	.31	281	2
0	18	.73	10	38	162	.27	1	.11	.09	.06	.85	.02	17	<1
0	18	.73	10	38	163	.28	<1	.09	.09	.06	.77	.02	13	<1
0	52	1.57	16	72	322	.5	0	.36	.02	.2	2.78	.02	57	0
0	309	13	245	944	3046	6.72	0	1.48	2.3	1.38	14	.57	363	0
0	18	.75	14	54	174	.38	0	.08	.13	.08	.8	.03	21	0
0	21	.91	17	66	214	.47	0	.08	.17	.09	.89	.04	20	0
0	300	13.2	118	1030	1770	3.27	1	1.54	3.44	1.81	15.8	.31	395	2
0	17	.75	7	59	101	.19	<1	.09	.2	.1	.9	.02	23	<1
0	17	.76	7	59	102	.19	<1	.07	.2	.09	.81	.02	18	<1
0	331	12.8	182	754	2996	5.18	2	1.97	2.51	1.52	17.3	.34	390	1
0	23	.91	13	53	211	.36	<1	.14	.18	.11	1.22	.02	27	<1
0	19	.74	10	44	174	.3	<1	.09	.15	.08	.9	.02	17	<1
0	572	15.8	209	627	2447	4.77	0	2.09	3	1.45	20.5	.49	204	0
0	26	.83	11	50	133	.26	0	.1	.14	.07	1.03	.02	19	0
0	26	.83	11	50	132	.26	0	.08	.14	.06	.93	.02	15	0
14	259	13.5	86	663	1629	2.91	100	1.84	4.95	1.74	16.3	.23	413	1
1	24	1.25	8	61	151	.27	9	.17	.46	.16	1.51	.02	38	<1
1	24	1.24	8	61	150	.27	8	.13	.46	.14	1.35	.02	12	<1
0	327	15	390	1144	2392	8.81	0	1.59	4.72	.93	17.4	.81	227	0
0	20	.92	24	71	148	.54	0	.1	.29	.06	1.08	.05	14	0
0	20	.93	24	71	148	.54	0	.08	.23	.05	.97	.04	9	0
0	64	2.18	24	148	1086	.56	162	.27	2.8	.21	2.96	.08	202	0
0	148	3.8	35	304	1069	.74	160	.39	2.78	.33	3.69	.12	39	4

Table A–1

Food Composition   (Computer code number is for West Diet Analysis program)   (For purposes of calculations, use "0" for t, <1, <.1, <.01, etc.)

Computer Code Number	Food Description	Measure	Wt (g)	H₂O (%)	Ener (cal)	Prot (g)	Carb (g)	Dietary Fiber (g)	Fat (g)	Fat Breakdown (g)		
										Sat	Mono	Poly
	BAKED GOODS: BREADS, CAKES, COOKIES, CRACKERS, PIES—Continued											
	Cakes, prepared from mixes using enriched flour and veg shortening, w/frostings made from margarine:											
	Angel food:											
371	Whole cake, 9 ¾" diam tube	1 ea	340	33	877	20	197	5	3	.4	.2	1.2
372	Piece, ¹⁄₁₂ of cake	1 pce	28	33	72	2	16	<1	<1	t	t	.1
373	Boston cream pie, ⅛ of cake	1 pce	123	45	310	3	53	2	10	3.1	5.4	1.2
	Coffee cake:											
374	Whole cake, 7¾ x 5⅛ x 1¼"	1 ea	336	30	1068	18	177	4	32	6.3	13	10.7
375	Piece, ⅙ of cake	1 pce	56	30	178	3	30	1	5	1	2.2	1.8
	Devil's food, chocolate frosting:											
376	Whole cake, 2 layer, 8 or 9" diam	1 ea	1021	23	3747	42	557	29	167	47.9	91.9	19.5
377	Piece, ¹⁄₁₆ of cake	1 pce	64	23	235	3	35	2	10	3	5.8	1.2
378	Cupcake, 2½" diam	1 ea	42	23	154	2	23	1	7	2	3.8	.8
	Gingerbread:											
379	Whole cake, 8" square	1 ea	603	33	1863	24	306	7	61	15.8	34	8.1
380	Piece, ⅑ of cake	1 pce	67	33	207	3	34	1	7	1.8	3.8	.9
	Yellow, chocolate frosting, 2 layer:											
381	Whole cake, 8 or 9" in diam	1 ea	1024	22	3880	39	567	18	178	49	99	21.4
382	Piece, ¹⁄₁₆ of cake	1 pce	64	22	243	2	35	1	11	3.1	6.2	1.3
	Cakes from home recipes w/enr flour:											
	Carrot cake, made with veg oil, cream cheese frosting:											
383	Whole, 9 x 13" cake	1 ea	1776	21	7743	82	838	21	469	86.8	116	242
384	Piece, ¹⁄₁₆ of cake, 2¼ x 3¼" slice	1 pce	111	21	484	5	52	1	29	5.4	7.2	15.1
	Fruitcake, dark:											
385	Whole cake, 7½"diam tube, 2¼"high	1 ea	1376	25	4458	40	848	51	125	15.4	57.4	44.6
386	Piece, ¹⁄₃₂ of cake, ⅔" arc	1 pce	43	25	139	1	26	2	4	.5	1.8	1.4
	Sheet, plain, made w/veg shortening, no frosting:											
387	Whole cake, 9" square	1 ea	774	24	2817	35	433	3	108	29.9	51.5	25.5
388	Piece, ⅑ of cake	1 pce	86	24	313	4	48	<1	12	3.3	5.7	2.8
	Sheet, plain, made w/margarine, uncooked white frosting:											
389	Whole cake, 9" square	1 ea	576	22	2148	20	339	2	83	13.8	35.5	29.5
390	Piece, ⅑ of cake	1 pce	64	22	239	2	38	<1	9	1.5	3.9	3.3
	Cakes, commerical:											
	Cheesecake:											
401	Whole cake, 9" diam	1 ea	960	46	3081	53	245	4	216	111	74.4	13.2
402	Piece, ¹⁄₁₂ of cake	1 pce	80	46	257	4	20	<1	18	9.2	6.2	1.1
	Pound cake:											
393	Loaf, 8½ x 3½ x 3"	1 ea	340	25	1319	19	166	2	68	38.1	19	3.7
394	Slice, ¹⁄₁₇ of loaf, 2" slice	1 pce	28	25	109	2	14	<1	6	3.1	1.6	.3
	Snack cakes:											
395	Chocolate w/creme filling, Ding Dong	1 ea	50	20	188	2	30	<1	7	1.6	2.7	2.1
396	Sponge cake w/creme filling, Twinkie	1 ea	43	20	157	1	27	<1	5	1.2	1.9	1.5
1677	Sponge cake, ¹⁄₁₂ of 12" cake	1 pce	38	30	110	2	23	<1	1	.3	.4	.2
	White, white frosting, 2 layer:											
397	Whole cake, 8 or 9" diam	1 ea	1136	20	4260	37	716	11	153	68.2	60.1	15.4
398	Piece, ¹⁄₁₆ of cake	1 pce	71	20	266	2	45	1	10	4.3	3.8	1
	Yellow, chocolate frosting, 2 layer:											
399	Whole cake, 8 or 9" in diam	1 ea	1024	22	3880	39	567	18	178	49	99	21.4
400	Piece, ¹⁄₁₆ of cake	1 pce	64	22	243	2	35	1	11	3.1	6.2	1.3
1332	Bagel chips	5 pce	70	3	298	6	52	6	7	1.2	2	3.4
2225	Bagel chips, onion garlic, toasted	1 oz	28		193	5	31	3	8	1.7	5.2	0
1035	Cheese puffs/Cheetos	1 c	20	1	111	2	11	<1	7	1.3	4.1	1
	Cookies made with enriched flour:											
	Brownies with nuts:											
403	Commercial w/frosting, 1½ x 1¾ x ⅞"	1 ea	61	14	247	3	39	1	10	2.6	5.1	1.6
1902	Fat free fudge, Entenmann's	1 pce	40	24	110	2	27	1	0	0	0	0

PAGE KEY: A–4 = Beverages   A–6 = Dairy   A–10 = Eggs   A–10 = Fat/Oil   A–14 = Fruit   A–20 = Bakery   A–28 = Grain   A–32 = Fish   A–34 = Meats   **A–23**
A–38 = Poultry   A–40 = Sausage   A–42 = Mixed/Fast   A–46 = Nuts/Seeds   A–50 = Sweets   A–52 = Vegetables/Legumes   A–62 = Vegetarian Foods
A–64 = Misc   A–66 = Soups/Sauces   A–68 = Fast   A–84 = Convenience   A–88 = Baby foods

A

Chol (mg)	Calc (mg)	Iron (mg)	Magn (mg)	Pota (mg)	Sodi (mg)	Zinc (mg)	VT-A (RE)	Thia (mg)	VT-E (a-TE)	Ribo (mg)	Niac (mg)	V-B6 (mg)	Fola (μg)	VT-C (mg)
0	476	1.77	41	316	2546	.24	0	.35	.34	1.67	3	.1	119	0
0	39	.15	3	26	210	.02	0	.03	.03	.14	.25	.01	10	0
45	28	.47	7	48	177	.2	28	.5	1.3	.33	.23	.03	18	<1
165	457	4.8	60	376	1414	1.51	134	.56	5.58	.59	5.11	.17	228	1
27	76	.8	10	63	236	.25	22	.09	.93	.1	.85	.03	38	<1
470	439	22.5	347	2042	3410	7.04	286	.28	17.3	1.36	5.89	.32	174	1
29	27	1.41	22	128	214	.44	18	.02	1.08	.08	.37	.02	11	<1
19	18	.92	14	84	140	.29	12	.01	.71	.06	.24	.01	7	<1
211	416	20	96	1453	2761	2.47	96	1.14	8.26	1.12	9.41	.23	60	1
23	46	2.22	11	161	307	.27	11	.13	.92	.12	1.05	.02	7	<1
563	379	21.3	307	1822	3450	6.35	276	1.23	27.6	1.61	12.8	.3	225	1
35	24	1.33	19	114	216	.4	17	.08	1.73	.1	.8	.02	14	<1
959	444	22.2	320	1989	4368	8.7	6819	2.42	74.9	2.77	17.9	1.35	213	19
60	28	1.39	20	124	273	.54	426	.15	4.68	.17	1.12	.08	13	1
69	454	28.5	220	2105	3715	3.72	261	.69	42.9	1.36	10.9	.63	261	5
2	14	.89	7	66	116	.12	8	.02	1.34	.04	.34	.02	8	<1
503	495	11.7	108	611	2322	2.74	372	1.24	11	1.39	10.1	.26	54	2
56	55	1.3	12	68	258	.3	41	.14	1.22	.15	1.12	.03	6	<1
323	357	6.16	35	305	1981	1.44	109	.58	10.9	.4	2.88	.2	156	1
36	40	.68	4	34	220	.16	12	.06	1.22	.04	.32	.02	17	<1
528	490	6.05	106	864	1987	4.9	1545	.27	10.1	1.85	1.87	.5	173	6
44	41	.5	9	72	166	.41	129	.02	.84	.15	.16	.04	14	<1
751	119	4.69	37	405	1353	1.56	530	.47	2.24	.78	4.45	.12	139	<1
62	10	.39	3	33	111	.13	44	.04	.18	.06	.37	.01	11	<1
8	36	1.68	20	61	213	.28	2	.11	1.01	.15	1.22	.01	14	<1
7	19	.55	3	39	157	.13	2	.07	.83	.06	.52	.01	12	<1
39	27	1.03	4	38	93	.19	17	.09	.17	.1	.73	.02	15	0
91	545	9.09	60	659	2658	1.76	368	1.14	20.4	1.48	10.2	.16	64	1
6	34	.57	4	41	166	.11	23	.07	1.28	.09	.64	.01	4	<1
563	379	21.3	307	1822	3450	6.35	276	1.23	27.6	1.61	12.8	.3	225	1
35	24	1.33	19	114	216	.4	17	.08	1.73	.1	.8	.02	14	<1
0	9	1.38	41	167	419	.9	0	.13	.46	.12	1.57	.19	58	0
0	0	2.52			490		0	.39	<.01	.24	3.5			0
1	12	.47	4	33	210	.08	7	.05	1.02	.07	.65	.03	24	<1
10	18	1.37	19	91	190	.44	12	.16	1.3	.13	1.05	.02	13	<1
0	0	1.08		90	140		0		.01					0

Table A–1

Food Composition (Computer code number is for West Diet Analysis program) (For purposes of calculations, use "0" for t, <1, <.1, <.01, etc.)

Computer Code Number	Food Description	Measure	Wt (g)	H_2O (%)	Ener (cal)	Prot (g)	Carb (g)	Dietary Fiber (g)	Fat (g)	Fat Breakdown (g) Sat	Mono	Poly
	BAKED GOODS: BREADS, CAKES, COOKIES, CRACKERS, PIES—Continued											
	Chocolate chip cookies:											
405	Commercial, 2¼" diam	4 ea	60	12	275	2	35	2	15	4.5	7.8	1.6
406	Home recipe, 2¼" diam	4 ea	64	6	312	4	37	2	18	5.2	6.7	5.4
407	From refrigerated dough, 2¼" diam	4 ea	64	13	284	3	39	1	13	4.5	6.5	1.3
408	Fig bars	4 ea	64	16	223	2	45	3	5	.9	2.6	.8
2052	Fruit bar, no fat	1 ea	28		90	2	21	0	0	0	0	0
2162	Fudge, fat free, Snackwell	1 ea	16	14	53	1	12	<1	<1	.1	.1	t
409	Oatmeal raisin, 2⅝" diam	4 ea	60	6	261	4	41	2	10	1.9	4.1	3
410	Peanut butter, home recipe, 2⅝"diam	4 ea	80	6	380	7	47	2	19	3.5	8.7	5.8
411	Sandwich-type, all	4 ea	40	2	189	2	28	1	8	1.7	4.7	1.1
412	Shortbread, commercial, small	4 ea	32	4	161	2	21	1	8	2	4.3	1
413	Shortbread, home recipe, large	2 ea	22	3	120	1	12	<1	7	4.5	2.1	.3
414	Sugar from refrigerated dough, 2" diam	4 ea	48	5	232	2	31	<1	11	2.8	6.2	1.4
1874	Vanilla sandwich, Snackwell's	2 ea	26	4	109	1	21	1	2	.5	.8	.2
415	Vanilla wafers	10 ea	40	5	176	2	29	1	6	1.4	2.4	1.5
416	Corn chips	1 c	26	1	140	2	15	1	9	1.2	2.5	4.3
	Crackers (enriched):											
417	Cheese	10 ea	10	3	50	1	6	<1	3	.9	.9	.5
418	Cheese with peanut butter	4 ea	28	4	135	4	16	1	6	1.4	3.4	1.2
	Fat free, enriched:											
2161	Cracked pepper, Snackwell	1 ea	15	2	60	2	12	<1	<1	.1	t	.1
2159	Wheat, Snackwell	7 ea	15	1	60	2	12	1	<1	.1	.1	.1
2075	Whole wheat, herb seasoned	5 ea	14	5	50	2	11	2	0	0	0	0
2077	Whole wheat, onion	5 ea	14	4	50	2	11	2	0	0	0	0
419	Graham, enriched	2 ea	14	4	59	1	11	<1	1	.4	.7	.2
420	Melba toast, plain, enriched	1 pce	5	5	19	1	4	<1	<1	t	t	.1
1514	Rice cakes, unsalted, enriched	2 ea	18	6	70	1	15	1	<1	.1	.2	.2
421	Rye wafer, whole grain	2 ea	22	5	73	2	18	5	<1	t	t	.1
422	Saltine-enriched	4 ea	12	4	52	1	9	<1	1	.3	.8	.2
1971	Saltine, unsalted tops, enriched	2 ea	6		25	1	4	0	<1	0	0	0
423	Snack-type, round like Ritz, enriched	3 ea	9	3	45	1	5	<1	2	.4	1	.7
424	Wheat, thin, enriched	4 ea	8	3	38	1	5	<1	2	.7	.8	.2
425	Whole-wheat wafers	2 ea	8	3	35	1	5	1	1	.2	.8	.2
426	Croissants, 4½ x 4 x 1¾"	1 ea	57	23	231	5	26	1	12	6.7	3.2	.7
1699	Croutons, seasoned	½ c	20	4	93	2	13	1	4	1	1.9	.5
	Danish pastry:											
427	Packaged ring, plain, 12 oz	1 ea	340	21	1349	19	181	1	65	13.5	40.9	6.4
428	Round piece, plain, 4¼" diam, 1" high	1 ea	88	21	349	5	47	<1	17	3.5	10.6	1.6
429	Ounce, plain	1 oz	28	21	111	2	15	<1	5	1.1	3.4	.5
430	Round piece with fruit	1 ea	94	29	335	5	45		16	3.3	10.1	1.6
	Desserts, 3 x 3" piece:											
1348	Apple crisp	1 pce	78	61	127	1	25	1	3	.6	1.2	.9
1353	Apple cobbler	1 pce	104	57	199	2	35	2	6	1.2	2.8	2
1349	Cherry crisp	1 pce	138	77	146	2	24	1	5	.9	2.5	1.8
1352	Cherry cobbler	1 pce	129	66	198	2	34	1	6	1.2	2.8	1.9
1350	Peach crisp	1 pce	139	75	155	2	27	2	5	.9	2.5	1.7
1351	Peach cobbler	1 pce	130	64	204	2	36	2	6	1.2	2.8	1.9
	Doughnuts:											
431	Cake type, plain, 3¼" diam	1 ea	47	21	198	2	23	1	11	1.8	4.5	3.8
432	Yeast-leavened, glazed, 3¾" diam	1 ea	60	25	242	4	27	1	14	3.5	7.8	1.7
	English muffins:											
433	Plain, enriched	1 ea	57	42	134	4	26	2	1	.2	.2	.5
434	Toasted	1 ea	52	37	133	4	26	2	1	.1	.2	.5
1504	Whole wheat	1 ea	66	46	134	6	27	4	1	.2	.3	.6
1414	Granola bar, soft	1 ea	28	6	124	2	19	1	5	2	1.1	1.5
1415	Granola bar, hard	1 ea	25	4	118	3	16	1	5	.6	1.1	3
1985	Granola bar, fat free, all flavors	1 ea	42	10	140	2	35	3	0	0	0	0
	Muffins, 2½" diam, 1½" high:											
	From home recipe:											
435	Blueberry	1 ea	57	39	165	4	23	1	6	1.4	1.6	3.1
436	Bran, wheat	1 ea	57	35	164	4	24	4	7	1.5	1.8	3.6
437	Cornmeal	1 ea	57	32	183	4	25	2	7	1.6	1.8	3.5

PAGE KEY: A–4 = Beverages A–6 = Dairy A–10 = Eggs A–10 = Fat/Oil A–14 = Fruit A–20 = Bakery A–28 = Grain A–32 = Fish A–34 = Meats A–38 = Poultry A–40 = Sausage A–42 = Mixed/Fast A–46 = Nuts/Seeds A–50 = Sweets A–52 = Vegetables/Legumes A–62 = Vegetarian Foods A–64 = Misc A–66 = Soups/Sauces A–68 = Fast A–84 = Convenience A–88 = Baby foods

A–25

Chol (mg)	Calc (mg)	Iron (mg)	Magn (mg)	Pota (mg)	Sodi (mg)	Zinc (mg)	VT-A (RE)	Thia (mg)	VT-E (a-TE)	Ribo (mg)	Niac (mg)	V-B6 (mg)	Fola (µg)	VT-C (mg)
0	9	1.45	21	56	196	.28	1	.07	1.74	.12	.97	.1	23	0
20	25	1.57	35	143	231	.59	105	.12	1.86	.11	.87	.05	21	<1
15	16	1.44	15	115	134	.32	11	.12	1.31	.12	1.27	.03	36	0
0	41	1.86	17	132	224	.25	3	.1	.45	.14	1.2	.05	17	<1
0	0	.36			95		0		.01					0
0	3	.29	5	26	71	.08	<1	.02	<.01	.02	.26	<.01		0
20	60	1.59	25	143	323	.52	98	.15	1.5	.1	.76	.04	18	<1
25	31	1.78	31	185	414	.66	125	.18	3.04	.17	2.81	.07	44	<1
0	10	1.55	18	70	242	.32	<1	.03	1.21	.07	.83	.01	17	0
6	11	.88	5	32	146	.17	4	.11	.98	.1	1.07	.01	19	0
20	4	.58	3	15	102	.09	67	.08	.18	.06	.64	<.01	2	0
15	43	.88	4	78	225	.13	5	.09	1.54	.06	1.16	.01	25	0
<1	17	.61	5	28	95	.16	<1	.05		.07	.69	.01		0
23	19	.95	6	39	125	.14	7	.11	.54	.13	1.24	.03	20	0
0	33	.34	20	37	164	.33	2	.01	.35	.04	.31	.06	5	0
1	15	.48	4	14	99	.11	3	.06	.1	.04	.47	.05	8	0
1	22	.82	16	69	278	.3	10	.11	1.24	.1	1.83	.42	25	0
<1	26	.73	4	19	148	.14	<1	.05		.06	.78	.01		<1
<1	28	.58	7	43	169	.21	<1	.04		.07	.73	.02		0
0	0				80		100							2
0	0	0			80		100							2
0	3	.52	4	19	85	.11	0	.03	.27	.04	.58	.01	8	0
0	5	.18	3	10	41	.1	0	.02	.01	.01	.21	<.01	6	0
0	2	.27	24	52	5	.54	1	.01	.02	.03	1.41	.03	4	0
0	9	1.31	27	109	175	.62	<1	.09	.44	.06	.35	.06	3	<1
0	14	.65	3	15	156	.09	0	.07	.2	.05	.63	<.01	15	0
0		.36		5	50				.1					
0	11	.32	2	12	76	.06	0	.04	.4	.03	.36	<.01	7	0
2	3	.28	5	16	70	.13	<1	.04	.02	.03	.34	.01	1	0
0	4	.25	8	24	53	.17	0	.02	.31	.01	.36	.01	3	0
43	21	1.16	9	67	424	.43	78	.22	.24	.14	1.25	.03	35	<1
1	19	.56	8	36	248	.19	1	.1	.32	.08	.93	.02	18	0
105	143	6.94	54	371	1261	1.87	20	.99	3.06	.75	8.5	.2	211	10
27	37	1.8	14	96	326	.48	5	.25	.79	.19	2.2	.05	55	3
9	12	.57	4	30	104	.15	2	.08	.25	.06	.7	.02	17	1
19	22	1.4	14	110	333	.48	24	.29	.85	.21	1.8	.06	31	2
0	22	.58	5	76	142	.12	24	.07		.06	.6	.03	4	2
1	21	.79	6	106	288	.16	76	.1	1.11	.09	.74	.04	3	<1
0	26	2.14	11	154	74	.15	150	.06	.93	.08	.6	.06	11	3
1	28	1.81	9	133	294	.2	135	.1	1.01	.11	.85	.05	9	2
0	20	.89	12	189	70	.19	108	.06	1.13	.05	1.06	.03	6	5
1	24	.91	10	159	291	.23	105	.09	1.16	.09	1.19	.03	6	3
17	21	.92	9	60	257	.26	8	.1	1.63	.11	.87	.03	22	<1
4	26	1.22	13	65	205	.46	6	.22	1.75	.13	1.71	.03	26	0
0	99	1.43	12	75	264	.4	0	.25	.07	.16	2.21	.02	46	<1
0	98	1.41	11	74	262	.39	0	.2	.07	.14	1.98	.02	38	<1
0	175	1.62	47	139	420	1.06	0	.2	.46	.09	2.25	.11	28	0
<1	29	.72	21	91	78	.42	0	.08	.34	.05	.14	.03	7	0
0	15	.74	24	84	73	.51	4	.07	.33	.03	.39	.02	6	<1
0	0	3.6			5		100							0
22	107	1.29	9	69	251	.31	16	.15	1.03	.16	1.26	.02	7	1
20	106	2.39	44	181	335	1.57	136	.19	1.31	.25	2.29	.18	30	4
26	147	1.49	13	82	333	.35	23	.17	1.08	.18	1.36	.05	10	<1

Table A-1

Food Composition (Computer code number is for West Diet Analysis program) (For purposes of calculations, use "0" for t, <1, <.1, <.01, etc.)

Computer Code Number	Food Description	Measure	Wt (g)	H_2O (%)	Ener (cal)	Prot (g)	Carb (g)	Dietary Fiber (g)	Fat (g)	Fat Breakdown (g) Sat	Mono	Poly
	BAKED GOODS: BREADS, CAKES, COOKIES, CRACKERS, PIES—Continued											
	From commercial mix:											
438	Blueberry	1 ea	50	36	150	3	24	1	4	.7	1.8	1.5
439	Bran, wheat	1 ea	50	35	138	3	23	2	5	1.2	2.3	.7
440	Cornmeal	1 ea	50	30	161	4	25	1	5	1.4	2.6	.6
1864	Nabisco Newtons, fat free, all flavors	1 ea	23		69	1	16		0	0	0	0
	Pancakes, 4″ diam:											
441	Buckwheat, from mix w/ egg and milk	1 ea	30	54	62	2	8	1	2	.6	.6	.8
442	Plain, from home recipe	1 ea	38	53	86	2	11	1	4	.8	.9	1.7
443	Plain, from mix; egg, milk, oil added	1 ea	38	53	74	2	14	<1	1	.2	.3	.3
1468	Pan dulce, sweet roll w/topping	1 ea	79	21	291	5	48	1	9	2	3.9	2.7
	Piecrust,with enriched flour, vegetable shortening, baked:											
444	Home recipe, 9″ shell	1 ea	180	10	949	12	85	3	62	15.5	27.4	16.4
	From mix:											
445	Piecrust for 2-crust pie	1 ea	320	10	1686	21	152	5	111	27.6	48.6	29.2
446	1 pie shell	1 ea	160	11	802	11	81	3	49	12.3	27.7	6.2
	Pies, 9″ diam; pie crust made with vegetable shortening, enriched flour:											
447	Apple: Whole pie	1 ea	1000	52	2370	19	340	16	110	21.1	59.4	20.9
448	Piece, ⅙ of pie	1 pce	167	52	396	3	57	3	18	3.5	9.9	3.5
449	Banana cream: Whole pie	1 ea	1152	48	3098	51	379	8	157	43.3	65.9	38
450	Piece, ⅙ of pie	1 pce	192	48	516	8	63	1	26	7.2	11	6.3
451	Blueberry: Whole pie	1 ea	1176	51	2881	32	394	16	140	34.3	60.2	36.2
452	Piece, ⅙ of pie	1 pce	196	51	480	5	66	3	23	5.7	10	6
453	Cherry: Whole pie	1 ea	1140	46	3078	32	439	17	139	34.1	60.5	37.1
454	Piece, ⅙ of pie	1 pce	240	46	648	7	92	4	29	7.2	12.7	7.8
455	Chocolate cream: Whole pie	1 ea	1194	63	2150	49	281	12	97	35.5	38.2	18.6
456	Piece, ⅙ of pie	1 pce	199	63	358	8	47	2	16	5.9	6.4	3.1
457	Custard: Whole pie	1 ea	630	61	1323	35	131	10	73	17.5	36.3	12.1
458	Piece, ⅙ of pie	1 pce	105	61	221	6	22	2	12	2.9	6	2
459	Lemon meringue: Whole pie	1 ea	678	42	1817	10	320	8	59	10.6	24.6	19.6
460	Piece, ⅙ of pie	1 pce	113	42	303	2	53	1	10	1.8	4.1	3.3
461	Peach: Whole pie	1 ea	1111	45	2994	26	443	16	130	31.1	55.7	37.4
462	Piece, ⅙ of pie	1 pce	139	45	375	3	55	2	16	3.9	7	4.7
463	Pecan: Whole pie	1 ea	678	19	2712	27	388	24	125	25.5	73.2	20.1
464	Piece, ⅙ of pie	1 pce	113	19	452	5	65	4	21	4.2	12.2	3.4
465	Pumpkin: Whole pie	1 ea	654	58	1373	25	179	18	62	13.2	32.8	10.5
466	Piece, ⅙ of pie	1 pce	109	58	229	4	30	3	10	2.2	5.5	1.7
467	Pies, fried, commercial: Apple	1 ea	85	40	266	2	33	1	14	6.5	5.8	1.2
468	Pies, fried, commercial: Cherry	1 ea	128	38	404	4	54	3	21	3.1	9.5	6.9
	Pretzels, made with enriched flour:											
469	Thin sticks, 2¼″ long	1 oz	28	3	107	3	22	1	1	.2	.4	.3
470	Dutch twists	10 pce	60	3	229	5	47	2	2	.4	.8	.7
471	Thin twists, 3¼ x 2¼ x ¼″	10 pce	60	3	229	5	47	2	2	.4	.8	.7
	Rolls & buns, enriched, commercial:											
472	Cloverleaf rolls, 2½″ diam, 2″ high	1 ea	28	32	84	2	14	1	2	.5	1	.3
473	Hot dog buns	1 ea	43	34	123	4	22	1	2	.5	1.1	.4
474	Hamburger buns	1 ea	43	34	123	4	22	1	2	.5	1.1	.4
475	Hard roll, white, 3¾″ diam, 2″ high	1 ea	57	31	167	6	30	1	2	.3	.6	1
476	Submarine rolls/hoagies, 11¼ x 3 x 2½″	1 ea	135	31	392	12	75	4	4	.9	1.3	1.4
	Rolls & buns, enriched, home recipe:											
477	Dinner rolls 2½″ diam, 2″ high	1 ea	35	29	112	3	19	1	3	.7	1.1	.7
	Sports/fitness bar:											
2043	Forza energy bar	1 ea	70	18	231	10	45	4	1			
2042	Power bar	1 ea	65		230	10	45	3	2			
2041	Tiger sports bar	1 ea	65	17	229	11	40	4	2			
478	Toaster pastries, fortified (Poptarts)	1 ea	52	12	204	2	37	1	5	.8	2.1	2
2132	Toaster strudel pastry—cream cheese	1 ea	54	32	188	3	24	<1	9	2.7		
2134	Toaster strudel pastry—french toast	1 ea	54	32	188	3	24	<1	9	2.9		

PAGE KEY: A–4 = Beverages A–6 = Dairy A–10 = Eggs A–10 = Fat/Oil A–14 = Fruit A–20 = Bakery A–28 = Grain A–32 = Fish A–34 = Meats A–38 = Poultry A–40 = Sausage A–42 = Mixed/Fast A–46 = Nuts/Seeds A–50 = Sweets A–52 = Vegetables/Legumes A–62 = Vegetarian Foods A–64 = Misc A–66 = Soups/Sauces A–68 = Fast A–84 = Convenience A–88 = Baby foods

A–27

A

Chol (mg)	Calc (mg)	Iron (mg)	Magn (mg)	Pota (mg)	Sodi (mg)	Zinc (mg)	VT-A (RE)	Thia (mg)	VT-E (a-TE)	Ribo (mg)	Niac (mg)	V-B6 (mg)	Fola (µg)	VT-C (mg)
23	12	.56	5	39	219	.19	11	.07	.7	.16	1.12	.04	5	<1
34	16	1.27	28	73	234	.57	15	.1	.75	.12	1.44	.09	8	0
31	37	.97	10	65	398	.32	22	.12	.75	.14	1.05	.05	5	<1
					77									
20	77	.56	17	70	160	.35	20	.05	.62	.08	.4	.04	5	<1
22	83	.68	6	50	167	.21	20	.08	.36	.11	.6	.02	14	<1
5	48	.59	8	66	239	.15	3	.08	.32	.08	.65	.03	3	<1
26	13	1.82	10	57	140	.35	87	.23	1.35	.21	1.98	.04	22	<1
0	18	5.2	25	121	976	.79	0	.7	9.94	.5	5.96	.04	121	0
0	32	9.25	45	214	1734	1.41	0	1.25	17.7	.89	10.6	.08	214	0
0	96	3.44	24	99	1166	.62	0	.48	8.83	.3	3.79	.09	19	0
0	110	4.5	70	650	2660	1.6	300	.28	16.5	.27	2.63	.38	220	32
0	18	.75	12	109	444	.27	50	.05	2.76	.04	.44	.06	37	5
588	864	12	184	1900	2764	5.53	806	1.6	16.9	2.38	12.1	1.53	311	18
98	144	2	31	317	461	.92	134	.27	2.82	.4	2.02	.25	52	3
0	82	14.5	94	588	2175	2.35	47	1.8	24.7	1.55	14	.4	270	8
0	14	2.41	16	98	363	.39	8	.3	4.12	.26	2.33	.07	45	1
0	114	21.1	103	878	2177	2.28	547	1.69	21.7	1.43	14.6	.39	308	11
0	24	4.44	22	185	458	.48	115	.35	4.56	.3	3.07	.08	65	2
109	1028	8.84	170	1705	2085	4.93	235	1.03	11.4	2.43	7.34	.37	59	6
18	171	1.47	28	284	348	.82	39	.17	1.9	.41	1.22	.06	10	1
208	504	3.65	69	668	1512	3.28	315	.25	7.5	1.31	1.84	.3	126	2
35	84	.61	12	111	252	.55	52	.04	1.25	.22	.31	.05	21	<1
305	380	4.14	102	603	990	3.32	353	.42	9.7	1.42	4.4	.2	88	22
51	63	.69	17	101	165	.55	59	.07	1.62	.24	.73	.03	15	4
0	59	12.1	79	1047	2025	1.88	386	1.41	26.2	1.19	15.3	.2	58	589
0	7	1.52	10	131	253	.23	48	.18	3.28	.15	1.91	.02	7	74
217	115	7.05	122	502	2874	3.86	319	.62	17.2	.83	1.69	.14	183	7
36	19	1.18	20	84	479	.64	53	.1	2.86	.14	.28	.02	30	1
131	392	5.17	98	1007	1844	2.94	3139	.36	10.5	1	1.22	.37	131	10
22	65	.86	16	168	307	.49	523	.06	1.75	.17	.2	.06	22	2
13	13	.88	8	51	325	.17	33	.1	.37	.08	.98	.03	4	1
0	28	1.56	13	83	479	.29	22	.18	.55	.14	1.83	.04	23	2
0	10	1.21	10	41	480	.24	0	.13	.06	.17	1.47	.03	48	0
0	22	2.59	21	88	1029	.51	0	.28	.13	.37	3.15	.07	103	0
0	22	2.59	21	88	1029	.51	0	.28	.13	.37	3.15	.07	103	0
<1	33	.88	6	37	146	.22	0	.14	.22	.09	1.13	.01	27	<1
0	60	1.36	9	61	241	.27	0	.21	.2	.13	1.69	.02	41	0
0	60	1.36	9	61	241	.27	0	.21	.2	.13	1.69	.02	41	0
0	54	1.87	15	62	310	.54	0	.27	.1	.19	2.42	.03	54	0
0	122	3.78	27	122	783	.85	0	.54	.1	.33	4.47	.05	40	0
13	21	1.04	7	53	145	.24	28	.14	.35	.14	1.21	.02	15	<1
0	300	6.3	160	220	65	5.25		1.5	20	1.7	20	2	400	60
0	300	5.4	140	150	110	5.25		1.5		1.7	20	2	400	60
	349	4.49	140	279	100		50	1.5	19.9	1.69	19.9	1.99	399	60
0	13	1.81	9	58	218	.34	149	.15	.97	.19	2.05	.2	34	<1
12	12	.97			217		17		1					0
12	12	.97			217		17		1					0

Table A–1

Food Composition

(Computer code number is for West Diet Analysis program)   (For purposes of calculations, use "0" for t, <1, <.1, <.01, etc.)

Computer Code Number	Food Description	Measure	Wt (g)	H₂O (%)	Ener (cal)	Prot (g)	Carb (g)	Dietary Fiber (g)	Fat (g)	Fat Breakdown (g) Sat	Mono	Poly
	BAKED GOODS: BREADS, CAKES, COOKIES, CRACKERS, PIES—Continued											
	Tortilla chips:											
1271	Plain	10 pce	18	2	90	1	11	1	5	.9	2.8	.7
1036	Nacho flavor	1 c	26	2	129	2	16	1	7	1.3	3.9	.9
1037	Taco flavor	1 pce	18	2	86	1	11	1	4	.8	2.6	.6
	Tortillas:											
479	Corn, enriched, 6" diam	1 ea	26	44	58	2	12	1	1	.1	.2	.3
480	Flour, 8" diam	1 ea	49	27	159	4	27	2	3	.6	1.4	1.4
1301	Flour, 10" diam	1 ea	72	27	234	6	40	2	5	.8	2.1	2
481	Taco shells	1 ea	14	4	63	1	9	1	3	.4	1.5	.6
	Waffles, 7" diam:											
482	From home recipe	1 ea	75	42	218	6	25	1	11	2.2	2.6	5.1
483	From mix, egg/milk added	1 ea	75	42	218	5	26	1	10	1.7	2.7	5.2
1510	Whole grain, prepared from frozen	1 ea	39	43	107	4	13	1	5	1.6	1.9	.9
	GRAIN PRODUCTS: CEREAL, FLOUR, GRAIN, PASTA and NOODLES, POPCORN											
484	Barley, pearled, dry, uncooked	1 c	200	10	704	20	155	31	2	.5	.3	1.1
485	Barley, pearled, cooked	1 c	157	69	193	4	44	6	1	.1	.1	.3
2009	Breakfast bars, fat free, all flavors	1 ea	38	25	110	2	26	3	0	0	0	0
	Breakfast bar, Snackwell:											
2165	Apple-cinnamon	1 ea	37	16	119	1	29	1	<1	.1	t	.1
2164	Blueberry	1 ea	37	16	121	1	29	1	<1	t	t	.1
2163	Strawberry	1 ea	37	16	120	1	29	1	<1	t	t	.1
	Breakfast cereals, hot, cooked w/o salt added:											
	Corn grits (hominy) enriched:											
486	Regular/quick prep w/o salt, yellow:	1 c	242	85	145	3	31	<1	<1	.1	.1	.2
487	Instant, prepared from packet, white	1 ea	137	82	89	2	21	1	<1	t	t	.1
	Cream of wheat:											
488	Regular, quick, instant	1 c	239	87	129	4	27	1	<1	.1	.1	.3
489	Mix and eat, plain, packet	1 ea	142	82	102	3	21	<1	<1	t	t	.2
1664	Farina cereal, cooked w/o salt	1 c	233	88	117	3	25	3	<1	t	t	.1
490	Malt-O-Meal, cooked w/o salt	1 c	240	88	122	4	26	1	<1	.1	.1	t
494	Maypo	1 c	216	83	153	5	29	5	2	.4	.7	.8
	Oatmeal or rolled oats:											
491	Regular, quick, instant, nonfortified cooked w/o salt	1 c	234	85	145	6	25	4	2	.4	.7	.9
	Instant, fortified:											
492	Plain, from packet	½ c	118	85	70	4	12	2	1	.2	.4	.4
493	Flavored, from packet	½ c	109	76	106	3	21	2	1	.2	.5	.5
	Breakfast cereals, ready to eat:											
495	All-Bran	1 c	62	3	160	8	46	20	2	.4	.4	1.3
1306	Alpha Bits	1 c	28	1	110	2	24	1	1	.1	.2	.2
1307	Apple Jacks	1 c	33	3	120	2	30	1	<1	.1	.1	.2
1308	Bran Buds	1 c	90	3	240	8	72	36	2	.4	.4	1.4
1305	Bran Chex	1 c	49	2	156	5	39	8	1	.2	.3	.7
1309	Honey BucWheat Crisp	1 c	38	5	147	4	31	3	1	.2	.3	.6
1310	C.W. Post, plain	1 c	97	2	421	9	73	7	13	1.7	6	4.7
1311	C.W. Post, with raisins	1 c	103	4	446	9	74	14	15	11	1.7	1.4
496	Cap'n Crunch	1 c	37	2	147	2	32	1	2	.5	.4	.3
1312	Cap'n Crunchberries	1 c	35	2	140	2	30	1	2	.5	.3	.3
1313	Cap'n Crunch, peanut butter	1 c	35	2	146	3	28	1	3	.7	1.1	.7
497	Cheerios	1 c	23	3	84	2	17	2	1	.3	.5	.2
1314	Cocoa Krispies	1 c	41	2	159	3	36	1	1	.7	.2	.2
1316	Cocoa Pebbles	1 c	32	2	131	1	27	1	2	1.1	.4	.1
1315	Corn Bran	1 c	36	3	120	2	30	6	1	.3	.3	.4
1317	Corn Chex	1 c	28	2	110	2	25	<1	<1	t	t	t
498	Corn Flakes, Kellogg's	1 c	28	3	100	2	24	1	<1	.1	t	.1
499	Corn Flakes, Post Toasties	1 c	24	3	93	2	21	1	<1	t	t	t
1340	Corn Pops	1 c	31	3	120	1	28	<1	<1	.1	.1	t
1318	Cracklin' Oat Bran	1 c	65	4	252	5	48	8	8	3.4	3.8	.9
1038	Crispy Wheat 'N Raisins	1 c	43	7	150	3	35	3	1	.1	.1	.2

PAGE KEY: A–4 = Beverages   A–6 = Dairy   A–10 = Eggs   A–10 = Fat/Oil   A–14 = Fruit   A–20 = Bakery   A–28 = Grain   A–32 = Fish   A–34 = Meats   A–38 = Poultry   A–40 = Sausage   A–42 = Mixed/Fast   A–46 = Nuts/Seeds   A–50 = Sweets   A–52 = Vegetables/Legumes   A–62 = Vegetarian Foods   A–64 = Misc   A–66 = Soups/Sauces   A–68 = Fast   A–84 = Convenience   A–88 = Baby foods

Chol (mg)	Calc (mg)	Iron (mg)	Magn (mg)	Pota (mg)	Sodi (mg)	Zinc (mg)	VT-A (RE)	Thia (mg)	VT-E (a-TE)	Ribo (mg)	Niac (mg)	V-B6 (mg)	Fola (µg)	VT-C (mg)
0	28	.27	16	35	95	.27	4	.01	.24	.03	.23	.05	2	0
1	38	.37	21	56	184	.31	11	.03	.35	.05	.37	.07	4	<1
1	28	.36	16	39	142	.23	16	.04	.24	.04	.36	.05	4	<1
0	45	.36	17	40	42	.24	6	.03	.04	.02	.39	.06	30	0
0	61	1.62	13	64	234	.35	0	.26	.62	.14	1.75	.02	60	0
0	90	2.38	19	94	344	.51	0	.38	.91	.21	2.57	.04	89	0
0	35	.36	15	34	25	.19	6	.04	.59	.02	.24	.04	7	0
52	191	1.73	14	119	383	.51	49	.2	1.73	.26	1.55	.04	34	<1
38	93	1.22	15	134	458	.35	19	.15	1.5	.19	1.23	.07	9	<1
39	84	.69	15	91	150	.45	25	.08	.53	.13	.75	.04	7	<1
0	58	5	158	560	18	4.26	4	.38	.26	.23	9.2	.52	46	0
0	17	2.09	34	146	5	1.29	2	.13	.08	.1	3.23	.18	25	0
0	20	.72			25		20							1
<1	17	5	6	68	103	3.88	260	.39		.44	5.2	.52		<1
<1	14	4.83	5	43	107	3.85	260	.39		.44	5.2	.52		<1
<1	14	4.82	6	47	102	3.83	260	.39		.44	5.2	.52		2
0	0	1.55	10	53	0	.17	14	.24	.12	.14	1.96	.06	75	0
0	8	8.19	11	38	289	.21	0	.15	.03	.08	1.38	.05	47	0
0	50	10.3	12	45	139	.33	0	.24	.03	0	1.43	.03	108	0
0	20	8.09	7	38	241	.24	125	.43	.02	.28	4.97	.57	101	0
0	5	1.17	5	30	0	.16	0	.19	.03	.12	1.28	.02	54	0
0	5	9.6	5	31	2	.17	0	.48	.03	.24	5.76	.02	5	0
0	112	7.56	45	190	233	1.34	633	.65	1.51	.65	8.42	.86	9	26
0	19	1.59	56	131	2	1.15	5	.26	.23	.05	.3	.05	9	0
0	109	4.2	28	66	190	.58	302	.35	.14	.19	3.65	.49	100	0
0	112	4.45	34	91	169	.66	306	.35	.14	.25	3.92	.51	100	<1
0	200	9	280	620	560	7.5	450	.75	1.14	.85	10	1	186	30
0	8	2.66	16	54	178	1.48	371	.36	.02	.42	4.93	.5	99	0
0	0	4.5	8	35	150	3.75	225	.38	.05	.43	5	.5	116	15
0	60	13.5	240	809	599	11.3	676	1.17	1.42	1.26	15	1.53	270	45
0	29	14	69	216	345	6.47	11	.64	.56	.26	8.62	.88	173	26
0	54	10.9	43	142	361	.68	913	.9	8.99	1.03	12.1	1.88	11	36
<1	47	15.4	67	198	167	1.64	1284	1.26	.68	1.46	17.1	1.75	342	0
<1	50	16.4	74	261	161	1.64	1363	1.34	.72	1.55	18.1	1.85	364	0
0	7	6.18	13	47	286	5.14	5	.51	.18	.58	6.85	.68	137	0
0	9	6.06	13	49	256	5.39	6	.5	.25	.57	6.72	.67	135	<1
0	3	5.85	24	80	264	4.87	5	.49	.19	.55	6.48	.65	130	0
0	42	6.21	25	68	218	2.88	288	.29	.16	.33	3.84	.38	77	11
0	0	2.38	11	79	278	1.97	298	.49	.19	.57	6.6	.66	123	20
0	5	2.02	13	53	180	1.7	424	.42	.04	.48	5.63	.58	113	0
0	27	10.1	19	75	338	5	5	.1	.19	.56	6.66	.67	134	0
0	3	8.01	4	23	306	.1	14	.36	.07	.07	4.93	.5	99	15
0	0	8.68	3	25	300	.17	225	.36	.03	.43	5	.5	99	15
0	1	.63	4	28	252	.07	318	.31	.06	.36	4.22	.43	85	0
0	0	1.86	2	25	120	1.55	225	.4	.03	.43	5.18	.5	109	15
0	26	2.41	79	305	226	1.95	299	.5	.43	.56	6.63	.66	181	20
0	54	3.52	33	180	223	.85	293	.29	.45	.33	3.91	.39	78	0

A

Table A–1

Food Composition   (Computer code number is for West Diet Analysis program)   (For purposes of calculations, use "0" for t, <1, <.1, <.01, etc.)

Computer Code Number	Food Description	Measure	Wt (g)	H₂O (%)	Ener (cal)	Prot (g)	Carb (g)	Dietary Fiber (g)	Fat (g)	Fat Breakdown (g) Sat	Mono	Poly
	GRAIN PRODUCTS: CEREAL, FLOUR, GRAIN, PASTA and NOODLES, POPCORN—Continued											
1319	Fortified Oat Flakes	1 c	48	3	180	8	36	1	1	.2	.3	.4
500	40% Bran Flakes, Kellogg's	1 c	39	4	121	4	32	7	1	.2	.2	.5
501	40% Bran Flakes, Post	1 c	47	3	152	5	37	9	1	.1	.1	.4
502	Froot Loops	1 c	32	2	120	2	28	1	1	.4	.2	.3
518	Frosted Flakes	1 c	41	3	159	2	37	1	<1	.1	t	.1
1320	Frosted Mini-Wheats	1 c	51	5	170	5	41	5	1	.2	.1	.6
1321	Frosted Rice Krispies	1 c	35	2	135	2	32	<1	<1	.1	.1	.1
1324	Fruit & Fibre w/dates	1 c	57	9	193	5	43	8	3	.4	1.3	1
1322	Fruity Pebbles	1 c	32	3	130	1	28	<1	2	1.4	.1	.1
503	Golden Grahams	1 c	39	3	150	2	33	1	1	.2	.4	.2
504	Granola, homemade	½ c	61	5	285	9	32	6	15	2.9	4.8	6.5
505	Granola, low fat	½ c	47	3	181	5	38	3	3	0		
1670	Granola, low fat, commercial	½ c	45	5	165	4	35	2	2	.6	.7	.9
505	Grape Nuts	½ c	55	3	196	7	45	5	<1	t	t	.1
1326	Grape Nuts Flakes	1 c	39	3	144	4	32	4	1	.6	.1	.2
1665	Heartland Natural with raisins	1 c	110	5	468	11	76	6	16	4	4.2	6.2
1327	Honey & Nut Corn Flakes	1 c	37	2	148	3	31	1	2	.3	.7	.6
506	Honey Nut Cheerios	1 c	33	2	126	3	27	2	1	.3	.5	.2
1328	HoneyBran	1 c	35	2	119	3	29	4	1	.3	.1	.3
1329	HoneyComb	1 c	22	1	86	1	20	1	<1	.2	.1	.1
1330	King Vitaman	1 c	21	2	81	2	18	1	1	.2	.3	.2
1039	Kix	1 c	19	2	72	1	16	1	<1	.1	.1	t
1331	Life	1 c	44	4	167	4	35	3	2	.3	.6	.8
507	Lucky Charms	1 c	32	2	124	3	27	1	1	.2	.4	.2
1323	Mueslix Five Grain	1 c	82	5	279	7	63	7	3	.5	1	1.2
508	Nature Valley Granola	1 c	113	4	510	12	74	7	20	2.6	13.3	3.8
1666	Nutri Grain Almond Raisin	1 c	40	6	147	3	31	3	2	.1	1	1.2
1336	100% Bran	1 c	66	3	178	8	48	19	3	.6	.6	1.9
509	100% Natural cereal, plain	1 c	104	2	462	11	71	8	17	7.5	7.5	2.3
1337	100% Natural with apples & cinnamon	1 c	104	2	477	11	70	7	20	15.5	1.8	1.3
1338	100% Natural with raisins & dates	1 c	110	3	496	12	72	7	20	13.6	3.7	1.7
510	Product 19	1 c	33	4	110	2	28	1	<1	t	.2	.2
1339	Quisp	1 c	30	3	121	2	25	1	2	.5	.4	.2
511	Raisin Bran, Kellogg's	1 c	61	9	200	6	47	8	1	.1	.1	.4
512	Raisin Bran, Post	1 c	56	9	172	5	42	8	1	.2	.1	.5
1667	Raisin Squares	1 c	71	9	241	6	55	7	2	.2	.2	.6
1041	Rice Chex	1 c	33	3	130	2	29	1	<1	t	t	t
513	Rice Krispies, Kellogg's	1 c	28	2	111	2	25	<1	<1	t	t	t
514	Rice, puffed	1 c	14	4	54	1	12	<1	<1	t	t	t
515	Shredded Wheat	1 c	43	5	154	5	35	4	1	.1	.1	.4
516	Special K	1 c	31	3	110	6	22	1	<1	t	t	.2
517	Super Golden Crisp	1 c	33	1	123	2	30	<1	<1	.1	.1	.1
519	Honey Smacks	1 c	36	3	133	3	32	1	1	.4	.1	.3
1341	Tasteeos	1 c	24	2	94	3	19	3	1	.2	.2	.2
1342	Team	1 c	42	4	164	3	36	1	1	.1	.2	.3
520	Total, wheat, with added calcium	1 c	40	3	140	4	32	4	1	.2	.2	.1
521	Trix	1 c	28	2	114	1	24	1	2	.4	.9	.3
1344	Wheat Chex	1 c	46	2	169	5	38	4	1	.2	.1	.5
1043	Wheat cereal, puffed, fortified	1 c	12	4	44	2	9	1	<1	t	t	.1
522	Wheaties	1 c	29	3	106	3	23	2	1	.2	.2	.1
523	Buckwheat flour, dark	1 c	120	11	402	15	85	12	4	.8	1.1	1.1
525	Buckwheat, whole grain, dry	1 c	170	10	583	23	122	17	6	1.3	1.8	1.8
526	Bulgar, dry, uncooked	1 c	140	9	479	17	106	26	2	.3	.2	.8
527	Bulgar, cooked	1 c	182	78	151	6	34	8	<1	.1	.1	.2
	Cornmeal:											
528	Whole-ground, unbolted, dry	1 c	122	10	442	10	94	9	4	.6	1.2	2
530	Degermed, enriched, dry	1 c	138	12	505	12	107	10	2	.3	.6	1
38041	Degermed, enriched, baked	1 c	138	12	505	12	107	10	2	.3	.6	1
	Macaroni, cooked:											
532	Enriched	1 c	140	66	197	7	40	2	1	.1	.1	.4
533	Whole wheat	1 c	140	67	174	7	37	4	1	.1	.1	.3

PAGE KEY: A–4 = Beverages A–6 = Dairy A–10 = Eggs A–10 = Fat/Oil A–14 = Fruit A–20 = Bakery A–28 = Grain A–32 = Fish A–34 = Meats A–38 = Poultry A–40 = Sausage A–42 = Mixed/Fast A–46 = Nuts/Seeds A–50 = Sweets A–52 = Vegetables/Legumes A–62 = Vegetarian Foods A–64 = Misc A–66 = Soups/Sauces A–68 = Fast A–84 = Convenience A–88 = Baby foods

A–31

A

Chol (mg)	Calc (mg)	Iron (mg)	Magn (mg)	Pota (mg)	Sodi (mg)	Zinc (mg)	VT-A (RE)	Thia (mg)	VT-E (a-TE)	Ribo (mg)	Niac (mg)	V-B6 (mg)	Fola (μg)	VT-C (mg)
0	68	13.7	58	228	220	2.54	636	.62	.34	.72	8.45	.86	169	0
0	0	10.9	81	229	309	5.03	505	.51	7.22	.58	6.71	.66	138	20
0	21	13.4	102	251	431	2.49	622	.61	.54	.7	8.27	.85	166	0
0	0	4.51	8	35	150	3.75	225	.37	.12	.43	5	.5	96	15
0	0	6.15	4	26	264	.2	298	.5	.05	.56	6.61	.66	123	20
0	0	15	60	170	0	1.5	0	.37	.46	.42	4.64	.5	102	20
0	0	2.42	8	27	256	.42	303	.49	.03	.56	6.72	.66	140	20
0	30	10.1	81	335	270	3.02	725	.75	1.32	.85	10.1	1	201	0
0	4	2.02	9	24	178	1.7	424	.42	.03	.48	5.63	.58	113	0
0	19	5.85	12	69	357	4.88	293	.49	.29	.55	6.51	.65	130	19
0	49	2.56	109	328	15	2.48	2	.45	7.87	.17	1.25	.19	52	1
0		2.71	36	143	90	5.64	226	.56	7.57	.64	7.52	.75	151	
0	15	1.35	30	127	101	2.84	169	.27	4.03	.31	3.74	.36	90	0
0	5	15.7	37	184	382	1.21	728	.71	.14	.82	9.68	.99	194	0
0	16	11.2	43	136	220	.78	516	.51	.1	.58	6.86	.7	138	0
0	66	4.02	141	415	226	2.83	7	.32	.77	.14	1.54	.2	44	1
0	0	3.03	3	40	249	.2	152	.26	.09	.3	3.37	.33	74	10
0	22	4.95	32	94	285	4.13	248	.41	.34	.47	5.51	.55	110	16
0	16	5.57	46	151	202	.9	463	.45	.81	.52	6.16	.63	23	19
0	4	2.09	7	25	124	1.17	291	.29	.09	.33	3.87	.4	78	0
0	3	5.92	18	58	176	2.65	212	.26	1.42	.3	3.53	.35	71	8
0	28	5.13	6	26	167	2.38	238	.24	.05	.27	3.17	.32	63	9
0	134	12.3	43	109	240	5.5	2	.55	.22	.62	7.35	.73	147	0
0	35	4.8	21	58	217	4	240	.4	.14	.45	5.34	.53	107	16
0	38	8.94	82	369	107	7.46	747	.75	8.94	.84	9.84	.99	197	1
0	85	3.53	107	375	183	2.27	0	.35	7.97	.12	1.25	.16	17	0
0	122	1.14	13	147	139	3.06	0	.32	4.38	.35	4.08	.41	80	0
0	46	8.12	312	652	457	5.74	0	1.58	1.53	1.78	20.9	2.11	47	63
1	100	3.11	109	457	28	2.5	1	.36	1.19	.17	1.84	.19	26	<1
0	157	2.89	72	514	52	2	6	.33	.73	.57	1.87	.11	17	1
0	160	3.12	124	538	47	2.11	7	.31	.77	.65	2.09	.16	45	0
0	0	19.8	18	55	308	16.5	248	1.65	24.4	1.88	22	2.21	429	66
0	6	5.1	15	40	216	4.26	4	.42	.15	.48	5.67	.56	113	0
0	40	4.5	80	350	390	3.75	225	.37	.56	.43	5	.5	122	0
0	26	8.9	95	345	365	2.97	741	.73	1.3	.84	9.86	1.01	198	0
0	0	21.7	54	335	4	1.99	0	.5	.38	.57	6.67	.64	142	0
0	5	9.44	8	38	276	.45	2	.43	.04	.01	5.81	.59	116	17
0	5	.7	12	27	206	.46	371	.52	.03	.59	6.92	.69	138	15
0	1	.41	4	16	1	.15	0	.06	.01	.01	.87	0	1	0
0	16	1.81	57	155	4	1.42	0	.11	.23	.12	2.26	.11	21	0
0	0	8.4	16	55	250	3.75	225	.53	.08	.59	7.01	.71	93	15
0	7	2.08	20	48	51	1.75	437	.43	.12	.49	5.81	.59	116	0
0	0	2.4	21	53	67	.4	300	.5	.18	.58	6.66	.68	133	20
0	11	6.86	26	71	183	.69	318	.31	.17	.36	4.22	.43	85	13
0	6	12	12	71	260	.58	556	.55	.1	.63	7.39	.76	7	22
0	344	24	43	129	265	20	500	2	31.3	2.27	26.8	2.67	533	80
0	30	4.2	3	16	184	3.5	210	.35	.56	.4	4.68	.47	93	14
0	18	13.2	58	173	308	1.23	0	.6	.17	.17	8.1	.83	162	24
0	3	.56	16	44	1	.37	<1	.05	.08	.03	1.43	.02	4	0
0	53	7.83	31	101	215	.68	218	.36	.36	.41	4.84	.48	97	14
0	49	4.87	301	692	13	3.74	0	.5	1.24	.23	7.38	.7	65	0
0	31	3.74	393	782	2	4.08	0	.17	1.75	.72	11.9	.36	51	0
0	49	3.44	230	574	24	2.7	0	.32	.22	.16	7.15	.48	38	0
0	18	1.75	58	124	9	1.04	0	.1	.05	.05	1.82	.15	33	0
0	7	4.21	155	350	43	2.22	57	.47	.82	.24	4.43	.37	31	0
0	7	5.7	55	224	4	.99	57	.99	.45	.56	6.94	.35	258	0
0	7	5.7	55	224	4	.99	57	.79	.5	.5	6.25	.32	181	0
0	10	1.96	25	43	1	.74	0	.29	.04	.14	2.34	.05	98	0
0	21	1.48	42	62	4	1.13	0	.15	.14	.06	.99	.11	7	0

Table A–1

Food Composition   (Computer code number is for West Diet Analysis program)   (For purposes of calculations, use "0" for t, <1, <.1, <.01, etc.)

Computer Code Number	Food Description	Measure	Wt (g)	H₂O (%)	Ener (cal)	Prot (g)	Carb (g)	Dietary Fiber (g)	Fat (g)	Fat Breakdown (g) Sat	Mono	Poly
	GRAIN PRODUCTS: CEREAL, FLOUR, GRAIN, PASTA and NOODLES, POPCORN—Continued											
534	Vegetable, enriched	1 c	134	68	172	6	36	2	<1	t	t	.1
535	Millet, cooked	1 c	240	71	286	8	57	3	2	.4	.4	1.2
	Noodles (see also Pasta and Spaghetti):											
1507	Cellophane noodles, cooked	1 c	190	79	160	<1	39	<1	<1	t	t	t
1995	Cellophane noodles, dry	1 c	140	13	491	<1	121	1	<1	t	t	t
537	Chow mein, dry	1 c	45	1	237	4	26	2	14	2	3.5	7.8
536	Egg noodles, cooked, enriched	1 c	160	69	213	8	40	2	2	.5	.7	.7
538	Spinach noodles, dry	3½ oz	100	8	372	13	75	11	2	.2	.2	.6
1343	Oat bran, dry	¼ c	24	7	59	4	16	4	2	.3	.6	.7
	Pasta, cooked:											
1418	Fresh	2 oz	57	69	75	3	14	1	1	.1	.1	.2
1417	Linguini/Rotini	1 c	140	66	197	7	40	4	1	.1	.1	.4
	Popcorn:											
539	Air popped, plain	1 c	8	4	31	1	6	1	<1	t	.1	.2
1042	Microwaved, low fat, low sodium	1 c	6	3	25	1	4	1	1	.1	.2	.3
540	Popped in vegetable oil/salted	1 c	11	3	55	1	6	1	3	.5	.9	1.5
541	Sugar-syrup coated	1 c	35	3	151	1	28	2	4	1.3	1	1.6
	Rice:											
542	Brown rice, cooked	1 c	195	73	216	5	45	4	2	.4	.6	.6
2215	Mexican rice, cooked	1 c	226		820	16	180	6	30	4	1	1
2216	Spanish rice, cooked	1 c	246	85	130	3	28	2	1			
	White, enriched, all types:											
543	Regular/long grain, dry	1 c	185	12	675	13	148	2	1	.3	.4	.3
544	Regular/long grain, cooked	1 c	158	68	205	4	45	1	<1	.1	.1	.1
545	Instant, prepared without salt	1 c	165	76	162	3	35	1	<1	.1	.1	.1
	Parboiled/converted rice:											
546	Raw, dry	1 c	185	10	686	13	151	3	1	.3	.3	.3
547	Cooked	1 c	175	72	200	4	43	1	<1	.1	.1	.1
1486	Sticky rice (glutinous), cooked	1 c	174	77	169	4	37	2	<1	.1	.1	.1
548	Wild rice, cooked	1 c	164	74	166	7	35	3	1	.1	.1	.4
1700	Rice and pasta (Rice-a-Roni), cooked	1 c	202	72	246	5	43	1	6	1.1	2.3	1.9
549	Rye flour, medium	1 c	102	10	361	10	79	15	2	.2	.2	.8
1044	Soy flour, low-fat	1 c	88	3	325	45	30	9	6	.9	1.3	3.3
	Spaghetti pasta:											
550	Without salt, enriched	1 c	140	66	197	7	40	4	1	.1	.1	.4
551	With salt, enriched	1 c	140	66	197	7	40	2	1	.1	.1	.4
552	Whole-wheat spaghetti, cooked	1 c	140	67	174	7	37	6	1	.1	.1	.3
1302	Tapioca-pearl, dry	1 c	152	11	544	<1	135	1	<1	t	t	t
553	Wheat bran, crude	1 c	58	10	125	9	37	25	2	.4	.4	1.3
554	Wheat germ, raw	1 c	115	11	414	27	60	15	11	1.9	1.6	6.9
555	Wheat germ, toasted	1 c	113	6	432	33	56	15	12	2.1	1.7	7.5
1669	Wheat germ, with brown sugar & honey	1 c	113	3	420	30	66	11	9	1.5	1.2	5.5
556	Rolled wheat, cooked	1 c	240	84	149	5	33	4	1	.1	.1	.5
557	Whole-grain wheat, cooked	1 c	150	86	84	4	20	3	<1	.1	.1	.2
	Wheat flour (unbleached):											
	All-purpose white flour, enriched:											
558	Sifted	1 c	115	12	419	12	88	3	1	.2	.1	.5
559	Unsifted	1 c	125	12	455	13	95	3	1	.2	.1	.5
560	Cake or pastry, enriched, sifted	1 c	96	12	348	8	75	2	1	.1	.1	.4
561	Self-rising, enriched, unsifted	1 c	125	11	443	12	93	3	1	.2	.1	.5
562	Whole wheat, from hard wheats	1 c	120	10	407	16	87	15	2	.4	.3	.9
	MEATS: FISH and SHELLFISH											
1045	Bass, baked or broiled	4 oz	113	69	165	27	0	0	5	1.4	1.5	2.3
1046	Bluefish, baked or broiled	4 oz	113	63	180	29	0	0	6	1.4	2	2.7
1686	Catfish, breaded/flour fried	4 oz	113	49	325	21	14	1	20	5	9	4.7
	Clams:											
563	Raw meat only	1 ea	145	82	107	19	4	0	1	.3	.4	.7
564	Canned, drained	1 c	160	64	237	41	8	0	3	.7	.9	1.5
1290	Steamed, meat only	10 ea	95	64	141	24	5	0	2	.4	.5	.9

PAGE KEY: A–4 = Beverages   A–6 = Dairy   A–10 = Eggs   A–10 = Fat/Oil   A–14 = Fruit   A–20 = Bakery   A–28 = Grain   A–32 = Fish   A–34 = Meats   A–38 = Poultry   A–40 = Sausage   A–42 = Mixed/Fast   A–46 = Nuts/Seeds   A–50 = Sweets   A–52 = Vegetables/Legumes   A–62 = Vegetarian Foods   A–64 = Misc   A–66 = Soups/Sauces   A–68 = Fast   A–84 = Convenience   A–88 = Baby foods

A

Chol (mg)	Calc (mg)	Iron (mg)	Magn (mg)	Pota (mg)	Sodi (mg)	Zinc (mg)	VT-A (RE)	Thia (mg)	VT-E (a-TE)	Ribo (mg)	Niac (mg)	V-B6 (mg)	Fola (µg)	VT-C (mg)
0	15	.66	25	41	8	.59	7	.15	.05	.08	1.43	.03	87	0
0	7	1.51	106	149	5	2.18	0	.25	.43	.2	3.19	.26	46	0
0	14	1	3	5	9	.23	0	.07	.06	0	.09	.02	1	0
0	35	3.04	4	14	14	.57	0	.21	.18	0	.28	.07	3	0
0	9	2.13	23	54	198	.63	4	.26	.07	.19	2.68	.05	40	0
53	19	2.54	30	45	11	.99	10	.3	.08	.13	2.38	.06	102	0
0	58	2.13	174	376	36	2.76	46	.37	.04	.2	4.55	.32	48	0
0	14	1.3	56	136	1	.75	0	.28	.41	.05	.22	.04	12	0
19	3	.65	10	14	3	.32	3	.12	.09	.09	.56	.02	36	0
0	10	1.96	25	43	1	.74	0	.29	.08	.14	2.34	.05	98	0
0	1	.21	10	24	<1	.27	2	.02	.01	.02	.15	.02	2	0
0	1	.14	9	14	29	.23	1	.02	.06	.01	.12	.01	1	0
0	1	.31	12	25	97	.29	2	.01	.03	.01	.17	.02	2	<1
2	15	.61	12	38	72	.2	3	.02	.42	.02	.77	.01	1	0
0	19	.82	84	84	10	1.23	0	.19	.53	.05	2.98	.28	8	0
0	300	9			2700		120							96
0		.72			1340									
0	52	7.97	46	213	9	2.02	0	1.07	.24	.09	7.75	.3	427	0
0	16	1.9	19	55	2	.77	0	.26	.08	.02	2.34	.15	92	0
0	13	1.04	8	7	5	.4	0	.12	.08	.08	1.45	.02	68	0
0	111	6.59	57	222	9	1.78	0	1.1	.24	.13	6.72	.65	427	0
0	33	1.98	21	65	5	.54	0	.44	.09	.03	2.45	.03	87	0
0	3	.24	9	17	9	.71	0	.03	.07	.02	.5	.04	2	0
0	5	.98	52	166	5	2.2	0	.08	.38	.14	2.12	.22	43	0
2	16	1.9	24	85	1147	.57	0	.25	.27	.16	3.6	.2	89	<1
0	24	2.16	76	347	3	2.03	0	.29	1.36	.12	1.76	.27	19	0
0	165	5.27	202	2261	16	1.04	4	.33	.17	.25	1.9	.46	361	0
0	10	1.96	25	43	1	.74	0	.29	.08	.14	2.34	.05	98	0
0	10	1.96	25	43	140	.74	0	.29	.38	.14	2.34	.05	98	0
0	21	1.48	42	62	4	1.13	0	.15	.07	.06	.99	.11	7	0
0	30	2.4	2	17	2	.18	0	.01	0	0	0	.01	6	0
0	42	6.15	354	686	1	4.22	0	.3	1.35	.33	7.89	.75	46	0
0	45	7.2	275	1025	14	14.1	0	2.16	20.7	.57	7.83	1.5	323	0
0	51	10.3	362	1070	5	18.9	0	1.89	20.5	.93	6.32	1.11	398	7
0	56	9.1	307	1089	12	15.7	11	1.51	24.9	.78	5.34	.56	376	0
0	17	1.49	53	170	0	1.15	0	.17	.48	.12	2.14	.17	26	0
0	9	.88	35	99	1	.73	0	.12	.3	.03	1.5	.08	12	0
0	17	5.34	25	123	2	.8	0	.9	.07	.57	6.79	.05	177	0
0	19	5.8	27	134	2	.87	0	.98	.07	.62	7.38	.05	193	0
0	13	7.03	15	101	2	.59	0	.86	.06	.41	6.52	.03	148	0
0	423	5.84	24	155	1587	.77	0	.84	.07	.52	7.29	.06	193	0
0	41	4.66	166	486	6	3.52	0	.54	1.48	.26	7.64	.41	53	0
98	116	2.16	43	515	102	.94	40	.1	.84	.1	1.72	.16	19	2
86	10	.7	47	539	87	1.18	156	.08	.71	.11	8.19	.52	2	0
92	41	1.44	34	376	598	1.05	33	.4	2.48	.18	3.37	.21	19	1
49	67	20.3	13	455	81	1.99	131	.12	1.45	.31	2.57	.09	23	19
107	147	44.8	29	1004	179	4.37	274	.24	3.04	.68	5.36	.18	46	35
64	87	26.6	17	597	106	2.59	162	.14	1.86	.4	3.18	.1	27	21

Table A-1

Food Composition

(Computer code number is for West Diet Analysis program)   (For purposes of calculations, use "0" for t, <1, <.1, <.01, etc.)

Computer Code Number	Food Description	Measure	Wt (g)	H₂O (%)	Ener (cal)	Prot (g)	Carb (g)	Dietary Fiber (g)	Fat (g)	Fat Breakdown (g) Sat	Mono	Poly
	MEATS: FISH and SHELLFISH—Continued											
	Cod:											
565	Baked	4 oz	113	76	119	26	0	0	1	.2	.1	.5
566	Batter fried	4 oz	113	67	196	20	8	<1	9	2.2	3.6	2.6
567	Poached, no added fat	4 oz	113	77	116	25	0	0	1	.2	.1	.3
	Crab, meat only:											
1048	Blue crab, cooked	1 c	118	77	120	24	0	0	2	.3	.3	.8
1049	Dungeness crab, cooked	1 c	118	73	130	26	1	0	1	.2	.3	.5
568	Blue crab, canned	1 c	135	76	134	28	0	0	2	.4	.3	.6
1587	Crab, imitation, from surimi	4 oz	113	74	115	14	11	0	1	.3	.2	.8
569	Fish sticks, breaded pollock	2 ea	56	46	152	9	13	<1	7	1.8	2.8	1.8
572	Flounder/sole, baked	4 oz	113	73	132	27	0	0	2	.5	.4	.9
1599	Grouper, baked or broiled	4 oz	113	73	133	28	0	0	1	.4	.4	.6
573	Haddock, breaded, fried	4 oz	113	55	264	22	14	1	13	3.2	5.4	3.3
1050	Haddock, smoked	4 oz	113	71	131	28	0	0	1	.3	.3	.5
	Halibut:											
17291	Baked	4 oz	113	72	158	30	0	0	3	.7	1	1.4
1051	Smoked	4 oz	113	64	203	34			4	.6	1.2	1.5
1054	Raw	4 oz	113	78	124	23	0	0	3	.7	.8	1.1
575	Herring, pickled	4 oz	113	55	296	16	11	0	20	4.4	11	4.8
1052	Lobster meat, cooked w/moist heat	1 c	145	76	142	30	2	0	1	.2	.2	.5
1687	Ocean perch, baked/broiled	4 oz	113	73	137	27	0	0	2	.4	1	.7
576	Ocean perch, breaded/fried	4 oz	113	59	249	22	9	<1	13	3.2	5.7	3.4
1056	Octopus, raw	4 oz	113	80	93	17	2	0	1	.3	.2	.3
	Oysters:											
577	Raw, Eastern	1 c	248	85	169	17	10	0	6	2	.9	2.8
578	Raw, Pacific	1 c	248	82	201	23	12	0	6	1.3	.9	2.2
	Cooked:											
579	Eastern, breaded, fried, medium	5 ea	73	65	144	6	8	<1	9	2.7	1.7	4.6
580	Western, simmered	5 ea	125	64	204	24	12	0	6	1.9	.9	2.7
581	Pollock, baked, broiled, or poached	4 oz	113	74	128	27	0	0	1	.3	.2	.6
	Salmon:											
582	Canned pink, solids and liquid	4 oz	113	69	157	22	0	0	7	1.7	2.1	2.3
583	Broiled or baked	4 oz	113	62	244	31	0	0	12	2.2	6	2.7
584	Smoked	4 oz	113	72	132	21	0	0	5	1.2	2.3	1.1
585	Atlantic sardines, canned, drained, 2 = 24 g	4 oz	113	60	235	28	0	0	13	1.9	4.4	6.4
586	Scallops, breaded, cooked from frozen	6 ea	93	58	200	17	9	<1	10	2	2.5	5.3
1588	Scallops, imitation, from surimi	4 oz	113	74	112	14	12	0	1	.1	.1	.3
1688	Scallops, steamed/boiled	½ c	60	76	64	10	1	0	2	.3	.7	.6
	Shrimp:											
587	Cooked, boiled, 2 large = 11g	16 ea	88	77	87	18	0	0	1	.2	.2	.5
588	Canned, drained	½ c	64	73	77	15	1	0	1	.3	.3	.7
589	Fried, 2 large = 15 g, breaded	12 ea	90	53	218	19	10	<1	11	2	3	5.9
1057	Raw, large, about 7g each	14 ea	98	76	104	20	1	0	2	.3	.3	1
1589	Shrimp, imitation, from surimi	4 oz	113	75	114	14	10	0	2	.3	.2	1
1053	Snapper, baked or broiled	4 oz	113	70	145	30	0	0	2	.4	.4	.7
1060	Squid, fried in flour	4 oz	113	64	198	20	9	0	8	2.1	3.1	2.4
1590	Surimi	4 oz	113	76	112	17	8	0	1	.2	.2	.6
1058	Swordfish, raw	4 oz	113	76	137	22	0	0	5	1.3	1.7	1
1059	Swordfish, baked or broiled	4 oz	113	69	175	29	0	0	6	1.7	2.2	1.3
590	Trout, baked or broiled	4 oz	113	70	170	26	0	0	7	1.8	2	2.2
	Tuna, light, canned, drained solids:											
591	Oil pack	1 c	145	60	287	42	0	0	12	2.2	4.3	4.2
592	Water pack	1 c	154	74	179	39	0	0	1	.4	.2	.5
1061	Bluefin tuna, fresh	4 oz	113	68	163	26	0	0	6	1.4	1.8	1.6
	MEATS: BEEF, LAMB, PORK and others											
	BEEF, cooked, trimmed to ½" outer fat:											
	Braised, simmered, pot roasted:											
	Relatively fat, choice chuck blade:											
593	Lean and fat, piece 2½ x 2½ x ¾"	4 oz	113	47	393	30	0	0	29	13	14.8	1.2
594	Lean only	4 oz	113	55	297	35	0	0	16	7.3	8.2	.7

PAGE KEY: A–4 = Beverages A–6 = Dairy A–10 = Eggs A–10 = Fat/Oil A–14 = Fruit A–20 = Bakery A–28 = Grain A–32 = Fish A–34 = Meats A–38 = Poultry A–40 = Sausage A–42 = Mixed/Fast A–46 = Nuts/Seeds A–50 = Sweets A–52 = Vegetables/Legumes A–62 = Vegetarian Foods A–64 = Misc A–66 = Soups/Sauces A–68 = Fast A–84 = Convenience A–88 = Baby foods

A–35

A

Chol (mg)	Calc (mg)	Iron (mg)	Magn (mg)	Pota (mg)	Sodi (mg)	Zinc (mg)	VT-A (RE)	Thia (mg)	VT-E (a-TE)	Ribo (mg)	Niac (mg)	V-B6 (mg)	Fola (μg)	VT-C (mg)
62	16	.55	47	276	88	.65	16	.1	.39	.09	2.84	.32	9	1
64	43	.92	36	443	124	.62	17	.13	.92	.13	2.58	.23	10	1
61	23	.54	41	496	69	.63	14	.09	.32	.08	2.48	.28	8	1
118	123	1.07	39	382	329	4.98	2	.12	1.18	.06	3.89	.21	60	4
90	70	.51	68	481	446	6.45	37	.07	1.33	.24	4.27	.2	50	4
120	136	1.13	53	505	450	5.43	2	.11	1.35	.11	1.85	.2	57	4
23	15	.44	49	102	950	.37	23	.04	.12	.03	.2	.03	2	0
63	11	.41	14	146	326	.37	17	.07	.77	.1	1.19	.03	10	0
77	20	.38	65	389	119	.71	12	.09	2.6	.13	2.46	.27	10	0
53	24	1.29	42	537	60	.58	56	.09	.71	.01	.43	.4	11	0
96	63	1.92	46	345	523	.59	33	.08	1.56	.14	4.49	.28	19	<1
87	55	1.58	61	469	862	.56	25	.05	.56	.05	5.73	.45	17	0
46	68	1.21	121	651	78	.6	61	.08	1.23	.1	8.05	.45	16	0
59	87	1.56	154	833	2260	.78	86	.11	1.11	.14	10.8	.64	22	0
36	53	.95	94	509	61	.47	53	.07	.96	.08	6.61	.39	14	0
15	87	1.38	9	78	983	.6	292	.04	1.81	.16	3.73	.19	3	0
104	88	.57	51	510	551	4.23	38	.01	2.1	.1	1.55	.11	16	0
61	155	1.33	44	396	108	.69	16	.15	1.84	.15	2.76	.3	12	1
71	136	1.57	38	323	431	.67	23	.14	2.41	.18	2.68	.24	15	1
54	60	5.99	34	396	260	1.9	51	.03	1.36	.04	2.37	.41	18	6
131	112	16.5	117	387	523	225	74	.25	1.98	.24	3.42	.15	25	9
124	20	12.7	55	417	263	41.2	201	.17	2.11	.58	4.98	.12	25	20
59	45	5.07	42	178	304	63.6	66	.11	1.66	.15	1.2	.05	23	3
125	20	11.5	55	378	265	41.5	183	.16	2.21	.55	4.53	.11	19	16
108	7	.32	82	437	131	.68	26	.08	.32	.09	1.86	.08	4	0
62	241	.95	38	368	626	1.04	19	.03	1.53	.21	7.39	.34	17	0
98	8	.62	35	424	75	.58	71	.24	1.42	.19	7.54	.25	6	0
26	12	.96	20	198	886	.35	29	.03	1.53	.11	5.33	.31	2	0
160	432	3.3	44	449	571	1.48	76	.09	.34	.26	5.93	.19	13	0
57	39	.76	55	310	432	.99	20	.04	1.77	.1	1.4	.13	34	2
25	9	.35	49	116	898	.37	23	.01	.12	.02	.35	.03	2	0
19	15	.15	32	168	246	.55	27	.01	.81	.04	.6	.08	7	1
172	34	2.72	30	160	197	1.37	58	.03	.66	.03	2.28	.11	3	2
111	38	1.75	26	134	108	.81	11	.02	.59	.02	1.77	.07	1	1
159	60	1.13	36	203	310	1.24	50	.12	1.35	.12	2.76	.09	7	1
149	51	2.36	36	181	145	1.09	53	.03	.8	.03	2.5	.1	3	2
41	21	.68	49	101	797	.37	23	.03	.12	.04	.19	.03	2	0
53	45	.27	42	590	64	.5	40	.06	.71	<.01	.39	.52	7	2
294	44	1.14	43	315	346	1.97	12	.06	2.09	.52	2.94	.07	16	5
34	10	.29	49	127	162	.37	23	.02	.28	.02	.25	.03	2	0
44	5	.91	30	325	102	1.3	41	.04	.56	.11	10.9	.37	2	1
56	7	1.18	38	417	130	1.66	46	.05	.71	.13	13.3	.43	3	1
78	97	.43	35	506	63	.58	17	.17	.57	.11	6.52	.39	21	2
26	19	2.02	45	300	513	1.31	33	.05	1.74	.17	18	.16	8	0
46	17	2.36	42	365	521	1.19	26	.05	.82	.11	20.5	.54	6	0
43	9	1.15	56	285	44	.68	740	.27	1.13	.28	9.77	.51	2	0
112	11	3.45	21	275	67	7.57	0	.08	.26	.27	3.54	.32	10	0
120	15	4.16	26	297	80	11.6	0	.09	.16	.32	3.02	.33	7	0

Table A-1

Food Composition   (Computer code number is for West Diet Analysis program)   (For purposes of calculations, use "0" for t, <1, <.1, <.01, etc.)

Computer Code Number	Food Description	Measure	Wt (g)	H₂O (%)	Ener (cal)	Prot (g)	Carb (g)	Dietary Fiber (g)	Fat (g)	Fat Breakdown (g) Sat	Mono	Poly
	MEATS: BEEF, LAMB, PORK and others—Continued											
	Relatively lean, like choice round:											
595	Lean and fat, pce 4⅛ x 2½ x ¾"	4 oz	113	52	311	32	0	0	19	8.5	9.7	.8
596	Lean only	4 oz	113	57	249	36	0	0	11	4.8	5.4	.5
	Ground beef, broiled, patty 3 x ⅝":											
597	Extra lean, about 16% fat	4 oz	113	54	299	32	0	0	18	8	9	.8
598	Lean, 21% fat	4 oz	113	53	316	32	0	0	20	8.9	10.1	.8
	Roasts, oven cooked, no added liquid:											
	Relatively fat, prime rib:											
601	Lean and fat, piece 4⅛ x 2¼ x ½"	4 oz	113	46	425	25	0	0	35	15.8	17.9	1.5
602	Lean only	4 oz	113	58	271	31	0	0	16	7	7.9	.7
	Relatively lean, choice round:											
603	Lean and fat, piece 2½ x 2½ x ¾"	4 oz	113	59	272	30	0	0	16	7.1	8.1	.7
604	Lean only	4 oz	113	65	198	33	0	0	6	2.9	3.3	.3
1701	Steak, rib, broiled, lean	4 oz	113	58	250	32	0	0	13	5.7	6.4	.5
	Steak, broiled, relatively lean,											
606	choice sirloin, lean only	4 oz	113	62	228	34	0	0	9	4	4.6	.4
	Steak, broiled, relatively fat,											
	choice T-bone:											
1063	Lean and fat	4 oz	113	52	349	26	0	0	26	11.8	13.3	1.1
1064	Lean only	4 oz	113	61	232	30	0	0	11	5.1	5.8	.5
	Variety meats:											
1086	Brains, panfried	4 oz	113	71	221	14	0	0	18	6.7	7	3.9
599	Heart, simmered	4 oz	113	64	198	32	<1	0	6	2.9	1.5	1.5
600	Liver, fried	4 oz	113	56	245	30	9	0	9	3.1	1.8	1.9
1062	Tongue, cooked	4 oz	113	56	320	25	<1	0	23	10.1	10.7	.9
607	Beef, canned, corned	4 oz	113	58	283	31	0	0	17	7.5	8.5	.7
608	Beef, dried, cured	1 oz	28	56	46	8	<1	0	1	.5	.5	.1
	LAMB, domestic, cooked:											
	Chop, arm, braised (5.6 oz raw w/bone):											
609	Lean and fat	1 ea	70	44	242	21	0	0	17	7.8	7.3	1.4
610	Lean only	1 ea	55	49	153	19	0	0	8	3.6	3.4	.6
	Chop, loin, broiled (4.2 oz raw w/bone):											
611	Lean and fat	1 ea	64	52	202	16	0	0	15	6.8	6.4	1.2
612	Lean only	1 ea	46	61	99	14	0	0	4	2.1	1.9	.4
1067	Cutlet, avg of lean cuts, cooked	4 oz	113	54	330	28	0	0	23	10.9	10.2	1.9
	Leg, roasted, 3 oz = 4⅛ x 2¼ x ½":											
613	Lean and fat	4 oz	113	57	292	29	0	0	19	8.7	8.1	1.6
614	Lean only	4 oz	113	64	216	32	0	0	9	4.1	3.8	.7
615	Rib, roasted, lean and fat	4 oz	113	48	406	24	0	0	34	15.7	14.7	2.8
616	Rib, roasted, lean only	4 oz	113	60	262	30	0	0	15	7	6.6	1.2
1065	Shoulder, roasted, lean and fat	4 oz	113	56	312	25	0	0	23	10.5	9.8	1.9
1066	Shoulder, roasted, lean only	4 oz	113	63	231	28	0	0	12	5.7	5.3	1.1
	Variety meats:											
1069	Brains, panfried	4 oz	113	76	164	14	0	0	11	4.4	3.7	1.9
1068	Heart, braised	4 oz	113	64	209	28	2	0	9	3.9	2.7	1.1
1070	Sweetbreads, cooked	4 oz	113	60	264	26	0	0	17	8.1	6.5	1.4
1071	Tongue, cooked	4 oz	113	58	311	24	0	0	23	8.8	11.6	1.4
	PORK, cured, cooked (see also Sausages and Lunch Meats)											
617	Bacon, medium slices	3 pce	19	13	109	6	<1	0	9	3.3	4.5	1.1
1087	Breakfast strips, cooked	2 pce	23	27	106	7	<1	0	8	2.9	3.8	1.3
618	Canadian-style bacon	2 pce	47	62	87	11	1	0	4	1.3	1.9	.4
	Ham, roasted:											
619	Lean and fat, 2 pces 4⅛ x 2¼ x ¼"	4 oz	113	64	201	25	0	0	10	3.4	5	1.7
620	Lean only	4 oz	113	68	164	24	2	0	6	2.1	3	.6
621	Ham, canned, roasted, 8% fat	4 oz	113	69	154	24	1	0	6	1.8	2.8	.5
	PORK, fresh, cooked:											
	Chops, loin (cut 3 per lb with bone):											
1291	Braised, lean and fat	1 ea	89	58	213	24	0	0	12	4.5	5.4	1
1292	Lean only	1 ea	80	61	163	23	0	0	7	2.7	3.3	.6
622	Broiled, lean and fat	1 ea	82	58	197	23	0	0	11	3.9	4.8	.8
623	Broiled, lean only	1 ea	74	61	149	22	0	0	6	2.2	2.7	.4

PAGE KEY: A–4 = Beverages A–6 = Dairy A–10 = Eggs A–10 = Fat/Oil A–14 = Fruit A–20 = Bakery A–28 = Grain A–32 = Fish A–34 = Meats **A–37**
A–38 = Poultry A–40 = Sausage A–42 = Mixed/Fast A–46 = Nuts/Seeds A–50 = Sweets A–52 = Vegetables/Legumes A–62 = Vegetarian Foods
A–64 = Misc A–66 = Soups/Sauces A–68 = Fast A–84 = Convenience A–88 = Baby foods

A

Chol (mg)	Calc (mg)	Iron (mg)	Magn (mg)	Pota (mg)	Sodi (mg)	Zinc (mg)	VT-A (RE)	Thia (mg)	VT-E (a-TE)	Ribo (mg)	Niac (mg)	V-B6 (mg)	Fola (μg)	VT-C (mg)
108	7	3.53	25	319	56	5.55	0	.08	.21	.27	4.21	.37	11	0
108	6	3.91	28	348	58	6.19	0	.08	.2	.29	4.61	.41	12	0
112	10	3.13	28	417	93	7.27	0	.08	.2	.36	6.61	.36	12	0
114	14	2.77	27	394	101	7.01	0	.07	.23	.27	6.75	.34	12	0
96	12	2.61	21	334	71	5.92	0	.08	.27	.19	3.8	.26	8	0
91	11	2.95	28	425	84	7.84	0	.09	.14	.24	4.64	.34	9	0
81	7	2.07	27	406	67	4.87	0	.09	.23	.18	3.92	.4	7	0
78	6	2.2	30	446	70	5.36	0	.1	.12	.19	4.24	.43	8	0
90	15	2.9	30	445	78	7.9	0	.11	.16	.25	5.42	.45	9	0
101	12	3.8	36	455	75	7.37	0	.15	.16	.33	4.84	.51	11	0
76	9	3.06	26	363	72	5.03	0	.1	.24	.24	4.46	.37	8	0
67	7	3.58	32	427	80	6	0	.12	.16	.28	5.23	.44	9	0
2254	10	2.51	17	400	179	1.53	0	.15	2.37	.29	4.27	.44	7	4
218	7	8.49	28	263	71	3.54	0	.16	.81	1.74	4.6	.24	2	2
545	12	7.1	26	411	120	6.16	12123	.24	.72	4.68	16.3	1.62	249	26
121	8	3.83	19	203	68	5.42	0	.03	.4	.4	2.43	.18	6	1
97	14	2.35	16	154	1136	4.03	0	.02	.17	.17	2.75	.15	10	0
12	2	1.26	9	124	972	1.47	0	.02	.04	.06	1.53	.1	3	0
84	17	1.67	18	214	50	4.26	0	.05	.1	.17	4.66	.08	13	0
67	14	1.49	16	186	42	4.02	0	.04	.1	.15	3.48	.07	12	0
64	13	1.16	15	209	49	2.23	0	.06	.08	.16	4.54	.08	11	0
44	9	.92	13	173	39	1.9	0	.05	.07	.13	3.15	.07	11	0
110	12	2.26	25	340	77	4.67	0	.12	.15	.32	7.48	.16	19	0
105	12	2.24	27	354	75	4.97	0	.11	.17	.3	7.45	.17	23	0
101	9	2.4	29	382	77	5.58	0	.12	.2	.33	7.16	.19	26	0
110	25	1.81	23	306	82	3.94	0	.1	.11	.24	7.63	.12	17	0
99	24	2	26	356	91	5.05	0	.1	.17	.26	6.96	.17	25	0
104	23	2.23	26	284	75	5.91	0	.1	.16	.27	6.95	.15	24	0
98	21	2.41	28	299	77	6.83	0	.1	.2	.29	6.51	.17	28	0
2308	14	1.9	16	232	151	1.54	0	.12	1.73	.27	2.79	.12	6	14
281	16	6.24	27	212	71	4.16	0	.19	.79	1.34	4.93	.34	2	8
452	14	2.4	21	329	59	3.03	0	.02	.78	.24	2.89	.06	15	23
214	11	2.97	18	179	76	3.38	0	.09	.36	.47	4.17	.19	3	8
16	2	.31	5	92	303	.62	0	.13	.1	.05	1.39	.05	1	0
24	3	.45	6	107	483	.85	0	.17	.08	.08	1.75	.08	1	0
27	5	.38	10	183	727	.8	0	.39	.15	.09	3.25	.21	2	0
67	9	1.51	25	462	1695	2.79	0	.82	.45	.37	6.95	.35	3	0
60	9	1.67	16	324	1359	3.25	0	.85	.29	.23	4.54	.45	3	0
34	7	1.04	24	393	1282	2.52	0	1.18	.29	.28	5.53	.51	6	0
71	19	.95	17	333	43	2.12	2	.56	.3	.23	3.93	.33	3	1
63	14	.9	16	310	40	1.98	2	.53	.3	.21	3.67	.31	3	<1
67	27	.66	20	294	48	1.85	2	.88	.27	.24	4.3	.35	5	<1
61	23	.63	20	278	44	1.76	2	.85	.31	.23	4.1	.35	4	<1

Table A-1

Food Composition   (Computer code number is for West Diet Analysis program)   (For purposes of calculations, use "0" for t, <1, <.1, <.01, etc.)

Computer Code Number	Food Description	Measure	Wt (g)	H₂O (%)	Ener (cal)	Prot (g)	Carb (g)	Dietary Fiber (g)	Fat (g)	Fat Breakdown (g) Sat	Mono	Poly
	MEATS: BEEF, LAMB, PORK and others—Continued											
624	Panfried, lean and fat	1 ea	78	53	216	23	0	0	13	4.7	5.5	1.5
625	Panfried, lean only	1 ea	63	59	152	16	0	0	10	3.2	3.9	1.2
626	Leg, roasted, lean and fat	4 oz	113	55	308	30	0	0	20	7.3	8.9	1.9
627	Leg, roasted, lean only	4 oz	113	61	233	35	0	0	9	3.2	4.3	.9
628	Rib, roasted, lean and fat	4 oz	113	56	288	31	0	0	17	6.7	7.9	1.4
629	Rib, roasted, lean only	4 oz	113	59	252	32	0	0	13	4.9	5.9	1
630	Shoulder, braised, lean and fat	4 oz	113	48	372	32	0	0	26	9.6	11.8	2.6
631	Shoulder, braised, lean only	4 oz	113	54	280	36	0	0	14	4.7	6.5	1.3
1088	Spareribs, cooked, yield from 1 lb raw with bone	4 oz	113	40	449	33	0	0	34	12.5	15.3	3.1
1095	Rabbit, roasted (1 cup meat = 140 g)	4 oz	113	61	223	33	0	0	9	4	2	3
	VEAL, cooked:											
632	Cutlet, braised or broiled, 4⅛ x 2¼ x ½"	4 oz	113	52	321	34	0	0	19	7.6	7.6	1.3
633	Rib roasted, lean, 2 pieces 4⅛ x 2¼ x ¼"	4 oz	113	60	258	27	0	0	16	6.1	6.1	1.1
634	Liver, panfried	4 oz	113	67	186	24	3	0	8	3.3	1.9	2.2
1096	Venison (deer meat), roasted	4 oz	113	65	179	34	0	0	4	1.4	1	.7
	MEATS: POULTRY and POULTRY PRODUCTS											
	CHICKEN, cooked:											
	Fried, batter dipped:											
635	Breast	1 ea	280	52	728	69	25	1	37	9.9	15.3	8.6
636	Drumstick	1 ea	72	53	193	16	6	<1	11	3	4.7	2.7
637	Thigh	1 ea	86	51	238	19	8	<1	14	3.8	5.9	3.3
638	Wing	1 ea	49	46	159	10	5	<1	11	2.9	4.5	2.5
	Fried, flour coated:											
639	Breast	1 ea	196	57	435	62	3	<1	17	4.9	7.1	3.8
1212	Breast, without skin	1 ea	86	60	161	29	<1	<1	4	1.1	1.5	.9
640	Drumstick	1 ea	49	57	120	13	1	<1	7	1.8	2.7	1.6
641	Thigh	1 ea	62	54	162	17	2	<1	9	2.5	3.7	2.1
1099	Thigh, without skin	1 ea	52	59	113	15	1	<1	5	1.4	2	1.3
642	Wing	1 ea	32	49	103	8	1	<1	7	1.9	2.9	1.6
	Roasted:											
643	All types of meat	1 c	140	64	266	40	0	0	10	2.9	3.8	2.4
644	Dark meat	1 c	140	63	287	38	0	0	14	3.7	5.2	3.2
645	Light meat	1 c	140	65	242	43	0	0	6	1.8	2.2	1.4
646	Breast, without skin	1 ea	172	65	284	53	0	0	6	1.8	2.2	1.3
647	Drumstick, without skin	1 ea	44	67	76	12	0	0	2	.7	.8	.6
1703	Leg, without skin	1 ea	95	65	181	26	0	0	8	2.2	2.9	1.9
648	Thigh	1 ea	62	59	153	16	0	0	10	2.7	3.9	2.1
1100	Thigh, without skin	1 ea	52	63	109	13	0	0	6	1.6	2.2	1.3
649	Stewed, all types	1 c	140	67	248	38	0	0	9	2.6	3.5	2.2
656	Canned, boneless chicken	4 oz	113	69	186	25	0	0	9	2.5	3.6	2
1102	Gizzards, simmered	1 c	145	67	222	39	2	0	5	1.5	1.3	1.5
1101	Hearts, simmered	1 c	145	65	268	38	<1	0	11	3.3	2.9	3.3
2300	Liver, simmered: Ounce	3 oz	85	68	133	21	1	0	5	1.6	1.1	.8
1098	Liver, simmered: Piece = 20 g	6 ea	120	68	188	29	1	0	7	2.2	1.6	1.1
	DUCK, roasted:											
1293	Meat with skin, about 2.7 cups	½ ea	382	52	1287	73	0	0	108	36.9	49.3	13.9
651	Meat only, about 1.5 cups	½ ea	221	64	444	52	0	0	25	9.2	8.2	3.2
	GOOSE, domesticated, roasted:											
1294	Meat only, about 4.2 cups	½ ea	591	57	1406	173	0	0	75	23.6	40.2	10.9
1295	Meat with skin, about 5.5 cups	½ ea	774	52	2360	195	0	0	170	53.2	80.5	24.8
	TURKEY:											
	Roasted, meat only:											
652	Dark meat	4 oz	113	63	211	33	0	0	8	2.7	1.8	2.5
653	Light meat	4 oz	113	66	177	34	0	0	4	1.2	.6	1
654	All types, chopped or diced	1 c	140	65	238	42	0	0	7	2.3	1.5	2
1103	Ground, cooked	4 oz	113	59	266	31	0	0	15	4.1	5.5	3.6
1106	Gizzard, cooked	2 ea	134	65	218	39	1	0	5	1.5	1	1.5
1107	Heart, cooked	4 ea	64	64	113	17	1	0	4	1.1	.8	1.1
1108	Liver, cooked	1 ea	75	66	127	18	3	0	4	1.4	1.1	.8

PAGE KEY: A–4 = Beverages A–6 = Dairy A–10 = Eggs A–10 = Fat/Oil A–14 = Fruit A–20 = Bakery A–28 = Grain A–32 = Fish A–34 = Meats A–39
A–38 = Poultry A–40 = Sausage A–42 = Mixed/Fast A–46 = Nuts/Seeds A–50 = Sweets A–52 = Vegetables/Legumes A–62 = Vegetarian Foods
A–64 = Misc A–66 = Soups/Sauces A–68 = Fast A–84 = Convenience A–88 = Baby foods

A

Chol (mg)	Calc (mg)	Iron (mg)	Magn (mg)	Pota (mg)	Sodi (mg)	Zinc (mg)	VT-A (RE)	Thia (mg)	VT-E (a-TE)	Ribo (mg)	Niac (mg)	V-B6 (mg)	Fola (μg)	VT-C (mg)
72	21	.71	23	332	62	1.8	2	.89	.32	.24	4.37	.37	5	1
52	14	.67	16	230	49	2.44	1	.46	.3	.23	2.8	.26	3	<1
106	16	1.14	25	398	68	3.34	3	.72	.34	.35	5.16	.45	11	<1
108	8	1.29	33	442	73	3.4	3	.91	.46	.4	5.56	.38	3	<1
82	32	1.06	24	476	52	2.33	2	.82	.41	.34	6.92	.37	3	<1
80	29	1.11	25	494	53	2.41	2	.86	.55	.36	7.25	.38	3	<1
123	20	1.82	21	417	99	4.72	3	.61	.5	.35	5.89	.4	5	<1
129	9	2.2	25	458	115	5.62	3	.68	.58	.41	6.71	.46	6	<1
137	53	2.09	27	362	105	5.2	3	.46	.52	.43	6.19	.4	5	0
93	21	2.57	24	433	53	2.57	0	.1	.96	.24	9.53	.53	12	0
133	32	1.23	27	316	90	4.1	0	.04	.45	.34	10.2	.29	16	0
124	12	1.1	25	333	104	4.62	0	.06	.4	.3	7.89	.28	15	0
634	8	2.96	21	232	60	10.8	9095	.15	.42	2.19	9.58	.55	858	35
127	8	5.05	27	379	61	3.11	0	.2	.28	.68	7.58	.42	5	0
238	56	3.5	67	563	770	2.66	56	.32	2.97	.41	29.4	1.2	42	0
62	12	.97	14	134	194	1.68	19	.08	.88	.15	3.67	.19	13	0
80	15	1.25	18	165	248	1.75	25	.1	1.05	.19	4.92	.22	16	0
39	10	.63	8	68	157	.68	17	.05	.52	.07	2.58	.15	9	0
174	31	2.33	59	508	149	2.16	29	.16	1.12	.26	26.9	1.14	12	0
78	14	.98	27	237	68	.93	6	.07	.36	.11	12.7	.55	3	0
44	6	.66	11	112	44	1.42	12	.04	.41	.11	2.96	.17	5	0
60	9	.92	15	147	55	1.56	18	.06	.52	.15	4.31	.2	7	0
53	7	.76	13	135	49	1.45	11	.05	.3	.13	3.7	.2	5	0
26	5	.4	6	57	25	.56	12	.02	.18	.04	2.14	.13	2	0
125	21	1.69	35	340	120	2.94	22	.1	.58	.25	12.8	.66	8	0
130	21	1.86	32	336	130	3.92	31	.1	.81	.32	9.17	.5	11	0
119	21	1.48	38	346	108	1.72	13	.09	.37	.16	17.4	.84	6	0
146	26	1.79	50	440	127	1.72	10	.12	.66	.2	23.6	1.03	7	0
41	5	.57	11	108	42	1.4	8	.03	.25	.1	2.68	.17	4	0
89	11	1.24	23	230	86	2.72	18	.07	.55	.22	6	.35	8	0
58	7	.83	14	138	52	1.46	30	.04	.35	.13	3.95	.19	4	0
49	6	.68	12	124	46	1.34	10	.04	.3	.12	3.4	.18	4	0
116	20	1.64	29	252	98	2.79	21	.07	.42	.23	8.57	.36	8	0
70	16	1.79	14	156	568	1.59	38	.02	.24	.15	7.15	.4	5	2
281	14	6.02	29	260	97	6.35	81	.04	2.29	.35	5.77	.17	77	2
351	28	13.1	29	191	70	10.6	13	.1	2.32	1.07	4.06	.46	116	3
536	12	7.2	18	119	43	3.69	4176	.13	1.45	1.49	3.78	.49	655	13
757	17	10.2	25	168	61	5.21	5895	.18	2.04	2.1	5.34	.7	924	19
321	42	10.3	61	779	225	7.11	241	.66	2.5	1.03	18.5	.69	23	0
197	26	5.97	44	557	144	5.75	51	.57	1.55	1.04	11.3	.55	22	0
567	83	17	148	2293	449	18.7	71	.54	9.16	2.3	24.1	2.78	71	0
704	101	21.9	170	2546	542	20.3	163	.6	13.5	2.5	32.3	2.86	15	0
96	36	2.63	27	328	89	5.04	0	.07	.94	.28	4.12	.41	10	0
78	21	1.53	32	345	72	2.31	0	.07	.12	.15	7.73	.61	7	0
106	35	2.49	36	417	98	4.34	0	.09	.59	.25	7.62	.64	10	0
115	28	2.18	27	305	121	3.23	0	.06	.45	.19	5.45	.44	8	0
311	20	7.29	25	283	72	5.57	74	.04	.27	.44	4.11	.16	70	2
145	8	4.41	14	117	35	3.37	5	.04	.13	.56	2.08	.2	51	1
470	8	5.85	11	146	48	2.32	2805	.04	2.41	1.07	4.46	.39	500	1

Table A–1

Food Composition

(Computer code number is for West Diet Analysis program)   (For purposes of calculations, use "0" for t, <1, <.1, <.01, etc.)

Computer Code Number	Food Description	Measure	Wt (g)	H₂O (%)	Ener (cal)	Prot (g)	Carb (g)	Dietary Fiber (g)	Fat (g)	Fat Breakdown (g)		
										Sat	Mono	Poly
	MEATS: POULTRY and POULTRY PRODUCTS—Continued											
	POULTRY FOOD PRODUCTS (see also items in Sausages & Lunchmeats section):											
1567	Chicken patty, breaded, cooked	1 ea	75	49	213	12	11	<1	13	4.1	6.4	1.6
659	Turkey and gravy, frozen package	3 oz	85	85	57	5	4	<1	2	.8	.8	.4
	Turkey breast, Louis Rich:											
1104	Barbecued	2 oz	56	72	58	12	2	0	<1	.2	.2	.1
1943	Hickory smoked	1 pce	80		80	16	2	0	1	0		
1947	Honey roasted	1 pce	80		80	16	3	0	1	.5		
1945	Oven roasted	1 pce	80		70	16		0	1	0		
661	Turkey patty, breaded, fried	2 oz	57	50	161	8	9	<1	10	2.7	4.3	2.7
662	Turkey, frozen, roasted, seasoned	4 oz	113	68	175	24	3	0	7	2.1	1.4	1.9
1704	Turkey roll, light meat	1 pce	28	72	41	5	<1	0	2	.6	.7	.5
	MEATS: SAUSAGES and LUNCHMEATS (see also Poultry Food Products)											
1072	Beerwurst/beer salami, beef	1 oz	28	53	92	3	<1	0	8	3.6	3.9	.3
1074	Beerwurst/beer salami, pork	1 oz	28	61	67	4	1	0	5	1.8	2.5	.7
1075	Berliner sausage	1 oz	28	61	64	4	1	0	5	1.7	2.2	.4
	Bologna:											
1297	Beef	1 pce	23	55	72	3	<1	0	7	2.8	3.2	.3
2115	Beef, light, Oscar Mayer	1 pce	28	65	56	3	2	0	4	1.6	2	.1
663	Beef & pork	1 pce	28	54	88	3	1	0	8	3	3.7	.7
2155	Healthy Favorites	1 pce	23		22	3	1	0	<1	0		
1298	Pork	1 pce	23	61	57	4	<1	0	5	1.6	2.2	.5
2114	Regular, light, Oscar Mayer	1 pce	28	65	56	3	2	0	4	1.6	2	.4
664	Turkey	1 pce	28	65	56	4	<1	0	4	1.4	1.3	1.2
1970	Turkey, Louis Rich	1 pce	28	67	57	3	<1	0	5	1.5	1.8	1.3
665	Braunschweiger sausage	2 pce	57	48	205	8	2	0	18	6.2	8.5	2.1
1073	Bratwurst, link	1 ea	70	51	226	10	2	0	19	6.9	9.3	2
666	Brown & serve sausage links, cooked	2 ea	26	45	102	4	1	0	10	3.4	4.4	1
1089	Cheesefurter/cheese smokie	2 ea	86	52	281	12	1	0	25	9	11.8	2.6
2157	Chicken breast, Healthy Favorites	4 pce	52		40	9	1	0	0	0	0	0
1556	Chorizo, pork & beef	1 ea	60	32	273	15	1	0	23	8.6	11	2.1
1090	Corned beef loaf, jellied	1 pce	28	69	43	6	0	0	2	.7	.7	.1
	Frankfurters:											
1077	Beef, large link, 8/package	1 ea	57	55	180	7	1	0	16	6.9	7.9	.8
1078	Beef and pork, large link, 8/package	1 ea	57	54	182	6	1	0	17	6.2	8	1.6
667	Beef and pork, small link, 10/pkg	1 ea	45	54	144	5	1	0	13	4.9	6.3	1.2
668	Turkey frankfurter, 10/package	1 ea	45	63	102	6	1	0	8	2.7	2.5	2.2
1968	Turkey/chicken frank 8/pkg	1 ea	43		80	6	1	0	6	2		
	Ham:											
669	Ham lunchmeat, canned, 3 x 2 x ½"	1 pce	21	52	70	3	<1	0	6	2.3	3	.7
670	Chopped ham, packaged	2 pce	42	64	96	7	0	0	7	2.4	3.4	.9
2156	Honey ham, Healthy Favorites	4 pce	52	73	55	9	2	0	1	.4	.8	.1
2113	Oscar Mayer lower sodium ham	1 pce	21	73	23	3	1	0	1	.3	.4	.1
673	Turkey ham lunchmeat	2 pce	57	71	73	11	<1	0	3	1	.7	.9
1091	Kielbasa sausage	1 pce	26	54	81	3	1	0	7	2.6	3.4	.8
1092	Knockwurst sausage, link	1 ea	68	55	209	8	1	0	19	6.9	8.7	2
1093	Mortadella lunchmeat	2 pce	30	52	93	5	1	0	8	2.8	3.4	.9
1097	Olive loaf lunchmeat	2 pce	57	58	134	7	5	<1	9	3.3	4.5	1.1
1952	Turkey breast, fat free	1 pce	28	77	22	4	1	0	<1	.1	.1	t
1080	Turkey pastrami	2 pce	57	71	80	10	1	0	4	1	1.2	.9
1969	Turkey salami	1 pce	28	72	41	4	<1	0	3	.9	1	.8
1081	Pepperoni sausage	2 pce	11	27	55	2	<1	0	5	1.8	2.3	.5
1094	Pickle & pimento loaf	2 pce	57	57	149	7	3	<1	12	4.5	5.5	1.5
1082	Polish sausage	1 oz	28	53	91	4	<1	0	8	2.9	3.8	.9
674	Pork sausage, cooked, link, small	2 ea	26	45	96	5	<1	0	8	2.8	4.1	.8
1079	Pork sausage, cooked, patty	4 oz	113	45	417	22	1	0	35	12.1	17.7	3.3
675	Salami, pork and beef	2 pce	57	60	143	8	1	0	11	4.6	5.2	1.1
677	Salami, pork and beef, dry	3 pce	30	35	125	7	1	0	10	3.7	5.1	1
676	Salami, turkey	2 pce	57	66	112	9	<1	0	8	2.3	2.6	2
	Sandwich spreads:											
1300	Ham salad spread	2 tbs	30	63	65	3	3	0	5	1.5	2.2	.8
678	Pork and beef	2 tbs	30	60	70	2	4	<1	5	1.8	2.3	.8
1296	Chicken/turkey	2 tbs	26	66	52	3	2	0	4	.9	.8	1.6

PAGE KEY: A–4 = Beverages A–6 = Dairy A–10 = Eggs A–10 = Fat/Oil A–14 = Fruit A–20 = Bakery A–28 = Grain A–32 = Fish A–34 = Meats
A–38 = Poultry A–40 = Sausage A–42 = Mixed/Fast A–46 = Nuts/Seeds A–50 = Sweets A–52 = Vegetables/Legumes A–62 = Vegetarian Foods
A–64 = Misc A–66 = Soups/Sauces A–68 = Fast A–84 = Convenience A–88 = Baby foods

Chol (mg)	Calc (mg)	Iron (mg)	Magn (mg)	Pota (mg)	Sodi (mg)	Zinc (mg)	VT-A (RE)	Thia (mg)	VT-E (a-TE)	Ribo (mg)	Niac (mg)	V-B6 (mg)	Fola (µg)	VT-C (mg)
45	12	.94	15	185	399	.78	22	.07	1.46	.1	5.04	.23	8	<1
15	12	.79	7	52	471	.59	11	.02	.3	.11	1.53	.08	3	0
25	14	.62	16	175	599	.59	0	.02		.06	5.35	.22	2	0
35	0	.72			1060		0							0
35	0	.72			940		0							0
35	0				910		0							0
35	8	1.25	9	157	456	.82	6	.06	1.36	.11	1.31	.11	16	0
60	6	1.84	25	337	768	2.87	0	.05	.43	.18	7.09	.3	6	0
12	11	.36	4	70	137	.44	0	.02	.04	.06	1.96	.09	1	0
17	3	.42	3	49	288	.68	0	.02	.05	.03	.95	.05	1	0
16	2	.21	4	71	347	.48	0	.15	.06	.05	.91	.1	1	0
13	3	.32	4	79	363	.69	0	.11	.06	.06	.87	.06	1	0
13	3	.38	3	36	226	.5	0	.01	.04	.02	.55	.03	1	0
13	4	.34	4	44	314	.53	0							0
15	3	.42	3	50	285	.54	0	.05	.06	.04	.72	.05	1	0
7		.18			255									
14	3	.18	3	65	272	.47	0	.12	.06	.04	.9	.06	1	0
15	14	.39	5	46	312	.45	0							0
28	23	.43	4	56	246	.49	0	.01	.15	.05	.99	.06	2	0
22	34	.45	5	51	242	.57	0	.01		.05	1.08	.05		0
89	5	5.34	6	113	652	1.6	2405	.14	.2	.87	4.77	.19	25	0
44	34	.72	11	197	778	1.47	0	.17	.19	.16	2.31	.09	3	0
16	2	.62	4	70	248	.3	0	.21	.06	.09	.96	.06	1	0
58	50	.93	11	177	931	1.94	33	.21	.27	.14	2.49	.11	3	0
25		.72			620		0							0
53	5	.95	11	239	741	2.05	0	.38	.13	.18	3.08	.32	1	0
13	3	.57	3	28	267	1.15	0	0	.05	.03	.49	.03	2	0
35	11	.81	2	95	585	1.24	0	.03	.11	.06	1.38	.07	2	0
28	6	.66	6	95	638	1.05	0	.11	.14	.07	1.5	.07	2	0
22	5	.52	4	75	504	.83	0	.09	.11	.05	1.18	.06	2	0
48	48	.83	6	81	642	1.4	0	.02	.28	.08	1.86	.1	4	0
40	60	1.08			480		0							0
13	1	.15	2	45	271	.31	0	.08	.05	.04	.66	.04	1	<1
21	3	.35	7	134	576	.81	0	.26	.11	.09	1.63	.15	<1	0
24	6	.7	18	144	635	1.02	0							0
9	1	.3	5	197	174	.42	0							0
32	6	1.57	9	185	568	1.68	0	.03	.36	.14	2.01	.14	3	0
17	11	.38	4	70	280	.52	0	.06	.06	.06	.75	.05	1	0
39	7	.62	7	135	687	1.13	0	.23	.39	.09	1.86	.12	1	0
17	5	.42	3	49	374	.63	0	.04	.07	.05	.8	.04	1	0
22	62	.31	11	169	846	.79	34	.17	.14	.15	1.05	.13	1	0
9	3	.34	8	59	387	.24	0							0
31	5	.95	8	148	596	1.23	0	.03	.12	.14	2.01	.15	3	0
21	11	.35	6	61	281	.65	0							0
9	1	.15	2	38	224	.27	0	.03	.02	.03	.55	.03	<1	0
21	54	.58	10	194	792	.8	4	.17	.14	.14	1.17	.11	3	0
20	3	.4	4	66	245	.54	0	.14	.06	.04	.96	.05	1	<1
22	8	.33	4	94	336	.65	0	.19	.07	.07	1.18	.09	1	<1
94	36	1.42	19	408	1462	2.84	0	.84	.29	.29	5.11	.37	2	2
37	7	1.52	9	113	607	1.22	0	.14	.12	.21	2.02	.12	1	0
24	2	.45	5	113	558	.97	0	.18	.08	.09	1.46	.15	1	0
47	11	.92	9	139	572	1.03	0	.04	.32	.1	2.01	.14	2	0
11	2	.18	3	45	274	.33	0	.13	.52	.04	.63	.04	<1	0
11	4	.24	2	33	304	.31	3	.05	.52	.04	.52	.04	1	0
8	3	.16	3	48	98	.27	11	.01	.57	.02	.43	.03	1	<1

Table A-1

Food Composition

(Computer code number is for West Diet Analysis program)   (For purposes of calculations, use "0" for t, <1, <.1, <.01, etc.)

Computer Code Number	Food Description	Measure	Wt (g)	H₂O (%)	Ener (cal)	Prot (g)	Carb (g)	Dietary Fiber (g)	Fat (g)	Sat	Mono	Poly
	MEATS: SAUSAGES and LUNCHMEATS—Continued											
1084	Smoked link sausage, beef and pork	1 ea	68	52	228	9	1	0	21	7.2	9.7	2.2
1083	Smoked link sausage, pork	1 ea	68	39	265	15	1	0	22	7.7	9.9	2.6
1085	Summer sausage	2 pce	46	51	154	7	<1	0	14	5.5	6	.6
1076	Turkey breakfast sausage	1 pce	28	60	64	6	0	0	5	1.6	1.8	1.2
679	Vienna sausage, canned	2 ea	32	60	89	3	1	0	8	3	4	.5
	MIXED DISHES and FAST FOODS											
	MIXED DISHES:											
1445	Almond Chicken	1 c	242	77	275	20	18	4	14	2	5.3	5.8
1981	Baked beans, fat free, honey	½ c	120	73	110	7	24	7	0	0	0	0
1454	Bean cake	1 ea	32	23	130	2	16	1	7	1	2.9	2.6
680	Beef stew w/ vegetables, homemade	1 c	245	82	218	16	15	2	10	4.9	4.5	.5
1109	Beef stew w/ vegetables, canned	1 c	245	82	194	14	17	2	8	2.4	3.1	.3
1116	Beef, macaroni, tomato sauce casserole	1 c	226	76	255	16	26	2	10	3.8	4.1	.5
2295	Beef fajita	1 ea	223	63	409	17	46	4	17	5.1	7.6	3.9
1265	Beef flauta	1 ea	113	49	360	16	13	2	27	4.9	11.6	9.1
681	Beef pot pie, homemade	1 pce	210	55	517	21	39	3	30	8.4	14.7	7.3
1898	Broccoli, batter fried	1 c	85	74	123	3	9	2	9	1.3	2.2	4.9
1462	Buffalo wings/spicy chicken wings	2 pce	32	53	98	8	<1	<1	7	1.8	2.8	1.6
1675	Carrot raisin salad	½ c	88	58	204	1	21	2	14	2	3.9	7.3
2248	Cheeseburger deluxe	1 ea	219	52	563	28	38		33	15	12.6	2
682	Chicken à la king, homemade	1 c	245	68	468	27	12	1	34	12.7	14.3	6.2
683	Chicken & noodles, homemade	1 c	240	71	367	22	26	2	18	5.9	7.1	3.5
684	Chicken chow mein, canned	1 c	250	89	95	6	18	2	1	0	.1	.8
685	Chicken chow mein, homemade	1 c	250	78	255	31	10	1	10	2.4	4.3	3.1
1266	Chicken fajita	1 ea	223	61	405	22	50	4	13	2.4	6	3.5
1264	Chicken flauta	1 ea	113	52	343	14	13	2	27	4.3	11.1	9.6
686	Chicken pot pie, homemade (⅓)	1 pce	232	57	545	23	42	3	31	10.9	14.5	5.8
1672	Chili con carne	½ c	127	77	128	12	11	2	4	1.7	1.7	.3
1112	Chicken salad with celery	½ c	78	53	268	11	1	<1	25	3.1	4.5	15.8
1382	Chicken teriyaki, breast	1 ea	128	67	176	26	7	<1	4	.9	1	.9
687	Chili with beans, canned	1 c	256	75	287	15	30	11	14	6	6	.9
1479	Chinese pastry	1 oz	28	46	67	1	13	<1	1	.2	.4	.8
688	Chop suey with beef & pork	1 c	220	63	425	22	31	3	24	5	8.6	9.3
690	Coleslaw	1 c	132	74	195	2	17	2	15	2.1	3.2	8.5
689	Corn pudding	1 c	250	76	273	11	32	4	13	6.4	4.3	1.8
1110	Corned beef hash, canned	1 c	220	67	398	19	23	1	25	11.9	10.9	.9
1255	Deviled egg (½ egg + filling)	1 ea	31	69	62	4	<1	0	5	1.2	1.7	1.5
	Egg foo yung patty:											
1467	Meatless	1 ea	86	78	113	6	3	1	8	1.9	3.3	2.1
1458	With beef	1 ea	86	74	129	9	3	<1	9	2.2	3.2	2.4
1465	With chicken	1 ea	86	74	130	9	4	<1	9	2.1	3.1	2.5
1602	Egg roll, meatless	1 ea	64	70	102	3	10	1	6	1.2	2.5	1.6
1550	Egg roll, with meat	1 ea	64	66	114	5	9	1	6	1.5	2.7	1.5
1113	Egg salad	1 c	183	57	586	17	3	0	56	10.6	17.4	24.2
691	French toast w/wheat bread, homemade	1 pce	65	54	151	5	16	<1	7	2	3	1.7
1355	Green pepper, stuffed	1 ea	172	75	229	11	20	2	11	5	4.9	.5
1487	Hot & sour soup (Chinese)	1 c	244	88	133	12	5	<1	6	2	2.5	1
2242	Hamburger deluxe	1 ea	110	49	279	13	27		13	4.1	5.3	2.6
1997	Hummous/hummus	¼ c	62	65	106	3	12	3	5	.8	2.2	2
	Lasagna:											
1346	With meat, homemade	1 pce	245	67	382	22	39	3	15	7.7	5	.9
1111	Without meat, homemade	1 pce	218	69	298	15	39	3	9	5.4	2.4	.6
1117	Frozen entree	1 ea	340	75	390	24	42	4	14	6.7	5.5	.8
1606	Lo mein, meatless	1 c	200	82	134	6	27	3	1	.1	.1	.3
1607	Lo mein, with meat	1 c	200	70	285	17	31	2	10	1.9	2.9	4.5
692	Macaroni & cheese, canned	1 c	240	80	228	9	26	1	10	4.2	3.1	1.4
693	Macaroni & cheese, homemade	1 c	200	58	430	17	40	1	22	8.9	8.8	3.6
1115	Macaroni salad, no cheese	1 c	177	60	461	5	28	2	37	4	6	25.5
1120	Meat loaf, beef	1 pce	87	63	182	16	4	<1	11	4.4	4.7	.5
1119	Meat loaf, beef and pork (⅓)	1 pce	87	60	205	15	4	<1	14	5.2	6.3	.9
1303	Moussaka (lamb & eggplant)	1 c	250	82	237	16	13	4	13	4.6	5.4	1.9

PAGE KEY: A–4 = Beverages   A–6 = Dairy   A–10 = Eggs   A–10 = Fat/Oil   A–14 = Fruit   A–20 = Bakery   A–28 = Grain   A–32 = Fish   A–34 = Meats
A–38 = Poultry   A–40 = Sausage   A–42 = Mixed/Fast   A–46 = Nuts/Seeds   A–50 = Sweets   A–52 = Vegetables/Legumes   A–62 = Vegetarian Foods
A–64 = Misc   A–66 = Soups/Sauces   A–68 = Fast   A–84 = Convenience   A–88 = Baby foods

Chol (mg)	Calc (mg)	Iron (mg)	Magn (mg)	Pota (mg)	Sodi (mg)	Zinc (mg)	VT-A (RE)	Thia (mg)	VT-E (a-TE)	Ribo (mg)	Niac (mg)	V-B6 (mg)	Fola (µg)	VT-C (mg)
48	7	.99	8	129	643	1.43	0	.18	.15	.12	2.2	.12	1	0
46	20	.79	13	228	1020	1.92	0	.48	.17	.17	3.08	.24	3	1
34	6	1.17	6	125	571	1.18	0	.07	.1	.15	1.98	.12	1	0
23	5	.51	6	75	188	.96	0	.03	.14	.08	1.4	.08	1	0
17	3	.28	2	32	305	.51	0	.03	.07	.03	.51	.04	1	0
35	81	2	59	551	615	1.54	75	.08	2.64	.19	8.59	.42	31	10
0	40	2.7			135		450							12
0	3	.67	6	57	55	.16	0	.07	1.14	.05	.55	.02	9	0
64	29	2.94	40	613	292	5.29	568	.15	.49	.17	4.66	.28	37	17
34	29	2.21	39	426	1006	4.24	262	.07	.34	.12	2.45	.2	31	7
39	26	2.7	40	522	862	3.14	97	.22	.57	.22	4.31	.29	20	14
26	76	3.69	38	427	850	2.38	52	.46	2.08	.3	4.73	.32	25	29
45	50	2.15	29	292	187	4.18	15	.07	4	.15	2.13	.25	10	14
44	29	3.78	6	334	596	3.17	519	.29	3.78	.29	4.83	.24	29	6
16	67	.94	20	242	62	.38	102	.08	2.1	.13	.75	.11	43	53
26	5	.4	6	59	61	.56	17	.01	.23	.04	2.06	.13	1	<1
10	26	.75	14	317	118	.19	1452	.08	5.03	.05	.64	.22	9	5
88	206	4.66	44	445	1108	4.6	129	.39	1.18	.46	7.38	.28	81	8
186	127	2.45	20	404	760	1.8	272	.1	.98	.42	5.39	.23	11	12
96	26	2.16	26	149	600	1.53	10	.05		.17	4.32	.19	10	0
7	45	1.25	14	418	725	1.3	28	.05	.05	.1	1	.09	12	12
77	57	2.5	28	473	718	2.12	50	.07	.75	.22	4.25	.41	19	10
41	83	3.7	51	532	439	1.77	55	.48	2.04	.37	6.64	.35	41	22
37	52	.97	27	243	189	1.18	21	.05	4.06	.1	3.21	.22	8	14
72	70	3.02	25	343	594	2	735	.32	3.25	.32	4.87	.46	29	5
67	34	2.6	23	347	505	1.79	84	.06	.81	.57	1.24	.16	23	1
48	16	.62	11	138	201	.79	31	.03	6.27	.07	3.28	.34	8	1
80	27	1.75	36	309	1866	1.94	16	.08	.35	.2	8.69	.46	13	3
43	120	8.78	115	934	1336	5.12	87	.12	1.88	.27	.92	.34	58	4
0	8	.51	6	28	3	.18	<1	.05	.25	<.01	.41	.02	1	0
46	39	4.16	54	515	818	3.52	134	.36	1.82	.37	5.63	.44	44	20
7	45	.96	12	236	356	.26	66	.05	5.28	.04	.11	.14	51	11
250	100	1.4	37	403	138	1.25	90	1.03	.52	.32	2.47	.29	63	7
73	29	4.4	36	440	1188	3.3	0	.02	.48	.2	4.62	.43	20	0
121	15	.35	3	36	94	.3	49	.02	.86	.14	.02	.05	13	0
184	31	1.04	12	118	310	.7	86	.04	1.57	.25	.44	.09	30	5
180	26	1.11	11	145	184	1.16	92	.05	1.79	.24	.74	.15	22	3
182	27	.86	12	144	187	.81	95	.05	1.87	.25	.96	.13	22	3
30	12	.76	9	98	306	.25	15	.08	.81	.1	.81	.05	12	3
37	13	.78	10	124	304	.46	14	.16	.78	.13	1.31	.1	9	2
574	74	1.8	13	180	665	1.42	260	.08	8.87	.66	.09	.46	62	0
76	64	1.09	11	86	311	.44	81	.13	.31	.21	1.06	.05	15	<1
34	16	1.77	20	233	201	2.28	44	.15	.75	.1	2.74	.3	17	55
23	29	1.83	27	351	1562	1.17	2	.19	.12	.22	4.58	.15	12	1
26	63	2.63	22	227	504	2.06	9	.23	.82	.2	3.69	.12	52	2
0	31	.97	18	108	151	.68	1	.06	.62	.03	.25	.25	37	5
56	258	3.22	50	461	745	3.25	158	.23	1.15	.33	3.97	.21	19	15
31	252	2.5	44	375	714	1.77	156	.22	1.07	.27	2.49	.17	17	15
55	263	3.44	64	752	823	3.7	248	.29	3.45	.39	5.07	.32	28	41
0	47	2.06	33	389	623	.92	130	.23	.35	.24	2.83	.19	49	12
30	25	2.11	39	246	276	1.63	6	.37	1.51	.24	4.25	.28	41	8
24	199	.96	31	139	730	1.2	73	.12	.14	.24	.96	.02	8	<1
42	362	1.8	37	240	1086	1.2	234	.2	.12	.4	1.8	.05	10	1
27	31	1.56	20	170	352	.53	44	.18	10.3	.1	1.43	.33	20	4
84	29	1.61	14	187	145	3.23	23	.05	.31	.22	2.61	.15	11	1
84	33	1.42	14	213	381	2.68	23	.19	.32	.22	2.68	.17	10	1
97	68	1.79	40	557	432	2.56	105	.15	.81	.31	4.14	.23	45	6

Table A-1

Food Composition   (Computer code number is for West Diet Analysis program)   (For purposes of calculations, use "0" for t, <1, <.1, <.01, etc.)

Computer Code Number	Food Description	Measure	Wt (g)	H₂O (%)	Ener (cal)	Prot (g)	Carb (g)	Dietary Fiber (g)	Fat (g)	Fat Breakdown (g)		
										Sat	Mono	Poly
	MIXED DISHES and FAST FOODS—Continued											
1899	Mushrooms, batter fried	5 ea	70	66	148	2	8	1	12	2.1	3	6.4
715	Potato salad with mayonnaise and eggs	½ c	125	76	179	3	14	2	10	1.8	3.1	4.7
1674	Pizza, combination, ¹⁄₁₂ of 12" round	1 pce	79	48	184	13	21		5	1.5	2.5	.9
1673	Pizza, pepperoni, ¹⁄₁₂ of 12" round	1 pce	71	46	181	10	20		7	2.2	3.1	1.2
694	Quiche Lorraine ⅛ of 8" quiche	1 pce	176	54	508	20	20	1	39	17.6	13.8	4.9
1449	Ramen noodles, cooked	1 c	227	82	156	6	29	3	2	.4	.5	.5
1671	Ravioli, meat	½ c	125	68	194	10	18	1	9	2.9	3.6	1
1597	Fried rice (meatless)	1 c	166	68	264	5	34	1	12	1.7	3	6.3
2142	Roast beef hash	½ c	117	66	230	9	11	1	16	7	5.8	3.2
	Spaghetti (enriched) in tomato sauce With cheese:											
695	Canned	1 c	250	80	190	5	38	2	1	0	.4	.5
696	Home recipe	1 c	250	77	260	9	37	2	9	2	5.4	1.2
	With meatballs:											
697	Canned	1 c	250	78	258	12	28	6	10	2.1	3.9	3.9
698	Home recipe	1 c	248	70	332	19	39	8	12	3.3	6.3	2.2
716	Spinach soufflé	1 c	136	74	219	11	3	3	19	9.5	5.7	2.2
1553	Sweet & sour pork	1 c	226	77	231	15	25	1	8	2.2	2.9	2.4
1263	Sweet & sour chicken breast	1 ea	131	79	117	8	15	1	3	.6	.8	1.5
1515	Three bean salad	1 c	150	82	139	4	13	3	8	1.2	1.9	4.9
717	Tuna salad	1 c	205	63	383	33	19	0	19	3.2	5.9	8.4
1121	Tuna noodle casserole, homemade	1 c	202	75	237	17	25	1	7	1.9	1.5	3.2
1270	Waldorf salad	1 c	137	58	408	4	13	2	40	4.1	7.3	27
	FAST FOODS and SANDWICHES (see end of this appendix for additional Fast Foods)											
699	Burrito, beef & bean	1 ea	116	52	255	11	33	3	9	4.2	3.5	.6
700	Burrito, bean	1 ea	109	52	225	7	36	4	7	3.5	2.4	.6
2106	Burrito, chicken con queso	1 ea	306	76	280	12	53	5	6	1.5		
701	Cheeseburger with bun, regular	1 ea	154	55	359	18	28		20	9.2	7.2	1.5
702	Cheeseburger with bun, 4-oz patty	1 ea	166	51	417	21	35		21	8.7	7.8	2.7
703	Chicken patty sandwich	1 ea	182	47	515	24	39	1	29	8.5	10.4	8.4
704	Corndog	1 ea	175	47	460	17	56		19	5.2	9.1	3.5
1922	Corndog, chicken	1 ea	113	52	271	13	26		13			
705	Enchilada	1 ea	163	63	319	10	28		19	10.6	6.3	.8
706	English muffin with egg, cheese, bacon	1 ea	146	49	383	20	31	1	20	9	6.8	2.1
	Fish sandwich:											
707	Regular, with cheese	1 ea	183	45	523	21	48	<1	28	8.1	8.9	9.4
708	Large, no cheese	1 ea	158	47	431	17	41	<1	23	5.2	7.7	8.2
709	Hamburger with bun, regular	1 ea	107	45	275	14	33	1	10	3.5	3.7	1.8
710	Hamburger with bun, 4-oz patty	1 ea	215	50	576	32	39		32	12	14.1	2.8
711	Hotdog/frankfurter with bun	1 ea	98	54	242	10	18		14	5.1	6.8	1.7
	Lunchables:											
2129	Bologna & American cheese	1 ea	128		450	18	19	0	34	15		
2130	Ham & cheese	1 ea	128		320	22	19	0	17	8		
2117	Honey ham & Amer. w/choc pudding	1 ea	176		390	18	34	<1	20	9		
2118	Honey turkey & cheddar w/Jello	1 ea	163		320	17	27	<1	16	9		
2131	Pepperoni & American cheese	1 ea	128		480	20	19	0	36	17		
2125	Salami & American cheese	1 ea	128		430	18	18	0	32	15		
2127	Turkey & cheddar cheese	1 ea	128		360	20	20	1	22	11		
712	Pizza, cheese, ⅛ of 15" round	1 pce	63	48	140	8	20	1	3	1.5	1	.5
	SANDWICHES: Avocado, chesse, tomato & lettuce:											
1276	On white bread, firm	1 ea	210	58	478	15	41	5	29	8.8	11.3	7.2
1278	On part whole wheat	1 ea	201	59	444	14	34	7	29	8.6	11.4	7.3
1277	On whole wheat	1 ea	214	58	468	16	40	8	30	8.7	11.6	7.5
	Bacon, lettuce & tomato sandwich:											
1137	On white bread, soft	1 ea	124	53	308	10	28	2	18	4.5	6.1	6.1
1139	On part whole wheat	1 ea	124	54	303	10	26	3	17	4.3	6.2	6.1
1138	On whole wheat	1 ea	137	53	328	12	32	4	18	4.4	6.4	6.3

PAGE KEY: A–4 = Beverages A–6 = Dairy A–10 = Eggs A–10 = Fat/Oil A–14 = Fruit A–20 = Bakery A–28 = Grain A–32 = Fish A–34 = Meats
A–38 = Poultry A–40 = Sausage A–42 = Mixed/Fast A–46 = Nuts/Seeds A–50 = Sweets A–52 = Vegetables/Legumes A–62 = Vegetarian Foods
A–64 = Misc A–66 = Soups/Sauces A–68 = Fast A–84 = Convenience A–88 = Baby foods

A–45

Chol (mg)	Calc (mg)	Iron (mg)	Magn (mg)	Pota (mg)	Sodi (mg)	Zinc (mg)	VT-A (RE)	Thia (mg)	VT-E (a-TE)	Ribo (mg)	Niac (mg)	V-B6 (mg)	Fola (µg)	VT-C (mg)
14	54	.76	8	180	121	.42	10	.07	.92	.22	1.65	.05	8	1
85	24	.81	19	318	661	.39	41	.1	2.33	.07	1.11	.18	8	12
20	101	1.53	18	179	382	1.11	101	.21		.17	1.96	.09	32	2
14	65	.94	9	153	267	.52	55	.13		.23	3.05	.06	37	2
205	201	1.9	27	271	549	1.66	243	.23	1.91	.44	4.71	.19	17	3
38	20	1.89	24	51	1349	.76	204	.22	.09	.1	1.75	.06	9	<1
84	32	2.03	20	259	619	1.67	94	.15	1.52	.22	2.95	.14	14	11
42	30	1.84	24	134	286	.89	21	.21	2.46	.11	2.25	.15	22	4
40	10	.9	22	362	695	2.99	0	.09		.12	2.33	.3	12	0
7	40	2.75	21	303	955	1.12	120	.35	2.13	.27	4.5	.13	6	10
7	80	2.25	26	408	955	1.3	140	.25	2.75	.17	2.25	.2	8	12
22	52	3.25	20	245	1220	2.39	100	.15	1.5	.17	2.25	.12	5	5
74	124	3.72	40	665	1009	2.45	159	.25	1.64	.3	3.97	.2	10	22
184	230	1.35	38	201	763	1.29	675	.09	1.22	.3	.48	.12	80	3
38	28	1.36	34	390	1219	1.46	28	.55	.62	.21	3.6	.41	10	20
23	16	.79	21	187	732	.66	20	.06	.39	.08	3.06	.18	6	12
0	35	1.42	25	224	514	.54	23	.07	1.96	.09	.4	.04	53	4
27	35	2.05	39	365	824	1.15	55	.06	1.95	.14	13.7	.17	16	5
41	34	2.3	30	182	772	1.2	13	.18	1.18	.15	7.78	.2	10	1
21	43	.88	39	270	236	.63	39	.1	8.67	.05	.36	.36	27	6
24	53	2.46	42	329	670	1.93	32	.27	.7	.42	2.71	.19	58	1
2	57	2.27	44	328	495	.76	16	.32	.87	.3	2.04	.15	44	1
10	40	.72			600		40							15
52	182	2.65	26	229	976	2.62	71	.32	1.34	.23	6.38	.15	65	2
60	171	3.42	30	335	1050	3.49	65	.35		.28	8.05	.18	61	2
60	60	4.68	35	353	957	1.87	31	.33	.55	.24	6.81	.2	100	9
79	102	6.18	17	263	973	1.31	37	.28	.7	.7	4.17	.09	103	0
64					668									
44	324	1.32	50	240	784	2.51	186	.08	1.47	.42	1.91	.39	65	1
234	207	3.29	34	213	784	1.81	158	.48	.6	.53	3.93	.16	47	1
68	185	3.5	37	353	939	1.17	97	.46	1.83	.42	4.23	.11	91	3
55	84	2.61	33	340	615	.99	30	.33	.87	.22	3.4	.11	85	3
43	51	2.46	22	215	564	2.05	13	.26	.43	.32	4.7	.13	52	3
103	92	5.55	45	527	742	5.81	4	.34	1.61	.41	6.73	.37	84	1
44	23	2.31	13	143	670	1.98	0	.23	.27	.27	3.65	.05	48	<1
85	300	2.7			1620		60							0
60	300	1.8			1770		80							
55	250	2.7			1540		40							
50	20	6			1360		80							
95	250	2.7			1840		60							
80	250	2.7			1740		60							
70	300	1.8			1650		60							
9	117	.58	16	110	336	.81	74	.18		.16	2.48	.04	35	1
34	294	3.06	54	581	550	1.71	140	.37	4.55	.39	3.77	.32	80	11
31	291	3.1	67	617	525	1.91	140	.35	4.55	.39	3.94	.35	80	11
31	281	3.53	102	679	593	2.68	140	.36	4.17	.37	4.29	.42	92	11
21	52	1.99	20	233	590	.96	31	.35	2.34	.19	3.2	.14	34	12
20	63	2.27	33	283	604	1.19	31	.36	2.7	.21	3.62	.17	37	12
20	55	2.68	64	342	670	1.9	31	.37	2.36	.2	3.97	.24	48	12

Table A-1

Food Composition   (Computer code number is for West Diet Analysis program)   (For purposes of calculations, use "0" for t, <1, <.1, <.01, etc.)

Computer Code Number	Food Description	Measure	Wt (g)	H₂O (%)	Ener (cal)	Prot (g)	Carb (g)	Dietary Fiber (g)	Fat (g)	Fat Breakdown (g) Sat	Mono	Poly
	MIXED DISHES and FAST FOODS—Continued											
	Cheese, grilled:											
1140	On white bread, soft	1 ea	119	37	400	18	30	1	24	13.2	7.5	2
1142	On part whole wheat	1 ea	119	37	396	18	28	3	24	13	7.6	2.1
1141	On whole wheat	1 ea	132	38	420	20	33	4	24	13.1	7.8	2.2
1596	Chicken fillet	1 ea	182	47	515	24	39	1	29	8.5	10.4	8.4
	Chicken salad:											
1143	On white bread, soft	1 ea	110	40	369	11	31	1	23	3.7	6.1	12.1
1145	On part whole wheat	1 ea	110	41	364	11	29	4	23	3.5	6.1	12.1
1144	On whole wheat	1 ea	123	41	387	13	34	5	23	3.6	6.3	12.2
1146	Corned beef & swiss on rye	1 ea	156	49	420	28	22	<1	26	9.4	7.4	6.3
	Egg salad:											
1147	On white bread, soft	1 ea	117	43	380	10	31	1	25	4.4	6.8	12
1149	On part whole wheat	1 ea	116	43	374	10	29	3	25	4.1	6.8	12
1148	On whole wheat	1 ea	130	43	400	12	35	5	25	4.3	7.1	12.2
	Ham:											
1279	On rye bread	1 ea	150	60	283	22	21	<1	13	2.4	3.8	6
1151	On white bread, soft	1 ea	157	55	334	22	30	1	14	3	4.4	5.9
1153	On part whole wheat	1 ea	156	55	328	22	28	3	14	2.7	4.4	6
1152	On whole wheat	1 ea	169	54	352	24	34	4	15	2.9	4.7	6.1
	Ham & cheese:											
1280	On white bread, soft	1 ea	157	49	403	23	30	1	22	8.4	5.9	6.1
1282	On part whole wheat	1 ea	156	50	397	23	28	3	21	8.1	5.9	6.2
1281	On whole wheat	1 ea	170	49	424	24	34	4	22	8.3	6.2	6.4
1150	Ham & swiss on rye	1 ea	150	54	339	22	22	<1	19	6.5	5.1	6
	Ham salad:											
1154	On white bread, soft	1 ea	131	47	362	11	37	1	20	4.8	6.7	7.4
1156	On part whole wheat	1 ea	131	48	357	11	35	3	20	4.5	6.8	7.5
1155	On whole wheat	1 ea	144	47	380	12	40	4	20	4.7	7	7.6
1157	Patty melt: Ground beef & cheese on rye	1 ea	182	46	561	37	22	3	37	13.2	11.7	8.4
	Peanut butter & jelly:											
1158	On white bread, soft	1 ea	101	26	351	12	47	3	15	3.1	6.7	3.9
1160	On part whole wheat	1 ea	101	27	346	12	45	5	15	2.9	6.7	4
1159	On whole wheat	1 ea	114	28	370	13	50	6	15	3	7	4.1
1161	Reuben, grilled: Corned beef, swiss cheese, sauerkraut on rye	1 ea	239	64	462	28	25	2	29	9.9	9.5	7.1
	Roast beef:											
713	On a bun	1 ea	139	49	346	21	33		14	3.6	6.8	1.7
1162	On white bread, soft	1 ea	157	46	404	29	34	1	17	3.4	4.2	8.2
1164	On part whole wheat	1 ea	156	47	398	29	32	3	17	3.2	4.3	8.3
1163	On whole wheat	1 ea	169	46	422	31	38	4	17	3.3	4.5	8.4
	Tuna salad:											
1165	On white bread, soft	1 ea	122	46	327	14	35	2	15	2.5	3.8	7.9
1167	On part whole wheat	1 ea	122	47	322	14	33	4	15	2.2	3.8	8
1166	On whole wheat	1 ea	135	46	346	16	39	5	15	2.3	4.1	8.1
	Turkey:											
1168	On white bread, soft	1 ea	156	54	346	24	29	1	15	2.4	3.2	8.3
1170	On part whole wheat	1 ea	155	54	338	24	27	3	14	2.1	3.2	8.3
1169	On whole wheat	1 ea	169	53	365	26	33	4	15	2.3	3.5	8.5
	Turkey ham:											
1272	On rye bread	1 ea	150	60	280	21	20	<1	14	2.5	2.8	6.9
1273	On white bread, soft	1 ea	156	55	331	21	29	1	14	3	3.4	6.8
1275	On part whole wheat	1 ea	156	56	326	21	28	3	14	3	4.2	5.6
1274	On whole wheat	1 ea	169	55	350	23	33	4	15	2.9	3.7	7
714	Taco	1 ea	171	58	369	21	27		20	11.4	6.6	1
	Tostada:											
1114	With refried beans	1 ea	144	66	223	10	26	7	10	5.4	3	.7
1118	With beans & beef	1 ea	225	70	333	16	30	4	17	11.5	3.5	.6
1354	With beans & chicken	1 ea	156	68	248	19	18	3	11	5.3	3.9	1.6
	NUTS, SEEDS, and PRODUCTS											
	Almonds:											
1365	Dry roasted, salted	1 c	138	3	810	22	33	19	71	6.7	46.2	14.9

A

Chol (mg)	Calc (mg)	Iron (mg)	Magn (mg)	Pota (mg)	Sodi (mg)	Zinc (mg)	VT-A (RE)	Thia (mg)	VT-E (a-TE)	Ribo (mg)	Niac (mg)	V-B6 (mg)	Fola (µg)	VT-C (mg)
55	399	1.81	25	154	1143	2.05	212	.24	1.13	.34	1.91	.06	24	<1
54	412	2.12	39	209	1160	2.31	212	.25	1.53	.36	2.39	.1	28	<1
54	402	2.54	73	271	1226	3.08	212	.26	1.14	.35	2.74	.17	40	<1
60	60	4.68	35	353	957	1.87	31	.33	.55	.24	6.81	.2	100	9
32	60	2.04	18	139	460	.8	24	.25	6.16	.18	3.68	.25	26	1
31	73	2.35	33	195	475	1.06	24	.26	6.57	.2	4.16	.29	30	1
30	63	2.79	68	259	543	1.85	24	.27	6.14	.19	4.5	.36	42	1
82	268	3.12	28	225	1392	3.65	81	.19	2.59	.33	2.72	.17	19	1
157	71	2.18	16	113	526	.74	76	.26	4.52	.31	1.98	.2	37	0
155	83	2.47	31	169	539	1	75	.27	4.91	.33	2.45	.23	41	0
155	73	2.92	66	234	611	1.8	76	.28	4.53	.32	2.82	.3	53	0
47	48	2.3	26	364	1566	2.11	8	.99	2.36	.31	5.45	.47	15	23
47	60	2.39	29	368	1619	2.04	8	1.02	2.34	.33	5.99	.47	24	22
45	71	2.68	43	421	1630	2.29	8	1.03	2.73	.35	6.45	.5	27	22
45	62	3.1	77	483	1696	3.05	8	1.04	2.34	.33	6.78	.57	39	22
61	232	2.28	30	315	1620	2.34	90	.76	2.53	.36	4.64	.36	25	15
59	244	2.58	45	368	1630	2.59	90	.77	2.93	.39	5.09	.39	28	15
59	236	3.02	79	432	1707	3.37	90	.78	2.56	.37	5.47	.46	40	15
57	258	2.25	29	344	1602	2.59	79	.72	2.52	.36	4.06	.35	16	15
30	56	2.07	19	159	921	1.06	8	.51	3.29	.22	3.25	.17	22	4
29	69	2.38	33	216	936	1.32	8	.52	3.69	.24	3.74	.21	26	4
29	59	2.81	69	279	1001	2.11	8	.53	3.29	.23	4.09	.28	38	4
113	222	4.19	36	391	701	7.11	123	.25	3.5	.46	6.14	.35	25	<1
2	60	2.25	56	245	293	1.07	<1	.27	.12	.17	5.33	.13	40	<1
0	72	2.55	70	299	308	1.32	<1	.28	.51	.19	5.8	.17	44	<1
0	63	2.97	104	361	375	2.09	<1	.29	.14	.17	6.14	.24	56	<1
80	288	4.24	38	361	1949	3.73	130	.21	4.46	.34	2.79	.27	38	13
51	54	4.23	31	316	792	3.39	21	.37	.19	.31	5.87	.26	57	2
45	60	3.98	28	432	1595	3.78	12	.29	3.3	.3	6.39	.39	30	12
43	72	4.27	42	485	1607	4.02	12	.31	3.69	.32	6.84	.42	34	12
43	62	4.7	77	547	1672	4.79	12	.31	3.31	.31	7.18	.49	45	12
14	60	2.24	22	161	567	.67	22	.25	2.72	.18	5.53	.11	25	1
13	73	2.55	37	217	582	.93	22	.26	3.12	.2	6	.16	29	1
12	63	2.98	72	280	649	1.72	22	.27	2.71	.19	6.34	.23	41	1
45	56	2	29	302	1585	1.33	12	.26	3.46	.23	8.97	.4	24	0
43	68	2.28	43	354	1589	1.57	11	.27	3.82	.25	9.38	.44	28	0
43	59	2.73	77	418	1665	2.35	12	.28	3.47	.23	9.77	.51	39	0
55	51	4.06	25	342	1185	3	8	.22	2.8	.33	4.3	.29	17	<1
55	62	4.09	28	346	1248	2.9	8	.27	2.75	.35	4.87	.28	25	0
53	74	4.37	42	400	1262	3.15	8	.28	3.57	.37	5.34	.32	29	0
53	65	4.81	76	462	1329	3.91	8	.29	2.76	.35	5.68	.39	41	0
56	221	2.41	70	474	802	3.93	147	.15	1.88	.44	3.21	.24	68	2
30	210	1.89	59	403	543	1.9	85	.1	1.15	.33	1.32	.16	43	1
74	189	2.45	67	491	871	3.17	173	.09	1.8	.49	2.86	.25	85	4
53	168	1.79	47	365	433	2.28	86	.11	1.87	.2	4.52	.32	53	3
0	389	5.24	420	1062	1076	6.76	0	.18	7.66	.83	3.89	.1	88	1

Table A–1

Food Composition   (Computer code number is for West Diet Analysis program)   (For purposes of calculations, use "0" for t, <1, <.1, <.01, etc.)

Computer Code Number	Food Description	Measure	Wt (g)	H₂O (%)	Ener (cal)	Prot (g)	Carb (g)	Dietary Fiber (g)	Fat (g)	Fat Breakdown (g)		
										Sat	Mono	Poly
	NUTS, SEEDS, and PRODUCTS—Continued											
	Almonds:											
718	Slivered, packed, unsalted	1 c	108	4	636	22	22	12	56	5.3	36.6	11.9
719	Whole, dried, unsalted	1 c	142	4	836	28	29	15	74	7	48.1	15.6
720	Ounce	1 oz	28	4	165	6	6	3	15	1.4	9.5	3.1
721	Almond butter:	1 tbs	16	1	101	2	3	1	9	.9	6.1	2
4572	Salted	1 tbs	16	1	101	2	3	1	9	.9	6.1	2
722	Brazil nuts, dry (about 7)	1 c	140	3	918	20	18	8	93	22.7	32.2	33.7
	Cashew nuts, dry roasted:											
723	Salted:	1 c	137	2	786	21	45	4	64	12.8	37.4	10.7
724	Ounce	1 oz	28	2	161	4	9	1	13	2.6	7.6	2.2
4621	Unsalted:	1 c	137	2	786	21	45	4	64	12.8	37.4	10.7
4621	Ounce	1 oz	28	2	161	4	9	1	13	2.6	7.6	2.2
725	Oil roasted:	1 c	130	4	749	23	37	5	63	12.6	36.9	10.6
726	Ounce	1 oz	28	4	161	5	8	1	13	2.7	7.9	2.3
4622	Unsalted:	1 c	130	4	749	21	37	5	63	12.6	36.9	10.6
4622	Ounce	1 oz	28	4	161	5	8	1	13	2.7	7.9	2.3
727	Cashew butter, unsalted	1 tbs	16	3	94	3	4	<1	8	1.6	4.7	1.3
4662	Cashew butter, salted	1 tbs	16	3	94	3	4	<1	8	1.6	4.7	1.3
728	Chestnuts, European, roasted (1 cup = approx 17 kernels)	1 c	143	40	350	5	76	7	3	.6	1.1	1.2
	Coconut, raw:											
729	Piece 2 x 2 x ½"	1 pce	45	47	159	2	7	4	15	13.5	.6	.2
730	Shredded/grated, unpacked	½ c	40	47	142	1	6	4	13	12	.6	.1
	Coconut, dried, shredded/grated:											
731	Unsweetened	1 c	78	3	515	6	19	13	50	45.1	2.1	.6
732	Sweetened	1 c	93	13	466	3	44	4	33	29.6	1.4	.4
733	Filberts/hazelnuts, chopped:	1 c	135	5	853	18	21	8	84	6.2	66.3	8.1
734	Ounce	1 oz	28	5	177	4	4	2	17	1.3	13.7	1.7
735	Macadamias, oil roasted, salted:	1 c	134	2	962	10	17	12	103	15.4	80.9	1.8
736	Ounce	1 oz	28	2	201	2	4	3	21	3.2	16.9	.4
1368	Macadamias, oil roasted, unsalted	1 c	134	2	962	10	17	12	103	15.4	80.9	1.8
	Mixed nuts:											
737	Dry roasted, salted	1 c	137	2	814	24	35	12	71	9.4	43	14.8
738	Oil roasted, salted	1 c	142	2	876	24	30	13	80	12.4	45	18.9
1369	Oil roasted, unsalted	1 c	142	2	876	27	30	14	80	12.4	45	18.9
	Peanuts:											
739	Oil roasted, salted	1 c	144	2	837	38	27	13	71	9.8	35.3	22.5
740	Ounce	1 oz	28	2	163	7	5	3	14	1.9	6.9	4.4
1370	Oil roasted, unsalted	1 c	144	2	837	38	27	10	71	9.8	35.3	22.5
741	Dried, salted	1 c	146	2	854	35	31	12	73	10.1	36.1	22.9
742	Ounce	1 oz	28	2	164	7	6	2	14	1.9	6.9	4.4
743	Peanut butter:	½ c	128	1	759	33	25	8	65	14.3	31.1	17.7
1371	Tablespoon	2 tbs	32	1	190	8	6	2	16	3.6	7.8	4.4
744	Pecan halves, dried, unsalted:	1 c	108	5	720	9	20	8	73	5.8	45.6	18
745	Ounce	1 oz	28	5	187	2	5	2	19	1.5	11.8	4.7
1372	Pecan halves, dry roasted, salted	¼ c	28	1	185	2	6	3	18	1.4	11.3	4.5
746	Pine nuts/piñons, dried	1 oz	28	6	176	3	5	3	17	2.6	6.4	7.2
747	Pistachios, dried, shelled	1 oz	28	2	162	6	7	3	14	1.8	9.2	2
1373	Pistachios, dry roasted, salted, shelled	1 c	128	2	776	19	35	14	68	8.8	45.7	10.2
748	Pumpkin kernels, dried, unsalted	1 oz	28	7	151	7	5	1	13	2.4	4	5.8
1374	Pumpkin kernels, roasted, salted	1 c	227	7	1184	75	30	9	96	18.1	29.7	43.6
749	Sesame seeds, hulled, dried	¼ c	38	5	223	10	4	3	21	2.9	7.9	9.1
	Sunflower seed kernels:											
750	Dry	¼ c	36	5	205	8	7	4	18	1.9	3.4	11.8
751	Oil roasted	¼ c	34	3	209	7	5	2	20	2	3.7	12.9
752	Tahini (sesame butter)	1 tbs	15	3	91	3	3	1	8	1.2	3.2	3.7
1334	Trail mix w/chocolate chips	1 c	146	7	707	21	66	8	47	9.3	19.8	16.5
753	Black walnuts, chopped:	1 c	125	4	759	31	15	6	71	4.8	15.9	46.9
754	Ounce	1 oz	28	4	170	7	3	1	16	1.1	3.6	10.5
755	English walnuts, chopped:	1 c	120	4	770	17	22	6	74	7.2	17	46.9
756	Ounce	1 oz	28	4	180	4	5	1	17	1.7	4	10.9

PAGE KEY: A–4 = Beverages A–6 = Dairy A–10 = Eggs A–10 = Fat/Oil A–14 = Fruit A–20 = Bakery A–28 = Grain A–32 = Fish A–34 = Meats A–38 = Poultry A–40 = Sausage A–42 = Mixed/Fast A–46 = Nuts/Seeds A–50 = Sweets A–52 = Vegetables/Legumes A–62 = Vegetarian Foods A–64 = Misc A–66 = Soups/Sauces A–68 = Fast A–84 = Convenience A–88 = Baby foods

Chol (mg)	Calc (mg)	Iron (mg)	Magn (mg)	Pota (mg)	Sodi (mg)	Zinc (mg)	VT-A (RE)	Thia (mg)	VT-E (a-TE)	Ribo (mg)	Niac (mg)	V-B6 (mg)	Fola (µg)	VT-C (mg)
0	287	3.95	320	791	12	3.15	0	.23	25.9	.84	3.63	.12	63	1
0	378	5.2	420	1039	16	4.15	0	.3	34.1	1.11	4.77	.16	83	1
0	74	1.02	83	205	3	.82	0	.06	6.72	.22	.94	.03	16	<1
0	43	.59	48	121	2	.49	0	.02	3.25	.1	.46	.01	10	<1
0	43	.59	48	121	72	.49	0	.02	3.25	.1	.46	.01	10	<1
0	246	4.76	315	840	3	6.43	0	1.4	10.6	.17	2.27	.35	6	1
0	62	8.22	356	774	877	7.67	0	.27	.78	.27	1.92	.35	95	0
0	13	1.68	73	158	179	1.57	0	.06	.16	.06	.39	.07	19	0
0	62	8.22	356	774	22	7.67	0	.27	.78	.27	1.92	.35	95	0
0	13	1.68	73	158	4	1.57	0	.06	.16	.06	.39	.07	19	0
0	53	5.33	332	689	814	6.18	0	.55	2.03	.23	2.34	.32	88	0
0	11	1.15	71	148	175	1.33	0	.12	.44	.05	.5	.07	19	0
0	53	5.33	332	689	22	6.18	0	.55	2.03	.23	2.34	.32	88	0
0	11	1.15	71	148	5	1.33	0	.12	.44	.05	.5	.07	19	0
0	7	.8	41	87	2	.83	0	.05	.25	.03	.26	.04	11	0
0	7	.8	41	87	98	.83	0	.05	.25	.03	.26	.04	11	0
0	41	1.3	47	847	3	.81	3	.35	1.72	.25	1.92	.71	100	37
0	6	1.09	14	160	9	.49	0	.03	.33	.01	.24	.02	12	1
0	6	.97	13	142	8	.44	0	.03	.29	.01	.22	.02	11	1
0	20	2.59	70	424	29	1.57	0	.05	1.05	.08	.47	.23	7	1
0	14	1.79	46	313	244	1.69	0	.03	1.26	.02	.44	.25	8	1
0	254	4.41	385	601	4	3.24	9	.67	32.3	.15	1.54	.83	97	1
0	53	.92	80	125	1	.67	2	.14	6.69	.03	.32	.17	20	<1
0	60	2.41	157	441	348	1.47	1	.28	.55	.15	2.71	.26	21	0
0	13	.5	33	92	73	.31	<1	.06	.11	.03	.57	.05	4	0
0	60	2.41	157	441	9	1.47	1	.28	.55	.15	2.71	.26	21	0
0	96	5.07	308	818	917	5.21	1	.27	8.22	.27	6.44	.41	69	1
0	153	4.56	334	825	926	7.21	3	.71	8.52	.31	7.19	.34	118	1
0	153	4.56	334	825	16	7.21	3	.71	8.52	.31	7.19	.34	118	1
0	127	2.64	266	982	624	9.55	0	.36	10.7	.16	20.6	.37	181	0
0	25	.51	52	191	121	1.86	0	.07	2.07	.03	4	.07	35	0
0	127	2.64	266	982	9	9.55	0	.36	10.7	.16	20.6	.37	181	0
0	79	3.3	257	961	1186	4.83	0	.64	10.8	.14	19.7	.37	212	0
0	15	.63	49	184	228	.93	0	.12	2.07	.03	3.78	.07	41	0
0	49	2.36	204	856	598	3.74	0	.11	12.8	.13	17.2	.58	95	0
0	12	.59	51	214	149	.93	0	.03	3.2	.03	4.29	.14	24	0
0	39	2.3	138	423	1	5.91	14	.92	3.35	.14	.96	.2	42	2
0	10	.6	36	110	<1	1.53	4	.24	.87	.04	.25	.05	11	1
0	10	.61	37	104	218	1.59	4	.09	.84	.03	.26	.05	11	1
0	2	.86	65	176	20	1.2	1	.35	.98	.06	1.22	.03	16	1
0	38	1.9	44	306	2	.37	6	.23	1.46	.05	.3	.07	16	2
0	90	4.06	166	1241	998	1.74	31	.54	8.26	.31	1.8	.33	76	9
0	12	4.2	150	226	5	2.09	11	.06	.28	.09	.49	.06	16	1
0	98	33.8	1212	1829	1305	16.9	86	.48	2.27	.72	3.95	.2	130	4
0	50	2.96	132	155	15	3.91	3	.27	.86	.03	1.78	.05	36	0
0	42	2.44	127	248	1	1.82	2	.82	18.1	.09	1.62	.28	82	<1
0	19	2.28	43	164	1	1.77	2	.11	17.1	.09	1.4	.27	80	<1
0	21	.95	53	69	<1	1.58	1	.24	.34	.02	.85	.02	15	0
6	159	4.95	235	946	177	4.58	7	.6	15.6	.33	6.44	.38	95	2
0	72	3.84	253	655	1	4.28	37	.27	3.28	.14	.86	.69	82	4
0	16	.86	57	147	<1	.96	8	.06	.73	.03	.19	.15	18	1
0	113	2.93	203	602	12	3.28	14	.46	3.14	.18	1.25	.67	79	4
0	26	.68	47	141	3	.76	3	.11	.73	.04	.29	.16	18	1

Table A–1

Food Composition

(Computer code number is for West Diet Analysis program)   (For purposes of calculations, use "0" for t, <1, <.1, <.01, etc.)

Computer Code Number	Food Description	Measure	Wt (g)	H₂O (%)	Ener (cal)	Prot (g)	Carb (g)	Dietary Fiber (g)	Fat (g)	Fat Breakdown (g)		
										Sat	Mono	Poly
	SWEETENERS and SWEETS (see also Dairy [milk desserts] and Baked Goods)											
757	Apple butter	2 tbs	36	52	66	<1	17	<1	<1	t	t	t
1124	Butterscotch topping	2 tbs	41	32	103	1	27	<1	<1	t	t	0
1125	Caramel topping	2 tbs	41	32	103	1	27	<1	<1	t	t	0
	Cake frosting, creamy vanilla:											
1127	Canned	2 tbs	39	13	163	<1	27	<1	7	1.9	3.4	.9
1123	From mix	2 tbs	39	12	165	<1	28	<1	6	1.3	2.6	2.2
	Cake frosting, lite:											
2061	Milk chocolate	1 tbs	16	18	58	<1	11	<1	1	.4		
2062	Vanilla	1 tbs	16	15	60	0	12	<1	1	.4		
	Candy:											
1128	Almond Joy candy bar	1 oz	28	10	131	1	16	1	7	4.8	1.8	.4
2069	Butterscotch morsels	¼ c	43	1	246	0	31	0	12	12.5	0	0
758	Caramel, plain or chocolate	1 pce	10	8	38	<1	8	<1	1	.7	.1	t
1961	Chewing gum, sugarless	1 pce	3		6	0	2		0	0	0	0
	Chocolate (see also #784, 785, 971):											
	Milk chocolate:											
759	Plain	1 oz	28	1	144	2	17	1	9	5.2	2.8	.3
760	With almonds	1 oz	28	1	147	3	15	2	10	4.8	3.8	.6
761	With peanuts	1 oz	28	1	155	5	11	2	11	3.4	5.1	2.5
762	With rice cereal	1 oz	28	2	139	2	18	1	7	4.4	2.4	.2
763	Semisweet chocolate chips	1 c	168	1	805	7	106	10	50	29.9	16.8	1.6
764	Sweet dark chocolate (candy bar)	1 ea	41	1	226	2	25	2	13	8.3	4.6	.4
765	Fondant candy, uncoated (mints, candy corn, other)	1 pce	16	7	57	0	15	0	0	0	0	0
1697	Fruit Roll-Up (small)	1 ea	14	11	49	<1	12	<1	<1	.1	.2	.1
766	Fudge, chocolate	1 pce	17	10	65	<1	13	<1	1	.9	.4	.1
767	Gumdrops	1 c	182	1	703	0	180	0	0	0	0	0
768	Hard candy, all flavors	1 pce	6	1	22	0	6	0	<1	0	0	0
769	Jellybeans	10 pce	11	6	40	0	10	0	<1	t	t	t
1134	M&M's plain chocolate candy	10 pce	7	2	34	<1	5	<1	1	.9	.5	t
1135	M&M's peanut chocolate candy	10 pce	20	2	103	2	12	1	5	2.1	2.2	.8
1130	Mars almond bar	1 ea	50	4	234	4	31	1	11	2.7	5.5	2.8
1129	Milky Way candy bar	1 ea	60	6	254	3	43	1	10	4.7	3.6	.4
1708	Milk chocolate-coated peanuts	1 c	149	2	773	19	74	7	50	21.8	19.4	6.4
1709	Peanut brittle, recipe	1 c	147	2	666	11	102	3	28	7.4	12.5	6.9
1132	Reese's peanut butter cup	2 ea	50	3	271	5	27	2	16	5.5	6.5	2.8
1133	Skor English toffee candy bar	1 ea	39	3	217	2	22	1	13	8.5	4.3	.5
1131	Snickers candy bar (2.2oz)	1 ea	62	5	297	5	37	2	15	5.6	6.5	3
1482	Fruit juice bar (2.5 fl oz)	1 ea	77	78	63	1	16	0	<1	t	0	t
771	Gelatin dessert/Jello, prepared	½ c	135	85	80	2	19	0	0	0	0	0
1702	SugarFree	½ c	117	98	8	1	1	0	0	0	0	0
772	Honey:	1 c	339	17	1030	1	279	1	0	0	0	0
773	Tablespoon	1 tbs	21	17	64	<1	17	<1	0	0	0	0
774	Jams or preserves:	1 tbs	20	29	54	<1	14	<1	<1	0	t	t
775	Packet	1 ea	14	34	34	<1	9	<1	<1	t	t	0
776	Jellies:	1 tbs	19	28	51	<1	13	<1	<1	t	t	t
777	Packet	1 ea	14	28	38	<1	10	<1	<1	t	t	t
1136	Marmalade	1 tbs	20	33	49	<1	13	<1	0	0	0	0
770	Marshmallows	1 ea	7	16	22	<1	6	<1	<1	t	t	t
1126	Marshmallow creme topping	2 tbs	38	18	118	1	30	<1	<1	t	t	t
778	Popsicle/ice pops	1 ea	128	80	92	0	24	0	0	0	0	0
	Sugars:											
779	Brown sugar	1 c	220	2	827	0	214	0	0	0	0	0
780	White sugar, granulated:	1 c	200		774	0	200	0	0	0	0	0
781	Tablespoon	1 tbs	12		46	0	12	0	0	0	0	0
782	Packet	1 ea	6		23	0	6	0	0	0	0	0
783	White sugar, powdered, sifted	1 c	100		389	0	99	0	<1	t	t	t
	Sweeteners:											
1711	Equal, packet	1 ea	1	12	4	<1	1	0	<1	0	t	t
1712	Sweet 'N Low, packet	1 ea	1		0	0	1	0	0	0	0	0

PAGE KEY: A–4 = Beverages A–6 = Dairy A–10 = Eggs A–10 = Fat/Oil A–14 = Fruit A–20 = Bakery A–28 = Grain A–32 = Fish A–34 = Meats A–38 = Poultry A–40 = Sausage A–42 = Mixed/Fast A–46 = Nuts/Seeds A–50 = Sweets A–52 = Vegetables/Legumes A–62 = Vegetarian Foods A–64 = Misc A–66 = Soups/Sauces A–68 = Fast A–84 = Convenience A–88 = Baby foods

Chol (mg)	Calc (mg)	Iron (mg)	Magn (mg)	Pota (mg)	Sodi (mg)	Zinc (mg)	VT-A (RE)	Thia (mg)	VT-E (a-TE)	Ribo (mg)	Niac (mg)	V-B6 (mg)	Fola (µg)	VT-C (mg)
0	2	.05	1	33	0	.02	0	<.01	.01	<.01	.03	.01	<1	1
<1	22	.08	3	34	143	.08	11	<.01	0	.04	.02	.01	1	<1
<1	22	.08	3	34	143	.08	11	<.01	0	.04	.02	.01	1	<1
0	1	.04		14	35	0	88	0	.79	<.01	<.01	0	0	0
0	4	.09	1	9	87	.04	42	.01	.79	.01	.13	<.01	0	0
0	1	.24			40		<1							0
0		.02			29		0							0
1	17	.39	18	69	41	.22	1	.01	.63	.04	.13	.02		<1
0	0	0		80	46		0	.03		.04	.03			0
1	14	.01	2	21	24	.04	1	<.01	.05	.02	.02	<.01	<1	<1
					0	0								
6	53	.39	17	108	23	.39	15	.02	.35	.08	.09	.01	2	<1
5	63	.46	25	124	21	.37	4	.02	.53	.12	.21	.01	3	<1
3	32	.52	34	150	11	.68	6	.08	1.3	.05	2.12	.04	23	0
5	48	.21	14	96	41	.31	3	.02	.35	.08	.13	.02	3	<1
0	54	5.26	193	613	18	2.72	3	.09	2	.15	.72	.06	5	0
<1	11	.98	47	139	3	.61	1	.01	.41	.1	.27	.02	1	0
0		.01		3	6	.01	<1	0	0	<.01	0	0	0	0
0	4	.14	3	41	9	.03	2	.01	.04	<.01	.01	.04	1	1
2	7	.08	4	17	10	.07	8	<.01	.02	.01	.02	<.01	<1	<1
0	5	.73	2	9	80	0	0	0	0	<.01	<.01	0	0	0
0		.02			2	<.01	0	0	0	0	0	0	0	0
0		.12		4	3	.01	0	0	0	0	0	0	0	0
1	7	.08	3	19	4	.07	4	<.01	.06	.01	.02	<.01	<1	<1
2	20	.23	12	69	10	.27	5	.03	.43	.04	.41	.02	7	<1
4	84	.55	36	163	85	.55	22	.02	.3	.16	.47	.03	9	<1
8	78	.46	20	145	144	.43	34	.02	.39	.13	.21	.03	6	1
13	155	1.95	134	748	61	2.8	0	.17	3.8	.26	6.33	.31	12	0
19	44	2.03	73	306	664	1.43	69	.28	2.41	.08	5.15	.15	103	0
2	39	.6	42	176	159	.7	9	.02	.66	.1	1.99	.04	27	<1
20	51	.02	13	93	108	.3	27	.01	.53	.13	.03	.01		<1
8	58	.47	42	209	165	.88	24	.13	.95	.1	2.26	.07	25	<1
0	4	.15	3	41	3	.04	2	.01	0	.01	.12	.02	5	7
0	3	.04	1	1	57	.04	0	0	0	<.01	<.01	<.01	0	0
0	2	.01	1	0	56	.03	0	0	0	<.01	<.01	<.01	0	0
0	20	1.42	7	176	14	.75	0	0	0	.13	.41	.08	7	2
0	1	.09		11	1	.05	0	0	0	.01	.02	<.01	<1	<1
0	4	.2	1	18	2	.01	<1	<.01	.02	.01	.04	<.01	2	<1
0	3	.07	1	11	6	.01	<1	0	0	<.01	<.01	<.01	5	1
0	2	.04	1	12	7	.01	<1	0	0	<.01	.01	<.01	<1	<1
0	1	.03	1	9	5	.01	<1	0	0	<.01	<.01	<.01	<1	<1
0	8	.03		7	11	.01	1	<.01	0	<.01	.01	<.01	7	1
0		.02			3	<.01	<1	0	0	0	<.01	0	<1	0
0	1	.08	1	2	17	.01	<1	0	0	0	.03	<.01	<1	0
0	0	0	1	5	15	.03	0	0	0	0	0	0	0	0
0	187	4.2	64	761	86	.4	0	.02	0	.01	.18	.06	2	0
0	2	.12	0	4	2	.06	0	0	0	.04	0	0	0	0
0		.01	0		<1	<.01	0	0	0	<.01	0	0	0	0
0		<.01	0		<1	<.01	0	0	0	<.01	0	0	0	0
0	1	.06	0	2	1	.03	0	0	0	0	0	0	0	0
0		<.01			<1	0	0	0	0	0	0	0	0	0
0	0	0			0		0							0

Table A–1
Food Composition

(Computer code number is for West Diet Analysis program)   (For purposes of calculations, use "0" for t, <1, <.1, <.01, etc.)

Computer Code Number	Food Description	Measure	Wt (g)	H₂O (%)	Ener (cal)	Prot (g)	Carb (g)	Dietary Fiber (g)	Fat (g)	Fat Breakdown (g) Sat	Mono	Poly
	SWEETENERS and SWEETS—Continued											
	Syrups, chocolate:											
785	Hot fudge type	2 tbs	43	22	149	2	27	1	4	2.4	1.6	1.4
784	Thin type	2 tbs	38	29	93	1	25	1	<1	.3	.2	t
786	Molasses, blackstrap	2 tbs	41	29	96	0	25	0	0	0	0	0
1710	Light cane syrup	2 tbs	41	24	103	0	27	0	0	0	0	0
787	Pancake table syrup (corn and maple)	2 tbs	40	24	115	0	30	0	0	0	0	0
	VEGETABLES and LEGUMES											
788	Alfalfa sprouts	1 c	33	91	10	1	1	1	<1	t	.1	t
1815	Amaranth leaves, raw, chopped	1 c	28	92	7	1	1	<1	<1	t	t	t
1816	Amaranth leaves, raw, each	1 ea	14	92	4	<1	1	<1	<1	t	t	t
1817	Amaranth leaves, cooked	1 c	132	91	28	3	5	2	<1	.1	.1	.1
1987	Arugula, raw, chopped	½ c	10	92	2	<1	<1	<1	<1	t	t	t
789	Artichokes, cooked globe (300 g with refuse)	1 ea	120	84	60	4	13	6	<1	t	t	.1
1177	Artichoke hearts, cooked from frozen	1 c	168	86	76	5	15	8	1	.2	t	.4
1176	Artichoke hearts, marinated	1 c	130	81	128	3	10	6	10	1.5	2.3	5.9
2021	Artichoke hearts, in water	½ c	100	91	37	2	6	0	0	0	0	0
	Asparagus, green, cooked:											
	From fresh:											
790	Cuts and tips	½ c	90	92	22	2	4	1	<1	.1	t	.1
791	Spears, ½" diam at base	4 ea	60	92	14	2	3	1	<1	t	t	.1
	From frozen:											
792	Cuts and tips	½ c	90	91	25	3	4	1	<1	.1	t	.2
793	Spears, ½" diam at base	4 ea	60	91	17	2	3	1	<1	.1	t	.1
794	Canned, spears, ½" diam at base	4 ea	72	94	14	2	2	1	<1	.1	t	.2
795	Bamboo shoots, canned, drained slices	1 c	131	94	25	2	4	2	1	.1	t	.2
1795	Bamboo shoots, raw slices	1 c	151	91	41	4	8	3	<1	.1	t	.2
1798	Bamboo shoots, cooked slices	1 c	120	96	14	2	2	1	<1	.1	t	.1
	Beans (see also alphabetical listing this section):											
1990	Adzuki beans, cooked	½ c	115	66	147	9	28	1	<1	t	t	t
796	Black beans, cooked	½ c	86	66	114	8	20	7	<1	.1	t	.2
	Canned beans (white/navy):											
803	With pork and tomato sauce	½ c	127	73	124	7	25	6	1	.5	.6	.2
804	With sweet sauce	½ c	130	71	144	7	27	7	2	.7	.8	.2
805	With frankfurters	½ c	130	69	185	9	20	9	9	3.1	3.7	1.1
	Lima beans:											
797	Thick seeded (Fordhooks), cooked from frozen	½ c	85	73	85	5	16	5	<1	.1	t	.1
798	Thin seeded (Baby), cooked from frozen	½ c	90	72	94	6	18	5	<1	.1	t	.1
799	Cooked from dry, drained	½ c	94	70	108	7	20	7	<1	.1	t	.2
1998	Red Mexican, cooked f/dry	½ c	112	70	126	8	23	9	<1	.1	.1	.2
	Snap bean/green string beans cuts and french style:											
800	Cooked from fresh	½ c	63	89	22	1	5	2	<1	t	t	.1
801	Cooked from frozen	½ c	68	91	19	1	4	2	<1	t	t	.1
802	Canned, drained	½ c	68	93	14	1	3	1	<1	t	t	t
1713	Snap bean, yellow, cooked f/fresh	½ c	63	89	22	1	5	2	<1	t	t	.1
	Bean sprouts (mung):											
806	Raw	½ c	52	90	16	2	3	1	<1	t	t	t
807	Cooked, stir fried	½ c	62	84	31	3	7	1	<1	t	t	t
808	Cooked, boiled, drained	½ c	62	93	13	1	3	<1	<1	t	t	t
1788	Canned, drained	½ c	63	96	8	1	1	<1	<1	t	t	t
	Beets, cooked from fresh:											
809	Sliced or diced	½ c	85	87	37	1	8	2	<1	t	t	.1
810	Whole beets, 2" diam	2 ea	100	87	44	2	10	2	<1	t	t	.1
	Beets, canned:											
811	Sliced or diced	½ c	79	91	24	1	6	1	<1	t	t	t
812	Pickled slices	½ c	114	82	74	1	19	2	<1	t	t	t
813	Beet greens, cooked, drained	½ c	72	89	19	2	4	2	<1	t	t	.1

PAGE KEY: A–4 = Beverages   A–6 = Dairy   A–10 = Eggs   A–10 = Fat/Oil   A–14 = Fruit   A–20 = Bakery   A–28 = Grain   A–32 = Fish   A–34 = Meats   A–38 = Poultry   A–40 = Sausage   A–42 = Mixed/Fast   A–46 = Nuts/Seeds   A–50 = Sweets   A–52 = Vegetables/Legumes   A–62 = Vegetarian Foods   A–64 = Misc   A–66 = Soups/Sauces   A–68 = Fast   A–84 = Convenience   A–88 = Baby foods

A–53

Chol (mg)	Calc (mg)	Iron (mg)	Magn (mg)	Pota (mg)	Sodi (mg)	Zinc (mg)	VT-A (RE)	Thia (mg)	VT-E (a-TE)	Ribo (mg)	Niac (mg)	V-B6 (mg)	Fola (µg)	VT-C (mg)
5	43	.52	21	92	56	.34	9	.01	0	.09	.09	.01	2	<1
0	5	5.17	25	183	58	.28	494	<.01	.01	.31	12.8	.01	2	<1
0	353	7.18	88	1021	23	.41	0	.01	0	.02	.44	.29	<1	0
0	68	1.76	100	376	6	.12	0	.03	0	.02	.08	.27	0	0
0		.04	1	1	33	.02	0	<.01	0	<.01	.01	0	0	0
0	11	.32	9	26	2	.3	5	.02	—	.04	.16	.01	12	3
0	60	.65	15	171	6	.25	82	.01	.22	.04	.18	.05	24	12
0	30	.32	8	85	3	.13	41	<.01	.11	.02	.09	.03	12	6
0	276	2.98	73	846	28	1.16	366	.03	.66	.18	.74	.23	75	54
0	16	.15	5	37	3	.05	24	<.01	.04	.01	.03	.01	10	1
0	54	1.55	72	425	114	.59	22	.08	.23	.08	1.2	.13	61	12
0	35	.94	52	444	89	.6	27	.1	.32	.26	1.54	.15	200	8
0	30	1.24	37	335	688	.41	21	.05	1.43	.13	1.06	.11	114	40
0	0	1.35		0	250		12							4
0	18	.66	9	144	10	.38	49	.11	.34	.11	.97	.11	131	10
0	12	.44	6	96	7	.25	32	.07	.23	.08	.65	.07	88	6
0	21	.58	12	196	4	.5	74	.06	1.13	.09	.94	.02	122	22
0	14	.38	8	131	2	.34	49	.04	.75	.06	.62	.01	81	15
0	11	1.32	7	124	207	.29	38	.04	.31	.07	.69	.08	69	13
0	10	.42	5	105	9	.85	1	.03	.5	.03	.18	.18	4	1
0	20	.75	5	805	6	1.66	3	.23	1.51	.11	.91	.36	11	6
0	14	.29	4	640	5	.56	0	.02	.8	.06	.36	.12	3	1
0	32	2.3	60	612	9	2.04	1	.13	.11	.07	.82	.11	139	0
0	23	1.81	60	305	1	.96	1	.21	.07	.05	.43	.06	128	0
9	71	4.17	44	381	559	7.44	15	.07	.69	.06	.63	.09	29	4
9	79	2.16	44	346	437	1.95	14	.06	.7	.08	.46	.11	49	4
8	62	2.25	36	306	559	2.43	19	.07	.61	.07	1.17	.06	39	3
0	19	1.16	29	347	45	.37	16	.06	.25	.05	.91	.1	18	11
0	25	1.76	50	370	26	.49	15	.06	.58	.05	.69	.1	14	5
0	16	2.25	40	478	2	.89	0	.15	.17	.05	.4	.15	78	0
0	42	1.86	48	369	240	.87	<1	.13	.08	.07	.37	.11	94	2
0	29	.81	16	188	2	.23	42	.05	.09	.06	.39	.03	21	6
0	33	.6	16	86	6	.33	27	.02	.09	.06	.26	.04	16	3
0	18	.61	9	74	178	.2	24	.01	.09	.04	.14	.02	22	3
0	29	.81	16	188	2	.23	5	.05	.18	.06	.39	.03	21	6
0	7	.47	11	77	3	.21	1	.04	.02	.06	.39	.05	32	7
0	8	1.18	20	136	6	.56	2	.09	.01	.11	.74	.08	43	10
0	7	.4	9	63	6	.29	1	.03	.01	.06	.51	.03	18	7
0	9	.27	6	17	88	.18	1	.02	.01	.04	.14	.02	6	<1
0	14	.67	20	259	65	.3	3	.02	.25	.03	.28	.06	68	3
0	16	.79	23	305	77	.35	4	.03	.3	.04	.33	.07	80	4
0	12	1.44	13	117	153	.17	1	.01	.24	.03	.12	.04	24	3
0	12	.47	17	169	301	.3	1	.01	.15	.05	.29	.06	30	3
0	82	1.37	49	654	174	.36	367	.08	.22	.21	.36	.09	10	18

Table A–1

Food Composition (Computer code number is for West Diet Analysis program) (For purposes of calculations, use "0" for t, <1, <.1, <.01, etc.)

Computer Code Number	Food Description	Measure	Wt (g)	H₂O (%)	Ener (cal)	Prot (g)	Carb (g)	Dietary Fiber (g)	Fat (g)	Fat Breakdown (g) Sat	Mono	Poly
	VEGETABLES and LEGUMES—Continued											
	Broccoli, raw:											
817	Chopped	½ c	44	91	12	1	2	1	<1	t	t	.1
818	Spears	1 ea	31	91	9	1	2	1	<1	t	t	.1
	Broccoli, cooked from fresh:											
819	Spears	1 ea	180	91	50	5	9	5	1	.1	t	.3
820	Chopped	½ c	78	91	22	2	4	2	<1	t	t	.1
	Broccoli, cooked from frozen:											
821	Spear, small piece	½ c	92	91	26	3	5	3	<1	t	t	.1
822	Chopped	½ c	92	91	26	3	5	3	<1	t	t	.1
1603	Broccoflower, steamed	½ c	78	90	25	2	5	2	<1	t	t	.1
823	Brussels sprouts, cooked from fresh	½ c	78	87	30	2	7	2	<1	.1	t	.2
824	Brussels sprouts, cooked from frozen	½ c	78	87	33	3	6	3	<1	.1	t	.2
	Cabbage, common varieties:											
825	Raw, shredded or chopped	1 c	70	92	17	1	4	2	<1	t	t	.1
826	Cooked, drained	1 c	150	94	33	2	7	3	1	.1	t	.3
	Cabbage, Chinese:											
1178	Bok choy, raw, shredded	1 c	70	95	9	1	2	1	<1	t	t	.1
827	Bok choy, cooked, drained	1 c	170	96	20	3	3	3	<1	t	t	.1
1937	Kim chee style	1 c	150	92	31	2	6	2	<1	t	t	.2
828	Pe Tsai, raw, chopped	1 c	76	94	12	1	2	2	<1	t	t	.1
1796	Pe Tsai, cooked	1 c	119	95	17	2	3	3	<1	t	t	.1
	Cabbage, red, coarsely chopped:											
829	Raw	1 c	89	92	24	1	5	2	<1	t	t	.1
830	Cooked, drained	1 c	150	94	31	2	7	3	<1	t	t	.1
831	Cabbage, savoy, coarsely chopped, raw	1 c	70	91	19	1	4	2	<1	t	t	t
1785	Cabbage, savoy, cooked	1 c	145	92	35	3	8	4	<1	t	t	.1
1896	Capers	1 ea	5	86		<1	<1	<1	<1			
	Carrots, raw:											
832	Whole, 7½ x 1⅛"	1 ea	72	88	31	1	7	2	<1	t	t	.1
833	Grated	½ c	55	88	24	1	6	2	<1	t	t	t
	Carrots, cooked, sliced, drained:											
834	From raw	½ c	78	87	35	1	8	3	<1	t	t	.1
835	From frozen	½ c	73	90	26	1	6	3	<1	t	t	t
836	Carrots, canned, sliced, drained	½ c	73	93	17	<1	4	1	<1	t	t	.1
837	Carrot juice, canned	1 c	236	89	94	2	22	2	<1	.1	t	.2
	Cauliflower, flowerets:											
838	Raw	½ c	50	92	12	1	3	1	<1	t	t	t
839	Cooked from fresh, drained	½ c	62	93	14	1	3	2	<1	t or t	t	.1
840	Cooked, from frozen, drained	½ c	90	94	17	1	3	2	<1	t	t	.1
	Celery, pascal type, raw:											
841	Large outer stalk, 8 x 1½"(root end)	1 ea	40	95	6	<1	1	1	<1	t	t	t
842	Diced	1 c	120	95	19	1	4	2	<1	t	t	.1
1789	Celeriac/celery root, cooked	1 c	155	92	39	1	9	2	<1	.1	.1	.2
1179	Chard, swiss, raw, chopped	1 c	36	93	7	1	1	1	<1	t	t	t
1180	Chard, swiss, cooked	1 c	175	93	35	3	7	4	<1	t	t	t
1855	Chayote fruit, raw	1 ea	203	94	39	2	9	3	<1	.1	t	.1
1856	Chayote fruit, cooked	1 c	160	93	38	1	8	4	1	.2	.1	.3
	Chickpeas (see Garbanzo Beans #854)											
	Collards, cooked, drained:											
843	From raw	½ c	95	92	25	2	5	3	<1	t	t	.1
844	From frozen	½ c	85	88	31	3	6	3	<1	.1	t	.2
	Corn, yellow, cooked, drained:											
845	From raw, on cob, 5" long	1 ea	77	73	72	2	17	2	1	.1	.2	.3
846	From frozen, on cob, 3½" long	1 ea	63	73	59	2	14	2	<1	.1	.1	.2
847	Kernels, cooked from frozen	½ c	82	77	66	2	16	2	<1	.1	.1	.2
	Corn, canned:											
848	Cream style	½ c	128	79	92	2	23	2	1	.1	.2	.3
849	Whole kernel, vacuum pack	½ c	105	77	83	3	20	2	1	.1	.2	.2
	Cowpeas (see Black-eyed peas #814-816)											
850	Cucumber slices with peel	7 pce	28	96	4	<1	1	<1	<1	t	t	t
1948	Cucumber, kim chee style	1 c	150	91	31	2	7	2	<1	t	0	.1

Chol (mg)	Calc (mg)	Iron (mg)	Magn (mg)	Pota (mg)	Sodi (mg)	Zinc (mg)	VT-A (RE)	Thia (mg)	VT-E (a-TE)	Ribo (mg)	Niac (mg)	V-B6 (mg)	Fola (μg)	VT-C (mg)
0	21	.39	11	143	12	.18	68	.03	.73	.05	.28	.07	31	41
0	15	.27	8	101	8	.12	48	.02	.51	.04	.2	.05	22	29
0	83	1.51	43	526	47	.68	250	.1	3.04	.2	1.03	.26	90	134
0	36	.65	19	228	20	.3	108	.04	1.32	.09	.45	.11	39	58
0	47	.56	18	166	22	.28	174	.05	.95	.07	.42	.12	28	37
0	47	.56	18	166	22	.28	174	.05	1.52	.07	.42	.12	52	37
0	25	.55	16	251	18	.39	5	.06	.23	.07	.59	.14	38	49
0	28	.94	16	247	16	.26	56	.08	.66	.06	.47	.14	47	48
0	19	.58	19	254	18	.28	46	.08	.45	.09	.42	.22	79	36
0	33	.41	10	172	13	.13	9	.03	.07	.03	.21	.07	30	22
0	46	.25	12	146	12	.13	19	.09	.16	.08	.42	.17	30	30
0	73	.56	13	176	45	.13	210	.03	.08	.05	.35	.14	46	31
0	158	1.77	19	631	58	.29	437	.05	.2	.11	.73	.28	69	44
0	145	1.28	27	375	995	.35	426	.07	.24	.1	.75	.34	88	80
0	58	.24	10	181	7	.17	91	.03	.09	.04	.3	.18	60	20
0	38	.36	12	268	11	.21	115	.05	.14	.05	.59	.21	63	19
0	45	.44	13	183	10	.19	4	.04	.09	.03	.27	.19	18	51
0	55	.52	16	210	12	.22	4	.05	.18	.03	.3	.21	19	52
0	24	.28	20	161	20	.19	70	.05	.07	.02	.21	.13	56	22
0	43	.55	35	267	35	.33	129	.07	.15	.03	.03	.22	67	25
0	2	.05			105		1							0
0	19	.36	11	233	25	.14	2025	.07	.33	.04	.67	.11	10	7
0	15	.27	8	178	19	.11	1547	.05	.25	.03	.51	.08	8	5
0	24	.48	10	177	51	.23	1914	.03	.33	.04	.39	.19	11	2
0	20	.34	7	115	43	.17	1292	.02	.31	.03	.32	.09	8	2
0	18	.47	6	131	177	.19	1005	.01	.31	.02	.4	.08	7	2
0	57	1.09	33	689	68	.42	2584	.22	.02	.13	.91	.51	9	20
0	11	.22	7	152	15	.14	1	.03	.02	.03	.26	.11	28	23
0	10	.2	6	88	9	.11	1	.03	.02	.03	.25	.11	27	27
0	15	.37	8	125	16	.12	2	.03	.04	.05	.28	.08	37	28
0	16	.16	4	115	35	.05	5	.02	.14	.02	.13	.03	11	3
0	48	.48	13	344	104	.16	16	.05	.43	.05	.39	.1	34	8
0	40	.67	19	268	95	.31	0	.04	.31	.06	.66	.16	5	6
0	18	.65	29	136	77	.13	119	.01	.68	.03	.14	.04	5	11
0	102	3.96	151	961	313	.58	550	.06	3.31	.15	.63	.15	15	31
0	34	.69	24	254	4	1.5	12	.05	.24	.06	.95	.15	189	16
0	21	.35	19	277	2	.5	8	.04	.19	.06	.67	.19	29	13
0	113	.44	16	247	9	.4	297	.04	.84	.1	.55	.12	88	17
0	179	.95	25	213	42	.23	508	.04	.42	.1	.54	.1	65	22
0	2	.47	22	193	3	.48	16	.13	.07	.05	1.17	.17	23	4
0	2	.38	18	158	3	.4	13	.11	.06	.04	.96	.14	19	3
0	3	.29	16	121	4	.33	18	.07	.07	.06	1.07	.11	25	3
0	4	.49	22	172	365	.68	13	.03	.11	.07	1.23	.08	57	6
0	5	.44	24	195	286	.48	25	.04	.09	.08	1.23	.06	52	9
0	4	.07	3	40	1	.06	6	.01	.02	.01	.06	.01	4	1
0	13	7.23	12	176	1531	.76	49	.04	.24	.04	.69	.16	34	5

Table A–1

Food Composition   (Computer code number is for West Diet Analysis program)   (For purposes of calculations, use "0" for t, <1, <.1, <.01, etc.)

Computer Code Number	Food Description	Measure	Wt (g)	H₂O (%)	Ener (cal)	Prot (g)	Carb (g)	Dietary Fiber (g)	Fat (g)	Fat Breakdown (g) Sat	Mono	Poly
	VEGETABLES and LEGUMES—Continued											
	Dandelion Greens:											
851	Raw	1 c	55	86	25	1	5	2	<1	.1	t	.2
852	Chopped, cooked, drained	1 c	105	90	35	2	7	3	1	.2	t	.3
853	Eggplant, cooked	1 c	99	92	28	1	7	2	<1	t	t	.1
1714	Endive, fresh, chopped	1 c	50	94	8	1	2	2	<1	t	t	t
856	Escarole/curly endive, chopped	1 c	50	94	8	1	2	2	<1	t	t	t
854	Garbanzo beans (chickpeas), cooked	1 c	164	60	269	14	45	12	4	.4	1	1.9
1939	Grape leaf, raw:	1 ea	3	73	3	<1	1	<1	<1	t	t	t
7914	Cup	1 c	14	73	13	1	2	2	<1	t	t	.1
855	Great northern beans, cooked	1 c	177	69	209	15	37	12	1	.2	t	.3
857	Jerusalem artichoke, raw slices	1 c	150	78	114	3	26	2	<1	0	t	t
1794	Jicama	1 c	120	90	46	1	11	6	<1	t	t	t
	Kale, cooked, drained:											
858	From raw	1 c	130	91	36	2	7	3	1	.1	t	.3
859	From frozen	1 c	130	90	39	4	7	3	1	.1	t	.3
860	Kidney beans, canned	1 c	256	77	218	13	40	16	1	.1	.1	.5
1181	Kohlrabi, raw slices	1 c	135	91	36	2	8	5	<1	t	t	.1
861	Kohlrabi, cooked	1 c	165	90	48	3	11	2	<1	t	t	.1
1183	Leeks, raw, chopped	1 c	89	83	54	1	13	2	<1	t	t	.1
1182	Leeks, cooked, chopped	1 c	104	91	32	1	8	1	<1	t	t	.1
862	Lentils, cooked from dry	1 c	198	70	230	18	40	16	1	.1	.1	.3
1288	Lentils, sprouted, stir-fried	1 c	124	69	125	11	26	5	1	.1	.1	.2
1289	Lentils, sprouted, raw	1 c	77	67	82	7	17	3	<1	t	.1	.2
	Lettuce:											
	Butterhead/Boston types:											
863	Head, 5" diameter	¼ ea	41	96	5	1	1	<1	<1	t	t	t
864	Leaves, inner or outer	4 ea	30	96	4	<1	1	<1	<1	t	t	t
	Iceberg/crisphead:											
867	Chopped or shredded	1 c	55	96	7	1	1	1	<1	t	t	.1
865	Head, 6" diameter	1 ea	539	96	65	5	11	8	1	.1	t	.5
866	Wedge, ¼ head	1 ea	135	96	16	1	3	2	<1	t	t	.1
868	Looseleaf, chopped	½ c	28	94	5	<1	1	1	<1	t	t	t
869	Romaine, chopped	½ c	28	95	4	<1	1	<1	<1	t	t	t
870	Romaine, inner leaf	3 pce	30	95	4	<1	1	1	<1	t	t	t
1930	Luffa, cooked (Chinese okra)	1 c	178	89	57	3	13	6	<1	.1	t	.1
	Mushrooms:											
871	Raw, sliced	½ c	35	92	9	1	2	<1	<1	t	t	.1
872	Cooked from fresh, pieces	½ c	78	91	21	2	4	2	<1	t	t	.1
1962	Stir fried, shitake slices	½ c	73	83	40	1	10	2	<1	t	t	t
873	Canned, drained	½ c	78	91	19	1	4	2	<1	t	t	.1
1951	Mushroom caps, pickled	8 ea	47	92	11	1	2	1	<1	t	t	.1
	Mustard greens:											
874	Cooked from raw	½ c	70	94	10	2	1	1	<1	t	.1	t
875	Cooked from frozen	½ c	75	94	14	2	2	2	<1	t	.1	t
876	Navy beans, cooked from dry	1 c	182	63	258	16	48	12	1	.3	.1	.4
	Okra, cooked:											
877	From fresh pods	8 ea	85	90	27	2	6	2	<1	t	t	t
878	From frozen slices	1 c	184	91	51	4	11	5	1	.1	.1	.1
1236	Batter fried from fresh	1 c	92	69	175	3	11	2	13	2.1	3.4	7.1
1930	Chinese, (Luffa), cooked	1 c	178	89	57	3	13	6	<1	.1	t	.1
	Onions:											
879	Raw, chopped	½ c	80	90	30	1	7	1	<1	t	t	t
880	Raw, sliced	½ c	58	90	22	1	5	1	<1	t	t	t
881	Cooked, drained, chopped	½ c	105	88	46	1	11	1	<1	t	t	.1
882	Dehydrated flakes	¼ c	14	4	49	1	12	1	<1	t	t	t
1934	Onions, pearl, cooked	½ c	93	87	41	1	9	1	<1	t	t	.1
883	Spring/green onions, bulb and top, chopped	½ c	50	90	16	1	4	1	<1	t	t	t
884	Onion rings, breaded, heated f/frozen	2 ea	20	28	81	1	8	<1	5	1.7	2.2	1
1917	Palm hearts, cooked slices	1 c	146	69	150	4	39	2	<1	.1	.1	t
885	Parsley, raw, chopped	½ c	30	88	11	1	2	1	<1	t	.1	t
888	Parsnips, sliced, cooked	½ c	78	78	63	1	15	3	<1	t	.1	t

Chol (mg)	Calc (mg)	Iron (mg)	Magn (mg)	Pota (mg)	Sodi (mg)	Zinc (mg)	VT-A (RE)	Thia (mg)	VT-E (a-TE)	Ribo (mg)	Niac (mg)	V-B6 (mg)	Fola (µg)	VT-C (mg)
0	103	1.71	20	218	42	.23	770	.1	1.38	.14	.44	.14	15	19
0	147	1.89	25	244	46	.29	1228	.14	2.63	.18	.54	.17	13	19
0	6	.35	13	246	3	.15	6	.07	.03	.02	.59	.08	14	1
0	26	.41	7	157	11	.39	103	.04	.22	.04	.2	.01	71	3
0	26	.41	7	157	11	.39	103	.04	.22	.04	.2	.01	71	3
0	80	4.74	79	477	11	2.51	5	.19	.57	.1	.86	.23	282	2
0	11	.08	3	8	<1	.02	81	<.01	.06	.01	.07	.01	2	<1
0	51	.37	13	38	1	.09	378	.01	.28	.05	.33	.06	12	2
0	120	3.77	88	692	4	1.56	<1	.28	.53	.1	1.21	.21	181	2
0	21	5.1	25	644	6	.18	3	.3	.28	.09	1.95	.12	20	6
0	14	.72	14	180	5	.19	2	.02	5.48	.03	.24	.05	14	24
0	94	1.17	23	296	30	.31	962	.07	1.11	.09	.65	.18	17	53
0	179	1.22	23	417	19	.23	826	.06	.23	.15	.87	.11	19	33
0	61	3.23	72	658	873	1.41	0	.27	.13	.22	1.17	.06	130	3
0	32	.54	26	473	27	.04	5	.07	.65	.03	.54	.2	22	84
0	41	.66	31	561	35	.51	7	.07	2.76	.03	.64	.25	20	89
0	52	1.87	25	160	18	.11	9	.05	.82	.03	.36	.21	57	11
0	31	1.14	15	90	10	.06	5	.03	.63	.02	.21	.12	25	4
0	38	6.59	71	731	4	2.51	2	.33	.22	.14	2.1	.35	358	3
0	17	3.84	43	352	12	1.98	5	.27	.11	.11	1.49	.2	83	16
0	19	2.47	28	248	8	1.16	4	.18	.07	.1	.87	.15	77	13
0	13	.12	5	105	2	.07	40	.02	.18	.02	.12	.02	30	3
0	10	.09	4	77	1	.05	29	.02	.13	.02	.09	.01	22	2
0	10	.27	5	87	5	.12	18	.02	.15	.02	.1	.02	31	2
0	102	2.7	48	852	48	1.19	178	.25	1.51	.16	1.01	.22	302	21
0	26	.67	12	213	12	.3	45	.06	.38	.04	.25	.05	76	5
0	19	.39	3	74	3	.08	53	.01	.12	.02	.11	.01	14	5
0	10	.31	2	81	2	.07	73	.03	.12	.03	.14	.01	38	7
0	11	.33	2	87	2	.07	78	.03	.13	.03	.15	.01	41	7
0	112	.8	101	570	420	.97	103	.23	1.22	.1	1.54	.33	81	29
0	2	.43	3	130	1	.26	0	.04	.04	.16	1.44	.03	7	1
0	5	1.36	9	278	2	.68	0	.06	.09	.23	3.48	.07	14	3
0	2	.32	10	85	3	.97	0	.03	.09	.12	1.1	.12	15	<1
0	9	.62	12	101	332	.56	0	.07	.09	.02	1.24	.05	10	0
0	2	.5	5	139	95	.28	0	.03	.05	.16	1.42	.03	6	1
0	52	.49	10	141	11	.08	212	.03	1.41	.04	.3	.07	51	18
0	76	.84	10	104	19	.15	335	.03	1.31	.04	.19	.08	52	10
0	127	4.51	107	670	2	1.93	<1	.37	.73	.11	.97	.3	255	2
0	54	.38	48	274	4	.47	49	.11	.59	.05	.74	.16	39	14
0	177	1.23	94	431	6	1.14	94	.18	1.27	.23	1.44	.09	269	22
15	104	.77	37	214	137	.5	43	.13	3.08	.1	.75	.13	37	10
0	112	.8	101	570	420	.97	103	.23	1.22	.1	1.54	.33	81	29
0	16	.18	8	126	2	.15	0	.03	.1	.02	.12	.09	15	5
0	12	.13	6	91	2	.11	0	.02	.07	.01	.09	.07	11	4
0	23	.25	12	174	3	.22	0	.04	.14	.02	.17	.13	16	5
0	36	.22	13	227	3	.26	0	.07	.19	.01	.14	.22	23	10
0	21	.22	10	154	218	.19	0	.04	.12	.02	.15	.12	14	5
0	36	.74	10	138	8	.19	19	.03	.06	.04	.26	.03	32	9
0	6	.34	4	26	75	.08	5	.06	.14	.03	.72	.01	13	<1
0	26	2.47	15	2636	20	5.45	10	.07	.73	.25	1.25	1.06	30	10
0	41	1.86	15	166	17	.32	156	.03	.54	.03	.39	.03	46	40
0	29	.45	23	286	8	.2	0	.06	.78	.04	.56	.07	45	10

Table A–1

Food Composition   (Computer code number is for West Diet Analysis program)   (For purposes of calculations, use "0" for t, <1, <.1, <.01, etc.)

Computer Code Number	Food Description	Measure	Wt (g)	H₂O (%)	Ener (cal)	Prot (g)	Carb (g)	Dietary Fiber (g)	Fat (g)	Fat Breakdown (g)		
										Sat	Mono	Poly
	VEGETABLES and LEGUMES—Continued											
	Peas:											
	Black-eyed, cooked:											
814	From dry, drained	½ c	86	70	100	7	18	6	<1	.1	t	.2
815	From fresh, drained	½ c	82	75	79	3	17	4	<1	.1	t	.1
816	From frozen, drained	½ c	85	66	112	7	20	5	1	.1	.1	.2
889	Edible pod peas, cooked	½ c	80	89	34	3	6	2	<1	t	t	.1
890	Green, canned, drained:	½ c	85	82	59	4	11	3	<1	.1	t	.1
5267	Unsalted	½ c	124	86	66	4	12	4	<1	.1	t	.2
891	Green, cooked from frozen	½ c	80	79	62	4	11	4	<1	t	t	.1
1786	Snow peas, raw	½ c	49	89	21	1	4	1	<1	t	t	t
1787	Snow peas, raw	10 ea	34	89	14	1	3	1	<1	t	t	t
892	Split, green, cooked from dry	½ c	98	69	116	8	21	8	<1	.1	.1	.2
1187	Peas & carrots, cooked from frozen	½ c	80	86	38	2	8	2	<1	.1	t	.2
1186	Peas & carrots, canned w/liquid	½ c	128	88	49	3	11	3	<1	.1	t	.2
	Peppers, hot:											
893	Hot green chili, canned	½ c	68	92	14	1	3	1	<1	t	t	t
894	Hot green chili, raw	1 ea	45	88	18	1	4	1	<1	t	t	t
1715	Hot red chili, raw, diced	1 tbs	9	88	4	<1	1	<1	<1	t	t	t
1988	Jalapeno, raw	1 ea	45	90	11	<1	2		<1			
895	Jalapeno, chopped, canned	½ c	68	89	18	1	3	2	1	.1	t	.3
1918	Jalapeno wheels, in brine (Ortega)	2 tbs	29		10	0	2		0	0	0	0
	Peppers, sweet, green:											
896	Whole pod (90 g with refuse), raw	1 ea	74	92	20	1	5	1	<1	t	t	.1
897	Cooked, chopped (1 pod cooked = 73g)	½ c	68	92	19	1	5	1	<1	t	t	.1
	Peppers, sweet, red:											
1286	Raw, chopped	½ c	75	92	20	1	5	1	<1	t	t	.1
1807	Raw, each	1 ea	74	92	20	1	5	1	<1	t	t	.1
1287	Cooked, chopped	½ c	68	92	19	1	5	1	<1	t	t	.1
	Peppers, sweet, yellow:											
1872	Raw, large	1 ea	186	92	50	2	12	2	<1	.1	t	.2
1873	Strips	10 pce	52	92	14	1	3	<1	<1	t	t	.1
898	Pinto beans, cooked from dry	½ c	85	64	116	7	22	7	<1	.1	.1	.2
1191	Poi, two finger	½ c	120	72	134	<1	33	<1	<1	t	t	.1
	Potatoes:											
	Baked in oven, 4¾"x2⅓" diam											
899	With skin	1 ea	202	71	220	5	51	5	<1	.1	t	.1
900	Flesh only	1 ea	156	75	145	3	34	2	<1	t	t	.1
901	Skin only	1 ea	58	47	115	2	27	5	<1	t	t	t
	Baked in microwave, 4¾"x 2⅓"dm:											
902	With skin	1 ea	202	72	212	5	49	5	<1	.1	t	.1
903	Flesh only	1 ea	156	74	156	3	36	2	<1	t	t	.1
904	Skin only	1 ea	58	63	77	3	17	3	<1	t	t	t
	Boiled, about 2½" diam:											
905	Peeled after boiling	1 ea	136	77	118	3	27	2	<1	t	t	.1
906	Peeled before boiling	1 ea	135	77	116	2	27	2	<1	t	t	.1
	French fried, strips 2–3½" long:											
907	Oven heated	10 ea	50	35	167	2	20	2	9	3	5.7	.7
908	Fried in vegetable oil	10 ea	50	38	158	2	20	2	8	1.9	4.7	.7
1188	Fried in veg and animal oil	10 ea	50	38	158	2	20	2	8	1.9	4.7	.7
909	Hashed browns from frozen	1 c	156	56	340	5	44	3	18	7	8	2.1
	Mashed:											
910	Home recipe with whole milk	½ c	105	78	81	2	18	2	1	.4	.2	.1
911	Home recipe with milk and marg	½ c	105	76	111	2	17	2	4	1.1	1.9	1.3
912	Prepared from flakes; water, milk, margarine, salt added	½ c	110	76	124	2	16	3	6	1.6	2.5	1.7
	Potato products, prepared:											
	Au gratin:											
913	From dry mix	½ c	123	79	114	3	16	1	5	3.5	1.5	.2
914	From home recipe, using butter	½ c	122	74	161	7	14	2	9	4.8	3.2	1.3
	Scalloped:											
915	From dry mix	½ c	122	79	113	3	16	1	5	3.2	1.5	.2
916	From home recipe, using butter	½ c	123	81	106	4	13	2	5	1.7	1.7	.9

PAGE KEY: A–4 = Beverages   A–6 = Dairy   A–10 = Eggs   A–10 = Fat/Oil   A–14 = Fruit   A–20 = Bakery   A–28 = Grain   A–32 = Fish   A–34 = Meats   A–38 = Poultry   A–40 = Sausage   A–42 = Mixed/Fast   A–46 = Nuts/Seeds   A–50 = Sweets   A–52 = Vegetables/Legumes   A–62 = Vegetarian Foods   A–64 = Misc   A–66 = Soups/Sauces   A–68 = Fast   A–84 = Convenience   A–88 = Baby foods

Chol (mg)	Calc (mg)	Iron (mg)	Magn (mg)	Pota (mg)	Sodi (mg)	Zinc (mg)	VT-A (RE)	Thia (mg)	VT-E (a-TE)	Ribo (mg)	Niac (mg)	V-B6 (mg)	Fola (μg)	VT-C (mg)
0	21	2.16	46	239	3	1.11	2	.17	.24	.05	.43	.09	179	<1
0	105	.92	43	343	3	.84	65	.08	.18	.12	1.15	.05	104	2
0	20	1.8	42	319	4	1.21	7	.22	.33	.05	.62	.08	120	2
0	34	1.58	21	192	3	.3	10	.1	.31	.06	.43	.11	23	38
0	17	.81	14	147	214	.6	65	.1	.32	.07	.62	.05	38	8
0	22	1.26	21	124	11	.87	47	.14	.47	.09	1.04	.08	35	12
0	19	1.26	23	134	70	.75	54	.23	.14	.08	1.18	.09	47	8
0	21	1.02	12	98	2	.13	7	.07	.19	.04	.29	.08	20	29
0	15	.71	8	68	1	.09	5	.05	.13	.03	.2	.05	14	20
0	14	1.26	35	355	2	.98	1	.19	.38	.05	.87	.05	64	<1
0	18	.75	13	126	54	.36	621	.18	.26	.05	.92	.07	21	6
0	29	.96	18	128	333	.74	739	.09	.24	.07	.74	.11	23	8
0	5	.34	10	127	798	.12	41	.01	.47	.03	.54	.1	7	46
0	8	.54	11	153	3	.13	35	.04	.31	.04	.43	.12	10	109
0	2	.11	2	31	1	.03	97	.01	.06	.01	.09	.02	2	22
			2	2			30		.37					53
0	16	1.28	10	131	1136	.23	116	.03	.47	.03	.27	.13	10	7
0				55	390		10		.2					21
0	7	.34	7	131	1	.09	47	.05	.51	.02	.38	.18	16	66
0	6	.31	7	113	1	.08	40	.04	.47	.02	.32	.16	11	51
0	7	.34	7	133	1	.09	428	.05	.52	.02	.38	.19	16	143
0	7	.34	7	131	1	.09	422	.05	.51	.02	.38	.18	16	141
0	6	.31	7	113	1	.08	256	.04	.47	.02	.32	.16	11	116
0	20	.86	22	394	4	.32	45	.05	1.28	.05	1.66	.31	48	342
0	6	.24	6	110	1	.09	12	.01	.36	.01	.46	.09	13	96
0	41	2.22	47	398	2	.92	<1	.16	.8	.08	.34	.13	146	2
0	19	1.06	29	220	14	.26	2	.16	.22	.05	1.32	.33	26	5
0	20	2.75	54	844	16	.65	0	.22	.1	.07	3.33	.7	22	26
0	8	.55	39	610	8	.45	0	.16	.06	.03	2.18	.47	14	20
0	20	4.08	25	332	12	.28	0	.07	.02	.06	1.78	.36	12	8
0	22	2.5	54	903	16	.73	0	.24	.1	.06	3.45	.69	24	30
0	8	.64	39	641	11	.51	0	.2	.06	.04	2.54	.5	19	24
0	27	3.45	21	377	9	.3	0	.04	.02	.04	1.29	.28	10	9
0	7	.42	30	515	5	.41	0	.14	.07	.03	1.96	.41	14	18
0	11	.42	27	443	7	.36	0	.13	.07	.03	1.77	.36	12	10
0	6	.83	11	270	307	.2	0	.04	.25	.02	1.34	.11	11	3
0	9	.38	17	366	108	.19	0	.09	.25	.01	1.63	.12	14	5
6	9	.38	17	366	108	.19	0	.09	.25	.01	1.63	.12	14	5
0	23	2.36	26	680	53	.5	0	.17	.3	.03	3.78	.2	10	10
2	27	.28	19	314	318	.3	6	.09	.05	.04	1.18	.24	9	7
2	27	.27	19	303	310	.28	21	.09	.31	.04	1.13	.23	8	6
4	54	.24	20	256	365	.2	23	.12	.77	.05	.74	.01	8	11
18	102	.39	18	269	540	.29	38	.02	1.48	.1	1.15	.05	8	4
18	145	.78	24	483	528	.84	46	.08	.64	.14	1.21	.21	13	12
13	44	.46	17	248	416	.3	26	.02	.18	.07	1.26	.05	12	4
7	70	.7	23	465	412	.49	23	.08	.4	.11	1.29	.22	13	13

Table A–1

Food Composition   (Computer code number is for West Diet Analysis program)   (For purposes of calculations, use "0" for t, <1, <.1, <.01, etc.)

Computer Code Number	Food Description	Measure	Wt (g)	H₂O (%)	Ener (cal)	Prot (g)	Carb (g)	Dietary Fiber (g)	Fat (g)	Fat Breakdown (g) Sat	Mono	Poly
	VEGETABLES and LEGUMES—Continued											
	Potato Salad (see Mixed Dishes #715)											
1192	Potato puffs, cooked from frozen	½ c	64	53	142	2	19	2	7	3.3	2.8	.5
918	Pumpkin, cooked from fresh, mashed	½ c	123	94	25	1	6	1	<1	t	t	t
919	Pumpkin, canned	½ c	123	90	42	1	10	4	<1	.2	t	t
1891	Radicchio, raw, shredded	½ c	20	93	5	<1	1	<1	<1	t	t	t
1894	Radicchio, raw, leaf	10 ea	80	93	18	1	4	1	<1	t	t	.1
920	Red radishes	10 ea	45	95	8	<1	2	1	<1	t	t	t
1793	Daikon radishes (Chinese) raw	½ c	44	95	8	<1	2	1	<1	t	t	t
921	Refried beans, canned	½ c	126	76	118	7	19	7	2	.6	.7	.2
1375	Rutabaga, cooked cubes	½ c	85	89	33	1	7	2	<1	t	t	.1
922	Sauerkraut, canned with liquid	½ c	118	92	22	1	5	3	<1	t	t	.1
923	Seaweed, kelp, raw	½ c	40	82	17	1	4	1	<1	.1	t	t
924	Seaweed, spirulina, dried	½ c	8	5	23	5	2	<1	1	.2	.1	.2
1866	Shallots, raw, chopped	1 tbs	10	80	7	<1	2	<1	<1	t	t	t
1557	Snow peas, stir-fried	½ c	83	89	35	2	6	2	<1	t	t	.1
925	Soybeans, cooked from dry	½ c	86	63	149	15	9	5	8	1.1	1.7	4.4
1996	Soybeans, dry roasted	½ c	86	1	387	34	28	7	19	2.7	4.1	10.6
	Soybean products:											
926	Miso	½ c	138	41	284	17	39	7	8	1.2	1.9	4.7
	Soy milk (see #144 and #2301 under Dairy)											
	Tofu (soybean curd):											
7540	Extra firm, silken	½ c	126	88	69	9	3	<1	2	.4	.4	1.3
7542	Firm, silken	½ c	126	87	77	9	3	<1	3	.5	.7	1.9
927	Regular	½ c	124	87	76	8	2	<1	5	.7	1	2.6
7541	Soft, silken	½ c	124	89	68	6	4	<1	3	.4	.6	1.9
	Spinach:											
928	Raw, chopped	½ c	28	92	6	1	1	1	<1	t	t	t
929	Cooked, from fresh, drained	½ c	90	91	21	3	3	2	<1	t	t	.1
930	Cooked from frozen (leaf)	½ c	95	90	27	3	5	3	<1	t	t	.1
931	Canned, drained solids:	½ c	107	92	25	3	4	3	1	.1	t	.2
5149	Unsalted	½ c	107	92	25	3	4	3	1	.1	t	.2
	Spinach soufflé (see Mixed Dishes)											
	Squash, summer varieties, cooked w/skin:											
932	Varieties averaged	½ c	90	94	18	1	4	1	<1	.1	t	.1
933	Crookneck	½ c	90	94	18	1	4	1	<1	.1	t	.1
934	Zucchini	½ c	90	95	14	1	4	1	<1	t	t	t
	Squash, winter varieties, cooked:											
	Average of all varieties, baked:											
935	Mashed	1 c	245	89	96	2	21	7	2	.3	.1	.6
936	Cubes	1 c	205	89	80	2	18	6	1	.3	.1	.5
937	Acorn, baked, mashed	½ c	123	83	69	1	18	5	<1	t	t	.1
1218	Acorn, boiled, mashed	½ c	122	90	41	1	11	3	<1	t	t	t
	Butternut squash:											
938	Baked cubes	½ c	103	88	41	1	11	3	<1	t	t	t
1219	Baked, mashed	½ c	103	88	41	1	11	3	<1	t	t	t
1193	Cooked from frozen	½ c	120	88	47	1	12	3	<1	t	t	t
1194	Hubbard, baked, mashed	½ c	120	85	60	3	13	3	1	.2	.1	.3
1195	Hubbard, boiled, mashed	½ c	118	91	35	2	8	3	<1	.1	t	.2
1196	Spaghetti, baked or boiled	½ c	77	92	22	<1	5	1	<1	t	t	.1
1189	Succotash, cooked from frozen	½ c	85	74	79	4	17	3	1	.1	.1	.4
	Sweet potatoes:											
939	Baked in skin, peeled, 5 x 2" diam	1 ea	114	73	117	2	28	3	<1	t	t	.1
940	Boiled without skin, 5 x 2" diam	1 ea	151	73	159	2	37	3	<1	.1	t	.2
941	Candied, 2½ x 2"	1 pce	105	67	144	1	29	3	3	1.4	.7	.2
	Canned:											
942	Solid pack	½ c	128	74	129	3	30	2	<1	.1	t	.1
943	Vacuum pack, mashed	½ c	127	76	116	2	27	2	<1	.1	t	.1
944	Vacuum pack, 3¾ x 1"	2 pce	80	76	73	1	17	1	<1	t	t	.1
1940	Taro shoots, cooked slices	1 c	140	95	20	1	4	1	<1	t	t	t
1941	Taro, tahitian, cooked slices	1 c	137	86	60	6	9	1	1	.2	.1	.4
	Tomatillos:											
1877	Raw, each	1 ea	34	92	11	<1	2	1	<1	t	.1	.1
1875	Raw, chopped	1 c	132	92	42	1	8	3	1	.2	.2	.6

Chol (mg)	Calc (mg)	Iron (mg)	Magn (mg)	Pota (mg)	Sodi (mg)	Zinc (mg)	VT-A (RE)	Thia (mg)	VT-E (a-TE)	Ribo (mg)	Niac (mg)	V-B6 (mg)	Fola (µg)	VT-C (mg)
0	19	1	12	243	477	.19	1	.12	.03	.05	1.38	.15	11	4
0	18	.7	11	283	1	.28	1330	.04	1.3	.1	.51	.05	10	6
0	32	1.71	28	253	6	.21	2713	.03	1.3	.07	.45	.07	15	5
0	4	.11	3	60	4	.12	1	<.01	.45	.01	.05	.01	12	2
0	15	.45	10	242	18	.5	2	.01	1.81	.02	.2	.05	48	6
0	9	.13	4	104	11	.13	<1	<.01	0	.02	.13	.03	12	10
0	12	.18	7	100	9	.07	0	.01	0	.01	.09	.02	12	10
10	44	2.09	42	336	377	1.47	0	.03	.39	.02	.4	.18	14	8
0	41	.45	20	277	17	.3	48	.07	.13	.03	.61	.09	13	16
0	35	1.73	15	201	780	.22	2	.02	.12	.03	.17	.15	28	17
0	67	1.14	48	36	93	.49	5	.02	.35	.06	.19	<.01	72	1
0	10	2.28	16	109	84	.16	5	.19	.4	.29	1.02	.03	8	1
0	4	.12	2	33	1	.04	125	.01	.01	<.01	.02	.03	3	1
0	36	1.73	20	166	3	.22	11	.11	.32	.06	.47	.13	28	42
0	88	4.42	74	443	1	.99	1	.13	1.68	.24	.34	.2	46	1
0	120	3.4	196	1173	2	4.1	2	.37	3.96	.65	.91	.19	176	4
0	91	3.78	58	226	5032	4.58	12	.13	.01	.34	1.19	.3	45	0
0	39	1.5	34	195	80	.76	0	.1	.18	.04	.3	.01		0
0	41	1.3	34	244	45	.77	0	.13	.24	.05	.31	.01		0
0	138	1.38	33	149	10	.79	1	.06	.01	.05	.66	.06	55	<1
0	38	1.02	36	223	6	.64	0	.12	.25	.05	.37	.01		0
0	28	.76	22	156	22	.15	188	.02	.53	.05	.2	.05	54	8
0	122	3.21	78	419	63	.68	737	.09	.86	.21	.44	.22	131	9
0	139	1.44	66	283	82	.66	739	.06	.91	.16	.4	.14	103	12
0	136	2.46	81	370	29	.49	939	.02	1.39	.15	.41	.11	105	15
0	136	2.46	81	370	29	.49	939	.02	1.39	.15	.41	.11	105	15
0	24	.32	22	173	1	.35	26	.04	.11	.04	.46	.06	18	5
0	24	.32	22	173	1	.35	26	.04	.11	.04	.46	.08	18	5
0	12	.31	20	228	3	.16	22	.04	.11	.04	.38	.07	15	4
0	34	.81	20	1070	2	.64	872	.21	.29	.06	1.72	.18	69	23
0	29	.68	16	896	2	.53	730	.17	.25	.05	1.44	.15	57	20
0	54	1.14	53	538	5	.21	53	.2	.15	.02	1.08	.24	23	13
0	32	.68	32	321	4	.13	32	.12	.15	.01	.65	.14	14	8
0	42	.62	30	293	4	.13	721	.07	.17	.02	1	.13	20	16
0	42	.62	30	293	4	.13	721	.07	.17	.02	1	.13	20	16
0	23	.7	11	160	2	.14	401	.06	.16	.05	.56	.08	20	4
0	20	.56	26	430	10	.18	725	.09	.14	.06	.67	.21	19	11
0	12	.33	15	253	6	.12	473	.05	.14	.03	.39	.12	11	8
0	16	.26	8	90	14	.15	8	.03	.09	.02	.62	.08	6	3
0	13	.76	20	225	38	.38	20	.06	.31	.06	1.11	.08	28	5
0	32	.51	23	397	11	.33	2487	.08	.32	.14	.69	.27	26	28
0	32	.85	15	278	20	.41	2574	.08	.42	.21	.97	.37	17	26
8	27	1.19	12	198	73	.16	440	.02	3.99	.04	.41	.04	12	7
0	38	1.7	31	269	96	.27	1936	.03	.35	.11	1.22	.3	14	7
0	28	1.13	28	396	67	.23	1013	.05	.32	.07	.94	.24	21	33
0	18	.71	18	250	42	.14	638	.03	.2	.05	.59	.15	13	21
0	20	.57	11	482	3	.76	7	.05	1.4	.07	1.13	.16	4	26
0	204	2.14	70	854	74	.14	241	.06	3.7	.27	.66	.16	10	52
0	2	.21	7	91	<1	.07	4	.01	.13	.01	.63	.02	2	4
0	9	.82	26	354	1	.29	14	.06	.5	.05	2.44	.07	9	15

A

Table A–1

Food Composition (Computer code number is for West Diet Analysis program) (For purposes of calculations, use "0" for t, <1, <.1, <.01, etc.)

Computer Code Number	Food Description	Measure	Wt (g)	H_2O (%)	Ener (cal)	Prot (g)	Carb (g)	Dietary Fiber (g)	Fat (g)	Fat Breakdown (g) Sat	Mono	Poly
	VEGETABLES and LEGUMES—Continued											
	Tomatoes:											
945	Raw, whole, 2 ⅗" diam	1 ea	123	94	26	1	6	1	<1	.1	.1	.2
946	Raw, chopped	1 c	180	94	38	2	8	2	1	.1	.1	.2
947	Cooked from raw	1 c	240	92	65	3	14	2	1	.1	.2	.4
948	Canned, solids and liquid:	1 c	240	94	46	2	10	2	<1	t	t	.1
5741	Unsalted	1 c	240	94	46	2	10	2	<1	t	t	.1
1879	Tomatoes, sundried:	1 c	54	15	139	8	30	7	2	.2	.3	.6
1881	Pieces	10 pce	20	15	52	3	11	2	1	.1	.1	.2
1885	Oil pack, drained	10 pce	30	54	64	2	7	2	4	.6	2.6	.6
2020	Tomato, raw	1 ea	123	94	26	1	6	1	<1	.1	.1	.2
949	Tomato juice, canned:	1 c	243	94	41	2	10	1	<1	t	t	.1
5397	Unsalted	1 c	243	94	41	2	10	2	<1	t	t	.1
	Tomato products, canned:											
950	Paste, no added salt	1 c	262	74	215	10	51	11	1	.2	.2	.6
951	Puree, no added salt	1 c	250	87	100	4	24	5	<1	.1	.1	.2
952	Sauce, with salt	1 c	245	89	73	3	18	3	<1	.1	.1	.2
953	Turnips, cubes, cooked from fresh	1 c	156	94	33	1	8	3	<1	t	t	.1
	Turnip greens, cooked:											
954	From fresh, leaves and stems	1 c	144	93	29	2	6	5	<1	.1	t	.1
955	From frozen, chopped	1 c	164	90	49	6	8	6	1	.2	t	.3
956	Vegetable juice cocktail, canned	1 c	242	93	46	2	11	2	<1	t	t	.1
	Vegetables, mixed:											
957	Canned, drained	½ c	81	87	38	2	7	2	<1	t	t	.1
958	Frozen, cooked, drained	½ c	91	83	54	3	12	4	<1	t	t	.1
1818	Water chestnuts, Chinese, raw	½ c	62	73	60	1	15	2	<1	t	t	t
	Water chestnuts, canned:											
959	Slices	½ c	70	86	35	1	9	2	<1	t	t	t
960	Whole	4 ea	28	86	14	<1	3	1	<1	t	t	t
1190	Watercress, fresh, chopped	½ c	17	95	2	<1	<1	<1	<1	t	t	t
	VEGETARIAN FOODS:											
7509	Bacon strips, meatless	3 ea	15	49	46	2	1	<1	4	.7	1.1	2.3
1511	Baked beans, canned	½ c	127	73	118	6	26	6	1	.1	t	.2
7526	Bakon crumbles	¼ c	7	16	28	2	1	<1	2			
7548	Chicken, breaded, fried, meatless	1 pce	57	70	97	6	3	3	7	1	2.9	2.5
7547	Chicken slices, meatless	2 ea	60	59	132	10	4	3	8	1.3	2	4.4
7557	Chili w/meat substitute	½ c	107	64	141	19	15	4	2	.3	.6	.9
7549	Fish stick, meatless	2 ea	57	45	165	13	5	3	10	1.6	2.5	5.4
7550	Frankfurter, meatless	1 ea	51	58	102	10	4	2	5	.8	1.2	2.7
7504	GardenBurger, patty	1 ea	45	53	87	5	13	3	1	.4	.3	.7
7505	GardenSausage, patty	1 ea	35	15	117	4	22	5	1	.7	.3	t
7551	Luncheon slice, meatless	1 sl	67	46	188	17	6	3	11	1.7	2.6	5.6
7560	Meatloaf, meatless	1 ea	71	58	142	15	6	3	6	1	1.5	3.3
1171	Nuteena	1 ea	55	58	162	6	6	2	13	5.1	5.8	1.7
7556	Pot pie, meatless	1 ea	227	59	524	15	41	5	34	9.5	12.6	9.8
7554	Soyburger, patty	1 ea	71	58	142	15	6	3	6	1	1.5	3.3
7562	Soyburger w/cheese, patty	1 ea	135	50	316	21	29	4	13	4.2	3.9	3.7
7564	Tempeh	1 c	166	55	330	31	28	9	13	1.9	2.9	7.2
7670	Vegan burger, patty	1 ea	78	71	75	11	6	4	<1	.1	.3	.2
	Vegetarian foods, Green Giant:											
7677	Breakfast links	3 ea	68	65	114	12	5	4	5	.7	1.2	3.1
7676	Breakfast patties	2 ea	57	65	95	10	5	3	4	.6	1	2.6
	Burger, harvest, patty:											
7673	Italian	1 ea	90	65	139	17	8	5	4	1.4	.3	.4
7674	Original	1 ea	90	65	137	18	8	5	4	1.3	.1	.4
7675	Southwestern	1 ea	90	65	135	16	9	5	4	1.4	.2	.4
	Vegetarian foods, Loma Linda											
7727	Chik nuggets, frozen	5 pce	85	47	245	12	13	5	16	2.5	4	8.8
7753	Chik-fried, frozen	1 pce	57	51	178	11	1	1	15	1.9	3.7	8.7
7744	Franks, big, canned	1 ea	51	59	110	10	2	2	7	1.1	1.7	3.8
7747	Linketts, canned	1 ea	35	60	72	7	1		4	.7	1.2	2.5
1173	Redi-burger, patty	1 ea	85	59	172	16	5		10	1.5	2.4	5.8
7755	Swiss steak w/gravy, canned	1 pce	92	71	120	9	8	4	6	.8	1.5	3.3

PAGE KEY: A–4 = Beverages A–6 = Dairy A–10 = Eggs A–10 = Fat/Oil A–14 = Fruit A–20 = Bakery A–28 = Grain A–32 = Fish A–34 = Meats A–38 = Poultry A–40 = Sausage A–42 = Mixed/Fast A–46 = Nuts/Seeds A–50 = Sweets A–52 = Vegetables/Legumes A–62 = Vegetarian Foods A–64 = Misc A–66 = Soups/Sauces A–68 = Fast A–84 = Convenience A–88 = Baby foods

A–63

Chol (mg)	Calc (mg)	Iron (mg)	Magn (mg)	Pota (mg)	Sodi (mg)	Zinc (mg)	VT-A (RE)	Thia (mg)	VT-E (a-TE)	Ribo (mg)	Niac (mg)	V-B6 (mg)	Fola (µg)	VT-C (mg)
0	6	.55	13	273	11	.11	76	.07	.47	.06	.77	.1	18	23
0	9	.81	20	400	16	.16	112	.11	.68	.09	1.13	.14	27	34
0	14	1.34	34	670	26	.26	178	.17	.91	.14	1.8	.23	31	55
0	72	1.32	29	530	355	.38	144	.11	.77	.07	1.76	.22	19	34
0	72	1.32	29	545	24	.38	144	.11	.91	.07	1.76	.22	19	34
0	59	4.91	105	1850	1131	1.07	47	.28	<.01	.26	4.89	.18	37	21
0	22	1.82	39	685	419	.4	17	.11	<.01	.1	1.81	.07	14	8
0	14	.8	24	470	80	.23	39	.06	.16	.11	1.09	.1	7	31
0	6	.55	13	273	11	.11	76	.07	.47	.06	.77	.1	18	23
0	22	1.41	27	535	877	.34	136	.11	2.21	.07	1.64	.27	48	44
0	22	1.41	27	535	24	.34	136	.11	2.21	.07	1.64	.27	48	44
0	92	5.08	134	2454	231	2.1	639	.41	11.3	.5	8.44	1	59	111
0	42	3.1	60	1065	85	.55	320	.18	6.3	.13	4.3	.38	27	26
0	34	1.89	47	909	1482	.61	240	.16	3.43	.14	2.82	.38	23	32
0	34	.34	12	211	78	.31	0	.04	.05	.04	.47	.1	14	18
0	197	1.15	32	292	42	.2	792	.06	2.48	.1	.59	.26	170	39
0	249	3.18	43	367	25	.67	1308	.09	4.79	.12	.77	.11	65	36
0	27	1.02	27	467	653	.48	283	.1	.77	.07	1.76	.34	51	67
0	22	.85	13	236	121	.33	944	.04	.49	.04	.47	.06	19	4
0	23	.75	20	154	32	.45	389	.06	.33	.11	.77	.07	17	3
0	7	.04	14	362	9	.31	0	.09	.74	.12	.62	.2	10	2
0	3	.61	3	83	6	.27	0	.01	.35	.02	.25	.11	4	1
0	1	.24	1	33	2	.11	0	<.01	.14	.01	.1	.04	2	<1
0	20	.03	4	56	7	.02	80	.01	.17	.02	.03	.02	2	7
0	3	.36	3	25	220	.06	1	.66	1.04	.07	1.13	.07	6	0
0	63	.37	41	376	504	1.78	22	.19	.67	.08	.54	.17	30	4
	8	.44	11	120	172	.25	0	.06		.02	.12	.02	7	0
0	13	.97	7	171	228	.37	0	.4	1.11	.27	2.68	.28	32	0
0	21	.78	10	198	474	.42	0	.66	1.61	.24	3.18	.42	46	0
0	53	4.24	36	362	527	1.26	78	.12	1.25	.07	1.21	.15	82	16
0	54	1.14	13	342	279	.8	0	.63	2.25	.51	6.84	.85	58	0
0	17	.92	9	76	219	.61	0	.56	.98	.61	8.16	.5	40	0
0	36	1.35		129	112		18	.05		.09		.06		<1
0	181	.33		307	78		3	.08		.2		.13		<1
0	27	1.54	15	188	576	1.07	0	.64	2.01	.37	7.37	.74	67	0
0	21	1.49	13	128	391	1.28	0	.64	1.23	.43	7.1	.85	55	0
0	9	.27	33	166	119	.46	0	.1		.35	1.04	.45	49	0
20	66	2.9	31	331	538	1.05	729	.65	4	.4	4.47	.31	40	10
0	21	1.49	13	128	391	1.28	0	.64	1.23	.43	7.1	.85	55	0
13	146	2.71	26	211	931	1.97	45	.77	1.43	.55	8.14	.86	69	1
0	154	3.75	116	609	10	3	115	.22	.03	.18	7.69	.5	86	0
0	79	2.66	15	398	351	.69	0	.23	.01	.51	3.78	.18	225	0
0	65	1.84			340	4.56	0	.18		.09	.27	.18		0
0	54	2			285	3.82	0	.15		.07	2.28	.15		0
0	74	2.61			374	6.93	3	.28		.14	4.05	.28		0
0	76	2.7			378	7.2	0	.29		.14	4.32	.29		0
0	71	2.52			371	6.66	3	.27		.13	4.05	.27		0
2	40	1.4		153	709	.43	0	.67		.3	2.89	.45		0
4	2	.63		76	503	.2	0	.98		.46	2.1	.35		0
2	8	.77		51	243	.89	0	.26		.46	1.98	.14		0
1	4	.39		29	160	.46	0	.13		.22	.64	.29		0
1	12	1.06	16	121	455	1.11	0	.14		.3	1.9	.51	21	0
2	24	.31		225	433	.41	0	1.25		.65	5.41	1		0

Table A-1
Food Composition

(Computer code number is for West Diet Analysis program)   (For purposes of calculations, use "0" for t, <1, <.1, <.01, etc.)

Computer Code Number	Food Description	Measure	Wt (g)	H₂O (%)	Ener (cal)	Prot (g)	Carb (g)	Dietary Fiber (g)	Fat (g)	Fat Breakdown (g) Sat	Mono	Poly
	VEGETARIAN FOODS:—Continued											
1174	Vege-Burger, patty	1 ea	55	71	66	10	2	2	2	.4	.6	.5
	Vegetarian foods, Morningstar Farms:											
7672	Better-n-burgers, svg	1 ea	78	71	75	11	6	4	<1	.1	.3	.2
7766	Better-n-eggs	¼ c	57	88	23	5	<1	0	<1	.1	.1	.1
57436	Breakfast links	2 pce	45	60	63	8	2	2	2	.5	.7	1.3
7752	Breakfast strips	2 pce	16	43	56	2	2	<1	4	.7	1.1	2.6
7725	Burger crumbles, svg	1 ea	55	60	116	11	3	3	6	1.6	2.3	2.5
7726	Burger, spicy black bean	1 ea	78	60	113	11	15	5	1	.2	.3	.4
7665	Chik pattie	1 ea	71	51	177	7	15	2	10	1.3	2.6	5.9
7724	Frank, deli	1 ea	45	52	109	10	3	3	7	1	2.1	3.5
7722	Garden vege pattie	1 ea	67	60	104	11	9	4	4	.5	1.1	2.2
7746	Grillers	1 ea	64	55	139	14	5	3	7	1.7	2.2	3
7664	Prime pattie	1 ea	64	64	94	16	4	3	2	.2	.4	.6
	Vegetarian foods, Worthington:											
7634	Beef style, meatless, frzn	3 pce	55	58	113	9	4	3	7	1.2	2.7	2.6
7732	Burger, meatless, patty	¼ c	55	71	60	9	2	1	2	.3	.5	1.1
1846	Chik slices, canned	2 pce	60	78	62	6	1	1	4	.6	.9	2.3
1833	Chili, canned	½ c	106	73	136	9	10	4	7	1.1	1.7	4.1
1835	Choplets, slices, canned	2 pce	92	72	93	17	3	2	2	.9	.3	.3
7608	Corned beef style, meatless, frzn	4 pce	57	55	138	10	5	2	9	1.9	4.1	3.1
1831	Country stew, canned	1 c	240	81	208	13	20	5	9	1.6	2.3	4.8
7632	Egg roll, meatless, frzn	1 ea	85	53	181	6	20	2	8	1.7	4.5	2.3
1838	Numete, slices, canned	1 pce	55	58	132	6	5	3	10	2.4	4.4	2.7
1839	Prime steaks, slices, canned	1 pce	92	71	136	9	4	4	9	1.4	2.9	4.9
1840	Protose, slices, canned	1 pce	55	53	131	13	5	3	7	1	3	2.4
7606	Roast, dinner, meatless, frzn	1 ea	85	63	180	12	5	3	12	2.2	5	5.2
1842	Saucette links, canned	1 pce	38	62	86	6	1		6	1.1	1.6	3.8
1844	Savory slices, canned	1 pce	28	66	48	3	2	1	3	1.2	1.3	.6
7735	Steaklets, frzn	1 pce	71	58	145	12	6	2	8	1.4	2.7	3.9
1847	Turkee slices, canned	1 pce	33	64	68	5	1	1	5	.8	1.9	2.1
	MISCELLANEOUS											
	Baking powders for home use:											
	Sodium aluminum sulfate:											
962	With monocalcium phosphate monohydrate	1 tsp	5	2	6	<1	2	0	0	0	0	0
963	With monocalcium phosphate monohydrate, calcium sulfate	1 tsp	5	5	3	0	1	<1	0	0	0	0
964	Straight phosphate	1 tsp	5	4	3	<1	1	<1	0	0	0	0
965	Low sodium	1 tsp	5	6	5	<1	2	<1	<1	t	0	t
1204	Baking soda	1 tsp	5		0	0	0	0	0	0	0	0
966	Basil, dried	1 tbs	5	6	13	1	3	2	<1	t	t	.1
2068	Cajun seasoning	1 tsp	3	5	6	<1	1	<1	<1			
961	Carob flour	1 c	103	4	185	5	92	41	1	.1	.2	.2
967	Catsup:	¼ c	61	67	63	1	17	1	<1	t	t	.1
968	Tablespoon	1 tbs	15	67	16	<1	4	<1	<1	t	t	t
1200	Cayenne/red pepper	1 tbs	5	8	16	1	3	1	1	.2	.1	.4
969	Celery seed	1 tsp	2	6	8	<1	1	<1	<1	t	.3	.1
1203	Chili powder:	1 tbs	8	8	25	1	4	3	1	.2	.3	.6
970	Teaspoon	1 tsp	3	8	9	<1	2	1	<1	.1	.1	.2
	Chocolate:											
971	Baking, unsweetened, square	1 oz	28	1	146	3	8	4	15	9.1	5.2	.5
	For other chocolate items, see Sweeteners & Sweets											
972	Cilantro/coriander, fresh	1 tbs	1	93		<1	<1	<1	<1	0	t	0
2287	Cinnamon	1 tsp	2	10	5	<1	2	1	<1	t	t	t
1197	Cornstarch	1 tbs	8	8	30	<1	7	<1	<1	t	t	t
2239	Curry powder	1 tsp	2	10	6	<1	1	1	<1	t	.1	.1
1202	Dill weed, dried	1 tbs	3	7	8	1	2	<1	<1	t	.1	t
975	Garlic cloves	1 ea	3	59	4	<1	1	<1	<1	t	0	t
2238	Garlic powder	1 tsp	3	6	10	<1	2	<1	<1	t	t	t
977	Gelatin, dry, unsweetened: Envelope	1 ea	7	13	23	6	0	0	<1	t	t	t
978	Ginger root, slices, raw	2 pce	5	82	3	<1	1	<1	<1	t	t	t

PAGE KEY: A–4 = Beverages A–6 = Dairy A–10 = Eggs A–10 = Fat/Oil A–14 = Fruit A–20 = Bakery A–28 = Grain A–32 = Fish A–34 = Meats A–38 = Poultry A–40 = Sausage A–42 = Mixed/Fast A–46 = Nuts/Seeds A–50 = Sweets A–52 = Vegetables/Legumes A–62 = Vegetarian Foods A–64 = Misc A–66 = Soups/Sauces A–68 = Fast A–84 = Convenience A–88 = Baby foods

A–65

Chol (mg)	Calc (mg)	Iron (mg)	Magn (mg)	Pota (mg)	Sodi (mg)	Zinc (mg)	VT-A (RE)	Thia (mg)	VT-E (a-TE)	Ribo (mg)	Niac (mg)	V-B6 (mg)	Fola (µg)	VT-C (mg)
0	8	.5	12	30	114	.58	0	.2		.25	.78	.31	15	0
0	79	2.66	15	398	351	.69	0	.23	.01	.51	3.78	.18	225	0
2	7	.63		68	90	.51	64	.01		.26	0	.11		0
1	15	2.14	16	59	338	.36	0	6.95		.22	5.19	.33	12	0
<1	7	.27		15	220	.05	0	.75		.04	.6	.07		0
0	40	3.2	1	89	238	.82	0	4.96	.34	.18	1.49	.27		0
1	56	1.84	44	269	499	.93	14	8.03	.36	.14	0	.21		0
1	11	1.02		163	536	.31	0	2.15		.16	1.51	.14		0
2	16	.26	4	50	524	.38	0	.14	1.26	.02	0	.01		0
1	34	.72	29	180	382	.59	20	6.47	.98	.1	0	0	29	0
2	43	1.16		127	256	.49	0	11.7		.24	2.99	.37		0
1	46	2.14		142	247	.74	0	.51		.25	.92	.41		2
0	4	2.63		44	624	.22	0	.89		.34	6.46	.56		0
0	4	1.73		25	269	.38	0	.13		.1	1.96	.24		0
1	9	.73		111	257	.26	0	.06		.05	.37	.08		0
0	20	1.49		195	523	.57	0	.02		.03	1.04	.31		0
0	6	.37		40	500	.65	0	.05		.05	0	.05		0
1	6	1.17		58	524	.26	0	10.6		.07	1.36	.3		0
2	51	5.09		270	826	1.03	216	1.85		.29	4.22	.86		0
1	15	.57		96	384	.31	0	1.22		.19	0	.03		0
0	10	1.12		155	272	.56	0	.08		.06	.54	.2		0
2	12	.38		82	445	.38	0	.12		.13	1.98	.38		0
<1	1	1.84		50	283	.7	0	.18		.13	1.34	.24		0
2	36	2.87		38	566	.64	0	2.13		.25	6.02	.6		0
1	9	1.15		25	205	.26	0	.59		.08	.09	.13		0
<1		.47		14	179	.08	0	.08		.06	.48	.1		0
2	49	.99		95	484	.5	0	1.51		.12	3.1	.26		0
1	3	.47		16	203	.11	0	1.13		.05	.39	.09		0
0	97	0		7	547	0	0	0	0	0	0	0	0	0
0	294	.55	1	1	530	<.01	0	0	0	0	0	0	0	0
0	368	.56	2		395	<.01	0	0	0	0	0	0	0	0
0	217	.41	1	505	4	.04	0	0	<.01	0	0	0	0	0
0	0	0	0	0	1368	0	0	0	0	0	0	0	0	0
0	106	2.1	21	172	2	.29	47	.01	.08	.02	.35	.06	14	3
				29	474									
0	358	3.03	56	852	36	.95	1	.05	.65	.47	1.96	.38	30	<1
0	12	.43	13	293	723	.14	62	.05	.9	.04	.84	.11	9	9
0	3	.1	3	72	178	.03	15	.01	.22	.01	.21	.03	2	2
0	7	.39	8	101	1	.12	208	.02	.24	.05	.43	.1	5	4
0	35	.9	9	28	3	.14	<1	.01	.02	.01	.06	.01	<1	<1
0	22	1.14	14	153	81	.22	279	.03	.08	.06	.63	.15	8	5
0	8	.43	5	57	30	.08	105	.01	.03	.02	.24	.06	3	2
0	21	1.77	87	233	4	1.12	3	.02	.34	.05	.31	.03	2	0
0	1	.02		5	<1	<.01	3	<.01	.02	<.01	.01	<.01	<1	<1
0	25	.76	1	10	1	.04	1	<.01	0	<.01	.03	<.01	1	1
0		.04			1	<.01	0	0	0	0	0	0	0	0
0	10	.59	5	31	1	.08	2	<.01	.01	.01	.07	.01	3	<1
0	53	1.46	13	99	6	.1	18	.01		.01	.08	.04		1
0	5	.05	1	12	1	.03	0	.01	0	<.01	.02	.04	<1	1
0	2	.08	2	33	1	.08	0	.01	0	<.01	.02	.08	<1	1
0	4	.08	2	1	14	.01	0	<.01	0	.02	.01	0	2	0
0	1	.02	2	21	1	.02	0	<.01	.01	<.01	.03	.01	1	<1

Table A–1

Food Composition

(Computer code number is for West Diet Analysis program)   (For purposes of calculations, use "0" for t, <1, <.1, <.01, etc.)

Computer Code Number	Food Description	Measure	Wt (g)	H₂O (%)	Ener (cal)	Prot (g)	Carb (g)	Dietary Fiber (g)	Fat (g)	Fat Breakdown (g) Sat	Mono	Poly
	MISCELLANEOUS—Continued											
1198	Horseradish, prepared	1 tbs	15	85	7	<1	2	<1	<1	t	t	.1
1997	Hummous/hummus	1 c	246	65	421	12	50	12	21	3.1	8.8	7.8
1909	Mustard, country dijon	1 tsp	5		5	<1	<1	0	0	0	0	0
2019	Mustard, gai choy Chinese	1 tbs	16	94	3	<1	1		<1			
979	Mustard, prepared (1 packet = 1 tsp)	1 tsp	5	80	4	<1	<1	<1	<1	t	.2	t
	Miso (see #926 under Vegetables and Legumes, Soybean products)											
980	Olives, green	5 ea	20	78	23	<1	<1	<1	3	.3	1.9	.2
981	Olives, ripe, pitted	5 ea	22	80	25	<1	1	1	2	.3	1.7	.2
26008	Onion powder	1 tsp	2	5	7	<1	2	<1	<1	t	t	t
2237	Oregano, ground	1 tsp	2	7	6	<1	1	1	<1	.1	t	.1
2236	Paprika	1 tsp	2	10	6	<1	1	<1	<1	t	t	.2
887	Parsley, freeze dried	¼ c	1	2	3	<1	<1	<1	<1	t	t	t
	Parsley, fresh (see #885 and #886)											
985	Pepper, black	1 tsp	2	10	5	<1	1	1	<1	t	t	t
	Pickles:											
986	Dill, medium, 3¾ x 1¼" diam	1 ea	65	92	12	<1	3	1	<1	t	t	t
987	Fresh pack, slices, 1½" diam x ¼"	2 pce	15	79	11	<1	3	<1	<1	0	0	t
988	Sweet, medium	1 ea	35	65	41	<1	11	<1	<1	t	t	t
989	Pickle relish, sweet	1 tbs	15	63	21	<1	5	<1	<1	t	t	t
	Popcorn (see Grain Products #539–541)											
917	Potato chips:	10 pce	20	2	107	1	11	1	7	2.2	2	2.4
44076	Unsalted	1 oz	28	2	150	2	15	1	10	3.1	2.8	3.4
1201	Sage, ground	1 tsp	1	8	3	<1	1	<1	<1	.1	t	t
1347	Salsa, from recipe	1 tbs	15	93	3	<1	1	<1	<1	t	t	t
2218	Salsa, pico de gallo, medium	1 tbs	15	92	2	0	1	<1	0	0	0	0
990	Salt	1 tsp	6		0	0	0	0	0	0	0	0
	Salt Substitutes:											
1205	Morton, salt substitute	1 tsp	6		0	0	<1		0	0	0	0
1207	Morton, light salt	1 tsp	6		0	0	<1		0	0	0	0
2067	Seasoned salt, no MSG	1 tsp	5	5	4	<1	1	<1	<1			
991	Vinegar, cider	½ c	120	94	17	0	7	0	0	0	0	0
2172	Balsamic	1 tbs	15	64	21	0	4	0	0	0	0	0
2176	Malt	1 tbs	15	90	5	0	<1	0	0	0	0	0
2182	Tarragon	1 tbs	15	95	3	0	<1	0	0	0	0	0
2181	White wine	1 tbs	15	89	5	0	<1	0	0	0	0	0
	Yeast:											
992	Baker's, dry, active, package	1 ea	7	8	21	3	3	1	<1	t	.2	t
993	Brewer's, dry	1 tbs	8	5	23	3	3	3	<1	t	t	0
	SOUPS, SAUCES, and GRAVIES											
	SOUPS, canned, condensed:											
	Unprepared, condensed:											
1210	Cream of celery	1 c	251	85	181	3	18	2	11	2.8	2.6	5
1215	Cream of chicken	1 c	251	82	233	7	18	<1	15	4.2	6.5	3
1216	Cream of mushroom	1 c	251	81	259	4	19	1	19	5.1	3.6	8.9
1220	Onion	1 c	246	86	113	8	16	2	3	.5	1.5	1.3
	Prepared w/equal volume of whole milk:											
994	Clam chowder, New England	1 c	248	85	164	9	17	1	7	2.9	2.3	1.1
1209	Cream of celery	1 c	248	86	164	6	14	1	10	3.9	2.5	2.6
995	Cream of chicken	1 c	248	85	191	8	15	<1	11	4.6	4.5	1.6
996	Cream of mushroom	1 c	248	85	203	6	15	<1	14	5.1	3	4.6
1214	Cream of potato	1 c	248	87	149	6	17	<1	6	3.8	1.7	.6
1213	Oyster stew	1 c	245	89	135	6	10	0	8	5	2.1	.3
997	Tomato	1 c	248	85	161	6	22	3	6	2.9	1.6	1.1
	Prepared with equal volume of water:											
998	Bean with bacon	1 c	253	84	172	8	23	9	6	1.5	2.2	1.8
999	Beef broth/bouillon/consommé	1 c	240	98	17	3	<1	0	1	.3	.2	t
1000	Beef noodle	1 c	244	92	83	5	9	1	3	1.1	1.2	.5
1001	Chicken noodle	1 c	241	92	75	4	9	1	2	.7	1.1	.6
1002	Chicken rice	1 c	241	94	60	4	7	1	2	.5	.9	.4
1208	Chili beef	1 c	250	85	170	7	21	9	7	3.3	2.8	.3
1003	Clam chowder, Manhattan	1 c	244	92	78	2	12	1	2	.4	.4	1.3

PAGE KEY: A–4 = Beverages A–6 = Dairy A–10 = Eggs A–10 = Fat/Oil A–14 = Fruit A–20 = Bakery A–28 = Grain A–32 = Fish A–34 = Meats A–38 = Poultry A–40 = Sausage A–42 = Mixed/Fast A–46 = Nuts/Seeds A–50 = Sweets A–52 = Vegetables/Legumes A–62 = Vegetarian Foods A–64 = Misc A–66 = Soups/Sauces A–68 = Fast A–84 = Convenience A–88 = Baby foods

A–67

Chol (mg)	Calc (mg)	Iron (mg)	Magn (mg)	Pota (mg)	Sodi (mg)	Zinc (mg)	VT-A (RE)	Thia (mg)	VT-E (a-TE)	Ribo (mg)	Niac (mg)	V-B6 (mg)	Fola (µg)	VT-C (mg)
0	8	.06	4	37	47	.12	<1	<.01	<.01	<.01	.06	.01	9	4
0	123	3.86	71	428	600	2.71	5	.23	2.46	.13	1.01	.98	146	19
0				10	120									
0	4	.1	2	6	63	.03	0	0	.09	0	0	<.01	0	0
0	12	.32	4	11	480	.01	6	0	.6	0	0	<.01	<1	0
0	19	.73	1	2	192	.05	9	<.01	.66	0	.01	<.01	0	<1
0	7	.05	2	19	1	.05	0	.01	<.01	<.01	.01	.03	3	<1
0	31	.88	5	33	<1	.09	14	.01	.03	.01	.12	.02	5	1
0	4	.47	4	47	1	.08	121	.01	.01	.03	.31	.04	2	1
0	2	.54	4	63	4	.06	63	.01	.06	.02	.1	.01	15	1
0	9	.58	4	25	1	.03	<1	<.01	.02	<.01	.02	.01	<1	<1
0	6	.34	7	75	833	.09	21	.01	.1	.02	.04	.01	1	1
0	5	.27	1	30	101	0	2	0	.02	<.01	0	<.01	0	1
0	1	.21	1	11	329	.03	5	<.01	.06	.01	.06	<.01	<1	<1
0	3	.12	1	30	107	.01	1	0	.02	<.01	0	0	0	1
0	5	.33	13	255	119	.22	0	.03	.98	.04	.77	.13	9	6
0	7	.46	19	357	2	.3	0	.05	1.37	.05	1.07	.18	13	9
0	16	.28	4	11	<1	.05	6	.01	.02	<.01	.06	.01	3	<1
0	1	.06	1	24	58	.02	22	.01	.04	<.01	.06	.01	2	5
0					130									
0	1	.02			2325	.01	0	0	0	0	0	0	0	0
	33			3018	<1									
	2		4	1560	1170									
			15	1542										
0	7	.72	26	120	1	0	0	0	0	0	0	0	0	0
	2	.07		10	3			<1	.07		.07	.07			<1
	2	.07		13	4			<1	.07		.07	.07			1
		.07		2	1			<1	.07		.07	.07			<1
	1	.07		12	1			<1	.07		.07	.07			<1
0	4	1.16	7	140	3	.45	<1	.16	.01	.38	2.79	.11	164	<1
0	17	1.38	18	151	10	.63	0	1.25		.34	3.03	.4	313	0
28	80	1.26	13	246	1900	.3	60	.06	.38	.1	.66	.02	5	<1
20	68	1.2	5	176	1972	1.26	113	.06	.33	.12	1.64	.03	3	<1
3	65	1.05	10	168	1736	1.19	0	.06	2.61	.17	1.62	.02	8	2
0	54	1.35	5	138	2115	1.23	0	.07	.57	.05	1.21	.1	30	2
22	186	1.49	22	300	992	.8	40	.07	.15	.24	1.03	.13	10	3
32	186	.69	22	310	1009	.2	67	.07	.97	.25	.44	.06	8	1
27	181	.67	17	273	1046	.67	94	.07	.24	.26	.92	.07	8	1
20	179	.59	20	270	918	.64	37	.08	1.34	.28	.91	.06	10	2
22	166	.55	17	322	1061	.67	67	.08	.1	.24	.64	.09	9	1
32	167	1.05	20	235	1041	10.3	44	.07	.49	.23	.34	.06	10	4
17	159	1.81	22	449	744	.29	109	.13	2.6	.25	1.52	.16	21	68
3	81	2.05	45	402	951	1.03	89	.09	.08	.03	.57	.04	32	2
0	14	.41	5	130	782	0	0	<.01	0	.05	1.87	.02	5	0
5	15	1.1	5	100	952	1.54	63	.07	<.01	.06	1.07	.04	19	<1
7	17	.77	5	55	1106	.39	72	.05	.07	.06	1.39	.03	22	<1
7	17	.75	0	101	815	.26	65	.02	.05	.02	1.13	.02	1	<1
12	42	2.13	30	525	1035	1.4	150	.06	.17	.07	1.07	.16	17	4
2	27	1.63	12	188	578	.98	98	.03	.73	.04	.82	.1	10	4

Table A-1

Food Composition   (Computer code number is for West Diet Analysis program)   (For purposes of calculations, use "0" for t, <1, <.1, <.01, etc.)

Computer Code Number	Food Description	Measure	Wt (g)	H₂O (%)	Ener (cal)	Prot (g)	Carb (g)	Dietary Fiber (g)	Fat (g)	Fat Breakdown (g) Sat	Mono	Poly
	SOUPS, SAUCES, and GRAVIES—Continued											
1004	Cream of chicken	1 c	244	91	117	3	9	<1	7	2.1	3.3	1.5
1005	Cream of mushroom	1 c	244	90	129	2	9	<1	9	2.4	1.7	4.2
1006	Minestrone	1 c	241	91	82	4	11	1	3	.6	.7	1.1
1211	Onion	1 c	241	93	58	4	8	1	2	.3	.7	.7
1007	Split pea & ham	1 c	253	82	190	10	28	2	4	1.8	1.8	.6
1008	Tomato	1 c	244	90	85	2	17	<1	2	.4	.4	1
1009	Vegetable beef	1 c	244	92	78	6	10	<1	2	.9	.8	.1
1010	Vegetarian vegetable	1 c	241	92	72	2	12	<1	2	.3	.8	.7
	Ready to serve:											
1707	Chunky chicken soup	1 c	251	84	178	13	17	2	7	2	3	1.4
	SOUPS, dehydrated:											
	Prepared with water:											
1299	Beef broth/bouillon	1 c	244	97	19	1	2	0	1	.3	.3	t
1376	Chicken broth	1 c	244	97	22	1	1	0	1	.3	.4	.4
1013	Chicken noodle	1 c	252	94	53	3	7	1	1	.3	.5	.4
1122	Cream of chicken	1 c	261	91	107	2	13	<1	5	3.4	1.2	.4
1014	Onion	1 c	246	96	27	1	5	1	1	.1	.3	.1
1217	Split pea	1 c	255	87	125	7	21	3	1	.4	.7	.3
1015	Tomato vegetable	1 c	253	93	56	2	10	<1	1	.4	.3	.1
	Unprepared, dry products:											
1011	Beef bouillon, packet	1 ea	6	3	14	1	1	0	1	.3	.2	t
1012	Onion soup, packet	1 ea	39	4	115	5	21	4	2	.5	1.4	.3
	SAUCES											
	From dry mixes, prepared with milk:											
1016	Cheese sauce	1 c	279	77	307	17	23	1	17	9.3	5.3	1.6
1017	Hollandaise	1 c	259	84	240	5	14	<1	20	11.6	5.9	.9
1018	White sauce	1 c	264	81	240	10	21	<1	13	6.4	4.7	1.7
	From home recipe:											
1206	Lowfat cheese sauce	¼ c	61	73	85	6	4	<1	5	2.1	1.9	.9
1019	White sauce, medium	¼ c	72	77	102	2	6	<1	8	2.3	3.2	2
	Ready to serve:											
2202	Alfredo sauce, reduced fat	¼ c	69		170	5	16	0	10	6		
1020	Barbeque sauce	1 tbs	16	81	12	<1	2	<1	<1	t	.1	.1
1706	Chili sauce, tomato base	1 tbs	17	68	18	<1	4	<1	<1	t	t	t
2126	Creole sauce	¼ c	62	89	25	1	4	1	1	.1	.2	.3
2124	Hoisin sauce	1 tbs	17	47	35	<1	7	0	1	0		
2199	Pesto sauce	2 tbs	16		83	2	1	0	8	1.8	5.4	.7
1021	Soy sauce	1 tbs	16	71	8	1	1	<1	<1	t	t	t
2123	Szechuan sauce	1 tbs	16	71	21	<1	3	<1	1	.1	.3	.4
1380	Teriyaki sauce	1 tbs	18	68	15	1	3	<1	0	0	0	0
	Spaghetti sauce, canned:											
1377	Plain	1 c	249	75	271	5	40	8	12	1.7	6.1	3.3
1378	With meat	1 c	250	74	300	8	37	8	14	2.7	7	3.2
1379	With mushrooms	½ c	123	84	108	2	13	1	3	.4	1.5	.8
	GRAVIES											
	Canned:											
1022	Beef	1 c	233	87	123	9	11	1	5	2.7	2.2	.2
1023	Chicken	1 c	238	85	188	5	13	1	14	3.4	6.1	3.6
1024	Mushroom	1 c	238	89	119	3	13	1	6	1	2.8	2.4
1025	From dry mix, brown	1 c	258	92	75	2	13	<1	2	.8	.7	.1
1026	From dry mix, chicken	1 c	260	91	83	3	14	<1	2	.5	.9	.4
	FAST FOOD RESTAURANTS											
	ARBY'S											
1402	Bac'n cheddar deluxe	1 ea	231	59	512	21	39	<1	31	8.7	12.7	10.1
	Roast beef sandwiches:											
1403	Regular	1 ea	155	47	383	22	35	1	18	7	8	3.5
1404	Junior	1 ea	89	48	233	11	23	<1	11	4.1	5.2	2.5
1405	Super	1 ea	254	58	552	24	54	1	28	7.6	12.2	8.4
1407	Beef 'n cheddar	1 ea	194	50	508	25	43		26	7.7	12	6.8
1408	Chicken breast sandwich	1 ea	204	52	445	22	52	1	22	3	9.7	10.1
1412	Ham'n cheese sandwich	1 ea	169	54	355	25	34	<1	14	5.1	5.8	3.8
1726	Italian sub sandwich	1 ea	297	58	671	34	47		39	12.8	15.7	8.5

Chol (mg)	Calc (mg)	Iron (mg)	Magn (mg)	Pota (mg)	Sodi (mg)	Zinc (mg)	VT-A (RE)	Thia (mg)	VT-E (a-TE)	Ribo (mg)	Niac (mg)	V-B6 (mg)	Fola (µg)	VT-C (mg)
10	34	.61	2	88	986	.63	56	.03	.2	.06	.82	.02	2	<1
2	46	.51	5	100	881	.59	0	.05	1.24	.09	.72	.01	5	1
2	34	.92	7	313	911	.73	234	.05	.07	.04	.94	.1	36	1
0	26	.67	2	67	1053	.61	0	.03	.29	.02	.6	.05	15	1
8	23	2.28	48	400	1006	1.32	45	.15	.15	.08	1.47	.07	3	2
0	12	1.76	7	264	695	.24	68	.09	2.49	.05	1.42	.11	15	66
5	17	1.12	5	173	791	1.54	190	.04	.32	.05	1.03	.08	10	2
0	22	1.08	7	210	822	.46	301	.05	.79	.05	.92	.05	11	1
30	25	1.73	8	176	889	1	131	.08	.18	.17	4.42	.05	5	1
0	10	.02	7	37	1361	.07	<1	<.01	.02	.02	.36	0	0	0
0	15	.07	5	24	1483	.01	12	.01	.02	.03	.19	0	2	0
3	33	.5	8	30	1282	.2	5	.07	.02	.06	.88	.01	18	<1
3	76	.26	5	214	1184	1.57	123	.1	.15	.2	2.61	.05	5	1
0	12	.15	5	64	849	.06	<1	.03	.1	.06	.48	0	1	<1
3	20	.94	43	224	1147	.56	5	.21	.13	.14	1.26	.05	39	0
0	8	.63	20	104	1146	.17	20	.06	.81	.05	.79	.05	10	6
1	4	.06	3	27	1018	0	<1	<.01	.01	.01	.27	.01	2	0
2	55	.58	25	260	3493	.23	1	.11	.42	.24	1.99	.04	6	1
53	569	.28	47	552	1565	.97	117	.15	.33	.56	.32	.14	13	2
52	124	.9	8	124	1564	.7	220	.04	.26	.18	.06	.5	22	<1
34	425	.26	264	444	797	.55	92	.08	1.58	.45	.53	.07	16	3
11	166	.25	10	100	389	.73	58	.03	.55	.14	.16	.03	4	<1
8	75	.21	9	100	82	.26	89	.05	.98	.12	.28	.03	4	0
30	150	0	8	80	600		80	0		.1	0			0
0	3	.14	3	28	130	.03	14	<.01	.18	<.01	.14	.01	1	1
0	3	.14	2	63	227	.05	24	.01	.05	.01	.27	.02	1	3
0	35	.31	9	187	339	.1	24	.03	.61	.02	.53	.07	9	0
0	0	0			250		0				0			0
4	64	.09	6	15	129	.29	39	.01		.03	0	.02	<1	0
0	3	.32	5	29	914	.06	0	.01	0	.02	.54	.03	2	0
0	2	.12	2	13	218	.02	10	<.01	.07	<.01	.1	.01	1	<1
0	4	.31	11	40	690	.02	0	<.01	0	.01	.23	.02	4	0
0	70	1.62	60	956	1235	.52	306	.14	4.98	.15	3.76	.88	54	28
15	68	1.94	60	952	1179	1.37	577	.13	5.91	.17	4.51	.87	52	26
0	15	1	15	332	494	.34	241	.08	1.35	.08	.93	.16	12	9
7	14	1.63	5	189	1304	2.33	0	.07	.15	.08	1.54	.02	5	0
5	48	1.12	5	259	1373	1.9	264	.04	.37	.1	1.05	.02	5	0
0	17	1.57	5	252	1356	1.67	0	.08	.19	.15	1.6	.05	29	0
3	67	.23	10	57	1075	.31	0	.04	.05	.08	.81	0	0	0
3	39	.26	10	62	1133	.32	0	.05	.05	.15	.78	.03	3	3
38	110	4.32		491	1094	3	40	.34		.46	9.6			11
43	60	4.86	16	422	936	3.75	0	.28		.48	11	.2	14	1
22	40	2.7	8	201	519	1.5		.18		.25	6.6	.1	7	
43	90	6.48	25	533	1174	3.75	30	.39		.58	12.4	.3	21	9
52	150	6.12		321	1166	3		.42		.63	9.8			1
45	60	2.88	30	330	1019	.15		.22		.54	9	.38	18	5
55	170	2.7	31	382	1400	.9	40	.82		.37	7.8	.31	26	24
69	410	4.32		565	2062		100	.91		.49	8.2			11

Table A-1

Food Composition   (Computer code number is for West Diet Analysis program)   (For purposes of calculations, use "0" for t, <1, <.1, <.01, etc.)

Computer Code Number	Food Description	Measure	Wt (g)	H₂O (%)	Ener (cal)	Prot (g)	Carb (g)	Dietary Fiber (g)	Fat (g)	Fat Breakdown (g) Sat	Mono	Poly
	FAST FOOD RESTAURANTS											
	ARBY'S—Continued											
1413	Turkey sandwich, deluxe	1 ea	195	69	260	20	33	<1	6	1.6	2.3	2.4
1680	Turkey sub sandwich	1 ea	277	62	486	33	46		19	5.3	6	7
	Milkshakes:											
1419	Chocolate	1 ea	340	74	451	10	76	<1	12	2.8	7	1.7
1420	Jamocha	1 ea	326	75	368	9	59	0	10	2.5	6.4	1.6
1421	Vanilla	1 ea	312	77	330	10	46	0	11	3.9	5.3	2.3
1728	Salad, roast chicken	1 ea	400	88	204	24	12		7	3.3	.9	.9
1729	Sports drink, Upper Ten	1 ea	358	88	169	0	42		0	0	0	0
	Source: Arby's											
	BURGER KING											
1423	Croissant sandwich, egg, sausage & cheese	1 ea	176	46	600	22	25	1	46	16		
	Whopper sandwiches:											
1425	Whopper	1 ea	270	58	640	27	45	3	39	11		
1426	Whopper with cheese	1 ea	294	57	730	33	46	3	46	16		
	Sandwiches:											
1629	BK broiler chicken sandwich	1 ea	248	59	550	30	41	2	29	6		
1432	Cheeseburger	1 ea	138	48	380	23	28	1	19	9		
1434	Chicken sandwich	1 ea	229	45	710	26	54	2	43	9		
1427	Double beef	1 ea	351	57	870	46	45	3	56	19		
1428	Double beef & cheese	1 ea	375	56	960	52	46	3	63	24		
1433	Double cheeseburger with bacon	1 ea	218	48	640	44	28	1	39	18		
1431	Hamburger	1 ea	126	48	330	20	28	1	15	6		
1437	Ocean catch fish fillet	1 ea	255	51	700	26	56	3	41	6		
1435	Chicken tenders	1 ea	88	50	230	16	14	2	12	3		
1439	French fries (salted)	1 svg	116	40	370	5	43	3	20	5		
1630	French toast sticks	1 svg	141	33	500	4	60	1	27	7		
1440	Onion rings	1 svg	124	51	310	4	41	6	14	2	8	4
1441	Milk shakes, chocolate	1 ea	284	75	320	9	54	3	7	4		
1442	Milk shakes, vanilla	1 ea	284	75	300	9	53	1	6	4		
1443	Fried apple pie	1 ea	113	47	300	3	39	2	15	3		
	Source: Burger King Corporation											
	CHICK-FIL-A											
	Sandwiches:											
69153	Chargrilled chicken	1 ea	150	54	280	27	36	1	3	1		
69152	Chicken	1 ea	167	61	290	24	29	1	9	2		
69155	Chicken salad	1 ea	167	55	320	25	42	1	5	2		
69154	Chicken salad club	1 ea	232	62	390	33	38	2	12	5		
	Salads:											
52139	Carrot and raisin	1 ea	76	53	150	5	28	2	2	0		
52136	Chicken plate	1 ea	468	85	290	21	40	6	5	0		
52134	Chicken garden, charbroiled	1 ea	397	89	170	26	10	5	3	1		
52135	Chick-n-strips	1 ea	451	86	290	32	21	5	9	2		
52138	Cole slaw	1 ea	79	70	130	6	11	1	6	1		
52137	Tossed salad	1 ea	130	85	70	5	13	1	0	0	0	0
15263	Chicken nuggets, svg	1 ea	110	51	290	28	12	60	12	3		
15262	Chicken-n-strips, svg	1 ea	119	59	230	29	10	0	8	2		
50885	Hearty breast of chicken soup, svg	1 ea	215	86	110	16	10	1	1	0		
7973	Waffle potato fries, svg	1 ea	85	28	290	1	49	0	10	4		
46489	Cheesecake, svg	1 ea	88	52	270	13	7	0	21	9		
49134	Fudge nut brownie, svg	1 ea	88	8	416	12	49	0	19	3.6		
20601	Icedream, svg	1 ea	127	74	140	11	16	0	4	1		
48214	Lemon pie, svg	1 ea	99	56	280	1	19	0	22	6		
	Source: Chick-Fil-A											

Chol (mg)	Calc (mg)	Iron (mg)	Magn (mg)	Pota (mg)	Sodi (mg)	Zinc (mg)	VT-A (RE)	Thia (mg)	VT-E (a-TE)	Ribo (mg)	Niac (mg)	V-B6 (mg)	Fola (µg)	VT-C (mg)
33	130	3.42	30	353	1262	1.5	40	.08		.41	15.4	.52	20	12
51	400	4.68		500	2033		20	13.2		.54	18.8			
36	250	.72	48	410	341	1.5	60	.12		.68	.8	.14	14	5
35	250	2.7	36	525	262	1.5	60	.12		.68	.8	.14	14	2
32	300	2.7	36	686	281	1.5	60	.12		.68	4	.14	37	2
43	170	1.98		877	508		485	.33		.54	5.6			51
0				0	40									
260	150	3.6			1140		80							0
90	80	4.5			870		100	.33		.41	7	.35		9
115	250	4.5			1350		150	.34		.48	7	.33		9
80	60	5.4			480		60							6
65	100	2.7			770		60							0
60	100	3.6			1400		0							0
170	80	7.2			940		100	.34		.56	10			9
195	250	7.2			1420		150	.35		.63	10			9
145	200	4.5			1240		80	.31		.42	6			0
55	40	1.8			530		20	.28		.31	4.89			0
90	60	2.7			980		20							1
35	0	.72			530		0							0
0	0	1.08			240		0							4
0	60	2.7			490		0							0
0	100	1.44			810		0							0
20	200	1.8			230		60	.13		.55	.13			0
20	300	0			230		60	.11		.57	.13			4
0	0	1.44			230		0							6
40					640									
50					870									
10					810									
70					980									
6					650									
35					570									
25					650									
20					430									
15					430									
0					0									
14					770									
20					380									
45					760									
5					960									
10					510									
36					773									
40					240									
5					550									

Table A-1
Food Composition

(Computer code number is for West Diet Analysis program)   (For purposes of calculations, use "0" for t, <1, <.1, <.01, etc.)

Computer Code Number	Food Description	Measure	Wt (g)	H₂O (%)	Ener (cal)	Prot (g)	Carb (g)	Dietary Fiber (g)	Fat (g)	Fat Breakdown (g) Sat	Mono	Poly
	FAST FOOD RESTAURANTS—Continued											
	DAIRY QUEEN											
	Ice cream cones:											
1446	Small vanilla	1 ea	142	63	230	6	38	0	7	4.5		
1447	Regular vanilla	1 ea	213	64	350	8	57	0	10	7		
1448	Large vanilla	1 ea	253	65	410	10	65	0	12	8		
1450	Chocolate dipped	1 ea	234	59	510	9	63	1	25	13		
1453	Chocolate sundae	1 ea	241	62	410	8	73	0	10	6		
1455	Banana split	1 ea	369	67	510	8	96	3	12	8		
1456	Peanut buster parfait	1 ea	305	51	730	16	99	2	31	17		
1457	Hot fudge brownie delight	1 ea	305	52	710	11	102	1	29	14	12	2
1459	Buster bar	1 ea	149	45	450	10	41	2	28	12		
1645	Breeze, strawberry, regular	1 ea	383	70	460	13	99	1	1	1	0	0
1460	Dilly bar	1 ea	85	55	210	3	21	0	13	7	3	3
1461	DQ ice cream sandwich	1 ea	61	46	150	3	24	1	5	2		
1463	Milk shakes, regular	1 ea	397	71	520	12	88	<1	14	8	2	2
1464	Milk shakes, large	1 ea	461	71	600	13	101	<1	16	10	2	2
1466	Milk shakes, malted	1 ea	418	68	610	13	106	<1	14	8	2	2
1470	Misty slush, small	1 ea	454	88	220	0	56	0	0	0	0	0
2250	Starkiss	1 ea	85	75	80	0	21	0	0	0	0	0
	Yogurt:											
1641	Yogurt cone, regular	1 ea	213	66	280	9	59	0	1	.5		
1643	Yogurt sundae, strawberry	1 ea	255	69	300	9	66	1	<1	.5	0	0
	Sandwiches:											
1481	Cheeseburger, double	1 ea	219	55	540	35	30	2	31	16		
1480	Cheeseburger, single	1 ea	152	55	340	20	29	2	17	8		
1474	Chicken	1 ea	191	56	430	24	37	2	20	4		
1647	Chicken fillet, grilled	1 ea	184	64	310	24	30	3	10	2.5		
1475	Fish fillet sandwich	1 ea	170	57	370	16	39	2	16	3.5		
1476	Fish fillet with cheese	1 ea	184	56	420	19	40	2	21	6	7	8
1477	Hamburger, single	1 ea	128	56	269	16	27	2	11	4.6	5.6	.9
1478	Hamburger, double	1 ea	212	62	440	30	29	2	22	10		
	Hotdog:											
1483	Regular	1 ea	99	57	240	9	19	1	14	5		
1484	With cheese	1 ea	113	55	290	12	20	1	18	8	8	2
1485	With chili	1 ea	128	61	280	12	21	2	16	6		
1489	French fries, small	1 ea	71	39	210	3	29	3	10	2	5	3
1490	French fries, large	1 ea	128	40	390	5	52	6	18	4	8	6
1491	Onion rings	1 ea	85	46	240	4	29	2	12	2.5		
	Source: International Dairy Queen											
	HARDEES											
1734	Frisco burger hamburger	1 ea	242		760	36	43		50	18		
1736	Frisco grilled chicken salad	1 ea	278		120	18	2		4	1		
1737	Peach shake	1 ea	345		390	10	77		4	3		
	Source: Hardees											
	JACK IN THE BOX											
	Breakfast items:											
1492	Breakfast Jack sandwich	1 ea	121	49	300	18	30	0	12	5	5	2.5
1494	Sausage crescent	1 ea	156	39	580	22	28	0	43	16		
1495	Supreme crescent	1 ea	153	40	530	23	34	0	33	10	18.9	7.8
1496	Pancake platter	1 ea	231	45	610	15	87	0	22	9	7.6	3.5
1497	Scrambled egg platter	1 ea	213	52	560	18	50	0	32	9	16.6	4.4
	Sandwiches:											
1654	Bacon cheeseburger	1 ea	242	49	710	35	41	0	45	15	15.7	8.7
1499	Cheeseburger	1 ea	110	41	330	16	32	0	15	6	5.9	2.3
1739	Chicken caesar pita sandwich	1 ea	237	59	520	27	44	4	26	6		
1655	Chicken sandwich	1 ea	160	52	400	20	38	0	18	4		
1656	Chicken sandwich, sourdough ranch	1 ea	225		490	29	45	1	21	6		
1505	Chicken supreme	1 ea	245	55	620	25	48	0	36	11	14.8	11.4
1583	Double cheeseburger	1 ea	152	44	450	24	35	0	24	12	11.6	3.1

PAGE KEY: A–4 = Beverages   A–6 = Dairy   A–10 = Eggs   A–10 = Fat/Oil   A–14 = Fruit   A–20 = Bakery   A–28 = Grain   A–32 = Fish   A–34 = Meats   A–38 = Poultry   A–40 = Sausage   A–42 = Mixed/Fast   A–46 = Nuts/Seeds   A–50 = Sweets   A–52 = Vegetables/Legumes   A–62 = Vegetarian Foods   A–64 = Misc   A–66 = Soups/Sauces   A–68 = Fast   A–84 = Convenience   A–88 = Baby foods

Chol (mg)	Calc (mg)	Iron (mg)	Magn (mg)	Pota (mg)	Sodi (mg)	Zinc (mg)	VT-A (RE)	Thia (mg)	VT-E (a-TE)	Ribo (mg)	Niac (mg)	V-B6 (mg)	Fola (µg)	VT-C (mg)
20	200	1.08		250	115		122	.05		.28				
30	300	1.8		390	170		150	.09		.38	.16	.13		2
40	350	1.8		451	200		200	.11		.4	.2			2
30	300	1.8		435	200		150	.09		.38	.16	.13		2
30	250	1.44		394	210		150	.08		.35	.4	.19		0
30	250	1.8		860	180		200	.15		.25	.4	.2		15
35	300	1.8		660	400		150	.15		.51	3	.22		1
35	300	5.4		510	340		80	.15		.68	.3	.18		1
15	150	1.08		400	280		80	.09		.17	3	.08		0
10	450	2.7		530	270		0	.13		.73				9
10	100	.36		170	75		60	.03		.14		.06		0
5	60	.72		105	115		40	.03		.25	.4	.05		0
45	400	1.44		570	230		80	.12		.59	.8	.19		<1
50	450	1.44		660	260		200	.15		.68	.8			<1
45	400	1.44		570	230		80	.12		.59	.8	.19		<1
0	0	0			20		0							0
0	0	0			10		0							0
5	300	1.8		285	170		0	.09		.38				2
5	300	1.8		352	180		0	.09		.49				6
115	250	4.5		426	1130		150	.29		.49	6.78			4
55	150	3.6		263	850		60	.29		.33	3.89			4
55	40	1.8		350	760		0	.37		.34	11			0
50	200	2.7		330	1040		0	.3		1.02	12			0
45	40	1.8		280	630		0	.3		.22	3			0
60	100	1.8		290	850		80	.3		.25	5			0
42	56	2.5		234	584		37	.27		.23	3.6			3
90	60	4.5		444	680		60	.32		.45	7.49			6
25	60	1.8		170	730		20	.22		.14	2			4
40	150	1.8		180	950		60	.22		.17	2			4
35	60	1.8		262	870		80	.23		.14	3			4
0	0	.72		430	115		0	.09		.03	2			5
0	0	1.44		780	200		0	.15		.07	3			9
0	0	1.08		90	135		0	.09		.05	.4			0
70					1280									
60					520									
25					290									
185	200	2.7		220	890		80	.47		.41	3			9
185	150	2.7		260	1010		100	.6		.51	4.6			0
210	150	3.6		270	930		150	.65		.54	4.2			12
100	100	1.8		310	890		80	.03		.85	7			6
380	150	4.5		450	1060		150			.66	5			9
110	250	5.4		540	1240		80	.24		.48	8.8	.39		9
35	200	2.7		200	510		60	.23		.23	3			1
55	250	2.7		490	1050		80							2
45	150	1.8		180	1290		40							0
65	150	1.8		340	1060									0
75	200	2.7		190	1520		100	.39		.32	11			2
75	250	3.6		320	900		100	.15		.34	6			0

Table A–1

Food Composition   (Computer code number is for West Diet Analysis program)   (For purposes of calculations, use "0" for t, <1, <.1, <.01, etc.)

Computer Code Number	Food Description	Measure	Wt (g)	H₂O (%)	Ener (cal)	Prot (g)	Carb (g)	Dietary Fiber (g)	Fat (g)	Fat Breakdown (g) Sat	Mono	Poly
	FAST FOOD RESTAURANTS—Continued											
	JACK IN THE BOX—Continued											
1651	Grilled sourdough burger	1 ea	223	48	670	32	39	0	43	16	17.8	7.9
1498	Hamburger	1 ea	97	42	280	13	31	0	11	4	4.9	2
1500	Jumbo Jack burger	1 ea	229	55	560	26	41	0	32	10	13	8
1501	Jumbo Jack burger with cheese	1 ea	242	55	610	29	41	0	36	12	15	9
1740	Monterey roast beef sandwich	1 ea	238	57	540	30	40	3	30	9		
1508	Tacos, regular	1 ea	78	57	190	7	15	2	11	4		
1509	Tacos, super	1 ea	126	59	280	12	22	3	17	6		
	Teriyaki bowl:											
1679	Beef	1 ea	440	62	640	28	124	7	3	1		
1668	Chicken	1 ea	440	62	580	28	115	6	1			
1516	French fries	1 ea	109	38	350	4	45	4	17	4		
1517	Hash browns	1 ea	57	53	160	1	14	1	11	2.5	6.8	.3
1518	Onion rings	1 ea	103	34	380	5	38	0	23	6	15.2	.9
	Milkshakes:											
1519	Chocolate	1 ea	322	72	390	9	74	0	6	3.5	2.1	
1520	Strawberry	1 ea	298	74	330	9	60	0	7	4	2	
1521	Vanilla	1 ea	304	73	350	9	62	0	7	4	1.8	
1522	Apple turnover	1 ea	110	34	350	3	48	0	19	4	10.6	1.5

Source: Jack in the Box Restaurant, Inc

Computer Code Number	Food Description	Measure	Wt (g)	H₂O (%)	Ener (cal)	Prot (g)	Carb (g)	Dietary Fiber (g)	Fat (g)	Fat Breakdown (g) Sat	Mono	Poly
	KENTUCKY FRIED CHICKEN											
	Rotisserie gold:											
1472	Dark qtr, no skin	1 ea	117	66	217	27	0	0	12	3.5		
1473	Dark qtr, w/skin	1 ea	146	62	333	30	1		24	6.6		
1513	White qtr with wing, w/skin	1 ea	176	65	335	40	1		19	5.4		
1525	White qtr with wing, no skin	1 ea	117	63	199	37	0	0	6	1.7		
	Original Recipe:											
1253	Center breast	1 ea	103	52	260	25	9	<1	14	3.8	7.8	2
1251	Side breast	1 ea	83	47	245	18	9	<1	15	4.2	8.8	2.2
1250	Drumstick	1 ea	57	54	152	14	3	<1	8	2.2	4.1	1.3
1252	Thigh	1 ea	95	49	287	18	8	<1	21	5.3	9.4	3.1
1249	Wing	1 ea	53	45	172	12	5	<1	12	3	6	1.8
	Hot & spicy:											
1451	Center breast	1 ea	125	48	360	28	13		22	5		
1452	Side breast	1 ea	120	43	400	22	16		28	6		
1430	Thigh	1 ea	119	47	370	24	10		27	6		
1471	Wing	1 ea	61	38	220	14	5		16	4		
	Extra crispy recipe:											
1261	Center breast	1 ea	118	48	330	26	14	<1	20	4.8	10.8	2.1
1259	Side breast	1 ea	116	40	400	21	19	<1	27	5.5	12.9	2.3
1258	Drumstick	1 ea	65	49	190	14	6	<1	12	3.4	7.7	1.7
1260	Thigh	1 ea	109	43	380	23	7	<1	30	7.7	16	4.2
1257	Wing	1 ea	59	34	240	13	8	<1	17	4.4	10.7	2.5
1390	Baked beans	½ c	167	70	200	8	36	6	3	1.5	.7	.4
1526	Breadstick	1 ea	33	30	110	3	17	0	3	0		
1388	Buttermilk biscuit	1 ea	65	28	234	5	28	<1	13	3.4	6.2	2.3
1391	Chicken little sandwich	1 ea	47	35	169	6	14	<1	10	2	4.7	3.4
1269	Coleslaw	1 svg	90	75	114	1	13	<1	6	1	1.7	3.4
1527	Cornbread	1 ea	56	26	228	3	25	1	13	2		
1268	Corn-on-the-cob	1 ea	151	70	222	4	27	8	12	2	1	1.5
1429	Chicken, hot wings	1 svg	135	38	471	27	18		33			
1386	Kentucky fries	1 svg	77	42	228	3	26	3	12	3.2		
1381	Kentucky nuggets	6 ea	95	48	284	16	15	<1	18	4		
1534	Macaroni & cheese	1 svg	114	71	162	7	15	0	8	3		
1387	Mashed potatoes & gravy	1 svg	120	80	103	1	16	<1	5	.4	.5	.2
1530	Pasta salad	1 svg	108	78	135	2	14	1	8	1		
1389	Potato salad	½ c	188	74	271	5	27	3	16	3	4.2	7.3
1383	Potato wedges	1 svg	92	59	192	3	25	3	9	3		
1535	Red beans & rice	1 svg	112	76	114	4	18	3	3	1		
1529	Vegetable medley salad	1 ea	114	77	126	1	21	3	4	1		

Source: Kentucky Fried Chicken Corp

PAGE KEY: A–4 = Beverages   A–6 = Dairy   A–10 = Eggs   A–10 = Fat/Oil   A–14 = Fruit   A–20 = Bakery   A–28 = Grain   A–32 = Fish   A–34 = Meats   A–38 = Poultry   A–40 = Sausage   A–42 = Mixed/Fast   A–46 = Nuts/Seeds   A–50 = Sweets   A–52 = Vegetables/Legumes   A–62 = Vegetarian Foods   A–64 = Misc   A–66 = Soups/Sauces   A–68 = Fast   A–84 = Convenience   A–88 = Baby foods

A

Chol (mg)	Calc (mg)	Iron (mg)	Magn (mg)	Pota (mg)	Sodi (mg)	Zinc (mg)	VT-A (RE)	Thia (mg)	VT-E (a-TE)	Ribo (mg)	Niac (mg)	V-B6 (mg)	Fola (µg)	VT-C (mg)
110	200	4.5		510	1140		150	.65		.48	8	.33		6
25	100	2.7		190	430		20	.15		.26	2			1
65	100	4.5		450	700		40	.36		.29	1.8			6
80	200	5.4		460	780		60	.36		.44	1.6			6
75	300	3.6		500	1270		80							5
20	100	1.08	35	240	410	1.2	0	.07		.17	1	.13		0
30	150	1.8	45	370	720	1.8	0	.12		.08	1.4	.18		2
25	150	4.5		430	930		1000							6
30	100	1.8		380	1220		1100							9
0	0	1.08		690	190		0	.18		.03	3.8			24
0	0	.36		190	310		0	.05			1			6
0	20	1.8		130	450		0	.29		.17	2.6			2
25	300	.72		680	210		0	.15		.6	.4			0
30	300	0		550	180		0	.15		.43	.4			0
30	300	0		570	180		0	.15		.34	.4			0
0	0	1.8		80	460		0	.2		.12	1.8			9
128	10	.18			772		15							1
163	10	.18			980		15							1
157	10	.18			1104		15							1
97	10	.18			667		15							1
92	30	.72			609		15	.09		.17	11.5			
78	68	1.2			604		15	.06		.13	6.9			
75	21	1.1			269		15	.05		.12	3.2			
112	40	1.08			591		31	.08		.3	5.5			
59	30	.54			383		15	.03		.08	3.7			
80	20	.72			750		15							6
80	40	1.08			850		15							6
100	20	1.08			670		15							6
65	20	.72			440		30							
75	33	.8			740		15	.11		.13	13.1			
75	20	.72			710		15	.09		.1	8.5			
65	20	.36			310		30	.06		.12	3.7			
90	49	1.2			520		30	.1		.21	6.5			
65	20	.36			320		30			.04	.06	3.3		
5	61	2.17	44	348	812	1.96	38	.09		.06	.76	.11	49	3
0	30	.18			15		0							0
3	43	1.92			565		28	.26		.2	2.77			
18	23	1.7			331		5	.16		.12	2.2			
5	30	.36			177		32	.03		.03	.2			27
42	60	.72			194		10							
0	0	.36			76		20	.14		.11	1.8			2
150	40	3.24			1230		15							6
4	11	.98			535		0							0
66	2	.1			865		15	.02		.02	1	.05		<1
16	120	.72			531		190							0
<1	20	.4			388		15			.04	1.2			
1	20	1.08			663		110							7
16	16	3.25	23	385	636	.44	120	.1		.03	.9	.29	11	
3					428		0							
4	10	.72			315									
0	20	.36			240		375							5

Table A–1

Food Composition (Computer code number is for West Diet Analysis program) (For purposes of calculations, use "0" for t, <1, <.1, <.01, etc.)

Computer Code Number	Food Description	Measure	Wt (g)	H₂O (%)	Ener (cal)	Prot (g)	Carb (g)	Dietary Fiber (g)	Fat (g)	Fat Breakdown (g)		
										Sat	Mono	Poly
	FAST FOOD RESTAURANTS—Continued											
	LONG JOHN SILVER'S											
1528	Chicken plank dinner, 3 piece	1 ea	399	56	890	32	101		44	9.5	24.8	9.4
1531	Clam chowder	1 ea	198	86	140	11	10	1	6	1.8	2.5	1.7
1532	Clam dinner	1 ea	361	46	990	24	114		52	10.9	31.4	9.9
	Fish, batter fried:											
1523	Fish & fryes (fries), 3 piece	1 ea	384	54	980	31	92		50	11.3	28.4	9.7
1524	Fish & fryes, 2 piece	1 ea	261	54	610	27	52		37	7.9	23.5	5.3
2240	Fish and lemon crumb dinner, 3 piece	1 ea	493	71	610	39	86		13	2.2	3.9	5.3
2241	Fish and lemon crumb dinner, 2 piece	1 ea	334	77	330	24	46		5	.9	1.6	1.2
1533	Fish & chicken dinner	1 ea	431	55	950	36	102		49	10.6	28.8	9.5
1537	Shrimp dinner, batter fried	1 ea	331	54	840	18	88		47	9.7	27.2	9.1
	Salads:											
1541	Cole slaw	1 ea	98	70	140	1	20	1	6	1	1.5	3.5
1539	Ocean chef salad	1 ea	234	89	110	12	13	2	1	.4	.4	.2
1540	Seafood salad	1 ea	278	79	380	15	12	2	31	5.1	8.2	17.5
1542	Fryes (fries) serving	1 ea	85	43	250	3	28	1	15	2.5	7.4	5.1
1543	Hush puppies	1 ea	24	40	70	2	10	<1	2	.4	1.3	.2
	Source: Long John Silver's, Lexington KY											
	McDONALD'S											
	Sandwiches:											
1221	Big mac	1 ea	216	53	510	25	46	3	26	9.3	7.5	4.1
1226	Cheeseburger	1 ea	122	46	318	15	36	2	13	5.6	3.8	1.1
1224	Filet-o-fish	1 ea	145	49	364	14	41	1	16	3.7	3.8	5.6
1225	Hamburger	1 ea	108	49	266	12	35	2	9	3.2	2.8	.9
1444	McChicken	1 ea	189	52	491	17	42	2	29	5.4	8.5	10.2
1591	McLean deluxe	1 ea	214	64	345	23	37	2	12	4.4	3.6	1.2
1438	McLean deluxe with cheese	1 ea	228	63	398	26	38	2	16	6.8	4.6	1.3
1222	Quarter-pounder	1 ea	171	52	415	23	36	2	20	7.8	6.7	1.3
1223	Quarter-pounder with cheese	1 ea	199	50	520	28	37	2	29	12.6	8.7	1.6
1227	French fries, small serving	1 ea	68	40	207	3	26	2	10	1.7	3.1	2.5
1228	Chicken McNuggets	4 pce	73	51	198	12	10	0	12	2.5	3.7	2.4
	Sauces (packet):											
1229	Hot mustard	1 ea	30	60	63	1	7	1	4	.5	1.1	2
1230	Barbecue	1 ea	32	58	53	<1	12	<1	<1	.1	.1	.2
1231	Sweet & sour	1 ea	32	57	55	<1	12	<1	<1	.1	.1	.3
	Low-fat (frozen yogurt) milk shakes:											
1232	Chocolate	1 ea	295		348	13	62	1	6	3.5	.1	.7
1233	Strawberry	1 ea	294		343	12	63	<1	5	3.4	.1	.6
1234	Vanilla	1 ea	293		308	12	54	<1	5	3.3	.1	.6
	Low-fat (frozen yogurt) sundaes:											
1237	Hot caramel	1 ea	182	56	307	7	62	1	3	2	.3	1
1235	Hot fudge	1 ea	179	60	293	8	53	2	5	4.7	.1	.4
1267	Strawberry	1 ea	178	65	239	6	51	1	1	.7	.1	.2
1238	Vanilla	1 ea	90	68	118	4	24	<1	1	.5	.2	t
1241	Cookies, McDonaldland	1 ea	56	3	258	4	41	1	9	1.7	6.4	.8
1242	Cookies, chocolaty chip	1 ea	56	3	282	3	36	1	14	3.9	4.4	1
1240	Muffin, apple bran, fat-free	1 ea	75	39	182	4	40	1	1	.2	.1	.3
1239	Pie, apple	1 ea	84	35	289	3	37	1	14	3.7	4.5	.3
	Breakfast items:											
1243	English muffin with spread	1 ea	63	33	189	5	30	2	6	2.4	1.5	1.3
1244	Egg McMuffin	1 ea	137	57	289	17	27	1	13	.7	4.5	1.6
1245	Hotcakes with marg & syrup	1 ea	222	44	557	8	100	2	14	2.4	4.6	5.8
1246	Scrambled eggs	1 ea	102	73	170	13	1	0	12	3.6	5.3	1.7
1247	Pork sausage	1 ea	43	45	173	6	<1	0	16	5.5	6.4	2.1
1248	Hashbrown potatoes	1 ea	53	55	130	1	13	1	8	1.3	2.3	1.9
1392	Sausage McMuffin	1 ea	112	42	361	13	26	1	23	8.3	8.2	2.8
1393	Sausage McMuffin with egg	1 ea	163	52	443	19	27	1	29	10	10.7	3.6
1394	Biscuit with biscuit spread	1 ea	76	32	260	4	32	1	13	3.8	3.7	.8

PAGE KEY: A–4 = Beverages   A–6 = Dairy   A–10 = Eggs   A–10 = Fat/Oil   A–14 = Fruit   A–20 = Bakery   A–28 = Grain   A–32 = Fish   A–34 = Meats   A–38 = Poultry   A–40 = Sausage   A–42 = Mixed/Fast   A–46 = Nuts/Seeds   A–50 = Sweets   A–52 = Vegetables/Legumes   A–62 = Vegetarian Foods   A–64 = Misc   A–66 = Soups/Sauces   A–68 = Fast   A–84 = Convenience   A–88 = Baby foods

A–77

A

Chol (mg)	Calc (mg)	Iron (mg)	Magn (mg)	Pota (mg)	Sodi (mg)	Zinc (mg)	VT-A (RE)	Thia (mg)	VT-E (a-TE)	Ribo (mg)	Niac (mg)	V-B6 (mg)	Fola (µg)	VT-C (mg)
55	200	4.5		1170	2000	3	40	.52		.51	16			9
20	200	1.8		380	590	.6	150	.09		.25	2			
75	200	4.5		910	1830	3	40	.75		.42	12			12
70	200	4.5		1120	1530	3	40	.45		.42	8			15
60	40	1.8		900	1480	1.2		.37		.34	8			9
125	200	5.4		990	1420	2.25	700	.75		.59	24			6
75	80	1.8		440	640	.9	1000	.3		.25	14			18
75	200	4.5		1280	2090	3	40	.6		.59	14			9
100	200	3.6		840	1630	3	40	.45		.42	9			9
15	60	.72		190	260	.6	40	.06		.07	2			
40	100	3.6		95	730	.3	500	.12		.14	3			21
55	150	4.5		130	980	.9	200	.15		.25	3			21
0	200	.72		370	500	.3	0	.09			1.6			6
	40	.72		65	25	.3		.06		.03	.8			
76	202	4.32	46	456	932	4.81	66	.49	1.01	.44	6.08	.25	49	3
42	134	2.73	27	281	766	2.62	64	.33	.46	.31	3.81	.15	24	2
37	123	1.85	32	266	708	.7	21	.32	1.52	.23	2.58	.07	30	0
28	126	2.73	24	260	531	2.25	22	.33	.23	.26	3.81	.14	21	2
52	128	2.5	32	319	797	1.06	29	.91	6.16	.24	7.74	.38	37	1
59	131	4.29	40	537	811	4.9	74	.42	.63	.34	7.16	.29	44	8
73	139	4.29	43	559	1046	5.26	115	.42	.85	.39	7.16	.3	47	8
70	127	4.33	33	405	692	4.66	33	.39	.36	.32	6.78	.24	27	3
97	143	4.5			1160		115	.39	.81	.43	6.78	.26	33	3
0	9	.53	26	469	135	.32	0	.05	.83	0	1.94	.24	26	8
42	9	.65	17	210	353	.69	0	.08	.96	.11	5.15	.21		0
3	7	.78		29	85		4	.01		.01	.15			0
0	4	0		51	277		0	.01		.01	.17			4
0	2	.16		7	158		74	0		.01	.08			0
24	372	1.04		543	241		46	.12		.51	.4	.1		3
24	366	.29		542	170		46	.12		.51	.4	.11		3
24	360	.29		533	193		45	.12		.51	.31			3
7	246	.15		344	197		18	.09		.34	.27			2
5	258	.59		441	190		7	.09		.34	.29			2
5	221	.25		325	115		6	.06		.34	.25			2
3	132	.23		175	84		4	<.01		.01	.23			1
0	10	1.73	11	62	267	.38	0	.24	.99	.16	2.01	.03		0
3	28	1.78	24	142	229	.4	0	.14	.92	.16	1.48			0
0	34	1.29	13	77	215	.33	0	.14	0	.14	1.32	.03	5	1
0	17	1.23	7	69	221	.23		.19	1.5	.12	1.55	.04	9	27
13	103	1.59	13	69	386	.42	33	.25	.13	.31	2.61	.04	57	1
234	151	2.44	24	199	730	1.56	100	.49	.85	.45	3.33	.15	33	2
11	108	1.98	27	285	746	.53	119	.24	1.2	.26	1.86	.09	<1	<1
424	50	1.19	10	126	143	1.06	168	.07	.92	.51	.06	.12	44	0
33	7	.5	7	102	292	.78	0	.18	.26	.06	1.7	.09		0
0	7	.3	11	213	332	.15	0	.08	.58	.02	.9	.08	8	3
46	132	2.07	22	191	751	1.51	48	.56	.66	.27	3.76	.14	16	0
257	156	2.8	26	251	821	2.07	117	.59	1.11	.49	3.79	.19	33	0
0	68	1.85	9	105	836	.3	2	.29	.81	.23	2.23	.03	5	0

Table A–1

Food Composition   (Computer code number is for West Diet Analysis program)   (For purposes of calculations, use "0" for t, <1, <.1, <.01, etc.)

Computer Code Number	Food Description	Measure	Wt (g)	H₂O (%)	Ener (cal)	Prot (g)	Carb (g)	Dietary Fiber (g)	Fat (g)	Fat Breakdown (g) Sat	Mono	Poly
	FAST FOOD RESTAURANTS—Continued											
	McDONALD'S—Continued											
1395	Biscuit with sausage	1 ea	119	37	433	10	32	1	29	8.6	10.1	2.8
1396	Biscuit with sausage & egg	1 ea	170	48	518	16	33	1	35	10.5	12.7	3.7
1397	Biscuit with bacon, egg, cheese	1 ea	152	46	450	17	33	1	27	8.7	8.9	2.3
	Salads:											
1398	Chef salad	1 ea	313	86	206	19	9	3	11	4.2	3	1.2
1400	Garden salad	1 ea	234	92	84	6	7	3	4	1.1	1.4	.7
1401	Chunky chicken salad	1 ea	296	87	164	23	8	3	5	1.3	1.6	1
	Source: McDonald's Corporation											
	PIZZA HUT											
	Pan pizza:											
1657	Cheese	2 pce	216	51	522	24	56	4	22	10	6.8	3.4
1658	Pepperoni	2 pce	208	49	531	22	56	4	24	8	9.9	3.7
1659	Supreme	2 pce	273	56	622	30	56	6	30	12	12	4.2
1660	Super supreme	2 pce	286	56	645	30	56	6	34	12		
	Thin 'n crispy pizza:											
1649	Cheese	2 pce	174	52	411	22	42	4	16	8	4.4	2.3
1623	Pepperoni	2 pce	168	48	431	22	42	2	20	8		
1622	Supreme	2 pce	232	57	514	28	42	4	26	10		
1620	Super supreme	2 pce	247	57	541	28	44	4	28	12		
	Hand tossed pizza:											
1619	Cheese	2 pce	216	53	470	26	58	4	14	7.9		
1618	Pepperoni	2 pce	208	51	477	24	58	4	16	8		
1648	Supreme	2 pce	273	56	568	32	60	6	24	10		
1617	Super supreme	2 pce	286	57	591	32	60	6	26	10		
	Personal pan pizza:											
1610	Pepperoni	1 ea	255	50	637	27	69	5	28	10	11.8	4.5
1609	Supreme	1 ea	327	57	721	33	70	6	34	12	14.7	5.6
	Source: Pizza Hut											
	SUBWAY											
	Deli style sandwich:											
69104	Bologna	1 ea	171	64	292	10	38	2	12	4		
69102	Ham	1 ea	171	69	234	11	37	2	4	1		
69103	Roast beef	1 ea	180	69	245	13	38	2	4	1		
69105	Seafood and crab:	1 ea	178	66	298	12	37	2	11	2		
69106	With light mayo	1 ea	178	68	256	12	37	2	7	2		
69108	Tuna:	1 ea	178		354	11	37	2	18	3		
69107	With light mayo	1 ea	178	67	279	11	38	2	9	2		
69101	Turkey	1 ea	180	69	235	12	38	2	4	1		
	Sandwiches, 6 inch:											
	B.L.T.:											
69135	On white bread	1 ea	191	67	311	14	38	3	10	3		
69136	On wheat bread	1 ea	198	65	327	14	44	3	10	3		
	Chicken taco sub:											
69131	On white bread	1 ea	286	70	421	24	43	3	16	5		
69132	On wheat bread	1 ea	293	69	436	25	49	4	16	5		
	Club :											
69117	On white bread	1 ea	246	73	297	21	40	3	5	1		
69118	On wheat bread	1 ea	253	71	312	21	46	3	5	1		
	Cold cut trio:											
69113	On white bread	1 ea	246	71	362	19	39	3	13	4		
69114	On wheat bread	1 ea	253	68	378	20	46	3	13	4		
	Ham:											
69115	On white bread	1 ea	232	73	287	18	39	3	5	1		
69115	On wheat bread	1 ea	239	71	302	19	45	3	5	1		

PAGE KEY: A–4 = Beverages A–6 = Dairy A–10 = Eggs A–10 = Fat/Oil A–14 = Fruit A–20 = Bakery A–28 = Grain A–32 = Fish A–34 = Meats A–38 = Poultry A–40 = Sausage A–42 = Mixed/Fast A–46 = Nuts/Seeds A–50 = Sweets A–52 = Vegetables/Legumes A–62 = Vegetarian Foods A–64 = Misc A–66 = Soups/Sauces A–68 = Fast A–84 = Convenience A–88 = Baby foods

Chol (mg)	Calc (mg)	Iron (mg)	Magn (mg)	Pota (mg)	Sodi (mg)	Zinc (mg)	VT-A (RE)	Thia (mg)	VT-E (a-TE)	Ribo (mg)	Niac (mg)	V-B6 (mg)	Fola (μg)	VT-C (mg)
33	75	2.35	15	207	1128	1.08	2	.48	1.07	.29	3.93	.12	5	0
245	100	2.95	20	271	1199	1.61	59	.51	1.53	.55	3.96	.18	27	0
238	103	2.6	20	245	1315	1.64	99	.39	1.49	.57	3.32	.13	30	0
179	157	1.81	40	605	727	2.16	1179	.33	1.45	.37	4.32	.36	100	22
139	52	1.34	24	407	61	.73	1114	.12	.95	.24	.65	.16	96	22
76	54	1.62	44	673	318	1.52	1973	.51	1.28	.21	8.46	.52	83	30
50	288	3	63	337	1002	4.32	211	.6		.64	5.48	.18		7
48	206	3.21	55	399	1140	4.14	190	.62		.48	5.31	.16	0	8
60	234	4.6	81	620	1529	6	195	.86		.84	6.4	.33		11
68	236	4.39	80	592	1649	5.99	201	.83		.73	7.13			12
50	291	2.06	56	307	1070	4.23	217	.46		.46	5.65	.18		6
50	208	2.2	51	330	1255	4.02	199	.48		.49	5.97			7
62	238	3.6	79	631	1591	5.4	197	.7		.57	6.27			12
70	238	3.41	73	563	1762	5.47	208	.72		.53	6.57			9
50	284	3	71	388	1242	4.6	198	.48		.48	5.3			10
48	202	3.21	84	610	1380	6.01	187	.72		.56	7.59			13
60	232	4.6	87	589	1769	5.48	192	.82		.66	8.45			14
68	232	4.39	89	607	1889	5.65	198	.84		.68	8.71			14
55	250	4	60	406	1338	3.8	233	.56		.66	8.16	.2		10
66	276	5.19	74	603	1757	4.69	240	.73		.82	9.91	.4		14
20	39	3			744		113							14
14	24	3			773		113							14
13	23	3			638		113							14
17	24	3			544		113							14
16	24	3			556		118							14
18	26	3			557		116							14
16	26	3			583		126							14
12	26	3			944		113							14
16	27	3			945		120							15
16	33	3			957		120							15
52	118	4			1264		209							18
52	124	4			1275		209							18
26	29	4			1341		120							15
26	35	4			1352		120							15
64	49	4			1401		130							16
64	55	4			1412		130							16
28	28	3			1308		120							15
28	35	3			1319		120							15

A

Table A–1

Food Composition   (Computer code number is for West Diet Analysis program)   (For purposes of calculations, use "0" for t, <1, <.1, <.01, etc.)

Computer Code Number	Food Description	Measure	Wt (g)	H₂O (%)	Ener (cal)	Prot (g)	Carb (g)	Dietary Fiber (g)	Fat (g)	Fat Breakdown (g)		
										Sat	Mono	Poly
	FAST FOOD RESTAURANTS—Continued											
	SUBWAY—Continued											
	Italian B.M.T.											
69139	On white bread	1 ea	246	66	445	21	39	3	21	8		
69140	On wheat bread	1 ea	253	64	460	21	45	3	22	7		
	Meatball:											
69129	On white bread	1 ea	260	70	404	18	44	3	16	6		
69130	On wheat bread	1 ea	267	67	419	19	51	3	16	6		
	Melt with turkey, ham, bacon, cheese:											
69127	On white bread	1 ea	251	70	366	22	40	3	12	5		
69128	On wheat bread	1 ea	258	68	382	23	46	3	12	5		
	Pizza sub:											
69133	On white bread	1 ea	250	66	448	19	41	3	22	9		
69134	On wheat bread	1 ea	257	65	464	19	48	3	22	9		
	Roast beef:											
69121	On white bread	1 ea	232	72	288	19	39	3	5	1		
69122	On wheat bread	1 ea	239	70	303	20	45	3	5	1		
	Roasted chicken breast:											
69125	On white bread	1 ea	246	70	332	26	41	3	6	1		
69126	On wheat bread	1 ea	253	68	348	27	47	3	6	1		
	Seafood and crab:											
69145	On white bread:	1 ea	246	69	415	19	38	3	19	3		
69147	With light mayo	1 ea	246	72	332	19	39	3	10	2		
69146	On wheat bread:	1 ea	253	67	430	20	44	3	19	3		
69148	With light mayo	1 ea	253	70	347	20	45	3	10	2		
	Spicy italian:											
69123	On white bread	1 ea	232	64	467	20	38	3	24	9		
69124	On wheat bread	1 ea	239	62	482	21	44	3	25	9		
	Steak and cheese:											
69119	On white bread	1 ea	257	68	383	29	41	3	10	6		
69120	On wheat bread	1 ea	264	67	398	30	47	3	10	6		
	Tuna:											
69141	On white bread:	1 ea	246	62	527	18	38	3	32	5		
69143	With light mayo	1 ea	246	70	376	18	39	3	15	2		
69142	On wheat bread:	1 ea	253	62	542	19	44	3	32	5		
69144	With light mayo	1 ea	253	68	391	19	46	3	15	2		
	Turkey:											
69111	On white bread	1 ea	232	73	273	17	40	3	4	1		
69112	On wheat bread	1 ea	239	71	289	18	46	3	4	1		
	Turkey breast and ham:											
69137	On white bread	1 ea	232	73	280	18	39	3	5	1		
69138	On wheat bread	1 ea	239	71	295	18	46	3	5	1		
	Veggie delite:											
69109	On white bread	1 ea	175	71	222	9	38	3	3	0		
69110	On wheat bread	1 ea	182	69	237	9	44	3	3	0		
	Salads:											
52128	B.L.T.	1 ea	276	91	140	7	10	2	8	3		
52124	B.M.T., classic Italian	1 ea	331	86	274	14	11	1	20	7		
52127	Chicken taco	1 ea	370	87	250	18	15	2	14	5		
52115	Club	1 ea	331	91	126	14	12	1	3	1		
52120	Cold cut trio	1 ea	330	89	191	13	11	1	11	3		
52123	Ham	1 ea	316	91	116	12	11	1	3	1		
52129	Meatball	1 ea	345	88	233	12	16	2	14	5		
52131	Melt	1 ea	336	88	195	16	12	1	10	4		
52121	Pizza	1 ea	335	86	277	12	13	2	20	8		
52126	Roast beef	1 ea	316	92	117	12	11	1	3	1		
52119	Roasted chicken breast	1 ea	331	89	162	20	13	1	4	1		
52117	Seafood and crab:	1 5	331	88	244	13	10	2	17	3		
52116	With light mayo	1 5	331	90	161	13	11	2	8	1		
52130	Steak and cheese	1 ea	342	87	212	22	13	1	8	5		
52122	Tuna:	1 ea	331	84	356	12	10	1	30	5		
52118	With light mayo	1 ea	331	89	205	12	11	1	13	2		
52114	Turkey breast	1 ea	316	92	102	11	12	1	2	1		
52125	With ham	1 ea	316	92	109	11	11	1	3	1		

Chol (mg)	Calc (mg)	Iron (mg)	Magn (mg)	Pota (mg)	Sodi (mg)	Zinc (mg)	VT-A (RE)	Thia (mg)	VT-E (a-TE)	Ribo (mg)	Niac (mg)	V-B6 (mg)	Fola (µg)	VT-C (mg)
56	44	4			1652		151							15
56	50	4			1664		151							15
33	32	4			1035		142							16
33	39	4			1046		142							16
42	93	4			1735		155							15
42	100	3			1746		156							15
50	103	4			1609		238							16
50	110	3			1621		238							16
20	25	4			928		120							15
20	32	3			939		120							15
48	35	3			967		123							15
48	42	3			978		123							15
34	28	3			849		121							15
32	28	3			873		131							15
34	34	3			860		121							15
32	34	3			884		131							15
57	40	4			1592		169							15
57	47	4			1604		169							15
70	88	5			1106		175							18
70	95	5			1117		176							18
36	32	3			875		125							15
32	32	3			928		146							15
36	38	3			886		126							15
32	38	3			940		146							15
19	30	4			1391		120							15
19	37	3			1403		120							15
24	29	3			1350		120							15
24	36	3			1361		120							15
0	25	3			582		120							15
0	32	3			593		120							15
16	24	1			672		273							32
56	41	2			1379		303							32
52	115	3			990		361							35
26	26	2			1067		273							32
64	46	2			1127		282							33
28	25	2			1034		273							32
33	30	2			761		295							33
42	90	2			1461		308							32
50	100	2			1336		390							33
20	23	2			654		273							32
48	32	2			693		276							32
34	25	2			575		273							32
32	25	2			599		284							32
70	86	3			832		328							35
36	29	2			601		278							32
32	29	2			654		298							32
19	28	2			1117		273							32
24	27	2			1076		273							32

Table A–1

Food Composition   (Computer code number is for West Diet Analysis program)   (For purposes of calculations, use "0" for t, <1, <.1, <.01, etc.)

Computer Code Number	Food Description	Measure	Wt (g)	H₂O (%)	Ener (cal)	Prot (g)	Carb (g)	Dietary Fiber (g)	Fat (g)	Fat Breakdown (g)		
										Sat	Mono	Poly
	FAST FOOD RESTAURANTS—Continued											
	SUBWAY—Continued											
52113	Veggie delite	1 ea	260	94	51	2	10	1	1	0		
	Cookies:											
47662	Brazil nut and chocolate chip	1 ea	48	12	229	3	27	1	12	3.5		
47655	Chocolate chip:	1 ea	48	14	209	2	29	1	10	3.5		
47658	With M&M's	1 ea	48	14	209	2	29	1	10	3		
47659	Chocolate chunk	1 ea	48	14	209	2	29	1	10	3.5		
47656	Oatmeal raisin	1 ea	48	15	199	3	29	1	8	2		
47657	Peanut butter	1 ea	48	13	219	3	26	1	12	2.5		
47660	Sugar	1 ea	48	11	229	2	28	0	12	3		
47661	White chip macademia	1 ea	48	12	229	2	28	1	12	2.5		
	Source: Subway International											
	TACO BELL											
	Breakfast burrito:											
1601	Bacon breakfast burrito	1 ea	99	48	291	11	23		17	4		
1627	Country breakfast burrito	1 ea	113	55	220	8	26	2	14	5		
1626	Fiesta breakfast burrito	1 ea	92	44	280	9	25	2	16	6		
1625	Grande breakfast burrito	1 ea	177	56	420	13	43	3	22	7		
1604	Sausage breakfast burrito	1 ea	106	49	303	11	23		19	6		
	Burritos:											
1544	Bean with red sauce	1 ea	198	58	380	13	55	13	12	4		
1545	Beef with red sauce	1 ea	198	57	432	22	42	4	19	8	6.7	.7
1546	Beef & bean with red sauce	1 ea	198	57	412	17	50	5	16	6	6.1	2.1
1569	Big beef supreme	1 ea	298	64	520	24	54	11	23	10		
1552	Chicken burrito	1 ea	171	58	345	17	41		13	5		
1547	Supreme with red sauce	1 ea	248	64	428	16	50	10	18	7.8		
1571	7 layer burrito	1 ea	234	61	438	13	55	11	19	5.8		
1538	Chilito	1 ea	156	49	391	17	41		18	9		
1549	Chilito, steak	1 ea	257	62	496	26	47		23	10		
	Tacos:											
1551	Taco	1 ea	78	58	180	9	12	3	10	4		
1554	Soft taco	1 ea	99	63	242	10	13	3	11	4.4		
1536	Soft taco supreme	1 ea	128	64	234	11	21	3	13	6.3		
1568	Soft taco, chicken	1 ea	128	63	212	15	22	2	7	2.6		
1572	Soft taco, steak	1 ea	100	63	180	12	16	2	8	1.9		
1555	Tostada with red sauce	1 ea	156	67	264	9	27	11	13	4.4		
1558	Mexican pizza	1 ea	223	53	578	21	43	8	35	10.1		
1559	Taco salad with salsa	1 ea	585	71	923	33	70	17	56	16.3		
1560	Nachos, regular	1 ea	106	40	343	5	36	3	19	4.3		
1561	Nachos, bellgrande	1 ea	287	51	708	19	77	16	36	10.1		
1562	Pintos & cheese with red sauce	1 ea	128	68	203	10	19	11	10	4.3		
1563	Taco sauce, packet	1 ea	9	94	2	<1	<1	<1	<1	0	0	0
1564	Salsa	1 ea	10	28	27	1	6		<1	0	0	0
1565	Cinnamon twists	1 ea	35	6	175	1	24	0	7	0		
1628	Caramel roll	1 ea	85	19	353	6	46		16	4		
	Source: Taco Bell Corporation											
	WENDY'S											
	Hamburgers:											
1566	Single on white bun, no toppings	1 ea	133	44	360	24	31	2	16	6		
1570	Cheeseburger, bacon	1 ea	166	55	380	20	34	2	19	7	10.3	1.4
1730	Chicken sandwich, grilled	1 ea	189	62	310	27	35	2	8	1.5		
	Baked potatoes:											
1573	Plain	1 ea	284	71	310	7	71	7	0	0	0	0
1574	With bacon & cheese	1 ea	380	69	530	17	78	7	18	4	10.7	3.3
1575	With broccoli & cheese	1 ea	411	74	470	9	80	9	14	2.5	8	2.5
1576	With cheese	1 ea	383	68	570	14	78	7	23	8	9.2	4.8

Chol (mg)	Calc (mg)	Iron (mg)	Magn (mg)	Pota (mg)	Sodi (mg)	Zinc (mg)	VT-A (RE)	Thia (mg)	VT-E (a-TE)	Ribo (mg)	Niac (mg)	V-B6 (mg)	Fola (µg)	VT-C (mg)
0	23	1			308		136							32
10	32	1.99			115		0							0
10	16	1.99			139		0							0
15	16	1			139		0							0
10	16	1			139		0							0
15	32	1			159		0							0
0	16	1			179		0							0
20	0	.72			179		0							0
10	16	1			139		0							0
181	80	1.8			652		310							
195	80	1.08			690		250							0
25	80	.72			580		150							0
205	100	1.8			1050		500							0
183	80	1.8			661		320							
10	150	2.7		495	1100		450	.04		2.02	1.98	.31		0
57	160	3.96		380	1303		530	.4		2.14	3.44	.32		1
32	170	3.78	50	442	1221	2.67	450	.49		.41	3.09	.59	38	1
55	150	2.7			1520		600							5
57	140	2.52			854		440							1
34	146	8.75	48	410	1196		486	.39		2.04	2.81	.34		5
21	165	2.98			1058		248							5
47	300	3.06			980		950							
78	200	2.7			1313		970							2
25	80	1.08		159	330		100	.05		.14	1.2	.12		0
27	88	1.19		211	363		110	.42		.24	2.95	1.08		0
31	90	1.62			532		135							3
37	85	1.52			571		63							1
19	62	1.13			797		31							0
13	132	1.59		401	573		441	.05		.17	.63	.26		1
46	253	3.65	80	408	1054	5.37	405	.32		.33	2.96	1.12	60	5
65	326	6.84		1048	1931	1.67	1736	.51		.76	4.8	.56	10	26
5	107	.77		160	610	1.68	64	.17		.16	.68	.19	10	0
32	184	3.31		674	1205		138	.1		.34	2.17			3
16	160	1.92	110	384	693	2.17	267	.05		.15	.43	.21	68	0
0	0	.07		9	75		30	0			.02			<1
0	50	.6		376	709		168	.02		.14	0			10
0	0	.45		27	238		50	.1		.04	.71	.04		0
15	60	1.44			312		330							4
65	110	4.14		296	580		0	.43		.38	6.71			0
60	170	3.42	38	375	850	5.9	80	.3		.31	6.43	.26	28	6
65	100	2.7			790		40							6
0	30	3.78	75	1187	25	.74	0	.31	.14	.12	4.3	.8	31	36
20	180	4.32	87	1498	1390	2.75	100	.24		.19	5.04	.94	36	36
5	210	4.5	93	1745	470	.97	350	.34		.29	4.5	.97	74	72
30	380	4.14	85	1510	640	.67	200	.25		.28	3.6	.88	36	36

Table A-1

Food Composition   (Computer code number is for West Diet Analysis program)   (For purposes of calculations, use "0" for t, <1, <.1, <.01, etc.)

Computer Code Number	Food Description	Measure	Wt (g)	H₂O (%)	Ener (cal)	Prot (g)	Carb (g)	Dietary Fiber (g)	Fat (g)	Fat Breakdown (g) Sat	Mono	Poly
	FAST FOOD RESTAURANTS—Continued											
	WENDY'S—Continued											
1577	With chili & cheese	1 ea	439	69	630	20	83	9	24	9		
1578	With sour cream & chives	1 ea	314	71	380	8	74	8	6	4		
1579	Chili	1 ea	227	81	210	15	21	5	7	2.5		
1582	Chocolate chip cookies	1 ea	57	6	270	3	36	1	13	6		
1580	French fries	1 ea	130	41	390	5	50	5	19	3	11.9	2.4
1581	Frosty dairy dessert	1 ea	227	68	330	8	56	0	8	5		
	Source: Wendy's International											
	CONVENIENCE FOODS and MEALS											
	BUDGET GOURMET											
1695	Chicken cacciatore	1 ea	312	80	300	20	27		13			
1692	Linguini & shrimp	1 ea	284	77	330	15	33		15			
1691	Scallops & shrimp	1 ea	326	79	320	16	43		9			
2245	Seafood newburg	1 ea	284	74	350	17	43		12			
1693	Sirloin tips with country gravy	1 ea	284	80	310	16	21		18			
1694	Sweet & sour chicken with rice	1 ea	284	72	350	18	53		7			
1689	Teriyaki chicken	1 ea	340	77	360	20	44		12			
1690	Veal parmigiana	1 ea	340	75	440	26	39		20			
1696	Yankee pot roast	1 ea	312	77	380	27	22		21			
	Source: The All American Gourmet Co.											
	HAAGEN DAZS											
1755	Ice cream bar, vanilla almond	1 ea	107		371	6	26		27	14.1	10	3
	Sorbet:											
1758	Lemon	½ c	113		140	0	35		0	0	0	0
1760	Orange	½ c	113		140	0	36		0	0	0	0
1759	Raspberry	½ c	113		110	0	27		0	0	0	0
	Yogurt, frozen:											
1753	Chocolate	½ c	98		171	8	26		4	2	2	0
1754	Strawberry	½ c	98		171	6	27		4	2	2	0
	Yogurt extra, frozen:											
1752	Brownie nut	½ c	101		220	8	29		9	4	4	1
1751	Raspberry rendezvous	½ c	101		132	4	26		2	1	1	0
	Source: Pillsbury											
	HEALTHY CHOICE											
	Entrees:											
2112	Fish, lemon pepper	1 ea	303	78	290	14	47	7	5	1		
1624	Lasagna	1 ea	383	76	390	26	60	9	5	2		
2111	Meatloaf, traditional	1 ea	340	79	320	16	46	7	8	4		
2104	Zucchini lasagna	1 ea	396	80	329	20	58	11	1	1		
2110	Dinner, pasta shells marinara	1 ea	340	74	360	25	59	5	3	1.5		
	Low-fat ice cream:											
973	Brownie	½ c	71	60	120	3	22	2	2	1	.3	.7
259	Butter pecan	½ c	71	60	120	3	22	1	2	1	.3	.7
650	Chocolate chip	½ c	71	62	120	3	21	<1	2	1	1	0
1608	Cookie & cream	½ c	71	62	120	3	21	<1	2	1.5	.5	0
650	Chocolate chip	½ c	71	62	120	3	21	<1	2	1	1	0
45	Rocky road	½ c	71	53	140	3	28	2	1	1	.5	0
1621	Vanilla	½ c	71	66	100	3	18	1	2	.5	1.5	0
391	Vanilla fudge	½ c	71	62	120	3	21	1	2	1.5		
	Source: ConAgra Frozen Foods, Omaha, NE											

PAGE KEY: A–4 = Beverages   A–6 = Dairy   A–10 = Eggs   A–10 = Fat/Oil   A–14 = Fruit   A–20 = Bakery   A–28 = Grain   A–32 = Fish   A–34 = Meats   A–38 = Poultry   A–40 = Sausage   A–42 = Mixed/Fast   A–46 = Nuts/Seeds   A–50 = Sweets   A–52 = Vegetables/Legumes   A–62 = Vegetarian Foods   A–64 = Misc   A–66 = Soups/Sauces   A–68 = Fast   A–84 = Convenience   A–88 = Baby foods

A–85

A

Chol (mg)	Calc (mg)	Iron (mg)	Magn (mg)	Pota (mg)	Sodi (mg)	Zinc (mg)	VT-A (RE)	Thia (mg)	VT-E (a-TE)	Ribo (mg)	Niac (mg)	V-B6 (mg)	Fola (µg)	VT-C (mg)
40	330	5.04	122	1745	770	4.15	200	.33		.29	4.5	.99	55	36
15	80	4.32	71	1438	40	.91	300	.23		.14	3.04	.8	32	48
30	80	2.9		501	800		80	.11		.15	2.66			4
30	10	1.8	13	89	120	.41	0	.05		.06	.36	.03	5	0
0	20	1.08	55	845	120	.62	0	.18		.04	3.6	.33	40	6
35	310	1.08	46	544	200	.97	150	.11		.47	.32	.13	17	0
60	150	1.8			810		40	.23		.51	5			21
75	10	3.6			1250		1000	.3		.17	3			2
70	150	.72			690		150			.26	3			12
70	100	.72			660		40	.23		.26	2			
40	60	.36			570		150	.15		.17	4	.28		2
40	60	.72			640		80	.12		.34	3			2
55	80	1.4			610		300	.15		.34	6			12
165	30	4.5			1160		1000	.45		.6	6			6
70	150	1.8			690		600	.15		.43	7			6
90	161	.38		221	85		161			.18				
0				30	20									7
0				80	20									20
0				60	15									7
40	147	.71		241	45		20			.17				
50	147			141	45		20	.03		.17				5
55	152	.73		250	60		20			.14				
20	81			97	25		0			.1				5
25	20	1.08			360		100							30
15	150	3.6		500	550		100	.3		.26	2			6
35	40	1.8			460		150							54
10	199	2.69			309		249							0
25	400	1.8			390		100							4
2	80	0		268	55		40							0
2	100	0		211	60		40							0
2	100	0		240	50		40							0
2	100			254	90		60	.03		.15				2
2	100	0		240	50		40							0
2	100	0		168	60		40	.03		.15				0
5	100	0		254	50		60	.05		.22				2
2	100	0		296	50		40							0

Table A-1
Food Composition

(Computer code number is for West Diet Analysis program)   (For purposes of calculations, use "0" for t, <1, <.1, <.01, etc.)

Computer Code Number	Food Description	Measure	Wt (g)	H₂O (%)	Ener (cal)	Prot (g)	Carb (g)	Dietary Fiber (g)	Fat (g)	Fat Breakdown (g) Sat	Mono	Poly
	CONVENIENCE FOODS and MEALS—Continued											
	HEALTH VALLEY											
	Soups, fat-free:											
2001	Beef broth, no salt added	1 c	240	98	18	5	0	0	0	0	0	0
2073	Beef broth, w/salt	1 c	240	98	30	5	2	0	0	0	0	0
2016	Black bean & vegetable	1 c	240	85	110	11	24	12	0	0	0	0
2017	Chicken broth	1 c	240	97	30	6	0	0	0	0	0	0
2018	14 garden vegetable	1 c	240	90	80	6	17	4	0	0	0	0
2015	Lentil & carrot	1 c	240	85	90	10	25	14	0	0	0	0
2014	Split pea & carrot	1 c	240	89	110	8	17	4	0	0	0	0
2013	Tomato vegetable	1 c	240	90	80	6	17	5	0	0	0	0
	Source: Health Valley											
	LA CHOY											
2100	Egg rolls, mini, chicken	1 svg	106	53	220	8	35	3	6	1.5		
2099	Egg rolls, mini, shrimp	1 svg	106	56	210	7	35	3	4	1		
	Source: Beatrice/Hunt Wesson											
	LEAN CUISINE											
	Dinners:											
1639	Baked cheese ravioli	1 ea	241	77	250	12	32	4	8	3	2	1
1632	Chicken chow mein	1 ea	255	81	210	13	28	2	5	1	2	1
1633	Lasagna	1 ea	291	79	270	19	34	5	6	2.5	1.5	.5
1634	Macaroni & cheese	1 ea	255	78	270	13	39	2	7	3.5	1.5	.5
1631	Spaghetti w/meatballs	1 ea	269	74	290	17	40	4	7	2	3	1.5
	Pizza:											
1636	French bread sausage pizza	1 ea	170	53	420	19	41	4	20	5	13.9	1.1
	Source: Stouffer's Foods Corp, Solon, OH											
	TASTE ADVENTURE SOUPS											
1905	Black bean	1 c	242		139	6	28	6	1			
1904	Curry lentil	1 c	241		138	6	30	5	1			
1906	Lentil chili	1 c	242		181	11	33	6	1			
1903	Split pea	1 c	244		140	5	27	5	1			
	Source: Taste Adventure Soups											
	WEIGHT WATCHERS											
	Cheese, fat-free slices:											
1978	Cheddar, sharp	2 pce	21	65	30	5	2	0	0	0	0	0
1980	Swiss	2 pce	21	65	30	5	2	0	0	0	0	0
1977	White	2 pce	21	65	30	5	2	0	0	0	0	0
1979	Yellow	2 pce	21	65	30	5	2	0	0	0	0	0
	Dinners:											
2029	Chicken chow mein	1 ea	255	81	200	12	34	3	2	.5		
1646	Oven fried fish	1 ea	218	78	230	15	25	2	8	2.5	5	2
1972	Margarine, reduced fat	1 tbs	14	49	59	0	0	0	7	1.5		
	Pizza:											
1653	Cheese	1 ea	163	48	390	23	49	6	12	4	3	1
1650	Deluxe combination pizza	1 ea	186	56	380	23	47	6	11	3.5	5	2
1652	Pepperoni pizza	1 ea	158	48	390	23	46	4	12	4	5	2
	Desserts:											
1644	Chocolate brownie	1 ea	182	75	190	6	35	4	4	1	2	1
2024	Chocolate eclair	1 ea	60	45	151	3	24	2	5	1.5		
2247	Chocolate mousse	1 ea	78	44	190	6	33	3	4	1.5		
1642	Strawberry cheesecake	1 ea	111	62	180	7	28	2	5	2	1	2

Chol (mg)	Calc (mg)	Iron (mg)	Magn (mg)	Pota (mg)	Sodi (mg)	Zinc (mg)	VT-A (RE)	Thia (mg)	VT-E (a-TE)	Ribo (mg)	Niac (mg)	V-B6 (mg)	Fola (µg)	VT-C (mg)
0				196	74						.98			
0	0	0		196	160		0				.98			5
0	40	3.6		676	280		2000	.34		.11	1.35	.22	135	9
0	20	1.8		147	170		0			.03	2.45			1
0	40	1.8		406	250		2000	.26		.08	2.25	.18	27	15
0	60	5.4		439	220		2000	.1		.16	5.63	.45	27	2
0	40	5.4		439	230		2000	.1		.16	5.63	.45		9
0	40	5.4		609	240		2000	.1		.08	2.25	.13	<1	9
5	20	1.44			460		20							0
5	20	1.44			510		20							0
55	200	1.08	42	400	500	1.5	150	.06		.25	1.2	.2	48	6
35	20	.36	30	300	510	1.1	20	.15		.17	5			6
25	150	1.8	44	620	560	2.9	100	.15		.25	3	.32		12
20	250	.72		170	550		20	.12		.25	1.2			0
30	100	2.7	47	480	520	2.5	80	.15		.25	3	.2		4
35	250	2.7	39	340	900	2.2	80	.45		.51	5	.07		6
				650	565									
				467	584									
				650	448									
				484	591									
0	99	0		64	306		56							0
0	99	0		74	276		56							0
0	99	0		64	306		56							0
0	99	0		64	306		56							0
25	40	.72		360	430		300							36
25	20	1.44		370	450		40	.09		.14	1.6			0
0	0	0		5	128		49							0
35	700	1.8		290	590		80	.3		.51	3	.06		6
40	500	3.6		370	550		150	.3		.51	3	.2		5
45	450	1.8		320	650		80	.23		.51	3			5
5	80	1.08		230	160		0	.06		.03	.2	.03		0
0	40	0		65	151		0							0
5	60	1.8		320	150		0							0
15	80	.36		115	230		40	.06		.07	1.6			2

Table A–1

Food Composition (Computer code number is for West Diet Analysis program) (For purposes of calculations, use "0" for t, <1, <.1, <.01, etc.)

Computer Code Number	Food Description	Measure	Wt (g)	H₂O (%)	Ener (cal)	Prot (g)	Carb (g)	Dietary Fiber (g)	Fat (g)	Fat Breakdown (g) Sat	Mono	Poly
	CONVENIENCE FOODS and MEALS—Continued											
	WEIGHT WATCHERS—CONTINUED											
2027	Triple chocolate cheesecake	1 ea	89	52	199	7	32	1	5	2.5		
	Source: Weight Watchers											
	SWEET SUCCESS:											
	Drinks, prepared:											
1776	Chocolate chip	1 c	265	81	180	15	30	6	3	1.6		
1777	Chocolate fudge	1 c	265	81	180	15	30	6	2			
1774	Chocolate mocha	1 c	265	81	180	15	30	6	1	1		
1778	Milk chocolate	1 c	265	81	180	15	30	6	2	1		
1775	Vanilla	1 c	265	81	180	15	33	6	1	.6		
	Drinks, ready to drink:											
2147	Chocolate mint	1 c	297	82	187	11	36	6	3	0		
2148	Strawberry	1 c	265	82	167	10	32	5	3	0		
	Shakes:											
1771	Chocolate almond	1 c	250	82	158	9	30	5	2	0	2.1	.2
1773	Chocolate fudge	1 c	250	82	158	9	30	5	2	0	2.1	.2
1768	Chocolate mocha	1 c	250	82	158	9	30	5	2	0	.6	1.8
1769	Chocolate raspberry truffle	1 c	250	82	158	9	30	5	2	0	2.2	.2
1770	Vanilla creme	1 c	250	82	158	9	30	5	2	0	2.1	.3
	Snack bars:											
1767	Chocolate brownie	1 ea	33	9	120	2	23	3	4	2	.5	.6
1766	Chocolate chip	1 ea	33	9	120	2	23	3	4	2	.4	.5
1921	Oatmeal raisin	1 ea	33	9	120	2	23	3	4	2		
1765	Peanut butter	1 ea	33	9	120	2	23	3	4	2	.6	.6
	Source: Foodway National Inc, Boise, ID											
	BABY FOODS											
1720	Apple juice	½ c	125	88	59	0	15	<1	<1	t	t	t
1721	Applesauce, strained	1 tbs	16	89	7	<1	2	<1	<1	t	t	t
1716	Carrots, strained	1 tbs	14	92	4	<1	1	<1	<1	t	t	t
1718	Cereal, mixed, milk added	1 tbs	15	75	17	1	2	<1	1	.3		
1719	Cereal, rice, milk added	1 tbs	15	75	17	<1	3	<1	1	.3		
1723	Chicken and noodles, strained	1 tbs	16	88	8	<1	1	<1	<1	.1	.1	t
1722	Peas, strained	1 tbs	15	87	6	1	1	<1	<1	t	t	t
1717	Teething biscuits	1 ea	11	6	43	1	8	<1	<1	.2	.2	.1

Chol (mg)	Calc (mg)	Iron (mg)	Magn (mg)	Pota (mg)	Sodi (mg)	Zinc (mg)	VT-A (RE)	Thia (mg)	VT-E (a-TE)	Ribo (mg)	Niac (mg)	V-B6 (mg)	Fola (μg)	VT-C (mg)
10	80	1.08		169	199		0							0
6	500	6.3	140	600	288	5.25	350	.52	7.05	.59	7	.7	140	21
6	500	6.3	140	750	336	5.25	350	.52	7.05	.59	7	.7	140	21
6	500	6.3	140	800	336	5.25	350	.52	7.05	.59	7	.7	140	21
6	500	6.3	140	750	336	5.25	350	.52	7.05	.59	7	.7	140	21
6	500	6.3	140	830	312	5.25	250	.52	7.05	.59	7	.7	140	21
6	470	5.94	131	526	226	5.05	329	.5	6.56	.56	6.53	.65	131	20
5	419	5.3	117	310	175	4.51	294	.45	5.86	.5	5.83	.58	117	17
5	396	5	110	443	190	4.25	277	.42	5.53	.47	5.5	.55	110	16
5	396	5	110	443	175	4.25	277	.42	5.53	.47	5.5	.55	110	16
5	396	5	110	403	175	4.25	277	.42	5.53	.47	5.5	.55	110	16
5	383	5	110	443	175	4.25	277	.42	5.53	.47	5.5	.55	110	16
5	396	5	110	293	175	4.25	277	.42	5.53	.47	5.5	.55	110	16
3	150	2.71	60	140	45	.59	150	.22	3.01	.25	3	.3	60	9
3	150	2.71	60	110	40	.59	150	.22	3.01	.25	3	.3	60	9
3	150	2.71	60		30	.59	150	.22	3.01	.25	3	.3	60	9
3	150	2.71	60	125	35	.59	150	.22	3.01	.25	3	.3	60	9
0	5	.71	4	114	4	.04	2	.01	.75	.02	.1	.04	<1	72
0	1	.03		11	<1	<.01	<1	<.01	.1	<.01	.01	<.01	<1	6
0	3	.05	1	27	5	.02	160	<.01	.07	.01	.06	.01	2	1
2	33	1.56	4	30	7	.11	4	.06		.09	.87	.01	2	<1
2	36	1.83	7	28	7	.1	4	.07		.07	.78	.02	1	<1
3	4	.07	1	6	3	.05	18	<.01	.04	.01	.07	<.01	2	<1
0	3	.14	2	17	1	.05	8	.01	.08	.01	.15	.01	4	1
0	29	.39	4	35	40	.1	1	.03	.05	.06	.48	.01	5	1

CONTENTS

RNI

Choice System for Meal Planning

Food Labels

Canada's Food Guide

Canada: Recommendations, Choice System, and Labels

Chapter 1 introduced recommended nutrient intakes, food guides, and food labels, and Chapter 20 introduced the U.S. exchange system (see Appendix C). This appendix presents details for Canadians. Appendix D includes addresses of

Table B-1
Recommended Nutrient Intakes for Canadians, 1990

| Age | Sex | Weight (kg) | Protein (g/day)[a] | Vitamins | | | | | |
				Vitamin A (RE/day)[b]	Vitamin E (mg/day)[c]	Vitamin C (mg/day)[d]	Iron (mg/day)	Iodine (µg/day)	Zinc (mg/day)
Infants (months)									
0–4	Both	6	12[e]	400	3	20	0.3[f]	30	2[g]
5–12	Both	9	12	400	3	20	7	40	3
Children (years)									
1	Both	11	13	400	3	20	6	55	4
2–3	Both	14	16	400	4	20	6	65	4
4–6	Both	18	19	500	5	25	8	85	5
7–9	M	25	26	700	7	25	8	110	7
	F	25	26	700	6	25	8	95	7
10–12	M	34	34	800	8	25	8	125	9
	F	36	36	800	7	25	8	110	9
13–15	M	50	49	900	9	30	10	160	12
	F	48	46	800	7	30	13	160	9
16–18	M	62	58	1000	10	40	10	160	12
	F	53	47	800	7	30	12	160	9
Adults (years)									
19–24	M	71	61	1000	10	40	9	160	12
	F	58	50	800	7	30	13	160	9
25–49	M	74	64	1000	9	40	9	160	12
	F	59	51	800	6	30	13[h]	160	9
50–74	M	73	63	1000	7	40	9	160	12
	F	63	54	800	6	30	8	160	9
75+	M	69	59	1000	6	40	9	160	12
	F	64	55	800	5	30	8	160	9
Pregnancy (additional amount needed)									
1st trimester			5	0	2	0	0	25	6
2nd trimester			20	0	2	10	5	25	6
3rd trimester			24	0	2	10	10	25	6
Lactation (additional amount needed)			20	400	3	25	0	50	6

NOTE: Recommended intakes of energy and of certain nutrients are not listed in this table because of the nature of the variables upon which they are based. The figures for energy are estimates of average requirements for expected patterns of activity (see Table B-2). For nutrients not shown, the following amounts are recommended based on at least 2000 kcalories per day and body weights as given: thiamin, 0.4 milligram per 1000 kcalories (0.48 milligram/5000 kilojoules); riboflavin, 0.5 milligram per 1000 kcalories (0.6 milligram/5000 kilojoules); niacin, 7.2 niacin equivalents per 1000 kcalories (8.6 niacin equivalents/5000 kilojoules); vitamin B$_6$, 15 micrograms, as pyridoxine, per gram of protein. Recommended intakes during periods of growth are taken as appropriate for individuals representative of the midpoint in each age group. All recommended intakes are designed to cover individual variations in essentially all of a healthy population subsisting upon a variety of common foods available in Canada.

SOURCE: Health and Welfare Canada, *Nutrition Recommendations: The Report of the Scientific Review Committee* (Ottawa: Canadian Government Publishing Centre, 1990), Table 20, p. 204.

[a]The primary units are expressed per kilogram of body weight. The figures shown here are examples.
[b]One retinol equivalent (RE) corresponds to the biological activity of 1 microgram of retinol, 6 micrograms of beta-carotene, or 12 micrograms of other carotenes.
[c]Expressed as δ-α-tocopherol equivalents, relative to which β- and γ-tocopherol and α-tocotrienol have activities of 0.5, 0.1, and 0.3, respectively.
[d]Cigarette smokers should increase intake by 50 percent.
[e]The assumption is made that the protein is from breast milk or has the same biological value as breast milk and that, between 3 and 9 months, adjustment for the quality of the protein is made.
[f]Based on the assumption that breast milk is the source of iron.
[g]Based on the assumption that breast milk is the source of zinc.
[h]After menopause, the recommended intake is 8 milligrams per day.

Canadian governmental agencies and professional organizations that may provide additional information.

RNI

As Chapter 1 mentioned, a major revision of the nutrient recommendations is underway in both the United States and Canada. The Dietary Reference Intakes (DRI) reports are replacing the 1989 RDA in the United States and the 1991 RNI in Canada. Recommendations from the DRI reports are presented on the inside front covers. For nutrients that do not yet have new values, the RNI will continue to serve health professionals in Canada (Tables B-1 and B-2).

Table B-2

Average Energy Requirements for Canadians

Age	Sex	Average Height (cm)	Average Weight (kg)	Requirements[a]					
				(kcal/kg)[b]	(MJ/kg)[b]	(kcal/day)	(MJ/day)	(kcal/cm)	(MJ/cm)
Infants (months)									
0–2	Both	55	4.5	120–100	0.50–0.42	500	2.0	9	0.04
3–5	Both	63	7.0	100–95	0.42–0.40	700	2.8	11	0.05
6–8	Both	69	8.5	95–97	0.40–0.41	800	3.4	11.5	0.05
9–11	Both	73	9.5	97–99	0.41	950	3.8	12.5	0.05
Children and Adults (years)									
1	Both	82	11	101	0.42	1100	4.8	13.5	0.06
2–3	Both	95	14	94	0.39	1300	5.6	13.5	0.06
4–6	Both	107	18	100	0.42	1800	7.6	17	0.07
7–9	M	126	25	88	0.37	2200	9.2	17.5	0.07
	F	125	25	76	0.32	1900	8.0	15	0.06
10–12	M	141	34	73	0.30	2500	10.4	17.5	0.07
	F	143	36	61	0.25	2200	9.2	15.5	0.06
13–15	M	159	50	57	0.24	2800	12.0	17.5	0.07
	F	157	48	46	0.19	2200	9.2	14	0.06
16–18	M	172	62	51	0.21	3200	13.2	18.5	0.08
	F	160	53	40	0.17	2100	8.8	13	0.05
19–24	M	175	71	42	0.18	3000	12.6		
	F	160	58	36	0.15	2100	8.8		
25–49	M	172	74	36	0.15	2700	11.3		
	F	160	59	32	0.13	1900	8.0		
50–74	M	170	73	31	0.13	2300	9.7		
	F	158	63	29	0.12	1800	7.6		
75+	M	168	69	29	0.12	2000	8.4		
	F	155	64	23	0.10	1500	6.3		

[a]Requirements can be expected to vary within a range of ±30 percent.
[b]First and last figures are averages at the beginning and end of the three-month period.
SOURCE: Health and Welfare Canada, *Nutrition Recommendations: The Report of the Scientific Review Committee* (Ottawa: Canadian Government Publishing Centre, 1990), Tables 5 and 6, pp. 25, 27.

Choice System for Meal Planning

The *Good Health Eating Guide* is the Canadian choice system of meal planning.[1] It contains several features similar to those of the U.S. exchange system including the following:

---

[1] The tables for the Canadian choice system are adapted from the *Good Health Eating Guide Resource,* copyright 1994, with permission of the Canadian Diabetes Association.

- Foods are divided into lists according to carbohydrate, protein, and fat content.
- Foods are interchangeable within a group.
- Most foods are eaten in measured amounts.
- An energy value is given for each food group.

Tables B-3 through B-10 present the Canadian choice system.

Table B-3

Canadian Choice System: Starch Foods

1 starch choice = 15 g carbohydrate (starch), 2 g protein, 290 kJ (68 kcal)

Food	Measure	Mass (Weight)
Breads		
Bagels	½	30 g
Bread crumbs	50 mL (¼ c)	30 g
Bread cubes	250 mL (1 c)	30 g
Bread sticks	2	20 g
Brewis, cooked	50 mL (¼ c)	45 g
Chapati	1	20 g
Cookies, plain	2	20 g
English muffins, crumpets	½	30 g
Flour	40 mL (2½ tbs)	20 g
Hamburger buns	½	30 g
Hot dog buns	½	30 g
Kaiser rolls	½	30 g
Matzo, 15 cm	1	20 g
Melba toast, rectangular	4	15 g
Melba toast, rounds	7	15 g
Pita, 20-cm (8") diameter	¼	30 g
Pita, 15-cm (6") diameter	½	30 g
Plain rolls	1 small	30 g
Pretzels	7	20 g
Raisin bread	1 slice	30 g
Rice cakes	2	30 g
Roti	1	20 g
Rusks	2	20 g
Rye, coarse or pumpernickel	½ slice	30 g
Soda crackers	6	20 g
Tortillas, corn (taco shell)	1	30 g
Tortilla, flour	1	30 g
White (French and Italian)	1 slice	25 g
Whole-wheat, cracked-wheat, rye, white enriched	1 slice	30 g
Cereals		
Bran flakes, 100% bran	125 mL (½ c)	30 g
Cooked cereals, cooked	125 mL (½ c)	125 g
Dry	30 mL (2 tbs)	20 g
Cornmeal, cooked	125 mL (½ c)	125 g
Dry	30 mL (2 tbs)	20 g
Ready-to-eat unsweetened cereals	125 mL (½ c)	20 g
Shredded wheat biscuits, rectangular or round	1	20 g
Shredded wheat, bite size	125 mL (½ c)	20 g
Wheat germ	75 mL (⅓ c)	30 g
Cornflakes	175 mL (⅔ c)	20 g
Rice Krispies	175 mL (⅔ c)	20 g

Table B-3 (continued)

Canadian Choice System: Starch Foods

1 starch choice = 15 g carbohydrate (starch), 2 g protein, 290 kJ (68 kcal)

Food	Measure	Mass (Weight)
Cheerios	200 mL (¾ c)	20 g
Muffets	1	20 g
Puffed rice	300 mL (1¼ c)	15 g
Puffed wheat	425 mL (1⅔ c)	20 g
Grains		
Barley, cooked	125 mL (½ c)	120 g
Dry	30 mL (2 tbs)	20 g
Bulgur, kasha, cooked, moist	125 mL (½ c)	70 g
Cooked, crumbly	75 mL (⅓ c)	40 g
Dry	30 mL (2 tbs)	20 g
Rice, cooked, brown & white (short & long grain)	125 mL (½ c)	70 g
Rice, cooked, wild	75 mL (⅓ c)	70 g
Tapioca, pearl and granulated, quick cooking, dry	30 mL (2 tbs)	15 g
Couscous, cooked moist	125 mL (½ c)	70 g
Dry	30 mL (tbs)	20 g
Quinoa, cooked moist	125 mL (½ c)	70 g
Dry	30 mL (2 tbs)	20 g
Pastas		
Macaroni, cooked	125 mL (½ c)	70 g
Noodles, cooked	125 mL (½ c)	80 g
Spaghetti, cooked	125 mL (½ c)	70 g
Starchy Vegetables		
Beans and peas, dried, cooked	125 mL (½ c)	80 g
Breadfruit	1 slice	75 g
Corn, canned, whole kernel	125 mL (½ c)	85 g
Corn on the cob	½ medium cob	140 g
Cornstarch	30 mL (2 tbs)	15 g
Plantains	⅓ small	50 g
Popcorn, air-popped, unbuttered	750 mL (3 c)	20 g
Potatoes, whole (with or without skin)	½ medium	95 g
Yams, sweet potatoes, (with or without skin)	½	75 g

Food	Choices per Serving	Measure	Mass (Weight)
NOTE: Food items found in this category provide more than 1 starch choice:			
Bran flakes	1 starch + ½ sugar	150 mL (⅔ c)	24 g
Croissant, small	1 starch + 1½ fats	1 small	35 g
Large	1 starch + 1½ fats	½ large	30 g
Corn, canned creamed	1 starch + ½ fruits and vegetables	12 mL (½ c)	113 g
Potato chips	1 starch + 2 fats	15 chips	30 g
Tortilla chips (nachos)	1 starch + 1½ fats	13 chips	20 g
Corn chips	1 starch + 2 fats	30 chips	30 g
Cheese twists	1 starch + 1½ fats	30 chips	30 g
Cheese puffs	1 starch + 2 fats	27 chips	30 g
Tea biscuit	1 starch + 2 fats	1	30 g
Pancakes, homemade using 50 mL (¼ c) batter (6" diameter)	1½ starches + 1 fat	1 medium	50 g
Potatoes, french fried (homemade or frozen)	1 starch + 1 fat	10 regular size	35 g
Soup, canned* (prepared with equal volume of water)	1 starch	250 mL (1 c)	260 g
Waffles, packaged	1 starch + 1 fat	1	35 g

*Soup can vary according to brand and type. Check the label for Food Choice Values and Symbols or the core nutrient listing.

Table B-4

Canadian Choice System: Fruits and Vegetables

1 fruits and vegetables choice = 10 g carbohydrate, 1 g protein, 190 kJ (44 kcal)

Food	Measure	Mass (Weight)
Fruits (fresh, frozen, without sugar, canned in water)		
Apples, raw (with or without skin)	½ medium	75 g
Sauce unsweetened	125 mL (½ c)	120 g
Sweetened	see *Combined Food Choices*	
Apple butter	20 mL (4 tsp)	20 g
Apricots, raw	2 medium	115 g
Canned, in water	4 halves, plus 30 mL (2 tbs) liquid	110 g
Bake-apples (cloudberries), raw	125 mL (½ c)	120 g
Bananas, with peel	½ small	75 g
Peeled	½ small	50 g
Berries (blackberries, blueberries, boysenberries, huckleberries, loganberries, raspberries)		
Raw	125 mL (½ c)	70 g
Canned, in water	125 mL (½ c), plus 30 mL (2 tbs) liquid	100 g
Cantaloupe, wedge with rind	¼	240 g
Cubed or diced	250 mL (1 c)	160 g
Cherries, raw, with pits	10	75 g
Raw, without pits	10	70 g
Canned, in water, with pits	75 mL (⅓ c), plus 30 mL (2 tbs) liquid	90 g
Canned, in water, without pits	75 mL (⅓ c), plus 30 mL (2 tbs) liquid	85 g
Crabapples, raw	1 small	55 g
Cranberries, raw	250 mL (1 c)	100 g
Figs, raw	1 medium	50 g
Canned, in water	3 medium, plus 30 mL (2 tbs) liquid	100 g
Foxberries, raw	250 mL (1 c)	100 g
Fruit cocktail, canned, in water	125 mL (½ c), plus 30 mL (2 tbs) liquid	120 g
Fruit, mixed, cut-up	125 mL (½ c)	120 g
Gooseberries, raw	250 mL (1 c)	150 g
Canned, in water	250 mL (1 c), plus 30 mL (2 tbs) liquid	230 g
Grapefruit, raw, with rind	½ small	185 g
Raw, sectioned	125 mL (½ c)	100 g
Canned, in water	125 mL (½ c), plus 30 mL (2 tbs) liquid	120 g
Grapes, raw, slip skin	125 mL (½ c)	75 g
Raw, seedless	125 mL (½ c)	75 g
Canned, in water	75 mL (⅓ c), plus 30 mL (2 tbs) liquid	115 g
Guavas, raw	½	50 g
Honeydew melon, raw, with rind	½	225 g
Cubed or diced	250 mL (1 c)	170 g
Kiwis, raw, with skin	2	155 g
Kumquats, raw	3	60 g
Loquats, raw	8	130 g
Lychee fruit, raw	8	120 g
Mandarin oranges, raw, with rind	1	135 g
Raw, sectioned	125 mL (½ c)	100 g
Canned, in water	125 mL (½ c), plus 30 mL (2 tbs) liquid	100 g
Mangoes, raw, without skin and seed	⅓	65 g
Diced	75 mL (⅓ c)	65 g
Nectarines	½ medium	75 g
Oranges, raw, with rind	1 small	130 g
Raw, sectioned	125 mL (½ c)	95 g

Table B-4 (continued)

Canadian Choice System: Fruits and Vegetables

1 fruits and vegetables choice = 10 g carbohydrate, 1 g protein, 190 kJ (44 kcal)

Food	Measure	Mass (Weight)
Papayas, raw, with skin and seeds	¼ medium	150 g
Raw, without skin and seeds	¼ medium	100 g
Cubed or diced	125 mL (½ c)	100 g
Peaches, raw, with seed and skin	1 large	100 g
Raw, sliced or diced	125 mL (½ c)	100 g
Canned in water, halves or slices	125 mL (½ c), plus 30 mL (2 tbs) liquid	120 g
Pears, raw, with skin and core	½	90 g
Raw, without skin and core	½	85 g
Canned, in water, halves	1 half plus 30 mL (2 tbs) liquid	90 g
Persimmons, raw, native	1	30 g
Raw, Japanese	¼	50 g
Pineapple, raw	1 slice	75 g
Raw, diced	125 mL (½ c)	75 g
Canned, in juice, diced	75 mL (⅓ c), plus 15 mL (1 tbs) liquid	55 g
Canned, in juice, sliced	1 slice, plus 15 mL (1 tbs) liquid	55 g
Canned, in water, diced	125 mL (½ c), plus 30 mL (2 tbs) liquid	100 g
Canned, in water, sliced	2 slices, plus 15 mL (1 tbs) liquid	100 g
Plums, raw	2 small	60 g
Damson	6	65 g
Japanese	1	70 g
Canned, in apple juice	2, plus 30 mL (2 tbs) liquid	70 g
Canned, in water	3, plus 30 mL (2 tbs) liquid	100 g
Pomegranates, raw	½	140 g
Strawberries, raw	250 mL (1 c)	150 g
Frozen/canned, in water	250 mL (1 c), plus 30 mL (2 tbs) liquid	240 g
Rhubarb	250 mL (1 c)	150 g
Tangelos, raw	1	205 g
Tangerines, raw	1 medium	115 g
Raw, sectioned	125 mL (½ c)	100 g
Watermelon, raw, with rind	1 wedge	310 g
Cubed or diced	250 mL (1 c)	160 g
Dried Fruit		
Apples	5 pieces	15 g
Apricots	4 halves	15 g
Banana flakes	30 mL (2 tbs)	15 g
Currants	30 mL (2 tbs)	15 g
Dates, without pits	2	15 g
Peaches	½	15 g
Pears	½	15 g
Prunes, raw, with pits	2	15 g
Raw, without pits	2	10 g
Stewed, no liquid	2	20 g
Stewed, with liquid	2, plus 15 mL (1 tbs) liquid	35 g
Raisins	30 mL (2 tbs)	15 g
Juices (no sugar added or unsweetened)		
Apricot, grape, guava, mango, prune	50 mL (¼ c)	55 g
Apple, carrot, papaya, pear, pineapple, pomegranate	75 mL (⅓ c)	80 g
Cranberry (see Sugars Section)		
Clamato (see Sugars Section)		

(continued on the next page)

Table B-4 (continued)

Canadian Choice System: Fruits and Vegetables

1 fruits and vegetables choice = 10 g carbohydrate, 1 g protein, 190 kJ (44 kcal)

Food	Measure	Mass (Weight)
Grapefruit, loganberry, orange, raspberry, tangelo, tangerine	125 mL (½ c)	130 g
Tomato, tomato-based mixed vegetables	250 mL (1 c)	255 g
Vegetables (fresh, frozen, or canned)		
Artichokes, French, globe	2 small	50 g
Beets, diced or sliced	125 mL (½ c)	85 g
Carrots, diced, cooked or uncooked	125 mL (½ c)	75 g
Chestnuts, fresh	5	20 g
Parsnips, mashed	125 mL (½ c)	80 g
Peas, fresh or frozen	125 mL (½ c)	80 g
Canned	75 mL (⅓ c)	55 g
Pumpkin, mashed	125 mL (½ c)	45 g
Rutabagas, mashed	125 mL (½ c)	85 g
Sauerkraut	250 mL (1 c)	235 g
Snow peas	250 mL (1 c)	135 g
Squash, yellow or winter, mashed	125 mL (½ c)	115 g
Succotash	75 mL (⅓ c)	55 g
Tomatoes, canned	250 mL (1 c)	240 g
Tomato paste	50 mL (¼ c)	55 g
Tomato sauce*	75 mL (⅓ c)	100 g
Turnips, mashed	125 mL (½ c)	115 g
Vegetables, mixed	125 mL (½ c)	90 g
Water chestnuts	8 medium	50 g

*Tomato sauce varies according to brand name. Check the label or discuss with your dietitian.

Table B-5

Canadian Choice System: Milk

Type of Milk	Carbohydrate (g)	Protein (g)	Fat (g)	Energy
Fat-free (0%)	6	4	0	170 kJ (40 kcal)
1%	6	4	1	206 kJ (49 kcal)
2%	6	4	2	244 kJ (58 kcal)
Whole (4%)	6	4	4	319 kJ (76 kcal)

Food	Measure	Mass (Weight)
Buttermilk (higher in salt)	125 mL (½ c)	125 g
Evaporated milk	50 mL (¼ c)	50 g
Milk	125 mL (½ c)	125 g
Powdered milk, regular	30 mL (2 tbs)	15 g
Instant	50 mL (¼ c)	15 g
Plain yogurt	125 mL (½ c)	125 g

Food	Choices per Serving	Measure	Mass (Weight)
NOTE: Food items found in this category provide more than 1 milk choice:			
Milk shake	1 milk + 3 sugars + ½ protein	250 mL (1 c)	300 g
Chocolate milk, 2%	2 milks 2% + 1 sugar	250 mL (1 c)	300 g
Frozen yogurt	1 milk + 1 sugar	125 mL (½ c)	125 g

Table B-6

Canadian Choice System: Sugars

1 sugar choice = 10 g carbohydrate (sugar), 167 kJ (40 kcal)

Food	Measure	Mass (Weight)
Beverages		
Condensed milk	15 mL (1 tbs)	
Flavoured fruit crystals*	75 mL (⅓ c)	
Iced tea mixes*	75 mL (⅓ c)	
Regular soft drinks	125 mL (½ c)	
Sweet drink mixes*	75 mL (⅓ c)	
Tonic water	125 mL (½ c)	
*These beverages have been made with water.		
Miscellaneous		
Bubble gum (large square)	1 piece	5 g
Cranberry cocktail	75 mL (⅓ c)	80 g
Cranberry cocktail, light	350 mL (1⅓ c)	260 g
Cranberry sauce	30 mL (2 tbs)	
Hard candy mints	2	5 g
Honey, molasses, corn & cane syrup	10 mL (2 tsp)	15 g
Jelly bean	4	10 g
Licorice	1 short stick	10 g
Marshmallows	2 large	15 g
Popsicle	1 stick (½ popsicle)	
Powdered gelatin mix		
(Jello®) (reconstituted)	50 mL (¼ c)	
Regular jam, jelly, marmalade	15 mL (1 tbs)	
Sugar, white, brown, icing, maple	10 mL (2 tsp)	10 g
Sweet pickles	2 small	100 g
Sweet relish	30 mL (2 tbs)	

Food	Choices per Serving	Measures	Mass (Weight)
The following food items provide more than 1 sugar choice:			
Brownie	1 sugar + 1 fat	1	20 g
Clamato juice	1½ sugars	175 mL (⅔ c)	
Fruit salad, light syrup	1 sugar + 1 fruits & vegetables	125 mL (½ c)	130 g
Aero® bar	2½ sugars + 2½ fats	1 bar	43 g
Smarties®	4½ sugars + 2 fats	1 box	60 g
Sherbet	3 sugars + ½ fat	125 mL (½ c)	95 g

Table B-7

Canadian Choice System: Protein Foods

1 protein choice = 7 g protein, 3 g fat, 230 kJ (55 kcal)

Food	Measure	Mass (Weight)
Cheese		
Low-fat cheese, about 7% milkfat	1 slice	30 g
Cottage cheese, 2% milkfat or less	50 mL (¼ c)	55 g
Ricotta, about 7% milkfat	50 mL (¼ c)	60 g
Fish		
Anchovies (see *Extras,* Table B-9)		
Canned, drained (e.g., mackerel, salmon, tuna packed in water)	(⅓ of 6.5 oz can)	30 g
Cod tongues, cheeks	75 mL (⅓ c)	50 g
Fillet or steak (e.g., Boston blue, cod, flounder, haddock, halibut, mackerel, orange roughy, perch, pickerel, pike, salmon, shad, snapper, sole, swordfish, trout, tuna, whitefish)	1 piece	30 g
Herring	⅓ fish	30 g
Sardines, smelts	2 medium or 3 small	30 g
Squid, octopus	50 mL (¼ c)	40 g
Shellfish		
Clams, mussels, oysters, scallops, snails	3 medium	30 g
Crab, lobster, flaked	50 mL (¼ c)	30 g
Shrimp, fresh	5 large	30 g
Frozen	10 medium	30 g
Canned	18 small	30 g
Dry pack	50 mL (¼ c)	30 g
Meat and Poultry (e.g., beef, chicken, goat, ham, lamb, pork, turkey, veal, wild game)		
Back, peameal bacon	3 slices, thin	30 g
Chop	½ chop, with bone	40 g
Minced or ground, lean or extra-lean	30 mL (2 tbs)	30 g
Sliced, lean	1 slice	30 g
Steak, lean	1 piece	30 g
Organ Meats		
Hearts, liver	1 slice	30 g
Kidneys, sweetbreads, chopped	50 mL (¼ c)	30 g
Tongue	1 slice	30 g
Tripe	5 pieces	60 g
Soyabean		
Bean curd or tofu	½ block	70 g
Eggs		
In shell, raw or cooked	1 medium	50 g
Without shell, cooked or poached in water	1 medium	45 g
Scrambled	50 mL (¼ c)	55 g

Food	Choices per Serving	Measures	Mass (Weight)
NOTE: The following choices provide more than 1 protein exchange:			
Cheese			
Cheeses	1 protein + 1 fat	1 piece	25 g
Cheese, coarsely grated (e.g., cheddar)	1 protein + 1 fat	50 mL (¼ c)	25 g
Cheese, dry, finely grated (e.g., parmesan)	1 protein + 1 fat	45 mL	15 g
Cheese, ricotta, high fat	1 protein + 1 fat	50 mL (¼ c)	55 g
Fish			
Eel	1 protein + 1 fat	1 slice	50 g

Table B-7 (continued)

Canadian Choice System: Protein Foods

1 protein choice = 7 g protein, 3 g fat, 230 kJ (55 kcal)

Food	Choices per Serving	Measures	Mass (Weight)
Meat			
Bologna	1 protein + 1 fat	1 slice	20 g
Canned lunch meats	1 protein + 1 fat	1 slice	20 g
Corned beef, canned	1 protein + 1 fat	1 slice	25 g
Corned beef, fresh	1 protein + 1 fat	1 slice	25 g
Ground beef, medium-fat	1 protein + 1 fat	30 mL (2 tbs)	25 g
Meat spreads, canned	1 protein + 1 fat	45 mL	35 g
Mutton chop	1 protein + 1 fat	½ chop, with bone	35 g
Paté (see *Fats and Oils* group, Table B-8)			
Sausages, garlic, Polish or knockwurst	1 protein + 1 fat	1 slice	50 g
Sausages, pork, links	1 protein + 1 fat	1 link	25 g
Spareribs or shortribs, with bone	1 protein + 1 fat	1 large	65 g
Stewing beef	1 protein + 1 fat	1 cube	25 g
Summer sausage or salami	1 protein + 1 fat	1 slice	40 g
Wieners, hot dog	1 protein + 1 fat	½ medium	25 g
Miscellaneous			
Blood pudding	1 protein + 1 fat	1 slice	25 g
Peanut butter	1 protein + 1 fat	15 mL (1 tbs)	15 g

Table B-8

Canadian Choice System: Fats and Oils

1 fat choice = 5 g fat, 190 kJ (45 kcal)

Food	Measure	Mass (Weight)	Food	Measure	Mass (Weight)
Avocado*	⅛	30 g	Nuts (continued):		
Bacon, side, crisp*	1 slice	5 g	Sesame seeds	15 mL (1 tbs)	10 g
Butter*	5 mL (1 tsp)	5 g	Sunflower seeds		
Cheese spread	15 mL (1 tbs)	15 g	Shelled	15 mL (1 tbs)	10 g
Coconut, fresh*	45 mL (3 tbs)	15 g	In shell	45 mL (3 tbs)	15 g
Coconut, dried*	15 mL (1 tbs)	10 g	Walnuts	4 halves	10 g
Cream, Half and half			Oil, cooking and salad	5 mL (1 tsp)	5 g
(cereal), 10%*	30 mL (2 tbs)	30 g	Olives, green	10	45 g
Light (coffee), 20%*	15 mL (1 tbs)	15 g	Ripe black	7	57 g
Whipping, 32 to 37%*	15 mL (1 tbs)	15 g	Pâté, liverwurst,	15 mL (1 tbs)	15 g
Cream cheese*	15 mL (1 tbs)	15 g	meat spreads		
Gravy*	30 mL (2 tbs)	30 g	Salad dressing: blue,	10 mL (2 tsp)	10 g
Lard*	5 mL (1 tsp)	5 g	French, Italian,		
Margarine	5 mL (1 tsp)	5 g	mayonnaise,		
Nuts, shelled:			Thousand Island	5 mL (1 tsp)	5 g
Almonds	8	5 g	Salad dressing,	30 mL (2 tbs)	30 g
Brazil nuts	2	10 g	low-calorie		
Cashews	5	10 g	Salt pork, raw	5 mL (1 tsp)	5 g
Filberts, hazelnuts	5	10 g	or cooked*		
Macadamia	3	5 g	Sesame oil	5 mL (1 tsp)	5 g
Peanuts	10	10g	Sour cream		
Pecans	5 halves	5 g	12% milkfat	30 mL (2 tbs)	30 g
Pignolias, pine nuts	25 mL (5 tsp)	10 g	7% milkfat	60 mL (4 tbs)	60 g
Pistachios, shelled	20	10 g	Shortening*	5 mL (1 tsp)	
Pistachios, in shell	20	20 g			
Pumpkin and	20 mL (4 tsp)	10 g			
squash seeds					

*These items contain higher amounts of saturated fat.

Table B-9

Canadian Choice System: Extras

Extras have no more than 2.5 g carbohydrate, 60 kJ (14 kcal)

Vegetables 125 mL (½ c)

Artichokes

Asparagus

Bamboo shoots

Bean sprouts, mung or soya

Beans, string, green, or yellow

Bitter melon (balsam pear)

Bok choy

Broccoli

Brussels sprouts

Cabbage

Cauliflower

Celery

Chard

Cucumbers

Eggplant

Endive

Fiddleheads

Greens: beet, collard, dandelion, mustard, turnip, etc.

Kale

Kohlrabi

Leeks

Lettuce

Mushrooms

Okra

Onions, green or mature

Parsley

Peppers, green, yellow or red

Radishes

Rapini

Rhubarb

Sauerkraut

Shallots

Spinach

Sprouts: alfalfa, radish, etc.

Tomato wedges

Watercress

Zucchini

Free Foods (may be used without measuring)

Artificial sweetener, such as cyclamate or aspartame	Lime juice or lime wedges
	Marjoram, cinnamon, etc.
Baking powder, baking soda	Mineral water
Bouillon from cube, powder, or liquid	Mustard
	Parsley
Bouillon or clear broth	Pimentos
Chowchow, unsweetened	Salt, pepper, thyme
Coffee, clear	Soda water, club soda
Consommé	Soya sauce
Dulse	Sugar-free Crystal Drink
Flavorings and extracts	Sugar-free Jelly Powder
Garlic	Sugar-free soft drinks
Gelatin, unsweetened	Tea, clear
Ginger root	Vinegar
Herbal teas, unsweetened	Water
Horseradish, uncreamed	Worcestershire sauce
Lemon juice or lemon wedges	

Condiments

Food	Measure
Anchovies	2 fillets
Barbecue sauce	15 mL (1 tbs)
Bran, natural	30 mL (2 tbs)
Brewer's yeast	5 mL (1 tsp)
Carob powder	5 mL (1 tsp)
Catsup	5 mL (1 tsp)
Chili sauce	5 mL (1 tsp)
Cocoa powder	5 mL (1 tsp)
Cranberry sauce, unsweetened	15 mL (1 tbs)
Dietetic fruit spreads	5 mL (1 tsp)
Maraschino cherries	1
Nondairy coffee whitener	5 mL (1 tsp)
Nuts, chopped pieces	5 mL (1 tsp)
Pickles	
unsweetened dill	2
sour mixed	11
Sugar substitutes, granular	5 mL (1 tsp)
Whipped toppings	15 mL (1 tbs)

Table B-10

Canadian Choice System: Combined Food Choices

Food	Choices per Serving	Measure	Mass (Weight)
Angel food cake	½ starch + 2½ sugars	1/12 cake	50 g
Apple crisp	½ starch + 1½ fruits & vegetables + 1 sugar + 1–2 fats	125 mL (½ c)	
Applesauce, sweetened	1 fruits & vegetables + 1 sugar	125 mL (½ c)	
Beans and pork in tomato sauce	1 starch + ½ fruits & vegetables + ½ sugar + 1 protein	125 mL (½ c)	135 g
Beef burrito	2 starches + 3 proteins + 3 fats		110 g
Brownie	1 sugar + 1 fat	1	20 g
Cabbage rolls*	1 starch + 2 proteins	3	310 g
Caesar salad	2–4 fats	20 mL dressing (4 tsp)	
Cheesecake	½ starch + 2 sugars + ½ protein + 5 fats	1 piece	80 g
Chicken fingers	1 starch + 2 proteins + 2 fats	6 small	100 g
Chicken and snow pea Oriental	2 starches + ½ fruits & vegetables + 3 proteins + 1 fat	500 mL (2 c)	
Chili	1½ starches + ½ fruits & vegetables + 3½ protein	300 mL (1¼ c)	325 g
Chips			
Potato chips	1 starch + 2 fats	15 chips	30 g
Corn chips	1 starch + 2 fats	30 chips	30 g
Tortilla chips	1 starch + 1½ fats	13 chips	
Cheese twist	1 starch + 1½ fats	30 chips	30 g
Chocolate bar			
Aero®	2½ sugars + 2½ fats	bar	43 g
Smarties®	4½ sugars + 2 fats	package	60 g
Chocolate cake (without icing)	1 starch + 2 sugars + 3 fats	1/10 of a 8" pan	
Chocolate devil's food cake (without icing)	2 starches + 2 sugars + 3 fats	1/12 of a 9" pan	
Chocolate milk	2 milks 2% + 1 sugar	250 mL (1 c)	300 g
Clubhouse (triple-decker) sandwich	3 starches + 3 proteins + 4 fats		
Cookies			
chocolate chip	½ starch + ½ sugar + 1½ fats	2	22 g
oatmeal	1 starch + 1 sugar + 1 fat	2	40 g
Donut (chocolate glazed)	1 starch + 1½ sugars + 2 fats	1	65 g
Egg roll	1 starch + ½ protein + 1 fat		75 g
Four bean salad	1 starch + ½ protein + 1 fat	125 mL (½ c)	
French toast	1 starch + ½ protein + 2 fats	1 slice	65 g
Fruit in heavy syrup	1 fruits & vegetables + 1½ sugars	125 mL (½ c)	
Granola bar	½ starch + 1 sugar + 1–2 fats		30 g
Granola cereal	1 starch + 1 sugar + 2 fats	125 mL (½ c)	45 g
Hamburger	2 starches + 3 proteins + 2 fats	junior size	
Ice cream and cone, plain flavour			
ice cream	½ milk + 2–3 sugars + 1–2 fats		100 g
cone	½ sugar		4 g
Lasagna			
regular cheese	1 starch + 1 fruits & vegetables + 3 proteins + 2 fats	3" x 4" piece	
low-fat cheese	1 starch + 1 fruits & vegetables + 3 proteins	3" x 4" piece	
Legumes			
Dried beans (kidney, navy, pinto, fava, chick peas)	2 starches + 1 protein	250 mL (1 c)	180 g
Dried peas	2 starches + 1 protein	250 mL (1 c)	210 g
Lentils	2 starches + 1 protein	250 mL (1 c)	210 g

* If eaten with sauce, add ½ fruits & vegetables exchange.

Table B-10 (continued)
Canadian Choice System: Combined Food Choices

Food	Choices per Serving	Measure	Mass (Weight)
Macaroni and cheese	2 starches + 2 proteins + 2 fats	250 mL (1 c)	210 g
Minestrone soup	1½ starches + ½ fruits & vegetables + ½ fat	250 mL (1 c)	
Muffin	1 starch + ½ sugar + 1 fat	1 small	45 g
Nuts (dry or roasted without any oil added)			
Almonds, dried sliced	½ protein + 2 fats	50 mL (¼ c)	22 g
Brazil nuts, dried unblanched	½ protein + 2½ fats	5 large	23 g
Cashew nuts, dry roasted	½ starch + ½ protein + 2 fats	50 mL (¼ c)	28 g
Filbert hazelnut, dry	½ protein + 3½ fats	50 mL (¼ c)	30 g
Macadamia nuts, dried	½ protein + 4 fats	50 mL (¼ c)	28 g
Peanuts, raw	1 protein + 2 fats	50 mL (¼ c)	30 g
Pecans, dry roasted	½ fruits & vegetables + 3 fats	50 mL (¼ c)	22 g
Pine nuts, pignolia dried	1 protein + 3 fats	50 mL (¼ c)	34 g
Pistachio nuts, dried	½ fruits & vegetables + ½ protein + 2½ fats	50 mL (¼ c)	27 g
Pumpkin seeds, roasted	2 proteins + 2½ fats	50 mL (¼ c)	47 g
Sesame seeds, whole dried	½ fruits & vegetables + ½ protein + 2½ fats	50 mL (¼ c)	30 g
Sunflower kernel, dried	½ protein + 1½ fats	50 mL (¼ c)	17 g
Walnuts, dried chopped	½ protein + 3 fats	50 mL (¼ c)	26 g
Perogies	2 starches + 1 protein + 1 fat	3	
Pie, fruit	1 starch + 1 fruits & vegetables + 2 sugars + 3 fats	1 piece	120 g
Pizza, cheese	1 starch + 1 protein + 1 fat	1 slice (⅛ of a 12")	50 g
Pork stir fry	½ to 1 fruits & vegetables + 3 proteins	200 mL (¾ c)	
Potato salad	1 starch + 1 fat	125 mL (½ c)	130 g
Potatoes, scalloped	2 starches + 1 milk + 1–2 fats	200 mL (¾ c)	210 g
Pudding, bread or rice	1 starch + 1 sugar + 1 fat	125 mL (½ c)	
Pudding, vanilla	1 milk + 2 sugars	125 mL (½ c)	
Raisin bran cereal	1 starch + ½ fruits & vegetables + ½ sugar	175 mL (⅔ c)	40 g
Rice krispie squares	½ starch + 1½ sugars + ½ fat	1 square	30 g
Shepherd's pie	2 starches + 1 fruits & vegetables + 3 proteins	325 mL (1⅓ c)	
Sherbet, orange	3 sugars + ½ fat	125 mL (½ c)	
Spaghetti and meat sauce	2 starches + 1 fruits & vegetables + 2 proteins + 3 fats	250 mL (1 c)	
Stew	2 starches + 2 fruits & vegetables + 3 proteins + ½ fat	200 mL (¾ c)	
Sundae	4 sugars + 3 fats	125 mL (½ c)	
Tuna casserole	1 starch + 2 proteins + ½ fat	125 mL (½ c)	
Yogurt, fruit bottom	1 fruits & vegetables + 1 milk + 1 sugar	125 mL (½ c)	125 g
Yogurt, frozen	1 milk + 1 sugar	125 mL (½ c)	125 g

B

Food Labels

Consumers can gather a lot of information from a nutrition label. Figure B-1 demonstrates the reading of a food label and Table B-11 defines terms.

Figure B-1

Example of a Food Label

WHAT YOU WILL FIND ON A LABEL:

Nutrition Claims
- in Canada, it is optional for a company to decide to use claims,
- when claims appear on a label, they must follow government laws

Nutrition Information
- gives detailed nutrition facts about the product, including serving size and core list
- does not have to appear by law on food products in Canada
- refers to the food as packaged, so if you add milk, eggs or other food, the nutritional content of the food you eat can be very different

Serving Size
- the amount of food for which the information is given
- check the serving size: the serving size on the label may not be the same as the serving size you would actually eat (for example, the serving size of cereal may be ¾ cup, much smaller than your regular serving

Core List
- the energy (in Calories and kilojoules), grams of protein, fat and carbohydrate for each serving
- some products break down fat into monounsaturates, polyunsaturates, saturates, and cholesterol (to find out what these mean, look at the Fats & Oils section)
- carbohydrates may include the amount of sugars, starch and fibre, or may list these items separately

Sodium and Potassium (in milligrams)

Vitamins and Minerals (as percent of your recommended daily intake)

Canadian Diabetes Association Food Choice Values and Symbols
- the Values and Symbols are tools to help you fit the food into your meal plan, they are not an endorsement by CDA
- it is up to the food company to decide if it wants its foods analyzed and assigned symbols
- when they are on a label, they have been assigned by a dietitian working for CDA, so you can be sure the information is correct

Ingredients
- must be found on all food labels by law
- ingredients are listed in decreasing order by weight, so what you see first is what you get the most of

Energy

kcalorie reduced: 50% or fewer kcalories than the regular version.

light: term may be used to describe anything (for example, light in colour, texture, flavour, taste, or kcalories); read the label to find out what is "light" about the product.

low kcalorie: kcalorie-reduced and no more than 15 kcalories per serving.

Fat and Cholesterol

low cholesterol: no more than 3 mg of cholesterol per 100 g of the food and low in saturated fat; *does not* always mean low in total fat.

low fat: no more than 3 g of fat per serving; *does not* always mean low in kcalories.

lower fat: at least 25% less fat than the comparison food; be aware that 80% fat-free still means the food is 20% fat.

Carbohydrates: Fibre and Sugar

carbohydrate reduced: not more than 50% of the carbohydrate found in the regular version; *does not* always mean the product is lower in kcalories because other ingredients such as fat may have increased.

source of dietary fibre: a product that provides 2–4 g of fibre.

high source of dietary fibre: a product that provides 4–6 g of fibre.

very high source of fibre: a product that provides 6 g (or more) of fibre.

sugar free: low in carbohydrates and kcalories; can be used as an extra food in the exchange system.

unsweetened or no sugar added: no sugar was added to the product; sugar may be found naturally in the food (for example, fruit canned in its own juice).

Canada's Food Guide

Canada's Food Guide to Healthy Eating, shown in Figure B-2, gives detailed information for selecting foods to meet the nutritional needs of all Canadians four years of age and older. Like the U.S. Daily Food Guide, Canada's Food Guide also takes a total diet approach, rather than emphasizing a single food, meal, or day's meals and snacks.

Figure B-2

Canada's Food Guide to Healthy Eating

Health and Welfare
Canada

Santé et Bien-être social
Canada

CANADA'S
Food Guide
TO HEALTHY EATING

Enjoy a variety
of foods from each
group every day.

Choose lower-
fat foods
more often.

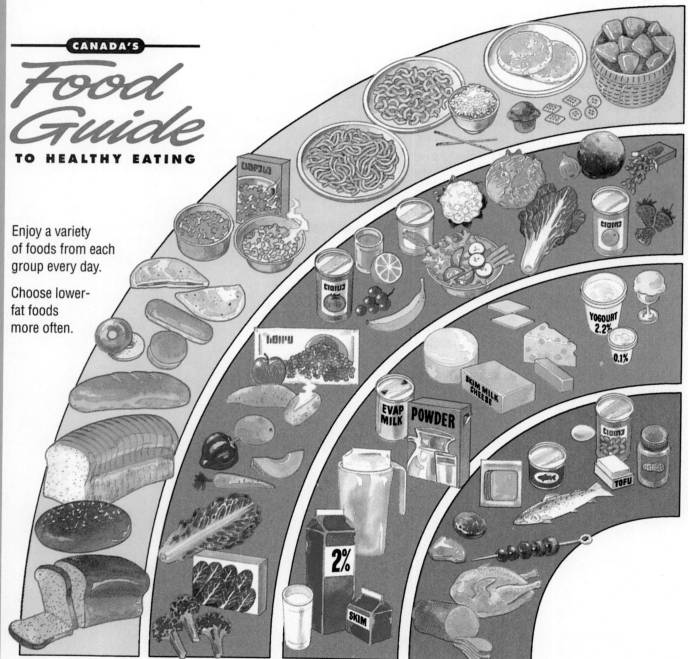

Grain Products
Choose whole grain
and enriched
products more
often.

Vegetables & Fruit
Choose dark green and
orange vegetables and
orange fruit more often.

Milk Products
Choose lower-fat
milk products more
often.

Meat & Alternatives
Choose leaner meats,
poultry and fish, as well
as dried peas, beans and
lentils more often.

CANADA'S

Food Guide

TO HEALTHY EATING

FOR PEOPLE FOUR YEARS AND OVER

Different People Need Different Amounts of Food

The amount of food you need every day from the 4 food groups and other foods depends on your age, body size, activity level, whether you are male or female and if you are pregnant or breast-feeding. That's why the Food Guide gives a lower and higher number of servings for each food group. For example, young children can choose the lower number of servings, while male teenagers can go to the higher number. Most other people can choose servings somewhere in between.

Grain Products

5–12

SERVINGS PER DAY

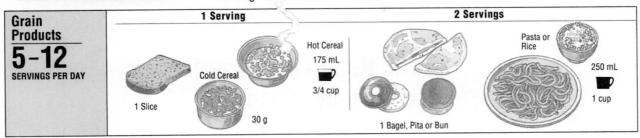

Vegetables & Fruit

5–10

SERVINGS PER DAY

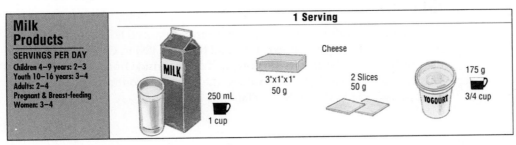

Milk Products

SERVINGS PER DAY

Children 4–9 years: 2–3
Youth 10–16 years: 3–4
Adults: 2–4
Pregnant & Breast-feeding Women: 3–4

Other Foods

Taste and enjoyment can also come from other foods and beverages that are not part of the 4 food groups. Some of these foods are higher in fat or Calories, so use these foods in moderation.

Meat & Alternatives

2–3

SERVINGS PER DAY

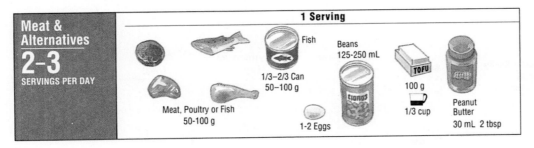

Enjoy eating well, being active and feeling good about yourself. That's VITALIT℮®

CONTENTS

RDA

Exchange Lists for Meal Planning

Diet and Health *Recommendations*

Nutrition Recommendations from WHO

United States: Recommendations and Exchanges

World Health Organization: Recommendations

Chapter 1 introduced the 1989 Recommended Dietary Allowances (RDA), and Chapter 20 introduced exchange systems; this appendix provides additional details. (See Appendix B for Canada's nutrition recommendations and exchange system.) Nutrition recommendations from the World Health Organization are also included.

RDA

As Chapter 1 mentioned, the Dietary Reference Intakes (DRI) reports are replacing the 1989 RDA in the United States and the 1991 RNI in Canada. Recommendations from the DRI reports are presented on the inside front covers. For nutrients that do not yet have new values, the RDA will continue to serve health professionals in the United States (Tables C-1, C-2, and C-3).

Table C-1

Estimated Safe and Adequate Daily Dietary Intakes of Additional Selected Minerals (United States)[a]

Age (yr)	Chromium (μg)	Molybdenum (μg)	Copper (mg)	Manganese (mg)
Infants				
0.0–0.5	10–40	15–30	0.4–0.6	0.3–0.6
0.5–1	20–60	20–40	0.6–0.7	0.6–1.0
Children				
1–3	20–80	25–50	0.7–1.0	1.0–1.5
4–6	30–120	30–75	1.0–1.5	1.5–2.0
7–10	50–200	50–150	1.0–2.0	2.0–3.0
11+	50–200	75–250	1.5–2.5	2.0–5.0
Adults	50–200	75–250	1.5–3.0	2.0–5.0

[a]Less information is available on which to base allowances for these nutrients. Therefore, they are not included in the main table of the RDA, and the figures provided here are in the form of ranges of recommended intakes; the toxic levels for many trace elements may be only several times usual intakes, so the upper levels for the trace elements given in this table should not be habitually exceeded.

SOURCE: *Recommended Dietary Allowances*, © 1989 by the National Academy of Sciences, National Academy Press, Washington, D.C.

Table C-2

Estimated Minimum Requirements of Sodium, Chloride, and Potassium

Age (yr)	Weight (kg)	Sodium[a] (mg)	Chloride (mg)	Potassium[b] (mg)
Infants				
0.0–0.5	4.5	120	180	500
0.5–1.0	8.9	200	300	700
Children				
1	11.0	225	350	1000
2–5	16.0	300	500	1400
6–9	25.0	400	600	1600
Adolescents	50.0	500	750	2000
Adults	70.0	500	750	2000

[a]Sodium requirements are based on estimates of needs for growth and for replacement of obligatory losses. They cover a wide variation of physical activity patterns and climatic exposure but do not provide for large, prolonged losses from the skin through sweat.
[b]Dietary potassium may benefit the prevention and treatment of hypertension, and recommendations to include many servings of fruits and vegetables would raise potassium intakes to about 3500 milligrams per day.
SOURCE: *Recommended Dietary Allowances*, © 1989 by the National Academy of Sciences, National Academy Press, Washington, D.C.

Table C-3

Median Heights and Weights and Recommended Energy Intakes (United States)

Age (yr)	Weight kg	Weight lb	Height cm	Height in	REE[a] (kcal/day)	Multiples of REE[b]	kcal/kg	kcal/day[c]
Infants								
0.0–0.5	6	13	60	24	320		108	650
0.5–1.0	9	20	71	28	500		98	850
Children								
1–3	13	29	90	35	740		102	1300
4–6	20	44	112	44	950		90	1800
7–10	28	62	132	52	1130		70	2000
Males								
11–14	45	99	157	62	1440	1.70	55	2500
15–18	66	145	176	69	1760	1.67	45	3000
19–24	72	160	177	70	1780	1.67	40	2900
25–50	79	174	176	70	1800	1.60	37	2900
51+	77	170	173	68	1530	1.50	30	2300
Females								
11–14	46	101	157	62	1310	1.67	47	2200
15–18	55	120	163	64	1370	1.60	40	2200
19–24	58	128	164	65	1350	1.60	38	2200
25–50	63	138	163	64	1380	1.55	36	2200
51+	65	143	160	63	1280	1.50	30	1900
Pregnant (2nd and 3rd trimesters)								+300
Lactating								+500

[a]REE (resting energy expenditure) represents the energy expended by a person at rest under normal conditions.
[b]Recommended energy allowances assume light-to-moderate activity and were calculated by multiplying the REE by an activity factor.
[c]Average energy allowances have been rounded.
SOURCE: *Recommended Dietary Allowances*, © 1989 by the National Academy of Sciences, National Academy Press, Washington, D.C.

Exchange Lists for Meal Planning

The U.S. exchange system groups together foods that have about the same amount of carbohydrate, protein, fat, and kcalories. Then any food on a list can be "exchanged" for any other food on that same list. Chapter 20 introduced the exchange lists and Tables C-4 through C-12 present the lists in detail.

Table C-4

U.S. Exchange System: Starch List

1 starch exchange = 15 g carbohydrate, 3 g protein, 0–1 g fat, and 80 kcal

NOTE: In general, a starch serving is ½ c cereal, grain, pasta, or starchy vegetable; 1 oz of bread; ¾ to 1 oz snack food.

Serving Size	Food	Serving Size	Food
Bread		½ c	Plantains
½ (1 oz)	Bagels	1 small (3 oz)	Potatoes, baked or boiled
2 slices (1½ oz)	Bread, reduced-kcalorie	½ c	Potatoes, mashed
1 slice (1 oz)	Bread, white (including French and Italian), whole-wheat, pumpernickel, rye	1 c	Squash, winter (acorn, butternut)
2 (⅔ oz)	Bread sticks, crisp, 40 x ½"	½ c	Yams, sweet potatoes, plain
½	English muffins	**Crackers and Snacks**	
½ (1 oz)	Hot dog or hamburger buns	8	Animal crackers
½	Pita, 6" across	3	Graham crackers, 2½" square
1 (1 oz)	Plain rolls, small	¾ oz	Matzoh
1 slice (1 oz)	Raisin bread, unfrosted	4 slices	Melba toast
1	Tortillas, corn, 6" across	24	Oyster crackers
1	Tortillas, flour, 7–8" across	3 c	Popcorn (popped, no fat added or low-fat microwave)
1	Waffles, 4½" square, reduced-fat	¾ oz	Pretzels
Cereals and Grains		2	Rice cakes, 4" across
½ c	Bran cereals	6	Saltine-type crackers
½ c	Bulgur, cooked	15–2" (¾ oz)	Snack chips, fat-free (tortilla, potato)
½ c	Cereals, cooked	2–5 (¾ oz)	Whole-wheat crackers, no fat added
¾ c	Cereals, unsweetened, ready-to-eat	**Dried Beans, Peas, and Lentils**	
3 tbs	Cornmeal (dry)	½ c	Beans and peas, cooked (garbanzo, lentils, pinto, kidney, white, split, black-eyed)
⅓ c	Couscous	⅔ c	Lima beans
3 tbs	Flour (dry)	3 tbs	Miso ✐
¼ c	Granola, low-fat	**Starchy Foods Prepared with Fat**	
¼ c	Grape nuts	**Count as 1 starch + 1 fat exchange.**	
½ c	Grits, cooked	1	Biscuit, 2½" across
½ c	Kasha	½ c	Chow mein noodles
¼ c	Millet	1 (2 oz)	Corn bread, 2" cube
¼ c	Muesli	6	Crackers, round butter type
½ c	Oats	1 c	Croutons
½ c	Pasta, cooked	16–25 (3 oz)	French-fried potatoes
1½ c	Puffed cereals	¼ c	Granola
½ c	Rice milk	1 (1½ oz)	Muffin, small
⅓ c	Rice, white or brown, cooked	2	Pancake, 4" across
½ c	Shredded wheat	3 c	Popcorn, microwave
½ c	Sugar-frosted cereal	3	Sandwich crackers, cheese or peanut butter filling
3 tbs	Wheat germ	⅓ c	Stuffing, bread (prepared)
Starchy Vegetables		2	Taco shell, 6" across
1/3 c	Baked beans	1	Waffle, 4½" square
½ c	Corn	4–6 (1 oz)	Whole-wheat crackers, fat added
1 (5 oz)	Corn on cob, medium		
1 c	Mixed vegetables with corn, peas, or pasta		
½ c	Peas, green		

✐ = 400 mg or more sodium per exchange.

Table C-5

U.S. Exchange System: Fruit List

1 fruit exchange = 15 g carbohydrate and 60 kcal

NOTE: In general, a fruit serving is 1 small to medium fresh fruit; ½ c canned or fresh fruit or fruit juice; ¼ c dried fruit.

Serving Size	Food	Serving Size	Food
1 (4 oz)	Apples, unpeeled, small	½ (8 oz) or 1 c cubes	Papayas
½ c	Applesauce, unsweetened	1 (6 oz)	Peaches, medium, fresh
4 rings	Apples, dried	½ c	Peaches, canned
4 whole (5½ oz)	Apricots, fresh	½ (4 oz)	Pears, large, fresh
8 halves	Apricots, dried	½ c	Pears, canned
½ c	Apricots, canned	¾ c	Pineapple, fresh
1 (4 oz)	Bananas, small	½ c	Pineapple, canned
¾ c	Blackberries	2 (5 oz)	Plums, small
¾ c	Blueberries	½ c	Plums, canned
⅓ melon (11 oz) or 1 c cubes	Cantaloupe, small	3	Prunes, dried
		2 tbs	Raisins
12 (3 oz)	Cherries, sweet, fresh	1 c	Raspberries
½ c	Cherries, sweet, canned	1¼ c whole berries	Strawberries
3	Dates	2 (8 oz)	Tangerines, small
1½ large or 2 medium (3½ oz)	Figs, fresh	1 slice (13½ oz) or 1¼ c cubes	Watermelon
1½	Figs, dried	**Fruit Juice**	
½ c	Fruit cocktail	½ c	Apple juice/cider
½ (11 oz)	Grapefruit, large	⅓ c	Cranberry juice cocktail
¾ c	Grapefruit sections, canned	1 c	Cranberry juice cocktail, reduced-kcalorie
17 (3 oz)	Grapes, small		
1 slice (10 oz) or 1 c cubes	Honeydew melon	⅓ c	Fruit juice blends, 100% juice
		⅓ c	Grape juice
1 (3½ oz)	Kiwi	½ c	Grapefruit juice
¾ c	Mandarin oranges, canned	½ c	Orange juice
½ (5½ oz) or ½ c	Mangoes, small	½ c	Pineapple juice
1 (5 oz)	Nectarines, small	⅓ c	Prune juice
1 (6½ oz)	Oranges, small		

Table C-6

U.S. Exchange System: Milk List

Serving Size	Food	Serving Size	Food
Fat-Free and Very Low-Fat Milk		**Low-Fat Milk**	
1 fat-free/low-fat milk exchange = 12 g carbohydrate, 8 g protein, 0–3 g fat, 90 kcal		1 reduced-fat milk exchange = 12 g carbohydrate, 8 g protein, 5 g fat, 120 kcal	
1 c	Fat-free milk	1 c	2% milk
1 c	½% milk	¾ c	Plain low-fat yogurt
1 c	1% milk	1 c	Sweet acidophilus milk
1 c	Fat-free or low-fat buttermilk	**Whole Milk**	
½ c	Evaporated fat-free milk	1 whole milk exchange = 12 g carbohydrate, 8 g protein, 8 g fat, 150 kcal	
⅓ c dry	Dry milk	1 c	Whole milk
¾ c	Plain fat-free yogurt	½ c	Evaporated whole milk
1 c	Fat-free or low-fat fruit-flavored yogurt sweetened with aspartame or with a nonnutritive sweetener	1 c	Goat's milk
		1 c	Kefir

Table C-7
U.S. Exchange System: Other Carbohydrates List

1 other carbohydrate exchange = 15 g carbohydrate, or 1 starch, or 1 fruit, or 1 milk exchange

Food	Serving Size	Exchanges per Serving
Angel food cake, unfrosted	1/12 cake	2 carbohydrates
Brownies, small, unfrosted	20 square	1 carbohydrate, 1 fat
Cake, unfrosted	20 square	1 carbohydrate, 1 fat
Cake, frosted	20 square	2 carbohydrates, 1 fat
Cookie, fat-free	2 small	1 carbohydrate
Cookies or sandwich cookies	2 small	1 carbohydrate, 1 fat
Cupcakes, frosted	1 small	2 carbohydrates, 1 fat
Cranberry sauce, jellied	1/4 c	2 carbohydrates
Doughnuts, plain cake	1 medium, (1 1/2 oz)	1 1/2 carbohydrates, 2 fats
Doughnuts, glazed	33/40 across (2 oz)	2 carbohydrates, 2 fats
Fruit juice bars, frozen, 100% juice	1 bar (3 oz)	1 carbohydrate
Fruit snacks, chewy (pureed fruit concentrate)	1 roll (3/4 oz)	1 carbohydrate
Fruit spreads, 100% fruit	1 tbs	1 carbohydrate
Gelatin, regular	1/2 c	1 carbohydrate
Gingersnaps	3	1 carbohydrate
Granola bars	1 bar	1 carbohydrate, 1 fat
Granola bars, fat-free	1 bar	2 carbohydrates
Hummus	1/3 c	1 carbohydrate, 1 fat
Ice cream	1/2 c	1 carbohydrate, 2 fats
Ice cream, light	1/2 c	1 carbohydrate, 1 fat
Ice cream, fat-free, no sugar added	1/2 c	1 carbohydrate
Jam or jelly, regular	1 tbs	1 carbohydrate
Milk, chocolate, whole	1 c	2 carbohydrates, 1 fat
Pie, fruit, 2 crusts	1/6 pie	3 carbohydrates, 2 fats
Pie, pumpkin or custard	1/8 pie	1 carbohydrate, 2 fats
Potato chips	12–18 (1 oz)	1 carbohydrate, 2 fats
Pudding, regular (made with low-fat milk)	1/2 c	2 carbohydrates
Pudding, sugar-free (made with low-fat milk)	1/2 c	1 carbohydrate
Salad dressing, fat-free	1/4 c	1 carbohydrate
Sherbet, sorbet	1/2 c	2 carbohydrates
Spaghetti or pasta sauce, canned	1/2 c	1 carbohydrate, 1 fat
Sweet roll or danish	1 (2 1/2 oz)	2 1/2 carbohydrates, 2 fats
Syrup, light	2 tbs	1 carbohydrate
Syrup, regular	1 tbs	1 carbohydrate
Syrup, regular	1/4 c	4 carbohydrates
Tortilla chips	6–12 (1 oz)	1 carbohydrate, 2 fats
Yogurt, frozen, low-fat, fat-free	1/3 c	1 carbohydrate, 0–1 fat
Yogurt, frozen, fat-free, no sugar added	1/2 c	1 carbohydrate
Yogurt, low-fat with fruit	1 c	3 carbohydrates, 0–1 fat
Vanilla wafers	5	1 carbohydrate, 1 fat

= 400 mg or more sodium per exchange.

Table C-8

U.S. Exchange System: Vegetable List

1 vegetable exchange = 5 g carbohydrate, 2 g protein, 25 kcal

NOTE: In general, a vegetable serving is ½ c cooked vegetables or vegetable juice; 1 c raw vegetables. Starchy vegetables such as corn, peas, and potatoes are on the starch list.

Artichokes	Mushrooms
Artichoke hearts	Okra
Asparagus	Onions
Beans (green, wax, Italian)	Pea pods
Bean sprouts	Peppers (all varieties)
Beets	Radishes
Broccoli	Salad greens (endive, escarole, lettuce, romaine, spinach)
Brussels sprouts	Sauerkraut
Cabbage	Spinach
Carrots	Summer squash (crookneck)
Cauliflower	Tomatoes
Celery	Tomatoes, canned
Cucumbers	Tomato sauce
Eggplant	Tomato/vegetable juice
Green onions or scallions	Turnips
Greens (collard, kale, mustard, turnip)	Water chestnuts
Kohlrabi	Watercress
Leeks	Zucchini
Mixed vegetables (without corn, peas, or pasta)	

= 400 mg or more sodium per exchange.

Table C-9

U.S. Exchange System: Meat and Meat Substitutes List

NOTE: In general, a meat serving is 1 oz meat, poultry, or cheese; ½ c dried beans (weigh meat and poultry and measure beans after cooking).

Serving Size	Food
Very Lean Meat and Substitutes	
1 very lean meat exchange = 7 g protein, 0–1 g fat, 35 kcal	
1 oz	Poultry: Chicken or turkey (white meat, no skin), Cornish hen (no skin)
1 oz	Fish: Fresh or frozen cod, flounder, haddock, halibut, trout; tuna, fresh or canned in water
1 oz	Shellfish: Clams, crab, lobster, scallops, shrimp, imitation shellfish
1 oz	Game: Duck or pheasant (no skin), venison, buffalo, ostrich
	Cheese with ≤ 1 g fat/oz:
¼ c	Fat-free or low-fat cottage cheese
1 oz	Fat-free cheese
	Other:
1 oz	Processed sandwich meats with ≤ 1 g fat/oz (such as deli thin, shaved meats, chipped beef, turkey ham)
2	Egg whites
¼ c	Egg substitutes, plain
1 oz	Hot dogs with ≤ 1 g fat/oz
1 oz	Kidney (high in cholesterol)
1 oz	Sausage with ≤ 1 g fat/oz
Count as one very lean meat and one starch exchange:	
½ c	Dried beans, peas, lentils (cooked)
Lean Meat and Substitutes	
1 lean meat exchange = 7 g protein, 3 g fat, 55 kcal	
1 oz	Beef: USDA Select or Choice grades of lean beef trimmed of fat (round, sirloin, and flank steak); tenderloin; roast (rib, chuck, rump); steak (T-bone, porterhouse, cubed), ground round
1 oz	Pork: Lean pork (fresh ham); canned, cured, or boiled ham; Canadian bacon; tenderloin, center loin chop
1 oz	Lamb: Roast, chop, leg
1 oz	Veal: Lean chop, roast
1 oz	Poultry: Chicken, turkey (dark meat, no skin), chicken white meat (with skin), domestic duck or goose (well-drained of fat, no skin)
	Fish:
1 oz	Herring (uncreamed or smoked)
6 medium	Oysters
1 oz	Salmon (fresh or canned), catfish
2 medium	Sardines (canned)
1 oz	Tuna (canned in oil, drained)
1 oz	Game: Goose (no skin), rabbit
	Cheese:
¼ c	4.5%-fat cottage cheese

Serving Size	Food
2 tbs	Grated Parmesan
1 oz	Cheeses with ≤ 3 g fat/oz
	Other:
1½ oz	Hot dogs with ≤ 3 g fat/oz
1 oz	Processed sandwich meat with ≤ 3 g fat/oz (turkey pastrami or kielbasa)
1 oz	Liver, heart (high in cholesterol)
Medium-Fat Meat and Substitutes	
1 medium-fat meat exchange = 7 g protein, 5 g fat, and 75 kcal	
1 oz	Beef: Most beef products (ground beef, meatloaf, corned beef, short ribs, Prime grades of meat trimmed of fat, such as prime rib)
1 oz	Pork: Top loin, chop, Boston butt, cutlet
1 oz	Lamb: Rib roast, ground
1 oz	Veal: Cutlet (ground or cubed, unbreaded)
1 oz	Poultry: Chicken dark meat (with skin), ground turkey or ground chicken, fried chicken (with skin)
1 oz	Fish: Any fried fish product
	Cheese with ≤ 5 g fat/oz:
1 oz	Feta
1 oz	Mozzarella
¼ c (2 oz)	Ricotta
	Other:
1	Egg (high in cholesterol, limit to 3/week)
1 oz	Sausage with ≤ 5 g fat/oz
1 c	Soy milk
¼ c	Tempeh
4 oz or ½ c	Tofu
High-Fat Meat and Substitutes	
1 high-fat meat exchange = 7 g protein, 8 g fat, 100 kcal	
1 oz	Pork: Spareribs, ground pork, pork sausage
1 oz	Cheese: All regular cheeses (American, cheddar, Monterey Jack, swiss)
	Other:
1 oz	Processed sandwich meats with ≤ 8 g fat/oz (bologna, pimento loaf, salami)
1 oz	Sausage (bratwurst, Italian, knockwurst, Polish, smoked)
1 (10/lb)	Hot dog (turkey or chicken)
3 slices (20 slices/lb)	Bacon
Count as one high-fat meat plus one fat exchange:	
1 (10/lb)	Hot dog (beef, pork, or combination)
2 tbs	Peanut butter (contains unsaturated fat)

= 400 mg or more sodium per exchange.

Table C-10

U.S. Exchange System: Fat List

1 fat exchange = 5 g fat, 45 kcal

NOTE: In general, a fat serving is 1 tsp regular butter, margarine, or vegetable oil; 1 tbs regular salad dressing. Many fat-free and reduced fat foods are on the Free Foods List.

Serving Size	Food
Monounsaturated Fats	
⅛ medium (1 oz)	Avocados
1 tsp	Oil (canola, olive, peanut)
8 large	Olives, ripe (black)
10 large	Olives, green, stuffed
6 nuts	Almonds, cashews
6 nuts	Mixed nuts (50% peanuts)
10 nuts	Peanuts
4 halves	Pecans
2 tsp	Peanut butter, smooth or crunchy
1 tbs	Sesame seeds
2 tsp	Tahini paste
Polyunsaturated Fats	
1 tsp	Margarine, stick, tub, or squeeze
1 tbs	Margarine, lower-fat (30% to 50% vegetable oil)
1 tsp	Mayonnaise, regular
1 tbs	Mayonnaise, reduced-fat
4 halves	Nuts, walnuts, English
1 tsp	Oil (corn, safflower, soybean)
1 tbs	Salad dressing, regular
2 tbs	Salad dressing, reduced-fat
2 tsp	Mayonnaise type salad dressing, regular
1 tbs	Mayonnaise type salad dressing, reduced-fat
1 tbs	Seeds: pumpkin, sunflower
Saturated Fats*	
1 slice (20 slices/lb)	Bacon, cooked
1 tsp	Bacon, grease
1 tsp	Butter, stick
2 tsp	Butter, whipped
1 tbs	Butter, reduced-fat
2 tbs (½ oz)	Chitterlings, boiled
2 tbs	Coconut, sweetened, shredded
2 tbs	Cream, half and half
1 tbs (½ oz)	Cream cheese, regular
2 tbs (1 oz)	Cream cheese, reduced-fat
	Fatback or salt pork†
1 tsp	Shortening or lard
2 tbs	Sour cream, regular
3 tbs	Sour cream, reduced-fat

= 400 mg or more sodium per exchange

*Saturated fats can raise blood cholesterol levels.

† Use a piece 1″ × 1″ × ¼″ if you plan to eat the fatback cooked with vegetables. Use a piece 2″ × 1″ × ½″ when eating only the vegetables with the fatback removed.

Table C-11

U.S. Exchange System: Combination Foods List

Food	Serving Size	Exchanges per Serving
Entrees		
Tuna noodle casserole, lasagna, spaghetti with meatballs, chili with beans, macaroni and cheese 🖊	1 c (8 oz)	2 carbohydrates, 2 medium-fat meats
Chow mein (without noodles or rice)	2 c (16 oz)	1 carbohydrate, 2 lean meats
Pizza, cheese, thin crust 🖊	¼ of 100 (5 oz)	2 carbohydrates, 2 medium-fat meats, 1 fat
Pizza, meat topping, thin crust 🖊	¼ of 10″ (5 oz)	2 carbohydrates, 2 medium-fat meats, 2 fats
Pot pie 🖊	1 (7 oz)	2 carbohydrates, 1 medium-fat meat, 4 fats
Frozen entrees		
Salisbury steak with gravy, mashed potato	1 (11 oz)	2 carbohydrates, 3 medium-fat meats, 3–4 fats
Turkey with gravy, mashed potato, dressing 🖊	1 (11 oz)	2 carbohydrates, 2 medium-fat meats, 2 fats
Entree with less than 300 kcalories 🖊	1 (8 oz)	2 carbohydrates, 3 lean meats
Soups 🖊		
Bean 🖊	1 c	1 carbohydrate, 1 very lean meat
Cream (made with water) 🖊	1 c (8 oz)	1 carbohydrate, 1 fat
Split pea (made with water) 🖊	½ c (4 oz)	1 carbohydrate
Tomato (make with water) 🖊	1 c (8 oz)	1 carbohydrate
Vegetable beef, chicken noodle, or other broth-type 🖊	1 c (8 oz)	1 carbohydrate
Fast Foods		
Burritos with beef 🖊	2	4 carbohydrates, 2 medium-fat meats, 2 fats
Chicken nuggets 🖊	6	1 carbohydrate, 2 medium-fat meats, 1 fat
Chicken breast and wing, breaded and fried 🖊	1	1 carbohydrate, 4 medium-fat meats, 2 fats
Fish sandwich/tartar sauce 🖊	1	3 carbohydrates, 1 medium-fat meat, 3 fats
French fries, thin	20–25	2 carbohydrates, 2 fats
Hamburger, regular	1	2 carbohydrates, 2 medium-fat meats
Hamburger, large 🖊	1	2 carbohydrates, 3 medium-fat meats, 1 fat
Hot dog with bun 🖊	1	1 carbohydrate, 1 high-fat meat, 1 fat
Individual pan pizza 🖊	1	5 carbohydrates, 3 medium-fat meats, 3 fats
Soft serve cone	1 medium	2 carbohydrates, 1 fat
Submarine sandwich 🖊	1 (60)	3 carbohydrates, 1 vegetable, 2 medium-fat meats, 1 fat
Taco, hard shell 🖊	1 (6 oz)	2 carbohydrates, 2 medium-fat meats, 2 fats
Taco, soft shell 🖊	1 (3 oz)	1 carbohydrate, 1 medium-fat meat, 1 fat

🖊 = 400 mg or more sodium per exchange.

Table C-12

U.S. Exchange System: Free Foods List

NOTE: A serving of free food contains less than 20 kcalories; those with serving sizes should be limited to 3 servings a day whereas those without serving sizes can be eaten freely.

Serving Size	Food	Serving Size	Food
Fat-Free or Reduced-Fat Foods		1 tbs	Cocoa powder, unsweetened
1 tbs	Cream cheese, fat-free		Coffee
1 tbs	Creamers, nondairy, liquid		Club soda
2 tsp	Creamers, nondairy, powdered		Diet soft drinks, sugar-free
1 tbs	Mayonnaise, fat-free		Drink mixes, sugar-free
1 tsp	Mayonnaise, reduced-fat		Tea
4 tbs	Margarine, fat-free		Tonic water, sugar-free
1 tsp	Margarine, reduced-fat	**Condiments**	
1 tbs	Mayonnaise type salad dressing, nonfat	1 tbs	Catsup
1 tsp	Mayonnaise type salad dressing, reduced-fat		Horseradish
	Nonstick cooking spray		Lemon juice
1 tbs	Salad dressing, fat-free		Lime juice
2 tbs	Salad dressing, fat-free, Italian		Mustard
¼ c	Salsa	1½ large	Pickles, dill
1 tbs	Sour cream, fat-free, reduced-fat		Soy sauce, regular or light
2 tbs	Whipped topping, regular or light	1 tbs	Taco sauce
Sugar-Free or Low-Sugar Foods			Vinegar
1 piece	Candy, hard, sugar-free		
	Gelatin dessert, sugar-free	Seasonings	
	Gelatin, unflavored	Flavoring extracts	
	Gum, sugar-free	Garlic	
2 tsp	Jam or jelly, low-sugar or light	Herbs, fresh or dried	
	Sugar substitutes	Pimento	
2 tbs	Syrup, sugar-free	Spices	
Drinks		Hot pepper sauces	
	Bouillon, broth, consommé	Wine, used in cooking	
	Bouillon or broth, low-sodium	Worcestershire sauce	
	Carbonated or mineral water		

= 400 mg or more sodium per exchange.

Diet and Health Recommendations

Not all diet recommendations apply equally to all diseases or to all people with a disease, but fortunately for the consumer, dietary recommendations do not contradict one another. In fact, they support each other. Most people can gain some disease-prevention benefits by dietary changes. To that end, *Diet and Health* recommendations describe the kinds of foods people should include, limit, or avoid to reduce chronic disease risks (see Table C-13).

Nutrition Recommendations from WHO

Like the Committee on Diet and Health in the United States, the World Health Organization (WHO) has also assessed the relationships between diet and the development of chronic diseases.[1] Its recommendations are expressed in average daily ranges that represent the lower and upper limits:
- Total energy: sufficient to support normal growth, physical activity, and body weight (body mass index = 20 to 22).
- Total fat: 15 to 30 percent of total energy.
  - Saturated fatty acids: 0 to 10 percent total energy.
  - Polyunsaturated fatty acids: 3 to 7 percent total energy.
  - Dietary cholesterol: 0 to 300 milligrams per day.

Table C-13

Diet and Health Recommendations

Nutrient and Energy Recommendations	Suggested Food Choices
Reduce total *fat* intake to 30% or less of kcalories. Reduce saturated fatty acid intake to less than 10% of kcalories and intake of cholesterol to less than 300 mg daily.	Reduce fat and cholesterol intake by substituting fish, poultry without skin, lean meats, and low-fat or fat-free dairy products for fatty meats and whole-milk products; by choosing more vegetables, fruits, cereals, and legumes; and by limiting fats, oils, egg yolks, and fried and other fatty foods.
Increase intake of starches and other *complex carbohydrates*.	Every day eat 5 or more servings of a combination of vegetables and fruits, especially green and yellow vegetables and citrus fruits, and six or more daily servings of a combination of breads, cereals, and legumes. The committee does not recommend increasing intake of added sugars, because their consumption is strongly associated with dental caries.
Maintain *protein* intake at moderate levels.	Meet at least the RDA for protein; do not exceed twice the RDA.
Balance food intake and physical activity to maintain appropriate *body weight*.	
For those who drink *alcoholic beverages,* the committee recommends limiting consumption to the equivalent of less than 1 oz pure alcohol in a single day. Pregnant women should avoid alcoholic beverages.	The committee does not recommend alcohol consumption. One ounce of pure alcohol is the equivalent of two cans of beer, two small glasses of wine, or two average cocktails.
Limit total daily intake of *salt* (sodium chloride) to 6 g or less.	Limit the use of salt in cooking, and avoid adding it to food at the table. Salty, highly processed salty, salt-preserved, and salt-pickled foods should be consumed sparingly.
Maintain adequate *calcium* intake.	
Avoid taking dietary *supplements* in excess of the RDA in any one day.	
Maintain an optimal intake of *fluoride,* particularly during the years of primary and secondary tooth formation and growth.	

SOURCE: Adapted from the National Academy of Sciences report *Diet and Health: Implications for Reducing Chronic Disease Risk* (Washington, D.C.: National Academy Press, 1989).

- Total carbohydrate: 55 to 75 percent total energy.
  - Complex carbohydrates: 50 to 75 percent total energy.
  - Dietary fiber: 27 to 40 grams per day.
  - Refined sugars: 0 to 10 percent total energy.
- Protein: 10 to 15 percent total energy.
- Salt: upper limit of 6 grams/day (no lower limit set).

Note

1. Diet, nutrition and the prevention of chronic diseases: A report of the WHO Study Group on Diet, Nutrition and Prevention of Noncommunicable Diseases, *Nutrition Reviews* 49 (1991): 291–301.

Nutrition Resources

Contents

Books

Journals

Addresses

People interested in nutrition often want to know where they can find reliable nutrition information. Wherever you live, there are several sources you can turn to:

- The Department of Health may have a nutrition expert.
- The local extension agent is often an expert.
- The food editor of your local paper may be well informed.
- The dietitian at the local hospital had to fulfill a set of qualifications before he or she became an RD (see Nutrition in Practice 1).
- There may be knowledgeable professors of nutrition or biochemistry at a nearby college or university.

In addition, you may be interested in building a nutrition library of your own. Books you can buy, journals you can subscribe to, and addresses you can contact for general information are given below.

Books

For students seeking to establish a personal library of nutrition references, the authors of this text recommend the following books:

- *Present Knowledge in Nutrition,* 7th ed. (Washington, D.C.: International Life Sciences Institute—Nutrition Foundation, 1996).

This 646-page paperback has a chapter on each of 64 topics, including energy, obesity, each of the nutrients, several diseases, malnutrition, growth and its assessment, immunity, alcohol, fiber, exercise, drugs, and toxins. Watch for an update; new editions come out every few years.

- M. E. Shils, J. A. Olson, M. Shike, and A. C. Ross, eds., *Modern Nutrition in Health and Disease,* 9th ed. (Baltimore: Williams & Wilkins, 1999).

This 1,951-page volume is a major technical reference book on nutrition topics. It contains encyclopedic articles on the nutrients, foods, the diet, metabolism, malnutrition, age-related needs, and nutrition in disease.

- Committee on Dietary Reference Intakes, *Dietary Reference Intakes for Calcium, Phosphorus, Magnesium, Vitamin D, and Fluoride* (Washington, D.C.: National Academy Press, 1997).
- Committee on Dietary Reference Intakes, *Dietary Reference Intakes for Thiamin, Riboflavin, Niacin, Vitamin B_6, Folate, Vitamin B_{12}, Pantothenic Acid, Biotin, and Choline* (Washington, D.C.: National Academy Press, 1998).

These two reports review the function of each nutrient, dietary sources, and deficiency and toxicity symptoms as well as provide recommendations for intakes. Watch for additional reports on the Dietary Reference Intakes for the remaining nutrients. Until they are published, you may need the following:

- Committee on Dietary Allowances, *Recommended Dietary Allowances,* 10th ed. (Washington, D.C.: National Academy Press, 1989).

The Canadian equivalent is *Nutrition Recommendations,* available by mail from the Canadian Government Publishing Centre, Supply and Services Canada, Ottawa, Ontario K1A OS9, Canada.

- Committee on Diet and Health, *Diet and Health Implications for Reducing Chronic Disease Risk* (Washington, D.C.: National Academy Press, 1989).

This 749-page book presents the integral relationship between diet and chronic disease prevention. Its nutrient chapters provide evidence on how diet influences disease development, and its disease chapters review the dietary patterns implicated in each chronic disease.

- E. M. N. Hamilton and S. A. S. Gropper, *The Biochemistry of Human Nutrition: A Desk Reference* (St. Paul, Minn.: West, 1987).

This 324-page paperback presents the biochemical concepts necessary for an understanding of nutrition. It is a handy reference book for those who have forgotten the basics of biochemistry or for those who are learning biochemistry for the first time.

We also recommend three of our own books that explore current topics in nutrition, health, and the life span:

- S. R. Rolfes, L. K. DeBruyne, and E. N. Whitney, *Life Span Nutrition: Conception through Life* (Belmont, Calif.: West/Wadsworth, 1998).
- E. N. Whitney and S. R. Rolfes, *Understanding Nutrition* (Belmont, Calif.: West/Wadsworth, 1999).
- C. B. Cataldo, L. K. DeBruyne, and E. N. Whitney, *Nutrition and Diet Therapy,* 5th ed. (Belmont, Calif.: West/Wadsworth, 1997).

Journals

Nutrition Today is an excellent magazine for the interested layperson. It makes a point of raising controversial issues and providing a forum for conflicting opinions. Six issues per year are published. Order from Williams and Wilkins, 351 West Camden Street, Baltimore, MD 21201-2436.

The *Journal of the American Dietetic Association,* the official publication of the ADA, contains articles of interest to dietitians and nutritionists, news of legislative action on food and nutrition, and a very useful section of abstracts of articles from many other journals of nutrition and related areas. There are 12 issues per year, available from the American Dietetic Association (see "Addresses," later).

Nutrition Reviews, a publication of the International Life Sciences Institute, does much of the work for the library researcher, compiling recent evidence on current topics and presenting extensive bibliographies. Twelve issues per year are available from Nutrition Reviews, P.O. Box 1897, Lawrence, KS 66044-8897.

Nutrition and the M.D. is a monthly newsletter that provides up-to-date, easy-to-read, practical information on nutrition for health care providers. It is available from Lippincott-Raven Publishers, 12107 Insurance Way, Hagerstown, MD 21740.

Other journals that deserve mention here are *Food Technology, Journal of Nutrition, American Journal of Clinical Nutrition, Nutrition Research,* and *Journal of Nutrition Education. FDA Consumer,* a government publication with many articles of interest to the consumer, is available from the Food and Drug Administration (see "Addresses," below). Many other journals of value are referred to throughout this book.

Addresses

Many of the organizations listed below will provide publication lists free on request. Government and international agencies and professional nutrition organizations are listed first, followed by organizations in the following areas: aging, alcohol and drug abuse, consumer organizations, fitness, food safety, health and disease, infancy and childhood, pregnancy and lactation, trade and industry organizations, weight control and eating disorders, and world hunger.

U.S. Government

- Federal Trade Commission (FTC)
  Public Reference Branch
  (202) 326-2222
  www.ftc.gov
- Food and Drug Administration (FDA)
  Office of Consumer Affairs, HFE 1
  Room 16-85
  5600 Fishers Lane
  Rockville, MD 20857
  (301) 443-1544
  www.fda.gov

- FDA Consumer Information Line
  (301) 827-4420
- FDA Office of Food Labeling,
  HFS 150
  200 C Street SW
  Washington, DC 20204
  (202) 205-4561; fax (202) 205-4564
  www.cfsan.fda.gov
- FDA Office of Plant and Dairy Foods and Beverages
  HFS 300
  200 C Street SW
  Washington, DC 20204
  (202) 205-4064; fax (202) 205-4422
- FDA Office of Special Nutritionals,
  HFS 450
  200 C Street SW
  Washington, DC 20204
  (202) 205-4168; fax (202) 205-5295
- Food and Nutrition Information Center
  National Agricultural Library,
  Room 304
  10301 Baltimore Avenue
  Beltsville, MD 20705-2351
  (301) 504-5719; fax (301) 504-6409
  www.nal.usda.gov/fnic
- Food Research Action Center (FRAC)
  1875 Connecticut Avenue NW,
  Suite 540
  Washington, DC 20009
  (202) 986-2200; fax (202) 986-2525
- Superintendent of Documents
  U.S. Government Printing Office
  Washington, DC 20402
  (202) 512-1071
  www.access.gpo.gov/su_docs
- U.S. Department of Agriculture (USDA)
  14th Street SW and Independence Avenue
  Washington, DC 20250
  (202) 720-2791
  www.usda.gov/fcs
- USDA Center for Nutrition Policy and Promotion
  1120 20th Street NW, Suite 200
  North Lobby
  Washington, DC 20036
  (202) 208-2417
  www.usda.gov/fcs/cnpp.htm
- USDA Food Safety and Inspection Service
  Food Safety Education Office,
  Room 1180-S
  Washington, DC 20250

(202) 690-0351
www.usda.gov/fsis
- U.S. Department of Education (DOE)
  Accreditation Agency Evaluation Branch
  7th and D Street SW
  ROB 3, Room 3915
  Washington, DC 20202-5244
  (202) 708-7417
- U.S. Department of Health and Human Services
  200 Independence Avenue SW
  Washington, DC 20201
  (202) 619-0257
  www.os.dhhs.gov
- U.S. Environmental Protection Agency (EPA)
  401 Main Street SW
  Washington, DC 20460
  (202) 260-2090
  www.epa.gov
- U.S. Public Health Service
  Assistant Secretary of Health
  Humphrey Building, Room 725-H
  200 Independence Avenue SW
  Washington, DC 20201
  (202) 690-7694

Canadian Government

Federal

- Bureau of Nutritional Sciences
  Food Directorate
  Health Protection Branch
  3-West
  Sir Frederick Banting Research
  Centre, 2203A
  Tunney's Pasture
  Ottawa, Ontario K1A 0L2
  www.hc-sc.gc.ca
- Canadian Food Inspection Agency
  Agriculture and Agri-Food Canada
  59 Camelot Drive
  Nepean, Ontario K1A 0Y9
  (613) 225-CFIA or (613) 225-2342
  www.agr.ca
- Nutrition & Healthy Eating Unit
  Strategies and Systems for Health
  Directorate, 1917C
  17th Floor—Jeanne Mance Bldg.
  Tunney's Pasture
  Ottawa, Ontario K1A 1B4
  www.hc-sc.gc.ca

- Nutrition Specialist
  Health Support Services
  Indian and Northern Health
  Services Directorate
  Medical Services Branch
  20th Floor—Jeanne Mance Bldg.,
  1920B
  Tunney's Pasture
  Ottawa, Ontario K1A 0L3
  www.hc-sc.gc.ca

Provincial and Territorial

- Population Health Strategies Branch
  Alberta Health
  23rd Floor, TELUS Plaza, North
  Tower
  10025 Jasper Avenue
  Edmonton, AB T5J 2N3
  www.health.gov.ab.ca

- Nutritionist
  Preventive Services Branch
  Ministry of Health
  1520 Blanshard Street
  Victoria BC V8W 3C8
  www.hlth.gov.bc.ca

- Executive Director
  Health Programs
  2nd Floor 800 Portage Avenue
  Winnipeg, MB R3G 0P4
  www.gov.mb.ca/health

- Project Manager
  Public Health Management Services
  Health and Community Services
  P.O. Box 5100
  520 King Street
  Fredericton, NB E3B 5G8
  www.gov.nb.ca/hcs-ssc

- Director, Health Promotion
  Department of Health
  Government of Newfoundland
  and Labrador
  P.O. Box 8700
  Confederation Building, West Block
  St. John's, NF A1B 4J6
  www.gov.nf.ca/health

- Consultant, Nutrition
  Health & Wellness Promotion
  Population Health
  Department of Health and Social
  Services
  Government of the Northwest
  Territories
  Centre Square Tower, 6th Floor
  P.O. Box 1320
  Yellowknife, NT X1A 2L9
  www.hlthss.gov.nt.ca

- Public Health Nutritionist
  Central Health Region
  201 Brownlow Avenue, Unit 4
  Dartmouth, NS B3B 1W2
  www.gov.ns.ca/health

- Senior Consultant, Nutrition
  Public Health Branch
  Ministry of Health, 8th Floor
  5700 Yonge St.
  North York, ON M2M 4K5
  www.gov.on.ca/health

- Coordinator, Health Information
  Resource Centre
  Department of Health and Social
  Services
  1 Rochford Street, Box 2000
  Charlottetown, PEI C1A 7N8
  www.gov/pe.ca/infopei/health

- Responsables de la santé cardio-
  vasculaire et de la nutrition
  Ministère de la Santé et des Services
  sociaux, Service de la Prévention
  en Santé
  3e étage
  1075, chemin Sainte-Foy
  Quèbec (Québec) G1S 2M1
  www.msss.gouv.qc.ca (French only)

- Health Promotion Unit
  Population Health Branch
  Saskatchewan Health
  3475 Albert Street
  Regina, SK S4S 6X6
  www.gov.sk.ca/govt/health

- Director, Nutrition Services
  Yukon Hospital Corporation
  #5 Hospital Road
  Whitehorse, YT Y1A 3H7
  www.hss.gov.yk.ca

International Agencies

- Food and Agriculture Organization
  of the United Nations (FAO)
  Liaison Office for North America
  2175 K Street, Suite 300
  Washington, DC 20437
  (202) 653-2400
  www.fao.org

- International Food Information
  Council Foundation
  1100 Connecticut Avenue NW,
  Suite 430
  Washington, DC 20036
  (202) 296-6540
  ificinfo.health.org

- UNICEF
  3 United Nations Plaza
  New York, NY 10017
  (212) 326-7000
  www.unicef.com

- World Health Organization (WHO)
  Regional Office
  525 23rd Street NW
  Washington, DC 20037
  (202) 974-3000
  www.who.org

Professional Nutrition Organizations

- American Academy of Nutritional
  Sciences
  9650 Rockville Pike
  Bethesda, MD 20814
  (303) 530-7050; fax (301) 571-1892
  www.nutrition.org

- American Dietetic Association
  (ADA)
  216 West Jackson Boulevard,
  Suite 800
  Chicago, IL 60606-6995
  (800) 877-1600; (312) 899-0040
  www.eatright.org

- ADA, The Nutrition Hotline
  (800) 366-1655

- American Society for Clinical
  Nutrition
  9650 Rockville Pike
  Bethesda, MD 20814-3998
  (301) 530-7110; fax (301) 571-1863
  www.faseb.org/ascn

- Dietitians of Canada
  480 University Avenue, Suite 604
  Toronto, Ontario M5G 1V2, Canada
  (416) 596-0857; fax (416) 596-0603
  www.dietitians.ca

- Human Nutrition Institute (INACG)
  1126 Sixteenth Street NW
  Washington, DC 20036
  (202) 659-0789
  www.ilsi.org

- National Academy of Sciences/
  National Research Council
  (NAS/NRC)
  2101 Constitution Avenue, NW
  Washington, DC 20418
  (202) 334-2000
  www.nas.edu

- National Institute of Nutrition
  265 Carling Avenue, Suite 302
  Ottawa, Ontario K1S 2E1
  (613) 235-3355; fax (613) 235-7032
  www.nin.ca

- Society for Nutrition Education
  7101 Wisconsin Avenue, Suite 901
  Bethesda, MD 20814-4805
  (301) 656-4938

Aging

- Administration on Aging
  330 Independence Avenue SW
  Washington, DC 20201
  (202) 619-0724
  www.aoa.dhhs.gov

- American Association of Retired
  Persons (AARP)
  601 E Street NW
  Washington, DC 20049
  (202) 434-2277
  www.aarp.org

- Canadian Association of
  Gerontology
  www.cagacg.ca

- National Aging Information Center
  330 Independence Avenue SW
  Washington, DC 20201
  (202) 619-7501
  www.aoa.dhhs.gov/naic

- National Institute on Aging
  Public Information Office
  31 Center Drive, MSC 2292
  Bethesda, MD 20892
  (301) 496-1752
  www.nih.gov/nia

Alcohol and Drug Abuse

- Al-Anon Family Group
  Headquarters, Inc.
  1600 Corporate Landing Parkway
  Virginia Beach, VA 23454-5617
  (800) 356-9996
  www.al-anon.alateen.org

- Alateen
  1600 Corporate Landing Parkway
  Virginia Beach, VA 23454-5617
  (800) 356-9996
  www.al-anon.alateen.org

- Alcohol & Drug Abuse Information
  Line
  Adcare Hospital
  (800) 252-6465

- Alcoholics Anonymous (AA)
  General Service Office
  475 Riverside Drive
  New York, NY 10115
  (212) 870-3400
  www.aa.org

- Narcotics Anonymous (NA)
  P.O. Box 9999
  Van Nuys, CA 91409
  (818) 773-9999; fax (818) 700-0700
  www.wsoinc.com

- National Clearinghouse for Alcohol
  and Drug Information (NCADI)
  P.O. Box 2345
  Rockville, MD 20847-2345
  (800) 729-6686
  www.health.org

- National Council on Alcoholism
  and Drug Dependence (NCADD)
  12 West 21st Street
  New York, NY 10010
  (800) NCA-CALL or (800) 622-2255
  (212) 206-6770; fax (212) 645-1690
  www.ncadd.org

- U.S. Center for Substance Abuse
  Prevention
  1010 Wayne Avenue, Suite 850
  Silver Spring, MD 20910
  (301) 459-1591 ext. 244; fax (301)
  495-2919
  www.covesoft.com/csap.html

Consumer Organizations

- Center for Science in the Public
  Interest (CSPI)
  1875 Connecticut Avenue NW,
  Suite 300
  Washington, DC 20009-5728
  (202) 332-9110; fax (202) 265-4954
  www.cspinet.org

- Choice in Dying, Inc.
  1035 30th Street NW
  Washington, DC 20007
  (202) 338-9790; fax (202) 338-0242
  www.choices.org

- Consumer Information Center
  Pueblo, CO 81009
  (888) 8 PUEBLO or (888) 878-3256
  www.pueblo.gsa.gov

- Consumers Union of US Inc.
  101 Truman Avenue
  Yonkers, NY 10703-1057
  (914) 378-2000
  www.consunion.org

- National Council Against Health
  Fraud, Inc. (NCAHF)
  P.O. Box 1276
  Loma Linda, CA 92354
  (909) 824-4690
  www.ncahf.org

Fitness

- American College of Sports
  Medicine
  P.O. Box 1440
  Indianapolis, IN 46206-1440
  (317) 637-9200
  www.acsm.org/sportsmed

- American Council on Exercise
  (ACE)
  5820 Oberlin Drive, Suite 102
  San Diego, CA 92121
  (800) 529-8227
  www.acefitness.org

- President's Council on Physical
  Fitness and Sports
  Humphrey Building, Room 738
  200 Independence Avenue SW
  Washington, DC 20201
  (202) 690-9000; fax (202) 690-5211
  www.indiana.edu/~preschal

- Shape Up America!
  6707 Democracy Boulevard,
  Suite 306
  Bethesda, MD 20817
  (301) 493-5368
  www.shapeup.org

Food Safety

- Alliance for Food & Fiber
  Food Safety Hotline
  (800) 266-0200

- Canadian Food Inspection Agency
  Agriculture and Agri-Food Canada
  59 Camelot Drive
  Nepean, Ontario K1A 0Y9
  (613) 225-CFIA or (613) 225-2342
  www.agr.ca

- FDA Center for Food Safety and
  Applied Nutrition
  200 C Street SW
  Washington, DC 20204
  (800) FDA-4010 or (800) 332-4010
  vm.cfsan.fda.gov

- National Lead Information Center
  (800) LEAD-FYI or (800) 532-3394
  (800) 424-LEAD or (800) 424-5323

- National Pesticide Telecommunications Network (NPTN)
Oregon State University
333 Weniger Hall
Corvallis, OR 97331-6502
(541) 737-6091
www.ace.orst.edu/info/nptn

- USDA Meat and Poultry Hotline
(800) 535-4555

- U.S. EPA Safe Drinking Water Hotline
(800) 426-4791

Health and Disease

- Alzheimer's Disease Education and Referral Center
P. O. Box 8250
Silver Spring, MD 20907-8250
(800) 438-4380
www.alzheimers.org

- Alzheimer's Disease Information and Referral Service
919 North Michigan Avenue, Suite 1000
Chicago, IL 60611
(800) 272-3900
www.alz.org

- Alzheimer Society of Canada
www.alzheimer.ca

- American Academy of Allergy, Asthma, and Immunology
611 East Wells Street
Milwaukee, WI 53202
(414) 272-6071; fax (414) 276-3349
www.aaaai.org

- American Cancer Society
National Home Office
1599 Clifton Road NE
Atlanta, GA 30329-4251
(800) ACS-2345 or (800) 227-2345
www.cancer.org

- American Council on Science and Health
1995 Broadway, 2nd Floor
New York, NY 10023-5860
(212) 362-7044; fax (212) 362-4919
www.acsh.org

- American Dental Association
211 East Chicago Avenue
Chicago, IL 60611
(312) 440-2800
www.ada.org

- American Diabetes Association
1660 Duke Street
Alexandria, VA 22314
(800) 232-3472 or (703) 549-1500
www.diabetes.org

- American Heart Association
Box BHG, National Center
7320 Greenville Avenue
Dallas, TX 75231
(800) 275-0448 or (214) 373-6300
www.amhrt.org

- American Institute for Cancer Research
1759 R Street NW
Washington, DC 20009
(800) 843-8114 or (202) 328-7744;
fax (202) 328-7226
www.aicr.org

- American Medical Association
515 North State Street
Chicago, IL 60610
(312) 464-5000
www.ama-assn.org

- American Public Health Association (APHA)
1015 Fifteenth Street NW, Suite 300
Washington, DC 20005
(202) 789-5600
www.apha.org

- American Red Cross
National Headquarters
8111 Gatehouse Road
Falls Church, VA 22042
(703) 206-7180
www.redcross.org

- Canadian Cancer Society
www.cancer.ca

- Canadian Diabetes Association
15 Toronto Street, Suite 800
Toronto, ON M5C 2E3
(800) BANTING or (800) 226-8464
(416) 363-3373
www.diabetes.ca

- Canadian Heart and Stroke Foundation
www.hsf.ca

- Canadian Public Health Association
400-1565 Carling Avenue
Ottawa, Ontario K1Z 8R1
(613) 725-3769; fax (613) 725-9826
www.cpha.ca

- Centers for Disease Control and Prevention (CDC)
1600 Clifton Road NE

Atlanta, GA 30333
(404) 639-3311
www.cdc.gov

- The Food Allergy Network
10400 Eaton Place, Suite 107
Fairfax, VA 22030-2208
(800) 929-4040 or (703) 691-3179
www.foodallergy.org

- Internet Health Resources
www.ihr.com

- National AIDS Hotline (CDC)
(800) 342-AIDS (English)
(800) 344-SIDA (Spanish)
(800) 2437-TTY (Deaf)
(900) 820-2437

- National Cancer Institute
Office of Cancer Communications
Building 31, Room 10824
Bethesda, MD 20892
(800) 4-CANCER or (800) 422-6237
www.nci.nih.gov

- National Diabetes Information Clearinghouse
1 Information Way
Bethesda, MD 20892-3560
(301) 654-3327
www.niddk.nih.gov

- National Digestive Disease Information Clearinghouse (NDDIC)
2 Information Way
Bethesda, MD 20892-3570
(301) 654-3810
www.niddk.nih.gov

- National Health Information Center (NHIC)
Office of Disease Prevention and Health Promotion
(800) 336-4797
nhic-nt.health.org

- National Heart, Lung, and Blood Institute
Information Center
P.O. Box 30105
Bethesda, MD 20824-0105
(301) 251-1222
www.nhlbi.nih.gov/nhlbi/nhlbi.htm

- National Institute of Allergy and Infectious Diseases
Office of Communications
Building 31, Room 7A50
31 Center Drive, MSC2520
Bethesda, MD 20892-2520
(301) 496-5717
www.niaid.nih.gov

- National Institute of Dental Research (NIDR)
  National Institutes of Health
  Bethesda, MD 20892-2190
  (301) 496-4261
  www.nidr.nih.gov

- National Institutes of Health (NIH)
  9000 Rockville Pike
  Bethesda, MD 20892
  (301) 496-2433
  www.nih.gov

- National Osteoporosis Foundation
  1150 17th Street NW, Suite 500
  Washington, DC 20036
  (202) 223-2226
  www.nof.org

- Office of Disease Prevention and Health Promotion
  odphp.osophs.dhhs.gov

- Office on Smoking and Health (OSH)
  www.americanheart.org/heart.org/
  Heart_and_stroke_A_Z_Guide/
  osh.html

Infancy and Childhood

- American Academy of Pediatrics
  141 Northwest Point Boulevard
  Elk Grove Village, IL 60007-1098
  (847) 228-5005
  www.aap.org

- Association of Birth Defect Children, Inc.
  930 Woodcock Road, Suite 225
  Orlando, FL 32803
  (407) 245-7035
  www.birthdefects.org

- Canadian Paediatric Society
  100-2204 Walkley Road
  Ottawa, ON K1G 4G8
  (613) 526-9397; fax (613) 526-3332
  www.cps.ca

- National Center for Education in Maternal & Child Health
  2000 15th Street North, Suite 701
  Arlington, VA 22201-2617
  (703) 524-7802
  www.ncemch.org

Pregnancy and Lactation

- American College of Obstetricians and Gynecologists Resource Center
  409 12th Street SW
  Washington, DC 20024-2188

  (202) 638-5577
  www.acog.org

- La Leche International, Inc.
  1400 N. Meacham Road
  Schaumburg, IL 60173
  (847) 519-7730
  www.lalecheleague.org

- March of Dimes Birth Defects Foundation
  1275 Mamaroneck Avenue
  White Plains, NY 10605
  (914) 428-7100
  www.modimes.org

Trade and Industry Organizations

- Beech-Nut Nutrition Corporation
  P.O. 618
  St. Louis, MO 63188-0618
  (800) 523-6633
  www.beechnut.com

- Borden Inc.
  180 East Broad Street
  Columbus, OH 43215
  (800) 426-7336

- Campbell Soup Company
  Consumer Response Center
  Campbell Place, Box 26B
  Camden, NJ 08103-1701
  (800) 257-8443
  www.campbellssoup.com

- Elan Pharma/Hi Chem Diagnostics
  2 Thurber Boulevard
  Smithfield, RI 02917
  (401) 233-3526
  www.hi-chem.com

- General Mills, Inc.
  Number One General Mills Boulevard
  Minneapolis, MN 55426
  (800) 328-6787
  www.generalmills.com

- Hoffmann-LaRoche, Inc.
  340 Kingsland Street
  Nutley, NJ 07110
  (973) 235-5000

- Kellogg Company
  P.O. Box 3599
  Battle Creek, MI 49016-3599
  (616) 961-2000
  www.kelloggs.com

- Kraft Foods
  Consumer Response and Information Center
  One Kraft Court

  Glenview, IL 60025
  (800) 323-0768
  www.kraftfoods.com

- Mead Johnson Nutritionals
  2400 West Lloyd Expressway
  Evansville, IN 47721
  (800) 247-7893
  www.meadjohnson.com

- Nabisco Consumer Affairs
  100 DeForest Avenue
  East Hanover, NJ 07936
  (800) NABISCO or (800) 932-7800
  www.nabisco.com

- National Dairy Council
  10255 West Higgins Road, Suite 900
  Rosemond, IL 60018-5616
  (847) 803-2000
  www.dairyinfo.com

- NutraSweet/KELCO
  P.O. Box 2986
  Chicago, IL 60654-0986
  www.equal.com

- Pillsbury Company
  Consumer Relations
  P.O. Box 550
  Minneapolis, MN 55440
  (800) 767-4466
  www.pillsbury.com

- Procter and Gamble Company
  One Procter and Gamble Plaza
  Cincinnati, OH 45202
  (513) 983-1100
  www.pg.com/info

- Ross Laboratories, Abbot Laboratory
  625 Cleveland Avenue
  Columbus, OH 43215
  (800) 227-5767
  www.abbot.com

- Sherwood Medical
  1915 Olive Street
  St. Louis, MO 63103
  (800) 428-4400

- Sunkist Growers
  Consumer Affairs
  Fresh Fruit Division
  14130 Riverside Drive
  Sherman Oaks, CA 91423
  (800) CITRUS-5 or (800) 248-7875
  www.sunkist.com

- United Fresh Fruit and Vegetable Association
  727 North Washington Street
  Alexandria, VA 22314
  (703) 836-3410

- USA Rice Federation
  4301 North Fairfax Drive, Suite 305
  Arlington, VA 22203
  Phone: (703) 351-8161
  www.usarice.com
- Weight Watchers International, Inc.
  Consumer Affairs Department/IN
  175 Crossways Park West
  Woodbury, NY 11797
  (516) 390-1400; fax (516) 390-1632.
  www.weightwatchers.com

Weight Control and Eating Disorders

- American Anorexia & Bulimia
  Association, Inc.
  165 West 46th Street #1108
  New York, NY 10036
  (212) 575-6200
  members.aol.com/amanbu
- Anorexia Nervosa and Related
  Eating Disorders (ANRED)
  P.O. Box 5102
  Eugene, OR 97405
  (541) 344-1144
  www.anred.com
- National Association of Anorexia
  Nervosa and Associated Disorders,
  Inc. (ANAD)
  P.O. Box 7

Highland Park, IL 60035
(847) 831-3438
members.aol.com/anad20/index.
html

- National Eating Disorder
  Information Centre
  200 Elizabeth Street, College Wing
  1-304
  Toronto, Ontario M5G 2C4
  (519) 253-7421; fax (519) 253-7545
- Overeaters Anonymous (OA)
  World Service Office
  6075 Zenith Court NE
  Rio Rancho, NM 87124
  (505) 891-2664; fax (505) 891-4320
  www.overeatersanonymous.org
- TOPS (Take Off Pounds Sensibly)
  4575 South Fifth Street
  P.O. Box 07360
  Milwaukee, WI 53207-0360
  (800) 932-8677 or (414) 482-4620
  www.tops.org

World Hunger

- Bread for the World
  1100 Wayne Avenue, Suite 1000
  Silver Spring, MD 20910
  (301) 608-2400
  www.bread.org

- Center on Hunger, Poverty and
  Nutrition Policy
  Tufts University School of Nutrition
  11 Curtis Avenue
  Medford, MA 02155
  (617) 627-3956
- Freedom from Hunger
  P.O. Box 2000
  1644 DaVinci Court
  Davis, CA 95617
  (530) 758-6200
  www.freefromhunger.org
- Oxfam America
  26 West Street
  Boston, MA 02111
  (617) 482-1211
  www.oxfamamerica.org
- SEEDS Magazine
  P.O. Box 6170
  Waco, TX 76706
  (254) 755-7745
  www.helwys.com/seedhome.htm
- Worldwatch Institute
  1776 Massachusetts Avenue NW,
  Suite 800
  Washington, DC 20036
  (202) 452-1999
  www.worldwatch.org

E

CONTENTS

Drug History: Nutrient Interactions

*Growth Charts and
Measures of Body Composition*

Laboratory Tests of Nutrition Status

NUTRITION ASSESSMENT: SUPPLEMENTAL INFORMATION

Chapter 13 described data from nursing assessments that help determine clients' nutrition status and nutrient needs. This appendix provides additional information to help evaluate clients' nutrition-related needs.

Drug History: Nutrient Interactions

Chapter 13 described nutrient-medication interactions and later chapters provided a series of diet-medication interactions boxes. Table E-1 shows where you can find examples of diet-medication interactions relevant to specific classes of medications.

Table E-1

Locating Examples of Diet-Medication Interactions

Medication	Page Number
Anabolic agents	579
Analgesics	398
Antacids	371
Antianxiety agents	398
Antibiotics	371, 549
Anticoagulants	507
Antidepressants	399
Antidiabetic agents	482
Antidiarrheals	399
Antiemetics, antinauseants	371
Antiflatuents	399
Anti-GERD, see *antisecretory agents*	371
Antigout agents	536
Antihypertensives	507
Anti-inflammatory agents	399
Antilipemics	507
Antineoplastic agents	579
Antisecretory agents	371
Antiviral agents	579
Appetite stimulants	549
Cardiac glycosides	507
Diuretics	507, 549
Exchange resins	
Phosphate binders	536
Potassium binders	536
Immunosuppressants	536, 549
Laxatives	399, 549
Sedatives	453
Other	
Interferon	549
Pancreatic enzyme replacements	399
Penicillamine	549
Sulfasalazine	399

Growth Charts and Measures of Body Composition

Growth charts, shown in Figures E-1 through E-6, allow health care professionals to evaluate the growth and development of children from birth to 18 years of age. The assessor follows these steps to plot a weight measurement on a percentile graph:

- Select the appropriate chart based on age and gender. (When length is measured, use the chart for birth to 36 months; when height is measured, use the chart for 2 to 18 years.)
- Locate the child's age along the horizontal axis on the bottom or top of the chart.
- Locate the child's weight in pounds or kilograms along the vertical axis on the lower left or right side of the chart.
- Mark the chart where the age and weight lines intersect. Read the percentile.

Revised growth charts will be available from the National Center for Health Statistics web site:
www.cdc.gov/nchswww

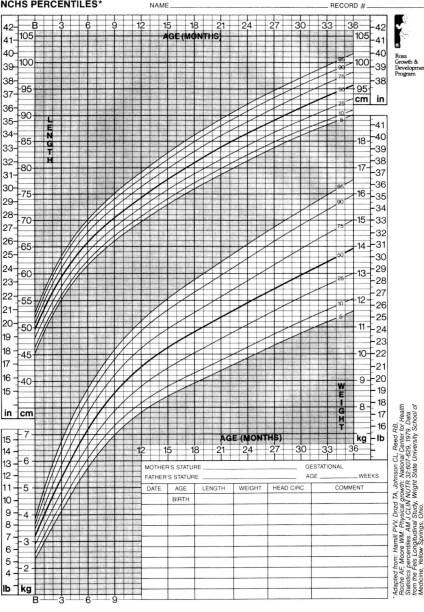

GIRLS: BIRTH TO 36 MONTHS
PHYSICAL GROWTH
NCHS PERCENTILES*

Figure E-1A

Girls: Birth to 36 Months Physical Growth National Center for Health Statistics NCHS Percentiles—Length and Weight for Age

Figure E-1B

Girls: Birth to 36 Months Physical Growth NCHS Percentiles—Head Circumference for Age and Weight for Length

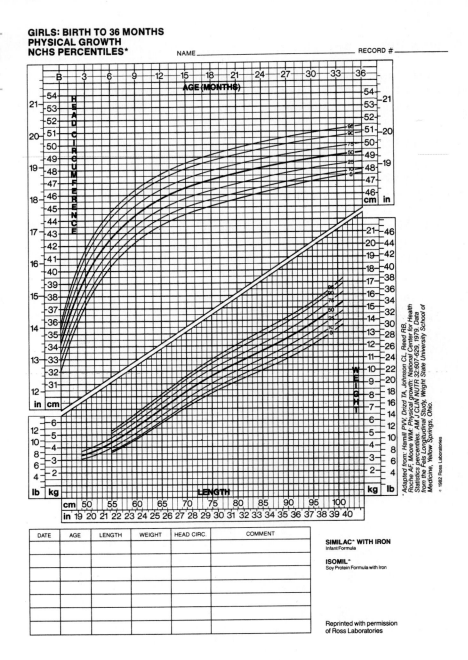

GIRLS: BIRTH TO 36 MONTHS
PHYSICAL GROWTH
NCHS PERCENTILES*

NAME_____ RECORD #_____

DATE	AGE	LENGTH	WEIGHT	HEAD CIRC.	COMMENT

SIMILAC® WITH IRON
Infant Formula

ISOMIL®
Soy Protein Formula with Iron

Reprinted with permission of Ross Laboratories

* Adapted from: Hamill PVV, Drizd TA, Johnson CL, Reed RB, Roche AF, Moore WM: Physical growth: National Center for Health Statistics percentiles. AM J CLIN NUTR 32:607–629, 1979. Data from the Fels Longitudinal Study, Wright State University School of Medicine, Yellow Springs, Ohio.

© 1982 Ross Laboratories

To plot length, height, or head circumference, the assessor follows the same procedure, using the appropriate chart. With height, weight, and head circumference measures plotted on growth percentile charts, a skilled clinician can begin to interpret the data (see p. 328).

Percentile charts divide the measures of a population into 100 equal divisions. Thus, half of the population falls above the 50th percentile, and half falls below. The use of percentile measures allows for comparisons among children of the same age and gender. For example, a six-month-old female infant whose weight is at the 75 percentile weighs more than 75 percent of the female infants her age.

**BOYS: BIRTH TO 36 MONTHS
PHYSICAL GROWTH
NCHS PERCENTILES***

NAME _____ RECORD # _____

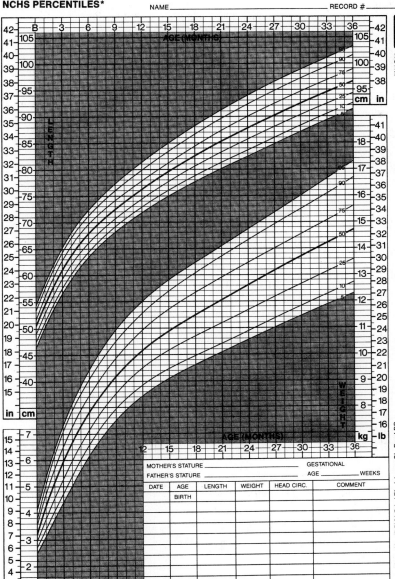

Ross
Growth &
Development
Program

*Adapted from: Hamill PVV, Drizd TA, Johnson CL, Reed RB, Roche AF, Moore WM: Physical growth: National Center for Health Statistics percentiles. AM J CLIN NUTR 32:607-629, 1979. Data from the Fels Longitudinal Study, Wright State University School of Medicine, Yellow Springs, Ohio.
© 1982 Ross Laboratories

MOTHER'S STATURE _____ GESTATIONAL
FATHER'S STATURE _____ AGE _____ WEEKS

DATE	AGE	LENGTH	WEIGHT	HEAD CIRC.	COMMENT
	BIRTH				

Figure E-2B

Boys: Birth to 36 Months Physical Growth NCHS Percentiles—Head Circumference for Age and Weight for Length

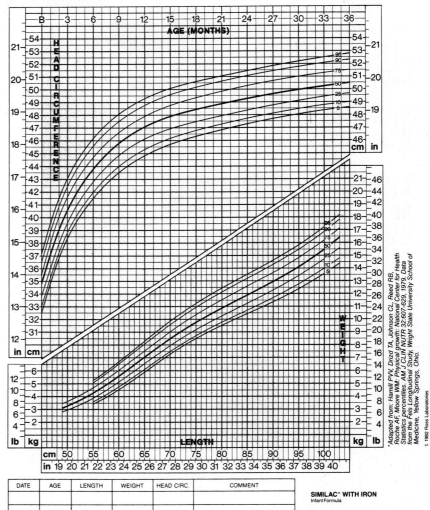

BOYS: BIRTH TO 36 MONTHS
PHYSICAL GROWTH
NCHS PERCENTILES*   NAME _____ RECORD # _____

*Adapted from: Hamill PVV, Drizd TA, Johnson CL, Reed RB, Roche AF, Moore WM: Physical growth: National Center for Health Statistics percentiles. AM J CLIN NUTR 32:607-629, 1979. Data from the Fels Longitudinal Study, Wright State University School of Medicine, Yellow Springs, Ohio.

© 1982 Ross Laboratories

DATE	AGE	LENGTH	WEIGHT	HEAD CIRC.	COMMENT

SIMILAC® WITH IRON
Infant Formula

ISOMIL®
Soy Protein Formula with Iron

Reprinted with permission
of Ross Laboratories

NUTRITION ASSESSMENT

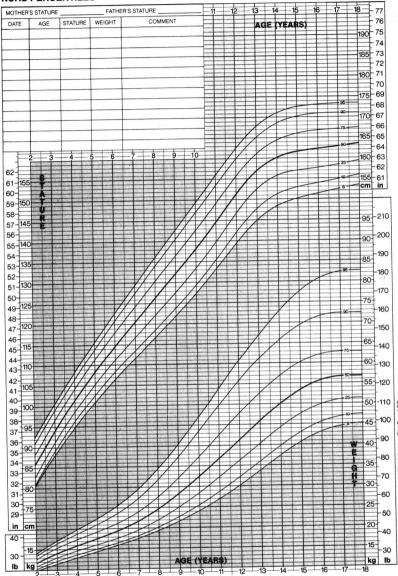

Figure E-4

Boys: 2 to 18 Years Physical Growth NCHS Percentiles—Height and Weight for Age

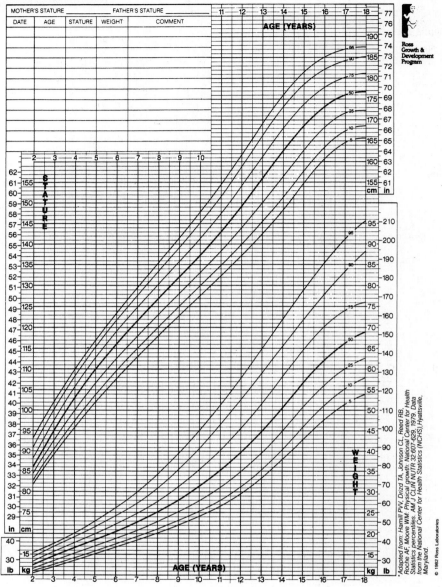

BOYS: 2 TO 18 YEARS
PHYSICAL GROWTH
NCHS PERCENTILES*

NAME _____ RECORD # _____

*Adapted from: Hamill PVV, Drizd TA, Johnson CL, Reed RB, Roche AF, Moore WM. Physical growth: National Center for Health Statistics percentiles. AM J CLIN NUTR 32:607-629, 1979. Data from the National Center for Health Statistics (NCHS), Hyattsville, Maryland

© 1982 Ross Laboratories

NUTRITION ASSESSMENT

Figure E-5

**Girls: Prepubescent Physical Growth
NCHS Percentiles—Weight for Height**

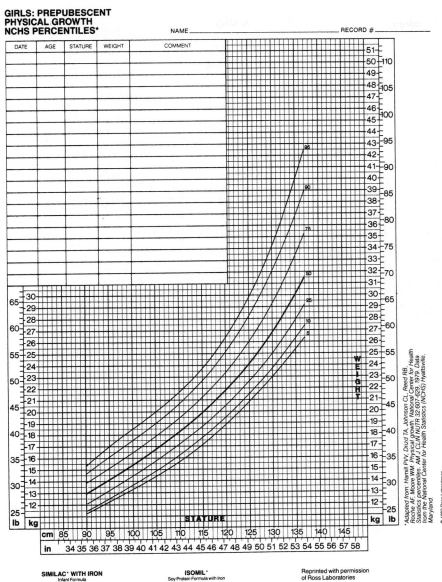

GIRLS: PREPUBESCENT
PHYSICAL GROWTH
NCHS PERCENTILES*

*Adapted from: Hamill PVV, Drizd TA, Johnson CL, Reed RB, Roche AF, Moore WM. Physical growth: National Center for Health Statistics percentiles. AM J CLIN NUTR 32:607-629, 1979. Data from the National Center for Health Statistics (NCHS) Hyattsville, Maryland.

© 1982 Ross Laboratories

SIMILAC® WITH IRON
Infant Formula

ISOMIL®
Soy Protein Formula with Iron

Reprinted with permission
of Ross Laboratories

Figure E-6

**Boys: Prepubescent Physical Growth
NCHS Percentiles—Weight for Height**

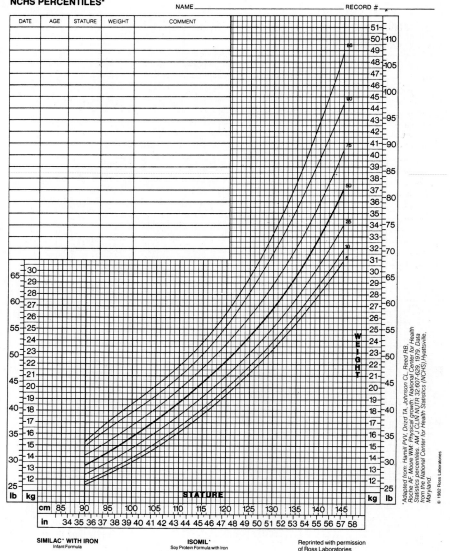

BOYS: PREPUBESCENT
PHYSICAL GROWTH
NCHS PERCENTILES*

NAME_____   RECORD #_____

SIMILAC* WITH IRON
Infant Formula

ISOMIL*
Soy Protein Formula with Iron

Reprinted with permission
of Ross Laboratories

*Adapted from: Hamill PVV, Drizd TA, Johnson CL, Reed RB, Roche AF, Moore WM. Physical growth: National Center for Health Statistics percentiles. AM J CLIN NUTR 32:607-629, 1979. Data from the National Center for Health Statistics (NCHS) Hyattsville, Maryland.
© 1982 Ross Laboratories

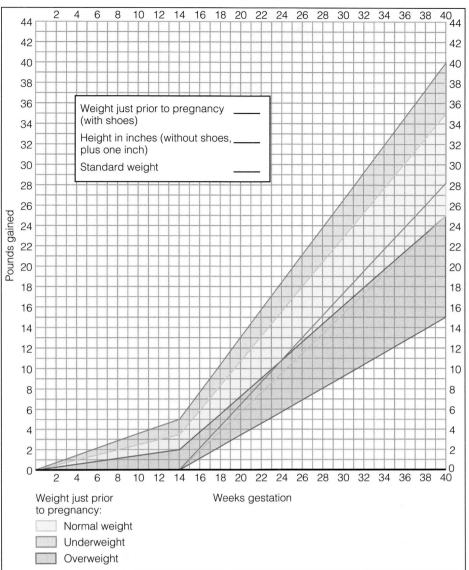

Figure E-7
Prenatal Weight-Gain Grid

Weight Gain during Pregnancy   Chapter 10 described desirable weight-gain patterns during pregnancy. Figure E-7 shows a prenatal weight-gain grid, used to plot the rate of weight gain during pregnancy. Normal-weight women should gain about 3½ pounds in the first trimester and just under 1 pound per week thereafter, achieving a total gain of 25 to 35 pounds by term; underweight women should gain about 5 pounds in the first trimester and just over 1 pound per week thereafter, achieving a total gain of 28 to 40 pounds by term; and overweight women should gain about 2 pounds in the first trimester and ⅔ pound per week thereafter, achieving a total gain of 15 to 25 pounds.

Figure E-8

How to Measure the Triceps Fatfold

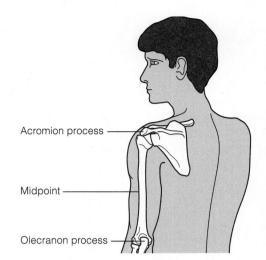
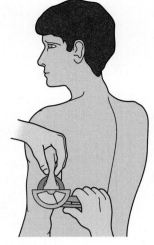

Acromion process

Midpoint

Olecranon process

A. Find the midpoint of the arm:

1. Ask the subject to bend his or her arm at the elbow and lay the hand across the stomach. (If he or she is right-handed, measure the left arm, and vice versa.)

2. Feel the shoulder to locate the acromion process. It helps to slide your fingers along the clavicle to find the acromion process. The olecranon process is the tip of the elbow.

3. Place a measuring tape from the acromion process to the tip of the elbow. Divide this measurement by 2, and mark the midpoint of the arm with a pen.

B. Measure the fatfold:

1. Ask the subject to let his or her arm hang loosely to the side.

2. Grasp the fold of skin and subcutaneous fat between the thumb and forefinger slightly above the midpoint mark. Gently pull the skin away from the underlying muscle. (This step takes a lot of practice. If you want to be sure you don't have muscle as well as fat, ask the subject to contract and relax the muscle. You should be able to feel if you are pinching muscle.)

Fatfold Measures Approximately half the fat in the body is located directly beneath the skin, and its thickness reflects total body fat. In parts of the body where fat is loosely attached, it can be pulled up between the thumb and forefinger to obtain a measure of fatfold thickness. Although such measurements can be taken from a variety of body sites, most clinicians measure the triceps fatfold (see Figure E-8), because the upper arm is easily accessible and its measurement involves little inconvenience to clients. Triceps fatfolds assist health care professionals in evaluating body fat. Triceps fatfold measures greater than 15 millimeters in men or 25 millimeters in women suggest excessive body fat and are consistent with a nursing diagnosis of altered nutrition: more than body requirements.

Bioelectrical Impedance Analysis Bioelectrical impedance analysis provides a good estimate of body composition using portable equipment that incurs minimal inconvenience to the client. To evaluate body composition using the bioelectrical impedance technique, an electrical current of very low intensity is briefly sent through the body by way of electrodes typically placed on the wrist and ankles. The measurement of electrical resistance is then used in a mathematical equation to estimate the total body water, lean body mass, and body fat. Since electrolyte-containing fluids, which readily conduct electrical current, are found primarily in lean tissues, the leaner the person, the less resistance to the current.

Laboratory Tests of Nutrition Status

As Chapter 13 pointed out, blood and urine tests provide valuable information about nutrition status. Table E-2 shows laboratory tests that help assess vitamin and mineral status.

Nutrition-Related Anemia   Some vitamin and mineral deficiencies, notably folate, vitamin B_{12}, and iron, can lead to anemia. Anemia, defined as a reduced number of red blood cells, can also result from many medical conditions unrelated to nutrition. Table E-3 on p. E-12 shows tests that help to define anemia and distinguish among its major nutrition-related causes. Tables E-4 through E-9 provide standards for hemoglobin, hematocrit, serum ferritin, serum iron, percent transferrin saturation, and folate concentrations.

Nitrogen Balance Studies   Nitrogen balance studies require that food intake be recorded and urine be collected during the same time period, often for 24 hours. Much effort is wasted if the proper techniques are not conscientiously followed or communications between health care professionals and clients are not clear. Nurses or assistants caring for clients must collect and save all urine samples and record food intake data or instruct clients on how to perform these tasks. Each urine sample is added to a collection container and refrigerated until the collection is complete. If even one urine sample is spilled or discarded, the test is invalid. Furthermore, any protein provided intravenously or by tube must be included as part of the protein intake. Nitrogen intake is calculated from the protein intake, and urine urea nitrogen (UUN) is measured from the urine sample. Nitrogen balance equals the nitrogen intake minus the nitrogen output.

In most cases, protein contains about 16 percent nitrogen. Thus, assessors multiply the protein intake by 16 percent or divide by 6.25—the mathematical equivalent—to calculate nitrogen intake. To determine nitrogen output, assessors must consider that in addition to nitrogen lost as UUN, the person also loses nitrogen through the skin, feces, and other sources of urinary nitrogen. To account for these losses, the assessor adds 4 grams of nitrogen to the UUN. Thus, nitrogen balance is calculated as follows:

$$\text{Nitrogen balance} = \frac{\text{24-hour protein intake (g)}}{6.25} - [\text{24-hour UUN (g)} + 4 \text{ g}].$$

For example, if a woman's 24-hour protein intake is 90 grams and her 24-hour UUN is 17 grams, nitrogen balance can be calculated as follows:

$$\text{Nitrogen balance} = \frac{90 \text{ g protein}}{6.25} - (17 + 4)$$

$$\text{Nitrogen balance} = 14.4 \text{ g} - 21 \text{ g} = -6.6 \text{ g}.$$

The woman in this example is in negative nitrogen balance; that is, she is losing about 6.5 more grams of nitrogen than she is consuming.

Table E-2

Biochemical Tests That May Be Used to Assess Vitamin and Mineral Status

Nutrient	Assessment Tests	Nutrient	Assessment Tests
Vitamins		**Minerals[b]**	
Vitamin A	Retinol-binding protein, serum carotene	Iron	Hemoglobin, hematocrit, serum ferritin, total iron-binding capacity (TIBC), transferrin saturation, protoporphyrin, mean corpuscular volume (MCV), serum iron
Thiamin	Erythrocyte (red blood cell) transketolase activity, urinary thiamin		
Riboflavin	Erythrocyte glutathione reductase activity, urinary riboflavin		
Vitamin B_6	Urinary xanthurenic acid excretion after tryptophan load test, urinary vitamin B_6, erythrocyte transaminase activity	Iodine	Serum protein-bound iodine, radioiodine uptake
Niacin	Urinary metabolites NMN (N-methyl nicotinamide) or 2-pyridone, or preferably both expressed as a ratio	Zinc	Plasma zinc, hair zinc
Folate	Free folate in the blood, erythrocyte folate (reflects liver stores), urinary formiminoglutamic acid (FIGLU), vitamin B_{12} status (folate assessment tests alone do not distinguish between the two deficiencies)		
Vitamin B_{12}	Serum vitamin B_{12}, erythrocyte vitamin B_{12}, urinary methyl-malonic acid synthesis or DUMP test (from the abbreviation for the chemical name of DNA's raw material, deoxyuridine monophosphate), Schilling test		
Biotin	Serum biotin, urinary biotin		
Vitamin C	Serum or plasma vitamin C,[a] leukocyte vitamin C, urinary vitamin C		
Vitamin D	Serum alkaline phosphatase		
Vitamin E	Serum tocopherol, erythrocyte hemolysis		
Vitamin K	Blood clotting time (prothrombin time)		

[a]Vitamin C shifts unpredictably between the plasma and the white blood cells known as leukocytes; thus, a plasma or serum determination may not accurately reflect the body's pool. The appropriate clinical test may be a measurement of leukocyte vitamin C. A combination of both tests may be more reliable than either one alone.
[b]Serum levels of the major minerals including sodium, potassium, phosphorus, magnesium, and chloride are routinely evaluated in hospitalized clients.
SOURCE: Adapted from A. Grant and S. DeHoog. *Nutritional Assessment and Support,* 3rd ed., 1985.

Table E-3

Laboratory Tests Useful in Evaluating Nutrition-Related Anemias

Test or Test Result	What It Reflects
General Tests for Anemia	
Hemoglobin (Hg)	Total amount of hemoglobin in the red blood cells (RBC)
Hematocrit (Hct)	Percentage of RBC in the total blood volume
Red blood cell (RBC) count	Number of RBC
Mean corpuscular volume (MCV)	RBC size; helps to determine if anemia is microcytic (iron deficiency) or macrocytic (folate or vitamin B_{12} deficiency)
Mean corpuscular hemoglobin concentration (MCHC)	Hemoglobin concentration within the average RBC; helps to determine if anemia is hypochromic (iron deficiency) or normochromic (folate or vitamin B_{12} deficiency)
Bone marrow aspiration	The manufacture of blood cells in different developmental states
Iron Deficiency	
↓ Serum ferritin	Early deficiency state with depleted iron stores
↓ Transferrin saturation	Progressing deficiency state with diminished transport iron
↑ Erythrocyte protoporphyrin	Later deficiency state with limited hemoglobin production
Folate Deficiency	
↓ Serum folate	Progressing deficiency state
↓ RBC folate	Later deficiency state
Vitamin B_{12} Deficiency	
↓ Serum vitamin B_{12}	Progressing deficiency state
Schilling test	Absorption of vitamin B_{12}

Table E-4

Standards for Hemoglobin Test Results

Age (yr)	Sex	Deficient (g/100 ml)	Acceptable (g/100 ml)
<2	M–F	<9.0	10.0 or >
2–5	M–F	<10.0	11.0 or >
6–12	M–F	<10.0	11.5 or >
13–16	M	<12.0	13.0 or >
	F	<10.0	11.5 or >
<16	M	<12.0	14.0 or >
	F	<10.0	12.0 or >
Pregnancy, 2nd trimester	F	<9.5	11.0 or >
Pregnancy, 3rd trimester	F	<9.0	10.5 or >

NOTE: To convert hemoglobin values (g/100 ml) to international standard units (g/L), multiply by 10.

Table E-5

Standards for Hematocrit Test Results

Age (yr)	Sex	Deficient (%)	Acceptable (%)
<2	M–F	<28	31 or >
2–5	M–F	<30	34 or >
6–12	M–F	<30	36 or >
13–16	M	<37	40 or >
	F	<31	36 or >
>16	M	<37	44 or >
	F	<31	38 or >
Pregnancy, 2nd trimester	F	<30	35 or >
Pregnancy, 3rd trimester	F	<30	33 or >

NOTE: To convert hematocrit values (%) to standard units, multiply by 0.01.

Table E-6

Standards for Serum Ferritin

Group	Deficient (ng/ml)
Children (3–14 years of age)	<10
Adolescents and adults	<12
Pregnant women	<10

Table E-7

Standards for Serum Iron

Age (yr)	Sex	Deficient (μg/100 ml)	Deficient (μmol/L)	Acceptable (μg/100 ml)	Acceptable (μmol/L)
<2	M–F	<30	<5.3	30 or >	5.3 or >
2–5	M–F	<40	<7.1	40 or >	7.1 or >
6–12	M–F	<50	<8.9	50 or >	8.9 or >
<12	M	<60	<10.7	60 or >	10.7 or >
	F	<40	<7.1	40 or >	7.1 or >

NOTE: To convert μg/100 ml to international units, multiply by 0.1791.

Table E-8

Standards for Percent Transferrin Saturation

Age (yr)	Sex	Deficient %	Acceptable %
<2	M–F	<15	15 or >
2–12	M–F	<20	20 or >
≥13	M	<20	20 or >
	F	<15	15 or >

Table E-9

Standards for Folate Concentrations

Measurement	Deficient (ng/ml)	Borderline (ng/ml)	Acceptable (ng/ml)
Serum folate	<3.0	3.0–6.0	>6.0
Erythrocyte folate	<140	140–160	>160

NOTE: To convert folate values (ng/ml) to international standard units (nmol/L), multiply by 2.266.

F AIDS TO CALCULATION

Contents

Conversion Factors

Percentages

Ratios

Weights and Measures

Many mathematical problems have been worked out as examples at appropriate places in the text. This appendix aims to help with the use of the metric system and with problems not fully explained elsewhere.

Conversion Factors

Conversion factors are useful mathematical tools in everyday calculations, including those encountered in the study of nutrition. A conversion factor is a fraction in which the numerator (top) and the denominator (bottom) express the same quantity in different units. One example used many times throughout this text is that 2.2 pounds (lb) and 1 kilogram (kg) are equivalent; they express the same weight. The conversion factor used to change pounds to kilograms or vice versa is:

$$\frac{2.2 \text{ lb}}{1 \text{ kg}} \text{ or } \frac{1 \text{ kg}}{2.2 \text{ lb}}.$$

Because both factors equal 1, measurements can be multiplied by the factor without changing the value of the measurement. Thus, the units can be changed.

To perform a conversion, use the factor with the unit you are seeking in the numerator (top) of the fraction. The following example illustrates the usefulness of conversion factors in the study of nutrition.

1. If you know that a 4-ounce hamburger contains 7 grams of saturated fat and you would like to determine how many grams (g) of saturated fat are contained in a 3-ounce (oz) hamburger, the conversion factor is:

$$\frac{7 \text{ g saturated fat}}{4 \text{ oz hamburger}}.$$

2. Multiply 3 ounces of hamburger by the conversion factor:

$$3 \text{ oz hamburger} \times \frac{7 \text{ g saturated fat}}{4 \text{ oz hamburger}} =$$

$$\frac{3 \times 7}{4} = \frac{21}{4} = 5 \text{ g saturated fat (rounded off to the nearest whole number).}$$

Percentages

A percentage is a comparison between a number of items (perhaps your intake of energy) and a standard number (perhaps the number of kcalories recommended for your age and sex—the energy RDA). The standard number is the number you divide by. The answer you get after the division must be multiplied by 100 to be stated as a percentage (*percent* means "per 100").

For example, if you wish to determine what percentage of the RDA your energy intake is, first find your energy RDA (see Appendix C). We'll use 2200 kcalories to demonstrate.

1. Total your energy intake for a day—for example, 1500 kcalories.
2. Divide your kcalorie intake by the RDA kcalories:

$$1500 \text{ kcal (your intake)} \div 2200 \text{ kcal (RDA)} = 0.68.$$

3. Multiply your answer by 100 to state it as a percentage:

$$0.68 \times 100 = 68 = 68\%.$$

In some problems in nutrition, the percentage may be more than 100. For example, suppose your daily intake of vitamin A is 3200 RE and your RDA (male) is 1000 RE. The following calculations show your vitamin A intake as a percentage of the RDA:

$$3200 \div 1000 = 3.2.$$

$$3.2 \times 100 = 320\% \text{ of RDA.}$$

Sometimes the comparison is between a part of a whole (for example, your kcalories from protein) and the total amount (your total kcalories). For example, suppose you wish to know what percentage of your total kcalories for the day come from protein, fat, and carbohydrate.

1. Find the total grams of protein, fat, and carbohydrate you consumed—for example, 60 grams protein, 80 grams fat, and 310 grams carbohydrate.
2. Multiply the number of grams by the number of kcalories from 1 gram of each energy nutrient (conversion factors):

$$60 \text{ g protein} \times \frac{4 \text{ kcal}}{1 \text{ g protein}} = 240 \text{ kcal.}$$

$$80 \text{ g fat} \times \frac{9 \text{ kcal}}{1 \text{ g fat}} = 720 \text{ kcal.}$$

$$310 \text{ g carbohydrate} \times \frac{4 \text{ kcal}}{1 \text{ g carbohydrate}} = 1240 \text{ kcal.}$$

$$240 + 720 + 1240 = 2200 \text{ kcal.}$$

3. Find the percentage of total kcalories from each energy nutrient:

- Protein: 240 ÷ 2200 = 0.109 × 100 = 10.9 = 11% of kcal.
- Fat: 720 ÷ 2200 = 0.327 × 100 = 32.7 = 33% of kcal.
- Carbohydrate: 1240 ÷ 2200 = 0.563 × 100 = 56.3 = 56% of kcal.
- 11% + 33% + 56% = 100% of kcal (total).

The percentages total 100 percent, but sometimes they total 99 or 101 because of rounding off. This is a reasonable error.

Ratios

A ratio is a comparison of two or three values in which one of the values is reduced to 1. A ratio compares identical units and so is expressed without units. For example, to find the potassium-to-sodium ratio of your diet, you first need to know how many milligrams of potassium and sodium you consumed, say, 3000 milligrams potassium and 2500 milligrams sodium.

1. Divide the potassium milligrams by the sodium milligrams:

   3000 mg potassium ÷ 2500 mg sodium = 1.2.

2. The potassium-to-sodium ratio is usually expressed as correct to one decimal point: 1.2.

The potassium-to-sodium ratio of your diet is 1.2:1 (read as "one point two to one" or simply "one point two"). A ratio greater than 1 means that the first value (in this case, milligrams of potassium) is greater than the second (sodium). When the second value is larger, the ratio is less than 1.

Weights and Measures

Length
1 inch (in) = 2.54 centimeters.
1 foot (ft) = 30.48 centimeters.
1 meter (m) = 39.37 inches.

To find degrees Fahrenheit (t_F) when you know degrees Celsius (t_C)*, multiply by 9/5 and then add 32:

$$9/5 \, t_C + 32 = t_F.$$

To find degrees Celsius (t_C) when you know degrees Fahrenheit (t_F), multiply by 5/9 after subtracting 32:

$$5/9 \, (t_F - 32) = t_C.$$

Volume
1 liter (L) = 1.06 quarts (qt) or 0.85 imperial quart.
1 liter = 1000 milliliters (ml).
1 milliliter = 0.03 fluid ounces.
1 gallon = 3.79 liters.
1 quart = 0.95 liter or 32 fluid ounces.
1 cup (c) = 8 fluid ounces.
1 tablespoon (tbs) = 15 milliliters.
3 teaspoons (tsp) = 1 tablespoon.
1 teaspoon = about 5 grams or 5 milliliters.
16 tablespoons = 1 cup.
4 cups = 1 quart.

Weight
1 ounce (oz) = approximately 28 grams (g).
16 ounces = 1 pound (lb).
1 pound = 454 grams.
1 kilogram (kg) = 1000 grams or 2.2 pounds.
1 gram = 1000 milligrams (mg).
1 milligram = 1000 micrograms (μg).

Energy units
1 kcalorie (kcal) = 4.2 kilojoules (kJ).
1 millijoule (mJ) = 240 kcalories.
1 kilojoule = 0.24 kcalories.
1 g carbohydrate = 4 kcal = 17 kJ.
1 g fat = 9 kcal = 37 kJ.
1 g protein = 4 kcal = 17 kJ.
1 g alcohol = 7 kcal = 29 kJ.

*Also known as *centigrade*.

Enteral Formulas

The staggering number of enteral formulas available allows health care professionals to meet a variety of their clients' medical needs, but also complicates the process of selecting an appropriate formula. The first step in narrowing the choice of formulas is to determine the client's ability to digest and absorb nutrients. Table G-1 lists examples of intact protein formulas for clients with the ability to digest and absorb nutrients, and Table G-2 provides examples of hydrolyzed formulas for clients with limited ability to digest and absorb nutrients. Each formula is listed only once, although the formula may have more than one use. A high-protein formula, for example, may also be a fiber-containing formula. Tables G-3 through G-5 list modular formulas. Although this appendix provides many examples, the list of formulas is not complete. The information reflects the manufacturer's literature and does not suggest endorsement by the authors. Be aware that formula composition changes periodically. Consult manufacturer's literature for updates. The following products are listed in this appendix:

- B. Braun-McGaw, Inc.[a]
  Hepatic-Aid® II
  Immun-Aid®

- Mead Johnson Nutritionals[b]
  Boost® with Fiber
  Boost® High Protein
  Boost® Plus
  Casec®
  Choice dm®
  Comply®
  Criticare HN®
  Deliver® 2.0
  Isocal®
  Isocal® HN
  Kindercal®
  Lipisorb®
  MCT Oil®
  Microlipid®
  Moducal®
  Magnacal® Renal
  Protain XL®
  Respalor®
  TraumaCal®
  Ultracal®

- Nestlé Clinical Nutrition[c]
  Crucial®
  Glytrol®
  Nutren®1.0
  Nutren® 1.0 with Fiber
  Nutren® 1.5
  Nutren® 2.0
  Nutren® Junior
  Nutren® Junior with Fiber
  NutriHep®
  NutriVent®
  Peptamen®
  Peptamen® Junior
  Peptamen® VHP
  ProBalance®
  Reabilan®
  Reabilan® HN
  Replete®
  Replete® with Fiber

- Novartis Nutrition Corporation[d]
  Compleat® Modified
  Compleat® Pediatric
  DiabetiSource®
  FiberSource®

FiberSource® HN
Impact®
Impact® 1.5
Impact® with Fiber
IsoSource® Standard
IsoSource® HN
IsoSource® VHN
IsoSource® 1.5
NovaSource® Renal
Resource® Diabetic
Resource® Standard
Resource® Plus
SandoSource® Peptide
Tolerex®
Vivonex® Pediatric
Vivonex® Plus
Vivonex® T.E.N.

- Ross Laboratories[e]
  Advera®
  Alitraq®
  Ensure®
  Ensure® Plus
  Ensure® Plus HN
  Ensure® with Fiber

Glucerna®
Jevity®
Jevity® Plus
Nepro®
Osmolite®
Osmolite® HN
Osmolite® HN Plus
Oxepa®
PediaSure®
PediaSure® with Fiber
Perative®
Polycose®
ProMod®
Promote®
Promote® with Fiber
Pulmocare®
Suplena®
TwoCal® HN
Vital® HN

---

[a]*Partners in Caring* (1998), B. Braun Medical Inc., Irvine, CA 92614.

[b]*Medical Nutritionals* (1999), Mead Johnson Nutritionals, Evansville, IN 47721.

[c]*Enteral Product Reference Guide* (1999), Nestlé Clinical Nutrition, Deerfield, IL 60015.

[d]*Enteral Products Pocket Guide* (1999), Novartis Nutrition, Minneapolis, MN 55440.

[e]*Ross Medical Nutritional System* (1999), Ross Laboratories, Columbus, OH 43215.

Table G-1

Intact Protein Formulas

Product	Form	Volume to Meet 100% RDI (ml)	Energy (kcal/ml)	Protein or Amino Acids (g/L)	Carbohydrate (g/L)	Fat (g/L)	Osmolality (mOsm/kg)	Notes
Lactose-Free, Low- to Moderate-Residue Formulas								
Compleat® Modified	liquid	1500	1.07	43	140	37	300	4.3 g fiber/L
Ensure®	liquid	948	1.06	38	169	25	555	
Isocal®	liquid	1890	1.06	34	135	44	270	20% fat from MCT
IsoSource Standard®	liquid	1165	1.20	43	170	41	360	50% fat from MCT
Nutren® 1.0	liquid	1500	1.00	40	127	38	315	25% fat from MCT
Osmolite®	liquid	1887	1.06	37	151	35	300	20% fat from MCT
ReSource® Standard	liquid	946	1.06	38	170	25	650	
Lactose-Free, Fiber-Containing Formulas								
Ensure® Fiber with Nutriflora FOS	liquid	1391	1.06	38	177	25	480	12 g fiber/L
FiberSource®	liquid	1165	1.20	43	170	39	490	10 g fiber/L
Impact® with Fiber	liquid	1500	1.00	56	140	28	375	10 g fiber/L, enriched with arginine, nucleic acids, and omega-3 fatty acids
Jevity®	liquid	1321	1.06	44	155	35	300	Isotonic, 14 g fiber/L
Nutren® 1.0 with Fiber	liquid	1500	1.00	40	127	38	320	Near-isotonic, 14 g fiber/L
ProBalance®	liquid	1000	1.20	54	156	41	350	10 g fiber/L
Promote® with Fiber	liquid	1000	1.00	63	138	28	370	14 g fiber/L
Replete® with Fiber	liquid	1000	1.00	63	113	34	310	Isotonic, 14 g fiber/L
Boost® with Fiber	liquid	1580	1.06	46	139	35	480	11 g fiber/L
Ultracal®	liquid	1180	1.06	44	123	45	310	Near isotonic, 14.4 g fiber/L

Table G-1

Intact Protein Formulas (continued)

Product	Form	Volume to Meet 100% RDI (ml)	Energy (kcal/ml)	Protein or Amino Acids (g/L)	Carbohydrate (g/L)	Fat (g/L)	Osmolality (mOsm/kg)	Notes
Lactose-Free, High-kCalorie Formulas								
Boost® Plus	liquid	1180	1.52	61	190	57	670	
Comply®	liquid	830	1.50	60	180	61	460	
Ensure® Plus	liquid	1420	1.50	55	200	53	690	
Ensure® Plus HN	liquid	947	1.50	63	200	50	650	
Deliver® 2	liquid	1000	2.00	75	200	102	640	
IsoSource® 1.5	liquid	933	1.50	68	170	65	650	8.0 g fiber/L
Nutren® 1.5	liquid	1000	1.50	60	169	67.5	430	50% fat from MCT
Nutren® 2.0	liquid	750	2.00	80	196	106	745	75% fat from MCT
Osmolite® HN Plus	liquid	1000	1.20	56	158	39	360	
ReSource Plus®	liquid	933	1.50	68	170	65	650	
TwoCal HN®	liquid	947	2.00	84	216	89	690	
Lactose-Free, High-Protein Formulas								
Boost® High Protein	liquid	1180	1.01	61	139	23	650	
FiberSource® HN	liquid	1165	1.20	53	160	39	490	7 g fiber/L
Isocal® HN	liquid	1180	1.06	45	124	45	270	Near-isotonic, low-residue, 20% fat from MCT
IsoSource® HN	liquid	1165	1.20	53	160	39	330	Low-residue
IsoSource® VHN	liquid	1250	1.00	62	130	29	300	Isotonic, 10g fiber/L
Jevity® Plus	liquid	1000	1.20	56	173	39	450	12.0 g fiber/L
Libisorb®	liquid	1180	1.35	57	161	57	630	85% fat from MCT
Osmolite® HN	liquid	1321	1.06	44	144	35	300	Lactose-free, low-residue, 20% fat from MCT
Promote®	liquid	1000	1.00	63	130	26	340	Fat primarily from polyunsaturated and monounsaturated sources, 20% fat from MCT
Special-Use Formulas: Pediatric (1 to 10 years)								
Compleat® Pediatric	liquid	900	1.00	38	126	39	N/A	Blenderized formula, 4.4 g fiber/L
Kindercal®	liquid	950	1.06	34	135	44	310	
Nutren Junior®	liquid	1000	1.00	30	128	42	350	
Nutren Junior® with Fiber	liquid	1000	1.00	30	128	42	350	6.0 g fiber/L
PediaSure®	liquid	1300	1.00	30	110	50	335	
PediaSure® with Fiber	liquid	1300	1.00	30	114	50	345	5.0 g fiber/L

Table G-1

Intact Protein Formulas (continued)

Product	Form	Volume to Meet 100% RDI (ml)	Energy (kcal/ml)	Protein or Amino Acids (g/L)	Carbohydrate (g/L)	Fat (g/L)	Osmolality (mOsm/kg)	Notes
Special-Use Formulas: Glucose Intolerance								
Choice dm®	liquid	1000	1.06	45	106	51	440	14 g fiber/L, 25% kcalories from monounsaturated fat
DiabetiSource®	liquid	1500	1.00	50	90	49	360	4.3 g fiber/L
Glucerna®	liquid	1422	1.00	42	96	54	355	14 g fiber/L, high in monounsaturated fat
Glytrol®	liquid	1400	1.00	45	100	48	380	15 g fiber/L
ReSource® Diabetic	liquid	1890	1.06	63	99	47	450	12 g fiber/L
Special-Use Formulas: Immune System Support								
Immun-Aid®	powder	2000	1.00	80	120	22	460	Enriched with arginine, glutamine, branched-chain amino acids, nucleic acids, omega-3 fatty acids, vitamins A, C, E, and B_6 and trace elements
Impact®	liquid	1500	1.00	56	130	28	375	Enriched with arginine, nucleic acids, and omega-3 fatty acids
Impact® 1.5	liquid	1250	1.50	84	140	69	550	Same as above
Special-Use Formulas: Renal Insufficiency								
Magnacal® Renal	liquid	1000	2.00	75	200	101	570	High in monounsaturated fat, 20% fat from MCT; intended for use once hemodialysis has been instituted
Nepro®	liquid	947	2.00	70	222	96	665	Lactose-free, high-calcium, low-phosphorus; intended for use once dialysis has been instituted
NovaSource® Renal	liquid	1000	2.00	74	200	100	700	Lactose-free, low in electrolytes; intended for use once dialysis has been instituted
Suplena®	liquid	947	2.00	30	255	96	600	Lactose-free, low in electrolytes; intended for use before dialysis is instituted

Table G-1

Intact Protein Formulas (continued)

Product	Form	Volume to Meet 100% RDI (ml)	Energy (kcal/ml)	Protein or Amino Acids (g/L)	Carbohydrate (g/L)	Fat (g/L)	Osmolality (mOsm/kg)	Notes
Special-Use Formulas: Respiratory Insufficiency								
NovaSource® Renal	liquid	1000	2.00	74	200	100	700	Enriched with arginine
NutriVent®	liquid	1000	1.50	68	100	94	330-450	Lactose-free, 55% kcal from fat, 40% fat from MCT
Oxepa	liquid	947	1.5	63	105.5	94	493	Lactose-free, 55% kcal from fat, enriched with antioxidant nutrients
Pulmocare®	liquid	947	1.50	63	106	93	475	Lactose-free, 55% kcal from fat, 20% fat from MCT, enriched with antioxidant nutrients
Respalor®	liquid	1420	1.52	76	148	71	580	Lactose-free; 41% kcal from fat; 30% fat from MCT; enriched with vitamins C and E, B-complex vitamins, zinc and trace elements
Special-Use Formulas: Wound Healing								
Protain XL®	liquid	1250	1.00	57	129	30	340	9 g fiber/L, 20% fat from MCT, enriched with vitamins A and C and zinc
Replete®	liquid	1000	1.00	63	113	34	300	Enriched with vitamins A and C and zinc
TraumaCal®	liquid	2000	1.50	82	142	68	560	Enriched with vitamins C, B-complex, and E and copper and zinc

Table G-2

Hydrolyzed Protein Formulas

Product	Form	Volume to Meet 100% RDI (ml)	Energy (kcal/ml)	Protein or Amino Acids (g/L)	Carbohydrate (g/L)	Fat (g/L)	Osmolality (mOsm/kg)	Notes
Special-Use Hydrolyzed Formulas: Heptatic Insufficiency								
Hepatic Aid® II	powder	N/A	1.20	44	169	36	560	Free amino acids, high in branched-chain amino acids, low in aromatic amino acids, no added vitamins or electrolytes.
NutriHep®	liquid	1000	1.50	40	290	21	690	Free amino acids, high in branched-chain amino acids, low in aromatic amino acids, contains vitamins and minimal electrolytes
Special-Use Hydrolyzed Formulas: HIV Infection or AIDS								
Advera®	liquid	1184	1.28	60	216	23	N/A	78% hydrolyzed and 22% intact protein, low-fat, fiber added, enriched with vitamins E, C, B_6, B_{12}, and folate
Special-Use Hydrolyzed Formulas: Immune System Support								
Alitraq®	powder	1500	1.00	53	165	16	575	47% free amino acids, 42% small peptides, enriched with glutamine and arginine
Crucial®	liquid	1000	1.50	94	135	68	490	Enriched with arginine and glutamine
Perative®	liquid	1155	1.30	67	177	37	385	Enriched with arginine and b-carotene
Vivonex® Plus	powder	1800	1.00	45	190	7	650	100% free amino acids, enriched with and E and copper and zinc

Table G-2

Hydrolyzed Protein Formulas (continued)

Product	Form	Volume to Meet 100% RDI (ml)	Energy (kcal/ml)	Protein or Amino Acids (g/L)	Carbohydrate (g/L)	Fat (g/L)	Osmolality (mOsm/kg)	Notes
Special-Use Hydrolyzed Formulas: Malabsorption								
Criticare HN®	liquid	1890	1.06	38	220	5	650	50% free amino acids, 50% small peptides
Peptamen®	liquid	1500	1.00	40	127	39	270	Near-isotonic
Peptamen VHP®	liquid	1500	1.00	63	105	39	300	70% fat from MCT
Reabilan®	liquid	2000	1.00	32	132	41	350	50% fat from MCT
Reabilan® HN	liquid	1500	1.33	58	158	54	490	50% fat from MCT
SandoSource® Peptide	liquid	1750	1.00	50	160	17	490	60% free amino acids and small peptides
Tolerex®	powder	1800	1.00	21	230	1.5	550	100% free amino acids
Vital® HN	powder	1500	1.00	42	185	11	500	87% hydrolyzed proteins, 13% essential amino acids
Vivonex® T.E.N.	powder	2000	1.00	38	210	3.0	630	100% free amino acids, enriched with branched-chain amino acids
Special-Use Hydrolyzed Formulas: Pediatric (1 to 10 years)								
Peptamin Junior®	liquid	1000	1.0	30	127.5	42	350	60% fat from MCT
Vivonex® Pediatric	powder	1000* 1170**	0.8	24	130	24	360	100% free amino acids

*Children 1 to 6 years old.
**Children 7 to 10 years old.

Product	Form	Major Protein Source	Energy (kcal/g)	Protein (g/100 g)
Casec®	powder	Calcium caseinate	3.7	88
Pro Mod®	powder	Whey protein	4.2	75.8

Table G-3
Protein Modules

Product	Form	Major Carbohydrate Source	Energy (kcal/ml or g)
Moducal®	powder	Hydrolyzed cornstarch	3.8 kcal/g
Polycose Liquid®	liquid	Hydrolyzed cornstarch	2.0 kcal/ml
Polycose Powder®	powder	Hydrolyzed cornstarch	3.8 kcal/g

Table G-4
Carbohydrate Modules

Product	Form	Major Fat Source	Energy (kcal/ml)	Fat (g/100 ml)
MCT Oil®	liquid	Derived from coconut oil	7.7	87
Microlipid®	liquid	Safflower oil	4.5	50

Table G-5
Fat Modules

Answers to Self Check Questions

Chapter 1

1. d	5. c	8. c
2. d	6. b	9. c
3. c	7. b	10. b
4. d		

Chapter 2

1. b	5. a	8. a
2. b	6. b	9. d
3. c	7. a	10. c
4. a		

Chapter 3

1. b	5. d	8. b
2. d	6. a	9. a
3. a	7. a	10. c
4. a		

Chapter 4

1. c	5. a	8. b
2. b	6. d	9. c
3. d	7. c	10. d
4. d		

Chapter 5

1. b	5. c	8. c
2. a	6. a	9. d
3. d	7. d	10. d
4. a		

Chapter 6

1. d	5. c	8. d
2. d	6. b	9. d
3. b	7. c	10. b
4. b		

Chapter 7

1. b	5. c	8. d
2. c	6. c	9. c
3. b	7. d	10. b
4. a		

Chapter 8

1. d	5. d	8. c
2. d	6. d	9. c
3. c	7. b	10. a
4. a		

Chapter 9

1. b	5. c	8. d
2. a	6. d	9. d
3. a	7. a	10. b
4. d		

Chapter 10

1. a	5. d	8. d
2. d	6. b	9. d
3. c	7. c	10. c
4. b		

Chapter 11

1. c	5. b	8. c
2. b	6. c	9. b
3. b	7. d	10. c
4. a		

Chapter 12

1. d	5. d	8. b
2. c	6. d	9. a
3. d	7. b	10. b
4. d		

Chapter 13

1. b	5. b	8. c
2. c	6. a	9. b
3. c	7. d	10. c
4. a		

Chapter 14

1. b	5. c	8. b
2. b	6. c	9. a
3. b	7. d	10. c
4. d		

Chapter 15

1. b	5. a	8. c
2. b	6. b	9. b
3. a	7. a	10. c
4. d		

Chapter 16

1. d	5. d	8. c
2. c	6. d	9. d
3. d	7. d	10. d
4. d		

Chapter 17

1. b	5. c	8. c
2. b	6. a	9. c
3. b	7. b	10. d
4. b		

Chapter 18

1. d	5. a	8. a
2. d	6. c	9. d
3. b	7. b	10. c
4. b		

Chapter 19

1. b	5. a	8. a
2. c	6. a	9. a
3. a	7. d	10. c
4. d		

Chapter 20

1. a	5. b	8. c
2. d	6. d	9. a
3. c	7. b	10. a
4. d		

Chapter 21

1. c	5. c	8. b
2. d	6. d	9. b
3. a	7. d	10. a
4. b		

Chapter 22

1. a	5. b	8. d
2. d	6. c	9. b
3. a	7. c	10. b
4. d		

Chapter 23

1. d	5. a	8. d
2. a	6. b	9. a
3. c	7. c	10. c
4. b		

Chapter 24

1. c	5. b	8. d
2. a	6. d	9. c
3. d	7. b	10. a
4. d		

GLOSSARY

abscess: an accumulation of pus that contains live microorganisms and immune system cells.

Acceptable Daily Intake (ADI): the amount of a sweetener that individuals can safely consume each day over the course of a lifetime without adverse effect.

acesulfame (AY-sul-fame) potassium: a 0-kcalorie artificial sweetener that tastes 200 times as sweet as sucrose; also known as acesulfame-K because K is the chemical symbol for potassium.

acetone breath: a distinctive fruity odor that can be detected on the breath of a person who is experiencing ketosis.

acetyl CoA (ASS-uh-teel or **uh-SEET-ul co-AY):** a 2-carbon compound (acetate, or acetic acid), to which a molecule of CoA is attached.

acid-base balance: the balance maintained between acid and base concentrations in the blood and body fluids.

acidosis (assi-DOE-sis): too much acid in the blood and body fluids.

acids: compounds that release hydrogen ions in a solution.

acquired immune deficiency syndrome (AIDS): the end stages of HIV infection, diagnosed when a person develops certain infections or severe wasting (AIDS-defining illnesses).

activities of daily living: a description of the client's lifestyle and ability to manage self-care, including patterns of food intake, urinary and bowel elimination, activity and exercise, and rest and sleep.

acupuncture (AK-you-PUNK-cher): a technique that involves piercing the skin with long thin needles at specific anatomical points to relieve pain or illness.

acute malnutrition: PEM that develops rapidly due to a sudden and dramatic demand for nutrients.

Adequate Intake (AI): the average amount of a nutrient that appears sufficient to maintain a specified criterion; a value used as a guide for nutrient intake when an RDA cannot be determined.

ADH (antidiuretic hormone): a hormone released by the pituitary gland in response to high salt concentrations in the blood. The kidneys respond by reabsorbing water.

adipose tissue: the body's fat, which consists of masses of fat-storing cells called adipose cells.

adolescence: the period of growth from the beginning of puberty until full maturity. Timing of adolescence varies from person to person.

advance directive: the means by which competent adults record their preferences for future medical interventions. A living will and durable power of attorney are types of advance directives.

aldosterone (al-DOS-ter-own): a hormone secreted by the adrenal glands that stimulates the reabsorption of sodium by the kidneys. Aldosterone also regulates chloride and potassium concentrations.

alitame (AL-ih-tame): a low-kcalorie artificial sweetener composed of two amino acids (alanine and aspartic acid) that tastes 2000 times as sweet as sucrose; FDA approval pending.

alkalosis (alka-LOE-sis): too much base in the blood and body fluids.

alpha-lactalbumin (lackt-AL-byoo-min): the chief protein in human breast milk, as **casein (CAY-seen)** is the chief protein in cow's milk.

alternative therapies: approaches to medical diagnosis and treatment that are not fully accepted by the established medical community. As such, they are not widely taught at U.S. medical schools or practiced in U.S. hospitals; also called *adjunctive, unconventional,* or *unorthodox therapies.*

alveoli (al-VEE-oh-lie): air sacs in the lungs. One sac is an *alveolus.*

Alzheimer's (ALTZ-high-merz) disease: a degenerative disease of the brain involving progressive memory loss and major structural changes in the brain's nerve cells.

amino (a-MEEN-oh) acids: building blocks of proteins. Each has a backbone, consisting of an amino group and an acid group attached to a central carbon, and also carries a distinctive side group.

amniotic (am-nee-OTT-ic) sac: the "bag of waters" in the uterus, in which the fetus floats.

amylase (AM-uh-lace): an enzyme that splits amylose (a form of starch).

anabolic steroids: drugs related to the male sex hormone, testosterone, that stimulate the development of lean body mass.

anabolism (an-ABB-o-lism): reactions in which small molecules are put together to build larger ones. Anabolic reactions consume energy.

anaphylactic (an-AFF-ill-LAC-tic) shock: a life-threatening whole-body allergic reaction to an offending substance.

aneurysm (AN-you-ris-um): a ballooning out of a portion of a blood vessel (usually an artery) due to weakness of the vessel's wall.

angina: a painful feeling of tightness or pressure in the area in and around the heart and often radiating to the back, neck, and arms; caused by a lack of oxygen to an area of heart muscle.

anorexia nervosa: an eating disorder characterized by a refusal to maintain a minimally normal body weight, self-starvation to the extreme, and a disturbed perception of body weight and shape; seen (usually) in adolescent girls and young women.

antibodies: large proteins of the blood and body fluids, produced in response to invasion of the body by unfamiliar molecules (mostly proteins).

antioxidant: a compound that protects other compounds from oxygen by itself reacting with oxygen.

anus (AY-nus): the terminal sphincter muscle of the GI tract.

appendix: a narrow blind sac extending from the beginning of the colon. The appendix stores lymphocytes.

appetite: the psychological desire to eat; a learned motivation that is experienced as a pleasant sensation that accompanies the sight, smell, or thought of appealing foods.

aroma therapy: a technique that uses oil extracts from plants and flowers (usually applied by massage or baths) to enhance physical, psychological, and spiritual health.

artery: a vessel that carries blood away from the heart.

arthritis: inflammation of a joint, usually accompanied by pain, swelling, and structural changes.

artificial feeding: parenteral and enteral nutrition; feeding by a route other than the normal ingestion of food.

artificial sweeteners: noncarbohydrate, 0- or low-kcalorie synthetic sweetening agents; sometimes called **nonnutritive sweeteners.**

ascites (ah-SIGH-teez): a type of edema characterized by the accumulation of fluid in the abdominal cavity.

ascorbic acid: one of the two active forms of vitamin C. Many people refer to vitamin C by this name.

aspartame (ah-SPAR-tame) or ASS-par-tame): a low-kcalorie artificial sweetener composed of two amino acids (phenylalanine and aspartic acid) that tastes 200 times as sweet as sucrose.

aspiration: the drawing in of food, gastric secretions, or liquid into the lungs.

atherosclerosis: the form of cardiovascular disease characterized by the buildup of plaques along the inner walls of the arteries, which narrow the lumen of the artery and restrict blood flow to the tissue supplied by the artery.

atrophic gastritis: a condition characterized by chronic inflammation of the stomach accompanied by a diminished size and functioning of the mucosa and glands.

ayurveda (EYE-your-VAY-dah): a traditional Hindu system of improving health by using herbs, diet, meditation, massage, and yoga to stimulate the body to make its own natural drugs.

Barrett's esophagus: changes in the cells of the esophagus associated with chronic gastroesophageal reflux that raise the risk of cancer of the esophagus.

basal metabolism: the energy needed to maintain life when a person is at complete rest after a 12-hour fast.

bases: compounds that accept hydrogen ions in a solution.

behavior modification: the changing of behavior by the manipulation of *antecedents* (cues or environmental factors that trigger behavior), the behavior itself, and *consequences* (the penalties or rewards attached to behavior).

beriberi: the thiamin-deficiency disease; characterized by loss of sensation in the hands and feet, muscular weakness, advancing paralysis, and abnormal heart action.

beta-carotene: a vitamin A precursor made by plants and stored in human fat tissue; an orange pigment.

bicarbonate: an alkaline secretion of the pancreas; part of the pancreatic juice. (Bicarbonate also occurs widely in all cell fluids.)

bifidus (BIFF-id-us, by-FEED-us) factors: factors in colostrum and breast milk that favor the growth of the "friendly" bacterium *Lactobacillus* (lack-toe-ba-SILL-us) *bifidus* in the infant's intestinal tract. This bacterium prevents other, less desirable intestinal inhabitants from flourishing.

bile: an emulsifier that prepares fats and oils for digestion; made by the liver from cholesterol, stored in the gallbladder, and released into the small intestine when needed.

biliary cirrhosis: severe liver damage from prolonged inflammation or obstruction of the bile ducts, which can lead to chronic pancreatitis and fat malabsorption.

bilirubin: a pigment in the bile whose blood concentration may rise as a result of some medical conditions, especially those that affect the liver.

binge eating disorder: an eating disorder similar to bulimia nervosa, excluding purging or other compensatory behaviors.

bioavailability: the rate and extent to which a nutrient is absorbed and used.

bioelectromagnetic medical applications: the use of electrical energy, magnetic energy, or both

to stimulate bone repair, wound healing, and tissue regeneration.

biofeedback: the use of special devices to convey information about heart rate, blood pressure, skin temperature, muscle relaxation, and the like to enable a person to learn how to consciously control these medically important functions.

biofield therapeutics: a manual healing method that directs a healing force from an outside source (commonly God or another supernatural being) through the practitioner and into the client's body; commonly known as "laying on of hands."

blind loop syndrome: the problems of fat malabsorption and vitamin B_{12} and folate deficiencies that result from the overgrowth of bacteria in a bypassed segment of the intestine.

body mass index (BMI): an index of a person's weight in relation to height; determined by dividing the weight (in kilograms) by the square of the height (in meters).

bolus (BOH-lus): the portion of food swallowed at one time.

bolus feeding: the delivery of about 250 to 300 ml of formula over 10 minutes or less.

bone marrow transplant: the replacement of diseased bone marrow in a recipient with healthy bone marrow from a donor; sometimes used as a treatment for breast cancer, leukemia, lymphomas, and certain blood disorders.

botulism: an often-fatal food poisoning caused by botulin toxin, a toxin produced by bacteria that grow without oxygen.

branched-chain amino acids: the amino acids leucine, isoleucine, and valine, which are present in large amounts in skeletal muscle tissue; falsely promoted as fuel for exercising muscles.

bronchitis (bron-KYE-tis): inflammation of the lungs' air passages.

brown sugar: refined white sugar crystals to which manufacturers have added molasses syrup with natural flavor and color; 91 to 96 percent pure sucrose.

buffers: compounds that can reversibly combine with hydrogen ions to help keep a solution's acidity or alkalinity constant.

bulimia (byoo-LEEM-ee-uh) nervosa: recurring episodes of binge eating combined with a morbid fear of becoming fat, usually followed by self-induced vomiting or purging.

caffeine: a natural stimulant found in many common foods and beverages, including coffee, tea, and chocolate.

calcium rigor: hardness or stiffness of the muscles caused by high blood calcium.

calcium tetany: intermittent spasms of the extremities due to nervous and muscular excitability caused by low blood calcium.

cancer cachexia (ka-KEKS-ee-ah) syndrome: a syndrome that frequently accompanies many types of cancer and is characterized by anorexia, loss of lean body mass, depletion of serum proteins, and debilitation.

cancers: diseases that result from the unchecked growth of cells.

capillary (CAP-ill-ary): a small vessel that branches from an artery. Capillaries connect arteries to veins. Oxygen, nutrients, and waste materials are exchanged across capillary walls.

capitation: prepayment of a set fee per client in exchange for medical services.

carbohydrates: energy nutrients composed of monosaccharides.

carcinogenic (CAR-sin-oh-JEN-ick): producing cancer.

cardiac cirrhosis: severe liver damage associated with the later stages of congestive heart failure.

cardiac sphincter (CARD-ee-ack SFINK-ter): the sphincter muscle at the junction between the esophagus and the stomach.

cardiovascular diseases (CVD): diseases of the heart and blood vessels.

cariogenic (KARE-ee-oh-JEN-ik): conducive to dental decay.

carnitine (CAR-ne-teen): a nonessential nutrient made in the body from lysine that helps transport fatty acids across the mitochondrial membrane. Carnitine supposedly "burns" fat and spares glycogen during endurance events, but in reality it does neither.

carrier: an individual who possesses one dominant and one recessive gene for a genetic trait, such as an inborn error of metabolism. When a trait is recessive, a carrier may show no signs of the trait but can pass it on.

cartilage therapy: the use of cleaned and powdered connective tissue, such as collagen, to improve health.

catabolism (ca-TAB-o-lism): reactions in which large molecules are broken down to smaller ones. Catabolic reactions usually release energy.

cataracts: thickenings of the eye lenses that impair vision and can lead to blindness.

cathartic (ca-THART-ic): a strong laxative.

CD4+ lymphocyte: a type of white blood cell that has a specific protein on its surface and is a necessary component of the immune system.

celiac (SEE-lee-ack) disease: a sensitivity to a part of the protein gluten that causes flattening of the intestinal villi and malabsorption; also called **gluten-sensitive enteropathy (EN-ter-OP-ah-thee)** or **celiac sprue.**

central obesity: excess fat on the abdomen and around the trunk of the body.

central total parenteral nutrition (central TPN): the provision of an IV solution that meets nutrient needs delivered through a central vein.

central veins: the large-diameter veins located close to the heart.

cerebral cortex: the outer surface of the cerebrum.

cerebral thrombosis: a blood clot that blocks the flow of blood through an artery that feeds the brain.

cesarean section: surgical childbirth, in which the infant is delivered through an incision in the woman's abdomen.

chelation therapy: the use of ethylene diamine tetraacetic acid (EDTA) to bind with metallic ions, thus healing the body by removing toxic metals.

chemotherapy: the use of drugs to arrest or destroy cancer cells. Drugs used for chemotherapy are called **chemotherapeutic** or **antineoplastic agents.**

chiropractic (KYE-roe-PRAK-tik): a manual healing method of manipulating vertebrae to relieve musculoskeletal pain suspected of causing problems with internal organs.

choline: a nonessential nutrient that can be made in the body from an amino acid.

chromium (CROW-mee-um) picolinate: a trace mineral supplement; falsely promoted as building muscle, enhancing energy, and burning fat. **Picolinate (pick-oh-LYN-ate)** is a derivative of the amino acid tryptophan that seems to enhance chromium absorption.

chronic diseases: degenerative diseases characterized by deterioration of the body organs; also called chronic, **noncommunicable diseases (NCD).** Examples include heart disease, cancer, and diabetes.

chronic malnutrition: PEM that develops from insufficient protein and energy intakes over long periods of time.

chronic obstructive pulmonary disease (COPD): one of several disorders, including emphysema and bronchitis, that interfere with respiration.

chronic or **congestive heart failure (CHF):** a syndrome in which the heart can no longer adequately pump blood through the circulatory system.

chronological age: a person's age in years from his or her date of birth.

chylomicrons (kye-lo-MY-crons): the lipoproteins that transport lipids from the intestinal cells into the body. The cells of the body remove the lipids they need from the chylomicrons, leaving chylomicron remnants to be picked up by the liver cells.

chyme (KIME): the semiliquid mass of partly digested food expelled by the stomach into the duodenum (the top portion of the small intestine).

cirrhosis (sih-ROW-sis): an advanced form of liver disease in which scar tissue replaces liver cells that have permanently lost their function.

clear liquids: foods that are liquid and transparent at body temperature.

clinical pathways, critical pathways, or **care maps:** charts or tables that outline a plan of care for a specific diagnosis, treatment, or procedure, with a goal of providing the best outcome at the lowest cost. The plan, developed by the health care team after a careful study of each facility's unique client population, is regularly reassessed and improved.

clinically severe obesity: a BMI of 40 or greater or 100 lb or more overweight for an average adult. A less preferred term used to describe the same condition is *morbid obesity.*

closed formula system: a system where formulas come prepackaged and ready to be attached to a feeding tube for administration.

CoA (coh-AY): a nickname for a small molecule (coenzyme A) that participates in metabolism.

coenzyme (co-EN-zime): a small molecule that works with an enzyme to promote the enzyme's activity. Many coenzymes have B vitamins as part of their structure.

cofactor: a mineral element that, like a coenzyme, works with an enzyme to facilitate a chemical reaction.

cognitive therapy: psychological therapy aimed at changing undesirable behaviors by changing underlying thought processes contributing to these behaviors. In anorexia nervosa, a goal is to replace false beliefs about body weight, eating, and self-worth with health-promoting beliefs.

collagen: the characteristic protein of connective tissue.

collaterals: small branches of a blood vessel that develop to divert blood flow away from the obstructed liver; also called **shunts.**

colon or **large intestine:** the last portion of the intestine, which absorbs water. Its main segments are the ascending colon, the transverse colon, the descending colon, and the sigmoid colon.

colostomy (ko-LOSS-toe-me): surgery that creates an opening from a portion of the colon through the abdominal wall and out through the skin.

colostrum (co-LAHS-trum): a milklike secretion from the breast that is rich in protective factors. It is present during the first day or so after delivery, before milk appears.

comatose: in a state of deep unconsciousness from which the person cannot be aroused.

competent: having sufficient mental ability to understand a treatment, weigh its risks and benefits, and comprehend the consequences of refusing or accepting the treatment.

complementary or **integrative therapy:** therapy that incorporates both traditional medical and alternative treatments.

complementary proteins: two or more proteins whose amino acid assortments complement each other in such a way that the essential amino acids missing from one are supplied by the other.

complete formulas: formulas designed to supply all needed nutrients when provided in sufficient volume.

complete protein: a protein containing all the amino acids essential in human nutrition in amounts adequate for human use.

complex carbohydrates: long chains of sugars arranged as starch or fiber; also called **polysaccharides**.

concentrated fruit juice sweetener: a concentrated sugar syrup made from dehydrated, deflavored fruit juice, commonly grape juice; used to sweeten products that can then claim to be "all fruit."

conditionally essential amino acid: an amino acid that is normally nonessential but must be supplied by the diet in special circumstances when the need for it becomes greater than the body's ability to produce it.

confectioners' sugar: finely powdered sucrose; 99.9 percent pure.

constipation: the condition of having infrequent or difficult bowel movements.

continuous feeding: slow delivery of formula in constant amounts over an 8- to 24-hour period.

continuous renal replacement therapy (CRRT): a method for slowly and continuously removing fluids and wastes from the blood during acute renal failure that relies on filtration, filtration plus replacement fluids, dialysis, or a combination of methods.

corn sweeteners: corn syrup and sugars derived from corn.

corn syrup: a syrup containing mostly glucose, produced by the action of enzymes on cornstarch.

cornea (KOR-nee-uh): the hard, transparent membrane covering the outside of the eye.

coronary heart disease (CHD): heart damage that results from an inadequate blood supply to the heart; also called **coronary artery disease (CAD).**

coronary thrombosis: a blood clot that blocks the flow of blood through an artery that feeds the heart.

counterregulatory hormones: hormones such as glucagon, cortisol, and catecholamines that oppose insulin's actions and promote catabolism.

creatine (KREE-ah-tine): a nitrogen-containing compound that combines with phosphate to form the high-energy compound creatine phosphate (or phosphocreatine) in muscles.

cretinism (CREE-tin-ism): an iodine-deficiency disease characterized by mental and physical retardation.

Crohn's disease: inflammation and ulceration along the length of the GI tract, often with granulomas.

cyclamate (SIGH-kla-mate): a 0-kcalorie artificial sweetener that tastes 30 times as sweet as sucrose; FDA approval pending in the United States; available in Canada in grocery stores but only as a tabletop sweetener, not as an additive.

cyclic parenteral nutrition: the continuous administration of central TPN solution for 8 to 12 hours a day with time periods when no nutrients are infused during the rest of the day.

cystic fibrosis: a hereditary disorder characterized by the production of thick mucus that affects many organs, including the pancreas, lungs, liver, heart, gallbladder, and small intestine.

cytokines (SIGH-toe-kynes): proteins that help regulate inflammatory responses. Some cause anorexia, fever, and discomfort.

Daily Food Guide: the USDA's food group plan for ensuring dietary adequacy that assigns foods to five major food groups.

Daily Values: reference values developed by the FDA specifically for use on food labels.

dawn phenomenon: early morning hyperglycemia that develops in response to elevated levels of counterregulatory hormones that act to raise blood glucose after an overnight fast. Without adequate insulin, the glucose cannot enter the cells and remains in the blood.

deamination: removal of the amino (NH_2) group from a compound such as an amino acid.

deficient: the amount of a nutrient below which almost all healthy people can be expected, over time, to experience deficiency symptoms.

dehydration: the loss of water from the body that occurs when water output exceeds water input. The symptoms progress rapidly from thirst, to weakness, to exhaustion and delirium and end in death if not corrected.

delusions (dee-LOO-shuns): inappropriate beliefs not consistent with the individual's own knowledge and experience.

dementia (dee-MEN-she-ah): irreversible loss of mental function.

denaturation (dee-nay-cher-AY-shun): the change in a protein's shape brought about by heat, acid, or other agents. Past a certain point, denaturation is irreversible.

dental caries: the gradual decay and disintegration of a tooth.

dental plaque (PLACK): a gummy mass of bacteria that grows on teeth and can lead to dental caries and gum disease.

dextrose: an older name for glucose and the name used for glucose in intravenous solutions.

dextrose monohydrate: a form of glucose that contains a molecule of water and is stable in IV solutions. IV dextrose solutions provide 3.4 kcal/g.

DHEA (dehydroepiandrosterone): a hormone made in the adrenal glands that serves as a precursor to the male hormone testosterone; falsely promoted as burning fat, building muscle, and slowing aging. Side effects include acne, aggressiveness, and liver enlargement.

diabetes (DYE-uh-BEET-eez) mellitus (MELL-ih-tus or mell-EYE-tus): a group of metabolic disorders of glucose regulation and utilization.

diabetic coma: unconsciousness precipitated by hyperglycemia, dehydration, ketosis, and acidosis in people with diabetes.

dialysate (dye-AL-ih-sate): a solution used during dialysis to draw wastes and fluids from the blood.

dialysis (dye-AL-ih-sis): removal of wastes and fluids from the blood using the principles of simple diffusion and osmosis through a semipermeable membrane.

dialyzer (dye-ah-LYES-er): the machine used for hemodialysis; also called an *artificial kidney.*

diarrhea: the frequent passage of watery bowel movements.

diet manual: a book that describes the foods allowed and restricted on a diet, outlines the rationale and indications for use of each diet, and provides sample menus.

diet order: a physician's written statement in the medical record of the client's diet prescription.

diet pills: pills that depress the appetite temporarily; often, physician-prescribed

amphetamines (speed). It is generally agreed that these drugs are of little value for weight loss and that their use can cause a dangerous dependency.

dietary adequacy: the characteristic of a diet that provides all the essential nutrients, fiber, and energy necessary to maintain health and body weight.

dietary balance: providing foods of a number of types in balance with one another such that foods rich in one nutrient do not crowd out foods that are rich in another nutrient.

Dietary Reference Intakes (DRI): a set of values for the dietary nutrient intakes of healthy people in the United States and Canada. These values are used for planning and assessing diets.

dietetic technician registered (D.T.R.): a professional who has earned an associate degree or higher; has completed a dietetic technician program approved by the American Dietetic Association (ADA); has passed a national registration exam; and assists in planning, implementing, and evaluating nutritional care.

dietetics: the practical application of nutrition, including the assessment of nutrition status, recommendation of appropriate diets, nutrition education, and the planning and serving of meals.

differentiation: the development of specific functions different from those of the original.

diffusion: movement of solutes from an area of high concentration to one of low concentration.

digestion: the process by which complex food particles are broken down to smaller absorbable particles.

dipeptide (dye-PEP-tide): two amino acids bonded together.

disaccharide (dye-SACK-uh-ride): a pair of sugar units bonded together.

diuretic (dye-yoo-RET-ic): a medication that promotes the excretion of water through the kidneys. Only some diuretics increase the urinary loss of potassium. Others, called potassium-sparing diuretics, result in potassium retention.

diuretic abuse: use of diuretics to promote water excretion by dieters who believe their weight excesses are due to water accumulation.

diverticula (dye-ver-TIC-you-la): sacs or pouches that develop in the weakened areas of

the intestinal wall (like bulges in an inner tube where the tire wall is weak).

diverticulitis: the condition of having infected diverticula.

diverticulosis: the condition of having diverticula.

dominant gene: a gene that has an observable effect on an organism. If an altered gene is dominant, it has an observable effect even when it is paired with a normal gene; see also *recessive gene.*

dumping syndrome: the symptoms that result from the rapid entry of undigested food into the jejunum: sweating, weakness, and diarrhea shortly after eating and hypoglycemia later.

duodenum (doo-oh-DEEN-um or doo-ODD-ah-num): the top portion of the small intestine (about "12 fingers' breadth" long, in ancient terminology).

durable power of attorney: a legal document in which one competent adult authorizes another competent adult to make decisions for her or him in the event of incapacitation.

dysentery (DIS-en-terry): an infection of the gastrointestinal tract caused by an amoeba or bacterium that gives rise to severe diarrhea.

dyspepsia: vague abdominal pain; also called indigestion.

dysphagia (dis-FAY-gee-ah): difficulty in swallowing.

dysuria (dis-YOU-ree-ah): painful or difficult urination.

eating disorder: a disturbance in eating behavior that jeopardizes a person's physical and psychological health.

eclampsia: following preeclampsia, a severe stage in which convulsions occur.

edema (eh-DEEM-uh): the swelling of body tissue that occurs when fluid leaks from the blood vessels and accumulates in the interstitial spaces.

electrolyte solutions: solutions that can conduct electricity.

electrolytes: salts that dissolve in water and dissociate into charged particles called ions.

embolism: the obstruction of a blood vessel by an **embolus (EM-boh-luss),** or traveling clot, causing sudden tissue death.

emetic (em-ETT-ic): an agent that causes vomiting.

emphysema (EM-feh-SEE-mah): a type of COPD in which the lungs lose their elasticity and the victim has difficulty breathing; often occurs along with bronchitis.

emulsifiers: substances that mix with both fat and water and that permanently disperse the fat in the water, forming an emulsion.

enamel: the hard, white, dense substance that forms a covering for the crown of the teeth.

end-stage renal disease: the severe stage of renal failure in which dialysis or a kidney transplant is necessary to sustain life.

energy metabolism: all the reactions by which the body obtains and spends the energy from food or body stores.

energy-yielding nutrients: the fuel nutrients, those that yield energy the body can use. The three energy-yielding nutrients are carbohydrates, proteins, and fats.

engorgement: overfilling of the breasts with milk.

enrichment: the addition of nutrients to a food to meet a specified standard. In the case of refined bread or cereal, five nutrients have been added: thiamin, riboflavin, niacin, and folate in amounts approximately equivalent to, or higher than, those originally present and iron in amounts to alleviate the prevalence of iron-deficiency anemia.

enteral formulas: liquid diets intended for oral use or for tube feedings.

enterostomy (EN-ter-OSS-toe-mee): a gastric or jejunal opening made surgically or under local anesthesia through which a feeding tube can be passed.

enzymes: protein catalysts that facilitate chemical reactions without being changed in the process.

EPA, DHA: omega-3 fatty acids made from linolenic acid. The full name for EPA is eicosapentaenoic (EYE-cosa-PENTA-ee-NO-ick) acid. The full name for DHA is docosahexaenoic (DOE-cosa-HEXA-ee-NO-ick) acid.

epigastric: the region of the body just above the stomach.

epiglottis (epp-ee-GLOT-tiss): a cartilage structure in the throat that prevents fluid or food from entering the trachea when a person swallows.

epithelial (ep-i-THEE-lee-ul) cells: cells on the surface of the skin and mucous membranes.

epithelial tissues: the layers of the body that serve as selective barriers between the body's interior and the environment (examples are the cornea, the skin, the respiratory lining, and the lining of the digestive tract).

erythrocyte (er-REETH-ro-cite) hemolysis (he-MOLL-uh-sis): rupture of the red blood cells, caused by vitamin E deficiency.

erythropoietin (eh-RITH-row-POY-eh-tin): a hormone secreted by the kidneys in response to oxygen depletion or anemia that stimulates the bone marrow to produce red blood cells.

esophageal ulcers: lesions or sores in the lining of the esophagus.

esophagus (e-SOFF-uh-gus): the food pipe; the conduit from the mouth to the stomach.

essential amino acids: amino acids that the body cannot synthesize in amounts sufficient to meet physiological need; also called *indispensable amino acids*. Nine amino acids are known to be essential for human adults:
- *Histidine* (HISS-tuh-deen).
- *Isoleucine* (eye-so-LOO-seen).
- *Leucine* (LOO-seen).
- *Lysine* (LYE-seen).
- *Methionine* (meh-THIGH-oh-neen).
- *Phenylalanine* (fen-il-AL-uh-neen).
- *Threonine* (THREE-oh-neen).
- *Tryptophan* (TRIP-toe-fane).
- *Valine* (VAY-leen).

essential fatty acids: fatty acids that the body requires but cannot make in amounts sufficient to meet its physiological needs.

essential nutrients: nutrients a person must obtain from food because the body cannot make them for itself in sufficient quantity to meet physiological needs.

Estimated Average Requirement (EAR): the amount of a nutrient that will maintain a specific biochemical or physiological function in half the people of a given age and sex group.

ethical: in accordance with moral principles or professional standards. Socrates described *ethics* as "how we ought to live."

ethnic diets: foodways and cuisines typical of national origins, races, cultural heritages, or geographic locations.

fad diets: diets based on exaggerated or false

theories of weight loss. Such diets are usually inadequate in energy and nutrients. Most fad diets, including the currently popular Zone Diet, advocate essentially the same high-protein, low-carbohydrate diet. Such diets may offer short-term weight-loss success to some who try them, but they fail to produce long-lasting results for most people. Furthermore, high-protein, low-carbohydrate diets are often high in fat and low in fiber, vitamins, and some minerals. Long-term use of such diets may produce adverse side effects such as nausea, fatigue, constipation, and low blood pressure. Some fad diets are more dangerous to health than obesity itself.

faith healing: healing by invoking divine intervention without the use of medical, surgical, or other traditional therapy.

fatfold measure: a clinical estimate of total body fatness in which the thickness of a fold of skin on the back of the arm (over the triceps muscle), below the shoulder blade (subscapular), or in other places is measured with a caliper. (The older, less preferred, term is **skinfold test.**)

fats: lipids that are solid at room temperature (70°F or 25°C).

fatty acids: organic compounds composed of a chain of carbon atoms with hydrogens attached and an acid group at one end.

fatty liver: an accumulation of triglycerides in the liver resulting from many conditions, including exposure to excessive alcohol, PEM, excessive weight gain, and diabetes mellitus; also called **hepatic steatosis, steatohepatitis,** and **fatty infiltration of the liver.**

female athlete triad: a potentially fatal triad of medical problems: disordered eating, amenorrhea, and osteoporosis.

fetal alcohol effect (FAE): partial abnormalities from prenatal alcohol exposure that are not sufficient for diagnosis with FAS, but are impairing to the child. Also called *alcohol-related birth defects (ARBD)* or *subclinical FAS.*

fetal alcohol syndrome (FAS) the cluster of symptoms seen in a person whose mother consumed excess alcohol during her pregnancy; includes mental and physical retardation with facial and other body deformities.

fetus (FEET-us): the developing infant from eight weeks after conception until its birth.

fibers: a general term denoting in plant foods the polysaccharides cellulose, hemicellulose, pectins, gums and mucilages, as well as the nonpolysaccharide lignins, that are not attacked by human digestive enzymes.

fibrocystic breast disease: a harmless condition in which the breast develops lumps; sometimes associated with caffeine consumption. In some, it responds to treatment by abstinence from caffeine; in others, it can be treated with vitamin E.

fistula (FIS-too-lah): an abnormal opening formed between two organs or between an internal organ and the skin.

fitness: the characteristics of the body that enable it to perform physical activity; more broadly, the ability to meet routine physical demands with enough reserve energy to rise to a sudden challenge; or the body's ability to withstand stress of all kinds.

flapping tremor: uncontrolled movement of the muscle group that causes the outstretched arm and hand to flap like a wing; occurs in disorders that cause encephalopathy; also called **asterixis (AS-ter-ICK-sis).**

flatus (FLAY-tuss): gas in the intestinal tract or the expelling of gas from the intestinal tract, especially through the anus.

fluid and electrolyte balance: maintenance of the necessary amounts and types of fluid and minerals in each compartment of the body fluids.

fluorapatite (floor-APP-uh-tite): the stabilized form of bone and tooth crystal, in which fluoride has replaced the hydroxy portion of hydroxyapatite.

fluorosis (floor-OH-sis): mottling of the tooth enamel from ingestion of too much fluoride during tooth development.

follicle (FOLL-i-cul): a group of cells in the skin from which a hair grows.

food allergies: adverse reactions to foods that involve an immune response; also called *food-hypersensitivity reactions.*

food aversion: a strong desire to avoid a particular food.

food consumption survey: a survey that measures the amounts and kinds of food people consume (using diet histories), estimates the nutrient intakes, and compares them with a standard such as the DRI.

food craving: a deep longing for a particular food.

food group plans: diet-planning tools that sort foods of similar origin and nutrient content into groups and then specify that people should eat certain numbers of servings from each group.

food intolerance: an adverse response to a food or food additive that does not involve the immune system.

food poisoning: illness transmitted to human beings through food and water, caused by infectious agents or by toxins that they produce.

fortification: the addition to a food of nutrients that were either not originally present or present in insignificant amounts. Fortification can be used to correct or prevent a widespread nutrient deficiency, to balance the total nutrient profile of a food, or to restore nutrients lost in processing.

free radical: a highly reactive chemical form that can cause destructive changes in nearby compounds, sometimes setting up a chain reaction.

fructose: a monosaccharide; sometimes known as fruit sugar. It is abundant in fruits, honey, and saps.

galactose: a monosaccharide; part of the disaccharide lactose.

galactosemia (ga-LAK-toe-SEE-me-ah): an inborn error of metabolism in which enzymes that normally metabolize galactose to compounds the body can handle are missing and an alternative metabolite accumulates in the tissues, causing damage.

gallbladder: the organ that stores and concentrates bile.

gangrene: death of tissue due to a deficient blood supply and/or infection.

gastrectomy (gas-TREK-tah-mee): surgery to remove all (total gastrectomy) or part (subtotal or partial gastrectomy) of the stomach.

gastric banding: a surgical means of producing weight loss by restricting stomach size with either a row of surgical staples or a constricting band; also called gastric partitioning or gastroplasty.

gastric bypass: surgery that reroutes food from the stomach to the lower part of the small intestine; creates a chronic, lifelong state of malabsorption by preventing normal digestion and absorption of nutrients.

gastric glands: exocrine glands in the stomach wall that secrete gastric juice into the stomach.

gastric juice: the digestive secretion of the gastric glands containing a mixture of water, hydrochloric acid, and enzymes. The principal enzymes are pepsin (acts on proteins) and lipase (acts on emulsified fats).

gastric motility: spontaneous motion in the digestive tract accomplished by involuntary muscular contractions.

gastritis: inflammation of the stomach lining.

gastroparesis: delayed gastric emptying.

gastrostomy (gas-TROSS-toe-mee): an opening in the stomach made surgically or under local anesthesia through which a feeding tube can be passed. The technique for creating a gastrostomy under local anesthesia is called **percutaneous endoscopic gastrostomy,** or **PEG** for short. When the feeding tube is guided from such an opening into the jejunum, the procedure is called **percutaneous endoscopic jejunostomy (PEJ),** a misnomer because the enterostomy site is in the stomach.

gatekeeper: with respect to nutrition, a key person who controls other people's access to foods and thereby exerts a profound impact on their nutrition. Examples are the spouse who buys and cooks the food, the parent who feeds the children, and the caregiver in a day-care center.

genes: the basic units of hereditary information, made of DNA, that are passed from parent to offspring in the chromosomes. A pair of genes codes for each genetic trait.

gestational diabetes: the detection of abnormal glucose tolerance during pregnancy.

GI tract: the gastrointestinal tract or digestive tract. The principal organs are the stomach and intestines.

gland: a cell or group of cells that secretes materials for special uses in the body. Glands may be *exocrine glands,* secreting their materials "out" (into the digestive tract or onto the surface of the skin), or *endocrine glands,* secreting their materials "in" (into the blood).

glomerular filtration rate (GFR): the rate at which the kidneys form filtrate. Normally, the GFR is about 120 ml/min. Falling GFR signifies deteriorating kidney function. When the GFR falls drastically, to about 10 ml/min, dialysis is usually initiated.

glomerulus (glow-MARE-you-lus): a cup-shaped membrane enclosing a tuft of capillaries within a nephron. (The plural is *glomeruli*.)

glucose (GLOO-kose): a monosaccharide, the sugar common to all disaccharides and polysaccharides; also called blood sugar or dextrose.

gluten (GLUE-ten): a protein found in wheat. **Gliadin (GLIGH-ah-din)** is the fraction of gluten that causes the toxic effects in celiac disease. Corresponding protein fractions in barley, rye, and possibly oats also have these effects.

glycemic effect: a measure of the extent to which a food raises the blood glucose concentration and elicits an insulin response, as compared with pure glucose.

glycerol (GLISS-er-ol): an organic compound, three carbons long, that can form the backbone of triglycerides and phospholipids.

glycogen (GLY-co-gen): a polysaccharide composed of glucose, made and stored by liver and muscle tissues of human beings and animals as a storage form of glucose.

glycogen loading: a regimen of exhausting exercise, followed by eating a high-carbohydrate diet, that enables muscles to temporarily store glycogen beyond their normal capacity; also called *carbohydrate loading or glycogen supercompensation.*

glycogen storage disease: one of several inherited disorders that can lead to severe liver damage. People with glycogen storage diseases lack one of the enzymes that allow them to utilize glycogen stores efficiently, and glycogen accumulates in the liver and other organs.

glycolysis (gligh-COLL-uh-sis): the metabolic breakdown of glucose to pyruvate.

glycosuria (GLY-ko-SUE-ree-ah) or **glucosuria (GLUE-ko-SUE-ree-ah):** glucose in the urine, which generally occurs when blood glucose exceeds 180 mg/100 ml.

goiter (GOY-ter): an enlargement of the thyroid gland due to an iodine deficiency, malfunction of the gland, or overconsumption of a thyroid antagonist. Goiter caused by iodine deficiency is *simple goiter.*

gout (GOWT): a metabolic disease in which crystals of uric acid precipitate in the joints.

granulated sugar: common table sugar, crystalline sucrose; 99.9 percent pure.

granuloma: a grainlike growth that contains live microorganisms and immune system cells enclosed within a fibrous coat.

HDL (high-density lipoprotein): the type of lipoprotein that transports cholesterol back to the liver from peripheral cells; composed primarily of protein.

health: a range of states with physical, mental, emotional, spiritual, and social components. At a minimum, *health* means freedom from physical disease, mental disturbances, emotional distress, spiritual discontent, social maladjustment, and other negative states. At a maximum, *health* means "wellness."

health maintenance organization (HMO): a form of managed care that limits the subscriber's choice of health care professionals to those affiliated with the organization and controls access to services by directing care through a primary care physician.

heart attack: sudden tissue death caused by blockages of vessels that feed the heart muscle; also called **myocardial infarction (MI), cardiac arrest,** or **acute heart failure.**

heartburn: a burning sensation felt behind the sternum caused by the presence of gastric juices in the esophagus; also called **pyrosis (pie-ROE-SIS).**

hematuria (HE-mah-TOO-ree-ah): blood in the urine.

hemochromatosis (HE-moe-CROW-mah-toe-sis): an inherited disorder in which people absorb too much iron from the intestine and store too much iron in the liver.

hemodialysis: removal of fluids and wastes from the blood by passing it through a dialyzer.

hemofiltration: removal of fluids and wastes from the blood using ultrafiltration and fluid replacement.

hemoglobin: the oxygen-carrying protein of the red blood cells.

hemorrhagic (hem-o-RAJ-ik) disease: the vitamin K–deficiency disease in which blood fails to clot.

hemosiderosis (heem-oh-sid-er-OH-sis): a condition characterized by the deposition of the iron-storage protein hemosiderin in the liver and other tissues.

hepatic coma: a state of unconsciousness that results from severe liver disease.

hepatic encephalopathy (en-SEF-ah-LOP-ah-thee): mental changes associated with liver disease that may include irritability, short-term memory loss, and an inability to concentrate.

herbal laxatives: laxatives containing senna, aloe, rhubarb root, castor oil, or buckthorn, which are commonly sold as "dieter's tea." Such products cause nausea, vomiting, diarrhea, fainting, and, in some users, possibly death.[a]

herbal medicine: the use of plants to treat disease or improve health; also known as *botanical medicine or phytotherapy.*

herbal products: substances extracted from plants to mimic drugs that suppress appetite. Such substances may have dangerous side effects. St. John's wort, for example, is often prepared in combination with the herbal stimulant ephedrine, extracted from the Chinese plant ma huang. Ephedrine has been implicated in several cases of heart attacks and seizures.

herpes virus: a virus that can lead to mouth lesions and may also affect the lower GI tract, causing diarrhea.

hiatal hernia: a protrusion of a portion of the stomach through the esophageal hiatus of the diaphragm.

high-fructose corn syrup (HFCS): the predominant sweetener used in processed foods today. HFCS is mostly fructose; glucose makes up the balance.

high-quality protein: an easily digestible, complete protein.

holistic (hoe-LIS-tik): to treat a person as a whole being rather than as a sum of many parts.

homeopathic (home-ee-OP-ah-thick) medicine: a practice based on the theory that "like cures like," that is, that substances that cause symptoms in healthy people can cure those symptoms when given in very dilute amounts.

homeostasis: the maintenance of constant internal conditions (such as chemistry, temperature, and blood pressure) by the body's control system.

honey: sugar (mostly sucrose) formed from nectar gathered by bees. An enzyme splits the sucrose into glucose and fructose. Composition and flavor vary, but honey always contains a mixture of sucrose, fructose, and glucose.

hormones: chemical messengers. Hormones are secreted by a variety of glands in the body in response to altered conditions. Each travels to one or more target tissues or organs and elicits specific responses to restore normal conditions.

human immunodeficiency virus (HIV): a virus that progressively hampers the function of the immune system and leaves its host defenseless against other infections and cancer and eventually causes AIDS.

hunger: the physiological need to eat, experienced as a drive for obtaining food; an unpleasant sensation that demands relief.

hydrochloric acid (HCl): an acid composed of hydrogen and chloride atoms; normally produced by the gastric glands.

hydrogenation: the process of adding hydrogen to unsaturated fat to make it more solid and resistant to chemical change.

hydrolyzed formula: a liquid diet that contains broken-down fragments of protein, such as amino acids and short peptide chains; also called **monomeric formula.**

hydrophilic colloids: laxatives composed of fibers that work like dietary fibers. They attract water in the intestine to form a bulky stool, which then stimulates peristalsis.

hyperactivity: inattentive and impulsive behavior that is more frequent and severe than is typical of others a similar age; professionally called **attention deficit hyperactivity disorder (ADHD).**

hypercalciuria (HIGH-per-kal-see-YOU-ree-ah): excessive urinary excretion of calcium.

hyperglycemia: elevated blood glucose. Normal fasting blood glucose is less than 110 mg/100 ml. Fasting blood glucose between 110 and 125 mg/100 ml suggests impaired glucose tolerance; values of 126 mg/100 ml or higher suggest diabetes.

hyperosmolar hyperglycemic nonketotic coma: coma that occurs in uncontrolled type 2 diabetes precipitated by the presence of hypertonic blood and dehydration.

hyperoxaluria (HIGH-per-OX-all-YOU-ree-ah): excessive urinary excretion of oxalate.

hypertension: higher-than-normal blood pressure.

hypertonic formula: a formula with an osmolality greater than that of blood serum.

hypnotherapy: a technique that uses hypnosis and the power of suggestion to improve health behaviors, relieve pain, and heal.

hypoglycemia: low blood glucose.

hypothalamus (high-poh-THALL-uh-mus): a part of the brain that helps regulate many body balances, including fluid balance.

idopathic (id-ee-oh-PATH-ic) cirrhosis: severe liver damage for which the cause cannot be identified.

ileocecal (ill-ee-oh-SEEK-ul) valve: the sphincter muscle separating the small and large intestines.

ileostomy (ILL-ee-OSS-toe-me): surgery that creates a stoma from the ileum through the abdominal wall and out through the skin.

ileum (ILL-ee-um): the last segment of the small intestine.

imagery: a technique that guides clients to achieve a desired physical, emotional, or spiritual state by visualizing themselves in that state.

inborn error of metabolism: an inherited flaw evident as a metabolic disorder or disease present from birth.

incomplete protein: a protein lacking or low in one or more of the essential amino acids.

indemnity insurance: traditional fee-for-service insurance.

indirect calorimetry: an indirect estimate of resting energy needs made by measuring the amount of carbon dioxide expired (carbon dioxide production) and the amount of oxygen inspired (oxygen consumption).

individualized: to treat a person as a unique being rather than as one of many others.

inflammatory bowel diseases (IBD): diseases characterized by inflammation of the bowel.

inflammatory response: the changes orchestrated by the immune system when tissues are injured by such forces as blows, wounds, foreign bodies (chemicals, microorganisms), heat, cold, electricity, or radiation.

inorganic: not containing carbon or pertaining to living things.

insoluble fibers: the tough, fibrous structures of fruits, vegetables, and grains; indigestible food components that do not dissolve in water.

insulin: a hormone secreted by the pancreas in response to high blood glucose. It promotes cellular glucose uptake and use or storage.

insulin resistance: the condition in which the cells fail to respond to insulin as they do in healthy people.

intermittent claudication: severe calf pain caused by inadequate blood supply. It occurs when walking and subsides during rest.

intermittent feeding: the delivery of about 250 to 400 ml of formula over 30 minutes or more.

intestinal flora: the bacterial inhabitants of the GI tract.

intestinal juice: the secretion of the intestinal glands; contains enzymes for the digestion of carbohydrate and protein and a minor enzyme for fat digestion.

intra-abdominal fat: fat stored within the abdominal cavity in association with the internal abdominal organs, as opposed to the fat stored directly under the abdominal skin (subcutaneous fat).

intravenous (IV): through a vein.

intrinsic: inside the system. Anemia that reflects a vitamin B_{12} deficiency caused by lack of intrinsic factor is known as **pernicious anemia.**

invert sugar: a mixture of glucose and fructose formed by splitting sucrose in a chemical process; sold only in liquid form, sweeter than sucrose. Invert sugar is used as an additive to help preserve food freshness and prevent shrinkage.

ions (EYE-uns): atoms or molecules that have electrical charges.

iridology: the study of changes in the iris of the eye and their relationships to disease.

iron deficiency: having depleted iron stores.

iron overload: toxicity from excess iron.

iron-deficiency anemia: a blood iron deficiency characterized by small, pale red blood cells; also called **microcytic hypochromic anemia.**

irritable bowel syndrome: an intestinal disorder of unknown cause. Symptoms include abdominal discomfort and cramping, diarrhea, constipation, or alternating diarrhea and constipation.

isotonic formula: a formula with an osmolality similar to that of blood serum (300 mOsm/kg).

IV catheter: a thin tube inserted into a vein. Additional tubing connects the IV solution to the catheter.

jaundice (JON-dis): a characteristic yellowing of the skin, whites of the eyes, mucous membranes, and body fluids resulting from the accumulation of bile pigments (bilirubin) in the blood.

jejunostomy (JEE-ju-NOSS-toe-mee): an opening in the jejunum made surgically or under local anesthesia through which a feeding tube can be passed. The technique for creating a jejunostomy under local anesthesia is called a **direct endoscopic jejunostomy (DEJ).**

jejunum (je-JOON-um): the first two-fifths of the small intestine beyond the duodenum.

Kaposi's (cap-OH-seez) sarcoma: a type of cancer rare in the general population but common in people with HIV infection.

kcalorie (energy) control: management of food energy intake.

kcalories: units by which energy is measured. One kcalorie (kcal) is the amount of heat necessary to raise the temperature of 1 kilogram (kg) of water 1°C.

keratin (KERR-uh-tin): a water-insoluble protein; the normal protein of hair and nails. In vitamin A deficiency, keratin-producing cells may replace mucus-producing cells.

ketonemia: ketones in the blood.

ketones (KEY-tones): acidic, fat-related compounds formed from the incomplete breakdown of fat when carbohydrate is not available; technically known as *ketone bodies.*

ketonuria: ketones in the urine.

kwashiorkor (kwash-ee-OR-core or **kawsh-ee-or-CORE):** a disease related to PEM; also called acute malnutrition because it develops rapidly.

lactation: production and secretion of breast milk for the purpose of nourishing an infant.

lactoferrin (lack-toe-FERR-in): a factor in breast milk that binds iron and keeps it from supporting the growth of the infant's intestinal bacteria.

lacto-ovo vegetarians: people who include milk or milk products and eggs, but omit meat, fish, shellfish, and poultry from their diets.

lactose: a disaccharide composed of glucose and galactose; commonly known as milk sugar.

lacto-vegetarians: people who include milk or milk products, but exclude meat, poultry, fish, shellfish, and eggs from their diets.

Laennec's (LAY-eh-necks) cirrhosis: severe liver damage related to alcohol abuse.

latent: the period in the course of a disease when the conditions are present but the symptoms have not begun to appear.

LDL (low-density lipoprotein): the type of lipoprotein derived from VLDL as cells remove triglycerides from them. LDL carry cholesterol and triglycerides from the liver to the cells of the body and are composed primarily of cholesterol.

lecithins: one type of phospholipid.

legal: established by law.

leptin: a protein produced by fat cells under the direction of the obesity gene; increases satiety and energy expenditure.

letdown reflex: the reflex that forces milk to the front of the breast when the infant begins to nurse.

levulose: an older name for fructose.

life expectancy: the average number of years lived by people in a given society.

life span: the maximum number of years of life attainable by a member of a species.

limiting amino acid: the essential amino acid found in the shortest supply relative to the amounts needed for protein synthesis in the body. It limits the amount of protein the body can make.

linoleic acid, linolenic acid: polyunsaturated fatty acids, essential for human beings.

lipids: a family of compounds that includes triglycerides (fats and oils), phospholipids, and sterols.

lipoprotein lipase (LPL): an enzyme mounted on the surface of fat cells (and other cells). It hydrolyzes triglycerides in the blood into fatty acids and glycerol for absorption into the cells. There they are metabolized or reassembled for storage.

lipoproteins: clusters of lipids associated with proteins that serve as transport vehicles for lipids in the lymph and blood.

living will: a document signed by a competent adult that specifically states whether the person wishes aggressive treatment in the event of terminal illness or irreversible coma from which the person is not expected to recover.

longevity: long duration of life.

low birthweight (LBW): a birthweight less than 5½ lb (2500 g); indicates probable poor health in the newborn and poor nutrition status of the mother during pregnancy. Normal birthweight for a full-term baby is 6½ to 8½ lb (about 3000 to 4000 g).

low-carbohydrate diets: diets designed to bring about metabolic responses similar to those of fasting. Without sufficient carbohydrate, the body cannot use its fat in the normal way, and ketosis results. Many physiological hazards accompany low-carbohydrate diets: high blood cholesterol, mineral imbalances, hypoglycemia, and more.

lymph (LIMF): the body fluid found in lymphatic vessels. Lymph consists of all the constituents of blood except red blood cells.

lymphatic system: a loosely organized system of vessels and ducts that conveys the products of digestion toward the heart.

macroangiopathies: disorders of the large blood vessels.

macrobiotic diet: a diet consisting of brown rice, miso soup, sea vegetables, and other traditional Japanese foods.

malnutrition: any condition caused by deficient or excess energy or nutrient intake or by an imbalance of nutrients.

maltose: a disaccharide composed of two glucose units; sometimes known as malt sugar.

managed care: a health care delivery system that aims to provide cost-effective health care by coordinating services and limiting access to services.

maple sugar: a sugar (mostly sucrose) purified from concentrated sap of the sugar maple tree. Maple sugar is expensive compared with other sweeteners.

marasmus (ma-RAZZ-mus): a disease that results from severe deprivation, or impaired absorption, of protein, energy, vitamins, and minerals; also called chronic malnutrition because it takes time to develop.

massage therapy: a healing method in which the therapist manually kneads muscles to reduce tension, increase blood circulation, improve joint mobility, and promote healing of injuries.

mastitis: infection of a breast.

mechanical soft diet: a diet that excludes only those foods that a person cannot chew; also called a **dental soft diet** or **edentulous diet.**

mechanical ventilator: a machine that "breathes" for the person who can't.

medical nutrition therapy: provision of a client's nutrient and nutrition education needs based on a complete nutrition assessment.

meditation: a self-directed technique of relaxing the body and calming the mind.

metabolism: the sum total of all the chemical reactions that go on in living cells.

microangiopathies: disorders of the capillaries.

microvilli (MY-cro-VILL-ee or MY-cro-VILL-eye): tiny, hairlike projections on each cell of every villus that can trap nutrient particles and transport them into the cells. The singular form is **microvillus.**

moderation: providing enough, but not too much, of a substance.

modified (or therapeutic diets): regular diets that are adjusted to meet special nutrient needs. Such diets can be adjusted in consistency, level of energy and nutrients, amount of fluid, or number of meals, or by the inclusion or elimination of certain foods.

modules: formulas or foods that provide primarily a single nutrient and are designed to be added to other formulas or foods to alter nutrient composition. They can also be combined together to create a highly individualized formula.

molasses: a thick brown syrup, left over from sugarcane juice during sugar refining. Blackstrap molasses contains iron, which comes from the machinery used to process it. This iron is not as well absorbed as the iron in meats and other foods.

molybdenum (mo-LIB-duh-num): a trace element.

monosaccharide (mon-oh-SACK-uh-ride): a single sugar unit.

monounsaturated fatty acid: a fatty acid that has one point of unsaturation; for example, the oleic acid found in olive oil.

mood disorders: mental illness characterized by episodes of severe depression or excessive excitement (mania) or both.

mouth ulcers: lesions or sores in the lining of the mouth.

mucous membranes: the membranes composed of mucus-secreting cells that line the surfaces of body tissues.

mucus (MYOO-cuss): a mucopolysaccharide (a relative of carbohydrate) secreted by cells of the stomach wall that protects the cells from exposure to digestive juices (and other destructive agents).

multiple daily injections: delivery of different types of insulin by injection three or more times daily.

muscular dystrophy (DIS-tro-fee): a hereditary disease in which the muscles gradually weaken. Its most debilitating effects arise in the lungs. This disease should not be confused with *nutritional* muscular dystrophy, a vitamin E–deficiency disease of animals characterized by gradual paralysis of the muscles.

mutation: an alteration in a gene such that an altered protein is produced.

mutual supplementation: the strategy of combining two protein foods in a meal so that each food provides the essential amino acid(s) lacking in the other.

myoglobin: the oxygen-carrying protein of the muscle cells.

nasoduodenal (ND): from the nose to the duodenum.

nasoenteric: from the nose to the stomach or intestine. *Nasoenteric feedings* include nasogastric, nasoduodenal, and nasojejunal feedings.

nasogastric (NG): from the nose to the stomach.

nasojejunal (NJ): from the nose to the jejunum.

naturopathic medicine: a system that integrates traditional medicine with botanical medicine, clinical nutrition, homeopathy, acupuncture, East Asian medicine, hydrotherapy, and manipulative therapy.

nephrons (NEF-rons): the working units of the kidneys. Each nephron consists of a glomerulus and a tubule.

nephropathy: a disorder of the kidneys.

neuron: a nerve cell; the structural and functional unit of the nervous system. Neurons initiate and conduct nerve transmissions.

neuropathy: a disorder of the nerves.

niacin equivalents (NE): the amount of niacin present in food, including the niacin that can theoretically be made from its precursor tryptophan present in the food.

night blindness: the slow recovery of vision after exposure to flashes of bright light at night; an early symptom of vitamin A deficiency.

nitrogen balance: the amount of nitrogen consumed (N in) as compared with the amount of nitrogen excreted (N out) in a given period of time.

nocturnal hypoglycemia: hypoglycemia that occurs while a person is sleeping.

nonessential amino acids: amino acids that the body can synthesize for itself; also called *dispensable amino acids* because they are dispensable from the diet.

nursing bottle tooth decay: extensive tooth decay due to prolonged tooth contact with formula, milk, fruit juice, or other carbohydrate-rich liquid offered to an infant in a bottle.

nursing process: is a systematic approach to identifying and addressing the nursing needs of clients, including nutrition. The nursing process consists of five steps:
1. Assessment.
2. Diagnosis.
3. Outcome identification and planning.
4. Implementation.
5. Evaluation.

nutrient density: a measure of the nutrients a food provides relative to the energy it provides. The more nutrients and the fewer kcalories, the higher the nutrient density.

nutrients: substances obtained from food and used in the body to provide energy and structural materials, as well as regulating agents to promote growth, maintenance, and repair. Nutrients may also reduce the risks of some diseases.

nutrition: the science of foods and the nutrients and other substances they contain, and of their ingestion, digestion, absorption, transport, metabolism, interaction, storage, and excretion. A broader definition includes the study of the environment and of human behavior as it relates to these processes.

nutrition assessment: the evaluation of the many factors that influence or reflect a client's nutrient and nutrition education needs.

nutrition screening: a tool for quickly identifying clients most likely to be at risk for malnutrition so that they can receive complete nutrition assessments.

nutrition status survey: a survey that evaluates people's nutrition status using nutrition assessment methods.

nutrition status: the state of the body's nutritional health.

nutritionist: a person who specializes in the study of nutrition. Some nutritionists are registered dietitians, but others are self-described experts whose training may be minimal or nonexistent. Some states make the term meaningful by allowing it to apply only to people who have master's (M.S.) or doctoral (Ph.D.) degrees from institutions accredited to offer such degrees in nutrition or related fields.

nutritive sweeteners: sweeteners that yield energy, including both the sugars and the sugar alcohols.

obesity: a chronic disease characterized by excessively high body fat in relation to lean body tissue.

oils: lipids that are liquid at room temperature (70°F or 25°C).

omega-3 fatty acid: a polyunsaturated fatty acid with its endmost double bond three carbons back from the end of its carbon chain; relatively newly recognized as important in nutrition. Linolenic acid is an example.

omega-6 fatty acid: a polyunsaturated fatty acid with its endmost double bond six carbons back from the end of its carbon chain; long recognized as important in nutrition. Linoleic acid is an example.

omega: the last letter of the Greek alphabet, used by chemists to refer to the position of the last double bond in a fatty acid.

open formula system: a system where formulas are decanted from their original packaging and then added to a feeding container before they can be administered through a feeding tube.

opportunistic illnesses: illnesses that would not normally occur or would cause only minor problems in the healthy population, but can cause great harm when the immune system is compromised.

oral antidiabetic agents: medications taken by mouth to lower blood glucose levels in people with type 2 diabetes.

oral rehydration therapy (ORT): the administration of a simple electrolyte solution to treat dehydration.

organic: carbon containing. The four organic nutrients are carbohydrates, fats, proteins, and vitamins.

orogastric: from the mouth to the stomach. This method is often used to feed infants because they breathe through their noses, and tubes inserted through the nose can hinder the infant's breathing. The tube is inserted before, and removed after, each feeding.

orthomolecular medicine: the use of large doses of vitamins to treat chronic disease.

osmolality (OZ-mow-LAL-eh-tee): a measure of the concentration of particles in a solution, expressed as the number of milliosmoles (mOsm) per kilogram.

osmosis: movement of water from an area of low solute concentration to one of high solute concentration.

osteoarthritis: a painful, chronic disease of the joints caused when the cushioning cartilage in a joint breaks down; joint structure is usually altered, with loss of function; also called *degenerative arthritis.*

osteomalacia (os-tee-o-mal-AY-shuh): a bone disease characterized by softening of the bones. Symptoms include bending of the spine and bowing of the legs. The disease occurs most often in adult women.

osteoporosis (oss-tee-oh-pore-OH-sis): literally, porous bones; reduced density of the bones, also known as *adult bone loss.*

overnutrition: overconsumption of food energy or nutrients sufficient to cause disease or increased susceptibility to disease; a form of malnutrition.

overt: out in the open, full-blown.

overweight: body weight above some standard of acceptable weight that is usually defined in relation to height (such as the weight-for-height tables).

oxalate: a nonnutrient found in significant amounts in rhubarb, spinach, beets, nuts, chocolate, tea, wheat bran, and strawberries.

ozone therapy: the use of ozone gas to enhance the body's immune system.

pancreas: a gland that secretes enzymes and digestive juices into the duodenum. (This is its exocrine function; it also has the endocrine function of secreting insulin and other hormones into the blood.)

pancreatic (pank-ree-AT-ic) juice: the exocrine secretion of the pancreas, containing enzymes for the digestion of carbohydrate, fat, and protein. Juice flows from the pancreas into the small intestine through the pancreatic duct. The pancreas also has an endocrine function, the secretion of insulin and other hormones.

pancreatic enzyme replacements: extracts of pork or beef pancreatic enzymes that are taken as supplements to aid digestion.

paranoia (PAR-ah-NOY-ah): mental illness characterized by delusions of persecution.

parenteral nutrition: delivery of nutrient solutions directly into a vein, bypassing the GI tract.

pathogens: disease-causing microorganisms.

pellagra (pell-AY-gra): the niacin-deficiency disease. Symptoms include the "4 Ds": diarrhea, dermatitis, dementia, and, ultimately, death.

pepsin: a protein-digesting enzyme (gastric protease) in the stomach. It circulates as a precursor, pepsinogen, and is converted to pepsin by the action of stomach acid.

peripheral parenteral nutrition (PPN): the provision of an IV solution that meets nutrient needs delivered through the peripheral veins.

peripheral veins: the small-diameter veins that carry blood from the arms and legs.

peripherally inserted central catheter (PICC): a catheter inserted into a peripheral vein and advanced into a central vein.

peristalsis (peri-STALL-sis): successive waves of involuntary muscular contractions passing along the walls of the GI tract that push the contents along.

peritoneal dialysis: removal of wastes and fluids from the body using the peritoneal membrane as a semipermeable membrane.

peritonitis: infection and inflammation of the membrane lining the abdominal cavity caused by leakage of infectious organisms through a perforation (hole) in an abdominal organ.

pernicious anemia: anemia caused by a lack of intrinsic factor and the consequent malabsorption of vitamin B_{12}.

persistent vegetative state: exhibiting motor reflexes but without the ability to regain cognitive behavior, communicate, or interact purposefully with the environment.

pH: the concentration of hydrogen ions. The lower the pH, the stronger the acid. Thus, pH 2 is a strong acid; pH 6 is a weak acid; pH 7 is neutral; and a pH above 7 is alkaline.

phospholipids: one of the three main classes of lipids; similar to triglycerides but with choline (or another compound) and a phosphorus-containing acid in place of one of the fatty acids.

physiological age: a person's age as estimated from her or his body's health and probable life expectancy.

phytates: nonnutrient components of grains, legumes, and seeds. Phytates can bind minerals such as zinc, iron, calcium, and magnesium in insoluble complexes in the intestine, and these complexes are then excreted unused.

phytochemicals: nonnutrient compounds in plant-derived foods that have biological activity in the body.

pica (PIE-ka): a craving for nonfood substances; also known as *geophagia (jee-oh-FAY-jee-uh)* when referring to clay-eating behavior.

pituitary (pit-TOO-ih-tary) gland: in the brain, the "king gland" that regulates the operation of many other glands.

PKU, phenylketonuria (FEN-el-KEY-toe-NEW-ree-ah): an inborn error of metabolism in which phenylalanine, an essential amino acid, cannot be converted to tyrosine. Alternative metabolites of phenylalanine (phenylketones) accumulate in the tissues, causing damage, and overflow into the urine.

placenta (pla-SEN-tuh): an organ that develops inside the uterus early in pregnancy, in which maternal and fetal blood circulate in close proximity and exchange materials. The fetus receives nutrients and oxygen across the placenta; the mother's blood picks up carbon dioxide and other waste materials to be excreted via her lungs and kidneys.

plaques (PLACKS): mounds of lipid material (mostly cholesterol) with some macrophages (a type of white blood cell) covered with fibrous connective tissue and embedded in artery walls. With time, plaques may enlarge and harden as the fibrous coat thickens and calcium is deposited in them.

platelets: tiny, disc-shaped bodies in the blood that play an important role in clot formation.

polydipsia (POLL-ee-DIP-see-ah): excessive thirst.

polypeptide: ten or more amino acids bonded together. An intermediate strand of between four and ten amino acids is an oligopeptide.

polyphagia (POLL-ee-FAY-gee-ah): excessive eating.

polyunsaturated fatty acid (PUFA): a fatty acid with two or more points of unsaturation.

polyuria (POLL-ee-YOU-ree-ah): excessive urine production.

portal hypertension: elevated blood pressure in the portal vein caused by obstructed blood flow through the liver.

postnecrotic cirrhosis: severe liver damage that develops as a complication of chronic hepatitis.

precursor: a compound that can be converted into another compound; with regard to vitamins, compounds that can be converted into active vitamins; also known as **provitamins.**

preeclampsia: a condition characterized by hypertension, fluid retention, and protein in the urine.

preferred provider organization (PPO): a form of managed care that encourages subscribers to select health care providers from a group that has contracted with the organization to provide services at lower costs.

preformed vitamin A: vitamin A in its active form.

pressure sores: the breakdown of skin and underlying tissues due to constant pressure and lack of oxygen to the affected area; also called **decubitus (dee-CUE-bih-tis) ulcers** or **bedsores.**

progressive diet: a diet that changes as a client's tolerances permit.

protein isolate: a protein that has been separated from a food. Examples include casein from milk and albumin from egg.

protein-energy malnutrition (PEM): a deficiency of protein and food energy; the world's most widespread malnutrition problem, including both marasmus and kwashiorkor.

protein-sparing effect: the effect of carbohydrate in providing energy that allows protein to be used for other purposes.

proteins: compounds composed of carbon, hydrogen, oxygen, and nitrogen atoms arranged into strands of amino acids. Some amino acids also contain sulfur atoms.

puberty: the period in life in which a person becomes physically capable of reproduction.

pyloric (pie-LORE-ic) sphincter: the sphincter muscle separating the stomach from the small intestine (also called *pylorus* or *pyloric valve*).

pyloroplasty (pie-LOOR-o-PLAS-tee): surgery that enlarges the pyloric sphincter.

pyruvate (PIE-roo-vate): pyruvic acid, a 3-carbon compound derived from glucose, glycerol, and certain amino acids in metabolism.

radiation enteritis: inflammation and scarring of the intestinal cells caused by exposure to radiation.

radiation therapy: the use of radiation to arrest or destroy cancer cells.

rancid: the term used to describe fats when they have deteriorated, usually by oxidation.

raw sugar: the first crop of crystals harvested during sugar processing. Raw sugar cannot be sold in the United States because it contains too much filth (dirt, insect fragments, and the like). Sugar sold domestically as raw sugar has actually gone through about half of the refining steps.

rebound hyperglycemia: hyperglycemia resulting from excessive secretion of counterregulatory hormones in response to excessive insulin and consequent low blood glucose levels; also called the **Somogyi (so-MOHG-yee) effect.**

recessive gene: a gene that has no observable effect on an organism. If an altered gene is recessive, it has no observable effect as long as it is paired with a normal gene that can produce a normal product. In this case, the normal gene is said to be *dominant.*

Recommended Dietary Allowances (RDA): the average daily amounts of nutrients considered adequate to meet the known nutrient needs of practically all healthy people; a goal for dietary intake by individuals.

rectum: the muscular terminal part of the GI tract extending from the sigmoid colon to the anus. The rectum stores waste prior to elimination.

refined grain: a product from which the bran, germ, and husk have been removed, leaving only the endosperm.

reflux esophagitis: (eh-sof-ah-JYE-tis): the backflow or regurgitation of gastric contents from the stomach into the esophagus, causing inflammation of the esophagus; also called **gastroesophageal reflux (GER)**.

registered dietitian (R.D.): a dietitian who has graduated from a university or college after completing a program of dietetics that has been accredited by the American Dietetic Association (or Dietitians of Canada). The dietitian must serve in an approved internship or coordinated program to practice the necessary skills, pass the association registration examination, and maintain competency through continuing education. Many states require licensing for practicing dietitians. Licensed dietitians (L.D.'s) have met all *state* requirements to offer nutrition advice.

renal colic: the severe pain that accompanies the movement of a kidney stone through the ureter to the bladder.

renal failure: failure of the kidneys to maintain normal function to a degree that requires dialysis or kidney transplantation for survival.

renal filtrate: the fluid that passes from the blood through the capillary walls of the glomeruli, eventually forming urine.

renal insufficiency: reduced renal function but not to the degree that requires dialysis or kidney transplantation.

renal threshold: the point at which a blood constituent that is normally reabsorbed by the kidneys reaches a level so high the kidneys cannot reabsorb it.

renin (REN-in): an enzyme secreted by the kidneys in response to a reduced blood flow that triggers the release of the hormone aldosterone from the adrenal glands.

requirement: the lowest continuing intake of a nutrient that will maintain a specified criterion of adequacy.

residue: the total amount of material in the colon; includes dietary fiber and undigested food, intestinal secretions, bacterial cell bodies, and cells shed from the intestinal mucosa.

respiratory failure: failure of the lungs to exchange gases.

retina (RET-in-uh): the layer of light-sensitive nerve cells lining the back of the inside of the eye; consists of rods and cones.

retinol-binding protein (RBP): the specific protein responsible for transporting retinol.

retinopathy: a disorder of the retina.

review of body systems: a brief review of clinical signs or symptoms associated with each organ system.

rheumatoid arthritis: a disease of the immune system involving painful inflammation of the joints and related structures.

rickets: the vitamin D–deficiency disease in children.

rooting reflex: a reflex that causes an infant to turn toward whichever cheek is touched, in search of a nipple.

saccharin (SAK-ah-ren): a 0-kcalorie artificial sweetener that tastes 500 times as sweet as sucrose; approved in the United States, but available in Canada only in pharmacies and only as a sweetener, not as an additive.

saliva: the secretion of the salivary glands. The principal enzyme is salivary amylase.

salivary glands: exocrine glands that secrete saliva into the mouth.

salt: a compound composed of a positive ion other than H^+ and a negative ion other than OH^-. An example of a salt is potassium chloride (K^+Cl^-).

saturated fatty acid: a fatty acid carrying the maximum possible number of hydrogen atoms (having no points of unsaturation). A saturated fat is a triglyceride that contains three saturated fatty acids.

schizophrenia (SKITz-oh-FREN-ee-ah): mental illness characterized by an altered concept of reality and, in some cases, delusions and hallucinations.

scurvy: the vitamin C–deficiency disease.

segmentation: a periodic squeezing or partitioning of the intestine by its circular muscles that both mixes and slowly pushes the contents along.

selenium (se-LEEN-ee-um): a trace element.

semivegetarians: people who include some, but not all, groups of animal-derived foods in their diets; they usually exclude meat and may occasionally include poultry, fish, and shellfish; also called *partial vegetarians*.

senile dementia: the loss of brain function beyond the normal loss of physical adeptness and memory that occurs with aging.

sepsis: the presence of infectious microorganisms in the bloodstream.

set-point theory: the theory that proposes that the body tends to maintain a certain weight by means of its own internal controls.

shock: a sudden drop in blood volume that disrupts the supply of oxygen to the tissues and organs and the return of blood to the heart. Shock is a critical event that must be corrected immediately.

short-bowel or **short-gut syndrome:** severe malabsorption that may occur when the absorptive surface of the small bowel is reduced, resulting in diarrhea, weight loss, bone disease, hypocalcemia, hypomagnesemia, and anemia.

simple carbohydrates (sugars): the monosaccharides (glucose, fructose, and galactose) and the disaccharides (sucrose, lactose, and maltose).

small intestine: a 10-foot length of small-diameter (1-inch) intestine that is the major site of digestion of food and absorption of nutrients.

soaps: chemical compounds formed between a basic mineral (such as calcium) and unabsorbed fatty acids.

soluble fibers: indigestible food components that readily dissolve in water and often impart gummy or gel-like characteristics to foods. An example is pectin from fruit, which is used to thicken jellies.

sphincter (SFINK-ter): a circular muscle surrounding, and able to close, a body opening.

standard (or **regular diets**): diets that include all foods and meet the nutrient needs of a healthy person.

standard formula: a liquid diet that contains complete molecules of protein; also called **intact** or **polymeric formula.**

starch: a plant polysaccharide composed of glucose and digestible by human beings.

steatorrhea (STEE-ah-toe-REE-ah): fatty diarrhea characterized by loose, foamy, foul-smelling stools.

sterile: free of microorganisms, such as bacteria.

steroid (STARE-oid): a medication used to reduce tissue inflammation, to suppress the immune response, or to replace certain steroid hormones in people who cannot synthesize them.

sterols: one of the three main classes of lipids; include cholesterol, vitamin D, and the sex hormones (such as testosterone).

stoma (STOH-ma): a surgically formed opening.

stress: the state in which a body's internal balance (homeostasis) is upset by a threat to a person's physical well-being (stressor).

stroke: an event in which the blood flow to a part of the brain is cut off; also called a **cerebral vascular accident (CVA).**

sucralose (SUE-kra-lose): a 0-kcalorie artificial sweetener that taste 600 times as sweet as sucrose.

sucrose: a disaccharide composed of glucose and fructose; commonly known as table sugar, beet sugar, or cane sugar.

sugar alcohols: sugarlike compounds; like sugars, they are sweet to taste but yield 2 to 3 kcal per gram, slightly less than sucrose. Examples are maltitol, mannitol, sorbitol, isomalt, lactitol, and xylitol.

syndrome X: the combination of insulin resistance, hyperinsulinemia, obesity, hypertension, elevated LDL and triglycerides, and reduced HDL that is frequently associated with type 2 diabetes and cardiovascular disease; also called **insulin-resistance syndrome** and **metabolic syndrome.**

systemic inflammatory response syndrome (SIRS): the complex of symptoms that occur as a result of immune and inflammatory factors in response to tissue damage that may progress to multiple organ failure.

teratogenic (ter-AT-oh-jen-ik): causing abnormal fetal development and birth defects.

terminal illness: a progressive, irreversible disease that will lead to death in the near future.

thermic effect of food: an estimate of the energy required to process food (digest, absorb, transport, metabolize, and store ingested nutrients).

thrombosis: the formation of a **thrombus,** or blood clot.

thrush: a fungal infection of the mouth and esophagus caused by *Candida albicans* that coats the tongue with a milky film, alters taste sensations, and causes pain on chewing and swallowing. The technical term for this infection is **candidiasis.**

tissue rejection: destruction of healthy donor cells by the recipient's immune system, which recognizes the donor cells as foreign; also called **graft-versus-host disease (GVHD).**

Tolerable Upper Intake Level (UL): the maximum amount of a nutrient that appears safe for most healthy people and beyond which there is an increased risk of adverse health effects.

tonsillectomy: (tawn-sill-ECK-tah-me): surgical removal of the tonsils.

toxins: poisons. Toxins produced by bacteria come in two varieties: *enterotoxins,* which act in the GI tract, and *neurotoxins,* which act on the nervous system.

trachea (TRAKE-ee-uh): the windpipe; the passageway from the mouth and nose to the lungs.

transferrin (trans-FERR-in): the body's iron-carrying protein.

transient ischemic attack (TIA): a temporary reduction in blood flow to the brain that causes temporary symptoms that depend on the part of the brain that is affected. Some common symptoms include light-headedness, visual disturbances, paralysis, staggering, numbness, or dysphagia.

transnasal: through the nose. A **transnasal feeding tube** is one that is inserted through the nose.

traveler's diarrhea: nausea, vomiting, and diarrhea caused by consuming food or water contaminated by any of several organisms, most commonly, *E. coli, Shigella, Campylobacter jejuni,* and *Salmonella.*

triglycerides (try-GLISS-er-rides): one of the three main classes of lipids; the chief form of fat in foods and the major storage form of fat in the body; composed of glycerol with three fatty acids attached.

tripeptide: three amino acids bonded together.

tubule: a tubelike structure that surrounds the glomerulus and descends through the nephron.

tumor: a new growth of tissue forming an abnormal mass with no function; also called a **neoplasm (NEE-oh-plazm).** Tumors that multiply out of control, threaten health, and require treatment are **malignant (ma-LIG-nant).** Tumors that stop growing without intervention or can be removed surgically and most often pose no threat to health are **benign (bee-NINE).**

turbinado (ter-bih-NOD-oh) sugar: raw (brown) sugar from which the filth has been washed; legal to sell in the United States.

type 1 diabetes: the less common type of diabetes in which the person produces no insulin at all.

type 2 diabetes: the more common type of diabetes that develops gradually and is associated with insulin resistance.

ulcer: an open sore or lesion. A **peptic ulcer** is an erosion of the top layer of cells from the mucosa of the lower esophagus, the stomach, or the small intestine. Ulcers may also develop in the mouth, upper esophagus, and other parts of the intestine, or on the skin.

ulcerative colitis (ko-LYE-tis): inflammation and ulceration of the colon.

ultrafiltration: a method of removing fluids and small to medium-sized molecules from the blood by using pressure to transfer the blood across a semipermeable membrane.

umbilical (um-BIL-ih-cul) cord: the ropelike structure through which the fetus's veins and arteries reach the placenta; the route of nourishment and oxygen into the fetus and the route of waste disposal from the fetus.

undernutrition: underconsumption of food energy or nutrients severe enough to cause disease or increased susceptibility to disease; a form of malnutrition.

unsaturated fatty acid: a fatty acid with one or more points of unsaturation where hydrogens are missing (includes monounsaturated and polyunsaturated fatty acids).

uremia (you-REE-me-ah): abnormal accumulation of nitrogen-containing substances, especially urea, in the blood; also called **azotemia (AZE-oh-TEE-me-ah).**

uremic frost: the appearance of urea crystals on the skin.

uremic syndrome: the many symptoms that may accompany the buildup of nitrogen-containing waste products in the blood including fatigue; diminished mental alertness; agitation; muscle twitches and cramps; anorexia, nausea, vomiting; inflammation of the mucous membranes of the mouth; unpleasant tastes in the mouth; dry, scaly, itchy skin; skin hemorrhages; gastritis and GI bleeding; and diarrhea.

uterus (YOO-ter-us): the womb, the muscular organ within which the infant develops before birth.

vagotomy (vay-GOT-o-mee): surgery that severs the nerves that stimulate gastric acid secretion.

varices (VAIR-ih-seez): blood vessels that have become twisted and distended.

variety (dietary): eating a wide selection of foods within and among the major food groups (the opposite of monotony).

vegans: people who exclude all animal-derived foods (including meat, poultry, fish, shellfish, eggs, cheese, and milk) from their diets; also called *strict vegetarians* or *total vegetarians.*

vein: a vessel that carries blood back to the heart.

villi (VILL-ee or VILL-eye): fingerlike projections from the folds of the small intestine. The singular form is **villus.**

vitamins: essential, noncaloric, organic nutrients needed in tiny amounts in the diet.

VLDL (very-low-density lipoprotein): the type of lipoprotein made primarily by liver cells to transport lipids to various tissues in the body; composed primarily of triglycerides.

voluntary activities: the component of a person's daily energy expenditure that involves conscious and deliberate muscular work—walking, lifting, climbing, and other physical activities.

waist circumference: a measurement used to assess a person's abdominal fat.

water balance: the balance between water intake and water excretion, which keeps the body's water content constant.

water intoxication: the rare condition in which body water contents are too high. The symptoms may include confusion, convulsion, coma, and even death in extreme cases.

weight cycling: repeated cycles of weight loss and subsequent regain that affect body composition and metabolism. With intermittent dieting, a person rebounds to a higher weight (and a higher body fat content) after each round. The weight-cycling pattern is popularly called the *ratchet effect* or *yo-yo effect* of dieting.

weight training (also called **resistance training**): the use of free weights or weight machines to provide resistance for developing muscle strength and endurance. A person's own body weight may also be used to provide resistance, as when a person does push-ups, pull-ups, or abdominal crunches.

wellness: maximum well-being; the top range of health states; the goal of the person who strives toward realizing his or her full potential physically, mentally, emotionally, spiritually, and socially.

white sugar: pure sucrose, produced by dissolving, concentrating, and recrystallizing raw sugar.

WHO (World Health Organization): an international agency that promotes cooperation for health among nations. WHO carries out programs to control and eradicate disease and strives to improve the quality of human life.

Wilson's disease: an inherited disorder in which people absorb too much copper from the intestine, and they have too little of the protein that transports copper from the liver to the sites where it is needed.

Zollinger-Ellison syndrome: marked hypersecretion of gastric acid and consequent peptic ulcers caused by a tumor of the pancreas.

This index lists primarily topics that received significant mention in the text. Inclusive pages (for example, 53–56) indicate major discussions; pages in **bold-face** refer to defined terms; pages in *italics* refer to figures, diagrams, or chemical structures; pages followed by a "t" refer to tables; pages followed by an "n" refer to notes; letter-number combinations (such as C-24) refer to appendixes.

A

Abdominal distention, tube feeding and, 419t
Abscess, **388, 449,** 450
Absorption, 105-107, *106. See also* Digestion; Malabsorption; *specific nutrients*
   alcohol effects on, 560, *560*
   calcium, 209
   diet-medication interactions and, 317-318, 318t
   formula selection and, 411
   intestinal resection and, 392-393, *393*
   iron, 188, 215-216
   nutrient release after, 107
   vitamin B_{12}, 185
Acarbose, 482
Accelerated metabolism, fever and wasting due to, 121
Acceptable Daily Intake. *See* ADI
ACE inhibitors, food interactions with, 507
Acesulfame potassium, **39,** 40
Acetone breath, **468**
Acetyl CoA, **122**
Achalasia, **359.** *See also* dysphagia
Acid-base balance, *79,* **80,** 202
   proteins and, 79-80
Acid indigestion, 103. *See also* indigestion, reflux esophagitis
Acidity, gastric, 362, 365
Acidosis, 79, **80,** 468, 529
Acids, **79**
Acquired immune deficiency syndrome (AIDS), **575-582.** *See also* Human immunodeficiency virus (HIV)
Activities of daily living, **323**
   in health history, 324-325
Activity. *See* Exercise
Acupuncture, 585, **586**
Acute gastritis, 364-365
Acute heart failure, **508**-509
Acute malnutrition, 82-84, **452**
   stress and, 452
Acute pancreatitis, 388
Acute phase (stress response), **451**

Acute renal failure, 525-528
Acute respiratory distress syndrome (ARDS), **511**
Acute respiratory failure, **511**-512
Acute stress, **449.** *See also* Stress
ADA. *See* American Dietetic Association
Adaptation, after intestinal resection, 392-393
Adaptive feeding devices, *377,* 377-378
Adaptive phase (stress response), **451**
Adaptogens, 195t
Addiction, alcohol, 559. *See also* Alcohol
Additives, in intravenous solutions, 432
Adenomas, **563**
Adequacy (dietary), **14**
Adequate Intakes. *See* AI
ADH. *See* Antidiuretic hormone
ADHD (attention deficit hyperactivity disorder), **270**
   amphetamines and, 372
ADI (Acceptable Daily Intake), **40**
Adipose tissue, **53.** *See also* Fat cells; Fat/lipids (in body)
Admixtures, TPN, 438
Adolescence, **275**
Adolescents, 275-281. *See also* Children
   with diabetes, 485-486
   drug use in, 278-279
   energy and nutrient needs, 276-277
   food choices and health habits of, 277-278, 278t, 279t
   growth and development, 276
   pregnancy in, 241. *See also* Pregnancy
   smoking in, 280
Adult(s)
   with cystic fibrosis, 390
   influence on adolescent eating, 278
   older. *See* Older adults
Adult bone loss. *See* Osteoporosis
Adult-onset diabetes, 465t. *See also* Noninsulin-dependent diabetes mellitus (NIDDM)
Adult rickets. *See* Osteomalacia
Advance directive, **445,** 446-447
Adverse reactions, food allergies vs., 270
African Americans. *See* Black Americans
AGA (appropriate for gestational age), **230**
Age/aging. *See also* Adolescents; Children; Infant(s); Older adults
   BMR and, 129t
   food choices and, 7
   physiological vs. chronological, 291
   slowing process of, 290-292
   U.S. population, 290, *290, 291*
Agriculture, U.S. Department of (USDA)
   Daily Food Guide, 18-19
   Elderly Nutrition Program, 354

   food assistance programs for pregnant women, 235, 354
   Food Stamp Program, 354
   Meat and Poultry Hotline, 115n
   School Lunch and Breakfast Programs, 354
AI (Adequate Intakes), **10**-11
   calcium, 208, 235
   fluoride, 220
   uses of, 12t
AIDS. *See* Acquired immune deficiency syndrome
AIDS-related complex (ARC), **576**
Alanine transaminase (ALT), 333t, 543
   in liver disease, 546t
Albumin, 333t, 334, 334t, 335t
   in liver disease, 546t
   in nephrotic syndrome, 523
Alcohol, 9
   abuse, 279-280, 558-561
   addiction to, 559
   adolescent abuse, 279-280
   amount constituting one drink, 472, 558
   in breast milk, 243
   cirrhosis and, 545, 552
   coronary heart disease and, 506
   diabetes and, 472
   diet and, 560-561
   excess energy from, 123
   folate and, 184, 560
   in food group plans, *19*
   hepatitis and, 545
   hypertension and, 506
   kcalories in, 559t
   long-term effects of abuse, 559
   metabolic effects of, 558-561
   moderate use of, 558
   nutrition status and, 558-561
   in pregnancy, 240-241
   sugar, 39
   tyramine and, 320t
   vitamin B_6 and, 560
   vitamin B_{12} and, 5394
   weight loss and, 151
Alcoholism, 559. *See also* Alcohol
Aldendronate sodium, nutrient effects on metabolism of, 320
Aldosterone, **201,** *546,* 547
   responses to severe stress, 450t
Alitame, **39,** 41
Alkaline phosphatase, in liver disease, 546t
Alkalosis, 79, **80**
All-in-one admixtures, 438
Allergy. *See* Food allergies
Allopurinol, 536
Aloe, 588t

Alpha-adrenergic blockers, 507
Alpha-glucosidase inhibitors, 482
Alpha-lactalbumin, **246**
Alprazolam, 398
ALT. *See* Alanine transaminase
Alternative sweeteners, 39-41, 472
Alternative therapies, 585-589, **586**, 588t
    anticoagulants and, 507
    cancer and, 570
    constipation and, 380
    examples of, 585, 585t
    health care professionals and, 589
    liver disease and, 544
    reasons for skepticism about, 586-587
    risks of, *587*, 587-589
    uses of, 585-586
Aluminum
    Alzheimer's disease and, 296
    in antacids, 371, 532
Alveoli, **511**
Alzheimer's disease, **296-297**, **340**
American Academy of Pediatrics
    on breastfeeding, 241
    on infant formulas, 249
American Dietetic Association (ADA)
    on breastfeeding, 241-242
    on supplement labeling, 197
Amiloride, 507
Amino acid(s), **75**, *75. See also* Protein; *specific*
      amino acids
    conditionally essential, **76**, 455, 462
    essential, **76**-77
    intravenous solutions, 431
    limiting, **86**
    nonessential, **76**
    sulfur-containing, 210
    supplements, 85-86
Amino acid metabolism, 123
    hepatic coma and, 547
    vitamin C and, 187
Ammonia, 546t
    elevated blood levels in cirrhosis, 547
    production in body, 547
Amniotic sac, **230**
Amoxicillin, food interactions with, 371
Amylase, **102**
    in digestion, 103
    serum, 333t
Anabolic agents, **578**, 579
Anabolism, **121**, *122*
Analgesics, lower GI tract and, 398
Anemia
    after gastrectomy, 369
    iron-deficiency, **212**-213, 265-266
    macrocytic, 185
    megaloblastic, 185
    microcytic hypochromic, **212**
    milk, 252
    nephrotic syndrome and, 524
    pernicious, **185**-186, **365**
    renal failure and, 529
    sports, 215
Anencephaly, **184**

Aneurysm, **496**, 498-499
Angina, **496**
Anorexia. *See also* Appetite
    cancer and, 567-568, 569t, 572
    HIV and, infection, 577, 578
    illness and, 314
    medications and, 315
Anorexia nervosa, 156, **160-163**
    characteristics of, 161-162
    diagnostic criteria for, 163t
    groups vulnerable to, 160-161
    treatment of, 162-163
Antacids, 103, 363
    renal failure and, 532
Anthropometric, **327**
Anthropometric measurements, 327t, 327-331,
      329t, 330t, E-9 to E-11
Antianxiety agents, 398
Antibiotics. *See also* Anti-infective agents
    for peptic ulcers, 371
Antibodies, **80**, 269
    proteins and, 80
Anticoagulants
    food interactions with, 507
    vitamin C and, 188
    vitamin K and, 318
Antidepressants, 399
Antidiabetic agents, oral, **472**, 482
Antidiarrheals, lower GI tract and, 399
Antidiuretic hormone (ADH), **200-201**
    responses to severe stress, 450t
Antiemetics, 399
    food interactions with, 371
Antiflatulents, 399
Antigens, 269
Antigout agents, 536
Antihistamine, vitamin C as, 187-188
Antihypertensives, 507
Anti-infective agents. *See also antibiotics and*
      antiviral
    food interactions with, 399, 536, 549, 579
    stress and, 453
Anti-inflammatory agents
    for inflammatory bowel disease, 399
    peptic ulcer and, 365
Antilipemics, food interactions with, 507
Antimicrobials, HIV infection and, 579
Antinauseants, food interactions with, 371
Antineoplastic agents, **568**, 579
Antioxidants, **169**. *See also specific antioxidants*
    beta-carotene, 169
    cancer and, 196, 197, 565
    cataracts and, 294
    coronary heart disease and, 506, 517-518
    recommendations, 10
    supplements, 196, 197
    vitamin C, 187
    vitamin E, 175
Antipromoters (cancer), **565-566**
Antisecretory agents, **363**
    food interactions with, 371
Antiulcer agents, **363**
Antiviral agents, **578**, 579

Anus, **96**
Apolipoprotein E, Alzheimer's disease
    and, 296n
Appendicitis, fiber and, 42
Appendix, **96**
Appetite. *See also* Anorexia
    activity and, 152
    in children, 262
    depression and, 303-304, 338-339
    stimulants, 549
    stress effects on, 456
Appropriate for gestational age (AGA), **230**
ARC (AIDS-related complex), **576**
ARDS (acute respiratory distress syndrome), **511**
Arginine, in stress formulas, 457
Ariboflavinosis, 191t
Aroma therapy, **586**
Arsenic, 221
Artery, **108**
    hepatic, 545
Arthritis, **294-295**, 295t
Artificial feeding, **445**. *See also* enteral and
      parenteral nutrition
Artificial sweeteners, 39-41
    recommendations, 41
Ascites, **547**
Ascorbic acid, **187**. *See also* Vitamin C
Aspartame, **39**, 40
Aspartate transaminase (AST), 333t, 543
    in liver disease, 546t
Aspiration, **359**
    dysphagia and, 359
    tube feeding and, 415, 419t
Aspirin
    folate and, 319-320
    food interactions with, 507
    in pregnancy, 240
Assessment. *See* Nutrition assessment
Assistants, skilled, 427
AST. *See* Aspartate transaminase
Asterixis, **547**
Atherosclerosis, **58**, **284**, 285, **496**. *See also*
      Cardiovascular disease; Cholesterol; Heart
      disease
    chronic renal failure and, 529
    coronary heart disease and, 496-502
    development of, 496-498, *497*
    HIV infection and, 576
    hypertension and, 499
Athletes. *See also* Exercise; Fitness
    eating disorders in, 160-161
    ergogenic aids used by, 90-94, 91t, 93t
    fluid balance in, 226t, 226-227, 227t
    iron deficiency in, 214-215
Atrophic gastritis, **300**, **365**
Attention deficit hyperactivity disorder
    (ADHD), **270**
Autoimmune disorders, **466**
Aversions
    pregnancy and, 235
    treatment related, 573
Avidin, **182**
Ayurveda, **586**

Azathioprine, 536
Azotemia, **526**

B

B vitamins, 178-186, 191t-193t. *See also* individual B vitamins
    Alzheimer's disease and, 297
    cirrhosis and, 552
    coronary heart disease and, 518
    deficiencies, 180-181, 340
    enrichment of foods with, 180
    interdependent systems and, 180
Bacteria. *See also* Food poisoning; Infection; Intestinal flora
    foods vulnerable to, 115-116
    kitchen safety and, 115-116
    peptic ulcer and, 365
    translocation, 462, 463
Bacterial overgrowth, malabsorption and, 391-392, *392*
Balance (dietary), **15**
Barrett's esophagus, **360**
Basal energy expenditure (BEE), Harris-Benedict equation and, 454t
Basal metabolic rate (BMR), 128-129
    activity and, 151-152
    factors affecting, 129t
Basal metabolism, **128**
Bases, **79**
Bedsores, **315**
BEE. *See* Basal energy expenditure
Bee pollen, 91t
Behavior
    iron deficiency in children and, 265-266
    malnutrition in children and, 267t
    strategies in feeding disabilities, 376t
    sugar and, 37
Behavior modification, **153**
    for weight loss, *153*, 153-155
Benign, **563**
Beriberi, **180**
Beta-blockers, 507
Beta-carotene, **168,** 170t, 178t. *See also* Vitamin A
    cancer and, 169
    conversion and toxicity, 171-172
    in foods, 172, *172*
    metabolic roles of, 169
Beverages. *See also* Water
    alcoholic. *See* Alcohol
    cola, 34, 35t
    sports drinks, 226-227
    tyramine-containing, 320t
Bicarbonate, **102**
Bifidus factors, **247**
Bile, **58, 102**
    in digestive process, 104, *104*
Biliary cirrhosis, **545**
Bilirubin, **431,** 543
    IV lipid emulsions and, 431-432
    in liver disease, 546t
Binge eating disorder, **160,** 164

Bioavailability, **184**
Biochemical analysis, 332-335, 333t-335t. *See also* Laboratory tests
Bioelectromagnetic medical applications, **586**
Biofeedback, **586**
Biofield therapeutics, **586**
Bioflavonoids, 187
Biotin, 182, 192t. *See also* B vitamins
    deficiency. *See* Biotin deficiency
    in foods, 182
    intestinal flora and, 104
    toxicity, 192t
Biotin deficiency, 182, 192t
Birth defects
    alcohol-related, 241
    folate deficiency and, 232
Birthweight, 230
    low, **230**
Black Americans, ethnic diet of, 4t
Black tea, coronary heart disease and, 507
Bladder cancer, 565t
Bland diet, 364-365, 365t
Blenderized formulas, 407. *See also* Pureed foods
Blind loop syndrome, **392**
Blindness
    night, **168**
    vitamin A deficiency and, 169-170
Blood
    albumin in nephrotic syndrome, 523
    ammonia levels, 547
    hemoglobin concentration, 213
Blood cholesterol. *See* Cholesterol
Blood clotting, *176*. *See also* Anticoagulants
    atherosclerosis and, 498
    vitamin K and, 176-177
Blood doping, 91t
Blood glucose. *See also* Glucose; Hyperglycemia; Hypoglycemia
    control in older adults, 488
    diabetes mellitus and, 466-467
    dumping syndrome and, 366
    fasting, 466
    monitoring, 478t, 484, 493
    monitoring during pregnancy, 487
    physical activity and, 480-481
Blood pressure. *See also* Hypertension
    atherosclerosis and, 496
    fluid exchange and, 498, *499*
    optimal, 498
    renal regulation of, 527
    vitamin E and, 506
Blood tests. *See* Laboratory tests; *specific type*
Blood urea nitrogen (BUN)
    in renal failure, 526
    tests, 333t
Blood vessels. *See* Cardiovascular system; Vascular system
BMI. *See* Body mass index
BMR. *See* Basal metabolic rate
Body
    energy budget, 123-130, *124, 125, 127,* 129t, 130t
    energy storage in, 9

fiber in, 35-36
Body composition, 140-141. *See also* Weight
    BMR and, 129t
    cancer and, 566
    HIV infection and, 576-577
    measures of, 327, E1-E10
    severe stess and, 450
Body fat. *See* Fat/lipids (in body)
Body fluids. *See* Fluid(s)
Body image, food choices related to, 7
Body mass index (BMI), **139-140,** *140,* 328
    disease risks and, 139t, 139-140, 501t
    pregnancy weight gain and, 237
    risk for diabetes, 466
Body temperature. *See also* Fever
    metabolism and, 120-121
Body weight. *See* Weight
Bolus, **97**
Bolus feeding, **415**
Bone
    calcium in, 206
    chronic renal failure and, 529
    disease after gastric surgery, 369
    malnutrition in children and, 267t
    minerals during pregnancy and fetal development of, 233-234
    nephrotic syndrome and, 524
    osteoporosis and, 22-23, 206-207, 207t
    prevention of loss, 206-207
    sodium and, 204
    structure of, *206*
    vitamin D and, 173
Bone marrow transplant, **568**-569
    nutrient losses and, 568
    oral diets after, 575
    parenteral nutrition after, 575
    tissue rejection after, 569
Bottled water, 117
Botulism, **113**
    infant, 116
Brain. *See also* Nervous system
    aging, 295-297
    alcohol effects on, 559
    cancer, 571t
    glucose and, 126
    nutrient deficiencies and, 297, 297t
Bread
    carbohydrate content, 45
    cariogenicity, 51t
    fats in, 63-64
    fiber content, *44,* 349t
    folate fortification, 184
    in food group plans, *18*
    protein content, 87t
Breast
    cancer, 565t
    fibrocystic disease, 175
Breast milk, *246,* 246-247. *See also* Breastfeeding
    alcohol in, 243
    drugs in, 244
    expressing, 259
    foods affecting flavor of, 243

Breastfeeding, 258-259, *259. See also* Breast milk; Lactation
  contraindications to, 243-244
  infants with cystic fibrosis and, 390
Breath, acetone, **468**
Bronchitis, **512**-514
Brown sugar, **38**
Buffalo hump, **576**
Buffers, **80**, 202
Bulimia nervosa, 156, **160**
  anorexia nervosa vs., 163-164
  consequences of, 164
Bulimia nervosa—continued
  diagnostic criteria for, 163t
  groups vulnerable to, 160-161
  treatment of, 164
BUN. *See* Blood urea nitrogen

C

Cachexia
  cancer, 566, *567*
  cardiac, **509**
CAD (coronary artery disease), **496**
Caffeine
  athletes and, 92
  effects on medication metabolism, 321
  lactation and, 243
  in pregnancy, 240
Calciferol, 178t. *See also* Vitamin D
Calcitonin, **206**
Calcitriol, **532**
Calcium, 205-209, 222t. *See also* Bone
  absorption of, 209
  adolescent needs, 277
  in antacids, 371
  in body fluids, 206
  deficiency. *See* Calcium deficiency
  in foods, 35t, *208*, 208-209, A
  hypertension and, 207-208, 506
  older adults and, 301
  osteoporosis and, 22-23, 206-207
  oxalate and, 386, 521, 522
  pregnancy and, 233-234, 234t
  recommendations, 208t, 208-209
  renal disease and, 522
  supplements in chronic renal failure, 530t, 532
  tests, 333t
  toxicity, 222t
  vegan diet and, 73
Calcium channel blockers, 507
Calcium deficiency, 206-207, 207t, 222t
  cirrhosis and, 552
  fat malabsorption and, *385*, 386
  gastric surgery and, 369
  lactose intolerance and, 384
  tooth development and, 50t
Calcium pangamate, 91t
Calcium rigor, **206**
Calcium tetany, **206**
Calculations, F-0 to F-1
Calculi, renal, **521**-523
Calories. *See* kCalories

Calorimetry, indirect, **454**
Camps, for children with diabetes, 494
Canada
  choice system, B-1 to B-13
  food labeling, B-14 to B-15
  Recommended Nutrient Intakes. *See* RNI
Canada's Food Guide to Healthy Eating, B-15 to B-17
Cancer(s), **563**-575. *See also specific location*
  alternative therapies, 570
  anorexia in, 567-568, 569t
  antioxidants and, 196, 197, 565
  beta-carotene and, 169
  development, 563-564, *564*
  dietary factors associated with, 565t, 565-566
  fat and, 24
  fiber and, 24, 42, 565
  fruits and vegetables and, 24, 565-566
  nutrition assessment in, 582
  nutrition consequences of, 567-568
  nutrition support in, 571-575
  omega-3 fatty acids and, 59
  prevention, 565t, 565-566
  therapy effects on nutrition status, 568, 569t
  treatment, 568-570
  vitamin C and, 188
Cancer cachexia syndrome, **566**, *567*
Candidiasis, **577**
Canned foods, home canning and, 113
Capillary, **108**
Capitation, **425**
Capsaicin, 170t
Carbohydrate(s) (in body), **32**-51. *See also* Fiber; Sugar(s)
  chemistry of, 32-41
  dental health and, 50-51
  excess of, 123
  stress effects on metabolism, 450
Carbohydrate(s) (in diet), 8, *32*, 45-47. *See also* Diet
  counting, 474, 481
  diabetes and, 471-472, 479t, 481
  in foods, A
  health and, 36-43
  hypertension and, 505-506
  intravenous solutions, 431
  kcalories in, 8
  needs during stress, 456
  older adults and, 299
  pregnancy and, 231
  protein-sparing effect, 87
  recommendations, 36, *37*, 45-47, *46*, 231
  weight loss and, 150-151
Carbon dioxide, tests, 333t
Carcinogen, **564**-565, 565t. *See also* Cancer
Carcinogenic, **564**
Carcinomas, **563**
Cardiac arrest, **508**-509
Cardiac cachexia, **509**
Cardiac cirrhosis, **545**
Cardiac glycosides, 509
  food interactions with, 507
Cardiac sphincter, **96**
  in achalasia, 359

  in reflux esophagitis, 362
  substances weakening, 360t
Cardiomegaly, **509**
Cardiovascular disease (CVD), **58, 284, 496**-518. *See also* Atherosclerosis; Cholesterol; Heart disease; Hypertension
  chronic renal failure and, 529
  diabetes and, 469-470
  fat and, 58
  nutrition assessment in, 514
  prevention in children, 284-287
Cardiovascular system. *See also* Cardiovascular disease (CVD); Heart; Vascular system
  malnutrition in children and, 267t
  metabolic function of, 120
  in nutrition assessment, 332t
Care maps, **343**-344
Care plan, 342-345, 343t
Caries. *See* Dental caries
Cariogenic, **50**
Carnitine, 91, 186
Carotenemia, **172**
Carotenoids, 170t
Carrier, **133**
Cartilage therapy, **586**
Casein, **246**
Catabolism, **121**-122, *122*
Cataracts, **294**
Catechin, 170t
Catecholamines, responses to severe stress, 450t
Cathartic, **160**, 164, **205**
Catheter
  complications related to, 436, 437t
  home TPN, 441-442
  insertion and care in TPN, 435-436
  IV, **434**
  peripherally inserted central, **435**
cc (cubic centimeter), 408
CD4+ lymphocyte, **576**
CDC. *See* Centers for Disease Control
Celiac disease, **393**-394
Celiac sprue, **393**-394
Cell(s). *See also* Red blood cells; White blood cells
  epithelial, **169**
  fat, *53*, 123, *125*, 143
  immune system, *461*, 461-462
Cell division, vitamin B$_{12}$ and folate and, 184-185
Cell salts, 91t
Cellulose, 35
  chemical structure of, *36*
Cellulose sodium phosphate, 536
Centers for Disease Control (CDC), 243
Central nervous system. *See* Brain; Nervous system
Central obesity, **140**-141
  diabetes and, 466
  heart disease and, 502
Central total parenteral nutrition, **434**-435, 436t. *See also* Total parenteral nutrition (TPN)
Central veins, **434**, *435*
Cereals. *See also* Bread; Grains
  cariogenicity, 51t
  in food group plans, *18*

protein content, 87t
Cerebral cortex, **296**
Cerebral thrombosis, **496**
Cerebral vascular accident (CVA), **510**
Cesarean section, **238**
Chamomile, 588t
Chaparral, 545, 588t
Charts. *See* Medical record
CHD. *See* Coronary heart disease
Cheese
    fat in, 64
    in food group plans, *19*
    protein content, 87t
    tyramine-containing, 320t
Chelation therapy, **586**
Chemical reactions, in body, 121-123
Chemistry
    of carbohydrates, 32-41
    of lipids, *54,* 54-58, *55,* 57t
    of proteins, 75
Chemotherapeutic agents, **568**
Chemotherapy, **568**
    effects on nutrition status, 568, 569t
    nutrient interactions, 579
Chewing difficulties, 357t, 357-359
    in cancer, 573
Chewing gum, sugar alcohols in, 39
CHF (chronic heart failure), **509**-510
Chicken. *See* Poultry
Child Care Food Program, 354
Children, 262-287. *See also* Adolescents; Infant(s)
    choking prevention in, 273
    with chronic renal failure, 529-530
    with cystic fibrosis, 390
    with diabetes, 485-486, 494
    energy needs, 262
    fluid balance in, 227, 228
    food allergies in, 269-270, 270t
    food assistance programs for, 265, 266, 354
    food choices and eating habits of, 263, *264,*
        271-275, 273t, 274t
    food dislikes of, 270
    growth and development of. *See* Growth
    growth charts, E-1 to E-8
    hyperactivity in, 270-271
    hypertension in, 285, 285t
    infusion pumps and, 416, 436
    iron deficiency in, 265-266
    lead poisoning in, 266-269, 267t, *268*
    malnutrition in, 264-266, 267t. *See also*
        Protein-energy malnutrition (PEM)
    mental effects of malnutrition in, 339-340
    obesity and disease development in,
        284-287, 285t
    PEM in. *See* Protein-energy malnutrition
        (PEM)
    television's influence on, 272-273
    vitamin D for, 174
    vomiting in, 363
Chinese Americans, ethnic diet of, 4t
Chiropractic, 585, **586**
Chlordiazepoxide, 398
Chloride, 204-205, 222t
    deficiency, 222t

tests, 333t
toxicity, 222t
Chlorpropamide, 482
Choking, prevention in children, 273
Cholecalciferol, 174, 178t. *See also* Vitamin D
Cholesterol (blood). *See also* Atherosclerosis;
        Cardiovascular disease (CVD); Fat; Heart
        disease
    biochemical tests, 333t
    in children and adolescents, 285, 285t,
        286-287
    coronary heart disease and, 496, 500, 501t
    diet for lowering, 505
    excretion of, 58
    monounsaturated fatty acids and, 59
    omega-3 fatty acids and, 59
    recycling in body, 58
    routes in body, 58
    saturated fat and, 59
    screening in children, 286-287
    synthesis of, 57-58
    transport of, 58
Cholesterol (dietary)
    in foods, 57, 63-64, A
    health and, 23, 60
Cholesterol-free (on labels), 25t
Cholestyramine, 507
Choline, **57,** 186
Chromium, 221, 223t
Chromium picolinate, 92
Chronic diseases, **13**
    childhood obesity and, 284-287, 285t
    obesity and, 145
Chronic gastritis, 365
Chronic heart failure (CHF), **509**-510
Chronic malnutrition, **452**
    stress and, 452
Chronic obstructive pulmonary disease (COPD),
        512-514
    consequences, 513
    treatment, 513-514
Chronic pancreatitis, 389
Chronic renal failure, 528-534
Chronic stress, **450.** *See also* Stress
Chylomicrons, **107**
    lipoproteins and, 107
Chyme, **97,** 101
Cigarette smoking. *See* Smoking
Cirrhosis, **544**-553
    biliary, **545**
    cardiac, **545**
    consequences of, 545-548, *546*
    diet therapy for, 550t, 550-553
    drug therapy for, 548-549
    idiopathic, **545**
    Laennec's, **545**
    nutrition assessment in, 555
    postnecrotic, **545**
    wasting in, 548, 548t
Cis-fatty acids, 67, *67*
Claudication, intermittent, **175**
Clear liquids, **346,** 347t
Client education
    in care plan, 343-344

in diabetes management, 494
Clients
    helping to accept oral formulas, 409
    helping to comply with renal diet, 533
    helping to cope with tube feedings, 412
    helping to eat in hospital, 349, 351
    helping to implement DASH diet, 504-505
Clinical pathways, 343-344
Clinically severe obesity, **148**
Clonidine, 507
Closed system (nutrition formula), **414**
Clotting. *See* Blood clotting
CoA, **122**
Coagulation. *See* Blood clotting
Cobalt, 221
Cocaine
    in adolescence, 279
    in pregnancy, 240
Coenzyme(s), **178**
    B vitamin, 178-179, *180*
Coenzyme Q10, 91t
Cofactor, **206**
Cognitive therapy, **160**
    for anorexia nervosa, 163
Cola beverages, 34, 35t
Colchicine, 536
Colestipol, 507
Colitis, ulcerative, 390-391
Collagen, **187**
    vitamin C in formation of, 187
Collaterals, **546**
Colloids, hydrophilic, **380**
Colon, **96,** 100
    cancer. *See* Colon cancer
    in digestive process, 104-105
    diverticular disease of, 394-395
    resection, 392-393, *393*
    residue, **348**-349, 408
Colon cancer, 565t, 571t
    dietary modifications, 571t
    fiber and, 42
Colorectal cancer, 565t. *See also* Colon cancer
Colostomy, *396,* **396**-397
Colostrum, **247**
Coma
    in diabetes, 468
    diabetic, **468**
    hepatic, **547**
    hyperosmolar hyperglycemic nonketotic, **468**
Comatose, **445**
Comfrey, 588t
Communication, professional, 350
Community nutrition, 353-355, 355t
Complementary proteins, **71,** 87
Complementary therapy, **586**
Complete formulas, **407**
Complete protein, **86**
Complex carbohydrates, **32,** 41
Computer web sites. *See* Web sites
Concentrated fruit juice sweetener, **38**
Conditionally essential amino acid, **76,** 455, 462
Confectioner's sugar, **38**
Congestive heart failure (CHF), **509**-510
Constipation, **380**-381

Constipation—continued
cancer and, 574
fiber and, 42
high-fiber formulas and, 408
HIV infections and, *574*
in pregnancy, 239
sedatives and, 453
tube feeding and, 419t
Contaminants
infant formula, 249
lead. *See* Lead poisoning
in water, 117
Continuous feeding, **415**
Continuous renal replacement therapy (CRRT),
526, 527, **539,** 540-541
Conversion factors, F-0
Cooking/food preparation
older adults and, 310-311
safety, 115-116
COPD. *See* Chronic obstructive pulmonary
disease
Copper, 219
deficiency, 219
in foods, 219
pregnancy and, 234t
recommendations, 219
toxicity, 219
Corn sweeteners, **38**
Corn syrup, **38**
Cornea, **168**
Coronary artery disease (CAD), **496**
Coronary heart disease (CHD), **284, 496.** *See also*
Heart disease; *specific risk factors*
antioxidants and, 506, 517-518
atherosclerosis and, 496-498
fat and, 24
fiber and, 24
physical activity and, 501
prevention and treatment, 500-507
risk factors, 499-500, 501t
vitamin E and, 518
Coronary thrombosis, **496**
Cortisol, **487**
responses to severe stress, 450t
Cost
diabetes in older adults and, 488
health care, 424-428, *426*
home nutrition support, 441
Coumadin. *See* Warfarin
Counterregulatory hormones, **449, 450**
Cow's milk. *See also* Milk and milk products
introduction in children, 249
CPK. *See* Creatine phosphokinase
Cramps, tube feeding and, 419t
Cranberry juice, urinary tract infections
and, 522
Cravings, pregnancy and, 235
Creatine, 92
Creatine phosphokinase (CPK), 333t
Creatinine, serum, 333t
Cretinism, **219**
Crib death, 240
Critical pathways, **343**-344
Crohn's disease, **390**-391

CRRT (continuous renal replacement therapy),
526, 527, **539,** 540-541
Cruciferous vegetables, 566
Cruzan, Nancy, 446
Cubic centimeter (cc), 408
Curcumin, 170t
CVA (cerebral vascular accident), **510**
CVD. *See* Cardiovascular disease
Cyanocobalamin. *See* Vitamin B_{12}
Cyanosis, **512**
Cyclamate, **39,** 41
Cyclic parenteral nutrition, **438**
Cyclosporine, 536, 549
Cysteine, sulfur in, 210
Cystic fibrosis, **389**-390
Cystine stones, 523
Cystinuria, 523
Cytokines, **449**
cancer cachexia and, 566
COPD and, 513

D

Daily Food Guide, **17**-20, *18-19*
for vegetarians, 73t
Daily Values, **21**-22
Dairy products. *See* Cheese; Eggs; Milk and milk
products
DASH (Dietary Approaches to Stop
Hypertension) diet, 502, 503t, 504-505
Dawn phenomenon, **468**
Deamination, **123**
Death, SIDS, 240
Debridement, **453**
Decubitus ulcers, **315**
Defecation, 100. *See also* Constipation; Diarrhea
Deficiencies. *See also specific nutrients*
B vitamin, 180-181, 340
biotin, 182, 192t
brain function and, 297, 297t
calcium, 206-207, 207t, 222t, 384, 552
in children, 265-266, 267t
chloride, 222t
chromium, 223t
community programs for prevention of, 354
copper, 219
fluoride, 220
folate, 192t-193t, 340, 369
after gastric bypass, 369
after gastric surgery, 368
iodine, 218-219, 223t
iron, 212-214, 223t, 265-266, 1189
lactase. *See* Lactose intolerance
magnesium, 210, 222t
niacin, 191t
pantothenic acid, 192t
potassium, 205, 222t
riboflavin, 191t
selenium, 218, 223t
sodium, 222t
thiamin, 191t
tooth development and, 50t
vitamin A, 169, 178t
vitamin B_6, 183, 192t

vitamin B_{12}, 186, 193t, 369
vitamin C, 188, 193t
vitamin D, 173, 178t, 384
vitamin E, 175, 179t
vitamin K, 176-177, 179t
zinc, 216-217, 223t
Degenerative diseases. *See also specific diseases*
prevention in older adults, 292-293, *293*
Dehydration, **200,** 226-228. *See also* Hydration
diabetes mellitus and, 467-468
diarrhea and, 381-383
physical signs of, 226, 226t, 331, 332t
tube feeding and, 419t
vomiting and, 363
7-Dehydrocholesterol, 173
Dehydroepiandrosterone. *See* DHEA
DEJ (direct endoscopic jejunostomy), **410**
Delavirdine, 579
Delusions, **340**
Dementia, **340**
AIDS and, 577
senile, **296**-297
Denaturation, 79, **80**
Dental caries, **38,** *50,* 50-51. *See also* Teeth
fluoride and, 220
prevention in children, 273
recommendations for control, 51t
sugar and, 38, 50, 51
Dental plaque, **50**
Dental soft diet, **357**
Deoxyribonucleic acid (DNA), 91t
Depression
food intake and, 303-304
nutrition status and, 338-339, 339t
Developing countries, diarrhea and dehydration
in, 227-228, 228t
Development. *See* Children; Growth
Dextrose, **38**
in central TPN solution, 434
monohydrate, **431**
in peripheral TPN solution, 433
DHA (docosahexaenoic acid), **56**
DHEA (dehydroepiandrosterone), supplements
in athletes, 93-94
Diabetes mellitus, **465**-494. *See also* Glucose;
Hyperglycemia
in children, 485-486, 494
complications, *467,* 467-470
control, 492-494
diagnosis, 466-467
fiber and, 42
gestational, **238,** 487
insulin-dependent, 465t, *467,* 468
intensive therapy, 492-493
monitoring and management, 484
noninsulin-dependent, 465t, *467,* 468
in older adults, 488
in pregnancy, 238, 486-488
sugar and, 37
traditional therapy, 492
treatment, 470-484, 492-494
types of, 465t, 465-466
Diabetic coma, **468**
Diadzein, 170t

Diagnostic tests. *See* Laboratory tests; *specific diagnosis or test*
Dialysate, **539**
Dialysis, **526, 539**-541, *540*. *See also* Renal failure
Dialzyer, **539**
Diaphragm, hiatal hernia and, 360-361, *361*
Diarrhea, **381**-383. *See also* Malabsorption
    breast milk and, 247
    cancer and, 573-574
    causes of, 382
    dumping syndrome and, 366
    fat malabsorption and, 385
    fiber and, 42
    fluid balance and, 227-228, 228t
    high-fiber formulas and, 408
    HIV infections, 578
    inflammatory bowel disease and, 390
    intractable, **382**
    irritable bowel syndrome and, 383
    osmotic, **382**
    ostomies and, 397
    secretory, **382**
    traveler's, **113**
    treatment of, 382
    tube feeding and, 417, 419t
Diary, weight loss and, 153, *153*
Diazepam, 398-399
Dicumarol, vitamin C and, 188
Didanosine, 579
Diet. *See also* Fasting; Foods; Medical nutrition therapy; Nutrients
    after acute stress, 456
    adjustment for chewing difficulties, 357-359, *358*
    alcohol and, 560-561
    bland, 364-365, 365t
    cancer risk and, 565t, 566
    diabetic, 471-480, 481, 486, 487-488
    edentulous, **357**-360
    ethnic, **3,** 4t-6t
    fad, **147**
    fat-restricted, 386-387, 387t, 388, 503
    fluid-restricted, 531
    gluten-free, 394, 395t
    high-fiber, 380, 383, 395
    high-kcalorie, 544
    high-protein, 544
    irritable bowel syndrome and, 383
    LDL-reduction, 502t, 503t
    liquid, 346-348, 347t
    low-carbohydrate, **147,** 240
    low-fat, 150-151
    low-fiber, 348-349, 349t, 391
    low-residue, 348-349
    low-sodium, 346, 525t
    macrobiotic, **586**
    mechanical soft, **357**-360, *358*
    modified, **345,** 347t, 348t
    oxalate-restricted, 386
    peptic ulcer and, 365
    postgastrectomy, 367-368, 368t
    progressive, **346**-349, 347t-349t, 382-383, 388, 438-439
    protein-restricted, 551

purine-restricted, 523
reflux esophagitis and, 360t, 362
regular, **345**
renal, 530-534
sodium-restricted, 506, 525, 525t, 550t, 551
soft, 348, 348t, 349
standard, **345**
therapeutic, **345**
tyramine-controlled, 320t
in ulcerative colitis, 390-391
vegetarian. *See* Vegetarian diets
very-low-kcalorie, 147-148
weight-loss, 147-148, 240
Diet diary, 153, *153*. *See also* Food record
Diet history, 324, 325-327
Diet manual, **346**
Diet-medication interactions, 302, **314,** 316-321, E-0
    antidiabetic agents and, 482
    cancer treatment and, 579
    cardiovascular disease treatment and, 507
    care plan and, 344
    classes of medications involved in, 317t
    folate and, 184
    food effects on medication absorption, 317, 318, 318t
    food effects on medication excretion, 318t, 321
    food effects on medication metabolism, 318, 318t, 320-321
    food intake and, 317, 318t
    formulas and, 417-418, 418t
    GI intolerance and tube feedings, 418
    health history and, *323,* 324
    HIV treatment and, 579
    inactive ingredients and, 321
    insulin and, 482, 483
    liver disease treatment and, 549-550
    lower GI tract disorders and, 398-399
    management strategies for, 319
    mechanisms of, 317, 318t
    medication effects on food metabolism, 318t, 318-320, 320t
    medication effects on nutrient absorption, 317-318, 318t
    medication effects on nutrient excretion, 318t, 321
    renal disease treatment and, 536
    risk factors for, 316
    stress effects on, 452, 453
    upper GI tract disorders and, 371
Diet order, **313,** 345-346
Diet pills, **147**
Diet planning
    food group plans, 17, *18-19*
    principles, 14-17
Diet therapy. *See also* Diet; Medical nutrition therapy; *specific diseases*
    after bone marrow transplant, 575
    in cancer, 570-574
    for celiac disease, 394, 395t
    for chronic heart failure, 510
    for chronic obstructive pulmonary disease, 513-514
    for chronic renal failure, 530-534

    for cirrhosis, 550t, 550-553
    for constipation, 38
    for Crohn's disease, 390-391
    for diabetes, 471-480, 481, 486, 487-488
    for dysphagia, 359-360
    after gastric partitioning, 370
    after heart attack, 508-509
    for hepatitis, 544
    in HIV infection, 580-581
    for hypertension, 502-506
    for kidney stones, 522-523
    after kidney transplant, 534, 534t
    for nephrotic syndrome, 524-525, 525t
    for PKU, 134-135
    for respiratory failure, 512
    after stroke, 510
Dietary adequacy, **14**
Dietary Approaches to Stop Hypertension (DASH) diet, 502, 503t, 504-505
Dietary balance, **15**
Dietary guidelines, 13, 14t
Dietary moderation, **15**
Dietary Reference Intakes. *See* DRI
Dietary variety, **16**
"Dieter's tea," 147
Dietetic technician registered (D.T.R.), **29**
Dietetics, **29**
Dietitians, 29, 313
    on nutrition support team, *426*
Differentiation, **168**
Diffusion, **539,** *540*
Digestibility, protein, 86
Digestion, **96**-112. *See also* GI (gastrointestinal) tract
    absorption and. *See* Absorption
    energy required for, 129-130
    in intestines, 103-105
    involuntary muscles and glands in, 99-102, *100, 101,* 103
    in mouth, 102
    nutrient transport and, 107-111, *108, 110*
    organs involved in, 96-99
    process of, 102-105, *104*
    rate of, 104
    in stomach, 103
Digestive juices, 103. *See also* Gastric juice; Pancreatic juice
Digitalis, 507
Digitoxin, 507
Digoxin, 507
Dihydroxy vitamin D, 173, 178t. *See also* Vitamin D
    cirrhosis and, 552
    fat malabsorption in, 385, 368
    renal failure and, 332
Dipeptide, **75**. *See also* Protein/amino acids (in body)
Direct endoscopic jejunostomy (DEJ), **410**
Disabilities. *See* Feeding disabilities
Disaccharides, **32,** 33t-34
Disease. *See also* Illness; *specific diseases*
    BMI values and risks of, 139t, 139-140
    chronic, **13**
    food label claims about, 24

Disease—continued
noncommunicable, **13**
Distention, tube feeding and, 419t
Disulfiram-like reaction, **482**
Diuretic(s), **205**
food interactions with, 507, 549
Diuretic abuse, **147**
Diuretic phase (acute renal failure), **527**
Diverticula, **394**-395
Diverticular disease, 394-395
Diverticulitis, **395**
Diverticulosis, **395**
fiber and, 42
DNA (deoxyribonucleic acid), 91t
Docosahexaenoic acid (DHA), **56**
Documentation, professional communication
and, 350
Dominant gene, **133**
DRI (Dietary Reference Intakes), **10**-12
uses of, 12t
Dronabinol, 549
food interactions with, 371
Drug(s). *See also* Medications; *specific drugs*
abuse by adolescents, 278-279
in breast milk, 244
diet pills, 147
inactive ingredients in, 321
in intravenous solutions, 432
for obesity, 146-147
older adults and, 302, 316
after organ transplant, 534
in peptic ulcer treatment, 365
peptic ulcers due to, 365
in pregnancy, 240
Drug-nutrient interactions. *See* Diet-medication
interactions
Dry mouth, 359
D.T.R. (dietetic technician registered), **29**
Dumping syndrome, **366**-368, *367*
postgastrectomy diet and, 367-368, 368t
Duodenum, **96**
Durable power of attorney, **445**, 446, 447
Dysentery, **83**
Dyslipidemia, **500**
Dyspepsia, **360**-363
Dysphagia, 357t, **359**-360
cancer and, 573
stroke and, 510
Dysuria, **521**

E

EAR (Estimated Average Requirements), 12t
Early satiety, cancer and, 567
Eating disorders, 156, **160**-167. *See also* Anorexia
nervosa; Bulimia nervosa
Echinacea, 588t
Eclampsia, **239**
Edema, 83-84
in chronic heart failure, 509
in chronic renal failure, 529
in cirrhosis, 546, 547
in nephrotic syndrome, 523
in pregnancy, 239

Edentulous diet, **357**-360
Eggs, fats in, 63
Eicosapentaenoic acid (EPA), **56**, 295
Elderly. *See* Older adults
Elderly Nutrition Program, 354-355
Electrolyte(s), **202,** 333t. *See also* Mineral(s);
specific electrolytes
acute renal failure and, 526-527
balance, 201-202
diarrhea and, 382
in intravenous solutions, 432
needs during stress, 454
tube feeding and, 419t
vomiting and, 363
Electrolyte solutions, **202**
Embolism, **496**
atherosclerosis and, 498
Embolus, **496**
Emetic, **160**, 164. *See also antiemetics.*
Emotions. *See also* Mental illness; Psychological
stress
diabetes management and, 493-494
food choices and, 6
nutrition status and, 303-304, 338-340
Emphysema, **512**-514
EMS. *See* Eosinophilia-myalgia syndrome
Emulsifiers, **57,** 104
Emulsion
bile and, 104
intravenous fat, 431-432, 437-438
Enamel, **50**
Encephalopathy, hepatic, **547**
End-stage renal disease, **529**
Endocrine glands, 100
Endometrial cancer, 565t
Endoscopic gastrostomy, percutaneous, **410**
Endoscopic jejunostomy
direct, **410**
percutaneous, **410**
Energy. *See specific diseases.*
for activities, 129
balance, 127-130, 129t
breaking down of nutrients for, 121-123
breast milk nutrients yielding, 246, *246*
budget, 123-130, *124, 125, 127,* 129t, 130t
cirrhosis and, 550, 550t
content in foods, A
diabetes and, 471
in diet for CHD prevention and
treatment, 502
estimation of day's output, 129
excess, 123-125, *125*
expenditure. *See* Energy expenditure
fat and, 47, 53, 63
foods dense in, 155-156
genetic influences on expenditure, 142
glycogen and, 47
measures of. *See* kCalories
needs during adolescence, 276
needs during lactation, 242
needs during stress, 454t, 454-455
needs in acute renal failure, 527
needs in chronic renal failure, 530t, 531
needs in cystic fibrosis, 389-390

needs in nephrotic syndrome, 524
needs in respiratory failure, 512
protein and, 47, 81
recommendations, 9
requirements, 129t, 129-130, 231
storage in body, 9
vegetarian diets and, 71-72
weight loss and, 150, 150t
Energy expenditure, 128
genetics and, 142
physical activity and, 151
Energy metabolism, **121**-123, *122, 124*
central pathways of, 123, *124*
energy budget and, 123-130, *124, 125, 127,*
129t, 130t
fasting and, 125-127, *127*
feasting and, 123-125, *125*
heat energy and, 120-121
Energy-yielding nutrients, **8,** 47-48
in breast milk, 246, *246*
children and, 262
in infant formulas, 248, *249*
older adults and, 297t, 298-299, 301t
polysaccharides, 34-35
Engorgement (breast), **258**
Enriched food, B vitamins in, 180
Enrichment, 180, **181**
Enteral formulas, **407**-413, *413*
Enteral nutrition, 407-424. *See also* Nutrition formulas; Parenteral nutrition; Tube feedings
after acute stress, 456-457, 462
in AIDS, 581
in cirrhosis, 552
in Crohn's disease, 391
formulas, **407**-413, *413,* G
in respiratory failure, 512
stress formulas, 456-457
transition from parenteral nutrition to,
438-439
Enteritis, radiation, **568**
Enteropathy
gluten-sensitive, **393**-394, 395t
HIV, **578**
Enterotoxins, 113
Environment
contaminants in. *See* Contaminants
obesity and, 143-144
Environmental temperature, BMR and, 129t
Enzyme(s), **77**-78, *78. See also specific enzymes*
serum, 333t
Enzyme replacements, **387,** 389, 399
Eosinophilia-myalgia syndrome (EMS),
tryptophan and, 85
EPA (eicosapentaenoic acid), **56,** 295
Ephedra, 588t
Ephedrine, 147, 589
Epigastric, **360**
Epiglottis, **96**
Epithelial cells, **169**
vitamin A in, 169
Epithelial tissue, **169**
Epoetin, 91t
Ergogenic aids, 90-94, 91t
Erythrocyte hemolysis, **175**

Erythropoietin, 520, **526**
Esophageal hiatus, **361**
Esophageal stricture, **360**
Esophageal ulcers, **360**
Esophageal varices, cirrhosis and, 546-547, 552
Esophagitis, reflux, 360t, **360**-363, *361*
Esophagus, **96**
    Barrett's, **360**
    cancer, 565t, 571t
    in digestive process, 97
    disorders, 359-363, 360t, *361*
Essential amino acids, **76**-77. *See also*
    Protein/amino acids; *specific amino acids*
Essential fatty acids, **55**-56
    deficiency in fat malabsorption, 385
Essential nutrients, **8**
Estimated Average Requirements. *See* EAR
Ethanol, 558. *See also* Alcohol
Ethical, **445**
Ethical issues, 444-447
Ethnic diets, **3,** 4t-6t
Ethnic heritage. *See* Race/ethnicity
Exchange resins, 528, 536
Exchange systems/lists
    Canadian system, B-1 to B-13
    diabetes and, 474-480, 476t, 479t, 480t, 481
    U.S. system, C-2 to C-9
Exercise. *See also* Fitness
    activity selection for, 152
    appetite control and, 152
    BMR and, 151-152
    to build muscles, 155
    childhood obesity and, 286
    coronary heart disease risk and, 501
    diabetes and, 480-481, 484
    diary of, 153, *153*
    energy RDA and, 130t
    hypertension and, 501
    insulin and, 484
    older adults and, 291-292
    pregnancy and, 237, 237t
    psychological benefits of, 152
    vitamins and minerals and, 214-215
    voluntary activities, 128, 129, 130t
    water and, 226t, 226-227, 227t
    weight loss and, 151-153
Exocrine glands, 100
Extra lean (on labels), 25t
Eyes. *See also* Blindness; Vision
    cataracts, 294
    malnutrition in children and, 267t
    in nutrition assessment, 332t

F

Face
    malnutrition in children and, 267t
    in nutrition assessment, 332t
    traits in fetal alcohol syndrome, 241
Fad diets, **147**
FAE (fetal alcohol effect), **241**
Faith healing, **586**
FANSA. *See* Food and Nutrition Science
    Alliance

FAO (Food and Agricultural Organization), 10
FAS (fetal alcohol syndrome), **240**-241
Fast foods, adolescents and, 278, 278t, 279t
Fasting. *See also* Anorexia nervosa
    BMR and, 129t
    hazards of, 126-127
    metabolic effects of, 125-127, *127*
    in pregnancy, 240
    responses to stress vs., 451, *451*
Fasting blood glucose, 466
Fasting triglycerides, coronary heart disease risk
    and, 501t
Fat (dietary), 8, 62-68, *66, 67*
    bile in digestion of, 104, *104*
    cancer risk and, 24
    cardiovascular disease and, 24, 58, 500,
        503-505
    children's intake, 286
    cirrhosis and, 551
    diabetes and, 472
    in eggs, 63
    excess of, 123-124
    excessive restriction of, 67
    in exchange lists, 474
    in fish, 63
    in food group plans, *19*
    in foods, A
    functions of, 62, 62t
    hard vs. soft, 55, *66*
    health and, 24, 58-60
    hidden, 63
    hydrogenated, 67
    hypertension and, 503
    intravenous emulsions, 431-432, 437-438
    kcalories in, 8, 65, *66*
    malabsorption and, *385, 386, 387*
    in meats, 63, 64
    melting temperature and saturation, 65, *66*
    in milk, 63, 64
    needs during stress, 456
    nephrotic syndrome and, 524, 525t
    older adults and, 299
    protein-sparing effect, 87
    rate of digestion, 104
    recommendations, 59-60, 64-68
    strategies for lowering intake, 64-68
    substitutes, 60-62
    weight loss and, 150-151
Fat cells, *53*
    development of, 143
    enlargement due to feasting, 123, *125*
Fat-free (on labels), 25t
Fat/lipids (in body), 53-70. *See also*
    Atherosclerosis; Body composition;
    Cardiovascular disease (CVD);
    Cholesterol; Heart disease; Lipoprotein
    chemistry of, *54,* 54-58, *55, 57*t
    coronary heart disease and, 496-497, 501t
    drugs for lowering, 506, *507*
    energy and, 47, 53
    functions of, 53-54, 54t
    ideal amount, 141-142
    intra-abdominal, **140**
    nephrotic syndrome and, 523-524

transport of, 109-111, *110*
Fat malabsorption, *385,* 385-392, 387t
Fat metabolism, 122-123
    stress effects on, 450
Fat-restricted diet, fat malabsorption, 386-387,
    387t, 388
    for heart disease, 502-503
Fat-soluble vitamins, 167t, 167-177, 178t-179t.
    See also Vitamin(s); *specific vitamins*
    fat malabsorption and, 385-386
    fat-restricted diets and, 386
Fat substitutes, 60-62, 388, 472
Fatfold measures, **141,** E-10
Fatigue, cancer and, 472, 567
Fatty acids, **54**-57, *55*
    chain length of, 54
    cis-, 67, *67*
    degree of saturation, 54-55
    essential, 55-56, 385
    health and, 58-60
    stress and, 456, 463
    trans-, *67,* 67-68, 504, 505
Fatty liver, **84,** 543-544
Fatty streaks, **284,** 285
FDA. *See* Food and Drug Administration
Feasting, 123-125, *125*
Fecal bulk. *See* Residue (in colon)
Feeding. *See also specific methods, e.g., Tube*
    feedings
    artificial, **445**
    bolus, **415**
    continuous, **415**
    evaluation, 374-376, 375t
    intermittent, **415**
    promotion of independent eating and,
        374-378
    transitional after parenteral nutrition, 439
Feeding devices, *377,* 377-378
Feeding disabilities, 374-378
    conditions leading to, 374t
    special implements and, *377,* 377-378
    stroke and, 510
Feeding tubes, 409-411. *See also* Tube feedings
Female athlete triad, **160,** 161
Ferritin, **212,** *E-12, E-13*
Ferulic-acid, 91t
Fetal alcohol effect (FAE), **241**
Fetal alcohol syndrome (FAS), **240**-241
Fetus, **230.** *See also* Infant(s); Pregnancy
    bone development, 233-234
    maternal PKU and, 135
Fever
    BMR and, 129t
    metabolism and, 121
Feverfew, 588t
Fewer (on labels), 25t
Fiber, **35**-36
    bacterial translocation and, 463
    in body, 35-36
    calcium binding by, 208
    cancer and, 565
    chronic heart failure and, 510
    diarrhea and, 383
    diet low in, 348-349, 349t

Fiber—continued
 diverticular disease and, 395
 in enteral formulas, 408
 food labels and, 24, 25t
 in foods, 36, 43, *44,* 47, A
 health effects of, 24, 42-43, 463
 hypertension and, 505
 inflammatory bowel disease and, 391
 irritable bowel syndrome and, 383
 older adults and, 299
 short chain fatty acids and, 463
 soluble vs. insoluble, 36, 43t
Fiberall, 380
Fibrocystic breast disease, **175**
Fibrous plaques, **284,** 285
Filtrate, renal, **526**
Fish
 fats in, 64
 protein content, 87t
 recommendations, 56
 safety, 116
Fish oil. *See* Omega-3 fatty acid
Fistula, **388**
 radiation therapy and, 568
Fitness, **13.** *See also* Exercise
 benefits of, 16t
 guidelines for, 13-14
Flapping tremor, **547**
Flatus, **381**
Flavonoids, 170t
Flora, intestinal, 104, 462
Fluid(s), 200-202. *See also* Dehydration; Fluid
  balance; Hydration; Water
 acute renal failure and, 527
 buildup in chronic renal failure, 531
 calcium in, 206
 chronic heart disease and, 509
 chronic renal failure and, 530t, 531
 cirrhosis and, 550t, 551
 diarrhea and, 382
 estimation of requirements, 416
 intake for weight gain, 156
 kidney stones and, 522
 needs after ostomy, 397
 needs during stress, 454
 respiratory failure and, 512
 vomiting and, 363
Fluid balance, 201-202. *See also* Water
 blood pressure and, 498, *499*
 proteins and, 78, *79*
 replacing losses and, 226t-228t, 226-228
Fluid-restricted diet, 531
Fluid retention, physical signs of, 331, 332t
Fluorapatite, **220**
Fluoridated water, 220
Fluoride, 220
 deficiency. *See* Fluoride deficiency
 pregnancy and, 234
 sources, 220, *220*
Fluoride deficiency, 220
 tooth development and, 50t
Fluorosis, **220**
Folacin. *See* Folate
Folate, 184, *185,* 192t-193t. *See also* B vitamins

 alcohol and, 184, 560
 aspirin and, 319-320
 bacterial overgrowth and, 391, 392
 chronic renal failure and, 530t, 532
 deficiency. *See* Folate deficiency
 drugs and, 184, 318-320, *320*
 in foods, 184, 233, A
 methotrexate and, 318-319, *320*
 older adults and, 300
 pregnancy and, 232-233, 233t, 234t
 recommendations, 232-233
 toxicity, 193t
 vitamin B$_{12}$ and cell division and, 184-185
Folate deficiency, 192t-193t
 birth defects and, 232
 depression and, 340
Folic acid. *See* Folate
Follicle, **170**
Follicular hyperkeratosis, **171**
Food(s). *See also* Diet; Nutrients
 adequacy of nutrients in, 196
 adverse effects of reactions to, 270
 aspiration, **359**
 associations influencing choices in, 2-3
 availability of, 6
 breast milk flavor and, 243
 caries-preventing, 51
 choices, reasons for, 2-7, 4t-6t
 choices during pregnancy and lactation,
   235, 235t
 choices for adolescents, 277-278, 278t, 279t
 choices for children, 263, *264,* 271-275,
   273t, 274t
 choices for infants, 252
 choices for older adults, 303-304
 choices for weight gain, 155-156
 choices for weight loss, 149-151
 composition, A
 convenience of, 6-7
 cooking. *See* Cooking/food preparation
 digestion of. *See* Digestion
 drug interactions with. *See* Diet-medication
   interactions
 economic factors in choices, 7
 energy-dense, 155-156
 energy nutrients in, 9, 128
 energy required to manage digestion,
   129-130
 enrichment of, 180, 181
 exchange lists. *See* Exchange system
 fat in, 62t, 62-68, *66, 67*
 fiber in, 36, 43, *44,* 47
 folate in, 184, 233
 gas-producing, 381, 381t, 397
 helping clients improve intake of, 349, 351
 intake affected by medications, 317, 318t
 intake affected by stroke complications, 510
 intake assessment in health history, 324,
   325-327
 intake improvement in cancer, 572-574
 introduction in infants, 249-252, 251t
 labeling. *See* Labeling
 in liquid diets, 346-348, 347t
 physical activity and intake in diabetes, 471

 portions for weight gain, 156
 preparation of. *See* Cooking/food preparation
 protein in, 86-87, 87t
 pureed, *358,* 358-359
 safety, 113-117
 soft, 348, 348t
 starchy, 34-35
 storage time in refrigerator, 115t
 stroke complications affecting intake, 510
 sugars in, 38-39
 thermic effect of, **130**
 transition from tube feeding to, 421-422
Food allergies, **269**
 in children, 269-270, 270t
 formula selection and, 413
 in infants, 252
Food and Agricultural Organization (FAO), 10
Food and Drug Administration (FDA)
 amino acid supplements and, 85-86
 artificial sweeteners and, 39, 40, 41
 fat substitutes and, 60-62
 food labeling, 21-23
 herbal remedies and, 588
 infant formulas and, 248
 seafood monitoring, 116
 supplement labeling, 197-198
 tanning lamps and, 174
Food and Nutrition Science Alliance (FANSA),
   28, 28t
Food assistance programs, 354-355, 355t
 for children, 265, 266, 354
 for older adults, 309, 309t, 354-355
 for pregnant women, 235
Food aversion
 cancer and, 573
 in children, 270
 pregnancy and, **235**
Food-borne infection
 bone marrow transplant and, 575
 HIV infection and, 581
 nutrition formula handling and, 414
Food consumption survey, **20**
Food cravings, **235**
Food frequency questionnaire, 325
Food group plans, **17**-20, *18-19*
Food Guide Pyramid, 17, *19*
Food-hypersensitivity reactions. *See* Food
   allergies
Food intolerance, **269**
Food poisoning, **113**
 prevention of, *114,* 114-116, 115t
Food records, 326
Food Stamp Program, 309t, 354
Foodservice, in hospital, 402-405, *404*
Formulas. *See also* Infant formulas; Nutrition
   formulas
 ORT, 227, 228, 228t, 382
Fortification, 180, **181**
 calcium, 208
 folate, 184
 vitamin D, 174
Fosamax. *See* Aldendronate sodium
Fosinopril, 507
Free (on labels), 25t

Free radical(s), **169**
 coronary heart disease and, 517-518
Fresh (on labels), 25t
Fructose, **33**
Fruits
 carbohydrate content, 45
 cariogenicity, 51t
 fiber content, *44,* 349t
 in food group plans, *18*
 health claims on labels and, 24
 in soft diets, 348t
 sugar vs., 34
Full-liquid diets, 34-348, 347t
Fulminant liver failure, 545

G

Galactose, **33**
Galactosemia, **133,** 135-136
Gallbladder, **96**
Gangrene, **468**
Garlic, 587, 588t
 coronary heart disease and, 507
Gas
 foods producing, 381, 381t, 397
 ostomies and, 397
Gastrectomy, *366,* **366**
 anemia after, 369
 bone disease after, 369
 diet after, 367-368, 368t
 dumping syndrome after, 366-368, *367*
 malabsorption after, 369
Gastric acidity, reflux esophagitis and, 362
Gastric banding, **148,** 366, 369-370
Gastric bypass, **148,** 366, 369-370
Gastric glands, **102**
Gastric juice, **102**
 in digestion, 103
Gastric motility, **99,** *100*
 nutrient delivery during stress and, 456-457
Gastric partitioning, **148,** 366, 369-370
Gastric residuals, 417
Gastric surgery, *366,* 366-370, *367,* 368t
Gastritis, **364**-365
Gastroduodenostomy, *366*
Gastroesophageal reflux (GER), **360**-363
Gastrointestinal tract. *See* GI tract
Gastrojejunostomy, *366*
Gastroparesis, **468,** 470
Gastroplasty, **148**
Gastrostomy, 366, 369-370, **410,** 411t
 percutaneous endoscopic, **410**
Gatekeeper, **271**
 children's eating habits and, 271
Gemfibrozil, 507
Genes, **133**
 dominant, **133**
 recessive, **133**
Genetics
 energy expenditure and, 142
 weight and, 142
Genistein, 170t
Geophagia, 214
GER (gastroesophageal reflux), **360**-363

Geriatrics. *See* Older adults
Gestational age, appropriate for, **230**
Gestational diabetes, **238,** 487
 prevention, 488
GFR. *See* Glomerular filtration rate
GHb (glycated hemoglobin), **484**
GI (gastrointestinal) tract, **96.** *See also* Digestion;
 specific organs
 anatomy of, 96-102, *98-99, 100, 101*
 factors in health of, 111
 formula selection and function of, 411
 immunity and, *461*
 infections in AIDS, 578
 lower GI tract disorders, 380-401
 metabolic function of, 119-120
 nutrient delivery during stress and, 456-457
 obstruction in cancer, 568
 specialization in, 107
 stress effects on function, 452, *461,* 461-463
 upper GI tract disorders, 357-373
Ginkgo, 588t
Ginseng, 91t, 587, 588t
Glands, **99.** *See also specific glands*
 in digestion, 99-102, *100, 101,* 103
 malnutrition in children and, 267t
 in nutrition assessment, 332t
Gliadin, **394**
Glimepiride, 482
Gliomas, **563**
Glipizide, 482
Glomerular filtration rate (GFR), **526**
Glomerulus, **520,** *520*
Glucagon, responses to severe stress, 450t
Gluconeogenesis, **123**
Glucose, **32-33.** *See also* Blood glucose; Diabetes
 mellitus; Hyperglycemia; Hypoglycemia
 chemical structure of, *32*
 metabolism of, 122
 renal threshold, 468
 stress response, 450, 451
 tests, 333t
 urine, 467, 484
Glucose metabolism, 122. *See also* Diabetes mel-
 litus; Hyperglycemia; Hypoglycemia
 fasting and, 126
Glucose tolerance, impaired, **466**
Glucose tolerance factor (GTF), **221**
Glucosuria, **468**
Glutamine
 bone marrow transplants and, 575
 in IV solutions, 431
 stress and, 462-463
Glutathione peroxidase, 218
Gluten, **394**
Gluten-free diet, 394, 395t
Gluten-sensitive enteropathy, **393**-394, 395t
Glyburide, 482
Glycated hemoglobin (GHb), 333t, **484**
Glycemic effect, **42**
Glycerol, **54**
Glycine, 91t
Glycogen, **35**
 energy and, 47
 fasting and, 126

Glycogen loading, 92
Glycogen storage disease, **545**
Glycolysis, **122**
Glycosuria, **468**
Glycosylated hemoglobin, **484**
Goals, in care plan, 343
Goiter, **218**
Goitrogen, **218**
Goldenseal, 588t
Good source (on labels), 25t
Gout, **188**
 treatment of, 536
Graft-versus-host disease, **569**
Grains. *See also* Bread; Cereals
 carbohydrate content, 45
 in celiac disease, 394, 395t
 in food group plans, *18*
 protein content, 87t
 refined, **180**
 in soft diets, 348t
Granisetron, food interactions with, 371
Granulated sugar, **38**
Granuloma, **390, 449,** 450
Grapefruit juice, effects on medication
 metabolism, 321, 507
Green tea, coronary heart disease and, 507
Growth
 in adolescence, 276
 BMR and, 129t
 chronic renal failure and, 529-530
 infant nutrition and, 244, *244*
 measures of, 327t, 327-331
 proteins and, 80-81
 vitamin A and, 169
Growth factors, intestinal cells and, 463
Growth hormone, administration during stress,
 456, 463
Growth hormone releasers, 91t
GTF (glucose tolerance factor), **221**
Guarana, 91t
Gum, chewing. *See* Chewing gum
Gums (mouth)
 malnutrition in children and, 267t
 in nutrition assessment, 332t

H

Habit. *See also* Lifestyle choices
 food choices and, 2
Hair
 malnutrition in children and, 267t
 in nutrition assessment, 332t
Harris-Benedict equation, energy needs during
 stress and, 454t
Hawthorn, coronary heart disease and, 507
HCl (hydrochloric acid), **102**
HDL (high-density lipoprotein), **109,** *110,* 500
 coronary heart disease risk and, 501t
 health implications of, 111
Head and neck cancer, 571t
 resection in, 571t
Head circumference, 327-328
Headache, tyramine and, 320
Health, **2**

Health care costs, 424-428, *426*
    changes related to reductions in, 427
    nutrition support team and, 426-427
    quality of care and, 425
    steps for reducing, 425
Health care professionals
    alternative therapies and, 589
    coordination and communication
        among, 426
    quality of care measurement and, 425
    role in diabetes control, 492, 493
    role in nutrition care, 29-30, 313-314
Health claims (on labels), 22-23, 24t-25t
    evaluating, 28-29
    supplements and, 197-198
Health history, in nutrition assessment, *323,*
    323-327
Health line, *3*
Health maintenance organization (HMO), **425**
Healthy People 2000, 21
Healthy People 2010, 21
    objectives of, 22t
Heart. *See also* Cardiovascular system
    metabolic function of, 120
Heart attack, **508**-509
    silent, **470**
Heart disease. *See also* Atherosclerosis;
        Cardiovascular disease (CVD);
        Cholesterol
    antioxidant supplements and, 196, 197
    coronary. *See* Coronary heart disease
    fiber and, 42
    prevention in children, 284-287
    selenium deficiency and, 218
    vitamin E and, 175
Heart failure
    acute, 508-509
    chronic, 509-510
Heartburn, **360**, *361*
    in pregnancy, 239
Heat energy, metabolism and, 120-121
Heavy metals. *See specific metals*
Height
    BMR and, 129t
    measurement, 328
    pregnancy weight gain and, 236, 237
    and weight tables/charts, 138, 138t, 139, 237,
        328-330, E-1 to E-8
Helicobacter pylori, peptic ulcer and, 365, 371
Hematocrit, **213,** 333t, *E-12, E-13*
Hematuria, **521**
Heme, **215**
Hemochromatosis, **214, 545,** 548
Hemodialysis, **539**
    nutrient needs in, 530t
    peritoneal dialysis vs., 540, 541
Hemofiltration, **539,** 540
Hemoglobin, **211,** 333t, *E-12, E-13*
    glycated, 333t, **484**
    glycosylated, **484**
    pregnancy and, 235
    structure of, 76, *77*
Hemoglobin concentration, **213,** *E-12*

Hemolysis, erythrocyte, 175
Hemorrhagic disease, **176**
Hemorrhoids, fiber and, 42
Hemosiderin, **212**
Hemosiderosis, **214**
Hepatic artery, 109, 545
Hepatic coma, **547**
    cirrhosis and, 547
Hepatic encephalopathy, **547**
Hepatic steatosis, **543**. *See also* Fatty liver
Hepatic vein, 109, 545
Hepatitis, 544
Hepatorenal syndrome, **548**
Herbal laxatives, **147**
Herbal medicine, **586,** 588t
    anticoagulants and, 507
    for cardiovascular disease, 507
    liver failure due to, 545
    risks of, 587-589
Herbal steroids, 91t
Herbs, sodium-free, 505
Heredity. *See* Genetics
Hernia, hiatal, 360-361, *361,* 363
Herpes virus, **577**
Hesperidin, 187
HFCS. *See* High-fructose corn syrup
Hiatal hernia, 360-361, *361,* 363
High (on labels), 25t
High blood pressure. *See* Hypertension
High-density lipoprotein. *See* HDL
High fiber (on labels), 25t
High-fiber diet
    adaptation to, 395
    constipation and, 380
    diverticular disease and, 395
    irritable bowel syndrome and, 383
High-fiber formulas, 408
High-fructose corn syrup (HFCS), **38**
High-kcalorie diet, hepatitis and, 544
High-protein diet, hepatitis and, 544
High-quality protein, **86**
    nonmeat mixtures containing, *72*
Hispanic Americans, ethnic diet of, 4t
Histidine, **76**
HIV. *See* Human immunodeficiency virus
HIV enteropathy, **578**
HMO (health maintenance organization), **425**
Holistic, **313**
Home canning, 113
Home nutrition support
    candidates for, 440
    in cystic fibrosis, 390
    health care costs and, 427-428
    specialized, 440-442
Homeopathic medicine, 585, **586**
Homeostasis, 32, **33**
Homocysteine
    Alzheimer's disease and, 297
    coronary heart disease and, 505-506, 518
Honey, **38**
    infant botulism and, 116
    sugar vs., 33-34
"Honeymoon phase" of insulin therapy, 483

Hormones, **80.** *See also specific hormones*
    counterregulatory, **449,** 450
    responses to chronic renal failure, 529
    responses to severe stress, 450t, 450-451
Hospital
    biotin deficiency in, 182
    foodservice in, 402-405, *404*
    helping client to eat in, 349, 351
    routine laboratory tests in, 332-334, 333t
Human immunodeficiency virus (HIV), **575**-582
    anorexia and inadequate nutrient intake
        in, 577
    breastfeeding and, 243-244
    nutrient losses in, 577-578
    nutrition assessment in, 582
    nutrition consequences of, 577-578
    nutrition support in, 581
    opportunistic illnesses, 576
    symptoms, 576
    wasting, 576-577
Human milk. *See* Breast milk
Hunger
    community programs for prevention of,
        354-355, 355t
    effects in children, 265
Hydration. *See also* Dehydration
    exercise and, 226t, 226-227, 227t
Hydrochloric acid (HCl), **102**
Hydrogenated fats, 67
Hydrogenation, **67**
Hydrolyzed formula, **407**-408. *See also* Nutrition
    formulas
    residue and, 408
Hydrophilic colloids, **380**
Hyperactivity, **270**-271
Hyperammonemia, **547**
    in cirrhosis, 547
Hypercalciuria, **521**
    treatment, 522
Hyperglycemia, **468,** 485t. *See also* Diabetes
        mellitus
    parenteral nutrition and, 436
    rebound, **468**
    symptoms, 467
    tube feeding and, 419t
Hyperglycemic nonketotic coma,
        hyperosmolar, **468**
Hyperkeratosis, **171**
Hyperlipidemia, **500**
Hypermetabolism, 451
Hyperosmolar hyperglycemic nonketotic
        coma, **468**
Hyperoxaluria, **522**
Hyperplastic obesity, **143**
Hypertension, **496,** 498-499. *See also* Blood
        pressure
    atherosclerosis and, 499
    calcium and, 207-208
    in children and adolescents, 285, 285t
    complications, 498-499
    coronary heart disease and, 499, 501t
    diet therapy, 502-506
    portal, **545**

potassium and, 205
preexisting in pregnant woman, 238
pregnancy-induced, 238n
prevention, 499
risk factors, 499
sodium and, 24, 204
transient, of pregnancy, 238, 487
Hypertonic formula, **408**-409
Hypertrophic obesity, **143**
Hypnotherapy, **586**
Hypochromic anemia, microcytic, **212**
Hypoglycemia, **468**, 485t
diabetes and, 469
nocturnal, **473**
symptoms, 469
treatment, 473
Hypoglycemic agents, **482**
Hyponatremia, **551**
Hypothalamus, **200**
Hypovitaminosis A, 178t. *See also* Vitamin A deficiency
Hypoxemia, **512**

I

IDDM. *See* Insulin-dependent diabetes mellitus
Ideal body weight, 329-330, 330t. *See also* Weight
Idiopathic cirrhosis, **545**
IGF-1. *See* Insulin-like growth factor-1
Ileocecal valve, **96**
Ileostomy, *396*, **396**-397
Ileum, **96**
Illness. *See also* Disease; Stress; *specific illness*
breastfeeding and, 243-244
in children, 284-287
conditions leading to feeding problems, 374t
diabetes treatment during, 473
food choices and, 7
nutrition and, 313, *313*, 314-316, 332t
opportunistic, **576**
in pregnancy, 237-239
prevention in older adults, 292-293, *293*
terminal, **445**
Imagery, **586**
Immobility, nutrition status and, 315
Immune system
breast milk and, 247
cells of, *461*, 461-462
intestines and, *461*
malnutrition and, 313, *313*
response to severe stress, 449-450, *451*
Immunity. *See also* Immune system
intestinal, *461*, 461-463, 462t
Immunosuppressants
kidney transplant and, 534, 536
liver transplant and, 549
Impaired glucose tolerance, **466**
Impaired swallowing, **359**
Inborn errors of metabolism, **133**-136, *134. See also specific disorders*
Incomplete protein, **86**
Indemnity insurance, **425**
Indigestion, 360-363

antacids and, 103
Indinavir, 579
Indirect calorimetry, **454**
Indispensable amino acids. *See* Essential amino acids; *specific amino acids*
Individual rights, 446
Individualized, **313**
Indoles, 170t
Infant(s), 244-255. *See also* Children; Pregnancy
birthweight, 230
breastfeeding. *See* Breastfeeding; Lactation
with cystic fibrosis, 390
with fetal alcohol syndrome, 240-241
first foods for, 249-252, 251t
fluid balance in, 227, 228
formulas for. *See* Infant formulas
honey and botulism in, 116
nutrient needs of, 244-245, *245*
nutrition assessment in, 254
SIDS in, 240
supplements for, 246t, 246-247
vitamin K for, 177
weight gain of, 244, *244*
Infant formulas, 247-249, *250*
cystic fibrosis and, 390
energy-yielding nutrients in human milk vs., *249*
transition to cow's milk from, 249
Infection. *See also specific infections*
formula preparation and, 414
TPN catheter and, 436
vitamin A and, 170-171
Inflammatory bowel diseases, **390**-391
Inflammatory response, **449**, 450
Infliximab, 399
Infusion pumps
children and, 416, 436
home TPN, 441-442
Initiators (cancer), **564**
Inorganic, **8**
Inorganic nutrients, 8
Inosine, 91t
Inositol, 186
Insoluble fibers, **36**, 43t
Insulin, **33**, 465. *See also* Blood glucose; Diabetes mellitus; Hypoglycemia
actions of different types, 483, *483*
acute renal failure and, 528
delivery, 483
diabetes treatment with, 483-484
food interactions with, 482, 483
forms, 483
physical activity and dosage of, 484
responses to severe stress, 450t
Insulin analogs, 482, 483
Insulin-dependent diabetes mellitus (IDDM), 465t. *See also* Type 1 Diabetes mellitus
Insulin-like growth factor-1 (IGF-1), 335
stress and, 456, 463
Insulin reaction, **469**
Insulin resistance, **466**
in cirrhosis, 547
Insulin-resistance syndrome, **469**

Insulin shock, **469**
Insurance, indemnity, **425**
Intact formula, **407**
Integrative therapy, **586**
Interferon, 549
Intermittent claudication, **175**
Intermittent feeding, **415**, 416
International unit (IU), **171**
Internet. *See* Web sites
Intestinal flora, **104**
immunity and, 461
vitamin K synthesis by, 176
Intestinal immunity, stress and, *461*, 461-463, 462t
Intestinal juice, **102**
Intestine. *See* Colon; Digestion; GI (gastrointestinal) tract; Small intestine
Intra-abdominal fat, **140**
Intractable diarrhea, **382**
Intrarenal (cause of acute renal failure), **526**
Intravenous (IV), **430**
Intravenous (IV) catheter, **434**
complications related to, 436, 437t
home TPN and, 441-442
insertion and care, 435-436
Intravenous (IV) nutrition, 430-438. *See also* Parenteral nutrition
abbreviations in solution names, 431
feeding methods, 430, *430*, 434-435, *435*, 436t
nutrient content calculation and, 433
simple solutions, 432
solutions for, 430-434
in stress, 456-457
techniques, 435-438
Intrinsic, **185**
Intrinsic factor, 185-**186**
loss of, 185-186, 365
Invert sugar, **38**
Involuntary muscles, in digestion, 99-102, *100, 101*
Iodine, 218-219, 223t
deficiency, 218-219, 223t
recommendations, 219
sources, 219
toxicity, 219, 223t
Ions, **201**
Iridology, **586**
Iron, 211-216, 223t
adolescent needs, 277
chronic renal failure and, 532
deficiency. *See* Iron deficiency
in foods, 215-216, A
foods providing for children, 266, 266t
foods providing for infants, 251
in infant formula, 249
older adults and, 300-301
pregnancy and, 234t, 234-235
recommendations, *215*, 215-216
toxicity, 214-215, 223t
vegan diet and, 72
vitamin C and, 188
Iron deficiency, **212**-214, 223t
causes of, 212

Iron deficiency—continued
  in children, 265-266, 266t
  gastric surgery and, 368
  prevalence of, 212
  in renal disease, 532
  self-diagnosis of, 214
  tests for, 212
  tooth development and, 50t
  in women athletes, 213-214
  world focus on, 266
Iron-deficiency anemia, 212-213
  in children, 265-266
Iron overload, 214, 223t, 545
Iron poisoning, 215
Irritable bowel syndrome, 383
Isoflavones, 170t
Isoleucine, 76
Isothiocyanates, 170t
Isotonic formula, 408
Italian Americans, ethnic diet of, 6t
IU (international unit), 171
IV. See Intravenous

J

Japanese Americans, ethnic diet of, 5t
Jaundice, 177, 544
  carotenemia vs., 172
JCAHO (Joint Commission on Accreditation of
    Healthcare Organizations), 425-426
Jejunostomy, 410, 411t
  direct endoscopic, 410
  percutaneous endoscopic, 410
Jejunum, 96
  dumping syndrome and, 366, 367
Jin bu huan, 588t
Joint Commission on Accreditation of Healthcare
    Organizations (JCAHO), 425-426
Juice, weight gain and, 156
Junk science, 28, 28t
Juvenile-onset diabetes, 465t. See also Type 1
    Diabetes mellitus (IDDM)

K

Kaposi's sarcoma, 577
kCalories, 8. See also Energy
  in alcoholic beverages and mixers, 559t
  breastfeeding and, 242
  counting in food intake data analysis, 326
  in enteral formulas, 408
  in intravenous fat emulsions, 431
  restriction and aging prevention, 292
  weight loss and, 147-148
Keratin, 169
Keratinization, 170
Keratomalacia, 170
Keshan disease, 218
Ketonemia, 126, 468
Ketones, 47
  monitoring in diabetes, 484
  protein breakdown and, 126
Ketonuria, 126, 468

Ketosis
  fasting and, 126, 240
  in type 1 diabetes, 468
Kidney(s)
  cancer, 571t
  diseases, 521-541
  metabolic function of, 120
  structure and function, 520, 520
  transplants, 484, 534, 534t
Kidney stones, 521t, 521-523
  conditions associated with, 521t
  fat malabsorption and, 386
Kilocalorie. See kCalories
Kilojoule (kJ), 8
Korean Americans, ethnic diet of, 5t
Kwashiorkor, 83-84. See also Protein-energy
    malnutrition (PEM)
  marasmus vs., 83-84

L

Labeling, 21, 21-23, 23
  Canadian, B-14 to B-15
  health claims in, 22-23, 24t-25t
  herbal remedies and, 588
  olestra and, 62
  package size and, 23
  required information, 21
  of supplements, 197-198
  terms used in, 24t-25t
Laboratory tests, 332-335, 333t-335t, E-11 to E-13
  in liver disease, 546t
  upper GI disorders and, 371
Lactase deficiency. See Lactose intolerance
Lactation, 242. See also Breastfeeding
  food choices during, 235t
  nutrient needs during, 232, 236t, 242-243
Lactic dehydrogenase (LDH), 333t
Lacto-ovo vegetarian diets, 72
Lacto-ovo vegetarians, 71
Lacto-vegetarians, 71
Lactoferrin, 247
Lactose, 34
  in medications, 321
Lactose intolerance, 34, 383-384
  formula selection and, 413
  gas and, 384
  liquid diets and, 347
Lactulose, 549
Laennec's cirrhosis, 545
Laetrile, 187, 588t
Lamivudine, 579
Large intestine, 96. See also Colon
Latent, 188
Laxatives, 549
  cathartic, 160, 164
  effects on nutrient absorption, 317-318, 399
  herbal, 147
  hydrophilic colloids, 380
LBW (low birthweight), 230
LCT (long-chain triglycerides), 386, 387
L.D. (licensed dietitian), 30
LDH. See Lactic dehydrogenase

LDL (low-density lipoprotein), 109, 110, 500
  coronary heart disease risk and, 501t
  diets for reduction of, 502t, 503t
  health implications of, 111
Lead poisoning, in children, 266-269, 267t, 268
Lean (on labels), 25t
Lecithins, 57
Length measurements, 327
Leptin, 143
Less (on labels), 25t
Less cholesterol (on labels), 25t
Less fat (on labels), 25t
Less saturated fat (on labels), 25t
Letdown reflex, 258, 259
Leucine, 76
Leukemias, 563
Levulose, 38
Liberal bland diet, 365t
Licensed dietitian (L.D.), 30
Licorice, effects on medication metabolism,
    321, 507
Life expectancy, 290
Life span, 290
Lifestyle choices
  coronary heart disease risk and, 500-501
  family of children with diabetes and, 486
  GI tract health and, 111
  after myocardial infarction, 509
  physiological age and, 291
Light. See also Sunlight
  riboflavin sensitivity to, 181-182
Light (on labels), 25t
Lignans, 170t
Limiting amino acids, 86
Limonene, 170t
Linoleic acid, 55-56
Linolenic acid, 55-56
Lipase, serum, 333t
Lipids, 53, 57t. See also Fat
Lipodystrophy, 576
Lipomas, 576
Lipoprotein lipase (LPL), 142-143
Lipoproteins, 107, 109-111, 110. See also HDL;
    LDL; VLDL
  Alzheimer's disease and, 296
Liquefying process, in digestion, 101
Liquid diets, 346-348, 347t. See also Nutrition
    formulas
Lispro, 482, 483
Liver
  cancer, 565t, 571t
  disorders, 543-557
  failure, 545
  fatty, 84, 543-544
  IV lipid emulsions and, 431-432
  metabolic function of, 119, 543
  in nutrient transport, 108, 108-109
  transplantation, 553-554
Living will, 445, 446-447
Loneliness, nutrition status and, 339, 339t
Long-chain triglycerides (LCT), 386, 387
Longevity, 290-294
Loop diuretics, 507

Lorazepam, 398-399
Lovastatin, 507
Low birthweight (LBW), **230**
Low-carbohydrate diets, **147**
    pregnancy and, 240
Low cholesterol (on labels), 25t
Low-density lipoprotein. *See* LDL
Low fat (on labels), 25t
Low-fat diet, 150-151
    irritable bowel syndrome and, 383
Low-fiber diet, 348-349, 349t
    in inflammatory bowel disease, 391
Low kcalorie (on labels), 25t
Low-residue diet, 348-349
Low-residue formulas, 408
Low saturated fat (on labels), 25t
Low sodium (on labels), 25t
Low-sodium diet, 346, 506, 525t, 550t, 551
Lower gastrointestinal tract, disorders, 380-401
LPL (lipoprotein lipase), **142**-143
Lungs. *See also* Pneumonia
    cancer, 565t
    disorders, 511-514
    vitamin E and, 175
Lycopene, 170t
Lymph, **107**
Lymphatic system, **107**
    in nutrient transport, 109
Lymphocytes, 334t, 335, 335t
    CD4+, **576**
Lymphomas, **563**
Lysine, **76**

M

Ma huang, 147, 588t
Macroangiopathies, **468**
Macrobiotic diet, **586**
Macrocytic anemia, **185**
Macrosomia, **487**
Magnesium, 210, 222t
    in antacids, 371, 532
    deficiency. *See* Magnesium deficiency
    in foods, 210, A
    hypertension and, 506
    pregnancy and, 233, 234
    recommendations, 210
    renal disease and, 527, 532
    tests, 333t
    toxicity, 222t
Magnesium deficiency, 210, 222t
    tooth development and, 50t
Major minerals, 203-211, 222t. *See also*
    Mineral(s); *specific minerals*
Malabsorption, 385-394. *See also* Absorption
    causes of malnutrition in, 386t
    in celiac disease, 393-394, 395t
    in Crohn's disease, 390-391
    in cystic fibrosis, 389-390
    fat, *385*, 385-388, 387t
    after gastrectomy, 369
    intestinal surgery causing, 392-393, *393*,
        396-397

    in pancreatitis, 388-389
    in ulcerative colitis, 390-391
Malignant, **563**
Malnutrition, **20**. *See also* Hunger;
        Malabsorption; Nutrition status
    acute, **452**
    BMR and, 129t
    cancer and, 566-568
    in children, 264-266, 267t, 339-340
    chronic, **452**
    in chronic heart failure, 509
    health problems leading to, 314-315, 332t
    HIV infection and, 576-578
    identifying risk for, 322t, 322-323
    illness and immunity and, 313, *313*
    in malabsorption, 386t
    mental effects in children, 339-340
    nursing diagnoses related to risk for,
        342, 343t
    older adults and, 301t, 304
    physical signs of, 331-332, 332t
    protein-energy. *See* Protein-energy
        malnutrition (PEM)
    respiratory failure and, 512
    senility vs., 340
Maltitol, 39
Maltose, **33**
Managed care, **425**
    nutrition services and, 427
Manganese, 219
Mannitol, 39
MAO inhibitors. *See* Monoamine oxidase
        inhibitors
Maple sugar, **38**
Marasmus, **83**. *See also* Protein-energy
        malnutrition (PEM)
    kwashiorkor vs., 83-84
Margarine
    cholesterol-lowering, 505
    trans-fatty acids in, 67-68
Marijuana
    in adolescence, 278-279
    in pregnancy, 240
Massage therapy, **586**
Mastication, **357**
    difficulties, 357-359
Mastitis, **258**
MCH, *333*
MCHC, *333*
MCT. *See* Medium-chain triglycerides
MCV. *See* Mean corpuscular volume
Meals/mealtime
    carbohydrate in, *46*
    child with diabetes and, 486
    child's involvement in planning, 273
    diabetes and, 473-474, 478-479, 480, 486
    for infants, 253
    for older adults, 303
    weight gain and, 156
Meals on Wheels, 354
Mean corpuscular hemoglobin (MCH), 333t
Mean corpuscular hemoglobin concentration
        (MCHC), 333t

Mean corpuscular volume (MCV), 333t
Measurements, anthropometric, 327-331,
        E-9 to E-11
Meats
    cariogenicity, 51t
    fats in, 64
    fiber content, *44*
    in food group plans, *18*
    low-fiber, 349t
    protein content, 87t
    safety, *114*, 115-116
    in soft diets, 348t
    tyramine-containing, 320t
Mechanical soft diet
    chewing difficulties and, 357-359
    dysphagia and, 359-360
Mechanical ventilator, **512**
Medical conditions. *See also* Illness; *specific
        conditions*
    food choices and, 7
    nutrition status and, 314-315
Medical history. *See* Health history
Medical nutrition therapy, 313, **314**. *See also
        specific disease*
Medical record, 350
    tube feeding and, 420-421
Medications. *See also* Drug(s)
    administration through feeding tubes,
        417-418, 418t
    lipid-lowering, 506
    nutrient interactions with. *See* Diet-
        medication interactions
    for type 2 diabetes, 482
Meditation, **586**
Medium-chain triglycerides (MCT), fat-
        restricted diets and, 386-387, 388
Megaloblastic anemia, **185**
Megestrol acetate, 549, 570, 579
Meglitol, 482
Melanomas, **563**
Mental illness. *See also* Emotions
    nutrition status and, 340
Menus
    clear-liquid diet, 347
    full-liquid diet, 348
    in hospitals, 402-404, *404*
    low-fiber/low-residue, 349
    PKU, 135
    selective, 402
Meprobamate, 399
Meridia (sibutramine), 147
Mesentery vein, 108
Metabolic rate, basal. *See* Basal metabolic rate
        (BMR)
Metabolic syndrome, **469**
Metabolism, **119**-136
    accelerated, 121
    alcohol effects on, 558-561
    basal, **128**
    beta-carotene and, 169
    diet-medication interactions and, 318t,
        318-319
    energy. *See* Energy metabolism

Metabolism—continued
  inborn errors of, **133**-136, *134*
  principal organs of, 119-120
  stress response effects on, 449-452, 450t, *451*
  TPN complications, 436, 437t
  untreated diabetes and, *467*
  vitamin A and, 168
  vitamin B$_6$ and, 183
  vitamin C and, 187-188
  vitamin D and, 173
Metamucil, 380
Metastasize, **564**
Metformin, 482
Methionine, **76**
  sulfur in, 210
Methotrexate, folate and, 318-319, *320*
Metoclopramide, food interactions with, 371
Metronidazole, food interactions with, 371
MFP factor, iron and, 215
MI (myocardial infarction), **508**-509
Microalbuminuria, **471**
Microangiopathies, **468,** 470
Microcytic hypochromic anemia, **212**
Microvillus/microvilli, **105**-107, *106*
Milk and milk products. *See also* Lactose
  breast. *See* Breast milk; Lactation
  calcium in, 208, 208t, 209
  carbohydrate content, 45
  cariogenicity, 51t
  fats in, 63, 64
  in food group plans, *19*
  gas and, 384
  introduction in children, 249
  iron and, 216
  lactose intolerance and, 384
  nutrients in, 209
  older adults and, 303, 309
  pregnancy and, 233
  protein content, 87t
  riboflavin in, 181
  in soft diets, 348t
  sugar in, 34
  vitamin D and, 174
  weight gain and, 156
Milk anemia, 252
Milliliter (ml), 408
Mineral(s), 8, *203,* 203t, 203-222. *See also*
    Electrolyte(s); *specific minerals*
  alcohol and, 560
  Alzheimer's disease and, 296
  in breast milk, 246
  cirrhosis and, 552
  exercise and, 214-215
  fat malabsorption and, 385-386
  fluid balance and, 78
  hypertension and, 506
  major, 203-211, 222t
  needs during stress, 456
  older adults and, 300-301
  in pregnancy, 233-235
  supplements, 197
  supplements during pregnancy, 233, 234t
  supplements in athletes, 91
  trace, 211-221, 212t, 223t

  vegetarian diet and, 72-73
  water and body fluids and, 200-202
Mineral oil, 399
Miscellaneous foods, in food group plans, 17, *19*
ml (milliliter), 408
Moderation (dietary), **15**
Modifiable risk factors, 499
Modified diets, **345,** 347t, 348t. *See also* Diet
    therapy
Modular formulas, 408
Modules (formula), **408**
Molasses, **38**
Molybdenum, **221**
Monoamine oxidase (MAO) inhibitors, 320
Monomeric formula, **407**-408
Monosaccharides, **32**-33, 33t
Monosodium glutamate (MSG), 270
Monoterpenes, 170t
Monounsaturated fatty acid, **55,** *55*
  blood cholesterol and, 59, 472, 503-504
Mood disorders, **340**
More (on labels), 25t
Morning sickness, 239
Motility
  gastric, **99,** *100*
  during severe stress, 456-457
Mouth. *See also* Teeth
  cancer, 571t
  candidiasis in HIV infection, 577
  in digestive process, 96-97, 102
  disorders, 357-360
  dry, 359
  ulcers, **359**
MSG (monosodium glutamate), 270
Mucosal block, **214**
Mucous membranes, **169**
Mucus, **102**
Multiple daily injections, **483**
Multiple organ failure, 450, 462
Muscle(s)
  malnutrition in children and, 267t
  in nutrition assessment, 332t,
    380, E-10
  strength training and, 155
Muscular activity. *See also* Exercise
  voluntary, **128,** 129, 130t
Muscular dystrophy, **175,** 374
Mutation, **133**
Mutual supplementation, **71, 87,** *87*
Myocardial infarction (MI), **508**-509
Myoglobin, **211**
Myrrh, 587

N

Nails
  malnutrition in children and, 267t
  in nutrition assessment, 332t
NANDA (North American Nursing Diagnosis
    Association), 342
Nandrolone, 579
Naphthoquinone, 179t. *See also* Vitamin K
Nasoduodenal (ND), **410,** 411t
Nasoenteric, **410**

Nasogastric (NG), **410,** 411t
Nasojejunal (NJ), **410,** 411t
National 5 a Day Better Health Program, 353
National Health and National Examination
    Survey (NHANES), 20
National Nutrition Monitoring and Related
    Research Act, 20
National School Lunch Program, 265, 274-275,
    275t, 354
Nationwide Food Consumption Survey (NFCS), 20
Native Americans, ethnic diet of, 6t
Naturopathic medicine, 585, **586**
Nausea, 363-364
  cancer and, 572-573, 567
  HIV infection and, 572-573
  in pregnancy, 239
  prevention, 364
  tube feeding and, 419t
ND. *See* Nasoduodenal
NE (niacin equivalents), **182**
Nelfinavir, 579
Neomycin, 549
Neoplasm, **563**
Nephron(s), **520,** *520*
Nephropathy, **468**
Nephrotic syndrome, 79, 523-525, *524,* 525t
Nervous system. *See also* Brain
  alcohol and, 559
  in diabetes, 469
  vitamin B$_{12}$ and, 185, 340
Neural tube defects, **184**
  folate and, 184, 232
Neuron, **296**
Neuropathy, **468,** 470
Neurotoxins, 113
Nevirapine, 579
Newborn. *See* Infant(s)
NFCS (Nationwide Food Consumption Survey), 20
NG (nasogastric), **410,** 411t
NHANES. *See* National Health and National
    Examination Survey
Niacin, 182, 191t. *See also* B vitamins
  deficiency, 180, 191t
  as drug, 182
  in foods, 182, A
  toxicity, 191t
Niacin equivalents (NE), **182**
Niacinamide. *See* Niacin
Nickel, 221
Nicotinamide. *See* Niacin
Nicotinic acid. *See* Niacin
NIDDM. *See* Type 2 Diabetes mellitus
Night blindness, **168**
Nitrogen balance, **81,** E-11
NJ (nasojejunal), **410,** 411t
No (on labels), 25t
Nocturnal hypoglycemia, **473**
Non-B vitamins, 186-187
Noncommunicable Diseases (NCD), **13.** *See also*
    Chronic diseases; *specific diseases*
Nonessential amino acids, **76**
Noninsulin-dependent diabetes mellitus
    (NIDDM), 465t. *See also* Type 2 Diabetes
    mellitus

Nonketotic coma, hyperosmolar hyperglycemic, **468**

Nonnucleoside reverse transcriptase inhibitors, **578,** 579

Nonnutritive sweeteners, **39**

Nonsteroidal anti-inflammatory agents, peptic ulcer and, 365

North American Nursing Diagnosis Association (NANDA), 342

NPO, **346**

Nucleoside reverse transcriptase inhibitors, **578,** 579

Nucleotides, **457**

Nurses
  in community nutrition education, 354
  hospital foodservice and, 404-405
  in nutrition care, 314
  on nutrition support team, *426*

Nursing assessments. *See also* Nutrition assessment
  nutrition status information in, 322, 322t
  physical examinations in, 327-336

Nursing bottle tooth decay, **249**

Nursing care plan. *See* Care plan

Nursing diagnoses
  nutrition-related, 342, 343t
  related to depression and loneliness, 339, 339t
  see also specific diagnoses

Nursing process, **313**
  nutrition care and, 313-352

NutraSweet (aspartame), 39, 40

Nutrients, 8-9. *See also specific nutrients*
  absorption after intestinal resection, 392-393, *393*
  absorption and release of, 107
  adequacy in foods, 196
  breaking down for energy, 121-123
  deficiencies. *See* Deficiencies
  density, **15**
  formula composition, 408
  inadequate intake in cancer, 567-568, 569t, 571t
  intravenous, 430-432
  losses in HIV infection, 577-578
  medical treatments and, 315. *See also* Diet-medication interactions
  needs during acute stress, 453-456
  needs during adolescence, 276-277
  needs during childhood, 262-263
  needs during infancy, 244-245, *245*
  needs during lactation, *232*, 236t, 242-243
  needs during late adulthood, 297-301
  needs during pregnancy, 230-235, *232*, 233, 234t, 236t
  needs in chronic renal failure, 530t, 530-532
  needs in cystic fibrosis, 389-390
  needs with dialysis, 540-541
  physiological vs. pharmacological effects of, 182
  recommendations, 10-13, *13*, B-0 to B-17, C-0 to C-13
  supplements. *See* Supplement(s)
  transport of, 107-111, *108, 110*

Nutrition, **2**

community, 353-355, 355t
cost-conscious health care and, 424-428, *426*
food choices and, 7
in health care, 313-323
quality of care and, 425

Nutrition assessment, **313,** E-11 to E-13
  cancer and, 582
  cardiovascular disease and, 514
  care plan development after, 342-345, 343t
  in children and adolescents, 281
  diabetes mellitus and, 489
  food intake data in, 325-327
  health history in, *323*, 323-327
  HIV infection and, 582
  in infants, 254
  kidney disease and, 535
  liver disorders and, 555
  lower GI disorders and, 398
  in older adults, 305
  ongoing, 344
  physical examinations and, 327-336
  pregnancy and, 254
  stress and, 458
  tube feeding and, 422
  upper GI disorders and, 370-371

Nutrition care, 313-323. *See also* Care plan
  medical record and, 350

Nutrition education
  in care plan, 343-344
  community, 353-355, 355t

Nutrition formulas, 409, *413*, G. *See also* Enteral nutrition; Formulas; Infant formulas
  administration of, 415-417
  availability of, 413
  blenderized, 407
  complete, **407**
  delivery techniques, 415
  diabetes and, 473
  drug interactions with, 417-418
  fiber in, 408
  helping clients to accept, 412
  high-fiber, 408
  hydrolyzed, **407**-408, 408
  hypertonic, 408-409
  intact, **407**
  isotonic, 408
  low-residue, 408
  medication interactions with, 418t
  modular, **408**
  monomeric, **407**-408
  nutrient composition, 408
  oral vs. tube feedings of, 409
  osmolality of, 408-409
  polymeric, **407**
  preparation for tube feeding, 414
  protein isolate, **407**
  safe handling of, 414-415
  selection of, 411-413, *413*
  volume and strength of, 415-416
  water in, 416-417

Nutrition guidelines, 9

Nutrition intervention, 342-352

Nutrition labeling. *See* Labeling

Nutrition Navigator, 28

Nutrition on the Net. *See* Web sites

Nutrition professionals, 29-30

Nutrition-related nursing diagnoses, 342, 343t

Nutrition resources, D-1 to D-7

Nutrition risk, screening criteria for, 323

Nutrition screening, **322,** 323

Nutrition status, **314.** *See also* Nutrition assessment
  cancer therapy effects on, 568, 569t, 571t
  in celiac disease, 394
  in chronic obstructive pulmonary disease, 513
  emotional/mental health and, 303-304, 338-340
  illness and, 314-316, 332t
  laboratory tests and, 332-335, 333t-335t. *See also* Laboratory tests
  nursing assessments and, 322, 322t
  stress effects on, 452, 452t

Nutrition status survey, **20**

Nutrition support, 407-424, 430-444. *See also* Diet therapy; Enteral nutrition; Feeding; Parenteral nutrition
  in acute renal failure, 528
  after acute stress, 452-457, 454t
  in cancer, 571-575
  in chronic renal failure, 532
  in cirrhosis, 552
  in Crohn's disease, 390-391
  ethical issues, 444-447
  health care costs and, 427-428
  HIV infection and, 581
  after intestinal surgery, 393
  specialized home support, 440-442
  in ulcerative colitis, 390-391

Nutrition support team, *426*, 426-427
  in home nutrition programs, 440

Nutrition surveys, 20-21, 353

Nutritional muscular dystrophy, 175

Nutritionist, **29**

Nutritive sweeteners, **39,** 472

O

Oatrim, 60

Obesity, 139, **142.** *See also* Weight (body)
  in adolescence, 276-277
  causes of, 142-146
  central, **140**-141, 466, 502
  in childhood, 284-287, 285t
  clinically severe, **148**
  coronary heart disease and, 499, 501t, 502
  definition of, 466
  diabetes and, 37, 466
  drug treatment of, 146-147
  health risks of, 145
  hypertension and, 500
  poor treatment choices for, 147
  pregnancy and, 236, 237
  prevention in children, 286
  sugar and, 37
  surgical treatment of, 369

Occupation, food choices and, 7

Octacosanol, 91t

Odansteron, 371
Odors
    acetone breath, 468
    ostomies and, 397
Oils, **53**
Oils—continued
    fats in, 66, 67
    MCT, 386-387
Older adults, 290-310
    arthritis in, 294-295, 295t
    brain in, 295-297
    cataracts in, 294
    depression in, 303-304, 339
    diabetes in, 488
    diet-medication interactions in, 302, 316. *See also* Diet-medication interactions
    energy needs, 297t, 298-299, 301t
    exercise in, 291-292
    food assistance programs for, 309t, 354-355
    food choices and eating habits of, 303-304
    mental health and nutrition status of, 340
    mineral needs, 300-301
    nutrition assessment of, 305
    strategies for, 302t, 309t, 309-311
    supplements for, 301
    vitamin needs, 299-300
    water requirements, 299
Olestra, 61-62, 472
Oliguria, 531
Oliguric phase (acute renal failure), **527**
Omega, **56**
Omega-3 fatty acid, **56**
    atherosclerosis and, 498
    blood cholesterol and, 59
    cancer and, 59
    rheumatoid arthritis and, 295
    supplements, 56, 57
Online resources. *See* Web sites
Open system (nutrition formula), **414**
Opportunistic illnesses, **575**
Oral antidiabetic agents, **472,** 482
Oral diets. *See* Diet; Foods
Oral formulas. *See* Nutrition formulas
Oral rehydration therapy (ORT), **227,** 228, 228t, 382
Organ(s). *See also specific organs*
    multiple failure of, 462
Organic, **8**
Organic nutrients, 8
Organosulfur compounds, 170t
Orlistat, 147
Orogastric, **410**
ORT. *See* Oral rehydration therapy
Orthomolecular medicine, **586**
Oryzanol, 91t
Osmolality, **227**
    formula, 408-409
Osmosis, **202, 539,** *540*
Osmotic diarrhea, **382**
Osteoarthritis, 294-295, 295t
Osteodystrophy, renal, **529**
Osteomalacia, **173.** *See also* Vitamin D deficiency
Osteoporosis, **173, 207**
    calcium and, 22-23, 206-207

prevention of, 206-207
    risk factors for, 207t
    sodium and, 204
    vitamin D and, 173
Ostomy, 396-397
Outcome identification, planning and, 343-344
Ovarian cancer, 565t
Overnutrition, **13**
Overt, **188**
Overweight, 138, **139,** *143. See also* Obesity; Weight
    pregnancy and, 236, 237
Oxalate, **208, 317, 386**
    foods high in, 522
Oxalate-restricted diet, 386
Oxalate stones, fat malabsorption and, 386
Oxandrolone, 579
Oxazepam, 398-399
Oxidation, 175
Oxidative stress, prevention, 518
Oxymetholone, 579
Oysters, raw, 116
Ozone therapy, **586**

P

PABA. *See* Para-aminobenzoic acid
Pain
    cancer and, 567
    heartburn, 360
    kidney stones and, 521
    mouth, 359
Pancreas, **96**
    cancer, 565t, 571t
    metabolic function of, 119-120
    transplant, 484
Pancreatic enzyme replacements, **387,** 399
Pancreatic juice, **102,** 103
Pancreatitis, 388-389
    acute, 388
    chronic, 389
Pantothenic acid, 182, 192t. *See also* B vitamins
    deficiency, 192t
    in foods, 182
    toxicity, 192t
Para-aminobenzoic acid (PABA), 186
Paranoia, **340**
Parathormone, **206**
Parenteral nutrition, **430-444.** *See also* Enteral nutrition; Intravenous (IV) nutrition; Total parenteral nutrition (TPN)
    cyclic, **438**
    monitoring during, 438t
    in respiratory failure, 512
    transitional feedings after, 439
Partial vegetarians, 71
Pathogens, **113**
Pathological stress, 449. *See also* Stress
Pediatrics. *See* Children; Infant(s)
PEG, 410
PEJ, 410
Pellagra, **180.** *See also* Niacin
PEM. *See* Protein-energy malnutrition
Penicillamine, 549-550

Pennyroyal, 371, 545, 588t
Pepsin, **102**
Peptic ulcer, **365**
Peptides. *See* Protein/amino acids (in body)
Percent fat free (on labels), 25t
Percentages, calculation, F-0 to F-1
Percutaneous endoscopic gastrostomy (PEG), **410**
Percutaneous endoscopic jejunostomy (PEJ), **410**
Peripheral parenteral nutrition (PPN), **434.** *See also* Total parenteral nutrition (TPN)
    infusion, 436, 437
Peripheral resistance, **498**
Peripheral veins, **433**
Peripherally inserted central catheter (PICC), **435**
Peristalsis, **100,** *100*
Peritoneal dialysis, **539**
    hemodialysis vs., 540, 541
    nutrient needs in, 530t
Peritonitis, **390**
Pernicious anemia, 185-186, **365**
Persistent vegetative state, **445**
    ethical issues, 445-446
Personal factors, in health history, *323, 324*
pH, 79, *79, 80*
Pharmacist
    medication information, 314, 319
    on nutrition support team, *426*
Phenolic acids, 170t
Phenylalanine, **76**
    PKU and. *See* Phenylketonuria (PKU)
Phenylketonuria (PKU), 77, **133,** 133-135, *134*
    aspartame and, 40
Phenytoin, tube feedings and, 418
Phlebotomy, 548
Phosphate binders, 536
Phosphate salt, 91t
Phospholipids, **57**
    roles of, 57
    structure of, 57
Phosphorus, 209-210, 222t
    chronic renal failure and, 530t, 532
    in foods, A
    foods high in, 532
    pregnancy and, 233, 234
    tests, 333t
    toxicity, 222t
Phosphorus deficiency, tooth development and, 50t
Phylloquinone, 179t. *See also* Vitamin K
Physical activity. *See* Exercise
Physical examinations, in nursing assessment, 327-336
Physical fitness. *See* Fitness
Physicians
    in nutrition care, 313
    on nutrition support team, *426*
Physiological age, **291**
Phytates, **208, 216, 317**
Phytic acid, 170t
Phytochemicals, **2,** 170t
Phytosterols, 91t, 170t
Pica, **214**
PICC (peripherally inserted central catheter), **435**
Picnics, safety, 116

formation of, 54
high levels and fat intake, 472
long-chain, 386, 387
medium-chain, 386-387
Trimesters, pregnancy, 231
Tripeptide, **75.** *See also* Protein/amino acids (in body)
Tryptophan, **76**
  niacin and, 182
  supplements, 85
Tube feedings, 409-421. *See also* Enteral nutrition
  administration of, 414-421
  advantages of, 409
  in cancer, 571
  candidates for, 409
  complications, 418-421, 419t, 420t
  diabetes mellitus and, 476
  in dysphagia, 360
  fiber and, 408
  helping clients cope with, 412
  at home, 441
  indications for, 409
  medical record, 420-421
  medication administration through, 417-418, 418t
  monitoring during, 418-419, 420t
  oral feedings vs., 409
  schedule for, 416
  transition to table foods from, 421-422
  tube features, 411
  tube placement, *410,* 410-411, 411t
  volume progression for, 416
Tuberculosis, HIV and, 577
Tubule, **520,** *520*
Tufts University, Nutrition Navigator, 28
Tumor, **563**
  formation, *564*
Turbinado, **38**
24-hour recall, 325
Two-gram sodium-restricted diet, 525t
Type 1 diabetes, 465t, **466.** *See also* Diabetes mellitus
  in children and teens, 485-486
  coma in, 468
  diet in, 471-480, 481
  insulin for, *483,* 483-484
  ketosis in, 468
  metabolic consequences if untreated, *467*
  physical activity in, 480-481
  weight loss in, 469
Type 2 diabetes, **37,** 465t, **466.** *See also* Diabetes mellitus
  metabolic consequences if untreated, *467*
  nonketotic coma in, 468
  treatment, 482
  weight changes in, 471
Tyramine, MAO inhibitors and, 320
Tyramine-controlled diet, 320t

U

Ubiquinone, 187
UL (Tolerable Upper Intake Levels), 11-12, 12t
  vitamin C, 188

Ulcerative colitis, **390**-391
Ulcers
  decubitus, **315**
  esophageal, **360**
  mouth, **359**
  peptic, **365**
Ultrafiltration, **539,** 540, *540*
Umbilical cord, **230**
Undernutrition, **13**
Underweight, 138. *See also* Weight
  dangers of, 155-156, 162
  pregnancy and, 236, 237
Unsaturated fats, food choices and, 64, 65, 66
Unsaturated fatty acid, **55,** *55*
Upper gastrointestinal tract, disorders, 357-373
Urea kinetic modeling, **531**
Uremia, **526**
Uremic frost, **529**
Uremic syndrome, **526,** 529
Uric acid
  kidney stones, 523
  tests, 333t
Urinary tract, *520. See also specific parts*
Urination, diabetes and, 467
Urine
  glucose in, 467, 484
  ketones in, 484
U.S. RDA. *See* RDA (Recommended Dietary Allowance)
USDA. *See* Agriculture, U.S. Department of
Usual body weight, 330t, 330-331
Usual intake, 325
Uterus, **230**

V

Vagotomy, **366**
Valerian, 587, 588t
Valine, **76**
Values, food choices and, 3, 6
Varices, esophageal, **546**-547, 552
Variety (dietary), **16**
Vascular system. *See also* Cardiovascular system
  diseases in diabetes, 470
  in nutrient transport, 108-109
Vegan(s), **71.** *See also* Vegan diets
Vegan diets, 71
  minerals in, 72-73
  vitamin B$_{12}$ deficiency and, 186
  vitamin D deficiency and, 174
  vitamins in, 72
Vegetables
  cancer and, 565-566
  cariogenicity, 51t
  coronary heart disease and, 505
  cruciferous, 566
  fiber content, *44,* 349t
  in food group plans, *18*
  health claims on labels and, 24
  protein content, 87t
  in soft diets, 348t
  starchy, 45
  tyramine-containing, 320t
Vegetarian diets, 71-73

diet planning, 71, 73t
food energy intakes in, 71-72
health advantages of, 73
minerals in, 72-73
protein in, 71, *72*
vitamins in, 72
zinc in, 217
Veins, **108**
  central, **434,** *435*
  feeding through. *See* Intravenous (IV) nutrition
  peripheral, **433**
  portal, 545
Ventilator, mechanical, **512**
Very-low-density lipoprotein (VLDL), **109,** *110*
Very-low-kcalorie diets (VLCD), 147-148
Very low sodium (on labels), 25t
Vietnamese Americans, ethnic diet of, 5t
Villus/villi, **105**-107, *106*
  immune cells protecting, 461, *461*
Vision. *See also* Eyes
  vitamin A in, 168, *169*
Vitamin(s), 8, 167t, **167**-198. *See also specific vitamins*
  alcohol effects on absorption, 560, *560*
  in breast milk, 246
  chronic renal failure and, 530t, 532
  cirrhosis and, 551-552
  fat malabsorption and, 385-386
  in intravenous solutions, 432
  needs during stress, 456
  non-B, 186-187
  older adults and, 299-300
  in pregnancy, 232-234, 233, 234t
  supplements, 196-198
  supplements during pregnancy, 233, 234t
  supplements in athletes, 91
  supplements in cystic fibrosis, 389-390
  vegetarian diet and, 72
Vitamin A, **168**-172, 178t
  in defense against infection, 170-171
  deficiency. *See* Vitamin A deficiency
  in epithelial cells, 169
  in foods, 35t, 172, *172,* A
  metabolic roles of, 168
  older adults and, 299-300
  preformed, **171**
  recommendations, 172
  toxicity, 171, 178t
  in vision, 168, *168*
Vitamin A deficiency, 169, 178t
  tooth development and, 50t
Vitamin B$_1$. *See* Thiamin
Vitamin B$_2$. *See* Riboflavin
Vitamin B$_3$. *See* Niacin
Vitamin B$_5$. *See* Pantothenic acid
Vitamin B$_6$, 183-184, 192t. *See also* B vitamins
  alcohol and, 560
  chronic renal failure and, 530t, 532
  deficiency, 183, 192t
  in foods, 184, A
  metabolic roles of, 183
  older adults and, 300
  pregnancy and, 234t

Vitamin B$_6$, continued
    protein intake and, 183
    toxicity, 183, 192t
Vitamin B$_{12}$, 184-186, 193t. *See also* B vitamins
    absorption of, 185
    alcohol and, 560
    bacterial overgrowth and, 391, 392
    chronic gastritis and, 365
    deficiency. *See* Vitamin B$_{12}$ deficiency
    folate and cell division and, 184-185
    in foods, 186
    nervous system and, 185
    older adults and, 300
    toxicity, 193t
    vegetarian diet and, 72
Vitamin B$_{12}$ deficiency, 186, 193t, 340
    in vegans, 186
"Vitamin B$_{15}$," 187
"Vitamin B$_{17}$," 187
"Vitamin B$_T$." *See* Carnitine
Vitamin C, 187-189, 193t
    deficiency. *See* Vitamin C deficiency
    dependency, 188-**189**
    in foods, *189,* 189-190, A
    foods providing for infants, 251
    iron absorption and, 188
    iron and, 215
    kidney stones and, 522
    metabolic roles of, 187-188
    pregnancy and, 234t
    recommendations, 188, 189
    safe limits of, 189
    special needs for, 189
    toxicity, 188-189, 193t
    withdrawal reaction from, 188-189
Vitamin C deficiency, 188, 193t
    tooth development and, 50t
Vitamin D, 173-175, 178t, **206**
    actions of, 173
    for adults, 174-175
    in bone, 173
    for children, 174
    cirrhosis and, 552
    deficiency. *See* Vitamin D deficiency
    metabolic conversions of, 173
    older adults and, 300
    pregnancy and, 234t
    from sun, 174
    supplements in chronic renal failure, 530t, 532
    toxicity, 173-174, 178t
    vegan diet and, 72, 174
Vitamin D deficiency, 173, 178t
    lactose intolerance and, 384
    tooth development and, 50t
Vitamin E, 175-176, 179t
    arthritis and, 295
    blood pressure and, 506
    coronary heart disease and, 518
    deficiency, 175, 179t
    in foods, 176, *176*
    myths about, 175
    supplements in athletes, 91

toxicity, 175-176, 179t
Vitamin K, 176, 179t
    deficiency, 176-177, 179t
    in foods, 177
    intestinal flora and, 104
    for newborns, 177
    parenteral nutrition and, 432
    toxicity, 177, 179t
    warfarin and, 318, 507
Vitamin P, 187
VLCD (very-low-kcalorie diets), 147-148
VLDL (very-low-density lipoprotein), **109,** *110*
Voluntary activities, energy expenditures and, **128,** 129, 130t
Vomiting
    in adults, 363-364
    cancer and, 572-573
    in children, 363
    in pregnancy, 239
    tube feeding and, 419t

W

Waist circumference, **141,** *141*
Warfarin (Coumadin)
    vitamin C and, 188
    vitamin K and, 318, 507
Waste products
    buildup in chronic renal failure, 531
    dialysis for removal of, 539-541
Wasting
    cancer and, 566
    cancer treatment and, 569t
    chronic renal failure and, 529, 530t
    cirrhosis and, 548, 548t
    HIV infection and, 576-577, 578
    metabolism and, 121
Water, 8. *See also* Fluid(s)
    balance, **200**-201, 201t. *See also* Fluid balance
    body fluids and, 200-202
    bottled, 117
    excretion regulation, 200-201
    fluoridation, 220, *220*
    in foods, A
    formula administration and, 416-417
    infant requirements, 244-245, 249-250
    intake regulation, 200
    intoxication, **200**
    lactation and, 242-243
    minimum requirement, 201
    in older adults, 299
    recommendations, 201, 416
    safety, 117
    sources, 201
    tube feeding and, 416-417
    weight loss and, 151
Water-soluble vitamins, 167t, 177-193, 191t-193t. *See also* Vitamin(s); *specific vitamins*
WBC. *See* White blood cells
Web sites
    on alternative therapies, 587
    on cancer, 584

    on carbohydrates, 49, 51
    on cardiovascular disease, 516, 517
    on care plans, 353
    on childhood and adolescence, 283, 284
    on cost-conscious health care, 424
    on diabetes mellitus, 491, 492
    on digestion and absorption, 112
    on eating disorders, 160
    on enteral nutrition, 424
    on ethical issues in nutrition care, 444
    extensions in names of, 28-29
    on fats/lipids, 70
    on feeding disabilities, 378
    on food safety, 113, 116, 117
    on HIV infection, 584
    on inborn errors of metabolism, 133, 136
    on kidney disease, 538, 539
    on lipids, 70
    on liver disease, 557
    on lower GI tract disorders, 401
    on metabolism and energy balance, 132
    on minerals, 225
    on nutrition assessment, 338
    nutrition information on, 27, 28-29
    on older adults, 307, 309
    on parenteral nutrition, 444
    on pregnancy and infancy, 256, 258
    on proteins and amino acids, 89
    on rehydration, 228
    reliability of information on, 28-29
    on severe stress, 460
    on supplements, 198
    on upper GI tract disorders, 373
    on vegetarian diets, 71
    on vitamins, 195
    on weight control, 158
Weight (body), 138-167. *See also* Obesity; Weight gain; Weight loss
    Alzheimer's disease and, 297
    arthritis and, 295
    artificial sweeteners and, 41
    control in type 2 diabetes, 486
    cystic fibrosis and, 390
    fiber and, 42
    genetics and, 142
    and height tables/charts, 138, 138t, 139, 237, 328-330, E-1 to E-8
    ideal, 329-330, 330t
    maintenance of, 153-154
    measurement, 327, 327t, 328-331, 329t, 330t
    prepregnancy, 230
    recommendations, 138t, 138-140
    upper GI disorders and, 370
    usual, 330t, 330-331
Weight cycling, **151,** *151*
Weight gain
    of infants, 244, *244*
    in pregnancy, 236-237, 237t, 331
    strategies for, 155-156, 574
Weight loss
    anorexia nervosa and. *See* Anorexia nervosa
    behavior modification in, *153,* 153-155
    in chronic heart failure, 510

PHOTO CREDITS